Lecture Notes in Civil Engineering 437

Lecture Notes in Civil Engineering (LNCE) publishes the latest developments in Civil Engineering—quickly, informally and in top quality. Though original research reported in proceedings and post-proceedings represents the core of LNCE, edited volumes of exceptionally high quality and interest may also be considered for publication. Volumes published in LNCE embrace all aspects and subfields of, as well as new challenges in, Civil Engineering. Topics in the series include:

- Construction and Structural Mechanics
- Building Materials
- Concrete, Steel and Timber Structures
- Geotechnical Engineering
- Earthquake Engineering
- Coastal Engineering
- Ocean and Offshore Engineering; Ships and Floating Structures
- Hydraulics, Hydrology and Water Resources Engineering
- Environmental Engineering and Sustainability
- Structural Health and Monitoring
- Surveying and Geographical Information Systems
- Indoor Environments
- Transportation and Traffic
- Risk Analysis
- Safety and Security

To submit a proposal or request further information, please contact the appropriate Springer Editor:

- Pierpaolo Riva at pierpaolo.riva@springer.com (Europe and Americas);
- Swati Meherishi at swati.meherishi@springer.com (Asia—except China, Australia, and New Zealand);
- Wayne Hu at wayne.hu@springer.com (China).

All books in the series now indexed by Scopus and EI Compendex database!

Stefano Gabriele · Amedeo Manuello Bertetto ·
Francesco Marmo · Andrea Micheletti
Editors

Shell and Spatial Structures

Proceedings of IWSS 2023

Springer

Editors
Stefano Gabriele
Department of Architecture
Roma Tre University
Rome, Italy

Amedeo Manuello Bertetto
Department of Structural, Geotechnical and Building Engineering
Politecnico di Torino
Turin, Italy

Francesco Marmo
Department of Structures for Engineering and Architecture
University of Naples Federico II
Napoli, Italy

Andrea Micheletti
Department of Civil and Computer Science Engineering
University of Rome Tor Vergata
Rome, Italy

ISSN 2366-2557 ISSN 2366-2565 (electronic)
Lecture Notes in Civil Engineering
ISBN 978-3-031-44327-5 ISBN 978-3-031-44328-2 (eBook)
https://doi.org/10.1007/978-3-031-44328-2

This Springer imprint is published by the registered company Springer Nature Switzerland AG
The registered company address is: Gewerbestrasse 11, 6330 Cham, Switzerland

Paper in this product is recyclable.

Preface

Shell and spatial structures represent highly efficient structural systems that blend optimized material utilization with effective structural forms. The ongoing advancement of analysis methods, design approaches, and construction techniques in this field has sparked growing interest among engineers, architects, and builders. The Italian Workshop on Shell and Spatial Structures (IWSS) focuses on a wide range of aspects related to shell and spatial structures. The second edition of the IWSS took place on June 26–28, 2023, at the Valentino Castle, an historic residence of the Royal House of Savoy and a UNESCO World Heritage Site since 1997.

The IWSS 2023 mascotte, the turtle *Shelly*, represents the feelings that the project of this workshop have brought to the organizers. In the collective imagination, the turtle is one of the most beloved animals, present in culture and myth, not only of western civilizations, but also in legends of Chinese, Indian, among Native Americans, and Pacific peoples. She is generally depicted as the one who holds up the world. Shelly thus gathers many meanings. Certainly, the appeal to the concept of the shell. Our turtle houses a real grid shell on its back. Shelly's accompanying motto *Shell we shape* is like a wordplay that invites participants to always find new ways for study and research without ever feeling stuck and never give up. In the myth, turtle is the point of contact and the boundary between sky and the Earth, between past and future and, in our case, it represents the undefined boundary between architecture and engineering, the limit between the beauty of a construction and whether it is really possible, buildable, realizable.

The IWSS 2023 represented an opportunity to comprehensive learning and sharing ideas, and participants had the chance to attend special keynote and plenary lectures of Italian and international speakers. Contributions were devoted to a broad range of topics related to experimental and theoretical studies, numerical methods and approaches for design, form finding, structural optimization, seismic analysis, sustainability analysis, stability analysis, manufacturing, testing, damage detection, structural health monitoring, maintenance, and historical reviews of all types of shell and spatial structures. The goal of the founders of the IWSS and chairmen of the 2023 event is to create a permanent network among researchers, professionals, and companies so as to make some structural engineering topics, which until now have been exclusively academic, closer to the world of work and business. This goal, in addition to making research topics current and usable, acts as a driving force for the development of new advances in science and research.

Participants to the IWSS 2023 had the opportunity to choose between two technical tours that demonstrate the legacy and the new perspective of shell and spatial structures. The first one, entitled *The Legacy of the Past*, consisted in a visit to Torino Esposizioni (Turin Exhibition Center) and Palazzo del Lavoro. The Turin Exhibition Center was conceived in the years immediately following World War II as a public exhibition space to host primarily the annual automobile show. The center houses some iconic concrete shell structures by Pier Luigi Nervi and Riccardo Morandi. They consist in three grand exhibition pavilions perfectly representing the structural art of these famous protagonists of contemporary engineering and architecture. The second tour, by the title of *The New Perspective*, allowed participants to visit the IAM@PoliTo, a mission-driven effort aimed at creating a comprehensive and multidisciplinary research platform for additive manufacturing. The overarching goal of this platform is to tackle the existing open challenges in the field of additive manufacturing, specifically in terms of machines, materials, and applications.

As part of the workshop, a permanent outdoor expo was held for the duration of the event, which featured the construction of two demonstrative structures: the Wings Pavilion and the Nested Tensegrity Balloons (Fig. 1). Both structures provided a valuable opportunity for joint work between research groups, faculties, researchers, and students. The first structure resulted from the collaboration between the groups of Giuseppe Fallacara and Amedeo Manuello, and it consisted of a saddle-shaped construction made from three arched structures built on the catenary principle and realized by steel chains and elastic cords. The second structure was designed by Andrea Micheletti and coworkers, in collaboration with the company Progetto Legno for the realization with modular wooden struts and marine ropes. The structural concept iterates the endoskeletal analogy of tensegrity systems by Kenneth Snelson, also known as the balloon analogy by Richard Buckminster Fuller, for which a stable tensegrity structure has the struts placed inside an outer network of cables. In this two-layer structure, two tensegrity balloons are placed one inside the other.

The IWSS 2023 collected more than 100 contributions which have been presented during the three days of conference, arranged into three parallel sessions and organized according to 14 main themes:

- *Form finding and optimization of shell and spatial structures*, organized by A. Manuello Bertetto, G. C. Marano and N. Lagaros
- *Multiscale structures for additive manufacturing*, organized by M. Galati
- *Structural theories and morphology: analyses and computational methods*, organized by A. Manuello Bertetto and J. Melchiorre
- *Historical vaulted structures: geometries and geometrical variations over time*, organized by C. Calderini and E. Piccoli
- *Analytical and numerical analysis of arches and vaults*, organized by S. Sessa and F. Marmo
- *Perspectives and advancements in construction of complex structures*, organized by S. Gabriele and A. Micheletti
- *Dynamics and stability of curved structures*, organized by A. Luongo
- *Space architecture: a multidisciplinary form finding*, organized by V. Sumini and M. Rossi
- *Monumental buildings and historical case studies: retrofitting and restoration*, organized by E. Lenticchia
- *Optimization in the analysis and design of shell and spatial structures*, organized by S. Gabriele and R. Cucuzza
- *Textile-hybrid structures, organized by A. Zanelli, Environmental impacts of shells and spatial structures*, organized by O. Iuorio
- *Realizations and case studies*, organized by G. Boller

Fig. 1. Wings Pavilion and Nested Tensegrity Balloons (top). Participants group photo (bottom).

- *Stability of dome structures and space frames*, organized by A. Manuello Bertetto and F. Bazzucchi

The parallel sessions of all days of conference were introduced by plenary lectures of prominent experts in the field of shell and spatial structures. The plenary lectures of the first day are entitled *Understanding Tensegrity with an Energy Function*, given by Simon Guest, and *Form Finding by Shape Optimization with Implicit Splines and*

Vertex Morphing, given by Kai-Uwe Bletzinger. During the second day, Valentina Sumini gave the lecture entitled *Space Architecture: A multidisciplinary form finding*, and Hugo Corres Peiretti gave the lecture entitled *Relationship between the past and the present of shells and spatial structures. From the teachings of Eduardo Torroja to more recent achievements*. Sigrid Adriaenssens connected online from Princeton to give the lecture entitled *How shell we build now?*, a title that also inspired the motto of the IWSS 2023. The contributions of the last day were introduced by the plenary lectures by Carlos Lázaro, entitled *Two lightweight structures with steel-fiber reinforced concrete*, and by Charis Gantes, entitled *Structural Challenges Encountered in the Design of Tubular Steel Wind Turbine Towers*.

Parallel sessions were introduced by keynote lectures regarding specific aspects of the design, analysis and construction of shell and spatial structures. Specifically, Fernando Fraternali gave the lecture entitled *Designing seismic isolators through lattice metamaterials*, Maurizio Angelillo discussed about *Membrane Equilibrium Analysis for the assessment of traditional masonry structures and the design of new sustainable structures*, Giuseppe Fallacara presented the *New frontiers of stereotomic stone architecture*, Angelo Luongo's lecture regarded *Perturbation methods for nonlinear continuous systems: statics, buckling, resonances and dynamic bifurcations*, Massimo Cuomo talked about *Variational formulations for the form finding of flexible structures*, and Gabriele Milani presented *Two novel lower and upper bound 3D distinct element limit analysis models for masonry domes and vaults*.

The IWSS also hosted the Pier Luigi Nervi Prize, in order to recognize talented young researchers, designers, and engineers working in the field of shell and/or spatial structures. The Nervi Prize Commission examined more than 50 submissions to the IWSS 2023 authored by young researchers and decided to assign the Pier Luigi Nervi Prize to Francesco Laccone from the Institute of Information Science and Technology "Alessandro Faedo" ISTI of the National Research Council of Italy (in Pisa) for the work entitled *Static- and fabrication-aware segmented concrete shells made of post-tensioned precast flat tiles*.

Due to the high number of participants and the high quality of presented contributions, the Nervi prize Commission also assigned six ex-aequo mentions of honor to (in alphabetic order) Gloria Rita Argento, for the work entitles *Isogeometric refinement for shape optimization with a tunable number of variables*, Fabio Bazzucchi for presenting *Snap n roll tuning and listening to the progressive buckling of reticulated ensembles*, Romane Boutillier for the contribution regarding *Topology generation of architectural meshes adapted to the support conditions*, Claudia Chianese, who presented *Form finding of membrane shells via isogeometric analysis*, Marta Lembo who talked about *Nervi's isostatic lines inspired floor slabs. Beyond the archetypal Gatti Woolen Mill in Rome*, and to Jonathan Melchiorre, who showed *Gridshell Multi-step Structural Optimization with Improved Multi-body Rope Approach and multi-objective genetic algorithm*.

In this book of proceedings, we are going to present the contributions to the workshop in the form of short papers illustrating the topics discussed during the IWSS 2023. To give higher prominence to the Nervi Prize recognition, we first report awarded contributions, while all others are included in order of submission.

The organizers of the IWSS 2023 wish to thank the institutions that supported the IWSS 2023, such as Regione Piemonte and the Ordine degli Ingegneri della Provincia di Torino. In addition, the workshop was endorsed by the Società Italiana di Scienza delle Costruzioni (SISCo - Italian Society of Solids and Structural Mechanics) and by the International Association of Shells and Spatial Structures (IASS). Finally, the financial contributions received from the IWSS 2023 sponsors listed below are gratefully acknowledged.

The IWSS 2023 Organizers

IWSS SHELL WE SHAPE 4S

July 2023

Amedeo Manuello Bertetto
Stefano Gabriele
Francesco Marmo
Andrea Micheletti

Conference Organization

Organizers

Amedeo Manuello Bertetto	Politecnico di Torino, Italy
Stefano Gabriele	Università degli Studi Roma Tre, Italy
Francesco Marmo	Università degli Studi di Napoli Federico II, Italy
Andrea Micheletti	Università degli Studi di Roma "Tor Vergata", Italy

Local Organizing Committee

Amedeo Manuello Bertetto	Politecnico di Torino, Italy
Giuseppe C. Marano	Politecnico di Torino, Italy
Federico Accornero	Politecnico di Torino, Italy
Valerio De Biagi	Politecnico di Torino, Italy
Erica Lenticchia	Politecnico di Torino, Italy
Fabio Bazzucchi	Politecnico di Torino, Italy
Gianfranco Piana	Politecnico di Torino, Italy
Laura Sardone	Politecnico di Torino, Italy
Costanza Anerdi	Politecnico di Torino, Italy
Marco Alforno	Politecnico di Torino, Italy
Giulia Marasco	Politecnico di Torino, Italy
Fiammetta Venuti	Politecnico di Torino, Italy
Jonathan Melchiorre	Politecnico di Torino, Italy
Marco Rosso	Politecnico di Torino, Italy
Riccardo Stevanato	Politecnico di Torino, Italy
Raffaele Cucuzza	Politecnico di Torino, Italy
Carla Serra	Politecnico di Torino, Italy
Gabriele Bertagnoli	Politecnico di Torino, Italy
Devid Falliano	Politecnico di Torino, Italy
Daniel David Suarez Riera	Politecnico di Torino, Italy
Renato Cafarelli	Politecnico di Torino, Italy

Secretariat and Marketing Communication

Yulia Shevchenko	Aristè, Torino, Italy

Scientific Committee

Sigrid Adriaensenss	Princeton University, USA
Marco Amabili	McGill University, Canada
Bruno Briseghella	Fuzhou University, China
Alberto Carpinteri	Politecnico di Torino, Italy
Bernardino Chiaia	Politecnico di Torino, Italy
Rosario Ceravolo	Politecnico di Torino, Italy
Pier Luigi D'Acunto	ETH Zurich, Switzerland
Giuseppe A. Ferro	Politecnico di Torino, Italy
Stefano Gabriele	Università degli Studi Roma Tre, Italy
Simon Guest	University of Cambridge, UK
Stefano Invernizzi	Politecnico di Torino, Italy
Ornella Iuorio	Leeds University, UK
Giuseppe Lacidogna	Politecnico di Torino, Italy
Nikos D. Lagaros	National Tech. Univ. of Athens, Greece
Carlos Lázaro	Technical University of Valencia, Spain
Amedeo Manuello Bertetto	Politecnico di Torino, Italy
Giuseppe C. Marano	Politecnico di Torino, Italy
Enzo Marino	Università degli Studi di Firenze, Italy
Francesco Marmo	Università degli Studi di Napoli Federico II, Italy
Marisela Mendoza Ramos	Nottingham Trent University, UK
Andrea Micheletti	Università degli Studi di Roma "Tor Vergata", Italy
Makoto Ohsaki	Kyoto University, Japan
Carlo Olivieri	Università degli studi di Salerno, Italy
Vittorio Paris	Università di Bergamo, Italy
Francesco Romeo	Università degli Studi di Roma "La Sapienza", Italy
Giuseppe Ruta	Università degli Studi di Roma "La Sapienza", Italy
Antonia Teresa Spanò	Politecnico di Torino, Italy
Edward Segal	Hofstra University, USA
Salvatore Sessa	Università degli Studi di Napoli Federico II, Italy
Francesco Tornabene	Università del Salento, Italy
Valerio Varano	Università degli Studi Roma Tre, Italy
Alessandra Zanelli	Politecnico di Milano, Italy

Honorary Committee

Guido Saracco	Dean of Politecnico di Torino, Italy
Carlos Lázaro	President of the Int. Association for Shell and Spatial Structures (IASS)
Bernardino Chiaia	President and CEO of InfraTo, Torino, Italy
Marco Nervi	President of PLN Project Foundation, Italy
Luciano Rosati	President of Società Italiana di Scienza delle Costruzioni (SISCo), Italy
Giuseppe A. Ferro	President of Ordine degli Ingegneri della Prov. di Torino (OIPT), Italy

Endorsements

international association for shell and spatial structures

Società Italiana di Scienza delle Costruzioni

PLN PROJECT

ADDOPTML Optimized 3D printed structures

CAMERA DI COMMERCIO INDUSTRIA ARTIGIANATO E AGRICOLTURA DI TORINO

REGIONE PIEMONTE

ORDINE DEGLI INGEGNERI DELLA PROVINCIA DI TORINO

Sponsors

SOFiSTiK

MARTINO Steel Engineering

Softing

LUNITEK

MAFFEIS engineering

LGA ENGINEERING investing in innovations

- H A N G A R -

ThalesAlenia Space a Thales / Leonardo company

MASERA ENGINEERING GROUP

Aristè marketing • branding

TORO&MORO CREA STAMPA INDOSSA

Contents

Static- And Fabrication-Aware Segmented Concrete Shells Made of Post-tensioned Precast Flat Tiles

Francesco Laccone[1(✉)], Sandro Menicagli[2], Paolo Cignoni[1], and Luigi Malomo[1]

[1] ISTI - CNR, via G. Moruzzi 1, Pisa, Italy
francesco.laccone@isti.cnr.it
[2] University of Pisa, Lungarno Pacinotti 43, Pisa, Italy

Abstract. This paper introduces a novel structural concept for freeform shells, in which the shape is decomposed into flat tiles to be assembled sequentially with the help of falseworks. Once the structure is completed, the tiles are post-tensioned to minimize the tension forces and avoid detachment. The entire design process, from an input shape to fabrication, is managed by an automatic pipeline. The input shape is segmented into a field-aligned quad mesh, computed from the principal stress of the thin shell. The flat tiles are obtained by extruding each face along the normal of the best-fitting plane per face. The contact between adjacent tiles is ensured only at their edge midpoints so the forces can mainly flow along the cross directions. The best configuration of cable paths and pre-loads is found by solving a constrained optimization problem exploiting a reduced beam model of the shell. All tiles can be prefabricated in the shop with an adaptable and reusable molding system. Once the structure is completed, the top surface is finally completed with an in situ cast that fills the gaps and activates the entire shell behavior. In contrast, the bottom surface maintains its jagged aesthetics.

Keywords: Concrete · Freeform surfaces · Computational design · Optimization · Digital fabrication

1 Concrete Shells: Fabrication and Design

The design of freeform concrete shells has regained popularity in recent times thanks to computational tools, which allow manipulating and exploring complex shapes interactively. Also, novel digital fabrication techniques shorten the distance between digital shape and fabricated structure [5,8]. However, in most cases, the actual fabrication of concrete shells, even on a small scale, still poses challenges of feasibility and cost.

The success of concrete shells lies in their ability to develop membrane behavior through their shape and the possibility of the fluid material being cast in any required geometry efficiently. Forming appropriate 3D shells requires pre-formed

S. Gabriele et al. (Eds.): IWSS 2023, LNCE 437, pp. 1–10, 2024.
https://doi.org/10.1007/978-3-031-44328-2_1

molds. Rigid continuous and dense formworks have made large and accurate concrete shells feasible since ancient times, from simple barrel vaults to form-found shapes. While in the past formworks were realized by skilled craftsmen that cut and bent wood or metal panels, additive and subtractive fabrication nowadays supports their realization as custom single-use pieces to be assembled. However, these are labor-intensive and costly technology, mainly because of the limited reusing degree. The use of dense formworks has gradually decreased in favor of simpler and faster construction processes. An alternative concept is the stay-in-place formwork, which provides knitted, wire mesh, bricks or other flexible/solid patches forming the mold and the surface finishing of the completed shell. In all these cases, concrete is cast or sprayed on site and the resulting structure is monolithic with few or no joints.

On the other hand, advances in manufacturing and analysis have provided new concepts based on segmentation into panels or tiles, which may be flat, single or double-curved. The molds for these tiles are fabricable with numerically controlled machines out of different materials. However, a single mold produces a unique element leading to a low prefabrication rate, especially in the case of variable curvature and unavoidable waste production. Flexible formworks are introduced as a viable alternative process to adapt their shape and form a variety of molds [1]. Segmented shells made of precast parts need to be assembled on site with the inherent formation of joints, which have to be stiff and resistant to not disturb the development of shell action.

Fig. 1. Conceptual view with technical details of our concrete shell: the shape is segmented into flat tiles that are jointed at the midpoint of the edges and post-tensioned; the outer surface is finished with wire patches and in situ casting.

In this paper, we combine the advantages of segmentation for prefabrication in the shop with the continuity provided by a final on-site casting as in Fig. 1. Such an approach was adopted in bold structural architectures from the past, such as the ribbed shell of Palazzetto dello Sport in Rome by Nervi. In the Nervi building system, curved precast ferrocement panels supported by falsework are finally finished with in situ concrete. Flat prismatic tiles are the best manufacturing option because they can be produced off-site with a master mold to be reused several times. High efficiency, precision and economy can be achieved. Moreover, the spreading of digital fabrication has made design variations for mass customization possible and cheap. However, freeform shapes can not be naturally split into flat tiles [2], and any discretization introduces weak lines that may alter the load path [6]. Therefore, we introduce a method for decomposing freeform shells into prismatic tiles, which are aligned to the principal stress directions for structural efficiency and flat for easy fabrication. These tiles share edge midpoint contact and are post-tensioned to favor a compression-only behavior. We develop an automatic pipeline that starts from a generic freeform shape and delivers tiles to the factory. We optimize the set of cable and their pre-load. We address all aspects of design and fabrication. Eventually the shell is completed with on-site casting.

2 Concept and Geometry of Segmented Shells

We propose a novel structural concept for freeform shells, in which the shape is decomposed into flat quadrilateral tiles to form right prisms (with sides and faces forming 90° angles), touching each other at the midpoint of their edges. The segmented shell is efficient as long as all tiles remain in touch, i.e. the shell is compressed. So, once assembled, the tiles are post-tensioned to minimize the resulting tension on the structure under service load. The outer surface is finally completed with an in situ cast that fills the gaps and activates the shell behavior. The bottom surface presents a jagged aesthetic due to gaps and misalignment at the seams.

Our pipeline to deliver the tiles is shown in Fig. 2. We convert the input shape (Fig. 2a) into a thin shell. Then, we run a static analysis under uniformly-distributed load, whose principal stress lines (Fig. 2b) yield a cross-field used

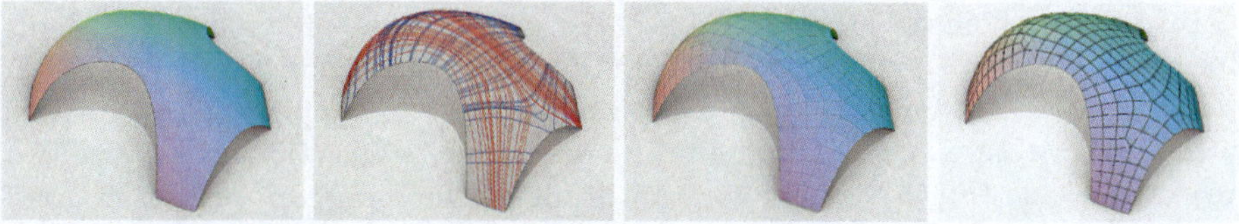

(a) (b) (c) (d)

Fig. 2. Pipeline: (a) input shape; (b) stress isolines from the continuum thin shell; (c) field-driven quad mesh; (d) disjointed flat tiles.

for producing a quad mesh (Fig. 2c). The obtained quad faces can have any curvature, however, we project each face on its best-fitting plane, obtaining a disjointed mesh of planar quad faces (Fig. 2d). The latter is very close to the checkerboard mesh associated with the starting quad mesh. The checkerboard mesh is a derivation of the Varignon theorem: a polygon (i.e., Varignon polygon) formed by the edge midpoint subdivision of a control face (quad in our case), which is planar whether the starting face is planar or not [7]. The resulting checkerboard mesh has 'black' planar rhomboid faces and arbitrary nonplanar 'white' faces. The closest points of two adjacent tiles are their edge midpoints as in Fig. 3a, which will be equipped with joints. Therefore, inter-tile forces can only flow through their edges' midpoint. Since the tiles are aligned with the principal stress directions, each tile is stressed mainly along its cross directions.

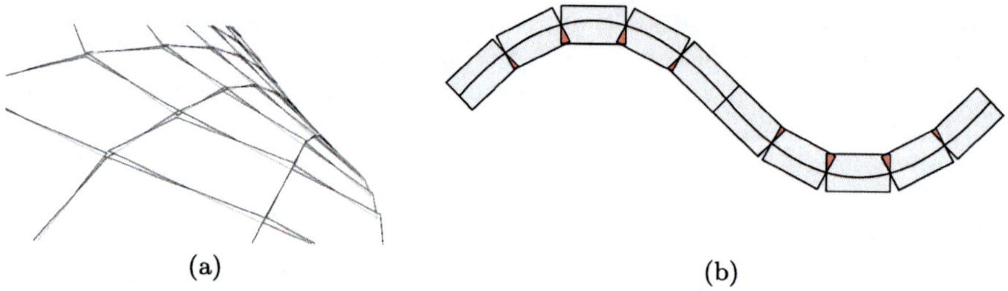

Fig. 3. (a) Flattening individual faces of a quad mesh on the best fitting plane through the vertices, obtaining a disjointed quad mesh; (b) the extrusion along the flat face normal produces overlaps and gaps among the tiles.

Practically, the tiles have a proper thickness, and the extrusion of the disjointed flat faces along their normals produces overlaps and gaps (Fig. 3b), which are a function of curvature and size of the tile (both edge size and thickness). While gaps are acceptable to a certain extent, a material intersections are not feasible. We start from the simple idea of having gaps only and extruding the planar face, which would have minimized the gaps. Working in the discrete prisms setup (as in Fig. 3b) would have required solving a complex iterative Boolean intersection.

Instead, we compute the mesh offset for both shell sides (yellow and light blue meshes in Fig. 4), which produces, in the general case, non-parallel and non-planar faces. Then, for each face of both offset meshes, we project the vertices on the best-fitting plane and compute the maximum intersection area of the two projected quads (Fig. 4). This operation leads to the minimum projected quad in the more straightforward case, i.e., in convex or concave areas. In contrast, in the general case, the intersection area can have multiple vertices (red polygon in Fig. 4) To conservatively restore the initial vertex number, vertices with incident segments closest to colinear are culled (green polygon). This straightforward approach remarkably preserves the number of vertices per face and is dependent on the selected thickness of the shell: the smaller, the closer the tiles.

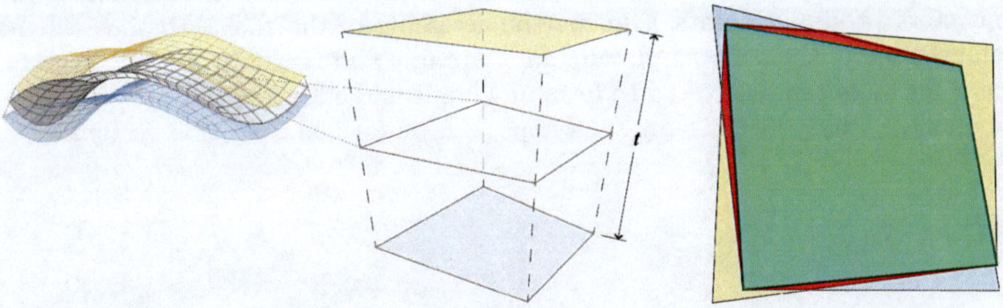

Fig. 4. Removing overlaps: the offset meshes are projected on a flat tile (extrusion thickness t is exaggerated for visualization purposes only).

Cables lie by design on the mid-surface of the shell to exert compression globally once post-tensioned. We build a large and distributed set of candidate cables, from which choosing the optimal subset within an optimization problem (Sect. 3). Candidate paths are found by chaining the segments linking pairs of opposite edge midpoints on a tile (Fig. 5a). We discard paths that do not end on the boundary (closed loops) or paths with significant kinks ($\alpha <$ *threshold* in Fig. 5a), to avoid localized shear on the surface (Fig. 5).

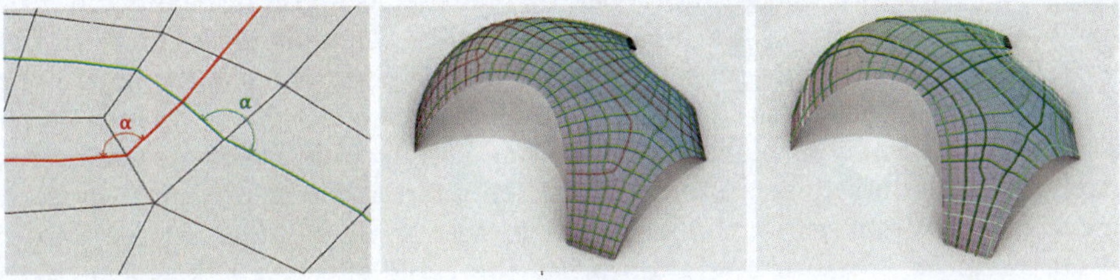

Fig. 5. Selection of cable paths: (a) closeup on candidate cable path (green) and discarded path (red); (b) all candidate and discarded paths; (c) selected cables after the optimization as per Eq. (1) (the darker, the higher the pre-load).

3 Structural Design, Modeling and Optimization

In the disjointed mesh, by design, the adjacent faces come (almost) in touch at the midpoints of each tile edge. These points are affected by a delicate structural task of transferring forces from tile to tile. C-profile steel segments are incorporated during the tile casting to avoid any possible local failure due to contacts and localized loads (Fig. 6a). These steel components serve two other functions: (a) restraining the ducts during the tiles' fabrication and (b) accommodating the connectors (bolts and plates) during the assembly of the shell. All tiles can be prefabricated in the shop employing an adaptable and reusable molding system. The mold walls can be altered to form any angle and edge length (Fig. 6b). Also, singular tiles, like triangles and polygons, can be formed with this system.

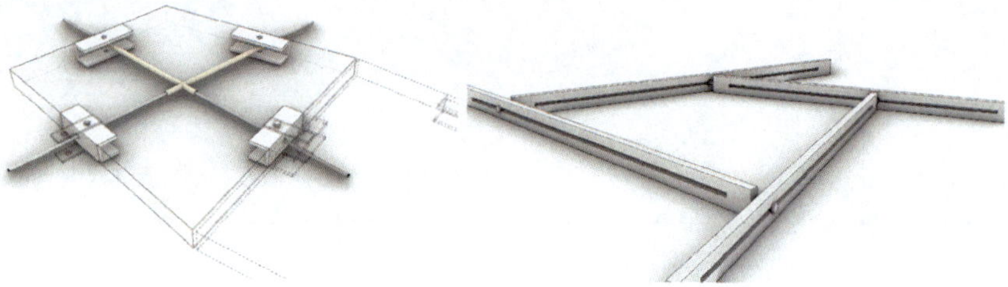

(a) (b)

Fig. 6. (a) Detailed design of the prismatic tile with embedded cables' ducts and steel C-profile segments at the contact interfaces. (b) Adaptable molding system.

Since the mesh faces are aligned with the principal stress directions, and the adopted detail enforces punctual force transfer, the tiles show a predominant strut-and-tie behavior along the two crossing directions (the tile shear is negligible). Therefore, the structure can be modeled as a grid of beams for optimization purposes (Fig. 7). Each cable is a polyline changing direction from tile to tile, overlapping a subset of the beams. It pre-loads the shell (namely the grid of beams) primarily at the kinks. We model the effect of the j-th cable on the structure as an equivalent load F_j acting at the cable kinks. Assuming for simplicity to be in a linear setup, the beams' internal forces produced by each F_j can be superimposed on the results of a linear analysis under service load.

Furthermore, because of the linearity, the cable effect can be scaled by a factor, i.e., its pre-load p_j. Similarly to [4], we select appropriate cables and their accompanying pre-load by solving a constrained minimization problem:

$$min(\sum f_t(p_j)) \tag{1}$$

$$s.t.\ \ 0 \leq p_j \leq p_{max} \tag{2}$$

In which $\sum f_t$ is the sum among all beams of tension forces only. All configurations (service load scenario and individual cables' equivalent-load scenarios) are pre-computed and linearly superimposed to find the minimum of the objective function. The equilibrium of the structure holds as a superimposition of multiple equilibrium conditions. Equation (2) defines a limitation on the allowable pre-stress on each cable.

4 Results and Discussion

We tested our pipeline on a quasi-membrane shape with large openings and non-trivial structural behavior. The service load adopted in the optimization is the sum of the structural weight and $\gamma_q q_k = 1.5 \times 0.5$ kN/m^2. The shape has a bounding box of $15.0 \times 13.0 \times 5.0$ m for a total projected area of 110 m^2, 269 prismatic faces of average edge length of 0.70 m and t = 0.06 m thickness. Our reduced model adopts beams cross sections equal to the web of the steel C-profile (0.06×0.18 m) and C28/35 concrete as material. The reduced model analysis at the starting condition (with no cables, Fig. 7a) confirms the effectiveness of the meshing and model reduction strategy, showing low bending moments and shear forces.

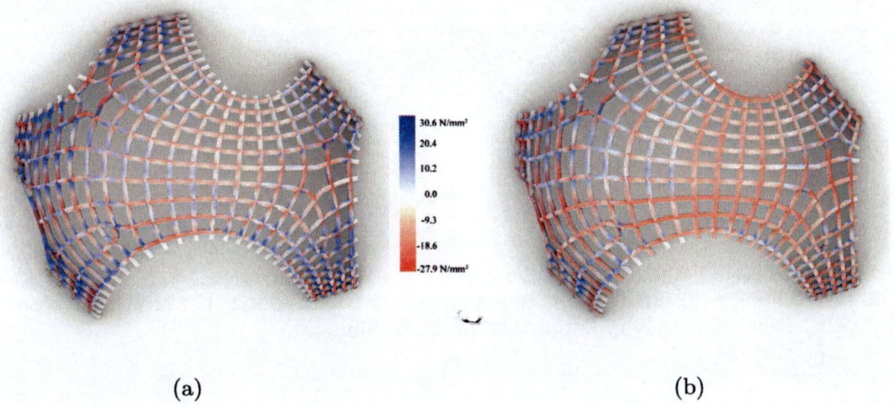

Fig. 7. Stress in the reduced model linear analysis: (a) starting shell under service load; (b) shell deploying optimized cables.

The entire cable set comprises 44 cables. 32 of them are selected in the optimization (see Fig. 5c). We adopt a 8 mm diameter cables for an allowable pre-stress p_{max} = 30 kN. The optimization problem of Eqs. (1) and (2) is solved using two algorithms from the NLopt library [3] in sequence: CRS2, for global evolutionary search, and BOBYQA, a derivative-free local refinement. The optimization leads to an significant reduction of the tension on the elements (Fig. 7b) and an overall improvement of the shell statics (Table. 1).

The pipeline results in a labeled set of prismatic tiles and cable segments, which are easily fabricable. The assembly (Fig. 8) requires falseworks to fulfill the stability of the tiles once they are sequentially moved to their target position. The shape curvature and the anchoring of connectors between the tiles gradually favor the formation of stable patches. Similarly, the connectors can transfer both tension and compression once the assembly is completed. The cables can be pushed into the ducts and post-tensioned from the boundary up to the desired

Table 1. Analysis results from the reduced model before (start) and after the optimization (opt).

	Unit	Value
Max displacement (start)	m	0.02
Max compression force f_c (start)	kN	35.30
Max tension force f_t (start)	kN	8.06
Sum of tension forces $\sum f_t$ (start)	kN	**582.54**
Max displacement (opt)	m	0.02
Max compression force f_c (opt)	kN	82.96
Max tension force f_t (opt)	kN	6.86
Sum of tension forces $\sum f_t$ (opt)	kN	**78.07**

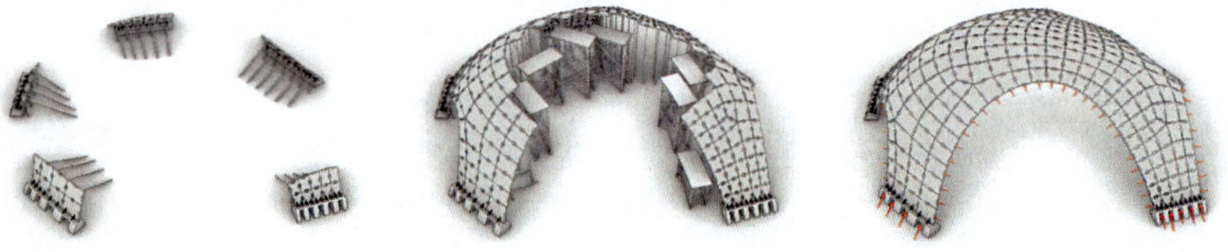

(a) (b) (c)

Fig. 8. Assembly: (a) laying the tiles starting from the supports; (b) sequential assembly with falseworks; (c) completion and post-tensioning.

pre-load (Fig. 8c). We tested the feasibility of the assembly routine on a scale model made of 3D-printed flat tiles as in Fig. 9.

To add more redundancy and to fill the gaps between the tiles while providing a finishing of the structure, an in situ cast can be arranged on the outer surface of the structure. This cast embeds a fiberglass or steel wire mesh. As a last step of our pipeline, we found a clustering method for the wire mesh (Fig. 10). The target double-curved surface is decomposed into developable flat patches having specific shapes and cuts so that these patches can be laid on the shell without significant distortions or wrinkles. As an additional fabrication constraint, we also require the patches to fit into the size of commercially-available sheets of material. Simultaneously, the ducts can be grouted so that the cables can be effectively coupled with concrete and prevent steel corrosion.

5 Conclusions

In this paper, we presented a design method that is an alternative to traditional concrete shells in both mechanics and fabrication and is expected to offer cost savings. We introduced a new structural concept and an automated pipeline to

Fig. 9. Assembly steps of a 3D printed demonstrator.eps

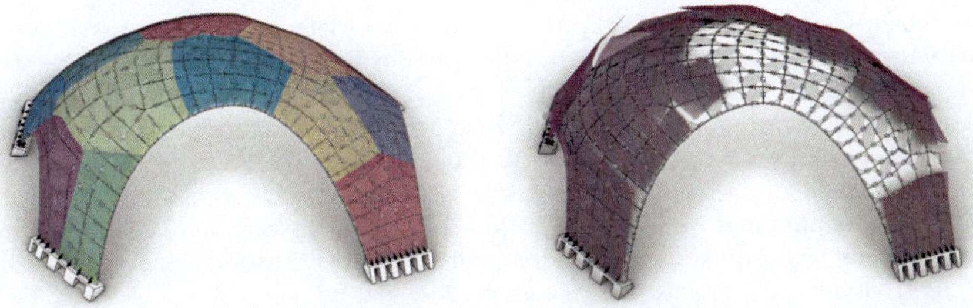

(a) (b)

Fig. 10. Minimum distortion patches of wire mesh to lay on the top face of the shell: (a) developable patches; (b) flat sheets of material.

deliver segmented shells made of prismatic tiles, which are easily fabricable and can be handily assemble to form a proper structure. To guarantee discrete tiles to develop a suitable load-bearing scheme, we adopt a meshing aligned to the principal stress directions, and to avoid detachment, we post-tension the tiles utilizing optimal embedded cables.

The method works for complex shapes and can handle several design variables to tune the stiffness and strength of the final solution. However, the efficiency of this system is higher in areas of double curvature where the geometry of the shape and the post-tensioning load provide the best contribution. Several topics deserve further research efforts. On the structural side, for instance, several redundancy layers are designed (post-tensioning, connectors, completion cast). However, their contribution to the statics and stability is to be investigated. We reasonably believe that the structural model adopted, as a grid of beams in accordance with the meshing and load transfer mechanism, could represent a lower-bound solution. Actually the structural response must be evaluated with more advanced simulations and testing. On the practical side, the essential topic is investigating fabrication tolerances and the effect of the size of the gaps between the tiles. Eventually, production and physical experiments could provide relevant insight to improve the design method.

References

1. Hawkins, W.J., et al.: Flexible formwork technologies - a state of the art review. Struct. Concr. **17**(6), 911–935 (2016). https://doi.org/10.1002/suco.201600117
2. Jiang, C., Wang, C., Tellier, X., Wallner, J., Pottmann, H.: Planar panels and planar supporting beams in architectural structures. ACM Trans. Graph. (2022). https://doi.org/10.1145/3561050
3. Johnson, S.G.: The NLopt nonlinear-optimization package (2021) https://github.com/stevengj/nlopt
4. Laccone, F., Malomo, L., Froli, M., Cignoni, P., Pietroni, N.: Automatic design of cable-tensioned glass shells. Comput. Graph. Forum **39**(1), 260–273 (2020). https://doi.org/10.1111/cgf.13801
5. Li, W., Lin, X., Bao, D.W., Xie, Y.M.: A review of formwork systems for modern concrete construction. Struct. **38**, 52–63 (2022). https://doi.org/10.1016/j.istruc.2022.01.089
6. Oval, R., Rippmann, M., Mesnil, R., Van Mele, T., Baverel, O., Block, P.: Feature-based topology finding of patterns for shell structures. Autom. Constr. **103**, 185–201 (2019). https://doi.org/10.1016/j.autcon.2019.02.008
7. Peng, C.-H., Jiang, C., Wonka, P., Pottmann, H.: Checkerboard patterns with black rectangles. ACM Trans. Graph. **38**(6), 1–13 (2019). https://doi.org/10.1145/3355089.3356514
8. Wangler, T., et al.: Digital concrete: opportunities and challenges. RILEM Tech. Lett. **1**, 67–75 (2016)

Isogeometric Refinement for Shape Optimization with a Tunable Number of Variables

Gloria Rita Argento[1](✉), Enzo Marino[2], Valerio Varano[1], and Stefano Gabriele[1]

[1] Department of Architecture, Roma Tre University, Rome, Italy
gloriarita.argento@uniroma3.it
[2] Department of Civil and Environmental Engineering, University of Florence, Florence, Italy

Abstract. The classical funicularity concept for shell structures has been extended defining the Relaxed Funicularity (R-Funicularity). A parameter called generalized eccentricity has been used for this purpose, following that a shell is R-Funicular when the generalized eccentricity doesn't exceed an admissibility limit. A shells' shape optimization process aiming at finding R-Funicular analytical shells is here modified at the geometry definition level by describing the shells' shape with spline surfaces and using the position of the control polygon vertices as optimization variables. An isogeometric (IG) refinement is applied in order to improve the local control of the shape in the optimization process. An advantage of this approach is that, since it introduces new vertices in the control polygon, the number of optimization variables becomes tunable. Namely, starting with the lowest number used to generate the initial geometry, the additional vertices can possibly be entered as new variables whenever a more accurate local control of the surface is needed. We present significant numerical examples.

Keywords: Shape optimization · NURBS · Generalized eccentricity · R-Funicularity · Isogeometric refinement

1 Introduction

Funicular shells are shaped to bear loads without introducing bending. They exploit the entire cross section, minimizing the material waste. For these reasons, funicular shells are an example of efficiency and many design approaches have been developed in order to find the "right" shape in accordance with the applied load [1].

Nevertheless, when bending stiffness is not negligible and/or the boundary conditions are not properly assigned, a funicular behaviour is not possible and bending arises. Accounting for this occurrence, the actual funicular behaviour of shells can be quantified by means of a method based on Relaxed Funicularity (R-Funicularity) [2]; the parameter used to quantify the funicularity is the generalized eccentricity. The amount of bending moments exploited by a shell to resist loads can be reduced by means of shape optimization, i.e. an inverse problem properly formulated to minimize an objective function somehow measuring the shell's bending moments. The minimization is pursued by

S. Gabriele et al. (Eds.): IWSS 2023, LNCE 437, pp. 11–21, 2024.
https://doi.org/10.1007/978-3-031-44328-2_2

updating some shell geometric features, i.e. variables, admitted to vary within a range. The goal of bending moments reduction is usually approached by minimizing the strain energy [3–5]; other objectives of shape optimization can be, for example, the minimization of the weight, the stress levelling [3, 4] or the minimization of displacements [6].

In [7], a shape optimization process aimed to find R-Funicular shells by minimizing the generalized eccentricity extrema has been developed. Geometry and mechanical behaviour of shells are strictly linked, thus the description of the first strongly influences the solution quality and the computational cost. Often, see for example [5, 6], when an R-Funicular shell is sought, the shape is described by means of analytical functions [7–9].

In this work, the geometry definition shown in [8] is modified by introducing B-spline basis functions and associated control points [10] to describe the shell mid-surface. Compared to existing works in which this approach was used (e.g., [5, 11]), here an isogeometric (IG) refinement is applied in order to improve the local control of the shape in the optimization process. An advantage of this approach is that, since it introduces new vertices in the control polygon, the number of optimization variables becomes tunable. This feature is exploited by performing a shape optimization where, starting with the lowest number of control points used to generate the initial geometry, additional vertices can be entered as new variables whenever a more accurate local control of the surface is needed. A similar approach has been employed in [12], where, in order to expand the feasible design space, a refinement of the initial geometry has been performed.

After a brief description of the concepts of R-Funicularity and generalized eccentricity $e(\alpha)$, we discuss the optimization strategy with a major focus on the geometry description. Two effective numerical examples are then proposed and the conclusions are drawn.

2 Generalized Eccentricity and R-Funicularity

The generalized eccentricity measure and its use to define the R-Funicularity concept are briefly introduced in this section. For a complete discussion, please refer to [2].

Let us consider a point P belonging to the surface S, endowed with its orthonormal basis $\boldsymbol{a}_1, \boldsymbol{a}_2, \boldsymbol{a}_3$, and a generic direction of the tangent plane represented by the unit vector $\boldsymbol{u} = (\cos\theta, \sin\theta)$ (Fig. 1).

The local tensors $\boldsymbol{N}$ and $\boldsymbol{M}$, respectively associated with the membrane forces and the bending moments, can be evaluated in each point P:

$$\boldsymbol{N} = \begin{pmatrix} N_{11} & N_{12} \\ N_{12} & N_{22} \end{pmatrix}, \boldsymbol{M} = \begin{pmatrix} M_{11} & M_{12} \\ M_{12} & M_{22} \end{pmatrix} \tag{1a,b}$$

Then, $N(\theta) = \boldsymbol{u}^T \boldsymbol{N} \boldsymbol{u}$ and $M(\theta) = \boldsymbol{u}^T \boldsymbol{M} \boldsymbol{u}$ are, respectively, the generalized membrane force acting along $\boldsymbol{u}$ and the generalized bending moment acting in the plane $(\boldsymbol{u}, \boldsymbol{a}_3)$. Using the multiple-angle and the power-reduction trigonometric formulae and setting $\alpha = 2\theta$, it has been possible to write the following explicit form for Eq. $N(\theta)$ and $M(\theta)$:

$$\begin{aligned} N(\alpha) &= \overline{N} + \hat{N} cos\alpha + N_{12} sin\alpha \\ M(\alpha) &= \overline{M} + \hat{M} cos\alpha + M_{12} sin\alpha \end{aligned} \tag{2}$$

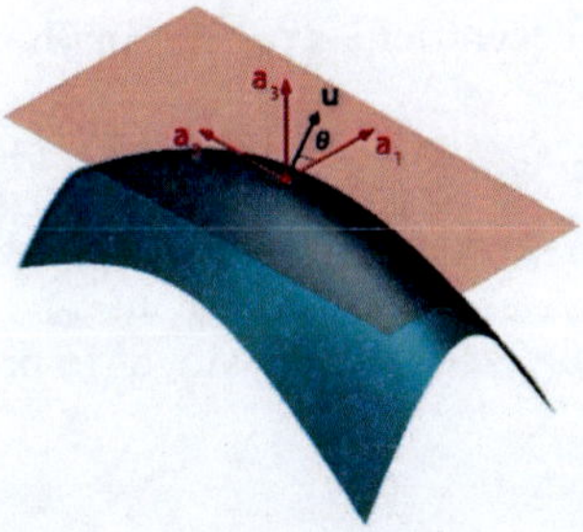

Fig. 1. Local orthonormal basis on P belonging to the surface S.

where $\overline{N} = \frac{N_{11}+N_{22}}{2}, \widehat{N} = \frac{N_{11}-N_{22}}{2}, \overline{M} = \frac{M_{11}+M_{22}}{2}, \widehat{M} = \frac{M_{11}-M_{22}}{2}$. Then, the generalized eccentricity is defined as the ratio:

$$e(\alpha) = \frac{M(\alpha)}{N(\alpha)} \tag{3}$$

Equation 2 is the equation of a parametric curve in the plane $(N, M) \in \mathbb{R}^2$, that is an ellipse when $\alpha \in [0,\ 2\pi]$, called *ellipse of eccentricity* or *Relaxed Funicularity (RF) ellipse* (Fig. 2).

It is well known that a surface structure is funicular, for a given load, if (4) is verified.

$$M(\alpha) = 0 \forall\alpha \tag{4}$$

The actual behaviour exhibited by shells with non negligible thickness shows that the nullity of bending internal forces is an ideal that never verifies. The aim of R-Funicularity is to define a criteria to consider a structure funicular in a relaxed manner even if Eq. 4 is not respected, when $e(\alpha)$ belongs to an admissible range set as $[-\lambda h,\ \lambda h]$, where h is the thickness of the surface and $\lambda \in \left[0,\ 1/2\right]$ an admissibility coefficient that depends on the applications.

In the light of above, the Relaxed Funicularity (RF) is defined as follow: "a shell is R-Funicular if, $e(\alpha) \in [-\lambda h,\ \lambda h]\ \ \forall\alpha \in [0, 2\pi]$, for each point of the surface". In practice the problem of the R-Funicularity is that of verifying:

$$[min(e(\alpha)), max(e(\alpha))] \subset [-\lambda h,\ \lambda h] \tag{5}$$

For completeness, we show the graphical representation of $e(\alpha)$ in the (N, M) plane. More details about its use are given in [2].

Considered the definitions below, the generalized eccentricity is the Rayleigh quotient: $e(\theta) = \frac{\boldsymbol{u}^T \boldsymbol{M} \boldsymbol{u}}{\boldsymbol{u}^T \boldsymbol{N} \boldsymbol{u}}$, and the maximum and minimum eccentricities (e_{max}, e_{min}) are found by solving the associated eigenvalue problem:

$$(\boldsymbol{M} - e\boldsymbol{N})\boldsymbol{u} = 0 \tag{6}$$

The solution of (6), discussed in [2], depends on the algebraic conditioning of $\boldsymbol{M}$ and $\boldsymbol{N}$ and gives the local extrema e_{max} and e_{min} shown in Eq. (17) of [2].

The global extrema of the eccentricities can be finally evaluated as follows:

$$min(e(\alpha)), max(e(\alpha)) = \begin{cases} -\infty, +\infty \; if \det(N) < 0 \\ e_{min}, e_{max} \; if \det(N) \geq 0 \end{cases} \tag{7}$$

The angles $\varphi_{max} = \text{atan}(e_{max})$ and $\varphi_{min} = \text{atan}(e_{min})$ in Fig. 2 have been used to define the functional to be minimized with the aim of improving the shells' mechanical behaviour.

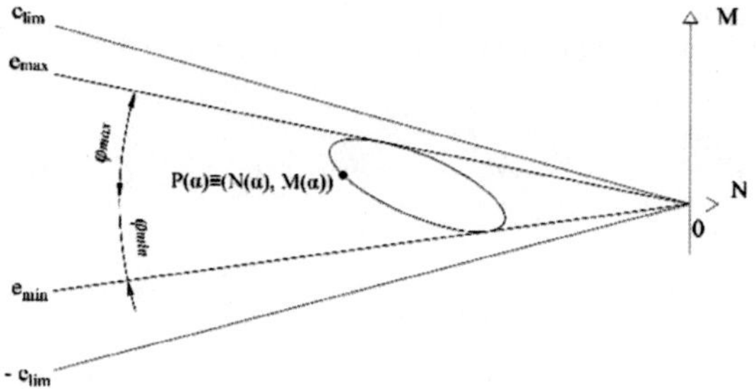

Fig. 2. RF ellipse, eccentricity extrema and limits e_{max} and e_{min}, e_{lim} and $-e_{lim}$, respectively equal to the slope of the upper and lower blue and red lines; eccentricity angles φ_{max}, φ_{min}

3 Shells' Shape Optimization for R-Funicularity

The main goal of the shells' shape optimization process proposed in [7] is to obtain a shape for the shell to be R-Funicular and is pursued by minimizing the generalized eccentricity extrema e_{max} and e_{min} all over the shell. Here the procedure is modified at the geometry definition level. Aiming at a wider design freedom, the shape is now described with B-spline surfaces and an IG refinement is applied to obtain a major local control of the shape; a MATLAB Toolbox [13] has been employed for this purpose.

3.1 Modelling of Geometry and Isogeometric Refinement

The shell geometry has been modelled by means of a non-rational B-splines surface, obtained by fixing a bidirectional net of control points $\mathbf{P}_{i,j}$, two knots vectors $\mathbf{U} = \{0, \ldots, 0, u_{p+1}, \ldots, u_{r-p-1}, 1, \ldots, 1\}$ and $\mathbf{V} = \{0, \ldots, 0, v_{q+1}, \ldots, v_{r-q-1}, 1, \ldots, 1\}$ and the products of the two univariate B-spline functions of degree p and q, $N_{i,p}(u)$ and $N_{j,q}(v)$:

$$\mathbf{S}(u, v) = \sum_{i=0}^{n} \sum_{j=0}^{m} N_{i,p}(u) N_{j,q}(v) \mathbf{P}_{i,j} \tag{8}$$

The vectors $\mathbf{U}$ and $\mathbf{V}$ contain, respectively, $r + 1$ and $s + 1$ knots, being:

$$\begin{aligned} r &= n + p + 1 \\ s &= m + q + 1 \end{aligned} \tag{9a,b}$$

The surface resulting from (8) is then manipulated by performing the isogeometric h-refinement that consists in the well-known algorithm of knots insertion. This type of refinement has multiple uses, among which: evaluating points and derivatives on curves and surfaces, subdividing curves and surfaces, adding control points in order to increase flexibility in shape control [10].

In order to understand the effect of knot insertion, let us consider a B-spline curve $\mathbf{C} = \sum_{i=0}^{n} N_{i,p}(u)\mathbf{P}_i$ defined on $\mathbf{U} = \{\mathrm{u}_0, \ldots, \mathrm{u}_m\}$, whose vector space is $\mathcal{V}_U$. If we insert $\overline{u} \in [u_k, u_{k+1})$ into U we obtain the new vector $\overline{\mathbf{U}} = \{\overline{u}_0 = u_0, \ldots, \overline{u}_k = u_k, \ldots, \overline{u}_{k+1} = \overline{u}, \overline{u}_{k+2} = u_{k+1}, \ldots, \overline{u}_{m+1} = u_m\}$, whose vector space $\mathcal{V}_{\overline{\mathrm{U}}}$ has dimension $\dim(\mathcal{V}_{\overline{\mathrm{U}}}) = \dim(\mathcal{V}_U) + 1$ and, since $\mathcal{V}_U \subset \mathcal{V}_{\overline{\mathrm{U}}}$, it is possible to represent $\mathbf{C}$ on $\overline{\mathbf{U}}$:

$$\mathrm{C(u)} = \sum_{i=0}^{n+1} \overline{N}_{i,p}(u)\mathbf{Q}_i \tag{10}$$

where $\mathbf{Q}_i$ are the new control points that can be obtained by exploiting the following equality:

$$\sum_{i=0}^{n} N_{i,p}(u)\mathbf{P}_i = \sum_{i=0}^{n+1} \overline{N}_{i,p}(u)\mathbf{Q}_i \tag{11}$$

and the property according to which in a given knot span $[u_j, u_{j+1})$, at most $p + 1$ $N_{i,p}$ are nonzero.

Omitting the intermediate mathematical steps, that can be found in [10] together with an exhaustive treatment of the topic, the new control points (Fig. 3) are:

$$\mathbf{Q}_\mathrm{i} = \alpha_i \mathbf{P}_i + (1 - \alpha_i)\mathbf{P}_{i-1} \tag{12}$$

where

$$\alpha_i = \begin{cases} 1 & i \le k - p \\ \frac{\overline{u} - u_i}{u_{i+p} - u_i} & k - p + 1 \le i \le k \\ 0 & i \ge k + 1 \end{cases}$$

and k is an index that locates the inserted knots in the vector $\overline{\mathbf{U}}$.

In this specific case, basis functions of degree 2 have been employed. Initially, a net of 3 × 5 control points and the following knots vectors $\mathbf{U} = \{000111\}$ and $\mathbf{V} = \{0001/32/3111\}$ have been considered. $\mathbf{U}$ and $\mathbf{V}$ contain, respectively 6 and 8 knots, accordingly to the Eqs. (9a,b). The resulting shape and the control net are depicted in Fig. 4.

The positions of the 9 inner control points (Fig. 5) have been updated in the optimization process by varying 4 different coefficients, each multiplying the z-coordinates of the points marked with the same colour in Fig. 5; the black ones are maintained fixed. Thus, each control point admitted to vary is defined as $\boldsymbol{P}_i \equiv (x_i, y_i, z_i c_i)$, where c_i are the optimization variables. The variables of the starting shape, denoted as c_{oi} are equal to 1.

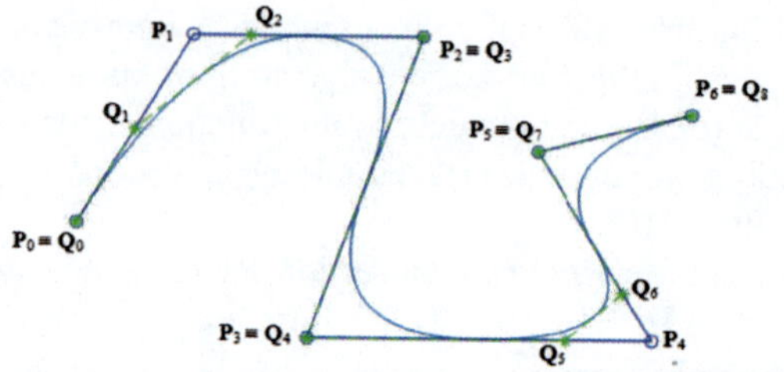

Fig. 3. Effect of knots insertion into a quadratic curve. Blue, solid line: control polygon before knot insertion; green, dashed line: control polygon after inserting $u = 0.1$ and $u = 0.7$ into $\mathbf{U} = \{0000.20.40.60.8111\}$.

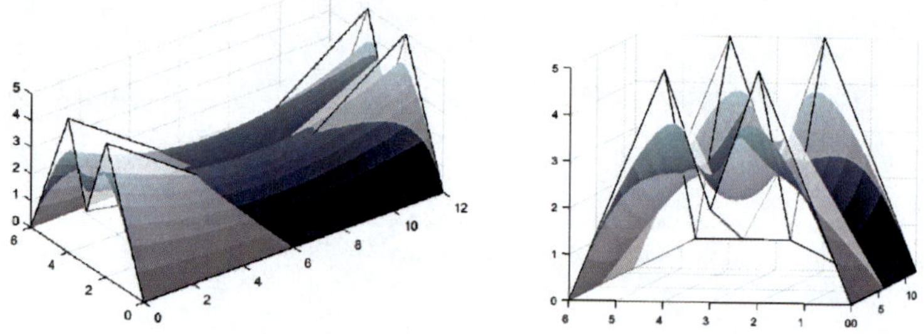

Fig. 4. Starting shape and control net

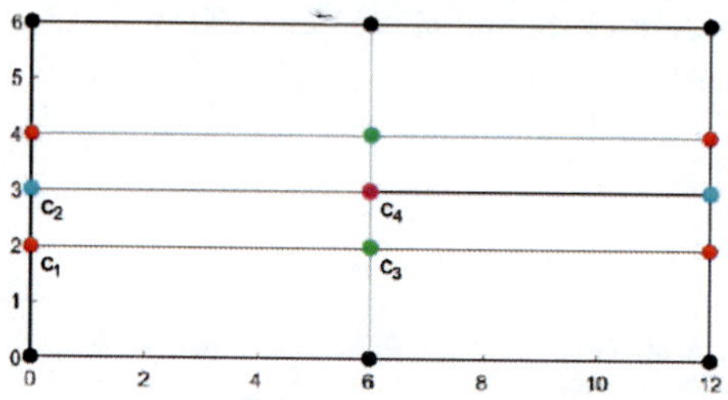

Fig. 5. Plane view of the initial control net. Black dots: control points maintained fixed in the optimization process; Coloured dots: control points updated in the optimization process.

The two knot vectors $\mathbf{U}_1 = \{1/32/3\}$ and $\mathbf{V}_1 = \{1/61/25/6\}$ have been inserted, respectively into $\mathbf{U}$ and $\mathbf{V}$, obtaining the new control net shown in Fig. 6.

Also in this case, the positions of the 30 inner control points (Fig. 7) have been updated in the optimization process by varying 9 different coefficients, each multiplying the z-coordinates of the points marked with the same colour in Fig. 7; the black ones are maintained fixed. The coordinates of the control points are defined analogously to the previous case.

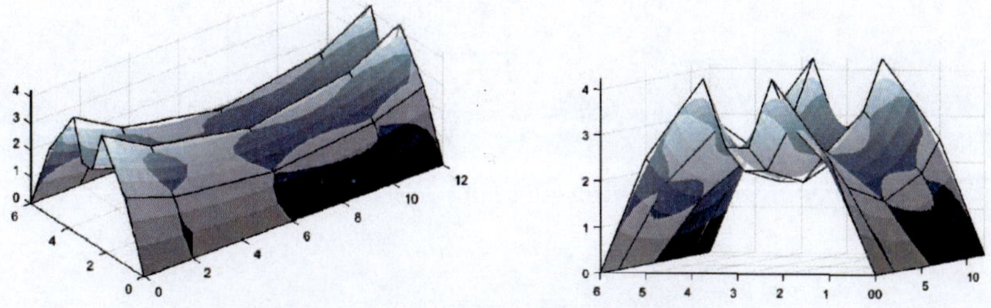

Fig. 6. Starting shape and new control net

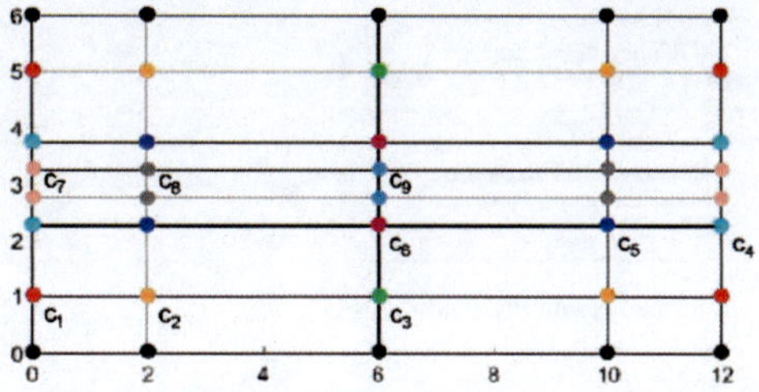

Fig. 7. Plane view of the refined control net. Black dots: control points maintained fixed in the optimization process; Coloured dots: control points updated in the optimization process.

3.2 Optimization Strategy

The shape optimization process is illustrated in Fig. 8. Firstly, a few general definitions should be provided: shell material, eccentricity limits and boundary conditions. The initial geometry is the B-spline surface described in Sect. 3.1 and shown in Fig. 4, properly discretized to build a Finite Element (FE) model. In the following step a linear static analysis of the structure subject to the self-weight load and an in plan uniform vertical load of $5\,\mathrm{kN/m^2}$ is performed and the internal moments and forces are stored for each i-th area element of the model. The minimum and maximum eccentricities of each i-th element are computed by using bending moments normalized with respect to the shell thickness and collected in the vectors $\boldsymbol{e}_{0,min}$, $\boldsymbol{e}_{0,max}$ respectively. In addition, the eigenvalues $\lambda_{1,i}$ and $\lambda_{2,i}$ of the tensor $\boldsymbol{N}_i$ (Eq. 1a) are evaluated and used to check if the ellipse of eccentricity of the i-th element crosses the M axis, that is $\lambda_{1,i}$ and $\lambda_{2,i}$ have different signs. Then, the values of eccentricity extrema $e_{min,\bar{i}} = 10000$ and $e_{max,\bar{i}} = 10000$ are imposed to the $\bar{i}$ areas with discordant eigenvalues. The eccentricities vectors that result from these computations, $\boldsymbol{e}_{min}$, $\boldsymbol{e}_{max}$, are used to evaluate the objective function to be minimized. If convergence is reached the process stops, otherwise the shape is updated by moving the position of the control points P or Q and the procedure is repeated until convergence.

Defining $\varphi_{M,i} = \max(|\varphi_{max,i}|, |\varphi_{min,i}|)$, where $i = 1, \ldots, N$, being N the number of shell elements in the FE model, the objective function is defined as the root mean square of $\varphi_{M,i}$:

$$f_{obj} = \sqrt{\frac{\sum_i \varphi_{M,i}^2}{N}} \tag{13}$$

In Table 1, the optimization settings are summarized.

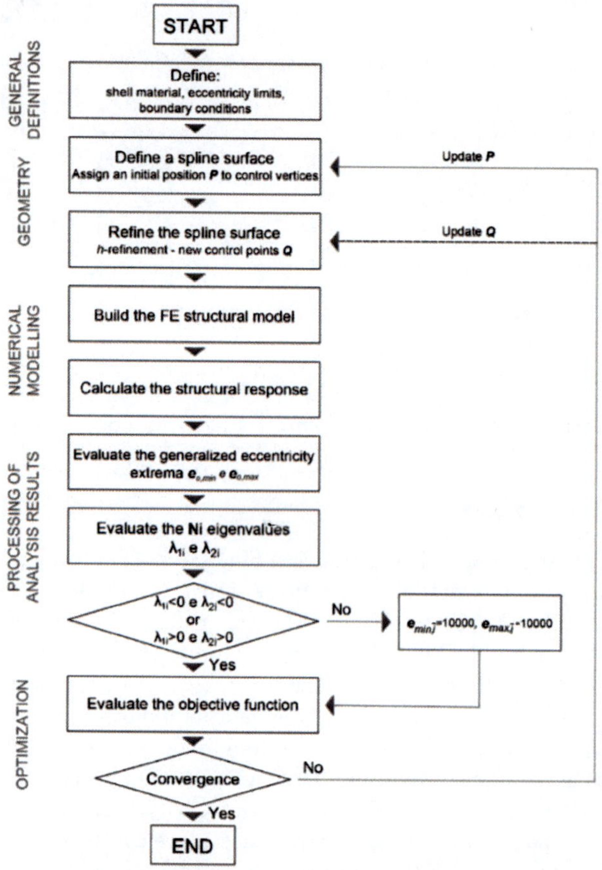

Fig. 8. Overview of the shape optimization process

4 Results and Conclusions

The shape optimization results are expressed in terms of shape variation and eccentricity distribution. As shown in Fig. 9, left, the starting shell is mostly non R-Funicular, being the eccentricity extrema larger than e_{adm} and the value of the objective function $f_{obj} = 1.28$. The image in the centre of Fig. 9 shows the results obtained by means of the first

Table 1. Optimization settings

Parameter	Value/Description
Material and thickness	Concrete: $w = 25\,\text{kN/m}^3; E = 3 \times 10^7\,\text{kN/m}^2$ --$h = 0.1\,\text{m}$
Admissible eccentricity	$e_{adm} = h\lambda$ with $\lambda = 1/6$ or $e_{adm} = \lambda$ when bending moments normalized with respect to h are used (middle third rule)
Starting shape	Non-rational B-spline surface. Control points multipliers:$c_{oi} = 1$
Mesh Boundary conditions Shell formulation	 Quadrangular mesh; pinned x edges; shell-thin element formulation
Objective function	$f_{obj} = \sqrt{\frac{\sum_i \varphi_{M,i}^2}{N}}$
Software	MATLAB (scripts for evaluation of eccentricities, optimization, shape update) SAP2000 (FE solver)
Optimization algorithm	Sequential quadratic programming (SQP): find local minimum of constrained nonlinear multivariable function
Variables constraints	$0.01 \leq c_i \leq 2$

shape optimization, where only 4 variables were updated to modify the control net and, consequently, the shell's shape. The optimization variables, starting from $c_{o1} = c_{o2} = c_{o3} = c_{o4} = 1$, turn $c_1 = 0.1, c_2 = 0.16, c_3 = 0.24, c_4 = 0.12$ after optimization. A notable improvement of the shell funicular behaviour has been obtained, especially in the central areas of the shell, also proved by the objective function that reaches the value $f_{obj} = 0.42$. The right image (Fig. 9) shows the results obtained by performing the shape optimization after knots insertion. Also in this case, the starting variables have been fixed to 1, the parameters after shape optimization are:

$c_1 = 0.59, c_2 = 0.53 c_3 = 0.39 c_4 = 0.57, c_5 = 0.49, c_6 = 0.33, c_7 = 0.91, c_8 = 0.76, c_9 = 0.47$. It can be observed that the increased local control and thus the major number of optimization parameters obtained by knot insertion allowed to obtain a shell mostly R-Funicular. The objective function, starting from 1.28, reduces to 0.32.

According to R-Funicularity, a good shell has been obtained in both examples. Nevertheless, by comparing the shapes and the eccentricity distribution, it can be stated that enriching the initial geometry description by means of IG refinement allowed for a better shape adjustment, also confirmed by the initial and the final f_{obj} in the two proposed cases. The objective function, in fact, providing the root mean square of the eccentricity angles (Fig. 2), gives a summary measure of how much the shell behaviour differs from the R-Funicular one, which is immediately evident by comparing the values assumed by f_{obj} before and after the optimization with $\text{atan}(e_{adm}) = 0.165$.

In the light of above, we believe that this approach could be useful when the accuracy of the shape local control should be modulated according to design and computational demands. Further studies about this approach could be devoted to observing the objective function trend as the number of optimization variables increase, in order to state whether if there is a threshold number over which the function no longer decreases. Also, attention should be paid to the numerical issues that could emerge when a large number of variables are considered.

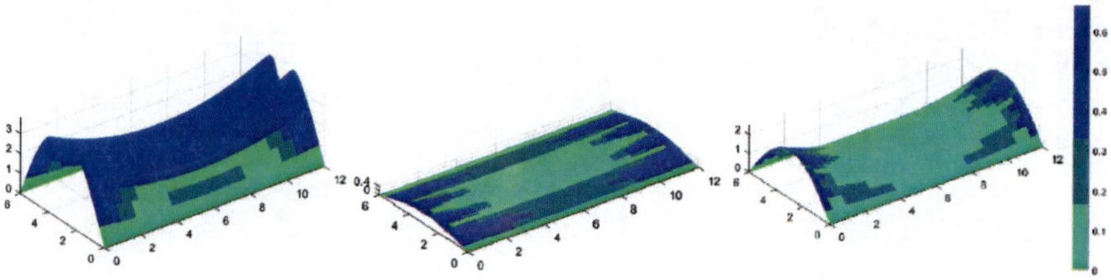

Fig. 9. Shapes and eccentricity distribution. The colorbar represents the values of the $e_{M,i} = max(|e_{max,i}|, |e_{min,i}|)$ that increases going from green to blue, the latter meaning that the eccentricity falls out of the cross section. The shell is R-Funicular in the light green areas. Left: before optimization, $f_{obj} = 1.28$; centre: after optimization with 4 variables, $f_{obj} = 0.42$; right: after optimization with 9 variables, $f_{obj} = 0.32$.

References

1. Adriaenssens, S., Block, P., Veenendaal, D., Williams, C. (eds.) Shell Structures for Architecture: Form Finding and Optimization. Routledge (2014)
2. Gabriele, S., Varano, V., Tomasello, G., Alfonsi, D.: R-Funicularity of form found shell structures. Eng. Struct. **157**, 157–169 (2018)
3. Ramm, E., Bletzinger, K.-U., Reitinger, R.: Shape optimization of shell structures. Revue Européenne des Éléments Finis **2**(3), 377–398 (1993)
4. Tomás, A., Martí, P.: Shape and size optimization of concrete shells. Eng. Struct. **32**, 1650–1658 (2010)
5. Marino, E., Salvatori, L., Orlando, M., Borri, C.: Two shape parametriza-tions for structural optimization of triangular shells. Comput. Struct. **166**, 1–10 (2016)
6. Robles, S.I., Ortega, N.F.: Study of volumetric displacements of shells. J. Int. Assoc. Shell Spat. Struct. **42**(137), 139–147 (2001)
7. Argento, G.R., Marmo, F., Varano, V., Gabriele, S.: Shells' shape optimiza-tion based on R-Funicularity. In: Behnejad, S.A., Parke, G.A.R., Samavati, O.A. (eds.) IASS Annual Symposium 2020/21 and the 7th International Conference on Spatial Structures – Inspiring the Next Generation, pp. 2548–2556 (2021)
8. Argento, G.R., Gabriele, S., Varano, V.: Different uses of the generalized eccentricity for shells' shape optimization: a comparison. In: Xue, S.-D., Wu, J.-Z., Sun, G.-J. (eds.) IASS 2022 Symposium affiliated with APCS 2022 conference – Innovation, Sustainability and Legacy, Beijing, China (2022)

9. Argento, G.R., Gabriele, S., Marmo, F., Varano, V.: Exploring the use of R-Funicularity in shells' shape optimization, In: IWSS2020 – 1st Italian Workshop on Shell and Spatial Structures, Web Meeting (2020)
10. Piegl, L., Tiller, W.: The NURBS Book. Springer, Heidelberg (2012). https://doi.org/10.1007/978-3-642-97385-7
11. Bletzinger, K.-U., Ramm, E.: Form finding of shells by structural optimization. Eng. Comput. **9**, 27–35 (1993). https://doi.org/10.1007/BF01198251
12. Nagy, A.P., IJsselmuiden, T.S., Abdalla, M.M.: Isogeometric design of anisotropic shells: optimal form and material distribution. Comput. Methods Appl. Mech. Eng. **264**, 145–162 (2013). https://doi.org/10.1016/j.cma.2013.05.019
13. Penguian: NURBS Toolbox by D.M. Spink (2022). https://www.mathworks.com/matlabcentral/fileexchange/26390-nurbs-toolbox-by-d-m-spink. MATLAB Central File Exchange, 22 December 2022

Check for updates

Snap'n'Roll: Tuning and Listening to the Progressive Buckling of Reticulated Ensembles

Fabio Bazzucchi[1,2](✉)

[1] Senseable City Lab, Massachusetts Institute of Technology, Cambridge, MA, USA
fbaz@mit.edu

[2] DISEG, Department of Structural Geotechnical and Building Engineering, Politecnico di Torino, Corso Duca Degli Abruzzi, 24, 10128 Turin, Italy

Abstract. Progressive snap through-buckling is a catastrophic collapse that is observed in slender and shallow structures such as reticulated roofs or gridshells [1]. It is usually triggered by a local instability and it spreads up to a portion or to the entire structure causing the static failure. To predict the extent of a local snap to the surrounding nodes, the mechanical consequences of local instabilities have to be investigated: apart from static force rearrangement, the impulse of the dynamic rebound can force a domino effect. In this work, each dynamic rebound has been tuned to an input frequency note to compose a simple music sequence with progressive collapses of Von Mises Truss ensembles. Local instabilities have been based on the interaction formulation [2] and fully elastic material and perfect connections have been considered for the whole deformation process. Finally, a comparison between perfect and imperfect systems is presented to translate imperfection sensitivity to musical performance.

Keywords: snap-through · progressive buckling · domes · musical composition · tuned buckling

1 Introduction

Progressive collapse represents the most severe type of instability mechanism for a reticulated roof [3,4]. Portions of the structural compartment buckle increasingly, causing, in the worst case, the total failure of the structure. This type of behavior, unpredictable and sudden, is a synonym of structural fragility. If redundancy and ductility are the key ingredients to reduce brittleness in common framed structures [5], structural engineers find it difficult to apply them to domes and gridshells. Redundancy is indeed limited by weight and architectural constraints, while ductility resources would not be engaged since buckling for slender elements usually happens significantly below the elastic limit. It is in fact a local instability, at element or group of element scale, to trigger progressive collapse [6]. It is a radical difference between latticed and continuous shells,

S. Gabriele et al. (Eds.): IWSS 2023, LNCE 437, pp. 22–30, 2024.
https://doi.org/10.1007/978-3-031-44328-2_3

where local dents create an energy dissipation area instead of propagating elastically [7]. Emblematic cases of these behaviors are tragically known after the collapse of the C.W. Post Dome or of the Bucarest Municipality dome [8]. The forensic analyses assessed that the whole structure collapsed (for the case of the Bucarest dome it did dramatically remain hanged and overturned) due to a local and asymmetrical accumulation of the snow. We can imagine a chain reaction-like collapse after the snap-buckling of the members directly loaded by the snow, that progressed through the entire structure. This implies that the design load needs to be assessed locally, and that it can be significantly lower than the total uniformly distributed load that the structure can bear statically (i.e. 1/3 for the case of the C.W. Post [9]).
Treating progressive collapse as a loadcase scenario, the AP (alternate path) method is the standard for structural analysis and design [10]. As the name suggests, it calculates the stability of the reticule once one (or more) buckled members have been removed from the frame. It provides sensitivity indexes as a measure of vulnerability, and it directly depends from a redundancy parameter that takes into account the maximum number or *removable* members before a global collapse. Intuitively, this has to deal with the redistribution and the balancing of the compressive forces inside the frame. Being form-resisting structures, non-isotropic fields translates into non-equilibrium. But static unbalance would not be sufficient to get the entire picture of how the progression advances. Snap-through has a dynamic dimension that directly depends from the motion that structure undergoes reaching the new stable position.
Considering the simplest mechanical model for snap-through in Fig. 1a) [11], such motion corresponds to the AB horizontal jump in the equilibrium path.

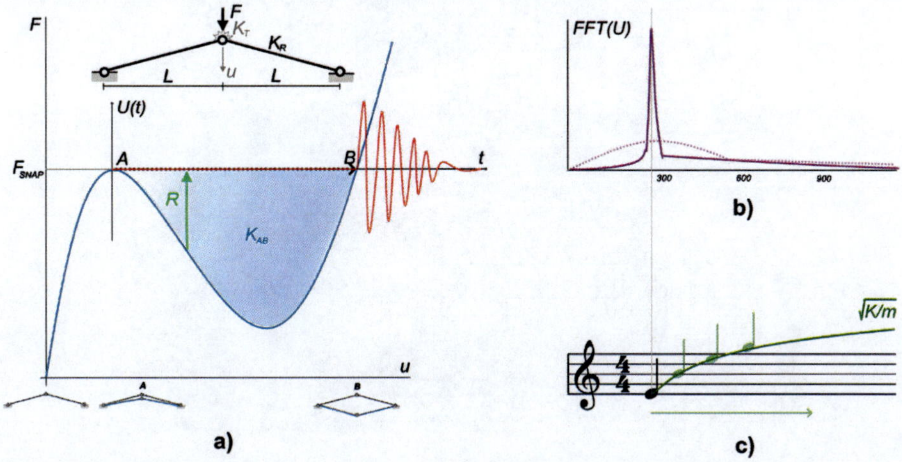

Fig. 1. a) Bellini [11] model for snap-through: equilibrium path (F-u) and post-snap vibrations U(t). b) FFT and frequency identification. c) corresponding FFT(U) note on the staff and linear increase of the stiffness composition

Because position B is a potential energy pit, the arch stops (apart from small oscillations around it) transmitting a dynamic force due to deceleration. The consequential impulsive action occurs also at the lateral restraints, that absorbs both the vertical forces and the vibrations. It is easy to understand that if this is a module of a reticulated roof, those lateral nodes could also snap due to the additional dynamic force on the buckled node, causing a domino effect to the entire structure. Some knockdown factors have been proposed to the AP method to take into account this effect [12], but considering the nonlinearity (and chaotic) nature of the problem, these represent more a blind safety factor than a proper physical counterpart account. Overall, the dynamic effect adds an energy layer to the problem by not treating a quasi-static load application, that has to be dissipated in some way once the system is in a new equilibrium position. Excluding material and connections damage by hypothesis, postbuckling vibrations are the solely dissipating apparatus of the system.
But if vibrations mean frequencies, and frequency means pitch, can we extract a note from the snap-buckling? And what these notes tell once the collapsing mechanism is turn into a musical composition? Examples of a mechanical musical arrangements that have a preset of instructions when externally loaded are the Kircher automata form the 16th century [13]. In Fig. 2, one of his designs for a mechanical organ, powered by steam [14]. Carillons works in the same principle, in which notes are produced by the vibrations of plucked bells or accorded foils. Tuned buckling systems have been instead designed for capacitors, sensors and MEMS and have popular and significant applications in current technology [15].

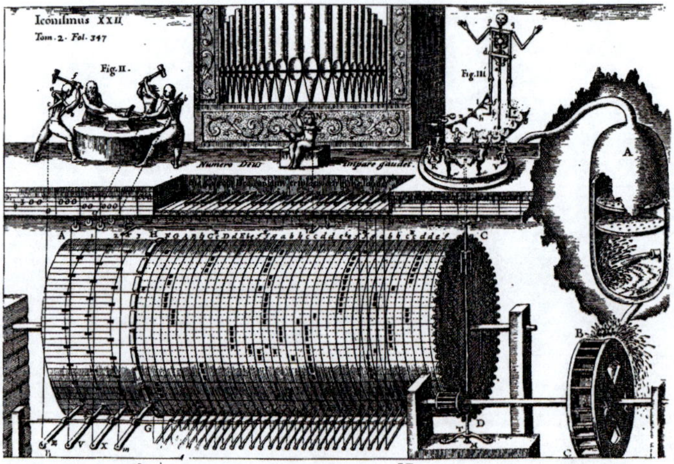

Fig. 2. Athanasius Kircher design for a steam powered mechanical organ [16]

2 Can We Play an Arch?

The structure in Fig. 1a) exhibits a snap-through along the vertical equilibrium path of the central node. Mechanically, the system is governed by the axial stiffness of the rods $K_R = \frac{EA}{L}$, and by the rotational spring stiffness at the crown K_T. If $K_T \neq 0$ the equilibrium path is not symmetrical along the horizontal axis and the snap-through occurs without sign inversion of the vertical reaction F, that can be expressed as follows:

$$F(u) = K_R u(\frac{H}{L}(1+u^2)^{-\frac{1}{2}} - 1) + 2K_T K_R \frac{(\frac{H}{L} - atan(u))}{1+u}. \tag{1}$$

The roots of (1) are the limit points of the path, from which the maximum value of the load F_{SNAP} can be derived. To follow the path after point A, a displacement control is necessary otherwise the system jumps to the next stable position B, at the same level of F_{SNAP} of the hardening branch. This means that for each value of $u \in [A, B]$, $R = F_{SNAP} - F(u)$ is the value of the unbalanced force acting on the system along the unstable curve. Therefore, the blue area in Fig. 1a) is the work energy W of the force R, i.e. the elastic energy that will be released by the snap.

$$W(u) = \int R du. \tag{2}$$

Excluding structural damages, this positive gap of energy is the kinematic quote K_{AB} that the system has in point B. Using momentum,

$$W(u) = \int R du = K_{AB} = \frac{1}{2} m(v_B^2 - v_A^2). \tag{3}$$

in where v are the velocities at points A and B and m is the equivalent mass system $m = \frac{F_{SNAP}}{g}$. Because the system is still at point A ($v_A = 0$), final speed at point B can be immediately derived by ,

$$v_B = \sqrt{\frac{2K_{AB}g}{F_{SNAP}}}. \tag{4}$$

Finally, using D'Alembert principal motion equation along vertical axis, oscillations of the system can be derived:

$$\begin{aligned} m\ddot{U} + c\dot{U} + kU = 0, \\ U(t) = \frac{v_B}{\omega} \sin \omega t + u_0 \cos \omega t. \end{aligned} \tag{5}$$

It is worth to note that condensing the structure at the crown node is a simplification allowed solely if this oscillations around position B are relatively small. This means that also that the total stiffness of the system k can be assumed as a constant is the same interval and not as a parametric excitation. Symmetry of the problem leads to

$$k = 2K_R \sin\frac{H}{L} + K_T - K_T = 2\frac{EA}{L}\sin^2\frac{H}{L}. \tag{6}$$

Therefore,

$$\omega = \sqrt{\frac{k}{m}} = \sqrt{\frac{kg}{F_{SNAP}}}. \tag{7}$$

and, as perfect duality with the string vibrations, a higher is the tensile stress (lower the compression), higher is the pitch.

3 Chords and Modules

Under this set of assumptions a SDOF system can only play one note[1] because F_{SNAP} is a property and not a control parameter. Thus, to play a sequence of notes additional *snapping nodes* have to be present inside the reticule, i.e. a progressive buckling. Analogously, a simultaneous buckling of multiple nodes would play a musical chord if the resulting frequencies are different. The structure in Fig. 3a) aptly describes this behavior. The two-rods archetype above is repeated along the X axis three times at equal distances (green). The central arch is positioned so that the central node (1) has double $Y-$height than the lateral ones in order to realize a double size arch on the perpendicular direction (red). Each node is connected with cylindrical hinges along the perpendicular beam axis. K_T is assumed equal to zero for every node. Built this way, the snap of the central loaded node causes a jump in the vertical forces of the adjacent nodes (2-3) that can consequently lead to their snap. The equilibrium paths of the nodes are shown in Fig. 3b). These have been obtained performing a GNIA (Geometrically Nonlinear Analysis with Imperfections) in 200 load increments in LUSAS FEM. Arc-Length control has been activated to follow the postbuckling branches [17]. For clarity, only a small portion after the limit point the equilibrium paths of nodes 2-3 are shown. Point A in the equilibrium path of node 1 represents the first snap-through of the system, causing the overturning of the central portion of the structure as shown by the deform (blue contour) in Fig. 3a). The symmetry of the system leads to a simultaneous snap-through of nodes 2-3, as reflected in the equivalent displacement of the equilibrium limit points. Using (7), the produced notes can be obtained as pictures in Fig. 3c). The higher pitch of node 2 has been realized by slightly increasing the E of the lateral member. Because the system is designed to not transmit bending moment, it is easy to tune (literally!) E,A or L parameters of the modules by forcing the frequency level on a reference k baseline. The total stiffness k of the system has been obtained by a linear static analysis imposing a unitary displacements to the corresponding notes in the local prebuckling configuration. Interestingly, since the second layer

[1] K_T and K_R are actually dependent by internal stress-strain state, thus controlling flexibility to produce large vibration would momentarily rise or low the pitch. Guitarists are use to this for what is called string *bending* technique.

of snaps occurs with a larger value of F_{SNAP} it produces a lower set of notes. This is intuitively consistent with the mechanic of the arrangement, that has a higher compression at the base level members.

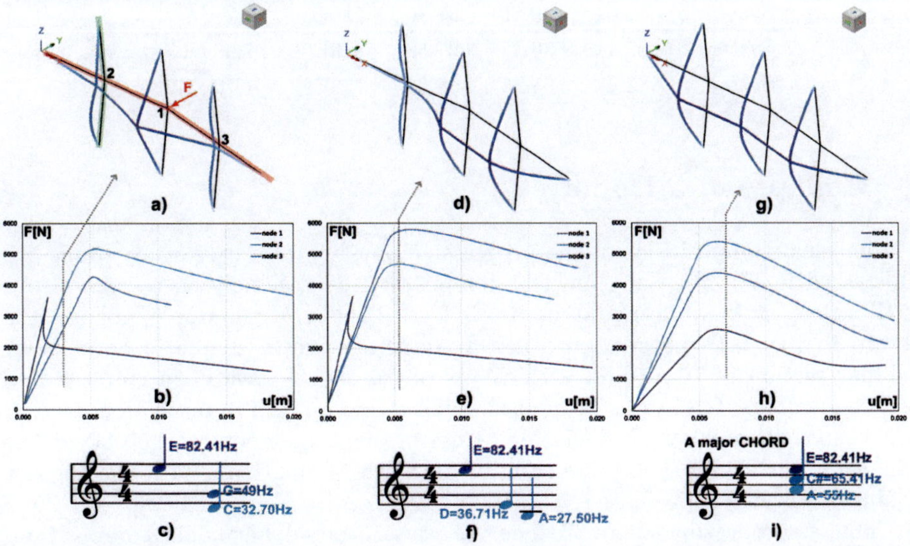

Fig. 3. a) 3 DOF structure, a 4×3 grid realized by 3 Von Mises arches in the Z direction and 1 in the X direction. Y is the vertical axis along acts the central node incremental Force F. In blue contour the deformed shape at the stage indicated by grey dotted line. b) structure equilibrium paths. c) musical composition. d) e) f) and g) h) i) constitutes the same result series.

Using the same logic, the system in Fig. 3d) has been tuned to create a *downward movement* musical composition. The deform shown in blue contours describe the vertical displacements field after the second snap-through has happened. A peculiar set of frequencies have instead been chosen for the system in Fig. 3g). Once a note is selected, is in fact possible to create a harmonic chord by tuning the notes to be *the fifth* and *the fourth* of *the fundamental.* According to harmonic progression (or Pythagorean) it is possible to build a chord by a frequency ratio of $\frac{3}{2}$ for the fifth and of $\frac{4}{3}$ for the forth in respect to the fundamental note. Table 1 reports the tuned elastic moduli of the members in order to obtain the chord and the simultaneous snap of the three nodes. It is worth to note that for a single-material structural arrangement the same tuning can be performed by varying cross section or length of the members (E has been used in this work for computational simplicity on the nonlinear analyses). It is worth to highlight that musical order correspond to an ordered (and very catastrophic) buckling mechanism for the whole portion. In Fig. 3g) the corresponding deformed has been pictured in blue contours.

Table 1. Tuned parameter (Young's modulus) to the corresponding desired notes for the structural arrangements in Fig. 3. System c) produces an A major chord (see the fundamental node of 55 Hz).

	ω_1 [Hz]	ω_2 [Hz]	ω_3 [Hz]	$E_1[GPa]$	$\frac{E_2}{E_1}$	$\frac{E_3}{E_1}$
System a)	82.41	49.00	31.70	210E9	1.2	1
System b)	82.41	36.71	27.50	210E9	1.06	0.94
System c)	82.41	65.41	55.00	173E9	1.20	1.23

4 Collapse of a Dome

The musical compositions of the progressive buckling of a case study dome are illustrated in Fig. 4b. Each musical composition decreasingly corresponds to a shallowness ratio, while the structure is loaded solely at the upper portion. As shown in Fig. 4a), the structure has a triangular mesh (diamatic dome) and tubular section of 10 cm and a thickness of 6 mm [18] have been employed to realize members. GNIA has been used and a random beam imperfection pattern has been applied by axis deviations from straight members of 1/1000 of the element length. In Fig. 4a) the equilibrium path of the dome with the highest shallowness ratio is presented.

This structure evidences in fact the most ordered harmonic composition. The simple metric of absolute distance in terms of frequencies from harmonic progression has been used to detect musical harmony. This result mirrors two structural behavior peculiarities: *(i)* being the flattest geometrical configuration, the meridian is highly susceptible to central loading. Consequently the parallel beam never reach buckling loads and the collapse progresses along the arch-like direction. Furthermore, (ii) in five snaps the dome is completely overturned and hanging from the lateral ring as pointed by point F in the equilibrium path. On the other hand, the higher configurations always present parallel beam collapse [18] and we may assume that they serve as impediments for the snap-through to progress uniformly.

However fascinating, it is necessary to discuss the methodological limitations of the presented results. Firstly, since the dome is not statically determined as the system in the previous Section, it is not possible to easily separate stiffness contributions and use Eq. 7. It has been instead numerically computed the area of the equilibrium path as previously presented in Fig. 1. Each produced frequency has been then derived by combining 4 and 7, assuming elastic prebuckling node stiffness as value for k. This could be not significantly distant from the actual value but further analyses should be performed to prove it. Moreover, considering a twelve tone scale, a difference in frequency higher than 8.3% produces a different note.

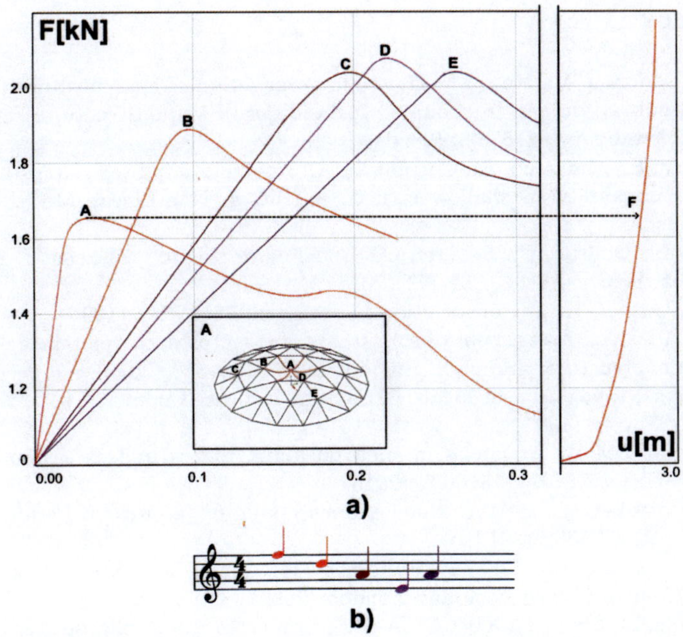

Fig. 4. a) Equilibrium paths of the diamatic dome in the small box. The dotted line indicates the snap-through of the crown node - at point F the dome is completely overturned [18]. b) Corresponding musical composition.

5 Conclusion

A musical flavor has been introduced to the classical problem of snap-through instability. In addition to the feasibility of a mechanical instrument, the results can serve as interpretation for snap patterns of progressive instability. Musical order is in fact highly coded by metrics and rules, therefore it may serve as reference model for structural performance or collapse identification. Concerning this, the interpretation of a progressive collapse case study has been presented. All limitations aside, the methodology presented grasped the most severe collapse mechanism using a simple musical disorder metric. Finally, a set of rules to tune a statically determined system to play notes and chords has been presented in Sect. 2.

Acknowledgments. This project has received funding from the European Union's Horizon 2020 research and innovation programme under the Marie Sklodowska-Curie grant agreement No 845064.

References

1. Ivan, A., Gioncu, V., Ivan, M.: Dynamic propagation of local instability for single layer reticulated domes. In: Coupled Instabilities in Metal Structures: CIMS 2000, World Scientific, pp. 515–522 (2000)
2. Bazzucchi, F., Manuello, A., Carpinteri, A.: Interaction between snap-through and Eulerian instability in shallow structures. Int. J. Non-Linear Mech. **88**, 11–20 (2017)
3. Lefevre, B., Douthe, C., Baverel, O.: Buckling of elastic gridshells. J. Int. Assoc. shell Spat. Struct. **56**(3), 153–171 (2015)
4. Feld, J., Carper, K.L.: Construction Failure, vol. 78. Wiley (1996)
5. Brett, C., Lu, Y.: Assessment of robustness of structures: current state of research, frontiers of structural and civil. Engineering **7**, 356–368 (2013)
6. Abedi, K.: Propagation of local instabilities in braced domes, Ph.D. thesis, University of Surrey (1997)
7. Thompson, J.M.T.: Advances in shell buckling: theory and experiments. Int. J. Bifurcat. Chaos **25**(01), 1530001 (2015)
8. Geis, J., Strobel, K., Liel, A.: Snow-induced building failures. J. Perform. Constr. Facil. **26**(4), 377–388 (2012)
9. Levy, M., Salvadori, M.: Why Buildings Fall Down: How Structures Fail. WW Norton Company, New York and London (2002)
10. Xianzhong, Z., Shen, Y., Yiyi, C.: A review on progressive collapse study for large-span space structures. J. Build. Struct. **34**(4), 1 (2013)
11. Bellini, P.X.: The concept of snap-buckling illustrated by a simple model. Int. J. Non-Linear Mech. **7**(6), 643–650 (1972)
12. Han, Q., Liu, M., Lu, Y., Wang, C.: Progressive collapse analysis of large-span reticulated domes. Int. J. Steel Struct. **15**, 261–269 (2015)
13. Chen, Y.-H., Ceccarelli, M., Yan, H.-S.: A historical study and mechanical classification of ancient music-playing automata. Mech. Mach. Theory **121**, 273–285 (2018)
14. Mazzola, G., Mannone, M., Pang, Y., et al.: Cool Math For Hot Music, Springer (2016). https://doi.org/10.1007/978-3-319-42937-3
15. Hu, N., et al.: Buckling-induced smart applications: recent advances and trends. Smart Mater. Struct. **24**(6), 063001 (2015)
16. Bedini, S.A.: The role of automata in the history of technology. Technol. Cult. **5**(1), 24–42 (1964)
17. FEA Ltd.: LUSAS user manual. Kingston-upon-Thames, UK (1999)
18. Bazzucchi, F., Manuello, A., Carpinteri, A.: Interaction between different instability phenomena in shallow roofing structures affected by geometrical imperfections. In: Proceedings of IASS Annual Symposia, vol. 2016, International Association for Shell and Spatial Structures (IASS), pp. 1–10 (2016)

Topology Generation of Architectural Meshes Adapted to the Support Conditions

Romane Boutillier[1,2](✉), Cyril Douthe[1], Laurent Hauswirth[2], and Olivier Baverel[1,3]

[1] Laboratoire Navier UMR8205, Ecole des Ponts, Univ Eiffel, CNRS, 77455 Champs-sur-Marne, France
romane.boutillier@enpc.fr

[2] Université Gustave Eiffel, Laboratoire d'Analyse et de Mathématiques Appliquées, UMR8050, Champs-sur-Marne, France

[3] GSA/ENS Architecture Grenoble, Grenoble, France

Abstract. The performances of gridshells depend on the topology of the grid but also on the geometry of the underlying surface. Topology and geometry are linked: the shape of a hanging chain model depends on the pattern of the suspended fabric. In this paper, a method is proposed that generates topologies by varying continuous parameters. The obtained grids are adapted to the support conditions of the considered surface: elements converge towards the supports and arrive perpendicular to the edges otherwise. Grids are then lifted: the spatial mesh is the result of an optimisation to have a funicular surface and flat panels. The spatial mesh is then mechanically evaluated and its performance according to different criteria is represented in a Pareto front. A non-intuitive result shows that a large number of different topologies are situated on the Pareto front. These results show the strength of the method to explore different topologies.

Keywords: Topology finding · Gridshell · Form finding

1 Introduction

1.1 Necessity of Topology Generation

Meshes are everywhere in free form architecture, especially when it comes to constructing surfaces. On so-called discrete surfaces, or gridshells, made of beams and panels, the mesh is clearly visible: the beams are the edges of the mesh, the panels are the faces. The topology of the grid, i.e. the connectivity between the different elements, is thus apparent.

The construction of the surfaces must meet several requirements, of a geometrical nature and relative to manufacturing (flat panels are less expensive and easier to implement, for example), of a mechanical nature (the engineer seeks to

S. Gabriele et al. (Eds.): IWSS 2023, LNCE 437, pp. 31–40, 2024.
https://doi.org/10.1007/978-3-031-44328-2_4

reduce deformations, to direct the forces towards the supports) or even aesthetic (one prefers an equal distribution of the elements and panels of regular shape)

The geometry of the grid, its shape, is one of the ways to meet these objectives. But before the geometry, it is necessary to choose the topology of the grid.

The topology of the grid can be derived from the geometry to be discretised. In the case of the Milan Trade Fair (Fig. 1), the abrupt change in curvature makes it necessary to introduce higher valence nodes (i.e. nodes that connect more elements, coloured on the image) in order to limit the distortion of the panels [16].

Fig. 1. Milan Trade Fair (Fuksas/SBP)

Fig. 2. Isler's hanging model [6]

But the topology of the grid can also control the resulting shape, in a process of form finding as illustrated by the image of a Heinz Isler model (Fig. 2). This shape was obtained by suspending a net, whose topology (the textile orientation) influences the deformation under gravity loading. The influence of the topology is also illustrated in [5] where different fabric patterns are lifted, and in Fig. 6 where a quad mesh with isolated singularities is submitted to gravity with a constraint on panel planarity.

In general, most form finding methods depend on the definition of an input grid topology, be it thrust network analysis [1], the Marionette method [10] or other historical methods developed by Frei Otto at the Institute for Lightweight Structures [12].

1.2 Existing Methods of Topology Generation

Topological generation has not been widely studied for architectural structures, but is a subject of research in the computer graphics community (see surveys [2,4,7,9] which are closer to the method described hereafter).

Topological generation is not to be confused with topological optimisation, which optimises the distribution of material in an element.

R. Oval [13,14] developed a method of topology generation based on the boundary of the surface to be meshed, as well as features such as points or curves chosen by the designer. The exploration of the different grids is then controlled by a grammar of elementary operations on strips.

The method presented here is a continuation of this approach called "topology finding". It proposes a new paradigm where the exploration of the grids is controlled by continuous parameters defined at the boundary. It is therefore intuitive, and starts from the support conditions which are a crucial element in the behaviour of the structure but also its integration in an architectural context.

1.3 Importance of Boundary Conditions

The support conditions are the starting point for the method presented in this article. Beyond aesthetics, the main issue of gridshell design is to carry the load towards the supports where forces are transmitted to the ground.

Fig. 3. Smithsonian Institute, Washington, DC (Foster and Partners/Buro Happold - 2001) [8].

Fig. 4. Cabot Circus, Bristol (Chapman Taylor with Nayan Kulkarni/SKM - 2008) [17]

On both surfaces in Figs. 3 and 4, the grid is formed of regular quadrilaterals. In the case of the Smithsonian Institute, the support conditions are punctual, and the topology of the grid is not changed at the columns. However, the thickness of the beams increases at this point where the forces are concentrated. In the case of Cabot Circus, the regular grid is supported by large edge beams, which interrupt the grid: at the boundary, the beams are of various lengths, and the connections may be unique.

The method, presented in the next section, generates topologies and grids based on parameters defined along the edge. The resulting grids have beams that converge at the supports, and beams that arrive perpendicular to the edge elsewhere. They are therefore optimal from a mechanical point of view.

2 Topology and Geometry Generation Method

2.1 Topology Generation

The topology finding process is detailed in [3]. It relies on the computation of minimal surfaces that are only an intermediate tool in the topology finding process, but are not connected to the architectural shape. However, the boundary of the minimal surface is chosen in relation to the supports of the real architectural surface, in order to find a topology and a grid that are adapted to the boundary conditions. One example is shown Fig. 5, the minimal surface boundary is plotted in red, its vertical components are aligned with the columns shown underneath.

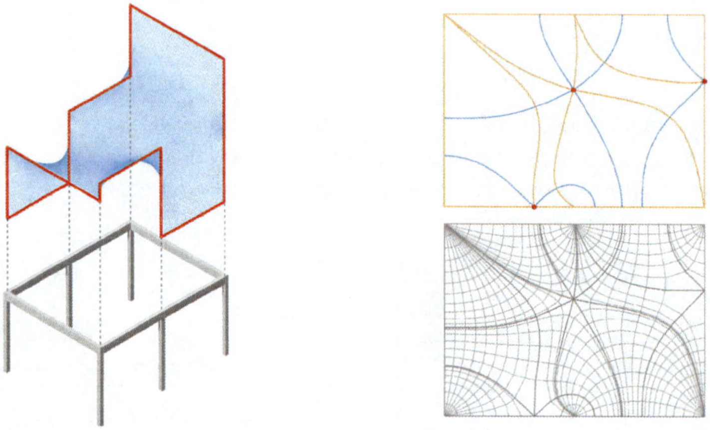

Fig. 5. Topology and grid generation process. A crenelated boundary (in red) is built over the real supports of the surface. This artificial boundary is controlled by height parameters. A minimal surface (in blue) is computed on this boundary. A topological skeleton is then retrieved (top right). It is made of singularities (red dots) of the height gradient field. Streamlines lines of the gradient are shown in blue, the level set is shown in orange. The skeleton defines patches that can be discretised regularly. The field is finally integrated to get patch subdivisions (bottom right). (Color figure online)

The grid is the discretisation of streamlines of a field, namely the gradient field and its conjugate set the level set, on minimal surfaces which boundary is made of horizontal curves and vertical lines. This choice provides thus a proper alignment of the grid to the boundary, and smooth streamlines which intersect at 90 degrees. Moreover, level lines converge at the location of vertical boundary lines, which we choose to locate at the supports. By doing so forces are channelled to the supports. This phenomenon is illustrated Fig. 5, where grey lines converge at six points on the boundary, where columns are, and arrive perpendicularly to the edge otherwise.

The boundary of the minimal surface can be modified: it is subdivided into slots whose heights can vary while keeping the same planar projection, and thus the resulting grid changes as well. This relation between the boundary and the grid is at the core of the generation process shown hereafter.

Generation of Grid Skeleton. In this study, grids with quad panels are generated. Some triangular panels appear at the boundary, and the boundary point they connect is called a pole. Other irregular points, located on the boundary or in the shape, are called singularities. At these points on the minimal surface, the gradient is null: singularities are saddle points. The topology finding process falls out into few parts: the determination of singularities (shown in red in Fig. 5) and the computation of a skeleton. The skeleton, represented with blue and orange curves Fig. 5, is made of streamlines that connect singularities between them and to the boundary. It defines patches: quadrilateral or triangular zones that can be discretised regularly.

2.2 Determination of the Geometry

Direct Method for Determining the Strip Subdivisions. The number of patch subdivisions is the same for all patches of a strip (a strip is a series of patches that are connected along their opposite edges) and is computed to get a target average edge length (of about 1.2 m in our example).

The resulting grid is made of quads panels (except at poles) that are not distorted because they come from the discretisation of lines that meet at right angle. There are overall distributed nicely in the mesh. However, concentration of lines at poles, and large faces at singularities can be noted. This could easily be remeshed locally, and the former grids are kept for the evaluation, despite this constructive remark.

Direct Method for Computing an Adapted Shape. In this study, in order to compare topologies, only one shape is computed for each topology. This shape depends on the topology, but is not the only one that is adapted to this topology. It is the result an optimisation under two major constraints: flat panels and a funicular shape. This optimisation is run with Kangaroo [15]. The funicular load is proportional to the tributary area in plane, therefore all grids, that have the same planar projection, are submitted to the same loading. Other goals are applied for achieving a smooth shape: mesh edge lengths are controlled, and a stiffness is provided at the ridge beams to smooth curvature. The optimised shape is rescaled to get a maximum target height that is the same for all grids (and corresponds to a ratio height/span of 0.2). Grid heights could also be normalised to get a constant enclosed volume. This lifting process is illustrated on two grids with different topologies in Fig. 6: because the shape depends on the topology, the resulting geometries vary.

As mentioned in the introduction, this method for generating topologies can also be used as a form finding method, as the resulting geometry is directly related to the topology of the grid. Therefore, each geometry is also evaluated as an isotropic shell, in order to compare their performance.

3 Case Study: Generation of Grids on a Hexagonal Boundary

The method is illustrated on a regular shape, that can be used to cover an interior space or as a building (like the restaurant of Felix Candela in Xochimilco

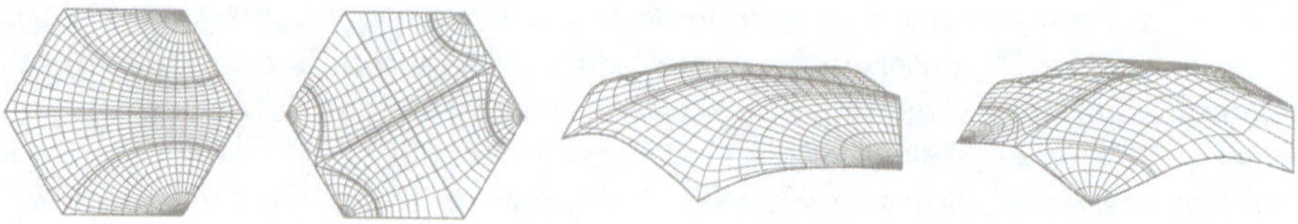

Fig. 6. Two grid examples that are lifted with a gravity load and constrained to have planar panels. The two lifted shapes differ, because grid topologies are different.

[11]). We consider a point supported structure, whose boundary projection is an hexagon, where supports are located at the discontinuities of the boundary. The crenellated contour is thus made of six slots, and five parameter heights are considered (when removing the rigid body motion). Some topologies that are found are shown Fig. 7.

3.1 Topologies that can be Reached with this Method

Even if this contour is very symmetrical, many topologies can appear. The types of topologies that can occur are controlled by the Poincaré Hopf theorem: the sum of the indices of the vertices must be equal to the Euler characteristic of the surface (equal to 1 in this case).

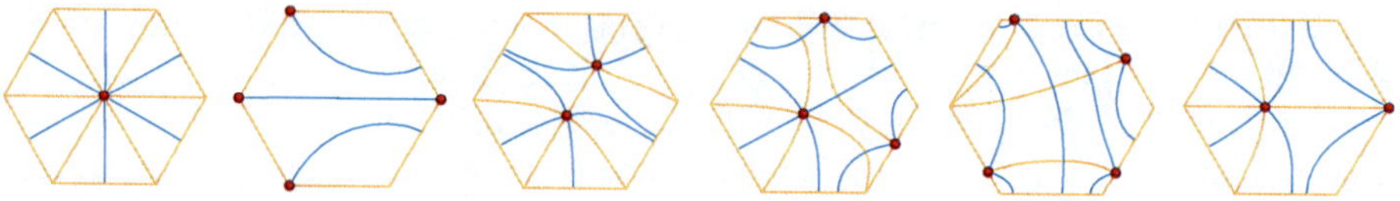

Fig. 7. Selection of topologies that can be achieved with the presented method. Various combinations of vertex indices can be created.

3.2 Mechanical Model

Structural Behaviour. Three structural types are tested: a shell (isotropic), a braced gridshell (anisotropic but with a membrane stiffness) and a gridshell without bracing (anisotropic). The gridshell without bracing serves as a reference for equivalent mechanical properties. The bracing section for the diagonals of the braced gridshell is taken as a tenth of the regular beams section. The shell properties are chosen in order to get a similar rigidity than the braced gridshell.

Applied Loads and Boundary Conditions. A symmetric gravity load is applied to mesh points and proportional to the tributary area. Hinged supports are set at the six corners.

Initialisation of the Model. Some parameters are chosen at the first iteration and kept constant for the rest of the generation process (same structure mass and section inertia in all grids). The only variables of the generation are those changing the topology, i.e. the slot heights of the boundary of the minimal surface.

Evaluation of the Grids. The grid geometry is thus given, and its properties can be evaluated. Among all the criteria that can be chosen, dimensionless ratios are built: the thrust divided by the vertical reaction, which can be a design criteria for structures on top of existing building for example, the ratio between axial and bending strain energies, which characterises deformations and forces in the structure, and deflection, which is a design factor for relatively low structures (deflection is kept as it is, since the height and span of the different structures are the same). Since several criteria are involved, the Pareto front of the generated grids is evaluated.

3.3 Results

In total, three structural topologies are tested. More than a hundred of grids are generated, but one cannot prove that it constitutes an exhaustive scan of the space of grids achievable with this method. Nevertheless, this sample allows to see the diversity of possible designs, and to compare grids between them (see Fig. 8).

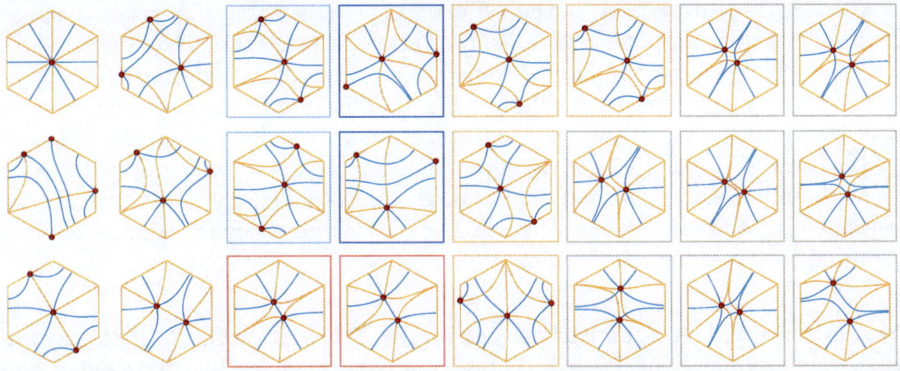

Fig. 8. Grids that belong to the Pareto front when evaluated for a symmetric loading. Different topologies appear, but also geometric variations of the same topology, which are framed with colours. (Color figure online)

Pareto Front. The Pareto front is calculated for the gridshell without bracing. Over the 138 tested grid, 24 of them belong to the Pareto front, and are represented Fig. 8. These grids can be grouped into topological families. Indeed, the generation method controls both the topology of the grid and the position of the elements in space. Some topologies are also equivalent (with one rotation) and are grouped together. Finally, eleven different topologies are present on the Pareto front.

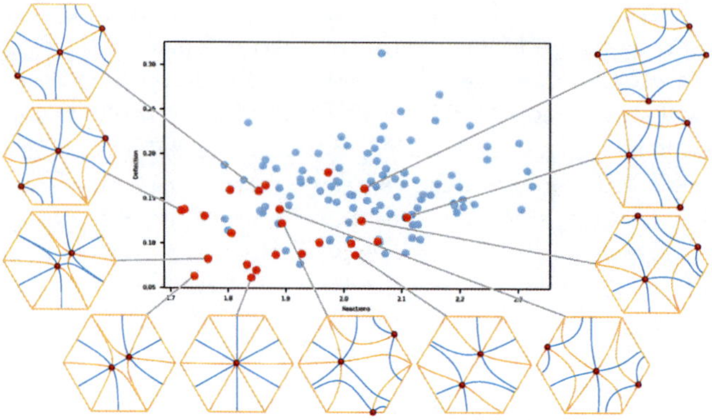

Fig. 9. Projection of the grids without bracing, in the plane of the support reactions and deflection criteria. The grids belonging to the Pareto front are shown in red, one representative of each topology family of the Pareto front is displayed on the side. (Color figure online)

These grids are plotted in the objective space, a three dimensional space whose axis are the supports reactions, the ratio of strain energies and deflection. The grids belonging to the Pareto front are represented in red among all other generated grids represented in blue in Fig. 9. The results are represented as a projection in the plane of two criteria (in this case the deflection on the ordinate and the ratio of horizontal to vertical reactions on the abscissa). A representative of each family is also displayed. This visualisation tool shows how the grids are positioned in relation to each other. It also allows the designer to make choices according to the criteria to be met.

Comparison of the Structure Typologies. Three structural typologies are tested and compared in Table 1. Values are computed for two topologies, the symmetrical one with one singularity in the center, and another topology with a singularity in the surface, two in corners, and no symmetry in the grid (these topologies are shown in Fig. 11 and the resulting 3D shapes are shown Fig. 10).

As expected, adding bracing improves drastically the behaviour of the grid, in particular with regard to strain energies. The relative performance depends on the type of structure considered: with bracing, the symmetrical topology works better than the asymmetrical topology. Deflection is much larger for the shell models than for the grids.

Table 1. Comparison of the performance of two grids, for three structural types (R. for reactions, E. for strain energy and d for deflection).

	3-symmetrical topology			asymmetrical topology		
	R	E	d (cm)	R	E	d (cm)
Gridshell	2.02	7.49	0.15	1.92	6.72	0.20
Braced gridshell	2.01	1.22	0.09	1.90	1.50	0.14
Shell	1.79	0.68	0.78	1.95	0.42	0.80

Fig. 10. Perspective views of the grid whose performances are compared for the different structural typologies: a symmetric topology with a central singularity (left) and an asymmetric topology with one singularity in the surface and two at corners (right).

The shell model serves as an evaluation of the geometry obtained with the grid topology. Stresslines can be computed on this shell, and compared to the grid that was used to create the geometry. Two examples are shown Fig. 11. The stress lines of the two funicular forms are completely different. However, the shape of the lines is close to the underlying grid topology. The deviations may be due to the added planarity objective, or to the method of determining these lines on imperfect meshes. This topic deserves to be explored in more detail.

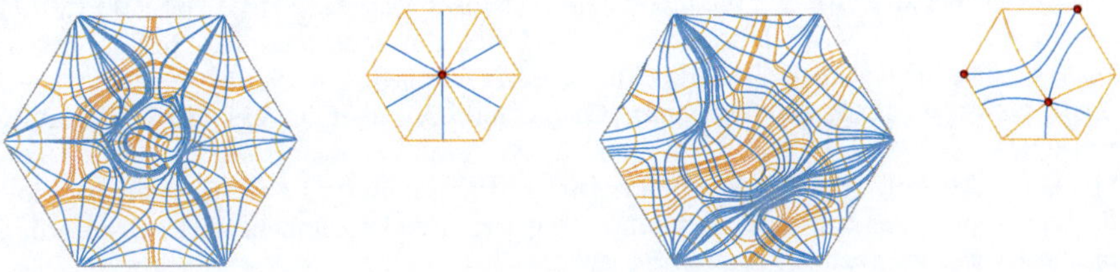

Fig. 11. Stress lines under gravity loading computed on the symmetric grid and on a grid with no symmetries and singularities located in the surface and on the boundary, with their associated topology.

4 Conclusion and Perspectives

The presented method allows the generation of a variety of grid topologies, all adapted to the support conditions, by varying parameters defined along the edge.

This variety is reflected in the Pareto front of the case study: a real choice is offered to the designer according to the chosen criteria. The topology generation tool can also be used upstream of many formfinding methods, for which the topology is a given. Local remeshing techniques and local optimisations could be added to improve grid performance. The link between topology and isotropic shells deserves further investigation.

References

1. Block, P., Ochsendorf, J.: Thrust network analysis: a new methodology for three-dimensional equilibrium. J. Int. Assoc. Shell Spat. Struct. **48**, 167–173 (2007)
2. Bommes, D., Levy, B., Pietroni, N., Puppo, E., Silva, C., Zorin, D.: Quad-mesh generation and processing: a survey. Comput. Graph. Forum **32**, 51–76 (2013)
3. Boutillier, R., Douthe, C., Hauswirth, L., Baverel, O.: Topology generation of structural grids for architectural surfaces (2023, to be published)
4. Campen, M.: Partitioning surfaces into quadrilateral patches: a survey. Comput. Graph. Forum **36**(8), 567–588 (2017)
5. Chilton, J., Chuang, C.C.: Rooted in nature: aesthetics, geometry and structure in the shells of Heinz Isler. Nexus Netw. J. **19**, 763–785 (2017)
6. Chilton, J., Isler, H.: Heinz Isler: The Engineer's Contribution to Contemporary Architecture. Thomas Telford (2000)
7. Dong, S., Kircher, S., Garland, M.: Harmonic functions for quadrilateral remeshing of arbitrary manifolds. Comput. Aided Geom. Des. **22**(5), 392–423 (2005)
8. Foster + Partners: Smithsonian Institution. https://www.fosterandpartners.com/projects/smithsonian-institution-courtyard. Accessed 25 May 2023
9. Jezdimirovic, J., Chemin, A., Reberol, M., Henrotte, F., Remacle, J.F.: Quad layouts with high valence singularities for flexible quad meshing. arXiv:2103.02939 [cs, math] (2021)
10. Mesnil, R., Douthe, C., Baverel, O., Leger, B.: Marionette Meshes: modelling free-form architecture with planar facets. Int. J. Space Struct. **32**(3), 184–198 (2017)
11. Moreyra Garlock, M.E., Billington, D.P.: Félix Candela: engineer, builder, structural artist. Princeton University Art Museum (2008)
12. Otto, F., et al.: IL 10 Gitterschalen/Grid Shells. Institut fur Leichte Flachentragwerke (IL) (1974)
13. Oval, R., Mesnil, R., Van Mele, T., Block, P., Baverel, O.: Rule-based topology finding of patterns towards multi-objective structural design. In: 2019 Proceedings of the IASS Symposium, Barcelona (2019)
14. Oval, R., Rippmann, M., Mesnil, R., Mele, T.V., Baverel, O., Block, P.: Feature-based topology finding of patterns for shell structures. Autom. Constr. **103**, 185–201 (2019)
15. Piker, D.: Kangaroo: form finding with computational physics. Archit. Des. **83**, 136–137 (2013)
16. Schlaich, J., Schober, H., Kürschner, K.: New trade fair in Milan-grid topology and structural behaviour of a free-formed glass-covered surface. Int. J. Space Struct. **20**, 1–14 (2005)
17. Schober, H., Justiz, S.: Cabot circus, bristol ebene vierecknetze für freigeformte glasdächer. Stahlbau **81**, 28–42 (2012)

Form-Finding of Membrane Shells via Isogeometric Analysis

Claudia Chianese(✉), Francesco Marmo, and Luciano Rosati

Department of Structures for Engineering and Architecture, University of Naples Federico II, 80138 Naples, Italy
claudia.chianese@unina.it
http://www.dist.unina.it

Abstract. A numerical form-finding procedure, based on Isogeometric Analysis (IgA), is proposed to determine the elevation of a shell having a prescribed covered area so as to carry applied loads in a pure membrane state of stress. The boundary-value problem of a membrane shell is described by Pucher's equation in terms of external load, Airy potential and shell mid-surface height. Membrane stress components at a point are computed via second derivatives of the Airy potential. Within the IgA framework, the equivalent weak form of the problem is derived and the resulting integral equation is discretized by approximating the relevant fields as a linear combination of control point values and B-Spline basis functions. Once the external load is prescribed, control point values of the Airy potential are computed via a nonlinear optimization routine aimed at minimizing thrusts at edge supports while ensuring a no-traction principal stress state fulfilling boundary conditions. The resulting shell form is then computed in terms of control point heights. Two numerical examples show the effectiveness of the proposed approach.

Keywords: Form finding · Membrane shells · Isogeometric analysis

1 Introduction

A shell surface resists applied loads by an internal stress distribution generally equivalent to three distinct internal actions, i.e. membrane (in-plane), out-of-plane shear and bending-twisting stress resultants. For an efficient employment of material in terms of strength and stiffness, the shell structural behaviour is deemed to be optimal if it experiences purely membrane stresses.

Membrane shells constitute inherently an attractive architectural solution for large lightweight structures of refined aesthetic elegance since they allow for wide interiors to be enclosed unobstructedly with minimal usage of material in the absence of bending and/or twisting moments. However, randomly curved shell forms are not free from moment actions. The design procedure aimed at

Supported by the University of Naples Federico II and the Compagnia di San Paolo, FRA grants, CUP E69C21000250005.

S. Gabriele et al. (Eds.): IWSS 2023, LNCE 437, pp. 41–50, 2024.
https://doi.org/10.1007/978-3-031-44328-2_5

determining the shell form possessing such moment-free property is called *form finding*. Existing form-finding algorithms are categorized in three groups [1].

Stiffness matrix methods do account for both geometric and material stiffness. Conventionally, the shell form is first determined via simulated physical models [2] starting from an initial geometry. Then, static analysis is performed via a finite-element scheme based on traditional shell theories, such as Kirchhoff-Love's [3] or Reissner-Mindlin's [4]. Recent unified approaches [5] enable simultaneous form finding and analysis procedures. Utility of these methods is undoubted as they allow for stress verification studies in form-found shells as well as in-depth buckling and large-displacement nonlinear investigations. However, including material properties in an inherently geometric form-finding problem is widely acknowledged as "unnecessary and computationally costly" [1]. **Geometric stiffness methods** constitute an ideal tool to address form finding of very thin (membrane) shells without unnecessary mathematical complications given their material independence. As the name suggests, only the stiffness attributed to the shell topology and internal force density is considered. An evident advantage of providing non-materialized solutions is that the resulting equilibrium shape is not affected by the material constitutive law in use. The Force Density Method [6,7], the Thrust Network Analysis [8] and form-finding algorithms [9] based on Pucher's membrane shell theory [10] belong to this group.

Dynamic equilibrium methods seek the shell static equilibrium in the form of the steady-state solution of a pseudo-dynamic problem of an equivalent spring-mass system. Dynamic Relaxation [11,12] and Particle-Spring System [13] are representatives of this group. Their strengths lie in an optimal convergence rate, obtained by a correct tuning of the fictitious damping, alongside a low computational cost, deriving from the non-necessity of assembling global stiffness matrices. As such, the versatility of dynamic methods justifies their use in form-finding of not only membrane shells [14], but also bending-active elastic models [15]. Latest works [5,16,18,19] combine mentioned approaches with recent Isogeometric Analysis [17] (IgA). The use of NURBS basis functions, endowed with high-order inter-element continuity, has proven to be ideally suited to model free-form shells as opposed to standard FEM segmented polynomial description. The strength of IgA is further revealed in iterative form-finding processes [20,21] featuring repeated communication between initial and current geometries. There, the integration between design and analysis is seen to significantly simplify the overall workflow, no longer requiring geometry conversions.

This paper presents a novel IgA-based form-finding procedure to determine the elevation of a shell of given planar footprint by solving the weak form of Pucher's membrane shell theory. This topic has already been discussed by Xia *et al.* [18], who dealt with masonry shells. Contrary to Xia *et al.* [18], our solution equation fully incorporates horizontal and vertical external loads, which may be of interest for applications in seismic zones. Furthermore, the field of the Airy potential is not simply user-prescribed, but computed via a nonlinear optimization routine aimed at minimizing thrusts at edge supports while ensuring a no-traction principal stress state fulfilling relevant boundary conditions.

The paper is organized as follows. In Sect. 2, the weak form of the membrane shell problem is derived. Section 3 describes the IgA computational model. Results are presented in Sect. 4. Eventually, Sect. 5 concludes the paper.

2 The Weak Form of the Membrane Shell Problem

According to Pucher's theory, the equilibrium of a shell infinitesimal element, subject to distributed loads and internal actions, reads:

$$\nabla^2\Phi\nabla^2 f - \nabla(\nabla\Phi)\cdot\nabla(\nabla f) = (\mathbf{q}^h\cdot\nabla f) + \mathbf{h}\cdot(\nabla\circ\nabla f) - q_z \tag{1}$$

being $\mathbf{h} = [\int q_x\,dx, \int q_y\,dy]^T$, $\mathbf{q} = [q_x, q_y, q_z] \equiv [\mathbf{q}^h, q_z]$ the force per unit projected[1] area, f the height of the shell mid-surface S, and Φ the Airy stress function associated with the projected membrane stress tensor, i.e. projected membrane stresses are obtained as suitable second derivatives of Φ. Supplemented with relevant boundary conditions, Eq. (1) is the strong form of the membrane shell problem.
Let $g(x,y)$ be an arbitrary variation function of f vanishing on the shell constrained boundary Γ_c. Applying the chain rule on the terms involving the second derivatives of f and eliminating similar addends, Eq. (1), pre-multiplied by g on both sides, becomes:

$$\begin{aligned} div\{\nabla^2\Phi\nabla f - \nabla f^T\nabla(\nabla\Phi) - \mathbf{h}\circ\nabla f\}g + [\nabla f^T\nabla(\nabla\Phi)]\cdot\nabla g - \\ +\nabla^2\Phi(\nabla f\cdot\nabla g) + (\mathbf{h}\circ\nabla f)\cdot\nabla g = -q_z g \end{aligned} \tag{2}$$

Integration of Eq. (2) over the shell projected domain Ω, combined with the divergence theorem, yields the weak form of the boundary-value problem:

$$\begin{aligned} \int_\Omega \Big[\nabla^2\Phi(\nabla f\cdot\nabla g) - [\nabla f^T\nabla(\nabla\Phi)]\cdot\nabla g - (\mathbf{h}\circ\nabla f)\cdot\nabla g\Big]\,d\Omega = \\ \int_{\Gamma_f} g\Big[\nabla^2\Phi\nabla f - \nabla f^T\nabla(\nabla\Phi) - (\mathbf{h}\circ\nabla f)\Big]\cdot\hat{\mathbf{n}}\,ds + \int_\Omega q_z g\,d\Omega \end{aligned} \tag{3}$$

3 The IgA Computational Model

The weak form is solved numerically by applying IgA discretization to approximate the shell mid-surface height $f(x,y)$. Thus, Eq. (3) is rewritten as a set of algebraic equations. To this end, the shell domain Ω is discretized into N^e sub-domains of a single B-Spline patch, identified by pairs of knot spans in the parametric directions ξ and η. Accordingly, the weak form reads:

$$\begin{aligned} \textstyle\sum_{e=1}^{N^e}\int_{\Omega^e}\nabla^2\Phi^e(\nabla f^e\cdot\nabla g^e) - [(\nabla f^e)^T\nabla(\nabla\Phi^e)]\cdot\nabla g^e \\ -(\mathbf{h}^e\circ\nabla f^e)\cdot\nabla g^e\,d\Omega^e = \textstyle\sum_{e=1}^{N^e}\int_{\Omega^e} q_z^e g^e\,d\Omega^e \end{aligned} \tag{4}$$

[1] Projections are intended to be taken on the (x,y) Cartesian plane.

In the previous equation, the integral on Γ_f has been omitted as null, provided that assigned membrane stresses fulfill static conditions (12) on free edges. Using IgA approach, the relevant fields f and g are computed as a linear combination of n^e_{cp} control point values and associated basis functions over an element Ω^e. In the present formulation, also the Airy stress function Φ is approximated through the same basis functions. Thus,

$$f^e(\xi,\eta) = S^e_k Z^e_k \Leftrightarrow f^e(\xi,\eta) = \left[R_1(\xi,\eta)...R_{n^e_{cp}}(\xi,\eta)\right]\left[f_1...f_{n^e_{cp}}\right]^T$$

$$f^e_{,j}(\xi,\eta) = B^e_{jk} Z^e_k \Leftrightarrow \begin{bmatrix} f_{,x}(\xi,\eta) \\ f_{,y}(\xi,\eta) \end{bmatrix} = \begin{bmatrix} R_{1,x}(\xi,\eta)...R_{n^e_{cp},x}(\xi,\eta) \\ R_{1,y}(\xi,\eta)...R_{n^e_{cp},y}(\xi,\eta) \end{bmatrix} \left[f_1...f_{n^e_{cp}}\right]^T$$

where a shorter notation of bivariate B-Spline basis functions, $R^{pq}_{ij}(\xi,\eta) \to R_i(\xi,\eta)$, is adopted to avoid index proliferation. The approximation scheme for g and Φ follows in an analogous way, differing only in the respective vectors - $\mathbf{W}^e$ and $\mathbf{A}^e$ - of control point values.
Similarly, $\nabla(\nabla\Phi^e)$ can be computed via a third-order tensor, containing the second derivatives of the basis functions times a vector of Φ-values at control points: $\Phi^e_{,ij}(\xi,\eta) = C^e_{ijm}(\xi,\eta) A^e_m$, where $i,j = \{x,y\}$ and $m = 1,..,n^e_{cp}$. Therefore:

$$\nabla^2\Phi(\xi,\eta) = L^e_p(\xi,\eta)\ P^e_p =$$
$$= [C_{111}(\xi,\eta)\dots C_{11n^e_{cp}}(\xi,\eta)|C_{221}(\xi,\eta)\dots C_{22n^e_{cp}}(\xi,\eta)][\Phi_1\dots\Phi_{n^e_{cp}}|\Phi_1\dots\Phi_{n^e_{cp}}]^T$$

Applying the previous discretization, Eq. (4) becomes:

$$\sum_{e=1}^{N^e}\left[\int_{\Omega^e}(\mathbf{L}^e\cdot\mathbf{P}^e)(\mathcal{B}^{e^T}\mathcal{B}^e) - \mathcal{B}^{e^T}(\mathbb{C}^e\mathbf{A}^e)\mathcal{B}^e - \mathcal{B}^{e^T}(\mathbf{h}^e\circ\mathcal{B}^e)d\Omega^e\right]\mathbf{Z}^e\cdot\mathbf{W}^e = \\ = \sum_{e=1}^{N^e}\left[\int_{\Omega^e} q^e_z\mathbf{S}^e d\Omega^e\right]\cdot\mathbf{W}^e \tag{5}$$

For clarity of reading, scalars, vectors, 2nd- and 3rd-order tensors are displayed in italic, bold, calligraphic bold and blackboard bold font styles, in turn.
Gauss-Legendre quadrature rule is employed for numerical integrations in Eq. (5). Because B-Spline basis functions are defined in the parametric space, an additional mapping from the Cartesian to the parametric domain is introduced, compared to standard FEM: $\hat{\phi} : \Omega^e \to \hat{\Omega}^e$. For numerical integrations, elements are then mapped to a regularly shaped parent element: $\tilde{\phi} : \hat{\Omega}^e \to \tilde{\Omega}^e$. The inverse mapping $\tilde{\phi}^{-1}$ to an element $\hat{\Omega}^e = [\xi_i, \xi_{i+1}] \otimes [\eta_j, \eta_{j+1}]$ is:

$$\xi = \frac{1}{2}[(\xi_{i+1} - \xi_i)\tilde{\xi} + (\xi_{i+1} + \xi_i)] \qquad \eta = \frac{1}{2}[(\eta_{j+1} - \eta_j)\tilde{\eta} + (\eta_{j+1} + \eta_j)]$$

where $\tilde{\xi}$ and $\tilde{\eta}$ denote the coordinates of Gaussian points in the parent element. Thus, the numerical integration of a generic integrand $f(x,y)$ over an element is performed as:

$$\int_{\Omega^e} f(x,y)\, d\Omega^e = \sum_{q=1}^{n_{gp}} f(\tilde{\xi}_q, \tilde{\eta}_q) w_{gp_q} |\mathcal{J}_{\hat{\phi}}||J_{\tilde{\phi}}|$$

where n_{gp} and w_{gp_q} are the number of Gaussian points and associated weights while the Jacobian matrices associated with the two mappings are defined as:

$$\mathcal{J}_{\hat{\phi}} = \begin{bmatrix} \frac{\partial x}{\partial \hat{\xi}} & \frac{\partial x}{\partial \hat{\eta}} \\ \frac{\partial y}{\partial \hat{\xi}} & \frac{\partial y}{\partial \hat{\eta}} \end{bmatrix} \qquad J_{\tilde{\phi}} = \frac{\partial \hat{\xi}}{\partial \tilde{\xi}} \frac{\partial \hat{\eta}}{\partial \tilde{\eta}}$$

By assembling over elements, Eq. (5) can be cast into a global standard form:

$$(\mathcal{K}\mathbf{Z}) \cdot \mathbf{W} = \mathbf{q} \cdot \mathbf{W} \tag{6}$$

where $\mathcal{K}$ takes the meaning of shell geometric stiffness matrix, according to the classification provided in Sect. 1.
Given the arbitrariness of $\mathbf{W}$, (6) can be written as:

$$\mathcal{K}\mathbf{Z} = \mathbf{F} + \mathbf{q} \tag{7}$$

being $\mathbf{F}$ the vector of control point residuals, which are nonzero only at constrained control points and equal to the reaction forces of supports.
To solve this algebraic system for the unknown components of $\mathbf{Z}$ and $\mathbf{F}$, these vectors are split into the sub-vector relevant to degrees of freedom where the shell height is prescribed (subscript P), and the sub-vector relevant to the degrees of freedom where the shell height is unknown (subscript U). Accordingly, Eq. (7) becomes:

$$\begin{bmatrix} \mathcal{K}_{PP} & \mathcal{K}_{PU} \\ \mathcal{K}_{UP} & \mathcal{K}_{UU} \end{bmatrix} \begin{bmatrix} \mathbf{Z}_P \\ \mathbf{Z}_U \end{bmatrix} = \begin{bmatrix} \mathbf{F}_P \\ \mathbf{F}_U \end{bmatrix} + \begin{bmatrix} \mathbf{q}_P \\ \mathbf{q}_U \end{bmatrix} \tag{8}$$

so that the unknown shell heights are computed as:

$$\mathbf{Z}_U = \mathcal{K}_{UU}^{-1}[\mathbf{F}_U + \mathbf{q}_U - \mathcal{K}_{UP}\mathbf{Z}_P] \tag{9}$$

and the constraint reactions are evaluated as:

$$\mathbf{F}_P = \mathcal{K}_{PP}\mathbf{Z}_P + \mathcal{K}_{PU}\mathbf{Z}_U - \mathbf{q}_P \tag{10}$$

Prior to solving Eqs. (9)-(10), a suitable Airy potential field must be provided in order to evaluate the entries of $\mathcal{K}$. Without loss of generality, the shell is assumed to operate in a no-traction stress state. Hence, denoting by $\mathcal{N}$ the projected membrane stress tensor, the no-traction assumption requires $\mathcal{N}$ to be negative semi-definite throughout the shell domain, i.e. $(\mathcal{N}\hat{\mathbf{v}})\hat{\mathbf{v}} \leq 0, \forall \hat{\mathbf{v}} \in \Omega$, that is:

$$\begin{aligned} tr\mathcal{N} &= N_x + N_y = (\Phi_{,yy} - h_x) + (\Phi_{,xx} - h_y) \leq 0 \\ det\mathcal{N} &= N_x N_y - N_{xy}^2 = (\Phi_{,yy} - h_x)(\Phi_{,xx} - h_y) - \Phi_{,xy}^2 \geq 0 \end{aligned} \tag{11}$$

Fulfillment of the static condition of null membrane stresses at control points on any free edge of outward unit vector $\hat{\mathbf{n}}$ is also required:

$$\mathcal{N}\hat{\mathbf{n}} = 0 \quad \text{on} \quad \Gamma_f \quad \rightarrow \quad \begin{cases} (\Phi_{,yy} - h_x)\hat{n}_x - \Phi_{,xy}\hat{n}_y = 0 \\ \Phi_{,yx}\hat{n}_x - (\Phi_{,xx} - h_y)\hat{n}_y = 0 \end{cases} \tag{12}$$

Being

$$T = |(\mathcal{N}\hat{\mathbf{n}})\hat{\mathbf{n}}| = |(\Phi_{,yy} - h_x)\hat{n}_x^2 - 2\Phi_{,xy}\hat{n}_x\hat{n}_y + (\Phi_{,xx} - h_y)\hat{n}_y^2|$$

the magnitude of the thrust transmitted to the support in the generic control point of the constrained boundary Γ_c, a suitable Φ field is computed via the MatLab© optimization routine `fmincon` by minimizing the objective function:

$$\text{obj} = |T_{max} - T_{threshold}| \tag{13}$$

subject to the nonlinear constraints in Eqs. (11)-(12). Convergence of the numerical procedure ensures that the corresponding distribution of membrane stresses is such that thrusts transmitted to edge supports do not exceed the user-defined limit value $T_{threshold}$.

4 Results

Two examples of fully compressed shells are illustrated in this section. The procedure starts from a flat surface having a planar footprint equal to that of the shell form to be found. In the first case, the covered area is a 4-meter-long square. In the second case, the shell projected domain is bounded by staggered arcs along the edges of outward normal -y and +y, connected by free-form curves to each other. The shell is parameterized as a bicubic B-Spline patch of uniform open knot vectors: $\Xi = H = [0, 0, 0, 0, 1/4, 2/4, 3/4, 1, 1, 1, 1]$. The order of the surface and the multiplicity of internal knots have been chosen so as to achieve an inter-element continuity of class C^2, suitable for smoothly modelling free-form shells. In the following examples, the shell is subject to a uniform projected vertical load of 2 kN/m^2, representative of its self-weight. In particular, the thrust upper limit value $T_{threshold}$ is set so that the resulting shell heights meet practical constructability criteria.

4.1 Shell on Squared Projected Domain

The shell is supported on three sides while the edge of outward normal $+x$ is free. According to Eqs. (11)-(12), the Φ distribution must be sought among the concave functions such that $\Phi_{,yy} = 0$ and $\Phi_{,xy} = 0$ on the free edge. A parabolic distribution, combined with a surface built by extruding a parabola arc along the y-axis into the first and fourth quadrants, is adopted as initial guess:

$$\Phi^{guess}(x, y) = \begin{cases} -150x^2 - 40y^2 & if \quad x < 0 \quad \forall y \in \Omega \\ -300x^2 & if \quad x \geq 0 \quad \forall y \in \Omega \end{cases}$$

Numerical convergence is attained by using the Sequential Quadratic Programming (SQP) algorithm with $T_{threshold} = 1\,kN$ and relative constraint and step tolerances equal to $1e^{-6}$ and $1e^{-10}$, respectively. Optimized Airy stress function and corresponding projected membrane stresses are visualized in Fig. 1, where the fulfillment of boundary conditions can be verified. An isometry of the resulting form-found geometry is reported in the same figure.

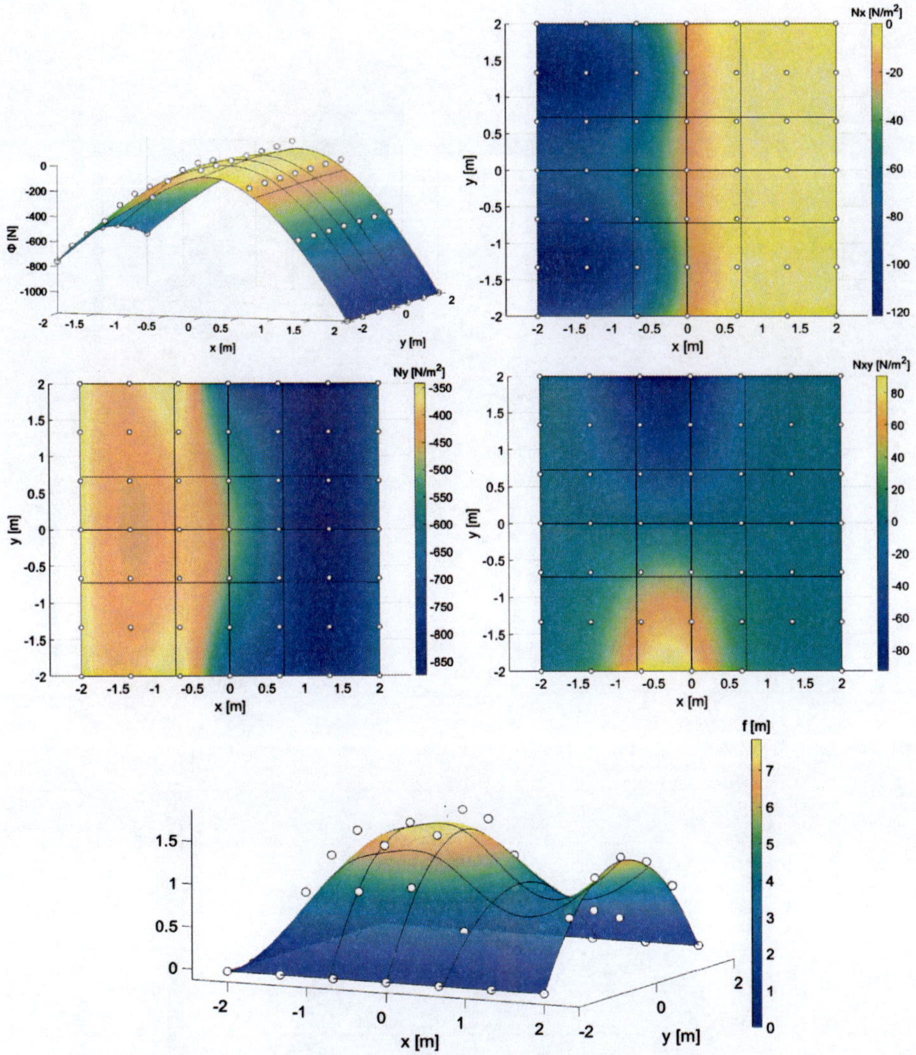

Fig. 1. Optimized Airy stress function (top left), projected membrane stresses (top right and middle row), and isometric view of form-found shell (bottom).

4.2 Shell on Curved Projected Domain

The shell is supported on two sides while the edges of outward normal $-x$ and $+x$ are free. Given the variability of the normal unit vector along the free edges, a polynomial surface of third order is prescribed as initial guess:

$$\Phi^{guess}(x,y) = -70 - 200x - 200y + 30x^2 + 15xy + 30y^2 + 10x^3 + 7x^2y + 7xy^2 + 10y^3 \qquad \forall (x,y) \in \Omega$$

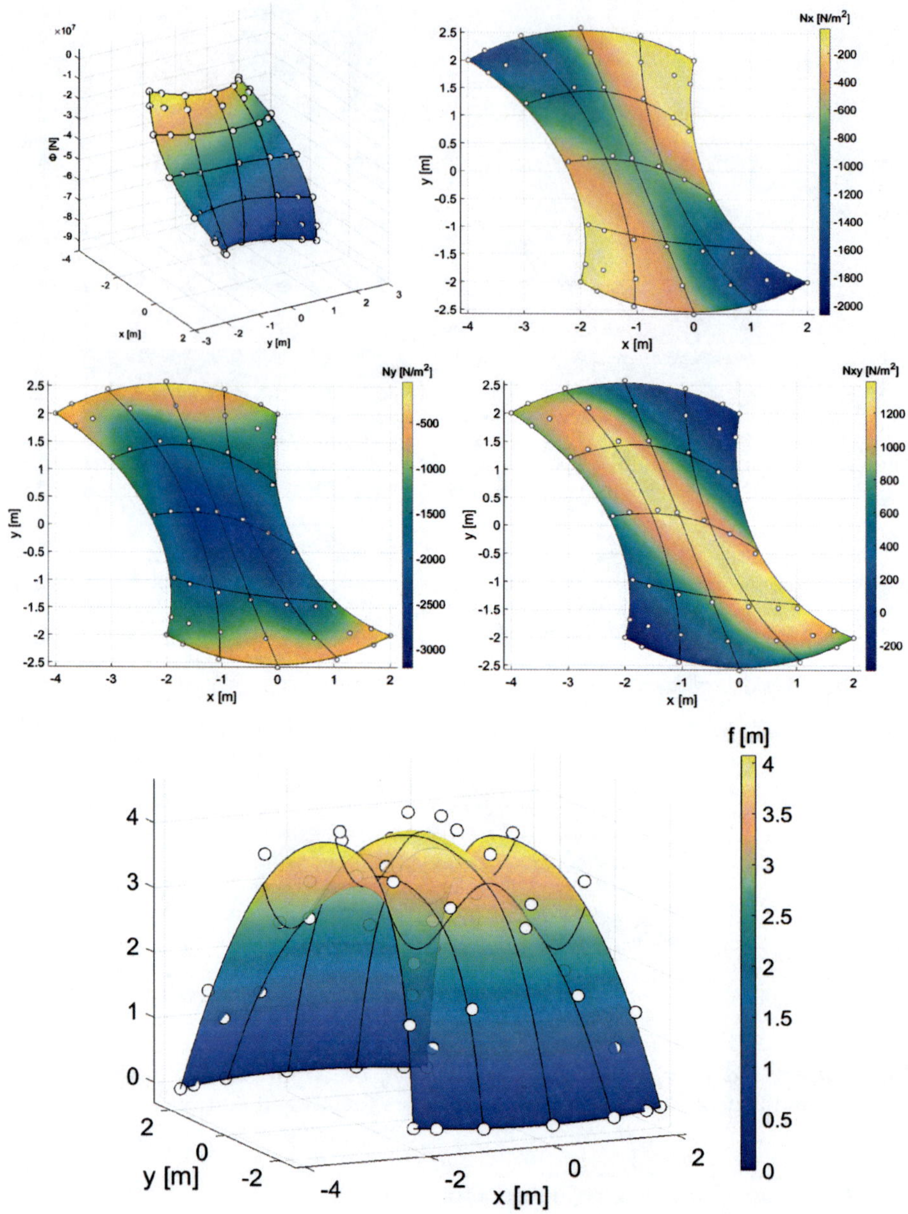

Fig. 2. Optimized Airy stress function (top left), projected membrane stresses (top right and middle row), and isometric view of form-found shell (bottom).

Numerical convergence of the SQP method is reached with $T_{threshold} = 1.55\,kN$ and relative constraint and step tolerances equal to $1e^{-5}$ and $1e^{-10}$, in turn. Corresponding optimized results are visualized in Fig. 2.

5 Conclusions

A numerical form-finding procedure, based on Pucher's theory, has been presented. The procedure finds the form of a shell of given planar footprint so as to equilibrate applied loads in a state of pure membrane stresses. Use of B-Spline basis functions, featuring Isogeometric Analysis, makes the approach applicable to shells of general free-form boundaries.
With predefined external loads, most of existing algorithms assign the Airy potential field as an input, then solve for the shell form. Besides concealing the subtle violation of boundary conditions, the arbitrary prescription of working stresses is often limiting, especially when the resulting configuration is not satisfactory and revising the input is not intuitive. Integration of the proposed optimization routine into the form-finding procedure proves to be a feasible tool not only to assist the user with the evaluation of an adequate Airy stress function, but also to conduct form finding instrumentally by satisfying a specific objective function, besides usual constraints.

Funding and Acknowledgments. The present research was supported by the University of Naples Federico II and the *Compagnia di San Paolo*, which are gratefully acknowledged by the authors, FRA grants, CUP E69C21000250005.

References

1. Veenendaal, D., Block, P.: An overview and comparison of structural form finding methods for general networks. Int. J. of Solids Struct. **49**, 3741–3753 (2012)
2. Li, Q., Su, Y., Wu, Y., Borgart, A.: Form-finding of shell structures generated from physical models. Int. J. Space Struct. **32**(3), 11–33 (2017)
3. Chen, L., Nguyen-Than, N., Rabczuk, T., Bordas, S., Limbert, G.: Explicit finite deformation analysis of isogeometric membranes. Comput. Meth. in Appl. Mech. Eng. **277**, 104–130 (2014)
4. Nguyen, N., Rabczuk, T., Nguyen-Xuan, H., Bordas, S.: A smoothed finite element method for shell analysis. Comp. Meth. Appl. Mech. Eng. **198**(2), 165–177 (2008)
5. Nguyen, N., Hien, T., Nguyen-Thoi, T., Lee, J.: A unified adaptive approach for membrane structures: form finding and large deflection isogeometric analysis. Comp. Meth. Appl. Mech. Eng. **369**, 113239 (2020)
6. Schek, H.J.: The force density method for form finding and computation of general networks. Comput. Methods Appl. Mech. Eng. **3**(1), 115–134 (1974)
7. Marmo, F., et al.: On the form of the Musmeci's bridge over the Basento river. Eng. Struct. **191**, 658–673 (2019)
8. Block, P., Ochsendorf, J.: Thrust network analysis: a new methodology for three-dimensional equilibrium. J. Int. Ass. Shell Spat. Struct. **48**(3), 8 (2007)

9. Marmo, F., Vaiana, N.: Form finding of shell structures by using membrane theory. In: Marmo, F., Sessa, S., Barchiesi, E., Spagnuolo, M. (eds.) Mathematical Applications in Continuum and Structural Mechanics. ASM, vol. 127, pp. 213–237. Springer, Cham (2021). https://doi.org/10.1007/978-3-030-42707-8_11
10. Pucher, A.: Über den spannungszustand in gekrümmten flächen. Beton Eisen **33**, 298 (1934)
11. Manuello, A.: Multi-body rope approach for grid shells: form-finding and imperfection sensitivity. Eng. Struct. **221**, 111029 (2020)
12. Manuello, A., Riberi, F.: Form-finding of pierced vaults and digital fabrication of scaled prototypes. Curved Layered Struct. **8**, 210–224 (2021)
13. Kilian, A., Ochsendorf, J.: Particle-spring systems for structural form finding. J. Int. Ass. Shell Spat. Struct. **46**(148), 77–84 (2005)
14. Ye, J., Feng, R., Zhuo, S.: The modified dynamic relaxation method for the form-finding of membrane structures. J. Comput. Theoret. Nanosci. **4**(8–10), 2845–2853 (2011)
15. Adriaenssens, S., Rhode-Barbarigos, L.: Form-finding and analysis of bending-active systema using dynamic relaxation. In: VI International Conference on Textile Composites and Inflatable Structures (2013)
16. Hilmersson, J., Larsson, F., Olsson, J., Ander, M.: Isogeometric analysis and form finding for thin elastic shells. In: Proceedings of IASS Annual Symposia: Contributions in Memory of Mike Barnes, pp. 1-8(8) (2019)
17. Cottrell, J.A., Hughes, T.J.R., Bazilevs, Y.: Isogeometric analysis: towards integration of CAD and FEA. Wiley (2009)
18. Xia, Y., Mantzaflaris, A., Juttler, B., Pan, H., Hu, P., Wang, W.: Design of self-supporting surfaces with isogeometric analysis. Comput. Methods Appl. Mech. Eng. **353**, 328–347 (2019)
19. Miki, M., Igarashi, T.: Parametric self-supporting surfaces via direct computation of airy stress functions. ACM Trans. Graph. **34**(4), 89 (2016)
20. Alic, V., Persson, K.: Form finding with dynamic relaxation and isogeometric membrane elements. Comput. Methods Appl. Mech. Eng. **300**, 734–747 (2016)
21. Milosevic, J., Kovacevic, S.: Form finding of shell structures based on isogeometric analysis. Izgradnja **70**, 353–360 (2016)

Nervi's Isostatic Lines' Inspired Floor Slabs. Beyond the Archetypal Gatti Woolen Mill in Rome

Marta Lembo[1](✉), Alberto Bologna[2], and Francesco Romeo[3]

[1] Department of Architecture, Construction and Design, Polytechnic of Bari, Bari, Italy
marta.lembo@poliba.it
[2] Department of Architecture and Design, Sapienza University of Rome, Rome, Italy
alberto.bologna@uniroma1.it
[3] Department of Structural and Geotechnical Engineering, Sapienza University of Rome, Rome, Italy
francesco.romeo@uniroma1.it

Abstract. In 1949 the construction company Nervi & Bartoli patented in Italy a construction system for two-dimensional structures with ribs bent in accordance with isostatic lines of bending moments or normal stresses, invented by an engineer from its staff: Aldo Arcangeli. In the years to come "isostatic lines" ribbed slab would become one of the main and distinctive archetypes used by Pier Luigi Nervi throughout the world. This is a cast-in-place reinforced concrete ribbed slab combining static efficiency with aesthetic and spatial qualities. The paper will present different case-studies, demonstrating (with a trans-disciplinary approach) how and when aesthetic, static and constructive issues overcome a rigorous mathematical morphogenesis. Case studies will show how the different patterns of the ribs are influenced both by structural instances (e.g. geometry, loads and constraints) and by the specificities imposed by each design and building site (e.g. use, craftmanship, economy, and replicability).

Keywords: Pier Luigi Nervi · Aldo Arcangeli · Isostatic lines · ribbed slab · Cuore Immacolato di Maria church Borgo Panigale · Cassa di Risparmio Venice

1 A Slab with Ribs Guided by Bending Moment Isostatic Lines

1.1 Aldo Arcangeli's Invention

The patent titled "Improvement in the construction of floors, vaults, domes, wall-beams and load-bearing structures in general with two or three dimensions, with the arrangement of the resistant ribs along the isostatic lines of bending

S. Gabriele et al. (Eds.): IWSS 2023, LNCE 437, pp. 51–61, 2024.
https://doi.org/10.1007/978-3-031-44328-2_6

moments or of normal strains" was filed in Rome on July 23, 1949 by the Nervi & Bartoli construction company: the inventor is Aldo Arcangeli [1,2]. Interestingly enough, on the same day Nervi filed the patent titled "Improvement in the construction of floors, slabs and other reinforced-cement structures" [3,4]. Arcangeli is an engineer who collaborates in those years with Pier Luigi Nervi, as employee of the Nervi & Bartoli construction company, where he works on the design of numerous structures; he also was Nervi's assistant at the Faculty of Architecture of Sapienza University of Rome until 1960, where he hold lessons on thin vaulted structures, on which he published the volume "Nozioni sulle volte sottili" [5]. The constructive success that leads to the definition of archetypal architectural elements capable of unequivocally marking Nervi's subsequent designs resulting from these two inventions is due to their simultaneous application. The first explains the theory of isostatic lines, identifies the potential applications and illustrates the opportunity to obtain economic savings through the construction of bended ribs; the second describes the method of realization of these structural elements using ferrocement formwork, i.e. it attributes feasibility to the theoretical concept.

1.2 The First Experiments: Gatti Woollen Mill

Arcangeli's patent and Nervi's studies on slabs with ribs bent in accordance with isostatic lines of the bending moments were applied in the design for the basement floor slab of the Gatti woollen mill in Rome: built in 1951, it immediately became famous for the geometry of its ribs thanks to an iconic photograph by Oscar Savio which was quickly spread around the world. Nervi created for the first time an extremely expressive floor slab which, through its ribs, showed the static forces acting inside the concrete, revealing them as something absolute through an architectural gesture characterised by strong aesthetic values. The isostatic lines express an harmonic flows of each structural element of the floor slab, linking their formal elegance to the geometry and the static stresses which they are subjected [6]. Mario Desideri, a long-time collaborator of Nervi, expressed soon some critical issues arising from the awareness of impossibility of building these ribs completely excluding twisting moments [7]. The ribs arranged along the isostatic lines cannot be exempt of twisting moments, since these lines belong to an infinite family of curves within a plate of constant thickness and, when the solid plate is transformed into a ribbed slab, they generate a variable section slab. It is therefore more appropriate to define these kind of structures as "ribbed slab inspired by isostatic lines": the application of physical principles is thus combined with the inspiration and interpretation of natural realities. In the years following the completion of the Gatti woollen mill, the isostatic-inspired ribbed floors became a distinctive archetype of Studio Nervi's production. However, it can be said that the success of these structural systems derives from the design ability not to exclusively express static issues, but to enrich the overall formal aspect of the building within which they were inserted. The panorama of works that Nervi designs and builds using this archetype is vast: however, some of them lend themselves to being studied better than others in order to understand the design methods through which Nervi was able to combine his distinctive expressive language with static and constructive issues.

2 Case Studies

The following case studies have been selected because they have common features that make a comparative analysis possible: the roofing slab of the Cuore Immacolato di Maria church in Borgo Panigale and the main hall ribbed slab of the Cassa di Risparmio in Venice. The case studies are comparable in size: 23 m is the diameter of the circular plan of the church in Borgo Panigale, 24 m is the side of the square plan of the hall of the Cassa di Risparmio in Venice. Both floors are supported by four isolated columns and have a geometric configuration with a double symmetry. The difference between these structures lies mainly in their static role. The roof of the Borgo Panigale church (Fig. 1a, Fig. 2a) consists of a 10 cm slab and a system of ribs with a height of 50 cm and a thickness of 15 cm. The supporting columns are 115 cm wide at the base and 40 cm wide at the top. The floor of the Cassa di Risparmio in Venice (Fig. 1b, Fig. 2b) is the roof of the main hall, but it is also designed to support the rest of the building above: it receives the loads from 24 columns on the side beams, from 12 columns in its central area and also supports the entire technical ceiling. The floor consists of a 20 cm slab and 200 cm high ribs with a depth varying from 25 to 70 cm: its supporting columns are 120 cm wide at the base and 380 cm wide at the top. The need for a different load-bearing capacity led to different construction and structural solutions.

Fig. 1. Photos of the roof structures. a) The church in Borgo Panigale (credit Sofia Nannini); b) the Cassa di Risparmio in Venice (creative commons https://www.flickr.com/photos/jaqueline_poggi/16705955686).

2.1 The Roofing Slab of the Cuore Immacolato di Maria Church in Borgo Panigale

The church in the Ina-Casa neighbourhood of Borgo Panigale, near Bologna, was designed by the architect Giuseppe Vaccaro and built between 1957 and 1962:

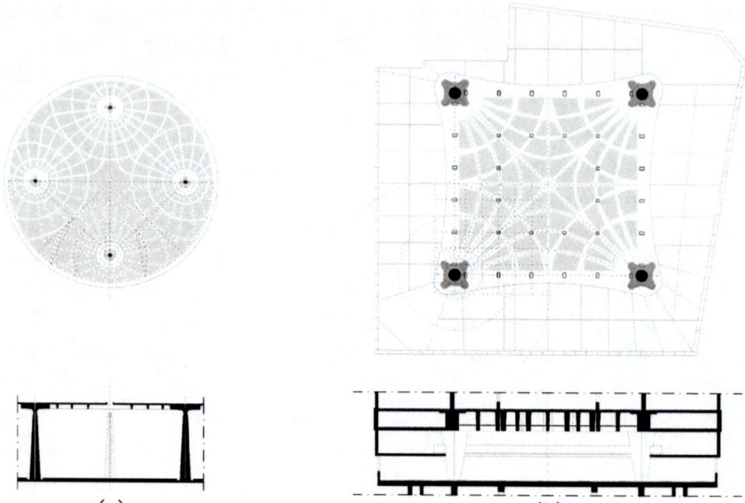

(a) (b)

Fig. 2. Plans of the roof structures. a) The church in Borgo Panigale; b) the Cassa di Risparmio in Venice.

Nervi was called upon to collaborate on the technical-constructive definition for the design and construction of the circular roof supported by four columns [8,9]. The church is characterised by a circular hall, covered by a ribbed slab, and an annular space, covered by a slab at a lower level with a smooth surface at the intrados. The liturgical space is defined solely by the presence of the four columns. The reinforced concrete columns have a cruciform cross-section, they are tapering upwards and have been realised using a formwork composed by wooden boards [10]. The connection to the roof slab is realised by means of an elegant mushroom-shaped element. For the construction of the roof floor slab, the characteristic tavelloni (on-site pre-fabricated thin concrete slabs), already used by Nervi for the construction of domes, were used. The design drawings (drawn up to explain the construction process to a construction company other than the Nervi & Bartoli) contain detailed explanations on the building method of the tavelloni and their geometry. Nervi envisages the negative forms of the tavelloni being made on the ground using masonry blocks finished externally with plaster, reproducing a single quadrant of the roof; on the top of these forms, the corresponding tavelloni were made one by one. In the joining of two tavelloni, the groove between the two elements remains visible: this groove thus keeps visible the construction process based on the assembly of previously pre-fabricated pieces. The design of the ribs (Fig. 3) of this floor is inspired by the isostatic moment lines of a single circular mushroom-shaped floor with a central constraint and two rows of beams: one is radial, the other is concentric. The whole floor slab is supported by 4 columns and is the result of the simple juxtaposition of 4 identical modules: the beams of each module are connected by curved joints and do not undergo any geometric change in the contact with the adjacent module. Finally, the dimension of the portion of the circular slab sector close to the capital incorporates the section of the structural elements close to the nodal point formed in correspondence with each column: the 20 radial beams and the first circular rib. The joining of the two outer circular beams of each module (resolved by the construction of another portion of solid slab) is foreseen in order to avoid the construction of 2 too narrow (and unfeasible) tavelloni and to follow the idea of the orthogonality of the families of isostatic lines.

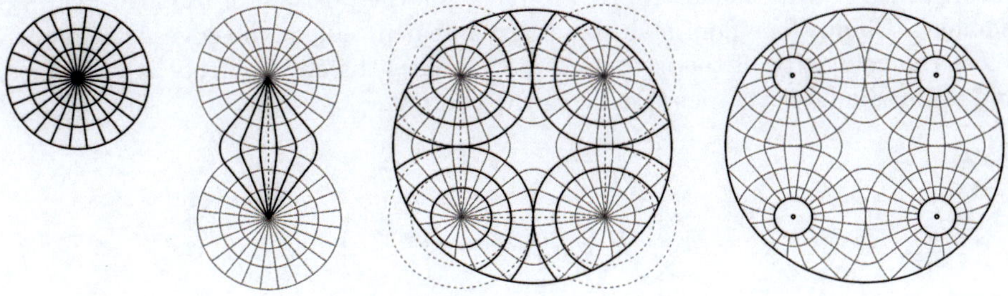

Fig. 3. Geometric genesis of the floor ribs of the church in Borgo Panigale.

2.2 The Main Hall Floor Slab of the Cassa di Risparmio in Venice

The building of the Cassa di Risparmio in Venice was built between 1963 and 1972. Architectural design is initially entrusted only to architect Angelo Scattolin, who later involved Nervi and the Nervi & Bartoli construction company [11]. Nervi's assignment is aimed at solving problems related to the requirement to minimise the footprint of ground floor structures and the fragile nature of Venice subsoil. For this reason, the design provided for the construction of a reinforced concrete foundation bulkhead with four walls onto which the columns of the ground floor are grafted: they support a huge technical floor slab supporting three upper floors. This design solution ensured total flexibility of the hall space on the ground floor, defined only by the presence of the large columns with variable cross-section: circular at the base, cruciform at the top. The curvilinear ribbed floor slab represents the emblem of this building: a single architectural element meets the requirements for extraordinary static capacity and at the same time defines the internal spatial quality [12]. The building site was opened in 1968: operations proved to be extremely laborious, also due to the absolute lack of space to organise a prefabrication site within the city of Venice. In an area located outside the city, a counter-mould corresponding of a quarter of the whole floor slab, with a thickness of 68 cm was first casted on the ground. On top of this, ferrocement formwork were then casted: they were then transported on motorboats and then assembled on-site for the casting of the cast-in-situ reinforced concrete floor slab [13, 14]. The design of the ribs (Fig. 4) of this floor slab is inspired by the isostatic lines of bending moments in a square floor constrained at the corners. However, the discretization of the lines is linked to the need to support columns with a regular square bay size of 3.48 m in the central area, in addition to those on the edge beams. For this reason, the axes of both rows of ribs pass over the axes of the upper columns. This constraint, linked to the structural bay of the entire building, was far more binding than any other design requirement and could not be disregarded. It can also be noted that the design of this floor slab can be traced back to the design of four mushroom-shaped modules with the central constraint in the main columns: there is only a misalignment of the concentric ribs to reconcile the position of the load-bearing elements. Finally, the floor slab has a curvilinear profile at the edges that reproduces the convexity of the isostatic lines, reduces the shear effects at the edges and accommodates the positioning of the column.

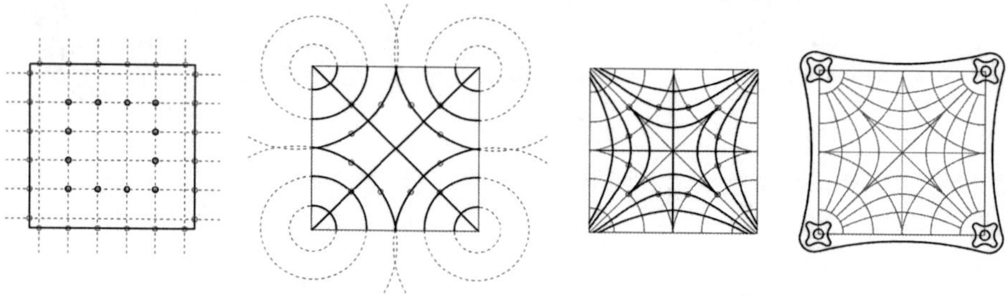

Fig. 4. Geometric genesis of the floor ribs of the Cassa di Risparmio in Venice.

3 Case Studies Comparative Static Analysis

The aim of this analysis is to make a comparison between ideal ribs following the isostatic lines of the bending moments and the built ones in the two case studies. The deviation between the two sets of ribs is measured in terms of twisting moments [15]. Similar steps were carried out for both case studies: slabs equivalent to the built ones in terms of external boundary geometry and mass were dimensioned; self-weight (concrete C25/30) and point hinge constraints were assigned; bending moment isostatic lines were numerically extracted from the static analysis; these lines were downsampled to match the actual number of ribs; loads and constraints were verified, according to the initial hypothesis; the difference between the twisting and bending moments along the selected curves has been evaluated.

3.1 The Roofing Slab of the Cuore Immacolato di Maria Church in Borgo Panigale

For the definition of the mesh, the position of the columns and other 208 equidistant and doubly symmetric points along the edge were considered as reference points. From the mesh analysis, the two families of bending moment isostatic lines were extracted (Fig. 5a). The observation of the obtained geometry allowed us to confirm the type of constraint (hinge). The family of lines 'concentric' to the column is not circular, while the orthogonal family of radial lines is sparser in the central zone, while in the outer zone tend to become parallel to the axis between two columns. It is also verified that the isostatic lines of moments run along the diagonals connecting the columns. Subsequently, an attempt was made to discretise the mesh points in relation to the realised shape in order to make a comparison on comparable geometries (Fig. 5b). Twenty concentric lines and 24 transverse lines were extracted. A comparison with the floor slab profile realized by Nervi shows that the first family of lines is not perfectly circular and the constructive simplification operated on the eight major ribs (two for each quadrant) at the intersection with the edge beam is almost irrelevant; the family of radial lines in the central area is similar to the built ribs, but does not have a change of direction towards the edge of the slab.

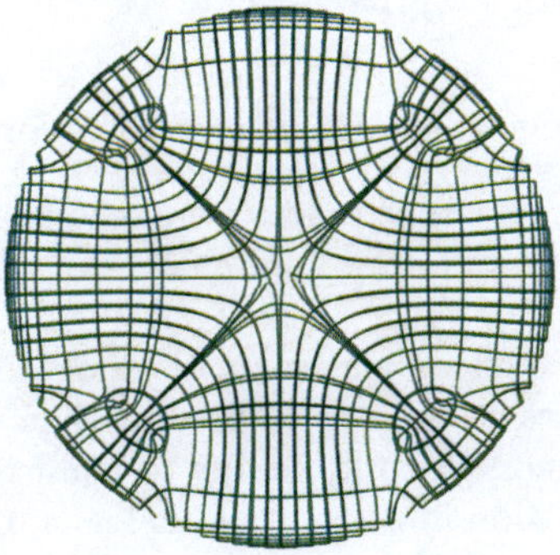

(a)

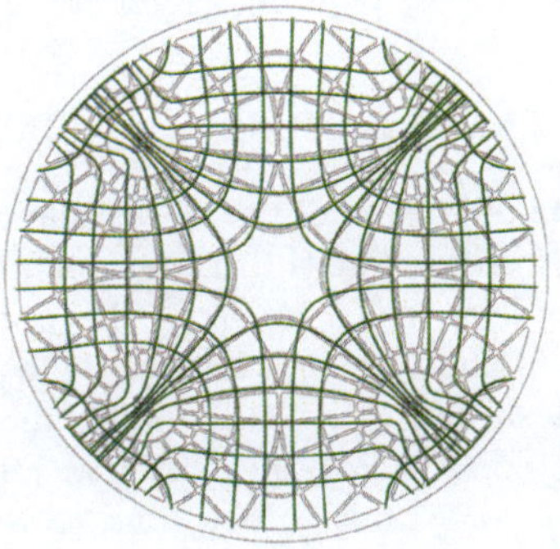

(b)

Fig. 5. Bending moment isostatic lines corresponding to equidistant points along the boundary. a) Numerical isostatic lines; b) Sampled isostatic lines superimposed to the realised ribs.

It was decided to compare the ideal geometry with the real one by analysing the twisting moment in the longest concentric ribs. This choice was prompted by the idea of focusing on the ribs exhibiting the greatest difference in geometry. The analysis (Fig. 6a) shows the twisting moments along the two curves. It is clear that the first case shows a negligible twisting moments. The results show that the maximum twisting moment of the ideal rib is only 11.23% of the maximum twisting moment of the actual rib. This result confirms that the ribs of the Borgo Panigale church are not derived from a strict structural optimisations,

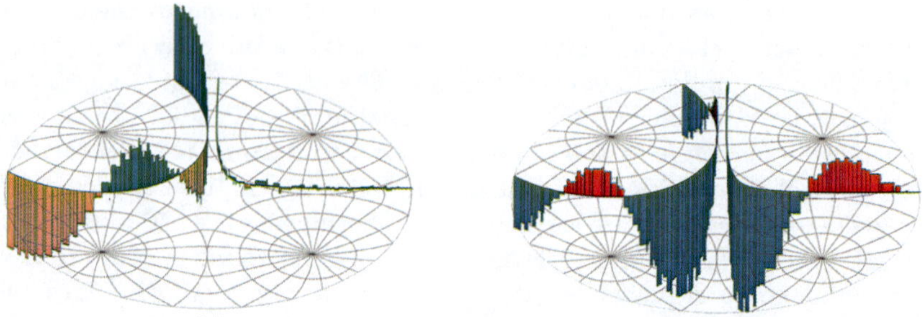

(a) (b)

Fig. 6. a) Diagrams of twisting moments along one of the built ribs (left) and along the corresponding ideal isostatic line (right); b) Diagrams of bending moments along one of the built ribs (left) and along the corresponding ideal isostatic line.

presenting a widespread geometric violation over a large part of the floor slab. Slightly larger bending moments are obtained along the ideal curves (Fig. 6b).

3.2 The Main Hall Floor Slab of the Cassa di Risparmio in Venice

For the definition of the mesh, the considered reference points were: the position of the columns at the corners of the structure and 46 points along the edge of each side in order to have the points with a distance equal to a submultiple (1/8) of the distance between the columns (3.48 m) acting on the floor. From the mesh analysis, the bending moment isostatic lines in both directions were extracted (Fig. 7a). The observation of this geometry allowed us to confirm the type of constraint (hinge) and to investigate the influence of point loads on the isostatic lines. The result is that, despite the pillar load being 78.51% of the slab mass, it is not possible to appreciate a real variation in the trajectories of the lines. This phenomenon is justified by the presence of punctual constraints that generally govern the flow of forces and by the homogeneous distribution of the 36 columns over the entire surface of the slab, which is therefore similar to a distributed load. Some considerations can be made about the two families of lines: the family of lines "concentric" to the column is not assimilated to a circumferential arc, but is orthogonal to the diagonal between the columns in the area close to the restraint and tapered towards the center; the family of lines arising from the column is sparser in the central area. Subsequently, an attempt was made to discretise the mesh points in relation to the built shape in order to make a comparison between similar geometries (Fig. 7b). Twelve concentric lines, 12 radial lines, 2 diagonal lines and 2 centerlines were extracted. The comparison with the profile of the floor slab realized by Nervi shows that the first family of lines has a variable and non-circular profile; it finds a significant coincidence with the second rib while the family of radial lines in the central area is similar to the built ribs. Given these results, it was decided to carry out a comparison between the ideal geometry and the actual geometry by analysing the twisting moments in two cases: the largest concentric rib, which has a more pronounced divergence, and the second concentric beam, which, on the contrary, is almost coincident. The analyses show the twisting moments stresses along the curves. From Fig. 8a it appears that for the non coincident case the maximum twisting moment of the ideal line is 10.77% of the maximum twisting moments of the real line. This figure confirms that, as assumed, this rib does not present a structural optimization with regard to the twisting moments. In any case, it is worth recalling that the position of this rib is constrained by the presence of two columns above it. This choice is also justified by the analysis of the bending moments. The maximum bending moment along the isostatic line of moments is 118.10% of the bending moment along the realized rib.

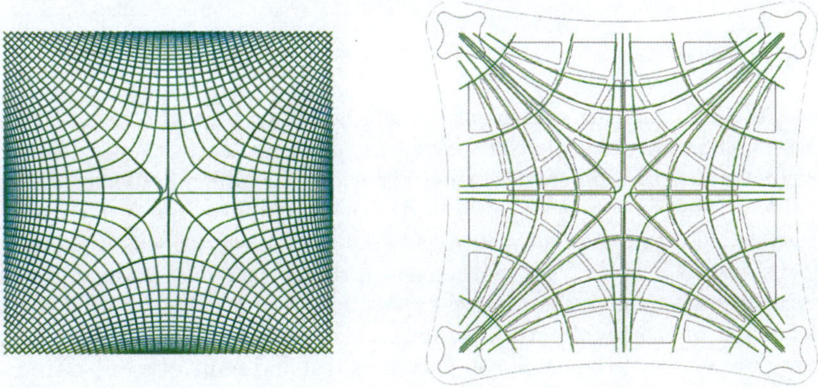

(a) (b)

Fig. 7. Bending moment isostatic lines corresponding to equidistant points along the boundary. a) Numerical isostatic lines; b) Sampled isostatic lines superimposed to the realised ribs.

In the second case, the maximum twisting moment of the ideal line is 14.42% of the maximum twisting moment of the real line at the fixed constraint and 93.24% at the centreline. This figure confirms that this rib has practically zero twisting moment at the centreline, deviating from the perfect solution only in the area where it is actually embedded in the edge beam. This coincidence is also confirmed in the analysis of the bending moments.

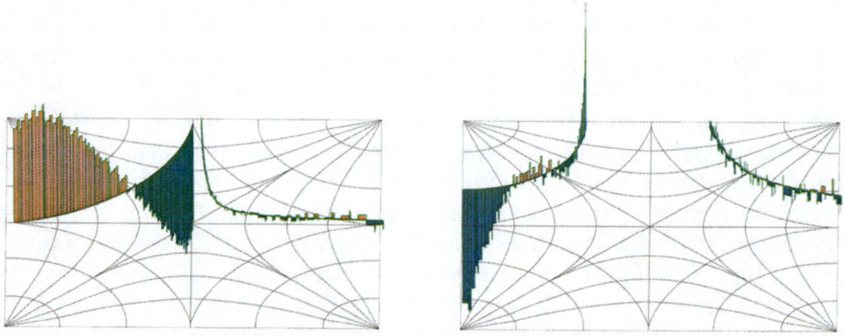

Fig. 8. Diagrams of twisting moments: a) Along one of the realized ribs (left) and along the corresponding ideal isostatic line (right); b) Along one of the realized ribs (left) and along the corresponding ideal isostatic line.

4 Conclusions

While it is clear that the success of Nervi's buildings were the result of a skilful agreement between spatial sensitivity, architectural expression, construction skill, structural truth and static-formal intuition, it is difficult to shed definitive light on how these factors were calibrated with each other in his design approach. The case studies show that the solutions of the realised works, with the exception of a few elements, are far from mathematical rigour: the ideal trajectories of the isostatic lines are a mere inspiration to the design and not a binding factor in the definition of the architectural form. By computing the self-weight twisting moment, a measure of the deviation between ideal and built ribs was introduced. It showed that the actual pattern of the ribs is independent of the theoretical isostatic line of bending moments. This happens as much due to construction issues related to the realization of ribs strictly following the ideal isostatic lines as to factors related to the aesthetic expectation that these floor slabs had to meet.

References

1. Iori, T.: Le plancher à nervures isostatiques de Nervi. In: Gargiani, R. (ed.) L'architrave le plancher la plate-forme: Nouvelle histoire de la construction, pp. 755–760. EPFL Press, Lausanne (2012)
2. Patent Archival Collection, Archivio Centrale dello stato, Roma (ACS), patent IT 455678, 23 July 1949
3. Greco, C.: Pier Luigi Nervi: dai primi brevetti al Palazzo delle Esposizioni di Torino 1917–1948, pp. 152–153. Heinz Wirz, Lucerna (2008)
4. Patent Archival Collection, Archivio Centrale dello stato, Roma (ACS), patent IT 455750, 23 July 1949

5. Arcangeli, A.: Nozioni sulle volte sottili. Agenzia Liber, Roma (1951)
6. Nervi, P.L.: Costruire correttamente: caratteristiche e possibilità delle strutture cementizie armate. Hoepli, Milano (1955)
7. Desideri, M.: In: Castelli, F.R., Del Monaco, A.I. (eds.) Pier Luigi Nervi e l'architettura strutturale, pp. 122–139. Edilstampa, Roma (2011)
8. Nervi, P.L.: Soluzione tecnico costruttiva della copertura per la chiesa dell'architetto Vaccaro a Borgo Panigale. Chiesa e Quartiere **4**, 32–35 (1957)
9. Nervi, P.L., Vaccaro, G.: Chiesa e complesso parrocchiale del Cuore Immacolato di Maria nel quartiere INA-Casa a Borgo Panigale. Chiesa e Quartiere **20**, 74–98 (1961)
10. Gargiani, R., Bologna, A.: The Rhetoric of Pier Luigi Nervi: Concrete and Ferrocement Forms, pp. 337–342. EPFL Press, Lausanne (2016)
11. De Magistris, A., Deambrosis, F.: Palazzo Nervi-Scattolin, pp. 36–73. Venezia. Skira, Milano (2021)
12. Collezione MAXXI Architettura. Archivio Pier Luigi Nervi (Maxxi - PLN) F3025
13. Collezione MAXXI Architettura. Archivio Pier Luigi Nervi (Maxxi - PLN) F3121
14. Collezione MAXXI Architettura. Archivio Pier Luigi Nervi (Maxxi - PLN) F3140
15. Halpern, A.B., Billington, D.P., Adriaenssens, S.: The ribbed floor slab systems of pier Luigi Nervi. J. Int. Assoc. Shell Spatial Struct. **54**, 176–177, 127–136 (2013)

Grid-Shell Multi-step Structural Optimization with Improved Multi-body Rope Approach and Multi-objective Genetic Algorithm

Jonathan Melchiorre[1(✉)], Stefanos Soutiropoulos[1], Amedeo Manuello Bertetto[1], Giuseppe Carlo Marano[1,2], and Francesco Marmo[3]

[1] Department of Structural, Geotechnical and Building Engineering, Politecnico di Torino, DISEG, Corso Duca Degli Abruzzi, 24, 10128 Turin, Italy
jonathan.melchiorre@polito.it

[2] College of Civil Engineering, Fuzhou University, Fuzhou 350108, China

[3] Department of Structures for Engineering and Architecture, University of Naples Federico II, Naples, Italy

Abstract. Grid-shell structures are popular in roofing for their aesthetic appeal and structural advantages, enabling the construction of lightweight, large roofs using slender elements. However, their widespread use is limited by the complexities involved in their construction. This paper presents an innovative solution to address the challenges associated with grid-shell construction. The proposed approach combines the enhanced Multibody Rope Approach (i-MRA) form-finding method with metaheuristic optimization algorithms to identify optimal design solutions from both structural and constructional perspectives. To facilitate practical implementation, a parametric code was developed using MATLAB, which was later converted to C# for integration with the parametric design software "Grasshopper". A multi-objective optimization problem was formulated to minimize the use of different structural elements, reduce material consumption, account for production waste, and ensure compliance with structural verification requirements. The optimization process yielded a Pareto front solution that enables the conceptual design of grid-shell structures. This approach provides an efficient and innovative solution to the complex construction of grid-shell structures, which could potentially pave the way for their widespread use in the future.

Keywords: Structural optimization · Form-finding · Pareto front · Gridshell · Multi-objective optimization

1 Introduction

Structural optimization plays a crucial role in achieving efficient and cost-effective designs in various fields of engineering [1–3]. By utilizing optimization techniques [4,5], engineers can enhance the performance of structures while

S. Gabriele et al. (Eds.): IWSS 2023, LNCE 437, pp. 62–72, 2024.
https://doi.org/10.1007/978-3-031-44328-2_7

minimizing material usage and construction costs [6]. In the realm of gridshell structures, which offer unique architectural possibilities, structural optimization becomes even more vital to strike a balance between form, function, ease of construction management and resource utilization [7]. In this paper, a comprehensive approach for the optimization of gridshell structures is presented. The method is specifically focused on the combination of form-finding techniques with meta-heuristic multi-objective optimization methods to identify a range of optimal solutions to guide the designer in the preliminary design phase. The proposed optimization methodology combines the improved Multi-body Rope Approach (i-MRA) method for form-finding of gridshell [8] with a multi-objective optimization genetic algorithm. The i-MRA method enables the determination of structurally optimal gridshell shapes by considering the given loading conditions, while the multi-objective optimization algorithm minimizes material usage while reducing the construction complexity by minimizing the number of types of elements to be handled on the construction site. The entire process and analyses are conducted using a self-made code in MATLAB [9], providing a versatile computational framework for optimization and analysis. To facilitate the practical application of the i-MRA form-finding method, a software component has been developed to integrate it within parametric design software Grasshopper, a tool that runs in Rhinoceros3D environment [10]. This component enhances accessibility by allowing structural designers and architects to utilize the i-MRA method in the design of gridshell structures. The primary objectives of this study are two-fold: first, to simplify the construction management of gridshell structures by reducing the number of different structural components required for their construction; and second, to minimize material consumption through the optimization process. Furthermore, the Cutting Stock Problem is incorporated to optimize the cutting pattern and minimize material waste during the production of structural elements. By achieving these objectives, the proposed approach aims to streamline the construction process, reduce costs, and promote sustainable practices in gridshell design. To demonstrate the effectiveness of the methodology, two case studies are presented. The first case study focuses on a square gridshell subjected to self-weight and various nodal forces. Optimization results are presented in terms of the number of different structural element types and the quantity of material used. The second case study involves the optimization of a more complex gridshell. Results are compared for different loading conditions, considering the number of different element typologies and both net material consumption and waste.

2 The MRA for Large Grid Shell: The Base Assumption and the Improved Version

The Multibody Rope Approach (MRA) [11,12] is a cutting-edge method for designing gridshell structures. MRA is specifically tailored for gridshell constructions that utilize free-forms and standardized building elements. By employing a dynamic model of falling bodies in space and time, MRA iteratively calculates the final equilibrium configuration for each node using D'Alembert's principle.

In the MRA the structural elements are modelled as ropes that connect the masses in the nodes. Specifically, the rope element exerts forces on the masses so that they do not move apart beyond the prescribed distance. The aim is to generate a geometry that is both structurally efficient and composed a reduced number of structural element typologies. The final equilibrium configuration of the structure is an inverted representation of the hanging net, resulting in a visually striking design.

In order to achieve an efficient construction of a gridshell, it is important to consider the number of required beam elements and their respective lengths. Thus, it is essential to determine the number of structural elements in the final configuration that possess a length equivalent to the target length l_{rope}. In practical applications, the complexity of gridshell assembly increases proportionally with the number of distinct elements needed for construction. Therefore, it is crucial to classify structural elements into three categories based on their length.

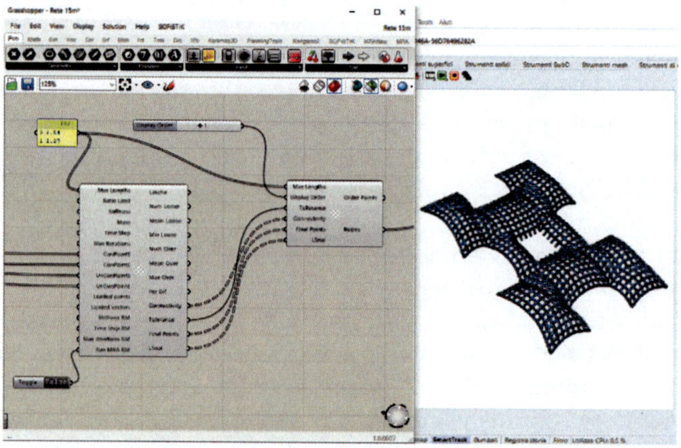

Fig. 1. Grasshopper i-MRA component.

The improved Multibody Rope Approach (i-MRA) method enhances the MRA approach by incorporating optimization techniques that facilitate the automation of the construction process. There are three main improvements to i-MRA:

- Multiple Orders MRA (MO-MRA) involves grouping together structural elements that share the same length, which reduces the number of distinct structural components required for construction.
- Repulsive Nodes MRA (RN-MRA) applies a repulsive force field to the dynamic model, allowing for minor geometry adjustments to further minimize the number of structural components needed.

- Variable time interval adjusts the time interval used in the dynamical system resolution, which reduces the number of computational iterations required to calculate the final equilibrium configuration, thus decreasing the computational effort involved.

The i-MRA method has been developed with the aim of optimizing the structural geometry of gridshells to streamline the construction process. To make the method easily accessible to designers, an original MATLAB [9] code was developed for its implementation. However, recognizing the need for a more user-friendly platform, the code was translated into a Grasshopper component [10], as depicted in Fig. 1.

3 The Optimization Strategy: The Multy-Subset Approach

In this paper, the i-MRA method is combined with a multi-objective genetic algorithm (MOGA) using the MATLAB [9] function "gamultiobj". The MOGA operates by creating a population of potential solutions and iteratively improving them over generations [13].

The objective functions defined in this work aim to minimize the number of different structural element types required for construction and the quantity of structural material needed to build the gridshell. The use of material is defined also considering the wastes due to cutting the industrially produced logs of 12 m in length. This approach allows for the creation of gridshells that are structurally optimised and efficient in terms of material use and construction management.

Figure 2 illustrates the workflow of the optimization process. In this process, the design variables are the parameters governing the form-finding process. Initially, a basic grid is employed, and the i-MRA is utilized to determine the structurally efficient shape under a given loading condition.

Subsequently, the Cutting Stock Problem (CSP) [14,15] is employed to calculate the minimum number of 12m-long billets required to obtain all the elements necessary for constructing the structure. This approach is adopted to account for the industrial production process of steel beams, which results in production waste from cutting the billets. The optimal cutting pattern can be determined by employing the algorithm to solve the CSP, minimizing waste during the production of structural elements.

Then, a finite element model is created based on the obtained structural shape, employing CHS-type (Circular Hollow Section) profiles made of S355 steel. Finite element analyses are conducted using a proprietary MATLAB code [16,17] to determine nodal displacements and stress characteristics acting on individual beams. These results are then used to verify if the generated structure meets the criteria for limit displacement, yield, and Eulerian buckling, as in the Eq. 1.

$$\delta_{max} < \delta_{lim} \qquad |\sigma_{max}| < f_y \qquad N_{max}^{Compression} < N_{Buckling} \tag{1}$$

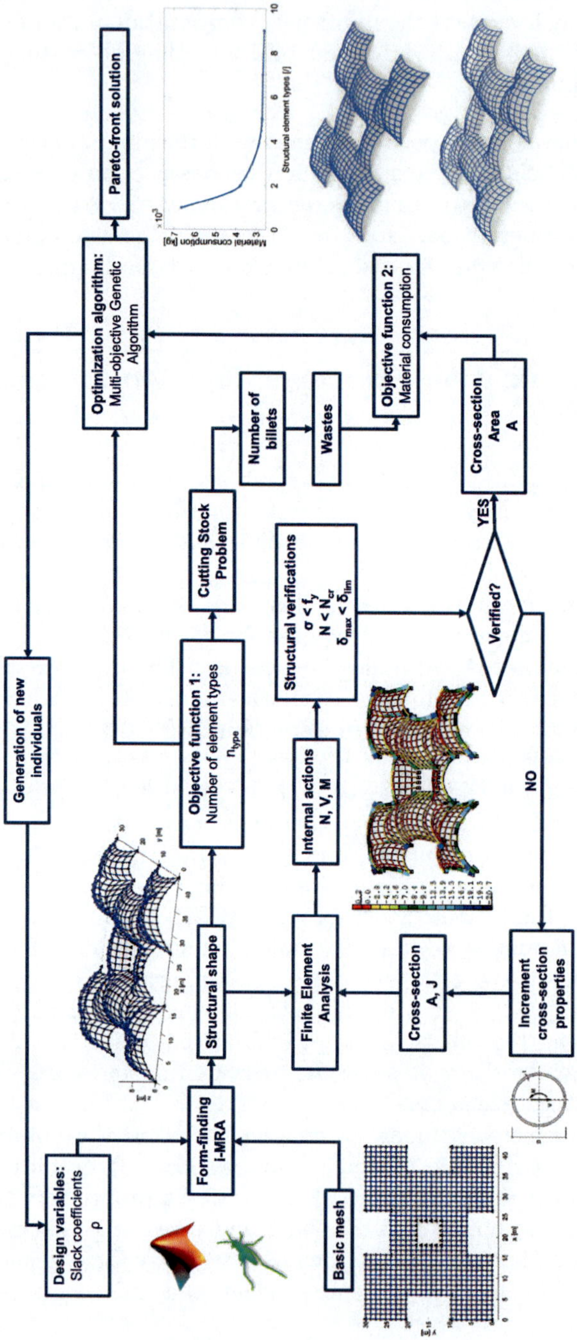

Fig. 2. Optimization procedure workflow.

where δ_{max} is the maximum nodal displacement experienced by the structure and δ_{lim} is the limit fixed to be $\frac{1}{250}$ of the maximum structural dimension without external constraints. $|\sigma_{max}|$ is the absolute maximum stress in the structure and should be less than f_y which represents the yielding stress of the employed steel. Finally, $N_{max}^{Compression}$ is the maximum compressive axial force and $N_{Buckling}$ is the critical eulerian buckling axial force.

The cross-section dimensions are increased iteratively until a CHS profile that enables the structure to pass the verifications is attained. This iterative process enables the determination of the cross-sectional area of the structural elements, which, when multiplied by the number of billets resulting from the CSP solution, determines the actual amount of material required for constructing the structure. Consequently, the fitness function evaluated by the MOGA considers two components: the number of different structural element types n_{type} of varying lengths and the amount of material used W. The process is iterated until a set of solutions is obtained, defining a Pareto front where each point represents an optimal structural geometry. This allows the designer to utilize the resulting curve to select the most suitable structure for the specific project, taking into consideration architectural, structural, economic, and management requirements.

4 Methods and Testing: Regular and Complex Shape Grid Shells

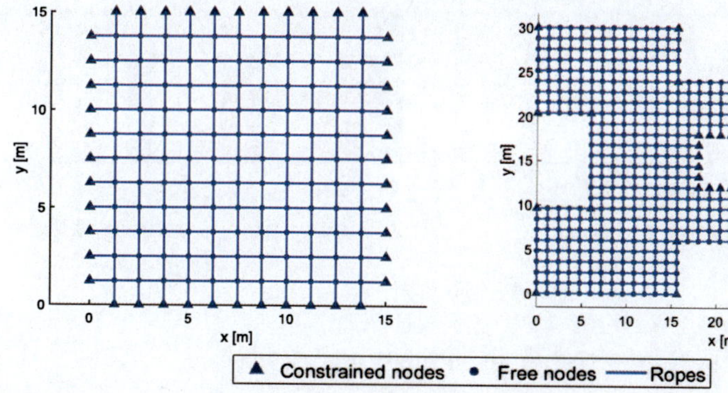

Fig. 3. Base mesh for the case studies

In this section, two examples of the application of the optimization process to form-finding are presented. The first case pertains to a square plan gridshell with a side length of 15m, as reported in Fig. 3(a). Figure 4 displays the optimization results for this structure subjected to self-weight and various nodal

Table 1. First case study optimization results in terms of use of material [kg].

		CSP				No CSP			
	n_{type}	**50 kN**	**10 kN**	**5 kN**	**1 kN**	**50 kN**	**10 kN**	**5 kN**	**1 kN**
1 var	**1**	7131	1432	1019	673	6641	1333	949	627
	2	3905	1019	/	/	3773	977	/	/
	3	/	/	956	549	/	963	857	505
	4	3330	/	900	/	/	/	823	/
	5	2683	/	/	/	2698	/	/	/
	7	2659	/	/	/	/	/	/	/
	9	2589	/	/	/	/	/	/	/
2 var	**1**	7131	1432	1019	673	6641	1333	949	627
	2	2552	1019	872	515	2492	970	820	504
	3	/	/	/	/	/	963	0	504

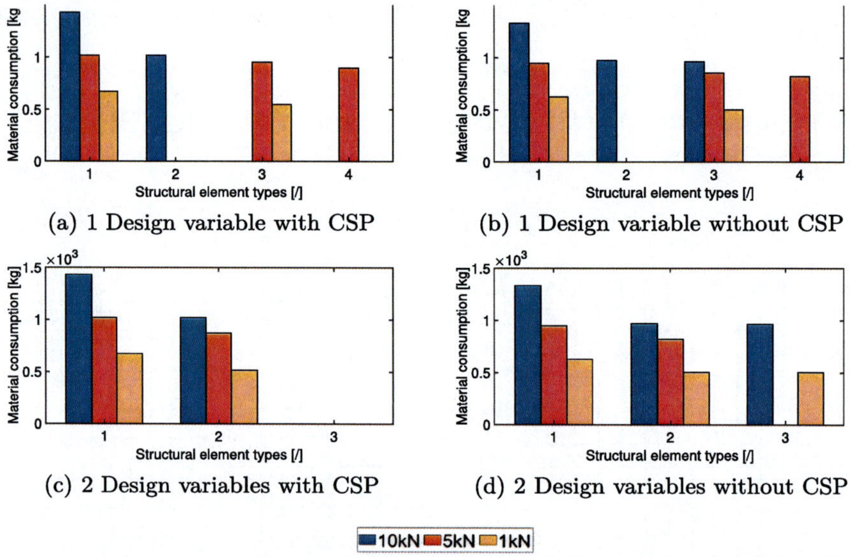

Fig. 4. First case study optimization results.

loads. Specifically, the results depict structures loaded at nodes with respective loads of 1kN, 5kN, and 10kN.

The reported results correspond to optimizations that were conducted considering waste using the CSP (Figs. 4(a) and (c)) or without considering waste, as in the cases of Figs. 4(b) and (d). Finally, the optimization was performed considering either one (Figs. 4(a) and (b)) or two (Figs. 4(c) and (d)) design variables represented by the slack coefficients assigned to the ropes for form-finding using the i-MRA method.

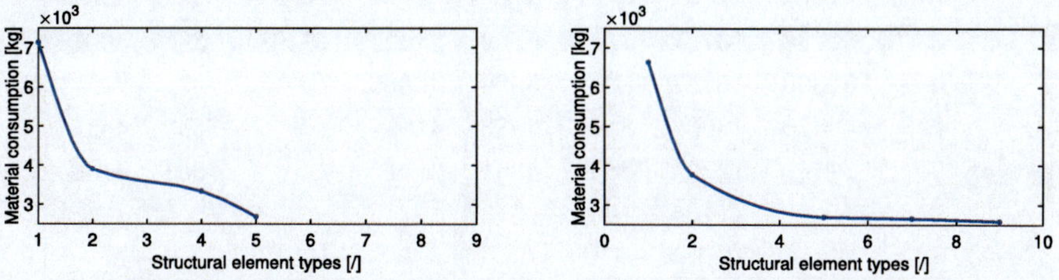

(a) CSP (b) No CSP

Fig. 5. 1st case study Pareto front solutions considering a nodal overload of 50kN.

The results are also presented in Table 1, which includes the case with a nodal load of 50 kN. It can be observed that the number of different structural element types is inversely proportional to the amount of material required for constructing the structure. Consequently, a range of optimal structural shapes that form a Pareto front is obtained, as depicted in Fig. 5.

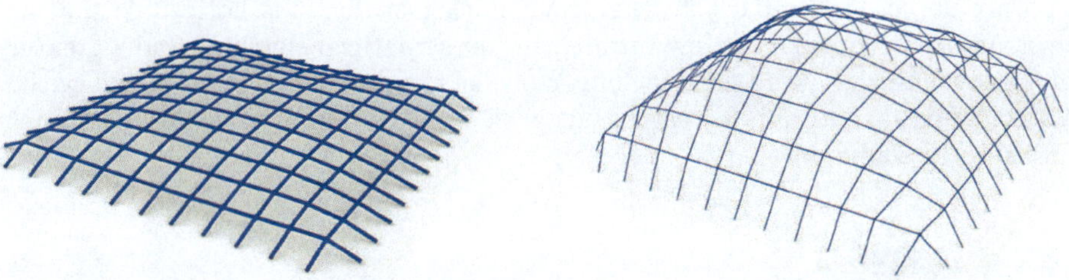

(a) $n_{type} = 1$ $m = 7.13ton$ (b) $n_{type} = 2$ $m = 2.55ton$

Fig. 6. Optimized structural shapes for case study 1 considering self-weight and a nodal overload of 50kN.

Notably, the lower-profile structural configurations tend to feature thicker structural elements due to the higher involved compression forces. However, they can be constructed using a smaller number of different structural element types. Conversely, the less low-profile configurations consist of structural elements of various types but allow for the utilization of slender elements.

Figure 6 presents the two optimal configurations for the first case study, considering both self-weight and a nodal overload of 50kN.

In the second case study, the presented procedure was applied to optimize the gridshell derived from the base plan depicted in Fig. 3(b). The optimizations

Table 2. Second case study optimization results in terms of use of material [kg].

		CSP				No CSP			
	n_{type}	**50 kN**	**10 kN**	**5 kN**	**1 kN**	**50 kN**	**10 kN**	**5 kN**	**1 kN**
1 var	**3**	56388	16797	10183	5399	52029	15498	9396	4903
	4	50390	15747	8662	5399	46865	/	/	/
	5	44670	13009	8149	4633	41934	/	7650	4416
	7	/	12224	6851	4361	/	/	6631	4189
	9	/	11464	/	/	/	/	/	/
	11	/	/	/	/	/	10513	/	/

were conducted using a design variable with the aim of minimizing both the number of structural elements of varying lengths and the material consumption, considering both waste and net quantities. The analyses were repeated for the same load cases as presented in the first case study, and the results are summarized in Table 2. Additionally, Fig. 7 provides visual representations of the material quantities used, both in net terms (Fig. 7(b)) and including the waste resulting from cutting industrial billets (Fig. 7(a)). It can be observed that the increased complexity of the studied structure leads to a higher number of different types of structural elements compared to the simpler case. Furthermore, in Fig. 8, the optimal structural forms for the case loaded with a nodal force of 50kN are depicted. Similarly to the previous case study, it is noticeable that as the structure becomes more low-profile, thicker structural elements and a greater amount of material are required. Conversely, as the structure becomes taller, the amount of used material decreases, but with a bigger number of different types of structural elements.

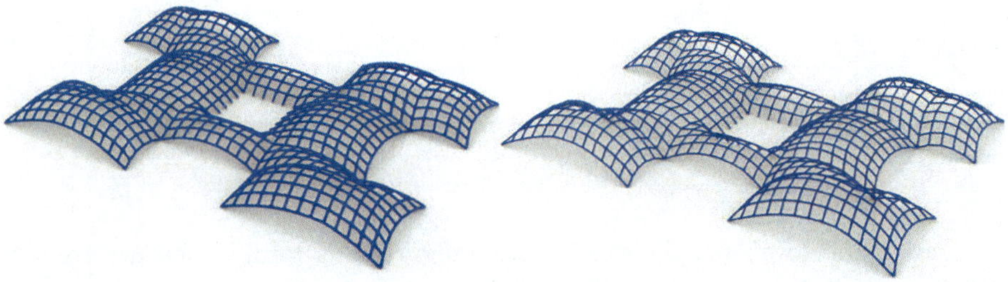

(a) $n_{type} = 3 \quad m = 52.0ton$ (b) $n_{type} = 5 \quad m = 41.9ton$

Fig. 7. Optimized structural shapes for case study 2 considering self-weight and a nodal overload of 50kN.

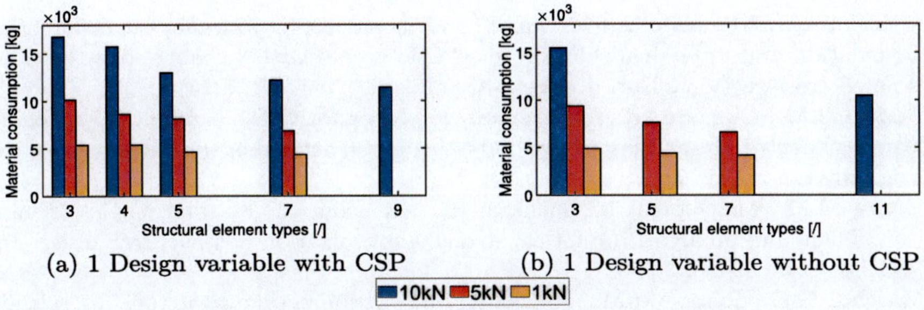

(a) 1 Design variable with CSP (b) 1 Design variable without CSP

Fig. 8. Second case study optimization results.

5 Conclusion

In this study, an optimization procedure combined with the form-finding process of gridshell structures is presented. Through the utilization of the i-MRA form-finding method, it was possible to optimize the structural form by minimizing the number of different types of structural elements and reducing material consumption. The utilization of the CSP allowed for the determination of the optimal cutting pattern, further minimizing waste during the production of the structural elements. The Pareto front obtained through the MOGA highlighted the trade-offs between different design criteria, such as structural performance, material usage, and aesthetic considerations. This provided designers with a range of optimal structural forms, enabling them to select the most appropriate solution based on specific project requirements. The presented case studies demonstrated how the optimization process could be applied to gridshell structures. The results indicated that the complexity of the structure influenced the number of different types of structural elements and the overall material consumption. Additionally, the relationship between structural height and material utilization was observed, with lower-profile configurations requiring thicker structural elements but a reduced number of element types, while taller configurations allowed for slimmer elements but involved a greater variety of element types. Finally, in order to make the i-MRA form-finding method used in this study accessible to professional designers, a component for the Grasshopper [10] parametric design environment was developed to facilitate the practical application of the proposed approach.

References

1. Cucuzza, R., Rosso, M.M., Aloisio, A., Melchiorre, J., Giudice, M.L., Marano, G.C.: Size and shape optimization of a guyed mast structure under wind, ice and seismic loading. Appl. Sci. **12**(10), 4875 (2022)
2. Melchiorre, J., Bertetto, A.M., Marano, G.C.: Application of a machine learning algorithm for the structural optimization of circular arches with different cross-sections. J. Appl. Math. Phys. **9**(5), 1159–1170 (2021)

3. Melchiorre, J., Manuello, A., Marmo, F., Adriaenssens, S., Marano, G.: Differential formulation and numerical solution for elastic arches with variable curvature and tapered cross-sections. Eur. J. Mech.-A/Solids **97**, 104757 (2023)
4. Rosso, M.M., Cucuzza, R., Di Trapani, F., Marano, G.C.: Nonpenalty machine learning constraint handling using PSO-SVM for structural optimization. Ad. Civil Eng. **2021**, 1–17 (2021)
5. Rosso, M.M., Melchiorre, J., Cucuzza, R., Manuello, A., Marano, G.C.: Estimation of distribution algorithm for constrained optimization in structural design. In: WCCM-APCOM 2022 900
6. Marano, G.C., Rosso, M.M., Melchiorre, J.: Optimization as a tool for seismic protection of structures. In: Cimellaro, G.P. (ed.) Seismic Isolation, Energy Dissipation and Active Vibration Control of Structures. WCSI 2022. Lecture Notes in Civil Engineering, vol. 309. Springer, Cham (2022). https://doi.org/10.1007/978-3-031-21187-4_8
7. Adriaenssens, S., Block, P., Veenendaal, D., Williams, C.: Shell structures for architecture: form finding and optimization. Routledge (2014)
8. Manuello, A., Melchiorre, J., Marano, G.C.: Improved multi-body rope approach for free-form grid shells. In: Italian Workshop on Shell and Spatial Structures (IWSS 2023) (2023)
9. MATLAB, version (R2022b). The MathWorks Inc., Natick, Massachusetts (2022). https://www.mathworks.com/products/matlab.html
10. McNeel, R., et al.: Rhinoceros 3D, version 7.0. (https://www.rhino3d.com/). Robert McNeel & Associates, Seattle, WA
11. Manuello, A.: Multi-body rope approach for grid shells: form-finding and imperfection sensitivity. Eng. Struct. **221**, 111029 (2020)
12. Manuello, A., Melchiorre, J., Sardone, L., Marano, G.C.: Multi-body rope approach for the form-finding of shape optimized grid shell structures. In: Proceedings of the 15th World Congress on Computational Mechanics (2022)
13. Deb, K.: Multi-objective optimisation using evolutionary algorithms: an introduction. In: Wang, L., Ng, A., Deb, K. (eds.) Multi-objective Evolutionary Optimisation for Product Design and Manufacturing. Springer, London (2011). https://doi.org/10.1007/978-0-85729-652-8_1
14. Ford, L.R., Fulkerson, D.R.: A suggested computation for maximal multi-commodity network flows. Manage. Sci. **5**(1), 97–101 (1958)
15. Gilmory, P., Gomory, R.: A linear programming approach to the cutting stock problem-part II. Oper. Res. **11**(6), 863–888 (1963)
16. Paradiso, M., Marmo, F., Rosati, L.: Consistent derivation of a beam model from the Saint Venant's solid model. Int. J. Solids Struct. **159**, 90–110 (2019)
17. Paradiso, M., Sessa, S., Vaiana, N., Marmo, F., Rosati, L.: Shear properties of isotropic and homogeneous beam-like solids having arbitrary cross sections. Int. J. Solids Struct. **216**, 231–249 (2021)

Free-Form Aluminum Grid Shell on Bolted Joints, Avenue Mall Shopping Center

Denis Gerasimov, Alexey Kashin, Yuri Lunev, Alexander Morozov(✉), Victoria Shmakova, Lili Bazygina, and Anna Frolova

Parametric Building Systems FZ-LLC, Ras Al Khaimah, United Arab Emirates
morozov@spacestructure.ru

Abstract. In 2015–2023, Parametric Building Systems LLC developed and erected free-form aluminum grid shells using only bolted assemblies. The materials used for all core and node elements are high—strength aluminum EN AW-6063 T5/6, EN AW-6082 T5/6. Beam elements and nodes are connected by high-strength bolts with a special coating without welding. The core elements of the grid shell frame are manufactured at the factory, the nodal elements are individually manufactured on 5-axis CNC machines.

The design is carried out using parametric and BIM design tools Rhino3D/Grasshopper/Kangaroo/Karamba3D, which ensures the design of mesh shells at the LOD 500 level of detail. One of the advantages of using aluminum is its low volumetric weight, less than three times that of steel, which makes it possible to install without the use of heavy cranes and other lifting equipment.

One example of such projects is the curved aluminum grid shell of the Avenue Mall shopping center. It is a shell of arbitrary shape and has dimensions of 55 m × 50 m, the maximum mark of the roof top from ground level is +25 m. The area of the coating, consisting of a metal frame and glass triangles, is 2750 m^2. The average length of the core elements is 2.5 m; the angles between them are as close as possible to the value of 60°.

Keywords: Architectural geometry · parametric modeling · aluminum shell · grid shell · Finite Elements Analysis · calculation of aluminum shells · aluminum structures · free-form constructions · Rhinoceros 3d

1 Introduction

One of the projects implemented by the company Parametric Building Systems LLC [1] is the curved aluminum grid shell [2] of the Avenue Mall in Dubai, UAE. It is a shell of arbitrary shape and has dimensions of 55 m × 50 m, the maximum elevation of the top of the roof from ground level is +25 m. The area of the coating, consisting of a metal frame and glass triangles, is 2750 m^2. The average length of the main elements is 2.5 m; the angles between them are as close as possible to the value of 60 degrees. One of the features of the project is the use of aluminum in the beam elements of the shell and in the nodal elements. In this case, all connections are bolted, without the use

S. Gabriele et al. (Eds.): IWSS 2023, LNCE 437, pp. 73–79, 2024.
https://doi.org/10.1007/978-3-031-44328-2_8

of welding. Supports for this design are free-standing three-, four-branch columns of circular cross section. The distance between the branches of the columns is from 5 to 10 m. Overhangs along the perimeter of the coating have a reach of up to 6 m. One of the features of the project is provides maximum horizontal transparency due to the absence of walls/beams along the perimeter of the structure, and a significant distance between the supports allows to achieve the effect of the maximum open space under the shell (see Fig. 1).

Fig. 1. Aluminum grid shell. General view of the finished structure.

2 Project Description

2.1 Loads on the Shell Structure

Another feature of the covering design is the fact that all its elements are in different working conditions; this is influenced by the following factors:

Another feature of the coating design is that all its elements are in different operating conditions; The following factors influence this:

- complex geometric shape of the shell
- standard wind load in different areas ranges from −30 to +20 kg/m^2;
- random arrangement of supports;
- distance from the supports, the span between adjacent supports, as well as the dimensions of the console.

To take into account all these factors, as well as to ensure the minimization of sections and the unification of structures, steel frame elements are designed with different sections:

- extruded rectangular aluminum sections of a specially designed type of section made of AW6082 T6 alloy EN.1999-1-1:2007;

The cross sections in the grid shell are different and are distributed according to the magnitude of the moments in individual elements. In the supporting areas of the shell of a larger section in the span areas of the shell of a smaller section. Rectangular sections of elements of a complex geometric shape, overall dimensions 250 * 56 mm, 300 * 56 mm (see Fig. 2).

Fig. 2. View of the structure from above, showing various profiles and nodal elements of the shell

2.2 Bolted Connections

Structural coating elements of various box-shaped sections are made to provide greater architectural expressiveness, as well as the perception of increased forces in the area of column support. The aluminum shell is made without welding. All nodal connections of the aluminum shell are made on bolted connections. The use of bolts in nodal connections made it possible to simplify the installation process of the shell, reduce the time, and not use crane equipment for its construction (see Fig. 3).

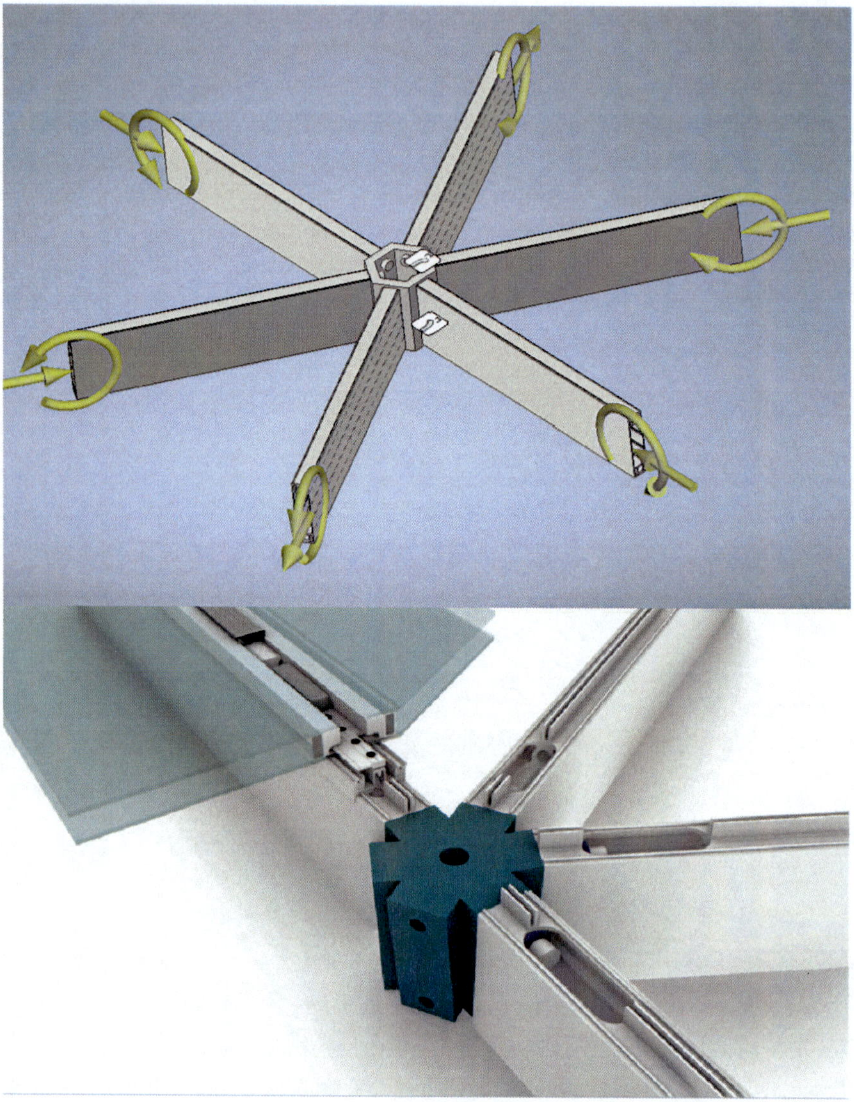

Fig. 3. Calculation model of the node in the program Inventor. Nodal Element Model

2.3 Parametric Model. Design Model

For the successful implementation of all the above tasks, special attention is paid to the construction of grid geometry at the design stage. An important issue was the creation of elements with maximum unification in length and area. This problem was solved using a parametric shell model in the Rhinoceros/Grasshopper [3, 4] program (see Fig. 4). Its feature is the ability to quickly change the geometric parameters of the structure (bar elements, nodes, panels, etc.) depending on the actual initial data and at the same time minimize manual routine work with the subsequent selection of the most optimal position of the rod. Elements and nodes. After optimizing and coordinating the shape of the coating, as well as the location of its rods and nodes, a strength calculation is performed for each element and node.

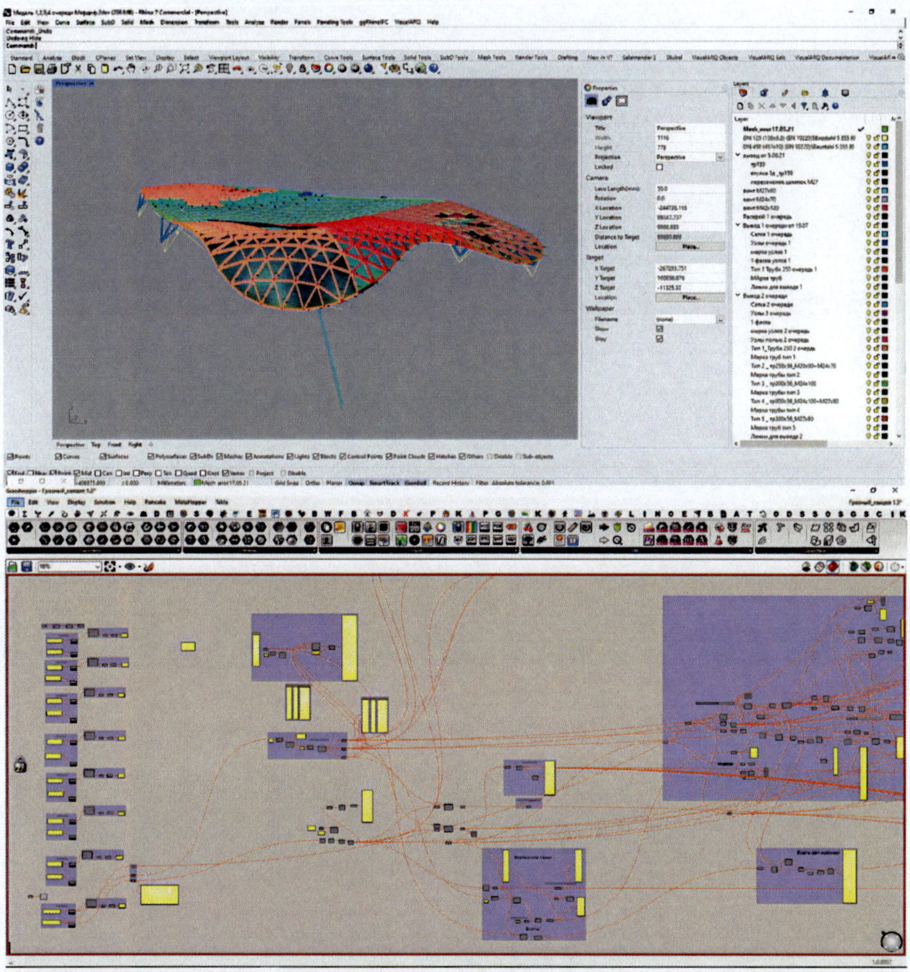

Fig. 4. Parametric model in Rhino and a script for building a detailed model in Grasshopper

The calculation scheme of the grid shell is a complex finite element model made in the Dlubal program with a rigid and elastic connection of the elements to each other (see Fig. 5). Wind loads on the mesh shell were determined using the Dlubal RWIND Simulation program. Based on the results of these studies, all sections that meet the requirements of the first and second groups of limit states are finally determined.

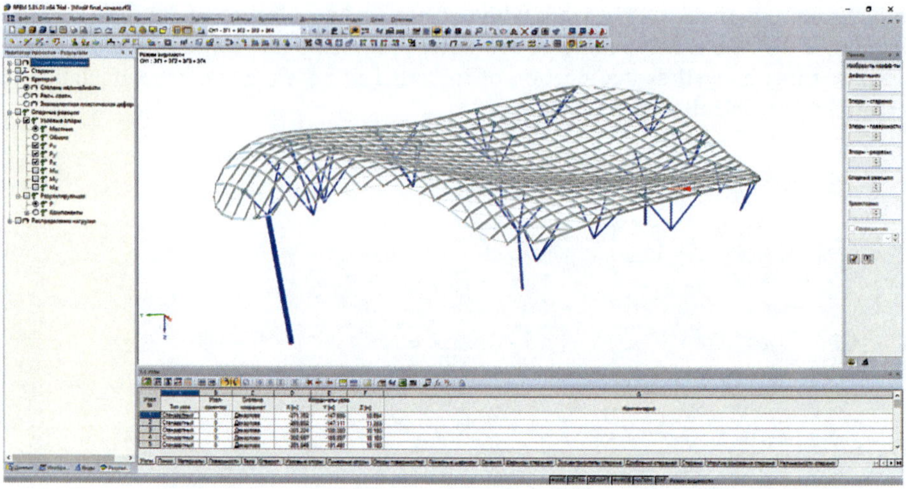

Fig. 5. Calculation model of grid aluminum shell in Dlubal

3 Shell Assembly

Fig. 6. Installation of the shell, the crane is visible, temporary supports blocks are installed on temporary supports and columns.

The installation of the mesh shell was carried out in large blocks. The shell was divided into 5 blocks of approximately the same weight. Each block was assembled on the ground. The supporting columns were previously installed in their design position and fixed with temporary supports (see Fig. 6).

After that, each block was lifted to the design mark using a crane. Each mounted block was placed on columns and temporary supports. After the complete assembly of all 5 blocks, temporary supports were dismantled.

4 Conclusions

Company Parametric Building Systems LLC carried out a large amount of work on the design and installation of the grid shell. The design was carried out entirely in parametric BIM models, and the working documentation was received in a very short time. The installation was carried out without the involvement of crane equipment in the shortest possible time.

References

1. Gerasimov, D.V., Kashin, A.V., Morozov, A.A., Shmakova, V.S., Lunev, Y.V., Averin, A.V.: Glass covering 'glass bark' of the natural park "zaryadye" in Moscow. J. Int. Assoc. Shell Spat. Struct. **59**(3), 225–228 (2018)
2. Schlaich, J., Schober, H., Kürschner, K.: New trade fair in Milan — grid topology and structural behaviour of a free-formed glass-covered surface. Int. J. Space Struct. **20**(1), 1–14 (2016)
3. Robert McNeel & Associates Rhino3D webpage. https://www.rhino3d.com/en/. Accessed 1 May 2023
4. Karamba3D Parametric Engineering Add-on webpage. https://www.karamba3d.com. Accessed 1 May 2023

Snap-Back Analysis of Fracture Evolution in Multi-cracked Masonry Arches

Renato Cafarelli[1](✉), Federico Accornero[2], and Alberto Carpinteri[1,2]

[1] Politecnico di Torino, Turin, Italy
renato.cafarelli@polito.it
[2] Shantou University, Shantou, People's Republic of China

Abstract. In the present work, the Cohesive/Overlapping Crack Model is extended to the study of masonry or plain concrete structures, which are subjected to off-center compression. Multi-cracking and multi-crushing damage phenomena are simulated by means of the Crack Length Control Scheme, in order to obtain a complete load history of the vaulted structure. This Nonlinear Fracture Mechanics model reveals a high capability in predicting the elastic-plastic-softening behaviour of arches and vaults as well as the local mechanical instabilities, such as snap-back and snap-through, occurring during the post-cracking regime.

In the first part of this paper, the Cohesive/Overlapping Crack Model and a description of the adopted multi-crack numerical procedures is provided. In the second part, some parametric analyses on scale effects in masonry arches are reported together with the results obtained for a real case-study. Finally, some future developments and applications of the model are presented.

Keywords: Cracking · Crushing · Scale effects · Cohesive Crack Model · Overlapping Crack Model

1 Introduction

1.1 Structural Theories for Masonry Arches

The design of masonry arch bridges during 19th and 20th centuries was based on Theory of Elasticity: the shape of the arch was chosen as close as possible to the line of thrust of the loads acting on the structure in order to reduce the tensile stresses generated by bending moment [1–3]. This design approach leads to the construction of very massive structures in which the variation of the thrust line due to live loads is negligible with respect to the position of the thrust line generated by dead loads [4, 5].

On the other hand, with the extension of Theory of Plasticity to masonry structures due to Heyman [6], it was provided that a limit condition is reached when the thrust line is tangent to the structure cross-section: in this case, a plastic hinge is formed and a variation in the internal stresses is obtained. The design of masonry arches by means of this approach leads to lighter structures as well as to material optimization since the thrust line is no longer forced to pass through the central kern of the cross-section [1, 7].

S. Gabriele et al. (Eds.): IWSS 2023, LNCE 437, pp. 80–87, 2024.
https://doi.org/10.1007/978-3-031-44328-2_9

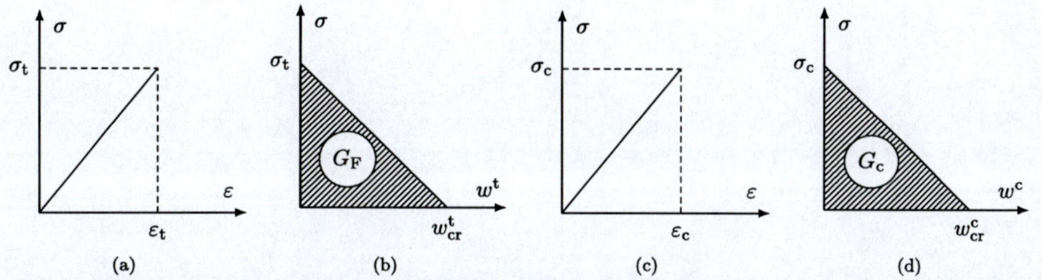

Fig. 1. Cohesive and Overlapping Crack Models: (a) Linear elastic stress – strain law in tension; (b) Post-peak σ-w^{t} cohesive relationship; (c) Linear elastic stress – strain law in compression; (d) Post-peak stress vs fictitious interpenetration relationship

In this context, although the two fundamental theorems of Theory of Plasticity may be used to define the limit condition and the loadbearing capacity of a structure, they are not able to clarify how this condition is reached. On the other hand, recent applications [1, 8] of Linear Elastic Fracture Mechanics (LEFM) have demonstrated that an increase in the bearing capacity of a masonry arch may be obtained by means of an increase in the cracking damage: due to cracking, a recentering of the thrust line is obtained and a fracturing benefit coefficient may be defined.

In the present paper, an extension of the Cohesive/Overlapping Crack Model (COCM) [9–15] is presented. This model is able to evaluate the elastic-plastic-softening loading history of masonry structures taking into account the damage evolution in tension as well as in compression together with scale effects.

1.2 The Cohesive/Overlapping Crack Model

COCM is a Nonlinear Fracture Mechanics model that integrates the Cohesive Crack Model to simulate the damaging process occurring in tension with the Overlapping Crack Model to simulate the damaging process occurring in compression. The Cohesive Crack Model [16, 17] is able to describe the damage evolution in the tension zone of a structure, assuming a linear elastic constitutive law (Fig. 1a) up to the first peak load, and a constitutive law in the form σ-w^{t}, σ being the applied stress and w^{t} the crack opening displacement, in the zone where strain localization in tension occurs [17]. More precisely, within the Cohesive Crack Model, a fictitious crack, longer than the real one, is introduced, providing a damage process zone where the material is still able to transfer tensile forces, albeit partially damaged. Thus, the residual bearing capacity of the structural element is simulated by means of the application of closing forces along the crack faces according to the constitutive law of Fig. 1b, where ${w^{\mathrm{t}}}_{\mathrm{cr}}$ is the threshold value of crack opening beyond which the closing forces vanish. The area subtended by the σ-w^{t} diagram represents the fracture energy, G_{F} (see Fig. 1b).

The Cohesive Crack Model has been effectively applied in the study of plain concrete or lightly reinforced/high strength concrete beams as well as in the nonlinear behaviour of masonry structures, not considering the potential crushing failure occurring in compression zones [18, 19]. In this framework, Carpinteri and co-workers [20] introduced

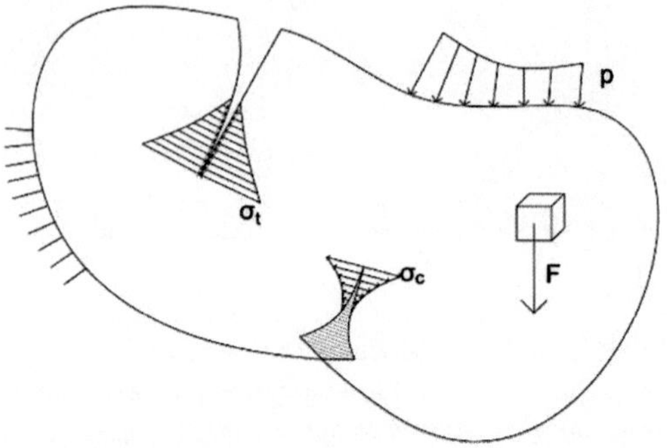

Fig. 2. Damaged body with process zones in tension and compression

the Overlapping Crack Model, which is able to describe the crushing damage of a quasi-brittle material by means of a fictitious interpenetration zone developing in the region where compressive strain localization takes place. The parameters involved in the Overlapping Crack Model are formally similar to those of the Cohesive Crack Model. The main variables entailed within this model are: the compressive strength of the material, σ_c, the threshold value of fictitious interpenetration beyond which the virtual opening compressive forces vanish, $w^c{}_{cr}$, and the crushing energy, G_c, which may be calculated as the area subtended by the σ-w^c diagram of Fig. 1d. Within the Overlapping Crack Model, the material behaves elastically until the compression strength σ_c is reached (Fig. 1c), then a constitutive law in the form σ-w^c (Fig. 1d) is adopted.

2 Numerical Algorithm

COCM may be integrated in a finite element method. In this context, for an undamaged elastic solid body, the weak form of equilibrium [21] leads to

$$\int_V \delta\{\varepsilon\}^{\mathrm{T}}\{\sigma\}\mathrm{d}V = \int_V \delta\{u\}^{\mathrm{T}}\{F\}\mathrm{d}V + \int_S \delta\{u\}^{\mathrm{T}}\{t\}\mathrm{d}S \tag{1}$$

$\{F\}$ being the vector of the forces acting on the volume of the body, $\{t\}$ the vector of the forces acting on the surface of the body, $\{u\}$ the displacement vector, and S the body surface. On the other hand, for a damaged solid having m cohesive cracks and n overlapping process zones (Fig. 2), the last term of Eq. (1) may be decomposed as

$$\begin{aligned}\int_S \delta\{u\}^{\mathrm{T}}\{t\}\mathrm{d}S = &\sum_j^m\left(\int_{S_{\mathrm{c},j}} \delta\{u\}^{\mathrm{T}}\{p_\mathrm{c}\}\mathrm{d}S\right)\\ &+\sum_j^n\left(\int_{S_{\mathrm{t},j}} \delta\{u\}^{\mathrm{T}}\{p_\mathrm{t}\}\mathrm{d}S\right) + \int_\Omega \delta\{u\}^{\mathrm{T}}\{p\}\mathrm{d}S\end{aligned} \tag{2}$$

with

$$\Omega = S - \sum_{j=1}^m S_{\mathrm{c},j} - \sum_{j=1}^n S_{\mathrm{t},j} \tag{3}$$

$\{p_c\}$ and $\{p_t\}$ being the force vectors of cohesive and overlapping process zones, respectively and $\{p\}$ the effective load vector acting on the body. Thus, if a finite element approach is used to solve Eq. (1), by means of the shape functions matrix, $[N]$, the matrix $[B]$ containing the derivatives of the shape functions and the elastic matrix $[D]$, it is possible to obtain

$$\left(\sum \int [B]^T [D][B] \mathrm{d}V - \sum \int [C][L][N] \mathrm{d}S - \sum \int [T][L][N] \mathrm{d}S\right)\{\overline{u}\} = \lambda \sum \int \{p\} \mathrm{d}S \tag{4}$$

$[C]$ and $[T]$ being the cohesive and the overlapping matrices set according to Fig. 1b and 1d, respectively; λ a multiplication factor for the external load vector and

$$[L] = \begin{bmatrix} -1 & 1 & 0 & 0 \\ 0 & 0 & -1 & 1 \end{bmatrix} \tag{5}$$

Equation (4) may be transformed as follows

$$([K] - [S_c] - [S_t])\{\overline{u}\} = \lambda \left\{\hat{f}_{ext}\right\} \tag{6}$$

$[S_c]$ and $[S_t]$ being the softening matrix in tension and compression, respectively and $\left\{\hat{f}_{ext}\right\}$, the reference external load vector. The equilibrium Eq. (6) represents an extension of the equilibrium condition initially proposed by Bocca et al. [22] and clearly suggests how in a damaged body, a loss of stability may occur similarly to Euler's buckling [17] since the effective stiffness matrix, $[K_{eff}]$, of the damaged solid is $[K_{eff}] = [K] - [S_c] - [S_t]$. Equation (6) may also be transformed as

$$\{f_{int}\} - \lambda \left\{\hat{f}_{ext}\right\} = \{0\} \tag{7}$$

$\{f_{int}\}$ being the internal force vector.

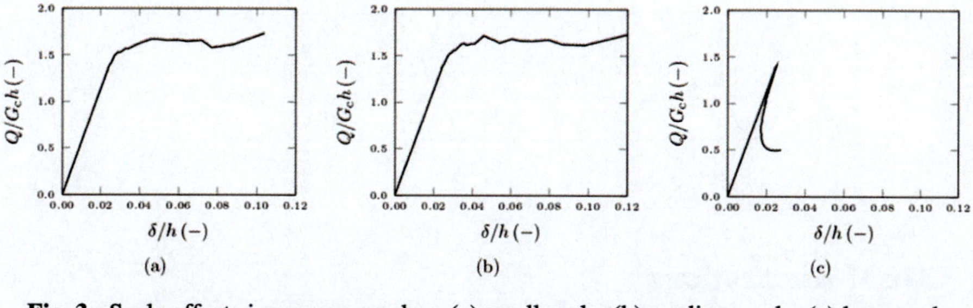

Fig. 3. Scale effects in masonry arches: (a) small scale; (b) medium scale; (c) large scale

Thus, by means of COCM an incremental steady-state analysis may be set according to Eq. (7). On the other hand, if we want to follow snap-back unstable branches in a stable manner, the multiplication factor, λ, must be determined according to a generalized path-following constraint [23] and, more precisely, assuming the crack length increment as controlling loading variable (Crack Length Control scheme). In this framework, at each computation step, it should be ensured that the internal force, f_i, at crack tip *i*, is equal to the ultimate force f_u set by the cohesive or overlapping constitutive laws (Fig. 1b, d). In this context, by means of Taylor expansion series, the load increment $\delta\lambda$ for each crack tip *i* between iteration *j* and $(j+1)$ may be written as

$$\delta\lambda_i^{j+1} = \frac{f_u - f_i^j - \{K_{\text{eff},i}\}^T [K_{\text{eff}}]^{-1} \{r^j\}}{\{K_{\text{eff},i}\}^T [K_{\text{eff}}]^{-1} \{\hat{f}_{\text{ext}}\}} \tag{8}$$

with $\{r^j\} = \lambda_i^j \{\hat{f}_{\text{ext}}\} - \{f_{\text{int}}^j\}$ and $\{K_{\text{eff},i}\}$ the *i*-th row of the effective stiffness matrix. At each computation step, Eq. (8) is solved for each crack tip *i* and the effective load increment is assumed as

$$\Delta\lambda = \min \left| \sum_j \delta\lambda_i^j \right| \tag{9}$$

In order to simulate the transferring stress capacity along cracks, *zero-thickness interface elements* [24] formed by an upper and a lower layer are employed. These layers have twin nodes sharing same coordinates and for which it is possible to calculate the relative displacement vector, $\{w\}$, as the displacement difference between each twin pairs (see Eq. (5)). Unlike many Fracture Mechanics models [25, 26], in COCM these special finite elements are not inserted in the mesh a priori but, at each computation step, a remeshing algorithm is adopted to modify mesh topology only in the zone ahead the crack tip that reaches the propagation condition. In this context, a reliable simulation of strain localization phenomena occurring in quasi-brittle materials may be performed.

Table 1. Geometrical parameters of the analysed arches

Scale	l (m)	r (m)	h (m)
Small	2.0	0.3	0.1
Medium	20.0	3.0	1.0
Large	200.0	30.0	10.0

3 COCM Applications

3.1 Scale Effects in Masonry Arches

In Fig. 3, numerical curves obtained by means of the application of COCM to masonry arches are presented. Masonry is assumed to have a compression strength $\sigma_c = 30$ MPa, and a tension strength $\sigma_t = 2.6$ MPa, whereas the fracture energy, G_F, and the crushing energy, G_c, are assumed equal to 0.14 and 30 N/mm, respectively. Three different scales have been considered according to a scale factor of 1, 10 and 100 for the arch span, l, the rise, r, and the section depth, h, as reported in Table 1. Each arch is analysed using the same mesh density and in order to have an accaptable computation time, the number of controlled sections along which strain localization phenomena may occur is fixed to 17.

In Figs. 3a and 3b the loading curves obtained for small and medium scales are reported. It is possible to observe that these two arches present a stable behaviour since once the elastic limit is reached, a pseudo-ductile plateau is described. Moreover, a small increase in the bearing capacity of the structure is detected in the post-elastic regime. The ultimate behaviour of these two scales is governed by damage in compression: COCM evidences that crushing takes place in all the 17 controlled sections along the arch span.

On the other hand, the large scale arch reported in Fig. 3c presents a completely unstable behaviour since after the peak load, a catastrophic snap-back is detected. For this scale COCM provides that the ultimate behaviour is mainly governed by cracking processes at the extrados sections close to springings.

3.2 Mosca Bridge Case Study

In this section, COCM will be applied to the investigation of the nonlinear behaviour of the Mosca bridge (Turin, Italy). The bridge was built in the first half of the XIX century and its span is formed by a unique 45 m arch made of local stone. The arch has a rise of 5.5 m and a section depth that varies from 2.0 m (close to abutments) to 1.5 m (close to keystone). The mesh is constructed in COCM following the abovementioned geometrical parameters and, as in the case of the numerical analyses presented in Sect. 3.1, the maximum number of sections along which cracking and crushing damage may occur is fixed to 17.

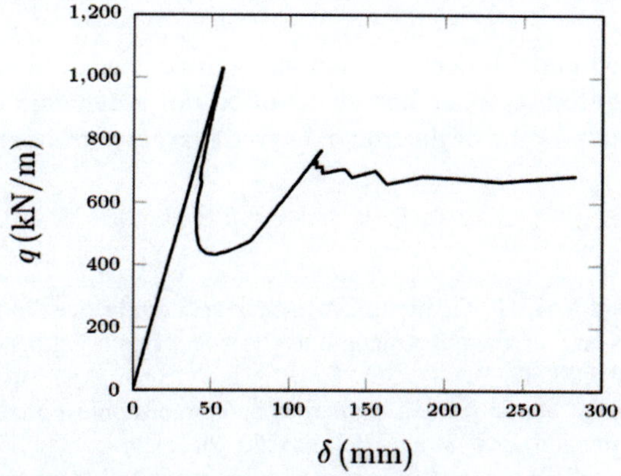

Fig. 4. Load-displacement curve for Mosca bridge

The analysis of the bridge is divided into two parts: in the first part, a parabolic load has been applied to simulate bridge construction phases and, in the second part an incremental analysis by means of COCM as presented in Sect. 2 is performed to investigate its ultimate behaviour.

For the construction phase analysis, it has been assumed a density for the arch granite equal to 27 kN/m^3 and a density of 23 kN/m^3 for the filling. The granite is supposed to have a compression strength $\sigma_c = 50$ MPa, and a tension strength $\sigma_t = 0.3$ MPa, whereas the fracture energy, G_F, and the crushing energy, G_c, are assumed equal to 0.02 and 50 N/mm, respectively. The application of COCM reveals that at the end of bridge construction, cracks are opened at the extrados of springings: these cracks were actually observed during centering removal [27]. On the other hand, with the application of live loads to the structure, these cracks close due to an increase in the compression force. The results of the incremental analysis performed by COCM is reported in Fig. 4 where the uniformly distributed load, q, is plotted against the keystone displacement, δ. It is possible to observe that the bridge has an almost linear behaviour until the peak load. After the peak load, a snap-back branch is detected due to granite crushing at keystone. Then, an increase in the loadbearing capacity is predicted due to cracking phenomena close to keystone and a redistribution of internal stresses: thus, a pseudo-ductile plateau is described until final failure.

4 Conclusions

In this paper, COCM has been integrated in a finite element method to simulate the elastic-plastic-softening loading history of masonry structures taking into account cracking and crushing damage phenomena together with scale effects. The numerical analyses within this paper clearly suggest the existence of a ductile-to-brittle transition in masonry arch bridges by increasing the scale. More precisely, COCM reveals that small scale arches present a pseudo-ductile behaviour similar to reinforced concrete beams: due to diffused damage along the arch span, an effective redistribution of internal stresses may be acknowledged.

The presented theoretical framework may be easily extended to include the effects of a possible FRP reinforcing plate and/or the effects of a dynamic load, providing a complete new approach for the evaluation of curved masonry structures.

References

1. Accornero, F., Lacidogna, G., Carpinteri, A.: Medieval arch bridges in the Lanzo Valleys, Italy: cases studies on incremental structural analysis and fracturing benefit. J. Bridge Eng. (ASCE) **23**(7), 05018005 (2018)
2. Manuello Bertetto, A., Marano, G.C.: Numerical and dimensionless analytical solutions for circular arch optimization. Eng. Struct. **253**, 113360 (2022)
3. Melchiorre, J., Manuello Bertetto, A., Marmo, F., Adriaenssens, S., Marano, G.C.: Differential formulation and numerical solution for elastic arches with variable curvature and tapered cross-sections. Eur. J. Mech.-A/Solids **97**, 104757 (2023)
4. Carpinteri, A., Lacidogna, G., Accornero, F.: Evolution of the fracturing process in masonry arches. J. Struct. Eng. (ASCE) **141**(5), 04014132 (2015)

5. Accornero, F., Lacidogna, G., Carpinteri, A.: Evolutionary fracture analysis of masonry arches: effect of shallowness ratio and size scale. Comptes Rendus Mécanique **344**(9), 623–630 (2016)
6. Heyman, J.: The Stone Skeleton, 1st edn. Cambridge University Press, Cambridge, UK (1995)
7. Lacidogna, G., Accornero, F.: Elastic, plastic, fracture analysis of masonry arches: a multi-span bridge case study. Curved Layered Struct. **5**, 1–9 (2018)
8. Accornero, F., Lacidogna, G.: Safety assessment of masonry arch bridges considering the fracturing benefit. Appl. Sci. **10**(10), 3490 (2020)
9. Carpinteri, A., Corrado, M., Mancini, G., Paggi, M.: Size-scale effects on plastic rotational capacity of reinforced concrete beams. Struct. J. (ACI) **106**(06), 887–896 (2009)
10. Carpinteri, A., Corrado, M., Paggi, M., Mancini, G.: New model for the analysis of size-scale effects on the ductility of reinforced concrete elements in bending. J. Eng. Mech. (ASCE) **135**(3), 221–229 (2009)
11. Carpinteri, A., Corrado, M.: Upper and lower bounds for structural design of RC members with ductile response. Eng. Struct. **33**(12), 3432–3441 (2011)
12. Accornero, F., Cafarelli, R., Carpinteri, A.: Cracking and crushing in prestressed concrete beams. Struct. J. (ACI) **118**(2), 101–109 (2021)
13. Accornero, F., Cafarelli, R., Carpinteri, A.: The cohesive/overlapping crack model for plain and RC beams: scale effects on cracking and crushing failures. Mag. Concr. Res. **74**(9), 433–450 (2022)
14. Accornero, F., Cafarelli, R., Carpinteri, A., Nanni, A.: Scale effects in GFRP-bar reinforced concrete beams. Struct. Concr. **24**(2), 2817–2826 (2023)
15. Carpinteri, A., Accornero, F., Cafarelli, R.: Scale effects in prestressed concrete structures: maximum reinforcement percentage to avoid brittle crushing. Eng. Struct. **225**, 113911 (2022)
16. Hillerborg, A., Modéer, M., Petersson, P.-E.: Analysis of crack formation and crack growth in concrete by means of fracture mechanics and finite elements. Cem. Concr. Res. **6**(6), 773–781 (1976)
17. Carpinteri, A.: Cusp catastrophe interpretation of fracture instability. J. Mech. Phys. Solids **37**(5), 567–582 (1989)
18. Belgin, C.M., Şener, S.: Size effect on failure of overreinforced concrete beams. Eng. Fract. Mech. **75**(8), 2308–2319 (2008)
19. Lourenco, P.B.: Computational strategies for masonry structures, PhD Thesis, Delft University of Technology, Netherlands (1996)
20. Carpinteri, A., Corrado, M., Mancini, G., Paggi, M.: The overlapping crack model for uniaxial and eccentric concrete compression tests. Mag. Concr. Res. **61**(9), 745–757 (2009)
21. Lanczos, C.: The Variational Principles of Mechanics, 4th edn. Dover Publication, New York, USA (1949)
22. Bocca, P., Carpinteri, A., Valente, S.: Mixed mode fracture of concrete. Int. J. Solids Struct. **27**(9), 1139–1153 (1991)
23. Ramm, E.: Strategies for tracing the nonlinear response near limit points. In: Wunderlich, W., Stein, E., Bathe, K.-J. (eds.) Nonlinear Finite Element Analysis in Structural Mechanics, pp. 63–89. Springer Berlin Heidelberg, Berlin, Heidelberg (1981). https://doi.org/10.1007/978-3-642-81589-8_5
24. Goodman, R.E., Taylor, R.L., Brekke, T.L.: A model for the mechanics of jointed rock. J. Soil Mech. Found. Div. (ASCE) **94**(3), 637–659 (1968)
25. Nguyen, V.P.: Discontinuos Galerkin/extrinsic cohesive zone modelling: implementation caveats and applications in computational fracture mechanics. Eng. Fract. Mech. **128**, 37–68 (2014)
26. D'Altri, A.M., et al.: Modeling strategies for the computational analysis of unreinforced masonry structures: review and classification. Arch. Comput. Methods Eng. **27**, 1153–1185 (2020)
27. Castigliano, A.: Théorie de l'Equilibre des Systémes Elastiques et ses Applications. 1st edn. Del Negro, Torino, Italy (1879)

Some Latest Research about the Messina Strait Bridge Crossing

Marco Peroni(✉), Giammarco Amore, and Shaik Maabu Subhani

Marco Peroni Ingegneria, Faenza, Ravenna, Italy
peroni@marcoperoni.it

Abstract. The construction of the suspension bridge over the Strait of Messina has been stopped due to political and economic problems and it seems also that the final approved project may have problems about his "serviceability" with respect to the wind action on the structure.

From since 2006 we have enterprises an intensive study of about an innovative solution for a design of an alternative solution for Messina based on a new type of suspended structure. In 2022 we have also performed a simpler solution applied on the design tender project based on the solution of the Italian engineer Sergio Musmeci.

Keywords: suspension bridge · tension structures · Messina Strait · hyperboloid · net ropes · wind actions · stabilizing cables · Sergio Musmeci

1 An Earlier Our New Design Proposal for Messina Strait Bridge

1.1 The Design Idea

Despite the great effort needing to cover the distance of the Messina Strait, all projects that have been presented until now, do not show a real new system for suspension: they are based on a classical suspension bridge solution with the research applied only on the deck aerodynamic performances. We want instead to propose two new innovative static systems: a first one is a new tension-structural system that not only allows to stabilize the bridge against the wind effects but that also is economically comparable with the tender design considered until now. A second, more simply solution, take in account the possibility to introduce a stabilizing rope under the tender deck of the bridge.

In 1969, for the Messina Strait crossing, architect Sergio Musmeci already proposed a "double effect" tensioning system, which for the first time in the design of infrastructures of such importance took into consideration the aero-elastics phenomena.

Musmeci's plan (Fig. 1), that won the international competition, consisted in a suspended bridge with, among other important innovations, an inferior stabilizing rope system that, with his intrinsic torsional damping, was contrasting the action of flutter and increases therefore the critical design speed wind. Even in the '800 there were some studies about stabilizing bridge systems.

S. Gabriele et al. (Eds.): IWSS 2023, LNCE 437, pp. 88–97, 2024.
https://doi.org/10.1007/978-3-031-44328-2_10

Fig. 1. Musmeci design project (1969) for Messina Strait.

Fig. 2. Le Ricolais experiments on hyperboloid structures.

Not in the last, even Le Ricolais has studied a bridge concept based on hyperboloid 3d surfaces (Fig. 2) and this studied confirm to me the necessity to develop this type of solution.

Developing the concepts proposed by Le Ricolais and Musmeci, my solution for the Messina Strait is a three-dimensional tensile structure with a hyperboloid shape: this consists of a 3D net with the ropes interlaced to each other to form a wicker basket containing inside the deck of the bridge.

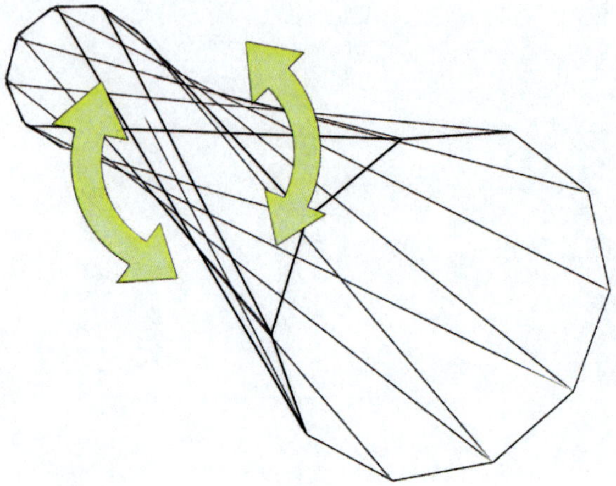

Fig. 3. Advantage of hyperboloid configuration.

1.2 Description of the Bridge Structure

The model that I propose for Messina is based on these guidelines: considerable in-crease in torsional and lift rigidity by means of the double-effect produced by the 3D cable net in hyperboloid configuration with anti-clastic opposite curvature (Fig. 3). Reduction of bridge weight using a less size diameter ropes distributed around an elliptic configuration in combination with a streamlined deck (like the Musmeci scheme) that allow to reduce the dead load of the bridge.

The principal net rope, beginning from two elliptic shaped towers at the extremities of the bridge 115 mt high above the sea level, has been designed with a series of twenty high steel strength cables each of 0.4 mt in diameter and is developed around an elliptic section that gradually reduces towards the midspan of the bridge (Fig. 4).

The particular interlaced cables conformation formed a "closed system" extremely stable with respect to the horizontal, vertical and torsional effects of the wind loads.

As a consequence of the shape designed for, all the ropes that build the bridge structural system have an extremely light weight in respect with the traditional suspension bridges and, thanks to their form, contribute together to the resistance and stability of the construction.

These new developments of the bridge design consist in a series of internal cable stayed bridge each 200 m long span (called sub-bridges), independent one to each other, hanging on the same frames that allow to build the cable nets.

In fact, the construction of the three-dimensional net is possible by assembling great steel frames that allow "to guide" the ropes that have already been assembled "in situ" with the technique of "areal spinning" to constitute the hyperboloid shape.

These frames should be at first assembled near the towers and then turned in position (from the two parts symmetrically) with the help of two temporary main ropes; at the same time the frames should be stabilized by two lower ropes.

Fig. 4. Model of our design proposal for the Messina Strait bridge crossing.

So, the structural philosophy is: the means of the internal bridges (Fig. 5) are made in a traditional and consolidated solution with a relative small span (max. 200 m). The innovation is in the external structural net-system that carried the complete structure.

This new stabilized structural scheme allow to reduce the deck torsion stiffness and design a lightness bridge deck: in our solution the deck is only 40 m width with the same road and rail lines of the classic design project. In consequence of this self-weight reduction, we have a total cables area of our solution (n°20 ropes of 0.40 m diameter) is about 2.52 mq against 4.80 mq of the tender winning solution.

We have also: 11 t/m deck weight in comparison with the 20 t/m deck weight of the traditional Messina scheme; 20 t/m of cable weight against 32 t/m and a cable total weight of 105000 t in comparison with 166000 t of the tender design project. This confirm the goodness of these configuration because this value is obtained with a structure more efficient in terms of such aspect than the traditional solution.

For about the towers we may considered the same steel quantity between our solution (440 m height in a unique elliptic configuration) and the classical tender solution (390 m height but with a means of transverse that connect together the two legs of the tower). In this new proposal the towers are rested on the same foundations designed for the classical solution.

1.3 The Computer Analysis

We may compare our results (Table 1) against those of the "classical" solution approved by Società Stretto di Messina (SM) as definitive design project (Table 2).

Table 1. FEM Analysis input data (new version).

Characteristics	Values
Diameter of the single main cable	0.40 m – n°20 ropes
Diameter of the secondary elliptic ropes	0.20 m

(*continued*)

Table 1. (*continued*)

Characteristics	Values
Max. Vertical deformation under total loads	5.5 m
Max. Lateral deformation under static wind load	4 m
First vibration mode (horizontal)	21 s
Max. Cable tension (static nonlinear analysis)	45000 kN (4500 t)

Table 2. Messina Strait crossing tender (SM version) design results.

Characteristics	Values
Diameter of the single main cable	1.20 m – n°4 ropes
Max. Vertical deformation under total loads	15 m
Max. Lateral deformation under static wind load	11 m
First vibration mode (horizontal)	32 s
Max. Cable tension (static nonlinear analysis)	810000 kN (81000 t)

A non-linear static analysis was performed at first to determinate the deformation of the bridge under permanent and live loads and under the horizontal static wind action.

The results showing a 5.5 m vertical deformation (Fig. 5) under overall live and permanent loads and a 4 m horizontal deformation (Fig. 6) under the static wind actions.

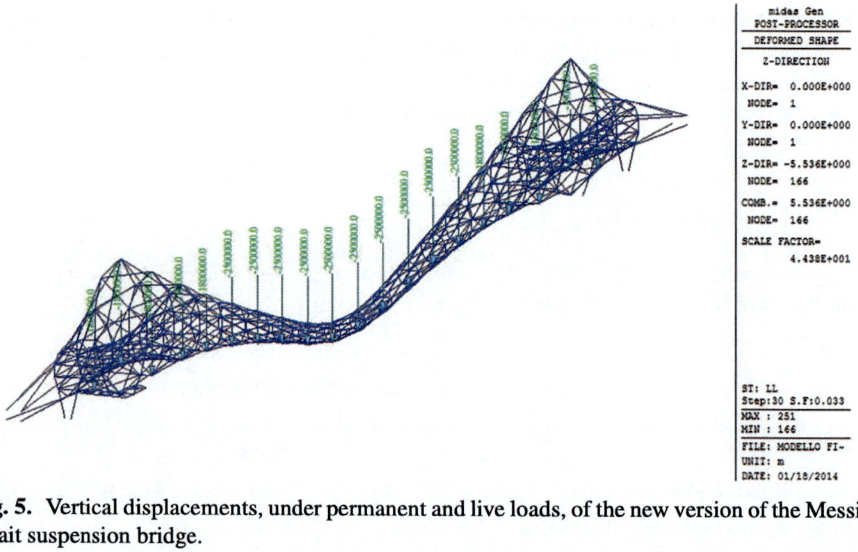

Fig. 5. Vertical displacements, under permanent and live loads, of the new version of the Messina Strait suspension bridge.

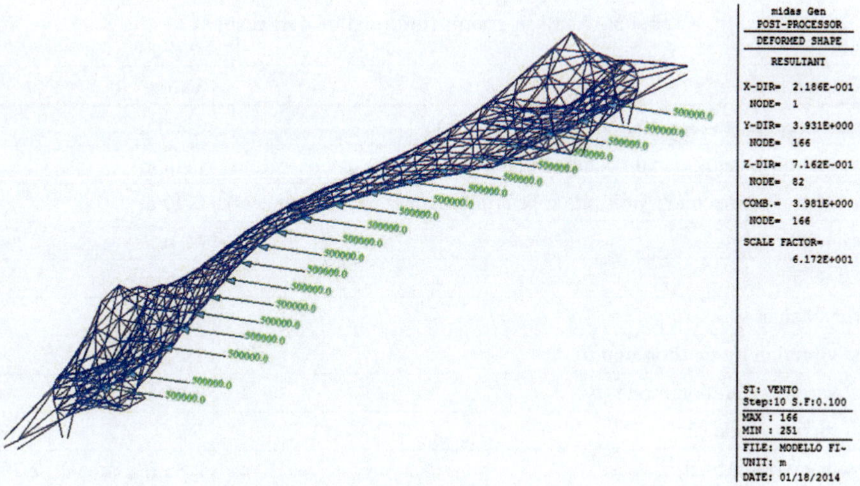

Fig. 6. Horizontal displacements, under the horizontal static wind action.

The maximum tensions in the cables of about 4500 t (Fig. 7).

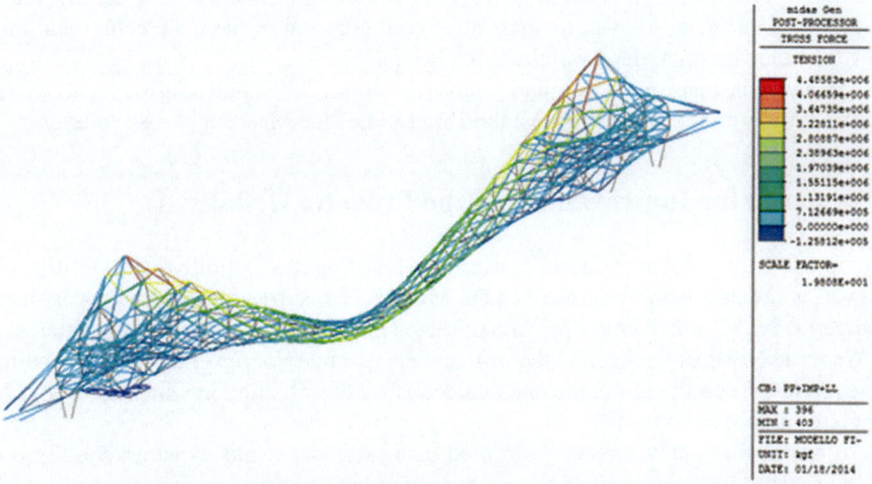

Fig. 7. Maximum tensions in the cables of the new version.

It was also performed an eigenvalue dynamic analysis to assess the modes of vibration of the structure. The first horizontal flectional mode is of about 13 s and the torsional one of about 4.70 s, showing a very stiffness structure and a great separation between the two principal modes as the flutter stability recommend (Table 3).

We know in fact as the rotational/vertical frequency ratio determines the wind response proprieties of the bridge for the most dangerous form of aeroelastic instability: the classic flutter which arises due to coupling of rotational and vertical modes. In

Table 3. Analysis results (information data results).

Characteristics	Values
Suspension bridge free span	3300 m
Diameter of the single main cable (external net ropes)	0.40 m
Diameter of the secondary elliptic ropes (internal net near the towers)	0.20 m
Deck width	40 m
Deck weight	11 t/m
Towers height	440 m
First vibration mode (horizontal)	19.5 s (Fig. 9)
First vertical vibration mode (fv)	14.5 s
First torsion mode (ft)	5 s
Frequency ratio (fv/ft)	2.85 (Storebealt: 2.79)
Total cables weight	110000 t
Total sub-bridges frames weight	20000 t

our solution this condition is made be possible without increase the deck aerodynamic and torsional stiffness but with an innovative configuration of the main cables and this allow to reduce the total dead weight of the bridge.

The analysis confirms a frequency ratio of about 2.85 that is close to those of Storebealt or Humber bridges against the 1.36 ratio of the Messina classic solution.

2 Proposal for Improvement of the Progetto Definitivo

Making use of the documentation concerning for Progetto Definitivo (2011) [9], the goal was to create a FEM model, using the MIDAS Gen software, with an approximate geometry (Fig. 8), which could provide reliable results for the analysis of our interest.

We started by introducing into the software the geometry of the bridge corresponding to the reference condition, i.e. the one that describes the situation in which the bridge is completed and open to traffic.

An eigenvalue analysis was performed using the static and dynamic equilibrium procedure valid for a suspension bridge, obtaining the first frequencies.

To understand if the masses provided by the MIDAS model, in term of weight of the structural steel deriving from the geometric model, could be plausible, a comparison was made with the information in possession (total weight of the steel for the for each structural element).

Furthermore, a further comparison was performed, this time in terms of frequencies: those obtained from the global model prepared by the design company COWI A/S and those present in the book [7] above were taken as a reference. For the purposes of our analysis, the frequencies of greatest interest are those corresponding to the following modes of vibration (Figs. 10, 11 and 12).

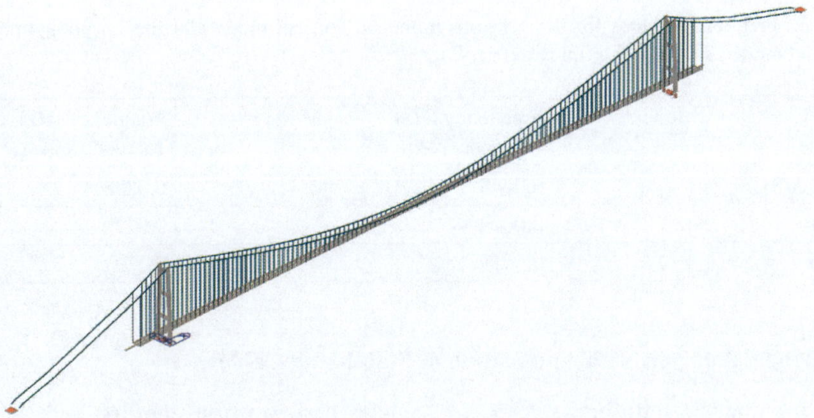

Fig. 8. MIDAS geometric model of the Messina Strait Bridge.

Fig. 9. 1st sym. Vertical mode (V.S.).

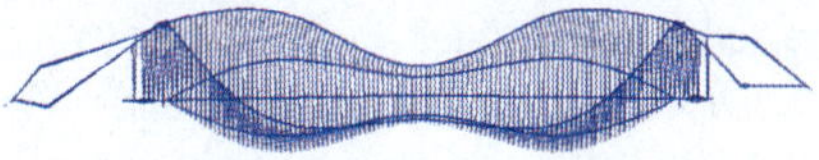

Fig. 10. 1st sym. Torsional mode (T.S.).

Fig. 11. 1st antisym. Vertical mode (V.A.).

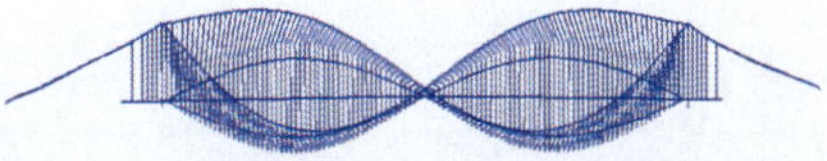

Fig. 12. 1st antisym. Torsional mode (T.A.)

Table 4. Frequency values for the 1st symmetric vertical mode and the 1st symmetric torsional mode as the models analyzed vary.

Model	Mode n°	Frequency [Hz] 1st mode V.S	Mode n°	Frequency [Hz] 1st mode T.S
Cowi A/S	7	0,0805	12	0,1003
Book	6	0,0805	9	0,0959
Midas	5	0,0828	11	0,1036

The comparison, therefore, can be summarized in the following Tables:

It is possible to note how the frequencies, corresponding to the first vertical and torsional modes, are similar between the different models, consequently it is legitimate to state that the MIDAS model is also validated under this aspect (Tables 4 and 5).

Table 5. Frequency values for the 1st antisymmetric vertical mode and the 1st antisymmetric torsional mode as the models analyzed vary.

Model	Mode n°	Frequency [Hz] 1st mode V.A	Mode n°	Frequency [Hz] 1st mode T.A
Cowi A/S	3	0,0648	8	0,0808
Book	3	0,0603	5	0,0799
Midas	3	0,0619	4	0,0791

2.1 Modal Analysis with Application of Stabilizing Cables

Knowing that the ratio between the frequencies deriving from the first vertical and torsional modes (symmetric and antisymmetric) determines the design of long-span suspension bridges, the objective is to increase this ratio by applying stabilizing cables in order to increase the wind speed at which the aeroelastic buckling at flutter occurs. The stabilizing cables were applied in 3 directions: inclined by about 45° to the vertical plane (Fig. 13), horizontal and vertical (it should be remembered that the arrangement of the latter two is difficult apply in reality).

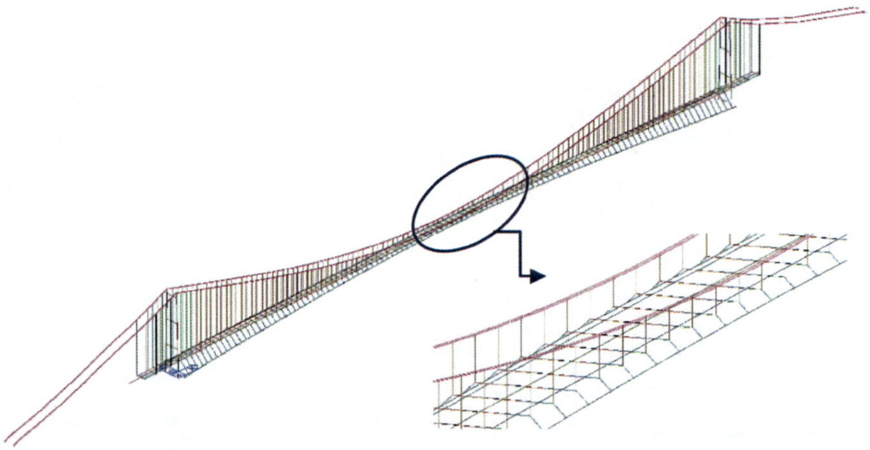

Fig. 13. Application of inclined stabilizing cables: view of the entire bridge and detail.

For each of these directions, then, the analyzes were carried out by varying the diameter of the cables (80-60-40 cm) and the assigned pretension. After performing the analyses, it was noted that, for any arrangement of these cables, significant results in terms of frequency ratios were not obtained as we would have expected, as the values obtained are slightly smaller than the starting value, i.e. the MIDAS model of the bridge without cables (T.S./V.S. = 1,251; T.A./V.A. = 1,278).

2.2 Modal Analysis with Application of Stabilizing Cables

We briefly report the results obtained from a non-linear analysis, of a geometric type, on an approximate model, by applying static forces equivalent to the wind along the deck, with and without the application of stabilizing cables (horizontal, vertical and inclined). Whatever the arrangement of the cables, a reduction of the lateral displacements, evaluated in the center of the main span, is shown as the diameter of the cables and the assigned pretension increase (in the best case, with inclined stabilizing cables of 80 cm and 100% of pretension, there is a reduction of about 50% of the lateral displacements, going from about 9,15 m to 4,3 m): in this way, the global behavior of the deck under the action of the wind tends to improve.

3 Conclusions

As we have shown in the previous results, the application of new innovative concepts (one completely redesigned and the other one simpler) on the design for the Messina Strait crossing is not also possible but even economically comparable with the tender winner design project that may be realized. In respect with the solution until now proposed by the tenders, these new ideas allow to obtain a better behavior in respect with the wind action and with the serviceability of the bridge about the horizontal, torsion and dynamic aspects.

Bibliography

1. Tadaki, K., Scott, R.: History of the Modern Suspension Bridge – Solving the Dilemma between Economy and Stiffness. ASCE PRESS (2010)
2. Borri, C., et al.: The aerodynamic advantages of a double-effect large span suspension bridge under wind loading. In: Space Structures, vol. 4. Thomas Telford, UK (1998)
3. Maeda, K., et al.: Applicability of CFRP cables to ultra long-span suspension bridges. In: Proceedings of IABSE Conference in Seoul, IABSE Report, vol. 84, pp. 224–225 (2001)
4. Scott, R.: In the Wake of Tacoma. ASCE PRESS (2001)
5. Petroski, H.: Design Paradigms: Case History of Error and Judgments in Engineering. Cambridge University Press, Cambridge (1994)
6. Ito, M.: Wind induced vibration on bridge, IABSE Report
7. Brancaleoni, F., et al.: The Messina Strait Bridge – A Challenge and a Dream. CRC Press, Taylor & Francis Group (2010)
8. Gimsing, N.: Cable Supported Bridges, 2nd edn. Wiley (1998)
9. Attraversamento stabile dello Stretto di Messina e collegamenti stradali e ferroviari sui versanti Calabria e Sicilia (2011). https://va.mite.gov.it/it-IT/Oggetti/Documentazione/1/2. Accessed 10 May 2023

The Role of Parametric Design in the Robotic Assembly of Dry-Constructed Shell Structures

Emil Korkis[1(✉)], Mehmet R. Dogar[1], and Ornella Iuorio[2]

[1] University of Leeds, Leeds LS2 9JT, UK
cn15ek@leeds.ac.uk
[2] Politecnico Di Milano, Milan, Italy

Abstract. Architecture is going through major changes as automation in construction is radically transforming standard processing technologies and could lead, in the long-term, to disruptive technologies such as 3D-printing and robotics being applied to improve construction processes. This paper describes the adoption of a parametric approach in the design, fabrication, and assembly of dry-constructed, lightweight, wooden shell structures.

The design of the structure in a parametric environment simplifies the process of running multiple kinematic simulations of the construction sequence. These simulations are used to determine the feasibility of each simulated sequence and detect collisions and errors in the robotic arm path. The instructions from the simulation stage are passed through a series of steps to a single collaborative robotic arm that assembles the structure with minimum human intervention. The simulated tests are carried out using a single KUKA lbr iiwa R800 robotic arm equipped with a 2-finger parallel jaw gripper.

The complexity of the design requires a multi-disciplinary effort in the development of suitable systems. This research aims to show the potential that robotics and 3D printing could offer in terms of enabling the adoption of more complex and efficient free-form shell structures with decreased need for formwork and temporary supports.

Keywords: Parametric Design · Robotic Assembly · Kinematic Simulation · Shell Structures

1 Introduction

Shell structures are thin, spatially curved structures that transmit external applied loads through membrane actions. Shells are found abundantly in nature such as eggshells, seashells, turtle shells, and many other notable examples. The efficiency of natural shells is also found in artificial shell structures developed in ancient times, and increasingly more in modern engineering disciplines of civil, architectural, mechanical, aeronautical, and marine engineering. Shell structures are extensively used in civil and architectural engineering, examples include large-span shell roofs, liquid retaining structures, silos, cooling towers, and arch dams.

S. Gabriele et al. (Eds.): IWSS 2023, LNCE 437, pp. 98–107, 2024.
https://doi.org/10.1007/978-3-031-44328-2_11

This paper addresses a solution for the robotic assembly and disassembly of a segmented dry constructed shell structure. Dry construction refers to the use of dry materials in the construction of the structure without the need for bonding agents like cement or mud. Dry structures can benefit from faster construction times, better insulation, better moisture resistance and fire safety. However, this type of assembly introduces a set of challenges that should be addressed in order to achieve a structurally stable, usable, and aesthetically appealing shell structure. Among those challenges is the need to connect the segmented modules in such a way that conserves the continuity of the structure and allow the shell to transmit the loads using membrane actions.

Membrane actions are the internal forces that act in compression and/or tension on the cross section of the shell. One of the significant advantages of shells is the ability to design a structure that minimizes bending moments and transmit external forces through the thin walls of the shell using membrane actions. Segmented shells offer an advantage for robotic assembly in that the size of the segments can be adjusted in the design stage to suit the capacity of the robotic system used. The process of creating a segmented surface is called tessellation. Tessellation is carried out in the design stage and results in segments that closely follow the original surface (Fig. 1).

Fig. 1. ICD/ITKE Research Pavilion, Landesgartenschau 2014 Exhibition Hall (Willmann et al. 2016)

2 Review

Robotics cannot replace skilled labour in the construction industry; however, robotics can complement the workforce. Faster and safer construction is a pressing need especially in an industry that have a high rate of injuries even in the most advanced and regulated economies. Robotics technologies have just started to realize commercial potential. A clear reason is that the construction tasks are notoriously difficult to automate. Research and development in this specific sector are much needed.

The use of robotics in the construction industry dates to the 80s of last century. Some processes were automated on or off-site, but the use of robotics was limited to experimental applications and the social and economic factors caused the progress to slow down (Bock 2015). More recently, and due to the rapid development in computer software and hardware, robotics use in the construction industry is becoming more viable. New methods are being developed for dry construction, fabrication, and assembly of structures; however, robotic systems are mostly being used in fabrication of shell structures rather than assembly (Fig. 2) (Li and Knippers 2015).

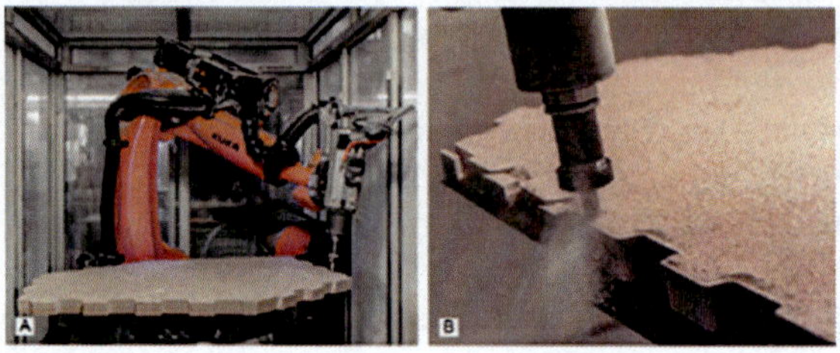

Fig. 2. Landesgartenschau Exhibition Hall in Schwäbisch Gmünd (Jian-Min Li 2015)

Robotic arms have been used in the dry stacking of wall sections, and simple structures. Two papers by (Thangavelu et al. 2018) and (Bonwetsch 2012) describe the vertical stacking of regular and irregular objects by a robotic arm, however, this method is only used in dry stacking where gravity provides the stability of the structure at all stages of construction (Fig. 3). This is not always the case for all types of structures and certainly not the case for most shell structures.

Fig. 3. "Structural Oscillation" Completed installation at Venice Biennale 2008 (Bonwetsch 2012)

3 Methodology

The approach to this paper in software is a CAD method called parametric design. Parametric design refers to linking the geometric design to parameters. Changing the parameters directly or indirectly changes the linked geometry and ultimately the overall design. Parametric design is an algorithm-based process that uses parameters to define relationships between a set of inputs and derives a result from this relationship. The outputs of this process remain tightly related to the parameters. Changing parameters at any stage of the process affects the resulting geometry. The magnitude of this effect depends entirely on the implementation of the parameter in the system being designed. The fundamental design approach of the research is illustrated in (Fig. 4) to highlight key features of the design.

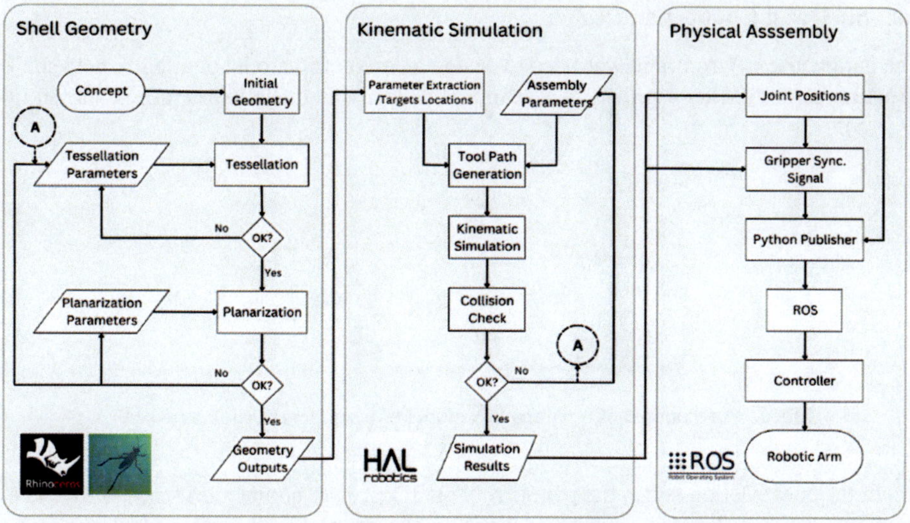

Fig. 4. The fundamental design approach proposed for this research.

3.1 Geometric Simulation

This approach contains three main conditional junctions where the design is evaluated. However, the actual design contains many more logical checks. The conceptual design is converted into a 3D design modelled in Rhinoceros/Grasshopper. The assembly of the shell structure is simulated in software using HAL Robotics plugin which offers great integration with Rhino3D and Grasshopper.

Setting up a simulation helps to identify problems and avoid most errors in the assembly sequence that could lead to crashes, unexpected behaviour of the robotic arm, damage, or injury. The simulation aims to investigate the possibility of automatically generating a path for the robot to follow, after the tessellation and planarization operations are completed. Extracting the parameters from the design will allow the code to adapt to different initial geometries and generate a new path for assembly. A robotic arm is used for this method, developed, and manufactured by KUKA robotics, the LBR iiwa collaborative robot is a 7-axis collaborative robotic arm is a lightweight manipulator with a very high repeatability and accuracy.

Geometric simulations are powerful tools to help designers make decisions on planning, analysis, and design in different areas of research and development. Different simulations are designed depending on the required model. For example, an assembler robot simulation is mainly concerned with the motion of the robot and the motion of the surrounding environment. A kinematic simulation is used in this case. On the other hand, a simulation of actuators mainly focuses on the dynamic models rather than kinematic models (Zlajpah 2008).

3.2 Shell and Connection Design

The parametric environment was used to design a parametric model of a segmented shell structure supported by 4 pillars. An example of segmented shell structure is shown in (Fig. 5).

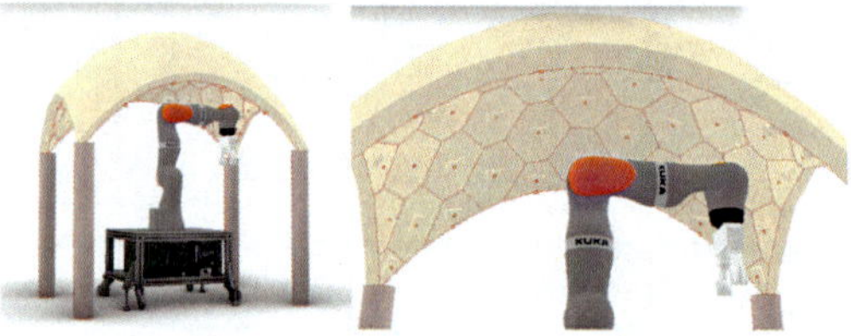

Fig. 5. An example of a parametric model of a segmented shell structure

In the construction stage, the panels rely on 3D-printed connection for stability. The segments can be affixed to each other using screws for stability. The connections were designed to act as temporary supports that minimize and possibly replace formwork.

The 3D-printed connections are standardized across the design, this provides excellent flexibility while exploring different solutions, given a practical limit of 25° between any two adjacent panels is maintained. A single connection is shown in (Fig. 6a) joining 3 panels at their intersection (Fig. 6b).

Fig. 6. A single connection (a) and a connection joining three panels (b)

The design of the shell structure forces catenary curves for the edge arches on all sides, which simplifies the assembly procedure by providing a self-supporting perimeter for the structure in the construction phase. The side arches' panels, however, differ from other panels of the shell in that they require a different shape for the connection to form the arch (in the construction phase).

3.3 Assembly Sequence

Due to the large number of segments, it is not practical to have a dedicated pick-up space for each segment, hence the pick-up location for all segments is unified, and after a segment is assembled, a new segment is placed in the same position. All segments have a rectangular shaped cut-out for the end-effector of the robot to attach to.

The assembly sequence starts with the arm at home position, which refers to the position where all axes of the robot are at the zero position. The arm then moves to a pick-up approach position, a specified distance over the pick-up location. The assembly operation is similar to the pick-up operation; the arm moves to the approach position then follows a predefined path to correctly assemble the panel into position. The connection is attached to the panel manually before assembly. Shown in (Fig. 7) are the steps to assemble a single panel.

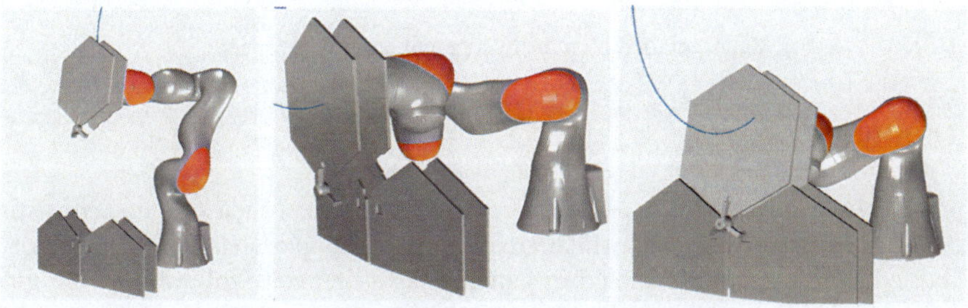

Fig. 7. Assembly of a single panel in position

The arm used is a 7-axis collaborative robotic arm. Due to the extra degree of freedom in the arm, it can reach an achievable goal in many different arrangements of joint positions. HAL robotics plugin for grasshopper handles motion planning, however, it only provides one solution for every run, which means that at every reachable target, only one solution is chosen. Some of the solutions provided are not optimal for an assembly operation, where a substantial manoeuvre is performed near targets. To get more favourable solutions, an intermediate axis-based goals are provided to force the arm to perform the needed manoeuvres away from the final target.

Each of the intermediate targets is shared between a number of panels. Panels are grouped into zones based on the intermediate target they share. Based on this approach, the assembly sequence of a single panel changes into the following: Home position, pick-up approach, pick-up target, back to pick-up approach, zone target, destination approach, and finally destination target. A detailed plan for N segments assembly procedure is found in (Fig. 8).

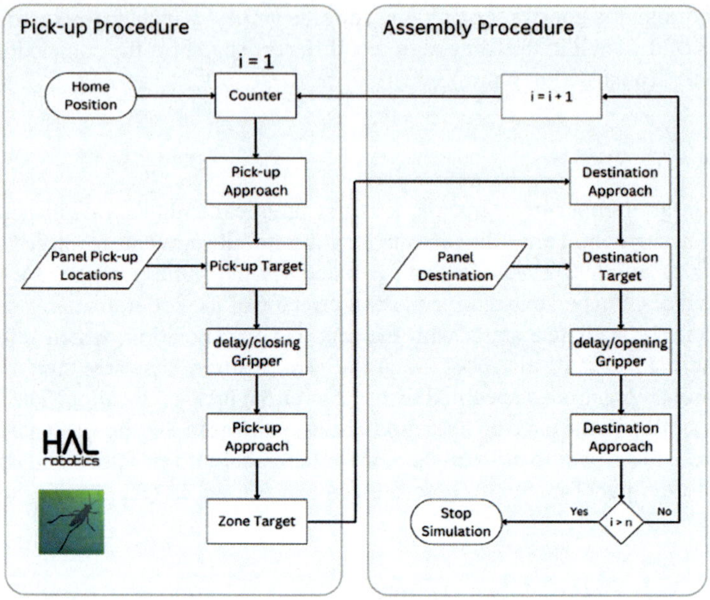

Fig. 8. A detailed plan for N-segment assembly procedure

4 Results

The parameters were adjusted to achieve a doubly symmetrical shell structure consisting of 61 hexagonal segments, 4 of which attach to the 4 supports (removed for clarity). The four edge arches are assembled in sequence as a first step, followed by the inner segments and finally the keystone (also known as the capstone). However, in this case, it was found that the keystone collides the structure due to the assembly from inside the shell. A possible solution was to use two segments as keystone (attached manually beforehand) which allows for twisting of the final segment to reach the approach location. The following (Fig. 9) and (Fig. 10) show the initial and final stages of the simulation respectively.

The overall simulation time is 1403 s (23 min 23 s) including the inserted delays for pick-up and place operations. The assembly can be examined visually or through a simulation dashboard and a timeline of joint movements, speeds, and accelerations. The visualised simulation is also available in all view ports. This gives an understanding of possible collisions or obstacles that can interfere with the assembly sequence.

Changing the parameters of the design changes the shell structure. A random set of parameters were chosen to present an example of a non-symmetrical shell (Fig. 11). The overall system worked as intended and a tool path was generated and simulated successfully.

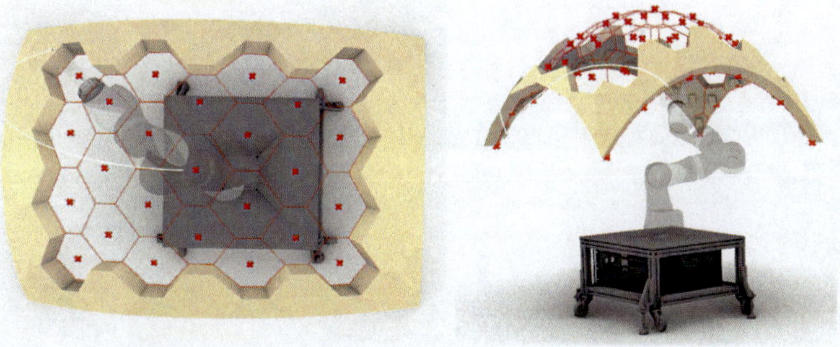

Fig. 9. Initial stages of the assembly procedure

Fig. 10. Final stages of the assembly procedure

The 3D-printed connection was tested with 3 panels of a shell structure. The connection was able to maintain the correct angles between the panels and provide stability for the structure in place of formwork (Fig. 12). Each panel was made out of two sheets of plywood, 5 cm apart. A larger number of panels need to be tested to check the ability of the connection system to replace formwork entirely and to assess the viability of this solution in the robotic assembly of shell structures.

Fig. 11. An example of a non-symmetrical shell assembly

Fig. 12. 3D-printed connection replacing supports.

5 Conclusion

This research can serve as step towards the automation of a wide array of construction needs. The efficiency of shell structures in load transfer and the structural capacity per unit weight compared to traditional construction methods is poised to be a major contributor to the exploration efforts in the future where the cost of transporting traditional construction materials can be prohibitively expensive.

The methodology followed is independent of the shell structure form. However, an investigation of how different form-finding techniques can influence the results is of a great importance, as well as the possibility of adapting the system for different types of shell structures, such as grid shells. This research tries to assess the viability of a collaborative robotic system to work independently or in collaboration with individuals in the assembly of a light-weight shell structure, and the benefits and drawbacks of such a system from the parametric design stage to the automatic path planning and the generation of tool paths.

References

Willmann, J., et al.: Robotic timber construction — expanding additive fabrication to new dimensions. Autom. Constr. **61**, 16–23 (2016)

Zlajpah, L.: Simulation in robotics. Math. Comput. Simul. **79**, 879–897 (2008)

Bock, T.: Construction robotics. Auton. Robot. **22**, 201–209 (2007)

Bock, T.: The future of construction automation: technological disruption and the upcoming ubiquity of robotics. Autom. Constr. **59**, 113–121 (2015)

Parascho, S., Han, I.X., Walker, S., Beghini, A., Bruun, E.P.G., Adriaenssens, S.: Robotic vault: a cooperative robotic assembly method for brick vault construction. Constr. Robot. **4**(3–4), 117–126 (2020)

Vasey, L., et al.: Behavioral design and adaptive robotic fabrication of a fiber composite compression shell with pneumatic formwork. Acadia, pp. 296–309 (2015)

Reality-Based FE Block Models of Masonry Cross Vaults: A Real Case Study

Marco Alforno[1,2(✉)], Alessia Monaco[1], Fiammetta Venuti[1], Chiara Calderini[2], Giulia Sammartano[1], Giacomo Patrucco[1], and Antonia Spanò[1]

[1] Department of Architecture and Design, Politecnico Di Torino, Torino, Italy
marco.alforno@edu.unige.it, marco.alforno@polito.it

[2] Department of Civil, Chemical and Environmental Engineering, Università Di Genova, Genova, Italy

Abstract. Nowadays, it is crucial to develop analysis methodologies to confidently simulate the structural behaviour of historical masonry constructions. From this perspective, the possibility of using reality-based models as starting data is becoming increasingly attractive. This methodology allows considering some aspects that are usually neglected in the analysis of masonry structures, e.g., the as-built geometry and the block arrangement. This study investigates the implications of applying this approach to a case study: the Palazzata di Vicoforte in Piedmont, Italy. Starting from dense point clouds obtained through LiDAR acquisitions and photogrammetric surveys, a very accurate NURBS modelling has been performed to develop Finite Element micro-Models of two cross vaults characterized by very similar macro-geometry, but different brick pattern. The models are numerically analysed under differential abutment settlements. Comparison of the results obtained from vaults of ideal geometry shows the importance of accurately modelling both macro and micro geometry.

Keywords: masonry vaults · 3D modelling · LiDAR · Digital photogrammetry · masonry pattern · FE block model

1 Introduction and Case-Study Description

In recent years, a major challenge for researchers operating in the framework of built heritage documentation, conservation and valorisation is represented by the high complexity that characterizes the invaluable assets witnessing our history. In this context, a multidisciplinary approach is essential, to deeply investigate historical buildings in their constructive complexity from different perspectives with the aim to deepen their knowledge. In the framework of the many processes connected to the knowledge of the built heritage structures, the instruments provided by Geomatics supporting Structural Engineering allow to carry out non-invasive investigations and analyses aimed at documentation or diagnostic purposes, and the fruitful synergy between these two disciplines is evidenced by several interdisciplinary studies (e.g., [1–3]). In fact, nowadays, thanks to the huge technological improvement in terms of acquisition techniques, sensors and

S. Gabriele et al. (Eds.): IWSS 2023, LNCE 437, pp. 108–117, 2024.
https://doi.org/10.1007/978-3-031-44328-2_12

processing algorithms, it is possible to obtain reality-based models characterized by very high accuracies and level of detail. Traditionally, the acquisition of point clouds – using range-based sensors such as laser scanning technologies, or image-based approaches following photogrammetric pipelines – represents the most efficient and effective solution for acquiring 3D metric data concerning historical buildings and masonries [4]. In particular, the integration of these methods allows the connection between 3D digital geometries characterised by very high spatial resolution and the radiometric information derived from image-based photogrammetric techniques, enabling an enhanced reading of elements that depend on both the geometry and radiometry of an architectural asset (e.g., material consistency or masonry texture) [5]. Reality-based models derived from dense point clouds can be used as primary data to develop structural models for the analysis of the surveyed object with different ad hoc solutions, usually based on the integration of different methods and techniques [6]. This methodology allows considering some aspects that are usually neglected in the structural analysis of masonry structures, e.g., the as-built geometry and the block arrangement [7].

This study investigates the implications of applying this approach to a case study: the Palazzata di Vicoforte in Piedmont, Italy, which is represented in Fig. 1. The *Santuario Regina Montis Regalis di Vicoforte* (also briefly known as *Santuario di Vicoforte*) is a monumental church located in Vicoforte, Piemonte, northern Italy. It is known for its elliptical dome, the largest in the world. The building is entirely constructed in brickworks, with the exception of stone pillars [8]. It consists of three main wings, a central wing and two lateral wings, built in different centuries, as shown in Fig. 1. The ground floor is characterized by a long portico facing the Sanctuary, obtained by a sequence of cross vaults of approximately 3.2 m of span. Most of the vaults do not have plasterwork at the intrados surface, hence the brick arrangement is visible. All vaults are built with bricks, regular in shape and arranged in a diagonal or radial pattern with respect to the head arches (Fig. 1).

Fig. 1. The Sanctuary of Vicoforte and the Palazzata (in color). View of two different vaults of the portico, with radial masonry pattern (right) and diagonal pattern (left).

First, a 3D metric survey campaign has been carried out on the masonry vaults belonging to the portico, exploiting the very high spatial resolution of the 3D model derived from LiDAR data as well as the radiometric resolution generated from the photogrammetric approach. Hence, it was possible to develop a hybrid strategy to achieve a 3D dense surface useful as the starting point for subsequent as-built NURBS discrete modelling. Finally, Finite Element models are developed of two cross vaults characterized by very similar macro-geometry, but different brick pattern. The models are tested under differential abutment settlements and results are compared with those obtained on vaults of ideal geometry. The scope is highlighting which is the influence of accurate real geometry, usually characterized by imperfections, in the assessment of the mechanical response of vaulted structures with different masonry block layouts.

The paper develops as follows: Sect. 2 illustrates the 3D metric survey and data post-processing; Sects. 3 and 4 are devoted to structural modelling and analysis, respectively, while conclusions are discussed in Sect. 5.

2 Geometric Survey and Data Processing

In this section, a data fusion strategy is proposed, based on the 3D models derived from LiDAR point clouds and the added-value metric products achieved following photogrammetric approach. These data represented the starting point for subsequent as-built NURBS discrete modelling and configuration of elements for structural analyses.

Geomatic survey of the Palazzata portico was performed according to the following procedure. First, a topographic network has been measured in order to provide a common reference system and a metric control to the subsequently acquired range-based and image-based data. Therefore, a set of points was measured using a total station to be used as control points. 12 LiDAR scans have been collected using a Faro Focus3D X 330 phase shift laser scanner, providing a high overlap between adjacent scans to facilitate the following registration procedures. Afterwards, a close-range photogrammetric survey has been carried out in order to collect high-resolution digital images of the selected cross vaults. A Digital Single Lens Reflex (DSLR) camera Canon EOS 5DSR, equipped with a Zeiss 25 mm lens, was set in nadiral configuration in relation to the vaults and mounted on a tripod. Then, two photogrammetric strips have been acquired from a shooting distance ≈3.2 m with an estimated overlapping of 90% and an average Ground Sample Distance (GSD) of 0.5 mm.

The LiDAR scans have been registered (oriented in the same reference system determined by the previously measured topographic network) using the following strategy:

- Initially, the scans were aligned using an ICP-based (Iterative Closest Point) approach, exploiting the high overlap between adjacent point clouds. The discrepancies observed among the registered scans at the end of this preliminary procedure were millimetre-level (approximately, a mean value of approximately 2 mm was achieved);
- Subsequently, the block of scans was rigidly rototranslated in the reference system using a set of measured control points (natural points or checkboard targets). In this case the observed accuracies are approximately 5 mm.

Subsequently, the acquired digital images have been processed following a standard SfM-based (Structure-from-Motion) pipeline [9]: Interior Orientation (determination of the intrinsic parameters of the camera); Relative Orientation of the images and generation of a sparse cloud of tie points; Absolute Orientation using a set of Ground Control Points (GCPs) homogeneously distributed on the surveyed surface and measured with the total station; accuracy evaluation using a set of Control Points (CPs). The observed Root Mean Square Errors (RMSEs) are millimetre-level for both GCPs and CPs, as reported in Table 1.

Table 1. Mean errors on GCPs and CPs.

	RMSE			
	X [m]	Y [m]	Z [m]	XYZ [m]
GCPs (n° 15)	0.003	0.002	0.002	0.004
CPs (n° 7)	0.002	0.002	0.004	0.005

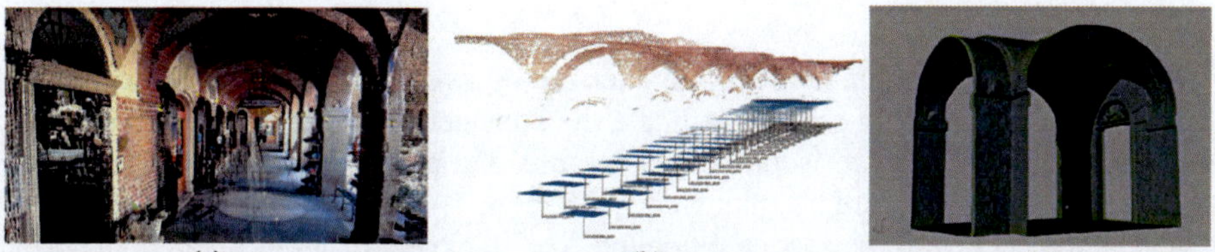

(a) (b) (c)

Fig. 2. (a) LiDAR point Cloud of the Palazzata; (b) Photogrammetric survey (oriented images and sparse cloud of ties points); (c) Mesh model of one vault in Rhinoceros.

Starting from the obtained reality-based data (Fig. 2 a, b) a very accurate NURBS modelling has been performed on Rhinoceros® platform. First, a dense mesh has been created from the point clouds (Fig. 2c). This densely meshed model represents the intrados surface of the vaults, and the surfaces of walls and pillars facing the interior of the portico. The mesh dimensions are sub-millimetric, hence it is possible to detect the mortar joints from the models. To improve the detection of the mortar joints, the radiometric information obtained from the photogrammetric survey has been projected onto the mesh, so as to have texturized models as a result.

The mesh of the geometric models is not suitable for FEM analyses, due to the fact that it is too irregular and extremely refined. Therefore, it was necessary to undertake a simplification process which consisted in transforming the intrados surface of the vaults into NURBS. This was done by projecting lines onto the mesh derived from the point clouds (blue lines in Fig. 3). These lines, which are complex polylines, have been described by simple regularized curves used to generate NURBS surfaces. In order to take into account the brick arrangements, the discretization of the intrados surface with regular curves has been done with lines that follow the bed-joints. Then, these lines have been translated normally to the intrados surface of a distance equal to the thickness of one brick (Fig. 4).

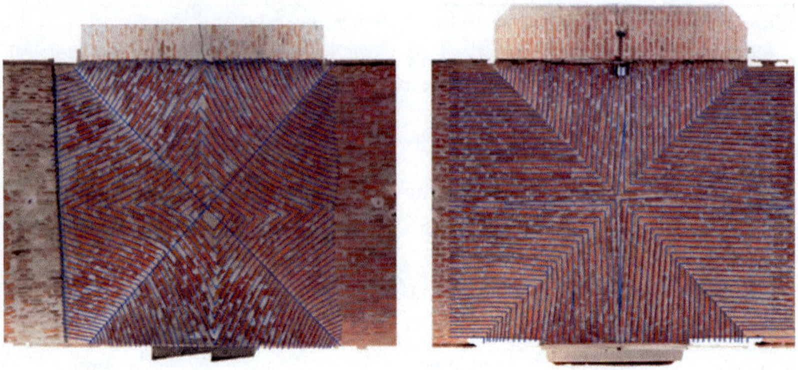

Fig. 3. Detection of bed joints on intrados surface of two vaults with diagonal (left) and radial pattern (right).

Two vaults have been modelled, characterized by two different brick arrangements. They have been chosen in order to validate the consideration made on vaults of ideal geometry in [7]. In order to isolate the two vaults and perform FE analyses on them, confinement structures were taken into consideration. The front arch and lateral arches, as well as the back wall, have been modelled. The thickness of the front arch has been easily measured from the survey of the facade, whereas the thickness of the lateral confinement arches has been estimated by observing the brick arrangements from the intrados: as a matter of fact, it is clearly visible that some bricks on the lateral arches are positioned vertically, resulting in the arch thickness of about the length of one brick (about 25 cm).

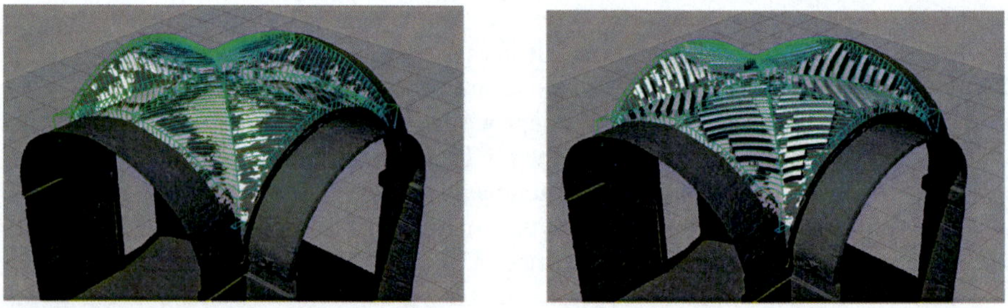

(a) (b)

Fig. 4. Construction lines for sweep-surfaces (a) and creation of a 3D volumes corresponding to the brick courses (b).

3 Reality-Based Finite Element Model

Figures 5a, b show a plan view of the two vaults that have been chosen as objects of the FE analyses, with radial and diagonal brick pattern, respectively. The detailed geometric models described in the previous section have been used to generate FE micro-models, in which every masonry unit is described as a closed three-dimensional volume that must be perfectly adjacent to subsequent units, in order to generate zero-thickness interfaces between blocks, according to the simplified micro-modelling approach [10].

The confining structures have been assumed as a rigid back wall and deformable front and lateral arches able to transfer the applied loads.

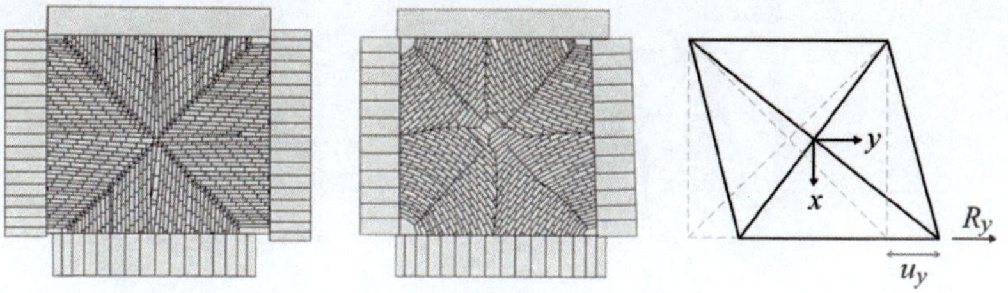

(a) (b) (c)

Fig. 5. Geometric model for FE analyses: radial (a) and diagonal (b). Scheme of the imposed shear displacement and measured reaction forces at the abutments (c)

The radial vault has a span of approximately 2.80 m and a rise of 0.81, whereas the diagonal vault spans over 2.95 m with the same rise. The discretization of the vault's geometry is performed with blocks of the size of the bricks and half the thickness of the mortar joints (about 6.5 × 13 × 25 cm). In the absence of data concerning the mechanical characterization of the existing masonry material, the mechanical properties of blocks are those suggested in [11] for historic brick masonry, i.e., $\rho = 1800$ kg/m^3, $E = 1200$ MPa, $\nu = 0.2$. Interfaces are characterized by frictional tangential behaviour, with friction coefficient $\mu = 0.5$, and a very high stiffness value $k_n = 5 \cdot 10^9$ N/m^3 in compression to simulate a quasi-rigid normal contact [7]. The mass of an infill 13 cm-thick at the vault's crown, with material density equal to 900 kg/m^3, is taken into account. The vaults are subject to their self-weight and to a horizontal settlement of two abutments in y direction, resulting in a shear mechanism (Fig. 5c). It is noteworthy to remark that the masonry pillars are not modelled, assuming the hypothesis that they are able to transfer the lateral loads from their base to their head without experiencing premature failure.

4 Results

The deformed shapes in Fig. 6 show that the two vaults experience quite different collapse shapes: however, in both cases only a partial collapse of the structure can be observed. In the case of the radial vault, a typical diagonal crack along the groins can be seen, as well as a hinge at the key of one cap, parallel to the orientation of the bed joints. The collapse of one cap can be observed. In the case of the diagonal vault, damage along the groins is not visible, and cracks develop normally to the groins. In this case, the local collapse of the structure involves two caps and is mainly concentrated along one groin. It is noteworthy to point out that the local collapses occur in quite different regions of the vaults, depending on the considered brick arrangement. The very same qualitative collapse shapes can be observed in the shear test of the vaults of ideal geometry described in [7], confirming that the masonry pattern strongly influences the structural behaviour of cross vaults. In fact, despite the differences in macro-geometry (overall shape and dimensions) of the two vaults in Vicoforte, the two different brick patterns possibly influence cracks propagation and therefore the generation of a collapse mechanism.

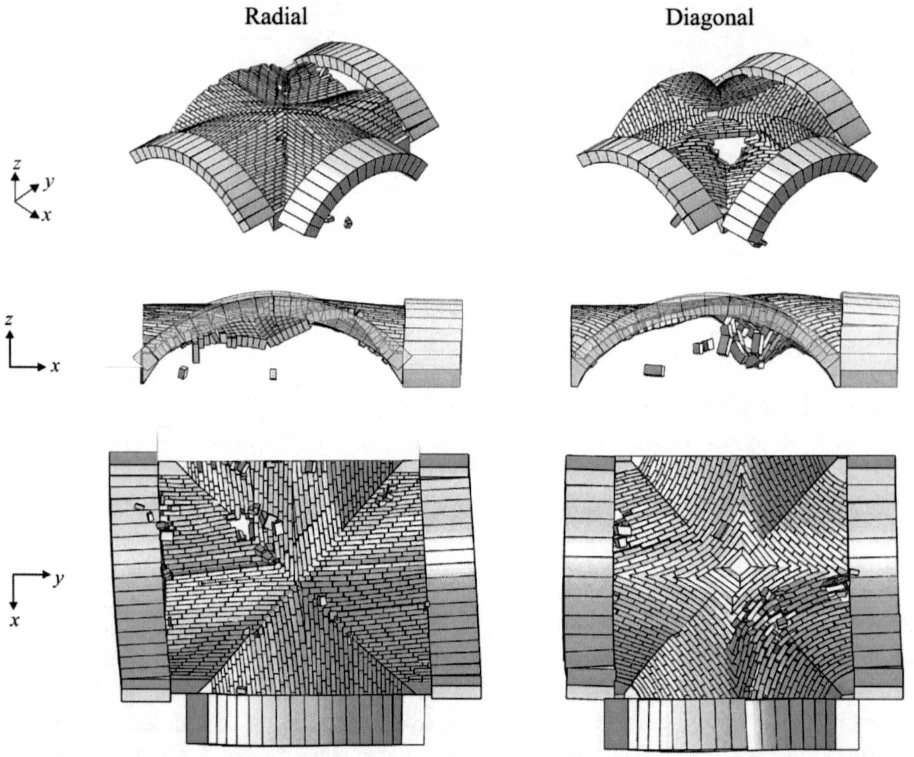

Fig. 6. Deformed shapes for shear settlement: axonometric view (first row), *x*-*z* plane view (second row), extrados plan view (third row).

Figure 7 plots the horizontal thrusts in the *y* direction at the abutments versus the horizontal displacement of the moving supports u_y for the Vicoforte vaults (Fig. 7a) and for two ideal vaults (Fig. 7b) subjected to shear settlement for comparison [7]. The two ideal vaults have the same macro geometry (span = 3.1 m, rise = 1.175 m), the same boundary condition as the real vaults, but differ in the static load, since the mass of infill was not taken into account in the ideal case. The two vaults in Vicoforte show similar values of peak force R_{max} and of initial stiffness K_{tan} (percentage differences around 20%).

Local collapses are highlighted on the load-displacement curves with black markers. In the radial vault the local collapse of one cap occurs in the post-peak branch in correspondence of an abrupt variation of the slope of the curve. Conversely, in the diagonal vault the local collapse, induced by sliding of few bricks, occurs when the reaction force reaches its peak value. If these local collapses are considered as a measure of the vault displacement capacity u_c, the radial vault has the largest displacement capacity being almost 6 times greater than the diagonal vault. Conversely, the two vaults show very similar ductile behaviour, being the ductility evaluated as the ratio between the imposed displacement corresponding to a decrease of 80% of the peak force value (u_{80}) and the displacement corresponding to 60% of R_{max} (u_{60}). Comparison of critical quantities is reported in Table 2.

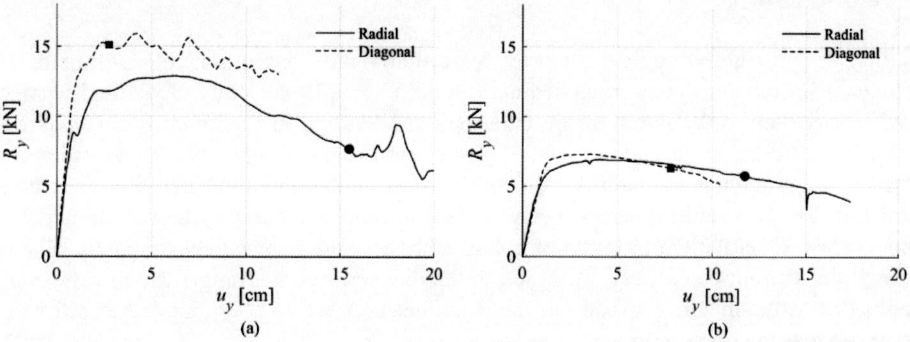

Fig. 7. Total reaction force vs. imposed displacement for Vicoforte (a) and ideal vaults (b).

Table 2. Table of critical quantities for Vicoforte and ideal vaults

Case study	Pattern	R_{max} [N]	u_c [m]	K_{tan} [N/m]	u_{80}/u_{60} [–]
Vicoforte	Radial	12891	0.16132	1217770	15.18
	Diagonal	15913	0.0276	1451315	17.35
	Δ = (D–R)/R [%]	23.44	–82.89	19.18	14.30
Ideal	Radial	6931	0.118	658815	18.82
	Diagonal	7314	0.078	581957	12.31
	Δ = (D–R)/R [%]	5.53	–33.90	–11.67	34.59

The comparison between the load-displacement curves and critical quantities obtained in ideal and real vaults (Fig. 7 and Table 2) allows to outline the following comments:

- The two ideal vaults only differ from each other for the brick pattern, therefore the differences in critical quantities between radial and diagonal vaults are certainly due to the micro-geometry;
- The two real vaults in Vicoforte have different brick arrangement and similar, but not equal macro-geometry, therefore differences in critical quantities cannot be attributed only to the brick pattern, but are also related to the influence of the overall vault dimensions (i.e., different rise-to-span-ratio and different self-weight);
- Despite a thorough comparison between ideal and real vaults is not possible, due to the different setups, the higher capacity of the real vaults with respect to the ideal ones can be attributed to the stabilizing effect of the infill (as already demonstrated in [7]), which was not considered in the ideal vaults.

5 Conclusions

In this paper, a methodological approach to develop reality-based finite element models of masonry vaults has been proposed and tested on a real case study. Specifically, two vaults belonging to the porch of the Palazzata di Vicoforte in Piedmont, Italy, characterized by similar macro-geometry but different masonry pattern, have been considered. The brick arrangement is carefully modelled, so as to take into consideration its structural role. The two vaults that have been studied show quite different behaviour under the same imposed settlement in terms of collapse shapes and displacement capacity. Even though this cannot be entirely attributed to the different brick arrangements of the two vaults (the actual macro-geometry of the structures comes into play), it is noteworthy to point out that the obtained results are consistent, from a qualitative point of view, with the results obtained in the study of ideal vaults in [7].

Further studies are needed to evaluate the possibility of reducing the great effort in the modelling process, e.g., by applying a discretization method that takes into consideration only the type of pattern (principal direction of bed joints) but doesn't exactly replicate the position of each block. By doing so, it would be possible to still take into consideration the role of micro and macro geometry, with a simplified approach.

References

1. Castellazzi, G., D'Altri, A.M., Bitelli, G., Selvaggi, I., Lambertini, A.: From laser scanning to finite element analysis of complex buildings by using a semi-automatic procedure. Sensors **15**, 18360–18380 (2015)
2. Korumaz, M., et al.: An integrated Terrestrial Laser Scanner (TLS), Deviation Analysis (DA) and Finite Element (FE) approach for health assessment of historical structures. A minaret case study. Eng. Struct. **153**, 224–238 (2017)
3. Angjeliu, G., Cardani, G., Coronelli, D.: A parametric model for ribbed masonry vaults. Autom. Constr. **105**, 102785 (2019)
4. Kersten, T.P., Lindstaedt, M.: Potential of automatic 3D object reconstruction from multiple images for applications in architecture, cultural heritage and archaeology. Int. J. Heritage Digit. Era **1**(3), 399–420 (2012)
5. Sànchez-Aparicio, L.J., Del Pozo, S., Ramos, L.F., Arce, A., Fernandes, F.M.: Heritage site preservation with combined radiometric and geometric analysis of TLS data. Autom. Constr. **85**, 24–39 (2018)

6. Spanò, A., et al.: Digital twinning for 20th century concrete heritage: HBIM cognitive model for Torino Esposizioni Halls. Sensors **23**(10), 4791 (2023)
7. Alforno, M., Venuti, F., Monaco, A., Calderini, C.: Numerical investigation of the influence of constructive aspects on the structural behaviour of masonry cross vaults. Int. J. Architect. Herit. (2021). https://doi.org/10.1080/15583058.2021.1992534
8. Beltramo, S.: Le botteghe dei pellegrini nel Piemonte Sabaudo tra XVI e XVII Secolo: i Santuari di Vicoforte e di Oropa. In: Calabi, D., Beltramo, S. (eds.) Il mercante patrizio: palazzi e botteghe nell'Europa del Rinascimento, pp. 257–282. Mondadori, Milano (2008)
9. Granshaw, S.I.: Photogrammetric terminology, 4th edn. Photogramm. Record **35**(170), 143–288 (2020)
10. Lourenço, P.B., Rots, J.G., Blaauwendraad, J.: Two approaches for the analysis of masonry structures – micro and macro-modeling. HERON **40**(4), 313–340 (1995)
11. Rossi, M., Calderini, C., Lagomarsino, S.: Experimental testing of the seismic in-plane displacement capacity of masonry cross vaults through a scale model. Bull. Earthq. Eng. **14**, 261–281 (2016)

Internal Force and Moment Surfaces for Shells

Emil Adiels and Chris J. K. Williams(✉)

Chalmers tekniska högskola, Gothenburg, Sweden
emil.adiels@chalmers.se, christopher.williams@chalmers.se

Abstract. There are 4 fundamental 2nd order tensors in shell theory, firstly the stress and internal moment tensors which are in equilibrium with the applied loads and loading couples, and secondly the rate of membrane strain and the rate of bending tensors which are compatible with the velocity of the shell as it deforms.

Shell theory is completed by the constitutive relationships which give the stress and moment in terms of the membrane strain and bending deformation.

The fundamental tensors are associated with a number of vectors including the applied loads, loading couples, velocity and angular velocity. We show that the 4 fundamental tensors can all be linearly assembled from vectors using the gradient and also a 3rd order tensor containing the surface permutation tensor and the unit normal.

Since a vector describes a surface in 3D space, all these quantities have a graphic significance which can be described using images and models - 3D graphic statics which is the application of differential geometry to statics.

We demonstrate that if the moment is described in an appropriate way, using 2 vectors, then internal moments are in equilibrium with zero imposed loads and loading couples, although boundary moments and forces are required. Thus internal moments in shells are 'redundant' in the nomenclature of structural theory. For flat plates, one of the redundant moments has to be replaced by a non-redundant moment.

Keywords: Shell structures · Differential geometry · Graphic statics

1 Introduction

Let us consider making an imaginary cut through the shell structure shown in Fig. 1 using the fret saw shown in Fig. 2. When we cut the shell we also cut through the reference surface **r** used to describe the shell, and we make no assumption that the shell is thin - that assumption is only necessary when analysing the stiffness and strength properties of the shell. **r** is arbitrary in the same way that the axis of a beam is arbitrary, the elastic centroid, plastic centroid and shear centre of a beam may all be different, and we choose a convenient line which we cut to make a reference point when we make a cross-section through

S. Gabriele et al. (Eds.): IWSS 2023, LNCE 437, pp. 118–128, 2024.
https://doi.org/10.1007/978-3-031-44328-2_13

the beam, which could be normal to the axis or in a vertical plane, or not even be planar. The orientation of the saw blade cutting the shell is also arbitrary, typically in shell theory the blade is effectively assumed to be normal to the reference surface, but again it could equally well be kept vertical, and it might even bend.

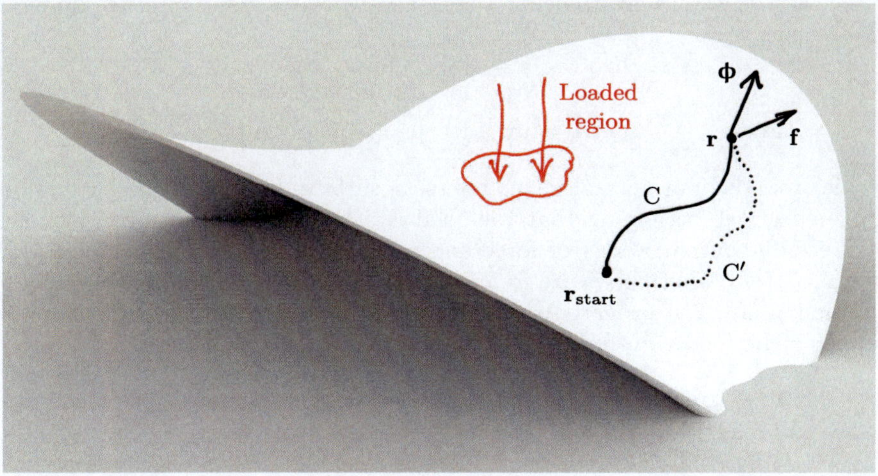

Fig. 1. A shell structure

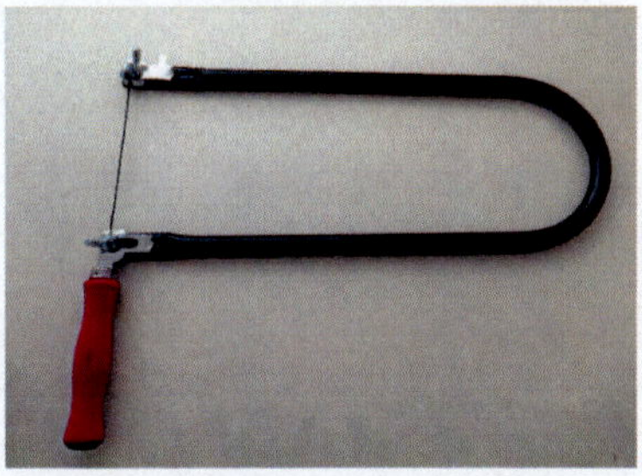

Fig. 2. A fret saw. Image: Gerd Fahrenhorst

Fig. 3. A kitchen sieve

There will be an element of force $d\mathbf{f}$ and moment $d\mathbf{m}$ crossing the element of cut $d\mathbf{r}$ through the reference surface. $d\mathbf{m}$ acts about the point $\mathbf{r}$. $d\mathbf{f}$ and $d\mathbf{m}$ will depend upon $d\mathbf{r}$, and also upon the orientation of the saw blade. For a uniform shell without individual members or ribs $d\mathbf{m}$ will be about an axis perpendicular to the blade, that is about axes tangential to the surface if the blade is normal to the surface or about the horizontal if the blade is vertical. The saw blade orientation clearly cannot influence the physical equilibrium of a shell, only our perceived values of moments. In the case of a sieve-like structure (Fig. 3) there will be moments about the normal to the reference surface due to the geodesic curvature of the members. These 'geodesic moments' about the normal might also be described as Cosserat moments [2].

We can use $d\mathbf{f}$ and $d\mathbf{m}$ to define the internal stresses $\boldsymbol{\sigma}$ and internal moments $\boldsymbol{\mu}$ in the shell by

$$d\mathbf{f} = (d\mathbf{r} \times \mathbf{n}) \cdot \boldsymbol{\sigma} = d\mathbf{r} \cdot (\mathbf{n} \times \boldsymbol{\sigma}) \tag{1}$$

and

$$d\mathbf{m} = (d\mathbf{r} \times \mathbf{n}) \cdot \boldsymbol{\mu} = d\mathbf{r} \cdot (\mathbf{n} \times \boldsymbol{\mu}) \tag{2}$$

in which $\mathbf{n}$ is the unit normal to the reference surface. In writing these equations we are effectively assuming that the shell is a continuum, although that does not preclude concentrations of force and moment in individual members. The equations apply whether we are in a loaded or unloaded region of shell.

$\boldsymbol{\sigma}$ and $\boldsymbol{\mu}$ are 2nd order tensors which are composed of the sum of tensor products. The tensor product $\mathbf{uv}$ of the vectors $\mathbf{u}$ and $\mathbf{v}$ is defined by

$$\mathbf{uv} \cdot \mathbf{w} = (\mathbf{uv}) \cdot \mathbf{w} = \mathbf{u}\,(\mathbf{v} \cdot \mathbf{w}) \quad \text{and} \quad \mathbf{w} \cdot \mathbf{uv} = \mathbf{w} \cdot (\mathbf{uv}) = (\mathbf{w} \cdot \mathbf{u})\,\mathbf{v}$$

where $\mathbf{w}$ is any vector and the $\cdot$ represents the inner or scalar product. The tensor product is sometimes written $\mathbf{u} \otimes \mathbf{v}$ and is also called the dyadic product or the outer product. Scalars are 0th order tensors, vectors are 1st order tensors and by adding repeated tensor products of vectors we can produce tensors of any order.

Thus we can use tensor products of the Cartesian base vectors $\mathbf{i}$, $\mathbf{j}$ and $\mathbf{k}$ to write the stress tensor in 3D as

$$\boldsymbol{\sigma} = \sigma_x \mathbf{ii} + \tau_{xy}\mathbf{ij} + \tau_{xz}\mathbf{ik} + \tau_{yx}\mathbf{ji} + \sigma_y\mathbf{jj} + \tau_{yz}\mathbf{jk} + \tau_{zx}\mathbf{ki} + \tau_{zy}\mathbf{kj} + \sigma_z\mathbf{kk}$$

which has units force per unit area. However $\boldsymbol{\sigma}$ in (1) is a force per unit length of cut and $\boldsymbol{\mu}$ in (2) is a moment per unit length of cut.

Since $d\mathbf{r} \times \mathbf{n}$ is perpendicular to $\mathbf{n}$ in (1) and (2) we can stipulate that

$$\mathbf{n} \cdot \boldsymbol{\sigma} = 0 \quad \text{and} \quad \mathbf{n} \cdot \boldsymbol{\mu} = 0 \tag{3}$$

without affecting the values of $d\mathbf{f}$ and $d\mathbf{m}$. In 3D a 2nd order tensor has $3 \times 3 = 9$ components. However (3) means that $\boldsymbol{\sigma}$ and $\boldsymbol{\mu}$ each only have 6 components, of which 4 are tangential to the surface and 2 are normal to the surface.

2 Equilibrium of an Unloaded Part of a Shell

In Fig. 1 there are 2 imaginary saw cuts from $\mathbf{r}_{\text{start}}$ to $\mathbf{r}$, C shown solid and C$'$ shown dotted. The saw blade orientation is the same for both cuts at the 2 ends, so that the cuts meet nicely.

If you travel from $\mathbf{r}_{\text{start}}$ to $\mathbf{r}$ along C and back from $\mathbf{r}$ to $\mathbf{r}_{\text{start}}$ along C′ you travel along a closed curve and the total force crossing this closed curve must be zero since the shell is unloaded in the region enclosed by the two curves. If you reverse your direction of travel you reverse the corresponding force based upon our definition of stress, (1). Thus the total force $\mathbf{f}$ crossing the paths from $\mathbf{r}_{\text{start}}$ to $\mathbf{r}$ must be the same along C or C′ or indeed any path which goes from $\mathbf{r}_{\text{start}}$ to $\mathbf{r}$ the same way round the loaded area.

If we move 1 of the end points, $\mathbf{r}$, keeping the other end, $\mathbf{r}_{\text{start}}$, fixed then $\mathbf{f}$ will vary and will be a function of $\mathbf{r}$ which we can represent as a surface.

If $\mathbf{r}_{\text{path}}$ is a typical point along C we can integrate (1) to give

$$\mathbf{f} = \int_{\mathbf{r}_{\text{start}}}^{\mathbf{r}} \left(d\mathbf{r}_{\text{path}} \times \mathbf{n}\left(\mathbf{r}_{\text{path}}\right)\right) \cdot \boldsymbol{\sigma}\left(\mathbf{r}_{\text{path}}\right) = \int_{\mathbf{r}_{\text{start}}}^{\mathbf{r}} d\mathbf{r}_{\text{path}} \cdot \left(\mathbf{n} \times \boldsymbol{\sigma}\right). \tag{4}$$

Now let us take moments about the point $\mathbf{r}$. The same argument tells us that moments due to internal stresses and internal moments crossing the two cuts C and C′ are equal and depend only on $\mathbf{r}$. Using (2) we can write this moment as

$$\boldsymbol{\phi} = \int_{\mathbf{r}_{\text{start}}}^{\mathbf{r}} d\mathbf{r}_{\text{path}} \cdot \left(\mathbf{n}\left(\mathbf{r}_{\text{path}}\right) \times \boldsymbol{\mu}\left(\mathbf{r}_{\text{path}}\right)\right) + \int_{\mathbf{r}_{\text{start}}}^{\mathbf{r}} \left(\mathbf{r}_{\text{path}} - \mathbf{r}\right) \times d\mathbf{f}_{\text{path}}. \tag{5}$$

Again we can represent $\boldsymbol{\phi}$ by a surface.

We can find $\boldsymbol{\sigma}$ in terms of $\mathbf{f}$ by taking the gradient of (4) to give $d\mathbf{r} \cdot \nabla\mathbf{f} = d\mathbf{r} \cdot (\mathbf{n} \times \boldsymbol{\sigma})$ and so $\nabla\mathbf{f} = \mathbf{n} \times \boldsymbol{\sigma}$ and

$$\boldsymbol{\sigma} = -\mathbf{n} \times \nabla\mathbf{f} = \boldsymbol{\epsilon} \cdot \nabla\mathbf{f}. \tag{6}$$

We need to explain the meaning of $\nabla\mathbf{f}$ and $\boldsymbol{\epsilon}$. Let us imagine that $\mathbf{Q}$ is some scalar, vector or higher order tensor which exists on $\mathbf{r}$ then the gradient of $\mathbf{Q}$, which is written $\nabla\mathbf{Q}$ is defined such that $d\mathbf{Q} = d\mathbf{r} \cdot \nabla\mathbf{Q}$ and $\mathbf{n} \cdot \nabla\mathbf{Q} = 0$ for any direction of $d\mathbf{r}$ on the surface. It is important to write the $d\mathbf{r}$ in front of the $\nabla\mathbf{Q}$ because $d\mathbf{r} \cdot \nabla\mathbf{Q}$ will not in general be the same as $\nabla\mathbf{Q} \cdot d\mathbf{r}$. $\nabla\mathbf{Q}$ is a tensor of order one higher than $\mathbf{Q}$, so that the gradient of a scalar is a vector, and so on. The gradient is a real physical quantity and it exists independent of any coordinate system, and so it is preferable to avoid coordinates in its definition.

The gradient of $\mathbf{r}$ itself is $\nabla\mathbf{r} = \mathbf{I} - \mathbf{nn}$ in which $\mathbf{I}$ is the unit tensor, which in Cartesian coordinates is $\mathbf{I} = \mathbf{ii} + \mathbf{jj} + \mathbf{kk}$. $\nabla\mathbf{r}$ satisfies $d\mathbf{r} = d\mathbf{r} \cdot \nabla\mathbf{r}$ and $\mathbf{n} \cdot \nabla\mathbf{r} = 0$.

The curvature tensor is minus the gradient of the normal, $\mathbf{b} = -\nabla\mathbf{n}$. It is a symmetric surface tensor, that is a tensor with no normal components, and as such it has two orthogonal principal directions and principal curvatures. The mean of the principal curvatures is $H = \text{tr}\,(\mathbf{b})\,/2$ and their product or Gaussian curvature $K = \left(4H^2 - \text{tr}\,(\mathbf{b} \cdot \mathbf{b})\right)/2$ where tr () is the trace. The components of $\mathbf{b}$ are known as the coefficients of the 2nd fundamental form [7].

The surface permutation tensor, $\boldsymbol{\epsilon} = \mathbf{n} \cdot \boldsymbol{\chi} = \boldsymbol{\chi} \cdot \mathbf{n}$, where the permutation tensor $\boldsymbol{\chi}$ is defined is defined in terms of the Cartesian base vectors, $\mathbf{i}$, $\mathbf{j}$, $\mathbf{k}$ by $\boldsymbol{\chi} = \mathbf{ijk} + \mathbf{jki} + \mathbf{kij} - \mathbf{kji} - \mathbf{ikj} - \mathbf{jik}$. The permutation tensor is useful in taking the vector product, so that if $\mathbf{u}$ and $\mathbf{v}$ are any vectors, $\mathbf{u} \times \mathbf{v} = \mathbf{v} \cdot \boldsymbol{\chi} \cdot \mathbf{u} = (\mathbf{uv}) : \boldsymbol{\chi}$.

Note the use of the double dot (:) notation, which in general for the vectors **a**, **b**, **c** etc. we take to mean $(\mathbf{ab....gef}) : (\mathbf{rgt....yz}) = (\mathbf{e} \cdot \mathbf{r})(\mathbf{f} \cdot \mathbf{g})(\mathbf{ab....gt....yz})$.

There are also the pure mathematical identities:

$$\begin{aligned} \boldsymbol{\epsilon} \cdot \boldsymbol{\chi} \cdot \mathbf{q} = \mathbf{q} \cdot (\mathbf{In} - \mathbf{nI}) = \mathbf{q} \cdot (\mathbf{I} - \mathbf{nn})\,\mathbf{n} - (\mathbf{q} \cdot \mathbf{n})(\mathbf{I} - \mathbf{nn}) \\ (\mathbf{I} - \mathbf{nn}) \cdot \boldsymbol{\chi} \cdot \mathbf{q} = \mathbf{q} \cdot (\boldsymbol{\epsilon}\mathbf{n} + \mathbf{n}\boldsymbol{\epsilon}) = -(\boldsymbol{\epsilon} \cdot \mathbf{q})\,\mathbf{n} + (\mathbf{q} \cdot \mathbf{n})\,\boldsymbol{\epsilon}, \end{aligned} \tag{7}$$

in which **q** is any vector with tangential and normal components,

$$\boldsymbol{\epsilon} : (\nabla(\nabla \mathbf{Q})) = \boldsymbol{\epsilon} : \nabla\nabla\mathbf{Q} = 0 \tag{8}$$

and

$$\nabla \cdot (\boldsymbol{\epsilon} \cdot \nabla \mathbf{Q}) = 0 \tag{9}$$

where **Q** is any scalar or higher order tensor quantity. (8) tells us that $\boldsymbol{\epsilon} : \nabla \mathbf{b} = 0$ which are the Peterson-Mainardi-Codazzi equations [7]. We explain the meaning of $\nabla \cdot$ below when we introduce the divergence, when perforce we have to use coordinates.

We can also take the gradient of (5) to give

$$\nabla \boldsymbol{\phi} = \mathbf{n} \times \boldsymbol{\mu} - (\mathbf{I} - \mathbf{nn}) \times \mathbf{f} \tag{10}$$

so that, after some manipulation,

$$\boldsymbol{\epsilon} \cdot \boldsymbol{\mu} = -\nabla \boldsymbol{\phi} + \mathbf{f} \cdot (\boldsymbol{\epsilon}\mathbf{n} + \mathbf{n}\boldsymbol{\epsilon}) \tag{11}$$

and

$$\mathbf{f} = -(\boldsymbol{\mu} - \boldsymbol{\epsilon} \cdot \nabla \boldsymbol{\phi}) \cdot \mathbf{n} + \frac{1}{2}(\operatorname{tr} \boldsymbol{\mu} + \boldsymbol{\epsilon} : \nabla \boldsymbol{\phi})\,\mathbf{n}.$$

The two surfaces **f** and $\boldsymbol{\phi}$ in (6) and (11) represent a state of stress and moment in a shell in equilibrium in a region where there are no applied loads, except, of course, forces and moments applied at the boundary of the shell. **f** and $\boldsymbol{\phi}$ can therefore be described as redundancies, which can produce any desired $\boldsymbol{\mu}$ unless the shell is a flat plate when one of the redundancies has to be replaced by a moment such as $\boldsymbol{\mu} = -M\boldsymbol{\epsilon}$ where M is a scalar. We have established the existence of **f** and $\boldsymbol{\phi}$ without considering the equilibrium equations. We shall see that (11) is unchanged even if the shell is loaded with forces or couples (unless it is a flat plate), but we shall have to add extra terms to (6).

The only moment that is usually included in shell and plate theory is $\boldsymbol{\mu} = \mathbf{M} \times \mathbf{n} = -\mathbf{M} \cdot \boldsymbol{\epsilon}$ where **M** is a symmetric 2nd order surface tensor. In plate theory $-\mu_{xy} = M_{xx}$ would be the sagging bending moment per unit width acting about the y axis. If we accept this condition we then have $\boldsymbol{\mu} \cdot \mathbf{n} = 0$ and $\operatorname{tr} \boldsymbol{\mu} = 0$ so that

$$\mathbf{f} = \boldsymbol{\epsilon} \cdot \nabla \boldsymbol{\phi} \cdot \mathbf{n} + \frac{1}{2}(\boldsymbol{\epsilon} : \nabla \boldsymbol{\phi})\,\mathbf{n} \tag{12}$$

and

$$\mathbf{M} = \frac{1}{2}\boldsymbol{\epsilon} \cdot \left(\nabla \boldsymbol{\phi} + (\nabla \boldsymbol{\phi})^{\mathrm{T}}\right) \cdot \boldsymbol{\epsilon} \tag{13}$$

in which $(\nabla \boldsymbol{\phi})^{\mathrm{T}}$ is the transpose of $\nabla \boldsymbol{\phi}$. However we shall continue assuming that we have all 6 components of $\boldsymbol{\mu}$, not just the 3 components of **M**.

3 Equilibrium of Forces Under Load

Let us suppose that the load applied to the shell per unit area of reference surface $\mathbf{r}$ is represented by the vector $\mathbf{p}$, which could include inertia forces. Then $\int_A \mathbf{p}dA$ is the total load applied to the portion of shell A which could be any part of, or indeed the whole shell.

Using (1), the total force crossing the cut represented by the closed curve C forming the boundary of A is $\oint_C d\mathbf{f} = \oint_C (d\mathbf{r} \times \mathbf{n}) \cdot \boldsymbol{\sigma}$ and thus the equation of overall equilibrium of forces is

$$\int_A \mathbf{p}dA + \oint_C (d\mathbf{r} \times \mathbf{n}) \cdot \boldsymbol{\sigma} = 0.$$

At this point we need the divergence theorem from pure mathematics, which for any vector or higher order tensor satisfying $\mathbf{n} \cdot \mathbf{Q} = 0$ on a curved surface can be written,

$$\int_A \nabla \cdot \mathbf{Q}dA = \oint_C (d\mathbf{r} \times \mathbf{n}) \cdot \mathbf{Q}.$$

Thus

$$\int_A \nabla \cdot \boldsymbol{\sigma}dA = \oint_C (d\mathbf{r} \times \mathbf{n}) \cdot \boldsymbol{\sigma}$$

and so

$$\int_A (\mathbf{p} + \nabla \cdot \boldsymbol{\sigma})\, dA = 0.$$

The boundary of the portion of shell is arbitrary which must mean that

$$\mathbf{p} + \nabla \cdot \boldsymbol{\sigma} = 0, \tag{14}$$

which is the equation of equilibrium of forces.

Unfortunately it is difficult to explain the meaning of the divergence of the stress, $\nabla\cdot\boldsymbol{\sigma}$, without introducing coordinates on the reference surface. If the x, y, z coordinates of $\mathbf{r}$ are known in terms of the surface coordinates or parameters θ^α where $\alpha = 1$ or $\alpha = 2$, then the covariant base vectors are

$$\mathbf{a}_\alpha = \mathbf{r}_{,\alpha} = \frac{\partial \mathbf{r}}{\partial \theta^\alpha} = \frac{\partial x}{\partial \theta^\alpha}\mathbf{i} + \frac{\partial y}{\partial \theta^\alpha}\mathbf{j} + \frac{\partial z}{\partial \theta^\alpha}\mathbf{k}.$$

$\mathbf{a}_\alpha$ are in the local tangent plane to the surface, perpendicular to $\mathbf{n}$. In general they are not unit vectors, nor are they mutually perpendicular. The contravariant base vectors $\mathbf{a}^\beta$ are also perpendicular to $\mathbf{n}$, and they are defined by

$$\mathbf{a}^\alpha = \nabla\theta^\alpha \quad \text{so that} \quad \mathbf{a}_\beta \cdot \mathbf{a}^\alpha = \delta^\alpha_\beta \begin{cases} = 0 \text{ if } \alpha \neq \beta \\ = 1 \text{ if } \alpha = \beta \end{cases}$$

in which δ^α_β are the Kronecker deltas. We can therefore write the gradient, $\nabla\mathbf{Q} = \mathbf{a}^\alpha\mathbf{Q}_{,\alpha} = \mathbf{a}^\alpha\partial\mathbf{Q}/\partial\theta^\alpha$ so that $d\mathbf{Q} = d\theta^\beta\mathbf{a}_\beta \cdot \mathbf{a}^\alpha\mathbf{Q}_{,\alpha} = d\theta^\alpha\mathbf{Q}_{,\alpha}$ as required. Note the use of the Einstein summation convention which implies that an index is summed from 1 to 2 if it is repeated as a subscript and superscript - it would be from 0 to 3 in relativity theory. This enables us to write any vector $\mathbf{q}$ as

$$\begin{aligned}
&\mathbf{q} = q_\alpha \mathbf{a}^\alpha + q\mathbf{n} = q^\alpha \mathbf{a}_\alpha + q\mathbf{n} \text{ in which}\\
&q_\alpha = \mathbf{a}_\alpha \cdot \mathbf{q} = a_{\alpha\beta} q^\beta \text{ and } q^\alpha = \mathbf{a}^\alpha \cdot \mathbf{q} = a^{\alpha\beta} q_\beta,\\
&a_{\alpha\beta} = \mathbf{a}_\alpha \cdot \mathbf{a}_\beta \text{ and } a^{\alpha\beta} = \mathbf{a}^\alpha \cdot \mathbf{a}^\beta,\\
&\text{and finally } \mathbf{I} - \mathbf{nn} = a_{\alpha\beta}\mathbf{a}^\alpha \mathbf{a}^\beta = a^{\alpha\beta}\mathbf{a}_\alpha \mathbf{a}_\beta = \mathbf{a}_\alpha \mathbf{a}^\alpha.
\end{aligned}$$

The distance $|d\mathbf{r}| = \sqrt{d\mathbf{r} \cdot d\mathbf{r}} = \sqrt{a_{\alpha\beta} d\theta^\alpha d\theta^\beta}$ and therefore $a_{\alpha\beta}$ are described as the components of the metric tensor, or the coefficients of the of the 1st fundamental form [7].

We can now define the divergence on the surface, $\nabla \cdot \mathbf{Q} = \mathbf{a}^\alpha \cdot \mathbf{Q}_{,\alpha}$ which is a tensor of order one less than that of $\mathbf{Q}$.

One might be tempted to write (14) in terms of components [4],

$$p^\alpha \mathbf{a}_\alpha + p\mathbf{n} + \mathbf{a}^\eta \cdot \left(\sigma^{\alpha\beta}\mathbf{a}_\alpha \mathbf{a}_\beta + \sigma^\alpha \mathbf{a}_\alpha \mathbf{n}\right)_{,\eta} = 0$$

in which $\sigma^{\alpha\beta}$ are the tangential components of $\boldsymbol{\sigma}$ and σ^α are the normal components. Then

$$\begin{aligned}
p^\beta + \partial_\alpha \sigma^{\alpha\beta} - \sigma^\alpha b^\beta_\alpha &= 0\\
p + \sigma^{\alpha\beta} b_{\alpha\beta} + \partial_\alpha \sigma^\alpha &= 0
\end{aligned} \tag{15}$$

where the *covariant derivative*, $\partial_\alpha \sigma^{\alpha\beta} = \sigma^{\alpha\beta}_{,\alpha} + \sigma^{\eta\beta}\Gamma^\alpha_{\eta\alpha} + \sigma^{\alpha\eta}\Gamma^\beta_{\eta\alpha}$ in which the Christoffel symbols of the 2nd kind, $\Gamma^\beta_{\eta\alpha} = \mathbf{a}^\beta \cdot \mathbf{a}_{\eta,\alpha}$. Clearly the form in (14) is to be preferred to that in (15) except under the exigency of performing calculations since 'les débauches d'indices masquent une realité géométrique souvent tres simple' [1]. It is important to note that (15) follows uniquely from (14).

It is also very important to note that the covariant derivative $\partial_\alpha \sigma^{\alpha\beta}$ does *not* include the normal component $\boldsymbol{\sigma} : \mathbf{b} = \sigma^{\alpha\beta} b_{\alpha\beta}$ which is vital for shell structures as it is the product of the membrane stress and the curvature, and is automatically included in $\nabla \cdot \boldsymbol{\sigma}$.

4 Equilibrium of Moments Under Load

The moment about the point $\mathbf{O}$ of any force $\mathbf{F}$ applied at the point $\mathbf{r}$ on the reference surface is $(\mathbf{r} - \mathbf{O}) \times \mathbf{F} = -\mathbf{F} \times (\mathbf{r} - \mathbf{O})$, acting clockwise looking along the vector $(\mathbf{r} - \mathbf{O}) \times \mathbf{F}$. Thus the moment of the loads applied to the surface about the point $\mathbf{O}$ is $\int_A (\mathbf{c} + (\mathbf{r} - \mathbf{O}) \times \mathbf{p})\, dA$ where $\mathbf{c}$ is the loading couple per unit surface area, perhaps caused by loads which are applied with an eccentricity to the shell reference surface. However the effects of such eccentricity are often negligible so that $\mathbf{c}$ is frequently taken as zero.

The total moment crossing the boundary cut due to internal stresses and moments is $\oint_C (d\mathbf{r} \times \mathbf{n}) \cdot (\boldsymbol{\mu} - \boldsymbol{\sigma} \times (\mathbf{r} - \mathbf{O}))$ and therefore the overall equation of equilibrium of moments is

$$\int_A (\mathbf{c} + (\mathbf{r} - \mathbf{O}) \times \mathbf{p})\, dA + \oint_C (d\mathbf{r} \times \mathbf{n}) \cdot (\boldsymbol{\mu} - \boldsymbol{\sigma} \times (\mathbf{r} - \mathbf{O})) = 0.$$

If we now apply the divergence theorem,

$$\mathbf{c} + (\mathbf{r} - \mathbf{O}) \times \mathbf{p} + \nabla \cdot (\boldsymbol{\mu} - \boldsymbol{\sigma} \times (\mathbf{r} - \mathbf{O})) = 0.$$

At this point we can use (14) to write, after some manipulation,

$$\mathbf{c} + \nabla \cdot \boldsymbol{\mu} + \boldsymbol{\sigma} : \boldsymbol{\chi} = 0. \tag{16}$$

This is the equation of equilibrium of moments.

5 Two Surfaces for Internal Forces Resisting Loads and Loading Couples

We will now introduce two new vectors or surfaces $\mathbf{u}$ and $\mathbf{s}$ to add to (6) and give

$$\boldsymbol{\sigma} = \boldsymbol{\epsilon} \cdot \nabla (\mathbf{f} + \mathbf{s}) + \mathbf{u} \cdot (\boldsymbol{\epsilon}\mathbf{n} + \mathbf{n}\boldsymbol{\epsilon})\,.$$

$\mathbf{f}$ is unchanged from (6). Let us write $\mathbf{u} = \mathbf{U} + u\mathbf{n}$ where $\mathbf{U} \cdot \mathbf{n} = 0$ so that

$$\boldsymbol{\sigma} = \boldsymbol{\epsilon} \cdot \nabla (\mathbf{f} + \mathbf{s}) + \mathbf{U} \cdot \boldsymbol{\epsilon}\mathbf{n} + u\boldsymbol{\epsilon}.$$

We know that $\boldsymbol{\phi}$ and $\mathbf{f}$ satisfy equilibrium of forces and moments when the load and loading couples are zero, and this can be confirmed using (9). Thus, the equation of equilibrium of forces (14) reduces to

$$\mathbf{p} = -\mathbf{a}^{\alpha} \cdot (\mathbf{U} \cdot \boldsymbol{\chi} \cdot \mathbf{n}\mathbf{n} + u\mathbf{n} \cdot \boldsymbol{\chi})_{,\alpha} = \boldsymbol{\epsilon} : \nabla \mathbf{U}\mathbf{n} + \mathbf{U} \cdot \boldsymbol{\epsilon} \cdot \mathbf{b} + \boldsymbol{\epsilon} \cdot \nabla u. \tag{17}$$

Writing $\mathbf{s} = \mathbf{S} + s\mathbf{n}$ where $\mathbf{S} \cdot \mathbf{n} = 0$ the equation of equilibrium of moments (16) becomes

$$\mathbf{c} + (\nabla \cdot \mathbf{S} - 2Hs)\,\mathbf{n} - \mathbf{S} \cdot \mathbf{b} - \nabla s + \mathbf{U} + 2u\mathbf{n} = 0.$$

We can solve (17) using

$$\boldsymbol{\epsilon} \cdot \mathbf{b} \cdot \boldsymbol{\epsilon} \cdot \mathbf{b} = \mathbf{a}_{\alpha} \epsilon^{\alpha\beta} b^{\lambda}_{\beta} \epsilon_{\lambda\mu} b^{\mu}_{\eta} \mathbf{a}^{\eta} = -K\,(\mathbf{I} - \mathbf{n}\mathbf{n}) = \mathbf{b} \cdot \boldsymbol{\epsilon} \cdot \mathbf{b} \cdot \boldsymbol{\epsilon}$$

to write
$$\mathbf{U} = \frac{1}{K}\,(\mathbf{p} - \boldsymbol{\epsilon} \cdot \nabla u) \cdot \mathbf{b} \cdot \boldsymbol{\epsilon} \tag{18}$$

so that
$$\boldsymbol{\epsilon} : \nabla \left(\frac{1}{K}\,(\mathbf{p} - \boldsymbol{\epsilon} \cdot \nabla u) \cdot \mathbf{b} \cdot \boldsymbol{\epsilon} \right) - \mathbf{p} \cdot \mathbf{n} = 0$$

which is a single equation in the 1 unknown u and we can find a single equation for s in a similar way once we have solved (18). Any solution for $\mathbf{u}$ is acceptable without regard to boundary conditions since boundary conditions are dealt with in the solution for $\mathbf{s}$.

Note that this approach fails when the Gaussian curvature, K, is zero in which case some other technique has to be adopted. The highest derivatives are of the form $(\boldsymbol{\epsilon} \cdot \mathbf{b} \cdot \boldsymbol{\epsilon}) : \nabla\nabla u$ and similarly for s which means that the differential equation is elliptic if the Gaussian curvature, $K > 0$, hyperbolic if $K < 0$ and parabolic if $K = 0$ [6], although as already noted the approach has to be modified if $K = 0$. In any event the issue is to try and simultaneously satisfy equilibrium and boundary conditions, which may not be possible without introducing bending moments.

It is possible to find $\mathbf{s}$ without first finding $\mathbf{u}$ by using the asymptotic directions on the reference surface in which there is no normal curvature [7]. The asymptotic directions are imaginary when the Gaussian curvature is positive, but they can still be used.

As an alternative, let us introduce the constant unit vector $\mathbf{k}$, often taken in the vertical direction. Then if we set $\boldsymbol{\mu} \cdot \mathbf{k} = 0$, $\boldsymbol{\phi} = \phi\mathbf{k}$ and $\mathbf{k} \cdot \mathbf{f} = 0$ in (10) we obtain $\boldsymbol{\sigma} = \boldsymbol{\epsilon} \cdot \nabla\left(\nabla\phi \cdot \left(\mathbf{I} - \mathbf{kn}/\left(\mathbf{k} \cdot \mathbf{n}\right)\right) \times \mathbf{k}\right)$ from (6), together with internal moments acting about horizontal axes. If we add a vertical stress and impose the condition $\boldsymbol{\sigma} \cdot \mathbf{n} = 0$, then we remove the internal moments and $\boldsymbol{\sigma} = \boldsymbol{\epsilon} \cdot \nabla\left(\nabla\phi \cdot \left(\mathbf{I} - \mathbf{kn}/\left(\mathbf{k} \cdot \mathbf{n}\right)\right) \times \mathbf{k}\right) \cdot \left(\mathbf{I} - \mathbf{nk}/\left(\mathbf{n} \cdot \mathbf{k}\right)\right)$, which is the symmetric surface tensor of membrane stress. ϕ is the Airy stress function [9] and the load, $\mathbf{p} = -\left(\boldsymbol{\sigma} : \mathbf{b}\right)\mathbf{k}/\left(\mathbf{n} \cdot \mathbf{k}\right)$, is in the direction of $\mathbf{k}$. This is Pucher's equation [3,8] which is written

$$\frac{\partial^2 z}{\partial x^2}\frac{\partial^2 \phi}{\partial y^2} - 2\frac{\partial^2 z}{\partial x \partial y}\frac{\partial^2 \phi}{\partial x \partial y} + \frac{\partial^2 z}{\partial y^2}\frac{\partial^2 \phi}{\partial z^2} = \text{downwards load per unit plan area}$$

in Cartesian coordinates. z is the height of the shell above some datum.

6 The Kinematics of Shell Structures

The meaning of the differential equations of equilibrium can be better understood by considering the kinematics of shells. Physically this is because shell structures are relatively flexible in bending and so we are primarily concerned about reducing bending deformation see, for example Harnach & Rothert [5].

Imagine a shell represented by a surface undergoing a velocity $\mathbf{v} = \partial\mathbf{r}/\partial t$ that produces a rate of membrane strain $\boldsymbol{\gamma}$, which is a symmetric surface tensor, and mean angular velocity vector $\boldsymbol{\omega}$. Then, using (7),

$$\nabla\mathbf{v} = \boldsymbol{\gamma} - \left(\mathbf{I} - \mathbf{nn}\right) \times \boldsymbol{\omega} = \boldsymbol{\gamma} + \boldsymbol{\omega} \cdot \left(\boldsymbol{\epsilon}\mathbf{n} + \mathbf{n}\boldsymbol{\epsilon}\right) \quad \text{and} \quad \boldsymbol{\gamma} = \frac{1}{2}\frac{\partial a_{\alpha\beta}}{\partial t}\mathbf{a}^\alpha\mathbf{a}^\beta.$$

$$\text{Thus} \qquad 0 = \nabla \cdot \left(\boldsymbol{\epsilon} \cdot \nabla\mathbf{v}\right) = \nabla \cdot \left(\boldsymbol{\epsilon} \cdot \boldsymbol{\gamma}\right) + \nabla \cdot \boldsymbol{\omega}\mathbf{n} - \nabla\boldsymbol{\omega} \cdot \mathbf{n}$$

$$\text{so that} \qquad \nabla\boldsymbol{\omega} \cdot \mathbf{n} = \nabla \cdot \left(\boldsymbol{\epsilon} \cdot \boldsymbol{\gamma}\right) \cdot \left(\mathbf{I} - \mathbf{nn}\right)$$

which is the compatibility equation relating $\boldsymbol{\omega}$ and $\boldsymbol{\gamma}$.

The rate of change of the coefficients of the 2nd fundamental form,

$$\frac{\partial b_{\alpha\beta}}{\partial t} = -\frac{\partial}{\partial t}\left(\mathbf{n}_{,\alpha} \cdot \mathbf{a}_\beta\right) = \mathbf{a}_\alpha \cdot \left(\mathbf{b} \cdot \boldsymbol{\gamma} + \nabla\boldsymbol{\omega} \cdot \boldsymbol{\epsilon}\right) \cdot \mathbf{a}_\beta$$

so that $\boldsymbol{\epsilon}\cdot\boldsymbol{\beta}\cdot\boldsymbol{\epsilon}=\boldsymbol{\epsilon}\cdot\nabla\boldsymbol{\omega}+\boldsymbol{\rho}\cdot(\boldsymbol{\epsilon}\mathbf{n}+\mathbf{n}\boldsymbol{\epsilon})$ where $\boldsymbol{\beta}=\frac{1}{2}(\mathbf{b}\cdot\boldsymbol{\gamma}+\boldsymbol{\gamma}\cdot\mathbf{b})-\frac{\partial b_{\alpha\beta}}{\partial t}\mathbf{a}^\alpha\mathbf{a}^\beta$ and $\boldsymbol{\rho}\cdot(\boldsymbol{\epsilon}\mathbf{n}+\mathbf{n}\boldsymbol{\epsilon})=\boldsymbol{\epsilon}\cdot\left(\frac{1}{2}(\mathbf{b}\cdot\boldsymbol{\gamma}-\boldsymbol{\gamma}\cdot\mathbf{b})\cdot\boldsymbol{\epsilon}+\nabla\cdot(\boldsymbol{\epsilon}\cdot\boldsymbol{\gamma})\cdot(\mathbf{I}-\mathbf{nn})\,\mathbf{n}\right)$.

The symmetric surface tensor $\boldsymbol{\beta}$ is the rate of bending tensor and we might call $\boldsymbol{\rho}$ the 'geodesic rate of bending vector'.

We can eliminate $\boldsymbol{\omega}$ using (9) to give $\nabla\cdot(\boldsymbol{\epsilon}\cdot\boldsymbol{\beta}\cdot\boldsymbol{\epsilon})=-\nabla\cdot(\boldsymbol{\rho}\cdot(\boldsymbol{\epsilon}\mathbf{n}+\mathbf{n}\boldsymbol{\epsilon}))$ which is the compatibility equation relating bending and stretching of the surface. It is equivalent to differentiating the Gauss-Codazzi equations (that is Gauss's Theorema Egregium [7] and the Codazzi equations) with respect to time.

7 Conclusions

Let us repeat the equations defining the 4 fundamental 2nd order tensors:

$$\boldsymbol{\gamma}=\nabla\mathbf{v}-\boldsymbol{\omega}\cdot(\boldsymbol{\epsilon}\mathbf{n}+\mathbf{n}\boldsymbol{\epsilon})\qquad \boldsymbol{\epsilon}\cdot\boldsymbol{\beta}\cdot\boldsymbol{\epsilon}=\boldsymbol{\epsilon}\cdot\nabla\boldsymbol{\omega}+\boldsymbol{\rho}\cdot(\boldsymbol{\epsilon}\mathbf{n}+\mathbf{n}\boldsymbol{\epsilon})$$
$$\boldsymbol{\epsilon}\cdot\boldsymbol{\mu}=-\nabla\boldsymbol{\phi}+\mathbf{f}\cdot(\boldsymbol{\epsilon}\mathbf{n}+\mathbf{n}\boldsymbol{\epsilon})\qquad \boldsymbol{\sigma}=\boldsymbol{\epsilon}\cdot\nabla(\mathbf{f}+\mathbf{s})+\mathbf{u}\cdot(\boldsymbol{\epsilon}\mathbf{n}+\mathbf{n}\boldsymbol{\epsilon})\,.$$

Space limitations preclude us from discussing these equations fully, but it is worth noting that $\boldsymbol{\gamma}$ and $\boldsymbol{\beta}$ are always symmetric surface tensors. $\boldsymbol{\sigma}$ is a symmetric surface tensor in the membrane theory and $\boldsymbol{\epsilon}\cdot\boldsymbol{\mu}$ is a symmetric surface tensor when $\boldsymbol{\epsilon}\cdot\boldsymbol{\mu}=-\boldsymbol{\epsilon}\cdot\mathbf{M}\cdot\boldsymbol{\epsilon}$ if $\mathbf{M}$ is symmetric. The equations on the left form one pair and the equations on the right are a second pair in which there is an $\boldsymbol{\epsilon}\,\cdot$ after the $=$ sign. Thus we can use a virtual bending of a surface to visualize a state of membrane stress, rather like a generalised Airy stress function, and a virtual stretching of a surface to visualize moments in a shell with or without load. Clearly they are all linked by the theorem of virtual power, that is virtual work using velocities instead of displacements.

Perhaps the most interesting results are in (12) and (13) which enable us to write the stresses and moments in an unloaded portion of shell in terms of the single vector $\boldsymbol{\phi}$. There are discontinuities in $\boldsymbol{\phi}$ if we follow a closed path around a loaded area and if we imagine that a free edge of a shell is extended into a zone with no stress or internal moment, then the corresponding virtual velocity is a rigid body translation and rotation in the extended zone.

References

1. Cartan, E.: Leçons sur la géométrie des espaces de Riemann, 2nd edn. Gauhier-Villars, Paris (1963)
2. Cosserat, E., Cosserat, F.: Théorie des corps déformables. Hermann et Fils, Paris (1909)
3. Flügge, W.: Stresses in Shells. Springer, Berlin (1960). https://doi.org/10.1007/978-3-662-01028-0
4. Green, A.E., Zerna, W.: Theoretical Elasticity, 2nd edn. Oxford University Press, Oxford (1968)

5. Harnach, R., Rothert, H.: On the theory of inextensional bending of shell structures. Int. J. Solids Struct. **12**(5), 359–376 (1976). https://doi.org/10.1016/0020-7683(76)90026-3
6. Pinchover, Y., Rubinstein, J.: An Introduction to Partial Differential Equations. Cambridge University Press, Cambridge (2005)
7. Struik, D.J.: Lectures on Classical Differential Geometry. Addison-Wesley, Reading (1961)
8. Timoshenko, S., Woinowsky-Krieger, S.: Theory of Plates and Shells, 2nd edn. McGraw-Hill, New York (1959)
9. Timoshenko, S.P., Goodier, J.N.: Theory of Elasticity, 3rd edn. McGraw-Hill, New York (1970)

Steel Frames Optimization Considering Beam-Column Joint Stiffness and Geometric Constraints

Marco M. Rosso[1(✉)], Stefanos Sotiropoulos[1], Rita Greco[2], and Giuseppe Carlo Marano[1,3]

[1] Politecnico di Torino, DISEG, Department of Structural, Geotechnical and Building Engineering, Corso Duca Degli Abruzzi, 24, 10128 Turin, Italy
marco.rosso@polito.it
[2] DICAR department, Politecnico di Bari, Bari, Italy
[3] College of Civil Engineering, Fuzhou University, Fuzhou 350108, China

Abstract. Structural optimization is an active research branch in engineering, especially dealing with complex and concomitant aspects likewise in seismic design. Capacity design criteria for seismic design and detailing must be respected, e.g. according to the "strong-column weak-beam" principle. In steel structures, the choice of a specific beam-column joint typology may strongly affect its behavior under horizontal actions. In this study, the authors investigated the role of beam-column joint stiffness within an optimization paradigm related to steel structure frames. Specifically, the authors adopted simplified modeling assumptions for analysis under lateral loads in the Python environment and Computer and Structures inc. SAP2000 finite element software. Indeed, the main focus hitherto is oriented toward the problem definition accounting for geometric constraints and beam-column rotational stiffness capacity. Future investigations will adopt more realistic modeling procedures accounting for the typical non-linearities involved during strong dynamic actions.

Keywords: steel structures · geometric constraints · beam-column joints · meta-heuristic algorithms · optimization

1 Introduction

In the civil engineering field, structural optimization procedures play a crucial and significant role nowadays with the final purpose of minimizing the various costs items involved in the design, construction, and maintenance phases [15,16,18]. Three main branches can be evidenced within the structural optimization field. Size optimization involves the definition of the optimal size of structural elements within the domain under investigation, whereas shape optimization studies the optimal shape of the structural members without varying

Supported by project ADDOPTML.

S. Gabriele et al. (Eds.): IWSS 2023, LNCE 437, pp. 129–138, 2024.
https://doi.org/10.1007/978-3-031-44328-2_14

their relative positions and structural layout [18,20]. The latter aspect is indeed investigated by the topology optimization branch [18,20].

With a growing consciousness and concerns for the environmental impact of the architecture and construction industry (AEC), in recent years life-cycle cost assessment posed the structural optimization area in a new perspective to globally account for optimizing cost-effective solutions considering the entire life-cycle of a structure [11,12]. However, this approach requires tremendous efforts to carefully evaluate many complex related aspects concurring with the definition of representative cost objective functions (OF) [1]. Nonetheless, as a first approximation, and in the absence of precise requirements and prescription, the weight minimization may still be successfully used as an indirect indicator of the construction cost, directly related to the minimum material consumption [6, 14,18,20]. Furthermore, the minimum self-weight optimization has other benefits besides the material consumption criterion, e.g. it may aid to reduce the inertial effects due to a reduction of the mobilized mass during earthquake conditions [17].

Mainly focusing on steel structures, one of the main aspects to account for is fulfilling capacity design criteria for seismic design and detailing, in which, e.g., the principle of "strong-column weak-beam" must be respected [5,21]. Furthermore, the technical choice of a specific beam-column joint typology may strongly affect the steel frame's behavior under horizontal actions. Numerous beam-column connection typologies exist and they affect the beam's ultimate rotational capacity behavior evaluated on the moment-rotation plane, thus permitting the formation of plastic hinges and safer ductile global mechanisms [3,7,8,10]. The stiffest connections are represented by the welded ones, but they entail increased complexities from a constructive point of view [23], therefore the less stiff bolted ones are often preferred. The specific technical solution adopted for the connection directly affects the rotational stiffness denoted as K_θ, thus determining a semi-rigid beam-column joint. In the present study, the authors investigated in a global and simplified manner how the various limit modeling approaches of the beam-column joint [9] affect the results of a size optimization problem of a steel plane frame. Since the current focus is oriented toward global frame optimization results, linear dynamic analyses were herein conducted. Future studies should further explore the rotational stiffness effects by adopting more detailed nonlinear analyses to account for the entire moment-rotation behavior of the specific technically adopted connections.

2 Steel Frame Case Study Description

The two-bay three-story plane steel frame analyzed in the current document is inspired by the steel frame example presented in [19], and it is illustrated in Fig. 1 (a). The black numbers represent the nodes' numbering system whereas the blue ones are referred to the structural finite element (FE) beams and columns members. The positions of semi-rigid joints are indicated with orange spirals, simulating the partial bending moment release due to the actual connection typology.

Using a steel S275 material grade, the FE model has been implemented in the Computer and Structures inc. SAP2000 commercial software [2]. Specifically, the dead loads are automatically computed by the software after choosing the specific cross-section profile for the beams and for the columns. These latter are selected directly from the available cross sections in SAP2000 according to the Europe area commercially available standard steel profiles. Specifically, 70 different profiles for the HE steel profiles are available for the columns (HE100A, HE100B, etc., until HE1000M), and 50 different IPE cross sections are available for the beams (IPE100, IPE120, etc., until IPE600V). In order to account for seismic conditions for the optimal design of the size of each structural element, the authors hypothetically located the steel frame in the Abruzzi region in Italy, precisely in L'Aquila city. The seismic analyses have been conducted with the dynamic linear modal analysis with elastic response spectrum for the life-safety limit state (SLV), which is depicted in Fig. 1 (b). The depicted response spectrum has been determined based on the geographical location of the structure, accounting for a ground typology B (typical of L'Aquila city), considering a commercial building class, with a nominal life of 50 years, an usage class III, thus a usage coefficient of $c_U = 1.5$, a topographical typology T1 for flat land, and considering a standard damping ratio of $\xi = 5\%$ [21]. The complete quadratic combination (CQC) has been performed to define the action effects envelope coming from the modal superposition of the first natural modes mobilizing a modal mass ratio greater than 85%. The mass considered in the modal analysis was accounted for quasi-permanent vertical load conditions, i.e. considering the dead load and the live load multiplied for a combinatorial factor $\psi_{2,1} = 0.6$ [21].

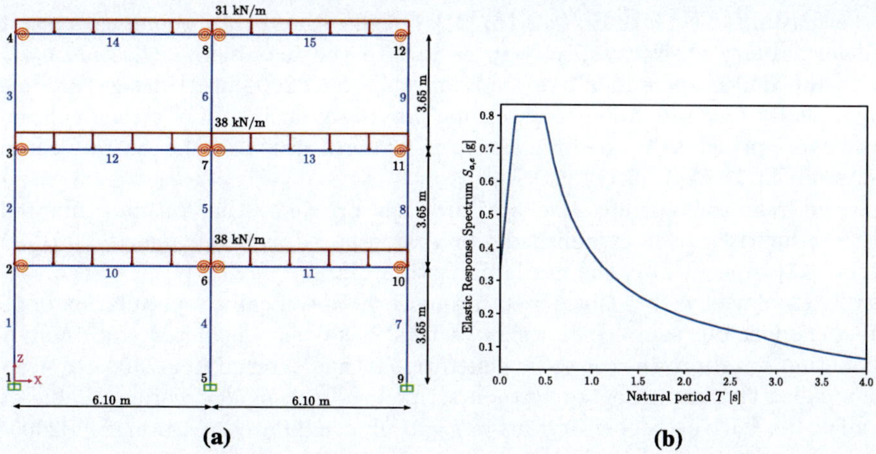

Fig. 1. View of the steel frame case study (a) and the adopted elastic response spectrum (b).

3 Discrete Size Optimization Strategy with Genetic Algorithm

The authors implemented a Python script to deal with the *SAP2000 OAPI* module [2] in order to parametrize and automatically run the FE model. Specifically, the size optimization of the steel frame can be stated as a constrained problem in which the OF $f(\boldsymbol{x})$ is represented by the dead load total base vertical reactions. The constraints are thus represented by the unitless ratios associated with the structural checks according to the NTC18 Italian structural design code regulations, which incorporate the EN1993 Eurocode 3 for steel structures. The assessment of the constraints has been performed with the *SAP2000 Steel Design* module [2]. The design vector $\boldsymbol{x}$ is a discrete-integer-value array whose components referred to the index of the steel profiles associated with the various structural members, thus spanning between 0 and 70 for selecting a specific HE profile, and varying between 0 and 50 for identifying a specific IPE profile.

$$\begin{cases} \min & f(\boldsymbol{x}) \\ s.t. & g_1(\boldsymbol{x}) = \sum_{i=1}^{15} \max\{r_{NMM,i}, 0\} \leq 1 \\ & g_2(\boldsymbol{x}) = \sum_{i=1}^{15} \max\{r_{V,i}, 0\} \leq 1 \\ & g_3(\boldsymbol{x}) = \delta_{max} \leq L/250 \\ & g_4(\boldsymbol{x}) = \delta_2 \leq L/300 \end{cases} \tag{1}$$

The constraint g_1 in Eq.(1) accounts for the combined bending and axial force cross-section checks ratios ($r_{NMM,i}$) according to Eq.(C4.2.37) of Method B of the NTC18 commentary [22]. This ratio also accounts for compression buckling and lateral torsional buckling checks. g_2 consider the shear design checks ratios ($r_{V,i}$) according to NTC18 Eq.(4.2.16) [21]. On the other hand, g_3 and g_4 account for deformability checks respectively related to the maximum deflection δ_{max} and to the deflection under live loads only δ_2. SAP2000 steel design module automatically accounts for special seismic provisions such as the "strong-column weak-beam" principle by defining an over-strength check of the beam-column joint, see NTC18 Eq.(7.5.11) [21].

To address the current size optimization problem, the authors adopted the meta-heuristic well-established population-based *genetic algorithm* (GA) method [13]. Specifically, the available Python library *pymoo* [4] has been successfully employed within the current framework. Specifically, a population of 50 individuals (various samples of design vectors) has been considered, and the GA termination criterion was set to 50 generations, thus determining 2500 OF evaluation calls. The implemented Python script has been made freely available at the following GitHub repository: https://github.com/marco-rosso-m/SAP2000-python-for-structural-optimization.

4 Results and Discussion

In order to explore the effects of various beam-column joint modeling strategies on the size optimization optimal solutions, the authors analyzed three structural configurations. The first configuration is related to a limit condition with perfect hinges between beams and columns, the second one is another limit condition with perfect fixity constraints between beams and columns. The third condition accounts for a more proper modeling condition referring to a realistic value of rotational stiffness K_θ. Based on the quite extensive experimental tests provided in the existing literature [3,7,8,10], the authors have identified that a reasonable range of K_θ values broadly spans between 10×10^3 (bolted connections, e.g. web cleats to column flanges) and slightly lower than 100×10^3 kNm/rad (welded connections, e.g. flush end plate to column web). Based on this quite wide range, the authors adopted a discrete range strategy to contain the search space complexity, i.e. by assuming only integer multiples of 10^3, e.g. 0×10^3, 10×10^3 20×10^3, etc.

The above-mentioned three modeling approaches have been analyzed into two main cases. In the first main case, a simplified optimization strategy has been performed, thus considering the same beams' and the same columns' profiles for every floor level. Therefore, in general, the design vector appears as a three-component array $\boldsymbol{x} = [x(1), x(2), x(3)]^T$. The first two components indicate respectively the steel profiles for the columns and for the beams for the whole elevation, whereas the third one considers a discrete possible value for K_θ for every beam-column joint. In the second main case, instead, a more complete optimization strategy has been implemented with an extended search space, i.e. based on adopting different profiles for both columns and beams at every floor level. In that case, in general, the design vector appears in general as a nine-components array. The first three components account for different columns' HE profiles for every level, the next three ones designated the different beams' IPE profiles for every level, whereas the latter three components identify the K_θ values for every single floor. Furthermore, two additional constraints have been introduced in order to account for geometric reasons. Indeed, when the column section varies with the elevation, it is typical that it tapers from the bottom to the top of the building. Therefore:

$$g_5(\boldsymbol{x}) = x(1) - x[0] \leq 0 \qquad \text{and} \qquad g_6(\boldsymbol{x}) = x(2) - x(1) \leq 0 \tag{2}$$

The summary of optimal results for all the analyzed cases is reported in Table 1.

4.1 Same Beams' and Columns' Profiles for Every Floor Case

In this section, the results of the first main case are presented, i.e. the ones with the same beams' and columns' profiles for every floor. In particular, the above-mentioned three modeling strategies sub-cases have been explored, as depicted in Figs. 2, 3 and 4 respectively. In all three sub-cases, the best optimal solution was found within the first 500 OF evaluations. Since the search space is composed of two design variables only, except when considering the optimization of the rotational stiffness, it is worthy to visualize how the GA explored the search space of possible combinations of steel profiles, evidencing the existence of a feasible region (blue) and an unfeasible one (blue). The feasible region appears noisy in Fig. 2 (b) and Fig. 4 (b), probably because of the imposed respectfulness of the special seismic provisions. Between the two limit case, the fixed one better exploited the load-carrying capacities redistribution, thus ensuring the minimum OF value. A similar finding was obtained in the K_θ sub-case, achieving an optimal remarkably large value of 90×10^3 kNm/rad.

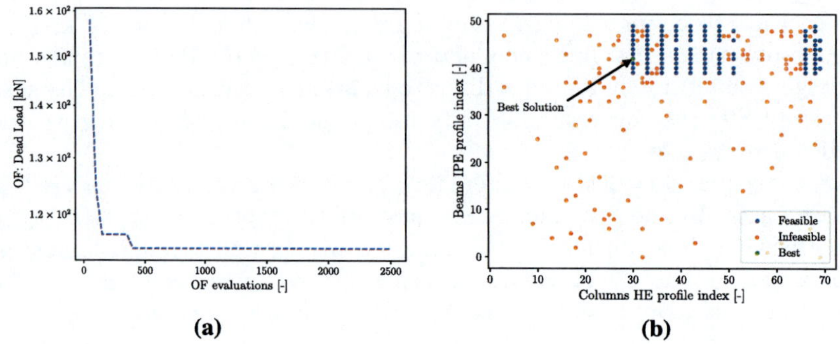

Fig. 2. Perfectly-hinged case with the same columns' and beams' profiles at every floor: OF history (a) and analysis of the GA's explored search space (b).

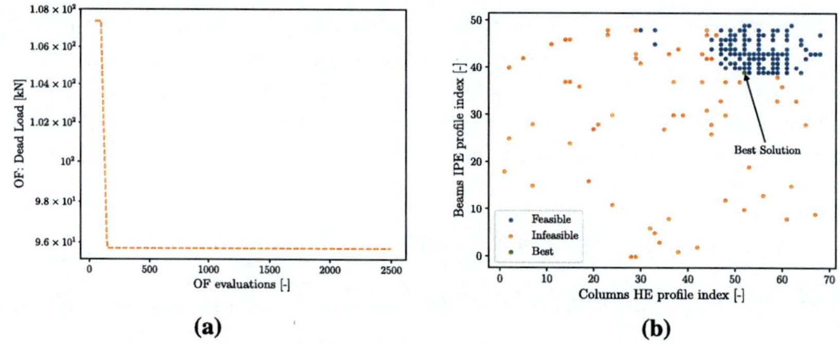

Fig. 3. Perfectly-fixed case with the same columns' and beams' profiles at every floor: OF history (a) and analysis of the GA's explored search space (b).

4.2 Different Beams' and Columns' Profiles for Every Floor Case

The results of the second main case are herein discussed, i.e. the one adopting different beams' profiles and different columns' profiles for every floor level. Specifically, the previously-mentioned three modeling strategies sub-cases have been analyzed, and their individual OF history during the optimization process

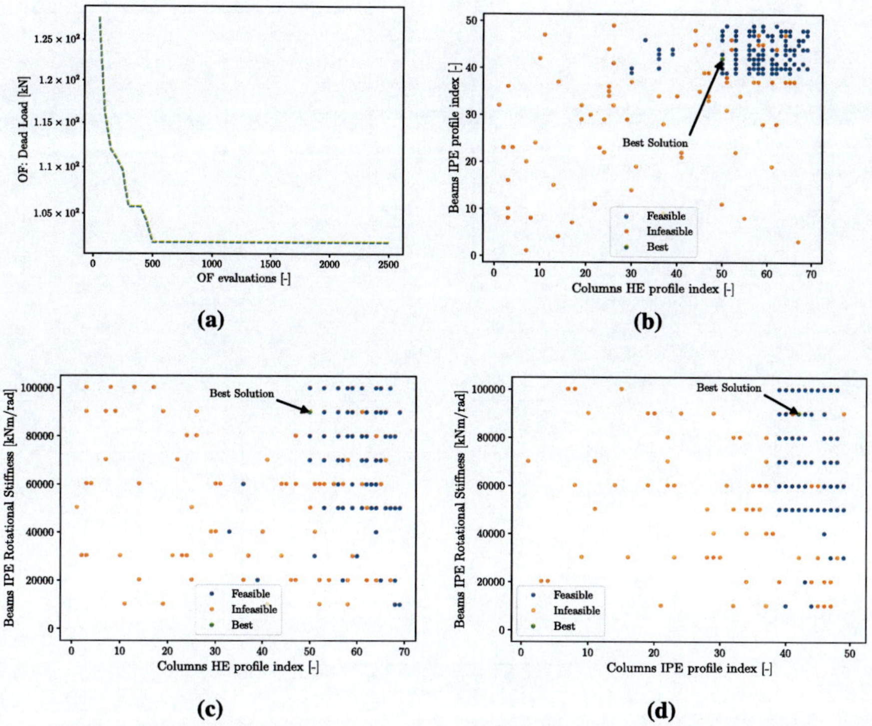

Fig. 4. Partially-released case with the same columns' and beams' profiles at every floor: OF history (a) and analysis of the GA's explored search space in terms of profiles (b), columns' profiles vs rotational stiffness (c), and beams' profiles vs rotational stiffness(d)

have been gathered in Fig. 5. Specifically, sub-figures (a) and (b) revealed that, during the initial 200 function evaluations, the entire population was unfeasible, and only from that point forward the GA founded the right feasible region and effectively started the actual optimization process. On the other hand, the semi-rigid connection K_θ sub-case revealed that, probably the optimization procedure should have been prolonged much more than 50 iterations, since the still decreasing trend of the OF. Finally, comparing the final results in Table 1, this latter second main case with a more refined and comprehensive size optimization problem delivered in virtually all the cases a considerable reduction of the OF. In this occasion, the three optimal solutions are quite comparable (around 80 kN), and it is worth noting that the semi-rigid case preferred an optimal solution with almost fixed connections at the first level (80×10^3 kNm/rad) and almost fully hinged to the other two floor levels (0 and 10×10^3 kNm/rad respectively).

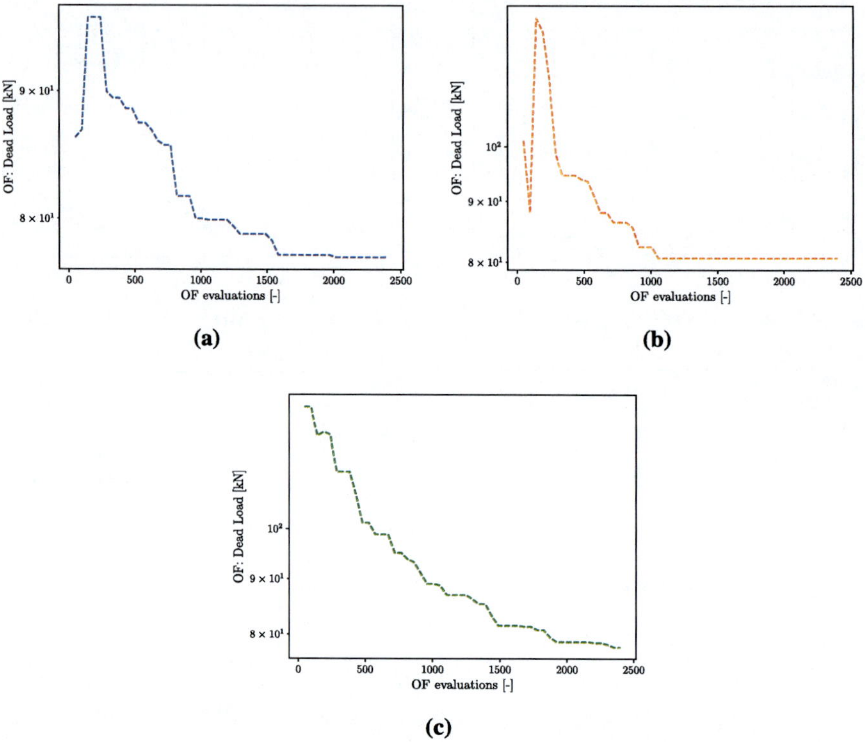

Fig. 5. Different columns' and beams' profiles at every floor case: OF history for perfectly-hinged beams (a), for perfectly-fixed beams (b), and for partially-released beams (c).

Table 1. Summary and comparison of the optimal results.

	Same profiles for every floor				Different profiles for every floor			
Elements	Hinged case profiles	Fixed case profiles	Partially released profiles	Rot.Stiff K_θ [kNm/rad]	Hinged case profiles	Fixed case profiles	Partially released profiles	Rot.Stiff K_θ [kNm/rad]
Columns:								
1-4-7					HE600A	HE600A	HE500A	
2-5-8	HE300M	HE600A	HE550B		HE400A	HE400A	HE450A	
3-6-9					HE280A	HE320A	HE280A	
Beams:								
10-11					IPE550	IPE550	IPE550	80×10^3
12-13	IPE550	IPE500O	IPE550	90×10^3	IPE550	IPE550	IPE500O	0
14-15					IPE450R	IPE550	IPE550	10×10^3
OF [kN]	114.366	95.741	101.976		77.181	80.652	77.925	

5 Conclusions

The current study permitted analyzing in a simplified and linear manner the effects of three beam-column joint modeling strategies and how it affects the size optimization of a steel frame under seismic loading. As expected, the results evidenced that optimal semi-rigid joints at lower-level floors should tend to be stiffer likewise welded joints, whereas at higher levels they may tend toward hinged connections. Future studies should further explore the rotational stiffness effects by adopting more detailed nonlinear analyses to account for the entire moment-rotation behavior of the specific technically adopted connections.

Supplementary material

The implemented code is made freely available at the following GitHub repository: https://github.com/marco-rosso-m/SAP2000-python-for-structural-optimization.

Acknowledgments. This research has been supported by the ADDOPTML project: "ADDitively Manufactured OPTi-mized Structures by means of Machine Learning" (No: 101007595) belonging to the Marie Skłodowska-Curie Actions (MSCA) Research and Innovation Staff Exchange (RISE) H2020-MSCA-RISE-2020. Their support is highly acknowledged.

References

1. Adeli, H., Sarma, K.C.: Cost Optimization of Structures: Fuzzy Logic, Genetic Algorithms, and Parallel Computing. Wiley, West Sussex (2006)
2. Berkeley, C.: CSI Analysis Reference Manual for sap2000. Computers and Structures Inc., Berkeley (2007)
3. Bing, G., Lei, W., Ying, W., Yan, S., Hailan, T.: Experimental study on rotational stiffness of steel frame beam-column connections. J. Build. Struct. **32**(10), 82 (2011)
4. Blank, J., Deb, K.: pymoo: multi-objective optimization in python. IEEE Access **8**, 89497–89509 (2020)
5. CEN: Eurocode 3-design of steel structures-. BS EN 1993-1 **1**, 2005 (2006)
6. Cucuzza, R., Rosso, M.M., Aloisio, A., Melchiorre, J., Giudice, M.L., Marano, G.C.: Size and shape optimization of a guyed mast structure under wind, ice and seismic loading. Appl. Sci. **12**(10), 4875 (2022)
7. Daniūnas, A., Urbonas, K.: Analysis of the steel frames with the semi-rigid beam-to-beam and beam-to-column knee joints under bending and axial forces. Eng. Struct. **30**(11), 3114–3118 (2008)
8. Davison, J., Kirby, P., Nethercot, D.: Rotational stiffness characteristics of steel beam-to-column connections. J. Constr. Steel Res. **8**, 17–54 (1987)
9. Díaz, C., Martí, P., Victoria, M., Querin, O.M.: Review on the modelling of joint behaviour in steel frames. J. Constr. Steel Res. **67**(5), 741–758 (2011)
10. Du, H., Zhao, P., Wang, Y., Sun, W.: Seismic experimental assessment of beam-through beam-column connections for modular prefabricated steel moment frames. J. Constr. Steel Res. **192**, 107208 (2022)

11. Gencturk, B.: Life-cycle cost assessment of RC and ECC frames using structural optimization. Earthq. Eng. Struct. Dyn. **42**(1), 61–79 (2013)
12. Guillen-Gosalbez, G., Caballero, J.A., Jimenez, L.: Application of life cycle assessment to the structural optimization of process flowsheets. Ind. Eng. Chem. Res. **47**(3), 777–789 (2008)
13. Jh, H.: Adaptation in natural and artificial systems. Ann Arbor (1975)
14. Kirsch, U.: Structural Optimization: Fundamentals and Applications. Springer, Heidelberg (2012)
15. Marano, G.C., Rosso, M.M., Melchiorre, J.: Optimization as a tool for seismic protection of structures. In: Cimellaro, G.P. (ed.) WCSI 2022. LNCS, pp. 100–113. Springer, Cham (2023). https://doi.org/10.1007/978-3-031-21187-4_8
16. Melchiorre, J., Manuello, A., Marmo, F., Adriaenssens, S., Marano, G.: Differential formulation and numerical solution for elastic arches with variable curvature and tapered cross-sections. Eur. J. Mech.-A/Solids **97**, 104757 (2023)
17. Plevris, V.: Structural seismic design optimization and earthquake engineering: formulations and applications: formulations and applications. IGI Global (2012)
18. Rao, S.S.: Engineering Optimization: Theory and Practice. Wiley, New York (2019)
19. Simoes, L.: Optimization of frames with semi-rigid connections. Comput. Struct. **60**(4), 531–539 (1996)
20. Spillers, W.R., MacBain, K.M.: Structural Optimization. Springer, Dordrecht (2009)
21. delle infrastrutture e dei Trasporti, M.: Norme tecniche per le costruzioni (ntc18). Decreto Ministero Infrastrutture. GU Serie Generale (42) (2018)
22. delle Infrastrutture e dei Trasporti, M.: Circolare 21 gennaio 2019 n. 7 cs ll. pp. istruzioni per l'applicazione dell'aggiornamento delle 'norme tecniche per le costruzioni' di cui al dm 17/01/2018. Suppl. Ord. Alla GU n. 35 Del 11/2/19 (2019)
23. Viana, D.D., Tommelein, I.D., Formoso, C.T.: Using modularity to reduce complexity of industrialized building systems for mass customization. Energies **10**(10), 1622 (2017)

Dynamic Analysis of Doubly-Curved Shells Made of Advanced Materials with Higher Order Theories and Generalized Differential Quadrature

Francesco Tornabene(✉), Matteo Viscoti, and Rossana Dimitri

Department of Innovation Engineering, University of Salento, Lecce, Italy
francesco.tornabene@unisalento.it

Abstract. In the present work, a refined theory with three-dimensional capabilities is proposed for the structural analysis of shell structures made of smart materials for advanced engineering applications. Principal curvilinear coordinates are used for the geometry definition of the structure. The kinematic description of the configuration variables is performed according to the Equivalent Single Layer (ESL) approach with higher order theories. A constitutive relation for elastic anisotropic laminates is considered. The fundamental equations, written in the strong form, are derived from the Hamilton principle together with the boundary conditions, and they are numerically solved by means of the Generalized Differential Quadrature (GDQ) method. The accuracy and the computational efficiency of the present theory is shown by means of different applicative examples. The numerical predictions of the present ESL model are compared to refined solutions coming from three-dimensional Finite-Element-based simulations. Furthermore, some parametric investigations are performed on a doubly-curved panel made of Variable Angle Tow (VAT) anisotropic materials in order to show the sensitivity of the VAT distribution on the dynamic response of the shell under consideration. The present formulation is very accurate if compared to refined models, despite its reduced computational demand, and it can be useful for design purposes of doubly-curved elements made of advanced materials.

Keywords: Anisotropic Materials · Doubly-curved Shells · Equivalent Single Layer · Generalized Differential Quadrature · Variable Angle Tow

1 Introduction

Structural components of complex shape made of advanced materials are extensively adopted in many engineering applications like mechanical systems, aeronautics, naval, automotive etc. In this context, laminated composite materials are widely used thanks to their high stiffness-to-weight ratio. An interesting application is the so-called Variable Angle Tow (VAT) composite [1], which consists in a tailored orientation of the reinforcing fibers of a classical composite material. In this way, the mechanical response of the

S. Gabriele et al. (Eds.): IWSS 2023, LNCE 437, pp. 139–149, 2024.
https://doi.org/10.1007/978-3-031-44328-2_15

structure is enhanced because some local buckling phenomena are prevented. In addition, the VAT technology orients the bending and vibrational response of the structural element in hand.

On the other hand, the design process of these advanced applications requires refined numerical predictions, which can be computationally demanding. Among two-dimensional methodologies, higher order theories, based on the Equivalent Single Layer (ESL) approach, can be a valid alternative to three-dimensional simulations [2, 3] since they provide very accurate results with a reduced number of Degrees of Freedom (DOFs). According to the ESL, a reference surface, located in the middle thickness of the structure, is used for the description of the shell, taking into account the main outcomes of differential geometry. Starting from classical two-dimensional theories like the First order Shear Deformation Theory (FSDT) and the Third order Shear Deformation Theory (TSDT), the description of the unknown field variable along the thickness direction can be made with a generalized approach, named as Higher order Shear Deformation Theory (HSDT). In this way, an arbitrary kinematic field assumption can be made depending on the particular application. When softcore laminates are studied, the higher order Layer-Wise (LW) approach [4] is usually adopted, because it provides a higher order description of the field variable for each layer of the stacking sequence and not for the entire laminate.

The solution of the fundamental equations requires a numerical tool, because closed-form analytical solutions are available only for a limited number of structures. The well-known Finite Element Method (FEM) is extensively adopted in literature and in practical applications for the derivation of an approximate solution. On the other hand, in advanced applications it can be very high computational demanding, therefore spectral collocation methods like the Generalized Differential Quadrature (GDQ) are valid alternatives, because they require a limited number of DOFs [5, 6].

In the present work, a two-dimensional ESL formulation, based on higher order theories, is presented for the dynamic analysis of doubly-curved shell structures made of anisotropic materials with general orientation. A continuous variation of the material orientation angle along the physical domain is modelled in each layer, taking into account a general power relation. The fundamental equations are derived from the Hamilton principle, and the numerical solution is found with the GDQ method. The present ESL formulation is applied to a doubly-curved panel for the evaluation of the mode frequencies and shapes. The material properties are taken from the database of the DiQuMASPAB research software, where the present formulation has been implemented [6].

In a first step, the proposed theory is validated starting from the results of a refined three-dimensional (3D) FEM model developed with a commercial software. Then, the vibrational response of the structure is investigated for different configurations of the VAT distribution, pointing out the influence of some significant governing design parameters. It is shown that the present higher order model can predict very accurately the mechanical response of doubly-curved structural elements for advanced applications with a reduced computational effort.

2 Theoretical Framework

According to the ESL approach, an arbitrary 3D shell element is described starting from a reference surface $\mathbf{r}(\alpha_1, \alpha_2)$, defined with a set of curvilinear principal coordinates α_1, α_2 varying in a rectangular physical domain $(\alpha_1, \alpha_2) \in [\alpha_1^0, \alpha_1^1] \times [\alpha_2^0, \alpha_2^1]$. As a consequence, the position vector $\mathbf{R}(\alpha_1, \alpha_2, \zeta)$ of an arbitrary point of the shell can be written from the following relation:

$$\mathbf{R}(\alpha_1, \alpha_2, \zeta) = \mathbf{r}(\alpha_1, \alpha_2) + \frac{h(\alpha_1, \alpha_2)}{2} z\mathbf{n}(\alpha_1, \alpha_2) \tag{1}$$

where $z = 2\zeta/h$ is a dimensionless coordinate introduced conveniently, whereas $h(\alpha_1, \alpha_2)$ denotes the thickness of the shell in each point of the physical domain, which can be computed as the sum of the thickness h_k of each k-th lamina with $k = 1, ..., l$.

At this point, the vector $\mathbf{U}^{(k)}(\alpha_1, \alpha_2, \zeta, t) = \begin{bmatrix} U_1^{(k)} & U_2^{(k)} & U_3^{(k)} \end{bmatrix}^T$ of the generalized field components $U_i^{(k)}$ with $i = 1, 2, 3$ is expanded thanks to the introduction of the so-called generalized thickness functions $F^{(k\tau)\alpha_i}$. In this way, the generalized displacement field vector $\mathbf{u}^{(\tau)}(\alpha_1, \alpha_2, t) = \begin{bmatrix} u_1^{(\tau)} & u_2^{(\tau)} & u_3^{(\tau)} \end{bmatrix}^T$ is introduced:

$$\mathbf{U}^{(k)}(\alpha_1, \alpha_2, \zeta, t) = \sum_{\tau=0}^{N+1} \mathbf{F}^{(k\tau)}(\zeta)\mathbf{u}^{(\tau)}(\alpha_1, \alpha_2, t) \tag{2}$$

where $\mathbf{F}^{(k\tau)}$ denotes the matrix of the thickness functions in hand. The previous relation can be expressed in a more expanded form as follows:

$$\begin{bmatrix} U_1^{(k)} \\ U_2^{(k)} \\ U_3^{(k)} \end{bmatrix} = \sum_{\tau=0}^{N+1} \begin{bmatrix} F^{(k\tau)\alpha_1} & 0 & 0 \\ 0 & F^{(k\tau)\alpha_2} & 0 \\ 0 & 0 & F^{(k\tau)\alpha_3} \end{bmatrix} \begin{bmatrix} u_1^{(\tau)} \\ u_2^{(\tau)} \\ u_3^{(\tau)} \end{bmatrix} \tag{3}$$

As far as the constitutive behaviour of each k-th lamina is concerned, a linear elastic relation is assumed, valid for generally anisotropic materials. Thus, the strain vector $\boldsymbol{\varepsilon}^{(k)}$ is related to the stress vector, denoted by $\boldsymbol{\sigma}^{(k)}$, by means of the stiffness matrix $\overline{\mathbf{E}}^{(k)}$, written in the reference system of the problem:

$$\boldsymbol{\sigma}^{(k)} = \overline{\mathbf{E}}^{(k)} \boldsymbol{\varepsilon}^{(k)} \tag{4}$$

The matrix $\overline{\mathbf{E}}^{(k)}$, defined in each k-th layer of the stacking sequence, is derived from the corresponding stiffness matrix $\mathbf{E}^{(k)}$ in the material reference system, taking into account a rotation matrix $\mathbf{T}^{(k)} = \mathbf{T}^{(k)}\left(\vartheta^{(k)}\right)$ depending on the material orientation angle $\vartheta^{(k)}$:

$$\overline{\mathbf{E}}^{(k)} = \mathbf{T}^{(k)}\mathbf{E}^{(k)}\mathbf{T}^{(k)T} \tag{5}$$

The matrix $\mathbf{E}^{(k)}$ can be written in a more expanded form as:

$$\mathbf{E}^{(k)} = \begin{bmatrix} E_{11}^{(k)} & E_{12}^{(k)} & E_{16}^{(k)} & E_{14}^{(k)} & E_{15}^{(k)} & E_{13}^{(k)} \\ E_{12}^{(k)} & E_{22}^{(k)} & E_{26}^{(k)} & E_{24}^{(k)} & E_{25}^{(k)} & E_{23}^{(k)} \\ E_{16}^{(k)} & E_{26}^{(k)} & E_{66}^{(k)} & E_{46}^{(k)} & E_{56}^{(k)} & E_{36}^{(k)} \\ E_{14}^{(k)} & E_{24}^{(k)} & E_{46}^{(k)} & E_{44}^{(k)} & E_{45}^{(k)} & E_{34}^{(k)} \\ E_{15}^{(k)} & E_{25}^{(k)} & E_{56}^{(k)} & E_{45}^{(k)} & E_{55}^{(k)} & E_{35}^{(k)} \\ E_{13}^{(k)} & E_{23}^{(k)} & E_{36}^{(k)} & E_{34}^{(k)} & E_{35}^{(k)} & E_{33}^{(k)} \end{bmatrix} \tag{6}$$

being $E_{ij}^{(k)}$ for $i, j = 1, ...7$ the generic component of the stiffness matrix $\mathbf{E}^{(k)}$, which can be the stiffness constants $C_{ij}^{(k)}$ or to the reduced elastic coefficients $Q_{ij}^{(k)}$, depending on the kinematic field assumption of Eq. (4).

Remembering the ESL kinematic field assumption, introduced in Eq. (3) and (4), a two-dimensional higher order constitutive relationship is written starting from the three-dimensional relation of Eq. (5), thus relating the generalized stress resultant vector $\mathbf{S}^{(\tau)\alpha_i}$ to the vector $\boldsymbol{\varepsilon}^{(\eta)\alpha_j}$ of the generalized strains, being $\mathbf{A}^{(\tau\eta)\alpha_i\alpha_j}$ the higher order generalized stiffness matrix:

$$\mathbf{S}^{(\tau)\alpha_i} = \sum_{\eta=0}^{N+1} \sum_{j=1}^{3} \mathbf{A}^{(\tau\eta)\alpha_i\alpha_j} \boldsymbol{\varepsilon}^{(\eta)\alpha_j} \tag{7}$$

for $\tau = 0, ..., N+1$ and $i = 1, 2, 3$.

The equilibrium equations are derived from the Hamilton principle, starting from the virtual variation $\delta \mathcal{T}$ of the kinetic energy and that $\delta \Phi$ of the elastic strain energy of the system:

$$\delta \int_{t_1}^{t_2} (\Phi - \mathcal{T}) dt = 0 \tag{8}$$

being $[t_1, t_2]$ an arbitrary time interval. If the generalized constitutive relation of Eq. (8) is considered, together with the kinematic relation of Eq. (3), the final set of the fundamental relations is derived. Furthermore, the boundary conditions of the problem are obtained. If $\mathbf{L}^{(\tau\eta)}$ and $\mathbf{M}^{(\tau\eta)}$ are the fundamental matrix and the mass matrix, respectively, defined for each $\tau, \eta = 0, ..., N+1$, the higher order fundamental equations can be written as follows:

$$\sum_{\eta=0}^{N+1} \mathbf{L}^{(\tau\eta)} \mathbf{u}^{(\eta)} = \sum_{\eta=0}^{N+1} \mathbf{M}^{(\tau\eta)} \ddot{\mathbf{u}}^{(\eta)} \quad \text{for } \tau = 0, ..., N+1 \tag{9}$$

The numerical solution of the previous relation, valid for a doubly-curved shell element, is derived on a non-uniform two-dimensional computational grid made of $I_N \times$

I_M discrete points, following the Chebyshev-Gauss-Lobatto (CGL) distribution. To this end, the GDQ method is extensively adopted. According to the GDQ, if $f = f(x)$ is an arbitrary smooth function, defined in a closed interval $[a, b]$ and sufficiently derivable, its n-th order derivative can be calculated in each discrete point $x_i \in [a, b]$ with $i = 1, ..., I_Q$ by means of the following relation, being $\varsigma_{ij}^{(n)}$ for $i, j = 1, 2, ..., I_Q$ the GDQ weighting coefficients [6]:

$$f^{(n)}(x_i) = \left.\frac{\partial^n f(x)}{\partial x^n}\right|_{x=x_i} \cong \sum_{j=1}^{I_Q} \varsigma_{ij}^{(n)} f(x_j) \quad i = 1, 2, ..., I_Q \tag{10}$$

3 Applications and Results

In this section, some numerical investigations are presented regarding a doubly-curved shell structure made of generally anisotropic layers. The ESL model is validated with respect to a 3D FEM solution, developed with a commercial software. Then, a continuous smooth variation of the material orientation angle is provided, and some useful parametric investigations are conducted.

3.1 Validation of the Higher Order ESL Model

In the present section, a validation of the present ESL formulation is performed. The first ten mode frequencies of an helicoidal panel are calculated with different higher order theories. The equation of the reference surface $\mathbf{r}(\alpha_1, \alpha_2)$ of the shell under consideration is written in principal coordinates α_1, α_2 as follows, setting $a = 1.2$ m:

$$\begin{aligned}\mathbf{r}(\alpha_1, \alpha_2) = & -a\cos(\alpha_1 + \alpha_2)\sinh(\alpha_1 - \alpha_2)\mathbf{e}_1 + \\ & -\sin(\alpha_1 + \alpha_2)\sinh(\alpha_1 - \alpha_2)\mathbf{e}_2 + a(\alpha_1 - \alpha_2)\mathbf{e}_3\end{aligned} \tag{11}$$

The lamination scheme consists in a central core of graphite-epoxy of thickness $h_1 = 0.1$ m and two external layers of a trigonal material of thickness $h_2 = h_3 = 0.005$ m. The orthotropic mechanical properties of the graphite-epoxy are provided in terms of the nine engineering constants reported below, whereas the material density is set equal to $\rho^{(k)} = 1450$ kg/m^3:

$$\begin{array}{lll} E_1^{(k)} = 137.9\,\text{GPa} & G_{12}^{(k)} = 7.1\,\text{GPa} & \nu_{12}^{(k)} = 0.30 \\ E_2^{(k)} = 8.96\,\text{GPa} & G_{13}^{(k)} = 7.1\,\text{GPa} & \nu_{13}^{(k)} = 0.30 \\ E_3^{(k)} = 8.96\,\text{GPa} & G_{23}^{(k)} = 6.21\,\text{GPa} & \nu_{23}^{(k)} = 0.49 \end{array} \tag{12}$$

For the derivation of the stiffness constants $E_{ij}^{(k)}$ of Eq. (7) starting from the orthotropic engineering constants of Eq. (13), the interested reader can refer to [6].

The constitutive matrix $\mathbf{E}^{(k)}$ of the trigonal material $\left(\rho^{(k)} = 2649\ \mathrm{kg/m^3}\right)$ looks as:

$$\mathbf{E}^{(k)} = \begin{bmatrix} C_{11}^{(k)} & C_{12}^{(k)} & C_{16}^{(k)} & C_{14}^{(k)} & C_{15}^{(k)} & C_{13}^{(k)} \\ C_{12}^{(k)} & C_{22}^{(k)} & C_{26}^{(k)} & C_{24}^{(k)} & C_{25}^{(k)} & C_{23}^{(k)} \\ C_{16}^{(k)} & C_{26}^{(k)} & C_{66}^{(k)} & C_{46}^{(k)} & C_{56}^{(k)} & C_{36}^{(k)} \\ C_{14}^{(k)} & C_{24}^{(k)} & C_{46}^{(k)} & C_{44}^{(k)} & C_{45}^{(k)} & C_{34}^{(k)} \\ C_{15}^{(k)} & C_{25}^{(k)} & C_{56}^{(k)} & C_{45}^{(k)} & C_{55}^{(k)} & C_{35}^{(k)} \\ C_{13}^{(k)} & C_{23}^{(k)} & C_{36}^{(k)} & C_{34}^{(k)} & C_{35}^{(k)} & C_{33}^{(k)} \end{bmatrix} = \begin{bmatrix} 86.74 & 6.99 & 0 & 0 & -17.91 & 11.91 \\ 6.99 & 86.74 & 0 & 0 & 17.91 & 11.91 \\ 0 & 0 & 39.88 & -17.91 & 0 & 0 \\ 0 & 0 & -17.91 & 57.94 & 0 & 0 \\ -17.91 & 17.91 & 0 & 0 & 57.94 & 0 \\ 11.91 & 11.91 & 0 & 0 & 0 & 107.20 \end{bmatrix} \mathrm{GPa} \tag{13}$$

In Table 1 the first ten mode frequencies are evaluated employing various higher order theories. Furthermore, some numerical predictions are provided with classical FSDT and TSDT theories. As can be seen, most accurate results is provided by the EDZ4 approach for all considered mode frequencies. As a consequence, next parametric investigations are conducted with this kinematic assumption. For the sake of completeness, the nomenclature for the identification of the selected thickness functions can be found in Ref. [6].

Table 1. Validation of the ESL kinematic field assumption with respect to a 3D FEM solution.

f [Hz]	3D FEM	FSDT	TSDT	ED3	EDZ3	ED4	EDZ4
DOFs	579411	5046	10092	10092	12615	12615	15138
1	62.55	62.20	61.32	61.53	61.48	61.18	61.16
2	79.46	81.30	79.10	79.28	79.21	78.88	78.83
3	179.76	183.86	178.87	179.28	178.87	178.68	178.29
4	229.72	244.87	235.92	236.90	236.30	235.76	235.17
5	300.69	309.74	301.11	303.28	302.19	302.38	301.35
6	347.68	359.64	353.69	354.17	353.40	353.67	352.91
7	463.84	477.06	465.68	467.70	466.17	466.12	464.50
8	469.96	483.88	467.92	472.08	470.38	468.13	466.45
9	626.91	658.46	628.04	630.40	628.99	629.30	628.08
10	644.43	698.44	658.33	660.58	659.39	658.90	657.67

$a = 1.2\,\mathrm{m}$, $\left[\alpha_1^0, \alpha_1^1\right] \times \left[\alpha_2^0, \alpha_2^1\right] = [0, 1] \times [0, 1]$, FFFC

(30/0/45), 1st layer: trigonal material, 2nd layer: graphite-epoxy, 3rd layer: trigonal material

3.2 Parametric Investigations

Once the accuracy of the higher order model is checked, a sensitivity analysis on the same doubly-curved structure is studied. To this end, the following smooth distribution of the material orientation angle $\vartheta^{(k)}(\alpha_1, \alpha_2)$ is considered in each k-th layer, remembering that $(\alpha_1, \alpha_2) \in \left[\alpha_1^0, \alpha_1^1\right] \times \left[\alpha_2^0, \alpha_2^1\right]$:

$$\begin{aligned}\vartheta^{(k)}(\alpha_1, \alpha_2) = c_1^{(k)}\left(\left(\vartheta_1^{(k)\alpha_1} - \vartheta_0^{(k)\alpha_1}\right)|2\overline{\alpha}_1 - 1|^{p_1^{(k)}} + \vartheta_0^{(k)\alpha_1}\right) \\ + c_2^{(k)}\left(\left(\vartheta_1^{(k)\alpha_2} - \vartheta_0^{(k)\alpha_2}\right)|2\overline{\alpha}_2 - 1|^{p_2^{(k)}} + \vartheta_0^{(k)\alpha_2}\right) + \overline{\vartheta}^{(k)}\end{aligned} \tag{14}$$

where $c_1^{(k)}, c_2^{(k)} = 1$ and $p_1^{(k)}, p_2^{(k)}$ are shape parameters of the distribution, whereas $\vartheta_0^{(k)\alpha_1}, \vartheta_0^{(k)\alpha_2}$ and $\vartheta_1^{(k)\alpha_1}, \vartheta_1^{(k)\alpha_2}$ are the material angles in the center and in the boundary region of the physical domain, respectively. In addition, $\overline{\alpha}_1, \overline{\alpha}_2$ are dimensionless coordinates defined as $\overline{\alpha}_i = \left(\alpha_i - \alpha_i^0\right)/\left(\alpha_i^1 - \alpha_i^0\right)$ with $i = 1, 2$. Finally, $\overline{\vartheta}^{(k)}$ is a constant shift value of the distribution.

In Table 2 the first five mode frequencies are reported for the helicoidal panel of Eq. (12) for different values of $p_1^{(k)} = p_2^{(k)} = p^{(k)}$ and $\vartheta_1^{(k)\alpha_1} = \vartheta_1^{(k)\alpha_2} = \vartheta_1^{(k)}$. On the other hand, a null value of $\vartheta_0^{(k)\alpha_1} = \vartheta_0^{(k)\alpha_2} = 0$ is considered.

Table 2. Modal analysis of an FFFC helicoidal panel made of anisotropic materials with a continuous variation of the material orientation along the physical domain.

Mode	$\vartheta_1^{(k)}$										
f [Hz]	0	10	20	30	40	45	50	60	70	80	90
$p^{(k)} = 1$											
1	61.16	54.85	49.08	45.45	43.22	42.44	41.88	41.51	42.49	45.14	49.07
2	78.83	82.34	81.92	79.78	77.06	75.57	74.09	71.73	71.19	72.23	74.45
3	178.29	167.73	153.55	141.11	131.81	128.59	126.31	124.56	126.86	133.10	142.07
4	235.17	232.66	224.32	214.95	206.77	203.48	200.81	198.28	201.41	208.23	212.94
5	301.35	302.26	299.31	294.84	295.03	296.96	298.06	291.42	288.93	293.32	301.95
$p^{(k)} = 2$											
1	61.16	56.94	52.26	48.70	46.32	45.46	44.76	43.82	43.66	44.47	46.07
2	78.83	80.44	79.38	77.43	75.51	74.64	73.82	72.51	72.02	72.46	73.89
3	178.29	171.21	160.43	150.18	141.85	138.60	135.98	132.61	131.76	133.33	136.70
4	235.17	232.06	225.40	218.72	213.00	210.60	208.54	205.83	206.05	209.57	214.16
5	301.35	302.09	301.16	298.16	295.73	295.50	296.10	299.84	298.90	296.65	298.53

(continued)

Table 2. (*continued*)

Mode f [Hz]	$\vartheta_1^{(k)}$ 0	10	20	30	40	45	50	60	70	80	90
$p^{(k)} = 3$											
1	61.16	58.05	54.20	50.94	48.54	47.63	46.86	45.72	45.17	45.44	46.32
2	78.83	79.62	78.38	76.50	74.64	73.80	73.02	71.76	71.15	71.28	71.85
3	178.29	172.87	164.08	155.45	148.19	145.23	142.73	139.16	137.50	137.60	138.81
4	235.17	232.07	226.20	220.43	215.52	213.42	211.58	208.87	208.02	209.62	212.91
5	301.35	301.64	300.80	298.75	296.60	295.96	295.80	297.23	300.92	301.30	300.42
$p^{(k)} = 4$											
1	61.16	58.73	55.48	52.53	50.20	49.27	48.47	47.20	46.45	46.42	47.04
2	78.83	79.20	77.92	76.13	74.33	73.49	72.70	71.37	70.61	70.48	70.60
3	178.29	173.84	166.34	158.87	152.44	149.76	147.45	143.99	142.16	141.64	141.77
4	235.17	232.23	226.94	221.74	217.25	215.31	213.57	210.80	209.38	209.78	212.05
5	301.35	301.32	300.29	298.49	296.61	295.95	295.60	296.19	298.79	303.09	304.02

$a = 1.2\,\text{m},\ \left[\alpha_1^0, \alpha_1^1\right] \times \left[\alpha_2^0, \alpha_2^1\right] = [0, 1] \times [0, 1]$

(30/0/45), 1st layer: trigonal material, 2nd layer: graphite-epoxy, 3rd layer: trigonal material

As can be seen from Fig. 1, the variation of $\vartheta_1^{(k)}$ leads to a minimum value of the mode frequency, whereas the second mode is characterized by a variation between a maximum and a minimum value, which depends on the selected value of $p^{(k)}$. The introduction of a smooth variation of the material angle leads to a reduction of the vibration frequency related to the fifth and the seventh mode. Finally, the value of the eighth mode is significantly influenced by the shape parameter $p^{(k)}$.

It should be specified that, for each simulation, the mode shape vectors have been evaluated. A three-dimensional representation of the first nine mode shape vectors of the helicoidal panel has been reported in Fig. 2, setting $p^{(k)} = 1$ and $\vartheta_1^{(k)\alpha_1} = \vartheta_1^{(k)\alpha_2} = 45$ deg in Eq. (14). Since the numerical simulations are performed with the EDZ4 theory, the modal shape vectors can contain both bending and stretching deformations, which can occur especially at higher frequencies. These vibrational effects are normally predicted only by three-dimensional formulations, whereas in classical FSDT and TSDT the stretching effects are neglected. Besides, the present two-dimensional model, thanks to the adoption of a higher order displacement field, allows one to predict complex effects with a reduced computational cost.

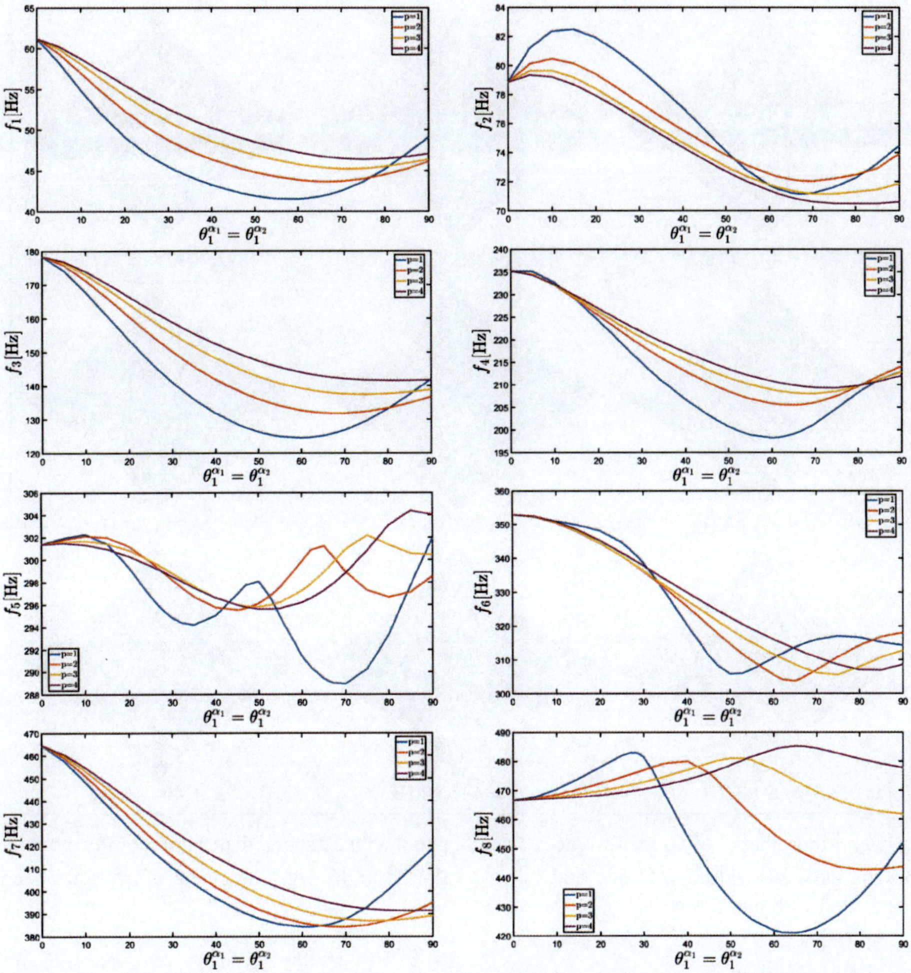

Fig. 1. Parametric investigation on the natural frequencies of an FFFC helicoidal panel for different values of the governing parameters of the adopted VAT distribution along the physical domain.

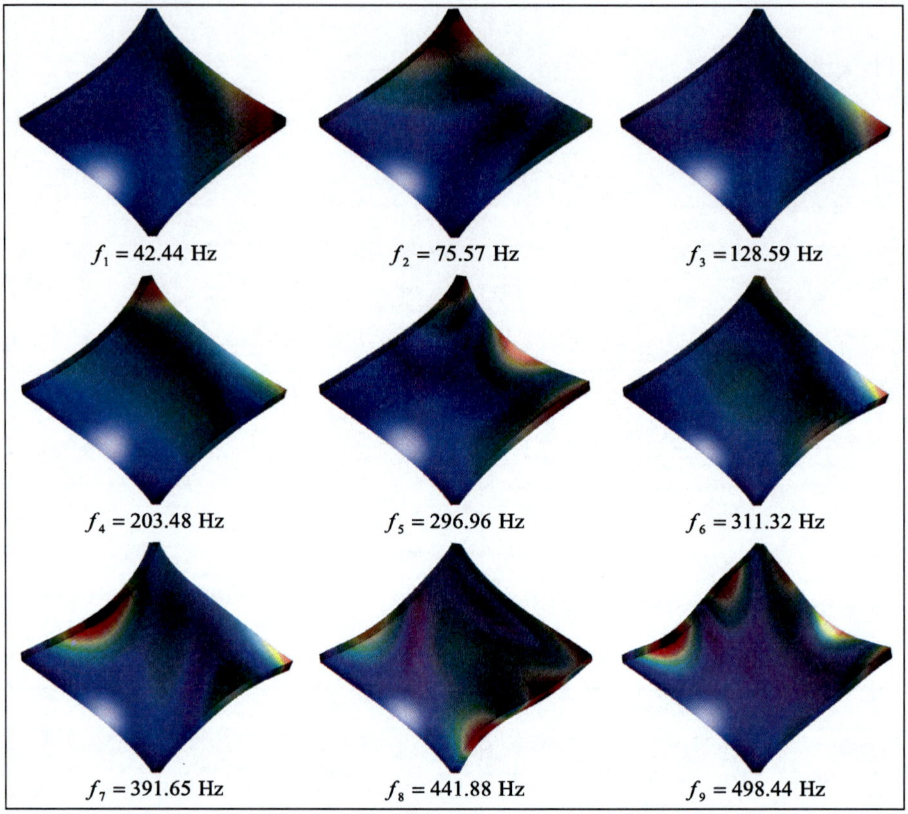

Fig. 2. Mode shapes of an anisotropic helicoidal panel with a general distribution of the material orientation angle, setting $p^{(k)} = 1$ and $\vartheta_1^{(k)\alpha_1} = \vartheta_1^{(k)\alpha_2} = 45$ deg. The modal analysis has been conducted with the EDZ4 theory.

4 Conclusion

In the present study the dynamic response of anisotropic doubly-curved shells have been studied, characterized by a continuous smooth variation of the material orientation. Following the ESL approach, a generalized expansion of the displacement field variables has been adopted, taking into account the Zig-Zag effects. In addition, a constitutive relation valid for generally anisotropic materials has been considered. The fundamental equations have been derived from the Hamilton principle, together with the boundary conditions of the problem. The GDQ method has been used for the discretization of the governing equations directly in the strong form. The formulation has been adopted for the vibration analysis of an helicoidal panel. A general relation has been provided for the variation of the material angle along the physical domain. Thanks to the computational efficiency of the formulation, a parametric investigation has been performed, showing the sensitivity of the governing parameters of the material angle distribution on the mode frequencies and shapes of the structure.

References

1. Tornabene, F., Viscoti, M., Dimitri, R.: Free vibration analysis of laminated doubly-curved shells with arbitrary material orientation distribution employing higher order theories and differential quadrature method. Eng. Anal. Boundary Elem. **152**(1), 397–445 (2023)
2. Tornabene, F., Viscoti, M., Dimitri, R., Reddy, J.N.: Higher order theories for the vibration study of doubly-curved anisotropic shells with a variable thickness and isogeometric mapped geometry. Compos. Struct. **267**, 113829 (2021)
3. Tornabene, F., Viscoti, M., Dimitri, R.: Static analysis of anisotropic doubly-curved shells with arbitrary geometry and variable thickness resting on a Winkler-Pasternak support and subjected to general loads. Eng. Anal. Boundary Elem. **140**, 618–673 (2022)
4. Tornabene, F., Viscoti, M., Dimitri, R.: Static analysis of anisotropic doubly-curved shell subjected to concentrated loads employing higher order layer-wise theories. Comput. Model. Eng. Sci. **134**(2), 1393–1468 (2023)
5. Tornabene, F., Viscoti, M., Dimitri, R.: Static analysis of doubly-curved shell structures of smart materials and arbitrary shape subjected to general loads employing higher order theories and generalized differential quadrature method. Comput. Model. Eng. Sci. **133**(3), 719–798 (2022)
6. Tornabene, F., Bacciocchi, M.: Anisotropic doubly-curved shells: higher-order strong and weak formulations for arbitrarily shaped shell structures. Esculapio, Bologna (2018)

Thrust Network Analysis of Masonry Arches and Domes of Any Stereotomy with Finite Compressive Strength: Multi-constrained Minimization Problem *versus* Stability Area Method

Danila Aita(✉), Matteo Bruggi, and Alberto Taliercio

Department of Civil and Environmental Engineering, Politecnico di Milano, Milano, Italy
{danila.aita,matteo.bruggi,alberto.taliercio}@polimi.it

Abstract. In this research, equilibrium analysis of curved structures is carried out via a computational approach based on the constrained force density method, with the aim of taking into account both the effects of a finite compressive strength of the material and any possible stereotomy of the voussoirs composing the structures. The procedure relies on independent sets of branches for networks with fixed plan projection. At each joint, a suitable set of local constraints is formulated to take into account the moment capacity. Then, a multi-constrained minimization problem is stated. In order to validate this approach, a benchmark case study is examined, i.e., a dome with arbitrary stereotomy. To this purpose, a classical semi-analytical graphical method, originally devised by the French scholar Durand-Claye is adopted. A re-visited version of this method, formulated in terms of the static theorem of Limit Analysis, is exploited in order to determine the complete set of admissible solutions with respect to both the equilibrium conditions and the strength requirements of the material.

Keywords: Funicular analysis · Masonry arches · Masonry domes · Stereotomy · Crushing

1 Introduction

In this contribution, a numerical approach based on the force density method is formulated to perform the funicular analysis of masonry arches and domes.

A peculiar aspect of this approach, with respect to classical and reformulated funicular and thrust network analysis methods [1–3], is the possibility of considering the limited compressive strength of masonry as well as the stereotomy of the voussoirs. The equilibrium of the loaded nodes of the network is imposed by assuming that they lie on the vertical straight lines passing through the centres of gravity of the blocks. Then, local constraints are set at each joint to take into account both strength requirements and stereotomy. By considering a fixed plan

S. Gabriele et al. (Eds.): IWSS 2023, LNCE 437, pp. 150–160, 2024.
https://doi.org/10.1007/978-3-031-44328-2_16

projection of the grid, a multi-constrained minimization/maximization problem arises, which is solved through sequential convex programming and provides the minimum/maximum values of the horizontal thrust. This method can handle networks of general topology and can be viewed as a development of the procedure presented in [4].

Based on the Static Theorem of Limit Analysis, the funicular method here proposed is not limited to Heyman's assumptions [5], since a more restrictive condition on the compressive stresses is imposed. A review of the scientific literature on the influence of the compressive strength in assessing the bearing capacity of masonry structures is out of the scope of this paper. To the aims of this research, it is sufficient to recall that Limit Analysis under Heyman's hypotheses can overestimate the bearing capacity of masonry arches and vaults, particularly for some typologies of masonry structures, among which segmental arches and domes with depressed profile. The effects of a bounded compressive strength on masonry structures is studied by different authors, as regards theoretical issues [6], walls [7], or arches [8,9]. Among the various methods, some of which are limited to the assumption of geometric criteria [8], the semi-analytical graphic procedure proposed by Durand-Claye [10,11], re-elaborated by [12–14], allows one to assess the bearing capacity of symmetric arches or domes of revolutions by considering in a simple, albeit rigorous, way the effects of the limited strength of masonry. As regards stereotomy, its influence on the equilibrium of the structure is very marked when a finite or zero friction is assumed [15]. However, the minimum and maximum crown thrust values are dependent on stereotomy even under the hypotheses of finite compressive strength, nil tensile strength, and infinite friction coefficient, as done in this research. To validate the funicular numerical method, a modern re-visitation of Durand-Claye's method is chosen [12–14]. A simple case study, that is a dome of revolution subjected to axisymmetric loads, is examined by considering both the stereotomy of the voussoirs and a limited compressive strength of the material. The results obtained via the two methods are compared and discussed.

2 Funicular Networks of Min/Max Thrust Using the Force Density Method and Mathematical Programming

2.1 Force Density Method

The "force density method" [16] is used to deal with the equilibrium of funicular networks. A spatial network is made of m branches, which undergo axial forces only, and $n_s = n + n_f$ nodes. The axes of the Cartesian reference system with origin O are denoted by x, y, and z. Hence, $\mathbf{x}_s$, $\mathbf{y}_s$, $\mathbf{z}_s$ collect the coordinates of the n_s nodes: the vectors $\mathbf{x}$, $\mathbf{y}$, $\mathbf{z}$ refer to the n unrestrained nodes, i.e. those subject to external loads; $\mathbf{x}_f$, $\mathbf{y}_f$, $\mathbf{z}_f$ collect the n_f restrained nodes, i.e. those where reactions arise. Let $\mathbf{C}_s$ be the connectivity matrix that defines the topology of the grid: it can be split into two submatrices $\mathbf{C}$, referring to the unrestrained nodes, and $\mathbf{C}_f$, referring to the restrained ones. The vectors $\mathbf{u}$, $\mathbf{v}$, $\mathbf{w}$ collect the difference in the coordinates of the nodes at the ends of each branch along the axis x, y, z, respectively. They can be expressed as:

$$\mathbf{u} = \mathbf{C}_s\mathbf{x}_s, \qquad \mathbf{v} = \mathbf{C}_s\mathbf{y}_s, \qquad \mathbf{w} = \mathbf{C}_s\mathbf{z}_s. \tag{1}$$

The force density vector $\mathbf{q} = \mathbf{L}^{-1}\mathbf{s}$ stores the force-to-length ratios for all the branches of the spatial network, being $\mathbf{s}$ the vector that collects the forces in the m branches. The lengths of the branches $l_i = \sqrt{u_i^2 + v_i^2 + w_i^2}$ are gathered in the vector $\mathbf{l}$, and $\mathbf{L} = \text{diag}(\mathbf{l})$. In this study, only gravity loads are considered. Vertical point forces are prescribed at the unrestrained nodes through vector $\mathbf{p}_z$. Because of the introduction of the force density vector, the equilibrium of the unrestrained nodes can be written by means of linear equations that are uncoupled along the three spatial directions. As discussed in [2,17,18], for networks with fixed plan projection subject to vertical forces only, the horizontal equilibrium of the loaded nodes reads:

$$\begin{bmatrix} \mathbf{C}^T\text{diag}(\mathbf{C}_s\mathbf{x}_{s0}) \\ \mathbf{C}^T\text{diag}(\mathbf{C}_s\mathbf{y}_{s0}) \end{bmatrix} \mathbf{q} = \begin{bmatrix} \mathbf{0} \\ \mathbf{0} \end{bmatrix}, \tag{2}$$

where the vectors $\mathbf{x}_{s0}$ and $\mathbf{y}_{s0}$ collect the fixed x and y coordinates of the nodes. Equation (2) implies that $m - r$ independent force densities $\overline{\mathbf{q}}$ can be retrieved, being r the rank of the coefficient matrix. The r dependent force densities $\widetilde{\mathbf{q}}$ can be expressed as:

$$\widetilde{\mathbf{q}} = \mathbf{B}\overline{\mathbf{q}} + \mathbf{d}, \tag{3}$$

where $\mathbf{B}$ and $\mathbf{d}$ are constant matrices whose entries can be derived by applying Gauss-Jordan elimination to Eq. (2), see [17]. Upon introduction of $\mathbf{Q} = \text{diag}(\mathbf{q})$, the equilibrium along the z axis, i.e.:

$$\mathbf{C}^T\mathbf{Q}\mathbf{C}\mathbf{z} + \mathbf{C}^T\mathbf{Q}\mathbf{C}_f\mathbf{z}_f = \mathbf{p}_z \tag{4}$$

can be solved to compute the vertical coordinates of the unrestrained nodes.

2.2 Forces and Eccentricities at the Joints

In Fig. 1, the i-th joint between two adjacent voussoirs is represented: C_i is the centroid of the rectangular section, whereas P_i is the point where the line of action of the funicular force $\mathbf{F}_i$ crosses the joint. Let $\boldsymbol{\xi}_i$, $\boldsymbol{\eta}_i$ denote the principal axes of inertia of the section and $\mathbf{n}_i$ its normal. The section has dimensions $l_{i,\xi} \times l_{i,\eta}$.

Denoting by $\mathbf{e}_x$, $\mathbf{e}_y$, $\mathbf{e}_z$ a triplet of unit vectors aligned with x, y, z, one has:

$$\mathbf{F}_i = s_i \left(\frac{u_i}{l_i} \mathbf{e}_x + \frac{v_i}{l_i} \mathbf{e}_y + \frac{w_i}{l_i} \mathbf{e}_z \right) = q_i \left(u_i \, \mathbf{e}_x + v_i \, \mathbf{e}_y + w_i \, \mathbf{e}_z \right). \tag{5}$$

Denoting by $\mathbf{N}_i$ the normal component of $\mathbf{F}_i$, its magnitude may be straightforwardly computed as $N_i = \mathbf{F}_i \cdot \mathbf{n}_i$. Its eccentricity with respect to $\boldsymbol{\xi}_i$ can be

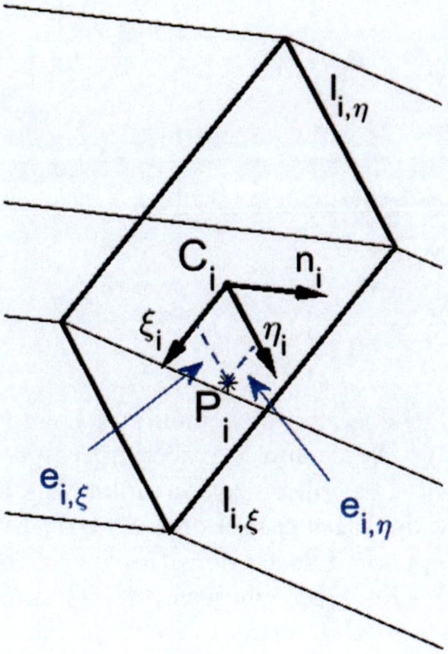

Fig. 1. The i-th joint between two adjacent voussoirs.

found by evaluating the moment of $\mathbf{N}_i$ about the same axis, $M_{i,\xi}$, and scaling by N_i. Recalling that the shear component of the force $\mathbf{F}_i$ does not provide any contribution to $M_{i,\xi}$, one may conveniently compute:

$$e_{i,\xi} = \text{abs}\left(\frac{M_{i,\xi}}{N_i}\right), \text{with } \frac{M_{i,\xi}}{N_i} = \frac{\boldsymbol{\xi}_i \cdot (\mathbf{r}_i \times \mathbf{F}_i)}{N_i}, \tag{6}$$

where $\mathbf{r}_i$ is the vector drawn from C_i to any point belonging to the line of action of $\mathbf{F}_i$ (in particular, any of the vertices of the funicular network lying in the two adjacent voussoirs), and abs$(\cdot)$ stands for the absolute value of the scalar argument. Similar relationships hold for $e_{i,\eta}$. They are not explicitly given here, since $e_{i,\eta} = 0$ for the numerical example considered next.

2.3 Optimization Problem

Within the framework of Limit Analysis it may be useful to study the range of admissible solutions, looking for both the minimum and maximum thrust. Aiming at minimizing/maximizing the overall thrust, the problem that follows is addressed:

$$
\begin{cases}
\min\limits_{\substack{\overline{\mathbf{q}} \le 0 \\ \mathbf{z}_f^{min} \le \mathbf{z}_f \le \mathbf{z}_f^{max}}} \quad \pm f = \sum_h^{n_f} \left(R_{x\,h}^2 + R_{y\,h}^2\right) & \text{(7a)} \\
\text{s.t. } \widetilde{\mathbf{q}} = \mathbf{B}\overline{\mathbf{q}} + \mathbf{d}, & \text{(7b)} \\
\mathbf{C}^T\mathbf{Q}\mathbf{C}\mathbf{z} + \mathbf{C}^T\mathbf{Q}\mathbf{C}_f\mathbf{z}_f = \mathbf{p}_z, & \text{(7c)} \\
\widetilde{q}_k \le 0 \quad \text{for } k = 1...r, & \text{(7d)} \\
\left(\dfrac{N_i}{\sigma_c\left(l_{i,\eta} - 2e_{i,\xi}\right) l_{i,\xi}}\right)^2 \le 1 \quad \text{for } i = 1...m. & \text{(7e)}
\end{cases}
$$

In the objective function of the above statement, R_{xh} and R_{yh} are the components of the reaction along the x and y direction, respectively, at the h-th restrained node, see e.g. [4]. The minimization unknowns consist of any reduced set of independent force densities $\overline{\mathbf{q}}$ and of the vertical coordinates of the restrained nodes $\mathbf{z}_f$. The dependent force densities $\widetilde{\mathbf{q}}$ are computed from the independent set $\overline{\mathbf{q}}$ by means of Eq. (7b), whereas Eq. (7c) enforces the equilibrium of the unrestrained nodes in the vertical direction. The occurrence of any positive independent force density is prevented by the side constraints on $\widetilde{\mathbf{q}}$, along with the set of local enforcements in Eq. (7d). Side constraints on $\mathbf{z}_f$ prescribe lower and upper bounds to the vertical coordinates of the n_f restrained nodes, which are $\mathbf{z}_f^{min}$ and $\mathbf{z}_f^{max}$ respectively. Denoting by $\sigma_c \le 0$ the strength in compression, Eq. (7e) is used to prevent crushing at the i-th no-tension joint. A uniform distribution of compressive stress is assumed to occur in a limited portion of the section of each joint, i.e. the area of size $(l_{i,\eta} - 2e_{i,\xi}) \times l_{i,\xi}$, to withstand the eccentric normal force $N_i \le 0$, see in particular [19] and [12]. By adopting $\sigma_c \to -\infty$ in Eq. (7e), the formulation in Eq. (7) retrieves networks of minimum/maximum thrust that comply with Heyman's assumptions. For finite values of σ_c, the assumption of infinite strength is removed at all joints.

As investigated in [4], the problem in Eq. (7) can be efficiently addressed through techniques of sequential convex programming that were originally conceived to handle problems of size optimization, see also the multi-constrained application in [20]. In more detail, the gradient-based Method of Moving Asymptotes (MMA) [21] is herein adopted.

3 Thrust Lines of Min/Max Thrust Using the Durand-Claye's Method

Durand-Claye's method was originally conceived for the analysis of symmetric masonry arches subjected to symmetric vertical loads [10]; then, it was extended to assess the equilibrium of masonry domes of revolution [11]. In [12–14] a modern version of this method has been proposed, by re-interpreting it in terms of internal forces, and considering the theoretical background of Limit Analysis. According to this method, for masonry arches the complete set of admissible solutions - with respect to both the equilibrium conditions and the strength capacity of the material - is represented by a region of the (f, e_0) plane, called stability area (Fig. 2a), where f denotes the horizontal thrust acting at point P_0 of the vertical cross section at the crown (Fig. 2b), while e_0 is the eccentricity of thrust f with respect to the cross section's center of gravity, C_0. When the friction is assumed to be infinite, as done to the purposes of the current contribution, the stability area coincides with the rotational domain. In symmetric masonry arches, at each joint i identified by a proper coordinate (for example by angle θ_i in Fig. 2c), imposing the equilibrium conditions to the portion of the arch comprised between the ideal vertical cross section and joint i (Fig. 2b, c) allows one for formally writing the expressions of the internal forces, $M_{i,\xi}$ and N_i in terms of f, e_0 and θ_i:

$$M_{i,\xi} = M_{i,\xi}(f, e_0, \theta_i) \text{ and } N_i = N_i(f, \theta_i). \tag{8}$$

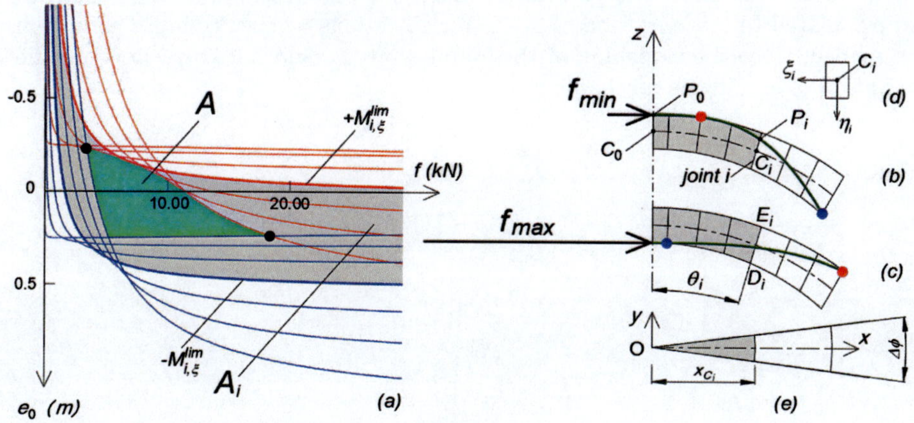

Fig. 2. Durand-Claye's method: (a) stability area; thrust lines corresponding to the minimum (b) or maximum (c) crown thrust; (d) generic section; (e) one of the lunes composing a dome, according to the slicing technique.

Moreover, at each joint i (Fig. 2b, d), the limit bending moment, $M_{i,\xi}^{lim}$ can be determined as a function of the normal force, N_i, of the dimensions of the cross-section, $l_{i,\xi}$ and $l_{i,\eta}$, and of the masonry compressive strength, σ_c, by assuming nil tensile strength and limited compressive strength:

$$M_{i,\xi}^{lim} = \frac{N_i\,(N_i - l_{i,\xi}\, l_{i,\eta}\, \sigma_c)}{2\, l_{i,\xi}\, \sigma_c}. \tag{9}$$

Thus, at generic i-th joint, the bending moment is bounded by the inequalities

$$-M_{i,\xi}^{lim} \le M_{i,\xi} \le +M_{i,\xi}^{lim}, \tag{10}$$

which define the region A_i in the plane (f, e_0) (Fig. 2a). This corresponds to the admissible solutions - in terms of equilibrium and strength requirements - for the portion of arch comprised between the vertical crown section and joint i.

By repeating the procedure for all the joints i, the intersection between all the areas A_i allows for obtaining the area of stability A for the entire arch.

If the stability area is a non-vanishing region, as the green region plotted in Fig. 2a, the arch is stable, and the values of the minimum and maximum thrusts, f_{min} and f_{max}, are determined as well as their eccentricity at the crown, e_0, by considering the coordinates (f, e_0) of the two black points in Fig. 2a. The related thrust lines are plotted in Fig. 2b, c, respectively. When the stability area shrinks to a single point, only one admissible solution (f, e_0) is found: the limit condition is attained, and the corresponding collapse mechanism or crushing condition is identified.

The method is adaptable also to masonry domes of revolution subjected to axi-symmetric loads, by applying the "slicing technique", i.e., by assuming that the cracked dome is divided into a number of lunes (Fig. 2e) behaving as independent half-arches [22,23]. The stability area method is first drawn for the single "lune" tackled as an arch of variable width; then, as observed in [14], the procedure related to masonry domes needs modifications when a collapse condition is searched, since the vanishing of the stability area could not correspond to the activation of a collapse mechanism.

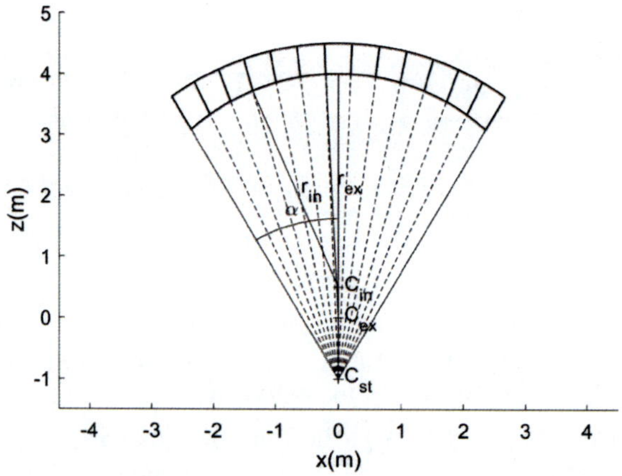

Fig. 3. Reference section of a dome of revolution.

4 Numerical Example

A dome of revolution is addressed under self-weight, see Fig. 3. The intrados lies along a circle with center at $C_{in} = (0, 0, 0.5)$ m and radius $r_{in} = 3.5$ m. The extrados lies along a circle having center at $C_{ex} = (0, 0, 0)$ m and radius $r_{ex} = 4.5$. One half of the angle of embrace is $\alpha = 30°$. A material with unit

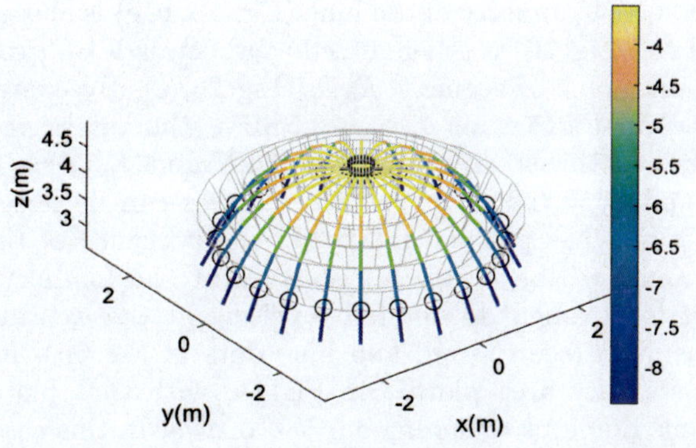

(a)

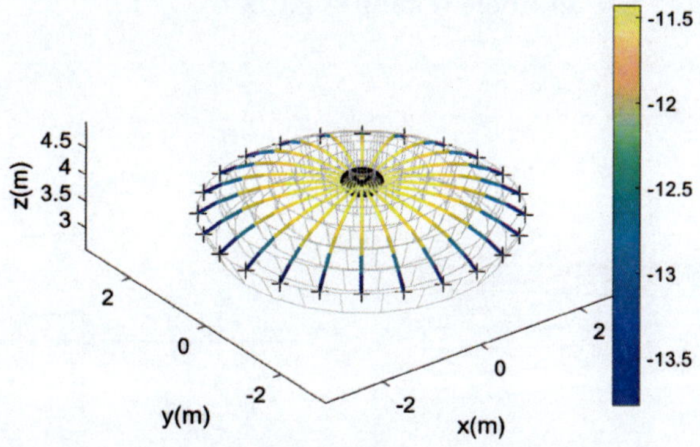

(b)

Fig. 4. Funicular networks for a dome with $\sigma_c = -0.5\,\text{MPa}$: (a) minimum thrust solution ($R_h = 3.49\,\text{kN}$); (b) maximum thrust solution ($R_h = 11.33\,\text{kN}$).

weight $\gamma_m = 15\ \text{kN/m}^3$ is considered. For the sake of simplicity, a network made of meridians only is considered. Excluding the keystone, each one of the twenty-four lunes is made of six voussoirs, whose stereotomy is defined by radial lines originating from the point $C_{st} = (0, 0, -1)\,\text{m}$. To test the method, a very small value of the compressive strength is herein considered. In Fig. 4, the funicular networks of minimum and maximum thrust found for $\sigma_c = -0.5\,\text{MPa}$ are given, along with the forces in the branches of the network in kN. In these picture, only the cross-section of the joints controlled within the optimization procedure are given (in grey). The symbols $\circ$ and $+$ stand for joint sections whose crossing branches activate a strength constraint. The former symbol is used when the branch intersects the cross-section of the joint below the centroid, the latter otherwise. The achieved results are in good agreement with those obtained via Durand-Claye's method, despite marginal discrepancies related to the geometry of blocks and joints. For the applications with Durand-Claye's method, an absolute Cartesian system $(O; x, y, z)$, with $O \equiv C_{st}$, and the (x, z) plane belonging to the vertical plane of symmetry of the lune (Fig. 2b, c, e) is chosen. The generic cross section (joint i, Fig. 2d) is schematized as a rectangle of area $l_{i,\eta} \times x_{C_i} \Delta\phi$, where C_i is the midpoint of segment D_iE_i (Fig. 2c, e). By examining the stability area plotted in Fig. 5a, for $\sigma_c = -0.5$ MPa, the values of the minimum and maximum crown thrust are $f_{min} = 3.58$ kN and $f_{max} = 11.75$ kN, with $e_0 = -0.181$ m and $e_0 = 0.0435$ m, respectively. As can be observed from the corresponding thrust lines plotted in Fig. 5b, the influence of the compressive strength is apparent, as the centers of pressure at the joints where the limit bending moment is attained do not fall precisely at the extrados/intrados of the joint, but inside it (see the red and blue dots in Fig. 5b). It is interesting to compare the stability area plotted in Fig. 5a, with that plotted in Fig. 2a, related to the same dome by assuming $\sigma_c = -10$ MPa. In this case the influence of the compressive strength is not evident as regards the position of the centers of pressure (see the red and blue dots in Fig. 2b, c).

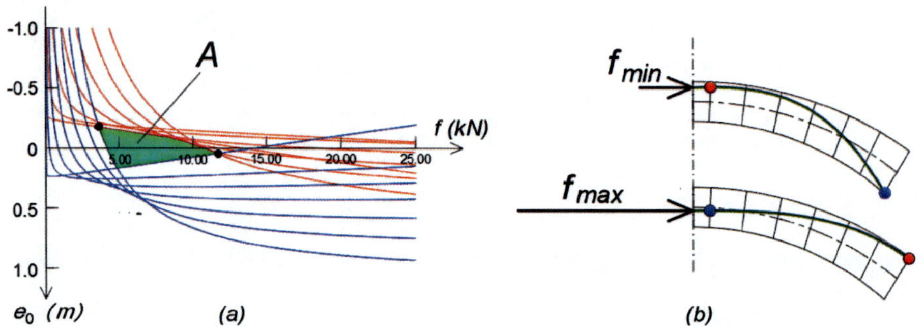

Fig. 5. Durand-Claye's method: (a) stability area of a dome with $\sigma_c = -0.5$ MPa; (b) thrust lines corresponding to the min/max values of the crown thrust.

5 Conclusions

A numerical method for the equilibrium analysis of masonry arches and domes has been proposed combining the constrained force density method with tools of mathematical programming, with the main aim of taking into account both the effects of a finite compressive strength of the material and any possible stereotomy of the voussoirs composing the curved structure. A dome with arbitrary stereotomy has been investigated modeling the lunes as independent arches by assuming two different values of the compressive strength. A re-visited version of the classical semi-analytical graphical method of Durand-Claye has been implemented for the assessment of the achieved preliminary results. The adopted slicing technique cannot account for the beneficial contribution of the hoop forces. Further investigations will address modeling the dome with a network made by both meridians and parallels, also extending the approach to problems of optimal reinforcement, see for instance [24].

References

1. O'Dwyer, D.: Funicular analysis of masonry vaults. Comput. Struct. **73**(1–5), 187–197 (1999)
2. Block, P., Lachauer, L.: Three-dimensional funicular analysis of masonry vaults. Mech. Res. Commun. **56**, 53–60 (2014)
3. Marmo, F., Rosati, L.: Reformulation and extension of the thrust network analysis. Comput. Struct. **182**, 104–118 (2017)
4. Bruggi, M.: A constrained force density method for the funicular analysis and design of arches, domes and vaults. Int. J. Solids Struct. **193–194**, 251–269 (2020)
5. Heyman, J.: The stone skeleton. Int. J. Solids Struct. **2**(2), 249–279 (1966)
6. Lucchesi, M., Padovani, C., Zani, N.: Masonry-like solids with bounded compressive strength. Int. J. Solids Struct. **33**(14), 1961–1994 (1996)
7. Pina-Henriques, J., Lourenço, P.B.: Masonry compression: a numerical investigation at the meso-level. Eng. Comput. **23**, 382–407 (2006)
8. Harvey, W.J.: Application of the mechanism analysis to masonry arches. Struct. Eng. **66**(5), 77–84 (1988)
9. Caporale, A., Luciano, R.: Limit analysis of masonry arches with finite compressive strength and externally bonded reinforcement. Compos. B Eng. **43**(8), 3131–3145 (2012)
10. Durand-Claye, A.: Note sur la vérification de la stabilité des voûtes en maçonnerie et sur l'emploi des courbes de pression. Ann. Ponts Chaussées **13**, 63–93 (1867)
11. Durand-Claye, A.: Vérification de la stabilité des voûtes et des arcs. Applications aux voûtes sphériques. Ann. Ponts Chaussées **19**, 416–440 (1880)
12. Aita, D., Barsotti, R., Bennati, S.: Equilibrium of pointed, circular, and elliptical masonry arches bearing vertical walls. J. Struct. Eng. **138**(7), 880–888 (2012)
13. Aita, D., Barsotti, R., Bennati, S.: Looking at the collapse modes of circular and pointed masonry arches through the lens of durand-claye's stability area method. Arch. Appl. Mech. **89**(8), 1537–1554 (2019)
14. Aita, D., Barsotti, R., Bennati, S.: Studying the dome of pisa cathedral via a modern reinterpretation of durand-claye's method. J. Mech. Mater. Struct. **14**, 603–619 (2019)
15. Aita, D.: Between stereotomy and mechanics: joints inclination and minimum thickness in frictionless pointed and circular arches. Int. J. Mason. Res. Innov. **7**(1–2), 61–88 (2022)
16. Schek, H.: The force density method for form finding and computation of general networks. Comput. Methods Appl. Mech. Eng. **3**(1), 115–134 (1974)
17. Liew, A., Avelino, R., Moosavi, V., Van Mele, T., Block, P.: Optimising the load path of compression-only thrust networks through independent sets. Struct. Multidiscip. Opt. **60**(1), 231–244 (2019)
18. Bruggi, M., Lógó, B., Deák, Z.: Funicular analysis of ribbed masonry vaults: a case study. Int. J. Archit. Herit. **16**(12), 1809–1823 (2022)
19. Fantin, M., Ciblac, T.: Extension of thrust network analysis with joints consideration and new equilibrium states. Int. J. Space Struct. **31**(2–4), 190–202 (2016)
20. Bruggi, M., Laghi, V., Trombetti, T.: Stress-based form-finding of gridshells for wire-and-arc additive manufacturing considering overhang constraints. Eng. Struct. **279**, 115654 (2023)
21. Svanberg, K.: The method of moving asymptotes-a new method for structural optimization. Int. J. Numer. Methods Eng. **24**(2), 359–373 (1987)

22. Heyman, J.: Equilibrium of Shell Structures. Oxford University Press, Oxford (1977)
23. Como, M.: Statics of Historic Masonry Constructions. Solid and Structural Mechanics, Springer, Cham (2017). https://doi.org/10.1007/978-3-319-54738-1
24. Bruggi, M., Taliercio, A.: Optimal strengthening of no-tension structures with externally bonded reinforcing layers or ties. Struct. Multidiscipl. Optim. **55**(5), 1831–1846 (2017)

Corrugated Shells: A Reinforced Concrete Roof Designed by Aldo Favini in Sesto S. Giovanni

M. Lai(✉), R. Argiolas, F. Musanti, D. Meloni, E. Reccia, and A. M. Cazzani

Dipartimento di Ingegneria Civile Ambientale e Architettura, Università degli Studi di Cagliari, via Marengo 2, Cagliari, Italy
matteolai@unica.it

Abstract. Aldo Favini, an exponent of the Italian structural school, was entrusted with the design of a fuel station in Sesto S. Giovanni for Aquila s.p.a. company. He envisaged a peculiar structure: a thin cylindrical vault made of reinforced concrete, whose external shape was sinuously corrugated with a thickness within the value of 60 mm. The corrugated roof is supported by a stiff hollow transverse beam, which rests on a V-shaped pillar and a slender column. Unfortunately, this structure has been demolished. The proposed corrugation shows an interesting enhancement "by shape" in structural behaviour since the shape increases the stiffness and the bearing capacity with respect to a plane surface, allowing for such a reduced thickness. Corrugated shells are also known for the iconic dome of the "Palazzetto dello Sport" in the Flaminio district of Rome, whose rim is gently corrugated and provides structural enhancements. This paper aims to give a new gaze to the tradition of shell construction in Italy, where the shape is wisely adapted to improve structural performances efficiently. The development of analytical and numerical methods to analyse such structures is crucial, especially when a safety assessment is required before a refurbishment to avoid disgraceful decisions like the demolition of the "Aquila" fuel station. Furthermore, although nowadays the building of corrugated concrete shells is decreasing, new construction techniques of 3d concrete printing make readily available the realization of sophisticated shapes, where corrugation should be fully included because of their natural predisposition to improve the mechanical behaviour of shape-resistant structures.

Keywords: Concrete shells · Corrugated shells · Shells · Favini

1 The Fuel Station Design

1.1 Introduction

Aldo Favini, an exponent of the Italian structural school [1] was commissioned to design a fuel station in Sesto S. Giovanni for the Aquila SpA company. An episode which occurred at a construction site triggered the idea for "Aquila" fuel station rooftop. Favini was strolling above a provisional corrugated metal sheet covering a hole; despite its thinness, the sheet was carrying much more load than its own self-weight. Since the materials' mechanical properties did not change, it was corrugation which raised the load-bearing capacity only because of its shape (Fig. 1).

S. Gabriele et al. (Eds.): IWSS 2023, LNCE 437, pp. 161–170, 2024.
https://doi.org/10.1007/978-3-031-44328-2_17

Fig. 1. External view of the fuel station in Sesto san Giovanni taken from [2].

The architecture of the "Aquila" fuel station in Sesto San Giovanni consists of two main parts: a closed, stereotomic, parallelepiped volume and an open-air portion on the front, covered by a light canopy supported by two pillars. The single-level rear building consists of three parallel self-contained spans, characterized by large glass windows at the front, overlooking the fuel distribution facilities and the car transit area. Even the lateral prospects have several full-height openings, but being less significant, show their secondary nature. Contrary to the stone solidity of the service block, the canopy is open and perceptively light. It is made of a particular undulated roof made of reinforced concrete, supported by two pillars that bear the hollow beam on which the roof stands. The beam on its outward side follows lengthwise the wavy shape of the roof. The corrugation is the same for the intermediate span, whereas in the cantilever part it has a continuous taper, enhancing the lightweight perception.

The two pillars placed parallel to the wide frontal window break the fuel pumps serially, are aesthetically autonomous thanks to their different shapes: the first is V-shaped, while the second has a circular cross-section. The character of the supporting system seems to resemble the name of the city —Sesto— in Roman cardinal numbers (VI). Two signs of the Aquila brand stand on the rooftop; its typeface is *sans serifs* (*carattere bastone*) and threadlike.

Favini envisaged a peculiar structure: a thin cylindrical vault made of reinforced concrete, whose exterior shape was sinuously corrugated. Its thickness is everywhere within the value of 60 mm. The shell is described in [3] by the designer himself and in [4] as an outstanding example of structural optimisation. Images of the building's exterior after its completion are shown in Figs. 2 and 3, courtesy of "Fondazione Favini" archives.

The original blueprints are now missing; nevertheless, some pieces of information can be found in Favini's own descriptions, and some historical memories are preserved in Favini's archives.

The method to replicate the wavy shape of the canopy is reported in [4]. The roof can be divided into two parts. The first span is clamped on one end, while it is supported by a transversal beam on the other one. It is a cylindrical surface where corrugation is kept constant with an amplitude of 1 m. The second span is instead overhanging, and corrugation is no longer constant but progressively varies until it disappears. The corrugated tapered shape was probably devised using cardboard models; nevertheless, a plot is reported in Fig. 7 and by using it, the points belonging to the shell are pinpointed in the cross sections and in the transversal section. Besides, the reinforced bars layout is also shown in Fig. 7.

The proposed corrugation shows attractive enhancements in the structural behaviour, like the increase in stiffness compared to a plane surface. Furthermore, enhanced capacity against horizontal loads like wind and seismic actions should be analysed appropriately. Moreover, the surface has been slightly tilted upwards, about $1.5 \div 2.0°$, which is beneficial for rainwater disposal and for the stress state. Unfortunately, the gasoline station was dismantled and destroyed in the 70's.

A comprehensive survey of the existing literature about the mechanical behaviour of shells can be found in [2, 5]; a particular attention to design of concrete shell buildings is drawn in [6–9]. An insight about mechanical behaviour of edge-corrugated dome, inspired by the work of Nervi, is addressed in [10–13].

1.2 Geometric Generation of the Model

A procedural modelling approach for the modelling of the carport designed by Favini is hereby presented, where the modeller Rhinoceros and its graphical development environment GrassHopper is used. This choice is dictated by the desire to carry out future analyses on the static capacities of this roof, even in the presence of dimensional variations, in order to identify the variation of resistance as a function of the various dimensional characteristics.

In fact, the procedural approach makes it possible, once the workflow of modelling operations to be carried out has been defined, to vary input parameters such as dimensions and shapes while leaving the shape generation process unchanged and avoiding the repetition of all manual modelling operations.

The first step for the development of the modelling algorithm is the definition of the main shape that characterises the surface, *i.e.*, the corrugation and its tapering at one edge.

An initial analysis of the available survey reveals a profile similar to a cosine function; in the cantilevered part, the amplitude of the waves decreases the farther away from the support until it reaches the linear configuration at the end. A second hypothesis is that the real form can be approximated by a module drawn using Bezier curves, with horizontal tangents at the end points.

In order to verify the correspondence between the real shape detected and the assumptions of the approximate shapes, the algorithm has been set up to generate the surface of the canopy from different profiles, selectable by the user; this way, not only it is possible to make a geometric comparison between the various profiles, but also to analytically study the variation in resistance given by the change in shape.

Fig. 2. External view of the fuel station in Sesto san Giovanni, courtesy of Favini's trust.

Fig. 3. External view of the fuel station in Sesto san Giovanni, courtesy of Favini's trust.

To make the three corrugations parameterizable and comparable, a unit module with a half period of length and amplitude equal to one unit has been assumed for the surface generation. In the case of the numerical data derived from the the actual form survey, the values are normalised, preserving the curve's profile but standardising its general dimensions; the points are then used as control points for an interpolation curve conforming to the actual geometry. Figure 4 shows the comparison between the three generated shapes, from which it is evident that the Bezier curve has a better fitting to the real shape.

The next modelling phase concerns the extrusion of the newly generated profiles, with constant shape in one direction and tapering in the opposite direction; the first extrusion, with no dimensional changes, is a simple horizontal extrusion, the extent of which is adjusted by a numerical parameter set by the user. As far as the tapering part is concerned, since this has a linear progression, it is sufficient to set a limited number of profiles with variable amplitudes, from the maximum to a null value; also in this case the user decides the extension of the cantilevered part, and the algorithm will generate the tapering by means of a *loft operation* based on a minimum of three values: maximum amplitude in the starting edge, half the amplitude in the middle of the cantilever and null amplitude in the opposite end.

The result is a half modulus that can be mirrored with respect to a vertical plane perpendicular to the plane of the curve, to obtain a complete modulus (period); this can then be repeated for a number of times set by the user. The surface obtained, corresponding to the surveyed intrados, can be extruded to generate the thickness of the roof, freely set by the user.

The process described can be schematised as in Fig. 5, while the modelling result is shown in Fig. 6.

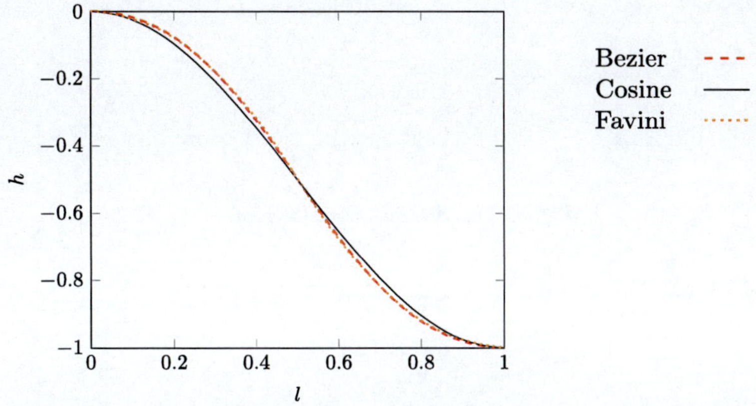

Fig. 4. Comparison of the generated curves: red dashed line, Bezier based profile, blue solid line, Cosine based profile and orange dotted line, Favini's profile. The axes are dimentionless.

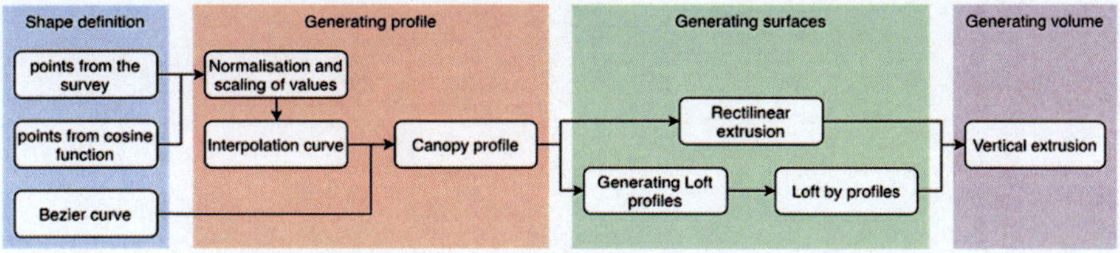

Fig. 5. Workflow for the procedural modelling of the canopy.

2 Simplified Structural Analysis of the Canopy

This section discusses the linear elastic analysis performed to address the structural assessment of Favini's canopy. The authors are fully aware that this kind of investigation cannot give a complete understanding of the structure behaviour; nevertheless, it can provide some practical considerations and perhaps an insight into conception and design. In fact, linear analyses were only possible in the 50s; besides, the complexity of shell design was often reduced by designers to hand-manageable simplifications. In this light, the following values for mechanical data are assumed: Young's modulus: 30 GPa, Poisson's modulus: 0.25, and volume density: 2500 kg/m^3.

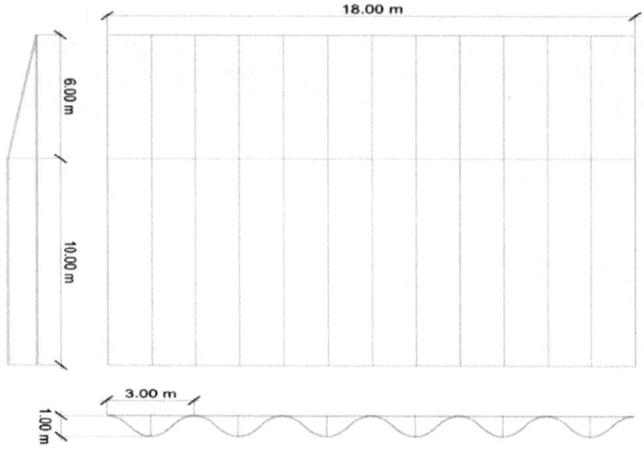

Fig. 6. Plan dimensions of the roof.

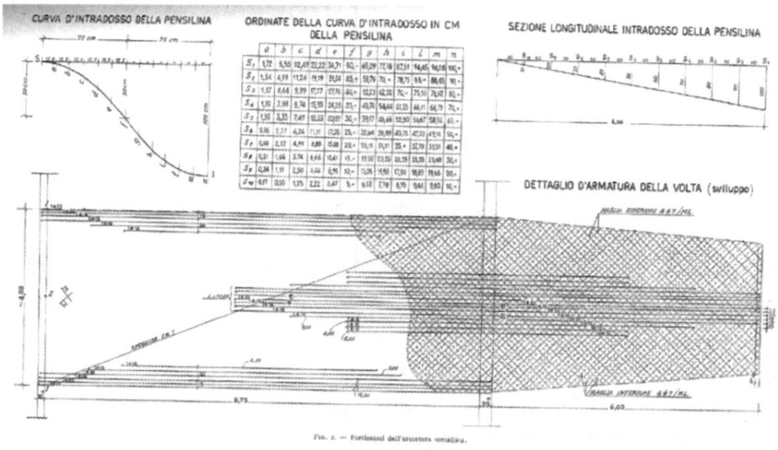

Fig. 7. Blueprint of the reinforcement bars positions, taken from [4].

For the sake of completeness, data on the construction materials (steel and concrete) used at the time of construction have been inferred through the statistical analyses of material tests carried out by that time in official University Laboratories ([14–17]).

The analysed roof has been perfectly clamped on one end, at the rear building, while the other edges are free. The intermediate support provided by the hollow beam has been modelled in a simplified way by distributed hinges. As a matter of fact, it seems reasonable that the designer conceived the shell as a continuous beam clamped on one end with an intermediate support, where only displacements are retained.

Favini reported the sketch of the reinforcement bars position in his original article see Fig. 7. A reinforcement mesh is placed equally over the entire surface area. Additionally, more reinforcement bars are placed where the tensile stresses occur, in the intrados and extrados part of the first span, while in the second additional bars are placed only at the extrados. By considering that the overhanging length is approximately equal to 6 m, the anchorage lengths are extended almost to the middle of the first span.

Let us consider now the stress positive if there is a compressive state and negative if there is a tensile state; then, the maximum value of the compressive stress in the longitudinal direction (x), disregarding stress concentrations due to restrains, is about -1.00 N/mm^2, which occurs nearby the support beam in the lower part of the waves; whereas the maximum tensile stress value, located near the support in the upper waves is about shell side with a value of 1.20 N/mm^2. Besides, the maximum compressive stress value in the transversal direction (y) is roughly –0.91 N/mm^2, whereas the maximum tensile value is 0.80 N/mm^2. Assuming the average compressive value of 20 N/mm^2, 1.20 N/mm^2 is an appropriate assumption for the maximum tensile stress. Where the maximum admissible tensile stress is roughly reached, cracks are possible but transversal reinforcement bars are able to limit the cracks opening. Figures 8 and 9 depict the contour plot of the stresses, and Fig. 10 and Fig. 11 show respectively the countour plot of the vertical displacements and the deformed shape of the roof when it is loaded with only a gravity load. It shows that the maximum vertical displacements are low, since the maximum deflection is about 2 mm at the cantilever end, which demonstrates the effectiveness of the cross-section stiffness and the balancing role of the rear span.

Regarding the snow load, this shape manifests one major drawback; in fact, due to the shape of the shell, snow between two waves cannot be easily removed; on the contrary, it makes it easier to accumulate and thereby represents a substantial increase in weight.

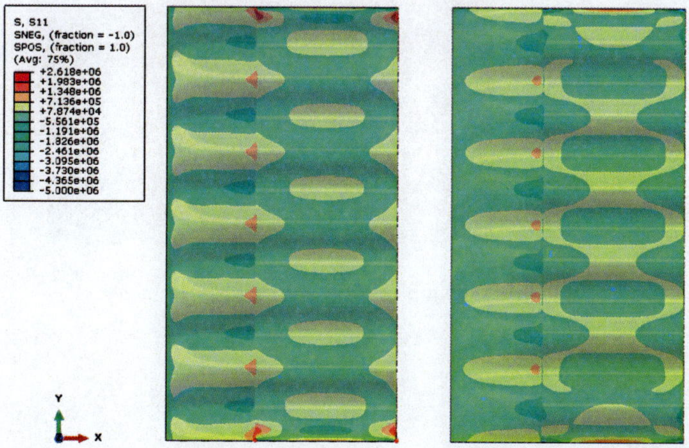

Fig. 8. Contour plot of the stress in the x-direction, (left) top view, (right) bottom view.

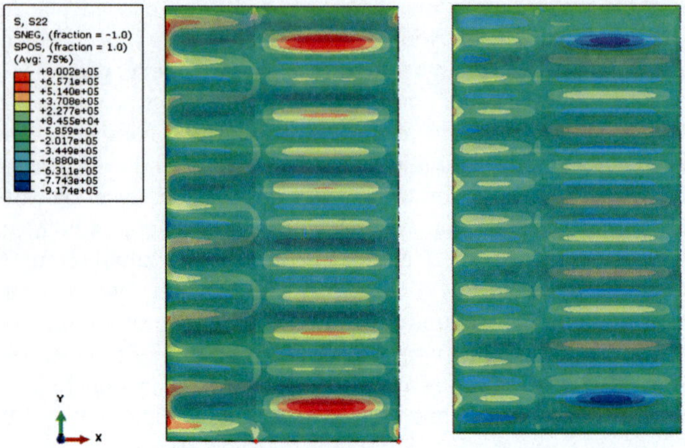

Fig. 9. Contour plot of the stress in the x-direction, intradosso

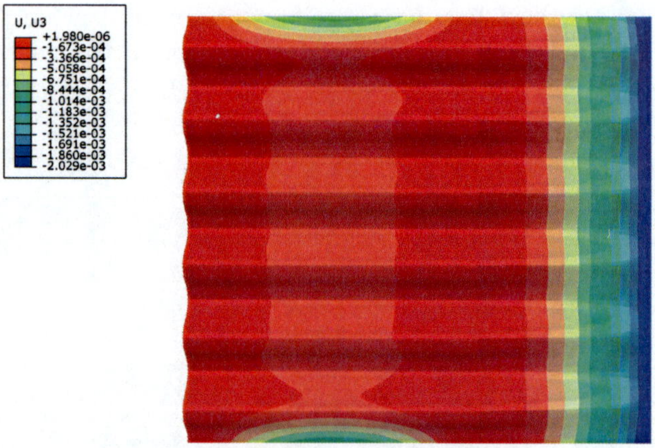

Fig. 10. Contour plot of the vertical displacement.

Fig. 11. Plot of the exaggerate deformed shape of the roof.

3 Conclusion

After some preliminaries elastic analyses, the Favini's design of Aquila fuel station rooftop has proved to be effective and properly designed with respect to stiffness and bearing capacity. In fact, the wavy shape of the structure surface has provided necessary stiffness to keep the deflection low under dead loads, even with a considerably small shell thickness. As regard the stress state, compressive and tensile stresses are within the admissibility limits with respect to compressive strength of medium-low concrete class and to occurrence of tensile cracks. Favini's design proved to be an outstanding example of well-conceinved structural optimisation.

Acknowledgment. The authors appreciate the financial support of "Aldo Favini" Foundation and thank the trust for the access to the historical archives. The copyright of Figs. 2, 3, 4 and 5 belongs to the Favini's Foundation. This research has been conducted at University of Cagliari — "Scuola di dottorato in Ingegneria Civile e Architettura" under the supervision of the scholars: A. M. Cazzani and E. Reccia.

References

1. Iori, T., Poretti, S. (eds.): SIXXI. Storia dell'ingegneria strutturale in Italia, vol. 1–5. Roma, Gangemi (2014)
2. Siegel, C.: Struttura e forma dell'architettura moderna. CELI, Bologna (1968)
3. Barazzetta, G. (ed.): Aldo Favini - Architettura e ingegneria in opera. CLUP, Milano (2004)
4. Favini, A.: Esempio di pensilina in volta sottile di cemento armato. G. Genio Civ. **9**, 533–536 (1950)
5. Pizzetti, G., Zorno Trisciuoglio, A.M.: Principi statici e forme strutturali. Utet, Torino (1980)
6. Haas, A.M.: Thin concrete shells, vol. 1. Wiley, New York (1962)
7. Billington, D.P.: Thin shell concrete structures. McGraw-Hill, New York (1965)
8. Haas, A.M.: Thin concrete shells, vol. 2. Wiley, New York (1967)
9. Gioncu, V.: Thin reinforced concrete shells: special analysis problems. Wiley, Bucarest (1979)

10. Lai, M., Reccia, E., Spagnuolo, M., Cazzani, M.A.: Structural behaviour of corrugated shells: A look at the Flaminio dome in Rome. In: 2nd fib Italy YMG Symposium on Concrete and Concrete Structures, pp. 87–94. Roma (2021)
11. Lai, M., Eugster, S.R., Reccia, E., Spagnuolo, M., Cazzani, A.: Corrugated shells: an algorithm for generating double-curvature geometric surfaces for structural analysis. Thin-Walled Struct. **173**, 109019 (2022). https://doi.org/10.1016/j.tws.2022.109019. Apr.
12. Lai, M., Zucca, M., Meloni, D., Reccia, E., Cazzani, M.A.: Thin corrugated-edge shells inspired by Nervi's dome: numerical insight about their mechanical behavior. Thin-Walled Struct., vol. to be submitted (2022)
13. Lai, M.: Edge-corrugation as structural enhancement of shallow shells. Nexus Netw. J. (2023). https://doi.org/10.1007/s00004-023-00716-1
14. Verderame, G.M., Manfredi, G., Frunzio, G.: Le proprietà meccaniche dei calcestruzzi impiegati nelle strutture in cemento armato realizzate negli anni 60. presented at the 10th ANIDIS conferece. Potenza-Matera, Italy (2001)
15. Verderame, G.M., Stella, A., Cosenza, E.: Le proprietà meccaniche degli acciai impiegati nelle strutture in c. a. realizzate negli anni '60'. presented at the 10th ANIDIS conferece. Potenza-Matera, Italy (2001)
16. Cosenza, E., Manfredi, G., Verderame, G.M.: La verifica di strutture in cemento armato progettate nel dopoguerra. presented at the Crolli e affidabilita'delle strutture civili. Istituto Universitario di architettura di Venezia (2001)
17. Fantilli, A., Ferraro, E., Chiaia, B.: La resistenza delle barre di armatura nei ponti in c.a. realizzati nel secolo scorso. In: Costruire in calcestruzzo: realizzazioni – ricerca, attualità e prospettive, pp. 817–824 (Apr. 2021)

Damage Detection in Heritage Vaults Through Geometric Deformation Analysis

Giulio Lucio Sergio Sacco(✉), Carlo Battini, and Chiara Calderini

Department of Civil, Chemical and Environmental Engineering, University of Genoa, Via Montallegro 1, 16145 Genoa, IT, Italy
giulioluciosergio.sacco@edu.unige.it

Abstract. Masonry vaults are common elements in heritage buildings and, due to their age, may accumulate damages and deformations due to various actions like overloads, support movements, earthquakes, etc. The use of terrestrial laser scanners and structure from motion photogrammetry survey to study these deformations can offer valuable information for structural assessment and conservation, especially when limited knowledge of buildings history is available. The analysis of the deviation between the surveyed point cloud and a reference geometry can reveal deformation patterns and aid the damage detection and action evaluation on the structure. This paper aims to explore the role of the reference geometry to provide a reliable and more accurate identification and quantification of deformations. Deviation analyses will be performed on the vaults of the main nave of two different churches with different reference geometries (i.e. geometries derived from the survey and geometrical primitives fitted on point-cloud) and the results are discussed comparing the deviation maps obtained and the already known deformation and caused.

Keywords: Damage Detection · Deformation analysis · Historic masonry vaults · Point clouds · Reference geometry

1 Introduction

In cultural heritage preservation, a comprehensive structural analysis of a building should be based on data from various sources. This includes knowledge about the history of the building and constructive techniques used in the different phases of its construction, as well as information gathered in the field about the materials, constructive details, detailed geometry etc. Geometric surveys are a critical part of this process and often serve as a starting point for further analyses. Moreover, the study of the geometry of a building plays a crucial role in many steps of the analysis, helping to identify structural damages on the structure and to understand the underlying causes.

The deformations observed in many masonry structures are the result of the cumulative effects of different actions such as overloads, foundation settlements, earthquakes, material degradation, creep, etc. These deformations can also be found in combination with other irregularities due to the construction techniques and subsequent modifications

S. Gabriele et al. (Eds.): IWSS 2023, LNCE 437, pp. 171–179, 2024.
https://doi.org/10.1007/978-3-031-44328-2_18

of a structure [1] increasing the difficulties in the damage recognition and interpretation. In this regard, the study of the deformations of arches and vaults is particularly interesting because they are relatively fragile and sensitive architectural elements that can shows specific damage pattern before and more clearly than other parts of a building. For these reasons, their study can help to identify different actions affecting the building and can help to direct the focus of further investigations.

This paper tries to investigate how to detect, distinguish and quantify the deformations of different vaulted structures starting from point clouds data. More specifically, it aims to assess the reliability to identify deformation of different approaches for generating reference geometries and their impact on the results of studying deformations analysing the deviation between point clouds and those reference geometries.

2 Deviation Analysis and Deformation Analysis

The investigation of geometrical irregularities and deformations in historical buildings has long been recognized as an important part of the knowledge process to reach a comprehensive understanding of the history and pathology of a structure [2]. It proved to be able to provide plenty of information useful in many fields from history of architecture to structural engineering [3]. With the technological advancements, surveying instruments have improved in accuracy and capability, enabling the collection of a greater amount of detailed metric data. Today, technologies like terrestrial laser scanners (TLS) and structure from motion photogrammetry (SfM) are a standard *de facto* to produce dense point clouds able to describe the entirety of a building and support the subsequent analysis, and are proved to be equally suitable for the study of the morphology of structures [4].

There is plenty of literature about the use of point cloud data to study the geometry, the state of conservation of the surfaces and damages of building. Many instances focus on the study of the surfaces to gather information about material degradation [5], crack recognition, historical stratigraphy [6], etc. Fewer examples are dedicated to the study of geometry from a structural point of view, with the aim of investigating causes of damage. In this regard, deformation maps obtained evaluating the distances of the surveyed point cloud from different reference geometries proved to be able to recognize in a qualitative way different kinds of damages. The effects of earthquakes was recognized in towers [7] and churches [8]; the deformation of arches and vaults was studied [9] and the mechanisms due to self-load [10] and foundation settlements [11] were identified. In those studies, different reference geometries were used to highlight different phenomena, but a discussion of the role of these choices was beyond the purpose of the research. Moreover, the interest in the role of the different reference geometries becomes more important as the geometry of the building and the deformation of the elements increase in complexity and further studies are required in this regard.

3 Description of the Case Studies

To investigate the possibilities offered by deviation analysis in the damage detection of masonry vaults, the main vaults of two already studied churches are taken as examples: the barrel vault of Nostra Signora della Bastia sanctuary and the lunette vault of San

Carlo Borromeo church, both located in the Liguria region (Italy). Both vaults underwent significant deformations due to different causes that have already been recognized.

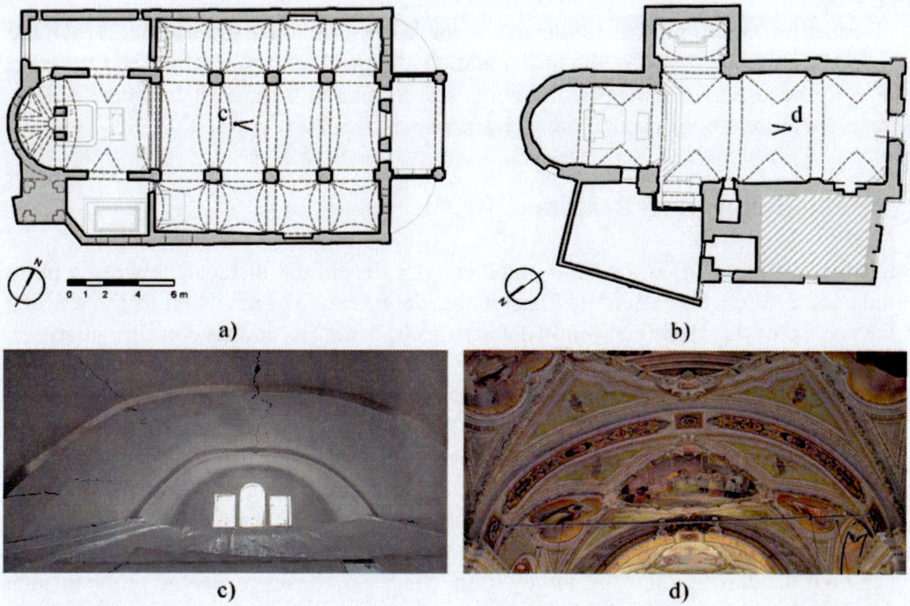

Fig. 1. Plans of the two churches and details of the annualized vaults: a) Nostra Signora della Bastia; b) San Carlo Borromeo; c) cracking pattern and deformation of the third, second and first bay of Nostra Signora della Bastia vault; d) deformation of the second arch of San Carlo Borromeo nave.

Nostra Signora della Bastia appears today as a three-nave church with a presbytery, an apse, a sacristy, a small chapel and a belltower (Fig. 1-a). The limited historical documentation and the evidence appeared during restoration works suggest that the first three bays from the façade are the most ancient part of the building while the remaining parts were made after the 1922. The church is a stone masonry structure except the vaults which are made of brick masonry. The barrel vault of the main nave is made of four bays divided by three arches. As will discussed in [12], the first three bays from the façade, which are also geometrically different in respect to the fourth, underwent a severe deformation (Fig. 1-c) compatible with a collapse mechanism due to the combination of piers displacement and the overloads applied on the crow on the arches and coming from the roof.

San Carlo Borromeo is a single nave church with an irregular plan developed in several constructive phases (Fig. 1-b). Today it consists in a three bays nave, a presbytery, an apse, two chapels, a sacristy and a clergy house. The oldest parts of the building are the presbytery and the apse, followed by the nave and the bell tower and then by all the other elements. All the church is made of stone masonry. The lunette vault of the nave is made of three bays divided by two arches. As found in [13] the vault present a significant rotation in the longitudinal direction due to a landslide acting in the area, a deflection due to the dead loads of the roof and the vault itself, and a local deformation in correspondence of the second arch due to a local settlement caused by the belltower (Fig. 1-d).

Further information about the structure, the history and the causes of damage of the two building can be found in the aforementioned papers alongside all the information related to the survey campaigns. In both cases, terrestrial laser scanners were used to capture the point clouds used in the analysis presented in this paper.

4 Methodology and Results

Deviation analysis [14] is a tool to graphically represent the distance between a point cloud and a reference geometry to assist in the identification of deviation patterns which in the context of this study can shows deformations, damages, constructive irregularities, and so on.

The deviation maps and all the operations on the point clouds presented in this work were performed in the Rhino/Grasshopper [15] visual programming environment reeling on Volvox [16] plugin. The signed distances between the point clouds representing the current state of the buildings and the different reference configurations were computed and plotted in deviation maps; negative distances are used for point below the reference geometry of vaults while positive values are used for point over it.

In both the case studies, the survey of the vault extradoses presented problems due to accessibility; as a result of this common occurrence, only the intrados of the vaults were studied. For each bay, the scan captured at its centre was extracted from the survey of the building and then was segmented to isolate the vault. This approach was chosen to limit the noise, reducing the need of post processing. The point clouds were therefore decimated with a voxel based spatial approach, obtaining datasets with a homogeneous density of one point every 5 cm (Fig. 2).

Fig. 2. Isometric view of the vaults studied: a) Nostra Signora della Bastia; b) San Carlo Borromeo.

The approaches adopted for the generation of the reference geometries needed for the deviation analysis can be divided in three groups: one (A) is based on the actual sections of the vault directly retrieved from the point clous, another one (B) uses idealized 2D primitives obtained from the same sections, and the last one (C) follows a more generalized 3D fitting approach.

In the first group (A), the generation of the reference geometry is based on the irregular geometry of the surveyed vaults as extracted from the point cloud. The least deformed sections were arbitrary chosen, taking care of their representativity of the original configuration of the vault, and extracted for the point clouds. Those sections are usually the ones in correspondence of the heads of a barrel vault and especially the one in correspondence of the façade, due to the higher rigidity usually observed in the areas. The reference geometry is then obtained extruding the least deformed section (AA') horizontally along the axis of naves, obtaining a portion of a horizontal cylinder with irregular base. The corresponding deviation maps are reported in Fig. 3-a, b. An alternative modelling technique is to retrieve two section lines (AA' and BB') in correspondence of the two heads of the vault and to generate the reference geometry connecting those lines through a loft operation. In this case the reference geometry obtained is usually no longer a cylinder and it can be tilted, also providing an estimation of the lining of the original vault. The corresponding deviation maps are reported in Fig. 3-d, c. In the case of Nostra Signora della Bastia (Fig. 3-b), the section BB' was taken in the third bay due to the fact that, as summarized in paragraph 3, the fourth one probably wasn't part of the original barrel vault of the nave and has a different geometry not representative of the original vault geometry.

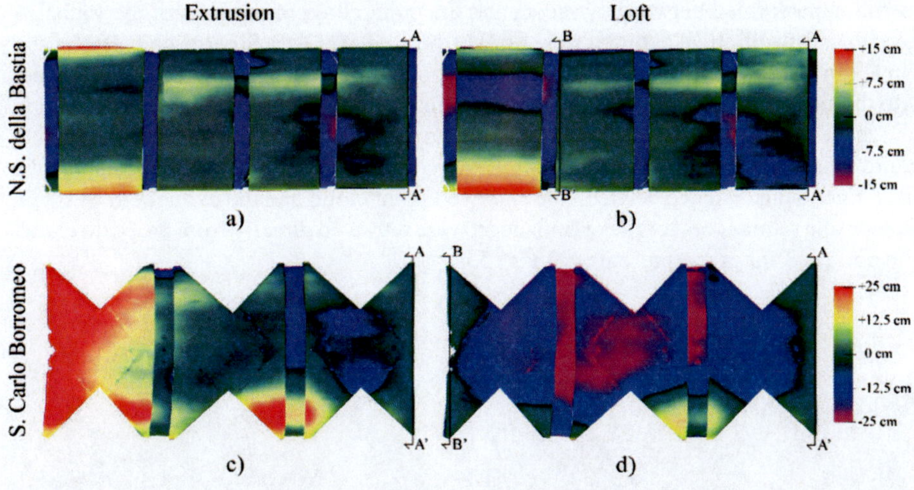

Fig. 3. Deviation maps of group A, obtained using the irregular geometries of the survey as references.

In the second group (B), the same sections already used for the group A are used but regular generatrix curves are derived from the surveyed data to obtain regular reference geometries. Circumferences and ellipses are fitted in the point cloud sections in the case of Nostra Signora della Bastia and San Carlo Borromeo respectively. As for group A,

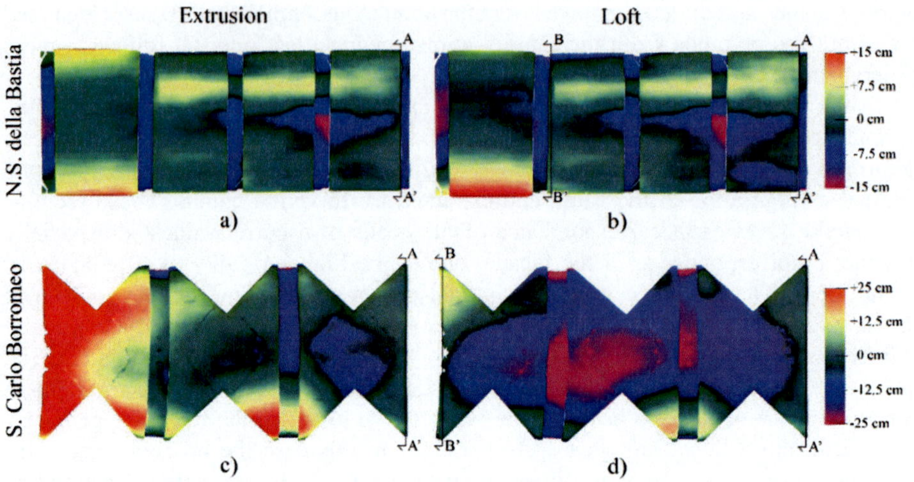

Fig. 4. Deviation maps of group B, obtained through fitting regular 2D generatrix curves to the survey data sections: a, b) circular generatrix; c, d) elliptical generatrix.

two sets of deviation maps are computed for each church following both the extrusion and the loft approach and are reported in Fig. 4.

In the third group (C), the reference geometry is directly obtained through a fitting operation performed between cylinders and the point cloud of the bays of the vaults. A circular and an elliptical cylinder were used in the case of Nostra Signora della Bastia and San Carlo Borromeo respectively. The fitting operation was performed in Grasshopper with the aid of the Goat [17] optimization plugin, following a local linear approximation approach. The optimization parameters were the ones needed to generate the ellipse (i.e. centre point, axis vector and radii) and the objective function to be minimized was the mean Euclidean distance between the surveyed point cloud and the cylinder to be fitted. The starting parameters for the optimization were retrieved directly from the point cloud. The deviation maps are presented in Fig. 5.

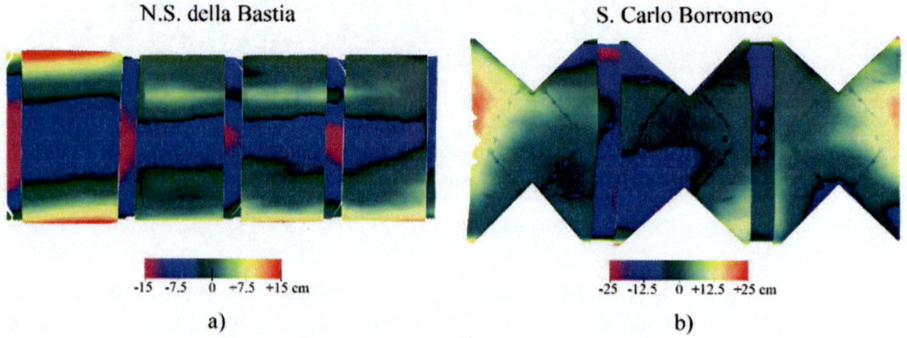

Fig. 5. Deviation maps obtained with the 3D fitting.

5 Discussion

Overall, the result of the analysis appears to be significantly sensitive to the reference geometry used generating the deviation maps. However, some significant phenomena can be qualitative observed in all cases. Regarding Nostra Signora della Bastia, all the deviation maps show the deflection of the crown of the first arch and the "rise" of the right portion of the vault, that combined could be a sign of the presence of the activation of a collapse mechanism. The maps also show the incoherence of the geometry of the fourth bay in respect to the other three, which could confirm the presence of different constructive phases. Regarding San Carlo Borromeo, which underwent more significant deformations, the combination of different phenomena leads to more variability in the deformation maps and the only occurrence present in all the analysis is the rise of the left part of the first arch.

In the first two groups of deviation maps (A and B, Fig. 3 and Fig. 4), the most significant difference can be observed between the maps obtained from the reference geometry extruded along the axis of the nave and the ones obtained with lofting operations. In the first case, in San Carlo Borromeo (Fig. 3-c and Fig. 4-c) the global rotation of the vault is evident, while in the second (Fig. 3-d and Fig. 4-d) no rotation can be seen but the deflection of the whole vault is recognizable. In this last case (group B), for both extruded and lofted reference geometry, the measurement of the deflection in the centre of the vault is in accordance with the previous analysis of about 20 cm [13]. Moreover, a smaller leaning of about 1.6°, compatible with the previously observed of about 2°, was present in the ellipses used as reference in these deviation maps. In Nostra Signora della Bastia the results are more homogeneous since just one main deformation mechanism is present. No significant rotation can be observed in the maps and the lining of the ellipse used to compute Fig. 4-b is of about 0.1°.

The deviation maps obtained from the fitted cylinders (Fig. 5) yielded mostly similar results compared to the previous one, although some phenomena appear in different ways. The deflection that can be observed along the crown of the vault of Nostra Signora della Bastia, of about 7.5 cm compared to a light of about 5 m, is probably due to physiologic deformation of vaults under their own weight [18]. Moreover it's interesting to notice that the outward displacement observed in the left side of the vault and mostly concentrated between the first and the second bay (bottom part of Fig. 5-a), could be an indication of support movement could partially justify the deflection of the corresponding arch, even if more investigations are needed. The tilting of San Carlo Borromeo vault is not visually apparent as it is inherently accounted for in the optimization parameters, ad resulted in about 1.5°. The deflection of the vault is difficult to recognize since the fitting process resulted in a rise in between the maximum and the minimum value of the height of the crown, as well as the local phenomena in the arches that would probably benefit from being analysed separately.

6 Conclusion

The use of deviation maps for damage detection in masonry vaults is a powerful tool but the difficulties and uncertainties in the definition of reference geometries poses a threat to its reliability, especially when multiple deformation are present at the same time.

The direct use of sections of the point cloud shown in the group A, despite not needing the introduction of additional assumptions, leads to the generation of irregular reference geometries which create significant difficulties in the interpretation of the deviation maps and could hide significant phenomena. This is even more relevant when all the sections of the vault under analysis are significantly deformed and one example of this can be seen in Fig. 3-d where all the deformation close to section BB' are not recognizable as well as the deformation pattern of the second arch of the nave. On the other hand, using regular primitives as reference geometries, as shown in the group B, even if it could be less representative of the deformations of the vault where the undeformed geometry is not known, leads to unambiguous deviation maps which are easier to analyse and help to better understand the deformation patterns. In the case studies, global rotation, deflection and damage mechanism were recognized in accordance with the previous studies.

The distances reported in the deviation maps, measured between the point cloud and a reference, cannot be considered a measure of deformation because the reference geometry is rarely a reliable representation of the original undeformed configuration of the structure which remain mostly unknown. However, the possibility to enhance direct three-dimensional fitting operations imposing additional constrains to control the coherence of the results, or the definition of parametric approaches to model a reference geometry more resembling the undeformed one could be a further step in the research to provide also a better estimation of deformations.

Acknowledgements. The authors would like to thank Arch. Andrea Mamone for the material provided on the case study of Nostra Signora della Bastia and the possibility to access the building during the renovation works.

References

1. Roca, P., Cervera, M., Gariup, G., Pelà, L.: Structural analysis of masonry historical constructions. classical and advanced approaches. Archi. Computa. Methods Eng. **17**, 299–325 (2010)
2. Roca, P., Cervera, M., Pelà, L., Clemente, R., Chiumenti, M.: Continuum FE models for the analysis of Mallorca Cathedral. Eng. Struct. **46**, 653–670 (2013)
3. Albuerne, A., Williams, M.S.: Structural Appraisal of a Roman Concrete Vaulted Monument: The Basilica of Maxentius. Int. J. Architect. Heritage 1–12 (2017)
4. Teza, G., Pesci, A., Ninfo, A.: Morphological analysis for architectural applications: comparison between laser scanning and structure-from-motion photogrammetry. J. Surv. Eng. (2016)
5. Teza, G., Pesci, A.: Geometric characterization of a cylinder-shaped structure from laser scanner data: Development of an analysis tool and its use on a leaning bell tower. J. Cult. Herit. **14**(5), 411–423 (2013)
6. Pesci, A., Bonali, E., Galli, C., Boschi, E.: Laser scanning and digital imaging for the investigation of an ancient building: palazzo d'Accursio study case (Bologna, Italy). J. Cult. Herit. **13**(2), 215–220 (2012)
7. Pesci, A., Casula, G., Boschi, E.: Laser scanning the garisenda and asinelli towers in bologna (Italy): detailed deformation patterns of two ancient leaning buildings. J. Cult. Herit. **12**(2), 117–127 (2011)

8. Bonali, E., Pesci, A., Casula, G., Boschi, E.: Deformation of ancient buildings inferred by terrestrial laser scanning methodology: the cantalovo church case study (northern italy)*: deformation of ancient buildings inferred by TLS methodology. Archaeometry **56**(4), 703–716 (2014)
9. Parrinello, S., De Marco, R.: Digital surveying and 3D modelling structural shape pipelines for instability monitoring in historical buildings: a strategy of versatile mesh models for ruined and endangered heritage. Acta IMEKO **10**, 84–97 (2021)
10. Quagliarini, E., Clini, P., Ripanti, M.: Fast, low cost and safe methodology for the assessment of the state of conservation of historical buildings from 3D laser scanning: the case study of Santa Maria in Portonovo (Italy). J. Cult. Herit. **24**, 175–183 (2017)
11. Ye, C., Acikgoz, S., Riley, E., DeJong, M.J., Pendrigh, S.: Mapping deformations and inferring movements of masonry arch bridges using point cloud data. Eng. Struct. **173**, 530–545 (2018)
12. Ferrero, C., Sacco, G.L.S., Ferrero, M., Battini, C., Calderini, C.: Coupled deformation and structural analysis for the damage assessment of cultural heritage buildings: the case of a masonry church exposed to slow-moving landslides. In: 13th International Conference on Structural Analysis of Historical Constructions (SAHC 2023) (2023). (Accepted)
13. Sacco, G.L.S., Ferrero, C., Battini, C., Calderini, C.: Combined use of deformation and structural analysis for the structural damage assessment of heritage buildings: A case study in the Liguria region (Italy). Eng. Fail. Anal (2023)
14. Anil, E.B., Tang, P., Akinci, B., Huber, D.: Deviation analysis method for the assessment of the quality of the as-is Building Information Models generated from point cloud data. Autom. Constr. 507–516 (2013)
15. Robert McNeel & Associates. Rhinoceros 7 (2020)
16. Zwierzycki, M., Evers, H.L., Leand, H.: Parametric architectural design with point-clouds - Volvox. Complexity & Simplicity: 34th eCAADe Conference (2016)
17. Rechenraum GmbH. Goat (2016)

Principal Lines on Shells: Which Are More Meaningful?

Arianna Venettoni(✉), Stefano Gabriele, Ginevra Salerno, and Valerio Varano

Roma Tre University, Via Madonna dei Monti, 40, 00184 Rome, Italy
arianna.venettoni@uniroma3.it
https://architettura.uniroma3.it/en/

Abstract. Several families of lines can be generated on a curved surface by plotting different properties. These lines can be classified as either geometrical, or strictly mechanical if the surface becomes a shell structure. Here, we plot these lines on shells and discuss the connection between them, as a preparatory investigation of the structural behaviour of curved surfaces. This analysis is performed by primarily using Rhinoceros [1], Grasshopper [2] and Karamba3D [3]. We also introduce a new typology of mechanical lines: the principal eccentricity lines. Finally, we propose a parametric analysis of different structural surfaces, in order to highlight the most relevant lines for the mechanical behaviour of shell structures and their possible use. Central to the premise of this work is the idea to explore the language of structural form, to the advantage of a structurally inspired design that can ensure the coexistence of structural and architectural design goals.

Keywords: Stream lines · General Eccentricity · Shell structure

1 Introduction

The performance of a structure is significantly influenced by geometry [4], and this is particularly true for shell structures. The form of a shell affects its structural behaviour more than any other factor. Many form finding approaches have been developed and deployed to obtain "optimal" structural geometry in static equilibrium with a specific load pattern [5,6]. Undoubtedly, structurally efficient designs can lower material consumption and lifespan costs, but it is fundamental to state that the development of innovative and intriguing architectural forms can and must be influenced by structural inputs. The process of building construction today often maintains a strict separation between architecture and engineering; in these circumstances, lines that show the flow of forces can be seen as a useful tool to help the visualization of the load paths by designers. The research undergone this paper aims to support architects and engineers in designing lighter shell and gridshell structures, and to enhance their understanding and mastery of the form [7].

S. Gabriele et al. (Eds.): IWSS 2023, LNCE 437, pp. 180–189, 2024.
https://doi.org/10.1007/978-3-031-44328-2_19

2 Generalized Eccentricity

The eccentricity is a useful parameter to define the condition of funicularity in arch structures. It is related to the normal force and the bending moment acting on a section. This parameter can be also generalized and used for shell structures. To the authors' knowledge, the first ones to use eccentricity as a parameter to quantify the funicularity of optimized shells are Marino et al. [8]. Marino estimates the eccentricity for a shell as the ratio of the membrane force matrix norm and moment matrix norm, noting that this value tends to zero when the membrane behaviour is prevalent. An earlier numerical method for masonry vault analysis by Lucchesi et al. [9] proposed the computation of the maximum modulus eccentricities surface (MMES), which was suggested in the context of no-tension materials as a tool for limit analysis.

Here, we briefly describe the concept of Generalized Eccentricity, according to which we evaluate the eccentricity lines. A complete discussion and further implications can be found in Gabriele et al. [10].

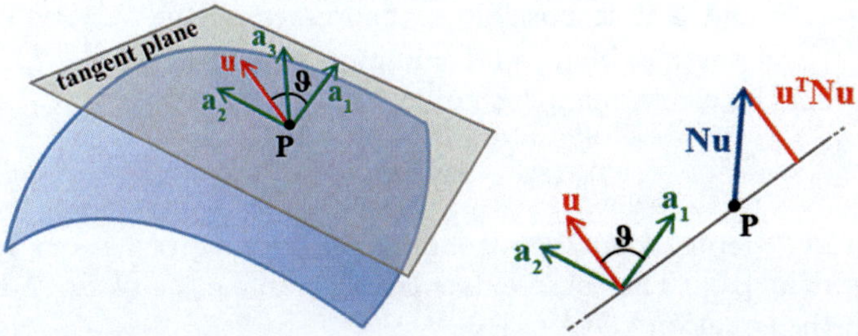

Fig. 1. Local basis and projection of internal forces [10]

Let's consider a point P belonging to the surface S, its orthonormal basis $\mathbf{a}_1, \mathbf{a}_2, \mathbf{a}_3$, and a generic direction $\mathbf{u} = (cos\theta, sin\theta)$ (see Fig. 1). The membrane force $N(\theta)$ acting along the direction $\mathbf{u}$ and the bending moment $M(\theta)$ acting in the plane $(\mathbf{u}, \mathbf{a}_3)$ are defined as follows:

$$N(\theta) = \mathbf{u}^T\mathbf{N}\mathbf{u} \tag{1}$$

$$M(\theta) = \mathbf{u}^T\mathbf{M}\mathbf{u} \tag{2}$$

where $\mathbf{N}$ and $\mathbf{M}$ are the local tensors, associated with the membrane forces and the bending moments respectively:

$$\mathbf{N} = \begin{pmatrix} N_{11} & N_{12} \\ N_{12} & N_{22} \end{pmatrix}$$

$$\mathbf{M} = \begin{pmatrix} M_{11} & M_{12} \\ M_{12} & M_{22} \end{pmatrix}$$

The generalized eccentricity is defined as the ratio between the generalized bending moment and membrane force:

$$e(\theta) = \frac{N(\theta)}{M(\theta)} \tag{3}$$

Using the multiple-angle, and the power-reduction trigonometric formulas and setting $\alpha = 2\theta$, the following explicit form can be written for Eq. 1 and Eq. 2:

$$N(\alpha) = \overline{N} + \widehat{N} cos\alpha + N_{12} sin\alpha \tag{4}$$

$$M(\alpha) = \overline{M} + \widehat{M} cos\alpha + M_{12} sin\alpha \tag{5}$$

where $\overline{N} = \frac{N_{11}+N_{22}}{2}$, $\widehat{N} = \frac{N_{11}-N_{22}}{2}$, $\overline{M} = \frac{M_{11}+M_{22}}{2}$, $\widehat{M} = \frac{M_{11}-M_{22}}{2}$. The generalized eccentricity is then reparametrized as:

$$e(\alpha) = \frac{N(\alpha)}{M(\alpha)} \tag{6}$$

From Eq. 1, 2 and 3 it is possible to express e as the Rayleigh quotient: $e(\theta) = \frac{\mathbf{u}^T\mathbf{M}\mathbf{u}}{\mathbf{u}^T\mathbf{N}\mathbf{u}}$, and the maximum and minimum eccentricities (e_{max}, e_{min}) are found by solving the eigenvalue problem in Eq. 7:

$$(\mathbf{M} - e\mathbf{N})\mathbf{u} = 0 \tag{7}$$

The solution of 7 depends on the algebraic conditioning of $\mathbf{M}$ and $\mathbf{N}$ and has been discussed in [10]. The local extrema e_{max} and e_{min} of Eq. 7 have been calculated in the equation (17a) of [10].

In this work, we evaluated the principal eccentricity directions by calculating the eigenvectors $\mathbf{v_1}$,$\mathbf{v_2}$ of the matrix $\mathbf{N}^{-1}\mathbf{M}$ using the analytical formula:

$$\mathbf{v_1} = \left(\frac{-(-e_{11} + e_{22} + \sqrt{e_{11}^2 + 4e_{12}e_{21} - 2e_{11}e_{22} + e_{22}^2}}{2e_{21}}, 1 \right)^T \tag{8}$$

$$\mathbf{v_2} = \left(\frac{-(-e_{11} + e_{22} - \sqrt{e_{11}^2 + 4e_{12}e_{21} - 2e_{11}e_{22} + e_{22}^2}}{2e_{21}}, 1 \right)^T \tag{9}$$

where e_{11}, e_{12}, e_{21} and e_{22} are the entries of the $\mathbf{N}^{-1}\mathbf{M}$ matrix.

In most cases, the values found for eccentricity provide a direct measurement of the extent to which internal moments affect the shell. The shell's capability to carry the imposed loads using only membrane forces will decrease as the eccentricity values increase. The shape of the shell can be regarded as entirely funicular if the eccentricity extrema have zero values. Essentially, this final goal is rarely met.

3 Implementation of a Script for Plotting Lines on Shells

Several families of lines can be generated on a curved surface. These lines can be defined as geometrical or mechanical, depending on the plotted properties. The goals of this research are to evaluate the line results on shells, to examine the relationship between them and to highlight the most relevant ones for the mechanical behaviour of shell structures. Here we also introduce a new typology of mechanical lines: the principal eccentricity lines, based on the principles introduced in Sect. 2. The necessity to visualize the eccentricity vectors field required a proper implementation, which is why the idea to plot the eccentricity lines on shells was crucial to this research.

3.1 Shape Definition

Before the analyses can occur, a parametrical family of shapes must be defined. The initial shape was chosen accordingly to its ease of both parametrization and definition in Grasshopper. Furthermore, this shape has already been investigated for shape optimization; see for istance [13,14] and [12]. As a result, it is seen to be a good candidate to evaluate the efficacy of the performed analysis. The following equation describes the chosen shape of a parabolic cylinder-shaped surface [15]:

$$z = 4fx\frac{l_x - x}{l_x^2} \qquad f = a_0 + a_1 y + a_2 y^2 \tag{10}$$

$$a_0 = c_1 \qquad a_1 = 4\frac{c_2 - c_1}{l_y} \qquad a_2 = 4\frac{c_1 - c_2}{l_y^2} \tag{11}$$

being l_y the surface span along y direction, c_1 the boundary rise and c_2 the midpoint rise. Parabolic cylinder

We set $l_x = 6$m and $l_y = 12$m and we investigate the design space generated by adjusting the c_1 and c_2 parameters. In this paper we choose to present the analysis of three sample shells, with Gaussian curvature equal to zero (shape A), positive (shape B) and negative (shape C) respectively (see Fig. 2).

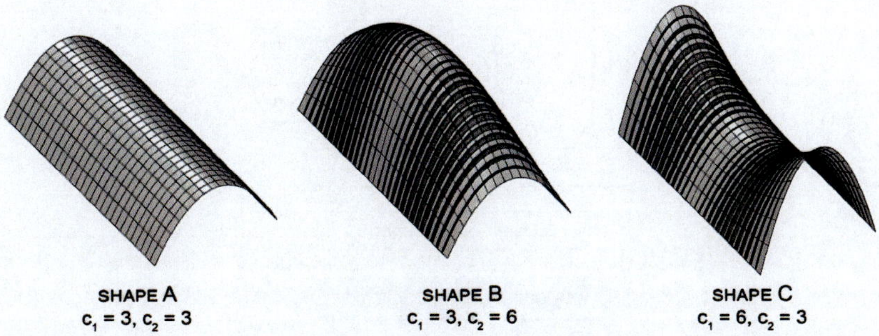

Fig. 2. Analyzed shapes.

3.2 Script Development

For each of the three shapes, we analysed both the option with pinned joints on the long side and the one with pinned joints on the short side, for a total of six shell structures. The shell thickness was set to 10 cm. A linear static analysis of the structure, subject to the self-weight load, was performed with Karamba 3D.

Once the parametric model was defined and analysed, a script was developed to plot the streamlines on the shells.

Usually related to fluids flow, *streamlines* are a family of curves whose tangent vectors constitute the velocity vector field of the flow. In a structure, these lines can show the flow of different force fields by connecting the projections of principal vectors directions (namely the eigenvectors) for a collection of points across the structural body. In this paper we generated three families of streamlines: *principal stress-lines*, that indicate the trajectories of internal normal forces; *principal-moment lines*, that follows the direction of principal bending moments, and *eccentricity-lines*, that shows the directions of maximum and minimum eccentricity.

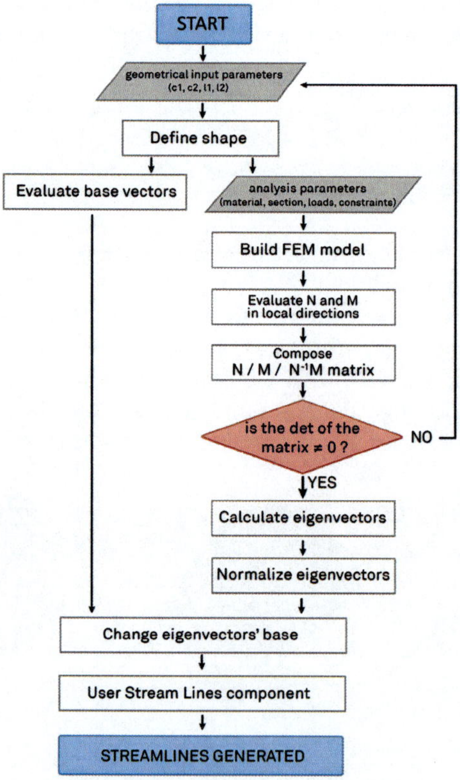

Fig. 3. Process of the developed script

We primarily used base Grasshopper tools and Karamba3D [3] *User Stream Lines* component. Karamba itself generates the principal stress and moment lines, but we developed our own path to have more control over the results: by using this script it is indeed possible to change the inputs and retrieve different lines at will. The script computes the eigenvectors using Eqs. 8 and 9. The process of the script is outlined in Fig. 3.

A similar method was adopted to plot the curvature lines; *a line of curvature* is a curve on a surface whose tangents are always in the direction of principal curvature, so its properties are strictly geometrical. In Fig. 4 we provide an example of each family of lines generated by the script, evaluated for shape B with constraints on the long side.

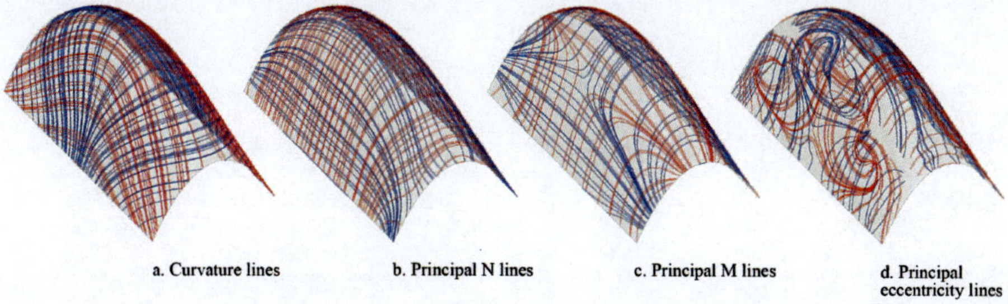

Fig. 4. Families of lines generated by the script.

3.3 Validation of Results

A comparison between moment lines generated with the Karamba component and the script respectively is shown in Fig. 5; the lines were evaluated for the shape A, with constraints on the short side.

The results are quite corresponding; the main differences can be found around the boundaries, and in the areas where both the script and the component present irregularities that still need be investigated.

Therefore, the shell structure was exported in SAP2000 and the same analysis was performed, in order to validate the achieved results. The coloured values of the principal stresses and moments were extracted, along with their directions. In Fig. 6, we can see a juxtaposition of the resultant vector of SAP2000 and our script, calculated for the previous example. A cross-reading of the plotted properties is quite effective for the comprehension of the natural load path of the forces in the system.

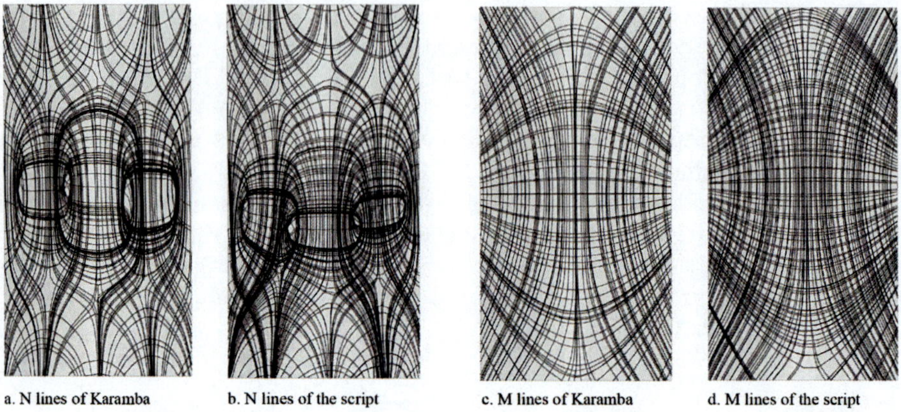

Fig. 5. Comparison between generated lines.

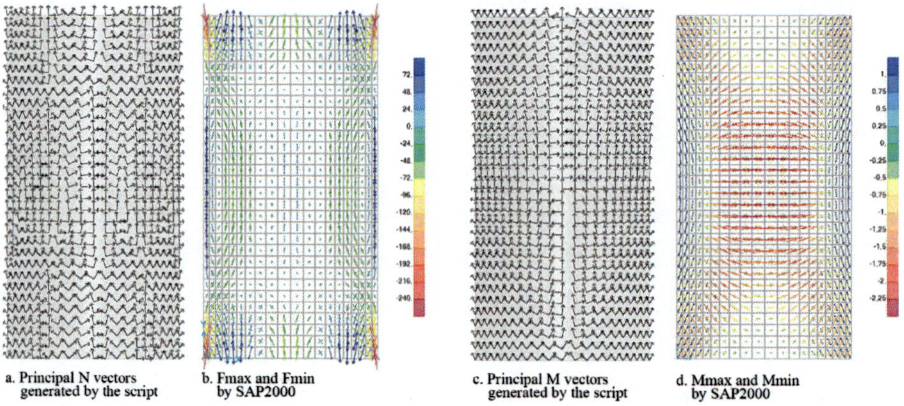

Fig. 6. Compared directions of principal normal forces and bending moments.

Finally, a Matlab script was implemented to plot the eccentricity values on the analysed surfaces. This script computes the maximum absolute value of the two eccentricity extrema $e_{M,i} = \max\left(\left|e_{max,i}\right|, \left|e_{min,i}\right|\right)$. In Fig. 7, the shape with negative Gaussian curvature is analysed. The colorbar represents the values of $e_{M,i}$ that increases going from green to blue, the latter meaning that the eccentricity is out of the cross section. This comparison was helpful to understand the numerical values associated with vectors and lines.

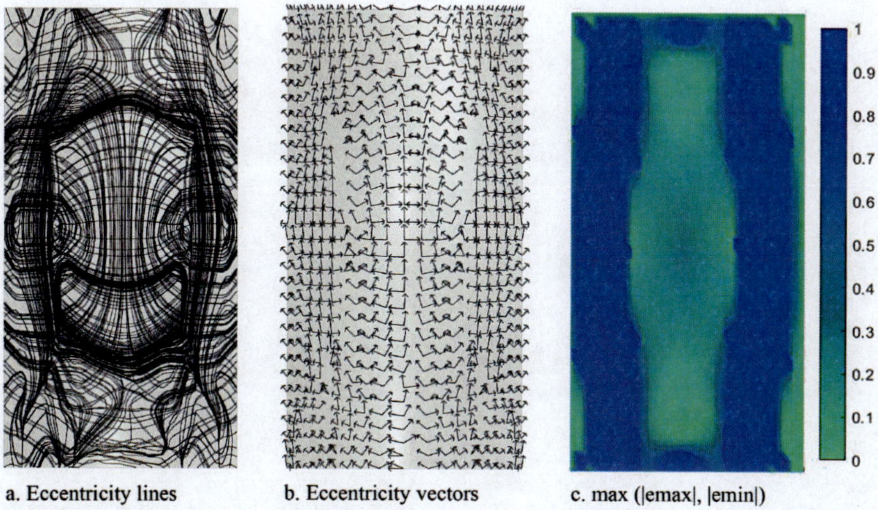

Fig. 7. Compared directions of principal eccentricities (Shape C, contraints on the long side).

4 Results and Conclusions

Some of the research results are presented here. We can see that mechanical lines are influenced by the geometrical ones, but don't trace their path. These families of lines resemble more when the shell works more like a membrane. In wider terms, global symmetries are preserved, but local ones are modified.

It is important to note that although $\mathbf{N}$ and $\mathbf{M}$ are symmetric, the products $\mathbf{N}^{-1}\mathbf{M}$ and $\mathbf{M}^{-1}\mathbf{N}$ are typically non-symmetric. This has two significant repercussions: Eqs. 8 and 9 can produce complex solutions, and real solutions will also typically have eigenvectors of $\mathbf{N}^{-1}\mathbf{M}$ and $\mathbf{M}^{-1}\mathbf{N}$ that aren't orthogonal. Therefore, the maximum and minimum eccentricity directions are typically not orthogonal.

As generally depicted in the figures, several inaccuracies can occur when plotting streamlines. Both the script and the software itself generate streamlines that suddenly shift direction, and present inconsistencies and oscillations. According to [16], this is because the principal stress vectors are eigenvectors, so their direction can be defined both in the positive or negative direction.

Another recurring inaccuracy verifies when the streamlines get closer to a singularity. After circling the singularity, the streamline will either stop or twist and cross over itself (Fig. 8).

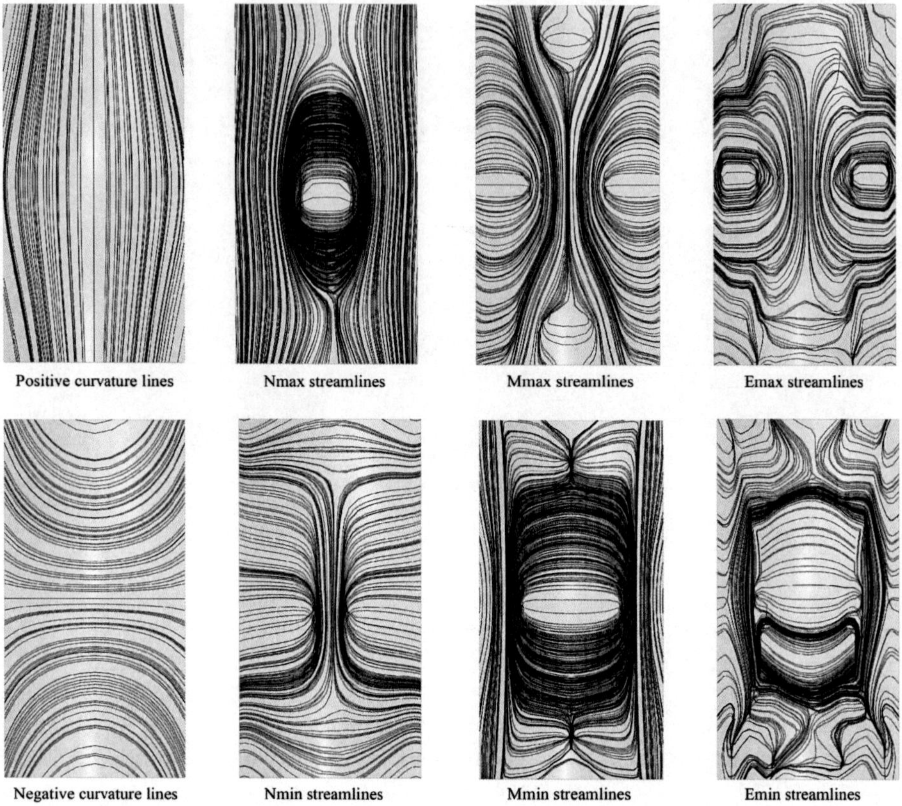

Fig. 8. Lines generated for the shape C with constraints on the long side.

In [16] is stated that these problems could be solved by developing a proper script using Python. Since Python is currently implemented in Grasshopper environment, the correction of the inaccuracies will be one of the research future goals.

From the analysis, it can be deduced that the significance of a typology of line over another depends on the design goals. Therefore, it is worth noting that the eccentricity lines are a significant synthesis of the path of the normal forces and the bending moment; hence, they're particularly useful to optimise the shape and find the most effective reinforcement and stiffening directions.

Taking into account its need for refinement, the script currently provides a better understanding of the general course of the lines. Principal stress lines are relevant to structural designers because they illustrate the natural load path of the forces in the system and the line of desired material continuity for a specific design domain. Furthermore, the possibility to generate not only principal stress lines, but several families of lines in one single environment enriches the designers' chance to explore and understand the form.

References

1. McNeel, R.: Rhinoceros 3D (2009)
2. McNeel, R.: Grasshopper 3D (2009)
3. Preisinger, C., Heimrath, M., Karamba? A toolkit for parametric structural design. Struct. Eng. Int. **24**(2), 217–221 (2014). https://doi.org/10.2749/101686614X13830790993483
4. Allen, E., Zalewski, W.: Form and Forces: Designing Efficient, Expressive Structures. Wiley, Hoboken (2010)
5. Ramm, E., Bletzinger, K.-U., Reitinger, R.: Shape optimization of shell structures. Revue Europenne des lments Finis **2**(3), 377–398 (1993)
6. Adriaenssens, S., Block, P., Veenendaal, D., Williams, C.: Shells Structures for Architecture: Form Finding and Optimization. Routledge, London (2014)
7. Kam-Ming, M.T.: Principal Stress Line Computation for Discrete Topology Design, Master Thesis, Massachusetts Institute of Technology (2015)
8. Marino, E., Salvatori, L., Orlando, M., Borri, C.: Two shape parametrizations for structural optimization of triangular shells. Comput. Struct. **166**, 1–10 (2015)
9. Lucchesi, M., Padovani, C., Pasquinelli, G., Zani, N.: Masonry Constructions: Mechanical Models and Numerical Applications. Springer, Cham (2008). https://doi.org/10.1007/978-3-540-79111-9
10. Gabriele, S., Varano, V., Tomasello, G., Alfonsi, D.: R-Funicularity of form found shell structures. Eng. Struct. **157**, 157–169 (2018)
11. Argento, G.R., Marmo, F., Varano, V., Gabriele, S.: Shells' shape optimization based on R-Funicularity. In: Behnejad, S.A., Park, G.A.R., Samavati, O.A. (eds.) Proceedings of the IASS Annual Symposium 2020/21 and the 7th International Conference on Spatial Structures–Inspiring the Next Generation, Guildford, UK, 23 – 27 August 2021, pp. 2548–2556 (2021)
12. Argento, G.R., Gabriele, S., Varano, V.: Different uses of the generalized eccentricity for shells' shape optimization: a comparison. In: Proceedings of IASS Annual Symposia, IASS/APCS 2022 Beijing Symposium: Future Challenges in the Design and Construction of Shell Structures for Low or Zero Carbon, pp. 1-11(11) (2022)
13. Bletzinger, K.-U., Ramm, E.: Form Finding of shells by structural optimization. Eng. Comput. **9**, 27–35 (1993)
14. Toms, A., Mart, P.: Shape and size optimization of concrete shells. Eng. Struct. **32**, 1650–1658 (2010)
15. Krivoshapko, S.N., Ivanov, V.N.: Encyclopedia of Analytical Surfaces. Springer, New York (2015)
16. Cobbs, M.: Investigating Principal Stress Lines: Optimization of Grid Shell Structures, Master Thesis, TU DELFT Repositories (2018)

Thrust Line Analysis for Reinforced Arches

Davide Pellecchia(✉), Luciano Rosati, and Francesco Marmo

Department of Structures for Engineering and Architecture, University of Naples Federico II, Naples, Italy
davide.pellecchia@unina.it

Abstract. The Thrust Line Analysis (TLA) is a methodology that can be employed to identify the position and shape of the thrust line in a masonry arch, providing information on the minimum and maximum thrust solutions that can exist within the arch. This paper presents an enhancement to the TLA that considers the influence of the Fiber-Reinforced Cementitious Matrix (FRCM) composite system. The numerical examples presented in this paper show that the proposed extension to the TLA can accurately forecast the behaviour of reinforced masonry arches and can offer valuable perspectives into the design and evaluation of such structures.

Keywords: Reinforced Masonry Arches · FRCM · Limit Analysis

1 Introduction

Frequently, in the context of seismic retrofitting for preexisting masonry structures, the reinforcement of arches and vaults, i.e. the elements that hold paramount importance in ensuring the overall structural stability of the building, is envisioned as a form of intervention. Among various techniques proposed in the past years, the prevailing methods currently employed involve the utilisation of high-strength fibers, such as glass or carbon fibers, embedded in an inorganic matrix composed of lime or cement mortar. This composite system is commonly known as Fabric Reinforced Cementitious Matrix (FRCM). The preference for utilising these composite materials over classical Fibre Reinforced Polymer (FRP) arises from the inorganic matrix which ensures a better grip between the reinforcement and the masonry support.

The scientific community has shown considerable interest in exploring novel applications of FRCMs for structural rehabilitation. Simultaneously, the distinctive characteristics of Italy's diverse building stock have captured the attention of numerous researchers in recent years [1–3].

Currently, a range of computational methods are accessible for the analysis of masonry structures. Among these, certain approaches are based on the shape optimisation of arches and vaults [4–6].

Experimental investigations have been extensively employed to examine the stability of masonry arches and vaults. Due to the challenges associated with

S. Gabriele et al. (Eds.): IWSS 2023, LNCE 437, pp. 190–198, 2024.
https://doi.org/10.1007/978-3-031-44328-2_20

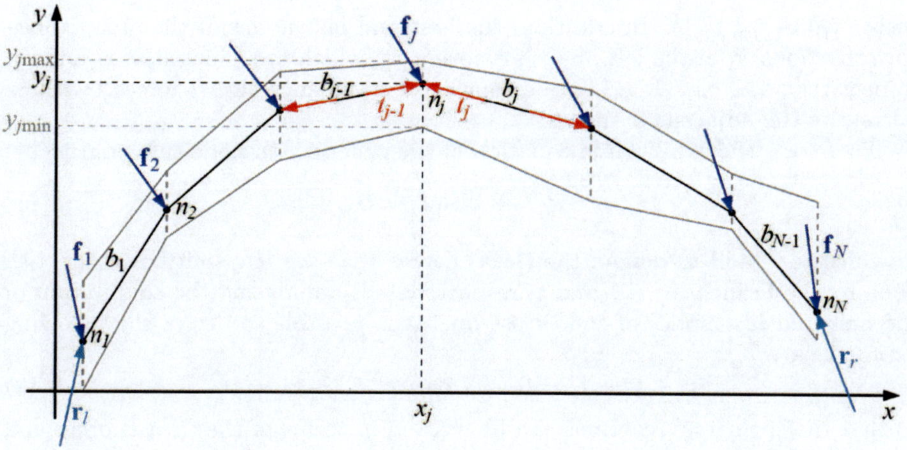

Fig. 1. Discretisation of thrust line.

conducting full-scale tests on arches and vaults, experimental campaigns are typically carried out on reduced-scale models [7].

In this paper, we propose an extension to the Thrust Line Analysis (TLA) [8] of masonry arches. The TLA is a specialised version of Thrust Network Analysis [9,10] for two-dimensional cases, and in our proposed extension, we include the effects of FRCM reinforcements. Specifically, this extension involves suitably expanding the geometric domain of the thrust line to incorporate the effects of the FRCM reinforcement. The size of the domain is evaluated iteratively, taking into account the factors that affect the strength of the reinforced arch, such as the limited compressive strength of masonry, tensile and detachment failure of the reinforcement, and the computed internal forces.

To demonstrate the feasibility of incorporating reinforcement within the TLA, two examples are presented in this paper. The results of these examples demonstrate that the proposed extension to the TLA can accurately predict the behaviour of reinforced masonry arches and can provide valuable insights into the design and assessment of such structures.

2 TLA for Reinforced Arches

The Thrust Line Analysis (TLA) is an equilibrium-based method used to determine admissible stress solutions in masonry arches. This analysis models internal stresses as a connected line with axial forces in branches and loads applied to its nodes. To discretise the thrust line, B branches and N nodes are used such that $B = N - 1$. An example of this discretisation can be seen in Fig. 1.

The position of the generic j-th node is identified, with respect to a Cartesian reference frame, by the two-dimensional vector $\mathbf{x}_j = [x_j, y_j]^T$. Each node can be loaded by external forces whose components are collected in the generic vector $\mathbf{f}_j = [f_{jx}, f_{jy}]^T$. In addition, the first and last nodes of the arch experience reactions from the left and right springers, which are denoted as $\mathbf{r}_l$ and $\mathbf{r}_r$, respectively, see, e.g., Fig. 1. Nodes may also experience thrust forces $\mathbf{t}_j$ transmitted by the adjacent branches.

Therefore, the equilibrium equation of the generic j-th node turns out to be:

$$\mathbf{t}_j^{j-1} + \mathbf{t}_j^j + \mathbf{f}_j = \mathbf{0} \tag{1}$$

in which $\mathbf{t}_j^{j-1}$ and $\mathbf{t}_j^j$ denote the thrust forces that are transmitted to the j-th node by the branches $j-1$ and j, respectively. By analysing the equilibrium of the first and last nodes of the thrust line, it is possible to derive the following expressions

$$\mathbf{r}_l + \mathbf{t}_1^1 + \mathbf{f}_1 = \mathbf{0}, \qquad \mathbf{r}_r + \mathbf{t}_N^{N-1} + \mathbf{f}_N = \mathbf{0} \tag{2}$$

so that the springers' reactions can be written in terms of the thrust forces and external loads.

In the framework of TLA, the horizontal position of the nodes and external forces are considered known, while the associated thrusts of the branches and the vertical position of all nodes are unknown. Hence, the set of equilibrium equations for all nodes consists of $2\,(N-1)$ equations with unknowns composed of B thrust forces and N vertical coordinates.

In accordance with Heyman's principles, the thrust line should be located within the limits of the arch thickness, meaning that the vertical coordinates of nodes must satisfy the inequalities $y_{j,\min} < y_j < y_{j,\max}$, where $y_{j,\min}$ and $y_{j,\max}$ are the heights of the intrados and extrados of the arch, respectively, see, Fig. 1. Additionally, the thrust force associated with the generic j-th branch of the thrust line must be a compressive force.

This set of equations and inequalities can have infinite, single, or no solutions. In the first scenario, one is focused on determining two extreme thrust line configurations corresponding to the minimum and maximum values of the horizontal thrust component at the springers. They are respectively called the solutions of minimum and maximum thrust. These solutions are also known as the deepest and shallowest solutions since they indicate the deepest and shallowest geometry of the thrust line.

Without delving into the mathematical details for determining these solutions, for which the reader is referred to the original publication [8], we shall now show the procedure for including FRCM reinforcement in the arch analysis. Specifically, this approach involves enlarging the boundary region of the arch, which is defined by its thickness, by including the effects of the FRCM.

The iterative evaluation of the domain size considers various factors that affect the reinforced arch's strength, including the limited compressive strength of masonry, the reinforcement tensile and detachment failure, and the computed internal actions. The iterative procedure starts by first computing the axial stress in the arch cross-sections that are intersected by the thrust line. This is achieved by projecting the thrust along the unit vectors tangent to the longitudinal axis.

Therefore, by means of the computation of the resisting bending moment in each cross-section, the expansions of the extrados and intrados boundaries of the

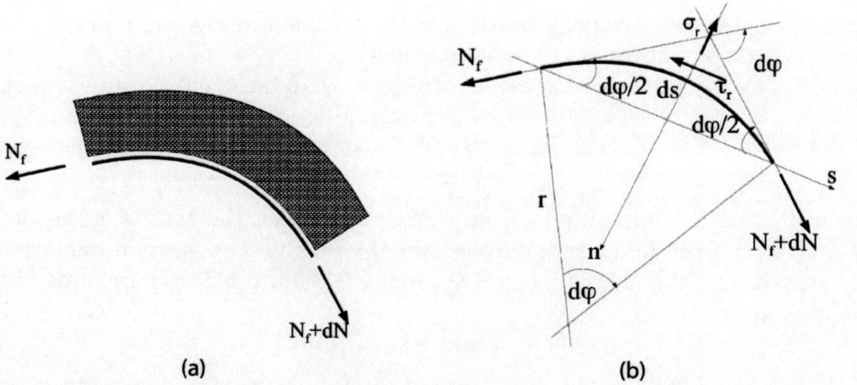

Fig. 2. Pictures taken from the Italian code CNR-DT 215/2018. (a) Reinforcement and the bond arch surface; (b) Curvilinear elementary section of length ds.

domain along the longitudinal axis, referred to as e^+ and e^- respectively, are determined as

$$e^+ = \frac{M_R^+}{N}, \qquad e^- = \frac{M_R^-}{N} \tag{3}$$

where N is the axial force and M_R^+ (M_R^-) is the positive (negative) resisting bending moment, according to whether it produces compression (tension) at the cross-section extrados.

The resisting bending moments M_R^+ and M_R^- are determined by a cross-section analysis, assuming that the FRCM in tension and the masonry in compression reach their ultimate strain values.

The ultimate strain of the masonry ε_{mu} is computed, following the guidelines of the Italian code CNR-DT 215/2018, by considering an elastic-plastic stress-strain law without accounting for tensile strength. The FRCM ultimate strain is determined by taking into account two collapse mechanisms: the tensile and detachment of the reinforcement. Specifically, the ultimate tensile strain of the reinforcement ε_{fut} is evaluated assuming an elastic stress-strain law without accounting for compressive strength. According to the Italian code CNR-DT 215/2018, the detachment failure between the reinforcement and the bond arch surface, (see, e.g., Fig. 2(a)), occurs due to the combined effect of the longitudinal stress τ_r induced in the reinforcement by the variation of the axial stress and the radial stress σ_r caused by the curvature of the bond arch surface, see Fig. 2(b). From the equilibrium along the radial and longitudinal direction of a curvilinear elementary section of length ds, one can arrive at

$$\sigma_r = \frac{N_f}{b_f\, r} \tag{4}$$

$$\tau_r = \frac{1}{b_f} \frac{dN}{ds} \tag{5}$$

where b_f is the reinforcement width, r is the radius of curvature and N_f is the axial stress on the fibres of the reinforcement.

Hence, by combining the previous stresses, the detachment failure is reached if

$$\frac{\sigma_r}{2} + \sqrt{\frac{\sigma_r^2}{4} + \tau_r^2} = \sigma_a \tag{6}$$

in which σ_a is the minimum tensile strength between the tensile strengths of the matrix and the bond arch surface. Finally, in the cross-section analysis for the evaluation of the resisting bending moment, the ultimate strain value of the FRCM is given by

$$\varepsilon_{fu} = \min(\varepsilon_{fut};\ \varepsilon_{fud}) \tag{7}$$

in which the details for computing the ultimate detachment strain of the reinforcement ε_{fud} will be explained in a forthcoming paper.

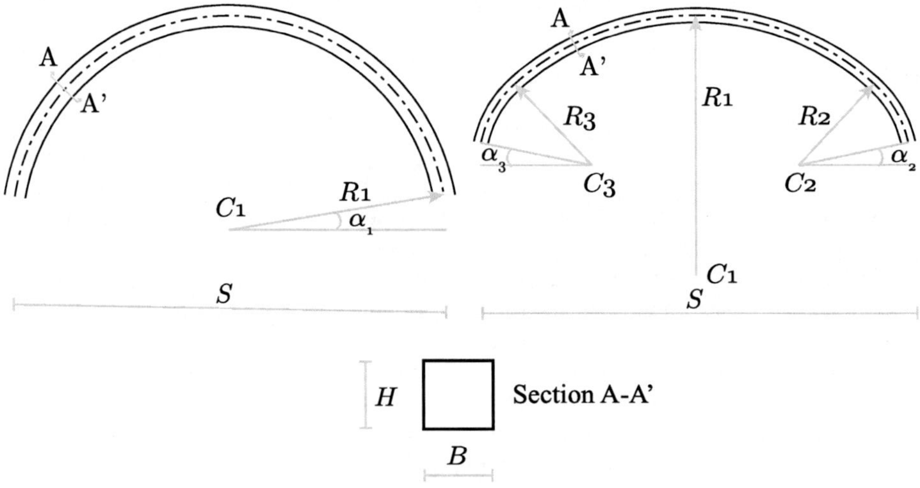

Fig. 3. Circular (left) and three-centred (right) arches.

3 Numerical Examples

Two numerical examples pertaining to circular and three-centred arches are here shown, see Fig. 3. The mechanical properties conform to the specifications provided in the Italian code CNR-DT 215/2018. In accordance with these guidelines, the Young modulus of the masonry is set at 1200 MPa, and the ultimate strength (strain) of the masonry is established as −2.4 MPa (−0.0035). Additionally, the Young modulus of the FRCM is set to 200 000 MPa, while its ultimate strength is established as 1200 MPa.

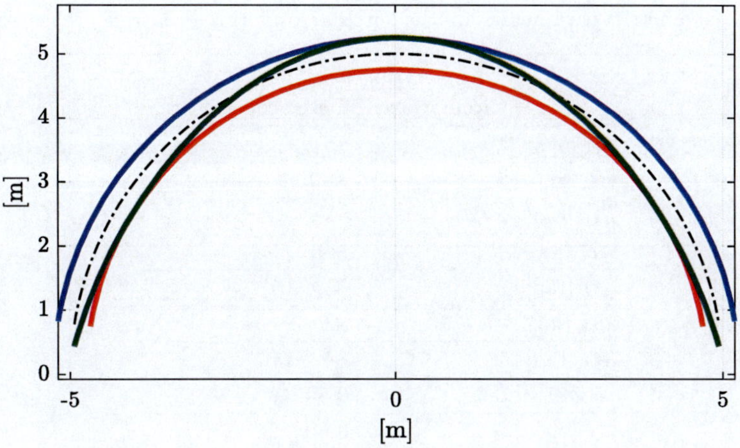

(a) *scale factor* $s = 1$

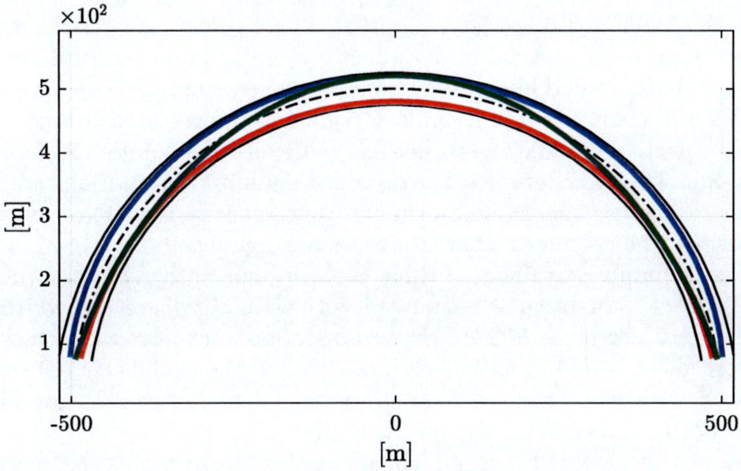

(b) *scale factor* $s = 100$

Fig. 4. Unreinforced circular arches and their deepest solutions. Blue line = upper limiting region; red line = lower limiting region; Blue line = upper limiting region; green line = thrust line.

In the first numerical example, an investigation is carried out to examine the scale effect of an unreinforced circular arch, taking into account the limited compressive strength of the masonry. When accounting for the limited compressive strength of masonry, it can be observed that the boundary of the safety region is smaller than the thickness of the arch. Consequently, the applicability of the law of proportions is no longer valid. An example of the above issue can be observed in Fig. 4(a), which displays a circular arch characterised by the geomet-

Table 1. Geometric parameters of the circular and three-centred arches shown in Fig. 3.

	Circular arch	Three-centred arch
S [m]	10	15
R_1 [m]	5	~ 9.21
α_1 [rad]	$\pi/20$	–
R_2 [m]	–	3.9
α_2 [rad]	–	$\sim \pi/15$
R_3 [m]	- -	3.9
α_3 [rad]	–	$\sim \pi/15$
B [m]	0.50	0.50
H [m]	0.50	0.50

ric properties listed in Table 1, and Fig. 4(b). This figure illustrates the arch with its geometric properties scaled up by a factor s of one hundred. Both arches are subjected to a distributed load proportional to their weight. The upper and lower boundaries of the equilibrated boundary region are represented by the blue and red lines, respectively, while the thrust line of the deepest solution is depicted by the green line. In both examples, the obtained equilibrated solutions are smaller than their respective thicknesses, although this disparity is more pronounced in Fig. 4(b) when the geometric characteristics are significant.

The last example pertains to a three-centred arch, with geometric properties listed in Table 1. This arch is reinforced with FRCM, where the width of the reinforcement is set to be 30% of the cross-section width, and the thickness of the reinforcement is 0.01 mm. The deepest (a) and the shallowest (b) solutions, equilibrated with a distributed load proportional to its weight, are shown in Fig. 5. In both solutions, an expansion of the boundary region is observed in comparison to the geometric region corresponding to the arch's thickness. However, at the imposts, the boundary domain is found to be smaller than at the voussoir and the keystone of the arch. This occurrence arises from the combined effect of higher axial stress values and lower detachment ultimate strain values due to the smaller radius of curvature.

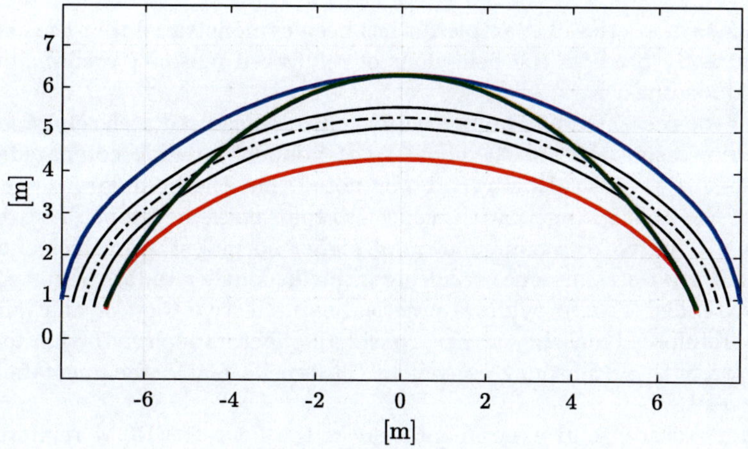

(a) *deepest solution*

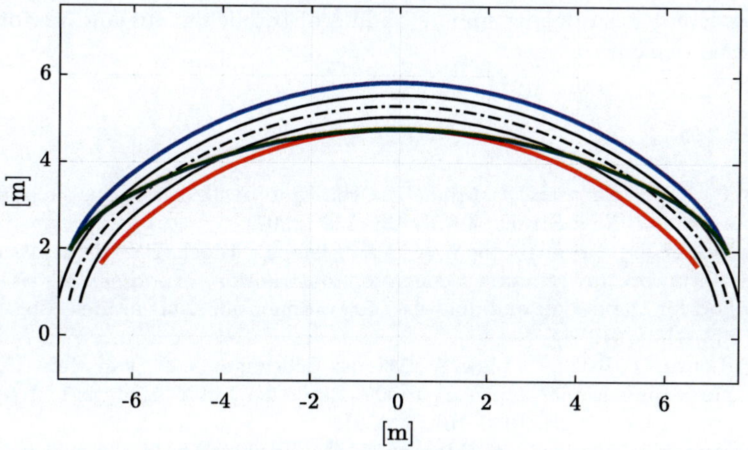

(b) *shallowest solution*

Fig. 5. Three-centred arch: deepest (a) and shallowest (b) solutions. Blue line = upper limiting region; red line = lower limiting region; Blue line = upper limiting region; green line = thrust line.

4 Conclusions

The extension of the Thrust Line Analysis (TLA) to incorporate the effects of Fabric Reinforced Cementitious Matrix (FRCM) reinforcement in masonry arches has been studied. The proposed extension successfully expands the geometric domain of the thrust line to consider the effect of FRCM reinforcement on arch behaviour.

By means of a series of examples, it has been demonstrated that the extended TLA accurately predicts the behaviour of reinforced masonry arches subjected to vertical loadings.

One of the presented examples examined a three-centred arch reinforced with FRCM, showcasing the enlargement of the boundary region compared to the arch's geometric region. However, it was noted that the boundary domain was smaller at the arch springers with respect to that at the keystone. This discrepancy was attributed to a combination of higher normal stress values and lower detachment limit strain values resulting from the smaller radius of curvature.

The extended TLA provides a practical and effective tool for analysing and designing reinforced masonry arches, considering factors such as the limited compressive strength of masonry as well as the tensile and detachment failure of reinforcement.

Further research and experimentation in the field of FRCM reinforcement and its integration with the TLA are certainly needed. The investigation of different arch geometries, reinforcement configurations, and loading conditions would contribute to a deeper understanding of the behaviour and performance of reinforced masonry arches.

References

1. Roca, P., López-Almansa, F., Miquel, J., Hanganu, A.: Limit analysis of reinforced masonry vaults. Eng. Struct. **29**(3), 431–439 (2007)
2. López López, D., Roca, P., Liew, A., Van Mele, T., Block, P.: Tile vaults as integrated formwork for reinforced concrete: construction, experimental testing and a method for the design and analysis of two-dimensional structures. Eng. Struct. **188**, 233–248 (2019)
3. López López, D., Roca, P., Liew, A., Méndez Echenagucia, T., Van Mele, T., Block, P.: A three-dimensional approach to the Extended Limit Analysis of Reinforced Masonry. Structures **35**, 1062–1077 (2022)
4. Melchiorre, J., Manuello, A., Marano, G.C.: Application of a machine learning algorithm for the structural optimization of circular arches with different cross-sections. J. Appl. Math. Phys. **9**(5), 1159–1170 (2021)
5. Manuello, A., Marano, G.C.: Numerical and dimensionless analytical solutions for optimization of circular arches. Eng. Struct. **253**, 113360 (2022)
6. Melchiorre, J., Manuello, A., Marmo, F., Adriaenssens, S., Marano, G.C.: Differential formulation and numerical solution for elastic arches with variable curvature and tapered cross-sections. Eur. J. Mech. - Solids A **97**, 104757 (2023)
7. Roselli, F., Alforno, M., Manuello Bertetto, A., Venuti, F.: Experimental investigation of the influence of brick patterns on the structural response of masonry arches and barrel vaults. Constr. Build. Mater. **368**, 130434 (2023)
8. Marmo, F.: ArchLab: a MATLAB tool for the thrust line analysis of masonry arches. Curved Layered Struct. **8**(1), 26–35 (2021)
9. Block, P., Ochsendorf, J.: Thrust network analysis: a new methodology for three-dimensional equilibrium. J. Int. Assoc. Shell Spat. Struct. **48**(3), 167–173 (2007)
10. Marmo, F., Rosati, L.: Reformulation and extension of the thrust network analysis. Comput. Struct. **182**, 104–118 (2017)

N'Albero, A Temporary Spatial Structure Made of Steel Scaffolding System: Loading Tests as Tool to Check Structural Design

Laura Giovanna Guidi(✉), Giuseppe Brandonisio, and Antonello De Luca

Università Degli Studi Di Napoli Federico II, Dipartimento Di Strutture Per L'Ingegneria E L'Architettura, Naples, Italy
lauragiovanna.guidi@unina.it

Abstract. The paper deals with design and construction stages of N'Albero, a temporary spatial steel structure built to be the main 2016 Christmas touristic attraction on Naples seafront. Opened to the public for three months, it was visited by more than 250.000 people, estimating the simultaneous presence of 750 visitors. N'Albero main structure consisted of 20.000 prefab steel pipe elements, adopting Allround System by LAYHER, a fast-assembly, safe and economic solution. The whole spatial system was about 45 m-high, with a squared base (30 m × 30 m), tapering gradually to the top through three sky terraces. Using light multidirectional scaffoldings, the main design aspects to consider were: stability, uplift or sliding of the structure; strength of components sufficient to withstand loads without failing; serviceability of a light structure under large deflections. Estimating a dead load of about 200 tons, a ballasting load of 130 tons was used to ensure global stability against wind force. For this deformable system, loading tests were performed during construction to check design, applying inclined (45°) concentrate forces at each terrace level: the effect of these forces, assigned at +13,00 m (F = 5t), +19,00 m (F = 5t) and +31,00 m (F = 2t), was obtained using steel tendons, connected to the structure through an additional steel truss, while the acting hoist was anchored to a crane as counterweight. During tests, all the applied forces were registered by a dynamometric cell. No visual damages or elements instabilities occurred in situ, validating design calculations. Furthermore, these tests revealed a high deformability of the structure, not readable from FEM model, requiring precautionary changes to the original solution. This makes loading tests a reliable way to check and improve structural design in the case of framework with high deformability.

Keywords: Temporary spatial structure · steel scaffolding · loading test · stability · ballasting system

1 A Temporary Landmark Along Napoli Waterfront

1.1 The Structure and Its Context: An Exemplary Proposal

The temporary installation "N'Albero", in Fig. 1, was placed along Napoli waterfront, at Rotonda Diaz, standing for about three months (December 2016 – March 2017). It became an amazing touristic attraction during 2016 winter holydays. This scaffolding

S. Gabriele et al. (Eds.): IWSS 2023, LNCE 437, pp. 199–209, 2024.
https://doi.org/10.1007/978-3-031-44328-2_21

Christmas tree was illuminated by 1,300,000 LED lamps and adorned with 900 pot plants of viburnum, later replanted in green area of the Parthenopean city.

Fig. 1. View of Via Caracciolo in Napoli: N'Albero at Rotonda Diaz.

The whole structure was about 45 m above ground. At the base, it was characterized by a bottom 7 m-high prism with squared base (30 m × 30 m), including services, ballast system to stabilize the light structure, and a gallery-shop area accessed for free. The upper 12 m-high central parallelepiped (15 m × 15 m) led to the second sky terrace at +19.00 m above the ground: it had a squared base of (21.00 m × 21.00 m), 15.00 m-large in central core and (3.00 m + 3.00 m) due to cantilever slabs. Progressively the central section reduced, until reaching a squared base (15 m × 15 m) at +31.00 m in correspondence of the third sky terrace (Fig. 2). At the top, the central core reduced to (2.50 m 2.50 m) squared base: it included stairs and lifts, making N'Albero an accessible structure free from any architectural barriers. Becoming a tourist attraction for more than 250 thousand people, N'Albero wanted to relaunch the image of Napoli on international scale, as occurred for previous similar temporary structures in Mexico City, with a record 110 m-high artificial tree, and in Rio De Janeiro, with the largest 84 m-high floating Christmas tree.

N'Albero was a very deformable structure made of scaffolding system: its design revealed the necessity to prioritize variable actions, firstly wind, adopting a monitoring system to guarantee users' safety. The same happened to "He" by Bam!(Fig. 3), a temporary installation which made its debut in the square of MAXXI in Rome on June 10, 2013, transforming the concrete facade of the Zaha Hadid-designed museum through a yellow, translucent and aerostatic prism.

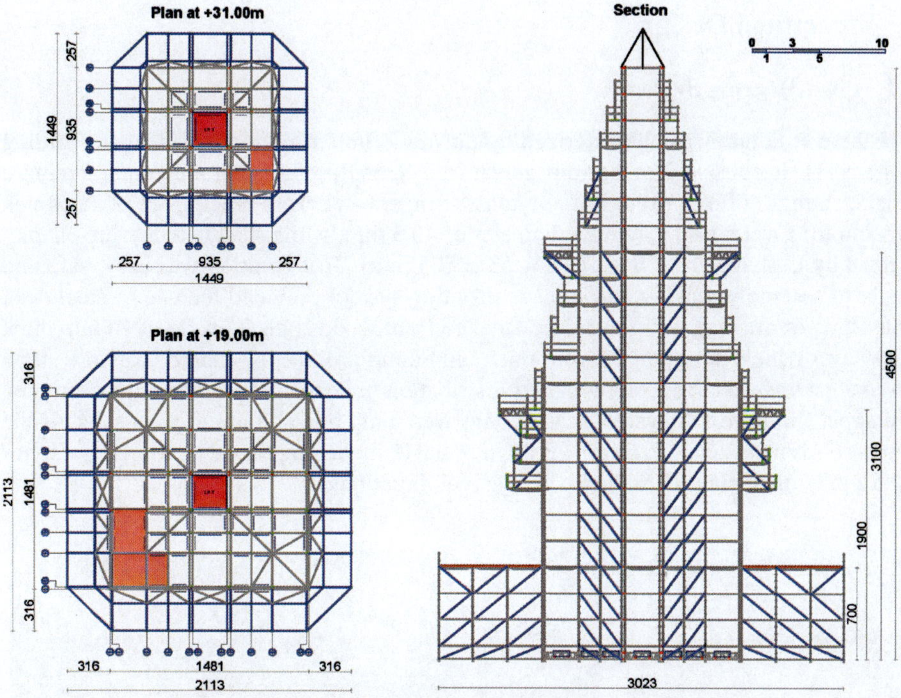

Fig. 2. N'Albero plans at (+19.00 m) (+31.00 m) and section

This diaphanous tensile structure provided an upper cable ring, using two fixing stays for each of four vertices. To preserve its squared shape, each lower vertex was connected to the ground by climbing stoppers. "He" was steadily monitored with anemometers, fixing limit to allowable wind velocity, as occurred for N'Albero.

Fig. 3. "He" by Bam!, MAXXI in Rome, 2013.

2 Structural Design

2.1 Load-Bearing System

N'Albero structure was characterized by *Layher Allround* multidirectional scaffolding system [1]. It consisted of hot-dip galvanized S235J prefab steel elements, having a tensile strength of 320 MPa. For columns, stringers and cross bracings 3,22 mm-thick pipe profiles were used, having a diameter of 48,3 mm, while steel beams were characterized by U-shape cross Sect. (48 × 53 × 2,5 mm). This solution was easy, safe and faster to assembly. *Layher Allround* scaffolding system provided the use of steel deck consisting of drawing hot-galvanized steel coil and wooden deck, made of 10 mm thick multilayer panels put upon 1,5 mm-thick aluminum profiles. N'Albero structure stood on road paving: to have a more regular distribution of concentrated loads transferred by the upper framework, the steel scaffoldings were put upon aluminum panels, as visible Fig. 4, ("Arena panels", 2,15 m × 3,00 m × 0,048 m, having a load bearing capacity of 1,86 t/m2), arranging them in two orthogonal directions.

Fig. 4. N'Albero construction: (left) ballasting plane and (right) internal core.

To guarantee a homogeneous load transfer on road surface, an adding dry foundation was adopted, introducing a 30 cm-high layer of mixed grain size below the aluminum panels. Considering the whole structure load (200 tons, automatically defined by FEM program), it could be valued a pressure of 900 kg/m^2, lower than carrying capacity of road surface. To avoid any overturning or sliding problems, a ballasting system was used: it consisted of 44 + 40 elements, 1,5 tons each one, for an overall stabilizing load of 126 tons. Meanwhile, to increase deck stiffness, plan bracings were introduced every two levels in correspondence of the core, as visible in Fig. 5.

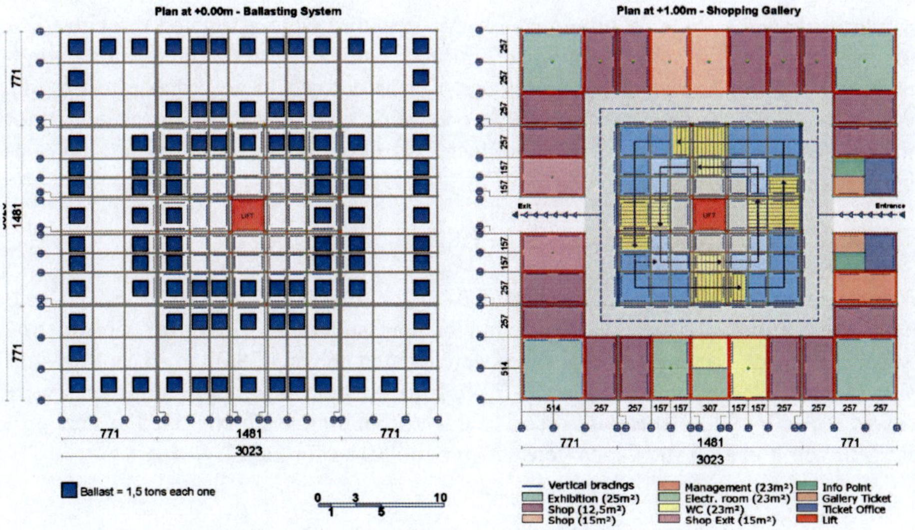

Fig. 5. N'Albero plans: ballast (+0,00 m) and functional distribution (+1,00 m).

2.2 Design Criteria and Loads Estimation

As temporary structure, N'Albero design referred to a *Working Life Category I* (nominal lifetime lower than 10 years), in accordance to UNI ENV 1993-1-1:2005 EUROCODE 3 [2] and Italian NTC08 [3]. Lasting less than 2 years, N'Albero did not require to be verified for seismic load. It was built by ITALSTAGE, according to architectural design by engineer A. Mazzarella and structural design by engineer F. Iasevoli. As a light tall building, the basic design criteria to be satisfied became [4–6]: *stability* against overturning, *uplift* or sliding of the structure as a whole; *strength* of the structural components of the building that was required to be sufficient to withstand imposed loading without failure; *serviceability* of a structure, whose inter-storey and overall deflections were expected to remain within acceptable limits.

Apart from dead loads (about 200 t), it was necessary to consider the most impactful live actions, as crowd and wind loads. N'Albero has been designed and verified to accommodate at most 747 people, 278 in the shopping gallery, 371in the restaurant at the lower terrace, 98 visitors for the other two sky terraces. For crowd load, variable actions were considered, changing with categories of occupancy at each level. In particular, for the first terrace, at +7,00 m, the category C1 (NTC08 or UNI-ENV 1993 -1-1-2005) was considered, having $Q_k = 300$ kg/m^2; for the upper terraces, at +19,00m and +31,00m, the category C3 was assumed, corresponding to $Q_k = 500$ kg/m^2.

Wind load was estimated according to CNR DT 207/2008 [7] ("*Istruzioni per la valutazione delle azioni e degli effetti del vento sulle costruzioni*"). For wind Zone 3 (that includes Napoli in Campania region), it could be considered a basic wind velocity $v_{b,0} = 27$ m/s. The corresponding design valued could have been reduced through a coefficient ($c_r = 0,75$) which considered the returned period of the event ($T_r <$ 1year). However, designers decided to adopt no wind load decrements, so the target velocity,

v_r, corresponded to $v_{b,0}$. Wind effects on this structure were estimated through FEM analysis, applying the most severe distribution of equivalent static forces on all wind-exposed surfaces. The peak wind pressure, p, was defined in accordance to Eq. (1), as function of: q_r, kinetic pressure; v_r, target velocity; $\rho = 1{,}25$ kg/m^{3}as density of air; c_p pressure coefficient, c_d dynamic factor, c_e exposure factor depending on c_t as topographic factor defined for a terrain *Category D,* $c_e = f(c_t)$.

$$p = q_r \cdot c_e \cdot c_p \cdot c_d = \left(0,5 \cdot \rho \cdot v_r^2\right) \cdot c_e \cdot c_p \cdot c_d \tag{1}$$

For steel uprights they have been assumed: $c_p = 2$, c_e (max at 45 m) = 3.73, $c_d = c_t = 1$. It has been estimated a mean value of wind load pressure of 343,16 kg/m^2. Considering N'Albero layout, no windproof surfaces were located above +29,00 m, except for steel pipe elements, stairs and lift. Form +7,00 m to +29,00 m windproof surfaces were characterized by 900 plants with roots, 1,20 – 1,50 m high, each one fixed to steel pipe element with three plastic bands (tensile strength: 50 kg for band).

Table 1. Synthesis of acting loads

Load type	Q (kg)	% TOT Vertical loads
Dead	200.000	**37,72%**
Super-imposed dead	44.000	**7,63%**
Live load- Terrace (1)	157.000	**27,25%**
Live load- Terrace (2)	112.500	**19,53%**
Live load- Terrace (3)	62.500	**10,85%**
Tot Live load	332.000	**57,63%**
Total Vertical load	576.000	**100%**
Total wind force	63.900	**11,09%**

The total wind force was approximately 10% of overall vertical loads (Dead + Super-imposed dead + Live), as visible in Table 1. N'Albero was monitored with an anemometer, an airspeed detection system put at the top of the structure: if wind velocity was greater than 20 m/s, the structure should have been fully evacuated.

2.3 FEM Analysis: Results and Check

The process of structural verification consisted in checking N'Albero behavior for Ultimate limit state (ULS) and for Serviceability limit state (SLS). A FEM model of the structure was analyzed, by using software SAP2000. The primary aspect to consider for this deformable steel structure was the global buckling. To this aim, an empirical method was adopted to evaluate buckling effects [8–10]. Fictitious forces as 1% of the overall vertical loads were applied, adequately distributed to each level: deformed shapes were defined, valuing a maximum drift (δ) at the top of 1,86 cm (δ/H = 1/2445), corresponding to a N-to-N_{crit} ratio of 1:10, as visible in Fig. 6.

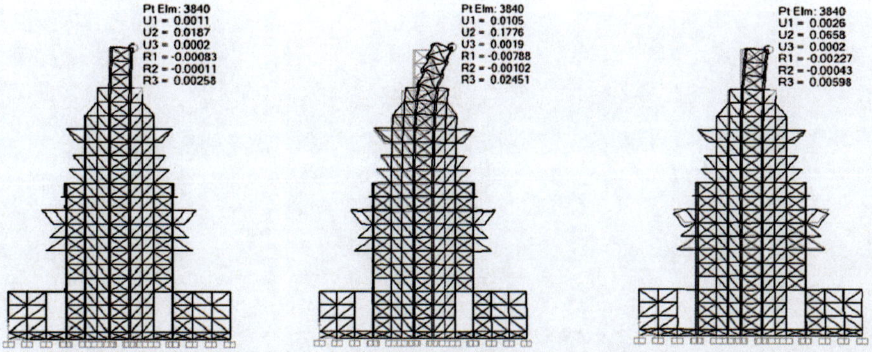

Fig. 6. FEM outputs: (left) buckling analysis, (center) effects wind fictitious uniformly distributed load and effects of wind effective load distribution (right).

To evaluate wind effects [11, 12], two different load assignments were compared. A fictitious uniformly distribution of total wind force for each level led to a maximum drift at the top (δ_1) of 17,76 cm ($\delta_1/H = 1/254$), about 10 time greater than the maximum value obtained for buckling. On the contrary, for the effective wind load distribution, a maximum displacement at the top, δ_2, of 6,58 cm ($\delta_2/H = 1/684$) has been estimated. In this case, the choice of eliminating windproof surfaces above +29,00 m, concentrating them at the bottom of the structure, led to reduce drift due to wind load at 1/3 of the expected value.

3 Loading Tests as Tool for Structural Design

3.1 Checking N'Albero Deformability *In Situ*

Considering that the structure was characterized by deformable connections, it was necessary to check and monitor its deformability also during construction. Three loading tests were performed, also to check the right installation of steel structural elements. Tests were carried out by CMG TESTING Srl, under the supervision of Prof. Eng. A. De Luca as inspector. The effect of the three concentrated forces, applied at +13,00 m, +19,00 m and +31,00 m respectively, was simulated using steel tendons, attached to the structure through an additional steel truss, visible in Fig. 7. To obtain the worst torsional effects on the structure, horizontal forces were inclined at 45° to the main plans direction (X and Y), parallel to the square base diagonal, as in Fig. 8. The hoist was anchored to a crane as counterweight. During tests, all the applied forces were registered by a dynamometric cell (DIGITAL DYNAMOMETER T1800 Mod.CR5/3.00). Drifts were defined through topographic measurements, using a total station "ZEISS S10". At each level, four control points were considered, as visible in Fig. 8: (P_1) is the point of application of concentrated force by tendons; (P_2) follows the first point along x-direction, staying aligned to the earliest frame of the stairwell; (P_3) is symmetrical on y-axis to (P_1); (P_4) is symmetrical on base-diagonal to (P_1). Thanks to the truss transferring system, the force was distributed among 12 uprights (six for direction), included in the pulling angle.

Fig. 7. Test at (+13,00m): connection to crane and tendons details

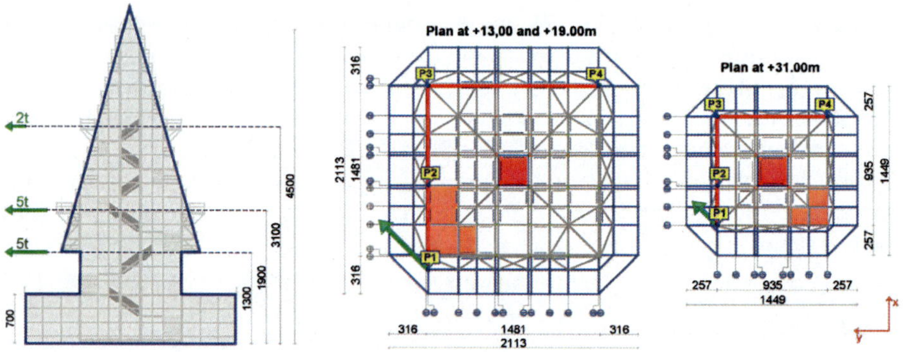

Fig. 8. Loading tests scheme: force layout

For Test 1, summarized in Table 2, and Test 2, summarized in Table 3**,** a maximum force of 5 tons was applied, respectively at +13,00 m and at +19,00 m. The highest tensile force was reached through eight loading steps, as: (S_1) early loading 0-2 tons; (S_2) 2-4 tons, (S_3) 4-5 tons; (S_4) early unloading 5-4 tons; (S_5) 4-2 tons; (S_6) 2-0 tons; (S_7) final loading 0-2 tons; (S_8) final unloading 2-0 tons.

Table 2. Test 1 (+13,00 m), $F_{max} = 5t - \delta_{lim} = H/500$ (=26mm)

Node	δ_{TOT} [mm]	δ_{FEM} [mm]	δ_{Res} [mm]	δ_{Exp} [mm]	$\delta_{TOT}/\delta_{FEM}$	$\delta_{TOT}/\delta_{Res}$	$\delta_{TOT}/\delta_{Exp}$	$\delta_{Exp}/\delta_{FEM}$
P1 δtot	30,21	2,43	10,38	19,83	**12:1**	3:1	2:1	**8:1**
P2 δtot	21,79	1,39	9,98	11,81	**16:1**	2:1	2:1	**8:1**
P3 δtot	22,32	1,98	3,85	18,47	**11:1**	6:1	1:1	**9:1**
P4 δtot	6,48	0,198	5,30	1,18	**33:1**	1:1	5:1	**6:1**

Table 3. Test 2 (+19,00 m), $F_{max} = 5t - \delta_{lim} = H/500$ (=38 mm)

Node	δ_{TOT} [mm]	δ_{FEM} [mm]	δ_{Res} [mm]	δ_{Exp} [mm]	$\delta_{TOT}/\delta_{FEM}$	$\delta_{TOT}/\delta_{Res}$	$\delta_{TOT}/\delta_{Exp}$	$\delta_{Exp}/\delta_{FEM}$
P1 δtot	51,74	11,46	16,48	35,26	**5:1**	3:1	1:1	**3:1**
P2 δtot	39.91	7,62	15,85	21,06	**5:1**	2:1	2:1	**3:1**
P3 δtot	36,61	7,10	15,38	21,23	**5:1**	2:1	2:1	**3:1**
P4 δtot	9,67	1,50	4,64	5,03	**6:1**	2:1	2:1	**3:1**

Table 4. Test 3 (+31,00 m), $F_{max} = 2t - \delta_{lim} = H/500$ (=62 mm)

Node	δ_{TOT} [mm]	δ_{FEM} [mm]	δ_{Res} [mm]	δ_{Exp} [mm]	$\delta_{TOT}/\delta_{FEM}$	$\delta_{TOT}/\delta_{Res}$	$\delta_{TOT}/\delta_{Exp}$	$\delta_{Exp}/\delta_{FEM}$
P1 δtot	38,89	19,12	16,81,	22,08	**2:1**	2:1	2:1	**1:1**
P2 δtot	30,51	14,66	12,75	17,76	**2:1**	2:1	2:1	**1:1**
P3 δtot	23,79	6,15	9,94	13,85	**4:1**	2:1	2:1	**2:1**
P4 δtot	6,32	2,08	3,96	2,36	**3:1**	2:1	3:1	**1:1**

Fig. 9. Tests at (+19,00 m) and at (+31,00 m): loading phase

During Test 3, visible in Fig. 9 and synthetized in Table 4, a maximum horizontal force of 2 tons was applied at +31,00 m. This peak value was reached by four steps: (S_1) loading 0-1 tons; (S_2) 1-2 tons; unloading (S_3) 2-1 tons; (S_4) 1-0 tons. All the aforementioned tables include: δ_{TOT}, drift measured when the maximum testing force was applied (final loading step); δ_{Res}, residual displacement at the end of each test (final unloading step), comparable to that one due to the backlash of connection; δ_{Exp}, maximum drift cleaned by residual displacements; δ_{FEM}, drift obtained as FEM analysis output; δ_{lim}, the maximum allowable displacement, set at H/500. Figure 10 shows the chart type to define δ_{Exp}.

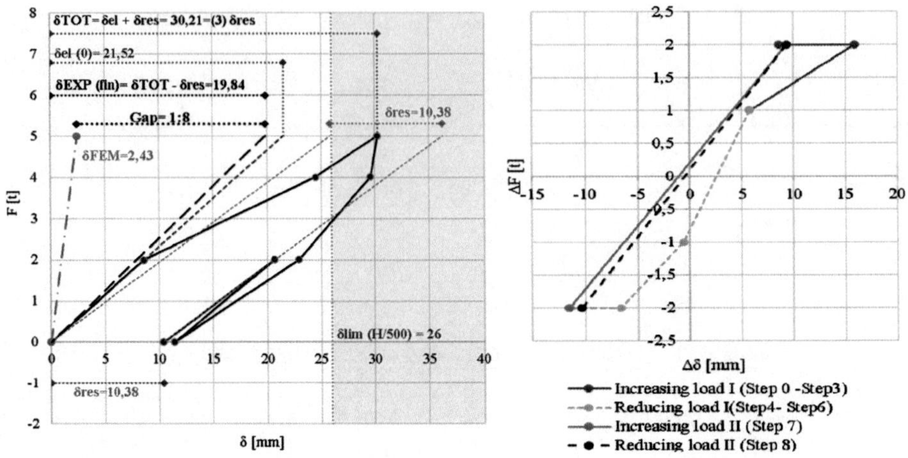

Fig. 10. Chart type to define δ_{Exp}: results from Test1 – Node P1- F-δ and ΔF- Δδ

Results show considerable local effects in correspondence of P1; furthermore, the applied forces result to be distributed in the whole structure. On the other hand, the evaluated drifts appeared so far from the expected values, obtained from FEM analysis. Loading tests checked a huge deformability of the structure, due to steel pipe connections mainly and showed a structural deformability about 10-time greater than that one obtained from the elastic theory.

4 Conclusions

Referring to a high deformable spatial structure as N'Albero, the paper shows as loading tests may be considered a reliable way to check and improve structural design. No visual damages or elements instabilities came out from FEM analyses of the case study. However, loading tests revealed a high deformability, non-deductible from modelling. Considering test results, structural changes have been made to improve structural stability, also complying architectural requirements. Tests in situ actually become the most efficient way to guarantee a correct design when structural behavior cannot be easily simulated through FEM analysis, being influenced by not moldable mechanical aspects (as bolt-hole clearance).

References

1. Layher.: Layher Allround. Il Sistema Polifunzionale per l'edilizia. (2019). https://www.layher.it/cataloghi/2019_Allround_Edilizia.pdf
2. UNI EN 1993-1-1:2005 Eurocode 3. Design of Steel Structures
3. Decreto Ministeriale 14 Gennaio 2008. Nuove Norme Tecniche per le costruzioni» (NTC08). GU Serie Generale n.29 del 04-02-2008 - Suppl. Ordinario n. 30. (ITA)
4. Smith, B.S., Coull, S.: Tall Building Structures: Analysis and Design. John Wiley & Sons (1991)
5. Sarkisian, M.: Designing Tall Buildings. Structure as Architecture. Routledge, Taylor & Francis Group. New York and London (2012)
6. Holmes, J.D., Bekele, S.: Wind Loading of Structures. 4th edn. Routledge, Taylor & Francis Group, New York and London (2021)
7. Consiglio Nazionale delle Ricerche (ITA). Istruzioni per la valutazione delle azioni e degli effetti del vento sulle costruzioni CNR DT 207/2008. (2009)
8. Timoshenko, S.P., Gere, J.M.: Theory of Elastic Stability. McGraw Hill, NY (1961)
9. Ballio, G., Mazzolani, F.M.: Theory and Design of Steel Structures. Chapman and Hall (1983)
10. Galambos, T.V., Surovek, A.E.: Structural stability of Steel: concepts and applications for structural engineers. John wiley & Sons, Inc. (2008)
11. Taranath, B.S.: Structural Analysis and Design of Tall Buildings. Steel and Composite Construction. CRC Press, Taylor & Francis Group (2012)
12. Simiu, E.: Design of Buildings for Wind: A Guide for ASCE 7–10 Standard Users and Designers of Special Structures. John Wiley & Sons Inc. (2011)

Structural Optimization Through Cutting Stock Problem

Raffaele Cucuzza(✉) and Giuseppe Carlo Marano

Department of Structural, Geotechnical and Building Engineering, Politecnico di Torino, Corso Duca degli Abruzzi, 24, 10129 Torino, Italy
raffaele.cucuzza@polito.it

Abstract. In this study, a novel optimization method has been applied to a geodesic dome inspired by real-world similar structures in which the environmental and cost impact has been minimized by reducing raw materials at the production stage. To achieve this goal, the Cutting Stock Problem (CSP) has been embedded inside the global optimization procedure of the entire structure. The CSP is one of the most famous combinatorial optimisation problems in the (one-dimensional) bin packing problems (BPP) class. The main objective is to produce d_j copies of each item type j (i.e. elements of the structures with the same cross-sectional Area) by employing the minimum number of bins such that the total weight in any bin does not exceed the capacity. In the civil engineering field, the traditional approach to structural optimization aims to improve the load-bearing capacity and the global performance of the structure itself. This includes, for instance, the maximization of the performance ratio through the minimization of the structure weight. However, this goal doesn't guarantee maximum efficiency in reusing structural elements and minimising waste during the industrial production phase. To overcome these limits, authors propose a stock-constrained structural optimization in which a heuristic search technique is adopted in order to find the best spatial arrangement of elements composing the structure whit the lowest cut-off waste. Finally, considerations have been discussed by comparing the solution obtained by the traditional weight-minimization approach and the stock-constrained one.

Keywords: Optimization · Steel · Genetic Algorithm · Cutting Stock Problem

1 Introduction

Since the beginning, researchers were focused on optimizing the weight or volume of specific structures by satisfying safety recommendations provided by specific standard regulations. In other words, the traditional approach which is largely adopted by the majority of the Scientific community relies on the minimum weight or cost design of structures. However, several authors demonstrated that

Supported by ADDOPTML Project, MSCA RISE 2020 Marie Skodowska Curie Research and Innovation Staff Exchange.

S. Gabriele et al. (Eds.): IWSS 2023, LNCE 437, pp. 210–220, 2024.
https://doi.org/10.1007/978-3-031-44328-2_22

optimal design derived by this approach leads to unrealistic and no practical technical solutions. Taking into account constructability issues or construction details in the optimization process assumes crucial importance if the feasibility of these optimality approaches has to be proven.

As demonstrated in this study, a considerable portion of construction expenses can be attributed to material waste resulting from the cutting process. Insufficient attention to meticulous cutting design and the minimization of waste during the construction phase can lead to suboptimal cost optimization. Additionally, the production phase may contribute to a higher cost and environmental impact expressed in terms of economic indices and CO2 emissions, respectively.

Construction and demolition activities are known to account for a significant proportion, approximately 23% (see [1]), of the total solid waste stream. This translates to an annual waste volume exceeding 100 million tonnes globally. Numerous surveys conducted worldwide, including those conducted in the United States, corroborate these estimates. Notably, a portion of the waste generated from stock reduction is preventable and stems from improper utilization of materials.

Efficient utilization of supplies could result in a reduction in the number of excess materials acquired, unnecessary craftsmanship, waste generation, as well as the associated costs of transportation and disposal. Therefore, emphasizing the importance of optimizing the cutting process in construction projects becomes crucial for minimizing waste and improving overall efficiency [1].

Indeed, optimizing resource utilization is not only beneficial to industrialists, but it also holds immense significance for the well-being of our planet. The disposal of waste resulting from stock-cutting operations can lead to pollution, while excessive wastage contributes to the depletion of our Earth's invaluable resources. By emphasizing the importance of cutting losses through optimization, we can effectively reduce our environmental impact and safeguard the future of our planet [2].

Cutting losses is possibly the most major source of steel waste. Cutting losses arise when normal steel lengths are shortened to fit the project's required lengths. Shahin et al. [3] pointed out the main causes deriving from cutting losses and provided some suggestions for optimal stock-cutting planning:

- dividing an order into separate, smaller orders typically results in more waste due to fewer cutting alternatives;
- using inefficient cutting patterns in the cutting schedule results in the generation of avoidable waste that could be avoided through better stock-cutting planning;
- using the optimum cutting patterns may result in unavoidable waste which is the minimum waste generated if the optimum cutting patterns are used

In this study, the potentiality of the cutting stock problem (CSP) is demonstrated and a novel optimization framework is developed by adopting a modified Genetic Algorithm (GA). Benefits derived from the proposed method are tested on a structural application case study inspired by a real-world structure.

2 Survey on CSP Techniques and Applications

In literature, Cutting and packing (C&P) problems appear under various specifications as cutting problems, knapsack problems, container and vehicle loading problems, pallet loading, bin packing, assembly line balancing, capital budgeting and changing coins [4]. All of these problems have essentially the same logical structure. One of the most prominent challenges in this domain is the bin packing problem (BPP), which addresses the optimization of item placement within a container. Practically, the goal is to minimize the number of containers (referred to as bins) required to accommodate a given inventory of goods. Specifically, the BPP can be defined as follows: given a set of n items, each characterized by an integer weight w_j $(j = 1, ..., n)$, and an unlimited supply of identical bins with an integer capacity c, the objective is to effectively allocate the items into as few bins as possible, while ensuring that the total weight packed into each bin does not exceed its specified limit [5].

Moreover, the cutting stock problem can be classified as a one-dimensional and two-dimensional problem [2–5].

The one-dimensional cutting stock problem (1D-CSP) involves the extraction of a specified set of order lengths from stock rods of fixed length. The primary objective is to minimize the number of rods (material input) required. On the other hand, the two-dimensional two-stage constrained cutting problem (2D-2CP) focuses on selecting the most valuable group of rectangular objects from a single rectangular plate of infinite length. It is worth noting that in the context of the two-dimensional cutting stock problem, shapes can be either regular or irregular, with the latter scenario referred to as nesting, posing a more challenging problem that demands a more intricate solution [4].

These types of problems fall under the purview of complex combinatorial optimization, constituting a strongly NP-hard problem from a mathematical standpoint. Consequently, numerous approaches have been proposed over the years, including linear programming techniques, heuristics, and metaheuristic methods, to tackle the complexities associated with these problems [6,7].

The initial attempts to tackle cutting and packing (C&P) problems can be attributed to Kantorovich in the 1960s [8], although his approach was limited in its applicability to small-scale cases. However, it played a crucial role in shaping problem understanding.

To address C&P problems, various deterministic and heuristic approaches have emerged, utilizing the technique of linear programming (LP) relaxation. These approaches involve solving the problem initially as an LP problem and subsequently converting the solution to an integer form. Relaxation results in removing the integrality constraints of variables, allowing the IP problem to be solved as an LP problem. This transformation converts the NP-hard optimization problem of integer programming into a polynomial-time solvable problem. By solving the related knapsack issue, Gilmore and Gomory [9] provided a novel method to identify the cutting patterns required to enhance the LP solution. They proposed a column generation approach inspired by the procedure developed by Dantzig and Wolfe [10] for decreasing stock and bin packing concerns

(BBP). Because enumerating all possible cutting patterns would take an inordinate amount of time, it reduces valid patterns repeatedly and adds them to the issue based on their contribution to the objective function. The column generation approach made large-scale cutting stock issues solvable in a reasonable amount of time.

In the last decade, soft computing techniques were largely used to solve these kinds of problems. Specifically, Genetic Algorithm [11], Evolutionary approach [13] and simulated annealing [14] still nowadays represent the most promising approach in terms of both computational effort and accuracy of the optimal result obtained.

Few applications have been recognized in the literature in which optimization methods and C&P formulation are adopted simultaneously to solve Civil Engineering problems or, by considering, optimal cutting stocks at the first stage of design and construction. Mainly, researchers focused on 2D structures like frames and trusses with the aim to minimize the waste during the fabrication phase or by reusing stock materials from other structures. Normally, the adopted objective functions were formulated taking into account structural cost parameters and environmental indices. In these works realized by Brutting et al. [15–17], comparisons between the 1-to-1 assignment of elements to positions in the truss and a cutting stock approach in which multiple members can be cut from individual elements are pointed out. The authors proposed a procedure to optimize the configuration of stock or kit-of-parts such that its elements can be reused in various structures. This last consideration allowed to spread of the stock of reusing items in many structures and the outcome leads to an ulterior minimization of the waste. The feasibility of the proposed approach was tested by the same authors in [18] for three different real-world application case studies. The proposed method includes form finding and digital fabrication and it applies to the design of trusses, gridshells, and space frames.

Finally, benefits derived from the proposed reusing procedure were illustrated by the same authors in [19] where the problem statement entirely based on a combination of Life Cycle Assessment (LCA) analysis and discrete optimization was introduced.

3 Combined CSP-GA Optimization Procedure

In this section, an optimization procedure for the minimum structural cost of spatial structures has been developed. The proposed method uses the potentiality of the Cutting Stock Problem for solving combinatorial problems and the effectiveness of the Genetic programming during the exploration and exploitation phase of the algorithm. Then, the performance of the proposed method will be checked in a real case study.

3.1 Problem Statement and Setting Parameters

The statement of the entire optimization process is the following:

$$\min f(x) = W(x) \tag{1}$$

$$\textbf{Where} \quad W(x) = \phi_1 \rho \sum_{g=1}^{g=k} n_g A_g L_g \tag{2}$$

$$\textbf{Subject to} \quad \frac{N_{ED}}{N_{t,RD}} \leq 1 \tag{3}$$

$$\frac{N_{ED}}{N_{c,RD}} \leq 1 \tag{4}$$

$$\frac{N_{ED}}{N_{b,RD}} \leq 1 \tag{5}$$

$$u_{max,x} \leq u_{lim,x} \tag{6}$$

$$u_{max,y} \leq u_{lim,y} \tag{7}$$

The goal of structural optimization is the minimization of the objective function $W(x)$ (see Eq. (2)) correlated to the stock mass e subjected to various structural and serviceability constraints. Specifically, $W(x)$ is the optimization objective function (OF) and represents the total mass evaluated as the sum of the purchased bars' weight for each group g. Specifically, k represents the total number of groups of elements with the same cross-sectional areas. n_g, A_g and L_g are the cardinality, the cross-sectional area and the length of bins belonging to the same group g of elements with the same cross-sectional area A_g, respectively. ρ is the material mass density assumed to be equal for all members composing the structure (in this case the steel density).

Equations (3)–(7) represent the structural constraints of the problem.

In detail, strength verifications about tensile stress (without any holes) and compression stress according to Eurocode 3 (EN 1993-12005 and EN 1993-2 2006) are introduced by Eqs. (3) and (4), respectively. Other constraints to satisfy is the maximum deflection along x and y directions (i.e. Eqs. (6) and (7)).

The mentioned-above constraints were considered in the optimization process through a suitable penalty coefficient ϕ_1. In this way, the unfeasible solutions have been penalized with respect to the feasible ones.

The resultant penalty, ϕ_1, applied to the OF is the sum of the single penalty related to each constraint's violation:

$$\phi_1 = \phi_{Nc} + \phi_{Nt} + \phi_{Nb} + \phi_{ux} + \phi_{uy} \tag{8}$$

More in detail, the violation functions are simply equal to the sum of the verification ratios

$$\phi_Q = \sum_{i}^{i=v} \frac{Q_{Ed}^i}{Q_{lim}} \tag{9}$$

where Q is the constraint parameter (stress or deflections) for an element of the truss i which does not satisfy the constraint condition. The overall penalty related to each solution, ϕ_Q, is the sum of the violation of all the violated items v. Particularly, the effective solicitation or deflection is at the numerator, Q_{Ed}, while the allowable value is at the denominator, Q_{lim}. The displacements of the free nodes in both directions had to be less than $L/300$ according to tumble recommendations largely adopted by practitioners.

The algorithm setting parameters are collected in Table 1.

Table 1. Optimization algorithm parameters set by the operator.

Parameter	Value
Maximum number of iterations	200
Number of individuals per population	200
Mutations' probability	1%
Stagnation condition	10 iterations

4 Case Study: Geodesic Dome

In this section, the results obtained by adopting the optimization framework described in the previous section will be shown.

4.1 Description of the Case Study

The geodesic dome is composed of 102 bars with 7 different lengths and inclinations and connected one each other with 43 joints. The reticulated dome is made of steel with a mass density of $\rho = 2.768\,\text{kg/m}^3$, elastic modulus $E = 275\,\text{GPa}$ and constrained by simple supports at each joint of the bottom level. More in detail, each level of the structure is realized by circular hoops, lying in the horizontal plane, and diagonal bars. Specifically, the first and second level is composed of 24 and 12 bars for the diagonals and the circular rings, respectively. Indeed, the third and four level of diagonals is composed of 18 and 6 bars, respectively, connected by a six-bar horizontal hoop. Finally, the kit-of-part is composed of tubular elements (CHS profile) connected with spherical nodes. In Fig. 1, an overview of the structure is shown and the theoretical ring which distinguishes the compressed ring from the tensioned one is reported in red.

The discrete design variables are chosen by selecting the cross-sectional areas from a standard list (EN 10210) with lower and upper bounds assumed to be equal to 182 and 24700 mm^2, respectively.

In the proposed method, the structure is subjected to only self-weight. Hence, the structural weight and, subsequently, the employed number of bins change iteration by iteration once the set of cross-sectional areas is selected by the optimizer.

As demonstrated in previous works of one of the authors (e.g. [20,21]), adopting a feasible grouping strategy is fundamental for reducing the computational effort and the global robustness of the entire optimization algorithm. Additionally, the feasible combinatorial solutions (i.e. number of bins), obtained by solving the CSP problem, are dramatically reduced. In this work, different cross-sectional areas are selected for each level of diagonals and horizontal rings. In Fig. 2, elements with the same colour are assigned to the same cross-sectional areas. Moreover, the grouping strategy is based on the level of stress in the structure: from the top to the bottom, horizontal rings and diagonals are always tensioned and compressed, respectively, with an increasing level of magnitude. For these reasons, different cross-sections (e.g. colours) were assigned to the diagonals and horizontal members at each level of the structure.

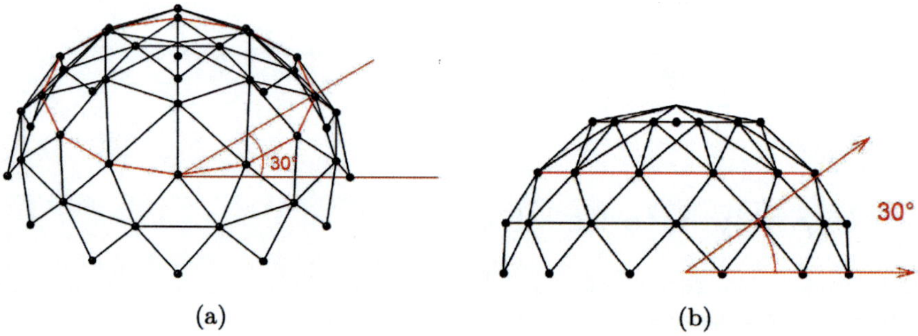

Fig. 1. Prospective (a) and perspective (b) view of the dome

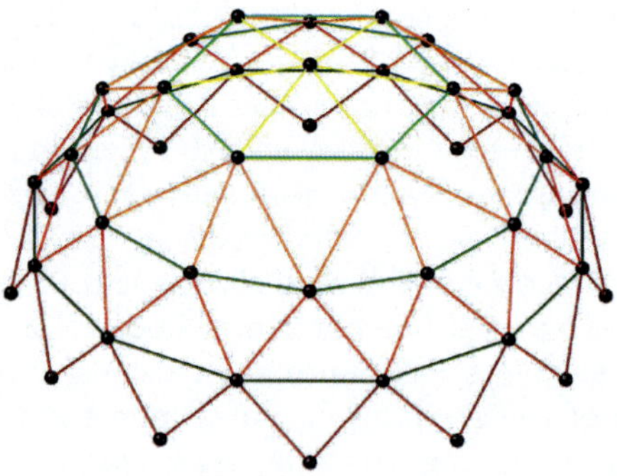

Fig. 2. Grouping strategy for the structural sizing. Each colour is representative of a specific cross-section property.

4.2 Results and Discussion

In this section, the results of the optimization process with specific regard to the investigated case study are pointed out. Specifically, two optimization scenarios have been performed:

- Scenario (a): optimization by considering CSP procedure (minimization of purchased steel bars);
- Scenario (b): optimization via traditional approach by minimizing the total weight of the structure without considering the CSP procedure (see [22, 23]).

Since the two approaches adopt different single-objective functions (minimum weight vs a minimum number of bins), the CSP procedure has been performed at the end of (b) such that the total number of bins requested for the assemblage of the optimized structure has been evaluated and a comparison with the optimal number of bins obtained by (a) would be possible. In this way, the benefit derived from incorporating the CSP procedure into structural optimization would be emphasized.

The problem statement described in the previous section has been adopted for the optimal design of the dome and 10 runs have been performed to assess the robustness and accuracy of the method. For simplicity, the authors will show the minimum value of the OF, among all runs, obtained for both scenarios.

Taking a look at Fig. 3(a)-(b), the advantages in terms of material saving derived from adopting the proposed method are quite evident. The optimal design of the dome obtained with a simple minimization of weight (scenario b) leads to significant material waste demonstrated by the no-optimal cutting pattern of pieces. Due to the adopted grouping strategy, 7 different cross-sectional areas have been selected by the optimizer aiming to maximize the structural performance of each group element. On the contrary, in scenario (a) where the CSP has been solved at each iteration of the optimization process, the total amount of bins requested at the production stage is dramatically reduced and the number of different cross-sections decreased to 4. The effect of coupled CSP with structural optimization becomes more evident through the comparison between the total amount of bins obtained by both the investigated scenarios. On the contrary, the structural mass derived by (b) results to be lower than the one obtained by (a). However, the gain in terms of mass waste evaluated in (a) seems to be more significant than the loss in terms of structural weight detected in (b).

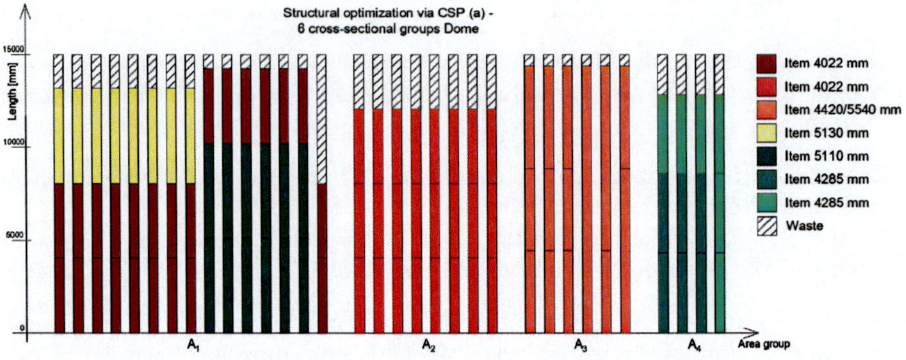

(a)

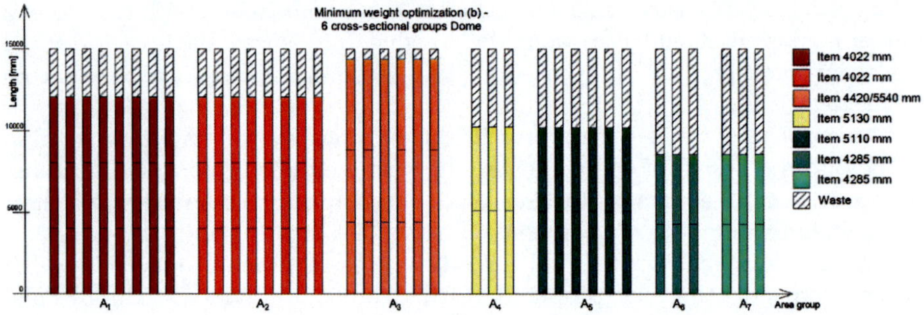

(b)

Fig. 3. Optimal cutting pattern obtained by scenario (a) and scenario (b)

5 Conclusions and Future Developments

In this work, a novel optimization procedure for the minimum structural waste has been introduced. The feasibility of the procedure has been tested and the convenience to consider constructability issues as cutting patterns during the production phase has been verified. The strategy to adopt the cutting stock procedure as an internal routine embedded in a well-known optimization algorithm, such as the Genetic algorithm, allows to achieve a significant reduction of the number of bins with a negligible increase in the structural weight. In future works, several improvements could be realized by considering geometry parameters as the total height of the dome or with new problem formulations entirely based on environmental impact indices.

References

1. Khalifa, Y., Salem, O., Shahin, A.: Cutting stock waste reduction using genetic algorithms. In: Proceedings of the 8th Annual Conference on Genetic and Evolutionary Computation, pp. 1675–1680 (2006)

2. Cheng, C.H., Feiring, B.R., Cheng, T.C.E.: The cutting stock problem-a survey. Int. J. Prod. Econ. **36**(3), 291–305 (1994)
3. Shahin, A.A., Salem, O.M.: Using genetic algorithms in solving the one-dimensional cutting stock problem in the construction industry. Can. J. Civ. Eng. **31**(2), 321–332 (2004)
4. Belov, G.: Problems, models and algorithms in one-and two-dimensional cutting (2003)
5. Delorme, M., Iori, M., Martello, S.: Bin packing and cutting stock problems: mathematical models and exact algorithms. Eur. J. Oper. Res. **255**(1), 1–20 (2016)
6. Martello, S., Toth, P.: Bin-packing problem. Knapsack problems: algorithms and computer implementations, pp. 221–245 (1990)
7. Cherri, A.C., et al.: The one-dimensional cutting stock problem with usable leftovers-a survey. Eur. J. Oper. Res. **236**(2), 395–402 (2014)
8. Kantorovich, L.V.: Mathematical methods of organizing and planning production. Manag. Sci. **6**(4), 366–422 (1960)
9. Gilmore, P.C., Gomory, R.E.: A linear programming approach to the cutting-stock problem. Oper. Res. **9**(6), 849–859 (1961)
10. Dantzig, G.B., Wolfe, P.: Decomposition principle for linear programs. Oper. Res. **8**(1), 101–111 (1960)
11. Onwubolu, G.C., Mutingi, M.: A genetic algorithm approach for the cutting stock problem. J. Intell. Manuf. **14**, 209–218 (2003)
12. Araujo, S.A.D., Poldi, K.C., Smith, J.: A genetic algorithm for the one-dimensional cutting stock problem with setups. Pesquisa Operacional **34**, 165–187 (2014)
13. Liang, K.H., Yao, X., Newton, C., Hoffman, D.: A new evolutionary approach to cutting stock problems with and without contiguity. Comput. Oper. Res. **29**(12), 1641–1659 (2002)
14. Leung, T.W., Yung, C.H., Troutt, M.D.: Applications of genetic search and simulated annealing to the two-dimensional non-guillotine cutting stock problem. Comput. Ind. Eng. **40**(3), 201–214 (2001)
15. Brütting, J., Fivet, C., Senatore, G.: Exploration of spatial structures made from reused elements and the design of optimal kits-of-parts. In: Structures and Architecture: Bridging the Gap and Crossing Borders, pp. 221–228. CRC Press (2019)
16. Brütting, J., et al.: Design of truss structures through reuse. In: Structures, pp. 128–137. Elsevier (2019)
17. Brütting, J., et al.: Optimum design of frame structures from a stock of reclaimed elements. Front. Built Environ. **6**, 57 (2020)
18. Brütting, J., Senatore, G., Fivet, C.: Design and fabrication of a reusable kit of parts for diverse structures. Autom. Const. **125**, 103614 (2021)
19. Brütting, J., Vandervaeren, C., Senatore, G., De Temmerman, N., Fivet, C.: Environmental impact minimization of reticular structures made of reused and new elements through life cycle assessment and mixed-integer linear programming. Energy Build. **215**, 109827 (2020)
20. Rosso, M.M., et al.: Nonpenalty machine learning constraint handling using PSO-SVM for structural optimization. In: Advances in Civil Engineering, pp. 1–17 (2021)
21. Cucuzza, R., Rosso, M.M., Marano, G.C.: Optimal preliminary design of variable section beams criterion. SN Appl. Sci. **3**(8), 745 (2021)

22. Cucuzza, R., Rosso, M.M., Aloisio, A., Melchiorre, J., Giudice, M.L., Marano, G.C.: Size and shape optimization of a guyed mast structure under wind, ice and seismic loading. Appl. Sci. **12**(10), 4875 (2022)
23. CUCUZZA, Raffaele, et al. Optimal strengthening by steel truss arches in prestressed girder bridges. In: Proceedings of the Institution of Civil Engineers-Bridge Engineering, pp. 1–21. Thomas Telford Ltd. (2021)

Integration of Digital Image Correlation in Flexural Tests for the Corrosion Evaluation of Ferrocement Plates

Erica Lenticchia(✉), Amedeo Manuello Bertetto, and Francesco Tondolo

Department of Structural, Building and Geotechnical Engineering (DISEG), Politecnico di Torino, Corso Duca degli Abruzzi, 24, 10129 Torino, Italy
erica.lenticchia@polito.it

Abstract. Ferrocement is a thin reinforced cementitious composites of hydraulic cement mortar with layers of wire steel mesh, which has been widely used in the roofing systems in major 20th-century architectures. However, few studies have analyzed its durability for preservation purposes and stiffens time-dependency evolution. In the present paper the results of a recent testing campaign on ferrocement specimens subjected to a corrosive environment are presented. The experimental campaign in monitored by 3D-Digital Image Correlation (DIC) signal processing. In particular, the external evidence obtained by the reconstruction of the strain field may be directly correlated with the 3-D localization of the microraking formation in the tested specimen obtained by AE detection. The aim was to assess the behavior of historical ferrocement (used by Pier Luigi Nervi in his constructions) and its durability for identifying the best procedures to protect and preserve it.

Keywords: Ferrocement · Digital Image Correlation · Corrosion · Cementitious composites · Fiber reinforced mortar

1 Introduction

As highlighted by the many international organization and experts [1–3], heritage of the 19th-20th centuries in the world is at risk. This is due to different reasons, such as the poor recognition and maintenance. Moreover, construction of reinforced concrete usually presents different fragilities connected to the durability of the materials employed. In fact, constructions in reinforced concrete have shown a marked vulnerability to aggressive environmental actions, mainly by carbon dioxide in the atmosphere, chlorides, and sulfates in water and soil. These vulnerabilities lead to material degradation often worsened by lack of maintenance of exposed concrete surfaces, and errors in the construction phase. These fragilities were emphasized by the experimental nature of numerous innovative structural and technological configurations adopted by the designers of the time. Ferrocement was one of the many experimentations in the construction history of the time. Although it was patented by J. L. Lambot as "*ferciment*" in 1855, the employment of ferrocement in civil structures began in the 1940s. In fact, it was usually employed

S. Gabriele et al. (Eds.): IWSS 2023, LNCE 437, pp. 221–230, 2024.
https://doi.org/10.1007/978-3-031-44328-2_23

to build hulls, silos, tanks, and roofs. Pier Luigi Nervi was one of the first designers to recognize the possible advantages of ferrocement not only for boats but also for civil structures applications [4, 5].

Nervi started experimenting with the material in the late thirties, by changing the percentage of cement and the amount of reinforcement in the chosen element. The aim was to develop a new construction solution by offering a light, user-friendly, low-cost material. By using ferrocement it was possible to manufacture very thin slabs of only a few centimeters, which were very ductile, elastic and crack-resistant, and not only possessed an impressive litheness that ensured easy transportation and installation, but also allowed shapes to be made in any desired configuration [6]. His experiments on boats and small-scale buildings (such as the shed in via della Magliana), culminated with different patents, but it was in the Torino Esposizioni pavilions, built between 1947 and 1953, that Nervi had the chance to apply his new material for the construction of a large structure [7]. After their construction, Nervi was consecrated as one of the most innovative designers of the world, and his legacy today is still admired and studied.

Due to its construction feature (the surface area of the reinforcement is very high) and the limited thickness of the concrete cover characterizing ferrocement, this material may present durability issues. This may be especially true for historical ferrocement, where the reinforcing wire meshes were not even galvanized or did not present any protective coating. Considering the lack of studies about historical ferrocement, especially in view of its preservation, the present paper investigates the durability of this material by carrying out an experimental campaign. In this sense, the paper includes the design of a specific experimental setup for testing very thin ferrocement elements built with Nervi's technique to subject them to a corrosive environment.

It is expected that the paper will contribute to a better understanding of the behavior of this material and apply the best preservation strategies.

The physical model has been subjected to different loading configurations and analyzed by 3D Digital Image Correlation (DIC) technique in order to recognize the pattern deformation during the increasing force application. DIC technique seems to be a very promising method to evaluate the stability and the collapse condition of these kind of structures [8, 9]. The rapid growth of computer technology that spurred continued increase of computational methods provided the foundation for the explosion of in vision-based full-field experimental measurement method: e.g. 2D-DIC for SEM, AFM and planar loading and surfaces, 3D-DIC for general motion and deformation of curved or planar surfaces, V-DIC or Digital Volume Correlation for interior deformation measurements in opaque solids. Today, the methods are used by scientists seeking to obtain full-field quantitative measurement of motions and deformations. The measurement of displacements and displacement gradients (strains) has always been an important topic in the evaluation of material properties, such as material strengths or fracture parameters and in experimental stress analysis. Optical techniques such as moiré interferometry [10], holography [11] and speckle interferometry [12] have been proven to be matured techniques to analyze macroscopic parameters and are being applied successfully in many different applications. However, all the interferometric techniques have stringent requirements for system's stability. Moreover, the processing of fringe patterns is laborious and time-consuming. This technical difficulty has raised many researchers' attention and computerized procedures [11, 13] have been developed to automate the processing of the data from the fringe patterns.

In the last decade, a non-contacting optical technique, digital image correlation (DIC), has been developed by Sutton et al. [14–17] and Bruck, et al. [11]. It was applied to measurement of displacements and strains. The applications include microscopic strain measurements in electronic packaging [18], strain fields in polyurethane foam plastic materials and evaluation of their mechanical properties [11], and evaluation of thermal strain in the solder joints. This methodology was even used for in situ evaluation of the state of conservation of mural frescoes [13]. The technique utilizes two similarly speckled images, which were captured by a solid state video camera, to represent the states of the object before and after deformation. By utilizing the concept of digitalization, one can characterize the image by the patterns of different levels of light intensity. Both digitized images are then correlated by an algorithm, based on their mutual correlation coefficient or other statistical functions, to find out the subtle differences between them. In the present paper, DIC analysis is used to recognize the typical deformation of the vault according to the loading process. 3D-DIC measurements were used to recognize in advance the critical condition of the vault before the collapse condition.

2 Experimental Investigations

2.1 Sample Preparation and Curing

Considering the historical value of the analyzed structures and since the ferrocement elements are extremely thin, to investigate the durability of this material, small-scale ferrocement mockups were built in the laboratory, starting from a small sample collected on-site. More specifically, the sample was collected from one of the ferrocement beams elements of Hall C, which elements are better described in [19, 20]. The sample was used to determine the matrix composition of the concrete (both chemical and petrographically) and identify the wire characteristics.

According to the results of chemical and petrographic analyses conducted on the sample, the mockups were reconstructed by using a premixed mortar based on pozzolanic cement. The pozzolanic component (in replacement of bentonite) confers to the mortar a thixotropic consistency, making it suitable to be set and modeled around the metallic net. Regarding the wire mesh, from the visual analysis of the original sample, the resulting steel meshes consist of woven steel wires spaced 10 mm apart. Thus, smooth steel bars and woven meshes with 1mm diameter and 10x10 mm spacing were selected for the mock-up reinforcements. The ferrocement samples have been realized following Pier Luigi Nervi's design and in accordance with ACI Standard PRC-549–18 [21, 22]. In the end, a total of ten samples were prepared to be aged to study the durability and their performance (30x7.5x3 cm), each sample included seven layers of metal mesh 270x55 mm (Fig. 1).

The S1, S2, S3, S4 and S5 samples were subjected to an accelerated weathering cycle of six months, to verify the effectiveness of the treatments in their durability. Sample S2 presents a layer of paint to simulate the possible presence of painting in actual structures. A solution with 3.5% NaCl solved in water is used to accelerate the corrosion process. The aging cycles run for a six-month period alternating wet and dry cycles,

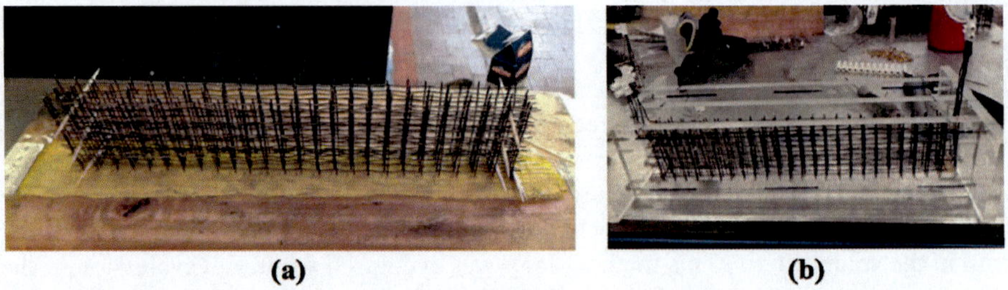

Fig. 1. **(a)** Steel net positioning. **(b)** Plexiglass formwork and layered wires.

where during the wetting day, samples were sprayed continuously for three hours. The drying occurred under natural conditions. After six months of aging, all samples showed evident deterioration on the surface. Table 1 reports the description of the treatments of each sample.

2.2 Digital Image Correlation

The 3D-Digital Image Correlation (DIC) method was employed to assess the four point tests under stress. The adopted correlation approach is based on monitoring the pattern of gray values in small local regions. Is **G** (x, y) the correlation algorithm reduced the total for the grey value of a pixel with the coordinates x and y inside the subset or facet (Fig. 2):

$$\sum_{x,y} (G_t(x_t, y_t) - G(x, y))^2 \tag{1}$$

where $G_t\,(x_t,y_t) = g_0 + g_1 G(x_t,y_t)$; $x_1 = a_0 + a_1 x + a_2 y + a_3 xy$ and $y_1 = a_4 + a_5 x + a_6\, y + a_7 xy$. By the variation of the illumination parameters $(g_0,\, g_1)$ and the parameters of the affine transformation $(a_0 \ldots a_7)$ an accuracy for the matching of better than 0.01 pixel can be achieved. The measurement and identification of the three-dimensional deformation of the object surface (deformed pattern) is the second step in digital three-dimensional correlation after the three-dimensional contour has been established (Fig. 3).

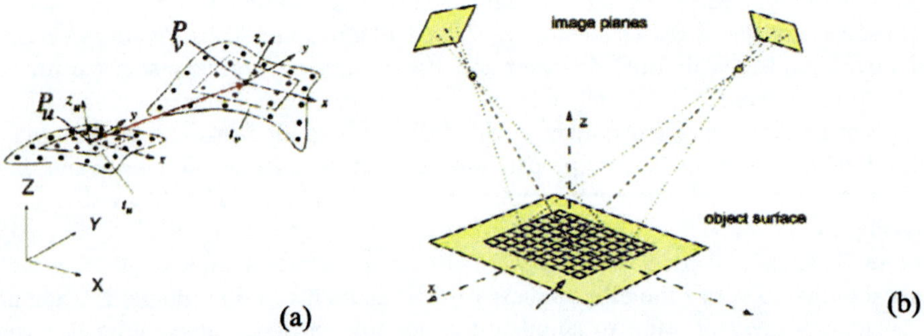

Fig. 2. The three-dimensional displacement vector (a) is determined. Principle of two-camera 3D image correlation (b).

Fig. 3. Region of interest (roi) for the four-point bending tests of ferrocement specimens. The target area (roi) was positions in the middle portion of the element.

2.3 Test Setups

Flexural tests were carried out with the configuration as required by [5, 22] with simply supported beam with third-point loading. Figure 4 shows the setup of the bending test. The results reported in Table 1 are used to evaluate the mechanical properties of the ferrocement samples by comparing the performance of the aged and non-aged ones. While Fig. 5 shows the results of the flexural tests of samples S4_SA_VF_01 and S6_M_01. During the experiments, additional measurements by acoustic emissions (AE) were performed [23].

(a) (b)

Fig. 4. **(a)** Setup of the flexural tests **(b)** Flexural test of aged sample S4_SA_VF_01 where it is possible to see the DIC pattern.

Table 1. List of the samples used in this study.

Sample ID	Treatment Type	Ultimate Load [kN]	∂_u [mm]
S1-MI-01	Aged	2.385	5.765
S3-PA-SA-01	Aged	2.352	5.788
S4-SA-VF-01	Aged	2.396	3.392
S5-VF-01	Aged	2.748	4.892
S6-M-01	No ageing applied	2.904	11.8
S6-M-02	No ageing applied	3.140	13.4
S6-M-03	No ageing applied	2.992	17.53
S6-M-04	No ageing applied	2.984	16.7
S6-M-05	No ageing applied	2.984	12.41
S6-M-06	No ageing applied	3.224	9.248

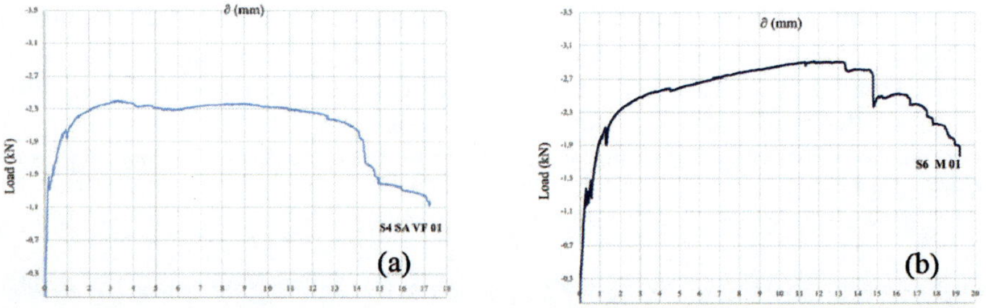

Fig. 5. (**a**) Flexural test of aged sample S4_SA_VF_01 (**b**) Non-aged sample S6_M_01.

3 Experimental Results and Discussion

Through the analysis of DIC, were especially analyzed specimens S4-SA-VF-01 and S6-M-01, in order to recognize possible variations in terms of displacements and deformations by comparing two cases of ferrocement elements with and without ageing applied. The figure below shows, in superposition, the grid recognized by the pattern recognition algorithm of the acquisition software. The superposition is made between the first step (state 0) and the last step of the test (step associated with the collapsed state). For the specimen S4-SA-VF-01 two major fractures can be recognized (Fig. 6a).

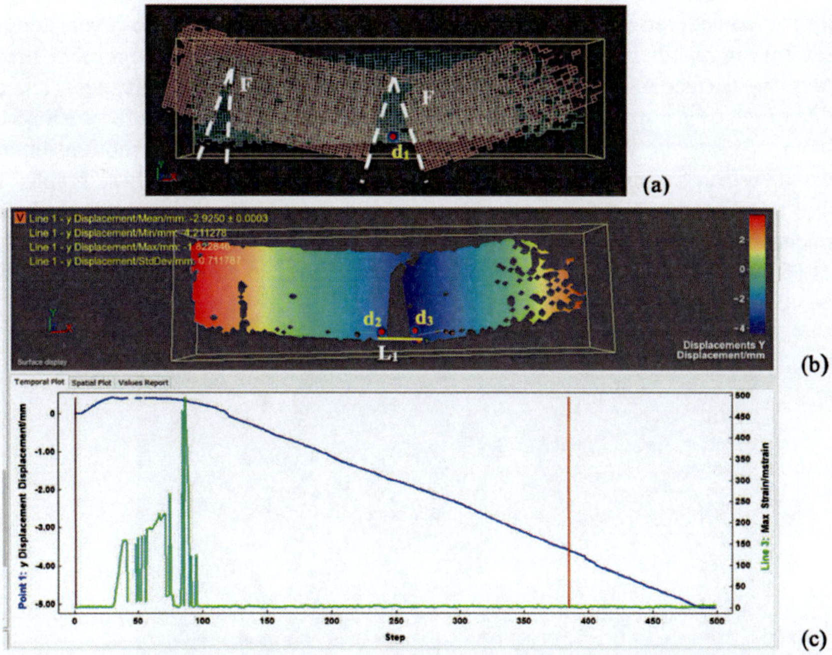

Fig. 6. Two main fractures F_1 and F_2 can be recognized initially (a). Later a third in the left portion will be recognized by recognizing symmetrical behavior. It was placed 3 smart displacement sensors d_1, d_2 and d_3 (b). An additional smart strain gauge (line gauge is positioned). The y displacement of d_1 over time is compared with the lagr. Deformation along x for L_1 (c).

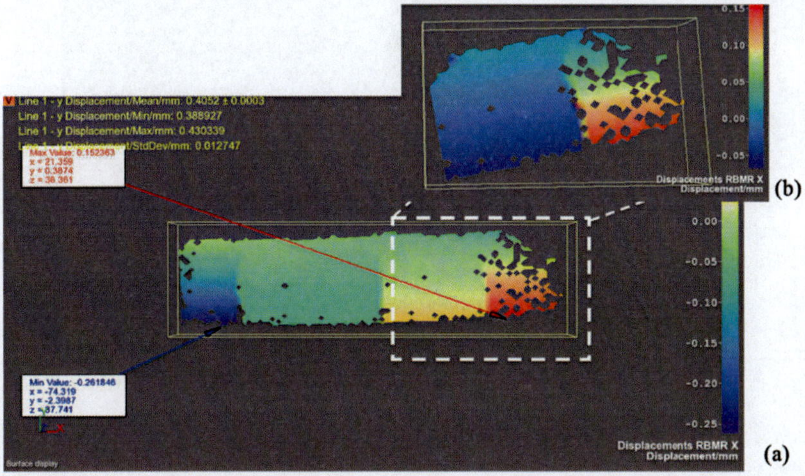

Fig. 7. RBMR Rigid body motion removed in x-direction for strain (a). It is possible to recognize a third macrocrack in the left portion with a symmetrical behavior of the specimen before failure (zoom in b).

Similar considerations can be made for specimen S6-M-01. It, however, does not present any aging phenomenon. In this case, it was possible to recognize 5 macro-cracks on the surface of the specimen and passing through the entire thickness. It is also possible to see by comparing the graphs at letters c in Figs. 6 and 8 that for the S6-M-01 specimen, a more ductile behavior is recognized RBMR Rigid body motion removed in the x-direction for strain (Fig. 7 and 9) can also be observed. By these graphs, it is possible to recognize in advance the formation of the crack pattern for both the two specimens. The crack mouth opening for specimen S4-SA-VF-01 (aging applied) was equal to 6.31 mm. The crack mouth opening for specimen S6-M-01 (ageing not applied) was equal to 6.08 mm.

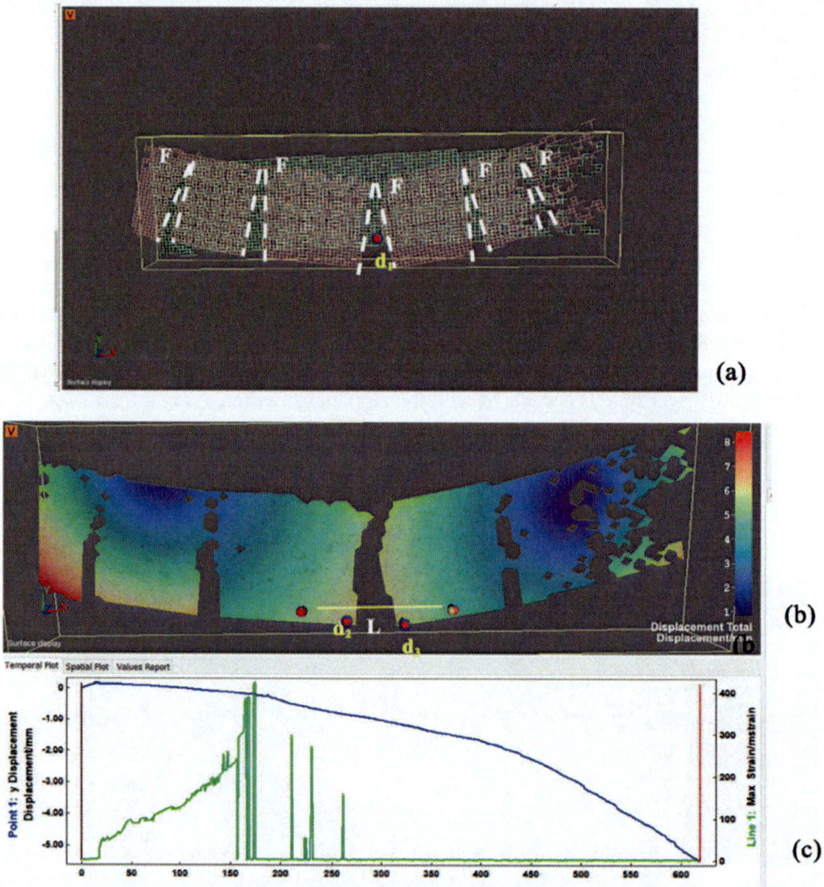

Fig. 8. Five main fractures F1-F5 can be recognized initially (a). 3 smart displacement sensors d1, d2 and d3 (b) were placed. An additional smart strain gauge (line gauge is positioned). The y displacement of d_1 over time is compared with the lagr. Deformation along x for L1 (c).

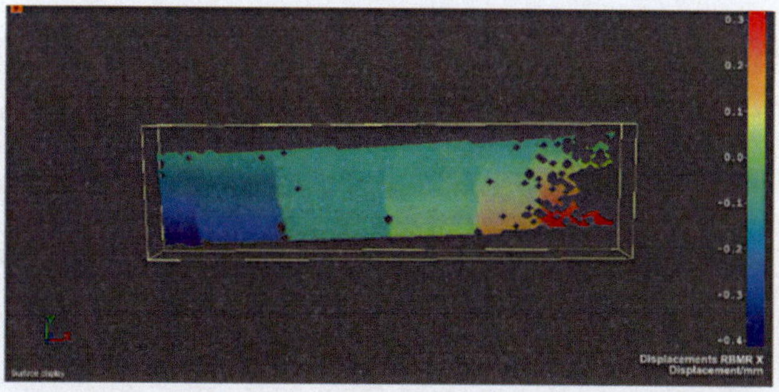

Fig. 9. RBMR Rigid body motion removed in x-direction for strain.

4 Conclusions

The study presented the behavior of ferrocement subjected to a corrosive environment; in particular, by reconstructing and analyzing the ferrocement patented by Pier Luigi Nervi. The study is of interest since, despite the widespread use of the material, few studies have analyzed its durability for preservation purposes. From Table 1, it can be inferred that the aged specimens showed a loss of capacity, which emerges both in terms of ultimate load and in the midspan deflection (∂_u), which is significantly lower in the aged samples (approximately 63% less). The results of the DIC tests confirm the loss of ductility. These results highlight the need to focus attention on the appropriate protection and preservation of this material.

Acknowledgements. The present work was supported with a Keeping it Modern grant awarded by The Getty Foundation of Los Angeles (USA). The authors would also like to acknowledge the City of Turin, owner of the buildings and partner of the project, Società Committenza Piemonte (SCR) for supporting the activities, and Buzzi Unicem for providing the material for the experimental campaign.

References

1. Harboe, G., Espinosa de los Monteros, F., Landi e, S., Normandin, K.C.: The Cádiz Document: InnovaConcrete Guidelines for Conservation of Concrete Heritage (2021)
2. ICOMOS International Committee on Twentieth Century Heritage (ISC20C), Approaches for The Conservation of Twentieth-Century Cultural Heritage (2017)
3. Macdonald e, S., Arato Gonçalves, A.P.: Conservation Principles for Concrete of Cultural Significance, Los Angeles: Getty Conservation Institute (2020)
4. Naaman, A.E., Evolution in Ferrocement and Thin Reinforced Cementitious Composites. Arab. J. Sci. Eng. **37**(2) (2012). https://doi.org/10.1007/s13369-012-0187-4
5. Naaman, A.: Ferrocement and Laminated Cementitious Composites., Vol. %1 di %2ISBN0-9674939-0-0, (http://www.technopress3000.com), p. 3000. Techno Press, Ann Arbor (2000)
6. Greco, A., Nervi, P.L.: Dai primi brevetti al Palazzo delle Esposizioni di Torino 1917–1948. Lucerne Quart Edizioni, Lucerna (2008)
7. Gargiani e, R., Bologna, A.: The rhetoric of Pier Luigi Nervi. Concrete and ferrocement forms. EPFL Press, Lausanne (2016)

8. Post, A.: Moiré Interferometry at VPI and SU. Experimental Mechanics **23**(2), 203–210 (1983)
9. Fottenburg, W.G.: Some Applications of Holographic Interferometry. Experimental Mechanics **8**, 281–285 (1969)
10. Bastawros e, A.F., Voloshin, A.S.: Thermal Strain Measurements in Electronic Packages through Fractional Fringe Moiré Interferometry. J. Elect. Packa. **112**(4), 303–308 (1990)
11. Bruck, H.A., McNeil, S.R., Sutton e, M.A., Peters, W.H.: Digital image correlation using newton-raphson method of partial differential correction. Experimental Mechanics **29**(3), 261–267 (1989)
12. Wang, Y.Y., Chen e, D.J., Chiang, F.P.: Material testing by computer aided speckle interferometry. Experimental Techniques **17**(5), 30–32 (1993)
13. Zhang, A., Zhang e, X., Cheng, G.: Compression strain measurement by digital speckle correlation. Experimental Mechanics **39**(1), 62–65 (1999)
14. Sutton, M.A., Wolters, W.J., Peters, W.H., Ranson e, W.F., McNeil, S.R.: Determination of Displacements using an improved Digital Correlation method. Image and Vision Computating **1**(3), 133–139 (1983)
15. Sutton, M.A., Cheng, M.Q., Peters, W.H., Chao e, Y.J., McNeill, S.R.: Application of an Optimized Digital Correlation Method to Planar Deformation Analysis. Image and Vision Computing **4**(3), 143–151 (1986)
16. Spagnolo, G., Paoletti, D.S., Ambrosini e, D., Guattari, G.: Electro-optic correlation for in situ diagnostics in mural frescoes. Pure & Applied Optics: Journal of the European Optical Society, Part A **6**(5), 557–563 (1997)
17. Sutton, M.A., Turner, J.L., Bruck e, H.A., Chae, T.A.: Full-field Representation of Discretely Sampled Surface Deformation for Displacement and Strain Analysis. Experimental Mechanics **31**(2), 168–177 (1991)
18. Sutton, M.A., McNeill, S.R., Jang e, J., Babai, M.: Effects of Subpixel Image Restoration on Digital Correlation Error. J. Opt. Eng. **27**(10), 870–877 (1988)
19. Lenticchia, A., Ceravolo e, R., Faccio, P.: Understanding the structures of Pier Luigi Nervi: a multidisciplinary approach. Vitruvio **8**, 66–75 (2023). https://doi.org/10.4995/vitruvio-ijats.2023.18862
20. Spanò, A., Patrucco, G., Sammartano, G., Perri, S., Avena, M., Fillia e, E., Milan, S.: Digital Twinning for 20th Century Concrete Heritage: HBIM Cognitive Model for Torino Esposizioni Halls. Sensors **23**(10) (2023). https://doi.org/10.3390/s23104791
21. ACI Committee 549: Guide for the Design, Construction, and Repair of Ferrocement. ACI Struct. J. **85**(3) (1988). https://doi.org/10.14359/3527
22. ACI Committee 549: Design Guide for Ferrocement. American Concrete Institute (2018)
23. Niccolini, G., Bosia, F., Carpinteri, A., Lacidogna, G., Manuello, A., Pugno, N.: Self-similarity of waiting times in fracture systems. Phys. Rev. E **80**(2), 026101 (2009)

Improved Multi-body Rope Approach for Free-Form Grid Shells

Amedeo Manuello Bertetto(✉), Jonathan Melchiorre, and Giuseppe Carlo Marano

Department of Structural, Geotechnical and Building Engineering, Politecnico di Torino, Corso Duca degli Abruzzi, 24, 10129 Turin, Italy
amedeo.manuellobertetto@polito.it

Abstract. Due to their ability to provide broad, lightweight roofs with slim main structural elements, grid-shell roofing systems are becoming more and more common in contemporary engineering and design. However, their widespread and extensive use outside of high-end projects or super architecture has been constrained by their complex and expensive fabrication. In order to solve this, it has been demonstrated that the Multi-body Rope Approach (MRA) is successful in calculating structurally efficient geometries that can decrease internal stresses, which are closely related to the structural shape of gridshells. Different MRA-improving strategies are put forth in this paper: The number of various structural components needed to build a gridshell can be significantly reduced thanks to Multiple Order MRA (MO-MRA) and Repulsive Nodes MRA (RN-MRA) strategy. Due to mass production, the suggested methodology offers lower manufacturing costs as well as improved building stage management. Finally, by a Matlab code and the new approaches were tested in various case-study structures.

Keywords: Multi-body Rope Approach · Multiple Order · Grid-shell · Form-finding · Light-weight Structures

1 Introduction

There has been an increase in demand in recent years for architectural structures that provide greater internal distribution flexibility. Double-curved shells and domes have emerged as a viable option for achieving both column-free areas and complicated designs[1]. Free-form shell buildings, often referred to as curved shell structures, have drawn a lot of interest because of their distinctive and attractive features [2]. Without the need of beams, columns, or walls, shells are thin, self-supporting structures with a single or double curve that can cover large spaces. The membranous behavior, which demonstrates superior structural efficiency, is the preferred. To avoid local (buckling) and non-local (snap-through)

Supported by organization x.

S. Gabriele et al. (Eds.): IWSS 2023, LNCE 437, pp. 231–240, 2024.
https://doi.org/10.1007/978-3-031-44328-2_24

instability issues, adequate bending stiffness must be given if compressive membrane states are present. A three-dimensional grid shell is constituted of straight elements connected by nodes makes up a gridshell, a type of free-form shell construction. Gridshells thus provide a variety of structural and aesthetically pleasing advantages. They are compact, self-supporting structures that can span wide areas with the least amount of material [3,4]. In the 1960s and 70s [5,6] innovative new methods for gridshell design were carried out. They produced intricate, double-curved structures using computer modeling and lightweight materials. Timber and other materials were frequently used to construct early gridshells [7] to create vast open spaces like exhibition halls, sports arenas etc. [8]. Gridshell structures are defined by the interaction between their shape and stress distribution. Because of this relationship between shape and forces, designing such structures directly, as in the case of conventional structures, is ineffective. In this scenario, the search for a suitable structural shape is critical, both for design aesthetics and for the structural efficiency. To achieve a feasible shape, it is necessary to balance the internal forces and loads acting on the structure and minimize the bending moment within the resisting elements [9]. Numerous form-finding strategies, including both physical and computational modeling techniques, have been developed over time to aid in the design of these complex structures [10]. Among the most popular form-finding methods are the force density method [11], thrust network analysis [12], dynamic relaxation method [13], particle-spring system [14], multi-body rope approach [15], and others [16–18]. Innovating shapes and free-form structures with structural optimization is the newest trend in structural design research [19–23]. The improved Multi-body Rope Approach (i-MRA), a novel advancement of the form-finding technique created exclusively for gridshells, is presented in this study. The Multi-body Rope Approach (i-MRA) is built upon by the MRA technique [15]. A net of masses connected by flexible ropes or cables serves as the MRA representation of a structure. As a result, the structure's geometry can be defined by reducing bending moments and stresses [24]. The MRA method minimizes the eccentricity of applied compression forces to design the structural geometry [25]. By including strategies that optimize the structural geometry for both structural and automation of the construction process, the i-MRA method outperforms MRA. The i-MRA has two main advancements: (i) Multiple Orders MRA (MO-MRA): reduces the variety of structural component types; (ii) Repulsive Nodes MRA (RN-MRA): creates a repulsive force field for the dynamic model. Different examples of increasing complexity were used to demonstrate the i-MRA method's ability to produce the best structural geometry. A structural analysis of the structures obtained by basic MRA and i-MRA was conducted in order to confirm that the geometries obtained by the new method proposed did not have any significant structural disadvantages (in terms of internal solicitations) compared to those calculated by pure form-finding methods. Finite element analyses were carried out for this purpose utilizing the SOFiSTiK [26].

2 Methods

2.1 The Basic MRA Method

The Multibody Rope Approach (MRA), developed by [15], is an original technique for figuring out the shape of gridshell structures, even for extremely complicated geometries and under any forming load. The D'Alembert's principle is used in MRA to iteratively determine the final equilibrium configuration for each node using a dynamic model of falling bodies in the space and time domains. An inverted image of the hanging net is the structure's final equilibrium configuration (also known as the funicular configuration). MRA makes the same assumption as particle-spring models on the localization of the nodes' self-weight and rope loads. In contrast to these strategies, MRA models the hanging network using ropes. Unlike the spring-particle (SP) and dynamic relaxation (DR) approaches, MRA's system of forces acting on individual nodes is separate from those in those techniques. No force is exerted if the distance between the ends is less than the rope's fixed length (l_rope). The forces F exerted on the end nodes can be represented as follows by defining l as the distance between the two ends of the rope and k as its axial stiffness:

$$\begin{cases} F_{rope} = 0 & \text{if } l < l_{rope} \\ F_{rope} = k(l - l_{rope}) & \text{if } l \geq l_{rope} \end{cases} \tag{1}$$

The MRA technique generally assumes very high stiffness levels in order to reduce axial deformations. Consider a generic node i in the nodes and ropes structural network that has a mass of m_i. The node i is roped together with n_i other nodes. The equilibrium equation looks like this if node i is subject to an external load p_i:

$$\vec{R}_i = \vec{p}_i + \sum_{j=1}^{n_i} \vec{F}_{rope,ji} + \vec{F}_i^I + \vec{F}_i^{II} = 0 \tag{2}$$

In this equation, the vector $\vec{Ri}$ represents the net force acting on node i, which is the result of the addition of various forces, such as the applied load p_i and the forces transmitted by the ropes holding the node in place. $\vec{F}rope, ji$, the inertial force $\vec{F}_i^{II}$, and the damping force $\vec{F}_i^I$. The magnitude of the inertial force $\vec{F}_i^{II}$ can be calculated as the product of the node's mass m_i and the magnitude of the acceleration vector $\vec{a}_i$, with the direction of the inertial force being opposite to the direction of node acceleration. Expressing the position of the generic node i as $\vec{u}_i = (x_i, y_i, z_i)$, the velocity and the acceleration can be obtained by deriving the position in time, as in the relations (3).

$$\vec{v}_i = \dot{\vec{u}}_i = (\dot{x}_i, \dot{y}_i, \dot{z}_i) \qquad \vec{a}_i = \ddot{\vec{u}}_i = (\ddot{x}_i, \ddot{y}_i, \ddot{z}_i) \tag{3}$$

Finally, the equilibrium equation can be projected in the three space dimensions, obtaining the system of Eq. (4).

$$\begin{cases} p_{ix} + \sum_{j=1}^{n_i} \left\{ \frac{(x_j - x_i)}{l_{ji}} \cdot F_{rope} \right\} - c_i \cdot \dot{x}_i - m_i \cdot \ddot{x}_i = 0 \\ p_{iy} + \sum_{j=1}^{n_i} \left\{ \frac{(y_j - y_i)}{l_{ji}} \cdot F_{rope} \right\} - c_i \cdot \dot{y}_i - m_i \cdot \ddot{y}_i = 0 \\ p_{iz} + \sum_{j=1}^{n_i} \left\{ \frac{(z_j - z_i)}{l_{ji}} \cdot F_{rope} \right\} - c_i \cdot \dot{z}_i - m_i \cdot \ddot{z}_i = 0 \end{cases} \tag{4}$$

The system may be resolved taking into account a time increment Δt. By knowing the position, velocity, and acceleration of each node at time t, these quantities can be determined at the next instant, $t + \Delta t$. A coefficient C_3 can is defined as a function of the known node positions at time t^*.

$$C_3 = \vec{p}_i + \sum_{j=1}^{n_i} \left\{ k \cdot \vec{F}_{rope,ji} \right\} \tag{5}$$

The coefficient C_3 depends solely on the position of the nodes at time t^*, and it defines the vector F_{rope}.

$$\ddot{\vec{u}} + \frac{c}{m} \dot{\vec{u}} = C_3 \tag{6}$$

In addition, the natural frequency of the system ω_n and critical damping ζ can be defined:

$$\omega_n = \sqrt{\frac{k}{m}} \tag{7}$$

$$\zeta = \frac{c}{2\omega_n m} \tag{8}$$

k is the stiffness, m is the mass, and c is the damping coefficient of the system. The natural frequency ω_n represents the frequency at which the system vibrates when it is not subjected to any external forces. The critical damping ζ is the damping coefficient value that results in the system being critically damped, which means that it returns to its equilibrium state as quickly as possible without oscillating.

$$\ddot{\vec{u}} + 2\omega_n \zeta \dot{\vec{u}} = C_3 \tag{9}$$

By adding the specific solution to the corresponding homogeneous differential equation, which is denoted by the following expression:

$$\vec{u}(t) = C_1 e^{-2\omega_n \zeta} + C_2 + \frac{C_3}{2\omega_n \zeta} t \tag{10}$$

The coefficients C_1 and C_2 can be calculated using the system's initial conditions. In this situation, they can be derived by using the nodes' locations and velocities at the previous instant $t - \Delta t$, as shown in the following equations:

$$C_1 = -\frac{2\omega_n \zeta \dot{\vec{u}}_{(t-\Delta t)} - C_3}{(2\omega_n \zeta)^2} \tag{11}$$

$$C_2 = -\frac{(2\omega_n\zeta)^2\vec{u}_{(t-\Delta t)} + 2\omega_n\zeta\dot{\vec{u}}_{(t-\Delta t)} - C_3}{(2\omega_n\zeta)^2} \tag{12}$$

The coefficients C_1, C_2, and C_3 in the solution rely on the nodes' locations, speeds, and accelerations at the preceding time instant.

$$\dot{\vec{u}}_t = \frac{\vec{u}_t - \vec{u}_{t-\Delta t}}{\Delta t} \tag{13}$$

Finally, the acceleration $\ddot{\vec{u}}_t$ can be calculated by subtracting the two velocities and multiplying the result by the time interval Δt.

$$\ddot{\vec{u}}_t = \frac{\dot{\vec{u}}_t - \dot{\vec{u}}_{t-\Delta t}}{\Delta t} \tag{14}$$

The suggested technique is made to determine the final configuration of a gridshell from its original mesh, which symbolizes the net's starting state. Until an equilibrium configuration is reached that represents the best structural geometry in relation to the applied force field $\vec{p}$, this process is repeated. The structural shape is a function of the nodal masses m, the system stiffness k, damping parameters c, the rope slack coefficient ρ, and the applied force field $\vec{p}$. The slack coefficient ρ is defined as the ratio of the initial distance between nodes to the target length of the ropes, as shown in the following equation.

$$\rho_{ij} = \frac{l_{rope}}{\vec{u}_i(0) - \vec{u}_j(0)} \tag{15}$$

Structural elements are divided into three groups according to their length: those with a length less than l_{rope} are called *loose elements*, those with a length greater than l_{rope} are called *over elements*, and those with a length equal to l_{rope} are referred to as *target elements*.

2.2 Shape and Constructability the Improved-MRA

To cut down on the variety of structural components obtained by MRA, the Multiple Order based on the MRA method is used, the new target in to reduce the number of *loose elements* in the final configuration. This is accomplished without materially changing the structural geometry by introducing additional families of shorter-length ropes that are positioned in the areas where the loose parts are found. The gridshell geometry is first calculated using traditional MRA with a goal length of $l_{rope,1}$. A new rope family with a shorter length will be created after the final equilibrium arrangement is achieved ($l_{rope,2} < l_{rope,1}$), it is introduced, and the geometric configuration is iteratively calculated.

$$\begin{cases} F_{rope} = 0 & \text{if } l < l_{rope,2} \\ F_{rope} = k(l - l_{rope,2}) & \text{if } l_{rope,2} < l \leq (l_{rope,1} - l_{rope,2}) + l_{rope,2} \\ F_{rope} = 0 & \text{if } \gamma(l_{rope,1} - l_{rope,2}) + l_{rope,2} < l < l_{rope,1} \\ F_{rope} = k(l - l_{rope,1}) & \text{if } l \geq l_{rope,1} \end{cases} \tag{16}$$

The γ coefficient can be modulated to obtain different geometric configurations by preferring ropes of length $l_{rope,1}$ or $l_{rope,2}$, and it was found that setting $\gamma = \frac{2}{3}$ yields sufficient results. It is possible to introduce new families of elements and set a target length of $l_{rope,3} < l_{rope,2}$ for any remaining *loose elements*. At the same time, the Repulsive Nodes MRA is a method designed to reduce the number of *loose elements* generated by applying the basic MRA. The basic idea is to introduce a repulsive force field $\vec{q}$ between the nodes of the geometric configuration obtained through the use of MRA. This force field $\vec{q}$ is introduced after the final equilibrium configuration is established using the MRA, which requires a new iterative computation process to determine the new equilibrium condition.

$$\vec{R}_i = \vec{p}_i + \vec{q}_i + \sum_{j=1}^{n_i} \left\{ k \cdot \vec{F}_{rope,ji} \right\} - c_i \cdot \vec{v}_i - m_i \cdot \vec{a}_i = 0 \tag{17}$$

Equation (17) introduces the repulsive force field $\vec{q}$ to each node i connected to slack ropes. This force field $\vec{q}$ is linearly proportional to the difference between the target length l_{rope} and the distance between nodes connected by a slack rope l_{ij}. To calculate the repulsive force field $\vec{q}$, Eq. (18) can be used, where k_{rep} is the proportionality constant that connects the modulus of the repulsive force to the distance between nodes i and j.

$$q_i = -k_{rep}(l_{rope} - l_{ij}) \tag{18}$$

The RN-MRA method has been observed to converge faster to an equilibrium configuration when a smaller value of k_{rep} is chosen relative to k. However, the designer has the flexibility to choose the value of k_{rep} based on the design situation. Ideally, RN-MRA should be applied to models with few *loose elements* relative to the number of tensioned ones, and where their length is close to the target length l_{rope}.

3 Application and Results

This section presents two main applications of the suggested technique to structural geometries with increasing levels of complexity. The comparison demonstrates that the i-MRA can lower gridshell structure realization costs while maintaining the structurally ideal solution obtained by a pure form-finding method (basic MRA). Overall, it is demonstrated that the suggested approach is successful in delivering structural solutions that are both practical and simple to build. The first application regards the structural geometry of a square gridshell defined using the MRA method applied to a square mesh with a side of 15 m and lines placed in an orthogonal grid linking nodes at a distance of $l_{ij} = 1.25$ m. The limitations are depicted as triangular points, and the construction is intended to be confined exclusively in the corners of the square plan. The second kind of application example seeks to assess the performance of the suggested strategy in a very broad scenario. When using the method, a base plan with a free-form curve as its definition is taken into account, and a parametric design tool is used to generate the mesh automatically. The quadrangular mesh and the mesh made up of hexagonal and pentagonal elements are the two types of meshes that are the subject of the analysis. As a result, this is the most complex application of MRA because the suggested goal is to produce equal structural pieces from a geometry made up of various elements. The edges in both of the meshes shown in this section have an average length of roughly 1.50 m (Fig. 1).

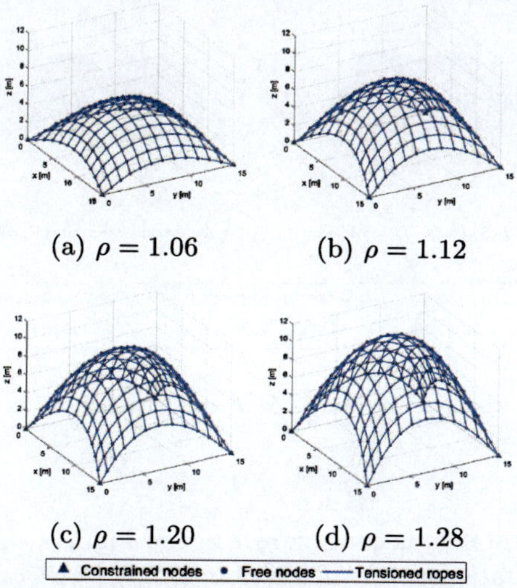

Fig. 1. MRA application on corner constrained square-plan structure for different slack coefficients ρ.

Nevertheless, the components of the quadrangular mesh in Fig. 2*a* range in length from approximately 0.80 m to 2.30 m, resulting in a significant amount of variety in the starting pieces. When structural elements' final lengths depart from the desired length, red is used to show them $L_{rope} = 1.78$ m (*loose ropes*). Due to the anomalies in the basic geometry, these components make up more than 13% of the 750 structural elements that make up the gridshell. To build the real construction, more than 100 distinct types of structural components are needed. Construction phase management may be more problematic than in the preceding examples, despite the fact that this may be doable on the construction site for simple and complicated structural shapes. However, Figures illustrate the structural configuration attained by utilizing the i-MRA described in this paper Figs. 2*b* and 2*d*. This example shows a substantial improvement. The method allowed for the calculation of a structural geometry that only required the use of 7 different lengths for structural component types $L_{rope} = [1.78; 1.58; 1.45; 1.35; 1.25; 1.15]m$. This amazing result shows how the methods described in Sect. 2 can significantly lessen building complexity. The

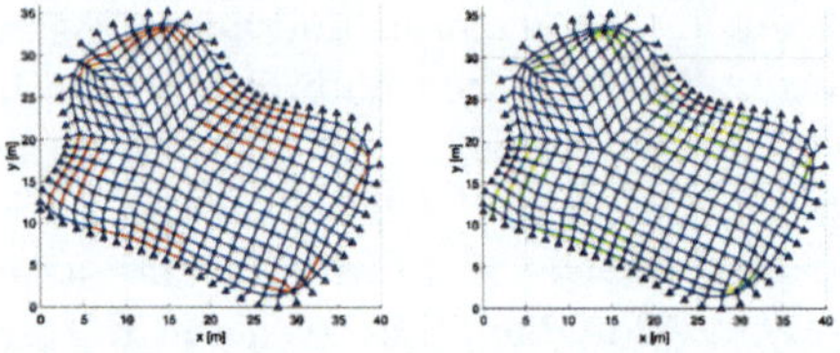

(a) MRA planimetric view (b) i-MRA planimetric view

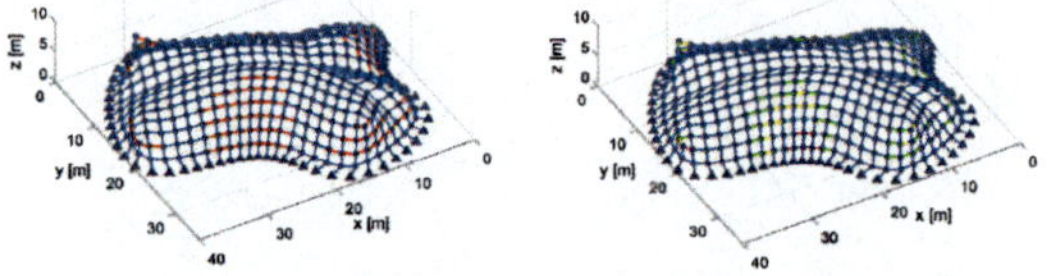

(c) MRA 3D view (d) i-MRA 3D view

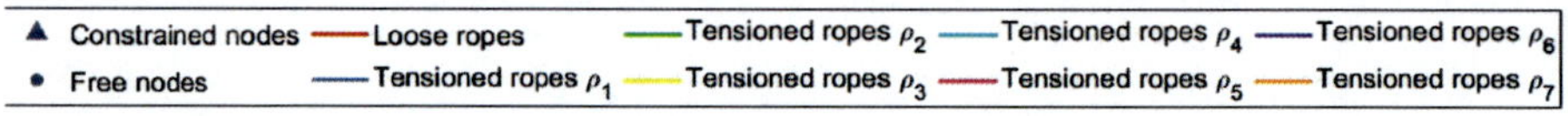

Fig. 2. Comparison between MRA and i-MRA on a free-form structure with quadrangular mesh.

results of the structural evaluations shown in the Figures show that the modifications to the structural geometry made possible by i-MRA did not significantly raise stresses. With the introduction of i-MRA, the number of structural elements used dropped from over 100 to just 6. The imposed Von Mises stresses increased by less than 0.7%, which is a minimal rise, despite the geometric alterations made. Additionally, the work's construction complexity—as measured by the various structural elements used—was reduced by nearly 20 times (Fig. 3).

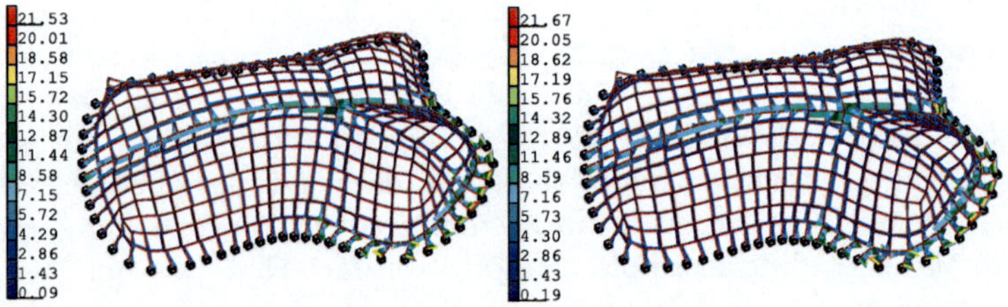

(a) MRA: $|\sigma^{VM}|_{max} = 21.53kPa$ (b) i-MRA: $|\sigma^{VM}|_{max} = 21.67kPa$

Fig. 3. Structural analysis performed on the structures obtained by applying basic MRA (left) and i-MRA (right) on the application example with quadrangular mesh. Comparison in terms of Von Mises stress [kPa].

4 Conclusions

In this paper, a form-finding technique for gridshell structures is presented. The approach expands on the Multi-body Rope Approach (MRA) created by [15]. Different improvement strategies are suggested for the fundamental MRA in order to further limit the complexity of structural construction management and lessen computing effort. The first method reduces the variety of structural elements needed to build the estimated structures by designating groups of structural parts with identical lengths. In the second method, a repulsive force field is applied to the dynamic model, allowing for minute geometry tweaks to reduce the amount of structural components that do not fall into the predefined groups. Four application examples with varying degrees of complexity were used to test the efficiency of the suggested strategy. The findings demonstrated that, particularly as the geometric complexity of the structure rises, the i-MRA technique significantly lowers the number of structural components needed.

References

1. Richardson, J.N., Adriaenssens, S., Coelho, R.F., Bouillard, P.: Coupled form-finding and grid optimization approach for single layer grid shells. Eng. Struct. **52**, 230–239 (2013)
2. Chilton, J., Chuang, C.-C.: Rooted in nature: aesthetics, geometry and structure in the shells of heinz isler. Nexus Netw. J. **19**(3), 763–785 (2017)
3. Tellier, X.: Bundling elastic gridshells with alignable nets. part ii: form-finding, Automation Constr. **141**, 104292 (2022)
4. Tomei, V., Grande, E., Imbimbo, M.: Design optimization of gridshells equipped with pre-tensioned rods. J. Build. Eng. **52**, 104407 (2022)
5. Happold, E.: Philosophy of design with particular respect to buildings. History and development, Structural Engineering (1997)
6. Addis, B., Walker, D.: Happold: The confidence to build. Taylor & Francis (2005)
7. Liddell, I.: Frei otto and the development of gridshells. Case Stud. Struct. Engineering **4**, 39–49 (2015)
8. Pan, W., Turrin, M., Louter, C., Sariyildiz, S., Sun, Y.: Integrating multi-functional space and long-span structure in the early design stage of indoor sports arenas by using parametric modelling and multi-objective optimization, Journal of Building. Engineering **22**, 464–485 (2019)
9. Gründig, L., Moncrieff, E., Singer, P., Ströbel, D.: A history of the principal developments and applications of the force density method in Germany 1970–1999. In: Proceedings of IASS-IACM 2000 Fourth International Colloquium on Computation of Shell & Spatial Structures, Chania-Crete, Greece (2000)
10. Mesnil, R., Douthe, C., Baverel, O., Gobin, T.: Form finding of nexorades using the translations method. Autom. Constr. **95**, 142–154 (2018)
11. Schek, H.-J.: The force density method for form finding and computation of general networks. Comput. Methods Appl. Mech. Eng. **3**(1), 115–134 (1974)
12. Otter, J.R.H., Cassell, A.C., Hobbs, R.E.: POISSON, dynamic relaxation. Proc. Inst. Civ. Eng. **35**(4), 633–656 (1966)
13. Block, P., Ochsendorf, J.: Thrust network analysis: a new methodology for three-dimensional equilibrium. J. Int. Assoc. Shell Spatial Struct. **48**(3), 167–173 (2007)

14. Kilian, A., Ochsendorf, J.: Particle-spring systems for structural form finding. J. Int. Assoc. Shell Spatial Struct. **46**(2), 77–84 (2005)
15. Manuello, A.: Multi-body rope approach for grid shells: form-finding and imperfection sensitivity. Eng. Struct. **221**, 111029 (2020)
16. Rian, I.M., Sassone, M., Asayama, S.: From fractal geometry to architecture: designing a grid-shell-like structure using the takagi-landsberg surface. Comput. Aided Des. **98**, 40–53 (2018)
17. Zhao, Z., Yu, D., Zhang, T., Zhang, N., Liu, H., Liang, B., Xian, L.: Efficient form-finding algorithm for freeform grid structures based on inverse hanging method. J. Build. Eng. **46**, 103746 (2022)
18. Huang, W., Wu, C., Hu, J., Gao, W.: Weaving structure: a bending-active gridshell for freeform fabrication. Autom. Constr. **136**, 104184 (2022)
19. Yamamoto, K., et al.: State-of-the-art for optimization of forms and strength for reticulated shells. In: IASS Annual Symposium; IASS-APCS, London, UK (2012)
20. Adriaenssens, S., Block, P., Veenendaal, D., Williams, C.: Shell structures for architecture: form finding and optimization, Routledge (2014)
21. Melchiorre, J., Manuello, A., Marmo, F., Adriaenssens, S., Marano, G.: Differential formulation and numerical solution for elastic arches with variable curvature and tapered cross-sections. Europ. J. Mech.-A/Solids **97**, 104757 (2023)
22. Bouhaya, L., Baverel, O., Caron, J.-F.: Optimization of gridshell bar orientation using a simplified genetic approach. Struct. Multidiscip. Optim. **50**, 839–848 (2014)
23. Marmo, F., et al.: On the form of the Musmeci's bridge over the Basento river. Eng. Struct. **191**, 658–673 (2019)
24. Manuello Bertetto, A., Riberi, F.: Form-finding of pierced vaults and digital fabrication of scaled prototype. Curved Layered Struct. **8**(1), 210–224 (2021)
25. Manuello, A., Melchiorre, J., Sardone, L., Marano, G.C.: Multi-body rope approach for the form-finding of shape optimized grid shell structures. In: Proceedings of the 15th World Congress on Computational Mechanics (2022)
26. SOFiSTiK, Text Editor 2023, Flataustr. 14, 90411 Nuremberg (2022). https://www.sofistik.com

Acoustic Emission Monitoring and Thrust Network Analysis of the Central Nave Vaults of the Turin Cathedral

Amedeo Manuello Bertetto[1(✉)], Francesco Marmo[2], and Jonathan Melchiorre[1]

[1] Department of Structural, Geotechnical and Building Engineering, Politecnico di Torino, Corso Duca degli Abruzzi 24, 10129 Torino, Italy
amedeo.manuellobertetto@polito.it

[2] Department of Structures for Engineering and Architecture, Universitá degli Studi di Napoli Federico II, Corso Umberto I 40, 80138 Naples, Italy

Abstract. Due to their age, elevation, and prolonged exposure to both static and dynamic loading conditions, historical constructions and old masonry structures such as medieval and Gothic cathedrals, bell towers, and other similar structures, are particularly vulnerable. The central nave vault of the Turin cathedral was subjected to an acoustic emission (AE) monitoring technique for structural integrity assessment. These findings are correlated with the evidence from the Thrust Network Analysis (TNA) performed on the cathedral's central nave vault, taking into account the additional elements added at the start of the XXth century to lessen horizontal forces. In this situation, the analysis's findings are strictly correlated to the vault's 3D AE localization, which was obtained by the triangulation method.

Keywords: Acoustic Emission · b-value analysis · AE source localization · Thrust Network Analysis · Masonry vaults

1 Introduction

Due to numerous coexisting circumstances operating on the bearing materials (thermal loading, static and dynamic loading, subsidence, fatigue, and creep), historical buildings commonly exhibit ubiquitous fracture patterns. We can evaluate the preservation status of these structures and their temporal evolution using non-destructive techniques [1–6]. In the past few years, numerous research have been carried out in historic masonry buildings using a control mechanism based on the spontaneous emission of elastic waves [7–9]. Through AE monitoring, wide-band piezoelectric (PZT) sensors capture the signals produced by microcracks (100–350 kHz), which are then post-processed by statistical and analytical analysis [10]. The authors gained extensive expertise in the monitoring of ancient structures using the AE technique, including masonry towers and monuments with stone bearing walls, as well as conventional and sack masonry

S. Gabriele et al. (Eds.): IWSS 2023, LNCE 437, pp. 241–249, 2024.
https://doi.org/10.1007/978-3-031-44328-2_25

systems [7,9]. The Syracuse cathedral's vertical bearing structure was monitored using the AE approach to manage the progression of structural damage brought on by pre-seismic and seismic activity. The Asinelli's Tower in Bologna, Central Italy, is an example of how the AE signals were employed simultaneously to evaluate the impact of recurring phenomena (motor traffic loads, wind effects) [8,9]. The results of AE monitoring for an array of eight pzt sensors installed on the central nave vault of the Turin cathedral were first analyzed in the current study. The examination of the AE cumulative number and several characteristics that can forecast how the damage behavior will change over time (such as the b-value and β^t). In particular, the b-value analysis reveals a declining trend toward values consistent with the expansion of localized micro and macro-cracks in the monitored structure's component around various locations identified in the vault's 3D model. These results are related to the numerical outcomes of the Thrust Network Analysis (TNA) performed on the vault, taking into account additional elements added at the beginning of the XX century to reduce the horizontal forces, along with supporting evidence from other AE parameters. The TNA is typically used to represent stresses in masonry vaults as a discrete network of forces in equilibrium with gravity loads, based on Heyman's ideas through a graphical interpretation. The 3D localisation is primarily connected with the outcomes of the study based on TNA. In 1694, the naves and the dome were finished. Carlo Filiberto of Savoy hosted the Holy Shroud at the end of the sixteenth century. In this regard, it has been made possible by the historical-critical analysis of the Cathedral of Turin to emphasize that the current structure is the result of a number of interventions that create a heterogeneous condition in terms of the materials used as well as the consolidating methods that were used. The first structural intervention was required in 1656 on the occasion of the partial collapse of the central nave's vault. The use of the aforementioned iron chains on the tangential side and the lower surface of the shutters allowed the vault to be stabilized until the successive interventions of the 20th century, which significantly altered the structure of the church attic and led to its current configuration. The vault was judged to be structurally unsafe in 1928. The lunette-adorned barrel-lowered vault was supported by steel beams that were positioned orthogonally to the vault's axis and hooked by iron tie-rods. In particular, there are 13 pairs of double T-shaped beams running the length of the vault extension. These beams are fixed in huge reinforced concrete blocks, 13 on each side, and are positioned in line with the pillars.

2 AE Monitoring Analysis and Methods

Eight piezoelectric (PZT) transducers are arranged in an array connected to a multi-channel system (each channel having a dedicated memory of 64 Mb), which automatically stores and processes important waveform parameters of the detected signal (the cumulative number of events, signal durations, peak amplitudes, and ring down counts), allowing in situ damage localization and quantification from the recorded parameters. The AE signal frequency connected to the damage evolution is between 200 and 50 times the sampling rate of the AE device (10 Ms/s) [11]. The threshold methods used in the various cases were implemented in the software of the device used and reported by the authors in several publications [9, 11]. The AE equipment automatically performs different kinds of analyses. The first parameter is represented by the cumulative number of AE signals N, detected during the monitoring time. In addition, the time dependence of the structural damage observed during the monitoring period, identified by parameter β^t, can also be correlated to the rate of the micro-cracks propagation in the time domain [6–10,24–26]. In the first localization step all the signals, recorded by different sensors, falling into a time intervals compatible with the formation of micro-cracks are recognized ad grouped. In the preliminary studies devoted to AE localization in concrete and concrete structures [12], it was usual to assume that the amplitude threshold of 100 mV of the non-amplified AE signal was appropriate to recognize the onset time of the AE P-wave. The AE localization problem in its traditional form can be written as:

$$d_i - d_j = v_p \Delta t_{ij} \tag{1}$$

where Δt_{ij} is the measured arrival time difference representing the difference of the signal arrival time at the positions of the sensor i and j. At the same time, the velocity v_p may be assumed as a material constant, or according to the damage evolution in the material, as a changing parameters correlated to the damage level. According to this last considerations the estimating variables of the AE localization procedure are the AE source coordinates (x, y, z) and the p-wave elastic propagation velocity. Different groups, composed by at least 5 sensors, must be recognized for each micro-crack to be localized. As far the velocity v_p is concerned the expression is obtained as,

$$v_p = \sqrt{E(1-\nu)/(\rho(1+\nu)(1-\nu))} \tag{2}$$

and for concrete and masonry material assuming $\nu = 0.2$:

$$E = 0.9(v_p^2 \rho) \tag{3}$$

These last assumption allowed to estimated the variation of E according to the evolving damage [13]. The instrumentation noise, the p-wave picking method utilized to assess the signal onset time, etc. can all be related to the accuracy of the AE localization. Regarding the last point, various techniques were employed to determine the best arrival time at each sensor. This suggests that the enhanced AIC picker approach can be applied. Here, the AE code used for the localization of the sources directly implements the improved AIC method, which is based on the Akaike information criterion. The signal is regarded as an AR model using this approach:

$$x_t = \sum_{k=1}^{N} a_m^i x_{j-m} + e_t^i \tag{4}$$

with t = 1, ..., k for interval 1 and t = k + 1, ..., n for interval 2. The model takes into account both the deterministic and non-deterministic components of the time series. The time series with no fixed order e_t^n or noise, is assumed to be Gaussian. The maximum likelihood estimation (MLE) of the AIC method is used to extract the non-deterministic part of the time series in intervals [1, k] and [k + 1, n] using Eq. (4), where k is the division point. We can express the approximate likelihood function L for the two non-deterministic time series in intervals [1, k] and [k + 1, n] as:

$$L(x; k, M, \Theta) = \prod_{i=1}^{2} \frac{1}{\sigma_i^2 2\pi} exp \frac{1}{\sigma_i^2} \sum_{k=1}^{N} ((a_m^i x_{j-m} + e_t^i)^2) \quad (5)$$

where $\Theta_i(\Theta(a_1^i..., a_M^i)$ represents the model parameters (σ_i^2 is dependent on k), and $p_1 = 1, p_2 = k+1, n_1 = k, n_2 = n-k$. By the maximization of the logarithmic form of the function L. Considering the AIC Criterion as: AIC=-ln(L)+2k and it can be rewritten considering the two portion of the signal the one characterized by the white noise and the second containing the signal:

$$AIC(k) = klog(\sigma_{1,max}^2) + (n-k)log(\sigma_{2,max}) + 2C \quad (6)$$

In this last relation the first term indicates the badness of the model fit and the second the unreliability. Point k where the joint likelihood is maximized, or AIC(k) in Eq. (6) is minimized, determines the optimal separation of the two time series:

$$AIC(k_w) = k_w log(var(R_w(1, k_w))) + (n_w - k_w) log(var(R_w(1 + k_w, n_w))), \quad (7)$$

where the subscript w denotes that not the whole time series is taken, but only the chosen window containing the onset time. n_w, is the last sample of the current time series, kw ranges from 1 to n_w. The term $var(R_w(1, k_w))$ is the variance function calculated from 1 to kw, while $var(R_w(1 + k_w, n_w))$ means that all samples ranging from $1 + k_w$ to n_w are considered. The global minimum of the AIC function defines the first P-wave onset time of the AE signal.

$$\sum_{t=k+1}^{10} |x_t|/10 \geq 4 \sum_{t=1}^{k} |x_t|/k \quad (8)$$

The fourfold mean amplitude of the interval of the time series ranging from 1 to k is compared with the mean amplitude of a shifting set of 10 data using this relation. The predetermined onset time (k_0) is chosen as the first value of k, which satisfies Eq. (8). Always localized after the actual onset time, the first estimation. In the AIC application, two different types of time frames with various time intervals are successively set taking into mind this fact. First, we apply the algorithm to the interval $[1, k_0]$ (window 1), which is used for a rough determination of the onset time by the AIC and gives k1 value. The second time window (2) is considered centered on the value k_1 with a length of $2\Delta k$. The value of Δk depends on the sample frequency. In our study, the sample frequency is 10 MHz, and Δk equal to 3000 samples gives us an ideal result. As the actual onset time of the AE signal under analysis, the time window (2) value $k_m in$ from the AIC-picker is taken into consideration. The outcomes of the AE localization process based on the modified AIC-picker approach [11] are reported in Fig. 1.

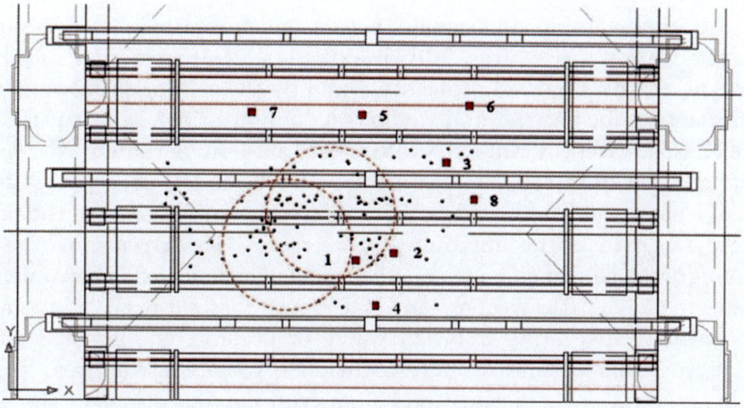

Fig. 1. Plan view of where EA sources are located.

3 Equilibrium Analysis of the Central Nave Vault

The stresses that cause a masonry arch or vault collapse are usually lower than those required for the material failure. Therefore, as Heyman pointed out in 1995, the stability of such structures depends mainly on their shape and self-weight distribution [14]. O'Dwyer contributed to its development in 1999 [15], but it was fully developed only in recent years by Block and coworkers (Thrust network analysis) [16,17]. Other approaches to modeling stresses in masonry vaults can be found in [18–20].

Marmo and Rosati [21] reformulated Block's version of the TNA by avoiding graphical interpretation of the method in order to avoid the employment of the so-called dual grid and focusing solely on the primal grid, resulting in a significant improvement in computational performance. This new version of the TNA incorporates horizontal forces in the analysis, as well as the inclusion of holes or free edges in the vault. Such approach has been successfully applied to a series of case studies including helical staircases [22] and fictile vaults [23]. In this paper it is be employed to analyse the central nave vaults of the Turin Cathedral.

3.1 Limit Conditions for the of Turin Cathedral Central Nave Vaults

For the equilibrium analysis of a masonry vault, internal forces can be modelled as a network of thrusts. These thrusts are compression forces that act within the structure, in equilibrium with the applied loads. This network is henceforth referred to as the *thrust network* in which thrust values associated to branches and the vertical coordinate of nodes are unknown.

Thrust values are required to be compressive, while nodal heights are required to be contained within the masonry thickness. Hence, equilibrium of the network is iused to evaluate unknown branch thrusts and nodal heights by employing coupled optimization procedures, fully described in [21], here omitted for brevity.

In general, two extreme solutions are found by the solving procedure, respectively representing the deepest and shallowest configurations of a thrust network, which are characterized by different values of a numeric parameter r, representing a scaling factor of forces applied to the network model. Let r_d (r_s) represent the value of r associated to the deepest (shallowest) configuration of the network. The ratio r_s/r_d reflects the difference between the two extremes of the equilibrated configurations. Since $r_d > r_s$, this ratio is always less than one. Lower r_s/r_d values mean that the deepest and shallowest configurations of the network are very distinct, suggesting a broad range of possible equilibrated solutions for the structure. Such values usually correspond to safer structures. When the r_s/r_d ratio equals one, the deepest and shallowest configurations of the network become indistinguishable, indicating a limit condition of the structure. At this point, only one equilibrated thrust network that fits within the geometric limits exists, and the structure's ultimate strength is reached.

The ratio $r_s/r_d = 1$ is utilized to determine an extreme geometry or a limit loading condition for the structure. This concept has practical applications, such as assessing the geometric safety factor of the vault [14], which involves determining the minimum thickness required for an equilibrated solution. The same approach can be used to determine the limit multiplier for a specific loading scenario. For instance, if the vault experiences a combination of vertical and horizontal forces, the maximum admissible multiplier of horizontal force can be determined by gradually increasing it until the ratio r_s/r_d equals 1.

3.2 Results of the Analysis

The geometry of the entire vault, particularly the intrados and extrados, was determined using the geometric survey of the central nave vault. This geometry is used to establish the node horizontal coordinates. The heights of the vault intrados and extrados are utilized as limit heights in the thrust network model. The initial configuration of the thrust network is a grid located on the mid-surface of the vault with a uniform horizontal spacing of $0.44, m \times 0.44, m$. Additional diagonal branches are included to simulate thrusts along the groins of each lunette.

Figure 2 presents a three-dimensional visualization of the thrust network's initial configuration, with the vault's intrados and extrados shaded in gray.

Nodes are loaded by the vault self weight, which is computed assuming a weight per unit volume of $18\,\mathrm{kN/m^3}$, and by a load of $5\,\mathrm{kN/m^2}$, uniformly distributed on the entire vault extrados. The additional weight of the fillings above the vault spingers and of the concrete blocks placed above the vault springers during a restoration work carried out at the beginning of the 20th century, each estimated to weight about 61.5 kN, was also considered. Following the approach described in [21], the TNA has been applied to evaluate the minimum thickness of the vault, for which the network configurations shown in Fig. 3 are obtained.

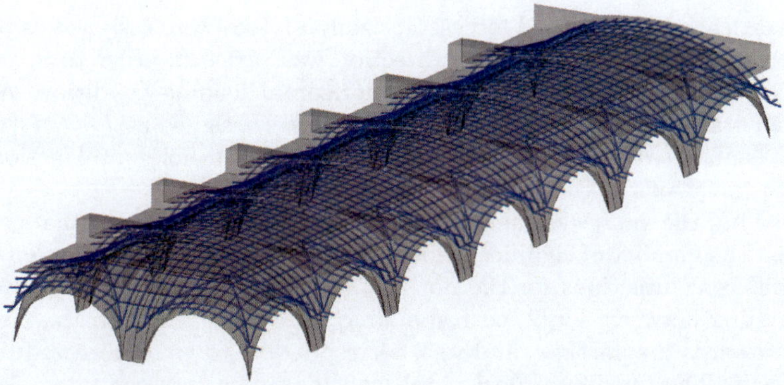

Fig. 2. Initial configuration of the thrust network employed for the equilibrium analysis of the central nave vault.

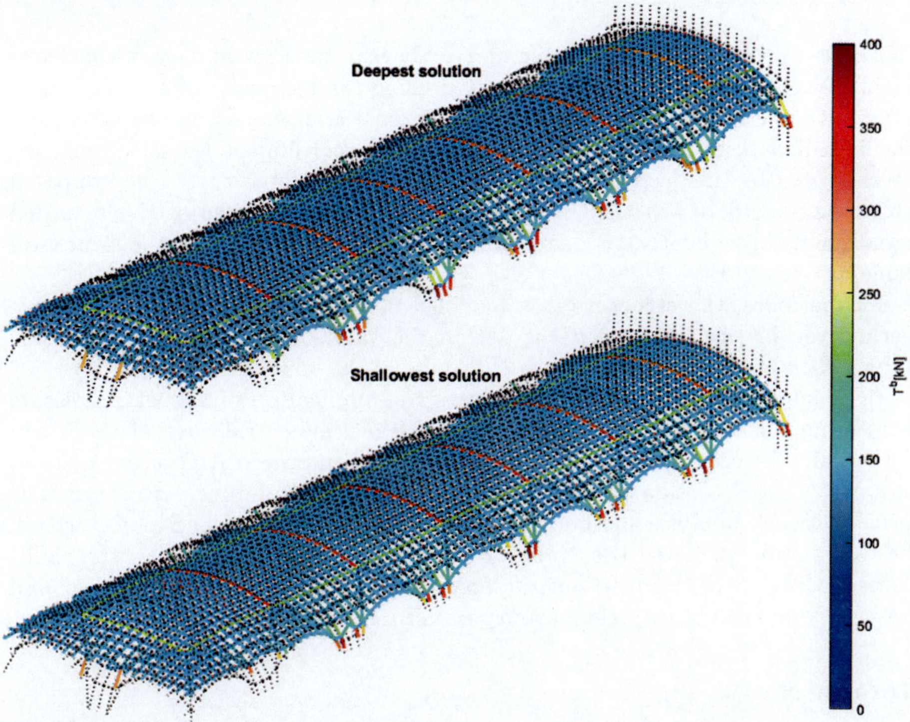

Fig. 3. Deepest and shallowest solutions for the vault having minimum thickness ($T_{min}/T = 0.63$, $R_s/R_d = 1$) subjected to vertical loads only.

In addition to the analyses regarding the vault subjected to vertical loads only, we also considered the case of the vault subjected to the combined effect of both vertical and horizontal forces. Horizontal forces were assigned as proportional to the vertical ones and their direction was varied in order to investigate the effect of different orientation of the horizontal loading condition. Accordingly, the horizontal components of nodal forces are assigned as $f_x = |f_z| \cos(\alpha)$, $f_y = |f_z| \sin(\alpha)$, where α is the angle formed by the horizontal action with respect to the x axis of the global reference frame.

Employing the vault symmetry about the longitudinal axis of the nave (i.e. the y axis) the number of significant loading cases is reduced by assuming a total of nine different directions for the horizontal loads, with angular increments of $\pi/8$, spanning between $-\pi/2$, corresponding to forces parallel to the nave and directed towards the narthex, and $\pi/2$, corresponding to forces directed towards the transept. The maximum horizontal load multiplier was computed for each direction of horizontal loads; such solutions are omitted here form brevity.

4 Conclusion

This article presents the monitoring and analysis of the central nave of the Turin Cathedral. The monitoring was conducted using the technique of acoustic emissions, with the installation of 8 different piezoelectric sensors on the structure. The installed instrumentation allowed for the recording of acoustic emission waves caused by the occurrence of fractures within the structure. The improved AIC picker method was then used to identify the onset time of the recorded signals and, consequently, to locate the sources of fractures within the structural elements.

Furthermore, the structure was modeled using the thrust network analysis method to identify the load paths within the vaults. Finally, the method was employed to evaluate the minimum thickness of the vault.

In conclusion, this study demonstrates the application of acoustic emission monitoring for assessing the structural health of the central nave of the Turin Cathedral. The use of piezoelectric sensors and the improved AIC picker method allowed for the detection and localization of fractures within the structure. The thrust network analysis method provided insights into the load paths within the vaults and facilitated the evaluation of the minimum thickness of the vault. These findings contribute to our understanding of the structural behavior and preservation needs of historic masonry structures.

References

1. Aggelis, D., Mpalaskas, A., Matikas, T.: Investigation of different fracture modes in cement-based materials by acoustic emission. Cem. Concr. Res. **48**(3), 1–8 (2013)
2. Simpson, D., Richards, P.: A probabilistic synthesis of precursory phenomena. In: Earthquake Prediction, pp. 566–574 (2013)
3. Bak, P., Christensen, K., Danon, L., Scanlon, T.: Unified scaling law for earthquakes. Phys. Rev. Lett. **88**, 178501-1–178501-4 (2002)
4. Ohtsu, M., Okamoto, T., Yuyama, S.: Moment tensor analysis of acoustic emission for cracking mechanisms in concrete. ACI Struct. J **95**(/), 87–95 (1998)

5. Rosso, M.M., Aloisio, A., Melchiorre, J., Huo, F., Marano, G.C.: Noise effects analysis on subspace-based damage detection with neural networks **54**, 23–37 (2023)
6. Marasco, G., et al.: Ground penetrating radar fourier pre-processing for deep learning tunnel defects automated classification. In: Iliadis, L., Jayne, C., Tefas, A., Pimenidis, E. (eds.) Engineering Applications of Neural Networks. EANN 2022. Communications in Computer and Information Science, vol. 1600, pp. 165–176. Springer, Cham (2022). https://doi.org/10.1007/978-3-031-08223-8_14
7. Anzani, A., Binda, L., Carpinteri, A., Lacidogna, G., Manuello, A.: Evaluation of the repair on multiple leaf stone masonry by acoustic emission **41**, 1169–1189 (2007)
8. Carpinteri, A., Lacidogna, G., Manuello, A., Niccolini, G.: A study on the structural stability of the asinelli tower in bologna. Struct. Control Health Monit. **23**, 659-667 (2015)
9. Lacidogna, G., Manuello, A., Carpinteri, A.: Acoustic emission monitoring of Italian historical buildings and the case study of the athena temple in syracuse. Archit. Sci. Rev. **58**(4), 290–299 (2015)
10. Melchiorre, J., Manuello Bertetto, A., Rosso, M.M., Marano, G.C.: Acoustic emission and artificial intelligence procedure for crack source localization. Sensors **23**(2), 693 (2023)
11. Carpinteri, A., Xu, J., Lacidogna, G., Manuello, A.: Reliable onset time determination and source location of acoustic emissions in concrete structures, vol. 34 (2012)
12. Shah, S.P., Li, Z.: Localizatin of microcracks in concrete under uniaxial tenssion. ACI Mater. J. **91**(4), 372–381 (1991)
13. Lenticchia, E., Manuello, A., Ceravolo, R.: Ae propagation velocity calculation for stiffness estimation in pier luigi nervi's concrete structures. Curved Layered Struct. **8**(1), 109–118 (2021)
14. Heyman, J.: The stone skeleton: Structural engineering of masonry architecture. Cambridge University Press (1995)
15. O'Dwyer, D.: Funicular analysis of masonry vaults. Comput. Struct. **73**, 187–197 (1999)
16. Block, P.: Thrust network analysis, Ph.D. thesis, Massachusetts Institute of Technology (2009)
17. Block, P., Lachauer, L.: Three-dimensional equilibrium analysis of gothic masonry vaults. Int. J. Archit. Heritage **8**, 1–24 (2014)
18. Angelillo, M., Cardamone, L., Fortunato, A.: A numerical model for masonry - like structures. J. Mech. Mater. Struct. **5**, 415–583 (2010)
19. Adriaenssens, S., Block, P., Veenendaal, D., Williams, C.: A thrust network approach for the equilibrium problem of unreinforced masonry vaults via polyhedral stress functions. Mech. Res. Commun. **37**, 198–204 (2010)
20. Tralli, A., Alessandri, C., Milani, G.: Computational methods for masonry vaults: a review of recent results. Open Civil Eng. J. **8**, 272–287 (2014)
21. Marmo, F., Rosati, L.: Reformulation and extension of the thrust network analysis. Comput. Struct. **182**, 104–118 (2017)
22. Marmo, F., Masi, D., Rosati, L.: Thrust network analysis of masonry helical staircases. Int. J. Archit. Heritage **12**(5), 828–848 (2018)
23. Marmo, F., Ruggieri, N., Toraldo, F., Rosati, L.: Historical study and static assessment of an innovative vaulting technique of the 19th century. Int. J. Archit. Heritage **13**(6), 799–819 (2019)

Adaptive Structures with Intrinsic Control: Exploratory Habitation-Vehicles for Space Environment

Andrea Manuello[1](✉), Amedeo Manuello Bertetto[2], and Giuseppe Carlo Marano[2]

[1] Department of Mechanical and Aerospace Engineering, Politecnico Di Torino, Corso Duca Degli Abruzzi, 24 - 10129 Torino, Italy
andrea.manuello@polito.it

[2] Department of Structural, Geotechnical and Building Engineering, Politecnico Di Torino, Corso Duca Degli Abruzzi, 24 - 10129 Torino, Italy

Abstract. A new type of habitation vehicle for future Mars colonization and exploration missions is proposed. The habitat vehicle is characterized by the movement of the habitat cell of varying sizes: from medium-small, for temporary colonization of near settlement, to larger with a greater autonomy, for the exploration purpose of unknown area. The vehicle consists of a pair of compass-shaped structures that support the habitat cell which, once the settlement position is defined, can be laid on the ground.

The mechanical and structural configuration allow the module to be raised from the surface, lowered, and transported to a constant altitude with horizontal layout. The module is kept at a constant altitude and horizontal trim during the motion of the vehicle. Furthermore, the function of being able to lower to the ground surface to allow access is ensured. The height and the constant trim during the motion and the lowering and ascent are achieved with a system of ropes wound on pulleys moved by the motion of the vehicle itself, with suitable transmission. The structural components it is previewing to exploits the principle of tensairity strucutres, with the aim of making maximum use of the material and reducing the mass of the structure facilitation the transportation from Earth to Mars.

Keywords: Space Architecture · Space habitat design · Tensearity · Space vehicle · Pulley System

1 Introduction

Interest in exploration and colonization has always been present in human history, and the next designated destination is space. This offers both technological, scientific as well as economic and sociological implications, such as the development of new construction techniques and the astrobiological and geological investigation of near-Earth planets, and escape valves for global overpopulation and the no longer sustainable exploitation of Earth's resources. Three-D printing methods for the creation of living quarters (or habitats) on extraterrestrial soil, new types of spacecraft and surface exploration craft have been developed, and many inventions have led to the advancement of technology on Earth, in areas as varied as the emergence of new diagnostic tools.

S. Gabriele et al. (Eds.): IWSS 2023, LNCE 437, pp. 250–258, 2024.
https://doi.org/10.1007/978-3-031-44328-2_26

Among the targets of this coming development is Mars. Its exploration by orbiters, landers and rovers began in the 1960s and continues to the present day, with growing interest in the realization of the first manned mission, a milestone for future colonization of the planet, which will require significant economic and technological effort, with development of sheltered housing isolated from the Martian environment and technologies for crew life support.

In the present paper, a new type of habitation vehicle for future Mars colonization missions is shown, with the aim of allowing the movement of a habitat cell of varying sizes: from medium-small, for exploration near the main settlement, to large, for exploration with greater range [1, 2].

The vehicle that is here presented consists of a pair of compass-shaped structures that support a habitat, allowing it to be raised from the surface, and then lowered, and transported with constant height e and horizontal set-up. The mechanism, that passively controls the set-up, consists of a multiplier and pulley system with a variable radius. The module is kept at a constant altitude and horizontal trim during the motion of the vehicle. Furthermore, the function of being able to lower to the ground surface to allow access is ensured. The height and the constant trim during the motion and the lowering and ascent are achieved with a system of ropes wound on pulleys moved by the motion of the vehicle itself, with suitable transmission. Cables ropes and flexible structural elements are often used in handling systems in aerospace projects, due to their high favorable strength/mass ratio [3, 4]. The vehicle moves forward or backwards using the legs of the compass resting on the ground which work as legs with a controlled movement at the base [5].

The structural components it is previewing to exploits the principle of tensairity structures, with the aim of making maximum use of the material and reducing the mass of the structure facilitation the transportation from Earth to Mars. The mode of the structure was carried out with the help of the ANSYS ambient.

2 Exploration and Settlement Environment

Mars is the fourth planet in the solar system, the last of the four terrestrial planets, along with Mercury, Venus and Earth, characterized by a solid surface altered by climate and meteorite impacts, including the intense late bombardment that occurred between 3.8 and 4.2 billion years ago. [6] Mars' orbit has a greater eccentricity, similar inclination and obliquity to Earth's, with nearly twice the period of revolution and a similar period of rotation. These orbital parameters, like those of any other planet, are subject to cyclical variations with periods of thousands to millions of years, called Milankovitch cycles, however their variation is greater for Mars than for Earth. For example, the tilt of the Earth's axis varies by 2.4° between a minimum of 22.1° and a maximum of 24.5° [7], while that of Mars in the order of tens of degrees, fluctuating between 0° and 80°, justifying the presence of ice deposits where their formation would be unstable today. [8] With the current tilt of the axis, the cadence of the seasons is similar to Earth's, although it has greater gaps in both the length of the seasons and temperatures due to the greater eccentricity, with spring one-third longer than winter [9–12]. In addition, Milankovitch cycles influence the seasons both by controlling the distribution of solar ra-dations at different latitudes, both as an annual and seasonal average, and by allowing exchanges of volatile elements between deposits and the atmosphere: the temperature at the poles increases relative to that at the equator as the axis tilt increases, resulting in the release of carbon dioxide.

Wind erosive action is also affected by these cycles. Unlike Earth, Mars has no active magnetic field although the presence of rocks in the Martian crust with residual magnetization, with an intensity at the surface one order greater than that found on Earth, suggest an active dynamo in early Mars. The core must have been partially liquid, rich in iron probably mixed with sulfides. The areas currently with residual magnetization are mostly in the southern hemisphere, of which the most intense is in Terra Cimmeria and Terra Sirenum (30°–90°S and 130–240°E), while the bases of craters created by impact grains show no magnetism, indicating that there was no longer a magnetic field at the time of impact [13].

The typical temperatures in summer at latitudes of 60° range between 180 K and 290 K, while at the equator of 210–220 K and at the poles during winter as low as 150 K, causing carbon dioxide caps to form over water deposits [14]. Atmospheric pressure varies from 14 mbar at the bottom of the Hellas Basin to 3 mbar at the top of the highest volcanoes, with seasonal variations on the order of 30 percent dependent on the sublimation of carbon dioxide at the poles along with some of the water vapor from the underlying glacier, which is transported from the north pole to the south pole, condensing along the way before reaching high southern latitudes. [12, 14]. The absence of an ozone layer allows very short ultraviolet radiation to reach the surface (particularly 190–300 nm, shielded on Earth), which results in the formation of free oxidant radicals that cause rapid decomposition of any organic material near the surface [14].

3 Vehicle-Habitat

3.1 The Equilibrium

To analyze the compass mechanism, let us consider the simplified diagram in the figure. The rods are assumed to be rigid bodies, the masses concentrated at the center of gravity of each rod, and the structure is studied as the following parameters change:

φ: angle at the base of the rods
φ_0: angle in the initial configuration of the rising phase
φ_1: angle in the final configuration of the ascent phase, coincident with the angle of the initial configuration of the opening phase at constant height
φ_2: angle of the final configuration of the opening phase at constant altitude
m_1: mass of rod 1
m_2: mass of rod 2
m_3: mass of rod 3 (middle rod)
La: length of the side rods, assuming they have the same length
d_1: distance of the center of mass of rod 1 from the end resting on the ground
d_2: distance of the center of mass of rod 2 from the end resting on the ground.
d_3: distance of the center of mass of rod 3 from the left end

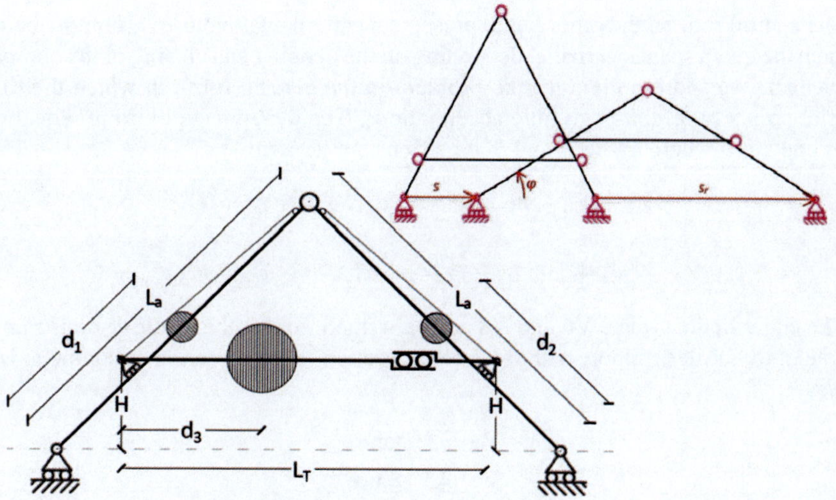

Fig. 1. Scheme of the compass mechanism showing the main geometric parameters defining the structure. The circle indicates the position of the material point equivalent to the mass of the side rods and the habitat to the central rod.

L_T: distance between the two constraints of rod 3 (center). It differs from the length of this rod in that L_T varies with the angle
H: elevation with respect to the ground of rod 3 (center)
s: displacement imposed at the base of the left constraint
s_r: displacement of the right-hand constraint

The compass mechanism can be divided into two parts: the side-rod assembly and the central rod. The first is connected to the outside by two constraints that can be hinges (in the case of quasi-stationary motion), a hinge and a roller support (in the case of moving one support at a time) or both roller supports (in the case of moving both.

Consistently V2 and V4 will take on the meaning of applied forces in the case where they are carriages, or of constraining forces, in the case where they are hinges (Fig. 2).

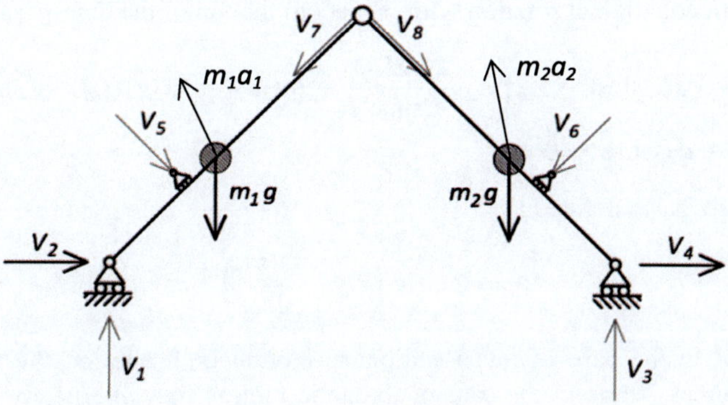

Fig. 2. Diagram of the self-weight and constraining forces applied on lateral beams.

The central rod, on the other hand, presents a pattern analogous to a support-barrel, in which the beam span is variable, depending on the opening and closing of the compass mechanism. We address the dynamic problem in the general form, in which the three material points have non-negligible accelerations. The dynamic equilibrium equations of the body as a whole in the X and Y directions are written:

$$V_2 + V_4 - m_1 a_{1x} - m_2 a_{2x} - m_3 a_{3x} = 0 \tag{1}$$

$$V_1 + V_3 - (m_1 + m_2 + m_3)g - m_1 a_{1y} - m_2 a_{2y} - m_3 a_{3y} = 0 \tag{2}$$

The equation that relate V6 and V8 can be written, so that the resulting constraint is equivalent to a roller support with the free translation in the direction along the axis of the legs:

$$\begin{aligned} V_8 &= V_6 \tan\varphi \\ V_7 &= V_5 \tan\varphi \end{aligned} \tag{3}$$

It is possible to write the dynamic equilibrium equations for rod 3 (center rod) in X and Y direction, and at rotation around the left end:

$$-V_5 \sin\varphi + V_7 \cos\varphi - m_3 a_{3x} = 0 \tag{4}$$

$$V_5 \cos\varphi + V_7 \sin\varphi + V_6 \cos\varphi + V_8 \sin\varphi - m_3 a_{3y} - m_3 g = 0 \tag{5}$$

$$(V_6 \cos\varphi + V_8 \sin\varphi) L_T - (m_3 a_{3y} + m_3 g) d_3 = 0 \tag{6}$$

The equilibrium equation at the rotation with respect to the top of the compass of the rod 1 is:

$$\begin{aligned} -V_1 L_a \cos\varphi + V_2 L_a \sin\varphi + V_5 \left(L_a - \frac{H}{\sin\varphi} \right) \\ +(m_1 a_{1y} + m_1 g)(L_a - d_1)\cos\varphi + \\ -m_1 a_{1x} (L_a - d_1)\sin\varphi = 0 \end{aligned} \tag{7}$$

The equilibrium equation at rotation with respect to the top of the com-pass of rod 2 is:

$$\begin{aligned} V_3 L_a \cos\varphi + V_4 L_a \sin\varphi - V_6 \left(L_a - \frac{H}{\sin\varphi} \right) - (m_2 a_{2y} + m_2 g)(L_a - d_2)\cos\varphi + \\ -m_2 a_{2x} (L_a - d_2)\sin\varphi = 0 \end{aligned} \tag{8}$$

Furthermore, L_T is defined as:

$$L_T = 2\cos\varphi \left(L_a - \frac{H}{\sin\varphi} \right) \tag{9}$$

V2 and V4 in the case of quasi-stationary motion both take on the meaning of constraining forces, while in the case of dynamic motion they alternately take on the meaning of constraint and action (null or nonzero).

Having defined the previous equations, they are reworked to obtain a system of 2 equations in the unknowns V2, V4, and s. To solve the problem, at least two other assumptions about the previous unknowns will be necessary: these assumptions can be, for example, quasi-stationary motion, for which inertial forces and displacement s are neglected; imposed motion, for which the function of displacement *s* and angle. φ, as in Fig. 1.

$$V_2 + V_4 - m_1 a_{1x} - m_2 a_{2x} - m_3 a_{3x} = 0 \tag{10}$$

$$(V_4 - V_2) + m_3 a_{3x}\left(1 - \frac{H}{L_a \sin\varphi}\right) + (m_3 a_{3y} + m_3 g)\frac{H\cos\varphi}{L_a \sin^2\varphi}$$
$$+\frac{(m_1 a_{1y} + m_1 g)}{\tan\varphi}\frac{d_1}{L_a} + \frac{(m_2 a_{2y} + m_2 g)}{\tan\varphi}\frac{d_2}{L_a} + m_1 a_{1x}\left(1 - \frac{d_1}{L_a}\right) - m_2 a_{2x}\left(1 - \frac{d_2}{L_a}\right) = 0 \tag{11}$$

Equation (1) and (11) allows the motion of the compass to be determined, once the value of either of the two forces V2 or V4 is known. In the case of an assumed motion, one can determine the hourly law of the force to guarantee that motion, simply by deriving the hourly law of the assumed motion and reinserting it within the formulas defining V2 and V4.

At the energy level, it is possible to write the total mechanical energy ***E***, assuming rigid bodies and absence of dissipative phenomena:

$$\begin{aligned} E = U + K &= \sin\varphi(m_1 g d_1 + m_2 g d_2 + m_3 g H) + m_1 g d_1 \\ &+ \frac{1}{2} m_1\left(v_{1x}^2 + v_{1y}^2\right) + \frac{1}{2} m_2\left(v_{2x}^2 + v_{2y}^2\right) + \frac{1}{2} m_3\left(v_{3x}^2 + v_{3y}^2\right) \end{aligned} \tag{12}$$

where ***K*** is the kinetic energy and ***U*** is the potential gravitational energy.

The variation of the mechanical energy ***E*** corresponds to the work of the external forces.

The accelerations of the three material points of the rods 1, 2 and 3, as in Fig. 1 with the masses considered to be concentrated in the center of gravity, are obtained by deriving the positions. The positions are given by the following equations, under the assumption that the central rod is held horizontally, under the general assumption that the height in-altitude H is also time-varying, and that the constraint in the lower left-hand corner moves by displacement ***s(t)***:

$$\begin{aligned} P_1 &= (d_1\cos\varphi + s;\ d_1\sin\varphi) \\ P_2 &= ((2L_a - d_2)\cos\varphi + s;\ d_2\sin\varphi) \\ P_3 &= (H\cot\varphi + d_1\cos\varphi + s;\ H) \end{aligned} \tag{13}$$

Deriving in time it is possible to write the velocities and the accelerations of the center of mass related to the rod 1, 2 and 3:

$$\begin{aligned} v_1 &= (-\dot{\varphi} d_1 \sin\varphi + \dot{s};\ \dot{\varphi} d_1 \cos\varphi) \\ v_2 &= (-\dot{\varphi}(2L_a - d_2)\sin\varphi + \dot{s};\ \dot{\varphi} d_2 \cos\varphi) \\ v_3 &= \left(-\frac{H\dot{\varphi}}{\sin^2\varphi} + \dot{H}\cot\varphi + \dot{s} - d_1\sin\varphi;\ \dot{H}\right) \end{aligned} \tag{14}$$

$$\begin{aligned}
\boldsymbol{a_1} &= \left(\boldsymbol{d_1}\left(-\ddot{\boldsymbol{\varphi}}\sin\boldsymbol{\varphi} - \dot{\boldsymbol{\varphi}}^2\cos\boldsymbol{\varphi}\right) + \ddot{\boldsymbol{s}}; \boldsymbol{d_1}\left(\ddot{\boldsymbol{\varphi}}\cos\boldsymbol{\varphi} - \dot{\boldsymbol{\varphi}}^2\sin\boldsymbol{\varphi}\right)\right) \\
\boldsymbol{a_2} &= \left((2\boldsymbol{L_a} - \boldsymbol{d_2})\left(-\ddot{\boldsymbol{\varphi}}\sin\boldsymbol{\varphi} - \dot{\boldsymbol{\varphi}}^2\cos\boldsymbol{\varphi}\right) + \ddot{\boldsymbol{s}}; \boldsymbol{d_2}\left(\ddot{\boldsymbol{\varphi}}\cos\boldsymbol{\varphi} - \dot{\boldsymbol{\varphi}}^2\sin\boldsymbol{\varphi}\right)\right) \\
\boldsymbol{a_3} &= \left(\frac{\boldsymbol{H}}{\sin^2\boldsymbol{\varphi}}\left(-\ddot{\boldsymbol{\varphi}} + 2\dot{\boldsymbol{\varphi}}^2\cot\boldsymbol{\varphi}\right) + \ddot{\boldsymbol{H}}\cot\boldsymbol{\varphi} + \ddot{\boldsymbol{s}} - \boldsymbol{d_1}\dot{\boldsymbol{\varphi}}^2\cos\boldsymbol{\varphi} - \boldsymbol{d_1}\ddot{\boldsymbol{\varphi}}\sin\boldsymbol{\varphi}; \ddot{\boldsymbol{H}}\right)
\end{aligned} \tag{15}$$

The compass system during quasi-stationary motion has no inertial forces. The equations in quasi-stationary motion are derived from the general Eqs. (1) and (11) by positing zero acceleration values:

3.2 Vehicle Ascent Mode

To analyze this mode, it is firstly necessary to define the function H(t), which must assume at the extremes of the time interval of the rising mode $H(t_0) = H_0$ and $H(t_1) = H_1$. It must be considered that the velocities of the centers of mass of the compass are zero at the initial instant $t = t_0$ and at the final instant $t = t_1$ since the rod (4) must stop when it reaches the height. For these conditions the velocity of the centers of mass in rods 1 (left), 2 (right) and 3 (middle) gives the values corresponding to different istant t, as well at the initial instant $t = t_0$ and at the final instant $t = t_1$. A step motion, for which $s = 0$ can be considered. To ensure that the velocities of rods 1 and 2 are zero, it is enough that:

$$\dot{\boldsymbol{\varphi}}(\boldsymbol{t_0}) = \dot{\boldsymbol{\varphi}}(\boldsymbol{t_1}) = 0 \tag{16}$$

For rod 3 it is necessary that $\dot{\boldsymbol{H}} = \mathbf{0}$

The latter equation is satisfied if condition (16) is verified:

$$\dot{\boldsymbol{H}} = \frac{\boldsymbol{dH}}{\boldsymbol{d\varphi}}\dot{\boldsymbol{\varphi}} = \boldsymbol{H}'\dot{\boldsymbol{\varphi}} \tag{17}$$

Condition (16) also must be verified at the initial instant $t = t_0$ and at the final instant $t = t_1$.

$$\boldsymbol{H}'(\boldsymbol{\varphi_1}) = \mathbf{0} \tag{18}$$

$$\boldsymbol{H}''(\boldsymbol{\varphi_1}) = \mathbf{0} \tag{19}$$

The function $H(\varphi)$ must satisfy previous equations and be worth $H(\varphi_0) = H_0$ and $H(\varphi_1) = H_1$. The simplest function that satisfies these conditions is a parabolical function having vertex at φ_1 (Fig. 3):

$$\boldsymbol{H}(\boldsymbol{\varphi}) = \frac{\boldsymbol{H_0} - \boldsymbol{H_1}}{(\boldsymbol{\varphi_0} - \boldsymbol{\varphi_1})^3}(\boldsymbol{\varphi} - \boldsymbol{\varphi_1})^3 + \boldsymbol{H_1} \tag{20}$$

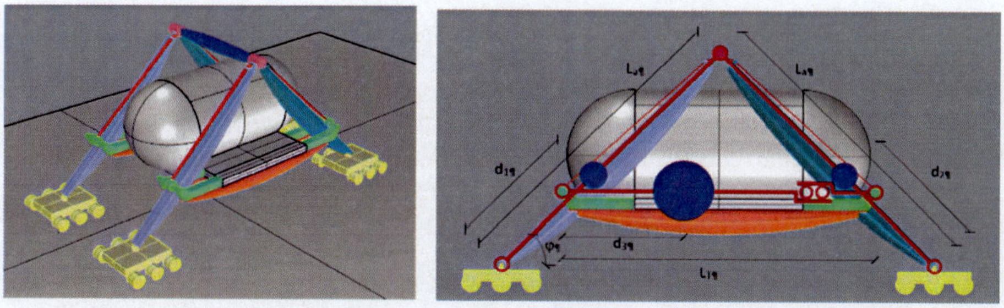

Fig. 3. Compass structure, one of the two constituents of the vehicle-habitat

4 Conclusions

The habitation vehicle described in this work has been analyzed in order to highlight the possibility of realizing the required performances: drive and loading and unloading phases, maintaining a constant trim and height of the habitation unit. The models described, albeit in the context of the hypotheses that have been declared, demonstrate the possibility of the vehicle to perform the tasks assigned to it. These models are, therefore, a reference that the authors deem useful for the design of the module.

The future developments of this work concern the study of the transversal stability of the vehicle during motion, outside the vertical plane in which the models have been developed in this work. Furthermore, a fundamental aspect will concern the methodologies for solving problems arising from the presence of friction in the mechanisms and rolling friction with the ground. Furthermore, the authors consider the possible motorized interlocking of the motion defined for the ropes to which the task of defining the height of the constraints of the transversal beam along the compass beams representing the legs of the system is an important aspect.

Acknowledgments. This paper is part of the project NODES which has received funding from the MUR–M4C2 1.5 of PNRR with grant agreement no. ECS00000036.

References

1. Drake, B.G.: Mars Architecture Steering Group, "Human Exploration of Mars–Design Reference Architecture 5.0", NASA document NASA. Technical report, SP-2009-566 (2009). 15, 17, 21, 23, 113, 114, 116, 125, 145G
2. Salotti, J.M.: Revised scenario for human missions to Mars. Acta Astronaut. **81**(1), 273–287 (2012). https://doi.org/10.1016/j.actaastro.2012.07.018
3. Khanakornsuksan, C., Wongratanaphisan, T.: Optimal design of a suspended cable driven system for locomotion device. Int. J. Mech. Control **21**(02), 63–71 (2020)
4. Fouad, I., Carbone, G., Ceccarelli, M.: Design and implementation of high level control modes for a 3D cable-driven-parallel robot. Int. J. Mech. Control **21**(02), 87–94 (2020)
5. Bonafede, Manuello Bertetto, A., Fiorini, P., Muscolo, G.G.: Compliant legs in biped robots for running optimization. Int. J. Mech. Control **23**(02), 53–60 (2022)
6. Cohen, M.M.: Mobile Lunar and Planetary Bases AIAA, 2003 (2003)

7. Seedhouse, E.: Bigelow Aerospace: Colonizing Space One Module at a Time. Springer, Svizzera (2015)
8. Ciardullo, C., Yashar M., Montes, J.: Martian Ice Habitats: Approaches to Additive Manufacturing with H2O Beyond Mars Ice House. ResearchGate (2016)
9. Yashar M., Ciardullo, C.: Mars X-House: Design Principles for an Autonomously 3D- Printed ISRU Surface Habitat, ResearchGate (2019)
10. Weissman, P.R.: The Solar System and Its Place in the Galaxy, in Encyclopedia of the Solar System. Elsevier (2014)
11. Buis, A.: Milankovitch (2020) (Orbital) Cycles and Their Role in Earth's Climate. https://climate.nasa.gov/news/2948/milankovitch-orbital-cycles-and-their-role-in-earths-climate/ (2020)
12. Barlow, N.: Chapter 6 – Atmospheric conditions and evolution. In: Mars: an introduction to its inferior, superior and atmosphere, Cambridge University Press (2008)
13. Catling, D.C.: Mars Atmosphere. In: Encyclopedia of the Solar System, pp. 343–357. Elsevier (2014). https://doi.org/10.1016/B978-0-12-415845-0.00016-5
14. Carr, M.H., Bell, J.F.: Mars: Surface and Interior, in fEncyclopedia of the Solar System. Elsevier (2014)

Multy Body Rope Approach and Funicular Prototype for a New Constructive System for Catenary Arches

Ilaria Cavaliere[1(✉)], Giuseppe Fallacara[1], Amedeo Manuello Bertetto[2], Jonathan Melchiorre[2], and Giuseppe Carlo Marano[2,3]

[1] Department of Architecture, Construction and Design, Politecnico di Bari, Bari, Italy
ilaria.cavaliere@poliba.it

[2] Politecnico di Torino, DISEG, Department of Structural, Geotechnical and Building Engineering, Corso Duca Degli Abruzzi, 24, 10128 Turin, Italy

[3] College of Civil Engineering, Fuzhou University, Fuzhou 350108, China

Abstract. The proposed method for the definition of the form is based on a multi-body rope approach (MRA) with masses connected by inextensible ropes characterized by a certain slack coefficient and by different constraint conditions. These parameters played a fundamental role in the definition of the shallowness ratio of the grid (rope), and therefore in the effect of the instability of the reversed shape under loading. In this paper, different MRA enhancing strategies were proposed and discussed. These approaches were used to determine the final shape of a funicular prototype for a pavilion inspired by Guarino Guarini's architecture. The prototype project is based on the principle of Gaudi's inverted catenary arch. It is composed of a series of arches, each one obtained hanging a metallic chain - which assumed the shape of a catenary curve - and then welding all the rings in order to fix its form. Different possible configurations will be described in this paper.

Keywords: Catenary arch · Form-finding · Funicularity · Multibody Rope Approac

1 Introduction

The aim of this paper is to describe an innovative method for the construction of catenary arches using iron chains. Over the years many optimization [1,2] and form-finding methods have been developed for the purpose of obtaining the best structural shape of arcs and curved structures. A catenary is a curve corresponding to the shape of a hanging string, whose ends are fixed to supports and which is left free to hang under its self-load [3]. In these conditions, the chain form is optimized for purely tensile stresses. If this shape is fixed and then reversed, it becomes a frame that works under a sheer compression load: this is the principle

S. Gabriele et al. (Eds.): IWSS 2023, LNCE 437, pp. 259–268, 2024.
https://doi.org/10.1007/978-3-031-44328-2_27

of the reversed catenary [4], which was hypothesized for the first time by Robert Hooke in 1675. The principle of the catenary was used in the past to verify the stability of arches since they can be considered stable if the line of force (which is a catenary) lies within the thickness of the arch itself [5]. Since a reversed catenary is shear-free, it is the ideal shape to build structures that work mostly in compression, such as masonry arches and vaults. One of the pioneers in the use of catenaries was the Spanish architect Antoni Gaudí, who was the first known architect to use a hanging model as an empirical optimization tool [6]. Several of Gaudí's architectures have been conceived using the catenary geometry as a design instrument: for example, the attic of Casa Milá is a vaulted space characterized by a sequence of diaphragm catenary arches, while the corridor of the School of Teresianas presents parabolic arched structures very similar to catenaries [7]. There are also some examples of the use of this peculiar shape in modern masonry structures. The St. George Orthodox Church, designed and built by Wallmakers in 2016, is a structure made of raw earth bricks and it is composed of a series of vaulted spaces whose form is based on the catenary geometry. The studio clearly stated that they were inspired by Gaudí's method of hanging ropes [8]. Another interesting application is represented by an experimental prototype of social housing realized by some researchers at the University of the Witwatersrand, who built a model of the house covered with a vaulted catenary-shaped ceiling. This allowed them to avoid steel reinforcements in the roof, to spare material [4]. A catenary arch is a complex curve, that is often created through an approximation and that can require the realization of a profile using CNC machines. The possibility to create a rigid catenary arch directly from a hanging chain would help to avoid the approximation and allows not necessarily use of CNC machines. These metal arches can be directly used to create architectural installations but can be also very useful in the realization of centrings for catenary vaulted spaces. In recent decades, advancements in computing technology have opened new avenues for research in structural engineering, particularly in the search for optimal structural forms. The emergence of digital tools, such as calculators and computer simulations, has greatly facilitated the exploration of various form-finding methods [9–12]. Among the diverse form-finding methods available, the Multibody Rope Approach (MRA) [13–15] has gained recognition for its ability to model structures as interconnected nodes connected by slack ropes. This approach treats the structure as a dynamic system and seeks to identify the funicular form that corresponds to the applied loading conditions. By considering the equilibrium of forces within the system, MRA determines the optimal configuration where the tension in the ropes is in balance, resulting in a structurally efficient and stable form. In this paper, the Multibody Rope Approach (MRA) is employed to determine the optimal structural form of a design inspired by the Chapel of the Holy Shroud by Guarino Guarini. The study focuses on the use of form-finding techniques to identify two distinct structural types: one composed of catenary chains and another representing a funicular form under the given loading conditions. After identifying these two structural configurations, the next step involves subjecting them to static self-weight loading analysis [16]. This analysis allows for the evaluation of the stress distribution within the structures and facilitates a comparison between the optimized funicular structure and the catenary-based structure, which is comparatively easier to construct. By quantifying the stress distances between the two structural types, the study aims to provide insights into the trade-off between structural performance and constructability. This analysis helps inform the decision-making process by highlighting the differences in stress distribution and potential structural implications of choosing one form over the other. Ultimately, the article contributes to the understanding of how form-finding techniques can assist in the design optimization process, considering both structural efficiency and practical feasibility.

2 The Construction Technique of the Catenary Frame

The constructive method presented has been conceived by prof. Giuseppe Fallacara to create rigid catenary arches directly "freezing" the shape of a hanging chain through welding. Some tests have been carried out using small pieces of chains, in order to evaluate the final outcome. The ends of the string were fixed to a horizontal bar and then the string itself was left hanging. It assumed naturally the configuration of a catenary curve, therefore all the joints were welded in order to fix the shape. After some successful results, an arch 3.10 m wide and 2.75 m tall has been realized. The experiment has been carried on using a galvanized iron chain, with a rod of 20 mm and rings 7 cm wide and 10.2 cm high (see Fig. 1). Two important considerations must be done, in order to explain the good quality of the prototype. First of all, the iron bar used as support has been placed in order to be perfectly horizontal, otherwise, the curve wouldn't have been a precise catenary. On the second hand, the chosen chain has a mass of 6.3 kg/m and the high weight prevents significant deformations during the welding phase.

3 The Assembly of the Catenary Arches

As previously written, these metallic catenary arches can be used to realize architectural installations. In this section, a first hypothesis will be described, together with some considerations linked to structural analysis.

3.1 A Configuration Inspired to Guarino Guarini

The first possible structure here described has been conceived as a tribute to the Italian architect Guarino Guarini, one of the main representatives of Italian Baroque. Guarini's vaults are generally characterized by lightness and transparency since they are usually formed by intersecting arches [17], and this principle is reprised in the design process of the described pavilion. The structure consists of three circular levels of catenary arches. The ground level is composed of six arched frames that intersect each other; the next level is as well composed

Fig. 1. The catenary arch obtained after the welding of the joints.

of six intersecting frames, that are rotated so that the bases of the arches are placed in correspondence with the intersection points of the previous level. The third level works the same way. The theme of staggered superimposed arches is inspired in particular by Guarini's dome of the Chapel of the Holy Shroud, which is characterized by a similar pattern (see Fig. 2). As a matter of fact, this dome appears on the inside as a sequence of superimposed elliptical arches. In the case of the designed self-bearing frame, in which each arch unloads its weight on the arches below, the catenary configuration is supposed to produce mainly compressive stresses, even if it is not a perfectly optimized structure anymore. Its actual static behaviour will be described in the next paragraph.

3.2 Analysis Through the Multi-body Rope Approach

The structural form depicted in Fig. 2, inspired by Guarino Guarini's Chapel of the Holy Shroud, was achieved through the application of the Multibody Rope Approach (MRA). In order to test the effectiveness of the method in defining catenary shapes, the MRA was initially applied to individual chains. By analyzing the behavior of these chains, the desired catenary shapes were obtained. Subsequently, these catenaries were superimposed to compose the complete geometry of the structure.

In addition to this approach, a comparative analysis was conducted by directly applying the MRA to the complete structure. This allowed for the determination of the funicular shape of the structure under the influence of self-weight

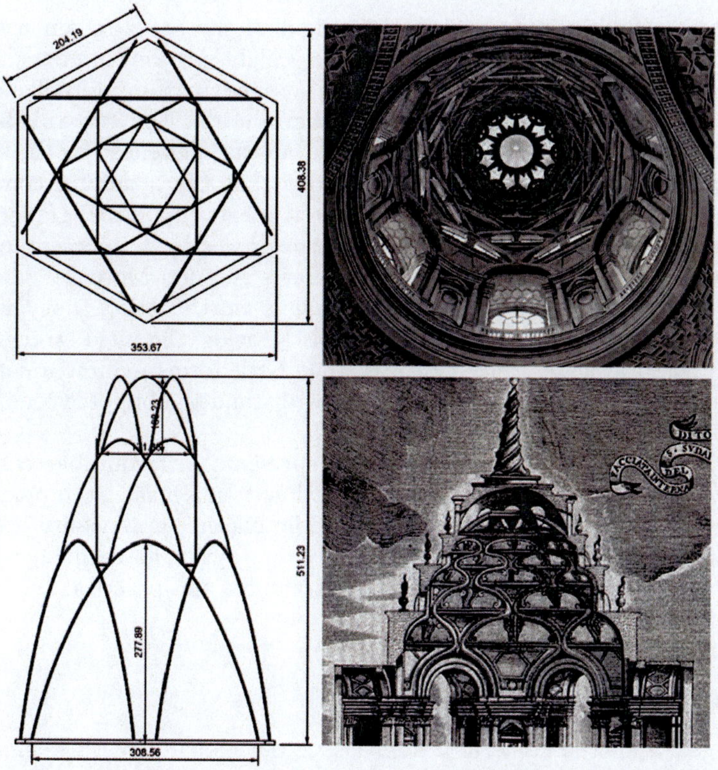

Fig. 2. A comparison between a scheme of the designed pavilion and the vault of the Chapel of the Holy Shroud.

loading. By considering the complete structure, the MRA was able to accurately define the optimal form of the structure in relation to the imposed loads. The resulting structures were subjected to analysis using the FEM SOFiSTiK software [16], assuming that they were composed of a solid steel cable with a diameter of 3 cm. The analysis was performed considering the self-weight of the structure. Based on this assumption, the calculated load per linear meter was determined to be 5.5 kg/m.

The analysis results, including internal actions and Von Mises stress, are presented in Fig. 3. It is evident from the results that the funicular structure proves to be the superior choice from a structural perspective. This is due to the fact that the funicular structure exhibits a bending moment of zero, which aligns with expectations for such a form. Consequently, the stress applied to the funicular structure is only 15% of that applied to the structure composed of catenaries. Taking into account the weight of the actual chain used in the construction of the structure, which has a weight per linear meter of 6.3 kg, an adjustment factor of 1.15 needs to be applied to the internal action and stress values shown in Fig. 3. Multiplying the values in Fig. 3 by this factor will provide the actual internal action and stress values experienced by the structure. Additionally, it is important to acknowledge that the structure composed of catenaries offers advantages from a constructional standpoint, aligning well with the technique described in Sect. 2. This form proves to be more practical and easier to construct compared to the funicular structure. Moreover, despite the structural efficiency of the funicular form, it is worth noting that the stresses applied to both structures remain significantly below the yield strength of any type of structural steel. This indicates that both forms exhibit a satisfactory level of structural integrity and can safely withstand the imposed loads without reaching the limit of their material strength.

In this context, the catenary form represents a favourable compromise between structural efficiency and constructional simplicity. It provides a satisfactory level of structural performance while offering practicality and ease of construction. This makes it a viable and attractive option for realizing the desired architectural design, considering both the technical and practical aspects of the construction process.

3.3 Other Possible Configurations

As emerged from the structural analysis, if the global structure consists of a composition of entwined catenary arches, it is no longer a frame optimized to work under sheer compression loads, since the optimal structure will be a funicular geometry. However, the installation previously described shows only one of the numerous possibilities offered by the assembly of catenary arches made of welded chains. First of all, the number of arches and levels can vary, producing more or less complex patterns. It is also possible to create totally different geometries, which don't refer necessarily to domes, but can be linked to other types of vaulted spaces. Figure 4 shows another possible configuration, consisting of a "winged" structure. It has only two footholds, where all the arches are connected. The central arch is disposed on a vertical plane, while the lateral ones lie on oblique planes, creating overhangs. The entire structure is balanced thanks to a network of elastic ropes with a diameter of 7 mm, that connect the three curved frames and generate two symmetrical saddle-shaped profiles. The ground connection is resolved through two triangular frames connected transversally by an iron bar, which is in charge of nullifying the pressure produced by the arches.

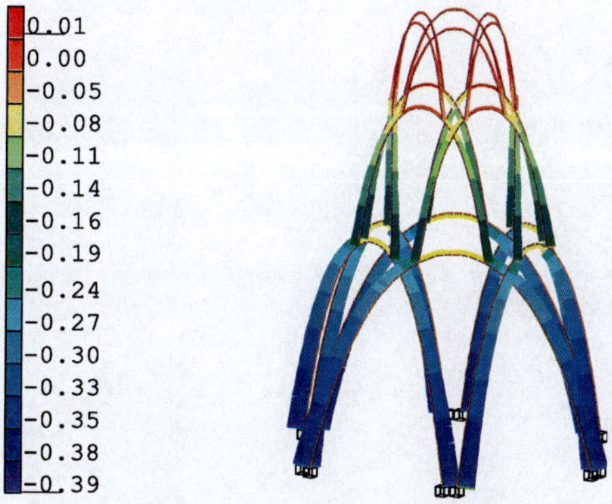

(a) Axial Force: $|N|_{max} = 0.39kN$

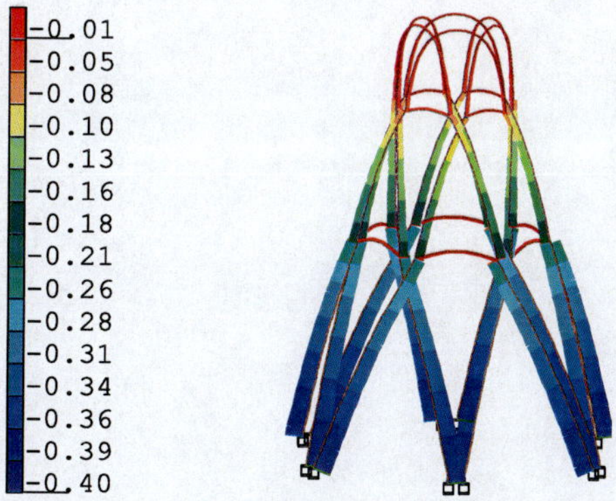

(b) Axial Force: $|N|_{max} = 0.40kN$

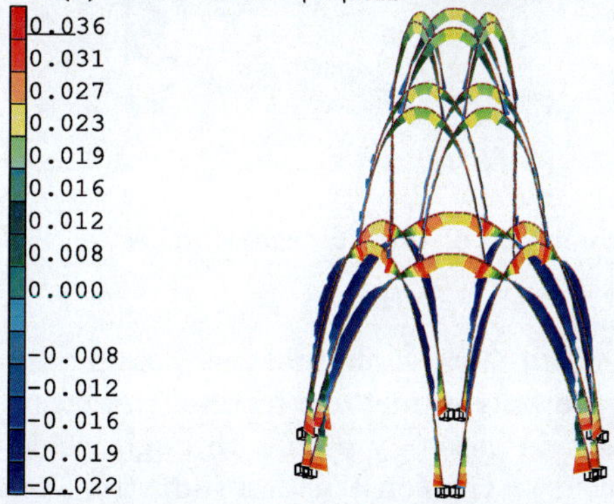

(c) Bending Moment: $|M|_{max} = 0.04kNm$

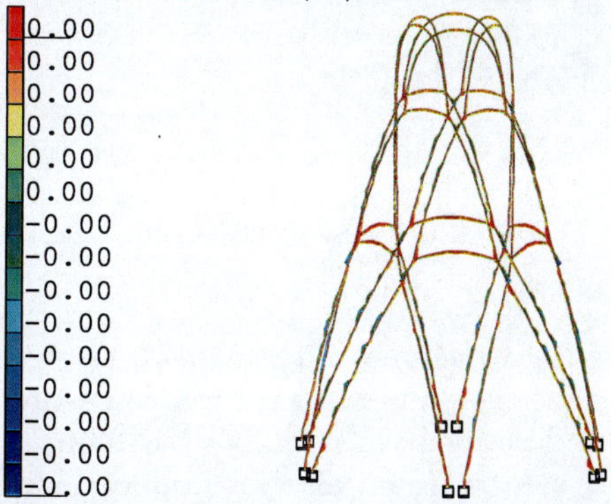

(d) Bending Moment: $|M|_{max} = 0kNm$

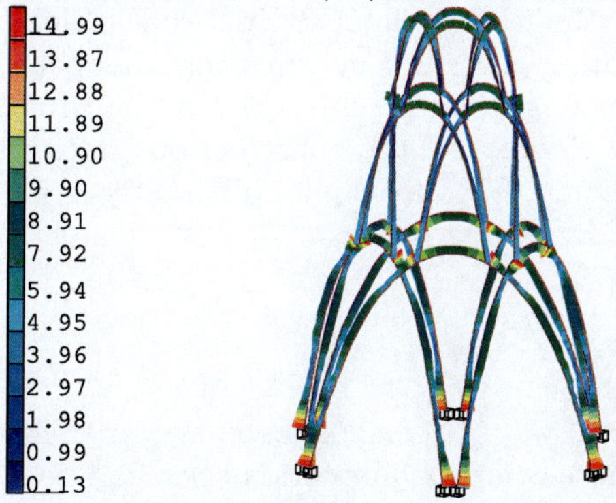

(e) V.M. Stress: $|\sigma^{VM}|_{max} = 14.99kPa$

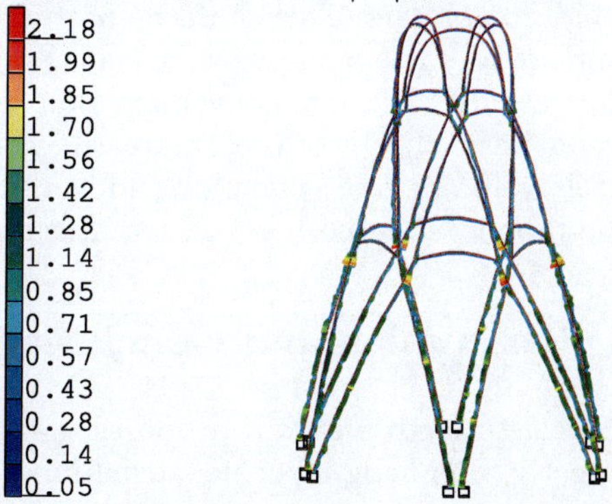

(f) V.M. Stress: $|\sigma^{VM}|_{max} = 2.18kPa$

Fig. 3. Results of the static analysis on the two structural typologies inspired by the Chapel of the Holy Shroud by Guarini. Catenary shape on the left and funicular shape on the right.

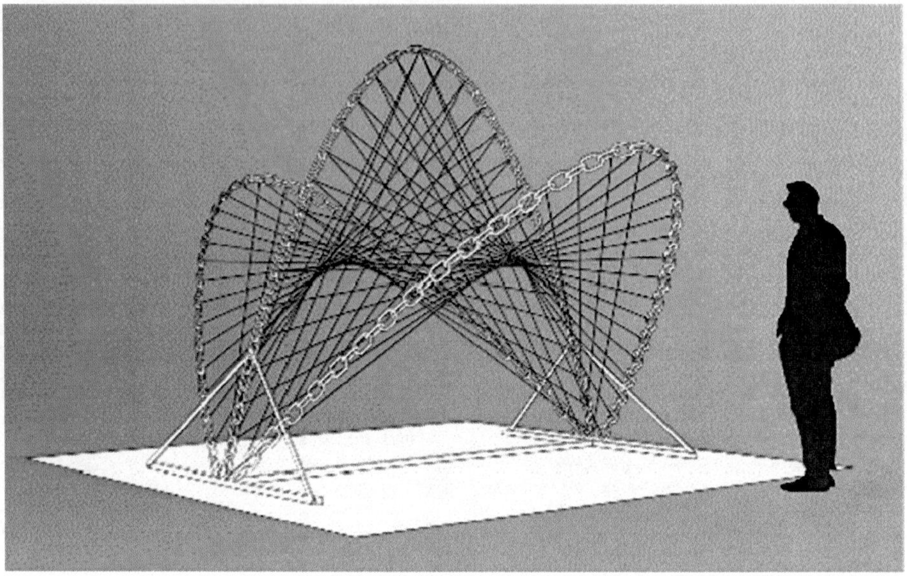

Fig. 4. 3D view of a winged structure composed of three catenary arches.

As it has been already said, there are much more configurations possible to realize this kind of structure, consisting essentially of metallic frames. However, the described constructive technique to obtain catenary arches through welding can be useful also for applications linked to the construction industry. For example, innovative centrings can be built using a series of arches connected by longitudinal bars. This could be useful for the realization of vaulted spaces that work only under compression loads. Creating centrings by "freezing" hanging chains would be more convenient than realizing wooden centrings, since cutting wood inevitably produces material waste. Moreover, the construction process itself would be more complex and would require CNC machines, differently from the proposed method, which requires only a welding phase.

4 Discussion and Conclusions

The described method proposes an easy way to realize catenary profiles directly from hanging chains, avoiding a more complex fabrication through CNC machines or using approximations. Such a flexible method would allow also to realize non-symmetrical arches, varying the reciprocal height of supports. These welded chains can be directly used or employed as profiles or centrings for masonry arches or vaults. Despite the advantages explained, it is necessary to highlight some limits of this technique. First of all, the welded arches can really work under sheer compression only if the global structure is simple: as a matter of fact, a catenary is a curve optimized only under the self-weight load. A composition of more catenaries entwined or superimposed does not correspond to an optimized structure, since every arch is subject to other loads and not only to gravity, therefore shear stresses will be present. Moreover, arches made of a welded chain are simple to produce, but difficult to transport, especially if they have large dimensions. Therefore this method would be more convenient only if applied on-site. Furthermore, in Sect. 3.2 of the study, it was observed that the shape obtained through the construction technique presented may not be the most structurally efficient one when compared to the funicular form obtained through the Multibody Rope Approach (MRA). However, it is important to highlight that the catenary form still represents a viable solution that balances both structural efficiency and constructive simplicity.

While the catenary form may not achieve the same level of structural optimization as the funicular form, it offers advantages in terms of construction feasibility and architectural considerations. The catenary shape aligns well with the construction technique discussed in Sect. 2 and can be more easily realized using available construction materials and methods.

Despite its slight compromise in structural efficiency, the catenary form remains a valid solution that takes into account both the structural requirements and architectural design intent. It strikes a balance between structural performance and practical construction considerations, making it a suitable choice for many real-world applications where both factors need to be considered.

By considering the architectural and construction aspects alongside structural efficiency, the catenary form provides a satisfactory solution that meets the overall objectives of the project.

A possible future development of this research can deal with a deeper investigation of the use of welded chains to create centring systems for vaulted ceilings or even in the use of weights hunged in specific points of the chains which allow the creation of funicular shapes using the presented techniques.

References

1. Melchiorre, J., Bertetto, A.M., Marano, G.C.: Application of a machine learning algorithm for the structural optimization of circular arches with different cross-sections. J. Appl. Math. Phys. **9**(5), 1159–1170 (2021)
2. Melchiorre, J., Manuello, A., Marmo, F., Adriaenssens, S., Marano, G.: Differential formulation and numerical solution for elastic arches with variable curvature and tapered cross-sections. Eur. J. Mech.-A/Solids **97**, 104757 (2023)
3. Gohnert, M., Bradley, R.: Catenary solutions for arches and vaults. J. Archit. Eng. **26**(2), 04020006 (2020)
4. Gohnert, M., Bulovic, I., Bradley, R.: A low-cost housing solution: earth block catenary vaults. In: Structures, vol. 15, Elsevier, pp. 270–278 (2018)
5. Block, P., DeJong, M., Ochsendorf, J.: As hangs the flexible line: equilibrium of masonry arches. Nexus Netw. J. **8**(2), 13–24 (2006)
6. Tomlow, J.: Gaudí's reluctant attitude towards the inverted catenary. In: Proceedings of the Institution of Civil Engineers-Engineering History and Heritage, vol. 164, no. 4, pp. 219–233 (2011)

7. Huerta, S.: Structural design in the work of Gaudi. Archit. Sci. Rev. **49**(4), 324–339 (2006)
8. St george orthodox church/wallmakers. https://www.archdaily.com/959657/st-george-orthodox-church-wallmakers?ad_source=search&ad_medium=projects_tab. Accessed 06 May 2023
9. Schek, H.-J.: The force density method for form finding and computation of general networks. Comput. Methods Appl. Mech. Eng. **3**(1), 115–134 (1974)
10. Otter, J.R.H., Cassell, A.C., Hobbs, R.E.: POISSON, Dynamic relaxation. In: Proceedings of the Institution of Civil Engineers, vol. 35, no. 4, pp. 633–656 (1966)
11. Block, P., Ochsendorf, J.: Thrust network analysis: a new methodology for three-dimensional equilibrium. J. Int. Assoc. Shell Spatial Struct. **48**(3), 167–173 (2007)
12. Kilian, A., Ochsendorf, J.: Particle-spring systems for structural form finding. J. Int. Assoc. Shell Spatial Struct. **46**(2), 77–84 (2005)
13. Manuello, A.: Multi-body rope approach for grid shells: form-finding and imperfection sensitivity. Eng. Struct. **221**, 111029 (2020)
14. Bertetto, A.M., Riberi, F.: Form-finding of pierced vaults and digital fabrication of scaled prototype. Curved and Layered Struct. **8**(1), 210–224 (2021)
15. Manuello, A., Melchiorre, J., Sardone, L., Marano, G.C.: Multi-body rope approach for the form-finding of shape optimized grid shell structures. In: Proceedings of the 15th World Congress on Computational Mechanics (2022)
16. SOFiSTiK. Text Editor 2023, Flataustr. 14, 90411 Nuremberg (2022). https://www.sofistik.com
17. Mazziotti, A., Brandonisio, G., Lucibello, G., De Luca, A.: Structural analysis of the basket dome in the chapel of the holy shroud by Guarino Guarini. Int. J. Architect. Heritage **11**(3), 324–338 (2017)

Numerical and Experimental Dynamic Analysis of 3D-Printed Pierced Vault

Raffaele Cucuzza(✉), Amedeo Manuello Bertetto, Marco Domaneschi, Raffaele Tarantini, Alessandro Cardoni, and Gian paolo Cimellaro

Department of Structural, Geotechnical and Building Engineering, Politecnico di Torino,
Corso Duca degli Abruzzi, 24, 10129 Torino, Italy
raffaele.cucuzza@polito.it

Abstract. In the present paper, a form-finding of shallow grid shells was introduced based on the multi-body rope approach (MRA) for the definitions of vaults with optimized shapes and different hole percentages. In order to obtain experimental validation, a physical model was reproduced at the laboratory scale performing ad hoc measurements to compare the observed respect to the simulated behaviour. A 3D printing procedure based on the Fuse Deposition Modeling (FDM) technique in polylactide (PLA) material was used to realize formworks of the cement-based blocks of the scaled prototype. Several static and dynamic load configurations are investigated, collecting the parameters into a sensitivity analysis, which mainly affects the structural behaviour. To compare the natural frequency obtained by experimental tests with those evaluated with a numerical model, an impulse test has been performed. Results show that an improvement in the modelling strategies adopted for the numerical model is necessary for the identification of the real structural dynamic response of the vault.

Keywords: Multi-body Rope Approach · Shape Optimization · Vaults · Dynamic Behaviour of Shells · Fuse Deposition Modeling · Experimental Dynamic Evaluation

1 Introduction

Several authors have developed novel applications for masonry structures over the past 20 years, integrating them into real-world architectural and engineering settings while using cutting-edge realizations and design methodologies (e.g. [1,2]). Due to the pressing environmental issue, innovative structural solutions that can balance structural efficiency and sustainability requirements are needed. Additionally, because of the development of optimization techniques (e.g. [3–8]), generative design and parametric analysis ([9]), masonry buildings are once again

Supported by ADDOPTML Project, MSCA RISE 2020 Marie Skodowska Curie Research and Innovation Staff Exchange.

S. Gabriele et al. (Eds.): IWSS 2023, LNCE 437, pp. 269–278, 2024.
https://doi.org/10.1007/978-3-031-44328-2_28

in the forefront of the architectural and technical world. The idea of structural design has been transformed by computer-based design and modelling processes that produce comprehensive methods for shape, material, and structures (e.g. [10,11]). In this area, the development of non-conventional and free-form structures was greatly aided by innovative shape-finding approaches based on automated tools and digital fabrication processes. Visual representation is necessary to give creative structural and architectural solutions (e.g. [12]) and real-time reactions (e.g. [1]) of buildings information have assumed a crucial role. Due to the durability of this specific structural topology and the high cost of raw materials for their chip extraction operation, there is a growing interest in studying ancient masonry vaults that have been built all over the world.

The issue of doubly-curved shells has emerged in recent decades and either compression-only vault design or curved surface design that takes into consideration the funicular's permanent structure is extensively treated by most publications (e.g. [1,13]). Additionally, early research in this area focused on hanging models and graphic statics to study dome-shaped buildings. Antoni Gaud was the first to use a hanging scaled model in the design of the Crypt of Colonia Güell at the turn of the 20th century, as was extensively covered in [1].

Recently, new optimization approaches have been created and applied to grid shells in an effort to determine the ideal topology, form, and size for the members that make it up (e.g. [14]). By selecting a number of variables and placing them under a number of restrictions, optimization is the process of determining the least or maximum value of a function.

Among these, the most promising form-finding methodologies, that seek to identify the optimal form and topological configuration, are the well-liked technique force density method [15] and the dynamic relaxation (DR) [16].

In the present paper, the shape of the vault model is principally obtained by a form-finding method (MRA technique) fine-tuned by one of the authors [13]. The uniqueness of this approach consists of its adaptation to situations involving free forms with standardised constituent parts. In fact, one of the goals of this study is to provide an assessment that takes into account the best option among the many patterns of the perforated configurations (disposition of the holes) in addition to the usefulness of the form-finding approach used.

The analysis of the effects of a few selected parameters, including the lowering degree (defined as the ratio between the maximum span D and h, defined as the height of the vault), the slack coefficient of the original hanging net, and the hole percentage (HP) of the covering surface, yields the final geometric configuration of the vault. The final shape and the pierced pattern were analysed by the MidasGen©software under static and seismic load pattern configuration once constraint conditions were fixed.

The manufacturing process, [17], was adopted with the aim of achieving a scaled prototype of the selected vault typology. Moreover, a geometric survey was necessary to obtain, with a certain level of confidence, the input geometric data for numerical analysis as *Mechanical Autocad*©3d Model of the scaled-prototype.

Finally, to compare the natural frequency obtained by experimental tests with those evaluated with a numerical model, an impulse test has been performed. Results show that an improvement in the modelling strategies adopted for the numerical model is necessary for the identification of the real structural dynamic response of the vault.

2 Case Study: Vault Scaled-Prototype

A scaled-prototype of the reverse suspended net was created in order to evaluate the form-finding process utilising the MRA technique and to perform the test.

The model measures 30 cm tall in total. In order to scale the previously described measurements to have a side equal to 1 m, a scaling method was performed using a scaling factor of 0.0625 (1/16). The scaled-prototype vault was 469 mm tall after this process, with a transverse section of 18.75 mm and a slenderness of 80. The appropriate bricks that made up the grid's rods were broken up into sections across the vault. For the final assembling step, a sequential numbering of the bricks was established, with each formwork corresponding to a particular brick. After this analysing the overall behaviour of the structure under a well-specified load pattern represents a crucial step for the success of the procedure.

The vault's form was divided into numerous dowels using fast prototyping toolkits. Since the bricks must be produced in mortar cement, a moulding architecture must be developed. The correctness of the formwork geometry must be guaranteed due to the tiny scale of the model. The geometry of each formwork was entered into Rhinoceros with this goal in mind using the parametric CAD program Grasshopper. Therefore, after obtaining the geometric characteristics of the moulds, they were sent as input data to a 3D Bio-printer using the Fuse Deposition Modeling (FDM) process and entirely eco-friendly PLA material. The following are the most important settings made for 3D printing: 1.75 mm filament diameter, 0.2 mm layer height, and 0.1 fill density of 30%. The printing temperature was 215 °C, and the print speed ranged from 80 to 150 mm/s. It is noted that the four basic pieces were produced using the FDM method for the purpose of simplicity. Every brick used in the shuttering was cast, and the cement mortar used had a water-to-cement ratio of 0.4.

3 Experimental Set-Up and Dynamic Test

Dynamic testing of structures can provide valuable information on the performance of structures (e.g. [18–20]). Modal parameters (e.g. natural frequencies, mode shapes, and damping) can be determined from the measured dynamic response that is caused by ambient or forced vibration. The performance of the structure can then be characterised using these identified parameters, and analytical models can subsequently be validated. This section outlines the dynamic tests performed on a shallow grid shell at the MASTRLAB laboratory (Politecnico di Torino, Italy), as well as the information regarding the instrumentation employed in the data acquisition processes.

3.1 Instrumentation

The instrumentation employed during the dynamic tests included a laptop equipped with LabVIEW software, a data acquisition device, and No.2 accelerometers, which allowed us to identify the structure's acceleration response. The TE Connectivity model 4030 triaxial MEMS DC accelerometers (see Fig. 1) were used. They have a $\pm 2g$ dynamic range with a nominal 0–200 Hz bandwidth. The sensors' characteristics are shown in Table 1.

Fig. 1. Model 4030 triaxial MEMS DC accelerometer.

Table 1. Accelerometer's specifications.

Range (g)	±2g
Sensitivity (mV/g)	1000
Frequency response (Hz)	0–200
Excitation Voltage (Vdc)	5–30
Cable	6 × 0.14 mm Conductors PVC Insulated, Braided Shield, PVC Jacket
Weight (grams)	50 (cable not included)

The data acquisition system consisted of a National Instruments (NI) USB-6211. It is a No.16 analogue input (16-Bit, 250 kS/s), No.2 analogue output (250 kS/s), No.4 digital input, No.4 digital output USB Multifunction I/O Device (see Fig. 2) which provides the accelerometers connection. An ad hoc LabVIEW software was developed to record the data.

3.2 Set-Up

As shown in Fig. 3, the structure was located at the floor level, inside the MASTRLAB laboratory. The plywood base was loaded with steel elements to increase its mass and avoid any unwanted movement. To analyse the response of the structure, a triaxial MEMS accelerometer was installed on the top of the grid

Fig. 2. National Instruments USB-6211.

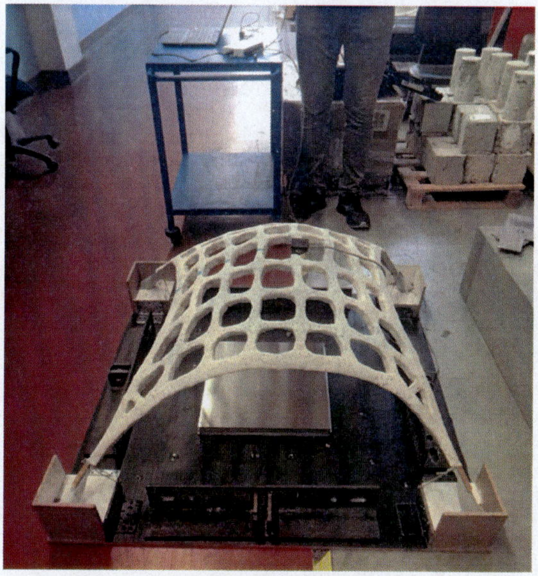

Fig. 3. Plywood base loaded with steel elements.

shell. A similar sensor was rigidly fixed to one of the four supports, close to the structure's impost (see Fig. 4). Both sensors were installed according to the same reference system. To provide a tight connection between the sensors and the structure, cable ties and double-sided adhesive tape were used to fix the sensor at the top, while screws firmly secure the other accelerometer.

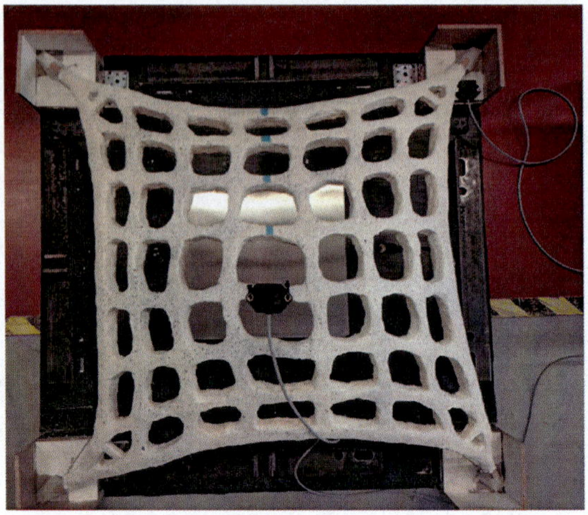

Fig. 4. Sensors layout.

3.3 Test and Experimental Methodology

Forced vibration testing refers to those techniques where the vibration is artificially induced, even if the excitation is not measured. The method can generally be classed as transient testing if the input force is not measured. This mean of forced excitation can be divided into impulse testing, used herein, and step relaxation [21]. The simplest means of delivering an impact to a structure is by using a dropped weight or a hammer. Two different types of dynamic tests were performed: the first one has been performed by exciting the structure with a steel marble dropped on five different structure's points (the top of the grid shell and the keystone of the transverse arches), while the other one has been carried out hammering the same five points. The experimental test consisted of ten acquisitions of about 30 s, using a sampling rate of 200 Hz.

Specifically, in order to investigate the response of the vault to different directions of the excitation source, the impact has been provided in $\pm X$, $\pm Y$ and $\pm Z$ axis corresponding to excitations along the two sides of the vault and the third one from the top to the bottom.

3.4 Results

In this section, the results obtained by the test performed with the hammer will be shown for clarity purposes. Additionality, the previous choice is justified by the significance of the results obtained during the acquisition phase. Due to the low weight of the steel marble, the resulting excitations on the structure were negligible and, consequentially, the accelerometer experienced difficulties in detecting the induced acceleration. For all these reasons, the authors decided to consider the acquisitions obtained by hammering the vault in five different positions.

Moreover, It has been recognized that the accelerations in the positive and negative sense of each axis show similar structural responses and comparable natural frequencies. Hence, the authors selected only the data obtained by hitting the vault according to the positive sense of the X and Y axis and in the negative sense along the Z axis.

As depicted in Figs. 5 (a)–(c)–(e), only the free vibration of the signal has been used for processing the FFT functions. For each direction, three curves have been plotted according to the output provided at the top of the vault, at the single support and the level of the shaking table.

The recorded data were processed through a Fast Fourier Transform (FFT) function to identify the fundamental frequencies of the structure.

By observing Figs. 5 (b)–(d), frequency domain outcomes highlight the first natural vibration frequencies with specific regard to the X and Y axis, respectively. For the accelerations along both X and Y, the peaks of the curves are observed for a value of the frequency equal to $15,77\,\text{Hz}$. As expected, due to the symmetry of the structure, the transversal natural frequencies along x and y axes assume about the same values. On the contrary, focusing on the FFT along the Z axis, which is the direction in which the structure is obviously most rigid, a clearly-defined peak of the function has not been recognized within the observed frequency range. This is also consistent with preliminary observations conducted with FEM modelling, despite the uncertainties on the parameters, where the first frequency in the transverse direction settled at about $210,71\,\text{Hz}$.

4 Comparison Between Numerical and Experimental Results

In this section, preliminary comparisons between the principal vibration modes obtained by the numerical model introduced in a previous work [22] and those derived by the experimental test will be shown. As reported in Table 2, a difference has been detected with regard to the natural frequency for the translation mode along X and Y directions, which represent the first and second vibration modes of the structure. More in detail, along the X and Y axis, corresponding to the first and second vibration modes respectively, the percentage variation between experimental and numerical results is equal to 39.5 and 46.2.

The inconsistency of these results could be addressed by the application of the mortar on the extrados and intrados sides of the bricks and, mainly, on the contact interface between two subsequent blocks. The intrinsic softness between the bricks could represent the main cause of the observed discrepancy. Due to the high frequency related to the translation mode along Z axis, no comparisons have been provided.

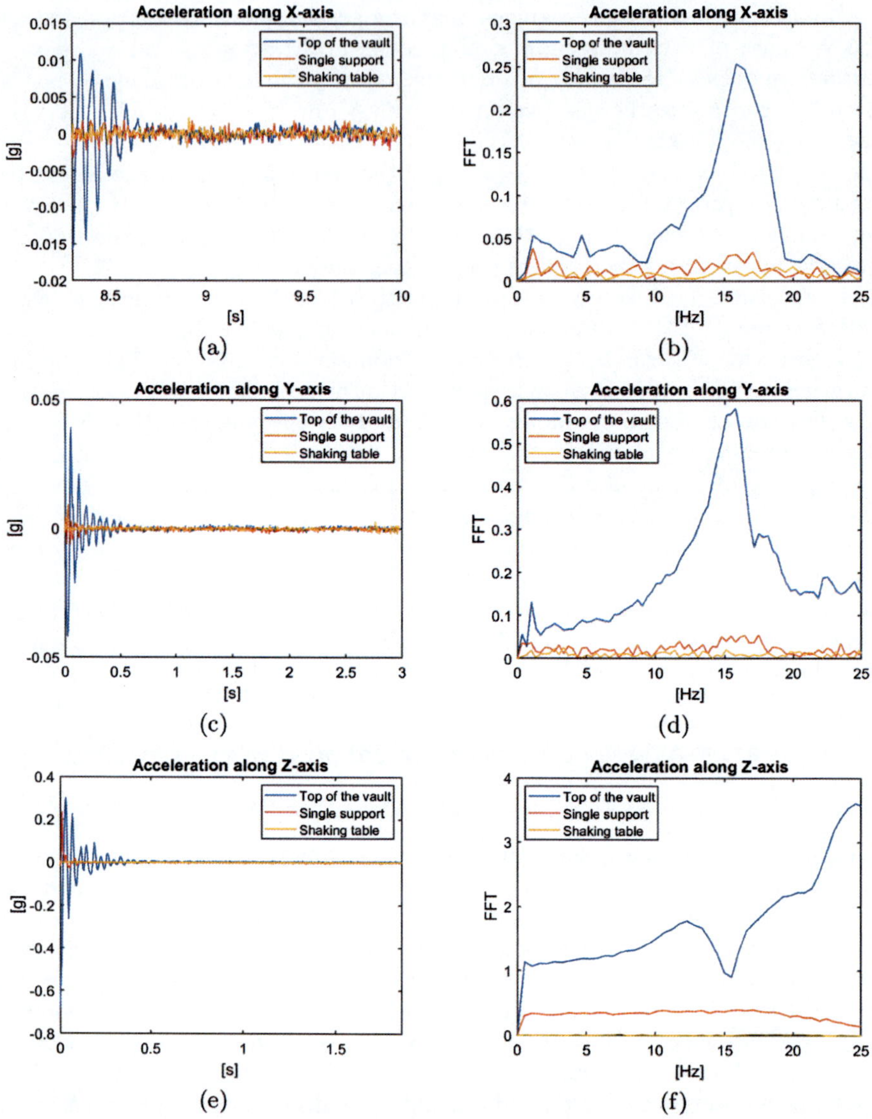

Fig. 5. Input acceleration (Figs. (a)–(c)–(e)) and corresponding FFT functions (Figs. (b)–(d)–(f)) along X, Y and Z axis, respectively

Table 2. Comparisons between experimental and numerical results

Mode	Numerical Frequency [Hz]	Experimental Frequency [Hz]	Variation [%]
1	22.15	15.88	39.5
2	23.05	15.77	46.2
3	210.71		

5 Conclusions and Future Developments

In this work, preliminary comparisons between numerical models and experimental tests have been shown. Specifically, an impulse test has been performed by hammering the structure or by adopting a steel marble dropped. The output obtained during the acquisition stage allow to point out the natural frequency of the translation modes along the X and Y axis. As expected, the vault exhibits similar transversal vibration mode frequencies. Unfortunately, the frequency range representative of the vertical mode has been not recognized since the law frequency samples set during the test. By performing an eigenvalue analysis with the numerical model, already introduced in [22], the natural vibration mode along the Z axis has been detected around a frequency value equal to 210, 71 Hz. This fact is demonstrated by the trend of the FFT function in which a well-defined peak of frequency has been not detected.

However, it seems clear that the discrepancy in terms of percentage variation between experimental and numerical results can be addressed by the modelling strategies adopted during the model realization. It does not take into account the contribution of the mortar applied to the extrados and intrados surface of the vault which leads to a softness behaviour.

In future works, the validation of the numerical model will be provided by adopting a homogenized section of each element composing the structure. In this way, the contribution of the mortar will be considered in the global behaviour of the structure. Moreover, the effect related to slip phenomena and the role of the bond's condition between mortar and concrete should be investigated.

Finally, dynamic tests will be realized by using a suitable shaking table and the structural response over time will be discussed.

References

1. Rippmann, M., Lachauer, L., Block, P.: Interactive vault design. Int. J. Space Struct. **27**(4), 219–230 (2012)
2. Oxman, R., Oxman, R.: New structuralism: design, engineering and architectural technologies. Archit. Des. **4**(80), 14–23 (2010)
3. Marano, G.C., Trentadue, F., Greco, R.: Optimum design criteria for elastic structures subject to random dynamic loads. Eng. Optim. **38**(7), 853–871 (2006)

4. Marano, G.C., Trentadue, F., Petrone, F.: Optimal arch shape solution under static vertical loads. Acta Mechanica **225**(3), 679–686 (2014)
5. Rosso, M.M., et al.: Enhanced multi-strategy particle swarm optimization for constrained problems with an evolutionary-strategies-based unfeasible local search operator. Appl. Sci. **202212**(5), 2285 (2014)
6. Cucuzza, R., Rosso, M.M., Marano, G.C.: Optimal preliminary design of variable section beams criterion. SN Appl. Sci. **3**(8), 745 (2021)
7. Cucuzza, R., et al.: Optimal strengthening by steel truss arches in prestressed girder bridges. In: Proceedings of the Institution of Civil Engineers-Bridge Engineering, pp. 1–21. Thomas Telford Ltd (2021)
8. Rosso, M.M., et al.: Structural optimization with the multistrategy PSO-ES unfeasible local search operator. In: Saraswat, M., Chowdhury, C., Kumar Mandal, C., Gandomi, A.H. (eds) Proceedings of International Conference on Data Science and Applications. LNNS, vol. 551, pp. 215–229. Springer, Singapore (2023). https://doi.org/10.1007/978-981-19-6631-6_16
9. Melchiorre, J., Bertetto, A.M., Marano, G.C.: Application of a machine learning algorithm for the structural optimization of circular arches with different cross-sections. J. Appl. Math. Phys. **9**(5), 1159–1170 (2021)
10. Cucuzza, R., et al.: Size and shape optimization of a guyed mast structure under wind, ice and seismic loading. Appl. Sci. **12**(10), 4875 (2022)
11. Rosso, M.M., et al.: Nonpenalty machine learning constraint handling using PSO-SVM for structural optimization. Adv. Civil Eng. **2021**, 1–17 (2021)
12. Kilian, A.: Design exploration through bidirectional modeling of constraints. Ph.D. thesis, Department of Architecture, Massachusetts Institute of Technology (2006)
13. Manuello, A.: Semi-rigid connection in timber structure: stiffness reduction and instability interaction. Int. J. Struct. Stabil. Dyn. **20**(07), 2050072 (2020)
14. Pedersen, P.: Optimal joint positions for space trusses. J. Struct. Div. **99**(12). 2459–2476 (1973)
15. Rasmussen, M.H., Stolpe, M.: Global optimization of discrete truss topology design problems using a parallel cut-and-branch method. Comput. Struct. **86**(13–14), 1527–1538 (2008)
16. Toan, V., Dalolu, A.T.: Optimization of 3D trusses with adaptive approach in genetic algorithms. Eng. Struct. **28**(7), 1019–1027 (2006)
17. Post, D.: Moiré interferometry at VPI & SU. Exp. Mech. **23**, 203–210 (1983)
18. De Mari, G., Domaneschi, M., Ismail, M., Martinelli, L., Rodellar, J.: Reduced-order coupled bidirectional modeling of the Roll-N-Cage isolator with application to the updated bridge benchmark. Acta Mech. **226**, 3533–3553 (2015)
19. Cimellaro, G., Domaneschi, M.: Stability analysis of different types of steel scaffolds. Eng. Struct. **152**, 535–548 (2017)
20. Domaneschi, M., Sigurdardottir, D., Glisic, B.: Damage detection on output-only monitoring of dynamic curvature in composite decks. Struct. Monit. Maintenance **4**, 1–15 (2017)
21. Salawu, O.S., Williams, C.: Review of full-scale dynamic testing of bridge structures. Eng. Struct. **17**(2), 113–121 (1995). https://doi.org/10.1016/0141-0296(95)92642-l
22. Cucuzza, R., Cardoni, A., Manuello, A., Domaneschi, M., Cimellaro, G.P., Marano, G.C.: Experimental investigation of the static and dynamic behaviors of 3D-printed shell structures. WCCM-APCOM **2022**, 900 (2022)

Sustainable Spatial Structures: A Design Approach Using Shape and Topology Optimization to Minimize Environmental Impact and Improve Buildability

Saaranya Kumar Dasari[1(✉)], Patrizia Trovalusci[1], Nicholas Fantuzzi[2], Marco Pingaro[1], and Roberto Panei[3]

[1] Sapienza University of Rome, 00197 Rome, Italy
saaranyakumar.dasari@uniroma1.it
[2] University of Bologna, Bologna, Italy
[3] Municipal Urban Hygiene Company, AMA, Rome, Italy

Abstract. The construction industry has a massive impact on climate change, and reducing its environmental impact is a critical challenge that requires innovative solutions. Shape and topology optimization methods can play a crucial role in addressing this issue by optimizing the structural geometry and material distribution. In this paper, we present a novel design approach for optimizing shape, topology, global warming potential and buildability. The proposed approach optimizes the shape and layout of structural elements to minimize the overall embodied energy and carbon emissions of the structure while ensuring that the structure is constructible. The methodology is demonstrated through case studies, the optimized design is evaluated based on the performance criteria and constraints that resulting in significant reduction of design cycle time as well as environmental impact while improving buildability. The research presented in this paper provides valuable insights for designers and engineers seeking to create environmentally sustainable yet elegant spatial structures with optimal buildability.

Keywords: Sustainability · Buildability · Structural Optimization · Global Warming Potential

1 Introduction

The construction industry is one of the largest contributors to global carbon emissions and has a significant impact on the environment. As a result, reducing the environmental impact of the construction industry has become a critical challenge that requires innovative and sustainable design approaches. Shape and topology optimization methods have emerged as a promising solution to this

Supported by PhD Program of Structural and Geotechnical Engineering, Sapienza University of Rome, Italy.

S. Gabriele et al. (Eds.): IWSS 2023, LNCE 437, pp. 279–288, 2024.
https://doi.org/10.1007/978-3-031-44328-2_29

challenge by optimizing structural geometry and material distribution. However, the current gap between sophisticated software models and the ability to realize designs within reasonable budgets remains a significant obstacle in structural architecture design. Bridging this gap requires consideration of Global Warming Potential (GWP) and Buildability in the design process. The design approach presented in this paper on Sustainable Spatial Structures aims to address these challenges by integrating form finding and topology optimization with the consideration of GWP and buildability. Form finding is the process of determining the optimal form of a structure, while topology optimization involves identifying the optimal material distribution within a structure to reduce its overall weight or volume. By minimizing the amount of material required and ensuring structural integrity, optimized spatial structures can significantly reduce the environmental impact of the construction industry. However, the success of such designs also relies on their practical feasibility and cost-effectiveness. To this end, this paper also examines the GWP and buildability of the optimized designs, ensuring that they can be efficiently constructed while minimizing waste. Overall, the design approach presented in this paper provides a comprehensive framework for creating environmentally sustainable and aesthetically elegant spatial structures with optimal buildability Fig. 1. The resulting insights can help designers and engineers in the construction industry to address the challenge of reducing the industry's environmental impact while improving efficiency and cost-effectiveness.

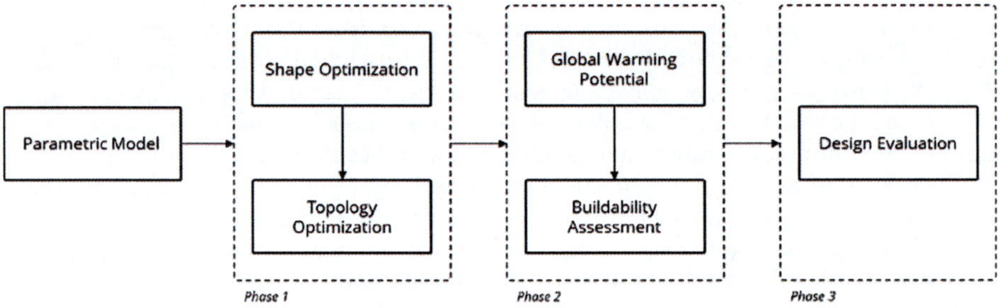

Fig. 1. The Framework of the Study

2 Methodology

In the field of structural architecture, the incorporation of minimal surfaces holds immense potential for designing highly efficient and structurally sound spatial structures. Differential geometry plays a crucial role in analysing minimal surfaces, it helps in understanding the relationship between geometry and topology, facilitating the creation of innovative designs [11]. Minimal surfaces are a special class of surfaces that have the property of minimizing their area for a given boundary. In other words, they are surfaces with the smallest possible surface area that encloses a specific boundary shape. These surfaces, which minimize their area for a given boundary, offer an unparalleled advantage in terms of material utilization and structural performance. For minimal surfaces, the mean curvature is zero, indicating that there is an equilibrium between the inward and outward curvatures, resulting in a surface with minimal area. The equilibrium configurations of minimal surfaces are intricately linked to the geometric and topological properties of the spatial structure (Fig. 2). The interplay between geometry, such as curvature and shape, and topology, including the arrangement of openings and handles, significantly influences the stability and behaviour of these surfaces. By harnessing this relationship, architects and engineers can unlock new possibilities for creating spatial structures that maximize structural integrity, optimize material utilization, and achieve visually captivating designs. Embracing minimal surfaces within the realm of geometry and topology paves the way for an innovative approach to design, enabling the realization of highly efficient and sustainable spatial structures [7].

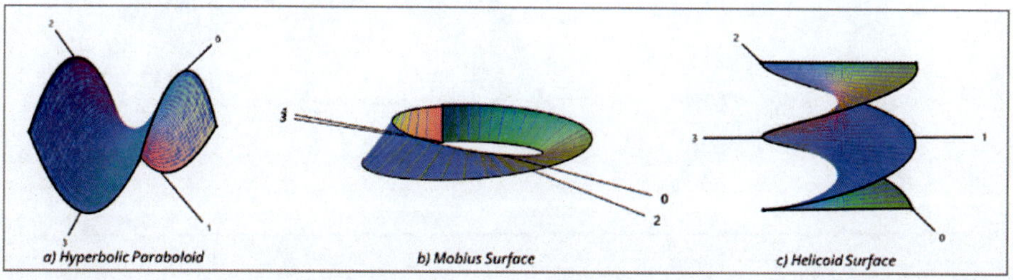

Fig. 2. Minimal Surfaces with their Topology

2.1 Shape Optimization

Strength through geometry is a key concept in structural design, and shape optimization techniques offer a powerful approach to achieve this objective. By leveraging the inherent properties of geometric forms, shape optimization enables the creation of structurally efficient and strong spatial structures. The goal of shape optimization is to find the optimal form or geometry of a structure that satisfies certain performance criteria, such as minimizing strain energy or achieving specific design objectives. In the context of minimal surfaces, shape optimization techniques can be employed to iteratively refine and evolve the shape of a structure until it converges to a minimal surface configuration. Shape optimization algorithms utilize mathematical optimization techniques, such as gradient-based methods or evolutionary algorithms, to iteratively update the design variables and improve the objective function value while satisfying the constraints [5,6]. The objective is to find the optimal shape that minimizes an objective function.

There are several other methods in the literature, some of them can be classified as gravity based or stiffness-based methods such as force density method, dynamic relaxation, particle spring systems.

Particle Spring Systems, for instance, simulate the behavior of particles interconnected by springs. By assigning specific properties to these particles, such as mass and elastic behavior, and subjecting them to forces, the system can be dynamically adjusted to find a shape. Through an iterative process of adjusting particle positions and optimizing the forces within the system, the structure gradually evolves towards a minimal surface configuration see an example in Fig. 3.

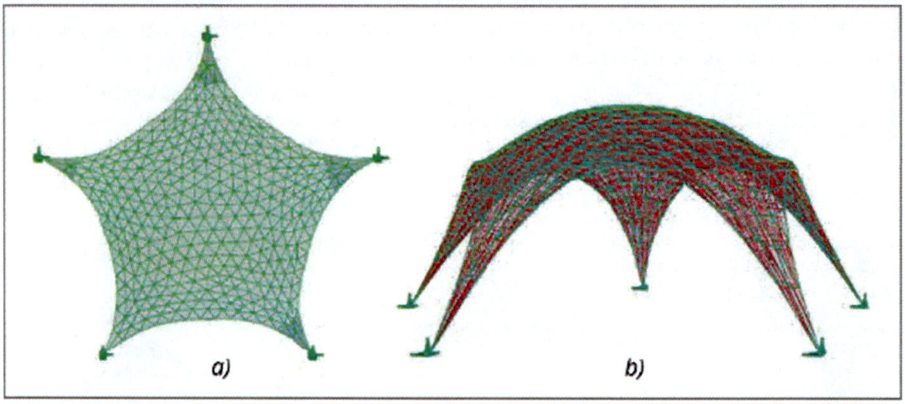

Fig. 3. Shell obtained from Kangaroo by using Particle string systems in Rhino/Grasshopper

2.2 Topology Optimization

We adopt topology optimization for finding the optimal material distribution that maximizes the stiffness of a structure by minimizing the compliance within a given design domain to achieve desired performance objectives while minimizing material usage. It involves identifying regions where material can be removed or redistributed to achieve the best structural efficiency. The Solid Isotropic Material with Penalization (SIMP) and the Bi-directional Evolutionary Structural Optimization (BESO) methods are widely used approaches for topology optimization.

The SIMP method formulates topology optimization as a material distribution problem, where the design domain is divided into finite elements. Each element is assigned a design variable denoting the material density, which represents the volume fraction of material in that element. SIMP method is based on penalization techniques. It assigns continuous material densities to each element in the design domain and penalizes intermediate densities to encourage either fully dense (solid) or void (empty) regions [14]. By iteratively adjusting the penalization parameter, SIMP optimizes the material distribution by either including or excluding elements based on their densities [3]. In SIMP topology optimization, several methods can be employed during the optimization process to iteratively improve the design such as Method of Moving Asymptotes (MMA), Optimal Criteria (OC) Methods and Level Set Methods [9]. These methods are not mutually exclusive, and different combinations or variations of these techniques can be employed depending on the specific requirements of the optimization problem. MMA, for instance, uses a series of approximations to the true constraint functions and solves sub-problems to find a feasible and optimal solution. MMA is known for its effectiveness in handling large-scale optimization problems [9]. To demonstrate, an example is shown in Fig. 4 we used tOpos [4] a topology optimization plugin for grasshopper, all though results are not so accurate we are using these plugins to demonstrate the idea presented in this paper.

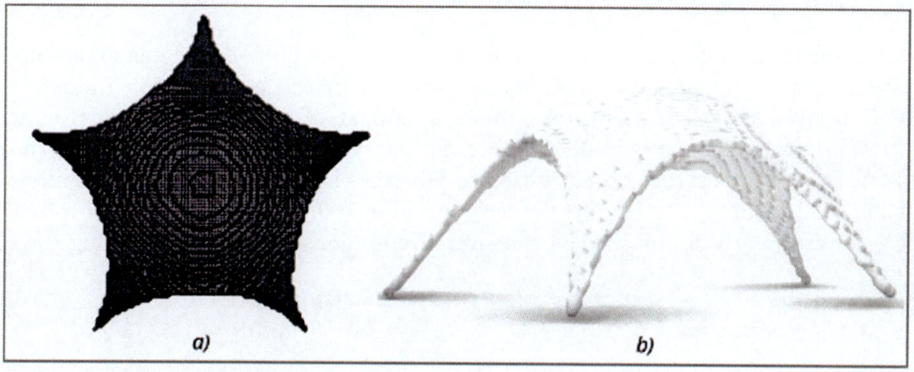

Fig. 4. Results obtained from tOpos plugin using SIMPin Rhino/Grasshopper

The BESO is a discrete version of the SIMP, it involves an iterative process of removing inefficient material and adding new material based on structural performance criteria [15]. It starts with an initial material distribution, typically a uniform distribution within the design domain. The algorithm then evaluates the structural response and identifies regions of high stress or strain. These regions are considered inefficient and marked for removal. See an example in Fig. 5, here we used Karamba 3D a finite element analysis plugin for grasshopper, the results are not considerable due to asymmetry in the mesh but to demonstrate the approach and the possibilities we show these example.

BESO focuses on removing inefficient material and redistributing the remaining material, while SIMP employs penalization techniques to control the presence or absence of material in each element. They both use the same filtering techniques either sensitivity or density filters.

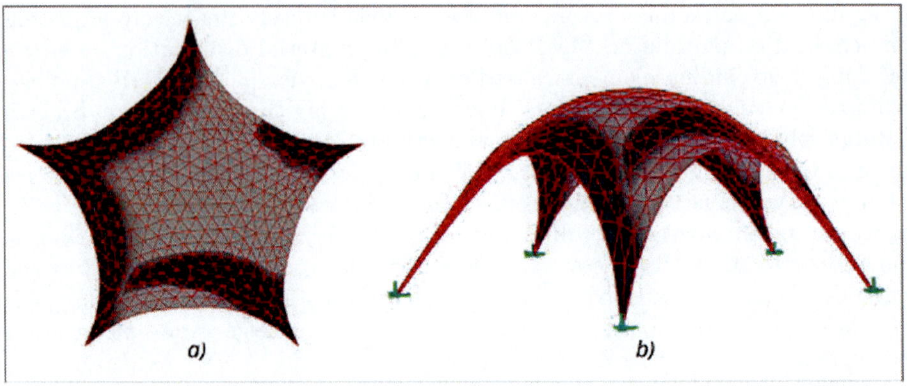

Fig. 5. Results obtained from Karamba3D using BESO in Rhino/Grasshopper

2.3 Global Warming Potential

Global warming potential (GWP) is a measure of the greenhouse gas emissions associated with a particular activity or process. In the construction industry, GWP is used to assess the environmental impact of building materials, construction methods, and overall building designs [8]. It helps in quantifying the contribution of construction activities to climate change. GWP[CO_2e]: Contribution to Greenhouse (GH) effect over a period of 100 years compared equivalent CO_2. Overall carbon footprint is the sum of emissions over entire building life.

$$\sum structural\ elements\big[Quantity\ (kg) * \\ CO_2\ e\ Material\ Factor(kgCO_2\ e/kg)\big] = GWP\ building\ (kgCO_2\ e)$$

2.4 Buildability Assessment

Buildability refers to the extent to which a building design facilitates ease of construction and the adoption of construction techniques and processes that enhance productivity levels. It underscores the importance of considering construction-related factors during the design phase to ensure smooth and efficient construction operations. By integrating buildability considerations into the design process, architects and engineers can optimize construction processes, minimize delays, reduce costs, and enhance overall project success. According to [2] buildability can be measured as Buidable Design Score in accordance with the Buildable Design Appraisal System (BDAS) as set out in the Code of Practice authorised by Singapore government. However rationalization techniques have also emerged as critical tools for complex geometries and streamlining the fabrication process.

Rationalization, a concept not exclusive to architecture but widely practiced in construction engineering and product design, has gained significance in response to the increasing complexity of architectural geometries [1]. However, the emergence of computational design tools has empowered architects to envision increasingly intricate geometries. Despite the potential offered by state-of-the-art digital fabrication processes, realizing such complex shapes often proves challenging and costly. Consequently, architects frequently encounter the need to modify these designs to align with the constraints imposed by the fabrication process. This process, known as architectural rationalization. By rationalizing the design, complex shell geometries can be simplified and optimized for efficient construction. One approach to achieve this is by leveraging machine learning algorithms [12], such as the Gaussian mixture algorithm, to cluster and group the panels of a shell structure.

3 Case Study

Los Manantiales is a famous structure designed by Mexican architect and engineer Felix Candela in the 1950s (Fig. 6). It is a hyperbolic paraboloid concrete shell structure that has become an iconic example of Candela's signature style. It is known for its elegant and lightweight design, which utilizes the inherent strength of the shell form to create a structurally efficient and visually stunning space. However, constructing such structures with heavy form work can be expensive and challenging. While Los Manantiales is already an impressive feat of engineering and design, it is possible to apply the design approach presented in this paper to further optimize the structure's environmental impact and buildability.

Fig. 6. a) Los Manantiales Restaurant, Xochimilco, Mexico City by Félix Candela 1958 b) Form work of Los Manantiales, From [www.archdaily.com]

The construction of hyperbolic paraboloid shell structures typically involves creating custom form work to shape and support the concrete during the casting process. The complex geometry and curvatures of these structures require intricate and precise form work, which adds to the overall cost and complexity of construction. The form work needs to be carefully designed, fabricated, and installed to achieve the desired shape and ensure the structural integrity of the concrete shell. According to [13], the cost attributed to formwork represents a substantial proportion, typically ranging from 40% to 60%, of the overall cost involved in the construction of a concrete structure (Fig. 7).

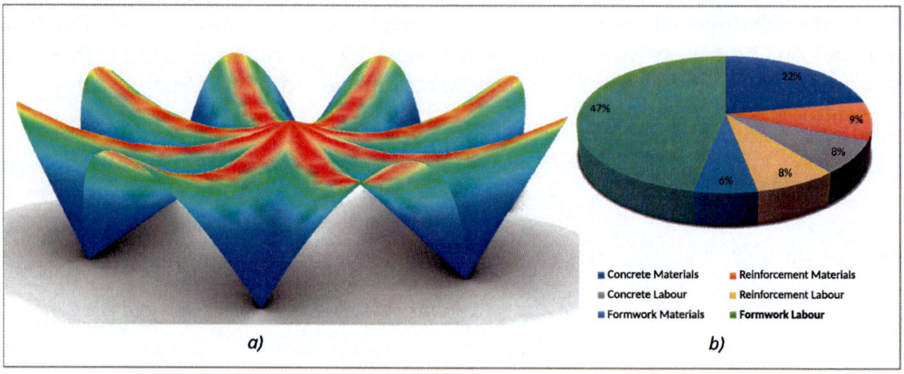

Fig. 7. a) Mean Curvature Analysis b) Cost of Formwork [13]

The high cost of constructing these structures with heavy form work can be attributed to several factors. Firstly, the labor and time required to create and assemble the form work are significant. Skilled labor is needed to construct the form work accurately, considering the complex geometry involved. Additionally, the materials used for the form work, such as timber or steel, add to the overall cost. Considering buildability and GWP in the design process is crucial to reduce the environmental impact associated with constructing such structures. To do

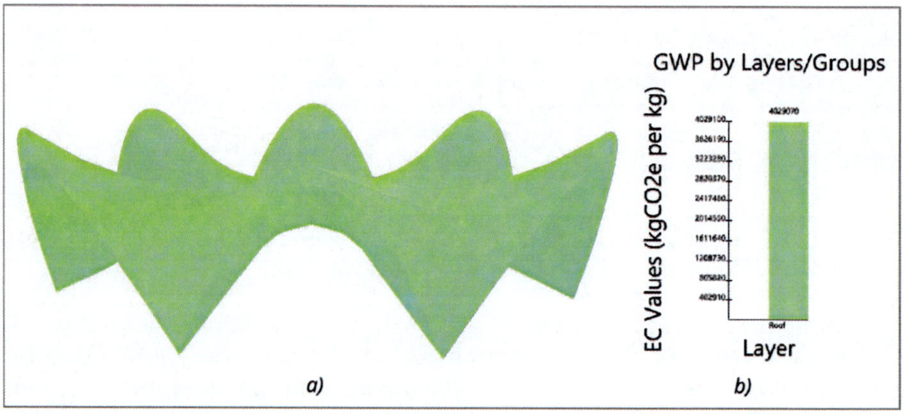

Fig. 8. GWP calculation in Rhino/Grasshopper a) Parametric Model b) GWP of the roof shell of Los Manantiales

this, we used Cardinal LCA plugin for grasshopper to calculate the GWP of the Los Manantiales considering the embodied carbon factors from the data base [8] (Fig. 8).

In the context of shell structures, rationalization involves identifying repetitive patterns and standardizing the design and fabrication process. By using the Gaussian mixture algorithm, which is a type of unsupervised learning algorithm, it becomes possible to analyze the geometric characteristics, dimensions, and connections of individual panels within a shell structure. The algorithm then clusters these panels into cohesive groups based on their similarities. To do this, we subdivide the roof surface into panels and group them into clusters for that we used Lunch Box ML plugin for Grasshopper [10].

The clustering process performed by the Gaussian mixture algorithm has several advantages for rationalization in shell structures. Firstly, it helps reduce the number of unique panel types that need to be fabricated. By grouping similar panels together, standard dimensions and connections can be established, enabling more efficient production and reducing the complexity of the fabrication process. This leads to cost savings and shorter manufacturing lead times. Secondly, the clustering of panels facilitates easier management and organization during transportation and installation (Fig. 9).

Fig. 9. Rationalization technique using Gaussian mixture algorithm by Lunch Box ML plugin in Rhino/Grasshopper

4 Conclusion

In conclusion, the use of shape and topology optimization methods in designing sustainable spatial structures is an innovative approach that prioritizes sustainability, buildability, and aesthetics. This design approach offers a promising solution to the challenge of achieving sustainable spatial structures that minimize environmental impact while improving buildability. Emphasizing buildability and GWP not only benefits the construction process but also contributes to creating more environmentally conscious and economically viable structures.

References

1. Austern, G., Capeluto, I.G., Grobman, Y.J.: Rationalization methods in computer aided fabrication: a critical review. Autom. Constr. **90**, 281–293 (2018)
2. BCA: Code of practice on buildability 2022 edn, pp. 1–105. https://www1.bca.gov.sg. Accessed 4 May 2023
3. Bendsøe, M.: Optimization of Structural Topology, Shape, and Material. Springer, Heidelberg (1995). https://doi.org/10.1007/978-3-662-03115-5
4. Białkowski, S.: tOpos GPGPUAccelerated Structural Optimisation Utility for Architects in - eCAADe, pp. 679–688 (2017)
5. Dasari, S.K., Fantuzzi, N., Trovalusci, P., Panei, R.: Computational approach for form-finding optimal design. Archit. Struct. Constr. **3**, 323–333 (2022). https://doi.org/10.1007/S44150-022-00077-2
6. Dasari, S.K., Fantuzzi, N., Trovalusci, P., Panei, R., Pingaro, M.: Optimal design of a canopy using parametric structural design and a genetic algorithm. Symmetry **15**, 142 (2023). https://doi.org/10.3390/sym15010142
7. Pellis, D., Kilian, M., Pottmann, H., Pauly, M.: Computational design of Weingarten surfaces. ACM Trans. Graph. **40**, 4, Article 114 (2021). https://doi.org/10.1145/3450626.3459939
8. Embodied Carbon Footprint Database - Circular Ecology. https://circularecology.com/embodied-carbon-footprint-database.htm
9. Svanberg, K.: The method of moving asymptotes - a new method for structural optimization. Int. J. Numer. Methods Eng. **24**, 3 (1987)
10. Miller, N., Stasiuk, D.: Negotiating structured building information data. In: B. Sheil, M. R. Thomsen, M. Tamke, S. Hanna (eds.) Design Transactions: Rethinking Information Modelling for a New Material Age, pp. 68–73 (2020)
11. Nourian, P.: Topology on topology and topological data models in geometric modeling of space (2018). https://doi.org/10.13140/RG.2.2.16572.74888
12. Ortner, F.P., Tay, J.Z., Exploring a circular economy solution space A comparative study to develop automated optimisation workflows supported by machine learning for circular design problems. Int. J. Archit. Comput. (2023)
13. STRUCTURE, Think Formwork - Reduce Costs: https://www.structuremag.org. Accessed 15 May 2023
14. Tuna, M., Trovalusci, P.: Topology optimization of scale-dependent non-local plates. Struct. Multidisc. Optim. **65**, 248 (2022). https://doi.org/10.1007/s00158-022-03351-5
15. Huang, X., Xie, M.: Evolutionary Topology Optimization of Continuum Structures: Methods and Applications. Wiley, Hoboken (2010)

Generative Design Process and Optimization of Geodesic Dome with Variable Frequency

Laura Sardone[1(✉)], Marco Martino Rosso[1], Jonathan Melchiorre[1], and Marco Pellegrino[2]

[1] DISEG, Department of Structural, Geotechnical and Building Engineering, Politecnico di Torino, Corso Duca degli Abruzzi, 24, 10129 Turin, Italy
{laura.sardone,marco.rosso,jonathan.melchiorre}@polito.it
[2] StruSoft AB, Hyllie Boulevard 53, 215 37 Malmö, Sweden
marco.pellegrino@strusoft.com

Abstract. A geodesic dome is a hemispherical thin-shell structure based on a geodesic polyhedron. The triangular elements of the hemisphere are structurally rigid and distribute the structural stress throughout the structure, making geodesic domes able to withstand heavy loads considering their size. After the first geodesic dome was built in Jena (Germany, 1922) on top of the Zeiss optics company as a projection surface for their planetarium projector, thanks to R. Buckminster Fuller, the geodesic domes have been explored far more thoroughly. As a result of their properties, these structures are used for many purposes: as residential modules, greenhouses, water reservoirs and as expositive pavilions. The present paper focuses on geodesic domes optimization, minimizing the overall volume of the frame structure and its connections. In the optimization phase, the base radius of the dome is considered constant, representing a shape constraint. The frequency variability allows for modification of the frame's number, thus varying the structural topology while acting on the shape of the sections (shape optimization). In the case study, self-weight and asymmetric load actions are considered while including the construction aspects for the assembly of a geodesic dome. The optimization phase involves evolutionary genetic algorithms (EAs) exploiting the results of a Finite Element Analysis (FEA).

Keywords: Generative Design · Domes · Structural Optimization · Size Optimization · Steel Structures · Risk Management

1 Introduction

Geodesic domes are unique architectural structures characterized by their curved, spherical, or partially spherical shapes, formed by a network of interconnected triangles or polygons. The concept of geodesic domes was popularized by the architect and engineer, R. Buckminster Fuller, in the mid-20th century (Fig. 1). A geodesic dome consists of a framework constituted by a series of struts or beams intersecting and forming triangular or polygonal panels. The panels distribute the structural stress evenly throughout the structure, making geodesic domes sturdy and capable of withstanding several environmental conditions and loads [1].

S. Gabriele et al. (Eds.): IWSS 2023, LNCE 437, pp. 289–298, 2024.
https://doi.org/10.1007/978-3-031-44328-2_30

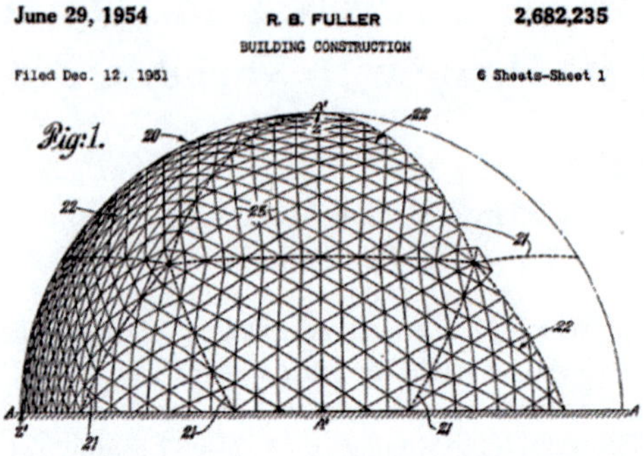

Fig. 1. Geodesic Dome representation by Fuller Richard Buckminster; Patent US261168A.

The structural integrity of geodesic domes is derived from the inherent strength of triangles. Triangles are inherently rigid and stable, which means that when multiple triangles are interconnected, they create a robust and stable structure. The more triangles incorporated into the dome's design, the stronger and more resilient it becomes [2].

Geodesic domes offer several advantages over traditional architectural forms. Firstly, their spherical or partially spherical shape allows for efficient distribution of stress and load-bearing capacity. This structural efficiency means that geodesic domes require fewer materials compared to conventional buildings of similar size, resulting in reduced construction costs. The unique shape of geodesic domes provides enhanced interior space utilization. With no internal columns or load-bearing walls, the interior can be easily adapted to various purposes, allowing for open and flexible floor plans. Additionally, the curved design of geodesic domes promotes efficient airflow and natural ventilation. These structures have found several applications, ranging from residential dwellings, recreational facilities, greenhouses, exhibition spaces, and even space exploration habitats. Their versatility and adaptability make them popular choices in various industries and contexts [3–7]. Considering the extensive use of geodesic domes in the Architecture, Engineering and Construction (AEC) industry and the continuous studies carried out on this type of structure, in this document the efficiency of geodesic domes is evaluated taking into account the self-weight and the action of the asymmetrical horizontal load (wind action). The FEM analysis is performed through *Alpaca 4D*, a Grasshopper plugin developed on top of *OpenSees*. Finally, a size optimization concerning the geodesic dome's frames is conducted using the meta-heuristic-based Galapagos tool. The results of the structural analyses are then processed through a Python script implemented using the *Iron-Python* component developed in the Grasshopper environment.

2 Current Implementation of a Geodesic Dome

2.1 Geodesic Dome Case Study Description

The geodesic dome under investigation presents a user-defined fixed base radius equal to 10.00 m, and it is depicted in Figs. 2 and 3. The dome was modelled in Grasshopper 3D, a visual programming language and environment that runs within the Rhinoceros 3D computer-aided design (CAD) application (Fig. 2) [8].

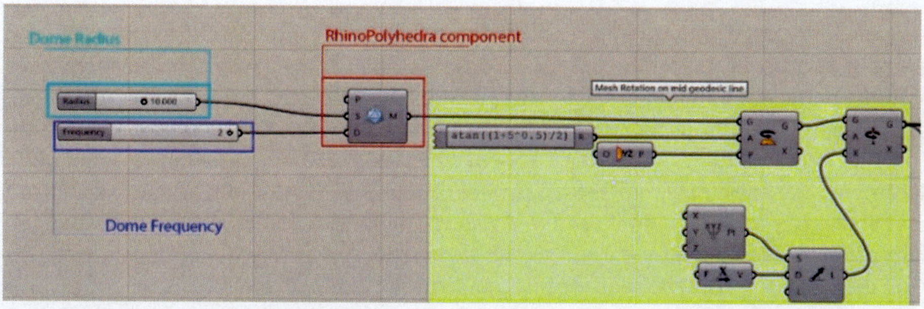

Fig. 2. Visual Programming environment and Geodesic Dome development with variable radius and frequency.

The spherical dome, implemented through the *RhinoPolyhedra* component, was rotated to find the preferential equatorial plane considering the following rotation angle:

$$\alpha \approx \operatorname{atan}\left(\frac{1+5^{0.5}}{2}\right) \approx 58^{\circ}.2825256. \quad (1)$$

The angle described in Eq. 1 permits the initial spherical dome to be rotated in order to obtain the cutting plane detecting the mid-cutting line of the hemisphere (Fig. 3).

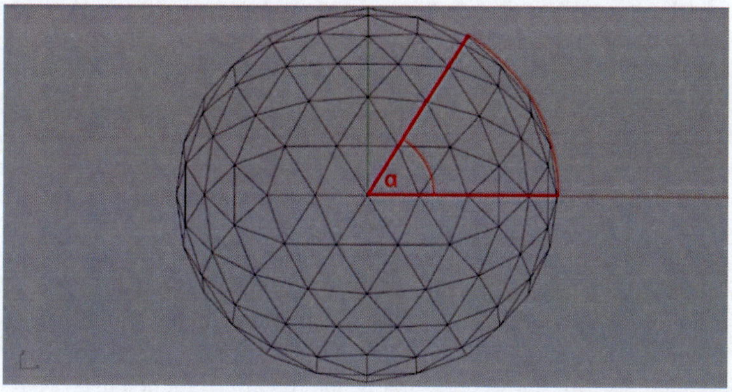

Fig. 3. Rotation angle α representation described by Eq. (1).

2.2 Structural Analyses and Actions

For structural analyses, the open-source plugin Alpaca4D has been adopted. It is based on the well-acknowledged Open System for Earthquake Engineering Simulation (OpenSees) software [9] developed by researchers at the University of California, Berkeley. The finite element (FE) model implements a steel material with grade S355 for the geodesic dome's structure using circular hollow core (CHS) section elements.

The boundary conditions of each member are fixed-fixed, and the entire dome presents global support restraints at the base. Two different load cases have been considered, the first one under ultimate limit state (ULS) and the second one under service limit state (SLS) design conditions, see EN 1990 Eurocode 0 "*Basis of structural design*" [10]. Specifically, the latter adopted the SLS characteristic load combination for maximum deformation assessments, i.e., the maximum displacement criterion, as discussed in the next section. The load actions in the structural models are related to the self-weight, automatically computed by the software based on the specific cross-section adopted for each member, a carried permanent load, empirically defined as 1.00 kN/M^2 of dome's surface accounting for the specific typology of the roofing system (e.g. glass panels and relative connection system to the supporting dome's frames), and a horizontal wind variable action. The wind action has been determined based on the technical procedure of the Italian regulation for structural design NTC2018 [11]. The wind pressure p is given by

$$p = q_r \cdot c_e \cdot c_p \cdot c_d \tag{2}$$

in which q_r is the kinetic pressure provided by the well-known Bernoulli theorem from hydraulics engineering $q_r = \frac{1}{2}\rho v_r^2$, i.e., depending on the air mass density assumed as 1.25 kg/m^3, and the wind reference velocity v_r of the specific geographical region considered. More precisely, the reference velocity $v_r = v_b \cdot c_r$ depends on a return coefficient c_r, set to unity in this case [11], and a reference base velocity $v_b = v_{b,0} \cdot c_a$, which in turn depends on an altitude coefficient c_a and an average base velocity $v_{b,0}$ measured in 10 min, at 10 m above plain ground, and with a return period of 50 years, and Since the roofing geodesic dome was hypothetically located in Turin (Italy) at an elevation of $z = 12$ m above the ground, thus in an urban area context denoted as Zone 1, the average base velocity is equal to $v_{b,0} = 25$ m/s, the altitude coefficient c_a can be set equal to the unity, determining that the reference velocity $v_r = 25$ m/s, and thus a kinetic pressure of $q_r = 15.625$ N/m^2. Furthermore, the geographical location of the dome and its elevation above the ground determines the value of the exposure coefficient c_e in Eq. (2), i.e., becoming equal to $c_e = 1.5$, thus characterizing an urban environment with a topographic coefficient c_t equal to unity and a ground roughness of class A [11]. The shape coefficient c_p depends on the inclination of the roof pitch. The geometry of the geodesic dome imposes considering an equivalent roof pitch with an inclination with respect to the horizontal plane forming an angle greater than +60°, imposing indeed a shape coefficient equal to $c_p = +0.8$. Finally, the dynamic coefficient c_d is assumed as unity in the absence of more precise indications or specific wind surveys. Therefore, the wind pressure for the dome becomes equal to $p = 18.75$ N/m^2. Eventually, considering the lateral surface under wind pressure, the wind global action that affects the dome is approximately equal to $P = 3$ kN. All the above – considered actions were distributed as acting as point loads in all the dome's nodes.

3 Size Optimization Problem Statement

The previously described geodesic dome undergoes an optimization process implemented in Rhinoceros Grasshopper using the meta-heuristic-based Galapagos tool. The size optimization problem has been defined according to the classical truss optimization problems [12, 13], in which the reduction of material consumption can be assumed as a reduction of the production costs of the structural part of the dome. The design vector $\boldsymbol{x}$ has been encoded with two components, i.e., the external diameter ϕ_{ext} in mm of the CHS adopted cross-section for every dome's member and the thickness t in mm of every tubular section,

$$\boldsymbol{x} = [x(1), x(2)]^T = [\phi, t]^T \tag{3}$$

For every component of the design vector, the authors have identified a reasonable interval in which to look for the optimal values, based on the admissible ranges of market availability of standard CHS profiles. Specifically, according to the EN 1993–1-1:2005 + AC2:2009 Sects. 6.2 & 6.3 (Eurocode standards for steel structures), the reasonable admissible intervals are represented by

$$21.3\,\text{mm} \leq \phi \leq 1219\ \text{mm} \tag{4}$$

$$2.3\,\text{mm} \leq t \leq 60\,\text{mm} \tag{5}$$

Therefore, the constrained size optimization problem statement can be expressed as follows:

$$\min V(x) = \sum_i A_i(\boldsymbol{x}) L_i \tag{6}$$

$$g_1(\boldsymbol{x}) = \frac{x(2)}{\frac{x(1)}{2}} - 1 \leq 0 \tag{7}$$

$$g_2(\boldsymbol{x}) = \frac{N_{Ed,t}(\boldsymbol{x})}{N_{Rd,t}(\boldsymbol{x})} - 1 \leq 0 \tag{8}$$

$$g_3(\boldsymbol{x}) = \frac{N_{Ed,c}(\boldsymbol{x})}{N_{Rd,b}(\boldsymbol{x})} - 1 \leq 0 \tag{9}$$

$$g_4(\boldsymbol{x}) = \frac{u_{max}}{u_{lim}} - 1 \leq 0 \tag{10}$$

In the previous equations, the $V(\boldsymbol{x})$ represents the steel volume of the entire dome, depending on the CHS cross-section $A_i(\boldsymbol{x})$ of each i-th adopted member and its length L_i. The first constraint $g_1(\boldsymbol{x})$ represents a geometrical feasibility condition that imposes that the chosen thickness must be lower than half of the CHS external diameter. The constraints $g_2(\boldsymbol{x})$ and $g_3(\boldsymbol{x})$ are referred to the respectfulness of the design criteria for ULS conditions for members under acting tensile forces $N_{Ed,t}(\boldsymbol{x})$ and acting compression forces $N_{Ed,c}(\boldsymbol{x})$. Specifically, $g_2(\boldsymbol{x})$ refers to the allowable tensile strength of each steel member, defined as

$$N_{Rd,t}(\boldsymbol{x}) = \frac{A_i(x) \cdot f_y}{\gamma_{M0}} \tag{11}$$

in which f_y is the steel strength equal to 355 MPa, unless the chosen thickness is greater than 40mm, in this case, the maximum strength is limited to 255 MPa. The coefficient γ_{M0}

is the partial safety factor accounting for neglected uncertainties of the semi-probabilistic calculation method, which may be assumed equal to 1.1. On the other hand, $g_3(\boldsymbol{x})$ represents the structural assessment of compressed members accounting for possible Eulerian buckling phenomena. Therefore, the compression-resisting value at ULS is defined as

$$N_{Rd,b}(\boldsymbol{x}) = \chi \frac{A_i(x) \cdot f_y}{\gamma_{M1}} \tag{12}$$

The reduction factor χ may be determined according to the Eurocode 3 formulation, which corresponds to the procedures prescribed in NTC18 [11]. For a hot-rolled CHS profile with steel S355, the instability curve is denoted as "curve a", which determines an imperfection factor of $\alpha = 0.21$. This latter accounts for the existing imperfections of real-world structural members, instead of the classic Eulerian buckling of ideal structural elements. Thereafter, a dimensionless slenderness parameter has to be determined

$$\overline{\lambda} = \sqrt{\frac{f_y \cdot A}{N_{cr}}} \tag{13}$$

which depends on the Eulerian critical buckling compression force

$$N_{cr} = \frac{\pi^2 \cdot E \cdot I(\boldsymbol{x})}{L_0^2} \tag{14}$$

being L_0 the effective length depending on the boundary conditions of each dome's compressed member. For the sake of conservativeness and for remaining on the safety side, the effective length has been considered equal to the actual length L_i of each structural element. Finally, the imperfection parameter has been defined as

$$\Phi = \frac{\left[1 + \alpha \cdot \left(\overline{\lambda} - 0.2\right) + \overline{\lambda}^2\right]}{2} \tag{15}$$

The imperfection parameter has been eventually adopted to define the reduction factor χ through the following relation

$$\chi = \frac{1}{\Phi + \sqrt{\Phi^2 - \overline{\lambda}^2}} \tag{16}$$

Eventually, the constraint $g_4(\boldsymbol{x})$ expresses the displacement limitation within a reasonable value under the SLS characteristic design condition, expressed in terms of the base radius R of the geodesic dome.

$$u_{lim} = \frac{1}{250} R \tag{17}$$

The constrained optimization problem stated in Eqs. (6)-(10) has been reformulated as an equivalent unconstrained optimization problem adopting the penalty function method [12–14]. Therefore, a new objective function is expressed considering an adaptive penalty term $\Pi(x)$ [14]:

$$\min \Phi(x) = V(x)\left(1 + k_p \cdot \Pi(x)\right) \tag{18}$$

A penalty factor k_p has been set equal to $k_p = 100$ [14]. The penalty term appears as an adaptive term since it will increase the original volume when one or more constraints are violated. Therefore, $\Pi(x)$ has a value greater than zero when the population individual is unfeasible, and it is zero otherwise:

$$\Pi(x) = \sum_{j}^{4} \max\{g_j(x), 0\} \tag{19}$$

4 Results and Discussion

The above-mentioned size optimization problem for the geodesic dome was implemented directly in Rhinoceros Grasshopper using the meta-heuristic-based Galapagos tool. Specifically, the structural analyses were conducted with Alpaca4D plug-in both under ULS and SLS conditions. Thereafter, the results of the structural analyses have been collected and processed through a Python script implemented in the Python plug-in of Grasshopper. The script was demanded to determine the objective function (volume) and the penalized volume assessing for the constraints and feasibility of violation. The plug-in Galapagos has been thus connected to the Python script component output, specifically to the penalized objective function, since it works as an unconstrained optimizer tool. The evolutionary solver has been adopted with a population size of 50 individuals. Since different search space topology is defined according to a specific value of the dome's frequency, the authors parametrized the frequency and herein discussed the results of frequency equal to 1 and equal to 2. In the following, the results related to the size optimization problem with the parametrization of the geodesic dome frequencies are presented.

4.1 Frequency Order 1

The optimization problem was solved by the Galapagos Evolutionary tool in Grasshopper software, and the exact optimal found solution respecting all the constraints after 100 generations for frequency 1 is the following:

$$\boldsymbol{x}_{\boldsymbol{opt\,frequency}=1} = \left[\phi_{opt}, t_{opt}\right]^T = [43.7\,\text{mm}, 2.4\,\text{mm}]^T \tag{20}$$

Figure 4 shows the 3D representation of the structural model with the optimal values of the design vector. The objective function history during the iterations is illustrated in Fig. 5 (a), whereas Fig. 5 (b) shows the bio-diversity ratio. This latter is a percentage value demonstrating that the Evolutionary tool is still exploring the search space, despite the best solution that has been already found in the early generations, leading to a final optimal steel volume of 0.06 m^3 at the exact design parameters. However, in order to provide a more realistic feasible structural solution, the exact design parameters were rounded up to the closest CHS steel profile already available on the market, according to the EN 1993-1-1:2005 + AC2:2009 Sects. 6.2 and 6.3. Therefore, the optimal technical solution for size optimization of the geodesic dome under investigation with the dome's frequency equal to 1 is $\boldsymbol{x}^*_{\boldsymbol{opt\,frequency}=1} = \left[\phi^*_{opt}, t^*_{opt}\right]^T = [48.3\,\text{mm}, 2.6\,\text{mm}]^T$ to which corresponds a slightly higher volume of 0.073 m^3.

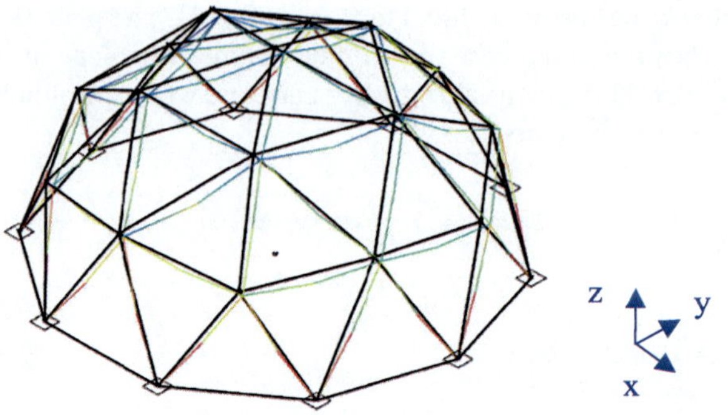

Fig. 4. 3D view of the optimal geodesic dome with frequency equal to 1 (black) and the deformed shape (colored).

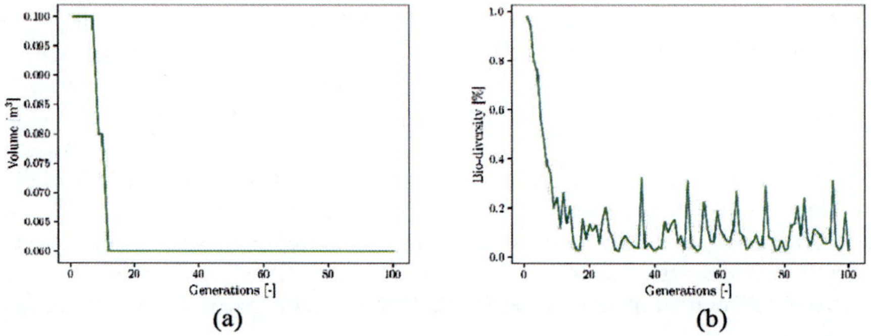

Fig. 5. Geodesic dome with frequency 1. (a) Best fitness value of the first element of the sorted population for every explored generation. (b) Bio-diversity ratio during the generations.

4.2 Frequency Order 2

On the other hand, the optimization process for frequency 2 leads to the following exact optimal solution:

$$x_{opt\,frequency=2} = \left[\phi_{opt}, t_{opt}\right]^T = [21.7\,\text{mm}, 3.4\,\text{mm}]^T \tag{21}$$

The above exact solution is respectful of all the imposed constraints, and it is associated with an optimal volume of 0.08 m^3. Figure 6 shows the 3D representation of the structural model with the optimal values of the design vector. The objective function history during the iterations is illustrated in Fig. 7 (a), whereas Fig. 7 (b) shows the bio-diversity ratio. Again, in order to provide a more realistic feasible structural solution, the exact design parameters were rounded up to the closest CHS steel profile already available on the market, according to the EN 1993-1-1:2005 + AC2:2009 Sects. 6.2 and 6.3. Therefore, the optimal technical solution for size optimization of the geodesic dome under investigation with the dome's frequency equal to 2 is $x^*_{opt\,frequency=2} = \left[\phi^*_{opt}, t^*_{opt}\right]^T = [26.9\,\text{mm}, 3.2\,\text{mm}]^T$ to which corresponds a slightly higher volume of 0.095 m^3. This solution is still respectful of all the imposed constraints.

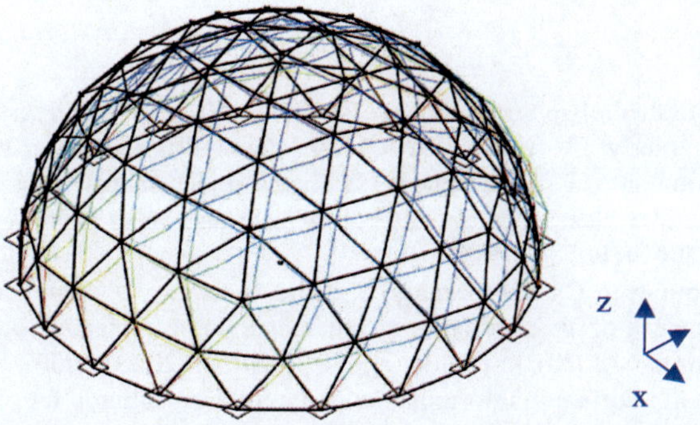

Fig. 6. 3D view of the optimal geodesic dome with a frequency equal to 2 (black) and the deformed shape (coloured).

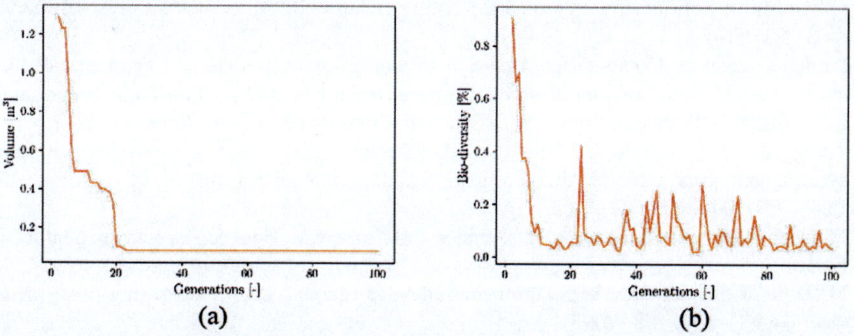

Fig. 7. Geodesic dome with frequency 2. (a) Best fitness value of the first element of the sorted population for every explored generation. (b) Bio-diversity ratio during the generations.

5 Conclusions

In this document, the efficiency of geodesic domes is evaluated considering the self-weight and the action of the asymmetrical horizontal load (wind action). The FEM analysis is then performed to retrieve the data necessary to achieve the optimal solution evaluating two different dome frequencies. The results show no violations of the imposed constraints while minimising the volume of structure and the computational costs, thanks to the high performance of the produced script. The entire workflow lays the foundations for a future study that includes seismic actions thanks to the connection between generative design – with the parameterization of geometric data – and OpenSees, used to simulate the performance of structural and geotechnical systems subjected to earthquakes. Although the application was performed on a geodesic dome, the workflow is suitable for several analyses and optimisation implementing different structures comprehending several structural systems.

Acknowledgement. This work is also part of the collaborative activity developed by Dr Laura Sardone within the framework of the "PNRR": VS3 "Earthquakes and Volcanos".

References

1. Šiber, A.: Icosadeltahedral geometry of geodesic domes, fullerenes and viruses: a tutorial on the t-number. Symmetry. **12**(4), 556 (2020). https://doi.org/10.3390/sym12040556
2. Rebelo, H.B., Amarante, F., dos Santos, C., Cismaşiu, D Santos: Exploratory study on geodesic domes under blast loads. Int. J. Protective Struct. **10**(4), 439–456 (2019). https://doi.org/10.1177/2041419618820540
3. Gythiel, W., Mommeyer, C., Raymaekers, T., Schevenels, M.: A comparative study of the structural performance of different types of reticulated dome subjected to distributed loads. Front. Built Environ. **6** (2020). https://doi.org/10.3389/fbuil.2020.00056
4. Melchiorre, J., et al.: Differential formulation and numerical solution for elastic arches with variable curvature and tapered cross-sections. Eur. J. Mech.-A/Solids **97**, 104757 (2023)
5. Marano, G.C., Rosso, M.M., Melchiorre, J.: Optimization as a tool for seismic protection of structures. In: Cimellaro, G.P. (ed.) Seismic Isolation, Energy Dissipation and Active Vibration Control of Structures: 17th World Conference on Seismic Isolation (17WCSI), pp. 100–113. Springer International Publishing, Cham (2023). https://doi.org/10.1007/978-3-031-21187-4_8
6. Sardone, L., et al.:Computational Design of Comparative models and geometrically constrained optimization of a multi-domain variable section beam based on Timoshenko model. In: Eurogen 2021 Proceedings, 14TH Eccomas Conference, Athens, Greece (2021)
7. Frangedaki, E., Sardone, L., Lagaros, N.D.: Design optimization of tree-shaped structural systems and sustainable architecture using bamboo and earthen materials. J. Architectural Eng. **27**(4), 04021033 (2021)
8. McNeel, R., et al.: Rhinoceros 3D, Version 7.0. Robert McNeel & Associates, Seattle, WA (2010)
9. Mazzoni, S., et al.: OpenSees command language manual. Pac. Earthq. Eng. Res. (PEER) Cent. **264**(1), 137–158 (2006)
10. EN 1990 (2002) (English): Eurocode – Basis of structural design [Authority: The European Union Per Regulation 305/2011, Directive 98/34/EC, Directive 2004/18/EC]
11. NTC 2018: DM del Ministero delle Infrastrutture e dei trasporti del 17/01/2018. Aggiornamento delle Norme Tecniche per le Costruzioni (in Italian) (2018)
12. MacBain, K.M., Spillers, W.R.: Structural Optimization. Springer, US (2009)
13. Rao, S.S.: Engineering Optimization Theory and Practice. Wiley (2019)
14. Rajeev, S., Krishnamoorthy, C.S.: Discrete optimization of structures using genetic algorithms. J. Struct. Eng. **118**(5), 1233–1250 (1992)

The Silberkuhl System (1958–1970): Exploring Form, Construction, and Structural Conception of HP-Shell by Parametric Modeling

Pio Lorenzo Cocco[1](✉), Ilaria Giannetti[2], and Martina Russo[3]

[1] Università Degli Studi Di Camerino, 63100 Ascoli Piceno, Italy
lorenzopio.cocco@unicam.it
[2] Università Degli Studi Di Roma "Tor Vergata", 00133 Rome, Italy
[3] Università Degli Studi Di Roma "La Sapienza", 00185 Rome, Italy

Abstract. The contribution focuses on the patents (1958–64) for modular thin vaults in reinforced concrete filed by the German engineer Wilhelm Johannes Silberkuhl (1912–1984). The thin modular vaults – shaped on narrow hyperbolic paraboloid (HP) sectors – are mainly applied for the roof of industrial buildings and were widely used in Germany and abroad in the 1960s and 1970s. Mainly known as *Silberkuhl System*, the vaults are entirely prefabricated and prestressed on-site. The study explores the relation between form, construction process, and structural conception of the Silberkuhl modular HP-shells, with the aid of a powered visualization framework based on parametric 3D modeling. The use of a computational modeling environment provides insight into the geometry and the structural behavior of the vaults. The application of the vaults for the Magazzino Merci Rinascente in Rome (1962), still existing, is presented as a paradigmatic case study for the spreading of the system in Italy.

Keywords: thin HP-shell · Silberkuhl System · Magazzino Merci Rinascente · Rome Italy · parametric 3D modeling

1 Introduction

During the 20th century, industrial buildings represented an *ideal laboratory* for experimentation and innovations in the building process. The architectural characteristics of the industrial buildings – such as large spans and modularity of the inner spaces – encouraged the development of standardized and industrialized structural elements, especially for roofing systems. In particular, the *shape-resistant* elements in reinforced concrete – such as shells and thin vaults applying the membrane theory – were widely applied from the 1920s to the 1970s, according to their convenience in terms of structural efficiency and economy of construction.

In the second half of the century, the so-called *Silberkuhl System* [1–3] stood out, counting on a broad international application that was comparable to the Zeiss & Dywidag system [4] in the first half of the century.

S. Gabriele et al. (Eds.): IWSS 2023, LNCE 437, pp. 299–308, 2024.
https://doi.org/10.1007/978-3-031-44328-2_31

The system, invented by the German engineer Wilhelm Johannes Silberkuhl (1912–1984) [5] in the 1950s, was widely applied to industrial buildings in Europe and abroad until the late 1970s. The spreading of the system was supported by a wide set of industrial patents filed throughout European countries, the US, and Japan, and the action of local dealers, such as Silberkuhl's own company Normko in Essen [6], and the noteworthy ATA Hyperbolic Shells in England [7].

The Silberkuhl Systems comprised two main shell-roofing typologies [2]: the ensemble of a steel lattice girder and a shell composed of individual precast slabs, representing a section of a circular cone; a doubly curved shells in precast prestressed concrete, representing in shape a sector of a hyperbolic paraboloid (HP), simply supported by lateral tympanum.

The present paper focuses on the second roof typology [1], the so-called *HP-shells*, following its local transfer in Italy. The case study of the Magazzino Merci Rinascente in Rome (1962) – still existing – is presented as a paradigmatic application [8]. The study explores the relation between form, construction process, and structural conception of the modular HP-shells, with the aid of a powered visualization framework based on parametric 3D modeling via the combined use of different platforms [9–13].

1.1 The Silberkuhl HP-Shell

In Italy, the Silberkuhl System was first registered in 1958 and, thus, widely applied to industrial buildings in the 1960s and 1970s via the joint action of national designers and construction firms [14]. The first industrial patents of the HP-shells system were filed by Silberkuhl in Germany in 1956 [15] and 1957 [16].

The patents described HP-shells as doubly curved precast units (20 m long, 2.4 m wide, and 5 cm thick) featured by a single skin hyperboloid shape. The shells formed modular roofs characterized by zenithal lighting systems obtained exploiting different high of the lateral tympanum.

Fig. 1. The three typologies of the Silberkuhl System for thin HP-shells [2].

The HP-shells were available in various forms [2]: the first type was lightly reinforced and characterized by a post-tensioned tie-rod that connected the ends points; the second type had not tie-rod but incorporated a joist running along the middle of the precast unit; the third type avoided the tie-rod and the joist, presenting a prestressing along its whole width (Fig. 1). This latter type, taking advantage of the generation of the HP as a ruled surface, presented straight prestressing wires (35 ⌀5), laid on the crossing generatrixes of the HP, and a light welded steel net.

A special device was designed to prefabricate the shells – on-site or off-site – by smoothing the upper surface of the element, poured in a mold, with the aid of a vibratory slider of arcuate configuration. For the setup of the shells, exploiting the small weight of the element, a single crane was used: compared with traditional cylindrical thin vaults [3], the HP-shells presented better stiffness in the two directions, with better performance to the load scenarios due to the setup process.

1.2 The Magazzino Merci Rinascente in Rome

A noteworthy example of the Silberkuhl System application in Italy was the roof of Magazzini Merci Rinascente in Rome, designed in 1961 by the engineer Aldo Molteni. The 250,000 square meters building presented a rectangular plan – 200 m long and 124.6 m large – with six spans of 20 m each. The roof of the building was composed of 400 HP-shell elements of the first type – a free span of 19.80 m, a wide of 2.50 m, and a thickness of 5 cm – simply supported by perimetral beams.

The geometry of the shell was generated through a translation surface: a circular arch with a radius of 3.30 m was shifted on a circular arch with a radius of 25 m. Each shell was equipped with a post-tensioned steel tie-rod connecting the two ends to counteract the horizontal trust; a light welded steel net characterized the reinforcement layout. The total weight of the steel used as reinforcement, counting the welded net plus the tie-rod, was 500 kg.

Fig. 2. Magazzino Merci Rinascente: construction steps of the HP-shell [8].

The execution process of the shells encompassed five main steps (Fig. 2): i) a mold composed of wooden board was set up on the ground; ii) a welded steel net was laid on the all whole surface and two steel plates, acting as anchoring devices for the tie-rod, were set at the ends points, together with four steel rings, functioning as lifting device; iii) concrete was poured and shaped by the action of a vibratory slider of arcuate configuration; iv) after 24-h curing, the shell was lifted from the mold, using a single bridge crane (the total weight of the single shell was 9,400 kg), and the tie-rod was tensioned; v) the shell was set up at the rood level through the support of the lateral beam (Fig. 3). Steps iv) and v) were carried out in 30 min and, thus, the mold was completely reusable for the cast of further shells; to support the 24-h curing process together with the workability of the material required in step iii), a specific mix design was prepared [8].

Fig. 3. Magazzino Merci Rinascente: the HP-shells assembling on the roof [8].

As the model presented in the patent, the assembled configuration of the roof was characterized by a zenithal lighting system, obtained by choosing different high of the lateral beams, serving as supports for the shells.

Today the building is still in function: despite the heavy functional and architectonic transformation, the roofing system is still visible in the commercial areas, showing the elegance of the HP-shells structural solution.

2 Digital Model of the *Magazzino Merci Rinascente* Shells System

To improve knowledge on both the design and the building details of the Silberkuhl System, a powered visualization framework was developed, exploiting a 3D modeling API. The customized workflow, shown in Fig. 4, was set up with the aim to provide the 3D geometric representation of each element of the HP-shells system, supported by augmented reality (AR) and cloud tools, and to export data for parametric structural analysis. The Magazzino Merci Rinascente HP-shell is selected as a case study to test the functionality of the proposed workflow.

2.1 Geometric Modeling

The process of modeling the elements of the HP-shell has been conducted using Grasshopper [9], a visual programming environment integrated within the 3D-modeling software Rhinoceros [10]. Using such tools makes it possible to generate and manage

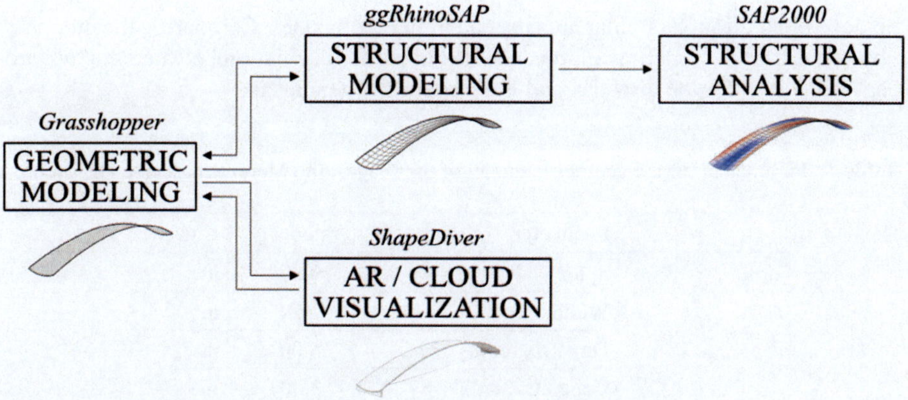

Fig. 4. The proposed workflow for the digital modeling of The Magazzino Merci Rinascente HP-shell; for each step is indicated the software or the plug-in chosen.

dynamic geometries that can be configured by varying parameters of different kinds. The HP-shell system under study has been divided into four elements: i) the shell, ii) the tympanums, iii) the steel net, and iv) the post-tensioning devices.

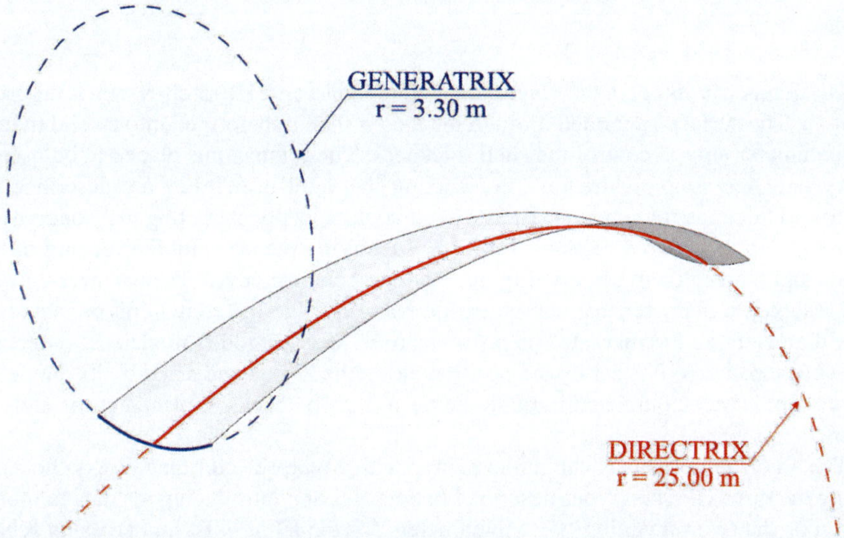

Fig. 5. The geometric construction of the Magazzino Merci Rinascente HP-shell used to generate the shape in Rhinoceros through Grasshopper parameters.

The geometric parameters of the analyzed shell are summarized in Table 1. The HP-shell surface is geometrically modeled by constructing the directrix and the generatrix of the HP, resulting from the combination of the span, the width of the vault, and the radii of the arcs (Fig. 5). As reported by Pietroboni [8], the geometry of the shell has been developed through circular arcs instead of parabolic ones. Comparing the outcome of the two different modeling approaches, the maximum deviation between the two arc types is 20.7 mm for the directrix and 2.3 mm for the generatrix.

Table 1. Parameters for the geometric model of the Magazzino Merci Rinascente HP-shell.

	Parameter	Value	Unit
Form	Span	19.80	m
	Width	2.50	m
	Directrix radius	25.00	m
	Generatrix radius	3.30	m
	Shell thickness	0.05	m
Construction	Concrete cover	50.00	mm
	Rebar diameter	10.00	mm
	Rebar spacing	200.00	mm
	Tie-rod diameter	20.00	mm
	Tympanum thickness	100.00	mm
	Tympanum height	150.00	mm

The geometric output of the Magazzino Merci Rinascente HP-shell system is reported in Fig. 6. The surface generated through the arcs is then transformed into a solid thanks to a parameter able to control the shell thickness. The tympanums placed at both ends of the vault are simple extrusions intersecting the vault itself; they are described by height and thickness parameters. The steel net is modeled by extracting the isocurves of the surface in their two directions, U and V, specifying the size and the spacing of the rebars, and placing them considering the reported concrete cover. The isocurves' net is then assigned a diameter that can be varied using a value list containing the standard rebar diameters on the market. The post-tensioned steel tie-rod is modeled connecting the two tympanums; it is set by the position along the Z-axis and sized by its diameter. The system is then completed through the tie-rod support rebars, placed at the ends of the shell.

The algorithm has also been implemented with a type-selector that allows choosing among the three HP-shell types described in Sect. 1.1 and introducing the data to model them. For the two remaining types, parameters for modeling joist and crossing rebars have been introduced in the code.

This process results in a Silberkuhl System HP-shell configurator, whose geometric outputs are dynamically adaptable to parameter changes and from which additional data (e.g., volume, weight) can be extracted.

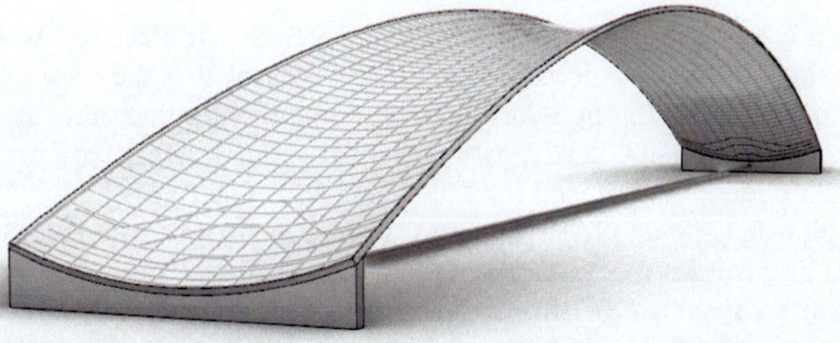

Fig. 6. The final geometric model of the Magazzino Merci Rinascente HP-shell as visualized in Rhinoceros; it is possible to recognize the four elements that characterize the system: the thin shell, the lateral support tympanums, the steel net, and the post-tensioning devices.

2.2 Structural Modeling

To analyze the structural behavior of the HP-shell without losing the ability to configure it parametrically, tools capable of interfacing with the computational environment should be used. For this purpose, the authors chose ggRhinoSAP [11], a plugin that allows Grasshopper to be linked to the FEM calculation software SAP2000 [12]. Implementing it in the algorithm, any changes made to the parametric model are also reflected in the structural analysis software.

It is necessary to set a series of operations to convert the continuous HP-shell model into a discretized model and then assign structural properties and computational settings. In particular, the mechanical model has been described by: i) the shell geometry, ii) the mechanical properties of structural materials, iii) the static load case, iv) the restraints system, and v) the settings for the conversion of the model into the SAP2000 environment. The continuous NURBS surface of the shell modeled into Rhinoceros is converted into a quadrangular mesh whose sizing is obtained by specifying a target dimension of the faces accordingly to the overall dimensions of the surface. Shell properties are then assigned by defining the orientation, the thickness, and the shell type (i.e., thin shell).

Table 2. Concrete mechanical properties of the Magazzino Merci Rinascente HP-shell.

Property	Value	Unit
Elastic modulus	30,000	mPa
Poisson ratio	0.15	–
Density (weight per unit volume)	3,596	kg/m^3
Specific weight	35.96	kN/m^3

The mechanical properties of the concrete are gained by Pietroboni's publication [8] and summarized in Table 2. Given the membrane behavior of this kind of shell, the contribution of steel rebars has not been considered for the static analysis.

The static load case considered at this stage is exclusively the self-weight of the shell. The restraints system locations are identified by extracting the vertices of the mesh at the ends of the shell, through sorting operations. Such points have been assigned the properties of fixed joints. Lastly, parameters for converting the parametric model into a SAP2000 model are introduced (i.e., units and solver settings).

The plug-in ggRhinoSAP generates a permanent link between Grasshopper and SAP2000 and allows the export of the analytical model in.sdb format (Fig. 7). The mechanical analysis can be run from the Grasshopper interface. Any change in shell parameters triggers the renewal of the analytical model export and calculation. Geometric variations and analysis results are visualized and reported in SAP as if the model had been built directly within it. The live connection between these tools grants interoperability between the parametric modeling process and the structural analysis, supporting the possibility of studying and highlighting how the structural performance varies at the variation of the geometric and mechanical features of the HP-shell.

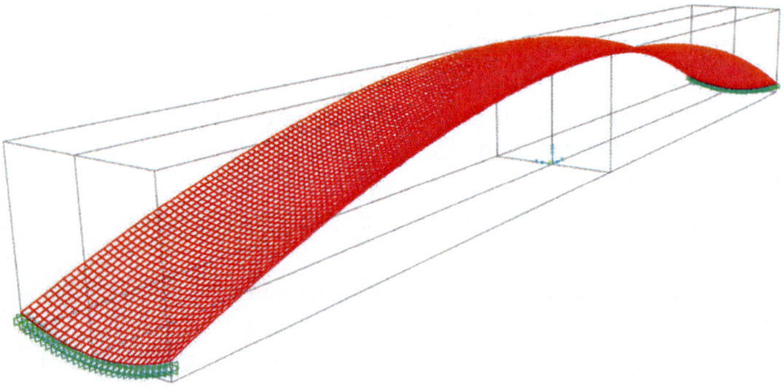

Fig. 7. The Magazzino Merci Rinascente HP-shell model as visualized in SAP2000.

2.3 Cloud Visualization

The HP-shell configurator programmed by the authors has been published on ShapeDiver [13], a cloud application that allows users to visualize and configure objects generated with Grasshopper through a browser. Uploading the file in.gh format to the platform, the latter will compute the Grasshopper definition in the cloud, combine the parameter set, and return the output in real time. In order to display and correctly compute the model on the platform, it should pay attention to the definition of some parameters, geometry type, and components optimization.

Regarding the parameters, ShapeDiver is able to automatically recognize them from the Grasshopper definition. The supported parameters can have different natures (e.g., sliders, booleans). If necessary, the parameters to be made non-editable by the user can be replaced by other types of components (i.e., text or numbers) or hidden during the following stage of platform configuration. The choice of editable parameters will also affect the usability of the interface on ShapeDiver, which should be as user-friendly as possible. Moreover, it is important to limit the range of parameters to suitable values to avoid errors and undesired results during the configuration phase made by the users. The authors excluded the parameters related to rebars and concrete cover, focusing on the parameters that determine the shell geometry.

While the HP-shell system has been modeled within Grasshopper as NURBS geometries, ShapeDiver returns geometries in the form of meshes, that are computed more quickly compared to NURBS. In order to limit the computation time, any kind of conversion from NURBS should be avoided, providing the geometric output for the visualization platform directly in the form of a mesh. Therefore, the strategy adopted was coding a second algorithm representing the same HP-shell system with the same parameters, which generates meshes as its geometric output. The geometries to be displayed in the configurator and their appearance are set through specific components. To make the configuration on ShapeDiver as smooth as possible, all components without a visual output (e.g., quantitative data extraction components) have been removed. In addition, the choice of the components was driven by the aim of reducing computation time, also making use of the *profiler widget* of Grasshopper.

Once the algorithm optimization operations for ShapeDiver have been carried out, the.gh file of the HP-shell system has been uploaded onto the ShapeDiver platform. Directly from the website, a description of the Silberkuhl System has been added, parameters have been grouped and sorted, and graphical settings have been specified. The interface for the three types of Silberkuhl Systems is shown in Fig. 8.

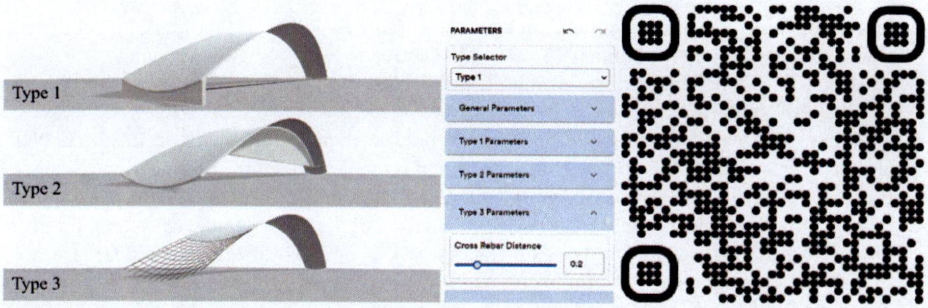

Fig. 8. Interactive view of the three types of the Silberkuhl shell obtained in ShapeDiver; using the QR code, it is possible to access to the digital model.

Once the various parameters have been configured, ShapeDiver provides users the ability to view the digital model in AR via a mobile device. For this purpose, the model scale of representation chosen is 1:100. The user still has the option of manually scaling the object with a pinch on the screen of the mobile device. Through the QR code shown in Fig. 8, it is possible to access the HP-shell configurator on ShapeDiver.

3 Conclusions

The present paper analyzed the Silberkuhl System – HP-shells in precast prestressed concrete – applied worldwide for industrial building roofs between the 1960s and 1970s. The case study of the Magazzino Merci Rinascente in Rome (1962) is presented as a significant application of the system in Italy. The relation between form, construction process, and structural conception of the HP-shells is explored, setting up a powered visualization framework based on parametric 3D modeling.

The study provides an insight into the Silberkuhl Systems, disclosing uncovered technical aspects: the acquired knowledge allows to frame the system in the history of thin shells in the 20th century, supporting the actual maintenance operations of the existing exemplars. From a methodological point of view, the study remarked the effectiveness of the presented modeling workflow to support the study of shell and spatial structures, featured by complex geometry and customized building details.

The showed 3D modeling API allows, at the same, the 3D visualization of the geometries and the building details, derived from the historical investigation, and the production of consistent data for structural analysis. In this way, the proposed workflow supported the integration of different research fields, such as construction history and structural mechanics, to improve knowledge of historical structures, facing maintenance operations and future design applications.

References

1. Siberkhul, W.J.: Vorgefertigte Schalenelemente, Eléments-voiles préfabriqués, Prefabricated shell elements. Bauen + Wohnen **15**, 428–430 (1961)
2. Joedicke, J.: Shell Architecture. Reinhold Publishing Corporation, New York, pp. 93–96, pp. 206–209 (1963)
3. Scheffler, T.: Development and application of precast hyperboloid shells in East and West Germany from the 1950s to the 1980s. In: Interfaces: Architecture, Engineering, Science, Proceedings of the IASS Annual Symposium 2017. Hamburg, Germany (2017)
4. May, R.: Shell sellers. The international dissemination of the Zeiss-Dywidag system. In: Fifth International Congress on Construction History, vol. 2, pp. 557–564. Chicago (2015)
5. Stiglat, K.: Bauingenieure und ihr Werk, p. 394. Ernst & Sohn, Berlin (2004)
6. Normko GmbH (1967) HP-Handbuch (4th ed.). Essen
7. Hyerbolic Shell Roof, A.T.A.: A new hyperbolic shell roof. Official Archit. Plan. **21**(1), 552 (1958)
8. Pietroboni, G.: Copertura con volte sottili prefabbricate. L'Industria italiana del Cemento **3**, 147–155 (1962)
9. Grasshopper. https://www.grasshopper3d.com/. Accessed 20 May 2023
10. Rhinoceros. https://www.rhino3d.com/. Accessed 20 May 2023
11. ggRhinoSAP. https://geometrygym.wordpress.com/. Accessed 20 May 2023
12. SAP 2000. https://www.csi-italia.eu/software/sap2000/. Accessed 20 May 2023
13. Shape Driver. https://shapediver.com/. Accessed 20 May 2023
14. Silberkuhl, W.J.: Costruzione perfezionata del tetto per strutture di tettoie o simil. Italian patent n. 656266, 24 June 1961
15. Silberkuhl, W.J.: Dachkonstruktion aus gekrümmten Shalen, insbesondere aus doppelt-gerkümmte Stahlbetonshalen. German patent n. 1052660, 9 April 1956
16. Silberkuhl, W.J.: Doppeltgerkümmte Shale aus Stahlbeton, insbesondere Dachshale. German patent n. 1053762, 5 Jan 1957

Vibration Analyses of an Hybrid Concrete and Cross-laminated Timber Building Case Study

Angelo Aloisio[1], Igor Gavrić[2], Marco M. Rosso[3(✉)], Dag P. Pasca[4], Roberto Tomasi[5], Massimo Fragiacomo[1], Giuseppe Carlo Marano[3,6], and Iztok Šušteršič[2]

[1] Università degli Studi dell'Aquila, via Giovanni Gronchi n.18, 67100 L'Aquila, Italy
[2] InnoRenew CoE, 6310 Izola, Slovenia
[3] Politecnico di Torino, DISEG, Department of Structural, Geotechnical and Building Engineering, Corso Duca Degli Abruzzi, 24, 10128 Turin, Italy
marco.rosso@polito.it
[4] Norsk Treteknisk Institutt, Børrestuveien 3, 0373 Oslo, Norway
[5] Norwegian University of Life Sciences, Ås, Norway
[6] College of Civil Engineering, Fuzhou University, Fuzhou 350108, China

Abstract. Nowadays, some innovative spatial structural typologies among others rely on timber-concrete hybrid solutions for designing modern buildings. However, the dynamic identification analysis may be more elaborate, and sometimes troublesome, due to the coupling effects of the different dynamic nature of cross-laminated timber and reinforced concrete members. In the current manuscript, the authors explore some preliminary results of the dynamic analysis of a hybrid timber concrete building case study. The operational modal analysis (OMA) based on output-only techniques has been employed, referring specifically to enhanced frequency-domain decomposition (EFDD) and the stochastic subspace identification (SSI) methods. The authors compared several ambient vibration OMA results with forced shaker-induced vibration responses highlighting the absence of nonlinearities during in-service operational conditions in two different moments.

Keywords: Timber Structure · Finite Element · Model Updating · Hybrid Timber Concrete Structures · Operational Modal Analysis

1 Introduction

The dynamic behaviour of timber buildings under service loads is a recently debated topic in the scientific literature, primarily thanks to the application of vibration-based identification methods. The growing interest in timber structures is confirmed by multiple initiatives and financed research projects involving the dynamic investigation of timber structures. Cross-laminated timber (CLT) structures are widespread among timber structures, especially tall buildings, due to

S. Gabriele et al. (Eds.): IWSS 2023, LNCE 437, pp. 309–318, 2024.
https://doi.org/10.1007/978-3-031-44328-2_32

their satisfactory structural performance and reduced environmental footprint. In detail, various building typologies have been introduced, with CLT as the central part of the load-bearing structure. A leading typology is represented by the timber-concrete hybrid buildings, where concrete elements are used in addition to the timber ones, e.g., for shear walls and elevator shafts. There is a lot of uncertainty in the as-built dynamic response of these buildings. Therefore, experimental vibration tests on timber buildings are precious since they provide insight into the as-built stiffness of different structural components, like shear walls and floors. As remarked by [21], in tall CLT buildings, the stiffness is not merely related to the properties of the CLT panels, but it is also affected by the connection system. Moreover, since timber structures are often lightweight and highly flexible, they may be more prone and susceptible to excessive vibrations [10].

Therefore, compared to other structural materials, many aspects of the dynamic response of timber buildings under service loads are still unknown. In particular, there are two important aspects which still deserve experimental investigations: the long-term dynamic performance of timber structures and the amplitude-dependent behaviour. Concerning the first aspect, as highlighted by Riggio and Dilmaghani [28], most of the recent investigations on the dynamic behaviour in timber structures have been conducted over a short monitoring period [2,25], and few studies hitherto involved long-term monitoring, e.g. Larsson et al. [22]. Regarding amplitude-dependent behaviour, most studies deal with laboratory tests, where structural assemblies or full-scale buildings are evaluated from shake table tests [9]. To the author's knowledge, no research attempts to detect possible amplitude-dependent phenomena in timber buildings from forced vibration tests on actual case structures nor by comparing forced and ambient vibration tests. Forced vibration tests were widespread in past years [5,14], when they were sole for reliably estimating the modal parameters [4,33]. However, since the pioneering studies of Trifunac in 1972 [30], the comparison of ambient and forced vibration proved that it is possible to determine with adequate accuracy natural frequencies, mode shapes and damping values from both data. Nonetheless, in forced vibration tests, the whipping of the floors might lead to modal interference, a consequence of the concentrated force excitation at a frequency only slightly different from the resonant frequency. This phenomenon may lead to discrepancies between the ambient and forced vibration tests predominantly at the top levels of a structure. Compared to ambient vibration tests, where stationary operational modal analysis can be used, input-output or nonstationary operational modal analyses must be carried out to correctly analyze the outcomes of forced vibration tests' results.

In this paper, the authors compare the stationary and nonstationary operational modal analysis results of an hybrid timber building located in Slovenia based on both ambient and forced vibration tests. The ambient vibration tests led to estimating the modal parameters using the classical stochastic subspace identification and enhanced frequency domain decomposition. The forced vibration tests were used to estimate the intrinsic modes using the Hilbert-

Huang Transform (HHT). Several methods have been proposed for estimating the time-varying natural frequencies of engineering structures based on measurements under ambient and forced vibrations [11,23,35]. Among these methods, the Hilbert transform (HT) technique, Teager energy operation (TEO) method, and their variants have been widely used, despite several drawbacks such as negative frequencies, ending effects, and fluctuations [6,34,36]. Besides, these two approaches are noise-sensitive, which will further cause distorted natural frequency estimation [34]. Recently, time-frequency representation (TFR) techniques have drawn increasing attention in field measurements of civil structures due to their advantage in concentrating signal energy along the instantaneous frequency curve while spreading the noise in the time-frequency plane [13,19,20]. Several time-frequency representation methods, including the short-time Fourier transform (STFT) [24,31], the wavelet transform (WT) [15], and the Wigner Ville distribution (WVD) [26], have been applied to analyze measured signals to obtain the instantaneous description of dynamic properties of civil structures. Still, the HHT [18] remains the most used algorithm among recent attempts. It is considered the most appropriate tool to deal with nonlinear and nonstationary signals since the uncertainty principle does not limit it, and its basis is adaptive [29].

This document has the following structure. Section 2 describes the case study under investigation, whereas Sect. 3 presents two subsections dedicated to ambient vibration tests and forced vibration tests respectively.

2 Case Study Description

The current case study building is an hybrid structure presented in Fig. 1 (a) with plan dimensions of 30 m × 20 m approximately, made of CLT, steel and concrete elements located in Slovenia. The main structural elements of the building's second, third and fourth floors are CLT shear walls, while the walls on the first floor are made of concrete. A concrete slab is placed between the first and second floor, while the rest of the floors consists of CLT panels. In the core of the building, there is a steel frame to support the floors. A concrete shaft is placed in one of the corners of the building. Additionally, some timber frame elements around the perimeter of the building support the CLT floor elements between the column-like structure of the outer walls.

The dynamic identification of the building is based on ambient and forced vibration tests, whose experimental accelerometers layout is depicted in Fig. 1 (b). The authors monitored only the third floor and the roof level to explore the dynamic behavior of the freely vibrating part of the structure, i.e. above the second floor, and disregarding the stiffening adjacent buildings effects. The experimental campaigns were executed in two different moments for the sake of comparisons, i.e. March 2021 and August 2021. The forced vibration tests were carried out using a shaker sweeping between frequencies from 2 to 22 Hz, at a rate of 5 Hz/h. Figure 2 shows the experimental setup with the shaker and acquisition system.

(a) **(b)**

Fig. 1. View of the hybrid timber building (a) and sensor placement (b).

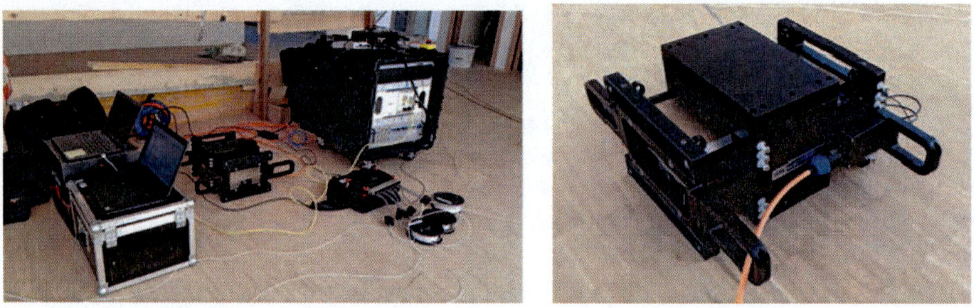

(a) **(b)**

Fig. 2. Forced vibration test's acquisition system (a) and the adottped shaker (b).

3 Dynamic Identification

3.1 Ambient Vibration Tests

The building was tested through two different moment in 2021 for the sake of comparison. The modal parameters are estimated with the covariance-based stochastic subspace identification (SSI-cov) [27] and the enhanced frequency domain decomposition (EFDD) [8]. Figure 3 reports the obtained stabilization diagram from SSI-cov (a) and the singular value decomposition (SVD) of the power spectral density (PSD) from EFDD. The best stabilization diagram, obtained by varying the time shift and the maximum model order, only shows three stable modes in the frequency range 0–25 Hz. The three modes at approximately 4.81, 6.27, and 7.08 Hz have been plotted in Fig. 4. The spatial discretization of the mode shapes does not allow a clear interpretation and distinction between translation and rotational modes. Each mode does not exhibit a prevailing direction of deformation. The mean of the root means square of the signals is approximate. 3.13×10^{-4}g, while the peak acceleration is 0.001g. These values are essential to understand the variation of the excitation level between ambient and forced vibration tests. Interestingly, there is a high modal component in the vertical direction compared to the translational ones. Although this fact is quite unusual, it may appear as measurement error. Nonetheless, this

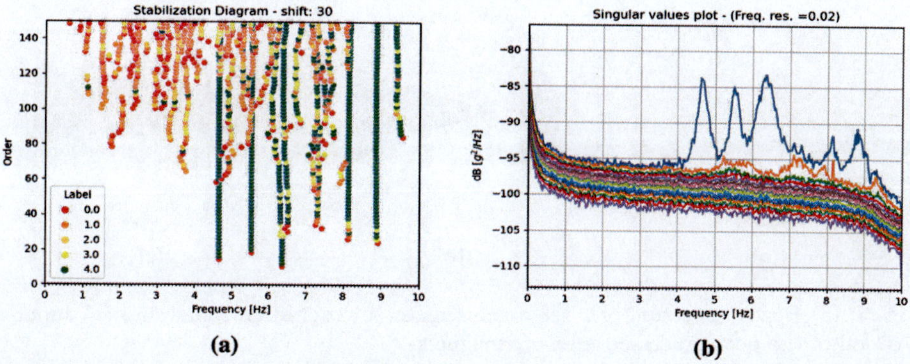

Fig. 3. Ambient vibration. SSI-cov stabilization diagram (a) and SVD of PSD for EFDD (b).

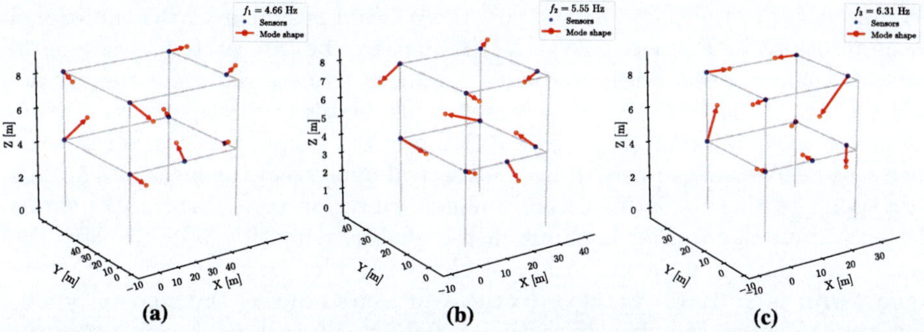

Fig. 4. Mode shapes for ambient vibrations, evidencing high vertical components.

outcome is conformed to a finite element model that the authors implemented for the sake of comparison, thus evidencing the existance of important modal components along the z axis.

3.2 Forced-Vibration Tests

The data were processed using the HHT to deal with non-stationary signals. The HHT was proposed by Norden E. Huang and initially published as a patent [16], and paper [18]. It consists of two steps, first the signal is decomposed by empirical mode decomposition (EMD), designed to decompose non-linear and non-stationary signals into a set of spectrally in-dependent oscillatory components, the intrinsic mode shapes (IMFs). Then, the IMFs are processed through the Hilbert Spectral Analysis (HSA) [12,32], resulting in the Hilbert spectrum, which preserves local properties in the time domain and provides information in the amplitude and frequency domain, favouring the identification of hidden local characteristics in the original signal [17]. The Hilbert spectrum is three-dimensional (time-frequency-energy) and can be represented in graphs, similar

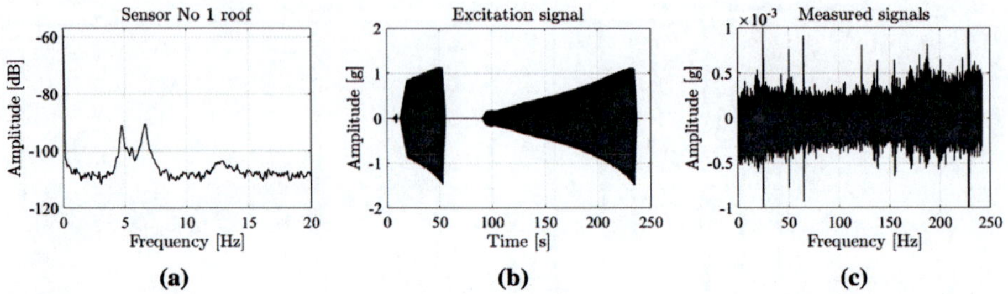

Fig. 5. (a) Power spectrum of a reference sensor, (b) excitation signal and (c) superposition of the post-processed measured signals.

to the wavelet spectrum. The EMD operates in sequences of local extremes, being performed by direct extraction of local energy associated with the intrinsic time scales of the signal itself, thus, the method is similar to the traditional decompositions of Fourier or Wavelet [1] and can be interpreted as a type of wavelet decomposition with sub-bands produced as needed during the extraction process of the IMFs, which represents the details of the signals on a given scale or frequency range [7]. Figure 5(a) shows the power spectrum of a reference sensor. The peaks revealed by the spectral analysis of the measured signals correspond to those identified from ambient vibration tests. Figure 5(b) shows the excitation signal used as input in the shaker, while Fig. 5(d) displays the superposed measured signals from all channels. The mean of the root mean square error is $1.15\times10^{-}4$g, close to those measured during the ambient vibration tests. The maximum acceleration is 0.011g, approximately ten times the peak observed during the operational response. The modes obtained from the HHT correspond to those estimated from ambient vibration tests, as reported in Table 1. Figure 6 plots the instantaneous frequencies and MAC between each mode estimate with the averaged mode shapes. The MAC plot proves that the averaged mode shapes cannot be considered physical modes, yielding a value close to 0.8 for all modes. Therefore, the averaged modes cannot be used for model updating. Nonetheless, the MAC variation is minimal, proving that the mode shapes are not very sensitive to the vibration amplitude. Higher sensitivity is shown in the instantaneous frequencies: In particular, the identification leads to higher discrepancies with the averaged ones when the vibration amplitude is lower and the signal is less informative. Nonetheless, the higher excitation level is not enough to activate possible nonlinear phenomena [3], as proven by Fig. 7. Figure 7(a)–(b) plots the dependence of the instantaneous frequencies on the response amplitude. The plots exhibit a higher scatter for lower vibration levels. However, increasing vibration levels do not trigger nonlinearities since the estimated values align along a vertical axis. This fact is further proven by cross-plotting the modal components of different sensors. The dots (see Fig. 7(c)) align along a line due to an almost linear structural response.

Table 1. Comparison between the ambient and forced vibration tests.

Mode	Ambient (March 2021)	Ambient (August 2021)	Forced vibration (August 2021)
1	4.81 Hz	4.58 Hz	4.65 Hz
2	6.27 Hz	5.80 Hz	5.77 Hz
3	7.08 Hz	6.71 Hz	6.67 Hz

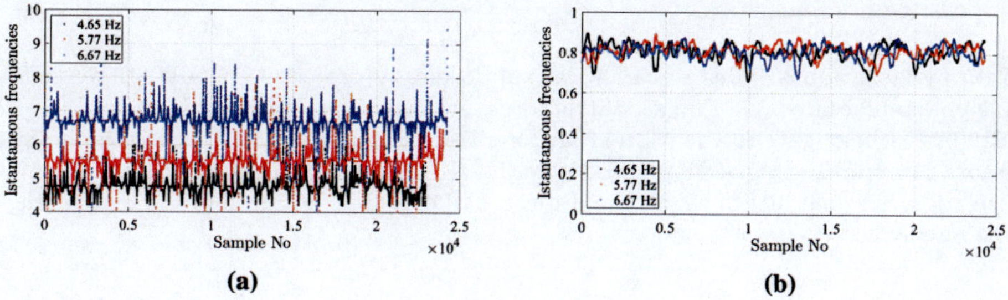

Fig. 6. (a) Instantaneous frequency values for the first three modes (b) MAC between the intrinsic modes and the averaged ones.

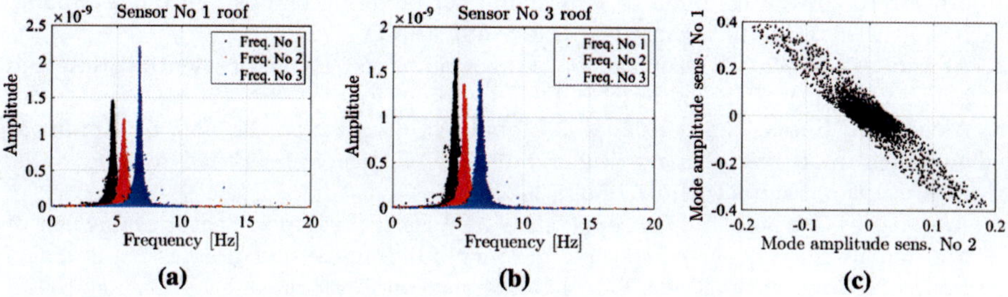

Fig. 7. Amplitude-frequency plots for two sensors and plot of the mode amplitudes between two sample mode components from two modes.

4 Conclusions

Experimental dynamic responses of a hybrid timber-concrete building located in Slovenia have been herein studied. The main objectives, major findings, novelty aspects, and future developments are summarized below:

- Comparisons between ambient and forced vibration tests have been conducted evidencing good modal parameters agreement;
- Estimation of the intrinsic modes from forced vibration tests using the Hilbert-Huang Transform (HHT) was employed to detect possible nonlinear phenomena;

- The intrinsic mode shapes (IMFs) obtained with the HHT and the modal components were in good agreement, however, the higher excitation level was not capable of triggering nonlinear phenomena;
- Therefore, in this case, the forced vibration tests are not more informative than the ambient ones.
- Immediate future extensions may involve automatic metaheuristic-based FE model updating and variance-based sensitivity analyses using the results both from forced or ambient vibration tests, comparing the final results to detect possible nonlinearities.

Acknowledgments. This research was supported by project ADDOPTML "ADDitively Manufactured OPTimized Structures by means of Machine Learning" (No: 101007595) belonging to the Marie Skłodowska-Curie Actions (MSCA) Research and Innovation Staff Exchange (RISE) H2020-MSCA-RISE-2020 - ADDOPTML (ntua.gr). The authors would like to thank the project ADDOPTML for funding and supporting this research.

References

1. Al-Baddai, S., Al-Subari, K., Lang, E., Ludwig, B.: Optimizing approach for sifting process to solve a common type of empirical mode decomposition mode mixing. Int. J. Electr. Comput. Eng. **11**(6), 678–681 (2017)
2. Aloisio, A., Pasca, D., Tomasi, R., Fragiacomo, M.: Dynamic identification and model updating of an eight-storey CLT building. Eng. Struct. (2020)
3. Aloisio, A., Rosso, M.M., Huang, D., Iqbal, A., Fragiacomo, M., Pei, S.: Nonlinear analytical modeling of mass-timber buildings with post-tensioned rocking walls. Bull. Earthq. Eng. **21**(1), 473–502 (2023)
4. Angotti, F., Vignoli, A., Giuseppetti, G., Panzeri, P.: Experimental evaluation of the dynamic behaviour of a stone masonry building causing progressive damage. In: Proceedings of the Tenth World Conference on Earthquake Engineering (1992). cited By 1
5. Beolchini, G., Vestroni, F.: Identification of dynamic characteristics of base-isolated and conventional buildings. In: Proceedings of 10th European Conference on Earthquake Engineering Conference, vol. 1, pp. 689–695 (1995). cited By 3
6. Bouchikhi, A., Boudraa, A.O.: Multicomponent am-fm signals analysis based on EMD-b-splines ESA. Sig. Process. **92**(9), 2214–2228 (2012)
7. Bouchikhi, A., Boudraa, A.O., Benramdane, S., Diop, E.H.: Empirical mode decomposition and some operators to estimate instantaneous frequency: a comparative study. In: 2008 3rd International Symposium on Communications, Control and Signal Processing, pp. 608–613. IEEE (2008)
8. Brincker, R., Zhang, L., Andersen, P.: Modal identification of output-only systems using frequency domain decomposition. Smart Mater. Struct. **10**(3), 441 (2001)
9. Casagrande, D., Pacchioli, S., Polastri, A., Pozza, L.: Influence of the rocking behavior of shearwalls on the fundamental period of CLT structures. Earthq. Eng. Struct. Dyn. **50**(6), 1734–1754 (2021)
10. Chapain, S., Aly, A.M.: Vibration attenuation in a high-rise hybrid-timber building: a comparative study. Appl. Sci. **13**(4), 2230 (2023)

11. Chen, J., Zhao, G.: Numerical and experimental investigation on parameter identification of time-varying dynamical system using Hilbert transform and empirical mode decomposition. Math. Probl. Eng. **2014** (2014)
12. Dhengyun, H., Ding, Y.: Marginal Hilbert spectrum based on EMD reconstruction and its application in fault diagnosis of cooling tower. In: 2013 Third International Conference on Intelligent System Design and Engineering Applications, pp. 926–929. IEEE (2013)
13. Feng, Z., Chen, X.: Adaptive iterative generalized demodulation for nonstationary complex signal analysis: principle and application in rotating machinery fault diagnosis. Mech. Syst. Sig. Process. **110**, 1–27 (2018)
14. Genovese, F., Vestroni, F.: Identification of dynamic characteristics of a masonry building. In: Proceedings of the 11th European Conference on Earthquake Engineering (1998). cited By 10
15. Hosseini, S.A., Amjady, N., Velayati, M.H.: A Fourier based wavelet approach using Heisenberg's uncertainty principle and Shannon's entropy criterion to monitor power system small signal oscillations. IEEE Trans. Power Syst. **30**(6), 3314–3326 (2014)
16. Huang, N.: Computer implicated empirical mode decomposition method. Apparatus, and article of manufacture. US Provisional Appl. **60**(23), 822 (1996)
17. Huang, N.E.: Introduction to the Hilbert–Huang transform and its related mathematical problems. In: Hilbert–Huang Transform and Its Applications, pp. 1–26. World Scientific (2014)
18. Huang, N.E., et al.: The empirical mode decomposition and the Hilbert spectrum for nonlinear and non-stationary time series analysis. Proc. Roy. Soc. Lond. Ser. A Math. Phys. Eng. Sci. **454**(1971), 903–995 (1998)
19. Khan, N.A., Ali, S.: A robust and efficient instantaneous frequency estimator of multi-component signals with intersecting time-frequency signatures. Sig. Process. **177**, 107728 (2020)
20. Khan, N.A., Mohammadi, M., Ali, S.: Instantaneous frequency estimation of intersecting and close multi-component signals with varying amplitudes. SIViP **13**(3), 517–524 (2019)
21. Kurent, B., Friedman, N., Ao, W.K., Brank, B.: Bayesian updating of tall timber building model using modal data. Eng. Struct. **266**, 114570 (2022)
22. Larsson, C., Abdeljaber, O., Bolmsvik, Å., Dorn, M.: Long-term analysis of the environmental effects on the global dynamic properties of a hybrid timber-concrete building. Eng. Struct. **268**, 114726 (2022)
23. Liu, J., Zheng, J., Wei, X., Ren, W.X., Laory, I.: A combined method for instantaneous frequency identification in low frequency structures. Eng. Struct. **194**, 370–383 (2019)
24. Meignen, S., Pham, D.H., Colominas, M.A.: On the use of short-time Fourier transform and synchrosqueezing-based demodulation for the retrieval of the modes of multicomponent signals. Sig. Process. **178**, 107760 (2021)
25. Mugabo, I., Barbosa, A., Riggio, M.: Dynamic characterization and vibration analysis of a four-story mass timber building. Front. Built Environ. **5** (2019). https://doi.org/10.3389/fbuil.2019.00086, cited By 14
26. Nikoo, A., Kahoo, A.R., Hassanpour, H., Saadatnia, H.: Using a time-frequency distribution to identify buried channels in reflection seismic data. Digit. Sig. Process. **54**, 54–63 (2016)
27. Peeters, B., De Roeck, G.: Reference-based stochastic subspace identification for output-only modal analysis. Mech. Syst. Sig. Process. **13**(6), 855–878 (1999)

28. Riggio, M., Dilmaghani, M.: Structural health monitoring of timber buildings: a literature survey. Build. Res. Inf. **48**(8), 817–837 (2020). https://doi.org/10.1080/09613218.2019.1681253, cited By 8
29. de Souza, U.B., Escola, J.P.L., da Cunha Brito, L.: A survey on Hilbert-Huang transform: evolution, challenges and solutions. Digit. Sig. Process. **120**, 103292 (2022)
30. Trifunac, M.D.: Comparisons between ambient and forced vibration experiments. Earthq. Eng. Struct. Dyn. **1**(2), 133–150 (1972)
31. Wang, C., Kong, F., He, Q., Hu, F., Liu, F.: Doppler effect removal based on instantaneous frequency estimation and time domain re-sampling for wayside acoustic defective bearing detector system. Measurement **50**, 346–355 (2014)
32. Yushan, Z., Jianwen, L., Yuxian, H.: Hilbert spectrum and intrinsic oscillation mode of dynamic response of a bilinear SDOF system: influence of harmonic excitation amplitude. Earthq. Eng. Eng. Vib. **4**(1), 17–26 (2005)
33. Zabel, V., Bucher, C., Riedel, J.: The dynamic behaviour of a historical bell tower - in-situ tests and numerical investigations, pp. 387–394 (2000). cited By 1
34. Zeng, M., Yang, Y., Zheng, J., Cheng, J.: Normalized complex teager energy operator demodulation method and its application to fault diagnosis in a rubbing rotor system. Mech. Syst. Sig. Process. **50**, 380–399 (2015)
35. Zhang, J., Shi, Z.: Instantaneous modal parameter identification of linear time-varying systems based on chirplet adaptive decomposition. Shock Vibr. **2019** (2019)
36. Zheng, J., Pan, H., Yang, S., Cheng, J.: Adaptive parameterless empirical wavelet transform based time-frequency analysis method and its application to rotor rubbing fault diagnosis. Sig. Process. **130**, 305–314 (2017)

Analytical and Numerical Instability Analysis of Corroded and Temperature-Varying Thin-Wall Shells

Cheng Huijuan Liu[1] and Giuseppe Lacidogna[2](✉)

[1] Department of Mathematics, Aberystwyth University, Aberystwyth SY23 3BZ, Ceredigion, Wales, UK

[2] Department of Structural, Geotechnical and Building Engineering, Politecnico di Torino, 10129 Turin, Italy

giuseppe.lacidogna@polito.it

Abstract. Thin-wall shells (steel plates, steel cylindrical shells, steel spherical shells, etc.) are widely used in many engineering fields such as construction, machinery, chemical industry, navigation, and aviation because of their light weight and high strength. Their failure modes under static pressure or impact dynamic load are mostly buckling instability, and the failure is very sudden, often causing structural failure or even catastrophic accidents without obvious symptoms. In this framework, the significance of this paper is that it considers the influence of external environment corrosion on steel shells' bearing capacity using plate and shell classical stability theory, and investigates the stable bearing capacity of thin-wall steel shells in view of corrosion impact. By this approach, a theoretical method for the time-varying stable bearing capacity of plate and shell thin-walled steel members under the simultaneous action of corrosion and temperature changes is obtained, providing a useful theory for complex engineering practices such as corrosion and temperature changes, including fire actions.

Keywords: Corrosion · Variable temperature · Fire · Thin-walled shell · Stable bearing capacity

1 Introduction

Thin shell structures (thin plates, thin cylindrical shells, thin spherical shells, etc.) often work in environments with corrosion and temperature changes. According to relevant research, most of their failure problems are controlled by stability. For their stable (unstable) behavior at this time, classical thin-walled shell theories, e.g., small or large deflection theories, can no longer be used directly. Gutman [1–3] proposed a series of calculation methods for cylindrical shells and spherical shells, mainly in corrosive environments. When dealing with temperature, he does not consider the temperature stress, but only the influence of temperature on the corrosion rate. Lacidogna also mentioned the problem of spherical shell stability after corrosion in literature [4]. However, it mostly focuses on the estimation of the stable bearing capacity based on the non-destructive

S. Gabriele et al. (Eds.): IWSS 2023, LNCE 437, pp. 319–328, 2024.
https://doi.org/10.1007/978-3-031-44328-2_33

testing data of the shell, and does not discuss the temperature stress in depth. Pronina in the literature [5, 6] took the spherical shell as an example, considered the influence of corrosion and temperature stress, and gave its stable bearing capacity calculation equation, and obtained a segmented analytical solution. However, as for the corrosion kinetic equation it uses comes from Gutman's metal immersion corrosion experiments and has not been verified to be applicable to the type of air corrosion. Also, during the derivation, it doesn't mention constraints (e.g.: elastic constraints). Although, this has little effect on spherical shells, which are usually considered free constraints. But once it is applied to cylindrical shells or other flat shells, the deficiency will appear. Because cylindrical shells or plate shells often have different levels of restraint at the ends. More importantly, in practice, temperature also changes the physical properties of the shell, such as elastic modulus, yield strength, and possibly even its mode of failure (e.g., stability control or strength control). And those things should be discussed. However, research on this type of content has yet to be found.

Among the factors that cause temperature changes, such as: temperature difference between day and night, seasonal temperature difference and fire, fire is more harmful. Therefore, as a special kind of variable temperature environment, its influence on the stability of corroded thin shells should also be discussed. In fact, no relevant research has been found so far.

Therefore, in this paper it is believed that it is necessary to use a better corrosion dynamic model, consider temperature stress, and different constraints to study the stability of shells.

2 Problem Description

2.1 Objectives of the Research

Our purpose is to study the calculation method of the stable bearing capacity of common thin shells (plates, cylindrical shells and spherical shells) in corrosion and variable temperature environments in the elastic range (Fig. 1).

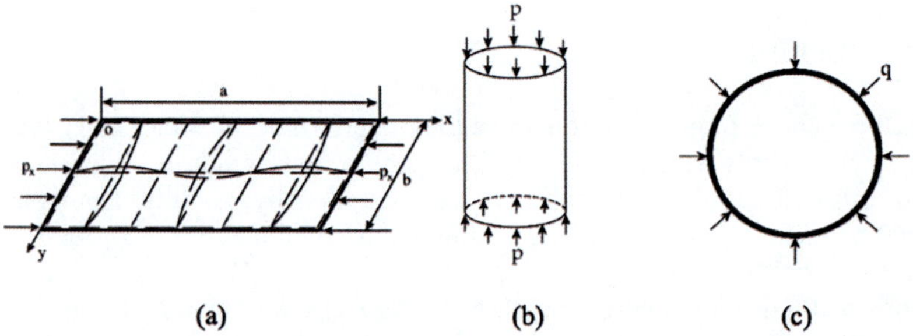

Fig. 1. Thin-wall shells: (a) plates, (b) cylindrical shells, (c) spherical shells

2.2 Stability Theory Under Known Ideal Environment

In this paper to carry out the research we start from the known small deflection and small deflection stability theory of part-thin shells [7], which is explained by the following equations.

2.2.1 Plate

Small deflection buckling:

$$\sigma_{x,cr}(t) = \frac{k\pi^2 E \cdot h^2(t)}{12(1-\mu^2)b^2}, \tag{1}$$

$$p_{x,cr}(t) = \sigma_{x,cr}(t) \cdot h(t) = \frac{k\pi^2 E \cdot h^3(t)}{12(1-\mu^2)b^2}. \tag{2}$$

Large deflection buckling:

$$\sigma_{xa} = \frac{4\pi^2 D}{hb^2} + \frac{E\pi^2 f^2}{8b^2} = \sigma_{cr} + \frac{E\pi^2 f^2}{8b^2}. \tag{3}$$

2.2.2 Cylindrical Shell

$$\begin{cases} \sigma_{cr}(t) = \begin{cases} \frac{E}{\sqrt{3(1-\mu^2)}}\frac{h(t)}{r}, \text{ small deflection buckling,} \\ \frac{1}{3}\frac{E}{\sqrt{3(1-\mu^2)}}\frac{h(t)}{r}, \text{ large deflection buckling,} \end{cases} \\ p_{cr}(t) = \sigma_{cr}(t) \cdot h(t), \\ N_{cr}(t) = \sigma_{cr}(t) \cdot 2\pi r \cdot h(t). \end{cases} \tag{4}$$

2.2.3 Spherical Shell

$$\sigma_{cr}(t) = \begin{cases} \frac{1}{\sqrt{3(1-\mu^2)}}\frac{Eh(t)}{R}, \text{ small deflection buckling,} \\ \frac{1}{4}\frac{1}{\sqrt{3(1-\mu^2)}}\frac{Eh(t)}{R}, \text{ large deflection buckling.} \end{cases} \tag{5}$$

$$q_{cr}(t) = \frac{2h(t)}{R}\sigma_{cr}(t) = \begin{cases} \frac{2E}{\sqrt{3(1-\mu^2)}}\frac{h(t)^2}{R^2}, \text{ small deflection buckling,} \\ \frac{E}{2\sqrt{3(1-\mu^2)}}\frac{h(t)^2}{R^2}, \text{ large deflection buckling.} \end{cases} \tag{6}$$

2.3 Other Required Unknown Conditions or Governing Equations

2.3.1 A More General and Improved Kinetic Corrosion Model

We propose an improved empirical model, considering the following equation, which has a different but optimized structure comparing the existing models [8]:

$$\frac{dh}{dt} = Ant^{n-1} exp\frac{V\sigma}{RT}. \tag{7}$$

2.3.2 Material Properties of Structural Steel at Varying Temperatures

The elastic modulus expression of steel at high temperature is the following [11]:

$$E(T_s) = \begin{cases} \frac{7T_s - 4780}{6T_s - 4760} E_s, & 20^\circ C \le T_s < 600^\circ C, \\ \frac{1000 - T_s}{6T_s - 2800} E_s, & 600^\circ C \le T_s < 1000^\circ C. \end{cases} \tag{8}$$

where, T_s is the temperature of steel; E_{Ts} is the elastic modulus of steel at high temperature (N/mm^2); E_{s} is the elastic modulus of steel at normal temperature, it can be taken as 206 GPa [10] according to the literature [11].

2.3.3 Relationship Between Temperature Stress and Constraints

Under arbitrary elastic constraints [12], the temperature stress equation of plates and cylindrical shells are the following.

Plate:

$$\sigma_{cr}(t_c, T_s) = \frac{k\pi^2 \cdot h^2(t_c) \cdot E(T_s)}{12(1-\mu^2)b^2} - k_r \alpha_T (T_s - T_{s0}) E(T_s). \tag{9a}$$

Cylindrical shell:

$$\sigma_{cr}(t_c, T_s) = \begin{cases} \frac{h(t_c)}{R} \frac{E(T_s)}{\sqrt{3(1-\mu^2)}} - k_r \alpha_t (T_s - T_{s0}) E(T_s), \\ \frac{1}{3} \frac{h(t)}{R} \frac{E(T_s)}{\sqrt{3(1-\mu^2)}} - k_r \alpha_t (T_s - T_{s0}) E(T_s). \end{cases} \tag{9b}$$

The spherical shell is considered free and unconstrained, so its temperature stress is:

$$\sigma_{cr}(t_c, T_s) = \begin{cases} \frac{h(t_c)}{r} \frac{E(T_s)}{\sqrt{3(1-\mu^2)}}, \\ \frac{1}{4} \frac{h(t)}{r} \frac{E(T_s)}{\sqrt{3(1-\mu^2)}}. \end{cases} \tag{9c}$$

2.3.4 Governing Equation of Fire Temperature Curve

Fire temperature curve [13] shows:

$$T_s = (\sqrt{0.044 + 5.0 \times 10^{-5} \alpha \frac{F_i}{V}} - 0.2)t + T_{s0} \; T_s \le 700^\circ\mathrm{C}. \tag{10}$$

In the equation, t is fire duration (s); T_{s0}-initial temperature of steel member before fire, which can be taken as 20 °C; F_{i} -fired surface area per unit length of steel member with fire protection (m^2); V-unit length of steel member volume (m^3); α -comprehensive heat transfer coefficient [W/ (m^2 °C)].

3 Stability Calculation Method for Thin Shells in Corrosion Environment

The stability conditions are deduced in this section according to Eqs. (1)–(10).

3.1 Plate

The stress of the plate can be expressed as,

$$\sigma(t) = p_0/h(t). \tag{11}$$

Substituting Eq. (11) into (7) gets:

$$\frac{dh}{dt} = -Ant^{n-1}exp\frac{V\sigma}{RT} = -Ant^{n-1}exp\frac{Vp_0}{RTh}. \tag{12}$$

We integrate it in the range of h_0 - h, t_0- t to get:

$$t = [-\frac{1}{A}\int_{h_0}^{h} exp(-\frac{Vp_0}{RTh})dh]^{\frac{1}{n}}. \tag{13}$$

In the equation $t = f(h) = [-1/A\int_{h_0}^{h} exp(-\frac{Vp_0}{RTh})dh]^{\frac{1}{n}}$, f expresses the corresponding rule between variable h and variable t.

Since the inverse operation exists, $h = f^{-1}(t)$, substituting it into Eq. (9a) we obtain:

$$\begin{cases} t = f(h) = [-\frac{1}{A}\int_{h_0}^{h} exp(-\frac{Vp_0}{RTh})dh]^{\frac{1}{n}}, \\ h = f^{-1}(t), \\ \sigma_{x,cr}(t) = \frac{k\pi^2 E}{12(1-\mu^2)}\frac{[f^{-1}(t)]^2}{b^2}, \\ p_{x,cr}(t) = \sigma_{x,cr}(t)\cdot h(t) = \frac{k\pi^2 E}{12(1-\mu^2)}\frac{[f^{-1}(t)]^3}{b^2}. \end{cases} \tag{14}$$

3.2 Cylindrical Shells

The stress of the cylindrical shell is expressed as:

$$\sigma(t) = \frac{N}{2\pi r\cdot h(t)}. \tag{15}$$

Considering Eq. (7), we have:

$$\frac{dh}{dt} = -Ant^{n-1}exp\frac{V\sigma}{RT} = -Ant^{n-1}exp\frac{VN}{2\pi rhRT}. \tag{16}$$

Integrating it over h_0 to h and 0 to t gives:

$$\int_{h_0}^{h} -\frac{1}{A}exp(-\frac{VN}{2\pi rhRT})dh = \int_{0}^{t} nt^{n-1}dt. \tag{17}$$

After integral operation, we get:

$$t = [-\frac{1}{A}\int_{h_0}^{h} exp(-\frac{VN}{2\pi rhRT})dh]^{\frac{1}{n}}. \tag{18}$$

Let $t = f(h) = [-1/A\int_{h_0}^{h} exp(-\frac{VN}{2\pi rhRT})dh]^{\frac{1}{n}}$, where f express the correspondence rule between variable h and variable t.

Substituting $h = f^{-1}(t)$ into above equations we obtain:

$$\begin{cases} t = f(h) = [-\frac{1}{A}\int_{h_0}^{h} exp(-\frac{VN}{2\pi rhRT})dh]^{\frac{1}{n}}, \\ h = f^{-1}(t), \\ \sigma_{cr}(t) = \begin{cases} \frac{E}{\sqrt{3(1-\mu^2)}}\frac{f^{-1}(t)}{r}, \text{ small deflection buckling,} \\ \frac{1}{3}\cdot\frac{E}{\sqrt{3(1-\mu^2)}}\frac{f^{-1}(t)}{r}, \text{ large deflection buckling.} \end{cases} \\ p_{cr}(t) = h(t)\cdot\sigma_{cr}(t), \\ N_{cr}(t) = 2\pi r\cdot h(t)\cdot\sigma_{cr}(t). \end{cases} \tag{19}$$

Among them, f^{-1} is the inverse of function f.

3.3 Spherical Shell

The stress of the spherical shell is expressed as:

$$\sigma(t) = \frac{qr}{2h(t)}. \tag{20}$$

Considering Eq. (7), we have:

$$\frac{dh}{dt} = -Ant^{n-1}exp\frac{V\sigma}{RT} = -Ant^{n-1}exp\frac{Vqr}{2hRT}, \tag{21}$$

integrating over h_0 to h and t_0 to t:

$$\int_{h_0}^{h} -\frac{1}{A}exp(-\frac{Vqr}{2hRT})dh = \int_{0}^{t} nt^{n-1}dt, \tag{22}$$

we find:

$$t = [-\frac{1}{A}\int_{h_0}^{h} exp(-\frac{Vqr}{2hRT})dh]^{\frac{1}{n}}. \tag{23}$$

In the equation $t = f(h) = [-1/A\int_{h_0}^{h} exp(-\frac{Vqr}{2hRT})dh]^{\frac{1}{n}}$, f expresses the corresponding rule between variable h and variable t.

Substituting $h = f^{-1}(t)$ it into above equations we obtain:

$$\begin{cases} t = f(h) = [-\frac{1}{A}\int_{h_0}^{h} exp(-\frac{Vqr}{2hRT})dh]^{\frac{1}{n}}, \\ h = f^{-1}(t), \\ \sigma_{cr}(t) = \begin{cases} \frac{E}{\sqrt{3(1-\mu^2)}}\frac{f^{-1}(t)}{r}, \text{ small deflection buckling,} \\ \frac{1}{4}\frac{E}{\sqrt{3(1-\mu^2)}}\frac{f^{-1}(t)}{r}, \text{ large deflection buckling,} \end{cases} \\ q_{cr}(t) = \frac{2h(t)\sigma_{cr}(t)}{r}. \end{cases} \tag{24}$$

4 Calculation Method of Stable Bearing Capacity of Thin Shell Under Corrosion Temperature Change

This section is based on the equations deduced in the Sect. 3, and then considers the influence of variable temperature (stress) (Eq. (9a), Eq. (9b), Eq. (9c)) to continue the derivation.

The relevant results are the following.

4.1 Plate

The critical stress expression when the plate loaded edge is subject to arbitrary elastic constraints is:

$$\begin{cases} \frac{dh}{dt_c} = Ant_c{}^{n-1} exp\frac{V\sigma}{RT_g}, \\ E(T_s) = \frac{7T_s-4780}{6T_s-4760}E_s, 20°C \le T_s < 600°C, \\ \sigma_{cr}(t_c, T_s) = \frac{k\pi^2 \cdot h^2(t_c)\cdot E(T_s)}{12(1-\mu^2)b^2} - k_r\alpha_T(T_s - T_{s0})E(T_s), \\ p_{cr}(t_c, T_s) = \sigma_{cr}(t_c, T_s)\cdot h(t_c). \end{cases} \tag{25}$$

4.2 Cylindrical Shell

When the two loaded sides of a cylindrical shell are subjected to arbitrary elastic constraints, its critical stress can be expressed as:

$$\begin{cases} \frac{dh}{dt_c} = Ant_c{}^{n-1} exp\frac{V\sigma}{RT_g}, \\ E(T_s) = \frac{7T_s-4780}{6T_s-4760}E_s, 20°C \le T_s < 600°C, \\ \sigma_{cr}(t_c, T_s) = \begin{cases} \frac{h(t_c)}{R}\frac{E(T_s)}{\sqrt{3(1-\mu^2)}} - k_r\alpha_T(T_s - T_{s0})E(T_s), \text{ small deflection buckling} \\ \frac{1}{3}\cdot\frac{h(t_c)}{R}\frac{E(T_s)}{\sqrt{3(1-\mu^2)}} - k_r\alpha_T(T_s - T_{s0})E(T_s), \text{ large deflection buckling} \end{cases} \\ p_{cr}(t_c, T_s) = h(t_c)\cdot\sigma_{cr}(t_c, T_s), \\ N_{cr}(t_c, T_s) = 2\pi r\cdot h(t_c)\cdot\sigma_{cr}(t_c, T_s). \end{cases} \tag{26}$$

4.3 Spherical Shell

The spherical shell is considered as a free structure in this paper. Therefore, its critical stress can be expressed as:

$$\begin{cases} \frac{dh}{dt_c} = Ant_c{}^{n-1} exp\frac{V\sigma}{RT_g}, \\ E(T_s) = \frac{7T_s-4780}{6T_s-4760}E_s, 20°C \leq T_s < 600°C, \\ \sigma_{cr}(t_c, T_s) = \begin{cases} \frac{h(t_c)}{r}\frac{E(T_s)}{\sqrt{3(1-\mu^2)}}, \text{small deflection buckling}, \\ \frac{1}{4}\frac{h(t)}{r}\frac{E(T_s)}{\sqrt{3(1-\mu^2)}}, \text{large deflection buckling}, \end{cases} \\ q_{cr}(t_c, T_s) = \frac{2\cdot h(t_c)\cdot\sigma_{cr}(t_c,T_s)}{r}. \end{cases} \quad (27)$$

5 Calculation Method of Stable Bearing Capacity of Corroded Thin Shells Under Variable Temperature in Fire

This section considers the fire control curve, Eq. (16), on the basis of the equations already derived in Sect. 4, while ignoring the change of wall thickness h during the fire, that is, $h(t_c) = h(t_0)$ (because usually the fire time is extremely short relative to the corrosion period), and the derivation is carried out.

We easily obtain its result as shown below.

5.1 Plate

$$\begin{cases} T_s = (\sqrt{0.044 + 5.0\times10^{-5}\alpha\frac{F_i}{V}} - 0.2)t_s + T_{s0}, T_s \leq 700^oC, \\ E(T_s) = \frac{7T_s-4780}{6T_s-4760}E_s, 20°C \leq T_s < 600°C, \\ \sigma_{cr}(t_c, T_s) = \frac{k\pi^2\cdot h^2(t_c)\cdot E(T_s)}{12(1-\mu^2)b^2} - k_r\alpha_T(T_s - T_{s0})E(T_s), 20°C \leq T_s < 600°C. \end{cases} \quad (28)$$

Among them, the third equation can also be written as:

$$\sigma_{cr}(t_c, t_s) = \frac{k\pi^2\cdot h^2(t_c)\cdot E(t_s)}{12(1-\mu^2)b^2} - k_r\alpha_T(T_s - T_{s0})E(t_s). \quad (29)$$

5.2 Cylindrical Shell

The calculation method for the stable bearing capacity of a thin cylindrical shell under variable temperature in fire is as follows:

$$\begin{cases} T_s = (\sqrt{0.044 + 5.0 \times 10^{-5}\alpha\frac{F_i}{V}} - 0.2)t + T_{s0}, T_s \leq 700^oC, \\ E(T_s) = \frac{7T_s - 4780}{6T_s - 4760}E_s, 20°C \leq T_s < 600°C, \\ \sigma_{cr}(t_c, T_s) = \frac{1}{3}\cdot\frac{h(t_c)}{R}\frac{E(T_s)}{\sqrt{3(1-\mu^2)}} - k_r\alpha_T(T_s - T_{s0})E(T_s), 20°C \leq T_s < 600°C. \end{cases} \quad (30)$$

The third equation can also be written as:

$$\sigma_{cr}(t_c, t_s) = \frac{1}{3}\cdot\frac{h(t_c)}{R}\frac{E(t_s)}{\sqrt{3(1-\mu^2)}} - k_r\alpha_T(T_s - T_{s0})E(t_s). \quad (31)$$

5.3 Spherical Shell

The calculation method of the stable bearing capacity of a thin spherical shell under fire temperature changes is as follows:

$$\begin{cases} T_s = (\sqrt{0.044 + 5.0 \times 10^{-5}\alpha\frac{F_i}{V}} - 0.2)t_s + T_{s0}, \ T_s \leq 700^oC \\ E(T_s) = \frac{7T_s - 4780}{6T_s - 4760}E_s, \ 20°C \leq T_s < 600°C, \\ \sigma_{cr}(t_c, T_s) = \frac{1}{4}\frac{h(t_c)}{r}\frac{E(T_s)}{\sqrt{3(1-\mu^2)}}, \ 20°C \leq T_s < 600°C. \end{cases} \quad (32)$$

The third equation can also be written as:

$$\sigma_{cr}(t_c, t_s) = \frac{1}{4}\frac{h(t_c)}{r}\frac{E(t_s)}{\sqrt{3(1-\mu^2)}}. \quad (33)$$

6 Conclusions

In this paper, a more general corrosion kinetic model is first proposed. Then, based on the model, starting from the basic stability theory, the calculation method of the stable bearing capacity of the thin shell under different corrosion temperature is deduced by combining the temperature stress and the change of steel physical properties under different constraints. Finally, considering the special variable temperature-fire control curve, the calculation method of the stable bearing capacity of the fire variable temperature corroded thin shell is deduced.

References

1. Gutman, E.M., Zainullin, R.S., Zaripov, R.A.: Kinetics of mechanical chemical failure and the life of constructional elements in tension in elastoplastic deformations. Soviet Mater. Sci. **20**, 101–103 (1984)
2. Gutman, E.M., Haddad, J., Bergman, R.: Stability of thin-walled high-pressure cylindrical pipes with non-circular cross-section and variable wall thickness subjected to non-homogeneous corrosion. Thin-Walled Struct. **43**, 23–32 (2005)
3. Gutman, E.M., Bergman, R.M., Levitsky, S.P.: Influence of internal uniform corrosion on stability loss of a thin-walled spherical shell subjected to external pressure. Corros. Sci. **111**, 212–215 (2016)
4. Liu, C.H., Lacidogna, G.A.: Non-destructive method for predicting critical load, critical thickness and service life for corroded spherical shells under uniform external pressure based on NDT data. Appl. Sci. **13** (2023)
5. Pronina, Y.G.: Analytical solution for decelerated mechanochemical corrosion of pressurized elastic-perfectly plastic thick-walled spheres. Corros. Sci. **90**, 161–167 (2015)
6. Pronina, Y., Sedova, O.: Analytical solution for the lifetime of a spherical shell of arbitrary thickness under the pressure of corrosive environments: the effect of thermal and elastic stresses. J. Appl. Mech. Trans. ASME **88** (2021)
7. Wu, L.: Plate and Shell Stability Theory, 1st edn. Huazhong University of Science and Technology Press, Wuhan (1996)
8. Liang, C., Hou, W.: Eight-year atmospheric exposure corrosion research on carbon steel and low alloy steel. Corros. Sci. Prot. Technol. **7**, 65–73 (1995)
9. Industry standard: Technical Code for Fire Protection of Building Steel Structures (GB51249-2017), China Planning Press, Beijing (2018)
10. Standard, N.: Technical Specifications for Steel Structures (GB 50017). China Architecture and Building Press, Beijing (2018)
11. Standard, N.: Code for Design of Steel Structures (GB50017-2003). China Planning Press, Beijing (2013)
12. Kang, J., Zhao, M., Jiang, Y., Jing, R.: Research on the relationship between degree of constraint and temperature stress. J. Hydraul. Archit. Eng. **12**, 21–25 (2014)
13. Technical Code for Fire Protection of Building Steel Structures (GB51249-2017). Beijing China Planning Press, Beijing (2016)

A Parametric Approach to the Morphological Structural Analysis of Thin Hyperbolic Paraboloid Shells

Martina Russo[1(✉)], Stefano Gabriele[2], Sigrid Adriaenssens[3], and Edoardo Currà[1]

[1] Sapienza Università Di Roma, Roma, Italy
martina.russo@uniroma1.it
[2] Università Degli Studi Di Roma Tre, Roma, Italy
[3] Princeton University, Princeton, NJ 08544, USA

Abstract. The undeniable fame achieved by thin concrete shells during the XX century is a well-known story. Their shapes represent the apex of the intersection between science and technology as well as a cultural distinctive character of modernity. The reasons for this enduring attraction on engineers and architects lay in the fact that the design of these geometry-based structures has a parametric nature. The identification of key parameters to control the pre-design phase has supported their design firstly in the realm of membrane theory and then, through the finite element calculation, in the bending theory, determining the extraordinary possibility of optimization that these structures still testify today. This paper illustrates the opportunity offered by the application of current parametric modeling and structural analysis tools to achieve a time-efficient structural evaluation of the influence of the morpho-typological choices make in the pre-design phase on the overall structural performance of hyperbolic paraboloid structures. Retracing the logic and processes of the past with new tools, the proposed analysis opens up new paths for future thin shell design applications.

Keywords: Thin concrete shells · Hyperbolic paraboloid · Geometry-based structures · Parametric modeling · Structural analysis

1 Introduction

The undeniable fame achieved by thin concrete shells during the 20th century is a well-known story. Their shapes represent the apex of the intersection between science and technology as well as a cultural distinctive character of modernity [1].

In recent years, thanks to the introduction of parametric modeling tools and their integration with algorithms for mechanical calculation, it is possible to study complex structures by visualizing, analyzing, and correlating quickly geometric choices with mechanical results.

Addressing the study and protection of the wide architectural heritage of thin concrete shells inherited from the 20th century, researchers cannot ignore the pivotal role of

S. Gabriele et al. (Eds.): IWSS 2023, LNCE 437, pp. 329–342, 2024.
https://doi.org/10.1007/978-3-031-44328-2_34

specific disciplines, such as mechanics and materials science, alongside the historical review, to achieve efficient strategies for their recovery and valorization. On the one hand, it is necessary to identify the role that these kinds of structures have had both in the history of architecture and construction and in the development of the history of engineering and science. On the other hand, it is essential to understand the specific calculations underlying their design, which have changed from decade to decade, with the evolution of the knowledge of their mechanical behavior and the mathematical tools available to engineers of the time.

Fig. 1. From top left, projects by: (a) Fernand Aimond, workshops for the School of Naval Mechanics in Rochefort, France, 1936; (b) Giorgio Baroni, warehouse for hemp storage in Tresigallo, Italy, 1939; (c) Felix Candela, Iglesia de la Medalla Milagrosa in Mexico City, Mexico, 1955; (d) Jack Christiansen, Mercer Island high school, Washington, USA, 1958; (e) Eduardo Catalano, Raleigh House, North Carolina, USA 1954; (e) Anton Tedesko, department store entrance hall in Denver, Colorado, USA, 1959.

As a consequence, approaching specific case studies of thin concrete shells is today a challenge to be carried out in partnership between historians and mechanical engineers, capable to retrace the historical context and control both archival research and analytical calculation. In this context, shells with hyperbolic paraboloid geometry (see Fig. 1) arouse a particular fascination, witnessed today by the recent numerous research focused on the beginning and diffusion of this constructive system, the main designers' bibliographies, and the stories of the most iconic construction and architectures [2–4].

The reasons for their enduring attraction on engineers and architects lay in the fact that the design of these geometry-based structures have a *parametric nature*. During the 20th century, the identification of key parameters to control the pre-design phase of hyperbolic paraboloid thin shell has supported their design firstly in the realm of membrane theory and then, through the finite element calculation, in the bending theory, determining the extraordinary possibility of optimization that these structures still testify today.

Starting from a long path of archival and historical literature research – undertaken by the authors from 2018 – both in the field of architectural and mechanical studies, this paper aims to widen the awareness of the complex structural behavior of hypar thin shells and the challenges that have been faced in achieving their understanding.

Combining the historical approach to structural analysis methodologies, this paper explores the opportunity offered by the application of current parametric modeling and structural analysis tools (i.e., Rhino, Grasshopper, SAP2000) to achieve a time-efficient structural evaluation of the influence of the morpho-typological choices make in the pre-design phase on the overall mechanical behavior. In particular, this work presents the pivotal role of historical research and findings in order to conduct a reliable and aware structural analysis on thin hypar shells.

Retracing the logic and processes of the past with new tools, the proposed analysis opens up new paths for future thin shell design applications and a more aware approach to their conservation and recovery.

2 Historical Studies and Findings

2.1 Geometry and Mechanics of Thin Hyperbolic Paraboloid Shell

A *shell* structure can be described geometrically as a curved three-dimensional surface. In mechanics, a shell is defined *thin* if one of its dimensions, the thickness t, is smaller than the radius of curvature r, so that the ratio between t/r is between $0.001 \leq t/r \leq 0.05$ m.

The mechanical model of the *membrane structure* can be applied to the geometric model of the *thin shell*. Based on the assumptions of the membrane theory, a structure following this model has no bending and torsional resistance. In this case, the shell structure will be characterized through its average surface and its thickness, and it will be considered able of carrying loads by tension, compression, and shear, neglecting its ability to carry loads by bending and torsion [5, 6].

This approximation allows the simplification of the equations for solving the mechanical problem of thin shells. The membrane theory applied to thin shells has opened up the possibility of analyzing not only the more classic three-dimensional forms – with single curvature (e.g., barrel vaults) and synclastic double curvature, (e.g., domes) – but also the more complex quadric geometries, such as surfaces with double anticlastic curvature.

This is the case of the hyperbolic paraboloid (also called *hypar*), a quadric surface with double anticlastic curvature – therefore with negative Gaussian curvature – which can be generated by translation. In fact, the hypar can be obtained by translating a parabola with downwards concavity along a parabola with upwards concavity; but also, by translating a straight line (i.e., the generatrix) along two other non-coplanar straight lines (i.e., the directrices, making it a so-called *ruled surface* [7].

Hypar structures can be classified by the way to cut and assemble part of the surface. The two main types can be considered the hypar bounded by parabolas and the one bounded by straight lines [7]. The latter has been considered a highly functional module that can originate numerous variations in the composition of shell roofs, thanks to the possibility of combining it in several different ways [1]. Among these, the *gable-type* hypar configuration was one of the most used – largely for industrial buildings –, due to the simplicity of the construction process needed and the high performance of the overall structure.

Being a ruled surface with double anticlastic curvature and being able to be analyzed using the simplifications of the membrane theory give the hypar evident constructive and mechanical advantages, which encouraged many 20th-century pioneering engineers to experiment with its application and design highly efficient structures of great span and extraordinary lightness.

2.2 The Evolution of the Theory for Thin Hyperbolic Paraboloid Shells Analysis

The structural analysis of thin shells has been tackled since the beginning of the 20th century in the membrane state and only starting from the seventies – with the advent of numerical calculation and finite element method [8] – in the bending state.

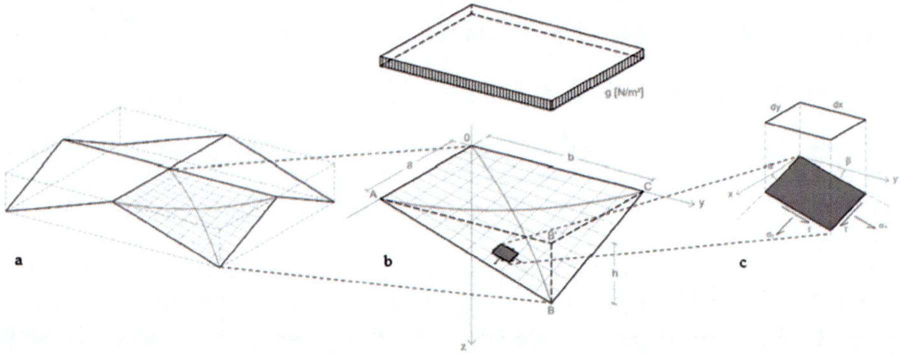

Fig. 2. The gable-type configuration of a shallow-shell roof, composed by four sections of hyperbolic paraboloid (a); each section can be defined by translating a straight line (A0) along two other non-coplanar straight lines (C0 and AB), with low inclination (b); membrane state on the shell with load uniformly distributed on the horizontal projection of the surface: bending and torsion are neglect (c).

The literature credits the French engineer Fernand Aimond (1902–1984) to be the first to have studied the behavior of a hypar structure, managing to define a mechanical model for the evaluation of stresses in self-supporting roofs with this shape [9, 10]. Considering a quadrilateral section of hypar (bordered by four straight lines, i.e., the generatrixes and directrices), with uniform thickness and uniformly loaded, Aimond was able to derive the equilibrium equations. In particular, in the hypothesis of a *shallow shell* (rice < span), Aimond observes that the stress at each point of the surface is proportional to the sine of the angle between the two generatrixes of the surface. In this case, in fact, the angle between the two generatrixes will have a very small variation, leading to the conclusion that there will be considerably homogeneous stress on it. Aimond thus demonstrated that hypar structures behave like a membrane structure, paving the way for a simplified approach to their calculation which determined their wide success (see Fig. 2).

In the same years as Aimond's research, the Austrian engineer Adolf Pucher (1902–1968), starting from his PhD obtained at the University of Graz in 1931, conducted in-depth research on the analysis of double-curvature geometries. Pucher investigated the possibility of finding the stress state of all kinds of doubly curved surfaces with a single calculation method.

In two articles published during the 1930s [11, 12], Pucher proposed the introduction of the Airy function within the equilibrium equations of an infinitesimal element part of a thin shell with double curvature. The Airy function would represent the membrane state of stress of an element in the case in which the load is limited only to the component along Z. In this way, the problem is reduced to the resolution of a single differential equation, in which the second unknown variable is the function that describes the shape of the surface. Once this function is known, the state of stress is obtained, so if the surface under examination is a known shape – such as the hypar – the differential equation can be solved easily.

It is interesting to note that Pucher's equation is symmetric with respect to the Airy stress function and the shape function. This makes it possible to solve both the direct problem (i.e., given the shape, find the stress with respect to the load condition) and the inverse problem (i.e., given the stress, find which shapes best perform with the given load conditions), which represents a first example of *form-finding* techniques, an analytics method to search for the optimal shape for structures [13, pp. 617–620].

Later studies extended the application of membrane theory. In the late forties, the German engineer K.G. Tester published a study, developed in 1944, on the analytical solution of a square-based hypar shell under multiple loading conditions, including wind, hydrostatic loading, loads of different shapes (i.e., triangular, trapezoidal, and parabolic), using the procedure presented by Pucher in his publications of 1934 and 1937 [14].

Since the end of the Second World War, after decades of supremacy by European designers in applying thin self-supporting shells (e.g. Aimond and Freyssinet in France, Torroja in Spain, Dischinger and Finsterwalder at the Dyckerhoff & Widmann in Germany, Giorgio Baroni, Dante Fornasir and Giulio Krall in Italy) the center of the architectural interest in this kind of structures and their mechanical behavior moved to America, while in Europe the interest shifts to prefabricated and prestressed systems [15].

The Spaniard architect, naturalized Mexican, Felix Candela (1910–1997), with his numerous and daring projects using hypar, definitively decreed the architectural success of these kinds of structures, behind their application for industrial constructions. In addition to his suggestive and innovative creations, Candela's contribution was also in the field of mechanical studies. In particular, among his numerous research, the publication on the calculation of *non-shallow* hypar allowed the application of the membrane theory to cases in which the sections of the hypar have a great inclination (rice > span) [16, 17]. For these *non-shallow* shell cases, Candela reformulates the equilibrium equations by introducing a correction coefficient for the load, since in this case it can no longer be considered as uniformly distributed on the horizontal projection of the surface, as Aimond did. In this regard, he formulated and presented the *K form factor*, as the ratio between rice r and the product of the two spans ($K = r / a * b$).

In the United States, engineer Alfred L. Parme (1909–1992), French by birth and naturalized American, thanks to his studies on the transmission of loads from the hypar shell sections to the edge beams [18, 19] and thanks to his work at the Portland Cement Association, decisively contributed to their widespread application.

Parme was part of the editorial board of the well-known hypar structures design guide [20], which also reports the first studies on the effects of secondary stresses, discussing the phenomenon called *degree of flatness*. The behavior variations in the hypar shell from what is predicted by the membrane theory are similar to what occurs in a two-hinged parabolic arch subjected to uniform loading if the rise/span ratio decreases. If the rise decreases, the bending moment of the arch becomes similar to that of a supported beam. The more the rise decreases, the less the membrane forces contribute to global statics.

2.3 From Membrane Theory to Bending Theory

Although numerous published studies have greatly contributed to clarifying the behavior of hypar and have supported engineers in their design, at the end of the 1960s there were still many unknown aspects of their behavior [21]. Considering the gable-type configuration (see Fig. 2a) – among the most frequently used –, although the presence of edge and ridge beams is necessary to transmit the forces to the edges of the shell, these introduce stresses that cannot be evaluated with membrane theory alone. Through their own weight, the stiffening beams introduce the bending moment at the edge of the shell, which is neglected in the membrane theory.

The membrane approach does not examine the interaction between the edge beams and the shell itself, considering the edge beams as a simple support for the shell. Although this assumption was considered realistic for a long period, the beam-shell system acts as a continuum in a structure of this kind. Investigating the occurrence of bending stresses at the geometrical discontinuity points is of primary importance. Furthermore, the bending state of a beam-shell system changes as the hypar configuration chosen varies. Specific support beams will be needed to design complex shapes. Each configuration of the support beams has a distinct effect on the overall behavior of the structure due to the specific bending moment introduced.

For these reasons, an important advance in the awareness of the behavior of thin hypar shells has been the introduction of *numerical calculation* and its implementation in the earliest software.

To solve the equations of the bending behavior of thin hypar shells bounded by straight lines [22, 23], starting from the beginning of the 1960s various numerical methods were attempted [24]. Among the methods that had the greatest impact on understanding the behavior of hypar shells, there were: the variational, finite difference, and finite element methods.

Starting from the theoretical studies of the Russian mathematicians Boris Galërkin and Leonid Kantorovič, the *variational methods* were useful for broadening the solutions to different boundary conditions and numerous rise/span ratios. Although they were efficient tools for the theoretical studies, the need to choose at the beginning of the analysis an appropriate function of the displacements with respect to the boundary condition of the specific case study made this approach not very competitive in the field of analysis with structural design purposes.

The *finite difference method* consists of replacing the derivatives that appear in the equations with the approximation of their finite difference. During the sixties, this method was applied to analyze a single hypar shell bounded by edge beams and then the inverted umbrella configuration, composed by it. Although this technique gave rise to several interesting studies for single-curved shells, it showed no convincing results for double-curved shells. Firstly, each structure requires its own formulation of the problem regarding the specific boundary conditions considered; as in the case of variational methods, this does not fit the needs of automatic calculation. Secondly, the accuracy of the results decreases progressively as the order of the derivatives increases. For these reasons, although this technique saw some use in the academies, it had limited application in practical design [24].

The *finite element method* (FEM) consists of splitting the shape of the shell (discretization phase) into smaller pieces, which approximate the overall shape of the surface into a mesh surface. This method demonstrated immediately the most accurate results for double-curved shells analysis and no specific limits in its implementation for automatic calculation. Starting from the mid-sixties, the application of this method became the most promising in the field of research on hypar thin shells.

It was immediately observed how the topology chosen for the discretization of the shape influenced the results. For this reason, numerous early studies were carried out using different types of mesh elements: triangular or quadrangular, flat or curved, with more or less control points [24].

Applying constant-strain flat triangular elements, the studies of Ray W. Clough (University of California, Berkeley) and C. Philip Johnson (University of Texas, Austin) have contributed significantly. First, they compared the limits and potential of the application of various types of elements. Then, they investigated the problem of compatibility with the edge beams, observing the need to intensify the mesh in the areas where the stresses showed rapid variations [25].

The curved elements were tested to check if the geometrical approximation of the shell to a mesh made up of flat elements led to an inaccurate solution. The results of their application were entirely compliant with those obtained with flat elements, nevertheless showed greater complexity of managing the calculation [26]. The curved elements had little application success compared to the flat elements with five degrees of freedom per node. Introducing an acceptable approximation, these elements offered a significant reduction in time for preparing the software input and for elaborating the calculation. Starting from the seventies, this kind of element was among those most used in the field of hypar shells and allowed to obtain significant results (see Sect. 3.3) [27, 28].

3 Parametric Modeling

It is possible to achieve a time-efficient structural analysis of hypar thin shells by taking advantage of their *parametric nature*. Considering the opportunities offered by software and tools currently available based on generative algorithms and parametric modeling, the authors set the methodology described in Fig. 3. The sequence of steps presented has been planned specifically for the analysis of gable-type hypar configuration (see Fig. 2a); nevertheless, it can be replicated for other hypar configurations.

The process provided here offers the possibility of rapidly obtaining a wide variety of geometric models to carry out the desired mechanical analyses for both ideal structures and specific case studies structures identified previously as of interest. Being able to compare the results on several different geometries, the algorithm elaborated for this

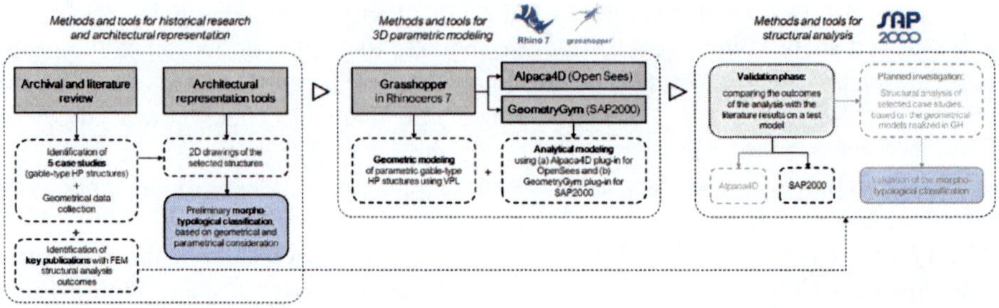

Fig. 3. Methodological workflow chart of the research process, describing the relation between disciplines, aims, software used, and results. The grey parts are not described in detail in this paper.

study contributes to carrying out morpho-typological studies that investigate how some key geometric parameters – underlying the pre-design phase of these structures – have determined their extraordinary possibility of optimization.

For the purpose of this paper, it will be presented just some phases of the overall workflow elaborated, highlighting the importance of the historical research and its findings for planning a reliable analysis of the hypar shells.

3.1 Historical and Architectural Research: Case Studies Selection and Analysis

The literature research was organized with two aims: 1. Identify historical publications that reported the results of FEM analysis that could be used to validate the results of the analyzes to be performed; 2. Choose some specific case studies among those realized during the 20th century using the gable-type hypar configuration with different geometrical characteristics (i.e., rise, span, shell thickness, and beams' sections dimensions) (see Fig. 4).

The collected data were organized in a database (see a sum in Table 1) and used to represent the chosen case studies through schematic drawings, highlighting some key geometrical characteristics for a preliminary morphological classification. Among them, Candela's *K form factor* was calculated for each case study.

3.2 Characteristics of the Geometric and Analytical Model

For the parametric modeling, the authors chose the visual programming language Grasshopper [29], implemented in McNeel's Rhinoceros 7 [30]. This environment is a graphical algorithm editor that allows programming the generation of geometrical shapes through control parameters.

Numerous plug-ins are available in Grasshopper to generate and solve mechanical models. For the purposes of this study, the authors explored the possibility offered by Alpaca4D [31] and Geometry Gym [32]. Alpaca4D is a plug-in exploits the advantages of the visual programming language to use the solution engine of OpenSees, an open-source software for finite element calculation, able to give reliable and accurate solutions for linear and non-linear shell analysis. Geometry Gym plug-in allows the connection of

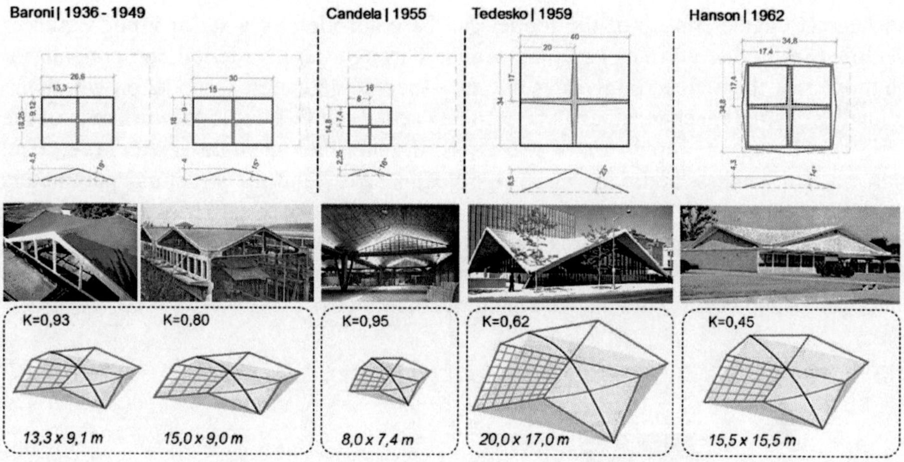

Fig. 4. The five projects chosen as case studies. From left to right: Felix Candela, Herdez factory warehouse, Mexico City, Mexico, 1955; Giorgio Baroni, after-work theater of the Vanzetti foundry, Milan, Italy, 1936; Giorgio Baroni, warehouse for the iron at Alfa Romeo, Milano, Italy, 1939; Thomas A. Hanson & Associates, Tucker's high school gym, Virginia, USA, 1962; Anton Tedesko, department store entrance hall, Denver, Colorado, USA, 1959.

Table 1. Summary of the geometrical characteristics of the 5 case studies chosen.

	Shell					Edge beams		Ridge beams	
	Span a	Span b	Rise	Thick	K	h	l	h	l
Baroni 1936	13.30	9.10	4.50	0.050	**0.93**	0.45	0.25	0.10	0.7
Baroni 1949	15.00	9.00	4.00	0.050	**0.80**	0.50	0.30	0.15	0.7
Candela 1955	8.00	7.40	2.25	0.066	**0.95**	0.61	0.31	0.10	0.71–0.41
Tedesko 1959	20.00	17.00	8.55	0.085	**0.63**	0.50–0.40	0.22	0.25	2.12–1.20
Hanson 1962	15.50	15.50	4.30	0.081	**0.45**	0.60–0.40	0.35	0.38	2.25–0.70

the parametric environment of Grasshopper with structural analysis platforms, such as SAP2000, to exploit the own features of each software. Using Alpaca4D the results can be visualized in the Rhinoceros environment and exported in Excel through TT Toolbox components; using Geometry Gym the results are visualized in SAP2000 and can be exported through it.

After the identification of suitable tools, it was prepared the algorithm for the geometric and mechanical modeling of a generic gable-type hypar. The geometrical and mechanical characteristics of the model can be controlled by a set of input variables identified in the literature as key parameters. Through the instruction components for the modeling, the algorithm gives as output a specific geometrical model on which the following desired mechanical analysis can be carried out (see Fig. 5). In the case of the gable-type hypar, the inputs are of two types: the morphological parameters (i.e., rise, span, shell thickness, and edge and ridge beams sizes) and the structural parameters (i.e., characteristics of the material, mesh subdivision, loads, supports, type of analysis). The results can include beams stresses (e.g., axial forces, bending moments), shell stresses (e.g., in-plane membrane forces, in-plane share forces), deformed model, normal deflection of the beams, and displacement of shell nodes.

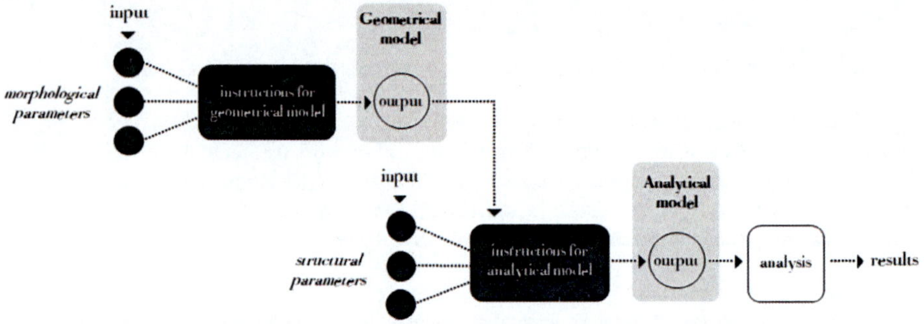

Fig. 5. Data flow of the parametric algorithm implemented in Grasshopper. The algorithm is composed of two sections: the first part controls the geometrical model through morphological parameters; the second applies mechanical parameters and solves the analytical model.

3.3 Validation of the SAP2000 Analytical Results

The outcomes of the mechanical analysis have been validated, comparing them with those published in the historical literature. Two papers published during the seventies applied the FEM to the same case study of gable-type hypar shell, obtaining consistent results with each other. These publications emerge for the quality, relevance, and extension of the analysis developed (see Table 2), which have been used by the authors to confirm the results acquired in SAP2000.

As an example, Fig. 6 shows the comparison between the results of axial forces in an edge beam with the same size in Schnobrich [27], Shaaban-Ketchum [28], and SAP2000. The same comparison has been accomplished for each analysis reported in Table 2.

The publications of Schnobrich and Shaaban-Ketchum also achieved important outcomes in terms of hypar behavior awareness, useful to plan the following phases of experimentation and morphological analysis of the present work.

Schnobrich's paper [27] deepens the understanding of stress development in edge and ridge beams. His publication highlights: 1. How the weight of the ridge and edge beams has a primary effect on the overall behavior of the roof; 2. How in the center of

Table 2. Analysis developed in Schnobrich 1972 and Shaaban-Ketchum 1976.

	Schnobrich, 1972	Shaaban and Ketchum, 1976
Beams	Normal deflections - Ridge Beams	
	Normal deflection - Edge Beams	
	Axial Forces - Ridge Beams	
	Axial Forces - Edge Beams	Axial Forces - Edge Beams
	Bending Moment - Ridge Beams	
	Bending Moment - Edge Beams	
Shell		In-plane membrane forces
	In-plane shear forces	In-plane shear forces
		Bending moment
		Direction of principal stresses in membrane
		Plan of deformed structures
		Principal compressive stresses in diagonal arch

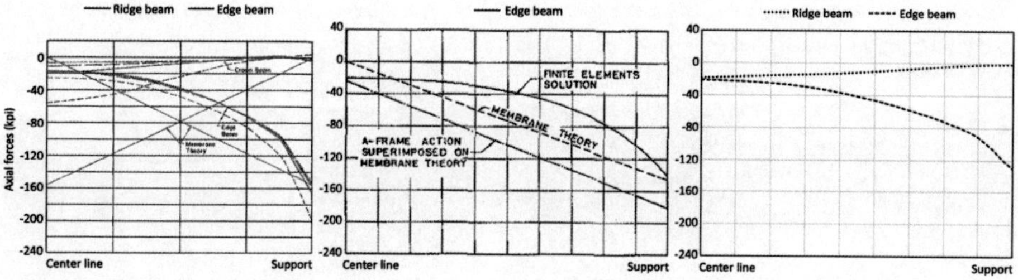

Fig. 6. Comparison between the results of axial forces in edge beams in (from left to right): Schnobrich, 1972; Shaaban and Ketchum, 1976; SAP2000.

the roof it is the shell that carries the loads of the ridge beams and not the contrary, as inevitably considered in the membrane theory.

The publication by Shaaban and Ketchum [28] investigates more in-depth the distribution of shell stresses. It observes: 1. How these stresses are not constant in the shell, as the membrane theory predicts; 2. How the action of the compressed arch is greater than that of the tension arch.

Billington [7] summarized and commented the results of the above papers, pointing out some key suggestions for the pre-design phase of these shells. About the overall morphology of the roof, he considers pivotal the geometry of the diagonal arch, since the gable-type hypar shell acts primarily as two diagonal arches crossing at the central area that transfer the entire load directly to the supports at the corners. In line with Parme results [20], the morphology of the compressed arches can predict the overall behavior of these roofs. About the edge and ridge beams, he strongly suggests to the designers to reduce their sizes – and consequence their loads – to the minimum necessary. Although the ridge beams are essential to stiffen the shell in the flat central area, they need to be light.

3.4 Planned Investigation

In view of these considerations could be of interest to explore the influence of the diagonal arch morphology on the overall mechanical behavior of this kind of shell to highlight if it can be identified the optimal value of the *K form factor* and if it depends on other geometrical features.

Using the parametric algorithm described, it will be possible to quickly control the effects of Billington's pre-design suggestions on the overall behavior of a gable-type hypar shell with a FEM structural analysis, checking how vary the principal compressive stress in the diagonal arch, the bending moment of ridge beams, and the deflection of the center of the roof to the variation of *K form factor* (see Fig. 7).

This process also contributes to the morphological classification of the case studies chosen, helping to understand how much their designers approached the optimal configuration using different design methodologies available during the 20th century.

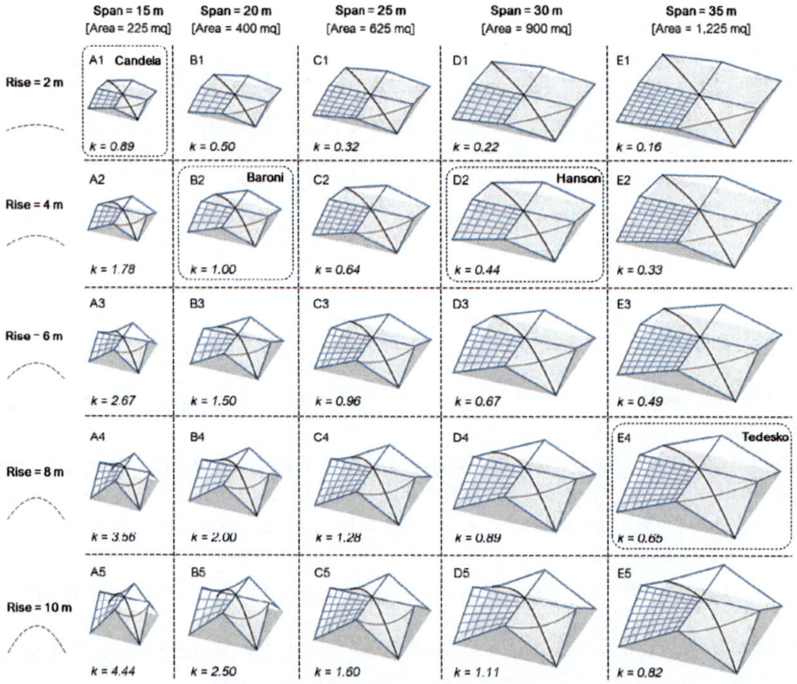

Fig. 7. Different morphological configurations of the gable-type hypar shell based on the K form factor; some of them approximate the configuration used by the designer of the case studies chosen.

4 Conclusion

The widespread diffusion of thin concrete shells during the XX century has left us an extensive architectural heritage to look after and a valuable example of structural optimization on which to reflect even today.

The present paper retraces the history of the studies for the mechanical calculation of these structures, firstly with the simplifications of the membrane theory, then – thanks to the numerical calculation and the advent of informatics – through the bending theory. By identifying the protagonists of this long adventure and their most important contributions, this research highlights some key studies to better understand the expected structural behavior of the numerous hypar roofs designed during the XX century, identifying knowledge and methods available at that time.

Acquainted with the parametric nature of these structures, the present work proposes a methodological process exploiting parametric modeling and contemporary structural analysis tools to retrace the genesis of the pre-design geometric choices in relation to mechanical behavior. The combination of the methodological approach of construction history and structural mechanics demonstrates how it is possible today, thanks to new tools and new awareness, better approach their study in order to correctly preserve the completed works and provide new ideas for future applications.

References

1. Joedicke, J.: Shell Architecture. Reinhold Publishing Corporation, New York (1963)
2. May, R.: Shell sellers. The international dissemination of the Zeiss-Dywidag system, 1923–1939. In: Friedman, D., Bowen, B., Leslie, T. (eds.) Proceeding of the 5th International Congress on Construction History (5ICCH, June 2015, Chicago, Illinois), pp. 557–565. Construction History Society of America (2015)
3. Espion, B.: Pioneering hypar thin shell concrete roofs in the 1930s. Beton und Stahlbetonbau **111**(3), 159–165 (2016)
4. Russo, M.: Costruire con leggerezza. Giorgio Baroni e la forma dell'ingegneria. LetteraVentidue, Siracusa (2023)
5. Flugge, W.: Stresses in Shells. Springer-Verlag, Berlin (1973)
6. Gould, P.L.: Analysis of Shells and Plates. Springer, New York (1988)
7. Billington, D.P.: Thin Shell Concrete Structures. McGraw-Hill, New York (1982)
8. Turner, M.J., Clough, R.W., Martin, H.C., Topp, L.C.: Stiffness and deflection analysis of complex structures. J. Aeronaut. Sci. **23**, 805–882 (1956)
9. Aimond, F.: Les voiles minces en forme de paraboloïde hyperbolique. Le Génie Civil **8**, 179–181 (1933)
10. Aimond, F.: Etude statique des voiles minces en paraboloïde hyperbolique travaillant sans flexion. In: International Association for Bridge and Structural Engineering, IABSE Proceedings, vol. 4, pp. 1–112 (1936)
11. Pucher, A.: Uber den Spannungszustand in Gekrümmten Flachen. Beton und Eisen **33**(19), 298–304 (1934)
12. Pucher, A.: Die Berechnung von doppelt gekrümmten schalen mittels Differenzengleichungen. Der Bauingenieur **18**(9–10), 118–123 (1937)
13. Benvenuto, E.: La scienza delle costruzioni e il suo sviluppo storico. Sansoni, Firenze (1981)
14. Tester, K.G.: Beitrag zur Berechnung der hyperbolischen Paraboloidschale. Ingenieur-Archiv **16**, 39–44 (1947)

15. Giannetti, I., Russo, M.: Gusci sottili in cemento armato: invenzioni e innovazioni tecnologiche per gli edifici industriali (1920–1970). Patrimonio Industr. **23** (2021)
16. Candela, F.: Discussion - warped surfaces: hyperbolic paraboloids. In: Proceedings of a Conference on Thin Concrete Shells (21–23 June 1954, MIT, Boston), pp. 91–98 (1954)
17. Candela, F.: General formulas for membrane stresses in hyperbolic paraboloidal shells. J. Am. Concr. Inst. **57**(16), 353–371 (1960)
18. Parme, A.L.: Hyperbolic paraboloids and other shells of double curvature. J. Struct. Div. **82**(5), 1–32 (1956)
19. Parme, A.L.: Shells of double curvature. Trans. Am. Soc. Civ. Eng. **123**(2951), 989–1013 (1958)
20. Portland Cement Association. Elementary Analysis of Hyperbolic Paraboloid Shells. Concrete Information, Chicago (1960)
21. Russo, M.: The collapse of the Tucker's gym: research impulses in the USA at the end of hypar shells era. In: Mascarenhas-Mateus, J., Pires, A.P. (eds.) History of Construction Cultures. Proceedings of the 7th International Congress on Construction History (7ICCH 2021, 12–16 July 2021, Lisbon, Portugal), vol. 1, pp. 392–399. Londra, CRC Press (2021)
22. Novozhilov, V.V.: The Theory of Thin Shells. Noordhoff Limited, The Netherlands (1959)
23. Apeland, K.: On the analysis of bending stresses in shallow hyperbolic paraboloidal shells. In: Proceedings, World Conference on Shell Structures (San Francisco, 1962), pp. 181–188. National Academy of Sciences, Washington DC (1964)
24. Schnobrich, W.C.: Analysis of hyperbolic paraboloid shells. In: Medwadowsky, S.J., Schnobrich, W.C., Scordelis, A.C. (eds.) Concrete thin shell, pp. 275–311. American Concrete Institute, Detroit (1971)
25. Clough, R.W., Johnson, C.P.: Finite element approximation for the analysis of thin shells. Int. J. Solids Struct. **4**(1), 43–61 (1968)
26. Clough, R.W., Johnson, C.P.: Finite element analysis of arbitrary thin shells. In: Medwadowsky, S.J., Schnobrich, W.C., Scordelis, A.C. (eds.) Concrete thin shell, pp. 333–363. American Concrete Institute, Detroit (1971)
27. Schnobrich, W.C.: Analysis of hipped roof hyperbolic paraboloid structures. J. Struct. Div. **98**(7), 1575–1583 (1972)
28. Shaaban, A., Ketchum, M.S.: Design of hipped HP shells. J. Struct. Div. **102**(11), 2151–2161 (1976)
29. Grasshopper. https://www.grasshopper3d.com/
30. Rhinoceros. https://www.rhino3d.com/
31. Alpaca4D. https://alpaca4d.github.io/index.html
32. Geometry Gym. https://geometrygym.wordpress.com/

On the Robustness of Tensegrity Systems Subjected to Local Damage

Valerio De Biagi[1](✉), Amedeo Manuello Bertetto[1], Andrea Micheletti[2], and Bernardino Chiaia[1]

[1] Department of Structural, Building and Geotechnical Engineering (DISEG), Politecnico Di Torino, Corso Duca Degli Abruzzi, 24, 10129 Torino, Italy
valerio.debiagi@polito.it

[2] Department of Civil and Computer Science Engineering, University of Rome Tor Vergata, Via Cracovia, 50, 00133 Roma, Italy

Abstract. Tensegrity systems are structures in equilibrium under a self-stress state prior to the application of external loads. Such self-stress state is the result of the superposition of self-stress basis vectors. Tensegrities have complex nonlinear behavior when subjected to external actions (loads or imposed displacements). In this work, the robustness of this particular structural typology is evaluated by numerical simulations. Robustness is the ability of a locally damaged structural system not to behave in a manner disproportionate to the original failure, i.e., the insensitivity to local damage. This concept and its insights have been thoroughly analyzed for several structural systems, as in the case of frame structures. To date, limited studies have been performed on the robustness of tensegrity systems to unexpected conditions. Among the potential damages, one can consider the settlement of one of the supports, the reduction of prestress and the failure of one of the cables. The present study reports the results of a set of damage simulations on a simple T-3 prismatic tensegrity structure. Several risk factors are considered to study the response of the structure to unexpected damage. In particular, the tensegrity configuration is subjected to different stress loss in the constituent elements. The results are obtained through simulations performed in MSC.visualNastran 4D (vN4D), which run dynamic simulations by merging CAD, motion, FEA, and controls technologies into a single functional modeling system. The models were established to obtain different equilibrium configurations and conditions of loss of equilibrium, with the corresponding large displacements measured for the most significant points of the structure, assumed as markers for the resulting effects on the structure.

Keywords: Tensegrity Systems · Robustness · Dynamic Simulations · Structural Damage · Prestress Reduction

1 Introduction

Buckminster Fuller and Snelson exhibited tensile structures with patents issued in 1962 [1] and 1965 [2], respectively. In the same period, Emmerich was studying similar structures [3]. In the beginning chapters of his 1974 book on geodesic dome geometry,

S. Gabriele et al. (Eds.): IWSS 2023, LNCE 437, pp. 343–350, 2024.
https://doi.org/10.1007/978-3-031-44328-2_35

the renowned literary critic Hugh Kenner [4] offered an insightful mathematical study of regular tensegrity prisms and spherical tensegrities. A companion book by Pugh [5] is essentially a how-to guide for constructing hundreds of regular tensegrity structures. Early findings on this topic are based on the works by Calladine, Tarnai, and Pellegrino [6–11], that discuss the basic mechanics of this kind of structures. The characterisation of rigid tensegrity configurations, which generalizes previous findings on the rigidity of pinned frames, has been the main focus of excellent mathematical research by Roth and Whiteley [12], Connelly [13], and Conneley and Whiteley [14]. There is also a corpus of research in cell biology, introduced by Ingber [15], that examines the mechanical behaviour of a tensegrity system that serves as the framework for the ctyoskeleton.

More recently, a few straightforward tensegrity structures, each of which has an analytical model, were analysed by Micheletti [16]. In this studies, a qualitative description of tensegrity system as a first-order mechanism were provided, together with static and dynamic analytical models (see also the literature referenced in [17]).

The focus of this paper's discussion will be on the findings regarding the robustness of simple tensegrity structures for applications in civil engineering and architecture.

In short, robustness is a structural property. According to Eurocode 1 – EN 1991–7 [18], "a localised failure due to accidental actions may be acceptable, provided it will not endanger the stability of the whole structure, and that the overall load-bearing capacity of the structure is maintained and allows necessary emergency measures to be taken". Hence, a robust structure is designed in order not to behave in a disproportionate way with respect to an initial localised failure. The concept, formulated after Ronan Point Building Accident in 1968 [19], has been applied to a large set of structures, mainly frame structures and bridges. The basic strategy for ensuring robustness in structures characterized by the presence of several elements that are continuously connected (such as in frame structures or in trusses) is to provide alternative load paths to ensure the transfer of forces when one or more elements are removed.

Although robustness studies have accounted for a multitude of structural types, so far, to the knowledge of the authors, extremely limited research has been performed on the robustness of tensegrity systems [20]. A tensegrity structure is a three-dimensional truss that has many tension members and few discontiguous compression members, that are in a pre-stressed condition at a certain geometry. While many patterns of connection may be used to assemble these structures, the simplest fundamental version has no more than one compression members and three tension members connected at each node.

The geometry we refer to as a T-3 prism has nine (tensile) cables and three (compressive) bars, making it the least complex of these spatial structures. There exist corresponding T-n prisms with $3n$ cables and n bars, and so on. It is a matter of fact that such a prism is a form-finding structure that becomes a stiff prestressed system as the last cable is pulled in tension. Our conventional use of Maxwell's rule does not adequately describe this structural type [6], and our traditional concepts of and statical determinacy, structural stiffness, and statical stability do not directly apply.

It has long been known that these prisms can be made to change their nodal geometry by appropriately adjusting the cable lengths, and it is because of this behavior that the prisms are compared to variable geometry truss manipulator in many applications.

We point out right away that there is no universal method for specifying the nodal locations that characterize a tensegrity state. For the situation of a right regular T-prism, a beginning state may be determined via symmetry and verified using straightforward statics. A 30 degree relative twist angle exists between the end faces of the T-3's two equilateral triangle end faces, which are parallel to one another and orthogonal to the axis connecting their centers. Every possible T-3 geometry can theoretically be achieved from that recognizable starting configuration by appropriately adjusting member lengths.

We shall demonstrate, however, that determining "suitable" (allowed) changes in member lengths is a difficult task in and of itself. We are motivated to investigate the appropriateness of tensegrity structures for robotic and civil engineering applications by a number of specific properties, including the following one: by tensioning the last cable, tenssegrity prisms may completely be erected from a low-volume bundle. This property is advantegeous for instance for erecting and dismantling temporary buildings.

A tensegrity tower might be quickly erected, designed to be strong and rigid enough to withstand the weight of the workers, who could then climb the tower and install more structural members to make a conventional tower framing. In an analogous way, a self-erecting prism could be remotely moved to a specific point inside a pipe system in its collapsed bundle condition, where it would subsequently take the self-stressed shape to create a crawling robot.

Because tensegrity structures typically include more cables, it is possible to obtain geometries with less total material weight than a traditional three-dimensional truss. Self-erecting capabilities and the anticipated weight economy point to applications for space constructions and robotics. Tensegrity structures present different families of shape-adaptable structures and corresponding machine kinematics when member lengths can be changed or when large elastic elongations are allowed. This has the advantage that actuation can be envisaged only for tension members, with simpler technological solutions than general bi-directional translational actuators.

The present study reports the results of a set of damage simulations on a simple prismatic tensegrity structure. Several risk factors are considered to study the response of the structure to unexpected damage. In particular, the T-3 structural configuration is subjected to different stress loss in the constituent elements. Results are obtained through simulations performed in MSC.visualNastran 4D (vN4D), which run dynamic simulations by merging CAD, motion, FEA, and controls technologies into a single functional modeling system. The models were established to obtain different equilibrium configurations and conditions of loss of equilibrium, with the corresponding large displacements measured for the most significant points of the structure, assumed as markers for the resulting effects on the structure.

2 Robustness of a Simple Tensegrity System to Local Damages

The first structure considered is the structure of the T-3 in its standard configuration. The realized model consists of rods 278 mm long with a diameter of 10 mm (elements a, b, c).

The top and base ropes have a length of 173 mm (u-1, u-2, u-3) and the diagonals have a length of 207 mm (d-1, d-2,d-3). These values correspond to a height *h* of the structure of 200 mm and a radius *r* of the circle circumscribed to the triagular bases of 200 mm. In the present analysis, we assume bars to be rigid and cables to be inextensible. The configuration shown in Fig. 1 below represents the stable equilibrium configuration characterized by a symmetrical loading condition at the vertices of rods A, B, C and pretension in all ropes u1–3, d1–3, f1–3. Ropes f1–3 take the same lengths of u1–3.

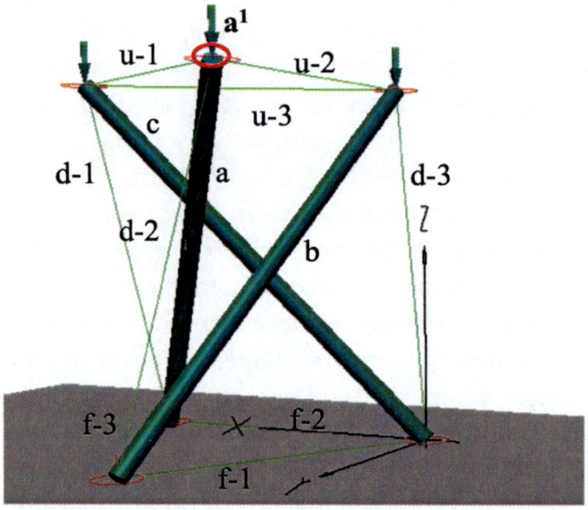

Fig. 1. Simple Tensegrity system T-3 configuration 3 rods and 9 ropes.

In order to be able to explore the collapse conditions as a function of reducing the degree of pretension of the ropes and also studying the relative level of associated slacking, we first proceed to increase the length of the rope d-1 passing to an increase equal to 1% from the initial length in pretension condition. The d-1 rope thus increases from a length of 207 mm to 209 mm. The effect is to achieve an equilibrium configuration close to the previous one that does not compromise the overall condition still far from the collapse condition.

By placing a position sensor on the vertex a^1 it is possible to obtains a new configuration with displacements of node a^1 (see Fig. 2) equal to 4% 3% and 2.5% of the height *h*, respectively for the x, y, z components. With an increase of the rope length d-1 by 2%, we still obtain a configuration without contact between the rigid elements (rods) a,b,c but with a displacement of node a^1 equal to 6% of 3.5% and 3% of the height *h*, respectively for the x,y,z components. The first configuration involving contact between the rigid rod elements is that correspondent to an increase in the length of the rope d-1 by 7% arriving at an overall length of 221 mm compared to the starting 207 mm. The collapse with overturning of the T-3 scheme occurs for an increase in rope length d-1 of 26% (see Fig. 2).

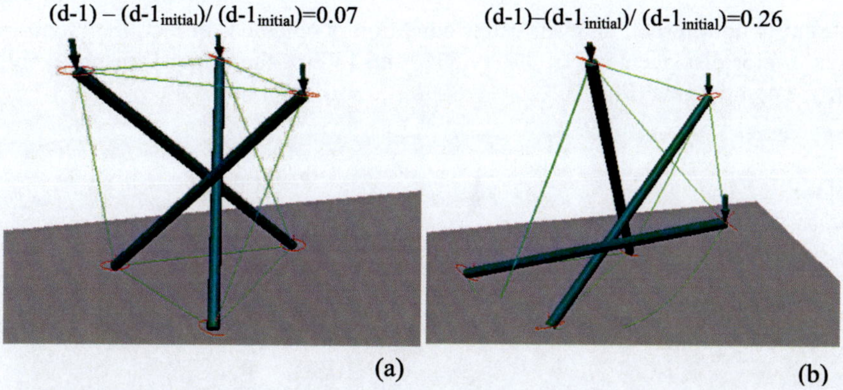

Fig. 2. Configuration of the T-3 basic system with one of the diagonal element d-1 with an increment of the length of 7% (a) and of 26% (b)

3 Robustness of a Reinforced Tensegrity Structure

In the configuration shown in Fig. 3, the stiffening ropes shown in red have been added. The effect of these additional ropes will be considered. In particular with respect to the (d-1)-(d-1initial)/ (d-1initial) = 0.26 configuration that resulted in the collapse condition in the basic version (Fig. 3a), a condition is obtained that does not result in the complete collapse of the system with the addition of the new ropes.

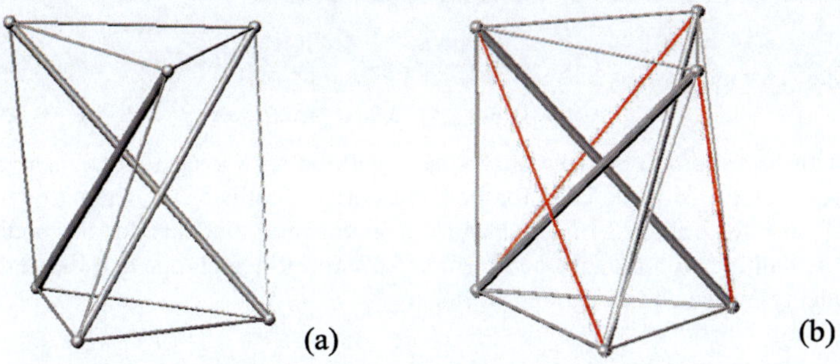

Fig. 3. Simple Tensegrity system T3 (a) and modified T3 with three additional diagonal ropes (b).

In particular, the stiffening ropes come into operation during the kinematics of the main mechanism, preventing tilting. Again, monitoring of the displacement of a^1, values of 5%, 24% and 16% of the height *h* are obtained respectively for the x,y,z components of the point a^1 with respect to the global reference system.

These results are obtained by considering the constituent bodies of the rods as bodies characterized by collision with each other and non-penetrating. This realizes an additional constraint of mutual support and friction (coulombian friction) between the three rods that come into contact at the central point of the system. In the case where the latter constraint is not present, an even better condition is obtained. In fact, the monitoring point a[1] yields displacements of 8% by 53% and 13% of the height h, respectively for the x, y, z components (Fig. 4).

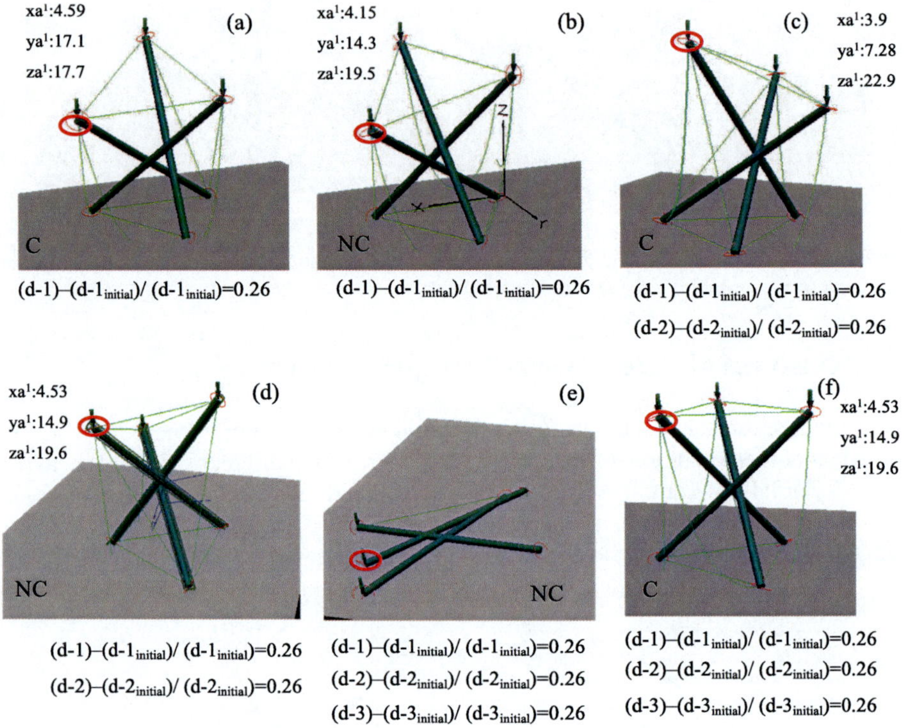

Fig. 4. T3 with three additional diagonal ropes: condition with a slack coefficient of 26% for principal rope d-1 in collision modality for the bodies (a), not collision (b). Condition with a slack coefficient of 26% for principal rope d-1 and d-2 in collision modality for the bodies (c), not collision (d). Condition with a slack coefficient of 26% for principal rope d-1, d-2 and d-3 in not collision modality for the bodies (e), and collision (f).

4 Conclusions

The present paper describes a preliminary study on the effects of cable slackening in a tensegrity structure. It is worth to note that the effectiveness of the load paths in a tensegrity structure is determined by the prestress force in the cables and the resulting geometrical configuration. If a cable has no stress it is no more able to transfer force. If a load path is made inactive, a reconfiguration of the elements occurs, possibly leading to the collapse. This is a direct result of the way tensegrity structures are made. Although the total number of elements composing them makes them a "mechanism", it results that under a specific elements configuration a self-stress state emerges making the mechanism stiff [19]. By changing the length of one of the elements, the self-stress state configuration is modified.

The positive influence of the additional stiffening cables emerges in second set of simulations. Here, the new connection elements provide an alternative path to transfer the forces that are no more balanced by the slacking cable, resulting in a more robust configuration. At present it is better to consider the addition of elements as an integrity-oriented technique to ensure the robustness of tensegrity systems. Additional studies are required to clarify their behavior and define the strategies to foster robustness in this type of structures.

Acknowledgements. The SISCON interdepartmental center (Safety for Infra-Structures and CONstructions), laboratory dedicated to the safety of both new and existing civil infrastructure, is gratefully acknowledged.

References

1. Fuller, R.B.: Tensile-integrity structures, U.S. Pat. 3.063.521 (1962)
2. Snelson, K.D.: Continuous tension, discontinuous compression structures. US Pat. **3**(169), 611 (1965)
3. Emmerich, D.G.: Construction de réseaux autotendants. French Patent No. 1,377,290 (1964)
4. Kenner, H.: Geodesic Math and How to Use It, University of California Press (1976)
5. Pugh, A.: An Introduction to Tensegrity, University of California Press (1976)
6. Calladine, C.R.: Buckminster Fuller's 'Tensegrity' structures and Clerk Maxwell's rules for the construction of stiff frames. Int. J. Solids Struct. **14**(2), 161–172 (1978)
7. Tarnai, T.: Simultaneous static and kinematic indeterminacy of space trusses with cyclic symmetry. Int. J. Solids Struct. **16**(4), 347–359 (1980)
8. Tarnai, T.: Problems concerning spherical polyhedra and structural rigidity. Struct. Topol. **4**, 61–66 (1980)
9. Pellgrino, S., Calladine, C.: Two-step matix analysis of prestressed cable nets. In: Proceedings Third Internal Conference on Space Structures, Guildford (1984)
10. Pellegrino, S., Calladine, C.R.: Matrix analysis of indeterminate frameworks. Int. J. Solids Struct. **22**(4), 409–428 (1986)
11. Calladine, C.R., Pellegrino, S.: First-order infinitesimal mechanisms. Int. J. Solids Struct. **27**(4), 505–515 (1991)
12. Roth, B., Whiteley, W.: Tensegrity frameworks. Trans. Am. Math. Soc. **265**(2), 419–446 (1981)
13. Connelly, R.: The rigidity of certain cabled networks and the second order rigidity of arbitrarily triangulated convex surfaces. Adv. Math. **37**(3), 272–299 (1980)
14. Connelly, R., Whiteley, W.: Second-order rigidity and prestress stability for tensegrity frameworks. SIAM J. Discrete Math. **9** (1996)
15. Ingber, D.E.: Cellular tensegrity: defining new rules of biological design that govern the cytoskeleton. J. Cell Sci. **104**(3), 613–627 (1993)
16. Micheletti, A.: Simple analytical models of tensegrity structures. In M. Frémond (a cura di), Novel Approaches in Civil Engineering, pp. 351–358 (2004)
17. Micheletti, A., Podio-Guidugli, P.: Seventy years of tensegrities (and counting) Archive of Applied Mechanics **92**, 2525–2548 (2022)
18. EN 1991–7: 2006. Actions on structures. Part 1–7: General actions - accidental actions (2006)

19. Pearson, C., Delatte, N.: Ronan point apartment tower collapse and its effect on building codes. J. Perform. Constr. Facil. **19**(2), 172–177 (2005)
20. Dong, W.: Topology, stability and robustness of tensegrity structures. Ph.D Thesis, Imperial College London, UK (2018)

Check for updates

Robustness and Reduced Bracing Condition of Tent-Structures: The Case Study of Campus Einaudi

Amedeo Manuello Bertetto(✉), O. Borla, Giuseppe Carlo Marano, and B. M. Chiaia

Department of Structural, Building and Geotechnical Engineering (DISEG), Politecnico Di Torino, Corso Duca Degli Abruzzi, 24, 10129 Torino, Italy
amedeo.manuellobertetto@polito.it

Abstract. The premises for Law and Political Science faculties are an important milestone in the building programme of the University of Turin. Not far from the historical seat of the University, the Luigi Einaudi Campus has transformed an abandoned industrial area into a new cultural forum for the city. The geometric complexity of the layout, which originates from the free volumes designed by Foster + Partners, has led to a sophisticated internal spatial differentiation and to innovative solutions developed to give formal unity due to the roof solution. The system of the light-weight tent structures of the roof is characterized by fifty-two reticulated arches anchored at the reinforced concrete slabs on the top of each architectural volumes. The arches are topped by waterproof teflon-coated fiberglass membrane with high mechanical strength and stability against weathering. In this paper, a study of some most representative portions of the structure is proposed considering the influence of connection stiffness reductions. In particular, the robustness under conditions of reduced bracing and various stiffness configurations of the steel nodes of the interconnected arches is an evident problem to be assessed in these kinds of buildings.

Keywords: Campus Einaudi · steel Structures · Tent-Structures · Robustness · Stiffness Reduction

1 Introduction

A significant milestone in the University of Turin's building plans is the establishment of new facilities for the colleges of Law and Political Science. The Luigi Einaudi Campus has converted an abandoned industrial area into a new cultural hub for the city, not far from the University's historical seat. The Foster + Partners design substitutes two groups of buildings connected by a single roof canopy for the proposed pavilions layout of the University's preliminary studies [1]. This project's distinct personality immediately became a "land mark" for its distinctiveness. The Law and Political Science faculties are to the left of the new four-story library, which is parallel to the River Dora on the northern side of the site. From the centre courtyard, there are separate entrances for each

S. Gabriele et al. (Eds.): IWSS 2023, LNCE 437, pp. 351–358, 2024.
https://doi.org/10.1007/978-3-031-44328-2_36

faculty. The layout's geometric intricacy, which derives from Foster + Partners' free volumes, has resulted in sophisticated internal spatial difference and creatively devised methods to provide formal unity. The new complex, developed on a triangular lot of about 45,000 m^2, consists of 7 large buildings of four floors above ground each (Fig. 2, view from the central courtyard), dedicated respectively to teaching (classrooms and departments) and to the library and attached services. It is considered a kind of "sapere" factory characterized by a deliberate coldness of the glass and metal facades, with a rectilinear course to the outside and a curvilinear course to the inside [1]. Each block, while representing a single building, is a kind of porous body broken and lightened through wide caesuras on the ground floor and suspended connections, generating a total of 7 structurally independent reinforced concrete volumes in terms of thermal and seismic responses. They echo, in their rounded forms, the profile of the gasometers on the former industrial site and faces the central courtyard, constituting the cohesive architectural element, a kind of extension of the interior spaces [1]. It boasts a surface area of about 16,700 m^2 and is made of structural membrane properly stretched over 54 tubular metal frames with an arcuate shape that generates a repetitive wave effect that is only apparently regular. In fact, the membrane surface is characterized by a constantly varying double curvature geometry in every part of the entire development. The arches consist of steel tubular members connected by bolted realized joints. Such joints may be subject to deterioration in connection stiffness (see Fig. 1).

This type of issue is particularly important and of considerable significance for roofs consisting of many elements, of lattice type, and consisting of hybrid joints or joints made by bolted connections [2]. Very often in order to consider a real condition, a particular pattern of imperfections (especially localized in the joints) must be assumed in order to assess the effect of deterioration in terms of lowering the load-bearing capacity of the structure and to recognize the early initiation of collapse phenomena due to local or non-local instability [2–5].

Fig. 1. The different arches are connected to the reinforced concrete slab and connected to the cantilevered bull-nose beam [1]. Each arch was preassembled on the ground and successively placed in place by hoisting [1]. All joints are of the bolted type.

For the joint connections in the case of timber domes with steel joints, a particular stiffness reduction coefficient was introduced in the recent paper [2]. The method chosen is based on the alteration of the moment of inertia connected to the joint section. This method aims to model both the connections' actual behaviour and how they deteriorate with time. In order to investigate the validity of the limit condition and other ways described in the literature, the model was also compared to other approaches. At other words, the evolution (depletion) of the stiffness at the nodes was taken into consideration when analysing the geometrically non-linear behaviour of specific existent arches of the structures.

2 Campus Luigi Einaudi Roof Structures

2.1 State of the Art

Each block, while representing a single building, is a kind of broken and lightened body through wide caesuras on the ground floor and suspended connections, generating a total of 7 reinforced concrete volumes that are structurally independent in terms of thermal and seismic responses [1]. They repeat, in their rounded forms, the profile of the gasometers on the former industrial site and face around a spacious green plaza about 80 m in diameter. The sinuous course of the tensile-structural roofing common to the 7 buildings, designed and shaped around the central courtyard, constitutes the architectural cohesion element, a kind of extension of the interior spaces. It boasts a surface area of about 16,700 m2 and is made of structural membrane properly stretched over 54 tubular metal frames with an arcuate shape that generates a repetitive wave effect that is only apparently regular. In fact, the membrane surface is characterized by a double curvature geometry changing throughout the entire development, a variable that conditioned the metal frames supporting the membrane [6–8].

The metal system supporting the structural membrane, is realized by means of 54 arch lattice frames, varying in span from 20 m to 52 m, consecutively arranged, for a total weight of the load-bearing carpentry of about 1200 tons. As already anticipated, the characteristic of geometric non-regularity, associated with the double curvature of the roof, inevitably led to a design characterized by complete heterogeneity of rods (in length) and nodes (in number of converging rods and angles of incidence at the node). All members (CHS series tubular and in angle profile) are in S355, the only exception being the accessory ceiling structures, which are made of IPE class S275 open profiles. It may be of interest to observe the mutual ground joint between the truss frames consisting of CHS 408 structural pipe arranged to take up the different members coming from the 2 adjoining trusses, and in specific in its sliding configuration placed at the constructive expansion joints straddling the 7 construction lots, where appropriate sliding support devices of the bidirectional type were inserted. The assembly and lifting of the trusses took place through the pre-assembly, on site, of the 108 semi-structures, which were then lifted and positioned to form the single arched module. Finally, the erection of the edge beam called "bull-nose" (Fig. 1), joining all the lattice tails, took place to define a perimeter, precisely in the shape of a "bull-nose," with an approximate length of 1000 m.

With regard to the design and construction of the structures, it is segned, leaving then to more detailed treatments any possible details, how the executive-constructive interaction has seen the implementation of an in-depth campaign of numerical investigations aimed at investigating and parameterizing the design resistances in chord failure (mechanical resistance of the walls of tube-lar nodes), due to the high number of converging connections on CHS series beams with thicknesses varying from 5 to 40 mm at non-standard geometry. The technical interaction, with subsequent executive update of the nominal sections of the beams, actually researched, by means of non-linear numerical analysis in the post-critical, the ultimate collapse resistances of the walls (ultimate resistance) to the actions of the afferent elements, thus being able to lead to the formulation of strength and stiffness abacuses used in the construction design for the adjustment of profile thicknesses where necessary.

The observation of plastic mappings during incremental loading sequences has proved to be of interest in this regard for some families of nodes, as it can often be associated with nodal stiffness losses. The typical phenomenon of propagation of differentially localized plastic zones on the walls of the bridleway, up to their conjunction (plastic bridges) and triggering of the degenerative mechanism of bearing capacity.

The membrane used for the roofing, called Sheerfill (Type I-HT and Type II-HT), consists of fiberglass (fiberglass) textile core with PTFE (polytetrafluoroethylene, jargon: Teflon) outer coating, and provides tensile strengths up to values of 170.8 and 157.8 kN/m along the warp and weft directions, respectively. The executive design provided the product stringing according to the two directions warp and weft (warp: arc-arc lay, weft: bullnose-bullnose lay), with values of 2.00 kN/m and 3.00 kN/m respectively. The provision of the aforementioned pre-stress state was carefully analyzed in construction design by the companies engaged in the design, construction and installation of the membrane, so as to mainly accommodate the issues related to the elevated (and often cantilevered) position of the tensile-structural edges.

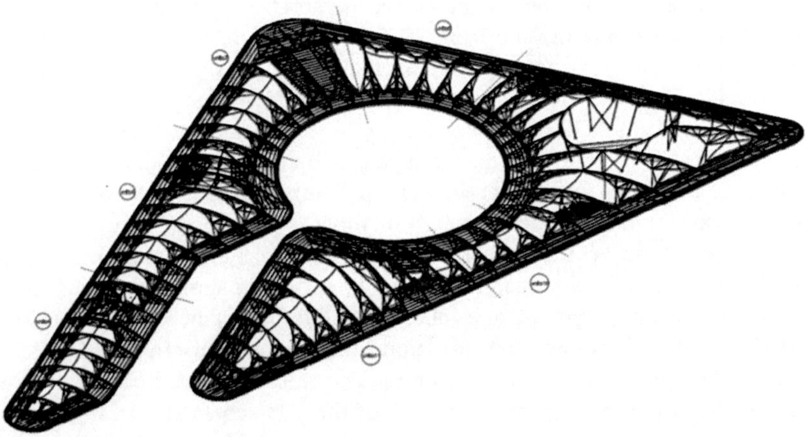

Fig. 2. The metal system supporting the structural membrane, is realized by means of 54 arch lattice frames, varying in span from 20 m to 52 m, consecutively arranged, for a total weight of the load-bearing carpentry of about 1200 tons.

3 Semi-rigid Nodes and Stiffness Reduction

Predicting the mechanical behavior of joints is a crucial step in the analysis of spatial systems with semi-rigid connections, as documented in several research. Numerous researches have recently been carried out to examine the mechanical behavior of joints in space constructions. [6–14]. López et al. [12], Ma et al. [15], and Kato et al. [16] verified that the joints rigidity is a crucial element that affects how a single-layer latticed dome behaves [15].

In this work, as reported in [2] First, we analyze a model that accounts for the connection's stiffness and the resulting rise in its yieldingness. The basic idea is to model the structure's rigidity deteriorating over time as a result of declining node efficiency. This effect is achieved by reducing the geometric stiffness of the node-forming section and, specifically, by assuming a drop in the section's inertia due to bending deformation. First, we consider the beam's rotational angle as a result of bending deformation:

$$\vartheta_x = \frac{M_x l}{EI_{xx}} \tag{1}$$

where M_x is the bending moment that stressed the beams, E is the elastic modulus of the material and I_{xx} is the moment of inertia respect to the x axis (barycentre x-axis of the section. we obtain a reduced $[I^{'}]$, where the reduced value is obtained for the component I_{yy} and I_{xy}. I_{yy}, in a compact form [2]:

$$[I][C'] - [K] + [K'] = [I'], \tag{2}$$

where: *[I]*, As previously stated, is the moment of inertia of the joint section free from any damage, stiffness decrease, section reduction, or fluctuation in term of yieldingness; *[C']* is a diagonal matrix in which the reduction coefficient for I_{xx} and I_{xy} are reported. *[K]* and *[K']* are the appropriate coefficient matrices that respectively were subtracted and added to the partially reduced contributions *[I*$^{''}$*]* [2]. According to the enhanced yieldingness of the joint, this tensor shows the decreased moment of inertia. In addition, considering Chen et al.'s model, the semi-rigid joint's yieldingness is expressed as [2, 6]:

$$\frac{L_b}{EI_{b,xx}} + \frac{2}{k} = \left(\frac{L_b k + 2E_b I_{b,xx}}{EI_{b,xx} k}\right), \tag{3}$$

where L_b is the beam of the grid, $I_{b,\mathrm{xx}}$ is its moment of inertia, E_b is the elastic modulus and *k/2* is the stiffness of the connections. Typically, L_b/L_n is greater than 10. From a general point of view, considering E_n and I_n respectively the elastic modulus of the material constituting the node and moment of inertia of the connection, it is possible to consider that $E_n I_{n,\mathrm{xx}} = E_b I_{b,\mathrm{xx}}$.

$$\frac{L_b}{E_b I_{b,xx}}\left(1 + \frac{2E_b I_{b,xx}}{kL_b}\right) = \frac{L_b + L_n}{E_b c I_{b,xx}} = \frac{L_b + L_n}{Eb\, c I_{b,xx}}, \tag{4}$$

where c is the reduction coefficient for $I_{b,xx}$. Taking onto account that in usual construction Lb $\cong L_b + L_n$. Equation (6) can be rewritten as:

$$1 + \frac{2E_b I_{b,xx}}{kL_b} = \frac{1}{c}. \tag{5}$$

It is possible to consider the relation between the semi-rigid law reported in Chen et al. [15] and the *reduction parameter c* in terms of connection stiffness:

$$c = \frac{kL}{kL + 2EI_{b,xx}}. \tag{6}$$

As previously noted, c has anticipated that $I_{b,xx}$ will be increased by the maximum value (100%) in the scenario when there has been no node section damage and no connection performance degradation. In the case, of $k \to 0$, considering Eq. (8) $c \to 0$. This last instance (severely damaged part) matched the node's greatest yieldingness and minimum stiffness. Using this method, the various elements of the reduction coefficients of *[I']* are.

$$c_x I_{xx}, \quad c_y I_{yy}, \quad c_{xy} I_{xy}. \tag{7}$$

4 Stiffness Reduction Strategies

In order to conduct the analyses, it is necessary to evaluate the different strategies that can be implemented for defining a plausible configuration for the pattern of imperfections located in the three-dimensional framework. First, it is necessary to assess that it might be possible to consider both the presence of imperfections of geometric type, along the constituent elements, of the structure (imperfections and deviations from the straight line of the axis line) and imperfections due to defects in the tightening of the bolts constituting the nodes (as reported and described in the previous section). Considering the typology of the elements constituting the arched type structure of the Einaudi Campus, it appears very indicative to consider the type of defect connected with the punctual stiffness reduction of some nodes of the structure.

This kind of analyses will be conducted by FEM models using specific element to evaluate the behavior of the nodes according to the previous Section.

As a preliminary step, an on-site assessment of the actual level of tightening of each of the connections constituting the different arches have to be carried out. Once such a survey has been performed, it will be possible to attribute to the structure a pattern of relative decrement in stiffness as closely as possible to actual condition. In other cases, in the absence of direct information a statistical investigation must be carried out considering the different possible patterns. In the light of a comparison made by evaluating in the preliminary model considering one of the arches facing the overhangs showed an increase in sagging of about 36% comparing a condition in which all connections were assumed to be perfectly rigid and one in which they were considered to have significantly increased yielding.

5 Conclusions

An on-site evaluation of the actual degree of tightness of each connections that make up the various arches must be done as a first step. After carrying out this survey, it will be feasible to assign the structure a pattern of relative stiffness reduction that corresponds as nearly as possible to actual state of conservation and in-service condition. In other situations, a statistical analysis must be conducted in the lack of direct evidence while considering the many potential patterns. Under the light of a comparison between a scenario in which all connections were supposed to be absolutely stiff and one in which they weren't, one of the arches facing the overhangs revealed an increase in sagging of roughly 36% comparing the assumption that all connections are absolutely stiff with the assumption that they have dramatically enhanced yielding.

Acknowledgements. The SISCON interdepartmental center (Safety for Infra-Structures and Constructions), laboratory dedicated to the safety of both new and existing civil infrastructure, is gratefully acknowledged for the data related to the monitored structures.

References

1. Zanchin, M., Mastropasqua, A.: The New Luigi Einaudi University Campus in Turin: an interesting roof created using a steel membrane tensile system. Costruzioni Metalliche. Gen-Feb. 30–40 (2013)
2. Manuello, A.: Semi-rigid connection in timber structure: stiffness reduction and instability interaction. Int. J. Struct. Stab. Dyn. **20**(7), 2050072 (2020)
3. Carpinteri, A., Bazzucchi, F., Manuello, A.: Nonlinear instability analysis of long-span roofing structures: the case-study of Porta Susa railway-station. Eng. Struct. **110**, 48–58 (2016)
4. Bazzucchi, F., Manuello, A., Carpinteri, A.: Interaction between snap-through and Eulerian instability in shallow structures. Int. J. of Non-Lin. Mech. **88**, 11–20 (2017)
5. Bazzucchi, F., Manuello, A., Carpinteri, A.: Instability load evaluation of shallow imperfection-sensitive structures by form and interaction parameters. Eur. J. of Mech. Solids A **66**, 201–211 (2017)
6. Chen, Z., Xub, H., Zhao, Z., Yan, X., Zhao, B.: Investigations on the mechanical behavior of suspend-dome with semi-rigid joints. J. Constr. Steel Res. **122**, 14–24 (2016)
7. Young-Lin, P., Trahair, N.S.: Three-dimensional nonlinear analysis of elastic arches. Eng. Struct. **18**(1), 49–63 (1996)
8. Kitipornchai, S., Kang, W., Lam, H.-F., Albermani, F.: Factors affecting the design and construction of Lamella suspended-dome systems. J. Constr. Steel Res. **61**(6), 764–785 (2005)
9. Huihuan, M., Feng, F., Gengbo, C., Zhenggang, C., Shizhao, S.: Numerical analyses of semi-rigid joints subjected to bending with and without axial force. J. Constr. Steel Res. **90**, 13–28 (2013)
10. Altuna Zugasti, A.M., Arancibia, A.L., Puente, I.: Influence of geometrical and structural parameters on the behavior of squared plan-form single-layer structures. J. Constr. Steel Res. **72**, 219–226 (2012)
11. López, A., Puente, I., Serna, A.: Direct evaluation of the buckling loads of semi-rigidly jointed single-layer latticed domes under symmetric loading. Eng. Struct. **29**(1), 101–109 (2007)
12. López, A., Puente, I., Serna, A.: Numerical model and experimental tests on single-layer latticed domes with semi-rigid joints. Comput. Struct. **85**(7), 360–374 (2007)
13. Loureiro, A., Goni, R., Bayo, E.: A one-step method for buckling analysis of single layer latticed structures with semi-rigid connections. In: Proceedings of the Fifth International Conference of Space Structures, Thomas Telford, London, pp. 1481–1490 (2002)
14. Ma, H., Fan, F., Shen, S.: Numerical parametric investigation of single layer latticed domes with semi-rigid joints. J. Int. Assoc. Shell Spat. Struct. **49**(2), 99–110 (2008)

15. Fan, F., Ma, H., Zhenggang, C., Shen, S.: A new classification system for the joints used in lattice shells. Thin-Walled Struct. **49**(12), 1544–1553 (2011)
16. Kato, S., Mutoh, I., Shomura, M.: Collapse of semi-rigidly jointed reticulated domes with initial geometric imperfections. J. Constr. Steel Res. **48**, 145–168 (1998)
17. Marmo, F., et al.: On the form of the Musmeci's bridge over the Basento river. Eng. Struct. **191**, 658–673 (2019)
18. Per il FEM (travi):
19. Paradiso, M., Marmo, F., Rosati, L.: Consistent derivation of a beam model from the Saint Venant's solid model. Int. J. Solids Struct. **159**, 90–110 (2019)
20. Paradiso, M., Sessa, S., Vaiana, N., Marmo, F., Rosati, L.: Shear properties of isotropic and homogeneous beam-like solids having arbitrary cross sections. Int. J. Solids Struct. **216**, 231–249 (2021)

Principal Symmetric Structures

Davide Pellis[1](✉) and Helmut Pottmann[2]

[1] ISTI-CNR, Pisa, Italy
davide.pellis@isti.cnr.it
[2] KAUST, Thuwal, Saudi Arabia

Abstract. We introduce a new class of quadrilateral gridshell structures in axial force equilibrium where beams run symmetrically to the principal stress directions of their limit membrane shell. This kind of structures have the property that, at each node, the axial forces in the four connected beams are approximately equal. This allows for a more homogeneous distribution of forces in the structure, particularly in shapes where stresses are significantly anisotropic, in which case a conventional gridshell typically results in numerous beams remaining nearly unloaded. In this work, we first discuss the properties of principal symmetric structures and evaluate their advantages relative to other types of gridshells. We introduce then a computational pipeline for the design of such structures based on a quadrilateral remeshing and a subsequent optimization, and show some results.

Keywords: Gridshells · Structural optimization · Architectural geometry

1 Introduction

Gridshell trusses offer an efficient and effective solution for constructing shell structures, combining strength with a lightweight design. By carrying loads through axial stresses in their beams, these structures make the most efficient use of material. Among all gridshells, quadrilateral ones are particularly well-suited for manufacturing, since disposing beams along the principal curvature lines of a surface leads to flat cladding panels and simplified joint connections.

A well-known approach to designing a quadrilateral gridshell truss is to follow the principal stress lines of a membrane shell. However, in cases where the principal stresses are significantly anisotropic, this can result in many beams remaining nearly unloaded in one direction, leading to a waste of structural material when using beams of constant cross-section.

In this paper, we introduce a new class of quadrilateral gridshell trusses that distribute axial forces highly homogeneously at each point, allowing for a more efficient use of structural material. It turns out that such gridshells must follow two specific directions that are symmetric to the principal stress directions of a tension-only or compression-only membrane.

S. Gabriele et al. (Eds.): IWSS 2023, LNCE 437, pp. 359–368, 2024.
https://doi.org/10.1007/978-3-031-44328-2_37

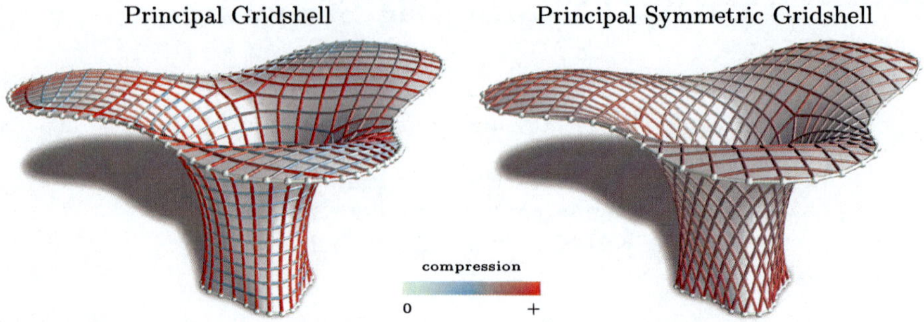

Fig. 1. On the *left*, a gridshell aligned with principal stress directions. Many beams remain almost unloaded. On the *right*, a principal symmetric gridshell. At each node, the distribution of axial forces is more homogeneous.

1.1 Related Work

This study falls within the broader context of gridshell design and optimization, which has been comprehensively reviewed in [1]. Our approach is based on decoupling horizontal and vertical equilibrium and on using the polyhedral Airy stress potential as a finite element discretization of a continuum, as described in [2,5]. Vouga et al. [14] have previously leveraged the resulting connections between statics and discrete differential geometry. Other studies have investigated the use of quadrilateral meshes aligned with principal stresses as statically-optimal gridshells, as in [6,8], and also aligned with principal curvatures, as discussed in [9]. Schling and colleagues [12] previously introduced structures symmetric to principal stress directions on shapes with rotational symmetry. The geometric properties and manufacturing implications of quadrilateral gridshells have been extensively studied in the field of architectural geometry, as surveyed in [11]. In this context, Pellis et al. [10] introduced gridshells that are symmetric to principal curvature directions, rather than principal stress directions, and investigated their potential for use in architectural panelization.

1.2 Overview and Contributions

We begin our exploration of principal symmetric structures in two dimensions with a brief overview of truss equilibrium at the limit of refinement in Sect. 2. Next, in Sect. 3, we generalize our findings to gridshell structures under vertical load. Finally, in Sect. 4, we outline a pipeline for the computational design of principal symmetric gridshells.

2 Principal Symmetric Trusses in 2D

In this section, we present the concept of principal symmetric structures in 2D. Our approach involves a refinement process of a gridshell truss, which eventually converges to a "truss-like continuum" where the mechanical state can be described using the Cauchy stress tensor. First, we provide an overview of the equilibrium of a truss at the refinement limit. Next, we analyze the problem in the continuous setting and then proceed to discretize the continuous solution to obtain a principal symmetric truss.

2.1 Discrete Equilibrium

Let us consider a two dimensional truss, with members corresponding to the edges of a mesh M with vertices $\mathbf{v}_i = (x_i, y_i)$, where loads and supports are applied at boundary vertices. Let $\mathbf{f}_{ij}$ be the force exerted by the oriented bar $\mathbf{v}_i - \mathbf{v}_j$ on the vertex $\mathbf{v}_i$. If the system is in axial equilibrium, at each unsupported vertex $\mathbf{v}_i$ we have

$$\sum_{j \sim i} \mathbf{f}_{ij} = 0, \tag{1}$$

where with $j \sim i$ we denote all the vertices j connected with the vertex i.

As shown by Maxwell [7], for such a truss we can construct a *force polyhedron* $\Phi = (x, y, \phi(x, y))$, whose edges and vertices coincide in the xy projection with the mesh M, in the following way. Let us suppose that each face f of Φ lies on a plane with gradient $\nabla\phi|_f$. Let f_l and f_r be the left and right faces of the oriented edge $\mathbf{v}_i - \mathbf{v}_j$, and let R be the counterclockwise 90° rotation matrix in the xy plane. If we set $\mathbf{f}_{ij} = R(\nabla\phi|_{f_r} - \nabla\phi|_{f_l})$, the equilibrium of Eq. 1 is ensured by the closure of the polyhedron Φ when turning around the vertex $\mathbf{v}_i$. This construction is uniquely defined up to vertical translations and shearing. For further details see [5,14].

2.2 Equilibrium at the Limit of Refinement

Let us now consider a refinement process that increases the density of a 2D truss. From a mechanical perspective, as the refinement limit is reached, the truss will approach a 2D continuum. Simultaneously, the force polyhedron representing the structure's equilibrium will converge to a continuous surface that represents the stresses of a 2D body, which is called the *Airy stress surface.* If the truss has the connectivity of a quadrilateral mesh, the edges will eventually converge to form a network of smooth curves.

Let D be a 2D continuum with loads and constraints applied on its boundary ∂D, and let it have a Cartesian reference system xy. At each point $p(x, y)$ within the continuum, the stress field can be described by the Cauchy stress tensor

$$S = \begin{pmatrix} \sigma_{xx} & \sigma_{xy} \\ \sigma_{xy} & \sigma_{yy} \end{pmatrix}. \tag{2}$$

The stress tensor S can be interpreted as follows: If we make an infinitesimal cut in the continuum at the point p along a vector $\mathbf{a}$, the force acting across the cut is

$$\mathbf{f}(\mathbf{a}) = SR\mathbf{a}. \tag{3}$$

The corresponding stress vector (force per unit length) is given by

$$\boldsymbol{\sigma}(\mathbf{a}) = \frac{\mathbf{f}(\mathbf{a})}{\|\mathbf{a}\|}. \tag{4}$$

Static equilibrium in the interior of the domain D is expressed by $\operatorname{div} S = \mathbf{0}$, where the divergence is applied to both columns separately. In a simply connected domain, this implies existence of an Airy potential $\phi(x, y)$ such that

$$H_\phi = \begin{pmatrix} \phi_{,xx} & \phi_{,xy} \\ \phi_{,xy} & \phi_{,yy} \end{pmatrix} = \begin{pmatrix} \sigma_{yy} & -\sigma_{xy} \\ -\sigma_{xy} & \sigma_{xx} \end{pmatrix} = \tilde{S}, \tag{5}$$

where H_ϕ is the Hessian of the function ϕ. Here we indicate partial derivatives by a comma and subscripts, and with a tilde the adjoint matrix operation given by $\tilde{S} = R^T S R$.

Let us consider now a network of curves on the domain D and its vertical projection on the Airy stress surface. At the limit of refinement, this network can be seen as a quadrilateral truss. The network should then span infinitesimal planar faces to form a force polyhedron. In such a parametrization, at a point $p = (x, y)$, let us consider the quadrilateral given by the points $\phi(p)$, $\phi(p + \mathbf{a}_1)$, $\phi(p + \mathbf{a}_2)$, and $\phi(p + \mathbf{a}_1 + \mathbf{a}_2)$. Such quadrilateral is planar if $\phi(p + \mathbf{a}_1 + \mathbf{a}_2) = \phi(p + \mathbf{a}_1) + \phi(p + \mathbf{a}_2)$. A Taylor expansion tells us that

$$\phi(p + \mathbf{a}_1 + \mathbf{a}_2) - \phi(p + \mathbf{a}_1) - \phi(p + \mathbf{a}_2) = 2\, \mathbf{a}_1^T H_\phi\, \mathbf{a}_2 + o(\mathbf{a}_1 + \mathbf{a}_2)^2.$$

At a point $p = (x, y)$, two directions $\mathbf{a}_1$, $\mathbf{a}_2$ are then in equilibrium at the limit of refinement if

$$\mathbf{a}_1^T H_\phi\, \mathbf{a}_2 = 0. \tag{6}$$

Such directions are said to be *conjugate* to H_ϕ. This result can also be interpreted informally as follows: Let us imagine to substitute the continuum at a point with two crossing beams $\mathbf{a}_1$ and $\mathbf{a}_2$. To insert the first beam, we have to make a cut $\mathbf{a}_1$ in the continuum. The resulting force across it is $SR\,\mathbf{a}_1$. For axial force equilibrium, this force should point in the direction of the second beam $\mathbf{a}_2$, therefore $\mathbf{a}_2^T R^T S R\, \mathbf{a}_2 = \mathbf{a}_1^T H_\phi\, \mathbf{a}_1 = 0$. This condition is symmetric and entails equilibrium also for the insertion of $\mathbf{a}_2$.

2.3 Stress Symmetric Networks in 2D

We look now for two directions $\mathbf{a}_1$, $\mathbf{a}_2$ such that the forces across these two cuts have the same magnitude, or rather $\|\boldsymbol{\sigma}(\mathbf{a}_1)\| = \|\boldsymbol{\sigma}(\mathbf{a}_2)\|$. To do this, let us first decompose the stress vector $\boldsymbol{\sigma}(\mathbf{a})$ into its normal and tangential components, given respectively by

$$\sigma_n(\mathbf{a}) = \frac{\boldsymbol{\sigma}^T(\mathbf{a}) R\,\mathbf{a}}{\|\mathbf{a}\|}, \quad \sigma_t(\mathbf{a}) = \frac{\boldsymbol{\sigma}^T(\mathbf{a})\mathbf{a}}{\|\mathbf{a}\|} \tag{7}$$

Let us now consider the eigenvalues of H_ϕ, denoted as σ_1, σ_2, and the corresponding eigenvectors $\mathbf{e}_1$, $\mathbf{e}_2$. Let then $\mathbf{a}(\theta)$ be a direction that makes an angle θ

with the eigenvector $\mathbf{e}_1$, and define $\boldsymbol{\sigma}(\theta) = \boldsymbol{\sigma}(\mathbf{a}(\theta))$. Applying the tensor transformation law, the normal and tangential components of the stress vector can be now expressed as functions of the angle θ as follows:

$$\begin{aligned} \sigma_n(\theta) &= \sigma_1 \cos^2\theta + \sigma_2 \sin^2\theta \\ \sigma_t(\theta) &= (\sigma_1 - \sigma_2)\sin\theta\cos\theta \end{aligned} \tag{8}$$

We can observe that the squared norm $\boldsymbol{\sigma}^2(\theta) = \sigma_n^2(\theta) + \sigma_t^2(\theta)$ is equal for couples of angles $\{\theta, -\theta\}$ or $\{\theta, \pi - \theta\}$ – that is for directions that are symmetric to the principal directions $\mathbf{e}_1$, $\mathbf{e}_2$.

If we desire a quad net in 2D which is principal symmetric and also in force equilibrium, it must fulfill the conjugacy Eq. 6. For that, let us consider the ellipse given by all vectors $\mathbf{a}$ such that $\mathbf{a}^T H_\phi\, \mathbf{a} = 1$. The principal semi-axes of this ellipse point along the directions $\mathbf{e}_1$ and $\mathbf{e}_2$, with corresponding radii $|\sigma_1|^{-\frac{1}{2}}$ and $|\sigma_2|^{-\frac{1}{2}}$. We denote the semi-axes as $\mathbf{e}_1^*$ and $\mathbf{e}_2^*$. Conjugate directions of H_ϕ correspond to conjugate diameters of the aforementioned ellipse. The only conjugate diameters that are symmetric to the principal axes are the so called *characteristic directions*, given by the diagonals of the principal axes rectangle. For these directions, the angle θ against the first principal direction satisfies

$$\tan\theta = \pm\sqrt{\frac{\sigma_1}{\sigma_2}}. \tag{9}$$

We observe that these directions, that we call *stress characteristic directions*, only exist if the stresses have equal sign, i.e., for a positive definite Hessian H_ϕ. Hence, principal symmetric quad structures in equilibrium must be either compression-only or tension-only structures.

Plugging Eq. 9 into Eqs. 8, we get

$$\boldsymbol{\sigma}^2 = \sigma_1\sigma_2. \tag{10}$$

We can see that, for a principal symmetric network, the forces are everywhere constant for stress fields where $\sigma_1\sigma_2$ = constant, or equivalently, $\det(H_\phi)$ = constant.

3 Principal Symmetric Gridshells

We extend now the previous results to surface-like gridshells in axial force equilibrium, loaded at joints with forces along a z axis. With a refinement process similar to the one described in the 2D case, such a structure will converge to a "truss-like membrane". If the gridshell has a quadrilateral connectivity, the beams will then converge to a network of curves on the membrane surface. The geometric and mechanical properties of this limit network can be analyzed with differential geometry and continuum mechanics. For a formal description of gridshells approaching membranes, see [8].

3.1 Equilibrium at the Limit of Refinement

For that, let us consider a membrane surface $\mathcal{S}$ given as the graph of a function $z(x, y)$, subject to a distributed load in z direction $p_z(x, y)$, expressed per unit xy area. Let I be the first fundamental form of the surface parametrization, given by

$$I = \begin{pmatrix} 1 + z_{,x}^2 & z_{,x} z_{,y} \\ z_{,x} z_{,y} & 1 + z_{,y}^2 \end{pmatrix}. \tag{11}$$

Let S be the membrane stress tensor, and let $\bar{S}$ be its projection on the xy plane, given by $\bar{S} = \sqrt{\Delta} S$, where $\Delta = \det(I)$. Since loads act only in z direction, the xy component of the equilibrium requires the projected stress tensor $\bar{S}$ to be divergence free. As in Sect. 2.2, this entails the existence of an Airy stress surface $\phi(x, y)$, representing the xy equilibrium, such that

$$\bar{S} = \tilde{H}_\phi = \begin{pmatrix} \phi_{,yy} & -\phi_{,xy} \\ -\phi_{,xy} & \phi_{,xx} \end{pmatrix}. \tag{12}$$

Being H_z the Hessian of the surface $z(x, y)$, the z component of the equilibrium requires

$$H_z : H_\phi + p_z = 0, \tag{13}$$

where with colon we indicate the matrix inner product (sum of element-wise multiplications). Note that the principal stresses on the surface are given by the eigenvalues of $\sqrt{\Delta} I^{-1} H_\phi$, as shown in [6].

3.2 Quadrilateral Gridshell Networks

Let us consider now a quadrilateral gridshell that at the limit of refinement converges to a network of curves on $\mathcal{S}$. At a surface point $z(x, y)$, let $\mathbf{a}_1$ and $\mathbf{a}_2$ be the tangent vectors of the two network curves passing through that point, and let $\bar{\mathbf{a}}_1$ and $\bar{\mathbf{a}}_2$ be the projections of these vectors in the xy plane. As in 6, horizontal equilibrium requires conjugacy of these directions to H_ϕ.

We can also ask the network to have additional properties to simplify manufacturing. For instance, we can look for a network with planar faces. As seen in Eq. 6, this requires conjugacy with respect to the surface Hessian H_z. Together with equilibrium, this lead to conditions

$$\bar{\mathbf{a}}_1^T H_z \, \bar{\mathbf{a}}_2 = 0; \quad \bar{\mathbf{a}}_1^T H_\phi \, \bar{\mathbf{a}}_2 = 0 \tag{14}$$

These conditions are both satisfied if $\bar{\mathbf{a}}_1$ and $\bar{\mathbf{a}}_2$ are pointing along the eigenvectors of $H_z^{-1} H_\phi$. For a given membrane in equilibrium, such a network always exists and it is uniquely determined. If we additionally ask for the orthogonality of $\mathbf{a}_1$ and $\mathbf{a}_2$, and therefore $\bar{\mathbf{a}}_1^T I \, \bar{\mathbf{a}}_2 = 0$, we have the condition that $\bar{\mathbf{a}}_1$ and $\bar{\mathbf{a}}_2$ must at the same time point along the eigenvectors of $I^{-1} H_z$ and $I^{-1} H_\phi$. This requires that the principal directions of stress and curvature must coincide. Such a problem is over constrained and have a solution only for a special class of membrane surfaces. A method for the design of such gridshells is described in [9].

3.3 Stress Symmetric Gridshell Networks

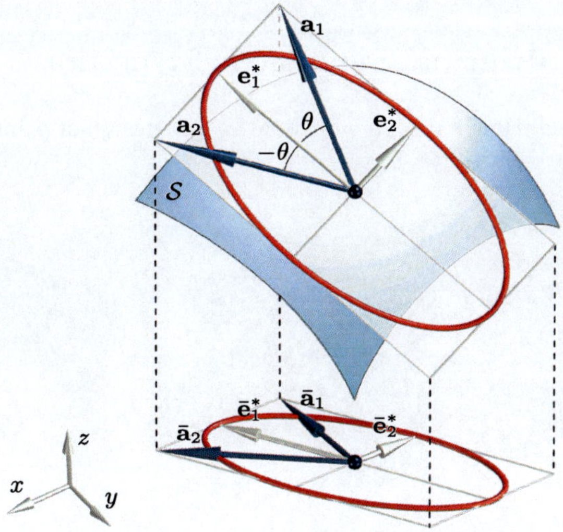

Fig. 2. Principal symmetric directions at a point of a membrane $\mathcal{M}$. In *red*, the Dupin indicatrix of the adjoint stress tensor $\sqrt{\Delta} I^{-1} H_\phi$ is shown.

We look now for two directions $\mathbf{a}_1$, $\mathbf{a}_2$ in equilibrium along which the membrane stress vector has the same magnitude. To do this, let us consider the tangent plane of the surface $\mathcal{S}$, and the Dupin indicatrix of the adjoint membrane stress on this plane. We have here the same conditions as described in Sect. 2.3, with characteristic directions being the diagonals of the principal axes rectangle, where now the principal stresses σ_1, σ_2 and the principal directions $\mathbf{e}_1$, $\mathbf{e}_2$ are the eigenvalues and eigenvectors of the adjoint membrane stress.

To compute these directions in our xy parametrization, let us consider the eigenvectors of the adjoint stress tensor $\sqrt{\Delta} I^{-1} H_\phi$, denoted as $\bar{\mathbf{e}}_1$, $\bar{\mathbf{e}}_2$. The xy projections of the semi-axes of the Dupin indicatrix are given by

$$\bar{\mathbf{e}}_1^* = \frac{\bar{\mathbf{e}}_1}{\sqrt{\bar{\mathbf{e}}_1^T H_\phi \bar{\mathbf{e}}_1}}, \quad \bar{\mathbf{e}}_2^* = \frac{\bar{\mathbf{e}}_2}{\sqrt{\bar{\mathbf{e}}_2^T H_\phi \bar{\mathbf{e}}_2}}.$$

As shown in Fig. 2, the xy projection of the characteristic directions can be computed as $\bar{\mathbf{a}}_{1,2} = \bar{\mathbf{e}}_1^* \pm \bar{\mathbf{e}}_2^*$. We observe that these directions are also conjugate to H_ϕ: This can be seen considering that $(\mathbf{a}+\mathbf{b})^T H_\phi (\mathbf{a}-\mathbf{b}) = \mathbf{a}^T H_\phi \mathbf{a} - \mathbf{b}^T H_\phi \mathbf{b}$. This network, as in the 2D case, is uniquely determined and exists only for tension-only or compression-only membranes.

A principal symmetric structure with planar faces requires the conjugacy of the directions $\bar{\mathbf{a}}_1$, $\bar{\mathbf{a}}_2$ with respect to the surface Hessian H_s, as in Eq. 14. These networks exists only on a special class of membrane surfaces. A particularly interesting structure would be one with constant forces in all the network. From Eq. 10, but considering now the membrane principal stresses, we observe that this happens for networks where $\det(\sqrt{\Delta} I^{-1} H_{\phi}) = \text{constant}$.

4 Design of Principal Symmetric Structures

In this section, we outline a design workflow for the design of principal symmetric gridshells, summarized in Fig. 3.

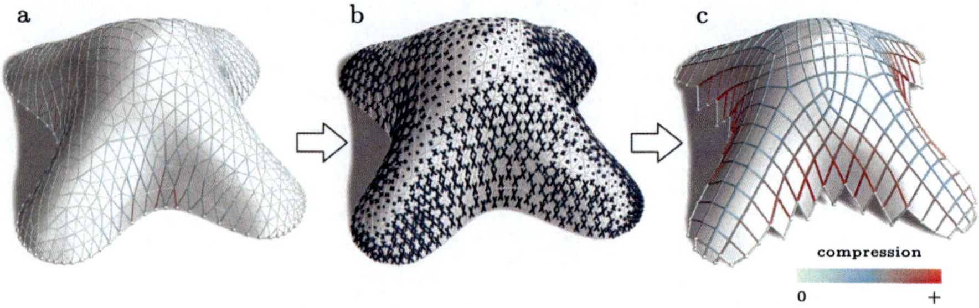

Fig. 3. Design pipeline. a) A membrane shell in funicular equilibrium is given as input. To compute it, we use a triangular gridshell in compression-only or tension-only equilibrium. b) We estimate the characteristic directions with Eq. 9. c) We compute a quadrilateral mesh aligned with characteristic directions and we optimize the resulting gridshell for equal forces at nodes.

Design of a Compression-Only (Tension-Only) Membrane in Equilibrium. As input, we need a compression-only or tension-only membrane in equilibrium. To compute it, we model the membrane as a triangular gridshell truss, with loads applied on its vertices. We impose then the equilibrium at each unsupported vertex asking for compression-only or tension-only axial forces. For that, we use the *force density method* with the implementation proposed by Tang et al. [13].

Computation of a Principal Symmetric Network. Once we have a membrane in equilibrium, we estimate the stress tensor at vertices from the forces of incoming edges, using the *extended shape operator* introduced by [4], as shown in [9]. From the estimated stress tensor, we compute the principal stresses and the corresponding principal directions. We compute then the principal symmetric directions with Eq. 9.

Quadrilateral Remeshing. From the resulting directions, we extract a quadrilateral mesh with *mixed integer quadrangulation* [3]. This method requires as input an orthogonal cross field. Since principal symmetric directions are not orthogonal, we first transform the triangular mesh by moving its vertices such that the computed directions become orthogonal through optimization. We apply then mixed integer quadrangulation on the transformed mesh and map back the result on the original shape through barycentric coordinates.

Post-optimization. We finally optimize the quadrilateral mesh for equilibrium under vertical load, additionally asking for equal forces at vertices. For that, at each unsupported vertex i with connected vertices j_1, j_2, j_3, j_4, we impose the constraint

$$\|\mathbf{f}_{ij_k}\| - \|\mathbf{f}_{ij_l}\| = 0, \quad (k,l) \in \{(1,2),(2,3),(3,4),(4,1)\}.$$

We solve this optimization problem using a Levenberg-Marquardt algorithm following the implementation of Tang et al. [13]. During the optimization, we also minimize the graph Laplacian energy of the mesh, with a small weight, to ensure smoothness of mesh polylines.

Results are shown in Figs. 1, 3, and 4. For all the presented examples, the structure is subjected to a constant area load along z direction, and it is supported by pinned joints at the boundary vertices.

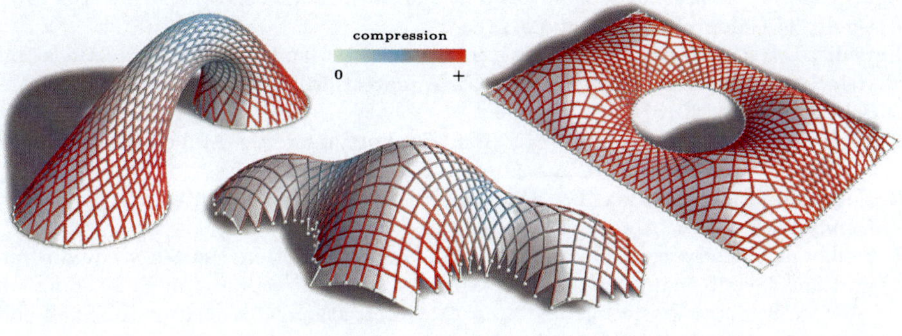

Fig. 4. Principal symmetric gridshells designed with our computational pipeline.

5 Limitations and Further Developments

Principal symmetric gridshells typically do not span planar faces, which can pose a challenge when they need to be covered with rigid panels. One solution to this problem is to enable the design of membranes that have stress characteristic directions that are also conjugate to curvature. In this case, a remeshing along these directions will result in a principal symmetric gridshell with planar faces. Another potential extension is the design of principal symmetric structures with constant forces in the beams, by computing membranes with constant stress determinant. From a theoretical perspective, an interesting direction lies in exploring the connections between principal symmetric structures and isotropic geometry.

References

1. Adriaenssens, S., Block, P., Veenendaal, D., Williams, C. (eds.): Shell Structures for Architecture. Taylor & Francis, Boca Raton (2014)
2. Block, P., Ochsendorf, J.: Thrust network analysis: a new methodology for three-dimensional equilibrium. J. Int. Assoc. Shell Spat. Struct. **48**(3), 167–173 (2007)
3. Bommes, D., Zimmer, H., Kobbelt, L.: Mixed-integer quadrangulation. ACM Trans. Graph. **28**(3), 77:1–77:10 (2009)
4. Cohen-Steiner, D., Morvan, J.M.: Restricted delaunay triangulations and normal cycle. In: Proceedings of the Nineteenth Annual Symposium on Computational Geometry, pp. 312–321. ACM, New York (2003)
5. Fraternali, F.: A thrust network approach to the equilibrium problem of unreinforced masonry vaults via polyhedral stress functions. Mech. Res. Commun. **37**(2), 198–204 (2010)
6. Kilian, M., Pellis, D., Wallner, J., Pottmann, H.: Material-minimizing forms and structures. ACM Trans. Graph. **36**(6), 173:1–173:12 (2017)
7. Maxwell, J.C.: On reciprocal figures, frames, and diagrams of forces. Trans. Roy. Soc. Edinburgh **26**, 1–40 (1872)
8. Mitchell, T.: A limit of economy of material in shell structures. Ph.D. thesis, University of California, Berkeley (2013)
9. Pellis, D., Pottmann, H.: Aligning principal stress and curvature directions. In: Hesselgren, L., et al. (eds.) Advances in Architectural Geometry 2018. Klein Publishing, Surrey (2018)
10. Pellis, D., Wang, H., Kilian, M., Rist, F., Pottmann, H., Müller, C.: Principal symmetric meshes. ACM Trans. Graph. **39**(4) (2020)
11. Pottmann, H., Eigensatz, M., Vaxman, A., Wallner, J.: Architectural geometry. Comput. Graph. **47**, 145–164 (2015)
12. Schling, E., Schikore, J., Oberbichler, T.: D-nets on rotational surfaces: equilibrium gridshell layout, symmetric to the principal stress directions. In: Proceedings of IASS Annual Symposia, pp. 1–13. no. 22. International Association for Shell and Spatial Structures (IASS) (2020)
13. Tang, C., Sun, X., Gomes, A., Wallner, J., Pottmann, H.: Form-finding with polyhedral meshes made simple. ACM Trans. Graph. **33**(4), 70:1–70:9 (2014)
14. Vouga, E., Höbinger, M., Wallner, J., Pottmann, H.: Design of self-supporting surfaces. ACM Trans. Graph. **31**(4), 87:1–87:11 (2012)

Elastic Stability of Shallow Bending-Active Arches

Gianfranco Piana[1(✉)], Fabio Bazzucchi[2], and Amedeo Manuello Bertetto[1]

[1] Department of Structural, Geotechnical and Building Engineering, Politecnico di Torino, Turin, Italy
gianfranco.piana@polito.it

[2] Senseable City Lab, Massachusetts Institute of Technology, Cambridge, MA, USA

Abstract. We present a numerical and experimental investigation on the nonlinear equilibrium paths and elastic stability of pinned shallow arches realized by buckling a straight strut and then subjected to a transverse central point load. Finite element analyses were run to compare the behavior of arches with and without initial pre-stress condition. The effects of initial shallowness ratio, axial-to-bending stiffness ratio, and initial geometric imperfections on the equilibrium path and bifurcation points were investigated. As an experimental verification, laboratory tests were conducted on a pinned-pinned shallow arch obtained by buckling a thin, slender steel bar. Static and dynamic measurements were carried out to determine the load-displacement curve, the subsequent static deformed configurations, the frequency-load curves, and the buckling load. The acquired data were used for buckling load predictions and to obtain stability domains in terms of form factors and imperfection-sensitivity parameters.

Keywords: Snap-through Buckling · Elastic Stability · Nonlinear Bifurcation · Shallow Arch · Experimental Test · Finite Element Analysis

1 Introduction

The behavior of pre-stressed arches intrigued for a long time the research in the field structural mechanics. Since the first experimental campaign by Roorda in 1965 [1, 2], the effect of pre-stressing on the buckling of arches put into shape by overcritical compression of a straight strut has proven difficult to determine in a direct way. The difficulties to overcome are many and of many species. From a theoretical standpoint (both in analytical and numerical frameworks), the geometrically nonlinear nature of the problem requires computational efforts in taking into account initial pre-stressing and imperfections while deriving the equilibrium path in different loading and geometrical conditions. Experimentally, imperfection sensitivity together with uncertainties regarding structure's stiffness, strength, geometry, actual constraint conditions, and loading transmission, pose another layer of complexity.

Simple closed-form solutions were found for the *linear* buckling load of pre-stressed arches, allowing to compare the load bearing capacity to that of the corresponding non

S. Gabriele et al. (Eds.): IWSS 2023, LNCE 437, pp. 369–378, 2024.
https://doi.org/10.1007/978-3-031-44328-2_38

pre-stressed arch [3]. However, investigating the *nonlinear* equilibrium path and buckling load requires approximate (analytical or numerical) solutions; furthermore, the above-mentioned difficulties and uncertainties make a direct comparison between experimental and numerical results very difficult; e.g. see [4].

As is well known, deep arches tend to loose stability by classic (linear) Euler buckling, whereas flat arches tend to loose stability by (non-linear) snap-through buckling. The former phenomenon is principally driven by the bending stiffness, while the latter is mainly governed by the axial stiffness. In the intermediate shallowness range, both axial and bending stiffness play a crucial role, so branching bifurcation of the nonlinear equilibrium path becomes possible, with a consequent reduction in loading capacity; initial (geometric or loading) imperfections further reduce the carrying capacity and favor the occurrence of the nonlinear limit-point (i.e. fold) bifurcation at lower loading levels. The phenomenon is governed by the shallowness ratio and by the axial-to-bending stiffness ratio of the arch. In this complicated scenario, the effect of a possible initial pre-stressing, introduced in the arch prior to the application of the additional transverse load, is not easy to determine.

In this contribution, we present part of the results of a work-in-progress about the matter. In particular, we show the results of an experimental campaign on the elastic stability of a slender, thin arch formed by a post-buckled straight beam and then subjected to a central point load. Load-displacement curves allowed to derive the equilibrium paths, thus reducing a continuous structure to a one-degree-of-freedom system. At the same time, a systematic geometry identification was carried out to determine the arch line configurations during the loading process and used to derive form factors for the buckling mechanism. High-frame rate cameras were also used to record the deflected shapes up to buckling. Moreover, the fundamental resonant frequency decay with the loading was measured, allowing for a buckling load prediction based on extrapolation [5]. Resting on this multi-level identification, we carried out a parametric ensemble of numerical analyses to provide stability domain predictions in terms of form factors and imperfection sensitivity. Results are in good accordance with experiments and enforce the importance of the problem, especially towards today's architectural trend of bending-active structures [6]. More results, with respect to what is shown here, will be presented at the Conference and in future contributions.

2 Problem Definition

Let us consider the plane problem of a pinned arch of span l_0 and rise f_0, obtained by buckling a slender straight beam of length l_n, under the action of an incremental central point load F (Fig. 1a). The arch (beam) is made of elastic material with Young's modulus E, shear modulus G, and is characterized by axial rigidity EA, bending rigidity EI, and shear rigidity GA_s, being A the area, I the principal bending moment of inertia, and A_s the shear-shape modified area of the cross-section ($A_s = A/t$, with t the shear factor). In particular, we consider a thin rectangular cross-section, of width b and thickness $h << b$. Thus: $A = bh$, $I = bh^3/12$, $A_s = 5/6\, A$; here, the effects of shearing deformability are small and can be neglected.

We call C_n the natural (i.e. stress-free) configuration, characterized by the strain state $\boldsymbol{\varepsilon}_n = \mathbf{0}$ and stress state $\boldsymbol{\sigma}_n = \mathbf{0}$. Applying an end shortening δ_0 to the straight beam, so that the beam results compressed by the post-critical force $P_0 > \pi^2 EI/l_n^2$, it reaches the pre-stressed configuration C_0, characterized by the strain state $\boldsymbol{\varepsilon}_0$ and stress state $\boldsymbol{\sigma}_0$ of the buckled beam (which is compressed and bent). Finally, applying an incremental central force F in the pre-stressed configuration C_0, the structure reaches the current (varied) configuration C, characterized by the strain state $\boldsymbol{\varepsilon}$ and stress state $\boldsymbol{\sigma}$. In Fig. 1b we have the corresponding arch where the initial curved shape coincides with the natural (stress-free) configuration, subject to the same incremental central point load F. In both cases, we call η the midpoint vertical displacement (different in the two cases) under the applied force F.

We are interested into investigating the elastic stability of the current configuration C, with respect to the applied force F, for the pre-stressed arch in Fig. 1a and for the corresponding non pre-stressed arch in Fig. 1b.

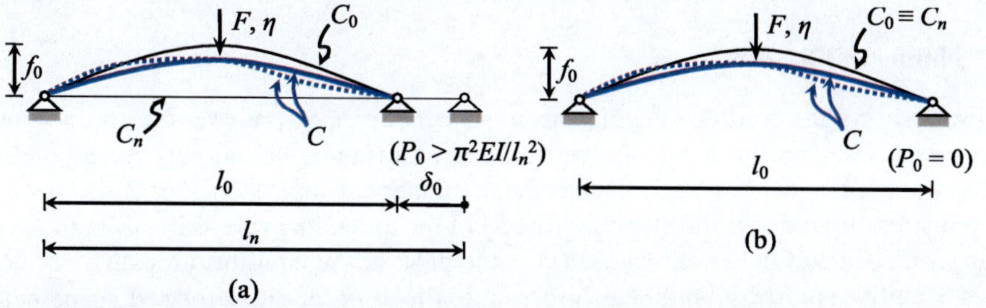

Fig. 1. (a) Pre-stressed arch made of a buckled strut, subject to incremental central point load and (b) corresponding non pre-stressed arch subject to incremental central point load. C_n is the natural (stress-free) configuration, C_0 is the pre-stressed configuration, and C is the current (varied) configuration.

3 Finite Element Analyses

Finite element analyses were run in LUSAS software (version 19.0-3 [7]). The arches were modelled by 20 nonlinear thick beam elements (BM131, a 3-node beam element with axial, bending and shearing deformability). Two different analysis typologies were conducted: linear eigenvalue (frequency and buckling) and nonlinear static. As for the eigenvalue analysis, full frontal solver was used, whereas for the nonlinear analysis a co-rotational formulation with arc-length was employed as the solution scheme. The nonlinear analyses used a prescribed displacement at the crown node to derive the equilibrium path, extracting the relevant vertical reaction (F). To initialize every analysis, a 1% of the total load factor was engaged. Nonlinear analyses were also used to apply the pre-stress by means of a prescribed support displacement. In such case, the equilibrium path with respect to the crown displacement was derived so to keep the internal stress/strain state introduced by the support displacement active. The same was done also for the frequency linear eigenvalue analysis, repeated at several load steps (i.e. different prescribed crown vertical displacements).

Several analyses were conducted to compare the behavior of arches with and without initial pre-stress. The effects of initial shallowness ratio, axial-to-bending stiffness ratio, and initial geometric imperfections on the equilibrium path and bifurcation points were investigated. Here, only the simulations related to the structure having the same features of the arch tested in the lab will be discussed. The input parameters adopted in the simulations are the following: rectangular cross-section with width $b = 20$ mm and thickness $h = 1.5$ mm ($A = 30\ \text{mm}^2$, $I = 5.625\ \text{mm}^4$, $A_s = 25\ \text{mm}^2$); isotropic, linearly elastic material with Young's modulus $E = 196{,}968.18$ MPa and Poisson's ratio $\nu = 0.3$ (see Subsect. 4.1); straight beam length $l_n = 1{,}114$ mm, prescribed end displacement (end "shortening") $\delta_0 = 14$ mm, producing the target values of arch span $l_0 = 1{,}100$ mm and arch rise $f_0 = 77$ mm. Hinged boundary condition was considered at the ends in all simulations.

The main results are summarized in the next subsection.

3.1 Numerical Results

Figure 2 shows the load (F) vs. displacement (η) curves of pre-stressed and non pre-stressed arches. The force F is the vertical reaction (numerical output) corresponding to the imposed crown vertical displacement (numerical input). In Fig. 2 we see that the pre-stressing reduces the slope (stiffness) of the ascending (pre-critical) branch and strongly diminishes the buckling load (i.e. the peak in the equilibrium path). For both cases, stability is lost by nonlinear bifurcation, the asymmetric deformed shape being the critical one. In the same figure, three deformed shapes are shown for the case with no pre-stress (load steps 1, 7, 20) and one for the case with pre-stress (load step 20). Still in Fig. 2, the linear buckling load is also marked: note that this value is the same for the two cases since the Euler critical value of the force F, obtained *via* a buckling linear eigenvalue analysis, is not affected by the pre-stress (the geometric stiffness matrix only considers the effect of the force F).

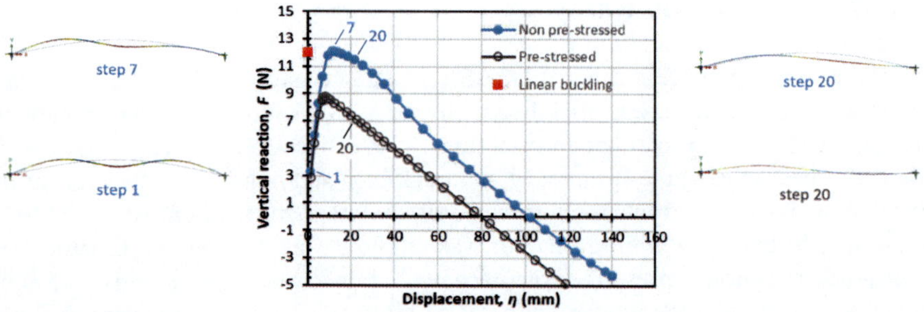

Fig. 2. Comparison between load vs. displacement ($F - \eta$) curves of pre-stressed and non pre-stressed arches.

Figure 3a shows the fundamental frequency (f_1) vs. load (F) curves for the arches with and without pre-stress. Again, the stiffness reduction induced by the pre-stressing is evident. Figure 3b shows the square of the fundamental frequency normalized to its initial value for $F = 0$ ($f_{1,0}$), vs. the force F for the two cases; note that these latter diagrams are linear. As is well known, the zero frequency condition can be used to predict the (static) buckling load [5, 8]; see Subsect. 4.3.

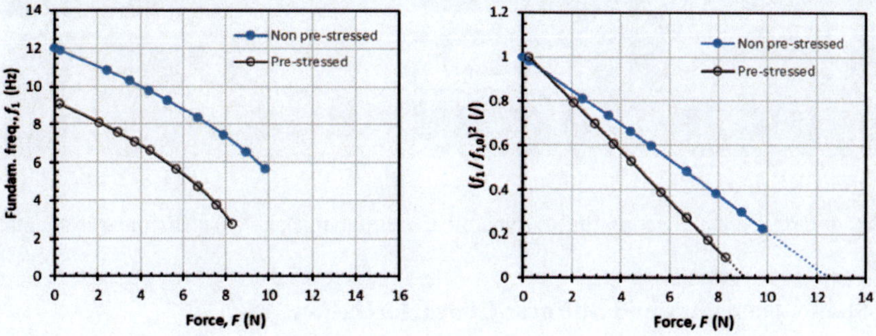

(a) (b)

Fig. 3. (a) Comparison between fundamental frequency vs. load ($f_1 - F$) curves of pre-stressed and non pre-stressed arches. Part (b) shows the normalized fundamental frequency squared vs. the force F.

4 Experimental Tests on a Pre-stressed Shallow Thin Arch

Laboratory tests were conducted on a pre-stressed, shallow thin steel arch. The arch had rise $f_0 = 77$ mm and span $l_0 = 1{,}100$ mm, obtained by axial buckling of a thin steel bar having length $l_n = 1{,}114$ mm. The arch was constrained in the vertical plane by cylindrical hinges at the ends (Fig. 4). The tested structure was characterized by the following non-dimensional parameters: slenderness ratio of the initially straight bar, $l_n/(I/A)^{0.5} = 2{,}573$; rise-to-span ratio, $f_0/l_0 = 0.07$; slenderness ratio of the arch (buckled bar), $EAl_0^2/EI = 6.45\text{E}+06$.

The tests were conducted by controlling (imposing) the vertical central force, realized in a quasi-static manner by filling a container with known quantities of fine sand, connected to the arch midpoint by a nylon line. Static displacement measurements were conducted by a Leica TCR1201 total station; the specimen was provided 11 targets along the arch line (Fig. 4b). The fundamental resonant frequency was extracted for increasing values of the applied force by analyzing the free response consequent to a small external impulsive excitation. The free dynamic response was measured at a point, located at about quarter-span, by a optoNCDT 1302-20 laser displacement transducer by Micro-Epsilon (Fig. 4a); this transducer has 10 μm resolution for dynamic acquisitions at 750 Hz top frequency, and 20 mm default measuring range. Two GoPro HERO 5 cameras were also employed to record the subsequent deflected shapes up to buckling (Fig. 4).

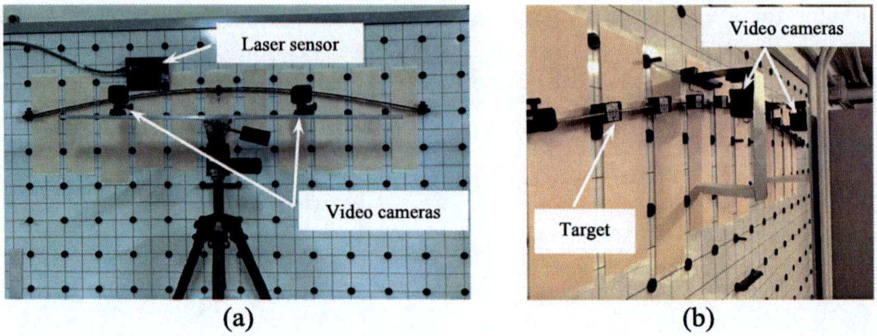

Fig. 4. Experimental set-up for (a) dynamic measurements and (b) static measurements.

4.1 Shape, Geometry and Stiffness Characterization

Given the uncertainties regarding the geometrical and mechanical features, a preliminary characterization was done.

In particular, the arch initial shape was measured both by means of the total station and of some strips of graph paper used as a reference (Fig. 4).

As regards the rectangular thin cross-section, the dimensions were measured by a digital caliper, obtaining the width $b = 20$ mm and thickness $h = 1.5 \pm 0.1$ mm (un uncertainty of $\pm$ 0.1 mm on the thickness was considered, in consideration of its effect on the bending moment of inertia I).

Lastly, direct bending tests were conducted to estimate the bending stiffness EI. The cantilever beam scheme was adopted, considering three free lengths equal to 200, 225, and 250 mm; each cantilever was tested under two different vertical tip forces, respectively equal to 0.87819 N and 1.87852 N. Under each combination of length and force, six reads of the tip deflection were done. The displacements were measured by the same laser sensor already mentioned. The resulting average bending stiffness was $(EI)_{ave} = 1{,}107{,}946$ Nmm^2.

4.2 Static Measurements: Load-Displacement Path

Figure 5a shows a comparison between experimental and numerical load vs. displacement curves. The experimental curves are limited to the pre-critical phase because the tests were conducted in force-controlled mode. Conversely, the numerical curve also shows the post-peak regime, having been obtained in displacement-controlled mode. As can be seen in Fig. 5a, the two sets of curves show a similar slope (stiffness) of the pre-critical branch. However, the experimental curves exhibited larger values of the peak load compared to the numerical prediction.

Figure 5b shows a sequence of static configurations, recorded for Test 1. It is clear how the deflection was symmetric at the early load levels, and then switched to the asymmetric configuration as the applied force was increased (compare to Fig. 2).

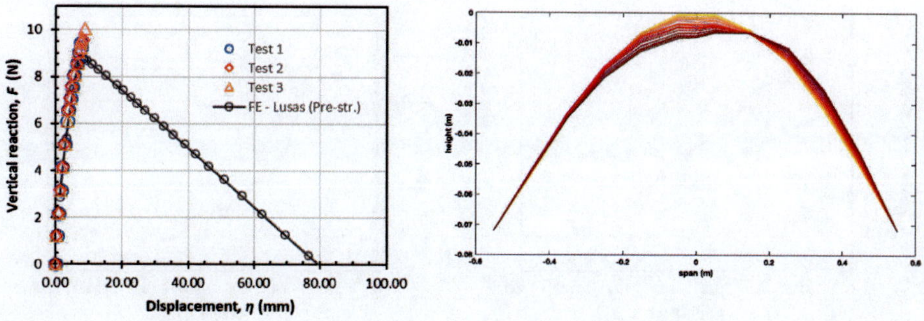

(a) (b)

Fig. 5. (a) Comparison between experimental and numerical load vs. displacement curves and (b) sequence of pre-critical static configurations for Test 1.

4.3 Dynamic Measurements: Fundamental Frequency Variation and Buckling Load Prediction

The decay of the fundamental frequency with the applied force was measured and used to predict the buckling load, according to a classic nondestructive method [5, 8]. The dynamic test was repeated three times (Tests 1 to 3). For each loading step, three acquisitions of the free response dynamic signal were done.

Figure 6a shows the fundamental resonant frequency, f_1, vs. the applied force, F, for the three tests and for the numerical simulation. We see that Tests 1 and 2 showed almost the same trend, while Test 3 exhibited a slightly different behavior, especially in the first phase; this can be ascribed to possible small changes in the initial boundary conditions, to which the system's response is pretty sensitive. Moreover, we observe that the experimental tests exhibited a stiffer response compared to the numerical one, possibly due to some friction in the hinges used in the lab, if not ascribed to an underestimated stiffness used in the computations. Extrapolating the frequency trend down to zero, provides an estimate of the expected buckling load (this estimate is quite rough).

Plotting the square of the normalized fundamental frequency vs. the applied force reduces the discrepancies between experimental and numerical curves and, also, improves the predictions of the buckling load (linear diagram in theory) [8]; see Fig. 6b ($f_{1,0}$ denotes the fundamental frequency for $F = 0$).

5 Stability Domains: Form Factors and Imperfection Sensitivity

Stability domains based on form factors are useful tools to separate the effect of different structural parameters on the overall behavior. Given the nonlinearity of the system, this separation does not produce a superimposable set, but if the system is not highly chaotic it produces monotonic trends [9, 10]. For flat arches, examples of these factors are the slenderness ratio and the shallowness ratio. Additionally, imperfection patterns or defects can be reduced to geometric factors or metrics. In this study, we made use of two form factors [10], related to the dangerous initial asymmetry of shallow arches. The

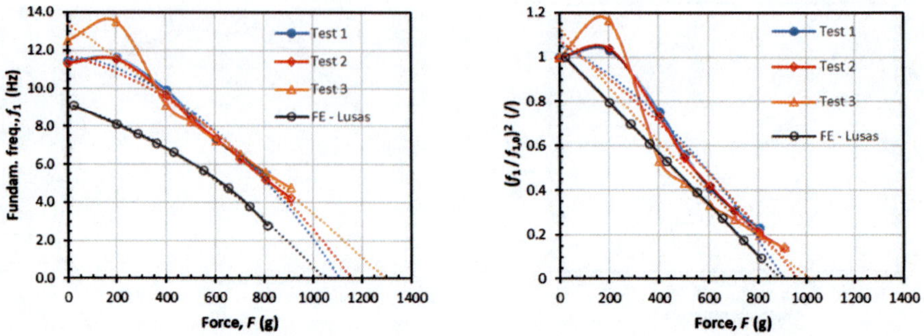

(a) (b)

Fig. 6. (a) Fundamental frequency vs. applied force and (b) normalized fundamental frequency squared vs. applied force: experimental and numerical results.

first one, Δ, is the maximum deviation in absolute value (maximum Euclidean distance) from the ideally perfect axis line (Fig. 7a), whereas the latter, Ω, is the absolute value of the difference between the maximum deviations Δ_l and Δ_r measured on the semi-spans and taken with sign (Fig .7b).

(a) (b)

Fig. 7. Form factors for an imperfect arch: (a) maximum deviation (in absolute value) from the ideal perfect arch, Δ; (b) absolute value of the difference between the maximum deviations measured, with sign, on the semi-spans, Ω.

The reference axis line (perfect) was taken as the sinusoid passing by the hinges and the crown node (with the values of l_0 and f_0 specified in Sect. 4). Form factors were calculated for every initial shape identified experimentally; an example is shown in Fig. 8 for Test 1 (compare with Fig. 5b). Interestingly, the maximum values of Δ and Ω correspond to the minimum value of the buckling load (Test 1). Anyway, if Ω acts as an ordering factor for the buckling loads, the intermediate value of Δ was observed for the highest buckling load (Test 3). Table 1 collects the discussed results.

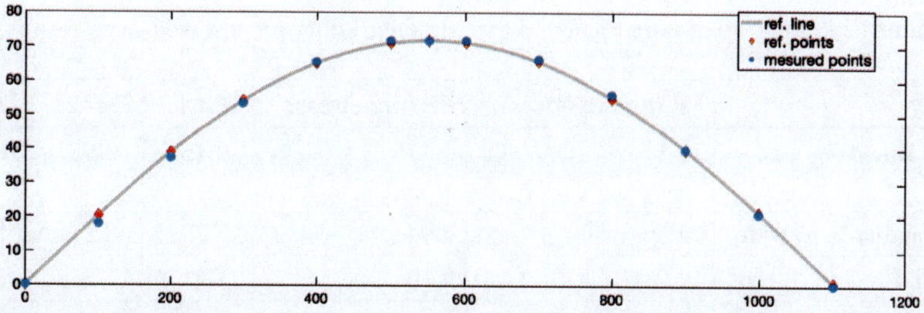

Fig. 8. Initial shape identified for Test 1 (imperfect arch) superposed to the ideal sinusoidal shape (perfect arch).

Table 1. Form factors and critical buckling loads for the three experimental tests.

	Δ (mm)	Ω (mm)	F_c (N)
Test 1	2.172	1.27	9.45
Test 2	1.830	1.12	9.55
Test 3	1.963	0.96	10.0

6 Conclusion

In connection with the problem of bending-active structures, we presented a numerical and experimental investigation on the nonlinear equilibrium paths and the elastic stability of pinned shallow arches obtained by buckling a straight strut and then subjected to a transverse central point load. Finite element analyses were conducted to compare the behavior of the pre-stressed arches with that of the corresponding arches with no initial pre-stress. As an experimental verification, laboratory tests were conducted on a pinned-pinned shallow arch obtained by buckling a thin, slender steel bar. The tested structure was characterized by the following non-dimensional parameters: slenderness ratio of the initially straight bar, $l_n/(I/A)^{0.5} = 2{,}573$; rise-to-span ratio, $f_0/l_0 = 0.07$; slenderness ratio of the arch (buckled bar), $EAl_0^2/EI = 6.45\text{E}{+}06$. The following main conclusions can be drawn.

It was found that the pre-stressing reduces both the stiffness of pre-critical branch and the buckling load of the arch. The instability occurred by nonlinear static bifurcation, with a switch from the symmetric to the asymmetric deformed shape. The dynamic criterion of stability, based on the zero-frequency condition, was effectively used to predict the (static) buckling load. A reasonable agreement was found between experimental and numerical results, although with some differences. Lastly, a suitable form factor, based only on the identified initial geometry, proved effective in ordering the buckling loads according to the initial geometrical imperfections of the tested arches. Table 2 summarizes the obtained results.

Further developments of the present work are in due course and will be presented in future contributions.

Table 2. Summary of numerical and experimental (only pre-stressed case) results.

Quantity	FE (no pre-stress)	FE (pre-stress)	Test 1	Test 2	Test 3
Nonlin. Buckling load, F_c (N)	12.1	8.77	9.45	9.55	10.0
Buckling displ., η_c (mm)	12.5	7.94	7.7	8.2	8.8
Fundam. Freq., $f_{1,0}$ (Hz)	12.0	9.10	11.4	11.3	12.5
Predicted buckl. Load, F_c^* (N)	12.5	9.10	9.1	9.7	10.2

Acknowledgments. We gratefully acknowledge Prof. Paola Antonaci for allowing us to perform the experimental tests at LaDiSSS lab of Politecnico di Torino – DISEG, as well as Geom. Mauro Perga for making the static measurements by the total station.

References

1. Roorda, J.: The instability of imperfect elastic structures. Ph.D. Thesis, University College, London (1965)
2. Roorda, J.: Stability of structures with small imperfections. J. Eng. Mech. Div. **91**(1), 87–106 (1965)
3. Thompson, J.M.T., Hunt, G.W.: On the buckling and imperfection-sensitivity of arches with and without prestress. Int. J. Solids Struct. **19**(5), 445–459 (1983)
4. Virgin, N.L., Guan, Y., Plaut, R.H.: On the geometric conditions for multiple stable equilibria in clamped arches. Int. J. Non-Linear Mech. **92**, 8–14 (2017)
5. Virgin, L.N.: Vibration of Axially Loaded Structures. Cambridge University Press, New York (2007)
6. Lienhard, J., Gengnagel, C.: Recent developments in bending-active structures. In: Mueller, C., Adriaenssens, S. (eds.) Proceedings of the IASS Annual Symposium 2018 Creativity in Structural Design, pp. 1–8. International Association for Shell and Spatial Structures (IASS) (2018)
7. LUSAS. http://www.lusas.com. Accessed 02 May 2023
8. Plaut, R.H., Virgin, N.L.: Use of frequency data to predict buckling load. J. Eng. Mech. **116**(10), 2330–2335 (1990)
9. Auslander, J., Yorke, J.A.: Interval maps, factors of maps, and chaos. Tohoku Math. J. Second Ser. **32**(2), 177–188 (1980)
10. Bazzucchi, F., Manuello, A., Carpinteri, A.: Instability load evaluation of shallow imperfection-sensitive structures by form and interaction parameters. Eur. J. Mech. A. Solids **66**, 201–211 (2017)

Postbuckling Echoes of iMRA Introduced Variation in Gridshells Mechanical Behaviour

Jonathan Melchiorre[1(✉)], Fabio Bazzucchi[2], Amedeo Manuello Bertetto[1], and Giuseppe Carlo Marano[1,3]

[1] DISEG, Department of Structural, Geotechnical and Building Engineering, Politecnico di Torino, Corso Duca Degli Abruzzi, 24, 10128 Turin, Italy
jonathan.melchiorre@polito.it
[2] Senseable City Lab, Massachusetts Institute of Technology, Cambridge, MA, USA
[3] College of Civil Engineering, Fuzhou University, Fuzhou 350108, China

Abstract. The collapse of a structure serves as a profound testament to its design quality. Forensic studies in structural engineering have revealed that collapses are intricate events with various interconnected metrics. The introduction of resilience and robustness concepts has revolutionized structural engineering, surpassing the conventional approach of designing for maximum load. This paradigm shift is especially crucial for shells and membranes. Postbuckling investigation emerged as a significant area of study since 1970, offering insights into shell design and stability. This paper focuses on the effect of the improved Multibody Rope Approach (i-MRA) for form finding on the postbuckling behaviour of gridshell structures. The i-MRA exhibits superior postbuckling behaviour, which directly relates to the method's underlying mechanics. The investigation centres around the mosque roofing in Dakar, Senegal. A comprehensive analysis was conducted, encompassing linear buckling analysis and geometrically nonlinear analysis with imperfections (GNIA) under incremental loading. The study successfully captured crucial insights into the behaviour of the structure throughout both the buckling and postbuckling phases. Additionally, a standard square-plan gridshell was analyzed to validate the findings and evaluate their generality. The research findings demonstrate the efficacy of the i-MRA and provide valuable insights for the design and stability assessment of gridshell structures.

Keywords: Form-finding · Instability · Gridshell · Snap-through · Buckling

1 Introduction

As karmic as it seems, the way of a structure collapses undoubtedly defines how well, or bad, it was designed in the first place. One of the biggest advancements, at least in cultural terms, in structural engineering has been the introduction

S. Gabriele et al. (Eds.): IWSS 2023, LNCE 437, pp. 379–389, 2024.
https://doi.org/10.1007/978-3-031-44328-2_39

of the concepts of resilience and robustness, overcoming the idea of designing towards a maximum load [1]. And if this is true, that a number expressed kilonewtons cannot solely describe a safe design, for common civil structures, it is a certainty for shells and membranes. Since 1970, postbuckling investigation became a hot topic in structural mechanics for its relevance to shell design [2,3]. Using an inverse logic, a design method that evidences a *better* postbuckling behaviour represents an improved procedure to conceive safer and more performing structures (in terms of stability) [4–6]. For the case of gridshells, hybrid little sisters of classical shells, we will show how the improved Multibody Rope Appoach (i-MRA) [7], an improved version of the Multibody Rope Approach (MRA) [8–10] for form finding, manifests a preferable postbuckling behaviour; and that this is strictly connected with the mechanics behind the method itself. The discovery of a qualitative improvement in the mechanisms of structural instability in structures generated using the i-MRA occurred while investigating the structural behaviour of the roofing of the mosque in Dakar, Senegal, with particular focus on the postbuckling phases. The roofing of the mosque, designed by the architectural firm "Fragomeli+Partners," serves as an intriguing case study. What sets this structure apart is its complex and captivating geometry, characterized by non-optimal curvatures [11], from a structural perspective. The structural analysis of the mosque's roofing was conducted through a comprehensive approach that encompassed both linear buckling analysis and geometrically nonlinear analysis with imperfections (GNIA) under incremental loading [12]. This methodology allowed for a thorough examination of the structure's behaviour, capturing crucial insights at different stages of instability [13]. The initial step involved a linear buckling analysis, which provided valuable information regarding the critical load multiplier for buckling. Building upon the linear buckling analysis, a geometrically nonlinear analysis with imperfections was performed using GNIA. This advanced technique enabled a more detailed assessment of the structural response, specifically focusing on the postbuckling phase. By applying incremental loading, the study measured the displacements corresponding to each load increment, capturing the progressive behaviour of the structure as it moved beyond its stable state. To simulate realistic conditions and account for imperfections that could influence structural stability, the imperfections imposed in the analysis were proportionate to the first mode of buckling. This comprehensive analysis provides valuable insights into the behaviour of the mosque's roofing throughout both the buckling and postbuckling phases. Lastly, to inspect the generality of the obtained conclusions, the proposed methodology was applied to investigate a standard square-plan gridshell. This additional study aimed to evaluate the consistency of the findings and ascertain if the observed trends hold true for a more common gridshell configuration.

2 The Postbuckling Behaviour of the Dakar Mosque Roof

In this section the postbuckling behaviour of the roof of a mosque planned for Dakar, Senegal, designed by the architectural firm "Fragomeli+Partners" in collaboration with "Wafai Architecture" is presented. The roof of the Dakar Mosque, illustrated in Fig. 1, showcases a unique geometry and is constructed using a gridshell system composed of tubular profiles CHS 60×3.6. The tubular profiles utilized in the gridshell are made from S355 structural steel and have a length of 2.00 m. This gridshell structure spans a rectangular area measuring 63×58 m without any intermediate supports. Consequently, it appears to be solely constrained at its edges, allowing for an open and expansive interior space.

Fig. 1. Dakar Mosque

The roof of the mosque represents an interesting case study as its structural form is defined by architectural design, which imposes constraints on the shapes achievable through traditional form-finding methods. These constraints have a tangible impact on both the quantitative and qualitative aspects of the structural response, indicating a distinct alteration in the structural stability characteristics. In fact, the roof exhibits a curvature that poses a structural weakness, resulting in a sub-optimal structural configuration. However, the use of MRA (Multibody Rope Approach) and i-MRA (improved Multibody Rope Approach) techniques allows for the definition of a form that, given the initial architectural design, achieves a structurally viable configuration while maintaining a harmonious geometry. The MRA method employs a modelling approach that represents the structure as a network comprising loose ropes and masses concentrated at the nodes. This representation allows for the depiction of the structural geometry as the inverted configuration of a hanging net. By adopting this approach, the MRA method facilitates the attainment of a funicular structural configuration, which minimizes bending actions on the structural elements. However, the quadrangular mesh of the suspended net leads to each mass being connected by four ropes, thereby constraining the three degrees of freedom associated with displacements in three-dimensional space. Consequently, certain constraints represented by the ropes become redundant, resulting in some of the ropes remaining slack in the final configuration. On the other hand, the i-MRA method aims to reduce the length of the slack ropes, ensuring that each rope is under tension in the final configuration. Regarding the specific case of the Dakar mosque, it is apparent that the geometries obtained using the two form-finding methods differ significantly. More specifically, the MRA-derived geometry ensures that only the ropes associated with the central archway along the longer side of the structure become tensioned. This establishes a structural hierarchy wherein the central arch supports the entire roof, constituting a secondary hierarchical structure suspended from the primary one. In the case of the geometry obtained through the i-MRA method, the final shape exhibits a smoother profile that closely resembles the curved surface of the architectural design. As a result, no prominent structural hierarchies are evident, and all elements appear to collaborate in bearing the loads The different in the structure geometry is evident in the linear buckling analysis, which reveals distinct values for the load multiplier for buckling between the two structural forms, as reported in Table 1.

Table 1. Dakar Mosque linear buckling eigenvalues considering symmetric and asymmetric load condition

		MRA		i-MRA		Difference %	
λ	Mode	Sym	Asym	Sym	Asym	Sym	Asym
100	**1**	82.3	69.9	63.1	40.8	23.3%	41.6%
	2	84.9	78.1	67.1	42.3	21.0%	45.8%
	3	90.7	86.2	68.0	44.9	25.1%	47.9%
150	**1**	15.1	12.4	12.2	8.3	19.5%	32.6%
	2	15.6	13.7	12.3	8.5	21.1%	38.2%
	3	16.4	14.8	12.5	8.9	23.8%	40.2%

Figure 2 presents the first buckling mode corresponding to the two different geometries, considering both a symmetrically distributed load on the roof nodes (Figs. 2(a) and 2(b)) and an asymmetrically distributed load on half of the roof (Figs. 2(c) and 2(d)). In the case of the MRA configuration, the observed buckling behavior closely resembles that of lowered arches. On the other hand, the i-MRA configuration exhibits a buckling mode that bears similarity to the typical buckling pattern observed in plates.

In order to examine the postbuckling behaviour of the structure, a geometrically nonlinear analysis with imperfections (GNIA) using incremental load analysis was conducted. Imperfections were introduced to account for potential defects

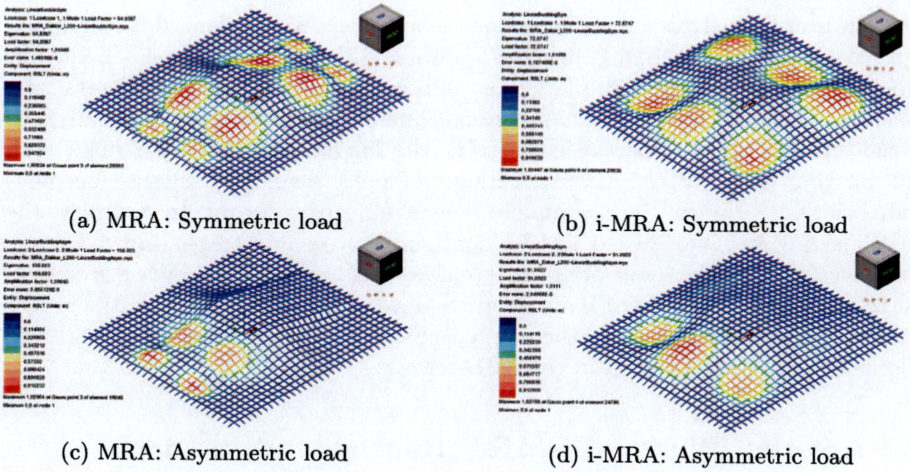

(a) MRA: Symmetric load
(b) i-MRA: Symmetric load
(c) MRA: Asymmetric load
(d) i-MRA: Asymmetric load

Fig. 2. Dakar Mosque: linear buckling mode comparison

that may occur during the production and assembly phases. These defects were modelled based on the deformation corresponding to the first mode of linear buckling, scaled to represent a maximum displacement within the accepted tolerance of 1 cm during assembly. Considering the impact of imperfections is essential for gridshell structures, as they can lead to significant deviations from ideal behaviour and affect structural stability. Figure 3 illustrates the displacement-load paths for the two geometries, considering both the symmetric and asymmetric load configurations. Notably, the disparity in postbuckling behaviour becomes more prominent when analyzing the asymmetrical loading case. Consequently, a detailed analysis focusing on this specific scenario is presented in Fig. 4.

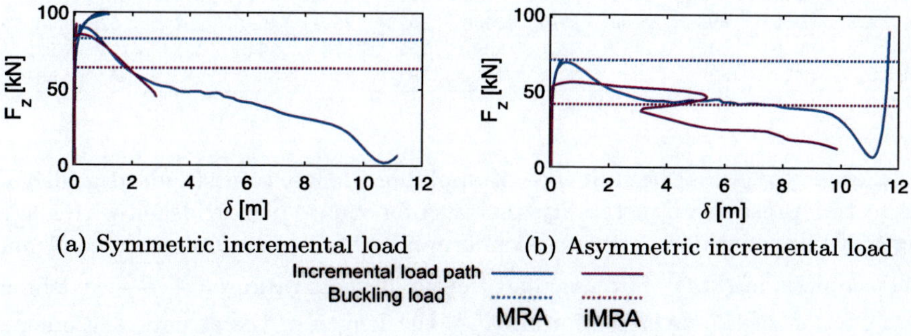

(a) Symmetric incremental load
(b) Asymmetric incremental load

Fig. 3. Displacement-load path of the central node of the Dakar mosque gridshell considering an incremental load geometric nonlinear analysis

The graph illustrates the displacements and stresses experienced by the structure along the load path, highlighting a notable difference in post-peak behavior. Specifically, for the MRA configuration, a behavior indicative of the interaction between linear buckling and snap-through phenomena is observed. Initially, the structure undergoes a partial collapse in the loaded section of the main arch during the early stages. After reaching the peak load, the collapse becomes catastrophic, resulting in a complete overturning of the roof. In contrast, the postbuckling behavior of the i-MRA configuration demonstrates improved performance, despite the lower peak load. In this case, the post-peak phase is characterized by a plateau, where the partial collapse experienced in the initial stages gradually expands. Notably, there is no sharp transition between partial and global collapse, as observed in the MRA case.

3 Are the Effects of i-MRA Relevant in Standard Gridshell Structures?

The case study of the roof covering of the Dakar Mosque has revealed that reducing slack through the application of i-MRA has a profound impact on the structural behavior, particularly during the post-buckling phase. It is worth noting that the analyzed case represents a structural anomaly in which the behaviour of the structure is heavily influenced by a combination of factors, including the architectural design, the big free-span, and variations in curvature along the roof covering profile.

Table 2. Mode 1: Euler linear buckling eigenvalue for square gridshell for both vertical symmetric and asymmetric load cases.

	Symmetric load $F_z\,[kN]$						Asymmetric load $F_z\,[kN]$					
	MRA			i-MRA			MRA			i-MRA		
$\lambda\backslash\eta$	**0.142**	**0.205**	**0.282**	**0.142**	**0.205**	**0.282**	**0.142**	**0.205**	**0.282**	**0.142**	**0.205**	**0.282**
50	2187	3861	6980	2189	3797	6956	1523	2671	4681	1524	2625	4656
100	146	256	440	146	252	439	101	176	294	101	173	292
150	26	51	87	26	50	87	18	35	58	18	34	58

In order to assess the influence of applying i-MRA to a standard gridshell, this section presents an instability analysis of a square plan gridshell with a side length of 15m, considering various slenderness ratios λ of structural elements and dimensionless heights η. In particular, the slenderness ratio $\lambda = \sqrt{\frac{AL^2}{J}}$ is defined as a function of the cross-section area A, the length of the structural elements L and the flexural inertia J while η is the ratio between the maximum height of the structure and the maximum base span. Specifically, the analysis involved calculating the linear buckling loads, as shown in Table 2. In this case, it is evident that the application of i-MRA does not introduce significant variations compared to the use of MRA, both geometrically and in terms of structural

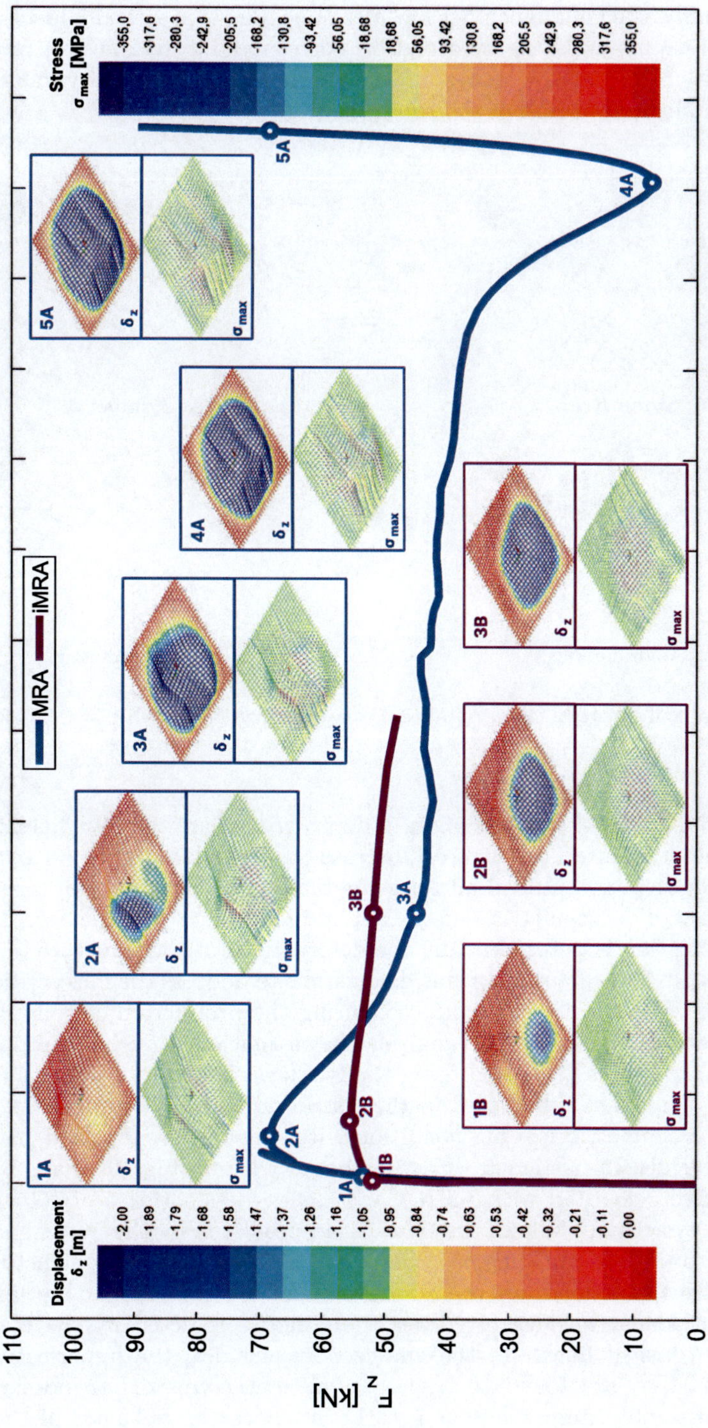

Fig. 4. Postbuckling behaviour of the Dakar Mosque

behaviour. Actually, the load multiplier for linear buckling is practically identical in both configurations, as is the deformation corresponding to the first mode of instability shown in Fig. 5, whether under symmetric loading or asymmetric loading applied only to one half of the roof covering.

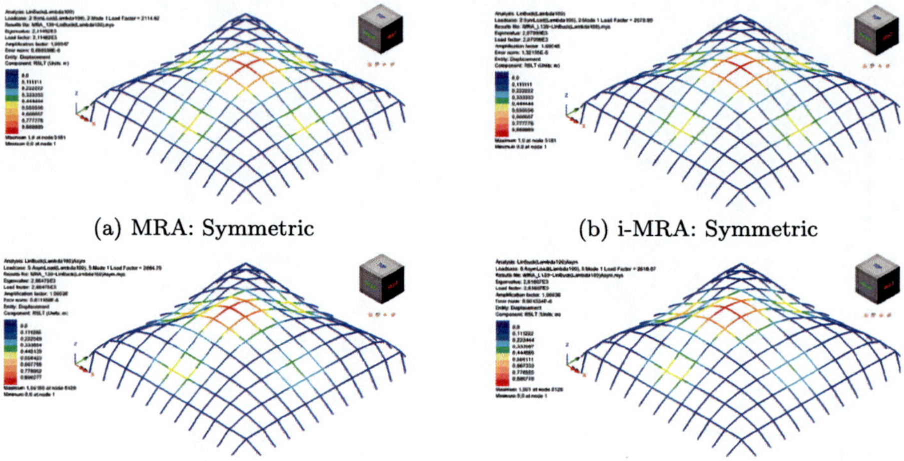

(a) MRA: Symmetric
(b) i-MRA: Symmetric
(c) MRA: Asymmetric
(d) i-MRA: Asymmetric

Fig. 5. Comparison of linear buckling mode 1 for square gridshell with $\eta = 0.205$ and $\lambda = 100$

Lastly, Fig. 6 displays the results of an in-depth analysis of the post-buckling behaviour of the structure. The figure illustrates the displacement-load paths obtained from an incremental load analysis conducted on the structure, considering the presence of a geometric defect that induces unstable behavior. The geometric imperfection is determined by the deformation of the structure in its first buckling mode. The inclusion of this defect aims to address the most critical scenario involving imperfections introduced during the production or construction of the structural elements. The analysis shows that even during the post-buckling phase, structures generated using the two form-finding methods do not appear to have significant differences in their structural behavior. Both under symmetric and asymmetric loading conditions, it is clear that the load paths exhibit similar trends. In particular, it can be observed from Figs. 6(d) and 6(f) that for structures generated with i-MRA and slenderness ratios λ of 100 and 150, the load path actually follows the bifurcation predicted by the linear buckling load. Confirming the presence of a load path bifurcation in this structure. When asymmetrically loaded, all the curves exhibit a peak followed by large displacements at loads considerably lower than those predicted by the buckling multipliers (dashed lines). In the analyzed geometries, the imperfections introduced by i-MRA are found to be less significant compared to potential defects during assembly. Imperfections greatly influence the behavior of these

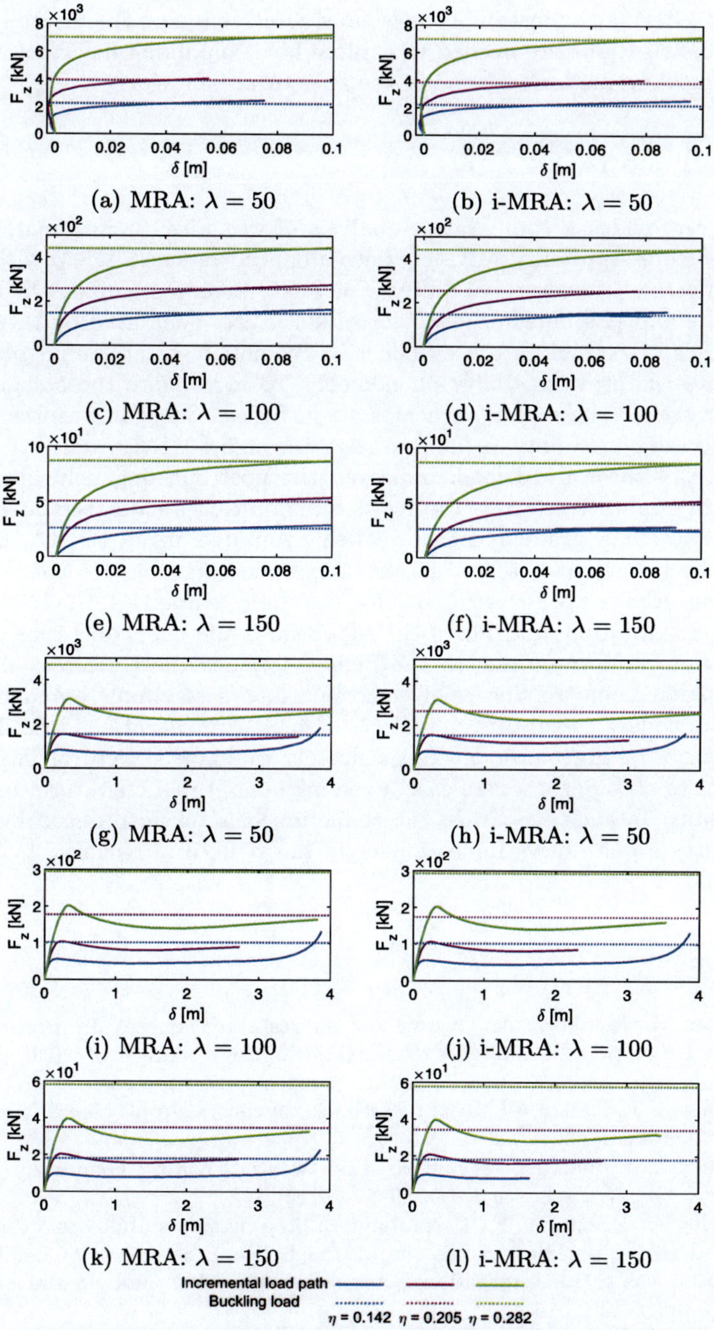

(a) MRA: $\lambda = 50$ (b) i-MRA: $\lambda = 50$

(c) MRA: $\lambda = 100$ (d) i-MRA: $\lambda = 100$

(e) MRA: $\lambda = 150$ (f) i-MRA: $\lambda = 150$

(g) MRA: $\lambda = 50$ (h) i-MRA: $\lambda = 50$

(i) MRA: $\lambda = 100$ (j) i-MRA: $\lambda = 100$

(k) MRA: $\lambda = 150$ (l) i-MRA: $\lambda = 150$

Fig. 6. Displacement-load path of the central node of the square gridshell considering an incremental load geometric nonlinear analysis structures, altering the instability mechanism and reducing the instability load. Further investigations are needed to explore how combining deformations from different buckling modes can modify the structural behavior.

4 Conclusions, So Far...

This paper provides a comparative analysis of two form-finding methods, the Multibody Rope Approach (MRA) and its enhanced version i-MRA, with a focus on their effects on structural stability. The study aims to compare the buckling mechanisms and postbuckling behavior of structures generated by these methods, specifically examining the influence of geometric variations introduced by i-MRA on structural instability phenomena. To investigate these aspects, the authors examined the roofing of a mosque in Dakar, Senegal. Surprisingly, the introduction of constraints in the geometry through i-MRA, even if resulted in a reduction of the critical load, improved the post-buckling behaviour of the structure in qualitative terms. Rather than exhibiting a catastrophic snap, the structure showed a gradual collapse when computed using i-MRA, contrasting with the behaviour observed in structures generated by the MRA method. Furthermore, the study revealed that for standard geometries, the introduction of geometric variations did not yield significant differences in terms of buckling behaviour. However, due to the highly nonlinear and complex nature of buckling phenomena, relying solely on the analysis of simple standard structures is inadequate to capture the effects of geometric variations introduced by different form-finding methods. To accurately evaluate structural integrity, it was crucial to consider a severe case involving suboptimal curvatures and other imperfections. In such scenarios, the reduction of slack introduced by i-MRA demonstrates a qualitative improvement in the structural response to buckling phenomena.

References

1. Bruneau, M., Reinhorn, A.: Overview of the resilience concept. In: Proceedings of the 8th US National Conference on Earthquake Engineering, vol. 2040, pp. 18–22 (2006)
2. Hutchinson, J., Koiter, W., et al.: Postbuckling theory. Appl. Mech. Rev. **23**(12), 1353–1366 (1970)
3. Elishakoff, I.: Probabilistic resolution of the twentieth century conundrum in elastic stability. Thin-Walled Struct. **59**, 35–57 (2012)
4. Manuello, A., Bazzucchi, F., Carpinteri, A.: Step-by-step stability analysis of shallow grid shells: buckling versus snap-through. In: Proceedings of IASS Annual Symposia, vol. 2016, International Association for Shell and Spatial Structures (IASS), pp. 1–10 (2016)
5. Bazzucchi, F., Manuello, A., Carpinteri, A.: Interaction between snap-through and Eulerian instability in shallow structures. Int. J. Non-Linear Mech. **88**, 11–20 (2017)

6. Carpinteri, A., Bazzucchi, F., Manuello, A.: Nonlinear instability analysis of long-span roofing structures: the case-study of Porta Susa railway-station. Eng. Struct. **110**, 48–58 (2016)
7. Manuello, A., Melchiorre, J., Marano, G.C.: Improved multi-body rope approach for free-form grid shells. In: Italian Workshop on Shell and Spatial Structures (IWSS 2023) (2023)
8. Manuello, A.: Multi-body rope approach for grid shells: form-finding and imperfection sensitivity. Eng. Struct. **221**, 111029 (2020)
9. Bertetto, A.M., Riberi, F.: Form-finding of pierced vaults and digital fabrication of scaled prototype. Curved Layer. Struct. **8**(1), 210–224 (2021)
10. Manuello, A., Melchiorre, J., Sardone, L., Marano, G.C.: Multi-body rope approach for the form-finding of shape optimized grid shell structures. In: Proceedings of the 15th World Congress on Computational Mechanics (2022)
11. Melchiorre, J., Manuello, A., Marmo, F., Adriaenssens, S., Marano, G.: Differential formulation and numerical solution for elastic arches with variable curvature and tapered cross-sections. Eur. J. Mech. A/Solids **97**, 104757 (2023)
12. Crisfield, M.A.: A fast incremental/iterative solution procedure that handles "snap-through. In: Computational Methods in Nonlinear Structural and Solid Mechanics, pp. 55–62. Elsevier (1981)
13. Thompson, J.: Instabilities, bifurcations and catastrophes. Phys. Lett. A **51**(4), 201–203 (1975)

Check for updates

About the Directions of Principal Eccentricities in Continuous Shells

Stefano Gabriele and Valerio Varano(✉)

Roma Tre University, via madonna dei monti, 40, 00184 Rome, Italy
valerio.varano@uniroma3.it
https://architettura.uniroma3.it/en/

Abstract. We recently extended the definition of funicularity for continuous shells, introducing the concept of Relaxed Funicularity (R-Funicularity or RF). Funicular shells are defined as shells whose static behavior is given by only local membrane actions. Extension to RF is needed as funicular shells [1] can only attain pure membrane behavior for very specific boundary conditions (bcs). The RF born also for quantifying the quality of the shape of a shell, for given load and bcs, including 'small' moments effects. Quantification of RF is made by defining a generalized eccentricity (GE) measure and verifying that the GE fall inside some eccentricity limits. Aim of this work is to discuss the nature of the GE and its associated eigenvalue problem, that allows to calculate principal eccentricities (PEs) and their directions. It is also shown how the directions of PEs are related to the relative angle between the principal directions of membrane and bending internal actions.

Keywords: Principal eccentricity directions · Relaxed funicularity · Continuous shells

1 Generalized Eccentricities and R-Funicularity

In this section we briefly describe the concept of Relaxed Funicularity, based on the generalized eccentricity; a complete discussion can be found in Gabriele et al. [2].

Let's consider a point P belonging to the surface S, endowed with its orthonormal basis $\mathbf{e}_1, \mathbf{e}_2, \mathbf{e}_3$, and a generic direction on the tangent plane represented by the unit vector $\mathbf{u} = (cos(\theta), sin(\theta))$ (see Fig. 1). The magnitudes of the generalized membrane force $N(\theta)$, acting along the direction $\mathbf{u}$ and the generalized bending moment $M(\theta)$, acting in the plane $(\mathbf{u}, \mathbf{a}_3)$ are defined as follows:

$$N(\theta) = \mathbf{u}^T \mathbf{N} \mathbf{u} \quad (1)$$

$$M(\theta) = \mathbf{u}^T \mathbf{M} \mathbf{u} \quad (2)$$

Supported by organization x.

S. Gabriele et al. (Eds.): IWSS 2023, LNCE 437, pp. 390–397, 2024.
https://doi.org/10.1007/978-3-031-44328-2_40

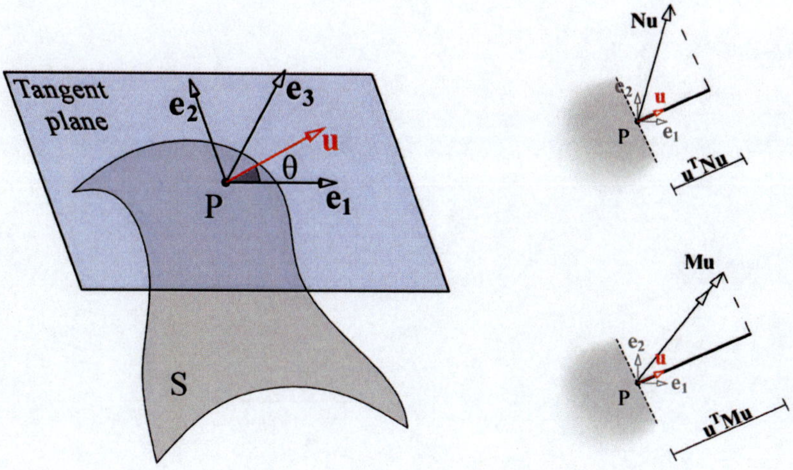

Fig. 1. Local basis and projection of internal forces [2]

being **N** and **M** the local tensors of the membrane forces and the bending moments, respectively:

$$\mathbf{N} = \begin{pmatrix} N_{11} & N_{12} \\ N_{12} & N_{22} \end{pmatrix}$$

$$\mathbf{M} = \begin{pmatrix} M_{11} & M_{12} \\ M_{12} & M_{22} \end{pmatrix}$$

The generalized eccentricity is defined as the ratio between the generalized bending moment and the generalized membrane force:

$$e(\theta) = \frac{N(\theta)}{M(\theta)} \tag{3}$$

Using the multiple-angle and the power-reduction trigonometric formulae and posing $\alpha = 2\theta$, it is possible to write the following explicit form for Eq. 1 and Eq. 2, respectively:

$$N(\alpha) = \overline{N} + \widehat{N}\cos\alpha + N_{12}\sin\alpha \tag{4}$$

$$M(\alpha) = \overline{M} + \widehat{M}\cos\alpha + M_{12}\sin\alpha \tag{5}$$

where $\overline{N} = \frac{N_{11}+N_{22}}{2}$, $\widehat{N} = \frac{N_{11}-N_{22}}{2}$, $\overline{M} = \frac{M_{11}+M_{22}}{2}$, $\widehat{M} = \frac{M_{11}-M_{22}}{2}$. Hence the reparametrized generalized eccentricity becomes:

$$e(\alpha) = \frac{N(\alpha)}{M(\alpha)} \tag{6}$$

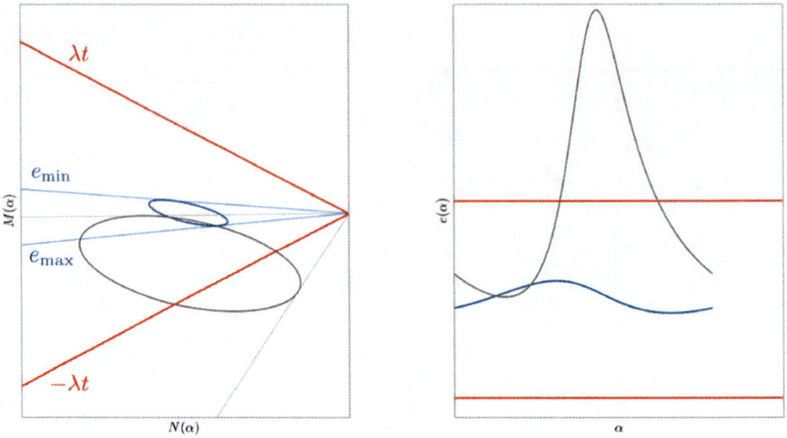

Fig. 2. (Left picture) Ellipse of eccentricity or RF ellipse, e_{max} and e_{min} are equal to the slope of the upper and lower blue lines respectively; the limits λh and $-\lambda h$ are equal to the slope of the red lines. (Right picture) Eccentricity trends w.r.t. α for the two depicted ellipses.

Equation 6 is the equation of a parametric curve in the plane $(N(\alpha), M(\alpha)) \in$ **R**, that is an ellipse when $\alpha \in [0, 2\pi]$, called *ellipse of eccentricity* or *Relaxed Funicularity (RF) ellipse*(see Fig. 2). The couples N, M represent the internal membrane forces and bending moments acting along the direction α. It is well known that a surface structure is strictly funicular if the following condition is verified.

$$M(\alpha) = 0 \quad \forall \alpha \tag{7}$$

The actual behaviour exhibited by continuous shells with non negligible thickness shows that the nullity of bending internal forces is an ideal state that never verifies. The aim of R-Funicularity is to define a criteria to consider a structure funicular, in a relaxed manner, even if Eq. (7) is not respected, when $e(\alpha)$ belongs to an admissible range, $[-\lambda h, \lambda h]$, where h is the thickness of the surface and $\lambda \in [0, \frac{1}{2}]$ an admissibility coefficient that depends on the applications. In the light of above, the Relaxed Funicularity (RF) is defined as follow: "a shell is R-Funicular if, $e(\alpha) \in [-\lambda h, \lambda h] \quad \forall \alpha \in [0, 2\pi]$, for each point of the surface". In practice, the problem of the R-Funicularity is that of verifying:

$$[\min(e(\alpha)), \max(e(\alpha))] \subset [-\lambda h, \lambda h] \tag{8}$$

For completeness, in Fig. 2 we show the graphical representation of $e(\alpha)$ in the (N, M) plane for two RF-ellipses (left picture) and the associated trends in the $(\alpha, e(\alpha))$ plane (right picture). More details and discussions about this point are given in [2] and here omitted for the sake of brevity.

From Eqs. 1, 2 and 3 it is possible to express e as the Rayleigh quotient: $e(\theta) = \frac{\mathbf{u}^T\mathbf{M}\mathbf{u}}{\mathbf{u}^T\mathbf{N}\mathbf{u}}$, and the maximum and minimum eccentricities (e_{max}, e_{min}) are found by solving the following generalized eigenvalue problem (see [3]):

$$(\mathbf{M} - e\mathbf{N})\mathbf{u} = 0 \tag{9}$$

The local extrema e_{max} and e_{min} of Eq. (9) have been calculated in the Eq. (17) of [2], based on the eigenvalue analysis performed on the *generalized eccentricity tensor*:

$$\mathbf{E} = \mathbf{N}^{-1}\mathbf{M} \tag{10}$$

the global extrema of eccentricity can be finally evaluated as follows:

$$\min(e(\alpha)), \max(e(\alpha)) = \begin{cases} -\infty, +\infty & if\ det(\mathbf{N}) < 0 \\ e_{min}, e_{max} & if\ det(\mathbf{N}) \geq 0 \end{cases} \tag{11}$$

2 Principal Directions of the Generalized Eccentricity Tensor

While in previous papers we focussed on the calculation of the eigenvalues of $\mathbf{E}$ in the present paper we want to discuss the existence and the features of the eigenvectors, i.e. the directions in which $e(\theta)$ assume local the minimum and maximum values. It is worth to recall that, although $\mathbf{N}$ and $\mathbf{M}$ are symmetric, in general the tensor $\mathbf{E}$ is non-symmetric, unless $\mathbf{N}$ and $\mathbf{M}$ commute, i.e. $\mathbf{NM} - \mathbf{MN} = 0$.

The two points to be discussed of the eigenvalue and eigenvector problem for a non symmetric tensor are:

P1. the eigenvalues can be complex,
P2. the eigenvectors are, in general, not orthogonal to each other.

It is easy to prove that the angle β between the two eigenvectors of a non symmetric 2×2 tensor $\mathbf{A}$ can be calculated by means of the following formula:

$$\cos(\Delta\beta)^2 = \frac{\|\mathrm{skw}(\mathbf{A})\|^2}{\|\mathrm{dev}(\mathbf{A})\|^2} = \frac{(\mathrm{A}_{12} - \mathrm{A}_{21})^2}{(\mathrm{A}_{11} - \mathrm{A}_{22})^2 + (\mathrm{A}_{12} + \mathrm{A}_{21})^2} \tag{12}$$

In order to understand how the geometric relationship between $\mathbf{N}$ and $\mathbf{M}$ affects the eigenvalues and eigenvectors of $\mathbf{E}$ we rewrite these by using the standard spectral decompositions:

$$\mathbf{N} = \begin{pmatrix} \cos\alpha_N & -\sin\alpha_N \\ \sin\alpha_N & \cos\alpha_N \end{pmatrix} \begin{pmatrix} N_1 & 0 \\ 0 & N_2 \end{pmatrix} \begin{pmatrix} \cos\alpha_N & \sin\alpha_N \\ -\sin\alpha_N & \cos\alpha_N \end{pmatrix} \tag{13}$$

$$\mathbf{M} = \begin{pmatrix} \cos\alpha_M & -\sin\alpha_M \\ \sin\alpha_M & \cos\alpha_M \end{pmatrix} \begin{pmatrix} M_1 & 0 \\ 0 & M_2 \end{pmatrix} \begin{pmatrix} \cos\alpha_M & \sin\alpha_M \\ -\sin\alpha_M & \cos\alpha_M \end{pmatrix} \tag{14}$$

where:

$$N_1 = \overline{N} - \sqrt{\overline{N}^2 - \det(\mathbf{N})}, N_2 = \overline{N} + \sqrt{\overline{N}^2 - \det(\mathbf{N})}$$

$$M_1 = \overline{M} - \sqrt{\overline{M}^2 - \det(\mathbf{M})}, M_2 = \overline{M} + \sqrt{\overline{M}^2 - \det(\mathbf{M})}$$

are the eigenvalues of **N** and **M** respectively, while

$$\alpha_N = \frac{1}{2}\arctan\left(\frac{N_{12}}{\widehat{N}}\right), \alpha_M = \frac{1}{2}\arctan\left(\frac{M_{12}}{\widehat{M}}\right)$$

are the angles of the principal directions of **N** and **M** with respect to $\mathbf{e}_1$.

By means of this notation the generalized eccentricity can be written as:

$$e(\theta) = \frac{2\overline{M} + \Delta M\cos(2(\alpha_M - \theta))}{2\overline{N} + \Delta N\cos(2(\alpha_N - \theta))} \tag{15}$$

where:

$$\Delta M = M_1 - M_2, \Delta N = N_1 - N_2,$$

that are the eigenvalues of the deviatoric part of **M** and **N** respectively.

In the special case in which **N** and **M** commute, then $\alpha_N = \alpha_M$.

We introduce the relative angle $\Delta\alpha = \alpha_M - \alpha_N$ between the eigenvectors of these two tensors. Following the introduced notation it is possible now to discuss the points P1 and P2, previously introduced, about the principal directions of the eccentricity tensor **E**.

P1: The existence condition for (distinct or coincident) real eigenvalues of **E** is:

$$\mathrm{Tr}(\mathbf{E})^2 \geq 4\det(\mathbf{E}) \tag{16}$$

By means of the polar representations (13) and (14) we can write:

$$\mathrm{Tr}(\mathbf{E}) = \frac{1}{2\det(\mathbf{N})}\left(4\overline{M}\,\overline{N} - \Delta M\Delta N\cos(2\Delta\alpha)\right) \tag{17}$$

$$\det(\mathbf{E}) = \frac{\det(\mathbf{M})}{\det(\mathbf{N})} \tag{18}$$

Then we distinguish between the following two cases:

case 1. $\det(\mathbf{N}) < 0$ & $\det(\mathbf{M}) < 0$. In this case the ellipse of eccentricity cross both the N and the M axis. Equation (16) is satisfied if and only if:

$$\cos(2\Delta\alpha) \leq \frac{4\overline{M}\,\overline{N} - 4\sqrt{\det(\mathbf{M})\det(\mathbf{N})}}{\Delta M\Delta N}$$
$$\text{or} \tag{19}$$
$$\cos(2\Delta\alpha) \geq \frac{4\overline{M}\,\overline{N} + 4\sqrt{\det(\mathbf{M})\det(\mathbf{N})}}{\Delta M\Delta N}$$

These ranges of values of $\Delta\alpha$ ensure that, even if the ellipse crosses both axes, it does not contain the origin.

case 2. $\det(\mathbf{N}) > 0$ & $\det(\mathbf{M}) < 0$ or $\det(\mathbf{N}) < 0$ & $\det(\mathbf{M}) > 0$ or $\det(\mathbf{N}) < 0$ & $\det(\mathbf{M}) > 0$. In this case the ellipse crosses at most one of the two axes and the Eq. (16) is satisfied $\forall \Delta\alpha \in R$.

P2: Let θ_1 and θ_2 be the principal directions of $\mathbf{E}$ and $\Delta\theta = \theta_2 - \theta_1$ the angle between the eigenvectors of $\mathbf{E}$, by means of Eq. (12):

$$\cos(\Delta\theta)^2 = \frac{\sin(2\Delta\alpha)\Delta M^2\,\Delta N^2}{4\overline{N}^2\,\Delta M^2 + 4\overline{M}^2\,\Delta N^2 - 8\overline{M}\,\overline{N}\Delta N\Delta M\cos(2\Delta\alpha)} \tag{20}$$

We can observe that the angle between the two eigenvectors of $\mathbf{E}$ depends directly on the angle between the directions of $\mathbf{M}$ and $\mathbf{N}$. In particular when eigenvectors of $\mathbf{N}$ and $\mathbf{M}$ are aligned $\Delta\alpha = 0$ implies that $\Delta\theta = \frac{\pi}{2}$ i.e. the eigenvectors of $\mathbf{E}$ are orthogonal and $\mathbf{E}$ is symmetric.

3 Minimum and Maximum Modulus Eccentricity Directions

In [3] the concept of maximum Modulus Eccentricity (ME) Surface has been introduced for shells made of no-tension materials, also used for shell's shape optimization techniques [4]. In this section we consider both the directions of max and min ME for a more general discussion.

While the sign of $N(\theta)$ has a mechanical meaning (tension vs compression), the sign of $M(\theta)$ is completely immaterial. Consequently the sign of $e(\theta)$, that indicates if the resultant of shell stresses pass above or below the middle surface of the shell, has not a role in the determination of the R-Funicularity. At authors' judgement, a more meaningful parameter is the ME:$=|e(\theta)|$. In Fig. 3 we can appreciate the difference between the global maxima and minima for the two parameters $e(\theta)$ and $|e(\theta)|$.

We can distinguish between four situations:

1. Both $\det(M) > 0$ and $\det(N) > 0$. In this case the eccentricity ellipse intersects neither the N nor the M axis (see Fig. 3.a3). The whole graph of the eccentricity lies above or below the N axis and there are not asymptotes. Both $e(\theta)$ and $|e(\theta)|$ have one local minimum and one local maximum. Local extrema are also global. (Figure 3.a1–a2)
2. $\det(M) \leq 0$ and $\det(N) > 0$. Now the eccentricity ellipse intersects only the N axis (see Fig. 3.b3), and there are one or two different values of θ such that $M(\theta) = 0 \rightarrow e(\theta) = 0$. Such values are given in the following Eq. (21) and correspond to the global minima for $|e(\theta)|$. (Figure 3.b1–b2)

$$\theta_M = \alpha_M \pm \frac{1}{2}\cos^{-1}\left(\frac{Tr(\mathbf{M})}{\Delta M}\right) \tag{21}$$

3. $\det(M) > 0$ and $\det(N) \leq 0$. In this case the eccentricity ellipse intersects only the M axis (see Fig. 3.c3), and there are two different value of θ such

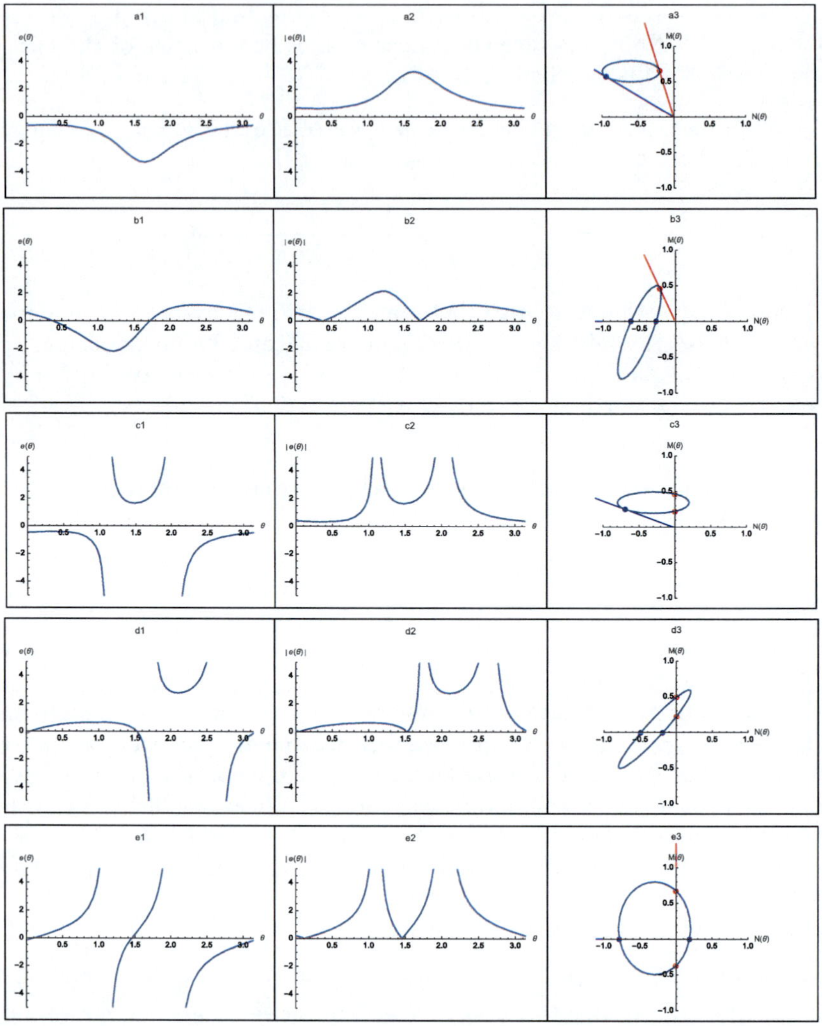

Fig. 3. Graphical representation of $e(\theta)$ (column 1), $|e(\theta)|$ (column 2) and R-Funicularity ellipse (Column 3). The parameters for each case are: case a: ($M1 = 0.5, M2 = 0.8, \alpha_M = \pi/4, N1 = -1, N2 = -0.2, \alpha_N = 0$), case b: ($M1 = 0.5, M2 = -0.8, \alpha_M = \pi/3, N1 = -0.8, N2 = -0.2, \alpha_N = -0.2$), case c: ($M1 = 0.2, M2 = 0.5, \alpha_M = \pi/4, N1 = -0.8, N2 = 0.2, \alpha_M = 0$), case d: ($M1 = -0.5, M2 = 0.6, \alpha_M = \pi/4, N1 = -0.8, N2 = 0.2, \alpha_N = \pi/5$), case e: ($M1 = -0.5, M2 = 0.8, \alpha_M = \pi/4, N1 = -0.8, N2 = 0.2, \alpha_N = 0$).

that $N(\theta) \rightarrow e(\theta) = \infty$. Such values correspond to two global maxima for $|e(\theta)|$, (Fig. 3.c1–c2).

$$\theta_N = \alpha_N \pm \frac{1}{2} \cos^{-1} \left(\frac{Tr(\mathbf{N})}{\Delta N} \right) \tag{22}$$

4. Both $\det(M) \leq 0$ and $\det(N) \leq 0$. In this last case the eccentricity ellipse intersects both the N and M axis, containing or not the axes origin (see Figs. 3.d3 and 3.e3). Now there are two different values of θ such that $e(\theta) = 0$ and two other values such that $e(\theta) = \infty$. These couples of θ values represent one or two global minima and one or two global maxima for $|e(\theta)|$. (Figure 3.d1–d2). When the ellipse contains the origin of the axes a very special case occurs, for which there are no local minima (Fig. 3.e1–e2).

4 Conclusions

In this work we presented a mathematical discussion about the directions of minima and maxima of generalized eccentricities and modulus eccentricities for a continuous shell, independent on the constitutive assumptions. The only assumption is that the internal fields of membrane forces and bending moments tensors are known in each point of the shell. This discussion is propaedeutic to applications, where the directions of eccentricity extrema could be considered as stiffening or reinforcement directions. It has been shown that, due to non symmetric nature of the eccentricity tensor $\mathbf{E}$, the principal eccentricity directions are, in general, not orthogonal and, in some cases, not well defined (more than one global maxima or minima). This property depends on the relative angle between the principal direction of $\mathbf{N}$ and $\mathbf{M}$ tensors. It has been also shown as the Modulus Eccentricity could be better used for interpreting the eccentricity directions of the global extrema (even when these eccentricities assume zero or infinite values).

References

1. Adriaenssens, S., Block, P., Veenendaal, D., Williams, C.: Shells Structures for Architecture: Form Finding and Optimization. Routledge, London (2014)
2. Gabriele, S., Varano, V., Tomasello, G., Alfonsi, D.: R-funicularity of form found shell structures. Eng. Struct. **157**, 157–169 (2018)
3. Lucchesi, M., Padovani, C., Pasquinelli, G., Zani, N.: The maximum modulus eccentricities surface for masonry vaults and limit analysis. Math. Mech. Solids **4**, 71–87 (1999)
4. Argento, G.R., Marmo, F., Varano, V., Gabriele, S.: Shells' shape optimization based on R-funicularity. In: Behnejad, S.A., Parke, G.A.R., Samavati, O.A. (eds.) Proceedings of the IASS Annual Symposium 2020/21 and the 7th International Conference on Spatial Structures–Inspiring the Next Generation, Guildford, UK, 23 – 27 August 2021, pp. 2548-2556 (2021)

The Use of Pneumatic Vacuum for the Assembly of Innovative Stereotomic Structures

Dario Costantino(✉), Giuseppe Fallacara, and Francesco Ciriello

Department of Architecture, Construction and Design, Politecnico di Bari, Bari, Italy
dario.costantino@poliba.it

Abstract. Bonding lithitic blocks using a pneumatic vacuum is a born by fortune research that started in 2006 at the exhibition Città di Pietra curated by prof. Claudio D'Amato Guerrieri at the X International Architecture Exhibition in Venice. In this occasion, professors D'Amato and Fallacara presented the Alexandros Obelisk. During its deconstruction, it was not possible to proceed with the separation of the ashlars due to the particular conformation of the aforementioned joints whose shapes had created a bonding through pneumatic vacuum. This situation inspired the idea of an innovative construction process of vaulted spaces without the use of centrings. A first attempt has been done by Giuseppe Fallacara in 2019 with a lithic oblique obelisk. This paper aims to illustrate the further research that is being carried on in order to apply the vacuum for structures made with innovative materials and technologies as 3D printing.

Keywords: Stereotomy · Architecture · 3D printing · Pneumatic vacuum

1 Introduction

Building practices are strictly linked to the use of connecting materials like mortars and temporary structures like centrings, without which it would often be impossible to work and build even the simplest and smallest architectures. Despite that, the evolution of technology, the study and updating of the legacy of tradition and the ceaseless flow of experiments are changing this situation and new methods and techniques are being developed in order to find new and faster ways to assemble manufacts.

This paper aims to focus on the analysis and description of one of these new assembling procedures, that is based on the use of pneumatic vacuum, combining the architectural theme of stereotomy with the scientific heritage of the Magdeburg hemispheres.

2 Research Background

2.1 Stereotomy

The word "stereotomy" derives from the Greek στερεός ("stereo", which means "firm, solid") and Τομή ("tomia", which means "cut") and it is used to mean the art of cutting solids in order to obtain the best configuration for a constructive system composed

S. Gabriele et al. (Eds.): IWSS 2023, LNCE 437, pp. 398–407, 2024.
https://doi.org/10.1007/978-3-031-44328-2_41

of smaller components. Adequately shaping ashlars leads not only to efficient static configurations, but also to material and time savings and eventually to obtain aesthetically interesting patterns respecting the Vitruvian triad: firmitas, venustas, utilitas. The stereotomic discipline was defined for the first time by Philibert Delorme in Le Première Tome de l'Architecture, in the SixteenthSeventeenth Century but it has been studied and improved during the years with the implementation of new technologies, becoming the center of a specific research field of the ArCoD department of the Polytechnic of Bari, started by prof. Claudio D'Amato Guerrieri and now carried on by prof. Giuseppe Fallacara [1, 2] (Fig. 1).

Fig. 1. Hyparvault, a stereotomic prototype by Giuseppe Fallacara and SNBR (2017).

2.2 Previous History

Aristotle believed that vacuum was physically impossible, since no region could be totally empty. This theory was confuted in the 17th century by the physicist Otto von Guericke, who demonstrated the existence of vacuum through the experiment of the so called "Magdeburg hemispheres". He used a spherical device composed of two brass hemispheres with a seal around the edges from which it was possible to extract air. It was demonstrated that once the entire sphere was assembled and the air was completely removed, it was almost impossible to separate them. This happens because the pressure inside the sphere is lower than the atmospheric pressure on the surface of the sphere.

Indeed, an illustration made by Gaspar Schott testifies the efficiency of this device showing the experiment that was made in 1654 consisting of using sixteen horses to pull the hemispheres in order to try to split the sphere. Even the power of the team of horses was not enough to separate the two halves [3, 4] (Fig. 2).

The assembly construction technique proposed is based on the experiment of the Magdeburg hemispheres, since it aims at facilitating the reciprocal adherence of ashlars

Fig. 2. Gaspar Schott's illustration of the vacuum experiment.

by removing air in correspondence of the joints, where an air chamber is purposely modeled.

2.3 Past Experience

Bonding lithic blocks using pneumatic vacuum is a born by fortune research that started in 2006 at the exhibition Città di Pietra curated by prof. Claudio D'Amato Guerrieri at the X International Architecture Exhibition in Venice. In this occasion, professors D'Amato and Fallacara presented the Alexandros Obelisk, which was a 15.5 m tall lithic monument split into six blocks built with the French group "Les Compagnon du Devoir du Tour de France", the Atelier Romeo and the Leopizzi enterprise [5]. Even though it was necessary to use cables and mortars in order to respect the safety rules, every block was cut following stereotomic principles, and so joints with a particular double-curvature shape were made in order to help ashlars get stuck in each other easily (see Fig. 3). This shaping led to an unexpected result: during the deconstruction of the obelisk, professors Fallacara and D'Amato's team noticed that it was almost impossible to proceed with the separation of the ashlars: as a matter of fact, the peculiar conformation of the aforementioned joints had led to the creation of a strong bonding through pneumatic vacuum. Only a fortuitous lateral movement of the building machine led to solve this problem and also to recognize the interstitial pneumatic vacuum [6].

This event inspired the idea of an innovative construction process of vaulted spaces without the use of centrings but using the vacuum instead. As a matter of fact, centrings

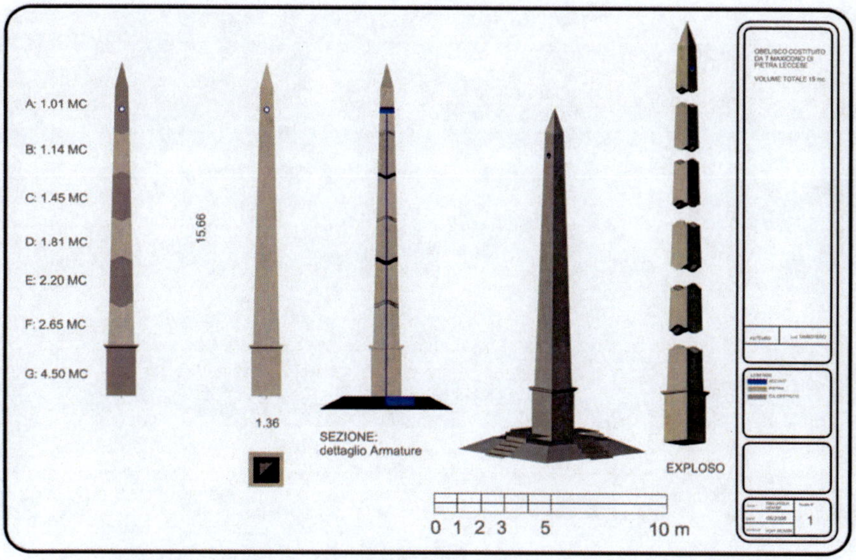

Fig. 3. Drawings of Alexandros Obelisk (© Giuseppe Fallacara)

are often expensive, they produce a great amount of material waste and their realization can be complex, especially if the vault to be built has not a regular shape.

A new focused attempt was made by Giuseppe Fallacara in 2019 with a new small lithic oblique obelisk (see Fig. 4). For this experiment four ashlars made of compact French limestone Anstrude were assembled using a simple pump that passed through the blocks, removing the air from the interstitial spaces carved in the contact surfaces. Despite the overhang, the ashlars of the obelisk remained reciprocally attached [6, 7].

Some difficulties in the realization of the system were linked to the necessity to have an extreme precision in the joints cut and not to have cracks, that could compromise the outcome of the experiment, but which can occur in stone.

The success of this experiment led to try the pneumatic vacuum again with new materials and technologies as 3D printing.

3 The Experiment

Stone has proven itself as an effective material for the vacuum, but how do other materials work? Moreover, how do new and unconventional materials work?

In order to answer these questions a new experiment has been conducted in 2023 focusing on the use of 3D printing technologies and PLA (polylactic acid) as a reference material. Therefore, a three-block flatbed has been 3D printed in order to try the pneumatic vacuum and see how plastic materials work.

Fig. 4. The pneumatic vacuum experiment made in 2019.

3.1 Phase 1: Modeling

The first step consisted in 3D modeling both the reference object and the auxiliary pieces (for example the coupling piece between pump and pipes) in Rhinoceros. The flatbed has been thought to be divided into three blocks and two small air chambers were designed in correspondence of the joints, creating lens-shaped inner air cavities to be linked to a pair of suction tubes. These lenses were the heart of the project, since they would have been the element that could allow the pneumatic vacuum to work subtracting the fluid (air) through the above-mentioned tubes.

A common vacuum cleaner modified for the occasion has been used for the suction. Since it was important not to lose air during the process, a particular junction has been modeled in order to join the vacum cleaner hose and the tubes.

3.2 Phase 2: 3D Printing

Once every piece was considered complete and ready to be printed (every model was verified to be a closed polysurface), they were sent to Simplify3D, that is a common slicing software (see Fig. 5). A FDM artillery X1 printer has been used.

Here a new check has been done before starting the printing phase using white PLA as material and a 0,4 mm nozzle. Each block was printed with a grid-shaped infill. The infill amount was 20% and the walls were 1,6 mm thick (see Fig. 6).

The printing process almost took 12 h for each block, whose final weight was about 0.35 kg.

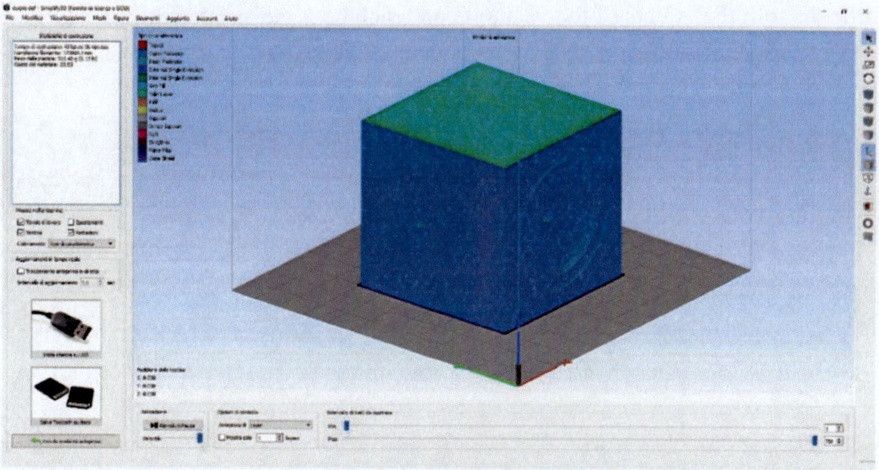

Fig. 5. One of the blocks inside the slicing software Simplify3D.

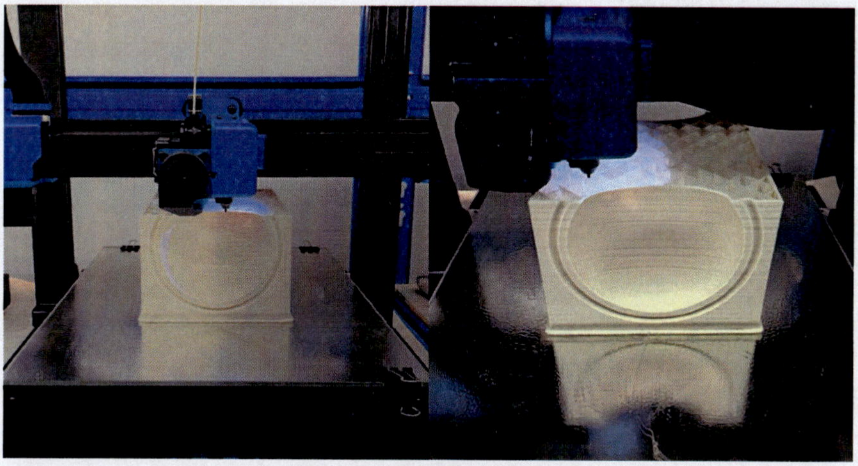

Fig. 6. The printing process.

3.3 Phase 3: Realization

Once the parts of the objects had been printed it was important to manually refine them in order to guarantee the vacuum to properly work. So, the lenses were bounded by circular gaskets that allowed the watertight seal for the air intake through a tube connected to a vacuum pump. This fluid was free to penetrate the lens through a diagonal hole made in the block near the joint.

Also, the pump had to be modified: as already mentioned it was necessary to find a way to adapt and link the vacuum and the 3D printed flatbed. To accomplish this task the 3D printed junction was fixed at the end of the hose and linked to two silicone tubes in order to split the effects of the vacuum and reach both the air chambers (see Fig. 8) (Fig. 7).

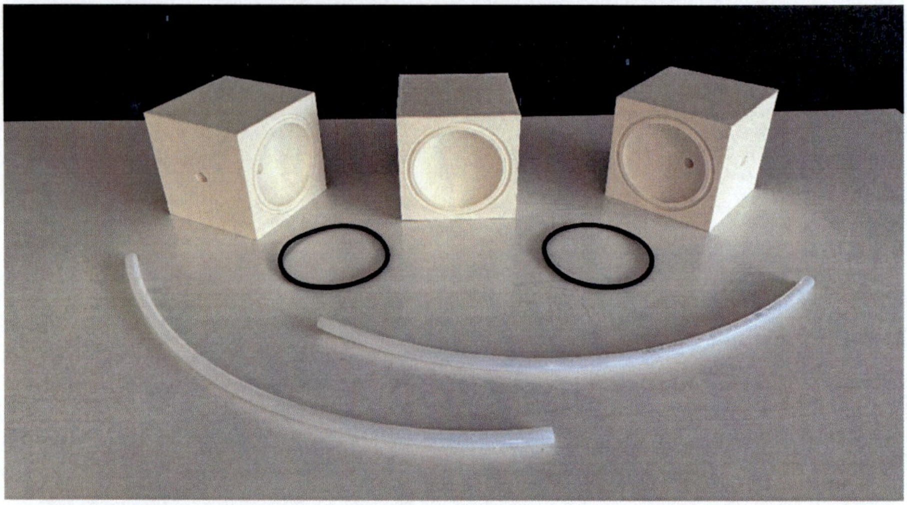

Fig. 7. Photo of all the components of the flatbed.

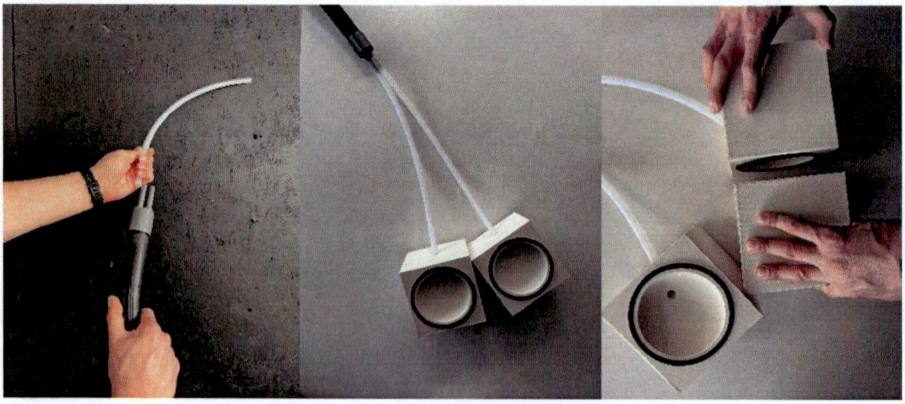

Fig. 8. Building process of the flatbed.

3.4 Phase 4: The Experiment

The experiment itself consisted in accomplishing a simple task: building the flatbed while the air kept being sucked by the vacuum and test its resistance putting some weight on it. So, once the three pieces were assembled, the entire flatbed was positioned with the ends rested on two tables and the central block free from any external support (see Fig. 9).

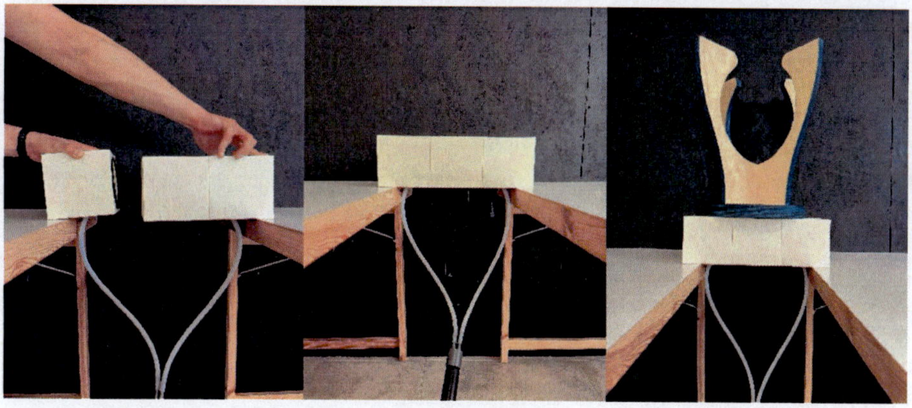

Fig. 9. Experiment.

Trials were made both with some items in the lab, and with precise weights put on the central block (see Fig. 10). In the end the flatbed weighted almost 1 kg and managed to bear a 5 kg stool.

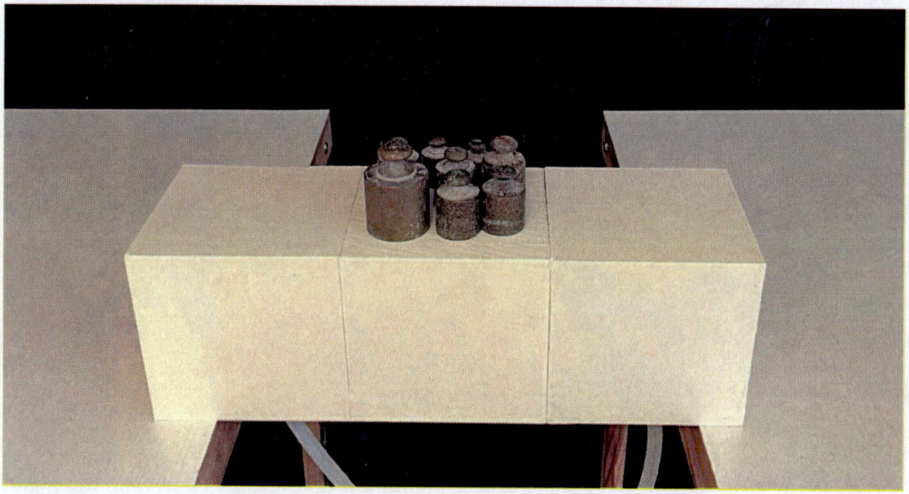

Fig. 10. 2 kg were put on the flatbed to test its endurance.

4 Discussion and Conclusions

Even if the experiment was conducted with small objects and overall light weights, the successful ending leads to the important acknowledgment that a different way of thinking the building procedures can be positively considered not only with ancient or traditional materials and geometries, but also with the new ones.

The possibility of building without the aid of centrings or other tools will considerably speed up the assembly and disassembly operations of different kinds of architectures, leading both to material and money savings; at the same time this technology could lead to build complex stereotomic architectures made of ashlars without planar faces.

Even if this construction process shows practical advantages in the realization of stereotomic structure, it is necessary to underline that it requires some technical limitations. As a matter of fact, the creation of vacuum implies that the two ashlars must perfectly fit each other, without any leak. Therefore, imperfection linked to fabrication are not allowed. Moreover, air leaks can be caused also by cracks, that usually don't happen with a plastic material, but can occur with a lithic material.

A future development of the research will deal with the realization of more complex structures: Fig. 11 shows a hypothesis made for a catenary arch, whose joints are conceived to be assembled through pneumatic vacuum. This would be the direct development of the experiment made for a flatbed.

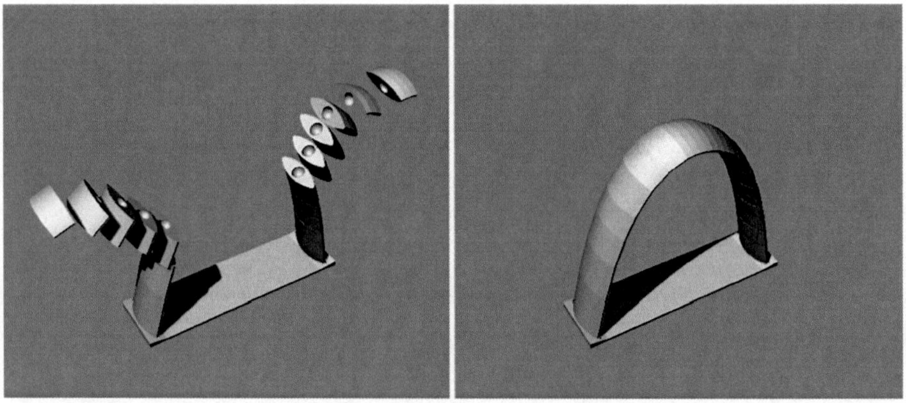

Fig. 11. Hypotesis of a catenary arch to be built through pneumatic vacuum.

References

1. Fallacara, G.: Verso una progettazione stereotomica. Nozioni di stereotomia, stereotomia digitale e trasformazioni topologiche: ragionamenti intorno alla costruzione della forma. Towards a stereotomic design. Notions of stereotomy, digital stereotomy and topological transformations: reasoning about the construction of the form. ARACNE editrice, Rome (2007)

2. Fallacara, G., Gadaleta, R.: Stereotomy: architecture and mathematics. In: Sriraman, B. (eds.) Handbook of the Mathematics of the Arts and Sciences, pp. 1–20. Springer, Cham (2019). https://doi.org/10.1007/978-3-319-70658-0_85-1
3. Boyer, T.H.: The classical vacuum. Sci. Am. **256**, 70–79 (1985)
4. Mona, C.: Magdeburg hemispheres. Phys. Today **74**(12), 68 (2021)
5. Fallacara, G.: Claudio D'Amato Guerrieri e la "scuola barese" di architettura. Storia e prospettiva dal basso. In: Fallacara, G., Restucci, A. (eds.) Claudio D'Amato Guerrieri e la "scuola barese" di architettura. A trent'anni dall'istituzione del Politecnico di Bari e della Facoltà di Architettura, pp. 3–26. Gangemi Editore, Roma (2020)
6. Aimar, F.: Le pietre sono unite con la forza dell'aria: l'esperimento al Politecnico di Bari. Teknoring (2020). https://www.teknoring.com/news/materiali-da-costruzione/stereotomia-politecnico-di-bari/. Accessed 19 May 2023
7. Cavaliere, I.: Obelisco obliquo con vuoto pneumatico. In: Fallacara, G. (ed.) My Archisign. Architettura e design in pietra. Pensieri, opere e visioni, pp. 120–123. Gangemi Editore, Roma (2022)

Stereotomic Fiber Architecture: Textile Architecture and Parametric Fibrous Structures

Alessandro Angione(✉)

Department of Architecture, Construction and Design, Polytechnic of Bari, Via Amendola 126/B, 70126 Bari, Italy
a.angione1@phd.poliba.it

Abstract. Parametric structures in composite material, such as fiber glass and carbon fiber with bicomponent epoxy resin, are usually constructed through the "robotic coreless filament winding" technique, which consist of weaving each fiber element according to a specific numerical parametric definition.

This paper illustrates various designs of pavilions in composite materials and discusses architectural implications of using this technology in terms of tectonics and space generated.

The topic of research for this paper is the analysis and the description of fiber pavilion designed using the methods of computational design and digital fabrication of spaces and structures.

The aim of this research project is to blend textile technology with the stereotomic knowledge of stone construction to parametric design and digital fabrication in order to prototype shell-like structure.

One of these is a tree-shaped structure with the stem and the crown which are joined together; or a fiber arch made by triangular blocks. They are a starter point of new structural projects for covered spaces.

New construction technology and typology to design new structural shapes are the two goals to demonstrate in lithic and stereotomic field, to transform them into complex but light structures.

The next stop of the research will be building a fiber tripod pavilion.

Keywords: Form Finding · Fibre · Composite Materials · Stereotomy · Shell · Computational Design · Digital Fabrication

1 Introduction

This research has the aim to join the tradition with the innovation in the field of fiber architecture [1], about the development of spatial structures or coverings of spaces designed through a parametric modelling with composite materials.

This paper starts with a small background of textile architecture to prove the origin of this study since the primitive architecture, to the innovative techniques of digital fabrication and computational design.

The new contribution of the research consists in studying the possibility of constructs parametric and computational structures [2], in carbon fiber and glass fiber, or basalt fiber

S. Gabriele et al. (Eds.): IWSS 2023, LNCE 437, pp. 408–415, 2024.
https://doi.org/10.1007/978-3-031-44328-2_42

respecting the architectural compositional rules of stereotomy and vaulted structures in cut stone.

The outputs of the research are two prototypes, the fiber arch in collaboration with the ICD University, and the fiber tree, a project bio inspired. They are developing from the analysis of two study cases: The Buga Fibre Pavilion and The Flux Pavilion.

This paper gives the contribution in studying the methods of design of parametric structures to generate prototypes, which validates the study carried out on the 3d model.

2 Background

2.1 A Bridge From Textile Architecture to Fiber Architecture

Architecture comes from the need to protect themselves from the forces of nature, and it is from the *Primitive Hut* [3] that begins the textile architecture. Huts have been built since the paleolithic period using stones for the walls, wood, straw and leaves for the ceiling, dung, mud, and vegetable fibers as binder. The first drawing, inspired by the Vitruvian description of the primitive hut, represents a structure consisting of four tree trunks with forked ends on which rests a gable roof. The French Marc-Antoine Laugier (1713–1769), in his treaty on architecture (*Essai sur l'architecture,1753*) wrote in Paris, states that a quality architecture must be legitimized by nature and reason. Laugier with this treatise argues that the primitive hut, consisting of an assembly of parts of trees that constitutes the origin of the classical temple and the house. Gottfried Semper followed as a great theorist of the 19th century. From 1850 he began to write his theories of architecture on the origin of it with a distinctly positivist character. The hut, as the first sign of human living after the caves, is found in many of his writings, especially in *Der Stil* [4], in which there is a radicalized idea of a hut. The hut is therefore made of wood, known as the first form of dwelling, the semperian model defines the Caribbean hut. It meets all the requirements of the primitive hut: a roof made of woven bamboo canes, a supporting structure tied in coir fiber ropes and palm leaves as cover (Fig. 1).

"The four elements of architecture seen as principles of living have shaped the hut: the production of heat and food - then the element of the hearth - of defense from the outside then the element of the fence - from the weather, then the roof - of defense from the floods then the embankment."

The structure of the walls supports this theory. In the hut, walls were not formed by masonry or wood but by carpets, then by woven fabrics that divided the infill of the house. This characteristic of interweaving highlights the origin of the research of the object of this paper since antiquity and from architectural and philosophical thought.

2.2 Stereotomic Shell of Stone Architecture

Stereotomy is a field of experimentation appropriate to demonstrate the accuracy of the new theoretical assumption, able to combine the architectural style with its constructive substance. To define stereotomic design [5], means to assume a theoretical-practical paradigm of reference suitable for the realization of architectures conceived, according to the use of systems or architectural elements resistant in their own shape through discrete elements/ashlars (Fig. 2).

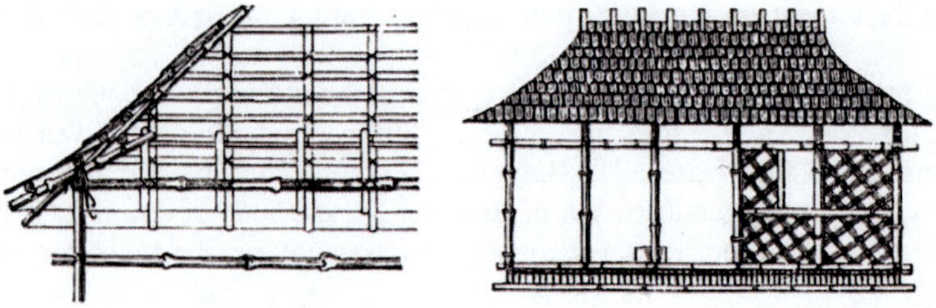

Fig. 1. Gottfried Semper, Caribbean hut, 1851

This design refers to the stone construction, where the concepts of "ligament" and "assembly" show the natural phenomena of composition of the material in a morphological way and in the architectural scale.

The design research is based on the traditional techniques of building stone from cutting, revisited in the modern computer technologies of computational design and digital fabrication [6], reformulating and updating the principles underlying the creation of the form for the structuring of a new stone architecture. This is in continuity with the traditional architecture of the stone construction.

The relationship between 3d modeling, cutting stone architecture, and design though digital fabrication is one of the peculiarities of the Bari school of architecture. In order to understand the cultural specificity of its territory of belonging.

Fig. 2. G Fallacara, M. Barberio, Portion of Infinity, exposed at the Garden of the Virgins,on the occasion of the 17th International Architecture Exhibition of the Venice Biennale, 2021.

2.3 Fiber Architecture (ICD)

One of the case studies analyzed is the Buga Fibre Pavilion [7]. The pavilion (Fig. 3) demonstrates how the combination of the latest state-of-the-art computational technologies with naturally occurring construction principles allows the development of truly new and digital construction systems. The pavilion structures were produced through a robotic set up with fiber filaments impregnated with resin. This structure is not only highly effective and exceptionally light, but also provides a unique architectural expression and an extraordinary spatial experience, just like a stereotomic domed space.

According to the study carried out by the project team with reference to biology, most of the supporting structures are made of composite fibers, such as cellulose, chitin or collagen, and a matrix material that supports them. The stunning performance and unparalleled efficiency of these carbon fiber and glass structures comes from these fibrous biological systems. This case study has been very useful for the deepening of the research on stereotomic fibrous structures that we will see in the next paragraphs. Especially for how much it looks for the set up and the textures of the ashlars.

Fig. 3. ICD/ITKE University of Stuttgart, Buga Fibre Pavilion, Heilbronn 2019.

3 Fiber Architecture Prototypes

3.1 Digital Fabrication and Computational Design for Project

A significant part is the implementation of computational design and digital fabrication in the architectural project of my research, with the creation of a parametrical code for the visualization of the study models [8].

Computational design is relevant to this research project primarily because it is for able to model a complex weaving pattern and define a weaving pathway.

The parametrical code (Fig. 4) is important to control the design with a computational model, but it is also important to make a little prototype to verify the project for the real future digital fabrication process.

The algorithm of weaving is generated through the construction of a code in Grasshopper. This plug-in is a visual scripting interface for 3D modelers like Rhinoceros, that allow you to create some parametrical code to obtain a specific 3d model. Another component that is very important for the creation of the code is Phyton, which is a programming language, and it is implemented in Grasshopper.

Prototyping takes place when the virtual model of project is completed. This can be done through robotic programming (Fig. 5) or through a traditional digital fabrication with numerical control machines.

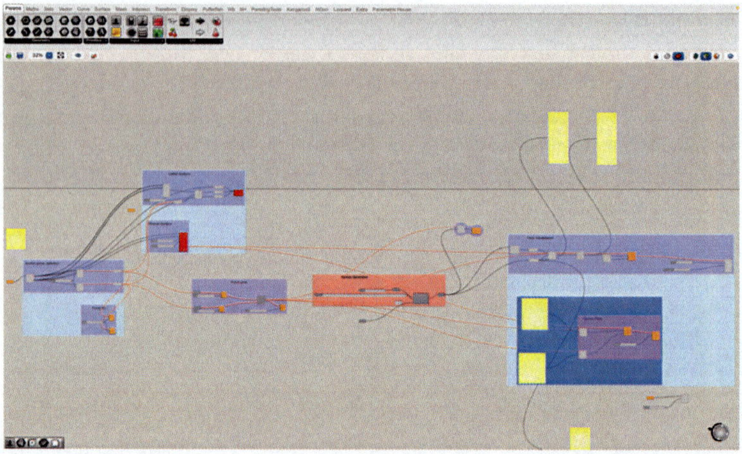

Fig. 4. Parametrical Code of winding 3d visualization, Fiber arch prototyping, 2022.

3.2 Fiber Arch

This technique allows to realize complex structures using a woven fabric of glass fibers and reinforced carbon fibers in epoxy resin baths with the consequent drying of the construction components.

It also allows to design and build large architectures through lightweight components and at the same time of great strength and refined aesthetic quality.

One of the first experiments was to manufacture a medium-scale demonstration structure, in particular a triangular arch section, which was exhibited at the cloister of the Hub of San Rocco, headquarters of the House of Emerging Technologies of the City of Matera.

The "core" of these activities was precisely to intertwine the stereotomic philosophy with that of the ICD University, in fact the stereotomic arch (Fig. 6) was made with the technique of fiber weaving.

For the prototyping we started with a simple arch with a triangular section, after we divided it into seven blocks to obtain the curves that define the two sides of the block.

The algorithm connects the array points of one face to the array points of the opposite face with a sequence of intertwined curves, that follow the shape of the minimal surface

Fig. 5. G. Fallacara, N. D'ambrosio, Fibre arch prototyping, 2022.

between the two sides, in order to simulate the interweaving with the glass fiber. Finally, the same operation was repeated for the carbon fiber that creates a woven pattern on the blocks, positioned corresponding to the tension lines on the block to make it more resistant, and give it an aesthetic touch. After the three-dimensional virtual study, the prototype was manufactured.

3.3 Fibre Tree

The experimentation that I advance in this paper is the design of a tree-shaped parametric-computational structure in composite material of carbon fiber and fiberglass impregnated in two-component epoxy resins. This structure was designed and then prototyped with digital fabrication using a three-dimensional modeling coupled to a parametric code, the same as the arch, to generate the weaving, but following another shape.

The focal point of the research is be able to develop complex shapes of structures with this technology and these innovative materials. This had already been developed in the lithic version [9], and the challenge was precisely to generate the same structure that can be used as a serial cover for fibrous material spaces. Below are the 3d model and its realization. (Fig. 7).

Fig. 6. G. Fallacara, N. D'ambrosio, Fibre arch, Matera, 2022.

Fig. 7. A. Angione, Stereotomic Fiber Tree, Polytechnic of Bari, 2023.

4 Conclusion

In conclusion, this experimentation is still only the beginning of a long research on materials, on their structural performance, and the possibility of creating new structures. The next goal, which I will try to achieve, is to generate a fiber tripod structure with blocks that create space arcs and spherical caps.

Finally, in addition to the formal research, the structural ones are already being considered through a precompression of the fiber joints.

References

1. Menges, A.: Material performance: fibrous tectonics & architectural morphology. Studio Report Fall 2013–2015. Harvard University Graduate School of Design Publication, Cambridge, MA, 02138 (2015)
2. Dunn, N.: Digital Fabrication in Architecture. Laurence King Publishing Ltd., 361–373 City Road, London (2012)
3. Laugier, M.A.: Essai sul l'architecture, Paris (1753)
4. Gottfried, S.: Der. Stil in den technischen und tektonischen kunten oder praktische Asthetik, 2 voll, Mäander Kunstverlag, Mittenwald, 1977, ed. Orig (1860)
5. Fallacara, G.: Verso una progettazione stereotomica. Nozioni di stereotomia, stereotomia digitale e trasformazioni topologiche: ragionamenti intorno alla costruzione della forma. ARACNE Editrice, Roma (2007)
6. Fallacara, G.: My ArchiSign. Architettura e design in pietra. Pensieri, opere e visioni. Gangemi Editore International, Roma (2022)
7. https://www.icd.uni-stuttgart.de/projects/buga-fiber-pavilion. Accessed 28 May 2023
8. Fallacara, G., Barberio, M.: An unfinished manifesto for stereotomy 2.0. Nexus Netw. J. **20**, 519–543 (2018). https://doi.org/10.1007/s00004-018-0390-z
9. Colella, M.: Stereotomy 2.0 and digital construction tools. Nexus Netw. J. **20**, 769–788 (2018). https://doi.org/10.1007/s00004-018-0407-7

3D Printing Stereotomy: Two Case Studies

Ilaria Cavaliere and Angelo Vito Graziano(✉)

Politecnico di Bari, 70126 Bari, Italy
angelovito.graziano@poliba.it

Abstract. Stereotomy – also known as "the art of cutting solids" – is a discipline that exploits rules of descriptive geometry to optimize the ashlars of a vaulted space, considering aesthetic, static and functional matters. Stereotomy and its innovation through the use of CNC machines have been investigated at the Polytechnic of Bari for twenty years. Even if this discipline is historically associated to the use of stone, the advent of 3D printing is stimulating research towards new possibilities and new directions.

This paper aims to present two case studies, consisting of two different prototypes of 3D printed stereotomy. The first one is a portion of a barrel vault composed of osteomorphic hollow blocks, printed using clay. The design process was inspired both by the ancient constructive system of fictile caroselli and by the shape of the Truchet's ashlars.

The second case study deals with the construction of a real scale tripod, composed of hollow ashlars printed using PLA.

Keywords: Stereotomy · 3D Printing · Additive Manufacturing · Digital Fabrication · Architecture · Design

1 Introduction

The word "stereotomy" – which was defined for the first time by Philibert Delorme in *Le Première Tome de l'Architecture*, in the Sixteenth-Seventeenth Century – derives from the Greek "stereo", which means "firm, solid" and "tomia", which means "cut". It is described as "the art of cutting solids", because this discipline consists in finding the best configuration of every single component of a constructive system, considering both static and aesthetic matters [1, 2].

Stereotomy is mainly associated with lithic vaulted spaces, whose blocks can be peculiarly shaped to optimize the efficiency of joints and to obtain decorative patterns.

In the past it was possible to design a specific configuration of a vaulted space applying the rules of descriptive geometry: trough complex projections the designer was able to represent every structural component, in order to convey its correct shape to stone manufacturers, who could realize it.

Nowadays this difficult method has been eased thanks to modern 3D modeling tools. The production process also has been improved by the introduction of CNC machines, that allow to obtain more complex shapes through an automated process.

S. Gabriele et al. (Eds.): IWSS 2023, LNCE 437, pp. 416–426, 2024.
https://doi.org/10.1007/978-3-031-44328-2_43

Moreover, an innovation of stereotomy is linked to additive manufacturing systems, as 3D printing, that works through material addition instead of subtraction.

The aim of this paper is to present two case studies of stereotomic prototypes realized using 3D printing technology.

2 3D Printing

2.1 An Overview of Additive Manufacturing

3D printing is an additive technology developed with the invention of Stereolithography and STL (Stereo Lithography interface format) by Chuck Hull in 1986. Stereolithography allows the creation of a three-dimensional object through the progressive solidification of liquid resin layers. Later, in 1988, Scott Crump invented Fused Deposition Modeling (FDM) technology. FDM is the most popular kind of 3D printing and involves the progressive extrusion of fused plastic layers in an additive process to create three-dimensional objects. 3D printing emerged early as a manufacturing technique, but it was not until the 2000s that it became more widespread. In fact, the advent of the Internet encouraged exchange among the large community interested in the technology, and the expiration of some patents favored the emergence of start-ups ready to put new models of printers on the market. Therefore, a real cultural and social phenomenon occurred and led this technology to be applied in all production sectors and to be used for educational purposes in all levels of education. The success of this technology in all manufacturing fields lies in the advancement it has brought to production processes and products. These improvements have been codified in what is now called Industry 4.0 and relate to the decentralization and mass customization of production [3]. The peculiarities of the various production sectors, including in the different materials used, has favored the constant updating of this technology, which is now able to use materials such as glass [4], cement [5], metal [6] as well as plastic polymers in additive processes. Interesting for the construction sector has been the development of LDM (liquid deposition modeling) technology, which allows the extrusion of material that is already liquid contained in special tanks and conveyed to the extruder through pumping systems.

2.2 3D Printing and Construction

In 2022, the construction sector dominated the market with a 72.8% share of global sales. It represents one of the countries' largest sources of employment and economic development. At the same time, the industry experiences low productivity determined by the onerous construction methodology. This aspect drives the industry to research automation methods to optimize processes. In recent times, various automation technologies, including 3D printing, have generated increasing interest in industry. Currently, the 3D printing for construction (3DCP) market has passed $1.5 million and it is expected to register a compound annual growth rate (CAGR) of 101.9% from 2023 to 2030 [6]. 3DCP is the result of the appropriation and reuse of technologies initially developed for other materials (plastics, resins) in the construction industry.

The possibility to transfer these technologies to the world of construction were first studied by Joseph Pegna in 1997 and then by Professor Behrokh Khoshnevis of the

University of Southern California in the early 2000s [7]. The Radiolaria Pavilion built in 2010 represents the first printing at the scale of architecture and consists of a complex free-form structure designed by Andrea Morgante and built with D-Shape technology by engineer Enrico Dini [8]. Nowadays the technology is under development, in fact, construction industries are the last to adopt 3D printing technology due to the singularity of each architectural project. Differences in the sizes and materials used have led to different 3D printing systems for the construction industry. The most popular are direct extrusion printing of viscous material (Extrusion System) and printing by injection of binder onto a bed of particles (Powder Bonding). In particular, the extrusion type leads the market and is characterized by a pumping system that pushes the viscous material (cement, concrete, gypsum, natural mortars and raw earth) toward the extruder, which, moving within the printing volume, deposits the material layer by layer until the finished product is realized.

Cementitious materials are clearly the most suitable for the construction field, but plastic materials and FDM processes are still applied in the design field, especially for the realization of temporary pavilions [9].

2.3 Stereotomy and 3D Printing

The use of Additive manufacturing is being explored also in the field of stereotomy. Various studies and research all over the world have led to the construction of prototypes as the Vulcan Pavilion, the DE: Stress Pavilion or the Striatus Bridge.

The Vulcan Pavilion. It is an eight-meter-wide temporary tripod, designed by the Studio LCD (Laboratory for Creative Design) for the 2015 Beijing Design Week. It was composed of 1086 3D printed blocks made of plastic. All the ashlars have a peculiar shape and show patterns that highlight both the flexibility of 3D printing processes in terms of customization and the high precision in terms of detail [10].

The DE: Stress Pavilion. It is the outcome of a research carried on by a team of the Ball State University in 2019. The study dealt with an innovative fabrication process, called "sub-additive concrete printing method", which mixes the use of subtractive manufacturing and additive manufacturing for the deposition of non-planar layers without supports. A reusable sand mould is carved using CNC machines, so that non-horizontal layers can be printed following the right profile. The pavilion is composed of 110 ashlars, each one different in terms of shape and weight [11, 12].

The Striatus Bridge. It is a prototype of a 3D printed pedestrian bridge designed by Zaha Hadid Architects and the Block Research Group in 2021. It consisted in a freeform structure composed of unreinforced concrete blocks. Each ashlar was fabricated using a robotic arm and the printing path was optimized to maximize the static performance by using non-parallels layers. As a matter of fact, every layer is oriented to be perpendicular to the main stresses [13] (Fig. 1).

Fig. 1. a) The Vulcan Pavilion (© Laboratory for Creative Design and Beijing Design Week). b) The DE: Stress Pavilion (© AntiStatics Architecture). c) The Striatus Bridge (© Chiara Becattini).

3 The Developed Prototypes

After an analysis of the state of the art of the use of 3D printing for vaulted structures, two different prototypes will be described. Both these two solutions aim at the creation of vaults through the use of hollow ashlars. The main inspiration derives from baroque coffered ceilings, for example the Borromini's vault of San Carlo alle Quattro Fontane, whose coffers help give the structure a sense of lightness and produce suggestive effects of light and dark [14].

Even if the baroque structures cannot be considered stereotomic, coffers can be transformed in structural elements that become part of the aesthetic component, too. Moreover, in this process of "updating" coffered ceilings, the sense of lightness can be taken to extreme and the structure can become transparent.

This type of vaults can be used as shading devices or also to realize spaces characterized by suggestive light effects. The outcome conceptually recalls the DE: Stress Pavilion, but a simpler technology has been used, since every block has been optimized to have a planar base. This way the fabrication processes haven't required complex machines as robotic arms or advanced parametric tools for non-planar slicing. Differently from the Striatus Bridge, whose layering was optimized to be always perpendicular to the stress direction, in these two case studies layers are parallel to forces. This implies a weaker behavior, that must be considered during the design process.

3.1 The Stereotomic Green Vault

The first prototype, the Stereotomic Green Vault, consists of a portion of a barrel vault composed of interlocking hollow ashlars made of clay.

The main inspiration has been the traditional constructive system of "caroselli", diffused especially in the Southern Italy. Caroselli are a specific type of fictile tubules, hollow terracotta bricks that were placed radially, following the direction of the vault itself and being kept together by mortar (see Fig. 2). Moreover, a good adherence to mortar was guaranteed by the horizontal signs typical of the use of lathe. This unevenness was sometimes emphasized manually, to improve the performance [15]. This has inevitably led to a parallelism with the layering that characterizes 3D printed objects, that could have a similar function.

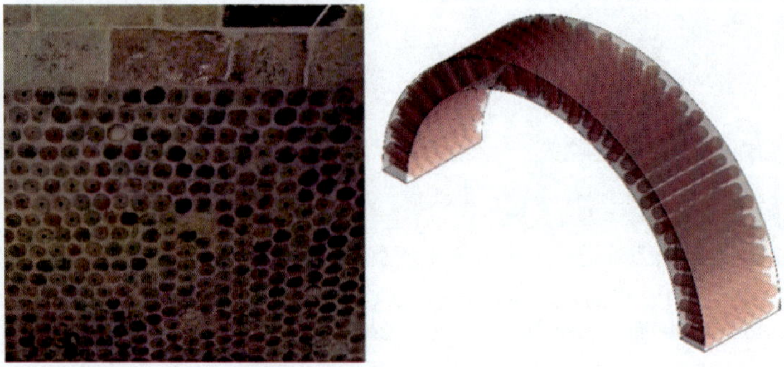

Fig. 2. A photo and a scheme of a vault built using caroselli.

A reflection upon the material has been conducted. Clay is a material widely used in the construction industry and it is printable. Clay 3D printing, which is made through LDM processes, does not require big and advanced machines. Moreover, if clay is not baked yet, it can be re-used. Therefore, printing errors don't necessarily lead to material waste.

This has led to the decision to design hollow components to be printed with clay, using a Wasp 40100 Clay printer, which is available at the FabLab Poliba in Bitonto.

Even if used for the construction of vaulted spaces, the constructive system of caroselli can't be considered stereotomic, since it consists of standardized components, that are not meant to have joints. Therefore, a rectangular pattern of interlocking ashlars was conceived (see Fig. 3), taking inspiration from the Truchet's ashlars (see Fig. 4). A similar geometry, which is characterized by curvilinear walls enclosing a cavity, is particularly suited to be realized through additive manufacturing instead of subtractive manufacturing, which would imply a great material waste and a complex fabrication process.

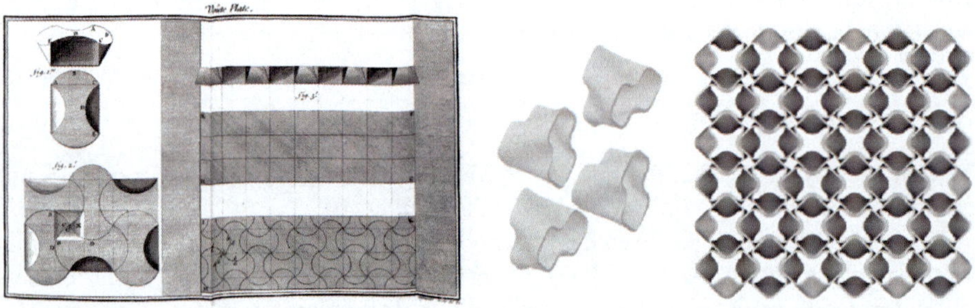

Fig. 3. Truchet's flat vault original patent and the designed pattern of interlocking ashlars.

The designed ashlars have been used to tessellate a portion of barrel vault 1,2 m wide and 25 cm high, so their shape has slightly changed due to the curvature, but their base has been kept planar, in order to maintain the printing process simple.

Through some printing tests it has been possible to calibrate the correct dimensions and proportions of the blocks. As a matter of fact, it is fundamental to keep in mind that print with supports is not recommended when printing with a fluid material, especially because their removal after the drying could imply fractures and imperfections. Therefore, it is necessary to consider some limits caused by the material itself: shapes with an excessive overhang tend to collapse, because clay remains fluid during the entire fabrication process. The shape of the ashlars has been kept more compact and the inclination of the walls has been reduced, since the first prototypes tended to deform excessively. The final ashlar was inscribed in a circle with a radius of 17 cm and was 10 cm high. A 6 mm nozzle has been used to print, and the fabrication of each block lasted eight minutes.

Fig. 4. The ashlars during the printing process phase and after the baking phase.

Some special blocks were designed in to guarantee the adherence of the arch to the wooden abutments.

The printed blocks required a few days to dry, and then a slight postproduction to remove imperfections was required. After each ashlar has been polished, they have been baked in a professional oven for pottery, available at the FabLab Poliba.

The assembly was carried on using a system of mobile centrings: given the interlocking shape of the ashlars, it would have been impossible to proceed from the bases of the vault and conclude the construction placing the key block. Therefore, the centring was divided into two halves, that could slide on wooden cylinders while placing the ashlars. One of the two abutments was fixed to the basis, while the other has been fixed once the assembly has been finished. Sylicon was used to fill excessive gaps between the ashlars, caused by deformations linked to material retirement during the drying and the baking phases (Fig. 5).

A load test has been performed on the model to test its global resistance. After having fixed the base to a panel in order to prevent its sliding, the structure was gradually loaded using bags with a mass of 25 kg each. The vault could bear a total mass of 375 kg (fifteen bags), that corresponds to a force of 3,75 kN. Because the weight of the clay structure was 0,25 kN, the vault was able to bear a load fifteen times bigger than its own weight.

Advantages and Limitations. The Stereotomic Green Vault experiment showed both pros and cons of LDM 3D printing applied to stereotomy. As previously said, the use of clay allows to reduce material waste thanks to the possibility to mix and reuse all the

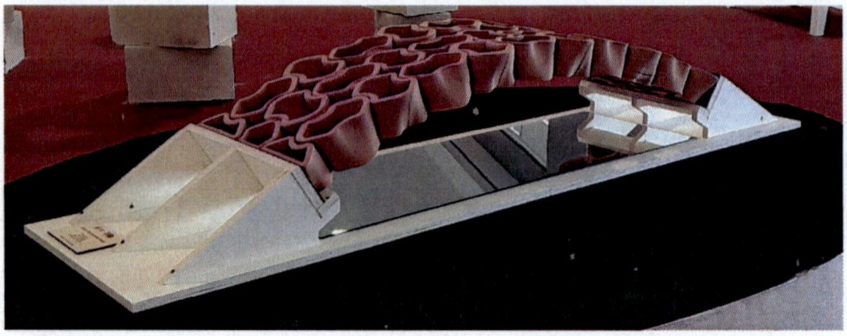

Fig. 5. The final model of the barrel vault.

material that has not been baked. This helps maintain the entire process more sustainable (hence the word "green" in the title of the project). Moreover, LDM 3D printing is a very fast process since a few minutes are enough to produce an ashlar.

On the other hand, a fluid material limits possibilities in terms of geometry because excessive overhangs cannot be realized. Moreover, it is often difficult to have a totally homogeneous mixture and variation in viscosity or the presence of air bubble can produce unpredictable deformations and printing errors. The last issue deals with the retirement of material during the drying and the baking process: the volume reduction can cause cracks and deformations that can lead to discard some printed objects.

3.2 The Stereotomic Tripod

The second prototype is a tripod conceived to be composed of hollow 3D printed ashlars with different shapes. The abutments are connected to pentagonal blocks, while the arches on the edges are composed of quadrangular ashlars; the corners of the tripod are filled with a triangular pattern, while the center is characterized by the presence of interlocking ashlars placed according to a hexagonal pattern.

The design process started from a catenary dome, which has been cut in order to have a triangular configuration, and which has been tessellated as previously described.

Because all the ashlars had to be 3D printed in a simple way, they were modeled to have a planar base.

The vault is inscribed inside a circle with a diameter of 5.5 m and the dimensions of the ashlars have been set in order to be included in a circle with a diameter of 1 m, which corresponds to the plate of the large-scale 3D printer Delta WASP 3 mt.

The entire structure has been thought to be assembled without the use of centrings. This method has already been used for the construction of the Flux Vault, a pavilion Designed by Giuseppe Fallacara and Maurizio Barberio in 2019. This prototype was conceived to have a system of metal sheets and screws in correspondence of the joints, so that every ashlar could be fixed to the previous one without the necessity of a centring.

This method is particularly innovative, since often custom complex wooden centrings have high costs and represent a large amount of waste. This way, on the contrary, a standardized scaffolding is sufficient to bear the components.

In this case, 3D printing allowed to directly realize holes to use screws in order to reciprocally connect the blocks.

A first prototype in scale 1:5 was printed using PLA with desktop printers available at the FabLab Poliba (Ultimaker 3 and Ultimaker S5).

This phase was important for a first verification of the model, especially because a construction process without centrings was simulated by gluing each ashlar to the near ones (Fig. 6).

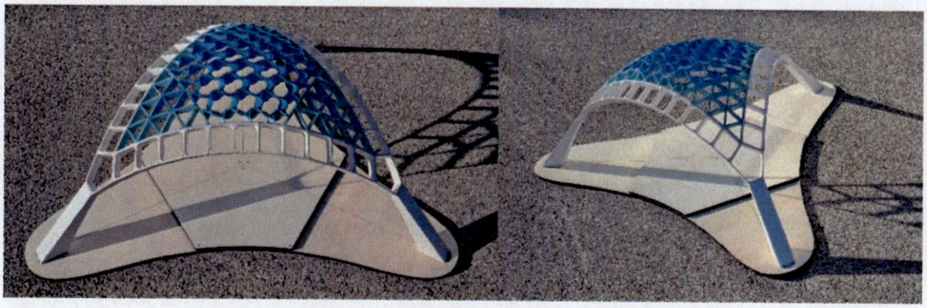

Fig. 6. 3D printed model of the tripod in scale 1:5.

The assembly of the small prototype showed some difficulties. First of all, even if each block was numbered, it was important to find the correct orientation to fit perfectly; moreover, the interlocking ashlars implied the same problem of the Stereotomic Green Vault: it was impossible to place the key block at the end. The strategy adopted consisted in dividing the structure into three thirds, which were assembled separately and united at the end. However, this process would require a system of mobile supports, that would be difficultly realized for a large-scale model.

For this reason, the full-scale prototype, that is now being realized at the FabLab Poliba (see Fig. 7), will be composed only of the arches and the triangular ashlars.

Blocks are being printed using the delta Wasp 3 mt and transparent PLA pellet is being used. A 3 mm nozzle has been chosen and each ashlar can require from eight to seventeen hours to be completed.

The pavilion will be exposed at the Marmomac Meets Academies in Verona in September 2023 and will have concrete abutments. Moreover, the pentagonal block at the corners of the triangular space will be covered with Apricena stone slabs, in order to simulate the possibility to create a coffered vault instead of a transparent one (see Fig. 8).

Advantages and Limitations. The Stereotomic Tripod was realized using a FDM process, which revealed a good level of precision but required also a higher time of fabrication. Moreover, even if PLA is an ecological plastic which is suitable to realize temporary pavilion, it is not appropriate for the construction field.

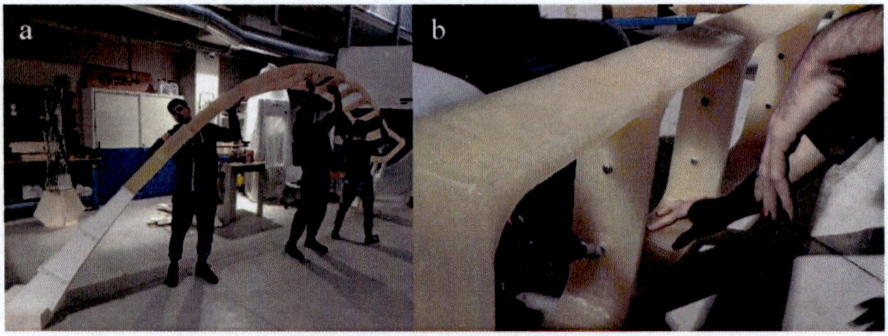

Fig. 7. a) an assembly test of an arch; b) a detail of the joints connected through screws.

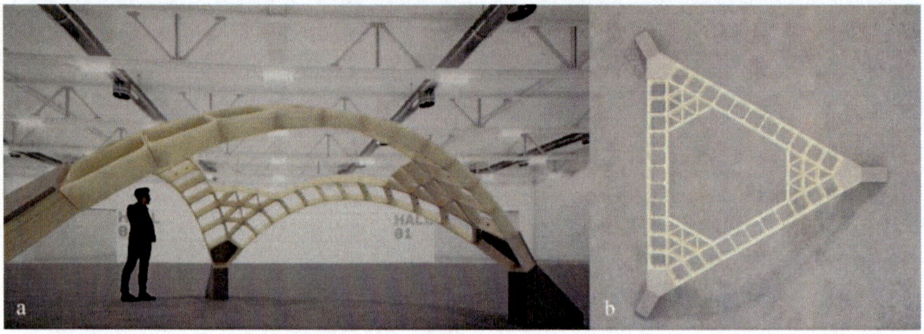

Fig. 8. a) rendering of the final pavilion; b) planimetry of the pavilion.

4 Discussion and Conclusions

In conclusion, 3D printing offers new important possibilities for the realization of vaulted spaces, since it keeps it possible to realize hollow ashlars to create transparent structures, which would be more difficult to realize through subtractive methods. Moreover, the system would offer further possibilities if ashlars were closed on the top or on the bottom: instead of totally diaphanous structures, coffered ceilings could be realized if the cavity is facing down. If the cavity is facing up it can be filled with vegetation, creating green vaults (see Fig. 9).

Both LDM and FDM processes showed pros and cons, that can be better investigated with further research.

A possible future development can deal with the possibility to print with stone powder, deriving from manufacturing waste from quarries.

Fig. 9. a) rendering of a coffered vault made of 3D printed ashlars; b) rendering of a green vault made of 3D printed hollow ashlars filled with vegetation.

References

1. Fallacara, G.: Verso una progettazione stereotomica. Nozioni di stereotomia, stereotomia digitale e trasformazioni topologiche: ragionamenti intorno alla costruzione della forma. Towards a stereotomic design. Notions of stereotomy, digital stereotomy and topological transformations: reasoning about the construction of the form. ARACNE editrice, Rome (2007)
2. Fallacara, G., Gadaleta, R.: Stereotomy: architecture and mathematics. In: Sriraman, B. (ed.) Handbook of the Mathematics of the Arts and Sciences, pp. 1–20. Springer, Cham (2019). https://doi.org/10.1007/978-3-319-70658-0_85-1
3. Barberio, M., Colella, M.: Architettura 4.0. Maggioli Editore, Santarcangelo di Romagna (2020)
4. Zhang, D., Liu, X., Qiu, J.: 3D printing of glass by additive manufacturing techniques: a review. Front. Optoelectron. **14**, 263–277 (2021)
5. Perrot, A.: 3D Printing of Concrete: State of the Art and Challenges of the Digital Construction Revolution. ISTE Ltd. and Wiley (2019)
6. Gardner, L., Kyvelou, P., Herbert, H., Buchanan C.: Testing and initial verification of the world's first metal 3D printed bridge. J. Constr. Steel Res. **172** (2020)
7. 3D Printing Construction Market Size, Share & Trends Analysis Report By Construction Method (Extrusion, Powder Bonding), By Material Type (Concrete, Metal), By End-user (Building, Infrastructure), And Segment Forecasts, 2023–2030. https://www.grandviewresearch.com/industry-analysis/3d-printing-constructions-market. Accessed 18 May 2023
8. Morgante, A.: Radiolaria pavilion. In: Glynn, R., Sheil, B. (eds.) Fabricate 2011. Making Digital Architecture, pp. 231–234. UCL Press (2017)
9. Naboni, R., Breseghello, L., Kunic, A.: Multi-scale design and fabrication of the Trabeculae Pavilion. Addit. Manuf. **27**, 305–317 (2019)
10. Xu, F.: VULCAN: closing the loop in 3D-printed architectural design. In: Fok, W.W., Picon, A. (eds.) Digital Property: Open-Source Architecture. Wiley (2016)
11. Battaglia, C., Miller, M., Verian, K.: DE: Stress Pavilion. Print-cast concrete for the fabrication of thin shell architecture. In: Doyle, M.S., del Campo, M., Ago, V., Slocum, B. (eds.) ACADIA 2020: Distributed Proximities/Volume II: Projects. Proceedings of the 40th Annual Conference of the Association of Computer Aided Design in Architecture, pp. 202–207. Acadia, Mexico City (2019)

12. Zivkovic, S., Battaglia, C.: Rough pass extrusion tooling. CNC post-processing of 3D-printed sub-additive concrete lattice structures. In: Anzalone, P., Del Signore, M., Wit, A.J. (eds.) Recalibration on Imprecision and Infidelity. Proceedings of the 38th Annual Conference of the Association for Computer Aided Design in Architecture (ACADIA), pp. 302–311. Universidad Iberoamericana, Mexico City (2018)
13. Bhooshan, S., Bhooshan, V., Dell'Endice, A., et al.: The striatus bridge. Archit. Struct. Constr. **2**, 521–543 (2022)
14. Portoghesi, P.: Borromini. Vita e Opere. Skira, Milan (2019)
15. Olivito, R.S., Scuro, C., Codispoti, R.: Experimental and numerical analysis on masonry arch built with fictile tubules bricks. Measurement **130**, 246–254 (2018)

Using Shape Optimization Settings in Comsol Multiphysics for Efficient Structural Design of Shells

Evgenia Ermakova(✉) and Marina Rynkovskaya

RUDN University, 6 Miklukho-Maklaya Street, Moscow 117198, Russian Federation
ermakova-ev@rudn.ru

Abstract. Shape optimization is one of the most important types of structural optimization for spatial coatings (shells), in which the boundaries, shapes and contours of the object under study are changed while maintaining its original structure-topology without creating any holes in the model itself. Correct adjustment of the components of the optimization process in the program allows to get a truly optimal result, both from a constructive point of view and from an aesthetic point of view. The article is devoted to the configuration and interaction of two important parameters, which are situated in Shape optimization settings in the Comsol Multiphysics program: maximum displacement (d_{max}) and filter radius (R_{min}) when optimizing the shape of two shells using the gradient method IPOPT (Interior Point OPTimizer). The purpose of the study is to select the most appropriate value of d_{max} and R_{min} for two shells for finding the effective shape optimization decision.

Keywords: Shape optimization · Comsol Multiphysics · Maximum displacement (d_{max}) · Filter radius (R_{min}) · IPOPT

1 Introduction

In Comsol Multiphysics the shape optimization technique works by deforming the mesh in existing geometry and in order to avoid too large deformations of the model elements it is necessary to adjust special settings: control variable-maximum displacement (d_{max}) and filtering-filter radius (R_{min}).

First parameter d_{max} is responsible for controlling the amount of optimization displacement. Parameter R_{min} controls the smoothness of the computational mesh on the model. Filter radius setting is well-established procedure in topology optimization to reduce the checker-board phenomenon [1, 2] and can also be used in the shape optimization process. The filtering techniques are discussed in [3, 4] and in article [5] is shown the study the influence of the filter radius (is varied between 0.0 and 6.0) in the geometry, manufacturability and mechanical performance of the design. Additionally, in article [6] truss bars are considered and filter radius (6, 3, 2.5, 2, 1.5) is used for different meshes. It is necessary to mention that no studies have yet been found that are dedicated

S. Gabriele et al. (Eds.): IWSS 2023, LNCE 437, pp. 427–435, 2024.
https://doi.org/10.1007/978-3-031-44328-2_44

to setting the maximum displacement and filter radius in Comsol Multiphysics. Therefore, it is of interest for us to investigate these parameters and the interaction between them. The d_{max} parameter shows how much the geometry maximally deviates from the existing one during the optimization process, and the R_{min} is responsible for the value of the bending radius that is used in the calculation [8]. Figure 1 shows the results that were obtained without detailed settings of the two parameters d_{max} and R_{min}. It can be observed that the shells have been changed so much that they no longer have a normal geometric appearance. To avoid such a final result, it is necessary to study in more detail the influence of parameters on the geometry of the model. The main purpose of the study is to find the optimal value of d_{max} and R_{min} during the optimization process of two shells.

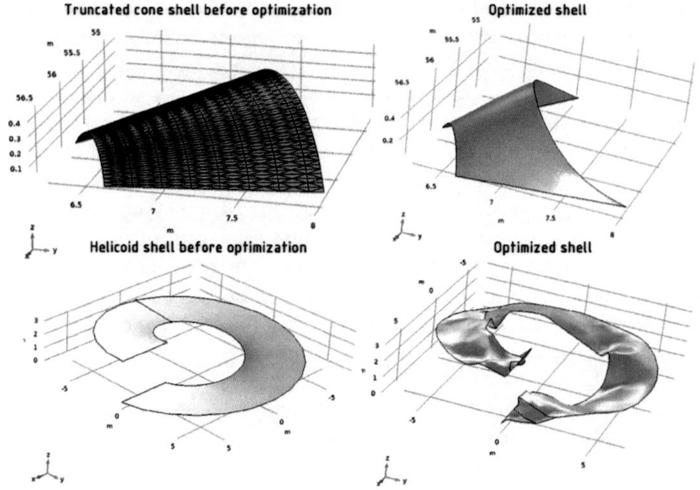

Fig. 1. Examples of optimized results without parameter settings

This paper is organized as follows: Sect. 2 describes the test models of shells and the preparatory process before optimization. Section 3 presents results, which were obtained during the optimization with d_{max} and R_{min}. Section 4 contains final remarks.

2 Materials and Methods

2.1 Setting Parameters for the Helicoid Shell and Truncated Cone Shell

As the test models for the calculation, a right helicoid and a truncated cone shell are taken. The right helicoid is a surface arising by moving a straight line along helicoidal line when the angle between the straight line and the axis of the surface is equal to 90° [7].

Among the exciting examples with the use of helicoid surface there are such buildings as the Ribbon Chapel (Hiroshima, Japan), the Bramante Staircase (Vatican, Italy), the project "Stairscaper" (New York, USA) and the Dali Museum (Florida, USA) (Fig. 2).

Fig. 2. Examples of the use of helicoid surfaces in the architecture

The right helicoid shell created in Comsol Multiphysics has inner radius – 3.1 m, outer radius – 7.1 m and the height of one loop – 3m. The shell consists of concrete with a volume weight of 2300 kg/m^3. The thickness of the shell is equal to 0.2 m. The size of finite mesh elements – "fine" (Fig. 3). The fixed constraint is installed on the initial and final faces of the shell.

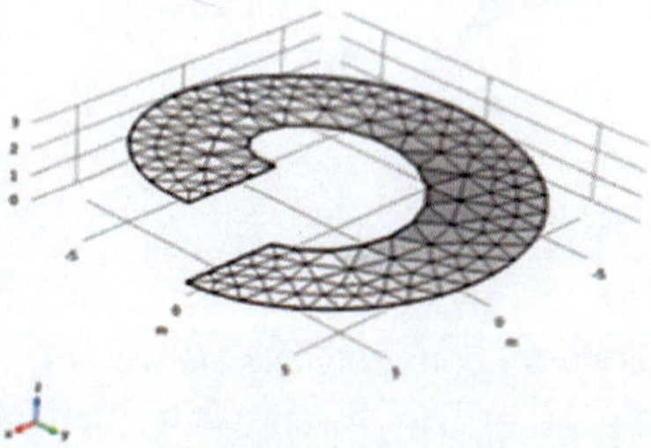

Description	Value
Maximum element size	1.14
Minimum element size	0.143
Predefined size	Fine
Triangle	264

Fig. 3. Mesh settings for the right helicoid

The truncated cone shell is taken as the second test example. A conical surface is a surface formed by the movement of a straight line (generatrix) passing through a given point (vertex of a conical surface) and intersecting a given line (guide) [10]. The surface of the cone is widely used in the architecture of buildings for various purposes: from technical to public buildings. Examples include buildings such as Grand Central Water Tower (Midrand, South Africa), Villa in Beekbergen (Beekbergen, Netherlands) and Catclife glass cone (Catcliffe, UK) (Fig. 4).

The truncated cone shell imported to Comsol Multiphysics has the following dimensions: length of two sides – 8 m, span – 11 m. The material, thickness and type of mesh are set the same as for the helicoid (Fig. 5). The lower base of the cone is completely fixed, and the other sides are free.

A detailed study of the optimization process of shells, including the helicoid, is contained in [8] and the description and static analysis of a truncated cone shell are shown in [11]. In this research we are interested in setting parameters in the Shape optimization module in Comsol Multiphysics. Maximum displacement (d_{max}) and filter radius (R_{min}) are located in Free Shape Boundary, which is added before the optimization calculation and set up control variables for all dimensions. By default, d_{max} is set to 5%

Fig. 4. Examples of the use of conoid surfaces in the architecture

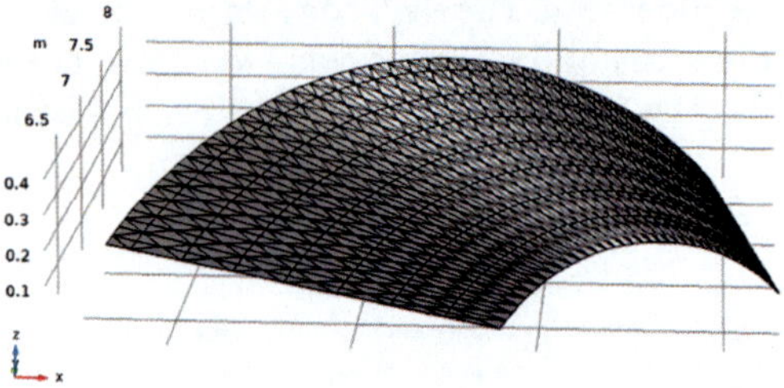

Description	Value
Maximum element size	0.191
Minimum element size	0.0344
Predefined size	Fine
Triangle	9673

Fig. 5. Mesh settings for the truncated cone

and R_{min} is set to 10% of the size of the geometry bounding box (BBox) [9]. The program automatically sets the following values: $d_{max} = 1$ m, $R_{min} = 2$ m (helicoid shell) and $d_{max} = 0.1$ m, $R_{min} = 0.3$ m (conoid shell). In [2] recommendations for setting a suitable R_{min} value are given and it should be noted that it is necessary to set the filter radius, which needs to be at least larger than the size of the smallest FE (finite element) used, in order to have an effect.

Thus, taking into account all of the above-mentioned, for this research a number of values are assigned: the values generated in the program and the values assigned manually according to the recommended percentage. Also, for a more visual result and subsequent comparison, additional values from outside of the recommended interval are taken.

The study considers the following parameter values for the helicoid shell: $d_{max} =$ 0.15 m, 0.2 m, 0.36 m, 1 m ($R_{min} = 0.3$ m, 0.71 m, 2 m, 5 m, 10 m). For the truncated cone shell: $d_{max} = 0.02$ m, 0.1 m ($R_{min} = 0.04$ m, 0.3 m, 0.8 m, 2 m, 5 m), $d_{max} = 0.4$ m ($R_{min} = 0.3$ m, 0.8 m, 2 m, 5 m), $d_{max} = 1$ m ($R_{min} = 0.8$ m, 2 m, 5 m).

Optimization is carried out using the IPOPT method, which is based on the interior-point algorithm with a filter line search method [12]. The total elastic strain energy is taken as the criteria of the system (objective function) and the type of optimization problem is the minimization of the function.

3 Results and Discussion

After carrying out the optimization process with different values of d_{max} and R_{min} the following results were obtained:

1. The maximum displacement value for the helicoid in the structural analysis before the optimization process is 53.5 mm. Setting the filter radius in the range of 0.3 m and 0.71 m for all researched values reduces the maximum displacement value by 90%. With $R_{min} = 2$ m there is an increase in the displacement to the initial value obtained during the structural analysis before optimization (Fig. 6).
2. For the truncated cone shell, the maximum displacement is 0.105 mm. The minimum value of the displacement can be observed in the range of $R_{min} = 0.04$ m and 0.8 m. In this interval, the value is reduced by 50–60% (Fig. 7).
3. The maximum value of the elastic strain energy for the helicoid before optimization is 18300 J and for the truncated cone shell the maximum value is 82.4 J. It should be noted that with an increase in the value of R_{min}, the value of the objective function also increases (Fig. 8, 9).
4. The greatest change in the shape of the helicoid shell is observed with $R_{min} < 5$ m. However, setting the $d_{max} \geq 0.36$ m and $R_{min} \leq 2$ m leads to large deformations of the shell, greatly changing its structure. It can be observed that at the value of $d_{max} = 1$ m and $R_{min} = 0.3$ m, 0.71 m the helicoid significantly narrowed, turning into a narrow strip, while with $R_{min} \geq 5$ m the shape of the helicoid is more smoothed, but gradually returns to the original geometry of the model. Therefore, it can be recommended to take $d_{max} < 1$ m and 0.71 m $< R_{min} \leq 5$ m for the researched cases (Fig. 10).

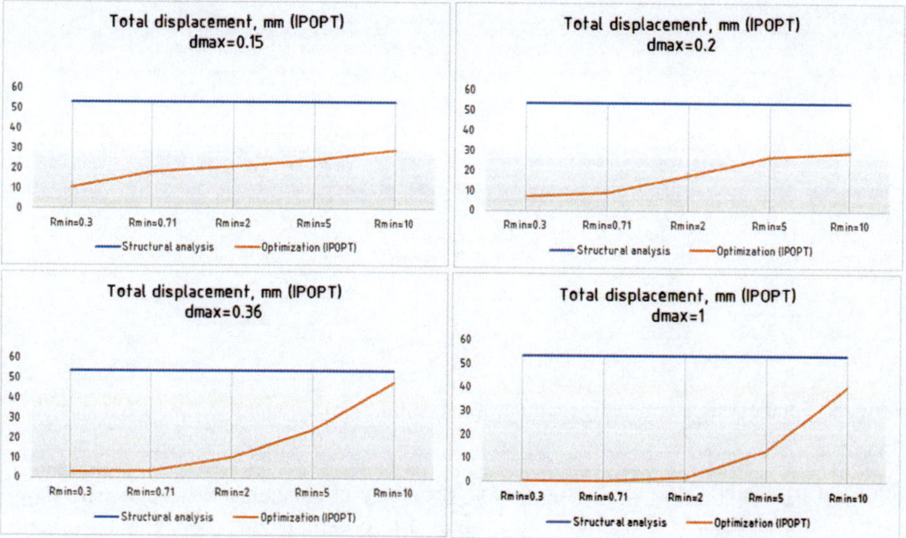

Fig. 6. Total displacement with different values of d_{max} and R_{min} (helicoid shell)

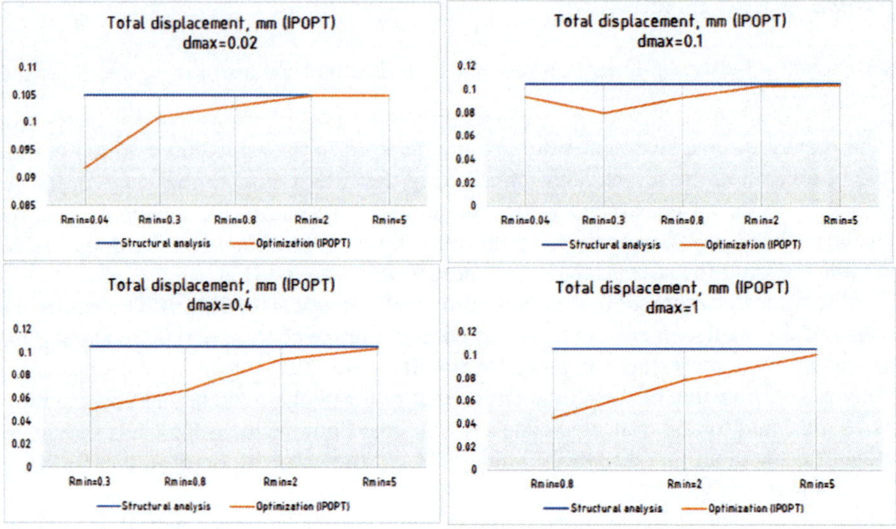

Fig. 7. Total displacement with different values of d_{max} and R_{min} (truncated cone shell)

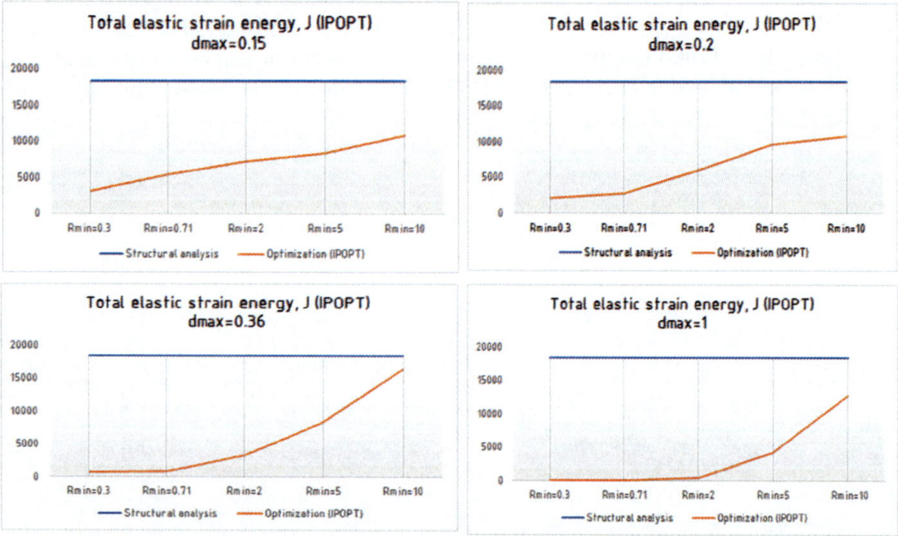

Fig. 8. Total elastic strain energy with different values of d_{max} and R_{min} (helicoid shell)

5. For the truncated cone shell, significant geometry changes are observed with d_{max} > 0.1 m and R_{min} < 2 m, but if the values are too small, the shell begins to bend strongly. When setting $d_{max} = 0.02$ m, the geometry of the shell practically does not change. The most optimal solution for a truncated cone shell can be considered with the 0.1 m ≤ d_{max} ≤ 0.4 m and 0.3 m < R_{min} < 2 m (Fig. 11).

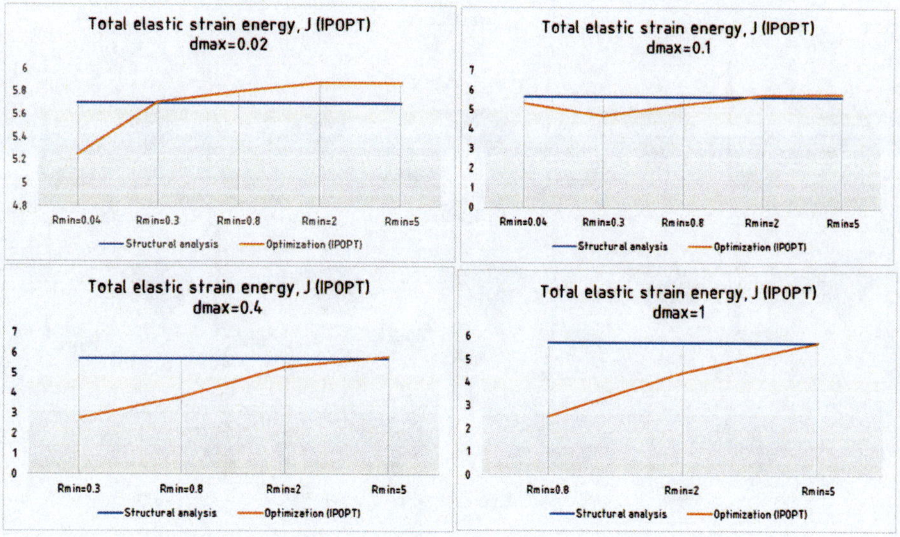

Fig. 9. Total elastic strain energy with different values of d_{max} and R_{min} (truncated cone shell)

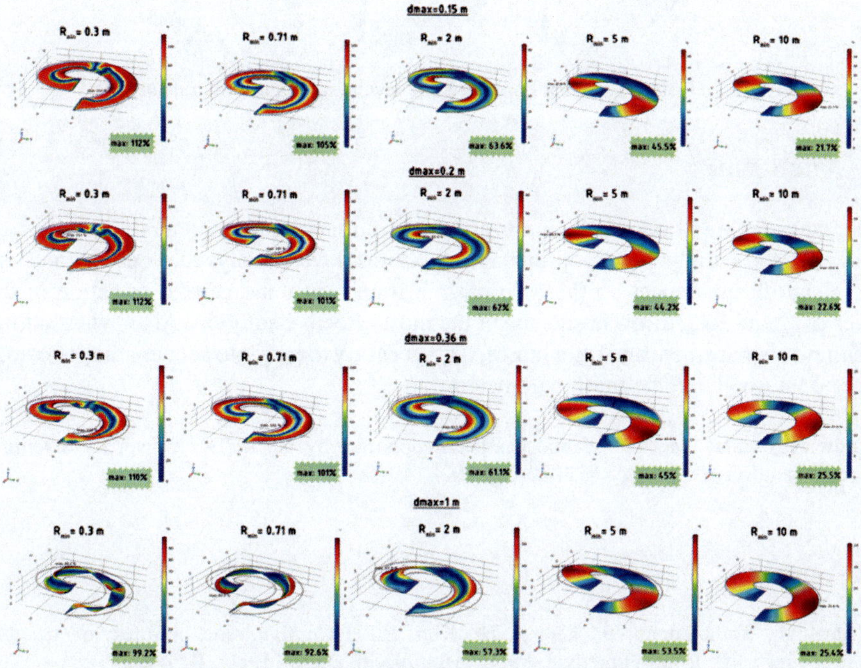

Fig. 10. Relative normal boundary displacement (helicoid shell)

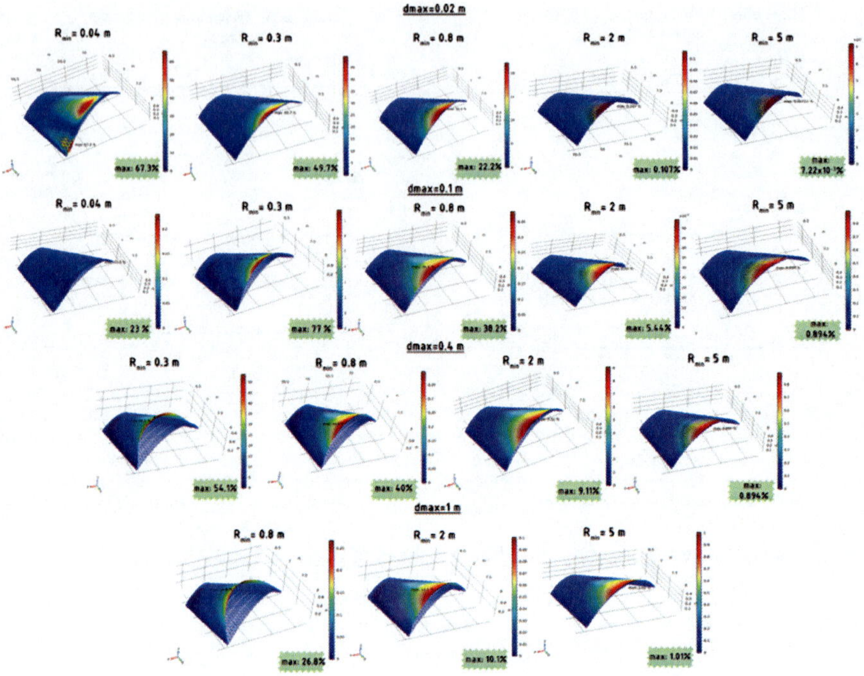

Fig. 11. Relative normal boundary displacement (truncated cone shell)

4 Conclusion

Using the example of two shells in the article, two important parameters d_{max} and R_{min} included in the settings before optimization in Comsol were analyzed. These parameters have a significant impact on the final optimization result and careful selection of the values d_{max} and R_{min} allows users to get the most effective solution. Also, when setting maximum displacement and filter radius, it is necessary to take into account the geometric shape of the shell and the mesh parameters.

Acknowledgments. This publication has been supported by the RUDN University Scientific Projects Grant System, project № 202247-2-000.

References

1. Jang, G., Kambampati, S., Chung, H., Kim, H.A.: Configuration optimization for thin structures using level set method. Struct. Multidiscip. Optim. **59**(1), 1–34 (2019)
2. Osvaldo, M.Q., Mariano, V., Alonso, C., Ansola, R., Marti, P.: Topology Design Methods for Structural Optimization, pp. 1–191 (2017)
3. Najian Asl, R., Bletzinger, K.: Implicit bulk-surface filtering method for node-based shape optimization and comparison of explicit and implicit filtering techniques. Struct. Multidiscip. Optim. **111**(66), 1–16 (2023)

4. Daoud, F., Bletzinger, K., Jurecka, F.: Filtering and regularization shape optimization techniques for preliminary design. In: 46th AIAA/ASME/ASCE/AHS/ASC Structures, Structural Dynamics and Materials Conference, pp. 1–8 (2005)
5. Montoya-Zapata, D., Acosta, D.A., Moreno, A., Posada, J., Ruiz-Salguero, O.: Sensitivity analysis in shape optimization using voxel density penalization. In: CEIG – Spanish Computer Graphics Conference, pp. 1–19 (2019)
6. Keshavarzzadeh, V., Kirby, R.M., Narayan, A.: Parametric topology optimization with multiresolution finite element models. Int. J. Numer. Methods Eng. **119**(7), 567–589 (2019)
7. Tupikova, E.M., Rynkovskaya, M.I.: Analytical approach to stress-strain analysis of right and oblique helicoid structures. Mag. Civ. Eng. **106**(6), 1–20 (2021)
8. Ermakova, E., Elberdov, T., Rynkovskaya, M.: Shape Optimization of a shell in comsol multiphysics. Computation **10**(4), 1–12 (2022)
9. Optimization Module User's Guide. https://doc.comsol.com/5.6/doc/com.comsol.help.opt/OptimizationModuleUsersGuide.pdf
10. Krivoshapko, S.N., Ivanov, V.N.: Encyclopedia of Analytical Surfaces. Springer, Cham (2015). https://doi.org/10.1007/978-3-319-11773-7
11. Ermakova, E., Rynkovskaya, M.: The use of shells in the architecture of the concert halls. Curved Layer. Struct. **8**(1), 61–69 (2021)
12. Yu, Y., Lyu, Z., Xu, Z., Martins, J.R.: On the influence of optimization algorithm and starting design on wing aerodynamic shape optimization. Aerosp. Sci. Technol. **75**, 183–199 (2018)

About Benefits and Limitations of Finite Elements and Thrust Network in Analyzing Homogeneous Masonry Domes: The Case Study of the Temple of Diana in Baiae

Salvatore Sessa(✉) and Francesco Marmo

Department of Structures for Engineering and Architecture, University of Naples, Federico II, 80138 Naples, Italy
salvatore.sessa2@unina.it
http://www.dist.unina.it

Abstract. The temple of Diana in Baiae represents one of the most important examples of masonry dome built in ancient times; presenting a very peculiar ogival shape, partially collapsed, is built in Opus Caementitium. In order to assess its structural vulnerability, the paper aims to investigate the performance of different analysis strategies: the Trust Network Analysis (TNA) and classical Finite Elements (FEM). While TNA proved to be very efficient for masonry structures presenting well-defined structural elements (rock blocks), the continuous nature of the roman concrete cast does not permit an univocal and effective discretization of the network. On the other hand, while FEM are generally very efficient in analyzing continuous elements, their solution is strongly influenced by external boundary conditions that, because of ground settlements, represent a critical action for the present case study. A numerical comparison between the outcomes of both strategies in presence of different external actions will present an insight on the limitations and the benefits of these analytical methodologies.

Keywords: Finite Element Method · Thrust Network Analysis · Masonry domes

1 Introduction

The Temple of Diana in Baiae was built at the beginning of the 3rd century AD. With diameter of about 30 m, it is the largest dome in antiquity after the Pantheon and its importance has been highlighted by several investigations available in the literature. One of the main features of Baiae – the *pusilla Roma* –

Supported by the University of Naples Federico II and the *Compagnia di San Paolo*, FRA grants, CUP E69C21000250005.

S. Gabriele et al. (Eds.): IWSS 2023, LNCE 437, pp. 436–445, 2024.
https://doi.org/10.1007/978-3-031-44328-2_45

consists in the fact that new construction techniques were hereby experimented before they were exported in Rome. In fact, Baiae domes date from the late Republican to the late Severian period and the Temple of Mercury is believed to be a pre-existing scaled model of the Pantheon.

The Temple of Diana is coherent with such a philosophy. In fact, one of its most important features is the shape of the top region of the shell. Contrarily to almost all roman domes, having circular shape, the radial section of the dome presents an ogival profile, often approximated by polycentric arches [17]. A recent investigation [24] pointed out that such a profile could be parabolic, rather than polycentric and this aspect would be of utmost importance since parabola is a very good approximation of a catenary.

Currently, the Temple of Diana is partially ruined. The dome presents an almost perfect fracture along the East-West diameter so that the southern half collapsed. Apparently, such a damage is due to asymmetric horizontal thrusts generated by different confinements of the outer walls.

Because of the importance of such a structure, and of its peculiar vulnerability, the present research aims to investigate the structural behavior of the dome's top region with respect to vertical loads. In particular, a geometric model, based on data available in the literature as well as on an on-field photo-geometric survey, is analyzed by the classic finite element method as well as by the thrust-network approach, based on limit analysis theories.

A comparison between the two methodologies is performed by computing at each integration point the eccentricity of the membrane generalized stress, computed by finite elements, since such a quantity is conceptually coherent with the location of the trust network. Moreover, the minimum thickness required to guarantee fully-compressive regimens in the whole dome is computed for both FEM and TNA. According to Heyman's computational approach [7], this provides a convenient safety factor of the analyzed model.

The paper is organized so that Sect. 2 describes the main geometric properties of the investigated model while Sects. 3 and 4 describe the analysis procedure with finite elements and thrust network, respectively. Finally, Sect. 5 provides a brief discussion about the results as well as the conclusions.

2 Geometric Model

The Temple of Diana in Baiae was built in *opus caementitium* and consists in a single room with circular shape. The external walls, follow an octagonal outline encompassing the inner room and forming a thick tambour supporting the proper dome, consisting in a 1.4 m thick shell.

The geometry of the dome radial section has been object of several researches including those of Fornaro [5] and Rakob [17] that determined a polycentric construction of the vault. More recently, Sinopoli and Aita [24] presented an essential investigation about the structural nature of the dome. One of their key findings consists in the fact that the upper portion of the shell, emerging from the tambour, is a self-sustaining structure built in order to avoid fractures typical of hemispherical domes.

Geometry was also investigated in this research by a photographic survey that permitted to reproduce with good accuracy the intradox and the thickness of the fractured front, as shown in Fig. 1. Thickness, which is $\delta = 1.4\,\text{m}$, turns out to be consistent with traditional design rules for vaults and domes suggesting a ratio between thickness and diameter of about 5%.

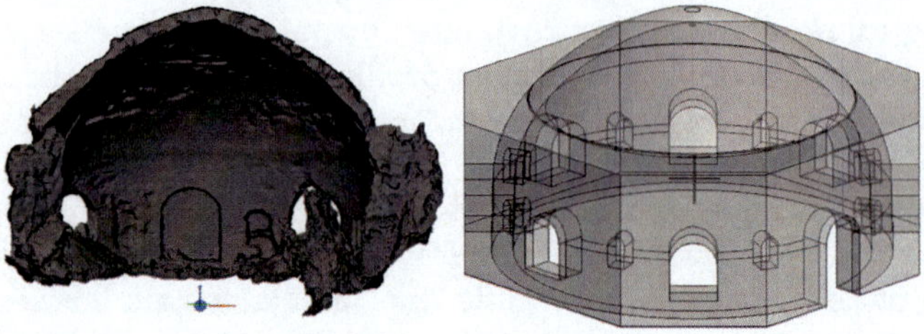

Fig. 1. Geometric survey of the investigated dome. Courtesy of M. Marmo [13].

3 Finite Element Model

The finite element model has been modeled in OpenSees; in particular, a triangular shell elements ShellDKGT, based on the theory of generalized conforming element [8], have been adopted to model the structure of interest. Such element presents 6 Degrees Of Freedom per node and 8 internal kinematic variables, relevant to membrane, flexural, and shear behavior. The model consists in 2129 nodes, fully constrained at the impost plane, and 4128 elements.

In order to define a suitable mechanical model, it is necessary to adopt constitutive parameters relevant to roman concrete. The adopted values, reported in Table 1, have been deduced by rare theoretical and experimental characterizations of roman concrete [3,19] that include fracture tests of significant volumes of authentic ancient core samples from the Great Hall of the Markets of Trajan (c.110AD) in Rome.

Besides such assessment is not relevant to the structure of interest, it is possible to assume analogous values for historical reasons. As a first consideration, lack of experimental data about roman concrete makes such a characterization [3,19] the most reliable information available in the literature. A second consideration concerns the historical development of Baiae since its buildings were often made as preliminary *testing models* for structures in Rome: it is reasonable to believe that the employed materials and the handcrafting techniques lead to similar constitutive parameters, at least in magnitude.

Since OpenSees does not provide any triaxial constitutive model capable of considering fracture, the standard Drucker-Prager elastic-plastic material [4] has been adopted since it represents one of the most reliable models capable

Table 1. Constitutive parameters of roman concrete.

Parameter:	unit	value	Parameter:	unit	value
Young modulus E	MPa	1600	Tensile strength σ_t	MPa	0.6
Poisson's ratio ν		0.2	Compressive Strength σ_c	MPa	−4.76
Density	Kg/m^3	1800	Tensile fracture energy	N/mm	0.04
			Compressive fracture energy	N/mm	8

of reproducing different behaviors for compressive and tensile loads, and conveniently used in modelling masonry [20,22,23]. Its parameters can be expressed as functions of the tensile and compressive strengths [21] as:

$$\sigma_y = \frac{2\,|\sigma_c|\,\sigma_t}{|\sigma_c| - \sigma_t}; \quad \rho_{dp} = \frac{|\sigma_c| + \sigma_t}{|\sigma_c| - \sigma_t}\sqrt{\frac{2}{3}} \tag{1}$$

Nonlinear analysis is performed by applying the vertical load, consisting in the self-weight of the dome, in ten incremental steps. In order to compare its outcomes with those of the thrust network analysis, it is necessary to determine the eccentricity of the membrane internal forces. In particular, membrane and bending generalized stresses of the ShellDKGT element can be represented in tensorial notation as:

$$\mathbf{N} = \begin{bmatrix} N_x & N_{xy} \\ N_{xy} & N_y \end{bmatrix}; \quad \mathbf{M} = \begin{bmatrix} M_x & M_{xy} \\ M_{xy} & M_y \end{bmatrix}; \tag{2}$$

where directions x and y represent the local axes belonging to the plane of the shell, hence N_x and N_y denote normal components while N_{xy} represents in-plane shear, M_x and M_y represent flexural contributions, and M_{xy} denotes the coupling moment.

A quantitative estimate of the membrane eccentricity [6] can be computed as:

$$e(\theta) = \frac{\bar{M} + \hat{M}\cos\theta + M_{xy}\sin\theta}{\bar{N} + \hat{N}\cos\theta + N_{xy}\sin\theta} \tag{3}$$

with:

$$\bar{M} = \frac{M_x + M_y}{2},\ \hat{M} = \frac{M_x - M_y}{2},\ \bar{N} = \frac{N_x + N_y}{2},\ \hat{N} = \frac{N_x - N_y}{2} \tag{4}$$

and with θ representing the angle between the direction along which the eccentricity is computed and the local reference of the element. At each integration point of the model, the membrane eccentricity is computed by maximizing the absolute value of $e(\theta)$ with respect to θ. Such a maximum value is then normalized by a half of the thickness so that values within the interval $[-1, 1]$ are relevant to fully-compressive regimens.

Membrane normalized eccentricity relevant to the original model of the dome, with thickness $\delta = 1.4\,\text{m}$, is reported in Fig. 2(a) while the relevant pressure

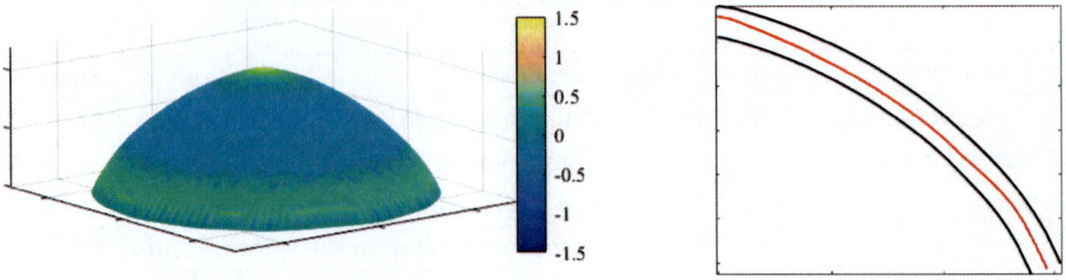

(a) Membrane eccentricity (b) Average pressure curve

Fig. 2. Results of the finite element analysis – real model $\delta = 1.4\,\mathrm{m}$.

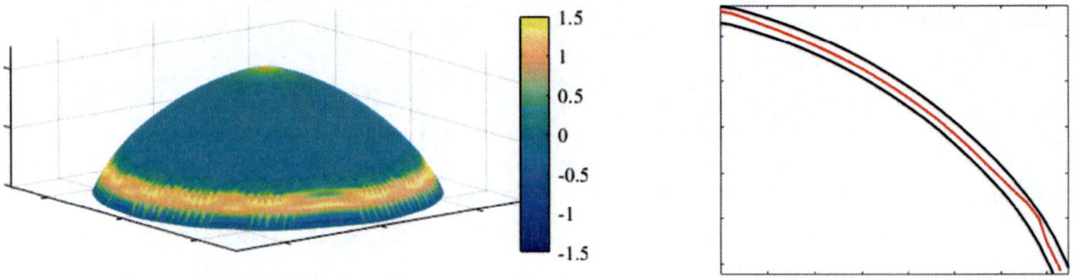

(a) Membrane eccentricity (b) Average pressure curve

Fig. 3. Results of the finite element analysis – FEM limit thickness $\delta = \delta_{FL} = 0.72\,\mathrm{m}$.

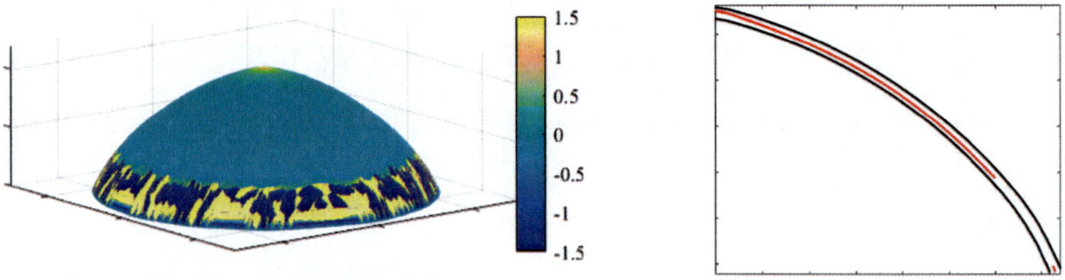

(a) Membrane eccentricity (b) Average pressure curve

Fig. 4. Results of the finite element analysis – TNA limit thickness $\delta = \delta_{FT} = 0.49\,\mathrm{m}$.

curve, obtained by computing the average of the membrane eccentricity along the dome's parallels, is reported in Fig. 2(b).

As a matter of fact, the model turns out to be completely compressed. This was expected, since the thickness is consistent with traditional design rules and is coherent with the results of the TNA reported in Sect. 4.

An important numerical indicator of structural safety for such kind of models is the so called geometric safety factor [7], which is the ratio between the actual thickness and its limit value. The latter is the minimum thickness for which the model results completely in compressive regimen under its self-weight.

Limit thickness has been computed by a constrained optimization strategy that converges at $\delta_{LF} = 0.72\,\mathrm{m}$. The relative membrane eccentricity relevant to the limit thickness and its average pressure curve are reported in Fig. 3. Indeed, both figures show the attainment of the limit condition clustered along a parallel nearby the impost plane.

A similar computation of the limit thickness by means of the Thrust Network Analysis is reported in Sect. 4 and yields $\delta_{LT} = 0.49\,\text{m}$, indeed smaller than δ_{LF}. A further finite element analysis adopting $\delta = \delta_{LT}$ has been performed and results are reported in Fig. 4.

Differently from the previous cases, FEM analysis does not achieve convergence. Moreover, the model presents a wide band close to the impost plane where computation of membrane eccentricity is unstable because of the presence of elements subject to purely flexural actions for which eccentricity is infinite.

This is possibly related to the attainment of plasticity condition for the whole thickness of the shell along the impost band; condition that induces the model to be non-statically determined.

4 Trust Network Analysis

The stresses that lead to failure in masonry arches or vaults are typically lower than those required for material failure. As a result, the stability of such structures is primarily determined by their shape and self-weight distribution [7].

Thrust network analysis (TNA) implements Heyman's principles by modelling stresses in masonry vaults as a discrete network of forces in equilibrium with applied loads. After the original formulation by O'Dwyer [16], it was fully developed in successive contributions [1,2,11].

In [11] a reformulation of the TNA is proposed, which incorporates horizontal forces in the analysis, as well as the inclusion of holes or free edges in the vault. Such approach has been successfully applied to a series of case studies including helical staircases [10] and fictile vaults [12].

According to such a formulation, the equilibrium analysis of a masonry vault is conducted by assuming that internal forces can be modelled as a network of thrusts, i.e., compressive forces that act within the structure, in equilibrium with the applied loads. This network is henceforth referred to as the *thrust network* and is described through the use of nodes connected by branches indicating the direction of the thrust forces, like funicular polygons.

The n-th node of the network has position $\mathbf{p}_n$ while the generic branch b connecting a pair of nodes is representative of the thrust force $t^{(b)}$ transmitted between such nodes. In addition, nodes are also subjected to an external force $\mathbf{f}^{(n)}$, whose magnitude depends on the region of influence of the node.

The horizontal positions of nodes are usually determined, while their height is unknown as well as the thrust values associated to branches. In basic applications of the method, horizontal and vertical equilibrium of nodes are used to compute the unknowns by solving linear optimization problems while complex cases require an iterative procedure [11].

The adopted thrust network is obtained by considering a net composed of meridian and parallel branches representing, respectively, the resultants of meridian and parallel components of the membrane force tensor. Nodes are loaded by

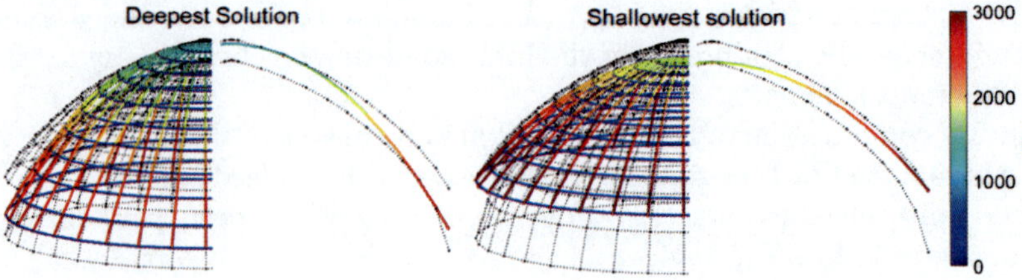

Fig. 5. Results of the thrust network analysis. Real model $\delta = 1.4\,\text{m}$.

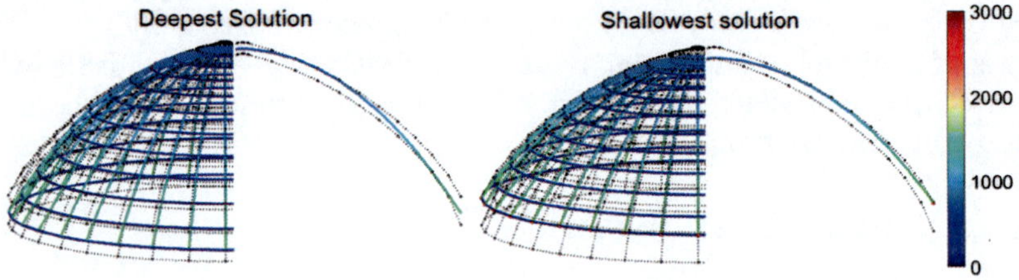

Fig. 6. Results of the thrust network analysis. FEM limit thickness $\delta = 0.72\,\text{m}$.

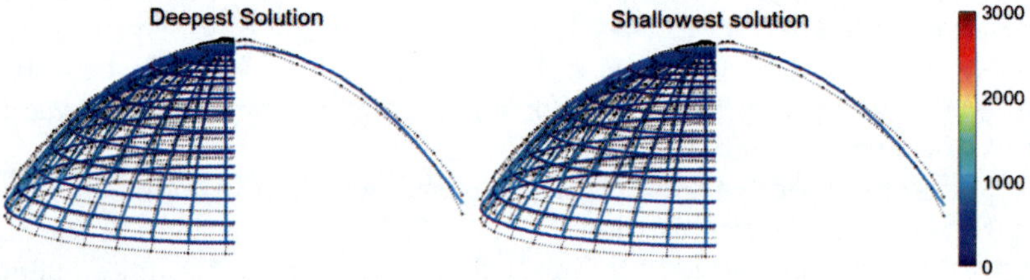

Fig. 7. Results of the thrust network analysis. TNA limit thickness $\delta = 0.49\,\text{m}$.

forces corresponding to the dome self weight, computed accounting for the region of influence of each node. The geometric domain for the vertical position of nodes corresponds, according to Heyman hypothesis, to the intrados and extrados of the dome (black dots in the figure).

The thrust network analysis of the dome furnishes two extreme configurations of the thrust network, commonly referred to as the deepest and shallowest configurations, respectively corresponding to the solutions of minimum and maximum horizontal thrust. The difference between these two configurations is quantified by the ratio $r_s/r_d \leq 1$. The deepest and shallowest solutions computed for the investigated dome having the real thickness $\delta = 1.4\,\text{m}$ and subjected to self weight are shown in Fig. 5, which correspond to $r_s/r_d = 0.69$. Lower r_s/r_d values mean that the deepest and shallowest configurations of the network are very distinct, suggesting a broad range of possible solutions for the structure. Such values usually correspond to safer structures.

Actually, when the dome thickness is reduced to the FEM limit thickness, that is $\delta_{FL} = 0.72$ m, the performance ratio increases slightly, reaching the value $r_s/r_d = 0.87$. The corresponding deepest and shallowest solutions are plotted in Fig. 6 and are shown to be slightly different one each other, meaning the equilibrium is still possible. Notice that the thrust values are also reduced due to the smaller self weight of the structure generated by the reduced thickness.

The dome thickness can be further reduced until the r_s/r_d ratio equals one, a value attained when $\delta_{FT} = 0.49$ m. In such a case the deepest and shallowest configurations of the network become indistinguishable, as shown in Fig. 7. The condition $r_s/r_d = 1$ indicates the attainment of a limit condition of the structure. At this point, only one thrust network fits the geometric limits, and the structure's ultimate equilibrium condition is reached.

In more general cases the ratio $r_s/r_d = 1$ is utilized to determine either an extreme geometry or a limit loading condition for the structure. This concept has practical applications, such as assessing the geometric safety factor of the vault [7], which, as done here, involves determining the minimum thickness required for the existence of a network configuration in equilibrium with applied loads and fulfilling the corresponding geometric limits. The same approach can be used in other applications to determine the limit multiplier for a specific loading condition.

5 Discussion and Conclusions

A static analysis of the top dome belonging to the Temple of Diana in Baiae has been presented. The analysis was performed by the classic Finite Element Method as well as by the Thrust Network Analysis in order to compare the results of such methodologies and to investigate their issues and advantages.

A significant outcome of such an investigation consists in the fact that results provided by the investigated strategies are mutually consistent. In fact, pressure curves determined by FEM turn out to be close to the shallowest solutions of the TNA nearby the impost region while they tend to the deepest solutions nearby the dome top. Moreover, the limit thickness, i.e. the minimum value for which the shell can reach equilibrium within an entirely compressive state, turns out to be similar, at least in magnitude, when computed with both the investigated strategies. Actually, FEM provide a limit thickness 50% greater than the value provided by TNA.

Such a difference is due to the fact that TNA neglects shear and compressive failures of masonry, while such phenomena are taken into account by the Drucker Prager plasticity model. Indeed, the constitutive characterization of materials represent the essential difference between TNA and FEM, and influences both benefits and drawbacks of these strategies. In fact, from one hand, finite elements provide a unique solution fulfilling both equilibrium and compatibility conditions. Its uniqueness permits to perform incremental analysis determining collapse mechanisms. Moreover, they are capable of accounting for constraint displacements. On the contrary, Thrust Network Analysis provides a set of solutions fulfilling equilibrium but they are not necessarily compatible. Moreover, constraint displacements should be taken into account by geometric perturbations of the model.

Nevertheless, the essential advantage of TNA consists in its independence from the constitutive behavior of the involved materials since its solutions depend only on geometry and load patterns. From this point of view, structural masses are far easier to be estimated with respect to parameters such as Young modulus, Poisson's ratio, yield stress etc. Indeed, for historical masonry it is complex even to determine an appropriate constitutive relationship, in fact the FEM analysis adopted the simplest model (Drucker Prager plasticity) permitting to account for compressive behaviors. Moreover, constitutive parameters of roman concrete were deduced by literature contributions reporting very rare experimental tests using original specimens coming from significantly distant sites and the model assumed the material to be homogeneous while brick patterns and non-homogeneities can strongly influence the behavior of the cast [18].

Indeed, both the investigated strategies present significant advantages as well as drawbacks and their conjunct use in structural analysis could be an effective strategy so that they can mutually overcome the respective issues. Possibly, exploitation of innovative solutions investigated for arches [9,14,15] could further enhance the synergy between differential and limit analysis approaches.

Funding and Acknowledgments. The present research was supported by the University of Naples Federico II and the *Compagnia di San Paolo*, which are gratefully acknowledged by the authors, FRA grants, CUP E69C21000250005. The authors wish also to acknowledge Arch. Mariano Marmo for his support about the geometric survey.

References

1. Block, P., Lachauer, L.: Three-dimensional (3D) equilibrium analysis of gothic masonry vaults. Int. J. Archit. Herit. **8**(3), 312–335 (2014)
2. Block, P.P.C.V.: Thrust network analysis: exploring three-dimensional equilibrium. Ph.D. thesis, Massachusetts Institute of Technology (2009)
3. Brune, P., Ingraffea, A., Jackson, M., Perucchio, R.: The fracture toughness of an imperial roman mortar. Eng. Fract. Mech. **102**, 65–76 (2013)
4. Drucker, D., Prager, W.: Soil mechanics and plastic analysis for limit design. Q. Appl. Math. **10**, 157–165 (1952)
5. Fornaro, D.: Le volte romane di Baia: Geometria, tecniche costruttive e modelli strutturali. Tesi di Laurea - Universitá degli Studi di Napoli Federico II, Napoli, Italy (1998)
6. Gabriele, S., Varano, V., Tomasello, G., Alfonsi, D.: R-funicularity of form found shell structures. Eng. Struct. **157**, 157–169 (2018)
7. Heyman, J.: The stone skeleton. Int. J. Solids Struct. **2**(2), 249–279 (1966)
8. Long, Y.Q., Xin, K.G.: Generalized Conforming Element Theory, pp. 89–100. Springer , Heidelberg (2009). https://doi.org/10.1007/978-3-642-00316-5_4
9. Manuello Bertetto, A., Marano, G.: Numerical and dimensionless analytical solutions for circular arch optimization. Eng. Struct. **253**, 113360 (2022)
10. Marmo, F., Masi, D., Rosati, L.: Thrust network analysis of masonry helical staircases. Int. J. Archit. Herit. **12**(5), 828–848 (2018)

11. Marmo, F., Rosati, L.: Reformulation and extension of the thrust network analysis. Comput. Struct. **182**, 104–118 (2017)
12. Marmo, F., Ruggieri, N., Toraldo, F., Rosati, L.: Historical study and static assessment of an innovative vaulting technique of the 19th century. Int. J. Archit. Herit. (2018)
13. Marmo, M.: Le cupole del Complesso archeologico delle Terme di Baia: la geometria degli spazi e dei fenomeni luminosi. Ph.D. thesis, Universitá degli Studi di Napoli Federico II (2018)
14. Melchiorre, J., Manuello Bertetto, A., Marano, G.: Application of a machine learning algorithm for the structural optimization of circular arches with different cross-sections. J. Appl. Math. Phys. **9**(5), 1159–1170 (2021)
15. Melchiorre, J., Manuello Bertetto, A., Marmo, F., Adriaenssens, S., Marano, G.: Differential formulation and numerical solution for elastic arches with variable curvature and tapered cross-sections. Eur. J. Mech. A/Solids **97**, 104757 (2023)
16. O'Dwyer, D.: Funicular analysis of masonry vaults. Comput. Struct. **73**(1–5), 187–197 (1999)
17. Rakob, F.: Römische Kuppelbauten in Baiae, Die Gewölbeprofile. Mitteilungen Des Deutschen Archäologischen Instituts, Römische Abteilung **95**, 257–301 (1988)
18. Roselli, F., Alforno, M., Manuello Bertetto, A., Venuti, F.: Experimental investigation of the effect of brick pattern on the structural response of masonry arches and barrel vaults. Constr. Build. Mater. **368**, 130434 (2023)
19. Santini, S., Baggio, C., Sabbatini, V.: Seismic assessment of roman concrete groin vaults through UAV, NDT and 3D analyses. Heritage **5**(1), 311–331 (2022)
20. Serpieri, R., Sessa, S., Rosati, L.: A MITC-based procedure for the numerical integration of a continuum elastic-plastic theory of through-the-thickness-jacketed shell structures. Compos. Struct. **191**, 209–220 (2018)
21. Sessa, S., Marmo, F., Vaiana, N., De Gregorio, D., Rosati, L.: Strength hierarchy provisions for transverse confinement systems of shell structural elements. Compos. B Eng. **163**, 413–423 (2019)
22. Sessa, S., Serpieri, R., Rosati, L.: A continuum theory of through-the-thickness jacketed shells for the elasto-plastic analysis of confined composite structures: theory and numerical assessment. Compos. B Eng. **113**, 225–242 (2017)
23. Sessa, S., Serpieri, R., Rosati, L.: Probabilistic assessment of historical masonry walls retrofitted with through-the-thickness confinement devices. In: AIMETA 2017 - Proceedings of the 23rd Conference of the Italian Association of Theoretical and Applied Mechanics, vol. 3, pp. 2324–2332 (2017)
24. Sinopoli, A., Aita, D.: The dome of the temple of Diana in Baiae: Opus caementicium, geometry and mechanics. Int. J. Archit. Herit. **16**(8), 1152–1183 (2022)

Structural Design of Palazzo dell'Edilizia of Alessandria

Matteo Buschini(✉), Andrea Alberto, Raffaele Todaro, and Pier Paolo Cairo

LGA Engineering S.r.l, 12038 Savigliano, Italy
ingegneria@lgaengineering.it

Abstract. The case study concerns the structural project of the "Palazzo dell'Edilizia of Alessandria", a multi-functional building that will host the activities of a public company managing building industry and construction workers issues. The early architectural concept design has been carried out by the well-known Architect Daniel Libeskind. LGA Engineering team by means of BIM approach developed the structural design in a Level Of Detail corresponding to a Definitive Project.

The building is organized into four above ground stories that house classrooms, offices, conference halls and multi-purpose spaces. Under structural point of view the main body is made of a central rigid concrete core and surrounding steel frames. The most iconic component of whole complex is the 50 m high tower made of steel frames interconnected each other and with the other main elements of the building.

During project development, the main structural challenges to deal with has been the design of long span beams, cantilever slabs and beams, out of plane structural elements and their mutual connection, and above all the managing of the complex geometry and the height of the tower steel frame.

Keywords: Steel · structure · BIM · tower · lattice girder · FEM analysis

1 Introduction

The building is organized into four above ground stories and one underground level. In addition to the main body, the structure has a 50 m high steel tower interconnected with the other main structural elements of the building (Fig. 1).

Due to the complex geometry of the building and to respect the architectural layout several structural configuration has been analyzed and the definitive solution adopted concerns a composite steel and concrete structure.

S. Gabriele et al. (Eds.): IWSS 2023, LNCE 437, pp. 446–453, 2024.
https://doi.org/10.1007/978-3-031-44328-2_46

Fig. 1. Side and front view of the BIM model of Palazzo dell'Edilizia of Alessandria.

2 Description of Building Components

Foundation structural elements are made of isolated concrete pads and concrete beams with underlying foundation piles. Most of the foundation elements were developed in a previous project phase and at the moment of the conception of the definitive structural design were already be realized (Fig. 2) therefore they have been considered as existing structures.

Fig. 2. Existing foundation elements and underground concrete structures.

The underground level hosts the car parking and has a main structure made of vertical concrete columns and concrete walls, horizontal elements of the floor composed by prestressed precast concrete beams and precast concrete slabs (Fig. 3a).

The ground floor is characterized by two large spaces designated to didactic laboratories, some accessory spaces and a coffee bar zone in the front side of the building.

Vertical main structural elements are both concrete walls and columns and steel columns, while the horizontal elements of the floor have a composite structure made of main steel beams and precast concrete slabs (Fig. 3b).

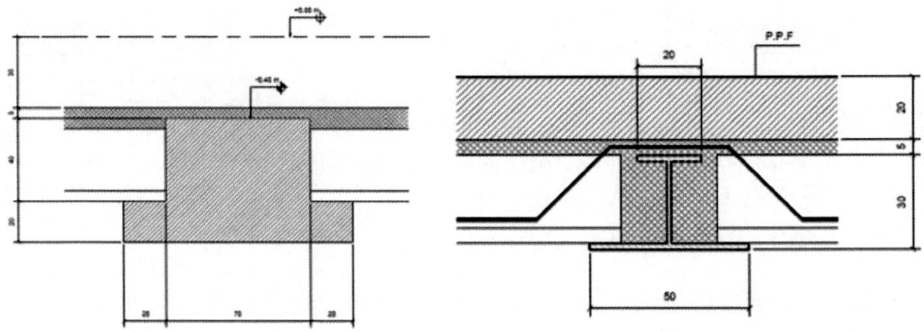

(a) (b)

Fig. 3. Schematic detail of connection between prestressed precast concrete beams and precast concrete slabs of underground level (a) and schematic detail of connection between steel beams and precast concrete slabs of upper floors (b).

Between the ground floor and the first floor there is an intermediate floor that takes up a reduced amount of the whole surface of the building and that house a dining hall and some accessory spaces. This floor is partially supported by vertical elements such as concrete walls and steel columns and in the forward part of the building is hung to the upper floor by means of steel tie-rods.

The first floor hosts classrooms and offices and is characterized by main structural elements made of steel lattice girders of height equal to the whole space between the two floors (Fig. 4a, b). The secondary structural elements are steel beams and precast concrete slabs.

The main space that is located in the second floor is the conference hall, placed in the front side of the building. The needs of ensure a large space without structural elements inside made the structural designers develop steel lattice girders connected with the ones of the floor below and supported by the four sloped concrete columns of the ground floor. Secondary structural elements, as for the first level, are made of steel beams and precast concrete slabs.

The roof structure is made of steel elements such as columns and beams and has been designed in order to be as light as possible. It is developed on a sloped plane with the upper edge corresponding to the forward part of the building and the lower one corresponding to the back part where the roof is partially open and are located some terraces accessible from the second floor.

Structural elements of the external walls are made of precast concrete panels with stiffener components strictly connected to the main structure at the level of the different floor by means of composite concrete and steel edge beams (Fig. 5). This layout let the whole structure be connected on the different levels and these wall elements works as windbrace system giving stability respect to wind and seisic horizontal actions.

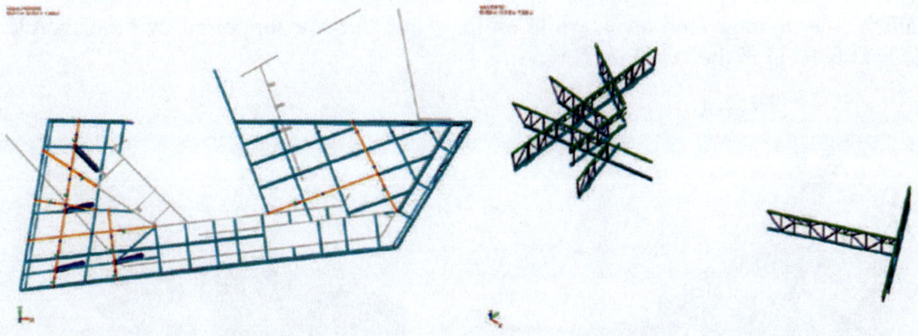

(a) (b)

Fig. 4. Location of lattice girders in first floor (a) and axonometric view of components of the steel frame (b).

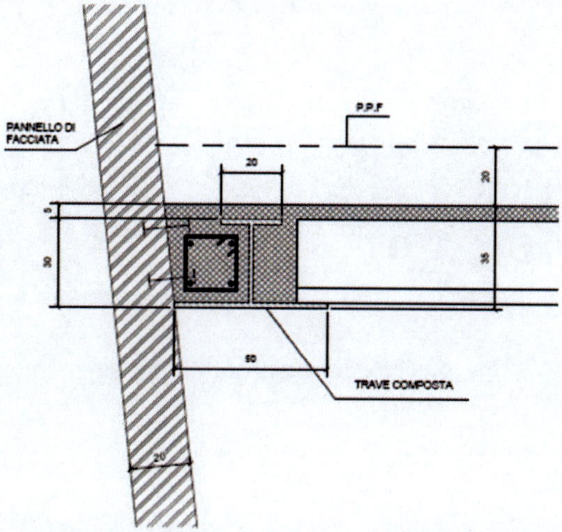

Fig. 5. Detail of the connection between concrete precast wall panels and edge composite beams of floor structure.

The central part of the main body of the building that hosts stairs and elevators is a concrete core made of concrete walls connected each other that provide vertical support to the tower steel structure and horizontal stiffness.

3 Description of Steel Tower Components

In addition to the main body of the building, the structure includes a steel tower with an overall height of 50 m which crosses the shape of the building and it is structurally connected to it. In particular the structure of the tower consist of steel frames that on the back part of the building provide supports to the floors and are supported by the concrete walls of the underground level, while on the front side are supported by the concrete core at the level of the last floor (Fig. 6).

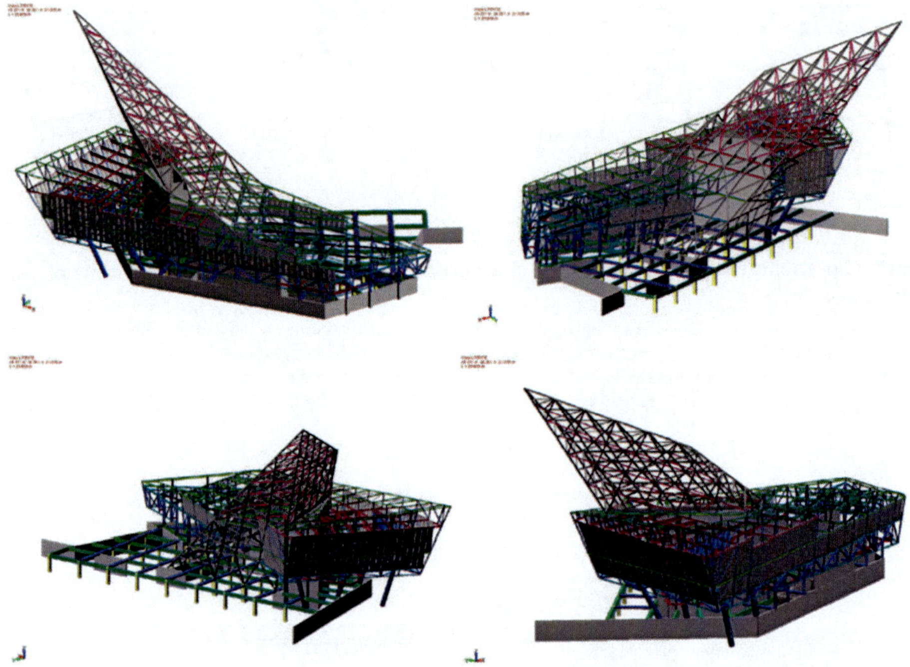

Fig. 6. Different views of FEM model.

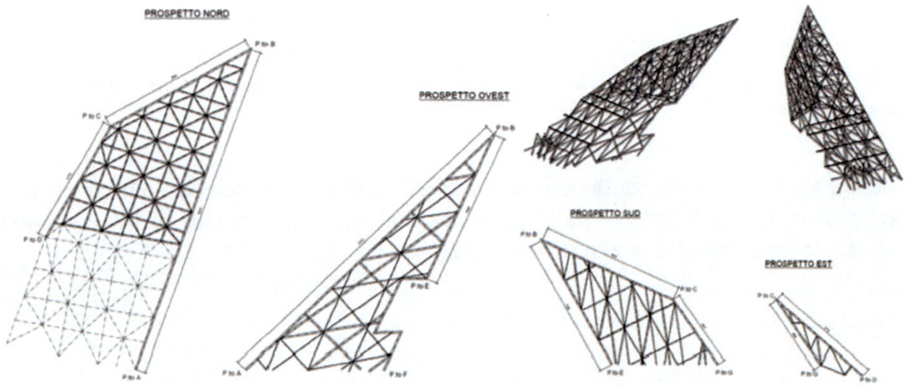

Fig. 7. Composition of steel tower frames.

Tower steel structure is subdivided into 4 plane frames realized with square hollow section steel elements coupled each other along the lateral side of the four faces (Fig. 7).

The back side of the tower hosts a photovoltaic plant with panels arranged on the steel structure up to the height of about 32 m, while the other three faces are open lattice structures without any cladding.

4 Structural Design Development and BIM Approach

Due to the complex configuration of the building, as described above, structural design was carried out by different steps starting from simple hand calculation until reaching advanced FEM analysis. Large use of BIM approach lets the project team manage the complex geometry of the building and simplify the developing of the structural conception of the components by means of software interoperability in exchanging information between Revit, Advance Design and Dolmen (Fig. 8).

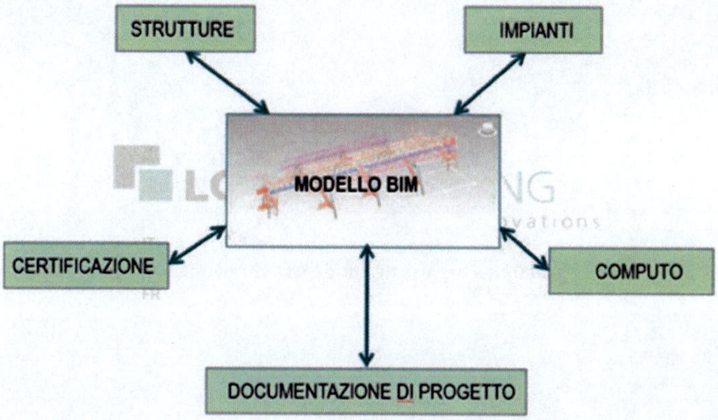

Fig. 8. Conceptual scheme of BIM interoperability between different project scenarios.

The most challenging task of the designers was to evaluate the behavior of the building in both static and dynamic configuration with specific regard to the tower steel structure response to wind and seismic action. In order to develop a complete structural design the early numerical evaluation has been the realization of two different FEM model of the main body of the building alone and the tower alone as well. This way to proceed let the designers perform simplified splitted analysis of the two main component of the complex in which the sizing of the main structural elements was carried out. After that the structural models merged in a unique FEM model and the detailed structural analysis was performed to develop all the detail of the structural configuration described above.

Complexity and importance of the project made the designers develop analogous FEM model of same components by means of different structural software such as Advance Design and Dolmen in order to compare and validate the results of the structural analysis performed (Fig. 9).

An in-depth numerical analysis has been carried out regarding the wind action on the steel lattice structure of the tower. In detail, evaluation of vortex shedding phenomena

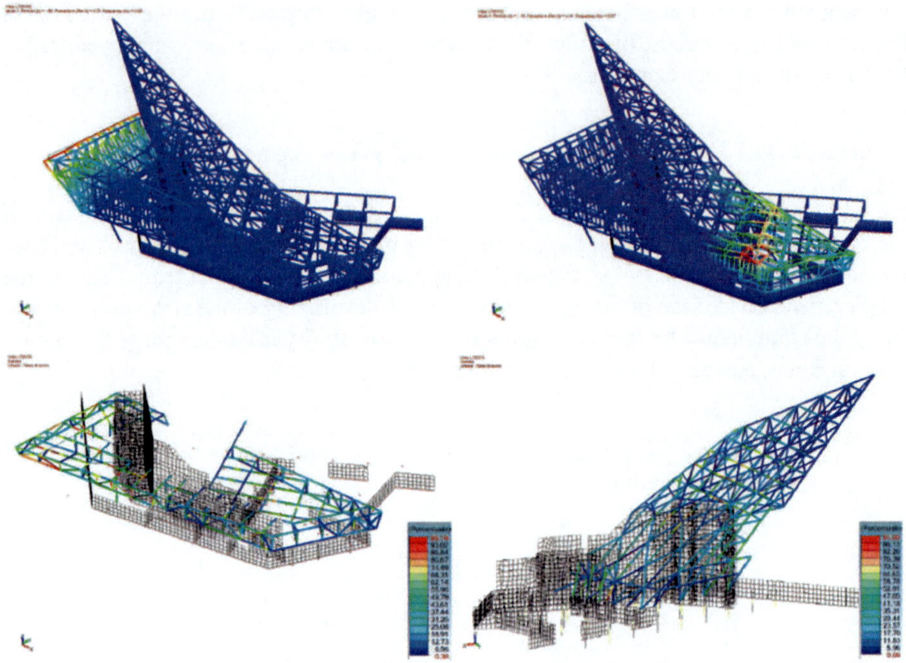

Fig. 9. Some results of FEM analysis regarding mode of vibrations and steel elements work ratio.

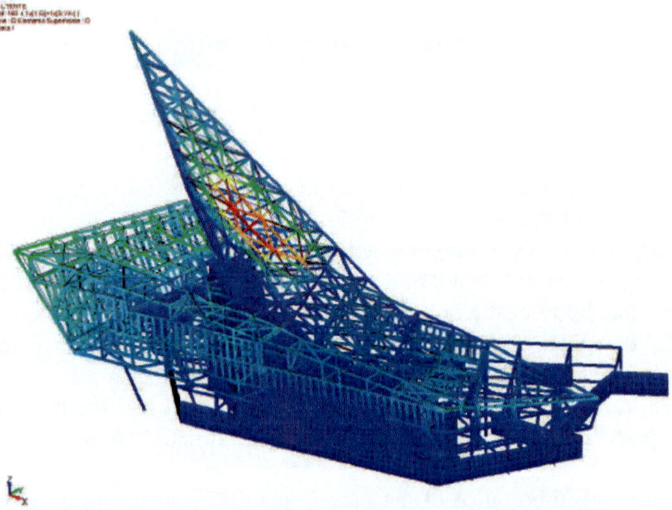

Fig. 10. Maximum displacements under wind action.

and occurrence of aeroelastic instabilities was performed according to EN 1991-1-4 [1] by means of determination of physical and dynamic characteristic of the structure, such as natural frequency of the flexural mode of cross-wind vibration, Strouhal number and critical wind velocity for bending vibration (Fig. 10).

More detailed analysis about interaction between wind and tower steel structure (e.g., CFD computation, wind tunnel test) shall be performed during next phase of executive structural design development.

Reference

1. EN 1991-1-4 "Action on structures – Part 1-4: General actions - Wind actions"

Characterisation of an Additively Manufactured 3D Cross-Based Fractal Structure for Impact Energy Absorption

Marco Viccica[1(✉)], Gabriel Ferreira Serra[2], Fábio Fernandes[2], Flaviana Calignano[1], Luca Iuliano[1], Ricardo Alves de Sousa[2], and Manuela Galati[1]

[1] Integrated Additive Manufacturing Center (IAM), Department of Management and Production Engineering (DIGEP), Politecnico di Torino, Turin, Italy
marco.viccica@polito.it
[2] Centre for Mechanical Technology and Automation (TEMA), Department of Mechanical Engineering, University of Aveiro, Aveiro, Portugal

Abstract. Designers propose using high-performance materials or cellular and bio-inspired structures in components with high strength-to-weight ratios, high heat transfer capacity, and energy absorption. An example is fractal geometries, which show highly complex 3D geometries that are unfeasible using conventional manufacturing processes. The work investigates the energy absorption performances of a 3D cross-based fractal structure (3D-CFS). The geometry is inspired by the mathematical 3D Greek cross geometry and designed for production using Additive Manufacturing (AM) powder bed fusion processes for polymers (PBF). The mechanical properties of Polyamide (PA12) and Thermoplastic Polyurethane (TPU) 3D-CFS structures designed with different volume fractions are evaluated using quasi-static and dynamic compression tests. The results show that the 3D-CFS structure is a good candidate for shock absorption applications such as personal protective equipment (PPE) applications.

Keywords: PPE · PBF-LB/P · PA12 · PBF-IrL · TPU · Fractal · shock absorption

1 Introduction

In the last few years, the Additive Manufacturing (AM) approach has led designers to explore more complex geometries capable of optimising and enhancing structural components in several engineering applications. Notably, increasing attention has been paid to personal protective equipment (PPE) for impact applications. The energy absorption properties of liner helmets in sports [1, 2] and military [3] fields have been insightfully investigated to develop more efficient and light devices.

Several studies have been carried out to evaluate the use of the periodic cellular structures, i.e., lattice structures [4], as an alternative absorber of the typical liner material, i.e., the expanded polystyrene (EPS). For this purpose, both strut-based [5, 6] and

S. Gabriele et al. (Eds.): IWSS 2023, LNCE 437, pp. 454–461, 2024.
https://doi.org/10.1007/978-3-031-44328-2_47

triply periodic minimal surfaces (TPMS) [7–9] lattice structures made via AM technologies have been demonstrated to be good candidates. However, more complex thin structures characterised by bio-inspired shapes [10, 11] and fractal geometry [12–14] have also been investigated over the last few years. Viccica et al. [15] have studied the energy absorption potentiality of a 3D Greek cross fractal [16] manufactured via AM in polyamide. The study results have shown that the developed fractal structure can absorb up to 77% of energy under compression loads than a typical EPS foam. However, the characterisation of the fractal object was limited to a single material tested under quasi-static compression loads.

This work aims to investigate the capabilities of the 3D cross-based fractal structure (3D-CFS) presented in Ref. [15] by comparing the energy absorption performance of the 3D-CFS produced with different polymeric materials and subjected to dynamic compression loads. In addition, the effects of the volume fraction variation were evaluated in terms of manufacturability and structural performance.

2 Materials and Methods

2.1 Greek Cross-Based Fractal Design

The 3D Greek cross-based fractal is a strut-based structure that follows the subsequent transformations of replication and scaling starting from a Greek cross initiator. This initiator (Fig. 1a left) consists of three equal struts linearly extruded along the axes of a cartesian tern. The strut has a circular cross-section with a D diameter and L length. The 3D-CFS is then properly transformed to get a cubic compression sample (Fig. 1b). The geometry of the 3D-CFS is parametric; therefore, its global dimension and volume fraction (VF) can be tailored by the two main geometric parameters, D and L.

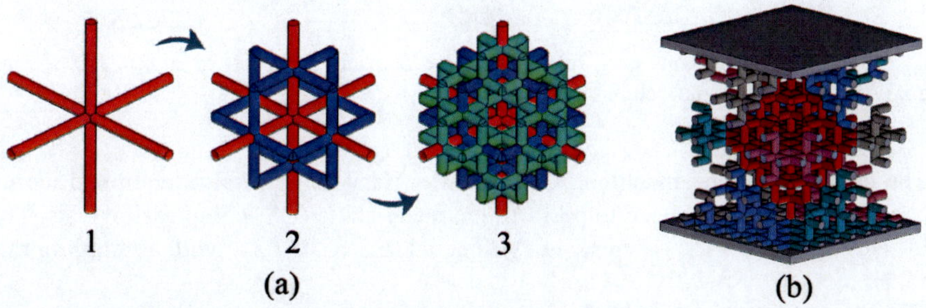

Fig. 1. (a) 3D-CFS iterations: in red, the first step; in blue, the added axes of the second; (b) cubic compression sample. (adapted from Ref. [15])

The diameter significantly affects the structure's stiffness, decreasing the strut's diameter [15]. In addition, from a process point of view, the diameter reduction induces some defects, contributing to the stiffness decrease. The variation of the VF must be yielded from both changing of L and D to consider also the material processability. Starting from the configuration analysed in the previous work [15], two more designs with a higher VF were investigated here. All designs are described by the variable L and D as follows: L40_D2.0 (L = 40 mm, D = 2 mm, VF = 9.1%) [15], L28_2.0 (L = 28mm, D = 2 mm, VF = 16.8%) and L28_D1.8 (L = 28 mm, D = 1.8 mm, VF = 14.1%). With this configuration, it is also possible to speculate on the possible effect of the sample height and diameter when one of these variables is fixed.

2.2 Production and Testing

The materials employed in this study are two different polymers: polyamide 12 (PA12) and thermoplastic polyurethane (TPU). The Formiga Velocis P110 (EOS GmbH) was used to fabricate the PA12 samples with the recycled PA2200 powder [17] (EOS trade-name of the PA12). At the same time, the ESTANE® 3D TPU M95A-545 powder [18] was processed via the HP® JetFusion 4200 MultiJetFusion machine to produce the TPU samples.

Three replicas for each configuration were produced and tested. Figure 2 reports the three configurations manufactured in PA12, TPU and compared to the nominal CAD.

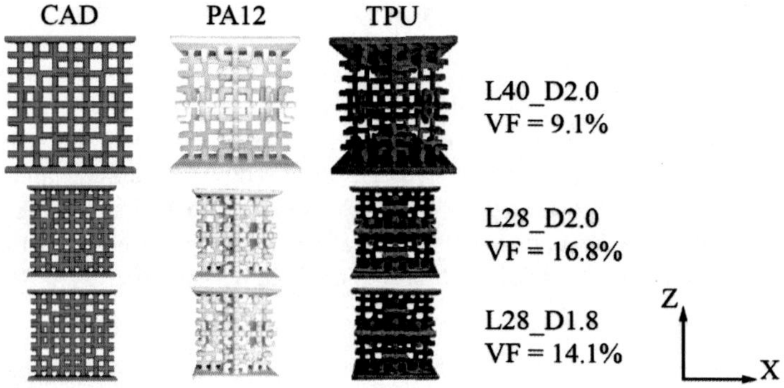

Fig. 2. 3D-CFS configurations with relative volume fraction: from left to right CAD version design, PA12 and TPU manufactured samples.

The universal testing machine, 3MZ Tensile (Easydur Srl, Italy), equipped with a load cell of 5 tonnes, was used to perform the quasi-static (QS) compression tests. The strain-rate selected for the compression was equal to 2×10^{-3} s^{-1} with a sampling rate of 100 Hz.

The impact tests were carried out by using a drop tower system. The total mass of the impactor was 20 kg. The flat diameter of the impactor (130 mm) guarantees a uniform application of the impact energy on the upper surface area of the samples. The displacement of the impactor and the response force generated during the impact was measured by the encoder and the load cell, respectively. The first sample (L = 40 mm) was used to perform a preliminary test under an impact energy quantity of 50 J, which was then decreased to 25 J for the other samples. The strain-rate related to the two high-speed (HS) impacts are 8.8 s^{-1} and 12.4 s^{-1}, respectively.

As an indicator of energy absorption properties, the energy per unit volume (W) was calculated considering the area under the stress-strain curve until the densification point (ε_d).

The specific energy absorption (SEA) was computed as the ratio between W and the material density to consider the actual absorber material density. The ε_d of the structure is localised in the strain point in which the absorber reaches the higher value of the Ideality (I) efficiency parameter (see Fig. 3). This parameter compares the actual material with the behaviour of an ideal absorber, which is characterised by constant strain for all strains [19].

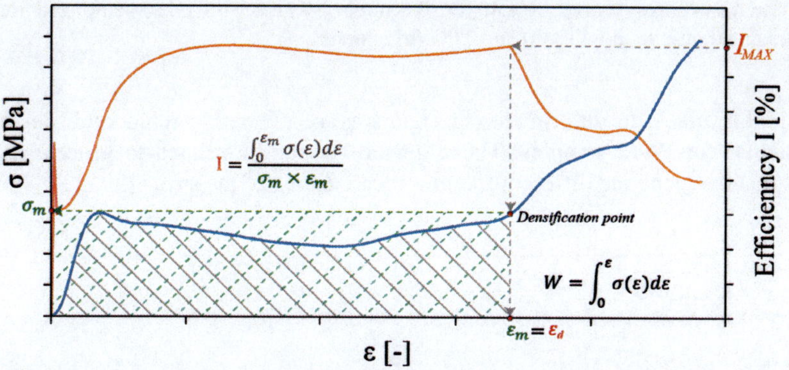

Fig. 3. Evaluation of the densification point through the Efficiency I curve (orange curve). The area under the stress-strain curve (blue light line) highlighted in grey represents W. The green area refers to the ideal amount of energy absorption up to the ε_d.

3 Results and Discussion

The results of the quasi-static and dynamic compression tests on the L40_D2.0 3D-CFS configuration are shown in Fig. 4. As mentioned above, as the first attempt of calibration, a 50 J impact energy was selected for this configuration. The engineering stress-strain curves of low and high strain-rate tests highlight the difference in strength of the PA12 material compared to the TPU [20]. In addition, as regards this configuration, both materials are strongly strain-rate dependent and, therefore, far from an ideal absorber. However, when tested at a high strain-rate, the TPU material's elasticity allows the structure to better accommodate the struts bending behaviour.

In contrast, the PA12 samples are dramatically affected by premature fractures of the struts due to the brittle behaviour of this material. This fracture mechanism leads to a significant decrease in the structure strength response. The PA12 sample showed more controlled deformation mechanisms at lower strain-rate compression loads. For this reason, the new structures were designed with higher VF and tested at lower impact energy (25 J).

The results of the impact tests on the L28_D2.0 and L28_D1.8 samples are reported in Fig. 5 and Fig. 6. Even if the stress and strain within the replicas of each configuration

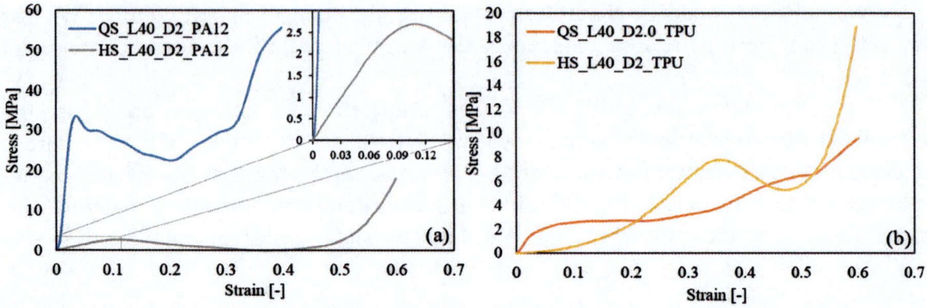

Fig. 4. Strain vs stress curve of L40_D2.0 samples tested with quasi-static and dynamic compression strain-rate in PA12 (a) and TPU (b) material.

show similar values, the deformation mechanisms are affected by remarkable deviations, in particular for PA12 samples. These deviations can be related to process-induced defects that can generate different fracture localisations of the struts [20, 21].

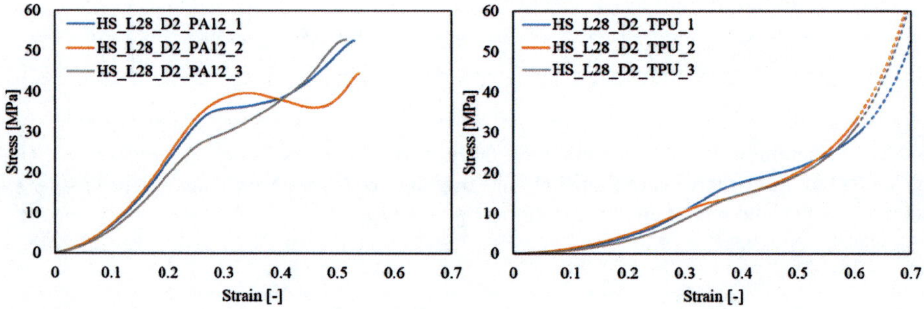

Fig. 5. Strain vs stress curve of all the replicas of L28_D2.0 samples tested with dynamic compression strain-rate in PA12 (a) and TPU (b) material.

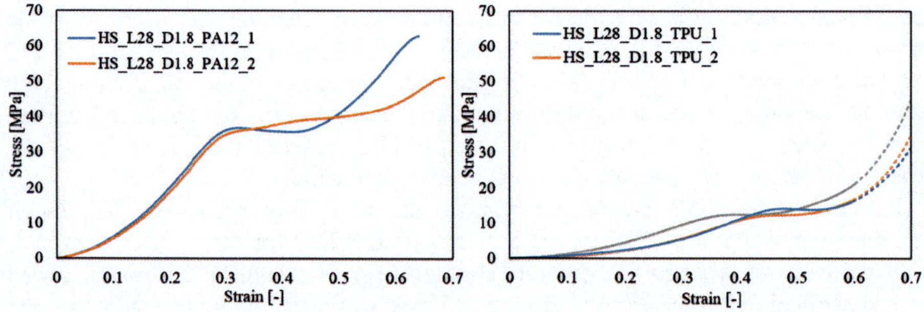

Fig. 6. Strain vs stress curve of all the replicas of L28_D1.8 samples tested with dynamic compression strain-rate in PA12 (a) and TPU (b) material.

Compared to the L40_D2.0 counterpart, both configurations and materials show higher energy absorption values with an appreciable plateau region. Because of the lower stiffness, energy absorption performances of TPU parts are slightly lower than PA12 samples. In contrast, the peak of stress on the first part of the impact is significantly lower.

Table 1 resumes the energy absorption properties extracted from the compression tests performed in QS and HS strain-rate. PA12 samples showed higher values of SEA and I than the TPU counterparts. Compared with the samples with L28, those with lower VF showed a slightly higher absorption efficiency because of the lower stress value at the densification point. On the other hand, the low VF of L40_D2.0 cannot guarantee acceptable energy absorption performance at high-impact loads.

Table 1. Energy absorption performance of the samples manufactured in PA12 and TPU, tested in quasi-static (QS) and dynamic (HS) loads.

ID	mean (SD)*					
	$\sigma_{\varepsilon d}$ [MPa]	ε_d [%]	$W_{\varepsilon d}$ [MJ/m^3]	SEA [kJ/kg]	I [%]	ρ [kg/m^3]
QS_L40_D2.0_PA12	33.8 (0.07)	33 (0.8)	8.78 (0.2)	8.7 (0.5)	78 (0.01)	930
QS_L40_D2.0_TPU	2.5 (0.2)	23 (1)	0.48 (0.05)	0.44 (0.05)	83 (0.02)	1100
HS_L40_D2.0_PA12	2.8	51	0.59	0.64	41	930
HS_L40_D2.0_TPU	7.9	55	2.17	1.98	51	1100
HS_L28_D2.0_PA12	38.6 (1.3)	44 (5)	9.7 (2.5)	10.4 (2.7)	56 (6)	930
HS_L28_D2.0_TPU	19.5 (2.6)	49 (3)	3.9 (0.8)	3.5 (0.7)	41 (2)	1100
HS_L28_D1.8_PA12	39.7 (1.3)	52 (3)	12.6 (1.2)	13.6 (1.3)	61 (0.3)	930
HS_L28_D1.8_TPU	13.9 (0.5)	53 (3)	3.4 (0.2)	3.1 (0.2)	46 (0.3)	1100

*The L40_D2.0 samples were tested with only a replica at the HS strain-rate to calibrate the impact energy level.

4 Conclusion

This study proposes an investigation of the energy absorption performance of a 3D cross-based fractal structure (3D-CFS) subjected to compression loads at different strain rates. The designed structures were manufactured with two different polymers, PA12 and TPU, by the powder bed fusion technology for polymers. The manufacturability and structural performance of the 3D-CFS were evaluated by analysing structures with different volume fractions. The testes results have shown that both PA12 and TPU samples are far from behaving as an ideal absorber because the response is significantly strain-rate dependent. Lower volume fraction can provide higher efficiency values in terms of energy absorption.

Although the PA12 samples can absorb more energy, the TPU counterparts guarantee relevant lower peak force impact, ideal for preventing brain injuries in head protection equipment.

References

1. Khosroshahi, S.F., Olsson, R., Wysocki, M., Zaccariotto, M., Galvanetto, U.: Response of a helmet liner under biaxial loading. Polym. Test. **72**, 110–114 (2018)
2. Gao, W., He, X., Wang, J., Feng, Y.T., Wang, C.: Numerical investigation of oblique impact behavior of a helmeted headform on a windshield considering failure. Thin-Walled Struct. **171**, 108722 (2022)
3. Rice University Rice tapped to develop 3D-printed 'smart helmets' for the military. https://news.rice.edu/news/2021/rice-tapped-develop-3d-printed-smart-helmets-military
4. Maconachie, T., et al.: SLM lattice structures: properties, performance, applications and challenges. Mater. Des. **183**, 108137 (2019)
5. Jhou, S.Y., Hsu, C.C., Yeh, J.C.: The dynamic impact response of 3D-printed polymeric sandwich structures with lattice cores: numerical and experimental investigation. Polymers (Basel) **13** (2021)
6. Sun, Z.P., Guo, Y.B., Shim, V.P.W.: Static and dynamic crushing of polymeric lattices fabricated by fused deposition modelling and selective laser sintering – an experimental investigation. Int. J. Impact Eng **160**, 104059 (2022)
7. Ramos, H., Santiago, R., Soe, S., Theobald, P., Alves, M.: Response of gyroid lattice structures to impact loads. Int. J. Impact Eng **164**, 104202 (2022)
8. Syam, W.P., Jianwei, W., Zhao, B., Maskery, I., Elmadih, W., Leach, R.: Design and analysis of strut-based lattice structures for vibration isolation. Precis. Eng. **52**, 494–506 (2018)
9. Miralbes, R., Higuera, S., Ranz, D., Gomez, J.A.: Comparative analysis of mechanical properties and energy absorption capabilities of functionally graded and non-graded thermoplastic sheet gyroid structures. Mech. Adv. Mater. Struct. 1–14 (2021)
10. Mehta, P.S., Solis Ocampo, J., Tovar, A., Chaudhari, P.: Bio-inspired design of lightweight and protective structures. SAE Tech. Pap. (2016). https://doi.org/10.4271/2016-01-0396
11. Najmon, J.C., Dehart, J., Wood, Z., Tovar, A.: Cellular helmet liner design through bio-inspired structures and topology optimization of compliant mechanism lattices. SAE Int. J. Transp. Saf. **6**, 217–235 (2018)
12. Wang, J., Zhang, Y., He, N., Wang, C.H.: Crashworthiness behavior of Koch fractal structures. Mater. Des. **144**, 229–244 (2018)
13. Zhang, Y., Wang, J., Wang, C., Zeng, Y., Chen, T.: Crashworthiness of bionic fractal hierarchical structures. Mater. Des. **158**, 147–159 (2018)
14. Zhang, Y., He, N., Song, X., Chen, T., Chen, H.: On impacting mechanical behaviors of side fractal structures. Thin-Walled Struct. **146**, 106490 (2020)
15. Viccica, M., Galati, M., Calignano, F., Iuliano, L.: Design, additive manufacturing, and characterisation of a three-dimensional cross-based fractal structure for shock absorption. Thin-Walled Struct. **181**, 110106 (2022)
16. Dickau, R.: Greek Cross Fractal (2008). https://www.robertdickau.com/greekcross.html. Accessed 10 Feb 2022
17. EOS GmbH - Electro Optical Systems: Material data sheet PA 2200 (2017)
18. Lubrizol: Estane 3D TPU datasheet (2021)
19. Miralbes, R., Ranz, D., Ivens, J., Gomez, J.A.: Characterisation of cork and cork agglomerates under compressive loads by means of energy absorption diagrams. Eur. J. Wood Wood Prod. **79**, 719–731 (2021)

20. Calignano, F., Giuffrida, F., Galati, M.: Effect of the build orientation on the mechanical performance of polymeric parts produced by multi jet fusion and selective laser sintering. J. Manuf. Process. **65**, 271–282 (2021)
21. Galati, M., Viccica, M., Minetola, P.: A finite element approach for the prediction of the mechanical behaviour of layered composites produced by Continuous Filament Fabrication (CFF). Polym. Test. **98**, 107181 (2021)

Designing with Technical Textiles on Mars: Material Properties and Sustainable-by-Design Principles for Enhancing Efficient Construction in Extreme Environments

Marta Rossi(✉), Valentina Sumini, Salvatore Viscuso, and Alessandra Zanelli

Politecnico di Milano, Milano, Italy
marta2.rossi@mail.polimi.it

Abstract. The flexibility and adaptability provided by technical textiles make them suitable for the aerospace field.

Specific requirements for space structures such as resilience, payload constraints, and economic costs can be tackled through a smart use of the newest technologies in material sciences. The future of space exploration has in store the building of human settlements beyond Earth; therefore, it is crucial to improve and perfect the technologies we use to go to space and sustain human life there.

For this reason, this paper focuses on studying textiles applied to a Martian habitat named E.L.L.E., an Extreme Livable Lightweight Environment, by proposing a new combination of materials for an inflatable system that allows to decrease the total weight of at least 30% than current space modules.

Experimenting with technical textiles in space will enable a deeper understanding of these technologies to encourage their application in extreme environments on Earth for a sustainable architecture.

Keywords: Technical Textiles · Inflatable · Lightweight · Extreme Environments

1 Introduction

Technical textiles and membranes can be employed within the aerospace field and their area of application is rather wide. The necessity of using these materials rises from the possibility to obtain lightweight structures which can withstand fatigue and stress and reduce fuel consumption for launches to space.

This kind of materials are suitable for the construction of various products, such as space suits, aircraft, or gossamer structures, which represent that category of ultra-lightweight structures which also includes inflatables.

Compared to traditional materials, technical textiles provide several advantages and help better manage certain critical points of the aerospace field, such as fuel consumption, economic costs, and habitability.

Studying technical textiles in extreme environments such as Mars allows to perfect these technologies due to the many challenges that such an environment poses, therefore bringing benefits to different fields of application that can also involve terrestrial contexts.

S. Gabriele et al. (Eds.): IWSS 2023, LNCE 437, pp. 462–471, 2024.
https://doi.org/10.1007/978-3-031-44328-2_48

2 Technical Textiles in the Aerospace Field

2.1 Gossamer Spacecrafts

The term "gossamer spacecraft" identifies a category of ultra-lightweight space structures which represent an example of the use of technical textiles and membranes in the aerospace field. Examples belonging to this category are space antennas, solar panels, inflatable structures, and solar shields [1]. The materials used for this kind of structures are usually thin and have a low level of stiffness. This type of technology allows to obtain both tensioned-planar and inflatable-curved conformations.

Gossamer spacecrafts will be more and more present in the context of space explorations, since their use facilitates the construction of larger and more efficient structures, thus improving the quality of exploration itself. Indeed, using a lower amount of resources would guarantee both a sustainable approach to space exploration and the possibility to widen the horizons and constraints of traditional space constructions [2].

Space Antennas. Space antennas can be considered part of the gossamer structures category. An example of this is brought by NASA and the Department of Defense of the United States who worked on the design of an inflatable antenna, called IAE - Inflatable Antenna Experiment. This is the first experiment of this genre and proved to be successful. The IAE was launched into orbit onboard a Space Shuttle in 1996, it inflated as expected and thus demonstrated that it was possible to have inflatable structures in space. The antenna consisted of a reflecting parabolic structure with a 14-m diameter with a transparent cap and a toroidal structure guaranteed the support of the reflector. The structure had a total size of 14 by 28 m once deployed, while during the launch phase it had been packaged in a much smaller volume [3].

Solar Panels. In order to support space exploration, it is necessary to design systems for energy production. Solar panels are particularly efficient and useful to generate the required energy to sustain the artificial probes sent to space. As research in the aerospace field progresses energy requirements increase too. The inflatable technology appears to be applicable to different types of structures, among which there are solar panels.

In 1993 a company named L'Garde, Inc. developed a project for an inflatable solar panel for DARPA (Defense Advanced Research Projects Agency) and Philips Laboratory. This structure was designed to be launched into orbit while packaged; upon arrival it deployed and stiffened.

LISA-T is another example of inflatable solar panel, its name is an acronym for Lightweight Integrated Solar Array and Transceiver. Its purpose is the increment of energy production to power satellites while using a lightweight structure [4].

Inflatable Structures. The inflatable technology could represent the starting point to implement and improve orbital habitats and to start building on other planets.

Future planetary settlements could be configured as clusters of inflatable shelters, which would be packed up before launch and deployed once they arrive at destination. These characteristics allow to obtain larger constructions, therefore providing habitats with more livable spaces if compared to current human space habitats, such as the International Space Station (ISS).

The TransHab (see Fig. 1) is an inflatable habitat designed in 1997 by a team of architects and engineers from the Johnson Space Center, headed by Doctor William Schneider.

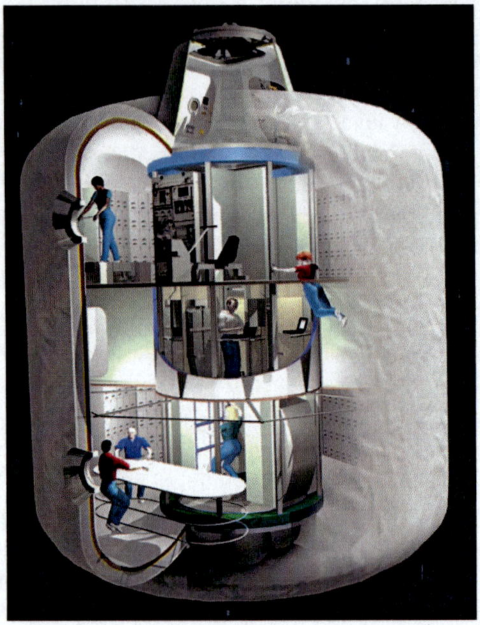

Fig. 1. View of the TransHab [5].

It is a hybrid space structure made of a rigid central core to which an inflatable outer shell is attached. It was the first project of an inflatable habitable module, and it was subjected to various tests, all brilliantly passed [5]. This project was later bought and further developed by the private space company Bigelow Aerospace. As a result, BEAM – Bigelow Expandable Activity Module was built and launched it into orbit, where it was connected to the ISS Tranquility Node [6].

Inflatable technologies also turn out to be useful for rovers, which could benefit from having inflatable wheels, therefore facilitating exploration by overcoming obstacles more easily and being faster and more efficient. An example of these inflatable rovers is Tumbleweed, invented by Jack Jones of NASA's Jet Propulsion Laboratory. It was designed to explore Mars and would take advantage of the planet's winds to move faster and explore a wider area [7].

Solar Shields. Solar shields are important to avoid overheating of space structures. An example is the James Webb Space Telescope, initially called the Next Generation Space Telescope, which was launched into orbit on December 25th, 2021, and is equipped with a main mirror with a 6.5 m diameter and a solar shield made with a five-layer structure of Kapton®. Each layer has a lower temperature than the previous one, this keeps the telescope at low temperatures which allows the observation of more distant points in the universe [8].

3 Traditional Space Habitats Technologies

3.1 Traditional Materials

There are several examples of space habitats that were built since the space exploration era began. One of the first of this kind is the Lunar Module (LM) which was created for the Apollo program in the 1960s, with the aim of bringing the first humans to the Moon. The LM ascent stage had an aluminum structural part, covered with layers of several materials whose purpose was to guarantee protection from heat and micrometeoroids. Above the aluminum skin there was a package of 25 layers of aluminized polyimide foil in Mylar® or H-film, each foil was covered with a very thin layer of aluminum on one side. These layers were then crimped to increase their thermal insulation capacity and the most exposed portions had additional layers of H-film or a sandwich layer made of two thin sheets of nickel, spaced out by a mesh made of Inconel®. The descent stage was also composed of a structure covered by a thermal and micrometeoroid shield similar to that of the ascent module. Here a Teflon® coated titanium shield was added, in order to protect against possible explosions during the ascent operation of the upper stage [9].

Skylab was the first station built by the United States, it was launched into orbit on April 14th, 1973, and remained operational until 1979. Its thermal protection device consisted of an internal layer of aluminum, covered with a Mylar® sheet which was attached to an outwards-facing layer of nylon with an adhesive [10].

Mir was launched in 1986 and was operational until 2001. The materials used for this space station included chemically milled 2 mm thick aluminum sheets welded to a 4 mm thick mesh. There also was a thermal blanket made of 25 layers of aluminized Mylar® coated with layers of materials with properties similar to those of Kevlar® [11].

The International Space Station is the first station to be built thanks to international cooperation between NASA, ESA, the Russian Federation (Roscosmos) and Japan (JAXA). It became operational in 1998 and hosted its first human mission in 2000. The external casing of the ISS modules is made with a type of light aluminum, above which there is a layer up to 10 cm thick that is composed of materials such as Kevlar® and ceramic fabrics, such as Nextel™, to guarantee protection from micrometeoroids and external temperatures [12].

The analysis of the materials used in past missions shows that aluminum has always been the most used element when building structures for space, since it is a lightweight metal. For this reason, space habitats are rigid and have several constraints in terms of volume, weight, and habitability.

3.2 Launch Vehicles and Payload Constraints

The current space scene has different stakeholders involved with launches of cargo and astronauts into space, such as NASA (National Aeronautics and Space Administration), ESA (European Space Agency), and CNSA (China Nation Space Administration). In these last few years, private space companies started emerging and infiltrating into a field that has always been led by government space agencies. Two examples of this are SpaceX, founded by Elon Musk in 2002, and Blue Origin, founded by Jeff Bezos in 2000. Therefore, collaboration between private and government space agencies has become crucial for the development of this field and to widen the horizons of space exploration. This cooperation brings novelties to launch systems since it is now customary for private space agencies to provide vehicles for government space missions. Falcon 9 (see Fig. 2) is an orbital reusable two-stage rocket developed by SpaceX and used by NASA to transport astronauts to the ISS. It is a 50 m rocket with a payload capability to low Earth orbit (LEO) of 22.8 tons [13].

Fig. 2. Falcon 9 rocket with Dragon capsule [https://www.nasa.gov/image-feature/falcon-9-rocket-with-dragon-spacecraft-vertical-at-launch-complex-39a].

Another SpaceX developed vehicle is Starship, which is a Super Heavy Rocket, that will be used to go to orbit, the Moon, or Mars. It can transport up to 100–150 tons of payload, therefore it will be especially useful for longer travels that require more cargo [14].

NASA's SLS (Space Launch System) is a super heavy lift launch vehicle that will be used to go beyond Earth's orbit [15]. The SLS Block 1 was employed to send Artemis 1 to the orbit of the Moon, therefore marking the first launch that combined SLS and the Orion capsule; this has a payload capability of 27 tons, while the other SLS Blocks go up to 46 tons in terms of cargo capacities [16].

Europe uses other types of space vehicles to send astronauts and cargo into space. ESA currently operates Ariane 5 and Vega-C from Europe's spaceport in French Guiana. Vega-C can transport 2.3 tons to orbit, while Ariane 6 has a payload capability of 21 tons [17].

4 E.L.L.E. an Extreme Livable Lightweight Environment

4.1 E.L.L.E.'s Technology

E.L.L.E. (see Fig. 3) is an inflatable module designed to accommodate six astronauts for 600 days on Mars.

Designing a project on Mars means that several challenges must be addressed: Mars has no magnetic field, a thin atmosphere mainly made of carbon dioxide, a reduced

Fig. 3. External view of E.L.L.E.

surface pressure, and low temperatures [18]. It is necessary to design a structure that can protect the internal environment against these external hazards, consequently a deep analysis on the material selection is required.

The inflatable membrane comprises different layers joined together using adhesives, with redundancy being a crucial consideration during its development [19]. The pneumatic system should be made of multiple layers to grant redundancy and protection against external hazards. A new combination of materials is proposed, each chosen based on prior research and technical properties [1]. The layers of the shell have specific functions, with the inner layer consisting of Nomex® to protect against fire, punctures, and other hazards, while also providing sound insulation and protection for the bladder layer. Nomex® is produced by DuPont and it is a flame-resistant meta-aramid synthetic fiber [20].

The bladder layer is composed of nylon laminate and ethylene vinyl alcohol and offers gas impermeability, flexibility, durability, and resistance to high temperatures. The strategy used to connect the bladder to the inner layer is an ultrasonic welding [1–19].

The restraint layer consists of MadFlex, which has vibration damping properties, is five times stronger than structural steel and has a flexible and a rigid side. This material was developed in 2015 by an Italian start-up called CoRe [21].

The protective layer protects against micrometeoroids and debris and is made of Vectran™ and Duocel®Foam, a foamy material that attenuates impacts. Vectran™ is a multifilament fiber obtained from a liquid crystal polymer manufactured by Kuraray. It provides qualities of resistance against heat, cut, and abrasion, it is an elastic material which is five times stronger than steel, therefore making it suitable for space applications [1–22].

The outer layer provides thermal insulation and radiation protection using Kapton® and Cermex. Kapton® is produced by DuPont and is a polyimide film, it provides a certain level of protection against radiation and Cermex has thermal insulation properties, to shield the internal environment from the low temperatures of Mars [23].

However, an adequate protection against galactic cosmic rays and solar flares can be achieved by adding an additional thick layer made of a high-density material, such as Martian regolith, this would reduce radiation absorption in long-term missions [18].

4.2 Mass of the Inflatable System

E.L.L.E. will be transported to Mars using Starship by SpaceX, which has a payload capability of over 100 tons. During transportation, the inflatable module will be deflated and packed.

The mass of the module can be easily determined because the project was entirely developed using Grasshopper3D, which is an algorithmic modeling software that allows for an easy and automatic calculation of the volume of the habitat.

As previously stated in paragraph 3.1 the inflatable shell is composed of Nomex®, nylon laminate, ethylene vinyl alcohol, Madflex, Duocel®Foam, Vectran™, Cermex, and Kapton®, each with a unique density and thickness.

In order to calculate the total mass of the system it is necessary to identify the density of each of the materials that constitute the external shell (Table 1).

Table 1. Thickness and density of the materials of E.L.L.E.'s shell.

Material	Thickness	Density
Nomex®	0.5 cm	1.37 g/cc
Nylon laminate	7 cm	1.15 g/cc
Ethylene vinyl alcohol	0.005 cm	1.14 g/cc
Madflex	0.5 cm	0.16 g/cc
Duocel®Foam	7 cm	1.4 g/cc
Vectran™	0.05 cm	1.4 g/cc
Cemex	5 cm	0.0961 g/cc
Kapton®	0.445 cm	1.42 g/cc

The next step consists in combining the densities of the different layers, to find the value of the density of the shell. Each density is converted into a percentile that describes the contribution of each material to the layers of the shell, these values are then multiplied by the densities. Table 2 shows the results of these calculations.

By adding up the percentile values multiplied by the densities we obtain the density of the shell which equals to 966.00 kg/m^3. A module with a radius of 7.5 m leads to a volume of 96.89 m^3, therefore the total mass of the system equals to 93,596.21 kg ($\approx$ 94 tons).

If the same structure was made of aluminum as if it was constructed with traditional technologies and materials, there would be some differences in terms of total mass of the system. Aluminum has an approximate density of 2,710.00 kg/m^3 [24], so if we consider the same volume of the shell and this density value the resulting mass is 262,571.90 kg ($\approx$ 263 tons).

Table 2. Percentile values of the materials.

Material	Percentile values	d x percentile
Nomex®	0.0244	33.41 kg/m^3
Nylon laminate	0.3415	392.68 kg/m^3
Ethylene vinyl	0.0002	0.28 kg/m^3
alcohol	0.0244	3.90 kg/m^3
Madflex	0.3415	478.05 kg/m^3
Duocel®Foam	0.0024	3.41 kg/m^3
Vectran™	0.2439	23.44 kg/m^3
Cemex	0.0217	30.82 kg/m^3
Kapton®		

This means that the payload capacity of Starship is respected when considering E.L.L.E.'s technology. Compared to traditional methods there is a decrement of approximately 64% of the mass, thus complying with the premises of the research.

4.3 Terrestrial Applications

The use of lightweight materials brings several advantages to the aerospace field, by ensuring a more sustainable way to build and manage space structures. However, the advancements that such research produces are not limited to the aerospace field, rather they can be extended to terrestrial environments.

Inflatables are particularly suitable for situations that require high levels of protection against external hazards or immediacy in construction, such as terrestrial extreme environments, like polar regions or oceans, or temporary emergency shelters. Therefore, E.L.L.E.'s technology could be repurposed, thus transferring its advantages to benefit other situations.

5 Conclusions

The purpose of the research consisted in demonstrating that technical textiles and membranes can be applied to a space architecture project to reduce the weight of the transportable cargo during a space mission to another planet.

E.L.L.E. was developed as a space architecture project that proposes an inflatable technology with a novel combination of materials that can fulfill the premises of the research. This study demonstrated that the obtained decrease in mass is equal to 64%, therefore exceeding initial expectations. This result shows that it is crucial to invest in lightweight materials within the aerospace field, since such technologies facilitate sustainable projects by decreasing the economic costs and the resources needed to travel to space.

Additionally, using an inflatable technology such as the one developed for E.L.L.E. allows to obtain larger livable volumes for the crew, therefore providing higher levels of comfort and habitability of space structure. This will improve the astronauts' lifestyles, which is particularly important point to consider for the longer missions to the Moon and Mars that will be seen during the next decades.

References

1. Jenkins, C.H.M.: Gossamer Spacecraft: Membrane and Inflatable Structures Technology for Space Applications. American Institute of Aeronautics and Astronautics Inc., Reston (2001)
2. Jenkins, C.H.M.: Compliant Structures in Nature and Engineering, 1st edn. WIT Press, Boston (2005)
3. Freeland, R., et al.: Inflatable Antenna Technology with Preliminary Shuttle Experiment Results and Potential Applications. Root (1996)
4. Lockett, T.R., et al.: Advancements of the lightweight integrated solar array an transceiver (LISA-T) small spacecraft system. In: 2015 IEEE 42nd Photovoltaic Specialist Conference (PVSC), pp. 1–6. IEEE, New Orleans (2015)
5. Kennedy, K.J., Adams, C.: ISS TransHab: an inflatable habitat. In: Johnson, S.W., Chua, K.M., Galloway, R.G., Richter, P.I. (eds.) Space 2000: Seventh International Conference and Exposition on Engineering, Construction, Operations, and Business in Space, pp. 89–100. American Society of Civil Engineering, Reston (2000)
6. NASA Extends Expandable Habitat's Time on the International Space Station. https://www.nasa.gov/feature/nasa-extends-beam-s-time-on-the-international-space-station. Accessed 13 May 2023
7. Kuhlman, K.R., et al.: Tumbleweed: a new paradigm for surveying the surface of mars for in-situ resources. In: Badescu, V. (eds.) Mars Prospective Energy and Material Resources, pp. 401–429. Springer, Heidelberg (2004). https://doi.org/10.1007/978-3-642-03629-3_14
8. James Webb Telescope Overview. https://www.nasa.gov/mission_pages/webb/about/index.html. Accessed 13 May 2023
9. Lunar Apollo Quick Reference Data. https://history.nasa.gov/alsj/LM04_Lunar_Module_ppLV1-17.pdf. Accessed 13 May 2023
10. Slemp, W.S.: Effects of simulated space environment on Skylab parasol material. In: NASA Technical Memorandum X-2965. National Aeronautics and Space Administration, Washington (1974)
11. Portree, D.S.F.: Mir Hardware Heritage. NASA Center for Space Information, Linthicum Heights (1995)
12. Reference Guide to the International Space Station. https://www.nasa.gov/pdf/508318main_ISS_ref_guide_nov2010.pdf. Accessed 13 May 2023
13. Falcon 9. https://www.spacex.com/vehicles/falcon-9/. Accessed 13 May 2023
14. Starship. https://www.spacex.com/vehicles/starship/. Accessed 13 May 2023
15. Space Launch System. America's Rocket for Deep Space Exploration. https://www.nasa.gov/exploration/systems/sls/fs/sls.html#:~:text=NASA%27s%20Space%20Launch%20System%2C%20or,Moon%20on%20a%20single%20mission. Accessed 13 May 2023
16. Space Launch System Lift Capabilities. https://www.nasa.gov/sites/default/files/atoms/files/sls_lift_capabilities_configurations_04292020_woleo.pdf. Accessed 13 May 2023
17. Europe's rockets. https://www.esa.int/ESA_Multimedia/Images/2019/06/Europe_s_rockets. Accessed 13 May 2023
18. Häuplik-Meusburger, S., Bannova, O.: Space Architecture Education for Engineers and Architects. Designing and Planning Beyond Earth. Springer, Switzerland (2016). https://doi.org/10.1007/978-3-319-19279-6
19. Pedley, M.D., Mayeaux, B.: TransHab selection material. In: Griffin, D.E., Stanley, D.C. (eds.) Proceedings of the 4th Conference on Aerospace Materials, Processes, and Environmental Technology (2001)
20. Nomex. https://www.dupont.com/brands/nomex.html. Accessed 13 May 2023
21. Tecnologia. https://www.composite-research.com/caratteristiche-tecniche. Accessed 13 May 2023

22. Roshan, P.: High Performance Technical Textiles. Wiley, Hoboken (2019)
23. DuPont™ Kapton®. https://www.dupont.com/content/dam/dupont/amer/us/en/ei-transformation/public/documents/en/EI-10142_Kapton-Summary-of-Properties.pdf. Accessed 13 May 2023
24. Density of Aluminum. https://www.thyssenkrupp-materials.co.uk/density-of-aluminium.html#:~:text=The%20density%20of%20aluminium. Accessed 13 May 2023

Methodology for Stability Assessment of Discretised Shell Structures During Robotic Assembly

Sam Wilcock[1(✉)], Mehmet R. Dogar[1], and Ornella Iuorio[2]

[1] University of Leeds, Leeds, UK
s.wilcock@leeds.ac.uk
[2] Politecnico di Milano, Milan, Italy

Abstract. Timber structure design can reduce embodied carbon for large span systems, by reducing material usage. The work in this paper presents the assessment of wood panel shell structures, focusing on the use of traditional joinery styles to produce self-supported structures. Key design criteria are to minimise external scaffolding to reduce falsework waste, to allow dry stacking without adhesive between panels for de-construction, and to be manufactured and assembled using digital processes. Focusing on a particular shell geometry, selected for its theoretical performance, a procedure is outlined for the definition of integral joints between planar panels. By modelling deflection using the coupled rigid-block analysis (CRA), different joint styles are assessed during and post-assembly, to compare their suitability and demonstrate the mitigation of falsework. Panels are both 3D printed and built as stacked plywood, validating the utility of CRA and finding the effect of scale to demonstrate its use as a structural design tool for intermediate assembly stages.

Keywords: Shell structures · rigid-block analysis · laser cutting · dry stacking · 3D printing · timber design · assembly sequencing

1 Introduction

The implication of building a form-found shell structure is that it is stable once fully assembled due to a membrane dominant stress profile, but it is not at all guaranteed during assembly, and this usually requires the use of temporary formwork. Computational techniques, such as the coupled rigid-block analysis technique (CRA) have been developed to test stability in block-like structures [1], working on the assumption that discrete elements act as rigid block elements with corresponding interaction forces, moments and friction driving parts to either static equilibrium or collapse. Additionally, the CRA method and its predecessor rigid block equilibrium (RBE) solver [2] have been integrated into ETH's COMPAS software package as COMPAS CRA [3], allowing easy conversion from CAD software Grasshopper's [4] solids and meshes into Python datastructures. Interfaces are found via a similar technique using a geometry package,

S. Gabriele et al. (Eds.): IWSS 2023, LNCE 437, pp. 472–482, 2024.
https://doi.org/10.1007/978-3-031-44328-2_49

and then the equilibrium coefficient matrices, friction and load vectors generated automatically. Displacements are additionally estimated by the coupling of the rigid block equations with kinematics.

The authors of this work study the wider field of robotic assembly of discrete shell structures. To this end, it is useful to be able to understand the stability of intermediate or local assembly stages as well as the global stability. For the sake of reducing material waste, and promoting reusability, it is also beneficial to mitigate formwork requirements traditionally associated with shell construction, and this presents questions about how to design shell panels which can temporarily support bending moments during assembly. Joint design is a key method by which discrete shell structures might be able to support these moments. Assessing specific joints designs for both stability and feasibility of assembly is a problem which must be integrated into the design process.

Feasibility of assembly for translational assemblies is an old computational problem [5]. Datastructures such as non-directional blocking graphs (NDBGs) can be used to describe blocking relationships between parts and can be integrated into design tools to verify the sequence and insertion of neighbouring elements [6].

Whilst the aforementioned CRA method was shown to be accurate at capturing the stability of the Armadillo [7] shell, which is characterised by both concave and convex geometries, it has not been previously exploited to assess the intermediate assembly process of shells or used for its displacement estimation. Therefore, the novelty of this work is to demonstrate CRA use as an early-stage design tool for assessing the stability throughout the assembly process, with a dual inspection of the parallel between stability and the local translational freedom of elements.

2 Methods

2.1 Design of the Shell Structure Surface and Discretisation

We propose a shell design that can be assembled using digital manufacturing tools. An initial base surface was designed and turned into a funicular form using the Kangaroo plugin's energy-based approach of minimising strain energy through a mesh. Panelling tools generally require the provision of untrimmed base NURBS surfaces due to how patterns are parameterised, and so by fitting lofts to a series of contour curves a close approximation surface to the shell mesh is obtained. The surface was tiled into hexagons which were then planarized and extruded along local normals at the hexagon corners [8] – in this way the panel edges between neighbours were kept colinear. Comparing to robot manipulator workspace analysis, the structure was checked to be a feasible assembly for a specific robot arm. Through sampling the surface at evenly spaced points at each step of the design, and comparing the distance of points to matching ones on previous surfaces using mean squared error, the relative changes to the design from each step could be quantified:

$$MSE = \frac{\Sigma_{i=0}^{n} d_i^2}{n} \tag{1}$$

where d_i is the Euclidean distance of the two nearest test points number i between surfaces at two design steps; and n is the number of test points.

2.2 Joint Design

The aim of the design was to ensure a locally stable structure, that is, one that could stand unaided during the assembly process. Of some standard woodworking joint styles, dovetails and finger joints were identified as being potentially viable and manufacturable.

Parameterisation of Base Joints. Basic parameterisation of both joint styles were initially created in Grasshopper (see Fig. 1). Joints were defined as planar curves to be extruded through mated panels, with a base line of the curve being drawn between co-ordinates (0, 0) and (1, 0) in the XY plane. Due to the use of the curves for extrusion, they were closed to allow boundary surfaces to be created. By allowing for the adjustment of the joint parameters, it would be possible to make comparisons between different percentages of edges covered by joints, as well as estimate the effect on deflection properties. For this work, the dovetail parameters were set as $w = 0.3, \theta = 70°$, whilst the finger joints parameters were set as $w = 0.2, h = 0.12$.

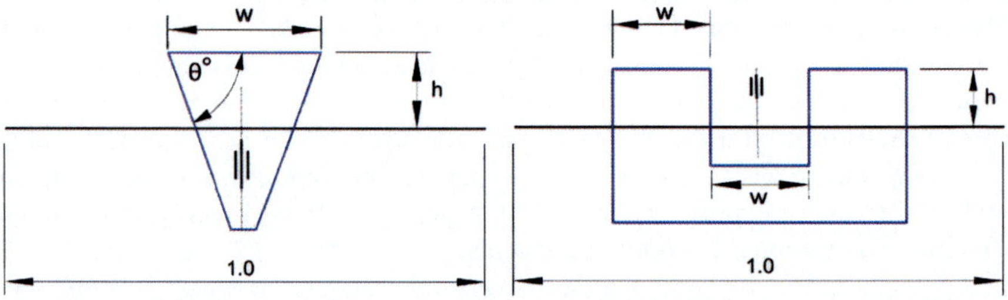

Fig. 1. Parameterisation of the two investigated joint styles, in blue, with vertical lines of symmetry. The reference line in black has length set as 1.0x the length of the edge it will be oriented onto; the w and h parameters are a percentage of this base length.

Calculating Panel Liaison Matrix. To reorient the scaled joints throughout the structure, it was required to find the topology of the shell structure. In our previous work on shell assembly sequencing, we demonstrated the use of Rhinoceros®' RhinoCommon method FindCoincidentBrepComponents to deconstruct a list of BREPs into their respective faces and find neighbouring faces within a certain tolerance [6]. Using this, it is possible to compactly describe the connections in an n-panelled structure as a binary $n \times n$ liaison matrix.

For this work, the liaison matrix was extended by additionally comparing panel centroids, and altering the matrix such that neighbours are assigned a $+1$ or -1 depending on whether they are higher or lower than the test panel (Fig. 2). The modified liaison matrix now allows further components in the definition to apply different rules to neighbours depending on their position relationships, for example, horizontal neighbours could have different joint styles than vertical neighbours to provide changing stiffness depending on whether joints are longitudinal or transverse to the main shell direction. Additionally, since the custom Grasshopper node is already testing for adjacent faces, it is also set to output a DataTree structure containing the face boundaries of these pairs of mating surfaces, without any major additional computing overhead.

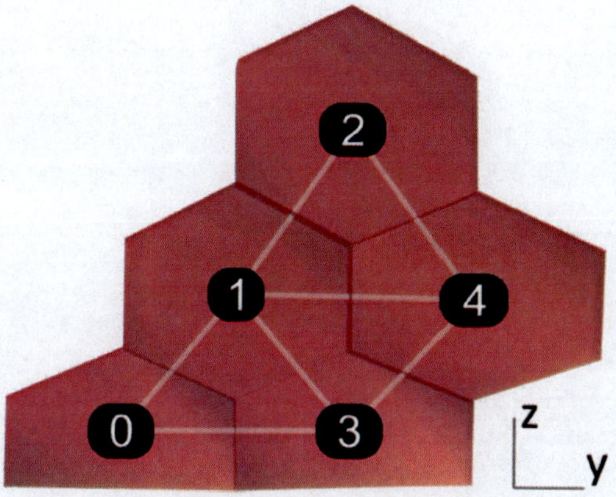

$$L = \begin{bmatrix} 0 & 1 & 0 & 0.5 & 0 \\ -1 & 0 & 1 & -1 & 0.5 \\ 0 & -1 & 0 & 0 & -1 \\ 0.5 & 1 & 0 & 0 & 1 \\ 0 & 0.5 & 1 & -1 & 0 \end{bmatrix}$$

Fig. 2. An example hexagon panel assembly with liaison graph overlay and associated extended liaison matrix. Note that the matrix could be made antisymmetric by encoding horizontal neighbours as +ve or −ve depending on relative y-location.

Propagating Joint Designs. Jiménez described the use of such liaison matrices for understanding precedence constraints in assembly sequencing [9]. In this work, they were applied both for understanding precedence in assembly sequencing and as a tool for deciding where to place sets of joints. Provided the face boundary structure and taking inspiration from Rezaei Rad's exploitation of RhinoCommon's native ChangeBasis Transforms [10], it is possible to take the previously designed joints in the world XY plane and reorient them throughout the structure along neighbouring panel's edges.

The parameterised joint curves were scaled to match the length of each top edge, before being reoriented into the plane of the "male" part that will be inserted into its neighbour. Using the liaison matrix, this is decided by propagating joints only to neighbours with relationships described by a −1 or 0.5, downwards or across. This ensures that each part is inserted male into female from above, implicitly enforcing the precedence requirement that parts are inserted once all lower neighbours are present.

Once the curves were propagated through the panel edges, they were then converted into planar surfaces and extruded through the thickness of the panel. To enable the addition of joint tolerance, scaled copies of each joint were made about centroids, with a scaling factor of $C_{joint} \geq 1.0$, where $C_{joint} = 1.0$ represents a perfect fit and increasing values give greater flexibility in the joint insertion. $C_{joint} = 1.07$ was used throughout.

The extrusion direction for the joints is along the panel conic vectors, meaning that the male surfaces also end up with the same conic shape as their panels and allowing panels to be fitted via translation into multiple slots simultaneously. By taking solid Boolean unions of the panels with the reoriented joint solids, and then taking solid Boolean differences of the panels with the scaled cutting solids, the panels and their integral joints are generated (Fig. 3).

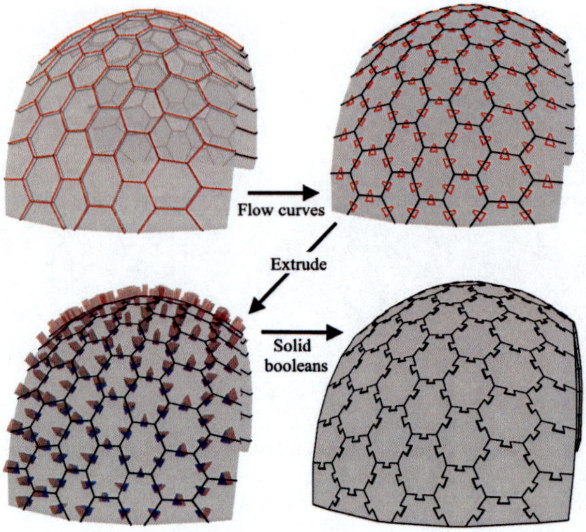

Fig. 3. Flowing and extruding joint curves through the shell geometry. First, top edges are found, joint designs are scaled and rotated to match male panel planes. The curves are then extruded (bottom left) with longer, larger red areas indicating cutting solids, blue indicating joints to union.

2.3 Coupled Rigid-Block Analysis and Comparison to 3DP Model

To validate the CRA method for such discretised shells, a subsection of the full shell was selected for manufacture. Testing was done by comparison of CRA to measurements of the deflected height for the central 3D printed panels (Fig. 4).

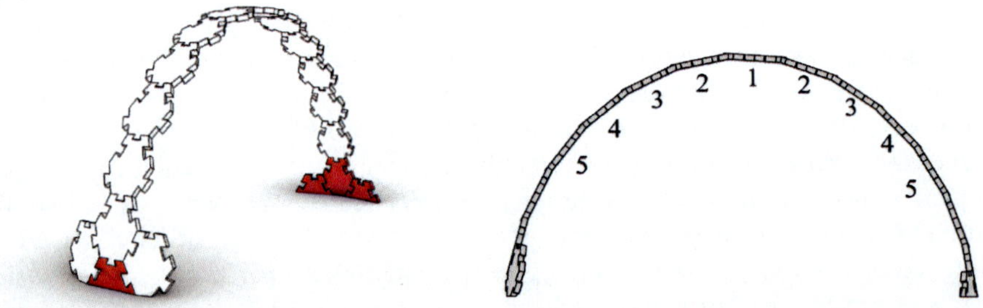

Fig. 4. **Left**) Subset of panels tested via CRA and through 3D printing. Red panels are those set as fixed base points for CRA modelling. **Right**) Numbered sequence of panels removed during the test.

By adding a small change to the code for COMPAS CRA, it also became possible to add external forces to the model and thus estimate the effects of additional dead loads on the structure. A custom ghPython component then allows quick export of models to JSON data describing the assembly, which were analysed for stability and displacement at various steps through the assembly process before being passed back to Grasshopper for viewing.

2.4 Assembly Sequencing Through Non-directional Blocking Graphs

Assembly sequencing allows a different insight into the assembly, allowing us to find a) if it is possible to assemble the arch using translational movements, and b) a sequence of parts and insertion directions. Previous work detailed the implementation of this process and the components written for Grasshopper [6], and is briefly summarised here.

The normal directions of all the planes comprising an interface surface, for panel i from neighbour j collected into a set, $X_{normal.i,j}$. Additionally, a set of test vectors of unit length, X_{test}, is generated, evenly sampling 3D space through a Fibonacci sphere. By comparing every interface vector with every test vector, the condition for a direction that the panel motion is blocked translationally by a neighbour is:

$$\vec{x}_{test} \cdot \vec{x}_{normal} > 0 \tag{2}$$

and these matching vectors are added to a set $X_{blockedi,j}$, whilst free directions are the set difference of test vectors and blocked vectors,

$$X_{freei,j} = X_{test} \backslash X_{blockedi,j}. \tag{3}$$

As panels are added to the structure, further neighbours become blocked, and so assembly sequences and potential free directions of insertion can be found using breadth-first search on the liaison graph and the sets of blocking relationships between panels, recalculating the total set of free panels at each step [11].

3 Results

3.1 Shell Design Step Effects

In Table 1, the different design steps are compared to demonstrate the amount of change that each step introduces to the design surface. Note particularly the order of the mean squared error calculation for the process of fitting a NURBs surface to the form-found mesh.

Table 1. Measured MSE across sampled points between two surfaces.

Test surface A	Test surface B	MSE (m^2)
Catenary mesh	Base surface	2.63×10^{-3}
Fitted loft	Catenary mesh	6.64×10^{-4}
Fitted loft	Base surface	2.45×10^{-3}
Planarized panelled	Catenary mesh	2.42×10^{-2}

Table 2. Comparison of predicted vs measured deflection heights and prediction error for 3DP arch vs. CRA model, with cases from Fig. 4 taking measurements on Left and Right.

#	Design height (mm) h_d		Predicted height (mm) h_P		Measured height (mm) h_m		Difference (mm) $h_m - h_p$		% Deflection $-\frac{h_m-h_d}{h_d}$		Error (%) $\varepsilon = \frac{h_m-h_p}{h_m}$		Average absolute error (%) $\frac{\lvert\varepsilon_L\rvert+\lvert\varepsilon_R\rvert}{2}$
	L	R	L	R	L	R	L	R	L	R	L	R	
0	250	246	241	239	243	240	2	1	3	2	1	0	1
1	250	246	187	201	179	190	−8	−11	28	23	−4	−6	5
2	239	228	198	201	202	194	4	−7	15	15	2	−4	3
3	213	197	190	183	190	179	0	−4	11	9	0	−2	1
4	178	156	165	150	166	148	1	−2	7	5	1	−1	1
5	133	108	129	108	129	105	0	−3	3	3	0	−3	1

3.2 Initial Validation of CRA Applied to 3DP Arch Model

There is significant droop in the structure (Table 2), which is most noticeable at case 1, with a maximum drop of 28% at the end point and least visible for the parts closest to the base. A relatively small sag is exhibited in the fully assembled arch, as would be expected since the arch has support from both ends and is based on a funicular.

The modified CRA Python script with added loads was tested with gradually increasing vertically downward dead loads on the top central panel, ultimately predicting a max load of 823 g for a small 3D printed model, spanning 510 mm with a max height of 250 mm and panel thickness 5 mm. This was tested physically, and the structure carried a 590 g measuring tape (Fig. 5), however it should be noted that it was precarious and seemed unlikely to take any more mass, as the panels were shifting and pushing outwards towards the base.

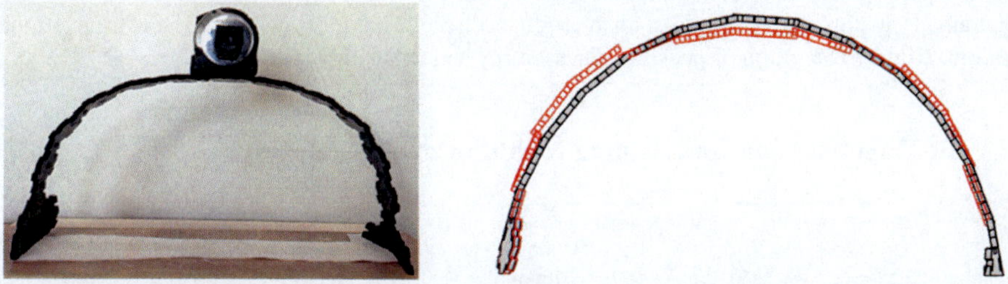

Fig. 5. Dead load testing. **Left)** The shape of the structure can be seen with a mass on the central panel. **Right)** The CRA model of the structure with its ultimate load, where the panels are predicted to separate (original design in black, failure mode in red).

3.3 Comparison of Joint Styles

Applying the same CRA case study to a finger jointed version, whilst the fully assembled arch was found to be stable as expected due to its funicular nature, only cases 4 and 5 were predicted stable from the set of reduced arches, with the mass of panels in longer cantilevered assemblies causing rotation about the joint interface and failure.

Part freedom testing also provided some differing results between the two joint styles (Fig. 6). The dovetail style is self-constraining, such that at any time, only the uppermost members without higher neighbours can be removed. On the other hand, for the finger-jointed structure, there is more freedom within the structure at any time – large areas of panels could be removed at once, as the panels are not blocked translationally by their higher neighbours. This means panels could also be removed from the middle of a structure, which while it may be of some benefit to assembly, would increase risk for sliding mode failure.

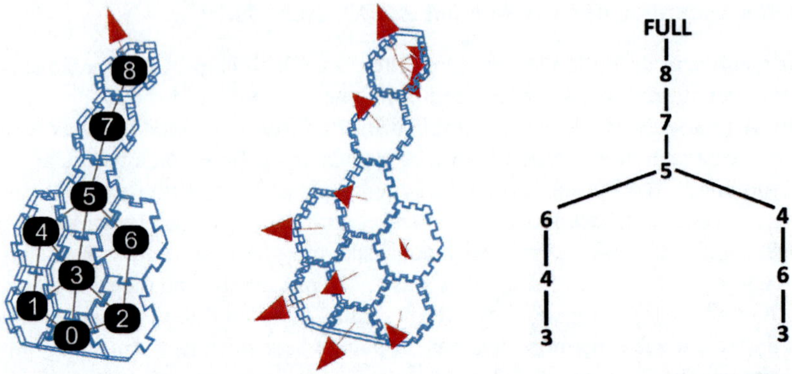

Fig. 6. **Left, Centre**) The results of the assembly sequencing/free direction testing for test structure of 9 panels. On the left, the dovetail joint system has only 1 panel which isn't completely constrained (the top panel #8), whilst on the right, each panel has at least 1 direction of translational freedom. **Right**) The graph of possible disassembly sequences for the dovetail set.

3.4 The Effect of Scale and Adding Neighbours on Modelling

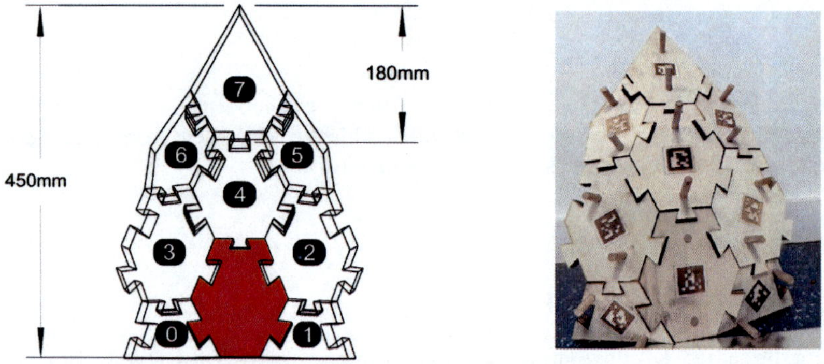

Fig. 7. A larger scale test set of panels, manufactured using a laser cutter and stacking of layers to approximate design. Panel numbers for testing are shown.

A larger scale test set was manufactured (Fig. 7) and had the deflection at the tip both measured and predicted, with a design height at tip of 450 mm. Tests were carried out with increasing numbers of neighbours supporting the central arch (Table 3).

Table 3. Tip heights and estimation error

Panels present	Predicted tip height (mm) h_p	Measured tip height (mm) h_m	Difference (mm) $h_m - h_p$	% Error $\frac{h_m - h_p}{h_m}$	% difference design to measured $\frac{h_m - h_d}{h_d}$
0, 1, 4, 7	443	415	−28	−7%	−8%
0–4, 7	444	433	−11	−3%	−4%
All	446	439	−7	−2%	−2%

4 Discussion

The loft fitting process produces an approximation of an order higher than the comparison between the base and catenary. The biggest source of change in the process is the planarization, and it could be worth comparing to the form-finding integral planarization technique of Contestabile [8], planarizing whilst applying form-finding instead of post-planarizing, to see if this provides a closer solution. Additionally, comparisons could be made to ensure funicularity, through a method such as relaxed funicularity [12].

The CRA method appears to provide a reasonable level of accuracy in the estimation of deflection for the styles of discrete panel shells studied, under the assumption that panels are suitably rigid. The effect of scaling appears to introduce more error, with the predicted deflection remaining the same despite the number of supporting elements and there are some speculations as to why. First, the error is more prominent in systems with less stiffening elements ensuring rigidity, and possibly the additional panels are aiding by providing greater friction to act against self-weight. Second, it is possibly an issue with the CRA implementation, and this will require further investigation.

The dovetail style joint is shown to be far better at holding temporary bending moment forces in the structure during assembly than finger joints, albeit with a large amount of sag within the structure that increases with longer cantilevers. Alternative designs might improve this sag, both of alternative parameters and styles.

Assembly sequences are shown to be a direct effect of the joint directions, and this makes implicit sense as the joint directions are driven by the liaison graph.

5 Conclusions

In this work, it has been shown that CRA is promising for the modelling of stability and deflection in discrete panel shell systems. It can be utilised for the analysis of assembly stages to look for early feedback on designing systems with reduced formwork requirements. The authors have demonstrated the potential for design of dry stacked discrete shell structures why may self-support during assembly.

In future work, data will be presented demonstrating the effect of joint parameter variation on performance alongside the effect of varying tolerance. A further investigation will be made on applying joints which will work in positive, negative and zero

Gaussian curvature regions, to work with a larger set of shells including hypars. Further investigation into the variables effecting the CRA method potentially also need to be made to ensure that it is generally suitable as a tool for local stability and deflection checking.

Research is ongoing into the use of the tools used here for design of shells for robotic assembly; future works will additionally demonstrate the manufacture design and assembly process for larger scale shell designs.

References

1. Kao, G.T.C., Iannuzzo, A., Thomaszewski, B., Coros, S., Van Mele, T., Block, P.: Coupled rigid-block analysis: stability-aware design of complex discrete-element assemblies. Comput.-Aided Des. **146**, 103216 (2022)
2. Frick, U., Mele, T.V., Block, P.: Decomposing three-dimensional shapes into self-supporting, discrete-element assemblies. In: Thomsen, M., Tamke, M., Gengnagel, C., Faircloth, B., Scheurer, F. (eds.) Modelling Behaviour, pp. 187–201. Springer, Cham (2015). https://doi.org/10.1007/978-3-319-24208-8_16
3. Kao, G.T.-C.: COMPAS CRA: Coupled Rigid-Block Analysis (CRA) for the COMPAS Framework (2022). https://blockresearchgroup.github.io/compas_cra
4. Rutten, D.: Grasshopper 3D (2007). https://www.grasshopper3d.com/
5. Wilson, R.: On geometric assembly planning. Department of Computer Science. Stanford University, Stanford University (1992)
6. Wilcock, S., Boyle, J., Dogar, M., Iuorio, O.: Automated robotics agents for assembly-aware design of shells. In: Hvejsel, M.F., Cruz, P.J.S. (eds.) Fifth International Conference on Structures and Architecture, pp. 1061–1068. CRC Press, Aalborg, Denmark (2022)
7. Van Mele, T., Popescu, M., Augustynowicz, E., Echenagucia, T.M.: The Armadillo Vault Computational design and digital fabrication of a freeform stone shell. In: Advances in Architectural Geometry 2016, pp. 344–363. vdf Hochschulverlag AG an der ETH Zürich, Zurich, Switzerland (2016)
8. Iuorio, O., Korkis, E., Contestabile, M.: Digital tessellation and fabrication of the ECHO shell. In: Proceedings of IASS Annual Symposia. 60th Anniversary Symposium of the International Association for Shell and Spatial Structures Structural Membranes 2019. International Association for Shell and Spatial Structures (IASS), Barcelona, Spain (2019)
9. Jiménez, P.: Survey on assembly sequencing: a combinatorial and geometrical perspective. J. Intell. Manuf. **24**, 235–250 (2013)
10. Rezaei Rad, A., Burton, H., Rogeau, N., Vestartas, P., Weinand, Y.: A framework to automate the design of digitally-fabricated timber plate structures. Comput. Struct. **244**, 106456 (2021)
11. Lee, S.: Backward assembly planning with assembly cost analysis. In: Proceedings 1992 IEEE International Conference on Robotics and Automation. IEEE Computer Society Press (1993)
12. Gabriele, S., Varano, V., Tomasello, G., Alfonsi, D.: R-Funicularity of form found shell structures. Eng. Struct. **157**, 157–169 (2018)

Check for updates

Heinz Isler's Form-Finding Models for his "HIB" Shells: Between Experiment and Design

Giulia Boller(✉), Philippe Block, and Joseph Schwartz

Department of Architecture, ETH Zurich, Zürich, Switzerland
boller@arch.ethz.ch

Abstract. The Swiss engineer Heinz Isler (1926–2009) is among the most prominent figures in shell design. Thanks to a form-finding approach based on the use of physical models, he designed and built many shell projects in reinforced concrete. His unconventional structures still represent an important source of inspiration for today's structural engineers. The paper reconstructs Isler's experimental method by looking at the multiple physical form-finding models he developed for his tennis hall shells. Designed for the first time in 1977, they became one of Isler's most successful shell typologies, promoted as "HIB" shells in Switzerland. Despite their apparently simple shape, Isler produced the largest number of physical form-finding models for this specific shell type. Their double symmetry challenged his design method: the highest precision was needed to avoid any irregularities in finding the appropriate geometry. By studying the original materials stored at the Heinz Isler Archive (gta Archives, ETH Zurich), details about Isler's experimental approach to the conceptual design of his shell structures will be revealed for the first time.

Keywords: Form finding · Reinforced concrete shells · Heinz Isler · Physical models · Conceptual Design

1 Introduction

Shells are among the most elegant structures. They combine their form with the most appropriate flow of forces. Thanks to their double curvature, they accommodate any load configuration through pure membrane action. In this context, form-finding methods help finding their appropriate equilibrium shape under boundary conditions, in compliance with external and internal forces. The foundations of computational form-finding tools were laid in early experiences with physical models [1]. Together with the German architect Frei Otto (1925–2015), and the Italian engineer Sergio Musmeci (1926–1981), the Swiss engineer Heinz Isler (1926–2009) is considered one of the masters of physical form-finding techniques in the second half of the 20th century [2]. This paper investigates Isler's experimental form-finding approaches by looking at the models that were fabricated for the tennis hall projects. Such shells exemplified Isler's ability in mastering his form-finding technique with physical models.

S. Gabriele et al. (Eds.): IWSS 2023, LNCE 437, pp. 483–491, 2024.
https://doi.org/10.1007/978-3-031-44328-2_50

2 Hanging Membranes

As advertised in the brochure of the tennis shell projects, Isler's design goal was to reproduce natural shapes through a physical form-finding experiment [3]. Thanks to multiple models, he controlled "new shapes for shells" [4], whose geometry could not be expressed with analytical equations at that time. The most complex shapes were found through the form-finding method of the "hanging membrane". Isler was not the first one who adopted it. The technique was based on Hooke's principle of the catenary line [5]. Early examples can be traced in Christopher Wren's studies for the dome of St. Paul Cathedral [1] and Antoni Gaudí's combinations of catenary curves [6]. Isler extended the principle to continuous shells: an isotropic rubber membrane hanging from a timber frame, on which gypsum was poured. During the experiment, the membrane found its equilibrium form due to the force of gravity [7]. Even though Isler adopted other form-finding techniques such as the "pneumatic membrane" or the "freely-shaped hill", the "hanging membrane" was considered "the best method for design" [4: 2]. Most of Isler's surviving models followed this principle. For the tennis projects, the highest number of models were made.

Fig. 1. Isler's "HIB" shells during construction, Heimberg, 1978 (217-0384, gta Archives, ETH Zurich).

In the tennis halls, Isler mastered the control of the relation between form and force in a system of juxtaposed buildings (Fig. 1). Their iconic form became the reference image for any tennis hall facility in Switzerland, as had happened for Isler's industrial buildings. The series was designated "HIB" shells, from the names of the three actors involved in their promotion (Haus + Herd), design (Isler), and construction (W. Bösiger AG). Between 1977 and 1993, fourteen projects were built following the "HIB" typology, for a total of sixty shells. After the industrial structures, they became the most prolific shell typology in Isler's oeuvre. The early studies looked back at his well-known shell series, either as a roof component – like the industrial bubble shells - or as a stand-alone building – like the garden center typology [8]. However, it became clear that the requirements of a tennis field – a longitudinal dimension of approximately 40 m, an inner height of a minimum of 5 m, a maximum of 9 m, and diffuse lighting conditions – asked for a new shell typology. Following Isler's statement that "the shell is the supporting structure and the space enclosure at the same time" [9: 149], the tennis shells guaranteed an elegant structural solution to the architectural program.

3 Form-Finding Apparatus

The "HIB" structures showed the purest application of Isler's hanging membrane form-finding method: a shell with a rectangular plan sitting on four supports at equal height. However, despite the shape's apparent simplicity, they represented Isler's most challenging project. After ten years of experience with hanging membrane models, extensive refinement of the form-finding technique became necessary. At first glance, this aspect is quite surprising. Other prominent projects by Isler, like the factory Sicli SA in Geneva (1968–69), asked for highly accurate apparatuses to find the unconventional doubly-curved form [10, 11]. However, it was easier to find a completely asymmetric shape than the doubly-symmetric one of the tennis shells. While formally-complex shells allowed tolerances in the form-finding phase, the "HIB" shells demonstrated the "need of highest accuracy" [12: 222] in the physical modeling technique. The equal thickness of the gypsum layer and symmetry in both directions required many iterations until the "[geometric] solution is found" [13]. The fifteen form-finding models for the first tennis shell project in 1977 underline these difficulties (Fig. 2). Isler worked for more than two months on them. It was the biggest exploration he ever did on form-finding methods for shell design. Through the twelve surviving models, the authors traced back his experimental procedure for the first time.

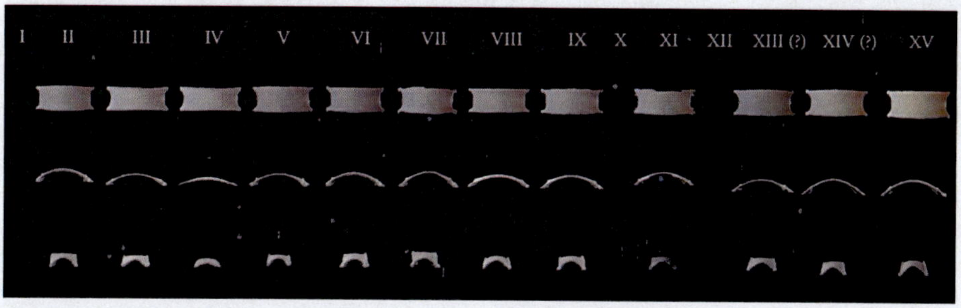

Fig. 2. Physical form-finding models for Isler's tennis shells, 1977. Surviving physical models ordered on the basis of (217-0364, S. 2/3, gta Archives, ETH Zurich).

Unlike Otto's research laboratory, where complex machines controlled his shape explorations [14], Isler's form-finding tools were constituted of economic materials borrowed from daily life [2]. However, despite the appearance, his engineering practice was the outcome of extremely accurate techniques that were refined over the course of his career [15]. The hand-crafted experimental devices allowed easy and fast repetitions of the same test. This became fundamental for the physical models of the "HIB" shells. At different stages, Isler worked on the position and prestressing of the hanging membrane, the gypsum mixture, the experimental equipment, and the overall methodology for achieving symmetrical results in a pure hanging membrane test.

3.1 Material

Even though Isler's public shows always displayed a hanging textile saturated with resin, the actual form-finding experiment he used for his projects employed a 0.3 mm-thick rubber membrane on which gypsum was poured. The isotropic material guaranteed a uniformly distributed stressed funicular shape, thanks to the even stiffness of the membrane and without the influence of the textile threads. Reference lines were drawn with black pencil and ruler, on the membrane side towards the poured material. For critical areas, a denser measurement grid was traced to provide a more precise picture of the membrane behavior. Such guidelines helped the evaluation of the most appropriate shape. Since they were automatically transferred on the model's gypsum surface during the form-finding process, they are still visible on the surviving objects. The membrane was cut according to the initial plan dimensions. Both its shape and elasticity influenced the found form: if too small and stiff, the resulting shape would be too flat; if too large and flexible, wrinkles developed. In the studies for the tennis shell projects, different cutting options were explored [16].

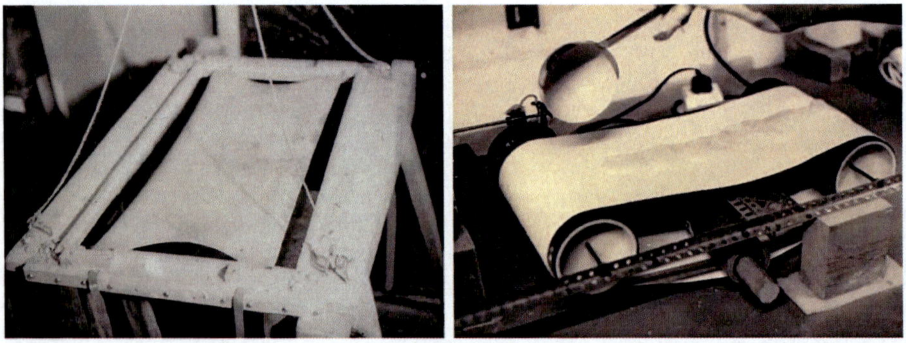

Fig. 3. Left: Rubber membrane and timber frame for the Düdingen shell project, 1977 (217-FX-2-77-18, gta Archives, ETH Zurich); Right: Setting test of the gypsum mixture, 1984 (217-FX-15-3, gta Archives, ETH Zurich).

Once the membrane was cut, it was anchored to the timber frame using nails and small timber elements (Fig. 3, left). It soon became clear that prestressing the membrane could help with controlling the form-finding process [17]. The applied prestressing force represented a design parameter since it highly influenced the final shape. Such studies are visible in the initial stages of the "HIB" form-finding explorations: the first ten physical models focused on this aspect. The more the membrane was prestressed, the more symmetric the shape was, but the lower the longitudinal curvature was. Isler explored tests with a range of membrane prestressing between 0 and 7% of the membrane's length. Experimental results showed that a prestressing of 2.8% helped achieve the most appropriate longitudinal curvature to meet the design requirements for the tennis shell projects [16].

If the choice of the rubber membrane was important for the form-finding study, the gypsum mixture was crucial for the success of the experiment. Unlike previous cases, where its quantity was expressed through the material thickness on the physical model

[18], for the "HIB" project the correct proportion between gypsum and water, together with the setting time, was investigated over the course of five different material tests in Isler's model workshop. A specific device – the "gypsum machine" – was manufactured to control the proportions of the mixture's components (Fig. 3, right) [19]. In a few tests, drops of lemon juice were even added to the mixture to slow down the setting time and avoid cracks on the surface caused by an early drying mechanism.

3.2 Equipment

Appropriate dimensions of the form-finding equipment were necessary for the experiment's success. If the device was too small, achieving the necessary precision for scaling up the results for the further design phases was difficult. If too big, the overall process could not be controlled by the two people who worked on it. In the "HIB" studies, the increase in model size from a scale of 1:100 to 1:75, with resulting longitudinal length of 47 and 63 cm, respectively, helped reduce incorrect outcomes [20].

As for the material choices, the form-finding equipment for the "HIB" experiments became more sophisticated than in Isler's early hanging studies. Each structure constituted a module that needed to be placed next to each other. For the prestressed edge beams of Isler's industrial shells to guarantee a perfect match between the modules while equilibrating the spatial structure [21], the tennis shells had to respect an accurate definition of both longitudinal and cross curvatures. The control of that connection detail on the small-scale physical form-finding model represented one of the most challenging studies. A few changes to the experimental device became necessary (Fig. 4) [22].

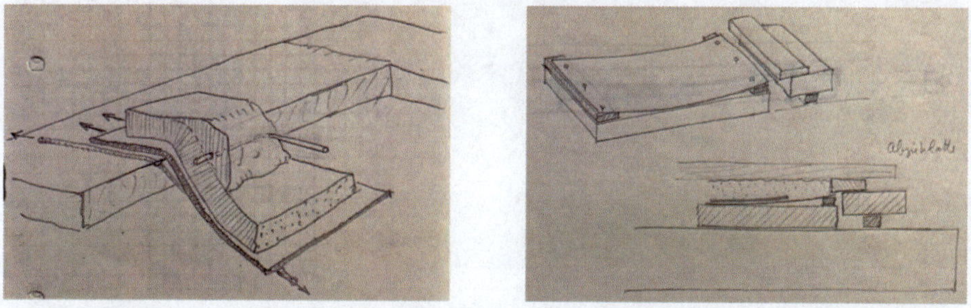

Fig. 4. Changes to the form-finding equipment for Isler's tennis shell projects, 1977. Left: Additional rubber bands along the longitudinal edges; Right: Curved bottom plate for the gypsum pouring process (217-0364, S. 2/3, gta Archives, ETH Zurich).

The longitudinal curvature was monitored through equal membrane prestressing thanks to a pre-tensioning mechanism in the timber frame. Along and close to the border of the longitudinal edges, two prestressed rubber bands were placed to control the symmetry of the cross curvature (Fig. 4, left). Additionally, a curved bottom plate was located right below the membrane and within the timber frame (Fig. 4, right). During the pouring process, the material distribution was not uniform due to the plate's transversal section: more gypsum in the central area increased the cross curvature of the hanging membrane.

4 Form-Finding Procedure

The form-finding experiment was based on a trial-and-error approach. If nowadays a computer program adopts a digital iterative procedure to find the most appropriate shape for a specific design problem, Isler's approach led to multiple physical models that were refined based on the experiment's observations. Each model was generated in approximately 20 min (preparation, pouring, and drying process). Therefore, it could be easily replicated multiple times. For the "HIB" shells, thirteen tests were conducted before the correct procedure for the form-finding study was defined [23]. The edges and the supporting points represented the most critical areas. They were carefully controlled. Additionally, the prestressing of the membrane and of the longitudinal bands was checked through visual observation and simple tensional tests [24]. After these preliminary steps, the experiment could start. Two timber trestle legs and a stool supported the hand-crafted apparatus (Fig. 5). The form-finding device consisted of the timber frame and the rubber membrane was placed on a timber board with curved bottom plate. The gypsum mixture was poured onto it. The material was distributed over the surface. After a few minutes, the membrane was hung from its supports and adapted its shape to the gypsum's load during the hardening phase. The timing was carefully controlled. If the operator had waited just a minute longer, cracks would have appeared on the surface. The experiment ended as soon as the material was fully hardened.

Fig. 5. Form-finding studies for the Düdingen tennis shell, 1977 (217-FX-2-77-18, gta Archives, ETH Zurich).

The free edges represented the most critical areas from the perspective of both symmetry and structural behavior. In early form-finding studies such as the one for the Grob shell (1968), a membrane reinforcement along the perimeter was adopted. This translated to a hidden edge beam in the real construction [17], which was discovered during the shell's demolition in 2021 [25]. Oppositely, the "HIB" studies showed that a more elegant alternative was possible. If the membrane was cut a bit bigger than the actual hole, the excess of flexible material folded over locally forming a negatively doubly-curved shape, which automatically resulted in a stiffened edge. Moreover, these upturned longitudinal edges, a reminiscence of Isler's garden center series, prevented the thin shell from buckling.

Once the gypsum hardened, the rubber membrane was removed. The multiple form-finding models were collected on a table. Their shapes were examined and compared. The one that accomplished the requirements of geometry, symmetry, structural behavior, and architectural appearance was chosen. The correlation between the model's form and the drawings for manufacturing the timber binders of the falsework made it possible to translate its geometry in full scale on the construction site.

5 Conclusion

Over his career, Isler consolidated an extremely accurate engineering practice that linked a form with its internal stress state. The fifteen physical models for the "HIB" shells embodied Isler's approach to the conceptual design of shell structures. Such physical modeling techniques highly inspired computational form-finding methods [26]. In the early 1990s, the geometry of the tennis hall shells was explored with digital tools at the University of Stuttgart [27]. Interestingly, Isler's physical models and these computational ones showed comparable results (Fig. 6, right). Even with different media, both were based on optimization processes linked to iterative procedures. The physical form-finding method of the hanging membrane considered gravity as the governing design load. On the contrary, computational tools combined different load cases to define the most appropriate shape. Only one-quarter of the shell was considered to run the optimization process digitally: the double symmetry helped reduce the calculation time and the complexity of the procedure. Isler adopted the same approach for the geometric studies and the consequent preparation of the timber falsework for the construction site. In this way, the risks of inaccurate results were reduced. However, the physical hanging model needed the full shape to produce the gypsum model during the experimental phase. While digital form-finding methods confirmed Isler's experimental results, recent FEM studies on the tennis hall shells highlighted that they helped find shapes with an appropriate structural behavior also from the point of view of stiffness and concrete allowable stresses [28]. Additionally, the shell's formal clarity reflected the shape's efficiency linked to the appropriate use of concrete as building material [29].

Isler's career took place during the generation shift from physical to digital approaches to designing structures. Isler understood the potential of digital methods

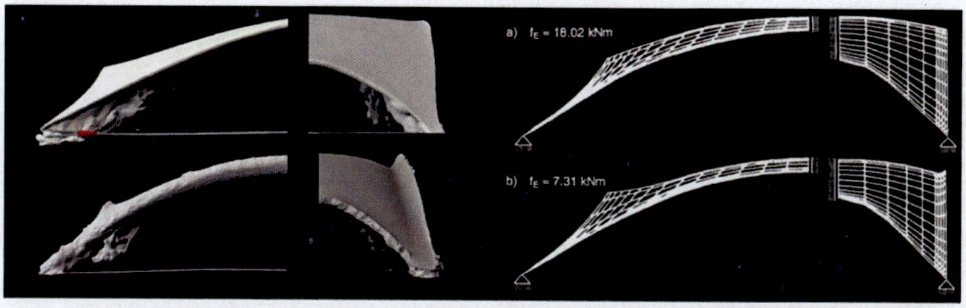

Fig. 6. Left, top: Physical model n. 2 (217-M-7-2:103, gta Archives, ETH Zürich); Left, bottom: Physical model n. 15 "Urform" (217-M-7-2:105, gta Archives, ETH Zürich); Right: [27]: 35.

to explore different design options in only a few seconds of calculation time. However, in his opinion, "the computer only answers the question we have asked it. It is not in a position to point out questions that we have not yet asked, if we have not yet recognized or experienced the problems. The comprehensive [physical] model, on the other hand, can do that." [30: 61, translation by the author]. Therefore, despite the recent discovery of Isler's use of early computational tools [15], his physical form-finding models remained the most important aspect of his unconventional engineering practice.

Acknowledgments. The authors would like to thank Hans Glanzmann, Isler's longtime collaborator in the office, and Heinz Bösiger, Isler's contractor, for sharing some memories about the "HIB" shells. Additionally, the authors thank Stefan Liniger for testing the archival findings by reproducing the procedures of Isler's "HIB" physical form-finding models at ETH Zurich.

References

1. Addis, B.: Physical modelling and form finding. In: Adriaenssens, S., Block, P., Veenendaal, D., Williams, C. (eds.) Shell Structures for Architecture, pp. 33–43. Routledge, New York (2014)
2. Boller, G., D'Acunto, P.: Structural design via form finding: comparing Frei Otto, Heinz Isler and Sergio Musmeci. In: Mascarenhas-Mateus, J., Pires Paula, A. (eds.) Seventh International Congress on Construction History, pp. 431–438. CRC Press, London (2021)
3. Isler, H.: Dreidimensionale Experimente. In: Noser, T. (ed.) Biologie und Bauen, pp. 176–188. Hochschule der Künste Berlin (1983)
4. Isler, H.: New Shapes for Shells. Bulletin of the IASS 8: Paper C-3 (1961)
5. Hooke, R.: A description of helioscopes and some other instruments. John Martin, London (1675)
6. Graefe, R.: The catenary and the line of thrust as a means for shaping arches and vaults. In: Addis, B. (ed.) Physical Models: Their Historical and Current Use in Civil and Building Engineering Design, pp. 79–126. Ernst & Sohn Verlag, Berlin (2020)
7. Isler, H.: Concrete shells derived from experimental shapes. J. IABSE **4**(3), 142–147 (1994)
8. Isler, H.: Early studies for the tennis shell typology. Document unpublished (217-01571, gta Archives, ETH Zuurich) (1969)
9. Isler, H.: Structural beauty of shells. IABSE Congr. Rep. **11**, 147–152 (1980)
10. Chuang, C., Chilton, J.: Design and modelling of Heinz Isler's Sicli shell. In: Proceedings of the IASS Annual Symposium, Tokyo (2016)
11. Delemontey, Y.: Un patrimoine d'exception: genèse et devenir d'une 'forme libre' en beton. L'usine SICLI à Genève, C. Hilberer architecte, H. Isler ingénieur, 1966–2017. In: Porrino M (ed.) Le béton armé. Histoire d'une technique et sauvegarde du patrimoine du 20e siècle, pp. 149–175. inFOLIO, Gollion (2019)
12. Isler, H.: Creating models. In: Proceedings of the 2. International Seminar on Structural Morphology, Stuttgart, pp. 221–230 (1994)
13. Isler, H.: Protocol sheet about the form-finding results of the physical model n. 15 for the Düdingen "HIB" project. Document Unpublished (217-0364 S. 2/3, gta Archives, ETH Zurich) (1977)
14. Otto, F.: Experiments (IL 25). Institut für Leichte Flächentragwerke, Stuttgart (1990)
15. Boller, G.: The model as a working method. Heinz Isler's experimental approach to shell design. PhD thesis, ETH Zürich (2022)

16. Isler, H.: Korrekturen an der Gummimembran. Document unpublished (217-0364, S. 2/3, gta Archives, ETH Zurich) (1977)
17. Boller, G., Chilton, J.: Heinz Isler's experimental approach to form finding. In: Beckh, M., del Cueto Ruiz-Funes, J.I., Ludwig, M., Schätzke, A., Schützeichel, R. (eds.) Candela, Isler, Müther. Positions on Shell Construction, pp. 98–109. Birkhäuser Verlag, Basel (2020)
18. Isler, H.: Protocol sheets about the form-finding results for the Grob AG project. Documents unpublished (217-0134 S. 3/3, gta Archives, ETH Zurich) (1968)
19. Isler, H.: Protocol sheet about the form-finding results of the physical model n. 10 for the Düdingen "HIB" project. Document unpublished (217-0364 S. 2/3, gta Archives, ETH Zurich) (1977)
20. Isler, H.: Protocol sheets about the form-finding studies for the Düdingen HIB project. Document unpublished (217-0364 S. 2/3, gta Archives, ETH Zurich) (1977)
21. Eigenraam, P., Borgart, A., Chilton, J., Li, Q.: Structural analysis of Heinz Isler's bubble shell. Eng. Struct. **210**, 110265 (2020)
22. Isler, H.: Protocol sheet about the form-finding results of the physical model n. 6 for the Düdingen "HIB" project. Document unpublished (217-0364 S. 2/3, gta Archives, ETH Zurich) (1977)
23. Isler, H.: Shell Düdingen. Bucket list for the form-finding experiment. Document unpublished 1977 (217-0364 S. 2/3, gta Archives, ETH Zurich) (1977)
24. Isler, H.: Gummikraft. Document unpublished (217-0364 S. 2/3, gta Archives, ETH Zurich) (1977)
25. Boller, G., Chilton, J., Ramm, E., Schwartz, J.: Demolition of Heinz Isler's free-form shell in Sargans. In: Proceedings of the IASS Annual Symposium, Beijing (2022)
26. Tomlow, J.: Designing and constructing the Olympic roof (Munich 1972). Int. J. Space Struct. **31**(1), 62–73 (2016)
27. Bletzinger, K.U., Ramm, E.: Form finding of shells by structural optimization. Eng. Comput. **9**(1), 27–35 (1993)
28. Baghdadi, A., Heristchian, M., Kloft, H.: Structural assessment of remodelled shells of Heinz Isler. Int. J. Adv. Struct. Eng. **11**, 491–502 (2019)
29. Isler, H.: Physical models in shell design. In: Proceedings of the 4th International Colloquium on Structural Morphology, Delft (2000)
30. Isler, H.: Moderner Schalenbau. Arcus **18**, 1–17 (1992)

Review of Lightweight Structures Use for Archaeological Shelters with a Proposal of the Design Improvement for Their Foundations

Milan Dragoljevic(✉), Marco Belloli, and Alessandra Zanelli

Politecnico di Milano, 20133 Milan, Italy
milan.dragoljevic@polimi.it

Abstract. Archeological sites are places of big importance for preserving history and culture (Iuorio 2016). However, because of their dimensions, their preservation is a challenging task. This paper aims to give a short overview of current requirements and how they are solved and identify the design points which can be improved further. One of them, which is in the focus of this paper is the issue of the foundations for lightweight structures in archeological use. The problem emerges because at the same time foundations need to be massive and able to resist uplift forces occurring in the lightweight structure while being limited by the archeological site importance which requires the least possible invasiveness to the soil (Zanelli 2013). Presented requirements are similar to the general requirements used in the design of lightweight structures: minimizing the size and the weight of the elements by isolating principal forces and loads and shaping the elements to respond to them. In this case, the observed forces and loads are at the foundation of the structure (Fig. 1). The new foundations' system is initially developed through virtual simulation that allows understanding the load directions and transfer of forces and also guides the design process. In the successive step, a series of form optimizations then leads to the improved design. Thanks to the virtual data collected through the simulations, it is possible to perform this data driven approach. In the end, these are tested and verified in the Politecnico di Milano wind tunnel, so the real-world data is collected and compared to the virtual one.

Keywords: Lightweight structure · archeological shelter · foundations design · membrane shelter

Archaeological sites are locations with concentrations of remains such as artifacts or structures, which reflect past activity and, subsequently, may be of interest for archaeological studies (Iuorio 2016). For research reasons, they are further examined and exposed to the access of the professional staff and later, when the conditions allow it, even the general public. In order to allow these activities, it is necessary to provide the safety of the remains as well as the safety of the persons visiting the site. Archaeological sites can be divided into three broad groups, which are often found in combination at a single location: unexcavated, above ground, and excavated. Unexcavated require the least work on their protection. Since the material of the artifacts is in the ground, it is

S. Gabriele et al. (Eds.): IWSS 2023, LNCE 437, pp. 492–502, 2024.
https://doi.org/10.1007/978-3-031-44328-2_51

protected from atmospheric impacts (such as rain, wind, sun, etc.). The other two cases, however, require additional protection, especially in the period of excavation when the transition between a stable soil environment and a variable exterior one is happening during a short period of time. A sudden change in the environment can cause damage to the material and cause deterioration of the remains. This is why the process of excavations needs to be very well planned and coordinated. In some special cases, the findings are buried back under the ground, so they are protected for further future research (Fig. 2).

Fig. 1. Neolithic temple of Mnajdra, Island of Malta

Fig. 2. Kourion archeological site shelter with natural ventilation

1 Shelter Requirements

The principal role of the shelter is to protect the site from possible damages. As presented by (Moghaddasi and Khajepour 2012) the potential damages can be divided into a couple of categories:

1) Human damages
   - Direct human damages: demolition, illegal excavations, vandalism, etc.
   - Indirect human damages: wars, destruction of natural resources, dam establishment, etc.
2) Damages through time
   - Long periods of time (millennia)
3) Natural damages
   - Climatic erosions: rain, snow, moisture, sun, etc.
   - Winds: potentially mixed with sand, dust, salts.
   - Animals and plants

All the mentioned potential sources of damage need to be addressed by the shelter, however natural ones are the most difficult to manage because of their variability. Rain and snow cause mechanical stress on the objects as well as can lead to the chemical dissolution of the binders. Moisture can cause damage through short evaporation and condensation cycles (because of salt crystallization and its spread on the surfaces and within the materials) (Palumbo and Teutonico 2002). Sun is affecting the site in two ways: through solar irradiation (causing thermal stress) and UV rays (which damage colors of the materials). Winds also lead to mechanical stress and when carrying particles can also cause abrasion.

The roof which protects against the direct sun, rain and other atmospheric conditions is almost always part of the shelter. However, addressing the wind, humidity, plants and animals could differ. Enclosing the area for a controlled climate condition is proposed by (Iuorio 2016) while (Zanelli 2013) opts for natural ventilation. Fully enclosed volume is good for control of the humidity, plants (Iuorio 2016) and animals access, while the natural ventilation reduces the need for active climate control (Zanelli 2013), therefore decreasing energy consumption and making the solution more sustainable.

The other requirements are more related to design and the way of use. Since it is possible that, during the research, new areas be discovered, it is necessary to adopt a design that allows the expansion of the covered area. Besides expansion, temporary removal and displacement could potentially be required. Also considering the structure itself, the foundations are usually the most challenging part. It is not known what could be under the surface (as other remains), so it is good to avoid massive foundations (Fig. 3).

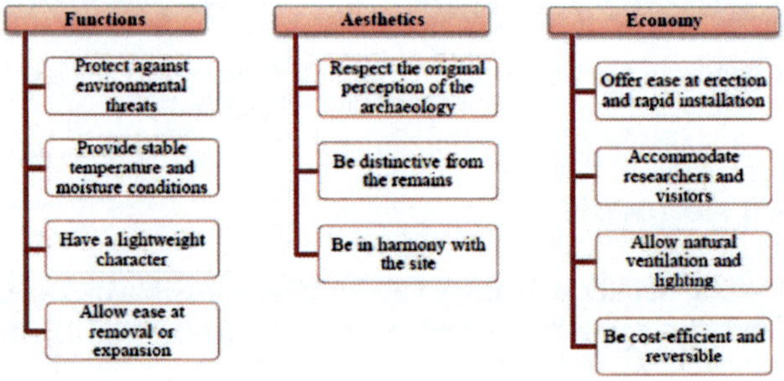

Fig. 3. Summary of needs and requirements for archeological shelter (Iuorio 2016)

2 Lightweight Solutions

Compared to the presented requirements and conditions, lightweight structures provide a very suitable response. Because of their weight, they are easy to construct and move on the site. This is a very important property since often it is necessary to construct the shelters in a short time or to reposition them according to the work on the site. Another positive side of lightweight structures is their aesthetical appearance. Current technology used for these structures is easily distinguishable from the archeological site, so the principle of clear separation between sheltering structures and historical remains is kept. Lightweight structures also create low visual impact. Moreover, since textile-based membranes come to the site as prefabricated elements (the same is true also for their additional structure), they have a very high level of reusability. After dismantling, the same elements could be used for another location. This is especially true if the elements are modular, so the area they cover can be variable.

The main unsolved tasks for temporary solutions are the appropriate anchorage to the soil and the adaptability to different working areas (Zanelli 2013). Also emphasized in a work of (Iuorio 2016) on the possibilities of use of grid shells as a special type of lightweight structure, prone to uplift due to wind loading, especially in open sites. Usually, the solution to overcome wind uplift phenomena is to implement heavier foundations or piles to increase the vertical resistance by either gravity loading or shear. In archaeological sites, however, the installation of bulky foundations should be prevented, as well as piles (Iuorio 2016).

3 Definition of the Problem

As already emphasized, one of the biggest challenges of lightweight archeological shelters is the design and construction of the foundations. Although the structure itself is designed in such an intelligent way that all the forces are absorbed with the minimum possible element size and material use, in the end, received forces need to be transferred to the ground. At the moment the foundations are designed to resist the resultant forces coming from the structure, while it could perhaps transfer them or use them to help the stability of the structure.

The greatest problem occurs in the case of the wind when there is also an uplifting force, so the structure needs to be kept attached to the ground. Even with the use of additional strings, the foundations need to be very massive and provide resistance to pulling upward.

4 Proposed Solution

As already emphasized, one of the biggest challenges of lightweight archeological shelters is the design and construction of the foundations. Although the structure itself is designed in such an intelligent way that all the forces are absorbed with the minimum possible element size and material use, in the end, received forces need to be transferred to the ground. At the moment the foundations are designed to resist the resultant forces coming from the structure, while it could perhaps transfer them or use them to help the stability of the structure.

The greatest problem occurs in the case of the wind when there is also an uplifting force, so the structure needs to be kept attached to the ground. Even with the use of additional strings, the foundations need to be very massive and provide resistance to pulling upward.

The idea of the project is to use similar diagonals which can fix the structure and resist its movement in a horizontal direction. An important part is the calculation of the forces in the support points of the structure. They can be calculated within the software, but if it is done with the use of the current standard, the results could be imprecise. The wind is applied following Eurocode EN 13782:2005. As can be noted in Fig. 7, the pressure force on the inclined surface depends on the angle of the surface, while the suction force is constant. In order to get more precise results, the experiment in the Wind tunnel is planned and performed. 3 structures are examined. The difference between

Fig. 4. Czech hedgehog with the tank in front of it

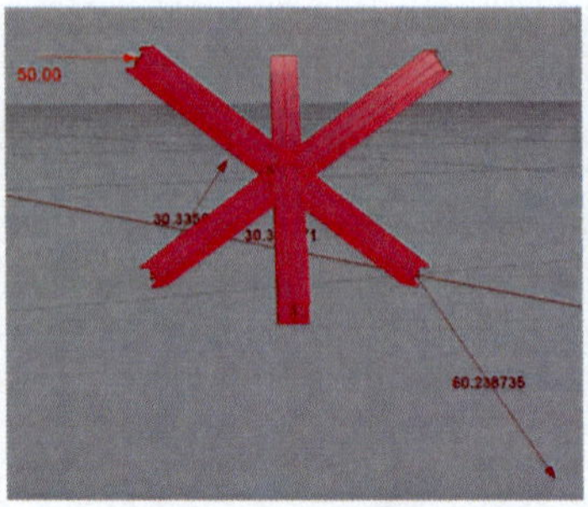

Fig. 5. Czech hedgehog: example of transfer of forces

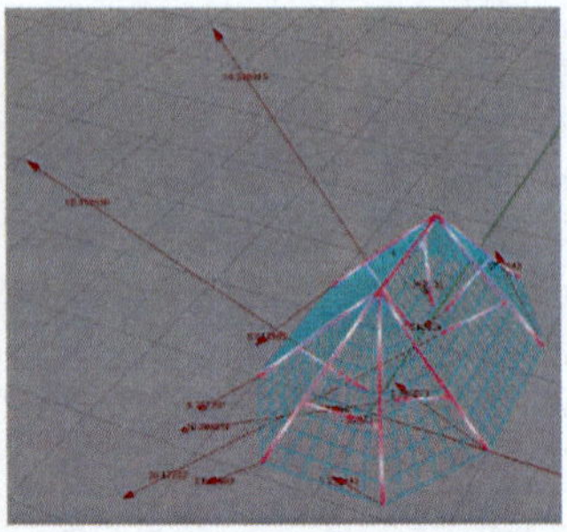

Fig. 6. Structural model of one of the options for the object

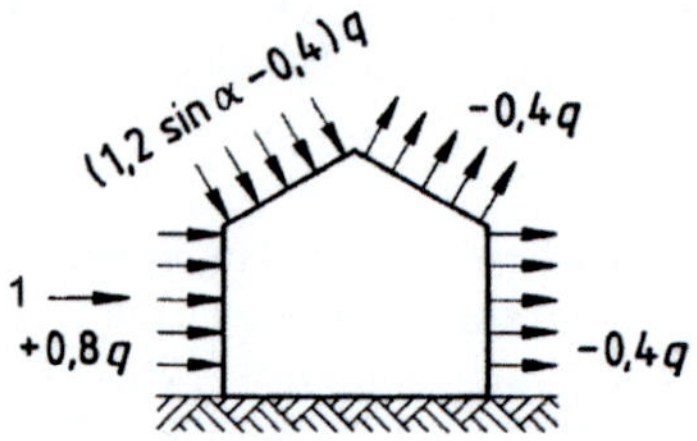

Fig. 7. Wind loads on the temporary structure by Eurocode EN 13782:2005

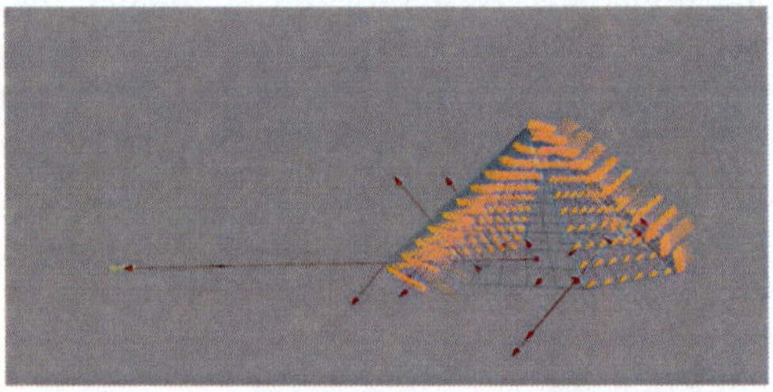

Fig. 8. Wind loads on the proposed structure

them is in the angle between the lateral surface and the ground. The 3 cases are 30°, 45° and 60° (Figs. 4, 5, 6 and 8).

The structure for examination is a simplified version of an archeological site cover. The triangular section of the shape is selected since the more refined shape can bring in too many variables. Also, the triangular form allows better control and understanding of the angles to the ground. Since the focus of the research is to understand the loads in the support points, for the foundations, the upper part of the structure is developed only on a conceptual level to understand the transfer of the loads to the ground. Considering this, the secondary structure is intentionally underdeveloped, since the load transfer within the structure is not of interest.

5 Wind Tunnel

A wind tunnel is a special facility used to simulate the wind conditions. For this experiment, the wind tunnel of Politecnico di Milano is used. It is a special closed-circuit wind tunnel, arranged in a vertical layout with two test rooms located on the opposite sides of the loop. The first one is located in the lower part of the loop and is suitable for Low Turbulence tests. The second one, bigger, is located in the upper part of the loop and is intended for civil engineering testing (the Boundary Layer Test Section). For this experiment, the Boundary Layer Test Section is used. It has dimensions: 13.84 m wide × 3.84 m high × 35 m long with a maximum wind speed of 16 m/s. As the setup, the terrain was defined as open-level country and the wind speed is defined as 25 m/s. Which is causing the pressure of 40 daN/m^2.

6 The Model

As already noted, the model is a simplified version of the archeological site cover, developed in order to test the wind impact and the load transfer to the ground. Certain dimensions are fixed for all of the models, so it is easier to compare the results. The variable is the angle of the side to the ground. For that, there are 3 options: 30°, 45° and 60°.

As already explained, the structural model for finite element analysis represents a simplified version of a structure. Since the elements dimensioning is not a scope of this project, the loads within the structure are directly transferred from the mesh to the primary linear structure (Figs. 9, 10, 11, 12, 13, 14, 15, 16, 17, 18, 19 and 20).

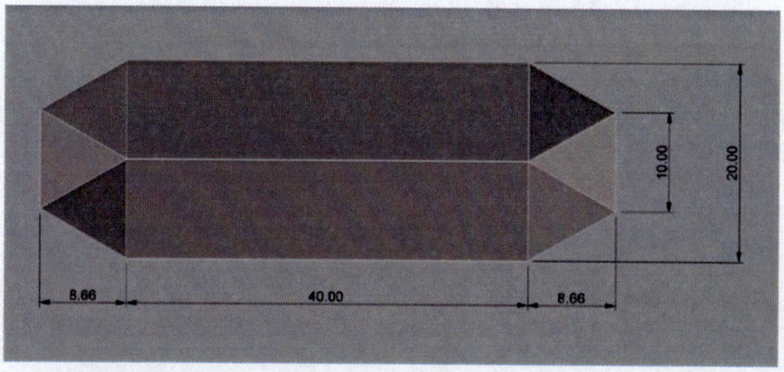

Fig. 9. Shared dimensions of the proposed structures

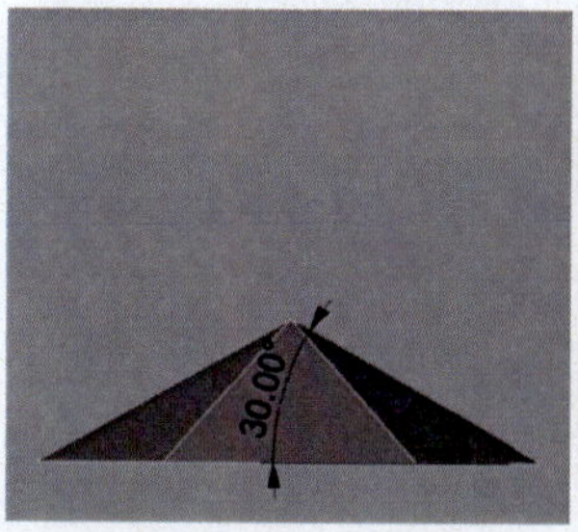

Fig. 10. Inclination 30° used for the first structure

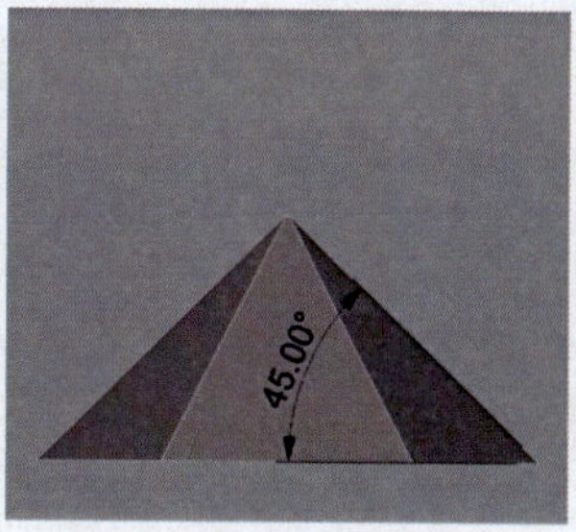

Fig. 11. Inclination 45° used for the second structure

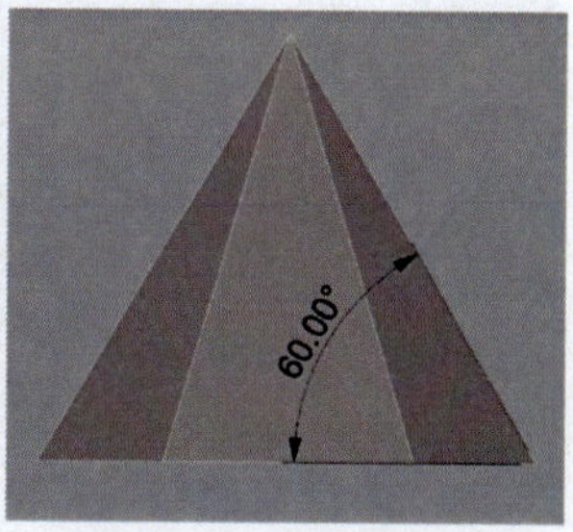

Fig. 12. Inclination 60° used for the third structure

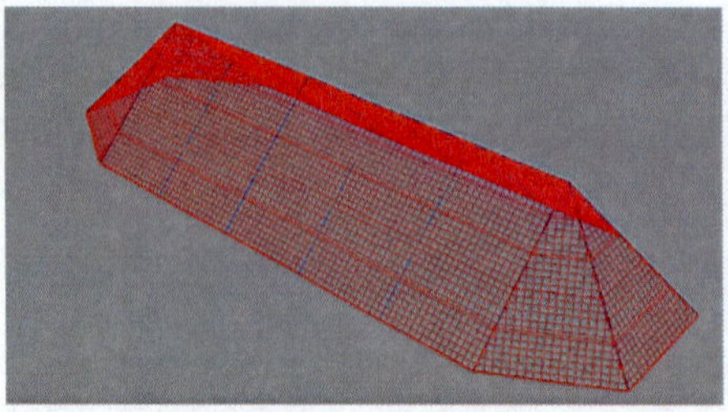

Fig. 13. Mesh fabric model for a 45° shelter

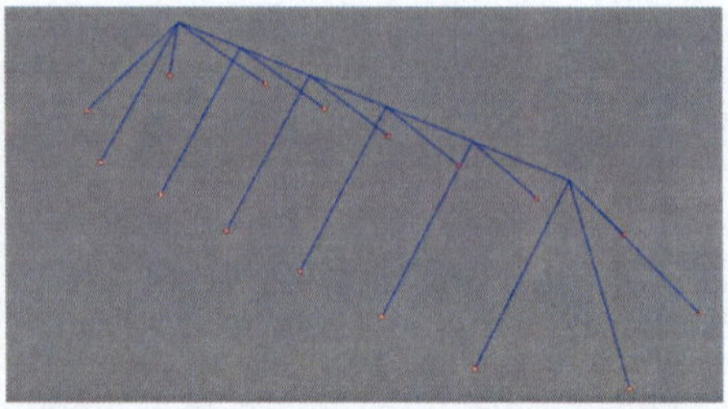

Fig. 14. Structural poles model for a 45° shelter

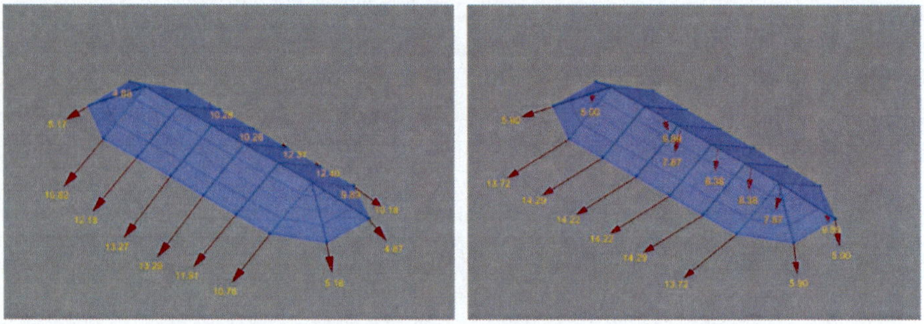

Fig. 15. Forces at the support points for a 30° shelter **a)** obtained through the Wind Tunnel experiment **b)** obtained through the application of Eurocode EN 13782:2005

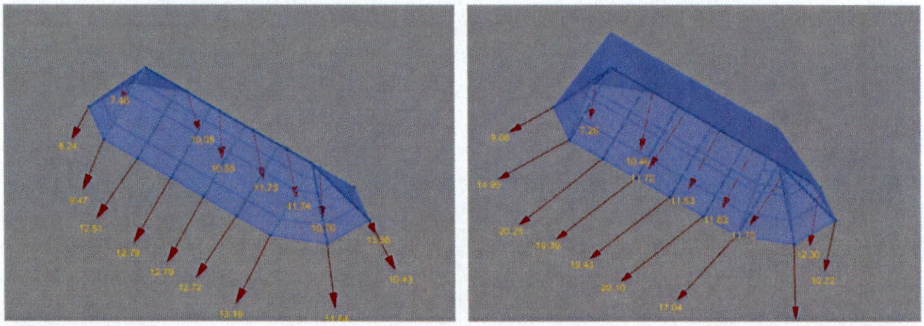

Fig. 16. Forces at the support points for a 45° shelter **a)** obtained through the Wind Tunnel experiment **b)** obtained through the application of Eurocode EN 13782:2005

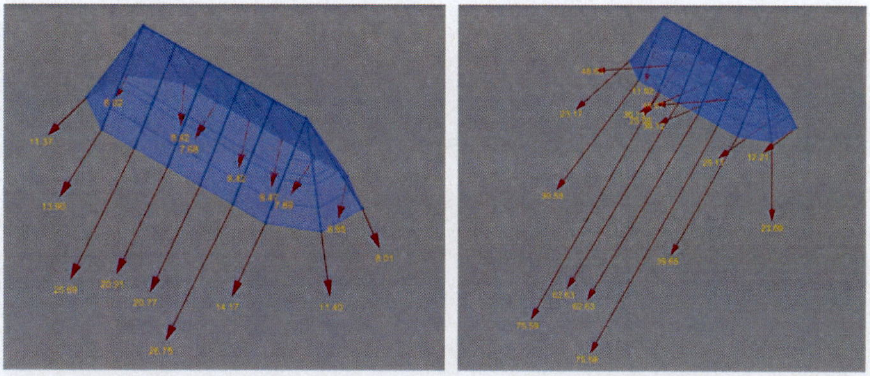

Fig. 17. Forces at the support points for a 60° shelter **a)** obtained through the Wind Tunnel experiment **b)** obtained through the application of Eurocode EN 13782:2005

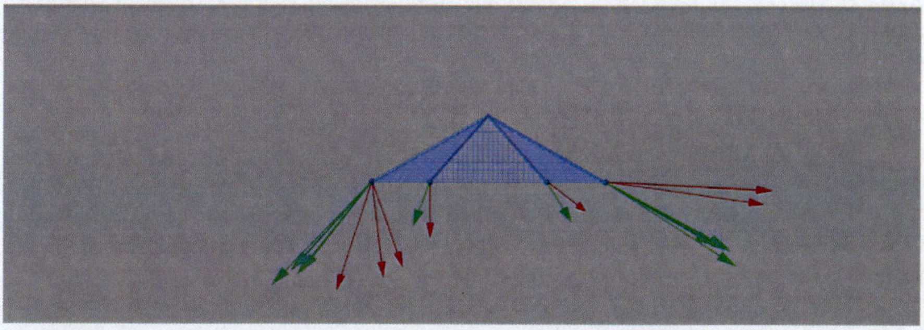

Fig. 18. Forces at the support points, obtained through the Wind Tunnel experiment (green) and the application of Eurocode EN 13782:2005 (red), for a 30° shelter (side elevation view)

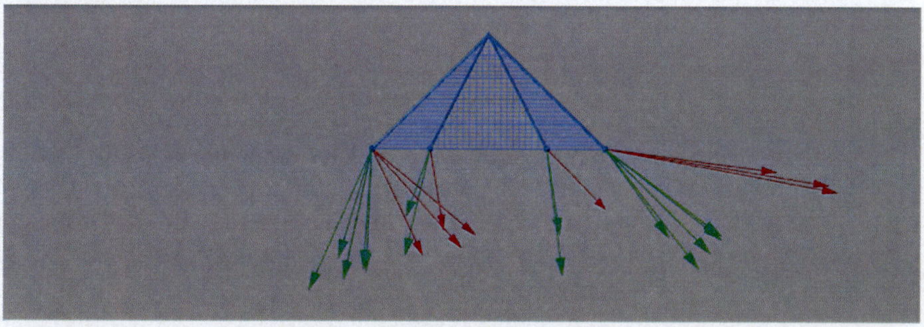

Fig. 19. Forces at the support points, obtained through the Wind Tunnel experiment (green) and the application of Eurocode EN 13782:2005 (red), for a 45° shelter (side elevation view)

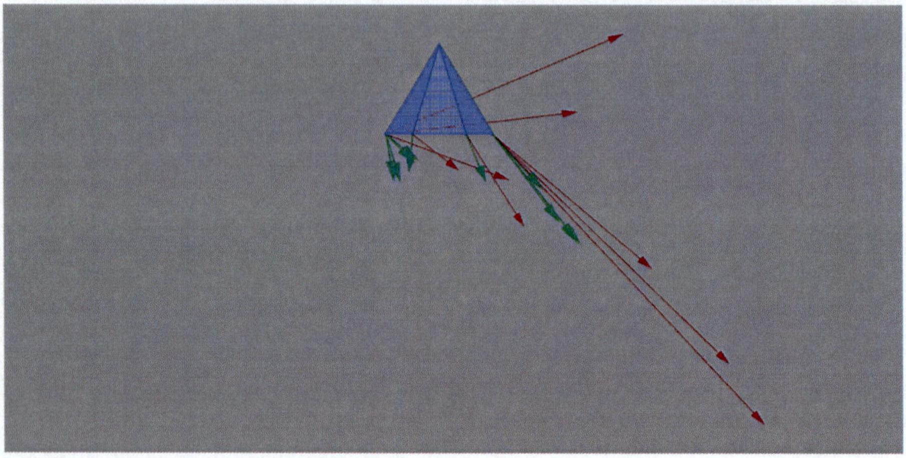

Fig. 20. Forces at the support points, obtained through the Wind Tunnel experiment (green) and the application of Eurocode EN 13782:2005 (red), for a 60° shelter (side elevation view)

7 Discussion of the Results

The conclusions for this research can be subtracted from the comparison of the FEM analysis based on the wind tunnel tests and the loads coming from Eurocode EN 13782:2005. It is noticeable that the reaction forces at the ground where foundations are positioned, have less intensity and are more oriented toward the ground in the case of the wind tunnel. This fact supports the idea that the use of the wind tunnel is bringing more precise and more plausible results for the improvement of the foundations' design.

Moreover, in Figs. 21, 22 and 23 it is noticeable how the reaction forces at the support points change the direction and intensity based on the shelter's side angle. The results support the idea that in all cases, the forces press the foundations to the ground and do not try to pull them out. Compared to the reaction forces coming out of the analysis based on the Eurocode EN 13782:2005, it is also noticeable that all the forces are oriented downward. It can also be concluded that with the increase of the angle, the forces increase as well.

This brings a conclusion that from a foundations design standpoint, the most feasible solution is the one with a sides angle of 60°. However, due to the increase in the elements' size and increased loads, it is in general not the most appropriate solution. This could further lead to redesigning the shelter structure in order to minimize these issues, while still keeping the idea of the side angle.

Future work should focus on including foundation design, above all the piles, into the process. The work of (Monfared et al. 2012) and the conclusions on the impact of horizontal loads on inclined piles could help to deepen the topic (Fig. 24).

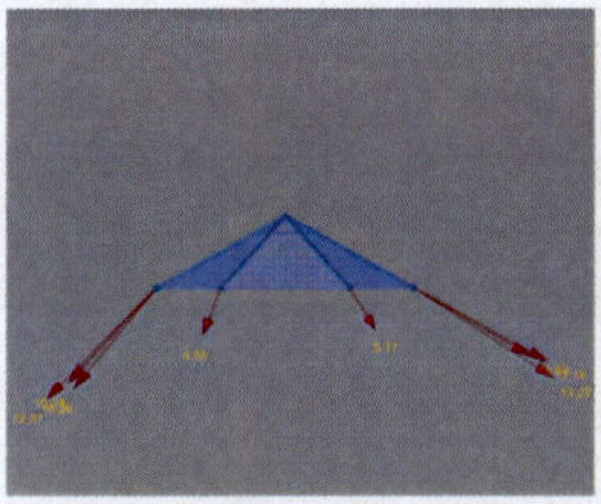

Fig. 21. Forces at the support points, obtained through the Wind Tunnel experiment, for a 30° shelter (side elevation view)

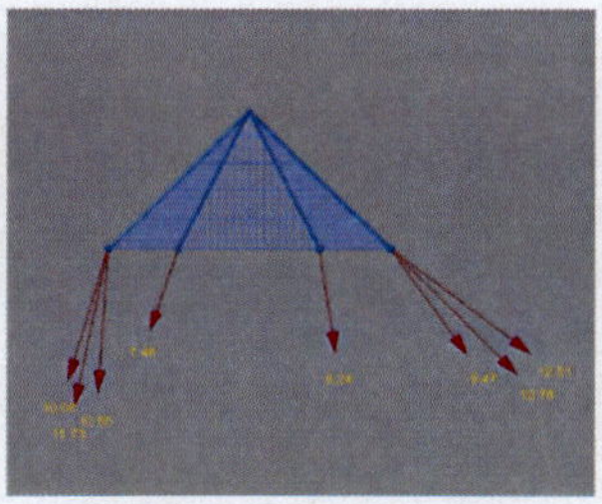

Fig. 22. Forces at the support points, obtained through the Wind Tunnel experiment, for a 45° shelter (side elevation view)

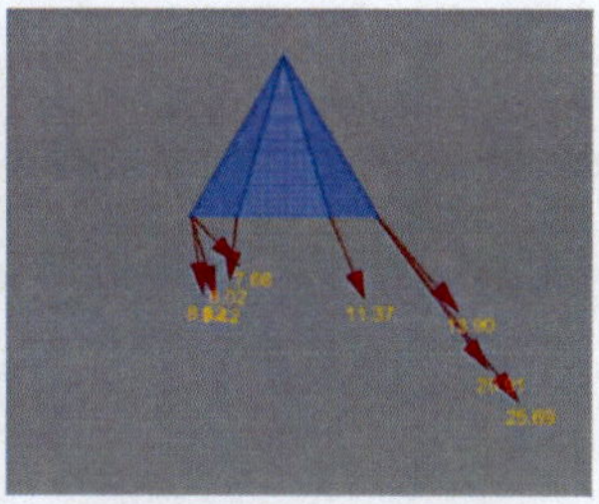

Fig. 23. Forces at the support points, obtained through the Wind Tunnel experiment, for a 60° shelter (side elevation view)

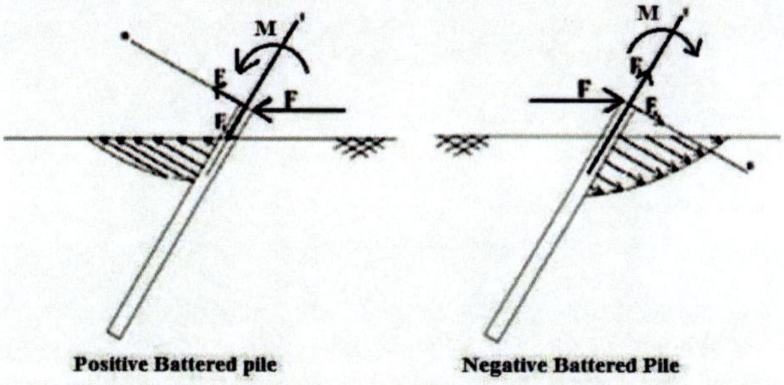

Fig. 24. Negative and positive battered piles (Monfared et al. 2012)

References

Iuorio, O., Homma, E.E., Tsavdaridis, D.K.: The application of free-form grid shells as protective shelters in archeological sites. In: International Association for Shell and Spatial Structures Symposium (IASS), Tokyo, Japan (2016)

Zanelli, A., Llorens, J.: Structural membranes for refurbishment of the architectural heritage. Proc. Eng. **155**, 18–27 (2016)

Zanelli, A., Rosina, E., Beccarelli, P., Carra, G.: Innovative solutions for ultra-lightweight textile shelters covering archaeological sites. In: Volume: Proceedings of the Second International Conference on Structures & Architecture ICSA 2013, pp. 1–8. CRC Press, Taylor & Francis Group (2013). ISBN: 9780415661959

Moghaddasi, A., Khajepour, M.: Typology of protective shelters in archeological sites; advantages and disadvantages. In: SOMA 2012: Identity and Connectivity; Proceedings of the 16th Symposium on Mediterranean Archaeology, Florence, Italy (2012)

Rosina, E., Zanelli, A., Beccarelli, P., Gargano, M., Romoli, E.: New procedures and materials for improving protection of archaeological areas. Mater. Eval. **69**(8) (2011)

Ertosun, I.A.: Evaluation of protective structures in archeological sites for in-situ conservation of architectural remains and artefacts. MSc Thesis at Middle East Technical University. Ankara, Turkey (2012)

Mollaert, M., De Laet, L., Verdonck, A., Lombardi, S., Beccarelli, P., Zanelli, A.: Textile shelters for archaeological or heritage areas: design references. In: STREMAH 2011, vol. 118 (2011)

Avrami, E., Guillaud, H., Hardy, M.: Terra Literature Review: An Overview of Research in Earthen Architecture Conservation. The Getty Conservation Institute, Los Angeles (2008)
Damavandinejad Monfared, S.: Numerical Study on the Behavior of Inclined Micropile (n.d.)

Operational Tunnel Model Generation Using Reinforcement Learning

Nicola Rimella[1](✉), Arianna Fonsati[2], and Anna Osello[1]

[1] Politecnico di Torino, Torino, Italy
nicola.rimella@polito.it
[2] KTH Royal Institute of Technology, Stockholm, Sweden

Abstract. Reinforcement Learning (RL) has emerged as a promising approach to solve complex problems in many different domains, including the Architecture, Engineering, Construction and Operation (AECO) industry. RL is a type of machine learning that focuses on training an agent to interact with an environment in order to maximize a reward signal.

In the AECO industry, RL has been used to optimize building design, construction planning and scheduling, and building energy management. This contribution presents an application of RL for the generation of an InfraBIM model of a tunnel built by mechanised excavation using a Tunnel Boring Machine (TBM). In particular, the present application was developed to minimise the distance between the theoretical layout and the operational one to be executed by the TBM, considering the geometry of the ring and the structural joints required for the solidity of the structure.

RL has the potential to improve efficiency, sustainability, and safety in the AECO industry by enabling intelligent decision-making and optimization across different phases of the construction process.

Keywords: InfraBIM · Tunnel Information Modelling (TIM) · Machine Learning (ML) · Reinforcement Learning (RL) · Tunnel Boring Machines (TBMs)

1 Introduction

The use of optimization techniques within the AECO sector increased significantly also due to the fast development of InfraBIM standards and processes. For instance, applications of computational design are currently used for the generation of as-built models of mechanized tunneling projects [1]. When specifically applied to management, simulations and visualizations in mechanized tunnelling projects the BIM methodology is defined Tunnel Information Modelling (TIM) [2].

Among the optimization techniques, the use of Machine Learning (ML) increased considerably in the last decades due to the development and implementation of enabling technologies that fostered its application. ML includes a series of algorithms developed between Seventies and early Nineties to analyse large datasets, and limited at the time by technologies constraints in data storage and processing.

S. Gabriele et al. (Eds.): IWSS 2023, LNCE 437, pp. 503–511, 2024.
https://doi.org/10.1007/978-3-031-44328-2_52

Such limitations have been overcome nowadays due to cloud-computing and Internet of Things (IoT) [3] and to the use of Big Data as data input for algorithms [4].

For the reasons mentioned above, the infrastructural domain, characterized by the integration of several different disciplines and heterogenous data-sources, represents a challenging field of application for ML techniques, integrated with digital information models produced within InfraBIM processes.

In particular, the present study addresses a critical aspect within the design of mechanized underground tunnels using Tunnel Boring Machines (TBM). TBMs are high-tech machines able to start from the design of a theoretical alignment, due to the TBM guidance system, and calculate at each step a new as-built alignment through a laser-based machine that deviates as little as possible from the theoretical one [2].

However, within this process the lack of information during the design phase causes a disruption due to the unknown locations of the rings before the construction phase. For this reason, the objective of this research was to optimize the design process by adopting an innovative approach based on automatic learning algorithms and its integration within InfraBIM processes.

2 Background

Machine learning is defined as the study of computer algorithms that can automatically detect patterns in data, and then use the uncovered ones to predict future data; the whole data prevision system improves automatically through experience [5, 6]. Indeed, learning algorithms for layered neural network models are inspired by the way that neurons communicate with one another and how they are modified by experience [3].

In terms of categories, machine learning is usually divided into the following types [7]: (i) **Predictive** or **Supervised** learning approach, in which the goal is to learn how to map inputs x into outputs y, given a labeled set of input-output pairs; (ii) **Descriptive** or **Unsupervised** learning approach, in which only inputs are provided to the machine whose goal is to find "interesting patterns" in the data in order to extract structures that link them; (iii) **Reinforcement learning** (RL), which is useful to learn how to behave when given occasional reward or punishment signals, according to the solution the algorithm proposes. In particular, RL finds applications in many branches of computer science and statistics, such as game theory, genetic algorithms and other areas [8–12].

The potentialities of ML techniques within the Construction industry have been widely exploited within previous researches, addressing several issues, such as design optimization [13, 14], data interpretation and standardization [15, 16], energy performance [17–21], asset and facility management [22]. However, applications involving an integration among ML, and specifically RL, and InfraBIM are still limited.

Within the tunnelling domain, Marcher et al. (2020) [23] carried out analyses of possible applications of ML in the tunnelling field, such as: (i) autonomous support installation, (ii) automatic rock (mass) classification, (iii) geological prognosis updating ahead of the tunnel face, (iv) overcoming limitations in the definition of constitutive behavior of soil and rock, (v) exploration of the applicability of reinforcement learning to fully automate different construction processes (self-driving TBMs). The authors described further possible phases to achieve the development of the previous applications: (i) use ML for design and construction optimization, as well as automation and operational guidance during maintenance; (ii) collect high quality data, data from sensors on TBMs as well as jumbos, drilling rigs and scanning and photos from the tunnel face; (iii) combined use of BIM and ML; (iv) standardize data input selection process and partitioning methods; (iv) standardize ANN (Artificial Neural Network) model architectures and performance measures of ANNs for tunnel engineering. An interesting attempt on the first step cited by Marcher et al. (2020) was developed by Gao et al. (2022) [24], who adopted RL to optimize the construction cost of a mountain railway alignment, but such application did not combine the use of ML and InfraBIM.

For this reason, the present study answers to previous shortcomings by presenting a possible integration between BIM and ML. The research scope is achieved through a methodological approach involving: (i) retrieving coordinates and elevations of the theoretical alignment and its combination with the data related to the geometry of rings from the TIM; (ii) the performance of simulations allowing the RL algorithm to understand, according to structural connections, the best positions and rotations for rings, in order to minimize the distance between the as-built and the theoretical alignment; (iii) the use of InfraBIM modelling tools to model the tunnel lining in order to enable future analyses.

3 Methodological Approach

Applying BIM methods and tools to the infrastructural domain is a topic of great interest nowadays. There are several reasons for this interest. Among these, the legislative compliance, both at international and national level, fostered the development of applications towards this direction. The intense research on BIM processes has shown that along with the advantages, there are also some criticalities arising from such applications. In this case, the aim of the paper is to implement machine learning algorithms as a supporting decision tool to evaluate the best combination of factors and determine a operational alignment the closest as possible to the theoretical one. In order to achieve such a result, the starting point of the process was the identification of the specific ML method to be applied; due to its characteristics the selected method was a RL algorithm. Reinforcement learning is a subcategory of machine learning based on the interaction of an agent with a known or unknown environment through actions and rewards, "constantly adapting and learning on giving points" [25]. This type of algorithm is perfectly suited to solve the proposed problem because RL is not based on dataset analysis but on simulations that reflect scenarios similar to reality.

The framework shown in Fig. 1**.** Includes the following five main methodological phases. **Data input for RL**: the input for the environment setup was the Tunnel Information Model (TIM), containing all relevant information on the project due to the adoption of the InfraBIM methodology during the design phase. Due to the TIM model, it was possible to analyze the structure of the rings and the possible axes increments at changing ring rotations and the coordinates of the theoretical layout. **Algorithm setup:** after having extracted the relevant data from the TIM model the environment and the brain have been set up in Python language, which represents a paramount phase in the definition of a high-performance RL algorithm. **Base line**: before performing the simulations using the RL algorithm, the operating alignment for the tunnel under examination was calculated using the deterministic method. In this case, the aim was to obtain a comparison alignment, thus evaluating the effectiveness of the defined reinforcement learning algorithm. **Simulations**: after having completed the first two phases, 18 different simulations were carried out, each one considering different value for the learning parameters. This analysis enabled the selection of the most effective settings for the resolution of the final operational alignment. **Data output**: the data output of the simulations has been used to update the TIM operational alignment. From the simulation carried out, the relative Q-table was extracted, containing all the information produced by the Q-Learning algorithm. Such data allowed the update of the operational alignment within the TIM environment, with a maximum deviation from the theoretical alignment of 1 cm.

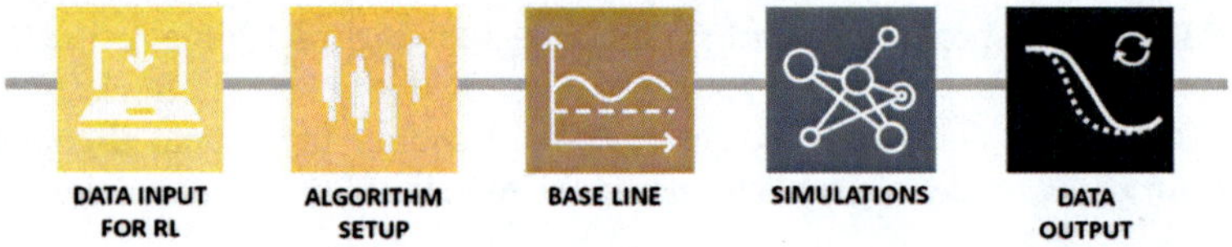

Fig. 1. Methodological framework

3.1 Environment Setup

The algorithm selected to solve the problem is based on Q-learning. This algorithm is based on Markov Decision Processes (MDP) [26] and it is a form of model-free RL. "It provides agents with the capability of learning to act optimally in Markovian domains by experiencing the consequences of actions, without requiring them to build maps of the domains." [27].

The RL problem consists of an environment and a single or multiple agents which, based on a chosen policy, take actions to interact with the environment. After each interaction, the agent receives a feedback (reward) from the environment and updates its state [28]. Within the research purpose, such a method allows to link the possible ring rotations with the operational alignment that will be created. This connection has been made through the Q-table. The compilation of this table takes place through simulations that, by testing different ring positions, associate to each position a reward based on the distance from the theoretical alignment, this score takes into account both the current reward, related to the current state, and the future rewards, related to the chosen one.

The Q-table consists of several columns equal to the possible actions that can be performed (rotations of the rings, depending on the positioning matrix), many rows equal to the positions that are investigated, and the Q-values that fill the different cells.

Learning of the algorithm is based on updating the Q-values using the formula:

$$q_{s_{t-1},a_{t-1}} = q_{s_{t-1},a_{t-1}} - \lambda\left(q_{s_{t-1},a_{t-1}} - r_{t-1} + \gamma max_{a' \epsilon A} q_{s_{t,a'}}\right) \quad (1)$$

where λ and γ are respectively the learning rate, which determines to what extent newly acquired information overrides old information, and discount factor, which determines the importance of future rewards.

Furthermore, the algorithm is improved with an additional ε-greedy parameter [29] that allows for the random exploration of new actions with a $(1 - \varepsilon greedy)$ probability.

The methodological approach followed is based on the analysis of several points. Firstly, the inputs of the problem were defined, i.e. the coordinates of the theoretical alignment and the geometries of the rings at the different possible rotations. Then, such a data has been imported into the calculation environment where the reward system was also defined, based on the formula:

$$r = e^{-\sqrt{(s_x - x^*)^2 + (s_y - y^*)^2 + (s_z - z^*)^2}} \quad (2)$$

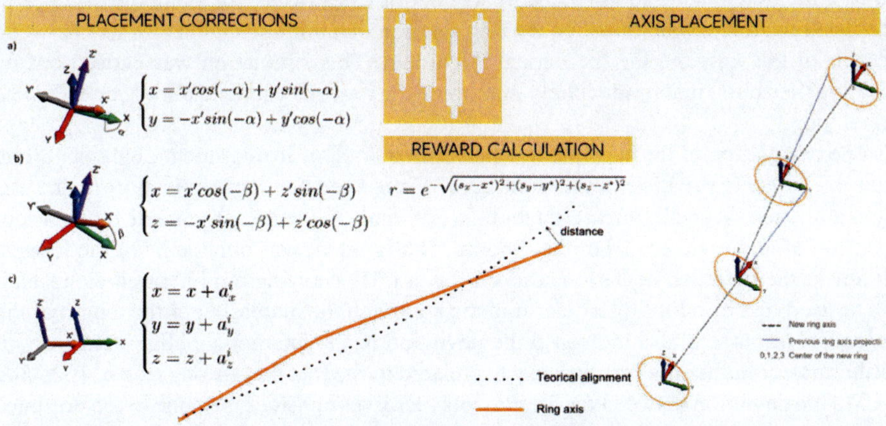

Fig. 2. Environment and agent setup

Equation (2) was designed to maximize the reward for minimum distances from the theoretical path (Fig. 2). In addition, a tolerance parameter was set to give negative rewards for actions that deviated more than 1 cm from the theoretical alignment.

Algorithm 1, shown below, shows the sequence of commands used to write the code in Python for the generation of the operating track.

Algorithm 1: Q-table learning: **ε**-Greedy

Input the number of episodes E, the horizon T, the learning rate λ
Initialize $q_{s,a} = 0$ for all $a \in A, s \in S$
for $e = 1,\ldots,E$ **do**
  Set the initial state s_0
  for $t = 1,\ldots,T$ **do**
    Choose action:
    $$a_{t-1} = \begin{cases} \arg max_{a' \in A} q_{s_{t-1},a'} & with\ prob.\ \varepsilon \\ sample\ uniformly\ from\ A & with\ prob.\ 1 - \varepsilon \end{cases}$$
    Input s_{t-1}, a_{t-1} in the environment
    Observe the new state s_t and the previous reward r_{t-1}
    Update the Q-table:
    $$q_{s_{t-1},a_{t-1}} = q_{s_{t-1},a_{t-1}} - \lambda\left(q_{s_{t-1},a_{t-1}} - r_{t-1} + \gamma max_{a' \in A} q_{s_{t,a'}}\right)$$

At the end, from the Q-table it was possible to extract the coordinates and rotations of the single rings to build a lining model. Through the use of visual programming in Dynamo and the BIM Revit modelling software, it was then possible to match the different data with the BIM elements. The final result is the Tunnel Information Model developed along the planned operational alignment as shown in Fig. 2.

4 Results

After having set the algorithm environment, agent, and brain, before proceeding with the simulations and its training, the operational alignment was calculated using a deterministic method to be used as a benchmark. In this case, instead of using the distance to calculate rewards, it was identified the ring rotation that minimized the distance between the axis of the ring and the theoretical alignment. This simulation was carried out by selecting the action that immediately maximize the reward described above for each new tunnel loop.

The calculation of the baseline was performed both on an alignment of about 400 m, made of 200 rings (which is the number of hives used for the application of the RL algorithm), and on an alignment longer than 2500 m. On the first alignment, a maximum deviation of about 1,8 cm from the theoretical alignment was obtained. On the longest alignment the recorded deviation reached 8,8 cm. By carrying out intermediate tests, it was noticed that by adopting a deterministic approach the increasing of the length of the theoretical alignment also increased the deviation of the operational alignment from it. Furthermore, another interesting aspect is related to the fact that among ring n. 1005 and n. 1040 the deviation between the alignments increases considerably, due to the presence of a curve and the impossibility of a deterministic approach to predict it. Since on the first 200 rings the baseline obtained a maximum deviation from the theoretical alignment of 1,8 cm, the tolerance parameter was set to 1 cm. As the number of placed rings increases, so does the distance between the alignments, because a deterministic approach is not able to self-correct through a reward system. In particular, it can be seen that there is a peak in the deviations after ring n. 1000. In order to evaluate the best combination of learning parameters to achieve the objective described in the article, 18 simulations of 10.000 episodes were initially carried out with different combinations of learning rate, discount factor and epsilon greedy (Fig. 3). The choice of these parameters can improve, or make worse, the learning capacity of the algorithm, measurable through the Eq. (3) that calculates the learning error as the difference between the estimated *q-value* for a state-action pair and the target *q-value*. When the values are equal, the value of $L(q_{s,a})$ is 0, so the learning curve flattens out and the algorithm stops learning.

$$L(q_{s,a}) = (q_{s,a} - r_t + \gamma \max_{a' \in A} q_{s_{t+1},a'})^2 \tag{3}$$

Among the different tested combinations of values (shown in Fig. 3), the combination n. 15 with λ, γ, and ε-greedy was selected because it guaranteed a more accurate learning (due to low value of λ) and a good approximation of Q values (due to medium γ and high ε-greedy).

After having identified the best combination of values to train the algorithm described above, the simulation was performed. It required one million eight hundred thousand

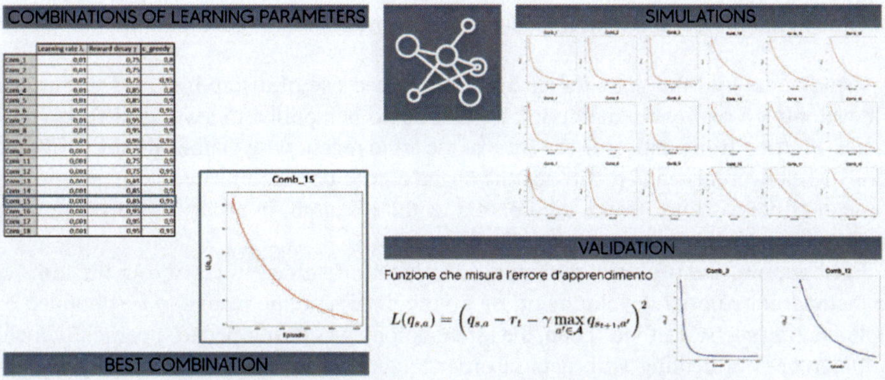

Fig. 3. Data validation and selection of the best combination

episodes to return a reliable output for the construction of the TIM model of the tunnel structure. The large number of simulations required is due to the high state-space dimension nature, which is difficult to manage by a method based on simple Q-learning. As the episodes continued, the algorithm managed to place an increasing number of rings, as it can be seen from the graphs in Fig. 4, meaning that the simulation was running correctly. Once the simulation was complete, the combinations of actions and positions with the highest Q-values have been extracted from the Q-table; these values represent the product of the accumulation of the assigned rewards, as previously described. Using this data as input, it was then possible to update the alignment and the structure of the tunnel assigning the correct ring geometry within the BIM platform using Visual Programming Language (VPL).

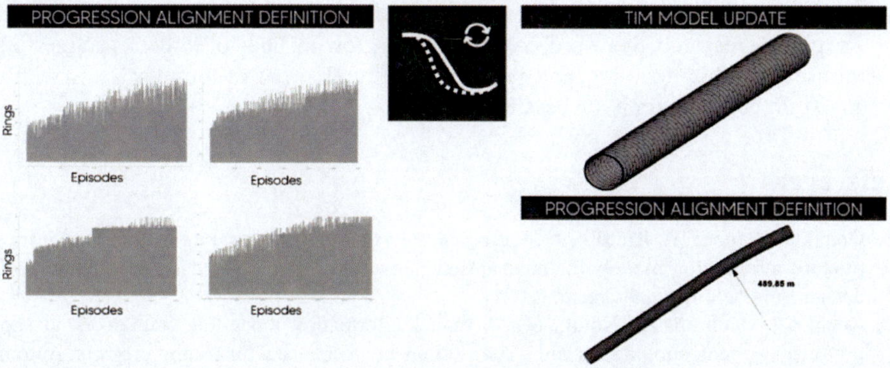

Fig. 4. Data output

5 Conclusion and Further Developments

This study analysed the potentiality of an integrated use of disciplines and techniques by developing a replicable methodology that could be applied to several different case studies. Starting from analysing the state of the art in regard with on the one side BIM and InfraBIM environment and processes and on the other side Machine Learning techniques, it was possible to better frame the context of the research, in order to deepen the real chances given by the proposed integration.

Furthermore, the problem statement identification helped in clarifying the aims of the methodology under development, by giving details on the results to be obtained by implementing such a method. Then, the methodology was developed to specifically focus on the process of dealing with data in order to achieve the solution to the highlighted issue, through the following process: (i) **data collection** and **input definition**, which included the necessity to develop a TIM model with all relevant information for the analysis; (ii) **data processing**, which involved several steps including the setting of the environment and the definition of a term of comparison to evaluate the performance of results from the RL approach, and to conclude (iii) **data validation**, selecting the best combination coming out from the simulations performed, and **update** the TIM model and its alignment on the basis of the output from the training of the RL algorithm.

In conclusion, the algorithm adopted was able to reconstruct an aligment composed of 200 rings with a maximum deviation from the theoretical alignment of 1 cm. However, the construction of such an alignment was computationally expensive because the number of states to explore in order to obtain the desired result are difficult to manage through a simple Q-learning method. For this reason, better results in terms of computational speed could be obtained by adopting algorithms using Artificial Neural Network (ANN). However, the results showed that due to the ease of extraction of the required data for the algorithm to operate, i.e. coordinates of the theoretical alignment and geometry of the ring, the method developed is efficient and easily replicable.

The present research, therefore, contributes to a growing body of evidence suggesting that the use of machine learning, and more specifically RL, to solve complex issues within the infrastructural domain is successful.

References

1. Getuli, V., Capone, P., Bruttini, A., Rahimian, F.P.: On-demand generation of as-built infrastructure information models for mechanised Tunnelling from TBM data: a computational design approach. Autom. Constr. (2021)
2. Koch, C., Vonthron, A., Konig, M.: A tunnel information modelling framework to support management, simulations and visualisations in mechanised tunnelling projects. Autom. Constr. **83**, 78–90, 2017
3. Sejnowski, T.J.: The Deep Learning Revolution. The MIT Press, Cambridge (2018)
4. Alpaydin, E.: Introduction to Machine Learning, 3rd edn.. The MIT Press, Cambridge (2014)
5. Mitchell, T.: Machine Learning, McGraw Hill (1997)
6. Kodratoff, Y.: Introduction to Machine Learning, Elsevier (1989)
7. Murphy, K.P.: Machine Learning : A Probabilistic Perspective. MIT Press (2012)
8. Kaebling, L., Littman, M., Moore, A.: Reinforcement learning: a survey. J. AI Res. **4**, 237–285 (1996)

9. Sutton, R., Barto, A.: Reinforcement Learning. MIT Press (1998)
10. Russel, S., Norvig, P.: Artificial Intelligence: A Modern Approach, 3rd edn. Prentice Hall (2010)
11. Szepesvari, C.: Algorithms for Reinforcement Learning, Morgan Claypool (2010)
12. Wiering, M., van Otterlo, M.: Reinforcement Learning: State of the Art. Springer, Heidelberg (2012). https://doi.org/10.1007/978-3-642-27645-3
13. Zhang, R., El-Gohary, N.: A machine learning approach for semantic matching of building codes and Building Information Models (BIMs) for supporting automated code checking. In: Recent Research in Sustainable Structures.Proceedings of the 3rd GeoMEast International Congress and Exhibition, Egypt 2019 on Sustainable Civil Infrastructures – The Official International Congress of the Soil-Structure Interaction Group in Egypt (SSIGE) (2019)
14. Troncoso-Pastoriza, F., Eguía-Oller, P., Díaz-Redondo, R.P., Granada-Álvarez, E.: Generation of BIM data based on the automatic detection, identification and localization of lamps in buildings. Sustain. Cities Soc. **36**, 59–70 (2018)
15. Croce, V., Caroti, G., De Luca, L., Jacquot, K., Piemonte, A., Véron, P.: From the semantic point cloud to heritage-building. Remote Sensing (2021)
16. Kim, M.K., Thedja, J.P., Chi, H.-L., Lee, D.E.: Automated rebar diameter classification using point cloud data based. Autom. Constr. 78–90 (2021)
17. Shapi, M.M., Ramli, N., Awalin, L.: Energy consumption prediction by using machine learning for smart building: case study in Malaysia. Developments in the Built Environment (2021)
18. He, Y., Wu, P., Li, Y. Wang, Y., Tao, F., Wang, Y.: A generic energy prediction model of machine tools using deep learning,. Appl. Energy (2020)
19. García-Martín, E., Rodrigues, C.F., Graham, R., Grahn, H.: Estimation of energy consumption in machine learning. J. Parallel Distrib. Comput. 75–88 (2019)
20. Seyedzadeh, S., Pour Rahimian, F., Oliver, S., Rodriguez, S., Glesk, I.: Machine learning modelling for predicting non-domestic buildings energy performance: a model to support deep energy retrofit decision-making. Appl. Energy (2020)
21. Amasyali, K., El-Gohary, N.: Machine learning for occupant-behavior-sensitive cooling energy consumption prediction in office buildings. Renew. Sust. Energy Rev. (2021)
22. McArthur, J.J., Shahbazi, N., Fok, R., Raghubar, C., Bortoluzzi, B., An, A.: Machine learning and BIM visualization for maintenance issue classification and enhanced data collection. Adv. Eng. Inform. **38**, 101–112 (2018)
23. Marcher, T., Erhater, G.H., Winkler, M.: Machine Learning in tunnelling – Capabilities and challenges. Geomech. Tunnel. **2**, 191–198 (2020)
24. Gao, T., et al.: A deep reinforcement learning approach to mountain railway alignment optimization. Comput. Aided Civ. Inf. **37**, 73–92 (2022)
25. Nandy, A., Biswas, M.: Reinforcement Learning. With Open AI, TensorFlow and Keras Using Python. Apress (2018)
26. Puterman, M.L.: Markov Decision Processes: Discrete Stochastic Dynamic Programming. Wiley, Hoboken (2005)
27. Christopher, P.D., Watkins, J.C.H.: Q-Learning. Mach. Learn. **8**, 279–292 (1992)
28. Amiri, R., Mehrpouyan, H., Fridman, L., Mallik, R., Nallanathan, A., Matolak, D.: A machine learning approach for power allocation in HetNets considering QoS. In: 2018 IEEE International Conference on Communications (ICC 2018) (2018)
29. Bradley, C., Levine, S., Abbeel, P.: Incentivizing exploration in reinforcement learning with deep predictive models (2016)

Algorithm-Aided Design and Analysis for the Comparative Models Study of Heinz Isler's Shells

Marco Pellegrino[1], Laura Sardone[2](✉), Marco Martino Rosso[2], Valeria Gozzi[2], and Bernardino Chiaia[2]

[1] StruSoft AB, Hyllie Boulevard 53, 215 37 Malmö, Sweden
marco.pellegrino@strusoft.com
[2] DISEG, Department of Structural, Geotechnical and Building Engineering, Politecnico di Torino, Corso Duca degli Abruzzi, 24, 10129 Turin, Italy
{laura.sardone,marco.rosso,valeria.gozzi, bernardino.chiaia}@polito.it

Abstract. Heinz Isler, known as the most famous contemporary shell designer, has employed physical modelling techniques for studying and constructing concrete shell structures. The Swiss engineer's work has spread over 50 years of continuous activity (1954 to 2008/9). The innovation in Isler's work stands by the pioneering form-finding methods based on physical structural models (expansion, inflation and hanging) of thin membranes. The strategy was adopted for many of his experimental structures, like (i) the Wyss Garden Centre in Solothurn (1962); (ii) COOP Warehouse Wangen in Olten (1960) and (iii) Deitingen Slid Service Station (1968).

In this document, the study focuses on the assessment of Isler's Shells employing new technologies and strategies based on Algorithm - Aided Design. The design steps comprehend the reparameterization of Isler's shells samples using Parametric Design techniques (geometry development in Rhinceros3D © + Grasshopper© add-on); the generative approach (parametric design + simulation) is conducted and achieved thanks to Alpaca4D©. The structural analysis relies on a Finite Element Analysis (FEA) carried on through the plug-in for Grasshopper© in Rhinoceros3D©. The results are retrieved to develop and validate the achieved results employing Alpaca4D© with a comparative model developed with the Karamba 3D© add-on and FEM-Design©, an advanced and intuitive structural analysis software.

Keywords: Generative Design · OpenSees · Alpaca4D · Benchmark Test · Parametric Development · Finite Element Analysis · Reverse Engineering

1 Introduction

Heinz Isler was a Swiss structural engineer and architect known for his innovative work in the field of shell structures. Born on June 19, 1926, in Zurich, Switzerland, Isler gave significant contributions to developing lightweight and efficient shell structures that challenged traditional architectural and engineering norms.

S. Gabriele et al. (Eds.): IWSS 2023, LNCE 437, pp. 512–521, 2024.
https://doi.org/10.1007/978-3-031-44328-2_53

Isler's approach to design was influenced by nature, specifically, the shapes and forms found in natural structures like shells, leaves, and flowers [1]. Isler's work builds upon the principles governing natural forms applied to create efficient and elegant structural designs [2].

One of Isler's notable achievements was his exploration of the double-curvature concrete shell structures. He developed a technique [3] involving the use of thin, reinforced concrete shells supported by a minimal number of slender columns or supports [4, 5]. This approach allowed for large-span structures that were both visually striking and structurally efficient. His work earned him international recognition and numerous awards throughout his career, including the Auguste Perret Prize in 1976 and the IASS (International Association for Shell and Spatial Structures) Gold Medal in 1993 [6].

Heinz Isler's innovative approach to shell structures revolutionized the field of architecture and engineering, and his designs continue to inspire and influence architects and engineers worldwide, showcasing the possibilities of creating elegant and efficient structures using natural principles.

Isler's portfolio includes a variety of projects such as roofs, shells, and pavilions. Some of his most renowned works include (i) The Wyss Garden Centre in Solothurn (1962); (ii) COOP Warehouse Wangen in Olten (1960) and (iii) Deitingen Slid Service Station (1968).

In this study, the attention will be focused on the latter three structures ((i), (ii), (iii)), validating Isler's model exploiting the Generative Design Approach. The models initially developed through the innovative parametric modelling techniques in Grasshopper 3D for Rhinoceros 3D will then be analyzed and validated in three different FEM solvers: (a) Karamba 3D, (b) Alpaca 4D and (c) FEM Design. The results obtained show good accordance between the employed software((a), (b), (c)), validating not only the structural strength of the shells modelled based on Isler's geometries but also the performance and computing power of Alpaca 4D© and FEM - Design©.

2 Isler's Models Development

The exceptional performance of Isler's shells within the four decades, and the importance of his approach to modelling, has encouraged many designers to re-evaluate his shells and get inspired by Isler's work [7]. Given the resonance of Isler's work concerning the shell structures and the studies produced in this field, three significant examples of Isler's design are considered. The following images (Fig. 1) describe the geometry of (from the top): (**a**) Wyss Garden Center (1961, Zuchwil, Solothurn, Switzerland); (**b**) Deitingen Service Station (1968, Deitingen, Solothurn, Switzerland); (**c**) Bürgi Garden Center (1973, Camorino, Switzerland).

The development of the geometries illustrated in the previous image (Fig. 1), was implemented in the Grasshopper environment. Grasshopper employs a visual programming paradigm, where users create and manipulate components called "nodes" within a canvas interface [8]. These nodes represent various operations, algorithms, and mathematical functions. For each geometry, the parameters have been set through design variables using the appropriate *number sliders* that allow the variation of the geometry and the consequent analysis in real-time (Fig. 2).

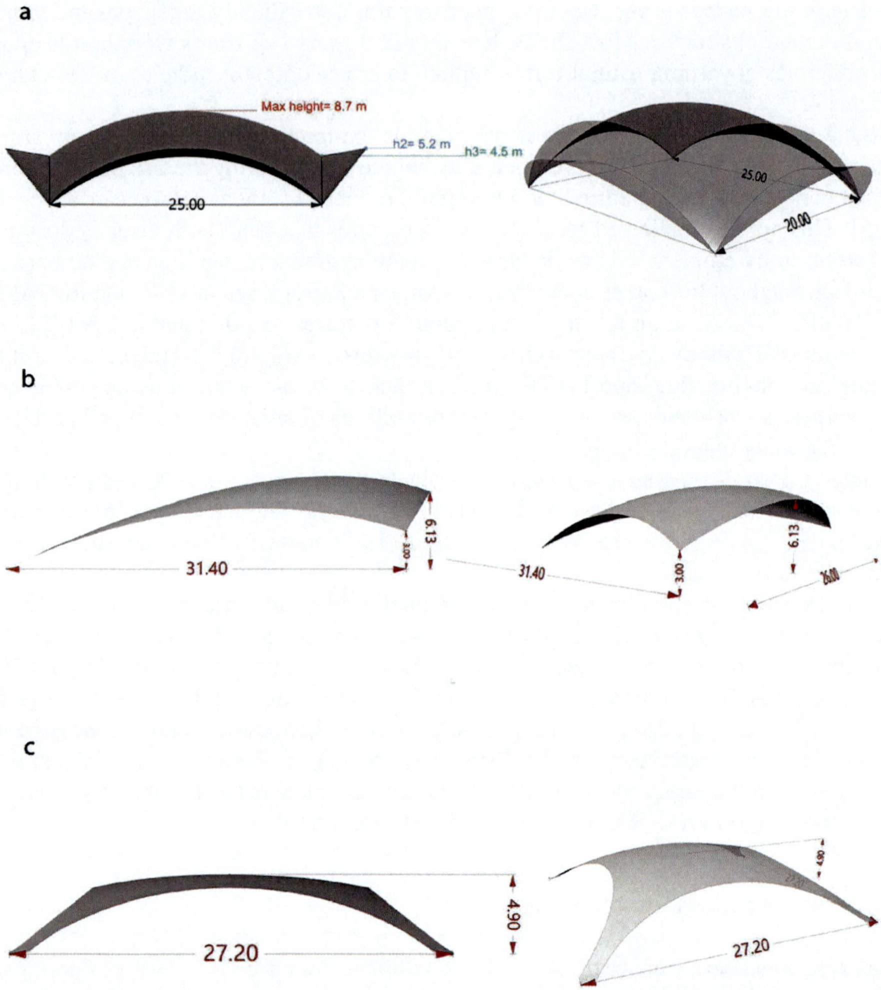

Fig. 1. (**a**) Wyss Garden Center (1961, Zuchwil, Solothurn, Switzerland); (**b**) Deitingen Service Station (1968, Deitingen, Solothurn, Switzerland); (**c**) Bürgi Garden Center (1973, Camorino, Switzerland).

Through the geometry definition, it was possible to obtain from the Boundary Representations (*BRep*) the mesh belonging to the three geometries to transform the structural object into Shell (*Mesh to Shell*).

Since the goal of the document is to study three of Isler's shells to achieve the software's comparison and benchmark, the next step is characterized by the geometry definitions using different finite element software.

To sum up, the work workflow for software testing through the definition of case studies is distributed as follows:

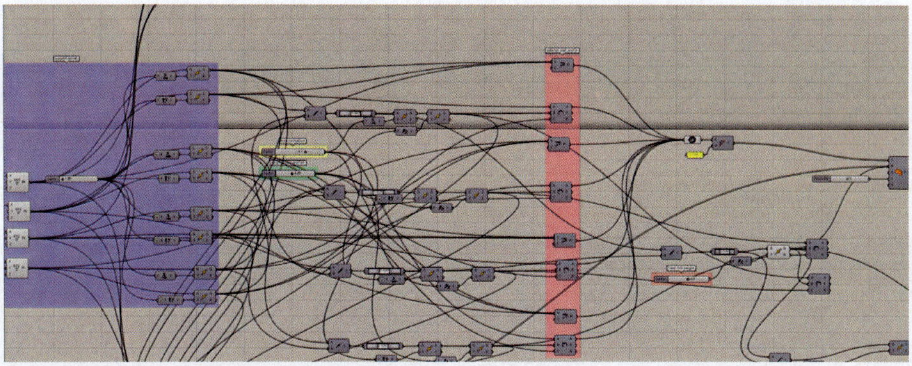

Fig. 2. Geometry development in Visual Language of the case study (**c**) Bürgi Garden Center

1. *Grasshopper development for parametric definition (Geometry)*
2. *FEM using Karamba - 3D plugin*
3. *Data transfer from Grasshopper to Alpaca 4D*
4. *Benchmark test using FEM - Design.*

Karamba3D is a parametric structural engineering tool which provides accurate analysis of spatial trusses, frames, and shells. Karamba3D is fully embedded in the parametric environment of Grasshopper which is a plug-in for the 3d modelling tool Rhinoceros (Rhino3d) [9]. In parallel to the Karamba 3D model, also Alpaca4d was used for the benchmark test. Alpaca 4D is a Grasshopper plugin which has been developed on top of OpenSees. It lets you analyse beam, shell, and brick elements through Static, Modal and Ground Motion Analysis. Alpaca4D is currently being developed by Marco Pellegrino and Domenico Gaudioso. Alpaca4d facilitates the interaction with OpenSees making it possible to model complex geometries within your parametric workflow and reducing the time spent in the modelling. The main idea of Alpaca4d is to provide an efficient and easy way to use OpenSees without writing any line of code. The library is mostly used by researchers and academia because of the not user-friendly interface even if the math at the core library is highly sophisticated [10].

To perform the analysis, the models have been evaluated under their self-weight with a structural uniformly distributed overload equal to 4 kN/m^2 for the simulation of the snow load, due to the location of the structural artefact. For all the parallel analyses was used the same linear material properties and definitions. The mechanical properties of the concrete material of Isler's shell constructions are not identical and different references have mentioned different values for it.

The typical concrete mixture of Isler's shells constitutes of cement percentage of a maximum (325 kgm^3), gravel stone with a maximum dimension of (15 mm) [7, 11] for the concrete of 28 MPa (C-28) compression strength. Since, the aim of this study is the comparison of the structural performance of different shells considering the benchmark test employing several finite element software, in all the shells, the same strength and material properties for concrete is assumed: C 45/55, E = 3600/3kN/cm^2, Specific weight (γ) = 25 kN/m^3, ν (Fig. 3).

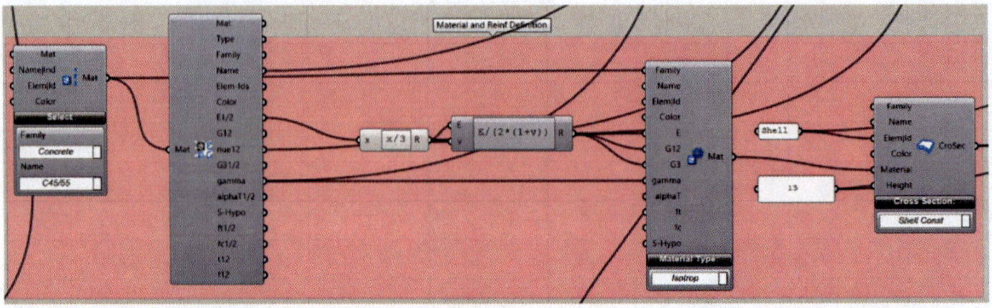

Fig. 3. Material Development and Shell Constant Cross Section in Visual Language

The data include the thickness, the span, and the height dimensions of shells. For the shells (a), (b), and (c), different heights were mentioned in the references. The shells' thicknesses vary, and the assumed values are based on ranges found in the relevant references. Considering the shells (a), (b), (c), the thickness (t) varied in the range between $6 \leq t \leq 10$ cm. However, for the analysis, the thickness of all shells $t = 15$ cm is assumed to develop comparative models. A uniform thickness is considered to acquire general results of the force flow and stress distribution.

3 Models Evaluation

In this section the retrieved results are described and compared in the following Tables 1, 2, 3. The main indicators of comparison between finite element analysis software (i.e., Alpaca4D©, Karamba3D© and FEM-Design©) are represented by *mass and displacement*. The trend of the vertical displacement is described through the following images for each of the shells analyzed through the different FE software (Figs. 4, 5, 6).

Table 1. Results comparison of the Test Case (a) Wyss Garden Center (1961).

Solver	Displacement [mm]	Mass [kg]
Karamba 3D	13.98	326186.93
Alpaca 4D	14.89	332618.58
FEM - Design	14.16	332416.48

The vertical deformation of the shells can be considered low considering their span; this is a clear indication that the load transmission work as in the membrane rather than in flexural form.

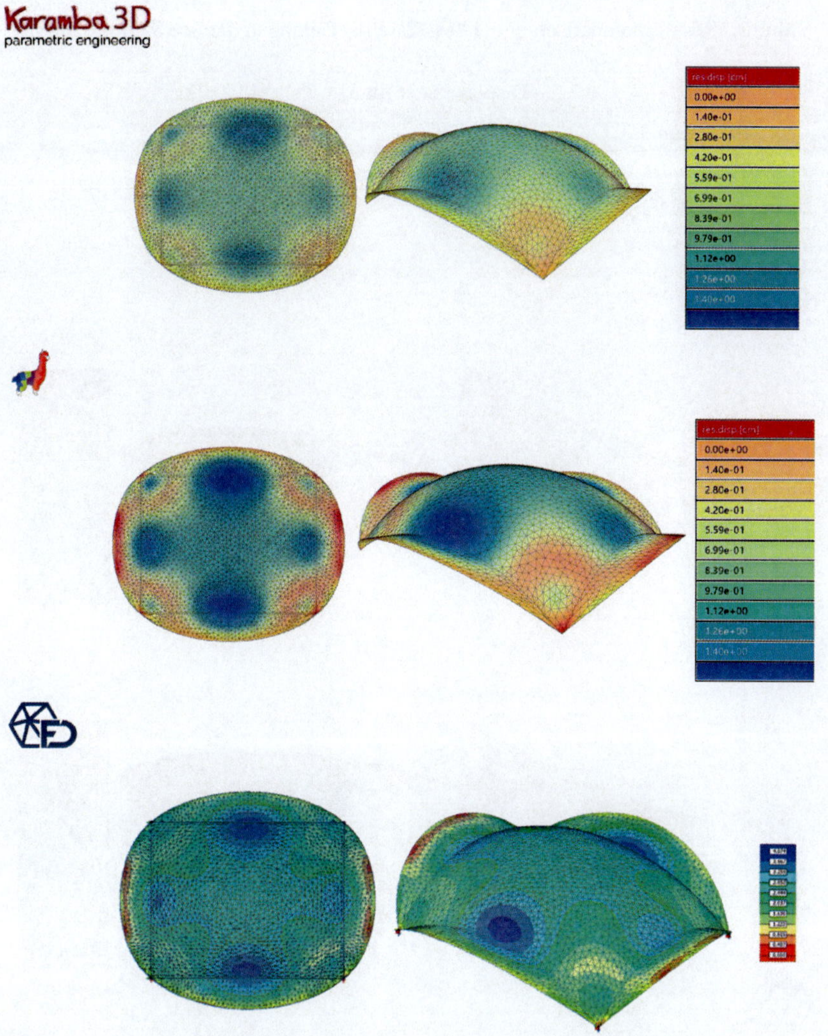

Fig. 4. Test case (a) Wyss Garden Center, 1961. Trend of the vertical displacement: Upper definition Karamba3D; Mid - definition Alpaca4D; Lower Definition: FEM-Design by StruSoft.

Table 2. Results comparison of the Test Case (b) Deitingen Service Station (1968)

Solver	Displacement [mm]	Mass [kg]
Karamba 3D	59.24	151877.17
Alpaca 4D	59.91	154871.83
FEM - Design	59.32	153275.16

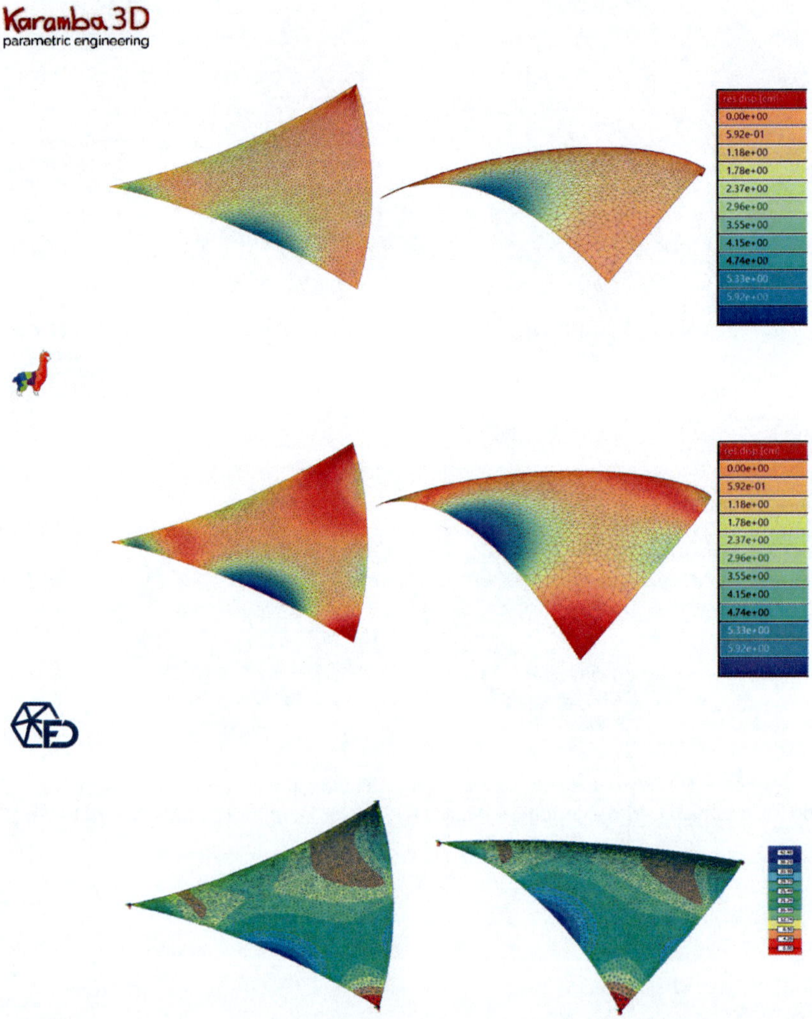

Fig. 5. Test case (b) Deitingen Service Station 1968. Trend of the vertical displacement: Upper definition Karamba3D; Mid - definition Alpaca4D; Lower Definition: FEM-Design by StruSoft.

Table 3. Results comparison of the Test Case (c) Bürgi Garden Center (1973)

Solver	Displacement [mm]	Mass [kg]
Karamba 3D	66.96	146495.55
Alpaca 4D	70.39	149384.10
FEM-Design	68.16	147548.61

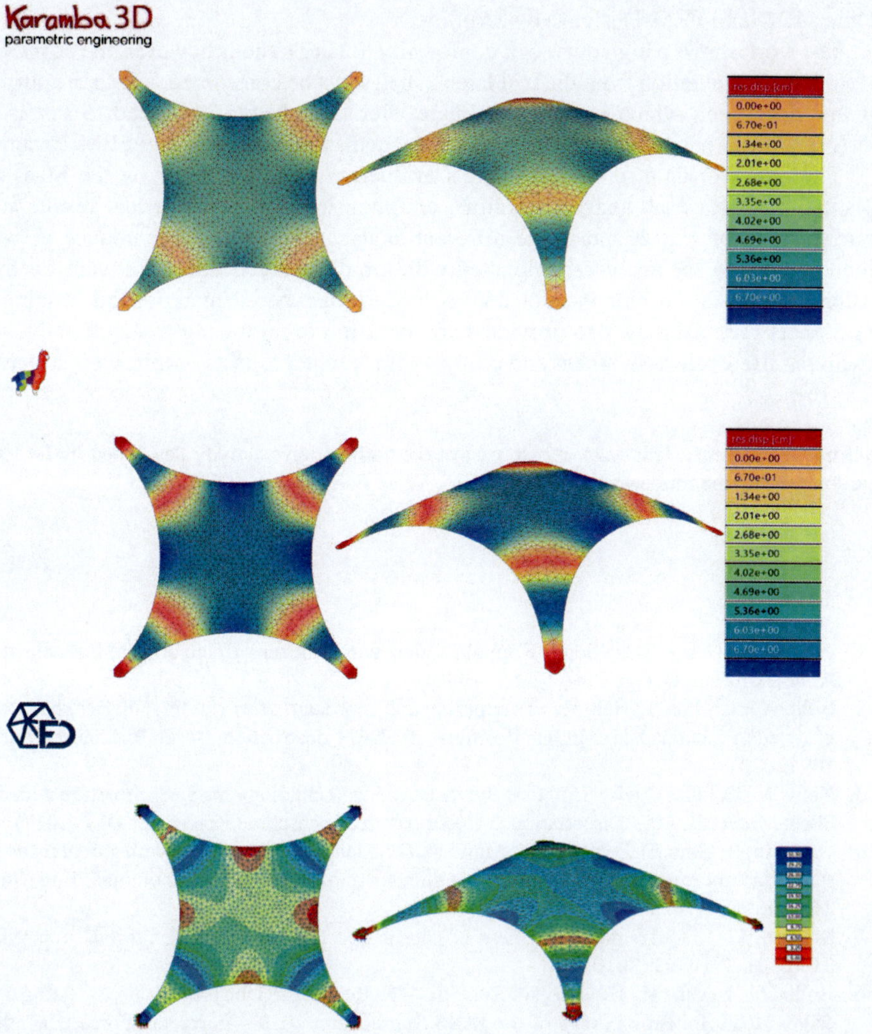

Fig. 6. Test case (c) Bürgi Garden Center, 1973. Trend of the vertical displacement: Upper definition Karamba3D; Mid - definition Alpaca4D; Lower Definition: FEM-Design by StruSoft.

4 Conclusions

In this research, three of Heinz Isler's shells are re-modelled and assessed through reverse engineering techniques. The modelling data for the development of the samples are retrieved by exploiting numerical and visual data from several references. The evaluation of the shell structures is processed within several software packages to estimate the accuracy of the results between the Finite Element Software employed within the Grasshopper© environment, a Visual Language environment, i.e., Karamba3D©, Alpaca4D© and FEM-Design© by StruSoft©.

The work shows a high degree of conformity to Isler's shells however, the possibility of a degree of deviation from the real Isler's shell must be considered, due to the purpose of this document, which is not to estimate Isler's result, but is instead to assess the performance of new production FE software, such as Alpaca 4D© and FEM-Design©.

The indicator data for the software's evaluation are represented by the Mass and Displacement for each analyzed sample, and, considering the numerical results summarized in Tables 1, 2, and 3, the different analyses show good accordance between them, validating the analyses conducted with the three independent software for three different samples. Current research on the frontiers of concrete material and sustainability aspects [12, 13] may also provide some near future promising research paths, also involving life-cycle assessment and optimization-related aspects of thin shell structures [14–16].

Acknowledgement. This work is also part of the collaborative activity developed by Dr Laura Sardone within the framework of the "PNRR": VS3 "Earthquakes and Volcanos".

References

1. Prerost, I.: Heinz Isler: Seine Kuppeln tönen wie Glocken, Hochparterre: Zeitschrift für Architektur und Design **3**(4), 52–55 (1990)
2. Boller, G., Chilton, J.: Heinz Isler's experimental approach to form finding, In: Beckh Matthias et al. (eds.) Candela Isler Müther. Positions on Shell Construction, Basel: Birkhäuser, pp. 98–109 (2020)
3. Xian, W.: Cellular Cavity Structure and its Building Technology for Shell Structure with Thin Sheet Materials. MSc Dissertation. TU Darmstadt. Hochschule Kennziffer D17 (2017)
4. Castaldo, P., Gino, D., Marano, G., Mancini, G.: Aleatory uncertainties with global resistance safety factors for non-linear analyses of slender reinforced concrete columns. Eng. Struct. **255**, 113920 (2022)
5. Marano, G.C., Quaranta, G.: A new possibilistic reliability index definition. Acta Mech. **210**(3–4), 291–303 (2010)
6. Boller, G., Beckh, M.: How to gain from loss – Heinz Isler's Haba project and its testing under critical load. In: Proceedings of the IASS Symposium 2019 – Form and Force, Barcelona, Spain (2019)
7. Baghdadi, A., Heristchian, M., Kloft, H.: Structural assessment of remodeled shells of Heinz Isler. Int. J. Adv. Struct. Eng. **11**, 491–502 (2019). https://doi.org/10.1007/s40091-019-00248-4

8. Tedeschi, A.: AAD_ALGORITHMS - AIDED DESIGN, Parametric Strategies using Grasshopper, pp 40–48. Le Penseur Publisher, English Edition (2014). ISBN 978-88-95315-30-0
9. SIMPLYRHINO.com. Karamba3D Overview. Article available on simplyrhino.com. Accessed 09 May 2023
10. Food4Rhino: ALPACA4D – OPENSEESGH. Available on food4rhino.com. Accessed 12 Apr 2023
11. Christian, L., David, B.: Autobahnraststätte Deitingen-Süd. Schalentragwerk von Heinz Isler (2012). docplayer.org. Accessed 27 May 2023
12. Fiore, A., Marano, G.C., Marti, C., Molfetta, M.: On the fresh/hardened properties of cement composites incorporating rubber particles from recycled tires. Advances in Civil Engineering,Article ID 876158 (2014). https://doi.org/10.1155/2014/876158
13. Gregori, A., Castoro, C., Marano, G.C., Greco, R.: Strength reduction factor of concrete with recycled rubber aggregates from tires. J. Mater. Civil Eng. **31**(8), 04019146 (2019). https://doi.org/10.1061/(ASCE)MT.1943-5533.0002783
14. Fiore, A., Quaranta, G., Marano, G.C., Monti, G.: Evolutionary polynomial regression–based statistical determination of the shear capacity equation for reinforced concrete beams without stirrups. J. Comput. Civil Eng. **30**(1), 04014111 (2016). https://doi.org/10.1061/(ASCE)CP.1943-5487.0000450
15. Sardone, L., Fiore, A., Greco, R., Moccia, C., Lagaros, N.D., De Tommasi, D.: Algorithm-aided structural-optimization strategies for the design of variable cross-section beams. In Proceedings of the International Fib Symposium on the Conceptual Design of Structures (2021)
16. Frangedaki, E., Sardone, L., Marano, G.C., Lagaros, N.D.: Optimisation-driven design in the architectural, engineering and construction industry. In: Proceedings of the Institution of Civil Engineers-Structures and Buildings (2022). https://doi.org/10.1680/jstbu.22.00032

Stability of Scaffoldless Dry Stone Vaults Under Microgravity and Seismic Loading

Peter Manos[1,2], Anjali Mehrotra[1,3], and Marina Konstantatou[4(✉)]

[1] Delft University of Technology, Delft, Netherlands
[2] Thornton Tomasetti, New York, NY, USA
PManos@thorntontomasetti.com
[3] Arup, London, UK
[4] Specialist Modelling Group, Foster + Partners, London, UK
mkonstantatou@fosterandpartners.com

Abstract. Vaults assembled from regolith-based components offer a compelling solution to the need for protective infrastructure in future off-Earth settlements. We have studied the stability and seismic performance of a novel pitched-brick vault geometry, which could shield various assets from radiation or other Space hazards. The dry stone scaffoldless vault was modeled as an assembly of mechanically interlocking components using geometric constraint solving. A novel methodology was developed to produce distinct element models with complex, non-convex rigid block geometries. The stability of these models was assessed in 3DEC, a distinct element modeling program, using quasi-static and dynamic time history analyses representative of expected loading in the Lunar context. For quasi-static pushover-type analysis, the proposed vault geometry was stable under its self-weight and exhibited favorable resistance to lateral acceleration. The vault's dynamic response to an artificial ground motion time history suggests that the structure effectively dissipated energy, and that progressive collapse is unlikely in this type of structure.

Keywords: Extraterrestrial design · Space Architecture · Interlocking masonry · Geometric constraint solving · Distinct element modeling · Seismic design

1 Introduction

Component-based solutions for extraterrestrial (ET) infrastructure offer several unique advantages, the most important being the potential for reconfiguration and reuse [4,5]. Furthermore, when considering the appreciable risk presented by shallow moonquakes [10], monolithic structures may be inadequate due to brittleness and low energy dissipation. Out of many infrastructure solutions needed to support permanent off-Earth human settlement, shielding structures are crucial for protecting habitats, vehicles, or other assets from space hazards [14]. Novel research investigates the use of discrete, brick-like interlocking components

S. Gabriele et al. (Eds.): IWSS 2023, LNCE 437, pp. 522–531, 2024.
https://doi.org/10.1007/978-3-031-44328-2_54

for the design of various structural assemblies, including vaults that can satisfy shielding requirements [5]. Amongst possible configurations, the pitched-brick vault typology found in vernacular North African and Middle Eastern architecture offers a particularly compelling shielding solution. This is because it can be constructed without the use of scaffolding, which is accomplished by building a series of inclined courses against an existing support [5,11]. This aspect is particularly appealing in extraterrestrial contexts, where dependency on temporary support structures can introduce additional points of failure. Furthermore, the use of discrete building units increases process robustness by decoupling fabrication from assembly. This decoupling is valuable for adapting to the uncertainties of manufacturing structural elements in-situ.

The component-based vault concept developed by [5] was based on limit state design principles, but its stability under moonquake loading was not studied. This paper presents a methodology for studying the stability of self-supporting dry-stone vaults proposed by [5], and provides preliminary insight into their structural performance under the combined influence of microgravity and seismic loading. A geometric and numerical modeling framework was first developed, which yielded a parametric model of a pitched-brick vault assembled from mechanically interlocking components. The geometry of these components is pictured in Fig. 1. Afterward, the models were subjected to quasi-static and dynamic seismic analyses. Rhino 7 was used to carry out the geometric modeling, which was based on the conceptual framework developed by [5]. The geometric models were converted into numerical models in 3DEC, a software package based on the distinct element method [3]. 3DEC was chosen because of its ability to accurately and efficiently simulate the behavior of systems characterized by nonlinear failure modes such as joint sliding or separation between elements [6].

2 A Methodology for Modeling Assemblies of Mechanically Interlocking Components

2.1 Component Geometry

Previous analyses of the vault concept from [5] were conducted in [8] to determine optimal component dimensions. The results showed that components with a length of 1.0 m, a width of 0.5 m, and a height of 0.1 m, would maximize seismic capacity while considering geometric limitations. Although [5] proposed components with three grooves, such as the one pictured in Fig. 1b, [8] increased the number of grooves to five while maintaining the same aspect ratio. This was done to deconstrain how the pins of one component could fit into the grooves of the components below. Since the component originally proposed by [5] was not directly compatible with 3DEC due to its non-convex geometry, the unit was idealized and modelled in Rhino as a group of several constituent components. Each of these constituents individually conformed to the topological requirements of 3DEC. Figure 1a illustrates how the component was initially subdivided in Rhino, after which it was joined into one rigid block in 3DEC

(Fig. 1b). For any given geometry, the component was broken down into a base, two end blocks, a series of groove walls, and two pins, as shown in Fig. 1a. The component geometry was defined by a parametric model where the dimensions of all parts of the block could be varied, as well as the number of grooves. Once the desired parameters were defined, the group of subcomponents could be imported into 3DEC and joined together to form a rigidly-linked collection of blocks that approximated the desired unit component geometry. For any given global geometry composed of these components, a script was used to iteratively assemble each group of elements into a single rigid block (Fig. 1b).

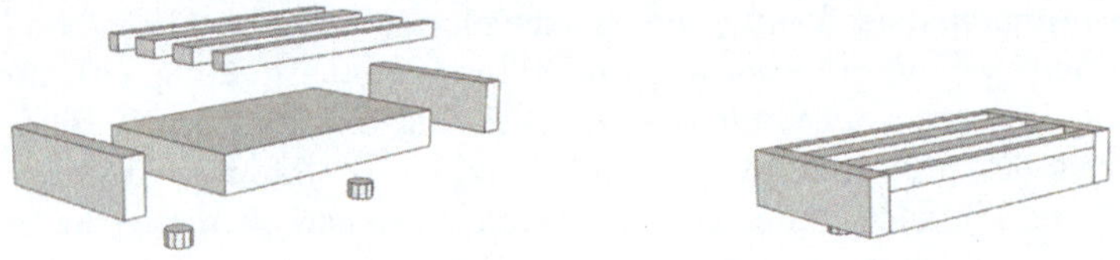

(a) Unit divided into convex solids (b) Joined unit

Fig. 1. Illustration of 3DEC-compatible component

2.2 Global Vault Geometry

For a given component definition, a global geometric model of the structure was created based on the conceptual modeling framework developed by [5]. The model definition was fully parameterized, so any combination of geometric parameters could be used to generate a global vault geometry which approximated how a collection of interlocking units could be assembled into a pitched-brick vault. The choice of geometric parameters such as rise and span of the vault were informed by [5], which proposed shielding structure dimensions that could accommodate an existing habitat module design concept. As such, the span of the vault was chosen to be 12 m, with a rise of 6 m and a length of 24 m. The influence of inclination and component dimensions are not presented in this work but were examined in previous research by [8], which indicated that an inclination of 45° maximized stability. Accordingly, it is assumed that a support with a 45° inclination can be constructed, possibly by utilizing an existing topographical feature.

2.3 Geometric Constraint Solving

The initial configuration of components in the global vault geometry did not satisfy interlocking requirements, so it needed to be input into a geometric constraint solving algorithm to reposition components in a way that allowed for mechanical interlocking to take place. This approach was taken because standard geometric logic could not be defined in Grasshopper to assign a component's pins to the grooves of components underneath it in a generic vault model. Since the arrangement of components in one course depended on the location of components below, attempts to use geometric logic in Grasshopper were too constrained to develop a fully parametric model. This necessitated the use of recursive geometric constraint solving through the Kangaroo 2 plugin by [12]. To this end, the main constraints defined for the Kangaroo 2 solver were: collision between adjacent components to prevent overlapping geometry; in-plane confinement to ensure that each course remained planar; and mechanical interlocking to ensure that components were connected between adjacent courses. Mechanical interlocking is necessary in this type of structure to recreate the self-supporting effect normally achieved in a pitched-brick vault with mortar.

2.4 Resulting Vault Model

After geometric constraint solving was implemented, a geometrically viable vault model was obtained which could be assembled from interlocking components. The resulting vault geometry is shown in Fig. 2. At this stage, each course did not have continuous load paths due to gaps that arose from relative rotations between neighboring components. To ensure that the vault was stable, wedge-like blocks were modelled to fill these gaps. These blocks are not pictured in Fig. 2. In reality, continuity may be enforced by filling the gaps with loose rocks or packed regolith, but it may also be possible to create close-fitting wedges from sintered regolith.

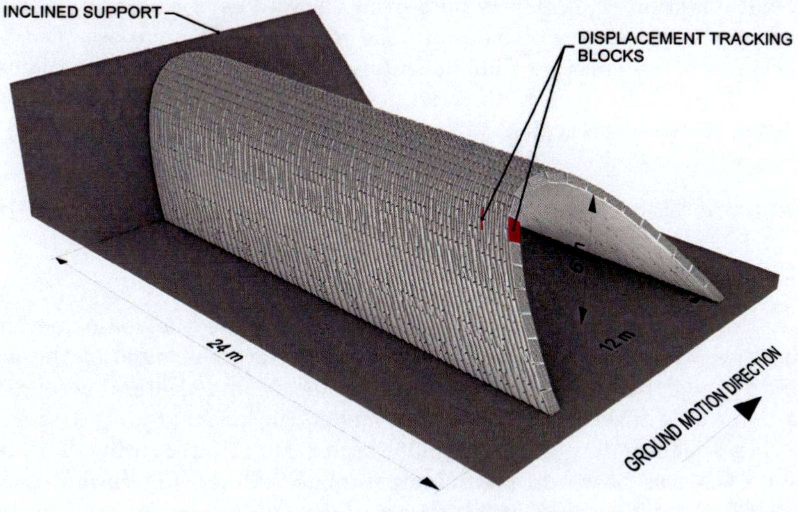

Fig. 2. Resulting vault model

2.5 Numerical Modeling

After a viable geometric definition of the vault was obtained in Rhino, the model needed to be translated into a 3DEC model for analysis. A rigid block approach was chosen because inter-block movements were expected to govern structural behavior and because of its advantage in computational efficiency [7]. To create a rigid block model of the vault, the geometry was exported as a VRML file. As previously shown in Fig. 1, each unit in a given vault model was represented by a group of constituent convex solids. In the VRML representation of the vault geometry, the order of the sub-components was consistent with the group-based definition for each component. This allowed for the geometry to be imported into 3DEC in one step as a collection of individual rigid blocks, which were iterated through using a Python script and joined into rigid assemblies corresponding to each unit. Additional model geometry such as the wedges required for continuity of load paths, as well as boundary supports, were exported from Rhino and directly imported into 3DEC as VRML files.

Modeling assumptions pertaining to boundary conditions and contact between blocks needed to be defined as well. It was assumed that the first and last blocks in each course were anchored to a rigid foundation. For contacts between blocks, a soft-contact approach was employed, whereby the extent of interpenetration between blocks was controlled by the interface stiffness. The tensile strength of the interface controlled the nonlinear response in the normal direction, while the Coulomb-slip model was assumed in the shear direction. For rigid block models, stiffness is lumped at the joints, and thus appropriate values for stiffness of contacts could be approximated by dividing the elastic stiffness parameters of the assumed material by the spacing between block centroids in the normal direction [1,6]. The results from [2] were used to define the fundamental material properties of blocks in the model, which were assumed to be manufactured using resistive radiant sintering of regolith simulant. Values of friction angle for sintered regolith could not be found in literature, so a value of 40° is assumed based on typical values for dry-stone masonry according to [9].

3 Seismic Performance of the Component-Based Vault

3.1 Quasi-Static Pushover-Type Analysis

Quasi-static pushover type analyses were conducted as the main method of assessing the seismic performance of the vault. The main aims of the analysis were to study the behavior of the vault geometry under lateral acceleration, and to determine whether it was stable under the peak ground acceleration (PGA) corresponding to possible ultimate limit state (ULS) loading. The design value for PGA was based on preliminary probabilistic seismic hazard analysis conducted by [13] corresponding to an area near the Apollo 17 landing site. In the absence of reliability guidelines for ET infrastructure, it was assumed that a moonquake with a 475-year return period could be used for preliminary ULS analysis. For this recurrence period, a PGA of $1.52\,\mathrm{m\,s^{-2}}$ was expected according to [13]. This value was conservatively rounded up to $1.6\,\mathrm{m\,s^{-2}}$ for quasi-static analysis in order to allow the pushover script to apply lateral acceleration in increments of $0.1\,\mathrm{m\,s^{-2}}$. The default value of 0.8 was used for local damping.

The model was first allowed to settle under the vertical component of gravity on the Moon, which is $1.62\,\mathrm{m\,s^{-2}}$. Afterward, the vault was subjected to a uniform lateral acceleration which increased in increments of $0.1\,\mathrm{m\,s^{-2}}$. After each new load step, the solver in 3DEC was cycled until equilibrium criteria were reached. This iteration was repeated until the model displayed significant permanent deformation, amounting to a maximum lateral displacement of 0.03 m. Given the preliminary nature of the analysis carried out, a precise understanding of the evolution of damage in the structure beyond the chosen cutoff displacement of 0.03 m was not economical to pursue. Therefore, the design level acceleration was subsequently applied in one load step in order to simply determine if the structure was stable under possible moonquake design loading. Once the model reached equilibrium under the design acceleration, the loading scheme was reversed, and the lateral acceleration was incrementally decreased to zero. This loading and unloading are reflected in the plot shown in Fig. 3, which is based on the lateral displacement reported for the centroid of a component in the outermost course of the vault model, which is consistent with the component highlighted in red in Fig. 2. In Fig. 3, acceleration and displacement were plotted on the Y and X-axes, respectively. The pushover plot indicates the component being tracked has a stiff response at first, followed by gradual stiffness degradation marked by decreasing slope. The mark placed at a displacement of 0.03 m indicates the discontinuity that arose when the design level acceleration was applied in one load step. At the final acceleration of $1.6\,\mathrm{m\,s^{-2}}$, the outermost course appears to experience significant deformation amounting to about 0.16 m, but the vault still reaches equilibrium at the design level acceleration. The unloading curve demonstrates that a residual deformation of about 0.12 m is present after the lateral acceleration is removed.

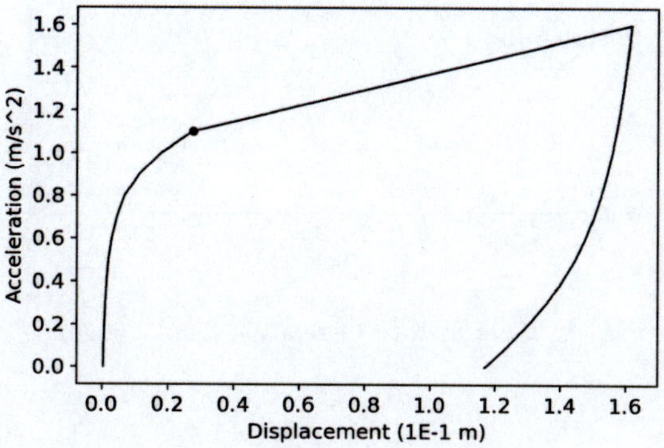

Fig. 3. Extended loading and unloading

3.2 Dynamic Time-History Analysis

To represent dynamic seismic actions on a given model, the velocity time history for a moonquake was applied to the inclined support, the ground support, and fixed base blocks in the model, which are colored in black in Fig. 5. Dynamic loading was only considered in the in-plane direction of the vault. Due to computational constraints, a single undamped dynamic time-history analysis was carried out.

The results of the time history analysis are presented in Fig. 4, which shows how displacement in the X-direction evolved over time as a moonquake ground motion was applied. Two points were tracked, which correspond to the block used for previous analysis, as well as a block with the same X and Z coordinates located ten courses behind the outermost course in the vault. The response of the vault relative to the ground was obtained by subtracting the displacement of the base, whose movement was defined by the ground motion time history, from the displacement of the tracking points. The input ground motion was truncated from 5 to 15 s, which corresponded to the most destructive part of the signal. Figure 4 shows that the displacement of the outer tracking block experienced a drastic increase soon after the 10-s mark. This corresponds to the outer courses of the vault exhibiting partial collapse, which is illustrated in Fig. 5. In this figure, a semi-transparent overlay shows the original, undamaged vault state. The displacement of the inside tracking block did not exhibit any irregular patterns and appeared to follow the base motion with no significant dynamic amplification. This is consistent with Fig. 5, which shows that only the first six courses collapsed.

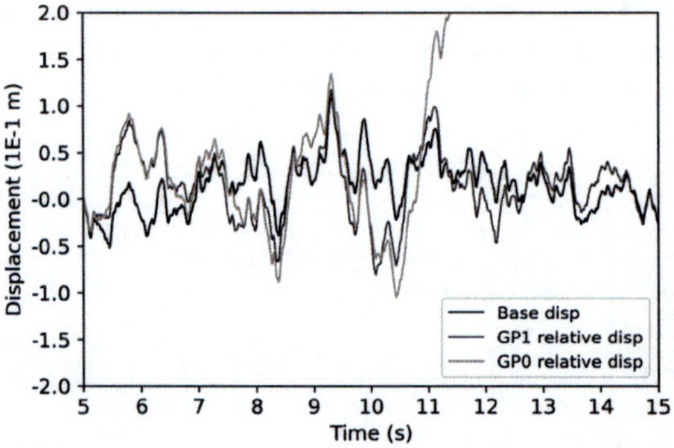

Fig. 4. Time history of relative displacement

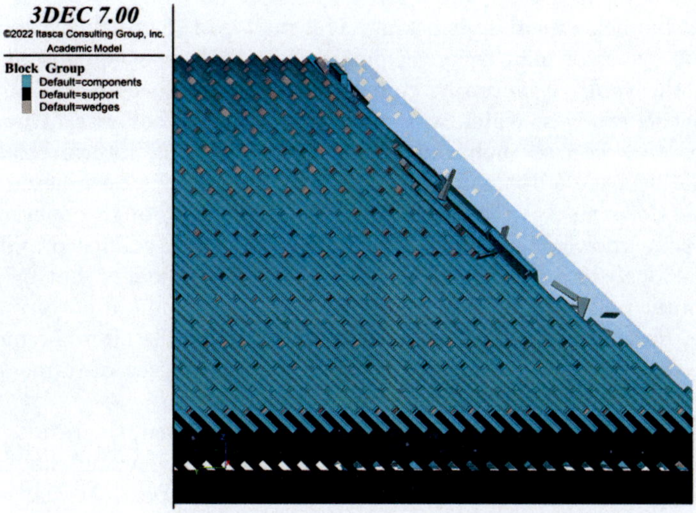

Fig. 5. Damage due to moonquake

4 Discussion

The pattern of gradual stiffness degradation observed in Fig. 3 is unlike conventional instability mechanisms in masonry, where a sudden onset of instability is typically expected. There appears to be an approximately elastic domain of loading up until an acceleration of $0.5\,\mathrm{m\,s^{-2}}$, after which the stiffness degrades slightly with each load step, but there is no sudden decrease in stiffness as otherwise seen in the pushover behavior of arches, for example. The gradual decrease of stiffness may be explained by the influence of the interlocking mechanism between components. Mechanical interlocking appears to result in redistribution of shear stresses by allowing force transfer to occur through the pins. In this way, the interlocking mechanism allows the vault to remain stable at a deformed state that would likely be unstable if the vault were assembled from components without mechanical interlocking.

The results from dynamic time-history analysis indicate that partial collapse of the outermost parts of the vault is possible when the assumed 475-year moonquake ground motion is applied in the direction of the vault span. The partial collapse observed in the end of the vault can be attributed to the lack of confinement in this region, which results in low frictional forces at the interface. Since the friction angle is lower than the angle of inclination, the stability of blocks in the outermost course is primarily defined by equilibrium considerations resulting from the vault geometry, in addition to the influence of mechanical interlocking. Interlocking, however, is only expected to influence the response once kinematic equilibrium is overcome. The low shear strength at the interface allows the blocks to displace with limited frictional resistance once kinematic equilibrium is overcome, though the outer components will be held in place to some degree by mechanical interlocking. This may lead to growing displacements with various cycles of load reversal, particularly if the wedge components separate from the vault. Eventually, component pins may become dislodged from their respective grooves, which will cause the course to collapse. The confining pressure of outer courses helps maintain the integrity of interior courses, but once the outermost course collapses, the blocks behind it becomes unconfined and are therefore susceptible to the same instability. Though concerning, it is important to acknowledge that less than 4% of the vault collapsed and that the damage was localized to the outermost courses. This indicates that by strengthening the outer parts of the vault, significant improvements in performance may be possible. Furthermore, the vault was shown to be stable after the moonquake simulation—aside from the partial collapse—so this degree of damage may be acceptable for ultimate limit state design.

Additionally, since the dynamic analysis was undamped, the results are likely to be conservative. Given that the presence of damping would help dissipate energy, it is possible that the local instabilities leading to partial collapse may not manifest in a damped analysis. The influence of out-of-plane motion in seismic loading was not studied in either the quasi-static analysis or the dynamic analysis, but is expected to impact the real dynamic performance of the vault. Additional uncertainties exist due to the use of a truncated ground motion because it does not capture the effect of the rise and decay of the signal, though this is not expected to have a very significant impact on the results. Despite these uncertainties, important conclusions can be drawn from the preliminary dynamic analysis results.

5 Conclusions

Novel scientific contributions resulted from this research, which address the gap in existing literature regarding the structural behavior of a dry-stone variant of the pitched-brick vault typology. Previous research into this type of structure indicated that it would likely be stable under its own weight based on limit design principles [5], but its behavior under seismic loading was not studied.

Based on the results, the vault may be able to safely resist moonquake loading, but further analysis is needed to account for various unconservative assumptions. For example, material nonlinearity must be accounted for to ensure that possible cracking of the components does not result in significantly diminished performance. Additionally, comprehensive dynamic analyses with ground motions in various directions are needed to more fully characterize the dynamic response of the vault. To address the limitations of the current analysis, further damped analysis is needed as well. If the additional analyses described above demonstrate significant potential for collapse, external reinforcement may be needed.

References

1. Gonen, S., Pulatsu, B., Erdogmus, E., Karaesmen, E., Karaesmen, E.: Quasi-static nonlinear seismic assessment of a fourth century A.D. Roman Aqueduct in Istanbul, Turkey. Heritage **4**(1), 401–421 (2021). https://doi.org/10.3390/heritage4010025
2. Indyk, S.J., Benaroya, H.: A structural assessment of unrefined sintered lunar regolith simulant. Acta Astronaut. **140**(September), 517–536 (2017). https://doi.org/10.1016/j.actaastro.2017.09.018
3. Itasca Consulting Group Inc.: 3DEC- Three-Dimensional Distinct Element Code, Ver. 7.0 (2019)
4. Konstantatou, M., Dall'igna, M., Wilkinson, S., Gallou, I., Piker, D.: Learning lessons from earth and space towards sustainable multi-planetary design. Spool **8**(2), 39–54 (2021)
5. Konstantatou, M., Navarro Perez, S.C., Piker, D., Dall'Igna, M., Gallou, I.: Off-Earth infrastructure assembly: a conceptual method for scaffoldless and mortarless component-based structures in static equilibrium. Int. J. Space Struct. **37**(3), 196–210 (2022). https://doi.org/10.1177/09560599221120032
6. Lemos, J.V.: Discrete element modeling of masonry structures. Int. J. Archit. Herit. **1**(2) (2007). https://doi.org/10.1080/15583050601176868
7. Lemos, J.V.: Discrete element modeling of the seismic behavior of masonry construction. Buildings **9**(2) (2019). https://doi.org/10.3390/buildings9020043
8. Manos, P.: Seismic and stability analysis of component-based extraterrestrial vaults. M.Sc. thesis, Delft University of Technology (2023)
9. Marino, M., Neri, F., Maria, A.D.E., Borri, A.: Experimental data of friction coefficients for some types of masonry and its correlation with an index of quality masonry (IQM). In: Second European Conference on Earthquake Engineering and Seismology, pp. 1–12 (2014)
10. Nakamura, Y., Latham, G.V., Dorman, H.J.: Apollo lunar seismic experiment - final summary. J. Geophys. Res. **87**(Supplement) (1982). https://doi.org/10.1029/jb087is01p0a117
11. Ochsendorf, J., Block, P.: Designing unreinforced masonry. In: Allen, E., Zalewski, W. (eds.) Form and Forces: Designing Efficient, Expressive Structures. Wiley, Hoboken (2009)
12. Piker, D.: Kangaroo: form finding with computational physics. Archit. Des. **83**(2), 136–137 (2013). https://doi.org/10.1002/ad.1569
13. Ruiz, S., Cruz, A., Gomez, D., Dyke, S.J., Ramirez, J.: Preliminary approach to assess the seismic hazard on a lunar site. Icarus **383**, 115056 (2022)
14. Sanders, G.B., Larson, W.E.: Progress made in lunar in situ resource utilization under NASA's exploration technology and development program. J. Aerosp. Eng. **26**(1), 5–17 (2013). https://doi.org/10.1061/(asce)as.1943-5525.0000208

Numerical Analysis of Additively Manufactured Polymeric Lattice Structures Under Tensile Load

Danilo Bruson(✉), Marco Viccica, Giovanni Rizza, Paolo Antonioni, Luca Iuliano, and Manuela Galati

Department of Management and Production Engineering, Politecnico di Torino, Torino, Italy
danilo.bruson@polito.it

Abstract. Owing to the challenging definition of an appropriate specimen design, the mechanical properties of lattice structures under tensile loads have rarely been investigated. Finite element analysis (FEA) may be used to analyze the deformation mechanisms and optimize the specimen design. In this work, a FEA is carried out on an ad-hoc designed specimen so that the fracture of the lattice specimen under tensile load is localized within the gauge length. The specimen is designed according to EN ISO 527 standard, in which the gauge length is replaced with a strut-based lattice structure. In addition, the lattice density is graded to control the stress distribution. The numerical analysis is experimentally validated using polymeric specimens produced by additive manufacturing (AM). The numerical result shows a deviation due to the geometrical specimen deviation produced by the AM process that, if compensated, enhanced the model's capability to forecast the lattice structure's mechanical behavior.

Keywords: Lattice structures · Tensile · FEA · PBF-LB/P · PA2200

1 Introduction

Reticular structures, also called lattices, are geometries well representative of the design freedom allowed by innovative processes such as additive manufacturing (AM). However, owing to such complexity, the mechanical behaviour is challenging to evaluate, especially with standard specimens. Because of that, lattice structures have mainly been studied under compression loads, while the resistance under tension is rarely addressed [1]. The main concern is due to the discontinuities in the stress distribution in the load introduction region [1–3] that cause a fracture out of the gauge length or at the interface between the lattice structure and the bulk ends [2]. When the fracture occurs with such mechanisms, the tensile properties are impossible to evaluate. As demonstrated by Drücker et al. [1], a smooth transition from the lattice to the bulk ends by increasing the strut diameter can distribute better the stress in the sample during the loading. This expedient realizes a density gradation that facilitates the control of fracture localization. However, the design of the density gradation region is difficult because it should

S. Gabriele et al. (Eds.): IWSS 2023, LNCE 437, pp. 532–539, 2024.
https://doi.org/10.1007/978-3-031-44328-2_55

consider the gradation pattern and its extension. The forecast of such fracture is difficult to assess during a tensile experiment. Simulation can be a helpful tool to predict fracture localization and analyze stress concentration. This information may be used to modify the specimen geometry and avoid stress concentration out of the gauge length. As regards the experimental mechanical characterization research on the mechanical simulation of polymeric lattice structures made by laser powder bed fusion, the literature focused mainly on compressive loads [4–7]. Porter et al. [8] also investigated the quasi-static tensile mechanical behavior of body-centered cubic and stochastic Voronoi lattices using experimental and FEA techniques. The resulting experimental and simulated force-displacement predictions for the structures differed up to 97%, highlighting the impact of in-process manufacturing dimensional errors on numerical precision, in agreement with Refs. [4, 8].

This work proposes a non-linear finite element analysis (FEA) to evaluate the effect of density gradation of the lattice cell on the load introduction and corresponding stress. The geometry under consideration adapts the ISO 527 standard for plastics. The gauge length is replaced with the lattice structure under consideration. Out of this region, an adaptation of the method proposed by Drücker et al. [1] is proposed. The specimens were produced in a polymeric laser powder bed fusion system and then tested under tensile loads To validate the simulations. X-ray computed tomography has been performed to investigate the geometric influence on the numerical simulations.

2 Material and Methods

The EN ISO 527-2:2019 [10] standard tensile specimen for plastics (Fig. 1) was chosen as a guide for designing the testing standard.

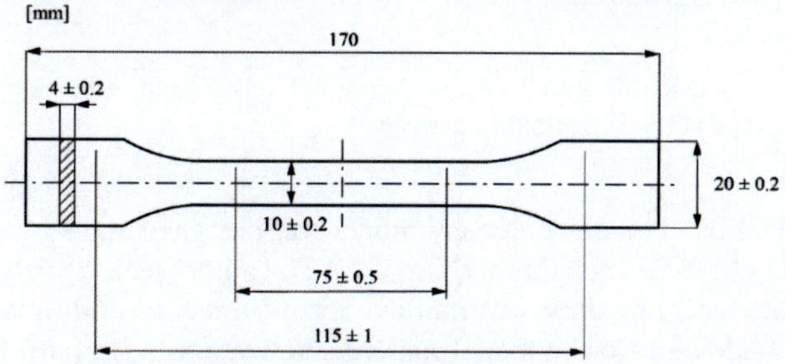

Fig. 1. EN ISO 527–2:2019: specimen type 1A [9]

The lattice fitting in the standard specimen geometry consisted of replacing the gauge length with the unit cells at constant relative density and achieving a smooth transition by the gradual increase of the strut diameter in the region between the end of the gauge length and the bulk tabs. In contrast to Drücker et al. [1], the hypothesi considers a complete lattice-bulk fusion with a strut diameter of the last step corresponding to a relative density of the graded cell of at least 90% bulk density. The gradation has been accomplished in three stages, each with a constant diameter increase of 44% (Fig. 2).

The progressively decreasing void percentage in the elementary cell led to a smooth transition from the lattice to the bulk ends. For experimental validation, a body-centered cell topology based on a cubic primitive BCC has been selected. According to the standard, a cell dimension of 5 mm and strut diameter of 1 mm (25% relative density) were chosen to achieve a width-thickness ratio of approximately 10:4. The thickness of the specimen has been designed to contain a single elementary cell. To accommodate the lattice cell, the standard specimen's width and thickness were marginally modified to 11 and 6 mm, respectively.

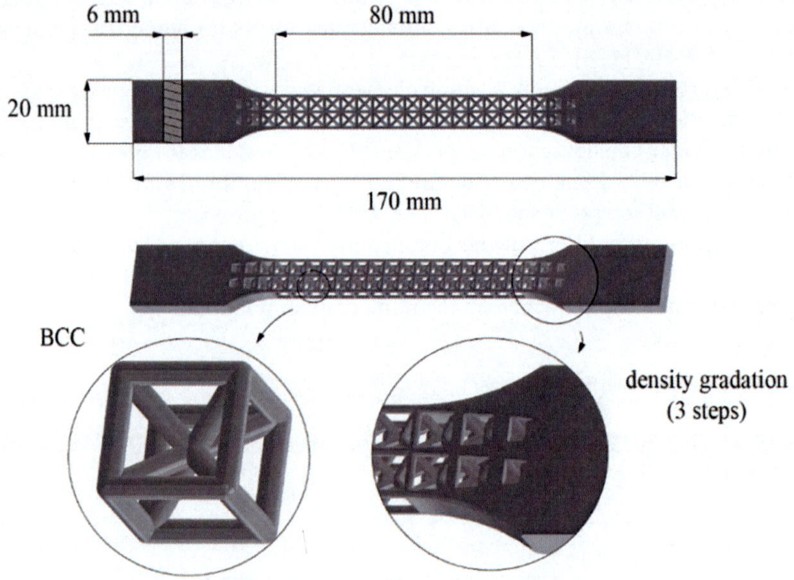

Fig. 2. Specimen dimensioning with the detailed view of the BCC unit cell topology and the three density gradation steps at the lattice-bulk interface

Before production, The finite element model was designed and solved via ANSYS Workbench 2021 (ANSYS Inc.; Canonsburg, PA). The model seeks to validate the gauge length region homogeneous stress distribution. Second order 10-node quadratic tetrahedrons have been adopted for the model finite discretization. Afterward a convergence analysis, the element size of the mesh was set to 0.20 mm, which also described the strut size. Figure 3b depicts a segment of the lattice specimen meshed cells. A joint constraint was applied to all nodes of the two surfaces of one end tab to simulate the clamp. A 3 mm longitudinal displacement was linearly applied on the opposite tab. Figure 3a is an illustration of the model. The behavior of polyamide 12 was modeled according to Porter et al. [10] using a multi-linear isotropic hardening (MISO) material model able to simulate the linear-elastic properties and the plasticity. The Young Modulus and the plasticity have been derived from the bulk tensile test. The Poisson ratio was set following Refs. [4, 10, 11].

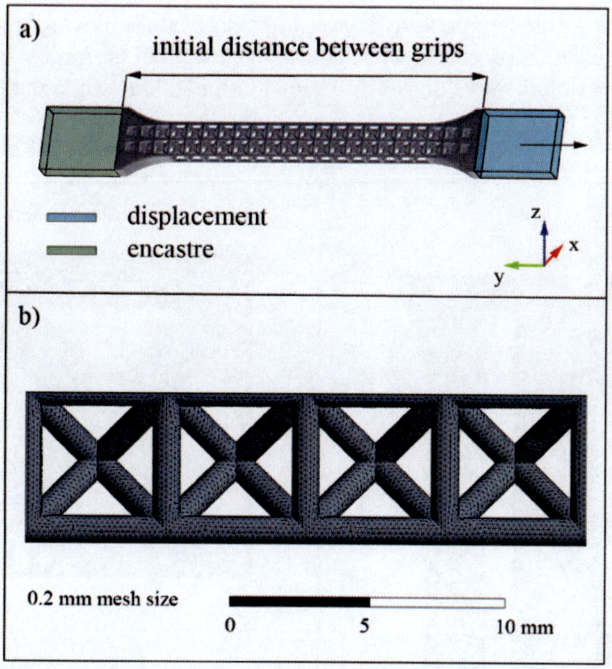

Fig. 3. a) Model used for the simulation of the tensile test. b) Segment of the lattice specimen meshed cells with a 0.2 mm mesh size.

The samples were manufactured using PA2200 powder (PA12 50% virgin, 50% recycled) processed by EOS Formiga P110 Velocis, a polymeric laser powder bed fusion (PBF-LB/P), with the standard process parameters extracted from Ref. [12]. The specimens were manufactured horizontally within the build volume, with the sample thickness perpendicular to the build orientation and the axis load along the x-axis. Five replicas have been manufactured and tested. As mentioned above, bulk specimens were also produced to calibrate the numerical model. All tensile tests were performed on an Easydur AURA 10 T equipped with a maximum load of 10 tons following EN ISO 527-1 [13] at a crosshead speed of 5 mm/min and an acquisition rate of 500 Hz until the sample failure. The dimensional and porosity analysis of the manufactured specimens were performed using a computed tomography system (CT-scan) (GE Phoenix v|tome|x s). The scans were conducted using 100 kV of voltage, 80 μA of current, and a voxel size of 16.74 μm. The 3D geometry was reconstructed Utilizing VGStudio Max 3.4.

3 Results and Discussion

The homogeneous stress distribution in the load introduction region without bulk-lattice interface perturbations has been assessed numerically. The Von Mises stress plot of the lattice specimen with a 3 mm of crosshead displacement is depicted in Fig. 4a. As recommended by the standards followed, the obtained results revealed high-stress regions (red zones) within the gauge length, with decreasing stress values moving toward the bulk ends. All failure locations occurred within the gauge length of the lattice specimens on a single cell row level, indicative of a uniform load introduction, validating the numerical analysis.

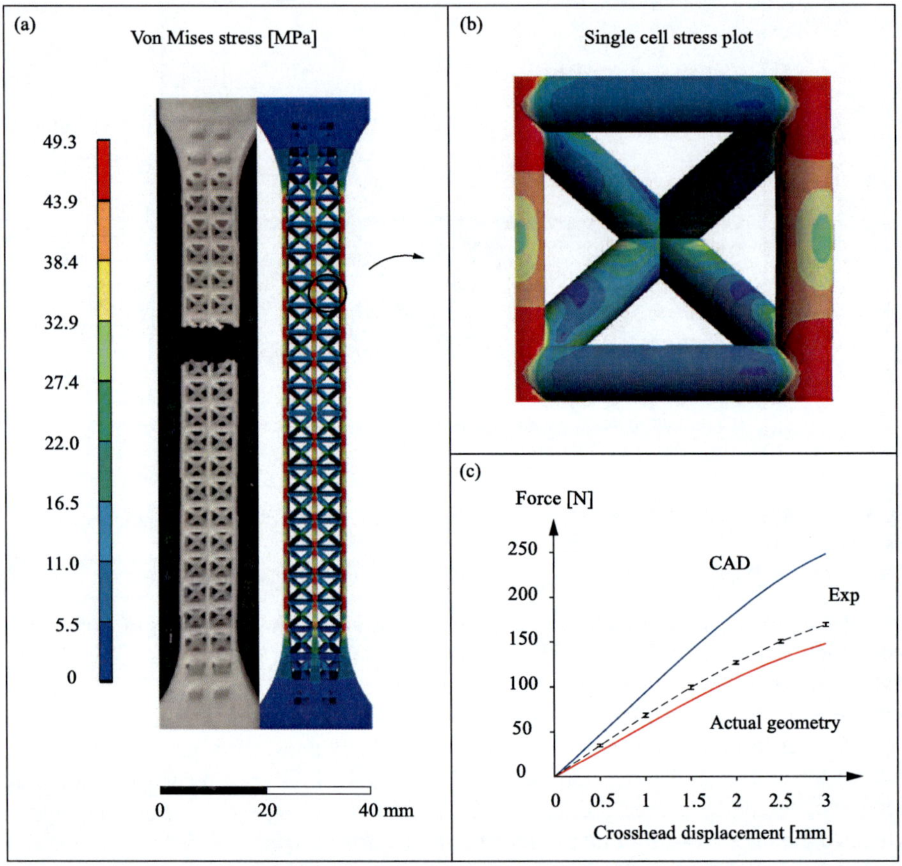

Fig. 4. a) Von Mises stress plot compared to the failed specimen after the tensile test. b) Single cell stress plot. c) Force-displacement plot comparison between numerical and experimental results.

As noticeable in the single cell stress plot magnification reported in Fig. 4b, the most stressed regions occurred in the strut oriented longitudinally to the tensile load axis, while the diagonal and perpendicular beams showed stress values mostly below half of the maximum stress magnitude. Remarkably, the reported stress plot highlighted stress concentration regions proximate to the nodes of the longitudinal beams, indicating a probable failure location at the single-cell level, which was experimentally confirmed (Fig. 4a). The load-bearing capacity of the analyzed topology is attributable primarily to the longitudinal struts and secondarily to the diagonal struts, in which the central node represents a higher stress region due to the CAD sharp re-entrant corners (cause for low stress convergence even at high mesh densities in agreement with the findings in Ref. [10]). Good agreement has been found between the forecasted stress concentration and the actual experimental failure of the structures that occurred in the longitudinal struts (Fig. 4a), validating the numerical model.

Comparing the forecasted force-displacement curves with the measured values, the numerical model behaves stiffer than the experimental counterpart (Fig. 4c). Considering a crosshead displacement of 3 mm, the predicted force was approximately 47% higher than the experimentally measured reaction load. This result could be attributed to some geometrical or material discrepancies between the experimental and numerical models [4, 8, 10]. In this regard, in accordance with the findings of Pavan et al. [14], the CT-scan measurements revealed that the struts oriented horizontally during the printing process resulted in an ovalized cross-section deviating from the circular one of the CAD model with a nominal strut diameter of 1 mm as reported in Table 1. The struts oriented according to the build direction and the diagonal struts exhibited circular cross-sections. For these orientations, the measured diameters were about 25% smaller than the nominal CAD. To account for this effect, the geometry of the numerical model has been modified based on the actual dimensions of the CT-scan reconstructed geometry listed in Table 1.

Table 1. BCC manufactured sample strut's measurements and standard deviations (SD).

BCC xy actual strut dimensions (SD) [mm]		
x	Major axis a	1.08 (0.03)
	Minor axis b	0.60 (0.04)
y	Major axis a	1.13 (0.05)
	Minor axis b	0.64 (0.02)
z	Diameter	0.76 (0.04)
diag	Diameter	0.73 (0.02)

The modified geometry reduced the deviation. Compared to previous results, the numerical-experimental force deviation at the maximum crosshead displacement decreased by a great extend and was about 12.5% (red line in Fig. 4c). However, the deviation is negative using the modified geometry, indicating that the new model is weaker than the experimental observation. At the base of this difference might be the material properties extracted from the bulk specimen, which could be microstructurally different from the material composing the thin struts. This finding is confirmed by the CT-scan analysis, which revealed differences between the bulk and lattice specimen in terms of porosity percentage content. In the gauge length, the lattice specimens showed a higher density with respect to the bulky counterpart (97.8% vs. 96.6%). Therefore, the total porosity is lower than the corresponding bulk specimen. According to Ref. [14], the amount of porosity in parts processed by L-PBF/P is correlated to the ratio between the inner area (hatching) and the contour of each component section. The higher the ratio is, the higher the porosity content. In the case of thin lattice struts, this ratio is relatively small because each layer is made of several small areas, including a perimeter and a relatively negligible inner area. In the bulk samples, the internal regions represent most of the area to be melted.

The obtained result suggests that small features inherent in lattice topologies display different mechanical responses compared to bulk samples [14], requiring a more detailed material model capable of accounting for local mechanical differences.

4 Conclusions

This study proposed a non-linear FEA to evaluate the mechanical response and the homogeneity of the stress distribution in the load introduction region of a novel specimen geometry based on the ISO 527 standard for the tensile characterization of lattice structures. The specimens were manufactured via PBF-LB/P and then tested under tensile loads. The obtained stress plot from the numerical analysis located the higher stress regions within the gauge length, assessing the efficacy of the performed density gradation. The FEA forecasted the failure locations within the gauge length of the lattice specimens on a single cell row level in agreement with the experimental failures, validating the numerical analysis. Remarkably, at a single cell level, the predicted most stressed regions are located in the struts oriented longitudinally to the load axis where the initial failure was observed experimentally. A discrepancy between the nominal CAD geometry and the manufactured geometry was observed via X-ray computed tomography. This finding explained the numerical reaction load, which was overestimated by 47% at a 3 mm crosshead. The inferior experimental mechanical responses compared to the results of the numerical analysis performed on the CAD nominal geometry can be explained by the ovalization of the xy struts and the smaller diagonal and z-oriented strut diameters. A CAD compensation obtained from the process-related dimensional deviations significantly improved the predictive capability of the numerical analysis. The numerical results deviated marginally from their experimental counterparts. The residual discrepancy may be attributed to the fact that the bulk-derived material model is suboptimal for representing the thin features of lattice structures, since the struts exhibited higher densities than the bulk material. Detailed experimental camping should be performed on single struts to compensate for this error and determine the correlation between the processed material and structure and the corresponding mechanical behavior.

References

1. Drücker, S., Inman, S., Fiedler, B.: Simulation and optimization of the load introduction geometry of additively manufactured lattice structure specimens. In: ECCM 2018 - 18th European Conference on Composite Materials (2020)
2. Alsalla, H., Hao, L., Smith, C.: Fracture toughness and tensile strength of 316L stainless steel cellular lattice structures manufactured using the selective laser melting technique. Mater. Sci. Eng. A (2016). https://doi.org/10.1016/j.msea.2016.05.075
3. Liu, W., Li, Y., Liu, B., Wang, G.: Development of a novel rectangular–circular grid filling pattern of fused deposition modeling in cellular lattice structures. Int. J. Adv. Manufact. Technol. (2020). https://doi.org/10.1007/s00170-020-05461-0

4. Neff, C., Hopkinson, N., Crane, N.B.:. Selective laser sintering of diamond lattice structures: Experimental results and FEA model comparison. In: Proceedings - 26th Annual International Solid Freeform Fabrication Symposium - An Additive Manufacturing Conference, SFF 2015 (2020)
5. Gümüs, S., et al.: Failure behavior of PA12 based SLS lattice structure with macro-porosity. In: MATEC Web of Conferences (2018). https://doi.org/10.1051/matecconf/201818803007
6. Prithvirajan, R., Balakumar, C., Arumaikkannu, G.: Effect of strut diameter on compressive behaviour of selective laser sintered polyamide rhombic dodecahedron lattice. In: Materials Today: Proceedings (2019). https://doi.org/10.1016/j.matpr.2020.09.684
7. Bai, L., et al.: Mechanical properties and energy absorption capabilities of functionally graded lattice structures: experiments and simulations. Int. J. Mech. Sci. (2020). https://doi.org/10.1016/j.ijmecsci.2020.105735
8. Viccica, M., Galati, M., Calignano, F., Iuliano, L.: Design, additive manufacturing, and characterisation of a three-dimensional cross-based fractal structure for shock absorption. Thin-Walled Struct. **181**, 110106 (2022). https://doi.org/10.1016/J.TWS.2022.110106
9. BS-EN-ISO-527-2: BS EN ISO 527-2 Plastics — Determination of tensile properties ,Part 2: Test conditions for moulding and extrusion plastics. Part, 1 (2012)
10. Porter, D.A., Di Prima, M.A., Badhe, Y., Parikh, A.R.: Nylon lattice design parameter effects on additively manufactured structural performance. J. Mech. Behav. Biomed. Mater. (2022). https://doi.org/10.1016/j.jmbbm.2021.104869
11. Stoia, D.I., Linul, E., Marsavina, L.: Influence of manufacturing parameters on mechanical properties of porous materials by selective laser sintering. Materials. https://doi.org/10.3390/ma12060871
12. Calignano, F., Giuffrida, F., Galati, M.: Effect of the build orientation on the mechanical performance of polymeric parts produced by multi jet fusion and selective laser sintering. J. Manufact. Processes (2021). https://doi.org/10.1016/j.jmapro.2021.03.018
13. British Standards Institute: ISO 527-–1 Plastics—Determination of tensile properties—Part 1: General principles. ISO standard
14. Pavan, M., Craeghs, T., Verhelst, R., Ducatteeuw, O., Kruth, J.P., Dewulf, W.: CT-based quality control of Laser Sintering of Polymers. Case Stud. Nondestruct. Test. Eval. **6** (2016). https://doi.org/10.1016/j.csndt.2016.04.004

Mixed Reality Holographic Models for the Interactive and Synergetic Exploration of Space Structures in Architectural Design and Education

Juan José Castellón González[1](✉), Anukriti Mishra[2], Guangyu Xu[1], Jose Daniel Velazco Garcia[3], and Nikolaos V. Tsekos[2]

[1] Rice University, Houston, TX 77005, USA
jcastellon@rice.edu
[2] University of Houston, Houston, TX 77004, USA
[3] Tietronix Software Inc, Houston, TX 77058, USA

Abstract. The purpose of this paper is to contextualize and introduce the academic potential of Augmented Reality (AR) models based on computer-generated holograms for the experimental exploration of spatial structures in architectural (interactive and immersive) design and education by multiple users. This work is driven by the potential impact of concurrently sharing and interacting with mixed reality (MR) scenes via holographic interfaces. With an interfaceable, scalable, and modular computational architecture, this platform is a powerful resource for multiple users, multiple formats (digital and physical), and multiple physical locations to design as well as assess structural and physical properties. The underlying didactic hypothesis is that holograph and MR offer an unparallel sense of immersion while enabling rapid virtual prototyping and project-driven synergy in combination with physical models. We present results on Augmented/Mixed reality as well as Virtual reality (VR) using devices such as the HoloLens and Vive head-mounted devices (HMD) and hand-held devices (HHD) such as the iOS iPad and iPhones and Android tablets all seeing the same time the same architectural holographic scene and interacting with it.

Keywords: Augmented Reality · Structural Design · Architecture · Structural Models · Experiential Learning

1 Introduction

The emergence of digital technologies has expanded the capacity to produce multiple models in different formats, both digital and analog. While physical models have traditionally helped students, engineers, and architects to understand static principles and material properties through hands-on experiments and through the implementation of empirical learning methods, digital models allow to explore multiple variations and simulate different loading conditions in a relatively short period of time. However, both models present different advantages and limitations.

S. Gabriele et al. (Eds.): IWSS 2023, LNCE 437, pp. 540–548, 2024.
https://doi.org/10.1007/978-3-031-44328-2_56

On the one hand, physical models are fully 3-dimensional objects that react to gravitational forces and applied loads in direct response to the proposed geometry, the specific materials, and assembly logic, as well as the imposed support conditions. Besides, the active interaction with the physical construct enhances an intuitive learning experience based on qualitative aspects following a trial-and-error methodology. However, this approach leads most of the time to a single solution in which the quantitative aspects based on the analysis and identification of the internal forces, the consequent evaluation of their dimensions, and the precise control of the relative interaction between the elements conforming to the spatial structure are missing. Also, a limited number of models can be constructed and evaluated secondary to time and budgetary constraints. From the educational perspective, students or professionals need to be physically present at the same location and time.

On the other hand, digital models in their more conventional forms are 2-dimensional representations (i.e., 2D projections on 2D displays or screens) of 3-dimensional objects in which applied loads, geometries, and material properties can be precisely defined, modeled, controlled, and quantified. Furthermore, the use of parametric software tools allows the generation of multiple solutions and complex formal configurations, as well as the possibility to simulate and analyze structural behavior to dimension the different parts of the structure accordingly. Nevertheless, the graphic representation of the structural behavior and the interaction with the model is constrained to the digital realm of 2D projections. Hence, the learning process relies on abstract hypothetical structural configurations that ultimately require verification through the construction of a physical model.

As an alternative to these two extremely different experimental approaches, Augmented Reality (AR) tools open a third path in the exploration of hybrid structural models that combine the advantages of physical and digital models. Moreover, this hybrid approach takes advantage of the experimental potential offered by the fruitful interaction between both formats and facilitates an interactive pedagogic model based on teamwork and collaborative learning processes. Furthermore, MR technologies work with perception by adding layers of virtual objects to the real environment, and this intended transformation of human perception through the senses is a key factor in architectural design and education. In fact, this approach is grounded on the tradition of some of the most influential avant-garde schools of architecture that emerged in the 20th Century such as the BAUHAUS in Germany or the WKhUTEMAS in the former Soviet Union. In both cases, courses and methodologies were strongly related to questions of perception and model-making.

One of the most relevant examples of this experimental approach was the Laboratory "Psychotechnical Laboratory" at Vkhutemas founded by Nikolay Ladovsky. Ladovsky was a Russian avant-garde architect and educator, leader of the rationalist movement in 1920s architecture, an approach emphasizing the human perception of space and form. According to his philosophy, architecture students were selected based on their visual and spatial skills and coordination. In 1927 Ladovski established his laboratory as a black room for testing spatial perception (of angles, volumes, linearity, etc.) using tools that he invented such as the *ligrazomer* (to test the perception of lines), *ploglazomer* (surfaces), *uglazomer* (angles), the *oglazomer* (volumes) or the most intriguing and sophisticated *prostrometr*, also called the space-meter [1]. The space-meter was conceived not only to experiment with mechanisms of perception but also to respond to the question of architectural talent as innate or acquired. For Ladovsky, "space, not stone, is the material of architecture". Nevertheless, although the pedagogic and design methods developed and applied in the Bauhaus and the VKhutemas were based on the integration of perception through the construction of physical tools and models in combination with other forms of representation, these methods were focused on compositional, aesthetic, psychological, or formal aspects neglecting analytical and quantitative questions related to structural behavior and the relationship between space, materials, and forces. Additionally, and from a more practical perspective, those pedagogical methods required students, educators, and professionals to be physically present. In this regard, Mixed Reality Holographic Models emerge as a powerful tool to develop innovative teaching and design methods combining the potential of physical and digital models and focusing on the experimental approach to these hybrid models to understand and visualize the implicit force-flow in a structure while interacting with physical models. This allows a fruitful interaction between the digital and physical realms enhancing and multiplying the effectiveness of the experiential learning process. Besides, this iterative process based on the production of mixed reality holographic models, can lead to new forms of collaborative work for the design and analysis of spatial structures in architecture, including the immersion into the three-dimensional space, as well as the interaction and immediate perception of the multiple outcomes; first the aesthetic and qualitative aspects of the space and structure and then the actual perception of the acting forces and their quantitative and parametric evaluation over time. Based on this theoretical framework, we present a computational platform that enables multiple users and multiple locations to access the same computer-generated holographic scene of spatial structures, interact with them, and share this experimental and collective learning experience.

2 System Design

In this first prototype, the primary design directive was to enable true real-time collaboration between users, i.e., as the example shown in Fig. 1(a), a user can alter the scene, and these changes are shared by the other users. The success of such bidirectional communication requires mimicking the actions that the user takes while collaborating in a real physical environment and communicating the same actions to all the other collaborating users. Moreover, we considered certain practical challenges. First, different users may have access to different devices; some to high-end head-mounted displays, others to tablets, or just a smartphone or laptop. And they run with different operating systems (iOS, Android, Windows, etc.). To accommodate that, we decided to develop a system that is independent of OS or platform. In developing the afore-discussed tool, another design directive was the ability to run appropriate software to model the physical world forces and interactions and dynamics. We also wanted practitioners to be able to modify the computation algorithms or add new modeling modules. This created yet another challenge: such high-quality software cannot run on a tablet or an HMD.

Aiming to develop a useful tool for practicing professionals, academicians as well as students and address those design directives and challenges, we decided to implement a computational net that linked all different involved entities based on the open-source framework by Velazco et al. [2]. This commutation fabric offered major flexibility. (A) All demanding computations were performed on a server (a PC with high computational power and memory). This is illustrated in Fig. 1(b). The devices were only used to perform renderings and as visualization interfaces. (B) Individual users were linked to the server rather than to each other, facilitating a highly efficient protocol. In addition, cybersecurity can be easily incorporated. (C) This framework allows for easier scaling between multiple platforms. Users can collaborate using immersive or non-immersive devices such as a desktop/laptop or augmented reality (AR) devices like smartphones, tablets, and head-mounted display (HMD) devices.

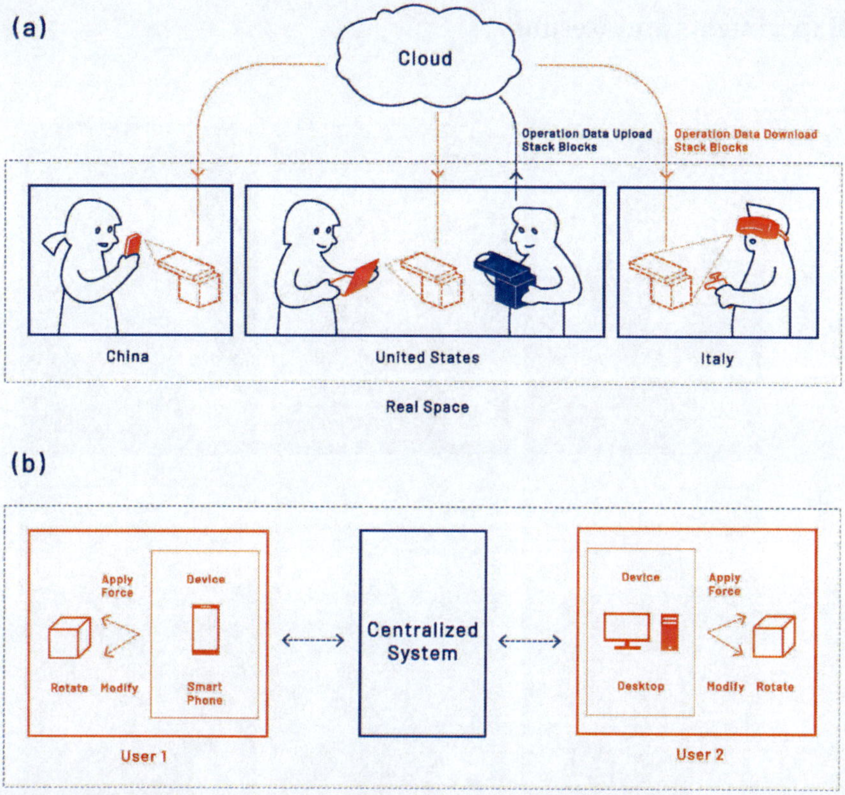

Fig. 1. (a) Example multi-user collaboration depicting the instructor stacking two blocks while multiple users at different locations using different devices (HMDs, tablets, or smartphones) experience on the fly those changes in the respective MR environment. Physical changes are tracked using stereo-cameras and appropriate spatial registration algorithms. (b) Illustration of the deployed framework set-up customized for Architectural immersion, underscoring the central control cyber-entity (server) and the two-way communication to the users.

The presented prototype was developed on Unity 3D engine [3] version 2021.3.1f1 in conjunction with Visual Studio 2019 for both AR and non-AR devices. Unity uses C# as its programming language and offers various built-in packages to implement AR/VR/MR functionalities across multiple devices. The ARFoundation package [4] facilitates the use of the device's augmented reality features by wrapping the APIs of ARKit [5] (AR library for iOS provided by Apple) and ARCore [6] (AR library for Android provided by Google). The server was implemented using the Photon Engine [7]. Experimental studies were performed using iPhone (Xs, iOS version 15.5, 64GB) as an AR platform device and an Apple desktop (Mini, OS version 13.3.1, 8GB). The application is compatible with iOS and Android for AR-enabled smartphones and tablets and both Windows and MacOS for non-AR desktops and laptops. The server for this prototype was setup up using the cloud.

3 Experiments and Results

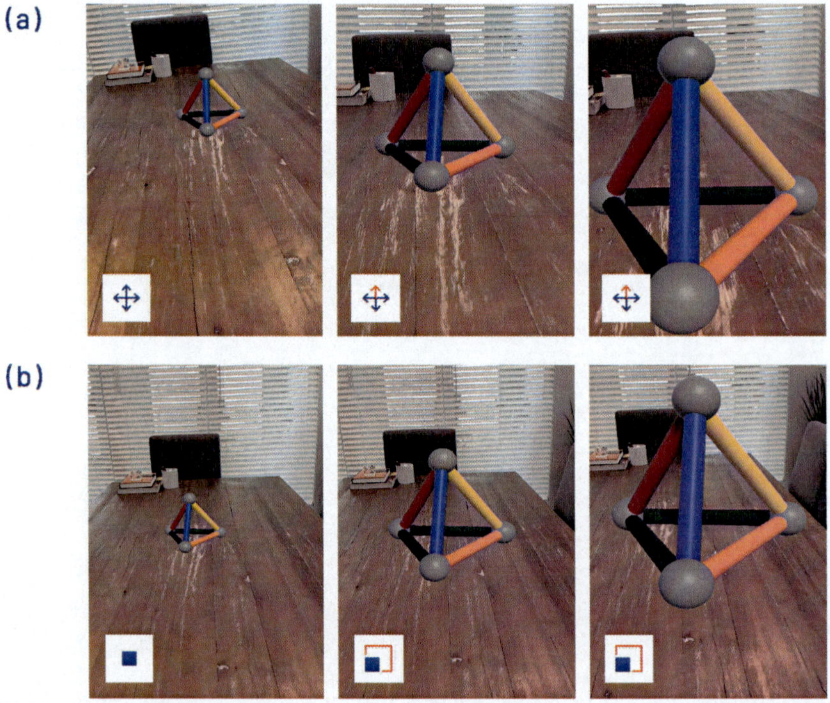

Fig. 2. iPhone screenshots of the holographic tetrahedron as the operator (a) physically moves closer to the object at an approximate distance of 1.3, 1.0, and 0.5 meters from it (respectively, from left to right) (b) enlarges the tetrahedron by setting its scale at 100%, 150%, and 200% of the original (respectively, from left to right)

Studies were performed to assess (a) the potential of computer-generated holograms to appreciate space and 3D structures, (b) multi-user collaboration, and (c) multi-platform compatibility. To test the multi-platform support, we ran the application on one AR device (iPhone XS) and one non-AR device (MacOS desktop). Figures 2 and 3 illustrate special immersion and provide an appreciation of 3D structures relative to the surrounding space (a computer-generated hologram of a tetrahedron) as seen by a user on an iPhone (these are screenshots made with the iPhone itself). As part of the immersion, by including real space, the holographic structure was placed by the user on the top of a conference table.

The three captures in Fig. 2(a) correspond to three positions as the user physically moves closer to the structure; as a reference, you may note the relative size of the chair and window shutters in the background. The panel in Fig. 2(b) illustrates one of the features incorporated in the application: the user can change the size of the holographic structure. This is performed by scaling it relative to its original position.

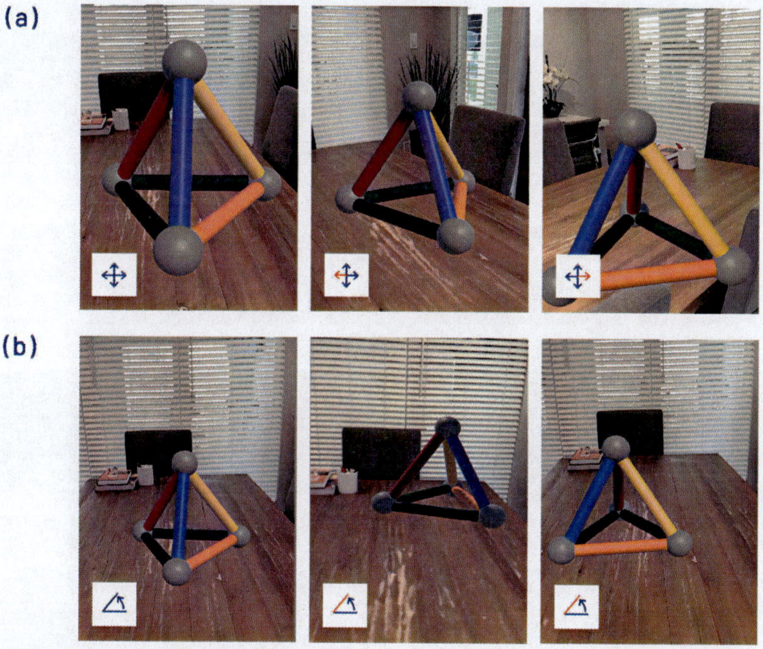

Fig. 3. Illustration of changing rotational perspectives of the tetrahedron with iPhone. (a) The user physically rotates around the fixed object while pointing the iPhone toward it. (b) The user stands at a fixed position and rotates the object.

Another illustration of the spatial immersion enabled by holographic can also be appreciated in Fig. 3(a), which shows iPhone captures as the user physically walks around the tetrahedron. Equally impactful to perception was moving and viewing the object from above or below when not placed on a table but just at any point in the air. Figure 3(b) shows rotations of the object by the user who stands at a fixed position as compared to a non-AR traditional 2D display. Holographic scenes via AR-enabled makes it easier to see the full 360-degree view of the virtual object whereas the standard display screen is not as effective.

To test the multi-user collaborative environment, we connect two users using different types of devices using the application to visualize the same space structure of a tetrahedron. User A connects using an AR-enabled device (iPhone), and User B connects using a non-AR device (desktop). The Users start with a tetrahedron composed of bars and nodes at equilibrium. We color-coded each bar to make it easier to identify the bars and nodes. Users then randomly remove one of the bars. The process is repeated until the tetrahedron collapses by its self-weight. Since both Users see the same virtual structure, information regarding each bar removal is relayed to both Users, and their virtual structure models are updated. Figure 4 shows step by step comparison of the User views after each bar removal. After each removal, we can see the status of the structure for the Users. For this experiment, the only force working on the space structure is gravity.

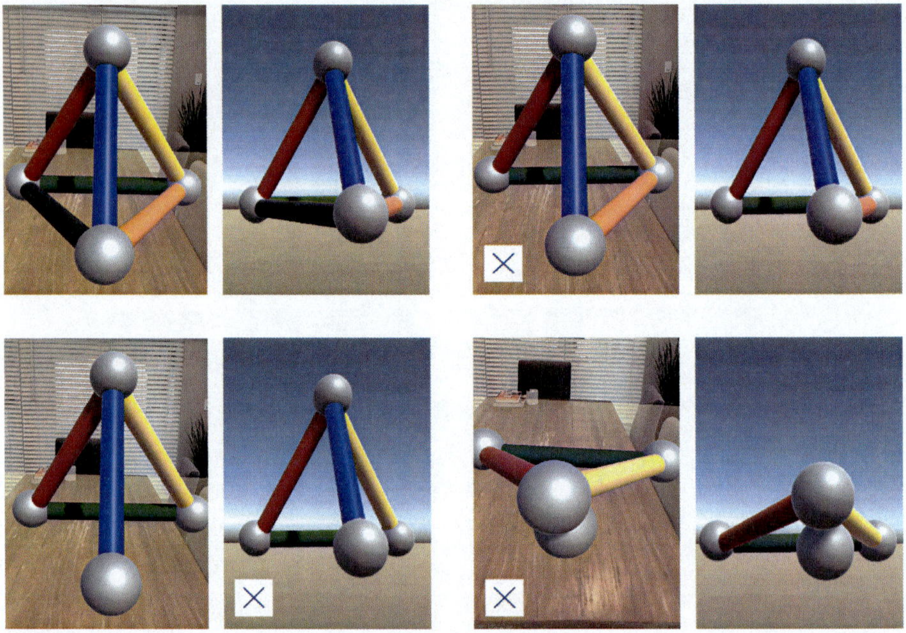

Fig. 4. Multi-User Collaboration: Corresponding view of the virtual space structure of tetrahedron after continuous bar removals until the structure falls as observed from: User A on AR-device (iPhone) and User B on Non-AR Device (Mac desktop). Sequence of removed bars: "Black" by User A, "Orange" by User B and "Blue" by User A.

It was observed that the time delay in communication between users was negligible. The time it took for the Users to remove the bar and both structures to reflect the change was instantaneous. As the communication is bi-directional, both Users took turns removing the bars, which further underscores the collaborative nature of the application.

While the prototype of the system described herein demonstrates the potential of computer-generated holography in architectural education, we have not incorporated a comprehensive range of physical properties into the simulated structure (material properties and forces other than gravity). Incorporation of such capabilities is a work in progress, and we feel that it does not limit the merit of this proof-of-concept prototype system. Notably, performing such demanding computations on servers and not on hand-held devices (tablets and smartphones) or tetherless head-mounted devices (like HoloLens) is a needed feature enabled by the used framework [2].

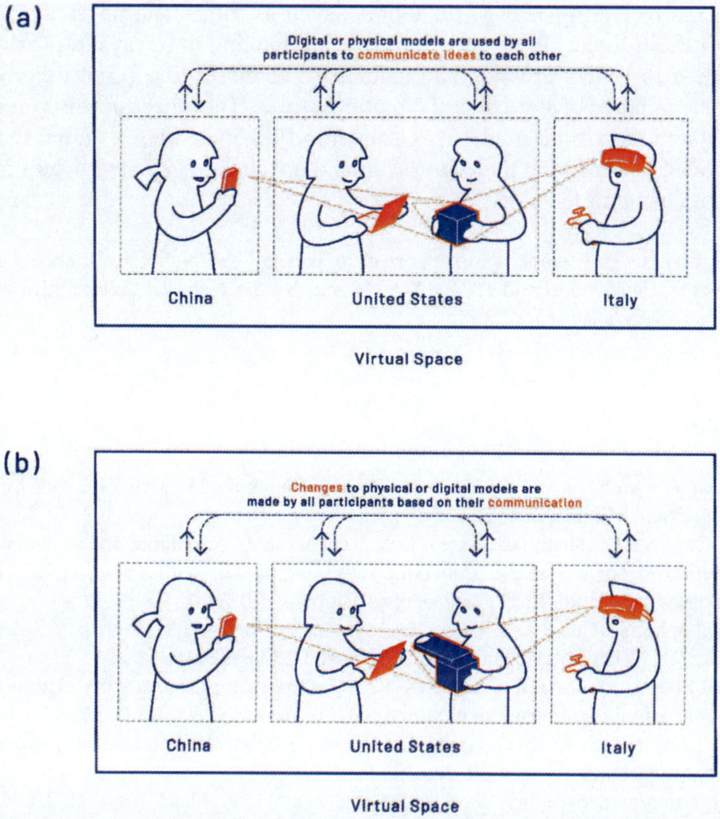

Fig. 5. Illustration of current works in fusing the physical (solid-blue object) and cyber worlds of computer-generated holograms (viewed by all users). Events in cyber and physical environments are concurrently tracked. (a) All users start with the same object, physical or computer-generated hologram. (b) A user changes the physical form of the real object (e.g., stretching force or adding new structural components). This physical change is tracked with the stereo cameras, the server performs all needed modeling and constraint calculations, and the updated data for rendering are sent to all other users updating their respective scenes on-the-fly.

4 Conclusions and Future Work

Experimental studies demonstrated the technical feasibility as well as functionalities of incorporating computer-generated holograms in architectural education and training. Although not possible to fully appreciate it on a page (in the 2D Fig. 2 and 3), of paramount value was proven the unparallel ability to appreciate 3D structures via the 3D spatial immersion experience. The modular and scalable nature of the deployed computational framework was proven effective in incorporating models to account for physical properties and laws of Physics (as shown in Fig. 4). Our team is currently pursuing methodologies towards enhancing the user experience and immersion. One thrust is in the realism of real-world actions, such as forces and physical forms, and the effect of these forces on the structure. We are looking into ways to overlay visual information, such as the direction and magnitude of forces like gravity, compression, tension, and simulation of any external directional force. The other thrust is in connecting the physical and the virtual models. As shown in Fig. 5, creating a digital twin of the physical model, and tracking them in real-time. Any change in that structure is tracked and replicated using immersion.

Acknowledgments. This work was supported in part by the National Science Foundation (NSF) awards CNS-1646566 and DGE-1746046 and by the National Aeronautics and Space Administration (NASA) award 222398.

References

1. Vronskaya, A.: Architecture of Life, Soviet Modernism & the Human Sciences. University of Minnesota Press, Minneapolis (2022)
2. Velazco-Garcia, J.D., Shah, D.J., Leiss, E.L., Tsekos, N.V.: A modular and scalable computational framework for interactive immersion into imaging data with a holographic augmented reality interface. Comput. Methods Programs Biomed. (2021)
3. Unity Technologies. Unity Real-Time Development Platform—3D, 2D VR and AR Visualizations [Internet] (2019). https://unity.com/. Accessed 23 Sept 2019
4. About AR Foundation—AR Foundation—4.2.7 [Internet]. https://docs.unity3d.com/Packages/com.unity.xr.arfoundation@4.2/manual/index.html. Accessed 30 Dec 2022
5. ARKit 6 - Augmented Reality - Apple Developer [Internet]. https://developer.apple.com/augmented-reality/arkit/. Accessed 30 Dec 2022
6. Build new augmented reality experiences that seamlessly blend the digital and physical worlds—ARCore—Google Developers. https://developers.google.com/ar
7. Photon Engine—PUN – [Internet]. https://www.photonengine.com/pun. Accessed 12 May 2022

A Geometry-Preserving Shape Optimization Tool Based on Deep Learning

Andrea Favilli[1,2](✉), Francesco Laccone[1], Paolo Cignoni[1], Luigi Malomo[1], and Daniela Giorgi[1]

[1] ISTI - CNR, via G. Moruzzi 1, Pisa, Italy
andrea.favilli@isti.cnr.it

[2] University of Pisa, Lungarno Pacinotti 43, Pisa, Italy

Abstract. In free-form architecture, computational design tools have made it easy to create geometric models. However, obtaining good structural performance is difficult and requires further steps, such as shape optimization, to enhance system efficiency and material savings. This paper provides a user interface for form-finding and shape optimization of triangular grid shells. Users can minimize structural compliance, while ensuring small changes in their original design. A graph neural network learns to update the nodal coordinates of the grid shell to reduce a loss function based on strain energy. The interface can manage complex shapes and irregular tessellations. A variety of examples prove the effectiveness of the tool.

Keywords: Design tool · Shape optimization · Graphical User Interface · Geometric learning

1 Shape Optimization of Shells and Grid Shells

In the structural design of shells and grid shells, shape optimization refers to methods and tools for maximizing the performance of an input surface. As building components, shells and grid shells are subject to unique contextual, functional, constructional or aesthetic constraints. Therefore, the relationship between shape and structural variables can differ case by case. Additionally, these structures are often shaped as free-form surfaces, which complicates the structural response.

A properly-designed shell or grid shell should possess a prevailing membrane behavior, avoid bending forces, and have good robustness [1]. In the ideal case, a 'natural' shape is sculpted by providing loading and boundary conditions to form-finding methods [2]. However, in the general case the shape is not naturally efficient and requires several design checks and improvements, which can

This work was supported by the NextGenerationEU programme under the funding schemes PNRR-PE-AI scheme (M4C2, investment 1.3, line on AI) FAIR (Future Artificial Intelligence Research).

S. Gabriele et al. (Eds.): IWSS 2023, LNCE 437, pp. 549–558, 2024.
https://doi.org/10.1007/978-3-031-44328-2_57

either rely on heuristics and experience, or adopt automated shape optimization methods. The objective functions usually involve structural quantities that are iteratively updated by moving the design variables within predefined bounds.

The literature offers a large variety of shape optimization methods, differing in the types of variables, objective functions, optimization methods etc. In [3], finite element mesh nodes are the variables updated through deformation gradient. The solution is enriched with surface regularization and distortion control. The authors of [4] emply a NURBS surface formulation, using control points and thickness as variables. The shape modifications result from finite element analysis and gradient evaluations, performed using Automatic Differentiation (AD). In [5], structural stiffness is maximized in a grid shell by iteratively updating nodal coordinates according to sensitivity information. Since the procedure is initialized with a displacement perturbation, which can lead to a jagged surface, a filtering scheme is adopted to normalize the non-smooth gradient fields. Other shape optimization methods are formulated as multi-criteria optimization. In [6], curves and surfaces can have discontinuities in tangent vectors and curvatures (creases). A multi-objective optimization problem is solved by the constraint approach to generate a trade-off design between smoothness and mechanical compliance.

In this paper, we implement a Graphical User Interface based on the method in [7], which modifies a grid shell shape using deep learning. The user can edit a set of input parameters, such as the beam characteristics and the solution speed. The graph neural network in the background is fed with geometric input features and the structural analysis of the grid shell. The variables are node coordinates. We showcase the potential of our tool with three representative examples. In two examples we perform shape optimization on free-form shells. Additionally, since shape optimization methods are often applied to solve form-finding problems [8], in a third example we solve a form-finding task on a flat surface input.

After briefly introducing the problem formulation and the learning model in Sect. 2, we describe how to reformulate the model for the tool implementation and a typical usage scenario in Sect. 3. Finally, Sect. 4 presents some example results that a user can achieve.

2 Geometric Deep Learning for Shape Optimization

Our tool is based on a graph neural network model [7] driven by the static analysis of grid shells. One of the main innovations is the introduction of a differentiable Euler-Bernoulli approach which fits usage in a learning model.

A triangular mesh $\mathcal{M} = (\mathcal{V}, \mathcal{E}, \mathcal{F})$ represents the input grid shell. $\mathcal{V}$ are the vertices (identified with the structural nodes), $\mathcal{E}$ are the edges (identified with the beams), and $\mathcal{F}$ are the mesh faces. The learning-based shape optimization acts on the vertices by imposing translations, while retaining the original mesh connectivity. In other words, the neural network model predicts an optimal translation vector $\delta_{\mathbf{v}} \in \mathbb{R}^3$ for each vertex $\mathbf{v} \in \mathcal{V}$. Fixed nodes can be constrained so that translations are null for a given subset of vertices $\overline{\mathcal{V}} \subseteq \mathcal{V}$.

The neural network model takes as input an augmented encoding of the original structure. We combine different pieces of information in an *input feature vector* $\mathbf{x_v} \in \mathbb{R}^{12}$, for each vertex $\mathbf{v} \in \mathcal{V}$. The input feature vector includes vertex coordinates, vertex normals, principal curvatures of the underlying free-form surface, and four distance and centrality measures of vertices with respect to the shell boundary $\partial\mathcal{V}$ and the fixed nodes $\overline{\mathcal{V}}$. Then, a deep sequence of layers transforms features until the final layer yields the prediction $\delta_{\mathbf{v}}$. At each layer, vertex features are updated by computing the weighted average of the features of the nearest-neighboring vertices. The network weights are determined by minimizing the *loss*, a target function accounting for structural compliance. The loss minimization follows a gradient descent method: starting from randomly initialized network weights, the weights are moved along the direction of maximum loss decrease (the opposite of the gradient) in an iterative procedure. At each step, the decrease direction vector is scaled by the *learning rate* parameter.

The learning model uses a two-node Euler-Bernoulli linear beam formulation. The loss function $\mathcal{L}(\mathcal{M})$ is defined as the mean strain energy over all edges:

$$\mathcal{L}(\mathcal{M}) := \frac{1}{|\mathcal{E}|} \sum_{e \in \mathcal{E}} E_e \tag{1}$$

with E_e the strain energy on a single edge e.

The loss in Eq. (1) is made differentiable with respect to vertex coordinates so that it can be used to determine the neural network weights in accordance with gradient descent minimization. The method relies on *automatic differentiation* (AD) to get punctual evaluations of gradients, thus avoiding the need to express analytically the partial derivatives of $\mathcal{L}$.

Our shape optimization task falls within the frame of *single-instance learning*. For each input mesh, we reset the network weights and we start a new loss minimization procedure. The task can be formalized as finding an optimal mesh $\mathcal{M}^* := T_{\boldsymbol{\theta}^*}(\mathcal{M})$ such that

$$\boldsymbol{\theta}^* \in \operatorname{argmin}_{\boldsymbol{\theta}} \mathcal{L}(T_{\boldsymbol{\theta}}(\mathcal{M})) \tag{2}$$

where the neural network is a function $T_{\boldsymbol{\theta}}$ of the weights $\boldsymbol{\theta}$.

While [7] adopts a scale-dependent stopping criterion for the iterative procedure, the graphical user interface proposed in this paper allows the user to enter the execution flow. The user can change the optimization parameters on the fly and visualize the shape modification from different perspectives. Moreover, the user can replay or restart the procedure until he/she founds a balance between performance metrics and amount of shape modification.

Finally, since a user interface needs a fast response for functional evaluation of the loss and its partial derivatives, we express the loss from Eq. (1) as a sequence of vectorized operations that support AD, without using code branching to build intermediate results. Our array-based implementation can benefit from GPU parallelization to speed up the process.

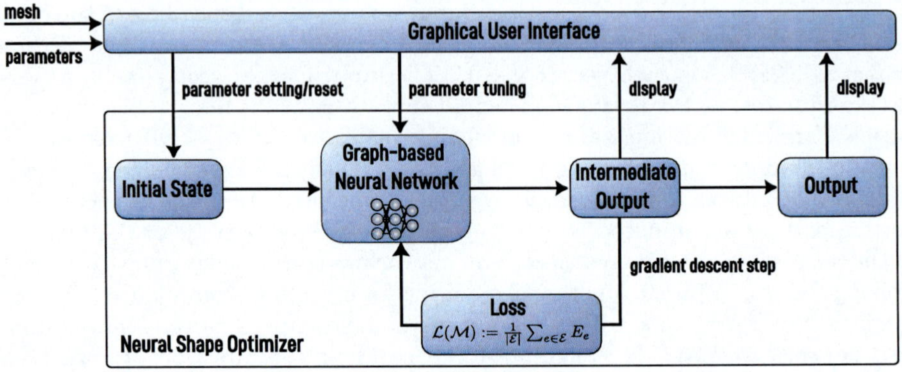

Fig. 1. The pipeline of our tool. The input is a grid shell encoded as a triangle mesh. The user can also input parameters concerning the structure (e.g., beam cross sections, Young's modulus, etc.). As the learning and optimization procedure starts, partial shape modifications are visualized in real time, so that the user can tune parameters on the fly, control the optimization flow or restart the procedure with different settings. The user can explore the outcomes until he/she reaches a good compromise between shape and performance.

3 Tool Design and User Interface

The tool is implemented in the Python programming language and is made of two main components: the Graphical User Interface and the Neural Shape Optimizer (Fig. 1). Each component runs on a different process, so computations can run asynchronously from the interface display task. This scheme implies that the user can explore the interface functionalities while the neural network modifies the shape. The Graphical User Interface detects parameter changes and updates the Neural Shape Optimizer process. Inter-process communication is implemented through two directed queues; one sends parameter changes from the Graphical User Interface to the Shape Optimizer, and the other sends back results to update the visualization.

We employed the PyTorch deep learning library [9] for the Neural Shape Optimizer. PyTorch offers an easy way to define the architecture of neural networks, with full support to Automatic Differentiation and a large variety of array-based operations. We also used PyTorch Geometric [10] to implement the graph-based network layers. To implement the Graphical User Interface we used Polyscope [11], a flexible 3D object viewer with a customizable user interface. Figure 2 shows the GUI.

3.1 Usage Scenario

To describe the usage and features of the user interface, we introduce Frank, a fictitious designer who can benefit from the tool in an early conceptual design phase. Even though Frank has a clear idea about the grid shell shape he wants

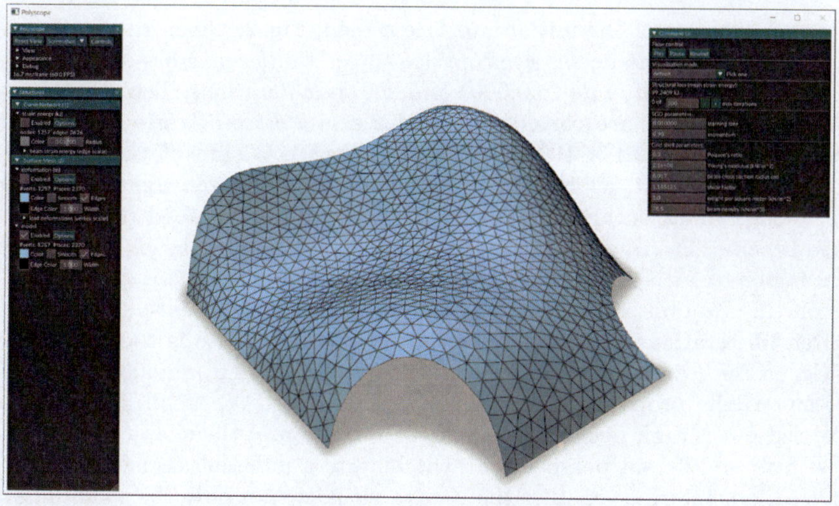

Fig. 2. Our interface at launching with an input grid shell. On the left, the user can find the default Polyscope tools (Polyscope and Structures boxes), where visualization settings (camera and shape appearance) can be modified. On the right (Command UI box), the user can control the shape optimization algorithm, check the structural performance in real time, and switch between different visualizations of metrics.

to design, he needs to produce a practically feasible solution. Thanks to our user interface, he can feed the optimization tool with a tentative shape concept, then drive the optimization towards a feasible solution from a structural point of view, while preserving his original design intent.

Frank loads the initial shell as a 3D model file and enters the optimization flow. He controls the shape optimization with a set of buttons recalling the videotape jargon: *Play*, *Pause*, *Rewind* and *Replay*. He knows almost nothing about Artificial Intelligence but was instructed about the only relevant parameter concerning the neural network he might want to tune: the learning rate. The interface proposes a default value for the learning rate, based on the dimension of the input structure. However, Frank might want to change the default parameter according to his needs, knowing that a lower learning rate would produce slow convergence of the gradient-based optimization, while a higher learning rate would lead to convergence in a smaller number of steps. On the other hand, Frank knows that the advantage of a lower learning rate is that it produces less distortion on shapes. The learning rate also depends on the initial loss, that is, on the energy of the input surface design: high energy inputs are likely to require lower rates to converge, while small energy inputs need higher rates to boost gradient descent towards a local optimum. In what follows, we describe Frank's interaction with the interface to refine his design.

Frank is in front of the initial interface screen, Fig. 2. Given the scale of the shell bounding box, he deems appropriate to set the *beam cross-section radius* field to 0.08 m, *learning rate* to 0.0001 and 200 *max iterations*. The other default *grid shell parameters* are considered valid, i.e. a *Poisson's ratio* of 0.3 and a *Young's modulus* of 2.1×10^8 kN/m^2. The Load is a sum of two terms: the structure's own weight, which is computed from the *beam density*, and an external face load, which is in the direction of gravity and is applied per node on a lumped scheme. Frank considers appropriate the default Load as a *weight per square meter* of 3 kN/m^2 and a *beam density* of 78.5 kN/m^3. At this point, he clicks on the *Play* button, and the shape optimization procedure starts.

After 50 iterations, Frank realizes that the loss decrease is too slow. Hence he clicks on the *Pause* button, increases the *learning rate* from 0.0001 to 0.0005, and resumes the procedure after clicking again *Play*. When all 200 iterations are carried out, Frank decides to check the evolution of the shape optimization process from a different perspective, switching to a different visualization.

Indeed, our interface enables the user to visualize not only the surface geometry and tessellation, but also the deformation under Service Load and the strain energy, mapped on the surface mesh in false colours (Fig. 3).

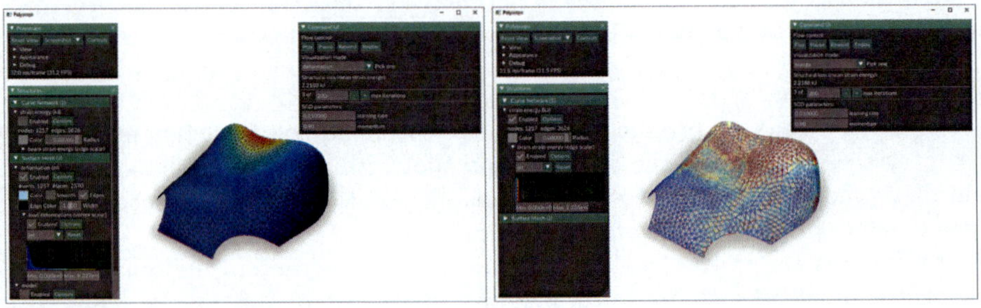

Fig. 3. Switching to different visualizations than geometry only: (Left) deformation under Service Load, (Right) strain energy visualization.

Frank selects the *energy* option from the dropdown menu, and clicks *Replay*. The *Replay* option is available only when the execution flow is paused in a state different from the initial. Finally, Frank can accept the outcome, perform some more iterations, or restart the procedure with different settings after clicking *Rewind*.

4 Results and Discussion

We analyze three different input shapes. The objective is to highlight the relation between user interaction inputs, parameter settings, and outcomes. Two examples shapes, Ice and Waving, are free-form grid shells with different structural compliance. The third example is a flat grid shell (Flat) to test the form-finding capabilities of our tool.

According to the size of the examples, we first use a learning rate of 0.01 (λ in Figs. 4, 5 and 6) and 200 gradient descent iterations (it) for shape optimization. However, the user can edit these parameters to alter the output shape. Ice and Waving show a different behavior for shape optimization in terms of design preservation. While both are complex surfaces designed sculpturally, Waving has a better initial static configuration, independently of the chosen beam cross-section. While static performance is the objective function, our tool tries to preserve the input shape. This happens with the settings mentioned above, as shown in Fig. 4(b), as well as with different input parameters set via the interface, as in Fig. 4(c), where the cross-section radius reduces to 0.04 m, the learning rate increases to 0.05 and the number of iterations to 500. For both cases, it can be seen that the loss goes down, while the overall design and surface features are preserved.

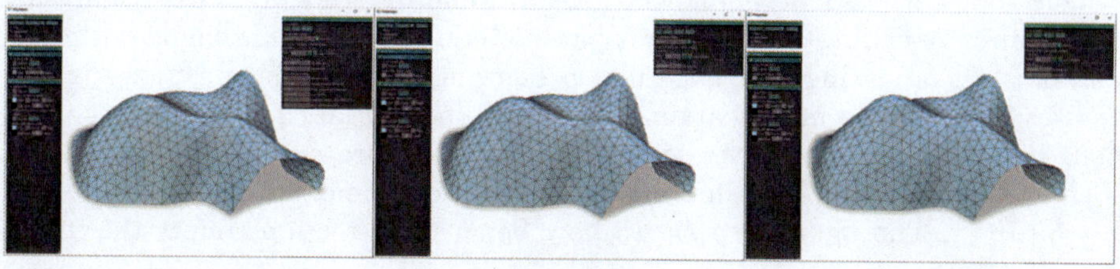

(a) | $r_{CS} = 0.08\ m$, $\lambda = 0.01$, 200 it (b) | $r_{CS} = 0.04\ m$, $\lambda = 0.05$, 500 it (c)

Fig. 4. Shape optimization on the Waving model: (a) the input structure ($\mathcal{L} = 0.0042$ kJ if $r_{CS} = 0.08$ m, $\mathcal{L} = 0.0097$ kJ if $r_{CS} = 0.04$ m); (b) the output for suitable parameters ($\mathcal{L} = 0.0038$ kJ); (c) shape modification is still slight if we reduce the cross-section radius, and we increase the learning rate and the number of iterations ($\mathcal{L} = 0.0033$ kJ).

Ice is more sensitive to the setting of the structural parameters, for the same gradient descent parameters $\lambda = 0.01$ and 200 iterations. In particular, as the size of the beam cross-section is closely involved with the edge strain energy, in Figs. 5(b) and (c), we compare the results obtained with two different solid circular cross-section radii (r_{CS}). On the one hand, a higher radius means a higher beam weight and, consequently, a higher load. On the other hand, larger beams have more bending stiffness and collect less stress due to external load. The case $r_{CS} = 0.08$ m (Fig. 5(b)) is less energized than the case $r_{CS} = 0.04$ m (Fig. 5(c)), leading to a lighter shape modification with respect to the input. The case $r_{CS} = 0.04$ m tends towards a membrane solution.

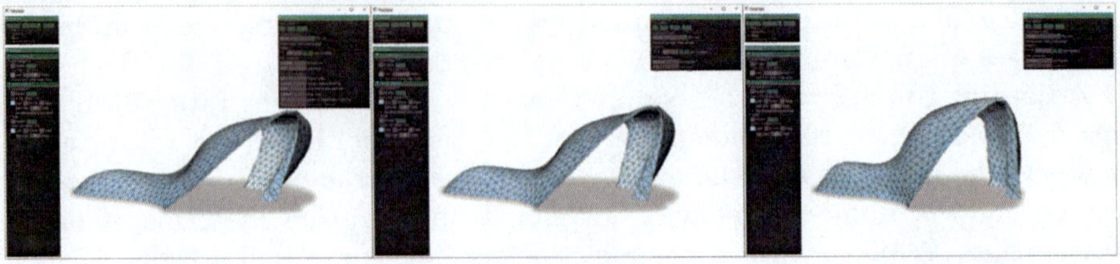

$r_{CS} = 0.08\ m$ $\lambda = 0.01$ 200 it $r_{CS} = 0.04\ m$ $\lambda = 0.01$ 200 it

(a) (b) (c)

Fig. 5. Shape optimization on the Ice model: (a) the input structure ($\mathcal{L} = 0.0466$ kJ if $r_{CS} = 0.08$ m, $\mathcal{L} = 0.3103$ kJ if $r_{CS} = 0.04$ m); (b) the output for suitable parameters ($\mathcal{L} = 0.0170$ kJ); (c) the output if the cross-section radius is reduced ($\mathcal{L} = 0.0117$ kJ).

Our tool can also be employed for form-finding. Dealing with a flat surface requires re-calibrating the input parameters. Indeed, while shape optimization aims at validating and hopefully preserving an input shape, form-finding requires heavy shape modification. The third model, Flat, is a high-energy grid shell. Hence, it requires lower learning rates to ensure convergence compared to the previous examples. Figure 6 reports the outcomes of form-finding with $\lambda = 5 \cdot 10^{-5}$. The figure also shows how beam cross-sections affect the initial

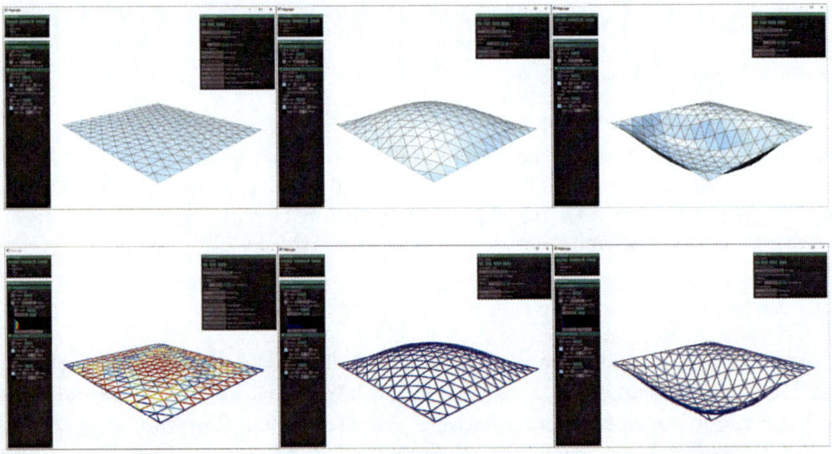

$r_{CS} = 0.1\ m$ $\lambda = 5 \cdot 10^{-5}$ 5000 it $r_{CS} = 0.08\ m$ $\lambda = 5 \cdot 10^{-5}$ 5000 it

(a) (b) (c)

Fig. 6. Form-finding starting from Flat grid shell. (a): the input structure ($\mathcal{L} = 15.7806$ kJ if $r_{CS} = 0.1$ m, $\mathcal{L} = 27.8012$ kJ if $r_{CS} = 0.08$ m); (b) and (c): two different local minima output ($\mathcal{L} = 0.0240$ kJ and $\mathcal{L} = 0.0901$ kJ respectively). The second row shows color-mapped strain energy on beams.

strain energy, and consequently the behaviour of the optimizer, which converges towards different minima: the classic funicular pillow shape in Fig. 6(b), or the symmetric membrane surface used in tension and compression in Fig. 6(c).

5 Conclusions

We developed a graphical interface tool to perform shape optimization and form finding based on deep learning. Our tool lets the user break up the linearity of a classical iterative procedure, and move the outcome back and forth as if the optimization task was a videotape playing.

This contribution is a first step towards the development of a complete AI-based software for assisted design. Indeed, architecture is for humans, and our tool puts humans at the center of the decision process. Automating the design process is not straightforward. However, we provide the designer with a tool to support the design space exploration. In real-time, different shape outcomes can be visualized after each parameter change.

This tool lays essential methodological foundations and can be expanded to include other feasibility/aesthetics metrics besides statics. A human-machine joint approach can enhance design quality, by accounting for both aesthetics and feasibility, and also benefit creativity.

References

1. Ramm, E.: Shape finding of concrete shell roofs. J. Int. Assoc. Shell Spat. Struct. **45**(1), 29–39 (2004)
2. Veenendaal, D., Block, P.: An overview and comparison of structural form finding methods for general networks. Int. J. Solids Struct. **49**(26), 3741–3753 (2012). https://doi.org/10.1016/j.ijsolstr.2012.08.008
3. Bletzinger, K.U., Firl, M., Linhard, J., Wüchner, R.: Optimal shapes of mechanically motivated surfaces. Comput. Meth. Appl. Mech. Eng. **199**(5–8), 324–333 (2010). https://doi.org/10.1016/j.cma.2008.09.009
4. Espath, L., Linn, R.V., Awruch, A.: Shape optimization of shell structures based on NURBS description using automatic differentiation. Int. J. Numer. Meth. Eng. **88**(7), 613–636 (2011). https://doi.org/10.1002/nme.3183
5. Wang, H., Chen, Z., Wen, G., Ji, G., Min Xie, Y.: A robust node-shifting method for shape optimization of irregular gridshell structures. Struct. **34**, 666–677 (2021). https://doi.org/10.1016/j.istruc.2021.08.003
6. Ohsaki, M., Ogawa, T., Tateishi, R.: Shape optimization of curves and surfaces considering fairness metrics and elastic stiffness. Struct. Multidiscip. Optim. **27**, 250–258 (2004). https://doi.org/10.1007/s00158-004-0382-3
7. Favilli, A., Giorgi, D., Laccone, F., Malomo, L., Cignoni., P.: Geometric deep learning for statics-aware 3D gridshells. Technical report, ISTI - CNR, July 2022. https://doi.org/10.32079/isti-tr-2022/016
8. Bletzinger, K.U., Ramm, E.: Form finding of shells by structural optimization. Eng. Comput. **9**, 27–35 (1993)

9. Paszke, A., et al.: PyTorch: an imperative style, high-performance deep learning library. In: Advances in Neural Information Processing Systems, vol. 32, pp. 8024–8035. Curran Associates, Inc. (2019). http://papers.neurips.cc/paper/9015-pytorch-an-imperative-style-high-performance-deep-learning-library.pdf
10. Fey, M., Lenssen, J.E.: Fast graph representation learning with PyTorch geometric (2019). arXiv arxiv:1903.02428
11. Sharp, N., et al.: Polyscope (2019). http://www.polyscope.run

New Directions of 3D Printed Earth Architecture

Nicola Parisi(✉)

Politecnico di Bari, Bari, Italy
nicola.parisi@poliba.it

Abstract. The aim of the research is to identify new implementation strategies for earth constructions using additive manufacturing techniques. The preliminary part of the study was about the analysis of traditional earth constructions in Abruzzo (Chieti province) with a reflection about coverage systems weaknesses. To provide a coverage issue solution, Nubian vault construction principles were rediscovered by the author. As a result, a massive organism completely made of earth has been designed: it is a 3D printed rural lamione and the application of it in widespread rural areas represents a museal complex itself. The study concludes with reflections about digital fabrication opportunities in the COVID and post-COVID era.

Keywords: Additive manufacturing · Earth materials · Nubian vault · Thrusting masonry

1 Introduction: Additive Manufacturing in Architecture with Earth Materials

Additive Manufacturing, or 3D printing, refers to all those processes that allow the creation of three-dimensional products from a digital model, in an additive manner, that is, by layering until the product is complete. There are several additive processes, the differences between them concern the material and the specific extrusion method. Among the best-known processes there are Stereolithography (SLA), Fused Deposition Modeling (FDM) and Selective Laser Sintering (SLS). For what concerns the large-scale additive processes, it is important to mention the Contour Crafting (CC) conceived by Dr. Behrokh Khoshnevis from the University of Southern California: it's a system characterized by the deposition of cement mixtures. The first project of a 3D printed house dates back to 1996 and it has been realized using this technology. The main characteristic of the Contour Crafting is the presence of two trowels that smooth the printed surfaces.

A new and simpler technology (that does not provide the presence of trowels) is called Liquid Deposition Modeling (LDM) and it has been created by the Italian 3D printers productor WASP (World's Advanced Saving Project) based in Massa Lombarda. The LDM Wasp Extruder is based on a pneumatic system in which a pressure generator allows the material to reach the deposition arm [1]. The LDM process is conceived to extrude viscous materials and in particular natural earth made materials. To realize the first architectural prototype ever made (Gaia), Wasp extruded an advanced mixture made

S. Gabriele et al. (Eds.): IWSS 2023, LNCE 437, pp. 559–568, 2024.
https://doi.org/10.1007/978-3-031-44328-2_58

of earth, straw and rice husk. The mixture is a composite and fibre-reinforced material, the fibers limit the shrinkage during drying phase.

The advantages of using additive manufacturing with the mentioned mixture are manifold. First of all, this technology provides freedom in design processes: it will be possible to realize a wide variety of new architectural forms. In addition, the mixture is a completely sustainable material: it produces no waste during the building disposal phase and can be reintegrate in nature, according to circular economy principles.In addition, the extruded material is very energy efficient: the selected mixture and the presence of isolation and ventilation chambers in the printed wall ensure excellent thermal transmittance values. The empty cavities can be also used for air circulation, air-conditioning and heating.

Other advantages concern the reduction of labor and material transportation costs (earth is available on-site) and the increasing of construction speed. Research is ongoing to improve the mechanical performance and the setting of the mixture extruded and to make lighter and more functional 3D printing systems. It is a technology designed to operate, as well as in industrialized countries, in world areas in which the housing emergency is very high and there are socio-economic disruptions or frequent natural disasters. In this sense, additive manufacturing with earth materials represents the suitable technology to satisfy basic human needs according to a sustainable development model [2].

Various experiments have already been made and among the research centres or architectural firms operating in this sector there are: World's Advanced Saving Project company WASP based in Massa Lombarda (pioneers of the LDM 3D printing process), the Institute of Advanced Architecture of Catalunya IAAC based in Barcelona and the architectural firm and research centre Emerging Objects based in San Francisco.

2 Aim of the Research: From Masonry Structures to Additive Manufacturing

The comparison between masonry constructive system and additive manufacturing system naturally emerged when the first large-scale manufacts were 3D printed. In his crucial article about concrete 3D printing Pegna [3] defines this technique "a new approach to masonry" and in the Contour Crafting inventor, Khoshnevis [4], mention the possibilities to print barrel vaults without supports.

As masonry is one of the first systems to be invented and additive manufacturing is the last, there are similarities between the two, particularly in terms of material behavior and layer arrangement, which lead us to study them more closely.

A masonry structure consists of discrete elements – the bricks – characterized by 3 dimensions (height h, length l and depth d). The bricks are connected to each other by mortar that ensures stability once the bricks are positioned.

A 3D printed architectural product, on the other hand, consists of continuous layers of material, characterized by two dimensions (height h and depth d), and the connection between them is provided by the extruded material itself. In general, the goal of large-scale additive manufacturing is to define a system in which robots can operate completely autonomously without human intervention, printing continuously. The continuity in the printing process would guarantee a perfect adhesion of the layers (we remind that the material extruded is viscous and subject to drying) and a reduction of discontinuities in the material (the cold joints in concrete structures).

The latest characteristic, the most relevant, that masonry structures and 3D printed structures share is the different behaviour that they have during and after the construction. Both the structures work well in compression, but poorly in tension. For this reason, often, in the masonry structures are used different coverage systems, realized with other materials [5].

The coverage systems for masonry constructions deserve a separate and more accurated discussion, but we can briefly affirm that load-bearing systems like arches or vaults would enable the construction of closing elements without the need of external supports or other materials. Anyway, when a certain clear span is exceeded, supports are anyway needed.

Historically existed a series of masonry structures built without scaffolds and they are successfully described by Cowan [6]. We cite some: corbels (historically used in the Treasury of Atreus in Mycenes, in some rural French constructions and in the Apulian trulli), the Nubian vault, the dome with inclined layers (according to Cowan, the true dome), the Catalan vault and the Persian vault.

In the past, it was necessary for several reasons, such as the unavailability of the materials needed for scaffolding, especially in desert areas or far from forests, the high cost of scaffolding (consider 30% to 60% of the total cost of concrete structures) and the technical difficulty of constructing scaffolding (just think of the construction of the dome of Santa Maria del Fiore in Florence, 45 m in diameter).

In this sense, transfering the knowledge acquired by the analysis of the existent thrusting masonry structures to the new additive manufacturing technique results very interesting. In this manner we could be able to generate possible closing solutions, since the solutions adopted so far are characterized by horizontal overlapping of the layers and provide other constructive systems for coverage. In addition, since one of the large-scale earth 3D printing intentions is to supply shelters in emergency conditions, it is really important to assume a prototype that could be built in areas in which the material needed for supports is unavailable or the transport and realization cost for them is too high.

The author is preparing to investigate the thrusting masonry systems realized without supports with the purpose to identify the constructive principles that could be applied in additive manufacturing processes. Aim of this paper is to present one of the possible applications of this method.

3 Additive Manufacturing: Evolutive Tool for Rural Architectural Heritage in Abruzzo

3.1 Rural Earth Architecture in Abruzzo

The earth architectural culture in Italy exists since ancient times and the first attestations date back to Etruscan period; the earth as construction material was massively used during Magna Grecia period. In Italy, the earth for constructions was almost used in rural areas where the peasants worked the land and knew qualities and characteristics of it. The availability of the material is the key to understanding the large presence of earth constructions in the area.

> ...l'abruzzese trova sul posto l'argilla per la sua pingiara come l'esquimese trova il ghiaccio per il suo igloo. La facilità e la semplicità intuitiva di manipolazione dell'argilla costituiscono altri motivi di sopravvivenza anacronistica della casa in terra [7].

There are other reasons for which earth constructions are so popular in rural areas in Abruzzo. One of them consists in the profound changes in agricultural sector that have taken place in the late XVIII – early XIX century. Due the significant population increase, people left the cities and moved to the countryside. The feudalism crisis had the effect to allow a new social class to have access to agricultural property. Earth construction was an obvious economic choice: the peasant draws the material directly from the ground, self-constructs his dwelling and personally provides for maintenance, there are no labour or materials costs.

In the late 1950s and early 1960s, economic growth and changing life needs led to phasing-out of traditional constructive techniques and to earth structures disuse. The existent earth architectures, in fact, were considered by the population a legacy of a poverty-stricken historical period. This degradation process caused the know-how loss and the disregard for recovery and restoration of existent earth structures [8].

Earth rural constructions are not characterized by a specific building typology but take their cues from the existent architectural heritage: the rural lamione.

There are two main types: (Fig. 1).

- The type 1, a schiera: the building is one-storeyed and the volumes are distributed along an axis.
- The type 2, italico: it consists of two levels, the level for dwelling and the level for work (commercial functions, shops...).

Both types derive from a square-plan basic block (cellula elementare) characterized by a side of 4 or 5 m [9]. The walls thickness is about 80–50 cm. This characteristic and the reduced size of windows minimize the heat dissipation, according to a conservative

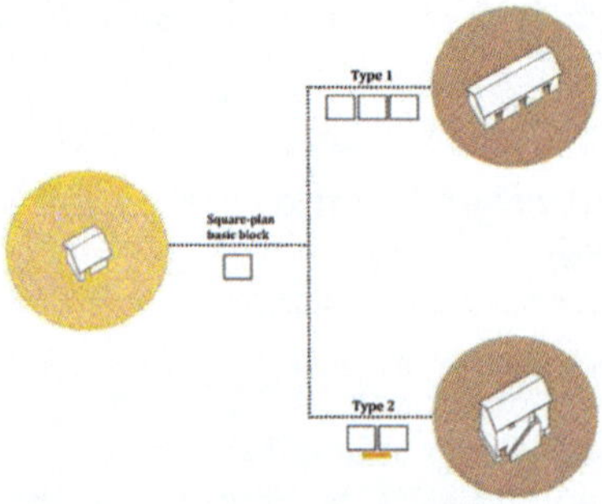

Fig. 1. Types scheme of traditional earth architectures in Abruzzo [10] © SelfMade Architecture V

energy model. The foundations were not very deep and sometimes were absent. Once the peasants had traced the construction perimeter, excavated the area to a depth of 50–100 cm, until they found a compact ground. The earth removed was placed in a pit and then they kneaded it with straw and water. Once the consistency of the mixture was good, the women produced massoni. Massoni (cob in English and bauge in French) are little clods of earth of the same dimensions. The massoni were brought to a flat area and amassed there. At this point the real construction began: the head of construction, the mastro, laid the massoni one over another all along the traced perimeter. The coverage was characterized by a timber roof truss and a roof covering made of brick tiles [10].

3.2 A 3D Printed Nubian Vault

The earthen lamione is an architectural typology that is no longer so common; it is disappearing for various reasons. First, we see the loss of the needed know-how for construction, there are no more expert workers able to manage all the construction phases and the different mixture typologies. Additionally, the earth, if combined with other materials – especially for what concerns the roof – is characterized by a poorer mechanical performance and there is no specific Italian regulation concerning earth construction. It is quite clear, in this sense, that massone principles recovery for new construction is a very anachronistic and unusual choice: it's hard this technique could really represent an environmentally friendly alternative continuously applicable in construction field. Consequently, the autor proposes the prototype development of a living module that embeds some architectural features of the traditional lamione but it's feasible with additive manufacturing.

For the prototype design has been essential using the basic constructive principles of the Nubian vault, since – as already explained – the aim is to define a completely made of earth and 3D printed architecture that incorporates thrusting masonry principles providing a coverage solution without supports.

The Nubian vault is an ancient technique that originated in Nubia, now Upper Egypt, and allows the construction of covered buildings made of clay bricks held together with mortar. The most ancient Nubian vaults are in Luxor and are still standing today, 3300 after years from their construction. Recently, in 1940, the Egyptian architect Hassan Fathy rediscovered this technique in the Nubian village of Abu al-Riche and applied it to the construction of a new village in Gourna. A traditional Nubian vault does not require the use of supports, even during construction. The bricks, measuring 24 × 12 × 4 cm, are laid diagonally, layer by layer, until they form a semicircular vault, except in the last part where they form a catenary. The mason lays each brick by hand, using an earth-based mortar (Fig. 2).

The key concept of this construction system concerns the fact that once the inclined arches are stable and the layers dimensions are constant, others inclined brick layers can be positioned without negative effect on the entire structure.

This system has been reproduced in a digital way and some tests were carried out to check the stability of it. The printing tests were conducted at the Digital Fabrication Laboratory Fablab Poliba based in Bitonto and a 3D printer WASP Delta 40100 was used. The material needed to simulate raw earth was industrial clay.

Fig. 2. Construction of a Nubian vault [11] ©AVN

In a first phase, a Nubian vault 3D model was realized using the modeling software Rhinoceros®, then multiple sections at an angle of 45° were made on it (Fig. 3) and a Grasshopper® G-Code was realized (Fig. 4). In particular, the G-code enabled the WASP machine to print inclined clay layers (not horizontal, as usually happens).

The printing tests demonstrated that Nubian vault is conceptually repeatable and can be adapted to additive manufacturing for constructions (Fig. 3).

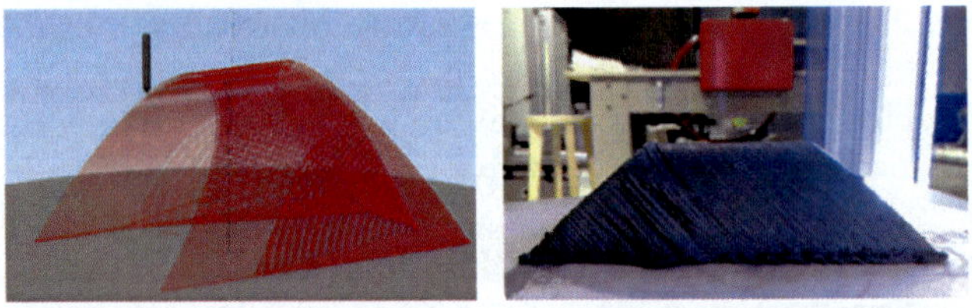

Fig. 3. The extrusor path and the 3D printed Nubian vault [10] ©Fablab Poliba

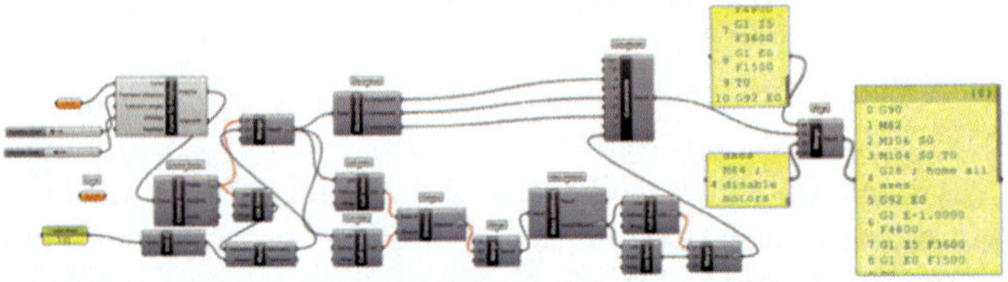

Fig. 4. Grasshopper® code for the G-Code realization [10] © Self Made Architecture V

3.3 The Lamione Prototype Development

The research is based on historical and technical considerations as shown in Fig. 5: the prototype development derives from traditional features of the lamione based in Abruzzo and traditional features of the Nubian vault. The prototype is completely made of earth, from the foundations to the coverage and it is characterized by two elements. The first is a massive solid made of horizontal layers with double function of access to the building and support for the inclined layers part. The second is the body of the prototype: it is made of inclined layers continuously printed from the base to the coverage. This element leans on the first part and can be extended according to the need. In this structure, as in traditional Nubian vault, external supports or scaffolds are not needed, even during the coverage construction. This happens because every new printed layer lies on the settled previous one.

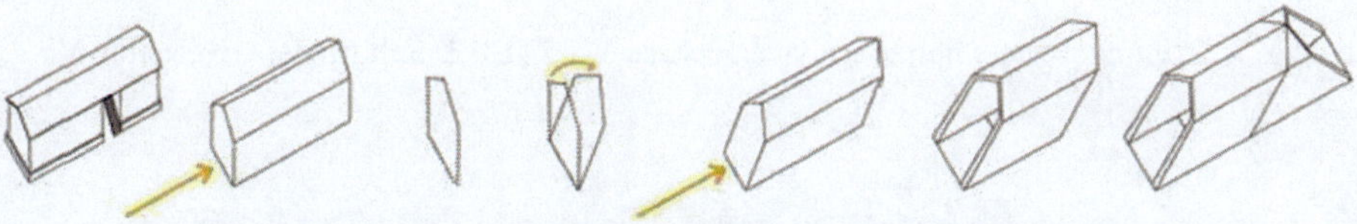

Fig. 5. Design process [10] © Self-Made Architecture V

Once the prototype was designed, printing laboratory tests were conducted, using clay as extrusion material. The tests clearly demonstrate that the most common horizontal layers extrusion has static limits for what concerns the coverage extrusion (Fig. 6). The clay layers, in fact, are not able to support their load during the coverage finalization. The tests conducted on the same model realized with inclined layers do not show the same static difficulties (Fig. 6). For this reason the author sustains this is the ideal strategy to print completely made of earth architecture without supports.

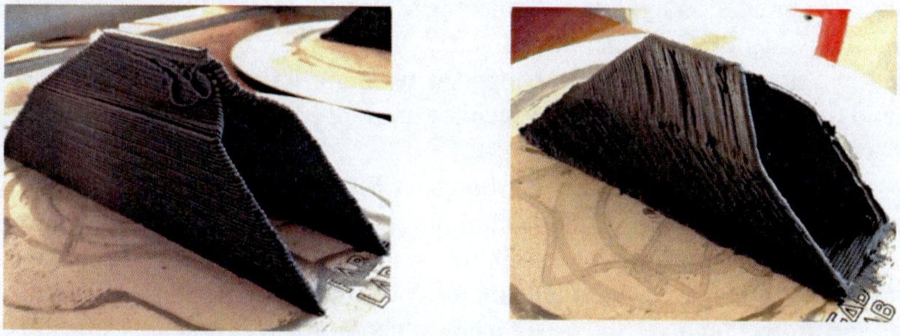

Fig. 6. Horizontal and inclined layers prototype [10] ©Fablab Poliba

The construction phases are described in Figs. 7 and 8. First, we see the positioning of two rails on the site, on which a six-axis robotic arm is mounted on a carriage. A disc cutter is mounted on the robotic arm to excavate the ground for the foundations. At the end of the process, the disc cutter is removed, and the extruder is mounted. An additional carriage is then placed on top of a tub filled with earth material. In the first step, the horizontal layer solid is formed as a support (Fig. 7), followed by the *body* of the structure, the inclined layer solid [10] (Fig. 8).

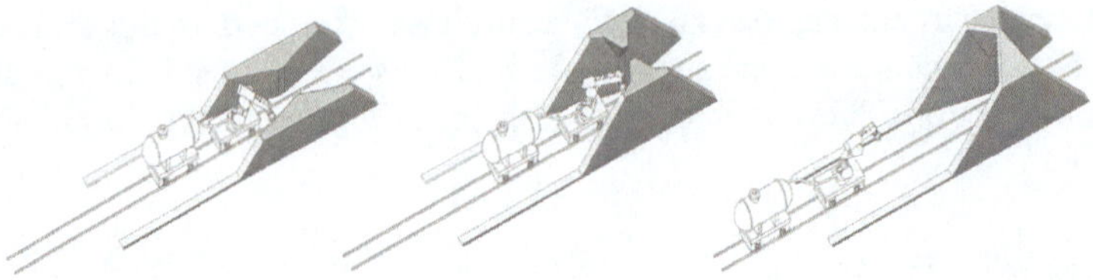

Fig. 7. Printing process horizontal layers, phase 5-6-7 [10] © Self Made Architecture V

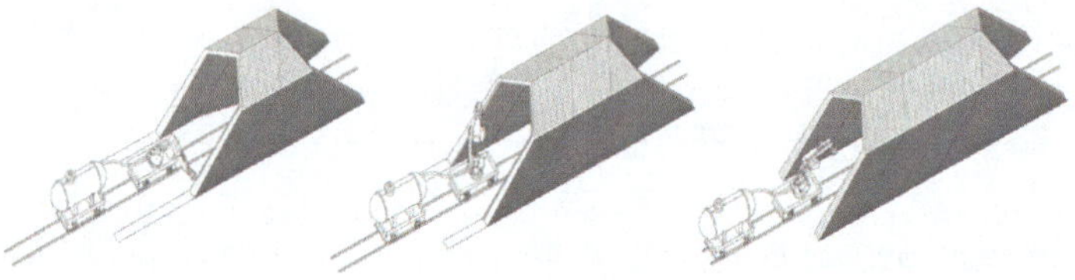

Fig. 8. Printing process inclined layers, phase 8-9-10 [10] © Self Made Architecture V

4 Countryside and Digital Fabrication: Architecture in the COVID and Post-COVID Era

The lamione prototype is designed to be realized in rural areas and in specific emergency situations. An application hypothesis concerns Malandra Vecchia area, a village located in the municipality of Casalincontrada (CH). In this area there are many raw earth constructions, now abandoned. As happened to many italian rural areas, during the Sixties and the Seventies, Malandra Vecchia experienced severe emigration, especially for work reasons and lack of services.

The prototypes are aggregated as shown in Fig. 9, although this is only one of the possible aggregation of them: the lamione prototype is thought as a single living module, repeatable according to need. In this sense, it would not be surprising placing the prototype in an isolated way, as the majority of traditional earth architectures are in the area. The project involves the use of a new technology in a rural area stuck in distant ages (Fig. 12). The basic idea is to transform earth materials in new, actual, testable materials.

The printed prototypes are single hall architectures that testify actual digital fabrication experiments and, in the future, could represent the first prototypes of a new architecture. In this sense, the entire complex – that is also characterized by the relationship with the traditional existent architectures in massone – could be an open-air museum with digital and rural features at the same time. The dislocation of the museum in the countryside generates a new and unexpected situation: the rural world encourages curiosity in citizens who, in contrast, do not really know it. The dislocation is in accordance with a series of actual considerations regarding the architectural world in general: the designer attention focuses on the countryside as a perfect project site for the future because the cities result in saturation more and more.

Rem Koohlaas, in the exhibition Countryside, The Future (2020) at the New York Guggenheim Museum, is very explicit in this regard: we should stop to neglect the 98%

Fig. 9. Malandra Vecchia project [10] © Self Made Architecture V

of the terrestrial surface that is not urban; the rural zones are also changing very fast in response to contemporary society transformations.

The still ongoing COVID-19 pandemic leads us to reflect about pandemic origins, infection spreading ways and some equilibrium we forgot. From this perspective, concepts as distance and isolation, since ever characteristic of the countryside and often considered by the citizen a penalization, assume positive meaning and become strength. In this sense, additive manufacturing technology is appropriate in pandemic emergencies, in situations where the mobility of craftsmen and building materials is limited, the application of additive manufacturing is possible through knowledge transfer alone. This museum-type approach lead us to reflect about the emergent coexistence between new technologies and human beings and it is really interesting how much in this process and in emergency conditions, the ego of the designer – as well specified by Rem Koolhaas [13] – it is absent. The purity of intention, undistorted by personal architectural visions, makes this research really attractive.

5 Conclusion and Future Research

The conducted study stimulates new research in different fields. The future additive manufacturing advances could concern the optimization of:

- the clay mixture, with the aim to improve mechanical characteristics of it and reduce drying time;
- the printing process, with the aim to make it continuous;
- the printer in general;

It would be interesting to continue putting in relation thrusting masonry and additive manufacturing with the aim to identify masonry principles that can be applied to 3D printing methods for architecture and to design new printable and repeatable coverage systems. The reiterated use of additive manufacturing in construction sector could contribute to important and irreversible society transformations in fields that are different from each other or still inexistent.

References

1. Rosenthal, M., Henneberger, C., Gutkes, A., Bues, C.-T.: Liquid deposition modeling: a promising approach for 3D printing of wood. Eur. J. Wood Wood Prod. **76**(2), 797–799 (2017)
2. WASP Site. https://www.3dwasp.com/stampante-3d-gigante-bigdelta-wasp-12mt/
3. Pegna, J.: Exploratory investigation of solid freeform construction. Autom. Constr. **5**(5), 427–437 (1997)
4. Khoshnevis, B.: Automated construction by contour crafting—related robotics and information technologies. Autom. Constr. **13**(1), 5–19 (2004)
5. Carneau, P., Mesnil, R., Roussel, N., Baverel, O.: Additive manufacturing of cantilever-from masonry to concrete 3D printing. Autom. Constr. **116**, 103184 (2020)
6. Cowan, H.J.: A history of masonry and concrete domes in building construction. Build. Environ. **12**(1), 1–24 (1977)
7. Baldacci, O.: L'ambiente geografico della casa di terra cruda in Italia. Rivista geografica italiana, LXV (1958)
8. Correia, M., et al.: Terra Europae. Earthen Architecture in the European Union. ETS, Pisa (2011)
9. Pedrini, D.: Ri–Abitare la Terra. (Unpublished master's thesis). Università di Firenze. College professors: Tonietti, U. (relatore), Conti, G., Catalano, R. (correlatori) (2017)
10. Carbonara, F.A., Delcuratolo, F.A., Lops, M., Tina, G.: Self Made Architecture V (Unpublished master's thesis). Politecnico di Bari. College professors: Parisi, N. (relatore), Ruggiero, F. (correlatore), Diaferio, M., Pastore, D., Scardigno, N (2021)
11. Chabaud, C., Hardy, M.: Manuel du maçon nubien. Livre Technique AVN (2013)
12. Salama, A.M.: Coronavirus questions that will not go away: interrogating urban and socio-spatial implications of COVID-19 measures. Emerald Open Res. **2020**(2), 14 (2020)
13. Koolhaas, R.: Museum in the countryside: aesthetics of the data centre. Archit. Des. **89**(1), 60–65 (2019)

Statics and Stability of Bending-Optimized Double-Layer Grid Shell

Francesco Laccone[1(✉)], Nico Pietroni[2], Maurizio Froli[3], Paolo Cignoni[1], and Luigi Malomo[1]

[1] ISTI - CNR, via G. Moruzzi 1, Pisa, Italy
francesco.laccone@isti.cnr.it
[2] School of Computer Science - UTS, 15 Broadway, Ultimo, NSW 2007, Australia
[3] University of Pisa, Lungarno Pacinotti 43, Pisa, Italy

Abstract. Grid shell structures are optimal when considering their aesthetics and lightness, but their efficiency is highly reduced when their shape deviates from a pure membrane. Many contemporary architectures possess a freeform shape, conceived mostly on aesthetics and functional criteria. In these cases, finding an efficient grid shell often requires substantial shape modifications. This work addresses a new kind of double-layer structure that aims to preserve the desired shape design. The structural system comprises a quad-meshed grid shell aligned to the target shape and enriched with an additional reinforcement layer that adds bending stiffness. This additional layer, going inward and outward of the main surface, presents variable height and discontinuous elements based on the required bending strength. The obtained structural system differs from both grid shells, as these latter may be very deformable in this setup, and from classic double-layer structures (space frames), which are heavier and redundant. In this paper, we show how the presented system compares with grid shell and double-layer competitors in terms of statics and stability. We highlight the pros and cons based on a systematic comparative analysis run on selected freeform shapes.

Keywords: Freeform surface · Space frame · Computational design · Structural design · Finite Element Model

1 From Grid Shells to Space Frames and Reinforced Structures

Grid shells are elegant and lightweight structures used in architecture to enclose large spaces. Their lightness is due to the structural efficiency in developing membrane actions to resist external loads, which flow along a grid of rods. This behavior relies on in-plane stiffness rather than on bending. It can be achieved only through careful shape design, i.e. form-finding, and properly selected topology of the grid and the structural scheme [1–4]. However, the shape of the grid

S. Gabriele et al. (Eds.): IWSS 2023, LNCE 437, pp. 569–578, 2024.
https://doi.org/10.1007/978-3-031-44328-2_59

shell is rarely determined by on site constraints, or it is simply sculpted by the architect. Therefore, a structural design that approaches a funicular shape to accomplish membrane behavior can strongly alter the initial design or can conflict with other requirements. Alternatively, while preserving the original design, bending forces inherently arise because transversal forces can not be balanced by in-plane stress. To address bending resistance demands and to increase the robustness, the alternative concept of double-layer structures or space frames can be used [5]. Two interconnected surfaces generate bending inertia from spacing the members of the two surfaces in the out-of-plane direction. These surfaces can be continuous or segmented, i.e., shaped as grids, plates, shells etc. The higher the distance between the layers, the larger the bending inertia.

Traditionally, a space frame is an assembly of linear elements arranged to transfer forces in 3D. Straight steel tubes, loaded mainly along the axes, and joints are the basic units of this system. A large variety of space frames exists, but their common features are to match the form of a surface in space and to be constructed of almost identical units arranged with a repeating pattern [6]. One of the two layers is usually filled or equipped with cladding panels, since space frames serve as roofs, envelopes or slabs. Double-layer plate structures are typically used in timber construction to realize dry-connected lightweight shells in the form of cassettes, hollow boxes or waffle structures [7–9]. This concept can be extended even towards less stiff materials [10].

An interesting design option is to vary the distance between the layers following the bending moment plot. This approach has inspired several design concepts for 2D structures such as truss bridges and frames. Some examples are the Berlin Central Station roof [11], designed by GMP Architekten and sbp, and the Waterloo station roof [12] in London, designed by Grimshaw Architects and YRM Anthony Hunt Associates. Both structures are translation surfaces where a transversal planar frame is replicated along a path. The main frame is an arch line, stiffened by cables and along-normal struts, oriented inward or outward according to the bending moment profile. This tension-stiffening system can be considered an added layer, and it is usually post-tensioned until bending forces are removed and/or the main layer results entirely compressed.

The approach of adding reinforcement-where-needed inspired research based on graphic statics aimed at transforming a generic free-form curve into a funicular shape. In the works [13,14], the funicularity of 2D and 3D shapes is achieved via an additional out-of-plane cable net using graphic statics. With a similar purpose, in [15], the funicular geometry is exploited to design a tension stiffening system shaped according to the bending moment in a 2D free-form curve.

To address the same issue in the continuum, the authors in [16] introduced 3D shell surfaces reinforced by ribs, whose height is proportional to the bending. The ribs form a grid aligned with a cross-field that considers the principal directions of the strain in the equivalent continuum. Several years before, in a similar fashion, Nervi patented a rib-stiffening scheme for isostatic concrete slabs that follows the principal bending directions [17]. Adding ribs and reinforcement [18,19] or aligning the beams of a grid along specific paths [20,21], i.e. principal

stress directions, is a well-known strategy that is inspired by the Michell-like optimum structures.
In this paper, we analyze the structural behavior of bending-optimized double-layer grid shells introduced in [22], which are quad-based grid shells reinforced with an out-of-plane discrete net of elements that is shaped to increase the bending strength. In this method, firstly, the input surface is tessellated, creating a main layer that aims at maximizing the resistance to out-of-plane bending; then, the location of an additional reinforcement layer is optimized using a novel energy formulation. We compare double-layer structures with state-of-the-art alternatives to prove that they have intermediate features of both the space frames and the grid shells.

2 Bending-Optimized Double Layer Grid Shells

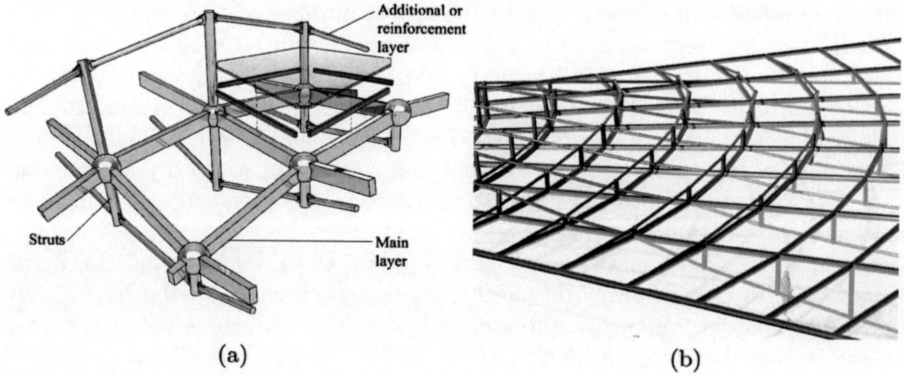

Fig. 1. Our bending-optimized double-layer grid ghells: (a) section view and nomenclature; (b) rendered closeup.

Our computational pipeline automatically designs a reinforcement structure optimized for a predefined load condition and capable of efficiently supporting out-of-plane loads. The details concerning the method and its implementation can be found in [22]. In the resulting structure, a quad-based grid shell is connected with a reinforcement structure having similar connectivity but whose nodes are placed at variable distances and orientations with respect to the original grid shell, as in Fig. 1. These two interconnected structures can generate bending inertia in 3D from the spacing between the two layers in the out-of-plane direction. To have always tension reinforcements (and compressed struts), we allow the additional layer to go inward or outward the base grid shell. The additional layer can be discontinuous if no bending strength is locally needed.

Given an initial surface, as a first step, we build a linear finite element analysis as a thick shell, and we use the strain tensor on corresponding top and bottom

Table 1. Dataset quantitative measures for the three structural cases: model name; size of the axis-aligned bounding box in meters; number of vertices #V; number of edges #E; steel volume of the structures V.

Model	Size	Grid shell			Ours			Space frame		
	$(m \times m \times m)$	#V	#E	V (m^3)	#V	#E	V (m^3)	#V	#E	V (m^3)
Rust	$80.35 \times 58.08 \times 19.15$	615	1194	46.43	1143	2460	66.10	1230	3003	122.03
Shell1	$51.11 \times 81.06 \times 23.56$	726	1398	50.62	1222	2587	70.01	1452	3522	136.90
Shell2	$90.59 \times 37.08 \times 18.33$	379	707	26.01	712	1460	38.55	758	1793	68.69
Vault	$42.75 \times 23.32 \times 12.37$	619	1184	20.18	1153	2352	30.21	1238	2987	65.91
Wave	$43.44 \times 44.01 \times 6.05$	431	818	24.94	1239	2626	58.87	1178	2865	77.10

nodes to create a cross-field. This vector field generates a quad mesh, in which its alignment to the cross-field is forced if the cross directions have higher anisotropy or higher magnitude. The quad mesh topology is chosen since it can effectively capture a planar principal stress with the lowest number of elements.

In the second step, we use the internal forces computed with a linear static analysis on the quad mesh to initialize the additional layer. Then, we compute the additional layer's optimal elevation, solving a constrained minimization problem in which the structural efficiency is improved. The optimization variables are the distances from the main layer. In the optimal setup, the stress due to bending would be constant in both layers. In the final step, the structure is modified to conform to fabrication constraints.

The main strength of this work is decoupling the meshing phase from the initialization/optimization of the additional layer, which is, to the best of our knowledge, the first method to automatically derive reinforcement structures for non-membrane stress in a discrete setting. Thus, any quad mesh can be used.

3 Experimental Setup and Analyses

We tested several input freeform surfaces having different features (Table 1). As we want to assess the structural performances of our bending-optimized double-layer grid shells, we set up a comparison with other state-of-the-art structures. For each model, we build and analyze three structural system cases:

- *Grid shell*, directly assembled from the quad mesh by converting edges into beams and vertices into structural nodes;
- *Our bending-optimized double-layer grid shells*, using a multi-chords approach [22];
- *Space frame*, by duplicating the grid shell at 2.5 m along the surface normal and connecting corresponding nodes.

The performance of discrete spatial structures, and of grid shells in particular, is a function of topology, grid spacing and the total weight of the elements [23,24]. For the sake of comparing different structural systems, we chose to keep fixed

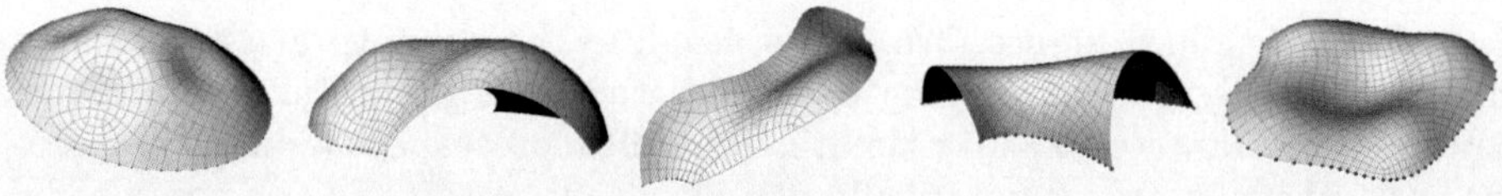

Fig. 2. Experimental dataset of surfaces, from left: Rust, Shell1, Shell2, Vault, Wave. All models are tessellated with the actual quad mesh used in the simulations; blue spheres indicate fixed supports. (Color figure online)

the grid spacing of the base mesh. So, the comparison criterion for the structure is to be initialized on the same quad mesh of Fig. 2. The quad mesh is computed using the approach of our bending-optimized double-layer grid shells. As a result, the topology is inherently different for the three structural systems, as so is the weight: our structures are heavier with respect to the grid shells (about 27 to 57% more) and can have nodes with various valence (larger or equal to 4); space frames are heavier than our structures (about 23 to 54% more) and than grid shells (about 62 to 70% more). The mode of node valence for space frames is 5.

All finite element analyses rely on a set of uniform assumptions:

- no stiffening systems are considered (i.e. provided by bracing cables), so all nodes are modeled as rigid;
- a subset of boundary nodes are supposed to be rigidly fixed (all 6 degrees of freedom are fixed);
- the base material is common S355 structural steel;
- the elements' cross-section are CHS of 0.2 m external diameter and 0.025 m thickness for the main beams, for the grid shell and for both layers of the space frame; solid circular sections 0.15 m diameter for the struts; solid circular sections of 0.1 m diameter for the added layer for our double-layer structures;
- the main layer of beams is kept fixed in all three models;
- the dead load is computed from the material density and the geometry of the cross sections;
- we impose a common uniform load $f_{ext} = 2\ \text{kN/m}^2$, representing a typical service load including non-structural load (e.g., cladding) and environmental load (e.g., snow), which is applied as a lumped load on the nodes of the grid shell and of the main layer of the other cases;
- material non-linearity is neglected as the analyses already involve many variables;
- each beam is modeled as a single finite element to reduce the computational time, even though this simplification prevents from detecting single member buckling.

This work aims not to solve real-world design cases but to evaluate our reinforcement strategy in bending-optimized double-layer structures compared to grid shells, which often need more stiffness, and space frames, which usually have more redundancy. We establish a two-level assessment: (a) serviceability limit state conditions (SLS) to check internal forces and displacements; (b) stability to control the failure modes. We do not analyze the strength at the ultimate limit state (ULS), even if yielding could sometimes be a realistic failure mode. Instead, we are interested in estimating the ultimate capacity on a comparative basis. We compute geometrically non-linear analyses (GNA) on commercial software [25]. Additionally, we compare the load factors obtained with the GNA with the first eigenvalue from a linear buckling analysis (LBA).

4 Results and Discussion

At the SLS, our double-layer grid shell has an intermediate stiffness between a grid shell and a space frame, as evidenced in Table 2. Some models, like Shell1 and Shell2, would not have enough stiffness as grid shells since they overshoot the usual permitted displacement. Therefore, it is at least recommended to change the structural system for these kinds of shapes. 'Well-shaped' surfaces like Vault are obviously performing well as grid shells. Concerning internal forces, the utilization of the elements is larger in grid shells, as expected. However, the less the shape is a membrane, the higher the risk of yielding. Additionally, it is worth converting grid shells to other structural systems despite the weight increase (Fig. 3a). Indeed, the load transfer mechanism changes and the elements are mainly utilized in the axial regime, which is more efficient. Similarly to a space frame, the bending moments is considerably reduced when using our double-layer grid shells (Fig. 3b). Remarkably, the utilization of the elements is more uniform in our case.

Table 2. Maximum displacement magnitude d at the SLS.

Model	Grid shell	Ours	Space frame
	d (m)	d (m)	d (m)
Rust	0.167	0.085	0.043
Shell1	0.445	0.235	0.100
Shell2	3.047	1.964	0.741
Vault	0.026	0.011	0.006
Wave	0.075	0.040	0.031

Under incremental load, the grid shell has the lowest load factor. As expected, our double-layer grid shells performance is in between grid shells and space frames (Table 3). Except for the Vault model, which has a more efficient shape, all grid shells do not exhibit a pure membrane behavior, and therefore they show a softening trend prior to collapse. Instead, our double-layer grid shells are often characterized by a post-critical stiffening, where the additional layer balances unstable displacements of the main layer. The critical shape of our double-layer grid shells and space frames is different. The stiffness and the load

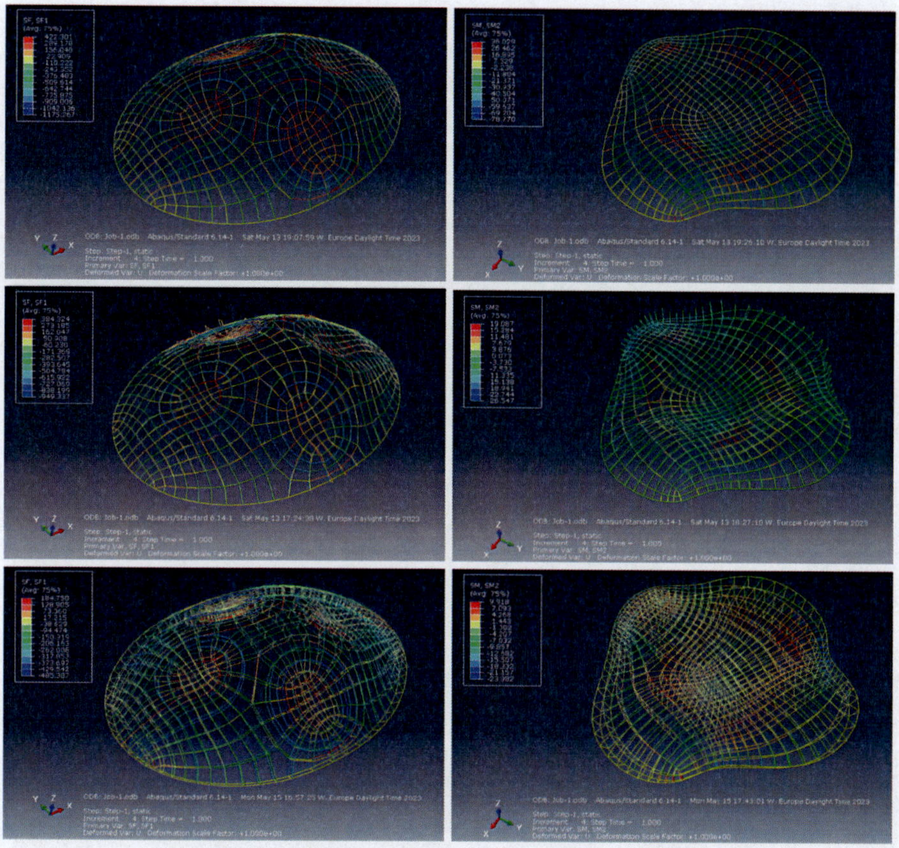

(a) (b)

Fig. 3. Internal forces for the three structural systems (from top: Grid shell, ours, space frame): (a) axial force, Rust model; (b) out-of-plane bending moment, Wave model.

transfer mechanisms of the two systems is quite different. In our double-layer grid shell, the additional layer uses a smaller cross section, it is mostly tensioned, and can be located inward or outward of the main layer. In space frames, the two layers build a more uniform stiffness and could be stressed either in tension or compression.

The collapse shapes obtained in LBA are very similar to GNA for our double-layer grid shells and space frames (Fig. 4). On the other hand, this is not the case for grid shells, in which probably a more complex combination of buckling modes occurs for the GNA.

Table 3. Comparison of load factors from the LBA (first eigenvalue) and from the GNA for the three structural cases (* indicates tension stiffening, - indicates not run).

Model	Grid shell		Ours		Space frame	
	LB	GNA	LB	GNA	LB	GNA
Rust	6.98	5.97	11.06	8.99	37.81	34.92
Shell1	3.89	*4.51	6.51	*7.52	40.89	*42.56
Shell2	0.63	0.39	1.35	1.03	18.67	14.49
Vault	75.27	46.99	149.36	162.20	722.72	-
Wave	22.92	18.71	52.40	35.29	121.71	*

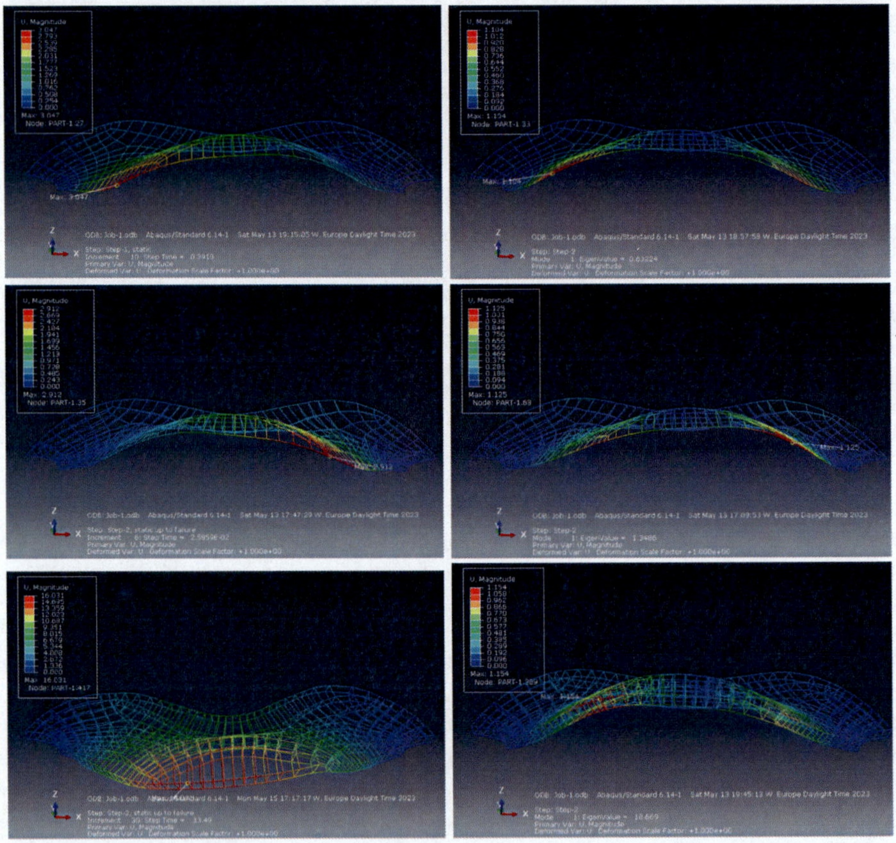

(a) (b)

Fig. 4. Buckling shape, Shell2 model (from top: Grid shell, ours, space frame): (a) GNA; (b) LBA.

5 Conclusions and Future Developments

This paper assesses the structural performance of bending-optimized double-layer grid shells introduced in [22]. These structures can represent a good alternative to classic lightweight spatial structures, such as grid shells and space frames. Space frames are more robust and can achieve a higher prefabrication rate. However, having all modular elements is nowadays not a primal request if digital fabrication tools and machines are employed. Hence, producing labelled pieces of any dimension comes at a low cost.

The role of imperfections in the GNA is still an open topic to investigate further. However, we adopted models and a meshing strategy that led to non-symmetric and non-regular structures. Thus, even in the case of grid shells, the membrane-to-bending strain energy ratio is lowered so that, in theory, structures would be less sensitive to imperfections. Another research direction could be introducing a discretization into multiple finite elements per beam to catch possible local buckling phenomena that may lower the load factors.

References

1. Richardson, J.N., Adriaenssens, S., Filomeno Coelho, R., Bouillard, P.: Coupled form-finding and grid optimization approach for single layer grid shells. Eng. Struct. **52**, 230–239 (2013). https://doi.org/10.1016/j.engstruct.2013.02.017
2. Mesnil, R., Douthe, C., Baverel, O.: Non-Standard Patterns for gridshell structures: fabrication and structural optimization. J. Int. Assoc. Shell Spat. Struct. **58**(4), 277–286 (2017). https://doi.org/10.20898/j.iass.2017.194.893
3. Grande, E., Imbimbo, M., Tomei, V.: Role of global buckling in the optimization process of grid shells. Eng. Struct. **156**, 260–270 (2018). https://doi.org/10.1016/j.engstruct.2017.11.049
4. Laccone, F., Malomo, L., Froli, M., Cignoni, P., Pietroni, N.: Automatic design of cable-tensioned glass shells. Comput. Graph. Forum **39**(1), 260–273 (2020). https://doi.org/10.1111/cgf.13801
5. IASS Working Group No. 8 on Spatial Steel Structures (Y. Tsuboi, chairman): Analysis, Design and Realization of Space Frames. Bulletin of the IASS XXV - 1/2(84/85) (1984)
6. Chilton, J.: Space Grid Structures. Routledge (2007). https://doi.org/10.4324/9780080498188
7. Robeller, C., Konaković, M., Dedijer, M., Pauly, M., Weinand, Y.: Double-layered timber plate shell. Int. J. Space Struct. **32**, 160–175 (2017). https://doi.org/10.1177/0266351117742853
8. Rezaei Rad, A., Burton, H., Rogeau, N., Vestartas, P., Weinand, Y.: A framework to automate the design of digitally-fabricated timber plate structures. Comput. Struct. **244**, 106456 (2021). https://doi.org/10.1016/j.compstruc.2020.106456
9. Wagner, H.J., Alvarez, M., Groenewolt, A., Menges, A.: Towards digital automation flexibility in large-scale timber construction: integrative robotic prefabrication and co-design of the BUGA wood pavilion. Constr. Robot. **4**(3–4), 187–204 (2020). https://doi.org/10.1007/s41693-020-00038-5
10. Laccone, F., Manolas, I., Malomo, L., Cignoni, P.: Exploratory study on a segmented shell made of recycled-HDPE plastic. In: Proceedings of IASS2020/21-SURREY (2021)

11. Gugeler, J., Havemann, K., Schober, H.: Lehrter Bahnhof Berlin: Das Nord-Süd-Dach. Stahlbau **75**(3), 194–202 (2006)
12. Hunt, A.J., Jones, A.C., Otlet, M., Dexter, D.I.: Waterloo international rail terminal trainshed roof structure. Struct. Eng. **72**, 123 (1994)
13. Todisco, L., Peiretti, H.C., Mueller, C.: Funicularity through external posttensioning: design philosophy and computational tool. J. Struct. Eng. **142**(2), 04015141 (2016). https://doi.org/10.1061/(ASCE)ST.1943-541X.0001416
14. Todisco, L., Corres-Peiretti, H., Mueller, C., et al.: Exploration of externally post-tensioned spatial structures using 3D graphic statics. In: Proceedings of IASS Annual Symposia, vol. 2018, pp. 1–8. International Association for Shell and Spatial Structures (IASS) (2018)
15. Van Mele, T., Lachauer, L., Rippmann, M., Block, P.: Geometry-based understanding of structures. J. Int. Assoc. Shell Spat. Struct. (IASS) **53**, 285–295 (2012)
16. Gil-Ureta, F., Pietroni, N., Zorin, D.: Reinforcement of general shell structures. ACM Trans. Graph. **39**(5), 1–19 (2020). https://doi.org/10.1145/3375677
17. Halpern, A.B., Billington, D.P., Adriaenssens, S.: The ribbed floor slab systems of Pier Luigi Nervi. In: Proceedings of IASS Annual Symposia, vol. 2013, pp. 1–7. International Association for Shell and Spatial Structures (IASS) (2013). ISSN 1996-9015
18. Li, W., Zheng, A., You, L., Yang, X., Zhang, J., Liu, L.: Rib-reinforced shell structure. Fabrication Des. **36**, 15–27 (2017). https://doi.org/10.1111/cgf.13268
19. Tam, K.M.M., Mueller, C.T.: Additive manufacturing along principal stress lines. 3D Printing Add. Manuf. **4**, 63–81 (2017). https://doi.org/10.1089/3dp.2017.0001
20. Kilian, M., Pellis, D., Wallner, J., Pottmann, H.: Material-minimizing forms and structures. ACM Trans. Graph. **36**(6), 1–12 (2017). https://doi.org/10.1145/3130800.3130827
21. Pietroni, N., Tonelli, D., Puppo, E., Froli, M., Scopigno, R., Cignoni, P.: Statics aware grid shells. Comput. Graph. Forum **34**, 627–641 (2015). https://doi.org/10.1111/cgf.12590
22. Laccone, F., Pietroni, N., Cignoni, P., Malomo, L.: Bending-optimized double-layer grid shells. SSRN (2023). https://doi.org/10.2139/ssrn.4313774
23. Gioncu, V.: Buckling of reticulated shells: state-of-the-art. Int. J. Space Struct. **10**(1), 1–46 (1995). https://doi.org/10.1177/026635119501000101
24. Tonelli, D., et al.: Stability of statics aware Voronoi grid-shells. Eng. Struct. **116**, 70–82 (2016). https://doi.org/10.1016/j.engstruct.2016.02.049
25. Dassault Systèmes: Abaqus analysis user's manual. Simulia Corp. Providence, RI, USA, vol. 40 (2014)

Space Architecture Goes Green: Investigating the Use of Hemp-Fibre Concrete for Martian Structures

Costanza Anerdi[1] (✉), Diego Gino[2], and Giuseppe Carlo Marano[2,3]

[1] DIMEAS, Department of Mechanical and Aerospace Engineering, Politecnico di Torino, 10128 Turin, Italy
costanza.anerdi@polito.it

[2] DISEG, Department of Structural, Geotechnical and Building Engineering, Politecnico di Torino, 10128 Turin, Italy

[3] College of Civil Engineering, Fuzhou University, Fuzhou 350108, China

Abstract. The establishment of sustainable human settlements in space requires the development of living and operational spaces capable to withstand the harsh environment of extra-terrestrial locations. To reduce the dependence from Earth resources, we should get used to introducing in-situ production of structural elements. This can include the construction of structural elements with a focus on the highest degree of recycling of materials and resources. Regolith, the loose soil found on the surfaces of Mars and other celestial bodies, presents a potential material for building structures due to its abundance and low cost.

This paper explores the use of a regolith simulant-based material with the addition of hemp fibres applied to a simple structure. We analyse the performance of a structure under various loading conditions and propose solutions that can improve its strength and durability. Our results demonstrate that the addition of hemp fibres in the regolith mixture enhances the thermo-mechanical properties of the material, making it a promising choice in structures where building efficiency is paramount. This research contributes to advancing sustainable and innovative solutions for space architecture and supports the goal of establishing a permanent human presence in space.

Keywords: ISRU (In-Situ Resource Utilization) · Martian concrete · Multi-planetary life · Hemp fibers · Regolith · Sustainability

1 Introduction

The colonization of Mars has become an increasingly compelling vision for future human space exploration [1, 2]. In order to establish sustainable and self-sufficient settlements on the Red Planet, the concept of In-Situ Resource Utilization (ISRU) has emerged as a crucial strategy. ISRU involves leveraging local resources to support human activities, including the construction of habitats, infrastructure, and other essential components [3, 4].

S. Gabriele et al. (Eds.): IWSS 2023, LNCE 437, pp. 579–587, 2024.
https://doi.org/10.1007/978-3-031-44328-2_60

The environment on Mars is significantly different from Earth, characterized by distinct atmospheric composition, lower atmospheric pressure, reduced gravity, variable temperatures, exposure to harmful radiation, and potential water resources [5].

The Martian atmosphere is composed mainly of carbon dioxide (CO_2), making up over 95% of its gas content. Compared to Earth's atmosphere, which consists mainly of nitrogen and oxygen, the high concentration of CO_2 on Mars results in a vastly different atmospheric composition. The average atmospheric pressure on Mars is extremely low, measuring at around 0.006 atmospheres. This pressure is equivalent to about 0.6% of Earth's atmospheric pressure. The thin atmosphere poses challenges for human exploration, as it affects aerodynamics, atmospheric flight, and the ability to breathe without proper life support systems [6].

Mars has a lower surface gravity compared to Earth, with Martian gravity approximately 0.38 times that of Earth's gravity. This reduced gravity can have significant effects on the human body, potentially leading to physiological changes, muscle and bone loss, and altered bodily functions. Long-term habitation on Mars would require adaptations to cope with the lower gravity environment.

The average temperature on Mars is quite cold, with temperatures around –63 °C. However, temperatures can vary considerably depending on the location, season, and time of day. While polar regions experience extremely cold temperatures, equatorial regions can see relatively warmer temperatures during the day.

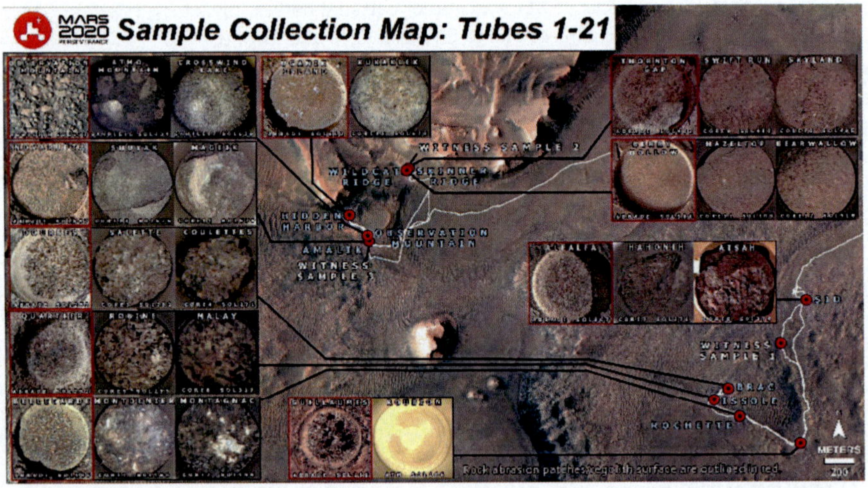

Fig. 1. A map of the samples collected by NASA's Perseverance Mars rover [2].

1.1 Martian Soil

A literature search was carried out on the composition of Mars soil, analyzing the data collected by NASA missions. The presence of a variety of minerals in Martian soil points to a history of significant geological processes, such as volcanic activity, impact events, and interactions with water. Over time, the soil has undergone extensive weathering, erosion, and sedimentation, leading to its wide-ranging characteristics and spatial distribution [7].

Polar regions present reserves of subsurface water ice, while equatorial regions may contain water molecules bound in minerals. Various methods can be employed to extract liquid water from the ice present in the ground, whether it is located at the surface or buried beneath the Martian surface. Water can also be acquired from the atmosphere or extracted from hydrated minerals. The discovery of water ice has profound implications for future construction methods, in particular when talking of concrete as it represents one of its fundamental components. Figure 1 shows a map of the 21 samples (containing rock, regolith, atmosphere, and witness materials) that have been collected over time by NASA's Perseverance Mars rover [2]. Martian regolith refers to the layer of loose, unconsolidated material covering the solid bedrock on the surface of Mars. It is composed of a mixture of dust, sand, small rocks, and other particles that have accumulated over time.

By comparing the results of NASA Mars rover measurements, it soon became apparent that the chemical composition of Mars' soil is not the same throughout the planet. In fact, it has a general basaltic composition, with very strong local variations in the concentration of silicates, iron sulphates, calcium sulphates, and iron minerals such as hematite. On the other hand, orbital observations show that the soil in the north-polar region consists mainly of gypsum and/or bassanite minerals [8].

1.2 Shielding

Without an active magnetic field, Mars lacks the protective shield provided by Earth's atmosphere against harmful cosmic and solar radiation. The Martian surface is exposed to intense radiation, which poses health risks to potential human explorers. Adequate shielding and protection would be crucial for ensuring the safety and well-being of humans on Mars [9]. It has been found that Martian regolith, especially when compacted and used as a building material for habitat construction, can serve as a radiation shield if applied with a proper thickness.

2 Material Selection

2.1 Martian Concrete

In its basic definition, concrete denotes a blend of binder and aggregate (inert phase). Typically, water is also a requisite component in the concrete formulation (although elemental sulfur stands out as an exception). Furthermore, additives or admixtures might be utilized to improve concrete properties [10]. Developing a viable composite material for use on Mars requires careful consideration of the components involved. One of the key factors is selecting an inert component that not only matches the grain size of the Martian regolith but also closely replicates its chemical composition. On Earth, Ordinary Portland Cement (OPC) is the most widespread binder, commonly referred to as "cement". However, any substance that fosters cohesiveness among the mixture components is recognized as a binder, and alternative binders to OPC could be suitable for concrete production on Mars [8]. In particular, gypsum has shown a low-energy intensity production process and has been selected as the target binder for our study.

2.2 Use of Hemp Fibers

While hemp fibers are not typically used as reinforcement for structural purposes, they can offer several benefits to the structural performance and improve certain properties of the concrete [11]. Hemp fibers can help control the formation and propagation of cracks in concrete, improving the overall durability. Moreover, natural fibers can assist in controlling drying shrinkage in concrete. Drying shrinkage occurs as the concrete loses moisture, leading to volume reduction and potential cracking. The fibers help mitigate this effect [12].

Of great interest are the thermal and thermo-hygrometric properties of these mixtures, assuming even more importance where the energetic building efficiency. In this context, hemp (Cannabis sativa) has received considerable attention because of its good thermal insulation properties, good mechanical properties, and rapid growth rate (only 3.5 months) [13]. Hemp is a fast-growing crop that requires minimal chemical inputs and can be cultivated with reduced environmental impact. It can be grown in greenhouses, directly on Mars [14].

According to previous study [15], it was possible to infer the mechanical properties of a material formed from Martian regolith and reinforced with hemp fibers, as reported in Table 1. These mechanical properties were used to define the material adopted in the finite element analysis (FEA) of a prototype shelter that could be constructed on the surface of Mars, as described below.

Table 1. Mechanical properties considered.

Material	fcm [Mpa]	ftm [Mpa]	E_{ci}[Mpa]	El_{ci}[Mpa]	GF [N/m]	ρ [Kg/m^3]
REF	6.2	3.6	219997.1	147270.7	48504	1800
FR (+hemp)	5.4	2.5	209966.2	97608.2	42104	1500

3 Structural Analysis

In this section, we present the preliminary design and structural analysis of a prototype shelter intended for human settlement on Mars. The shelter takes the form of a barrel with a radius of 6 m and a length of 20 m. It is proposed to have a thickness of 30 cm. This specific shape is particularly suitable as it allows for the full utilization of the internal space without the need for columns [16, 17]. By doing so, it maximizes the available volume for the inhabitants who will spend a significant portion of their time indoors due to the hostile Martian environment.

The structural integrity and stability of a construction, considering both short and long-term perspectives, must be evaluated through valid and appropriate analysis methods, considering probable static and dynamic structural loads.

In this study, probable Martian structural loads were thoroughly investigated and applied to a simple structure. Valid data from previous studies, as well as information gathered from landers, rovers, and orbiters, were utilized to explore potential vertical and horizontal loads that structures on Mars may encounter.

Specifically, the analysis considered self-weight loads, calculated according to the lower gravity on Mars. Moreover, an internal pressure of 101,325 kPa was taken into consideration, representing Earth's atmospheric pressure at sea level. It is essential to recreate Earth's atmospheric pressure inside the habitat, making it suitable for human life. However, unlike the Earth, the internal pressure is not balanced by the external one, being Martian atmospheric pressure far lower than the terrestrial one. As a result, a net pressure difference exists between the interior and exterior of the habitat.

Indeed, the analysis also considered the effect of wind on the shelter's structural integrity. However, due to the thin atmosphere of Mars, the wind action is considerably lower than the one usually experienced on Earth [18].

For the sake of completeness, we applied a thermal load that takes into consideration the temperature gradient between the shelter's intrados and extrados. Inside the shelter, a temperature of 20 °C was maintained for the inhabitants, while outside the structure a temperature of –62 °C has been taken into account.

3.1 Results and Discussion

Structural analysis has been conducted based on the results obtained through a 3D numerical model of the vault performing a linear elastic analysis. The numerical model is realized through 8 nodes bi-dimensional elements. Each node has three degrees of freedom allowing translations in the x, y, and z directions. The concrete slabs have been modelled with quadrilateral 200 x 200 mm shell elements with a uniform thickness of 300 mm.

In Fig. 2, a three-dimensional FE model is presented. The model describes an arched shelter with applied loads including self-weight, internal pressure, and temperature gradient.

The results of the finite element analysis in terms of membrane stresses have been presented in Fig. 3. The distribution of stresses exhibits a unique and entirely different qualitative pattern compared to that observed in equivalent structures on Earth.

Indeed, it can be observed that a portion of the structure experiences tensile stresses, not only compressive stresses as one would typically expect in an arched structure. This peculiar stress distribution is a result of the unique configuration of applied loads, which is entirely different from what is normally encountered in such structures on Earth.

Generally, gravitational loads play a dominant role in the behavior of these structures on Earth. However, in this case, the reduced Martian gravity combined with the significant difference between internal and external pressures results in a stress configuration that subjects the structure to tensile stresses.

Finally, Fig. 4 shows the vertical displacements experienced by the structure concerning the applied load configuration. Once again, a unique qualitative configuration

Fig. 2. Tridimensional model with applied loads: self-weight, internal pressure, temperature gradient.

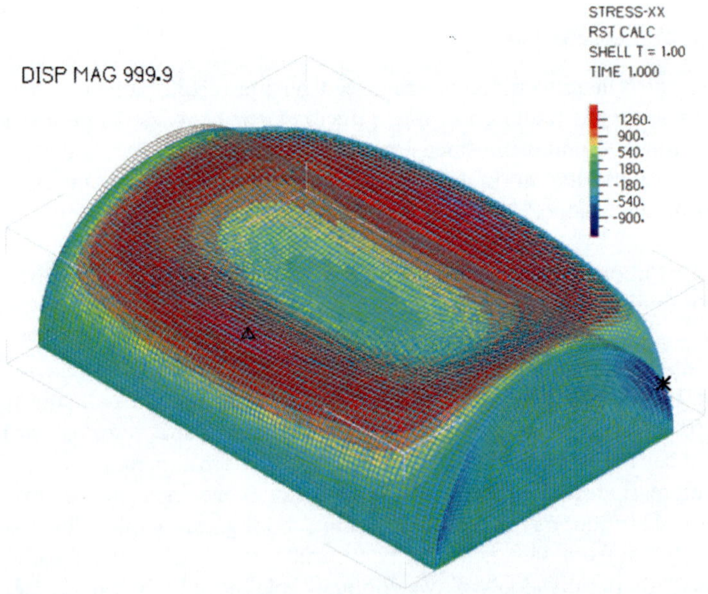

Fig. 3. Membrane stresses due the applied loads. The maximum tensile value is 1.54 MPa, lower than the design value.

can be observed. In quantitative terms, the vertical displacements are minimal, with a maximum value of 1.8 mm. This is due to the fact that the vertical loads are low because of Martian gravity.

The atypical behavior in terms of stress distribution and vertical displacements highlights the importance of considering the specific environmental conditions on Mars when designing shelters or structures for future human habitation. The analysis provides valuable insights into the structural behavior of the shelter under Martian conditions and helps in optimizing its design to ensure safety and stability in the challenging Martian environment.

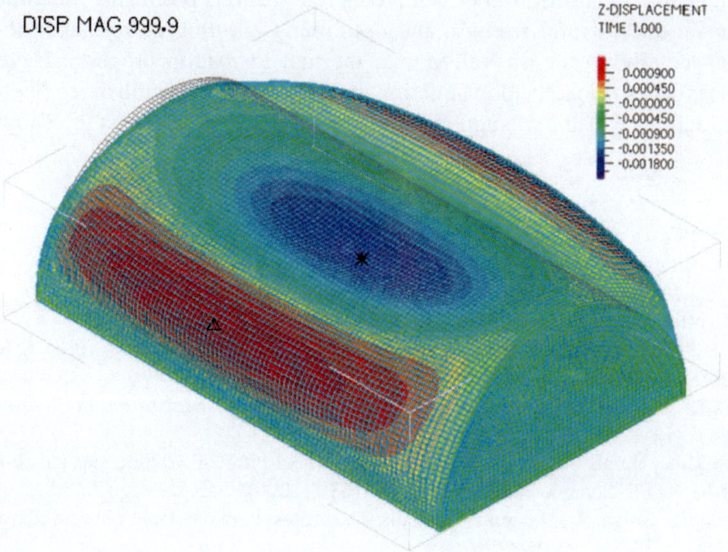

Fig. 4. Vertical displacement due the applied loads. The maximum displacement in the center vault is 1.8 mm.

4 Conclusions

The pre-sizing analysis of a Martian concrete vault structure provided results indicating tensile stress levels below the design threshold. The influence of Martian loads, particularly the internal pressure and the temperature gradient, resulted in a prominent tensile response, while the dead load alone proved insufficient in countering the radial pressure exerted on the structure.

Introducing gypsum hemp concrete has shown a significant improvement in flexural strength, allowing for a significant reduction in compressive strength but maintaining a crucial residual tensile strength, making it an ideal choice for the challenging Martian conditions. However, further refinement is necessary to establish a precise constitutive law, necessitating an essential experimental campaign to facilitate accurate non-linear finite element analyses.

Moreover, it is essential to explore alternative structural designs compared to the conventional ones used on Earth. The disparity in applied loads on Mars necessitates reconsidering the traditional forms designed primarily to withstand gravitational forces. Such forms may not be the most efficient or effective in the unique Martian environment. As a result, it becomes decisive to study and develop innovative structural configurations that can adapt to the specific conditions on Mars. This could involve utilizing novel materials and designs that optimize structural strength and efficiency in the Martian setting.

In conclusion, the utilization of hemp concrete exhibits promising potential for fortifying tensile and flexural strength and optimizing building performance in Martian habitats, especially as for its well-known thermal insulation property. Nevertheless, additional rigorous research and validation are imperative to establish resilient and sustainable structures capable of supporting future human settlement on the enigmatic red planet.

References

1. Musk, E.: Making life multi-planetary. New Space **6**, 2–11 (2018)
2. NASA. http://mars.nasa.gov/mars2020. Accessed 03 May 2023
3. Starr, S., Muscatello, A.: Mars in situ resource utilization: a review. Planet. Space Sci. **182**(104824) (2020)
4. Liu, J., Li, H., Sun, L., Guo, Z., et al.: In-situ resources for infrastructure construction on Mars: a review. Int. J. Transp. Technol. **11**, 1–16 (2022)
5. Bandfield, J., Hamilton, V., Christensen, P.: A global view of Martian surface compositions from MGS-TES. Science **287**(5458), 1626–1630 (2000)
6. Banfield, D., Spiga, A., Newman, C., et al.: The atmosphere of Mars as observed by InSight,» Nat. Geosci. **13**(3), 190–198 (2020)
7. Hu, T.S., Muqiu, C., et al.: Research progress on lunar and Martian concrete. Construct. Build. Mater. **343**(128117) (2022)
8. Reches, Y.: Concrete on mars: options, challenges, and solutions for binder-based construction on the Red planet. Cement Concr. Compos. **104**(103349) (2019)
9. Ramstad, R.: Global mars-solar wind coupling and ion escape. J. Geophys. Res. Space Phys. **8**(122), 8051–8062 (2017)
10. Bertagnoli, G., Tondolo, F., Canonico, F., Anerdi, C., Buzzi, L., Mancini, G.: Calcium Sulfoaluminate based concrete: a mechanical characterization. IOP Conf. Ser. Mater. Sci. Eng. **471**, 032006 (2018)
11. Sassoni, E., Manzi, S., Motori, A., Montecchi, M., Canti, M.: Novel sustainable hemp-based composites for application in the building industry: physical, thermal and mechanical characterization. Ener. Build. **77**, 219–226 (2016)
12. Zampori, L., Dotelli, G., Vernelli, V.: Life cycle assessment of hemp cultivation and use of hemp-based thermal insulator materials in buildings. Environ. Sci. Technol. **47**(13), 7413–7420 (2013)
13. NASA. https://www.nasa.gov/feature/can-plants-grow-with-mars-soil. Accessed 03 May 2023
14. Awwad, E., Mabsout, M., Hamad, B., Talal Farran, M., Khatib, H.: Studies on fiber-reinforced concrete using industrial hemp fibers. Construc. Build. Mater. **35**, 710–717 (2012)

15. Parcesepe, E., De Masi, R.F., Lima, C., Mauro, G.M., Pecce, M.R., Maddaloni, G.: Assessment of mechanical and thermal properties of hemp-lime mortar. Materials **14**(882) (2021)
16. Melchiorre, J., Manuello Bertetto, A., Marano, G.C.: Application of a machine learning algorithm for the structural optimization of circular arches with different cross-sections. J. Appl. Math. Phys. **9**(5) (2021)
17. Manuello, B.A., Gabriele, S., Marmo, F., Micheletti, A.: Shell and spatial structures: between new developments and historical aspects. Curved Layered Struct. **7**(1), 186–187 (2020)
18. Wilson, C.F.: Measurement of Wind on the Surface of Mars, Linacre College (2003)

Exploring the Potential of Fibrous Tectonics for Facade Application: A Form-Finding, FEM Analysis, and Optimization study

Amirhossein Ahmadnia(✉), Salvatore Viscuso, Carol Monticelli, and Alessandra Zanelli

DABC, Politecnico di Milano, Via Ponzio 31, 20133 Milano, Italy
Amirhossein.Ahmadnia@polimi.it

Abstract. Fiber-reinforced polymer composites have garnered considerable interest in the field of structural engineering due to their promising mechanical properties. However, their application at the building scale still requires further efficiency. Current design approaches often rely on traditional methods used for conventional structural elements; in contrast, this paper focuses on the concept of fibrous tectonics. While thermoset matrices are commonly employed in this composite, our study focuses on developing a comprehensive computational workflow encompassing design, structural analysis, and optimization specifically tailored for fiber-reinforced thermoplastic polymer composite roving in facade applications. The primary objective is to minimize the length of the fiber roving used while ensuring acceptable levels of utilization and deflection, in addition to proposing possible architectural functions for using FRP at building scale, e.g., shading façades.

Keywords: Fiber-reinforced thermoplastic polymer composite · Finite-element analysis · Multi-objective optimization · Structural element · Geometrical form-finding · Fibrous construction

1 Introduction

Fiber-reinforced polymer (FRP) composites are composed of a polymer matrix embedding a fiber reinforcement, which can be a thermoplastic, elastomer, or thermoset material (Bhargava 2004). FRP composites possess notable characteristics such as high strength, lightweight, excellent fatigue resistance, a high elastic modulus, low thermal expansion, and corrosion resistance (Liu et al. 2015). Additionally, they exhibit anisotropic behavior, meaning they demonstrate varying properties in different directions (Waimer et al. 2013). Considering their composition, which consists of approximately 60% fiber reinforcement and 40% polymer matrix (Liu et al. 2015; Bhargava 2004), the mechanical properties of FRP primarily depend on the fiber direction, resulting in relatively high strength and modulus, while in the direction perpendicular to the fiber axis, the properties are influenced by the polymer and resin, leading to relatively lower strength and modulus (Liu et al. 2015). Also, they have some disadvantages like poor performance at high temperatures and expensive material and manufacturing costs (Viscuso et al. 2023).

S. Gabriele et al. (Eds.): IWSS 2023, LNCE 437, pp. 588–598, 2024.
https://doi.org/10.1007/978-3-031-44328-2_61

Although FRP composites have been used since the 1930s, they are now gaining popularity in the construction and structural industries (Bakis et al. 2002). They find applications in various industries in place of metal, including construction, military, energy, transportation, chemical, and electric power (Xue et al. 2022). FRPs can be used as structural elements, bridge decks, internal or external reinforcements for concrete structures, and for repair, strengthening, and pre-fabricated shapes (Bakis et al. 2002). However, there is still a need to optimize their design and fabrication processes for use on a building scale, as they have been successfully used in the aerospace and automotive industries (Waimer et al. 2013).

Despite the introduction of FRP as a novel material, traditional architectural and civil engineering design approaches rooted in conventional materials like wood, concrete, and steel continue to dominate architectural practices (Moskaleva et al. 2021). Achim Menges and Jan Knippers present an innovative biomimetic design approach in "Fibrous Tectonics" (2015) that was inspired by fibrous structures, which are found in most load-bearing natural systems. This paper also considers the use of FRP roving, considering the concept of fibrous tectonics, as a load-bearing structural element to create more efficient and sustainable structures, considering the unique properties and behavior of tailor-made continuous composites.

2 Research Aims and Methodology

Considering the costs, the traditional way of using FRP, and the lack of methodology to functionally use them in the building sector, this paper aims to present a computational workflow for the design, analysis, and optimization of fibrous construction as a facade structure. The investigation focused on the anisotropic behavior of tailor-made fibrous construction, which requires a different methodology compared to mesh-based orthotropic structures presented by Viscuso et al. (2023). The research project and investigation criteria can be summarized in the following list:

(i) A uniaxial lab test to determine the mechanical properties of the FRP;
(ii) Form-finding process and random fibrous pattern generation;
(iii) FEA analysis of the generated pattern to calculate the utilization by Karamba3D;
(iv) Multi-objective optimization for finding the best results based on equilibrium between material usage and structural efficiency;

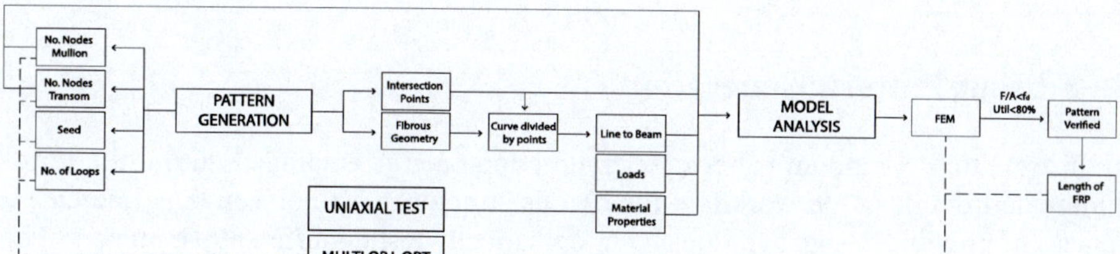

Fig. 1. Research overview

3 Uniaxial Lab Test of FRP Roving

The aim of a uniaxial lab test is to determine the mechanical properties of FRP roving, which refers to a bundle of straight strands wound onto spools. While thermoset resins are commonly used in fiber-reinforced polymer composites, this study focuses on a transparent elastomer thermoplastic resin. In this research project, the mechanical test was conducted on six specimens of pre-impregnated fiber-reinforced thermoplastic polymer composite with a 1200 TEX. The specimens had a cross-section area of 3.5 x 0.4 mm and a density of 2.5 g/cm^3.

The test involved applying a controlled load in a single direction to the glass fiber-reinforced thermoplastic polymer (GFRTP) composite sample until it reached the point of failure, allowing for the measurement of its breaking strain. The test procedure followed the guidelines outlined in Standard UNI ISO EN 1421:2017 and ISO 4606:1995. To ensure consistency, an initial strain of 10 N and a test speed of 100 mm/min were set for the experiment, as suggested by Besnard et al. (2006). The obtained test results provide crucial information about the linear anisotropic mechanical properties of the material, including tensile strength, elastic modulus, and other relevant parameters. These properties play a vital role in the design and analysis of the material for various structural applications, such as beams. Figure 2. Shows the uniaxial test on GFRTP samples.

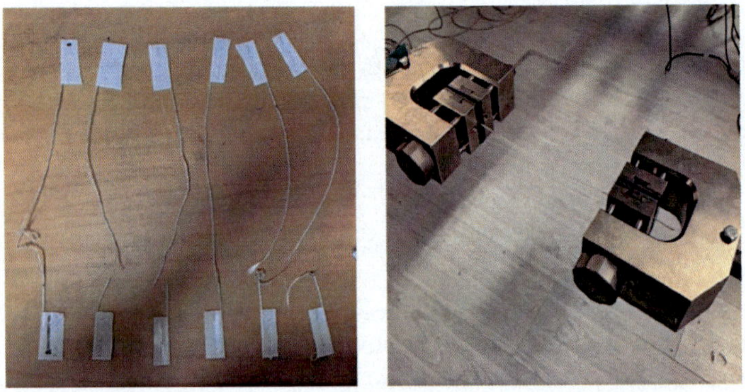

Fig. 2. Uniaxial Lab test of FRP roving done by TEXTILESHUB laboratory of POLIMI

4 Fibrous Pattern Generation

In this research, a random pattern generation approach is employed during the form-finding and optimization workflow for fibrous structures. The objective is to create a data set of possible design solutions by systematically testing different parameters. The random pattern generation is implemented using Grasshopper, a visual programming language integrated with Rhinoceros 3D. The authors have developed a custom code within GhPython for this purpose. The process begins with the existing frame serving as the base for the fibrous structure. The frame is divided into consecutive segments consisting of two mullions and two transoms. The number of nodes on each mullion and transom is determined by the input parameters X and Y. These nodes act as support points for the fibrous structure and are strategically placed on both mullions and transoms to facilitate the FRTP winding process. Certain constraints are applied to ensure the workability of the fibrous pattern; for instance, consecutive points cannot reside on the same mullion or transom.

The nodes within each segment are organized into four separate lists, each with distinct indexes. In each iteration of the fibrous structure, the code generates four random numbers within the range of 0 to (X and Y inputs minus 1), one for each list. These random numbers represent the index of the selected node from each list. By connecting these four selected nodes together, a loop of fibrous structure is formed. Each loop is represented as a closed polyline, resulting from connecting the four lines. To illustrate the concept, Fig. 3 provides a simplified depiction of the code generating one loop of the fibrous facade pattern.

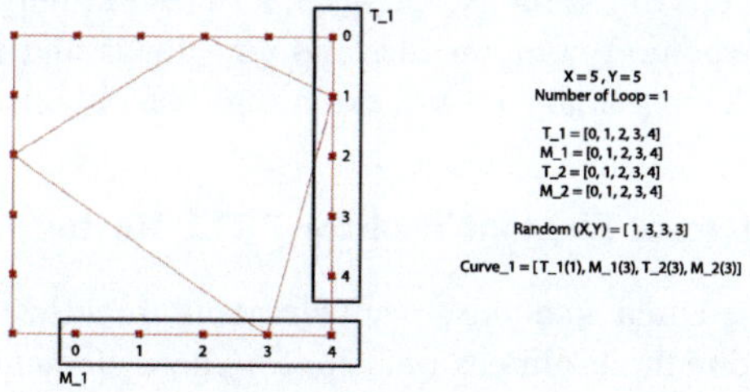

Fig. 3. Random Pattern Generation concept

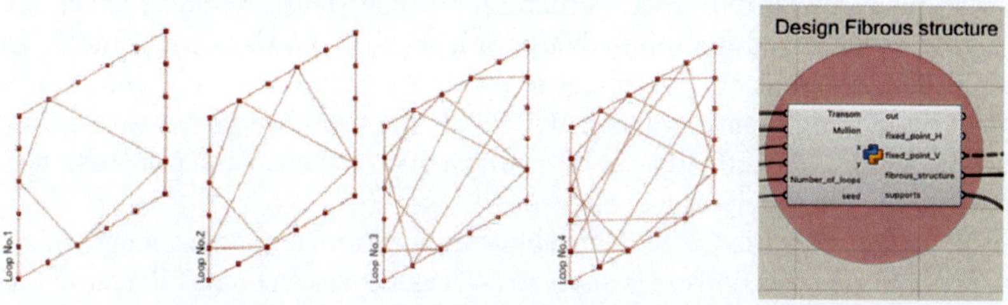

Fig. 4. Loop Generations and scripted code for generation fibrous pattern

Figure 4 illustrates the inputs and outputs of the scripted component used for random fibrous pattern generation. The 'seed' input serves as a pseudorandom number generator, influencing the generation of the pattern. On the other hand, the 'number_of_loops' input determines the quantity of loops involved in generating the fibrous pattern.

The outputs of the component are the supporting points and the fibrous structure itself. Both outputs play essential roles in the subsequent structural analysis. The values of the inputs, including 'X,' 'Y,' 'Number_of_loops,' and 'seed,' are variables that can be modified to produce different results. By adjusting these input values, a new fibrous pattern with distinct characteristics can be generated, enabling exploration and evaluation of various design possibilities.

5 Finite-Element Modeling (FEM) Analysis

The results obtained from the uniaxial mechanical test served as the basis for conducting finite-element modeling (FEM) simulations using Rhinoceros software. The previously generated fibrous patterns can now undergo structural analysis, taking into account the applied loads and assigned mechanical properties. The FEM analysis is conducted using the Karamba plugin for Grasshopper, an interactive and parametric finite element program (Preisinger et al. 2014). To analyze the behavior of FRTP roving in Karamba 3D, the first crucial step is to determine the appropriate structural element type for simulation. One of the key questions is whether the FRTP roving should be modeled as a beam, truss, or cable element. Once the choice is verified, the analysis process can proceed by following the typical steps of assembly, analysis, and result interpretation. These steps involve defining the structural components, applying loads and boundary conditions, performing the finite element analysis, and examining the obtained results.

5.1 Selection of Structural Element Type for FRTP Roving

In the context of finite-element analysis, beam elements are utilized as one-dimensional finite elements oriented in three-dimensional space. These elements require a minimum of two nodes placed at their ends to define their geometric properties. Beam elements are particularly well-suited for representing structural elements where one dimension dominates over the other two. In the case of FRTP roving, while it can be initially defined as a beam, it exhibits a bending moment of zero. This characteristic aligns more closely with the behavior of a 'truss' structural element. Truss members are assumed to be connected by idealized pin joints that permit rotational movement while restricting translation. The pin joints enable trusses to primarily bear axial loads, either in tension or compression, along their members. For FRTP, the presence of pre-tensioning must also be considered. Consequently, FRTP roving can be regarded as a pre-tensioned truss structural element or simply as a cable.

To validate this condition, a FRTP roving structural element with a length of 5 m and a cross-sectional area of 0.056 cm is analyzed. The analysis accounts for a point load of 1 kN applied at the element's center and two fixed translation supports at its ends. Manual calculations suggest a maximum displacement of approximately 91 (Fig. 5–3); however, the finite-element analysis yields different results. When considering the FRTP roving structural element as a continuous element, the calculated maximum displacement is zero (Fig. 5–1), indicating an incorrect calculation. To address this discrepancy, the FRTP roving structural element is divided into smaller elements (in this case, five pieces), which are connected by fixed translation joints (Fig. 5–2). The obtained result aligns with the manual calculation, confirming the accuracy of the simulation and the suitability of the truss structural element type for FRTP roving.

The results of the finite-element analysis (FEA) have been obtained using the Karamba 3D algorithm based on second-order theory. In this analysis, the displacements are directly related to linear deformations and tensions within the structural elements. By employing this approach, a more accurate representation of the behavior of the FRTP roving can be achieved.

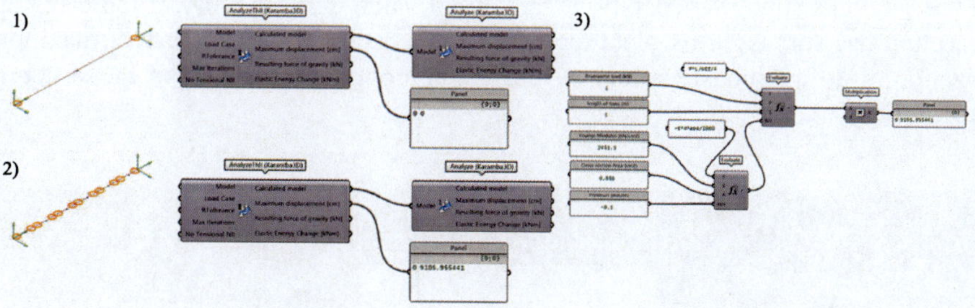

Fig. 5. Verification of Structural Element Type for FRTP Roving

5.2 Analysis Process: Assembly, Analysis, and Result Interpretation

During the assembly stage, several components such as "LineToBeam," "Support," "Load," "MaterialProperties," and "Cross-Section" are utilized to feed the "AssembleModel" component. The polylines obtained from the pattern generation result are connected to the "LineToBeam" component to define the geometry and shape of the fibrous structure. The nodes generated in the pattern generation process are designated as "supports" with zero degrees of freedom in translation (to represent the pin joints). The "load" component incorporates various load conditions. The initial strain is defined to account for the pre-tensioning of the fibrous beams. The wind load is calculated based on EN 1991 wind action on buildings for Milano, Italy, and is applied as a linear distributed mesh load. Additionally, the effect of gravity is considered by applying a "gravity" load. The "temperature load" is also taken into account as a relevant load condition. The inputs for the "MaterialProperties" component are determined based on the results of the uniaxial mechanical test. While FRTP roving behaves as an isotropic material, it is important to note that FRP itself is an anisotropic material. The cross-section is simplified as a filled trapezoidal section with a height of 0.04 cm and a width of 0.35 cm for a single layer of FRTP roving. The number of roving layers can influence the height of the section, which affects the moment of inertia and maximum deflection of the beam. In the final step, all these components are connected to the "AssembleModel" component, which collects the necessary data and initiates the analysis process.

Finally, the model is ready to be analyzed using the 'AnalyzeThII' component, which employs second-order theory. This component calculates the maximum vertical displacement in the fibrous structure, a critical parameter for verifying its performance. According to the EN-14351-1 standard, the deflection limit under wind load should not exceed the length divided by 200 or 15 mm. This criterion is a benchmark to assess the maximum vertical displacement obtained in the analysis. The 'ReactionForce' and

'BeamForce' components provide valuable data on reactions and internal forces, respectively. The reaction forces are instrumental in designing the structural nodes, while the axial forces enable accurate dimensioning of the FRTP roving.

The 'utilization' component offers insights into the relationship between stresses and the section's strength capacity. If the output value exceeds unity, it indicates that the section has surpassed its yield tension. In this research, a conservative maximum utilization of 80% is considered to ensure the appropriate utilization of the structural elements (Fig. 6).

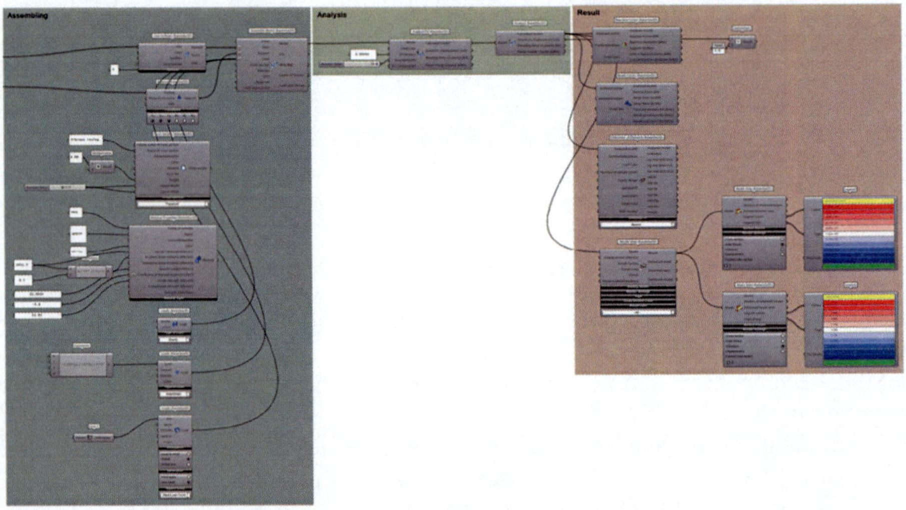

Fig. 6. Loop Generations and scripted code for generation fibrous pattern

6 Multi-objective Optimization (MOO)

The combination of random fibrous pattern generation and FEM using Karamba 3D offers the opportunity to employ multi-objective optimization (MOO). In the field of computer-aided architectural design (CAAD), the form-finding process has evolved beyond a single objective of structural form-finding. Designers are now guided by multiple objectives to achieve the best or near-best results (Brown 2016). The MOO process (Gunantara 2018) enables the identification of optimal fibrous patterns that are structurally verified and possess the minimum possible length. Table 1 and Table 2 present the variables and objectives involved in the optimization process. To perform multi-objective optimization, the research utilizes Wallacei, a plugin for Grasshopper developed by Makki et al. (2019). Wallacei incorporates the NSGA-2 (Non-dominated Sorting Genetic) algorithm (Deb et al. 2000) as the primary evolutionary algorithm, empowering users to run comprehensive evolutionary simulations.

Table 1. Table of Design Space Variable

Symbol	Variables	Font size and style
X1	Number of anchor points on mullion	x > 5
X2	Number of anchor points on transom	x > 5
X3	Number of loops	1 < x < 20 (Assumption)
X4	Random Selection of anchor points (seed)	x > 1
X5	Number of FRTP roving layers	1 < x < 5 (Assumption)

Table 2. Table of Objective Function Assumptions

Objectives	Metric (condition, unit)	Evaluation method
Max Deflection	Max deflection in model (= < 15, mm)	FEA
Max Utilization	Utilization of all active elements (< 80, %)	FEA
Structural Efficiency	FRP's length (min, m)	Geometric measurement

7 Results and Visualization

After setting up the fibrous pattern generation, finite-element analysis, and defining the variables (genes) and objectives, the optimization process can be initiated. The search space consists of 3.1 x 10^6 potential results; however, for this study, the researchers selected a population size of 8000 individuals in 400 generations, with 20 individuals per generation. While the first generation was generated randomly, subsequent generations are produced through the mutation of the previous generation, introducing further variation and exploration. For result visualization, Fig. 7–1 presents a projection of all the results onto a 3D tri-objective plot, offering a comprehensive visual representation of the optimization outcomes. The Pareto front are indicated with yellow rectangles. Additionally, Fig. 7–2 displays parallel coordinate plots depicting each generation based on objectives. In the plots, earlier generations are represented by the red color, while the blue color represents the latest generations. Figure 8 showcases a catalog of results, featuring the 20 results from the Pareto front and an additional selection of 34 results chosen by the authors.

The final decision and selection of the optimal solution can be influenced by various factors, including the prioritization of design variables (Brown 2018), aesthetic considerations, ease of construction, or other measurable variables like the calculation of shading to enhance comfort within internal spaces. While the decision-making process is not extensively explored in this research, Fig. 9 provides an illustration of axial forces, utilization forces, and deflection for one of the potential solutions.

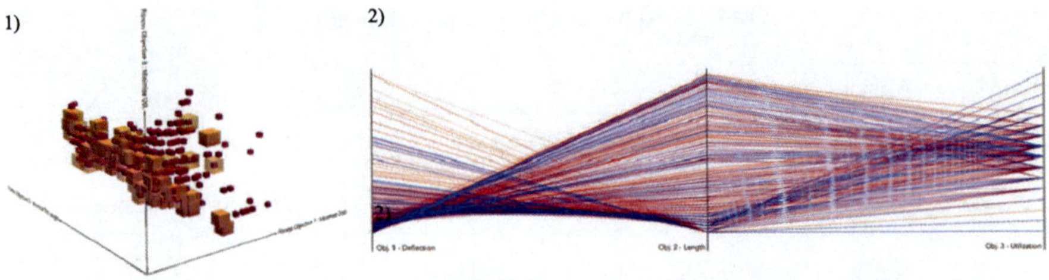

Fig. 7. Visualization of data on the tri-objective and parallel coordinate plots

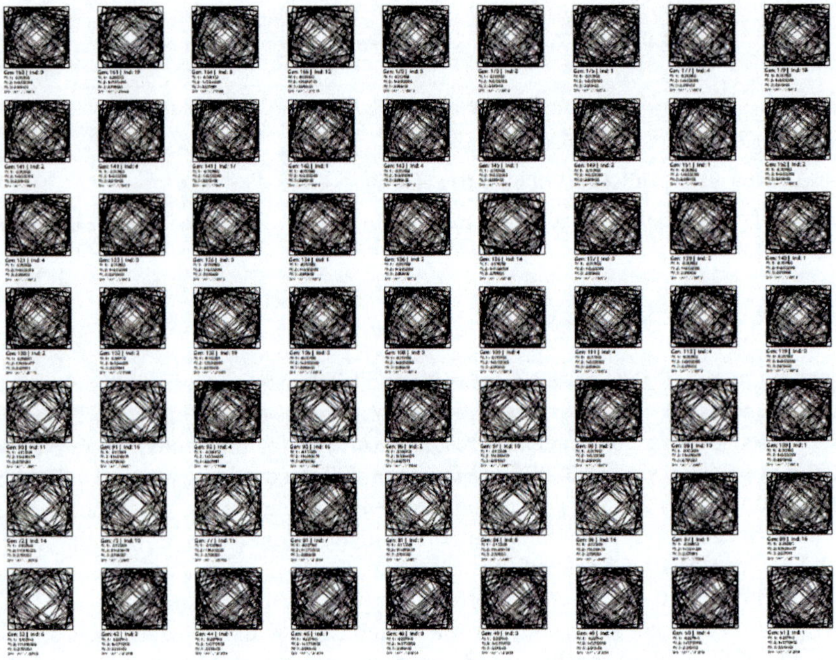

Fig. 8. Catalogue from design space sampling

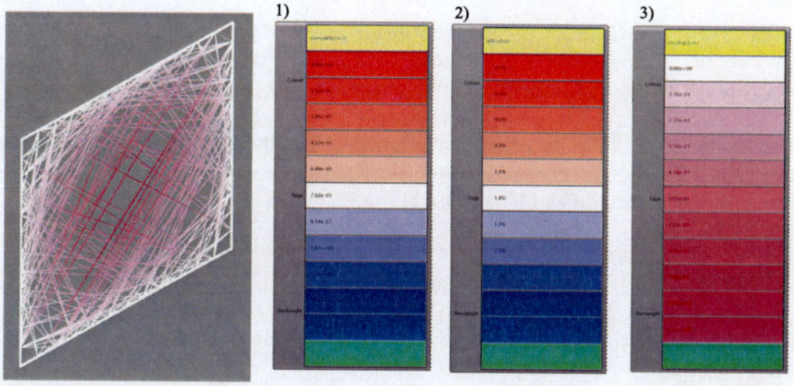

Fig. 9. FEA of one possible solution (1-Axial Stress, 2-Utilization, 3-Deflection)

8 Conclusion

This paper presents a methodology that integrates structural analysis into the architectural design environment for fibrous structures. The computational workflow enables designers to seamlessly design, analyze, simulate, and optimize their structures by exchanging data between Rhino Grasshopper, Karamba, and Wallacei. This integration offers benefits such as improved efficiency, enhanced design exploration, and informed decision-making. The interactive visualization of optimized solutions provides designers with a clear understanding of the structural verification process. This methodology has the potential to significantly advance the field of fibrous structures and transform the design process, allowing for more efficient and informed design decisions.

We would like to express our sincere gratitude to i-mesh company for providing the materials used in this research. Additionally, we extend our thanks to the TextilesHUB lab at POLIMI for conducting the uniaxial tests on the materials. Their expertise and assistance have greatly contributed to the success of this research.

References

Bakis, C.E., et al.: Fiber reinforced polymer composites for construction; state-of-the-art review. J. Compos. Constr. **6**(2), 73–87 (2002)

Besnard, G., Hild, F., Roux, S.: Finite-element displacement fields analysis from digital images: application to Portevin - Le Châtelier bands. Exp. Mech. **46**, 789–803 (2006)

Bhargava, A.K. 2004. Engineering Materials: Polymers, Ceramics and Composites. New Delhi, India: Prentice-Hall of India Pvt. Ltd.

Brown, N.: Multi-Objective Optimization for the Conceptual Design of Structures. Massachusetts Institute of Technology, Cambridge (2016)

Deb, K., Agrawal, S., Pratap, A., Meyarivan, T.: A Fast Elitist Non-Dominated Sorting Genetic Algorithm for Multi-Objective Optimization: NSGA-II. In: International Conference on Parallel Problem Solving From Nature, pp. 849–858. France, Springer, Paris (2000)

Gunantara, N.: A review of multi-objective optimization: methods and its applications. Cogent Eng. **5**, 1 (2018). https://doi.org/10.1080/23311916.2018.1502242

Liu, Y., Zwingmann, B., Schlaich, M.: Carbon fiber reinforced polymer for cable structures—a review. Polymers **7**, 2078–2099 (2015). https://doi.org/10.3390/polym7101501

Makki, M., Showkatbakhsh, M., Song, Y.: Wallacei Primer 2.0. (2019). https://www.wallacei.com/

Menges, A., Knippers, J.: Fibrous tectonics. Archit. Des. **85**, 40–47 (2015)

Moskaleva, A., Safonov, A., Hernández-Montes, E.: Fiber-reinforced polymers in freeform structures: a review. Buildings **11**, 481 (2021)

Preisinger, C., Moritz, H.: Karamba - a toolkit for parametric structural design. Struct. Eng. Int. **24**(2), 217–221 (2014)

Viscuso, S., Monticelli, M., Zanelli, A., Fiorenzi, A.: Development and validation of an experimental methodology for the characterization and FEM analysis of fibre-reinforced architectural meshes. In: Tensinet Symposium 2023, Nantes, France (2023)

Waimer, F., La Magna, R., Reichert, S., Schwinn, T., Menges, A., Knippers, J.: Rethink prototyping design modeling symposium. In: Gengnagel, C., (ed.) Proceedings, pp. 277–90. Verlag der Universität der Künste Berlin, Berlin 2013, pp. 277–290 (2013)

Xue, X., Liu, S.Y., Zhang, Z.Y., Wang, Q.Z., Xiao, C.Z.: A technology review of recycling methods for fiber-reinforced thermosets. J. Reinf. Plast. Compos. **41**(11–12), 459–480 (2022)

Spatial Plates Structures Behavior: The Case Study of the Control Tower of Swiss Railways in Pollegio – CH

Valeria Gozzi[1(✉)], Cristina Zanini Barzaghi[2], Thomas Brunelli[3], and Bernardino Chiaia[4]

[1] Politecnico di Torino – SUPSI, Zanini Gozzi Sagl, Tessin, Switzerland
valeria.gozzi@supsi.ch
[2] Zanini Gozzi Sagl, former Borlini & Zanini SA, Paradiso, Switzerland
[3] Borlini & Zanini SA, Collina d'Oro, Switzerland
[4] Politecnico di Torino, Turin, Italy

Abstract. The proposed case study is the control tower of Swiss Railways in Pollegio - the railway sector with the 50 km long Gotthardtunnel through the Alps. The structure is the result of a competition held in 2006, the construction has been completed in 2014 and it is until now under deformation control.

The control tower's periscope shape, with an asymmetrical upper box and a solid lower shaft, is the direct result of the load-bearing structure being made of prestressed reinforced concrete plates. The shape resulted from specific choices made in the early design stages together with the architects.

The structure is a complex spatial system that works three-dimensionally and required many different and detailed studies.

In this paper, we will present the overall design process. That process began with the traditional analysis methods based on graphic statics and tension fields refined with finite element analysis that confirmed the important interactions between the multiple structural elements. Those interactions were ultimately validated by the results obtained from continuously monitoring the structure from the construction phase to the present day.

Keywords: spatial plate structure · graphic static · tension fields · monitoring · correlation of testing/analysis

1 Introduction and Context

To understand the design method proposed in this paper, contextualizing it within Switzerland's architectural and engineering culture [1] is essential. The Swiss approach to design has traditionally been a dialogue between architecture and structure [2]. This approach emerged from a historical and academic context that fostered numerous infrastructure projects in the country since the 19th century.

The establishment of the ETH Zurich in 1855 aimed to train professionals capable of designing the infrastructure needed to connect different parts of the country. Karl

S. Gabriele et al. (Eds.): IWSS 2023, LNCE 437, pp. 599–608, 2024.
https://doi.org/10.1007/978-3-031-44328-2_62

Culmann's "Die Graphische Statik" [3] in 1864 promoted a geometric and graphic method for analyzing structures, distinct from other European schools' analytical and mechanical approaches. The Swiss school's focus on drawing-led analysis emphasized the formal and aesthetic aspects of artifacts [4].

Another significant aspect of the design process in this paper is the role of materials, particularly concrete. Switzerland's abundant quarries of crushed stone, gravel, and limestone have fostered a strong tradition of concrete construction. Engineer R. Maillart's work on concrete bridges between the two world wars revolutionized the use of the material [5] [6]. His work facilitated the design later by engineers, trained at ETH in the wake of its innovative experiences. More recently, architect Flora Ruchat [7] issued the architectural guidelines for AlpTransit railway line, which is part of Swiss Federal Railways systems including base tunnel and plain lines from Lugano to Flüelen, where the Pollegio control tower is located.

This contextual background sets the stage for the conception of the "Periscope," a control tower at the southern portal of the Gotthard Base Tunnel, resulting from a design competition held in 2006, that encouraged collaboration between architects and engineers (Fig. 1).

Fig. 1. Landscape insertion of the structure, photographer Hélène Binet

2 Case Study

The Berlin-based architecture firm, Bruno Fioretti Marquez Architekten, proposed a sculptural entrance to a tunnel inspired by the Basque artist Chillida. The structure resembles an ever-changing object depending on the perspective of passing train passengers. It also serves as an observant eye for the landscape and infrastructure. Initially conceived as a periscope shape, it evolved into a box rotated on a slanted stem with asymmetrical overhangs and a west-facing window [8, 9].

The structural language comes from different reference buildings that were the products of intense collaborations between an architect and an engineer: the Voltastrasse school in Basel (2000, architect Miller Maranta and engineer Conzett Bronzini) [10] and the Forsterstrasse residential building in Zurich (2003, architect C. Kerez and engineer J. Schwarz) [11].

The 33 m high building combines architectural aesthetics and static requirements with a lower massive shaft and an upper box-like section box with asymmetrical overhangs of 16 m and 9 m. Overall stability is ensured by the balance given by two jutting bodies of different sizes but similar mass and the resultant vertical loads of the building act on the barycenter of the shaft. Reinforced concrete with internal insulation is used as the primary material for the load-bearing structure. The lower floors handle technical activities with loads evenly distributed along the perimeter walls. At the upper floors' transverse wall beams, special steel flag profiles with connectors are inserted to ensure the passage of the large, concentrated forces from the upper box to the shaft (Fig. 2).

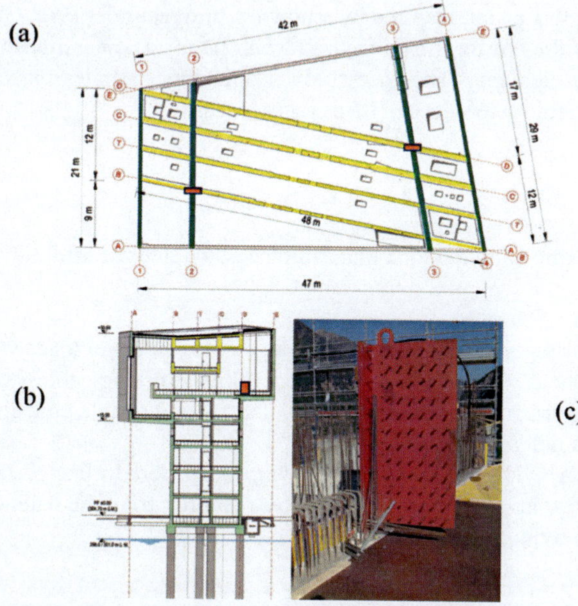

Fig. 2. Upper box structural scheme (a), structure vertical section (b) showing steel connection flag positions and picture (c)

The control room's roof, spanning approximately 33 m, resembles bridge-like box girders, supported by four prestressed longitudinal beams of varying height (1.80 m to 3.30 m). The design features a two-pitch structure with a compluvium, aiding rainwater drainage. The 6th-floor office slab is suspended from the roof, ensuring the 5th-floor room, where railway traffic is controlled, remains column-free and transparent. The cantilever part's loads are borne by four reinforced transverse wall beams with prestressing cables, and the arrangement of openings is carefully designed to meet plant engineering requirements.

3 Spatial Structures Design

The analysis method proposed in this paper has its origins in Switzerland's academic environment. In 1899, Karl Wilhelm Ritter developed the concept of truss models embedded in concrete, with tension elements made of reinforcement and compression elements of concrete [12]. This theory was further extended and studied by Emil Mörsch [13], Leonhardt and Walther [14], leading to the definition of design basis in the CB FIB Model code in 1978 [15].

Later, Schlaich and others [16] applied the theory of plasticity to the strut-and-tie model, and ETHZ's research on plastic failure analysis, led by Prof. Bruno Thürlimann [17], was introduced in the Swiss SIA codes in 1989 [18]. These approaches were further developed, including the stress field method by Muttoni [19], whose research at ETHZ culminated in the book "Design of Concrete Structures with Stress Fields" published in 1997 [20].

The authors of this paper, inspired by renowned bridgebuilder Prof. Christian Menn [21, 22], extended the tension fields method from 2D to 3D, decomposing monolithic structures into 2D elements. The case study in this paper applies these fundamental concepts and academic references in its design process.

4 Method

The calculation method applied took place in successive steps accordingly to the project phases.

a. Competition phase.

From the early stage, the concept of carrying on the cantilever loads with full-height wall beams reinforced with horizontal and vertical prestressing cables was elaborated. The analysis was performed with manual calculations and simple conceptual drawings showing the stress fields in the wall beam.

Foundations were immediately planned with deep bored piles of large diameter, starting from the tensile and compression forces resulting from the balance of forces in the plane of the individual wall beams (Fig. 3).

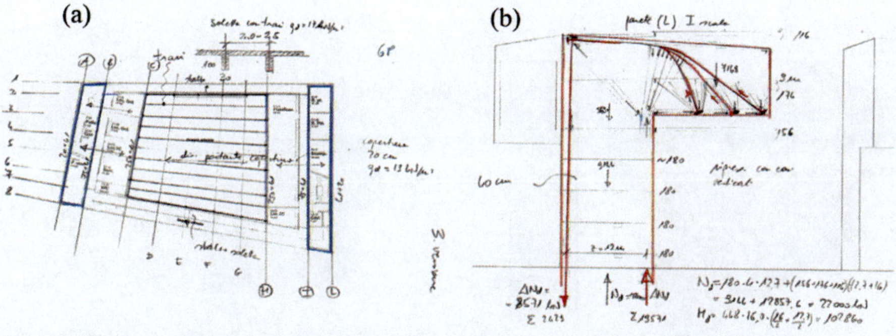

Fig. 3. Loads pattern analysis for competition, in plan (a) and in section (b)

b. Final design.

b1 Refinement of competition design.

A detailed analysis of the load transmission and decomposition of horizontal (slab-plates) and vertical (wall slabs and beams-wall) elements was carried out. Then, a comparison with two-dimensional calculations FEM program (software Cedrus) was conducted to calibrate the position of the openings and set the concept of prestressing and introduction of loads concentrated on the underlying walls of the massive shaft.

b2 Quickly moved to the three-dimensional calculation.

At the same time, it has been investigated the interaction between the different elements of the building, using a 3D model (software AXIS).Since that:

- the connections are monolithic and,
- horizontal plates can be activated against the in-plane horizontal loads as slabs.

This analysis shows that the internal wall beams (n.2 and n.3), theoretically more loaded, bear only 60% of the vertical loads foreseen during the competition design refinement (b1 phase). This is explained because the two pairs of wall beams interact with the slabs and behave like bent tubes with a compressed flange below and a tensioned flange above (Figs. 4 and 5).

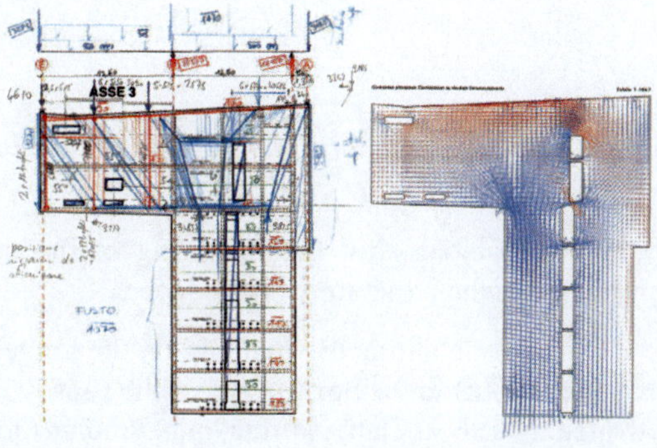

Fig. 4. Comparison of bidimensional analysis with software Cedrus stress-fields

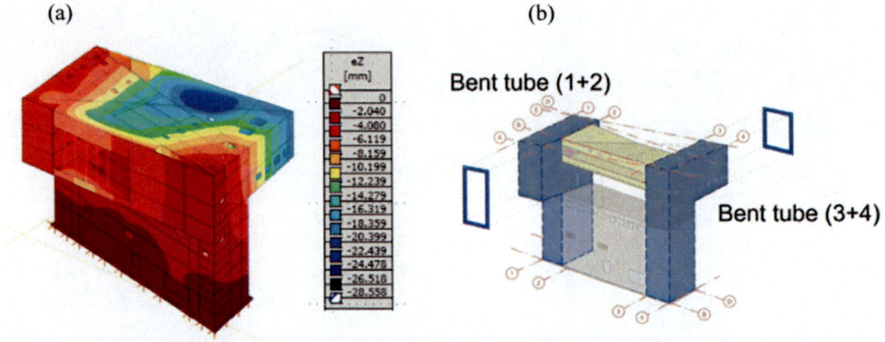

Fig. 5. 3D Model with software AXIS: vertical displacements (a) and bent tubes, stem and roof structure (without façades A and E) (b)

c. Executive project.

In the executive project, we deepened how the different structural elements interact to bear the torsion between the two cantilever parts in an asymmetrical way.

Numerous 3D models have been elaborated, simulating different stiffness conditions considering fissured or not fissured behavior for individual elements. In addition, a simplified model has been developed to make it easier to compare the results (FIg. 6).

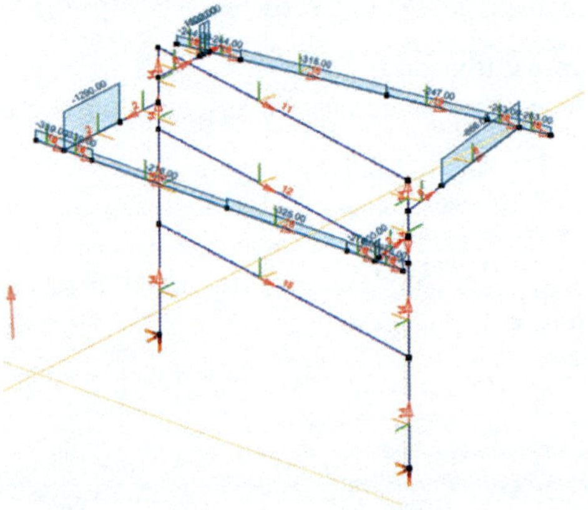

Fig. 6. Simplified structural model for the comparison of different stiffness

With these analyses and comparisons, the following main equilibria of torsion recovery due to the antimetry of the structure were deduced:

- the two antimetric tubes activate antimetric bending on the longer façades (n.A and n.E), and tubular torsion (Bredt) in the box girders of the roof.
- the roof slab is activated as a shear slab and transmits tubular torsion (Bredt) in the two vertical cores at the ends of the building, which are continuous over the entire height. The torsion is then distributed throughout the perimeter of the stem (Fig. 7).

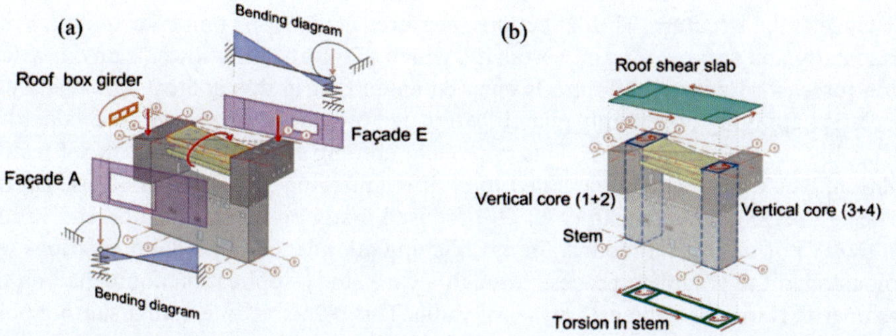

Fig. 7. Façade bending diagram and roof box girder (a), roof shear slab diagram and torsion in the cores and stem (b)

This overlaying of equilibriums ensures that the centre of gravity of the vertical loads, despite the different geometric shapes of the two cantilevered parts, corresponds to the centre of mass of the building: the stem helps to distribute the stresses and loads uniformly on the piles, making the foundation system efficient.

Prestress concept allows the mobilization of these different equilibrium systems, through the creation of ties along the connections of the elements to activate the maximum forces transmission in the plane of the various elements (bent plates).

Dimensioning was then made considering different possible equilibriums and load patterns, to have a certain redundancy as well as robustness.It must be noted that, in this project, in advance of international codes and regulations, robustness concepts have been included and can be related to monitoring results.

5 The Importance of Monitoring

Because of uncertainties involved in modeling, construction, and measurement systems, the assessment of the FE model validation must be conducted based on stochastic measurements to provide designers with confidence for further applications [23]. The objective of model validation is to refine and confirm the analysis model of structure by using reference data obtained from experimental tests and monitoring can thus be used for design optimization.The complexity of the analyzed structure as well as the difficulty to model in a uniquely precise and exhaustive model the overall structural behavior induced the engineer to perform a monitoring concept.The following monitoring has therefore been planned:

- Control of deformations and building overhangs, in directions x y and z
- Control of foundation displacements.
- Crack openings in concrete.
- Post tensioning efficiency (electrical control from cable heads, as in bridges, but not usual in construction)

The deformations have therefore been measured during and after the construction site until today. The monitoring allowed us to verify the plausibility of the static calculation and to ascertain that the real behavior is strongly influenced by the torsional resistance of the structure. The deflections measured at the worst point are approx. 6 mm horizontally and approx. 12 mm vertically, which is significantly lower than our calculation forecast of approx. 40 mm. It must be noted that in this context a test–analysis correlation applying uncertainty quantification and propagation to monitored values has not applied at the design time. This approach, coming from structural model testing tradition, was only recently extended to civil engineering structures analysis. During the design period, Atamturktur et al. [24] focused firstly on the verification and validation (V&V) of numerical models for establishing confidence in model predictions and demonstrated the complete process through a case study application completed on the Washington National Cathedral masonry vaults.This offers an interesting starting point to continue the research and evaluate the degree of uncertainty of monitoring to have greater certainty on the reliability of the calculation model and extend any considerations to future applications (FIg. 8).

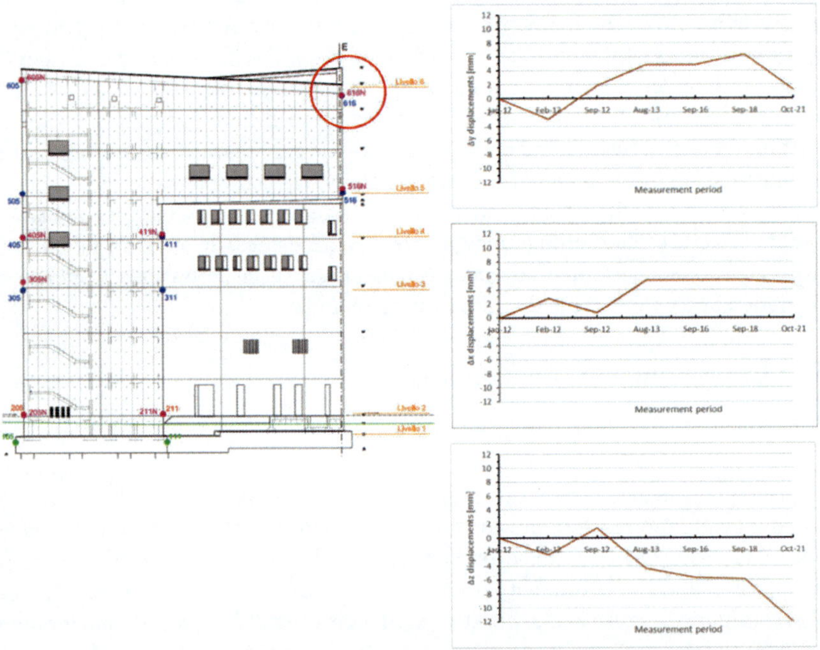

Fig. 8. Monitored points on façade East and displacements evolution of most sensitive point 616

The first measurements of the monitoring have allowed to confirm that the building has an important torsional behavior, so much so that the displacements are more accentuated as the rotation of the upper box than box cantilever displacements, which were initially imagined as a greater risk.

The detailed survey of the trend of the cracks on the facades, of imperceptible openings, confirms the important effect of the torsion of the entire lower shaft.

6 Conclusions

In this paper, the design process of a spatial plate structure is discussed, emphasizing the significance of cultural and technical context in the approach. Collaboration between engineer and architect from the early design stages, along with the use of materials and technical knowledge, is crucial in proposing and implementing structural solutions. The main conclusions are as follows:

1. An efficient structural system should be defined from the early stages of design.
2. The statical solution raise from the specific cultural context in which the engineer operates (see chapt. 3)
3. The spatial systems can be simplified in bidimensional problems using manual analysis performed with strut-and-tie and stress fields methods that are very efficient for concrete plate design.
4. 3D models can be used to validate the first assumptions and analysis and to estimate overall behavior.
5. The different proposed analysis models can be used according to the different project stages.

The importance of refining the method through iterative steps to deepen the understanding of structural behavior is evident. The presented case study evolves from geometric and visual methods to analytical approaches using FEM models.

Complex structures require merging different methods and simplifying the system into manageable sub-parts to comprehend overall behavior and ensure redundancy and robustness. Monitoring plays a vital role and including it in the early design stages is crucial as fundamental design requirements.

References

1. Davos Declaration, International Conference of Ministers of Culture, Davos 20–22, January 2018. https://davosdeclaration2018.ch/. Accessed 20 May 2023
2. Gozzi, V., Boller, G.: Ontologia della costruzione: progettazione concettuale delle strutture tra passato e presente. Archi **1**(2022), 11–14 (2022)
3. Culmann, K.: Die Graphische Statik, Meyer & Zeller, ETH-Bibliothek Zürich, Rar 4547: TEX ex. A. Zürich, Switzerland (1866)
4. Billington, D.P.: The Art of Structural Design: A Swiss Legacy. Princeton University Art Museum, Princeton (2003)
5. Billington, D.P.: Robert Maillart and the Art of Reinforced Concrete. The MIT Press, Cambridge (1990)
6. Aprea, S., Navone, N., Stalder, L., Et al,: Concrete in Switzerland. Histories from the Recent Past, EPFL Press, Lausanne, Switzerland (2021)
7. Maffioletti, S., Navone, N., Toson, C.: Un dialogo ininterrotto. Studi su Flora Ruchat-Roncati, Il Poligrafo. Padova, Italia (2018)
8. Zanini Barzaghi, C.: Periscopio Alptransit Pollegio: la struttura portante di una scultura in calcestruzzo, quaderno speciale trilingue (ita, de, fr), Espazium. Zurigo, Svizzera (2015)
9. Zanini Barzaghi C.: The Periscope in Pollegio: the prestressed structure of a concrete sculture. In: Structural Concrete in Switzerland, 5. FIB Congress Proceedings, Melbourne, Australia, pp. 12–17 (2018)

10. Conzett, J., Mostafavi, M.: Structure as Space: Engineering and Architecture in the Works of Jürg Conzett and His Partners. AA Publications, London (2006)
11. Kerez, C.: Mehrfamilienhaus forsterstrasse. In: Ebner, P., Gerstenberg, F., Höllbacher, R. (eds.) Housing Moves On. Springer. Wien (2009)
12. Ritter, W.: Die Bauweise Hennebique. Schweizerische Bauzeitung **33**(7), 41–61 (1899)
13. Mörsch, E.: Der Eisenbetonbau, seine Theorie und Anwendung, third edition, Verlag von Konrad Wittwer. Hamburg, Germany, 1908, p. 376 (1908)
14. Leonhardt, F., Walther, R.: Wandartige Träger, Deutscher Aus schuss für Stahlbeton. Berlin, Germany 159 p. (1966)
15. CEB-FIP Model Code for Concrete Structures: Comité Euro-International du Béton (CEB). Lausanne, Switzerland (1978)
16. Schlaich, J., Weischede, D.: Ein praktisches Verfahren zum methodischen Bemessen und Konstruieren im Stahlbetonbau, Bulletin d'Information No. 150, Comité Euro-International du Béton. Lausanne, Switzerland, pp. 163 (1982)
17. Thurlimann, B., Marti, P., Pralong, J., et al.: Application of the Plasticity Theory to Reinforced Concrete, Institut für Baustatik und Konstruction, ETH. Zürich, Switzerland, p. 252 (1983). (in German)
18. Swiss Society of Engineers and Architects: Code 162 for Concrete Structures (1989)
19. Muttoni, A.: The Applicability of the Theory of Plasticity in the Design of Reinforced Concrete, Institut für Baustatik und Konstruktion, Bericht, No. 176, ETH. Zürich, Switzerland, 159 pp. (1989). (in German)
20. Muttoni, A., Schwartz, J., Thürlimann, B.: Design of Concrete Structures with Stress Fields, Birkhaüser. Basel, Boston and Berlin, Switzerland (1997)
21. Menn, C., Brühwiler, E.: Stahlbetonbrücken. Springer, Wien, Österreich (2003)
22. Menn, C., Schärer, C.: Christian Menn– Bridges, Scheidegger & Spiess. Zürich, Switzerland (2015)
23. Xiaosong, L., Zhouhong, Z., Niu, J.: Finite element model validation of bridge based on structural health monitoring—Part II: Uncertainty propagation and model validation. J. Traffic Transp. Eng. (2015)
24. Atamturktur, S., Hemez, F., et al.: Verification and validation under uncertainty applied to Finite Element models of historic masonry monuments. In: Conference Proceedings of the Society for Experimental Mechanics Series (2009)

A Parametric BIM Approach for Bridge Design and Analysis

Valentina Boretti[1], Luis Bohórquez[1], Pietro Palumbo[1(✉)], Rebecca Asso[2], and Davide Masera[1]

[1] Masera Engineering Group S.r.l., Turin, Italy
pietro.palumbo@masera-eg.com

[2] Department of Structural, Geotechnical and Building Engineering DISEG, Politecnico di Torino, Turin, Italy

Abstract. Bridges are complex structures in several aspects, from design to construction, mainly due to their complicated and constrained geometry, high design loads and unique boundary conditions. Moreover, the narrow schedules and the demand for efficiency in the civil engineering industry require a new approach. The Parametric BIM approach removes repetitive time-consuming tasks by introducing parameters and numerical relationships between them. This enhances the design process, increasing efficiency and reducing time and effort. This paper presents a methodology for the development of the geometrical model of different bridge typologies through the modification of input parameters. The approach is applied to composite and concrete box girders bridges, enabling efficient analysis of different structural configurations to explore a range of innovative cost-effective solutions. Thus, the workflow is proposed as a supportive tool for decision-making in the design of new complex structures.

Keywords: Parametric Design · BIM · Structural Analysis · 3D Modelling

1 Introduction

Civil infrastructures support human activities, providing shelter and essential services and enabling economic development and social interactions. They are known as society's lifelines. Bridges are one of the most important and complex infrastructures and have a significant impact on society by favoring territorial connection [1].

Bridges are geometrically complex civil engineering structures, and, in the case of curved bridges, the design complexity is greatly increased by the numerous dependencies between the geometric entities and the complex 3D curved surfaces [2]. These projects usually exhibit significant geometric variations between the structures and the geometric dependencies can hardly be managed using a conventional 2D design approach. Moreover, the designs must integrate different requirements to meet the needs and the structure's context must be considered during design. Therefore, there is a low degree of reuse of the models in similar projects.

S. Gabriele et al. (Eds.): IWSS 2023, LNCE 437, pp. 609–618, 2024.
https://doi.org/10.1007/978-3-031-44328-2_63

Nowadays, advancements in architectural and engineering design tools have broadened the scope of incorporating innovative and impressive curves and structural details. However, with the infrastructure industry embracing digitalization to enhance productivity and to be updated with new technologies and innovations, there is a framework of increasing costs, sustainability issues, and short deadlines. Consequently, an increasing number of civil engineers are turning to parametric design and BIM-based workflows as a response to these limitations.

1.1 Building Information Modeling

Building information modelling (BIM) is a process that builds, analyses, documents, and evaluates virtually a digital representation of a facility's physical and functional attributes before making iterative revisions until the ideal model is documented. The BIM model is a virtual database that contains management data in addition to graphic designs [3].

BIM can be implemented in different types of building projects; however, its use has been limited in bridge construction. Currently, BIM has been customized to bridge projects and it is known as Bridge Information Modeling (BrIM). BrIM considers bridges' physical and functional characteristics and provides information for their entire lifecycle. A perfect graphical and functional representation of a bridge and the outcomes of its design are provided by intelligent and physical 3D bridge models, which enhance design quality, constructability, and teamwork.

1.2 Parametric Modeling

The implementation of parametric modeling enhances the infrastructure design process, resulting in increased efficiency through time and effort reduction. This approach involves the creation of a flexible geometric model that incorporates parameters and numerical relationships among them. It is implemented using computer programming codes, such as scripts, tailored for design purposes.

In a parametric model, the user interacts directly on the geometry to generate the structural components, as opposed to predefining them like in direct modeling [4]. As a result, a parametric model becomes an optimal framework that offers multiple solutions based on the values imposed for each of the parameters without the need to redraw the model [5].

Therefore, Building Information Modelling (BIM) with parametric modelling provides every tool for a complete workflow, from design to construction and maintenance, enabling data reuse and decreasing costs. This can be carried out thanks to direct connections between BIM software and visual programming tools like Grasshopper, a pre-installed plug-in for Rhinoceros 7, a 3D computer graphics and computer-aided design (CAD) application, structural engineers can implement a parametric workflow without prior programming experience [6].

1.3 Bridging the Gap Between BIM and Parametric Design

The workflow proposed in the present report aims to bridge the gap between BIM and parametric design. To accomplish this, parametric structural elements that are used to assemble projects in BIM environments are further connected and related to one another. In contrast to manual modelling and static positioning, the construction of objects is guided by rules and recipes.

The workflow is based on the definition of input parameters such as coordinates, dimensions, or curves and visually script rules that act on those parameters, generating the desired geometry or output in Grasshopper, which can then be applied directly to live objects using parametric BIM software. Repetitive operations and change updates are automated by integrating the parametric design into a 3D parametric modelling software, Tekla Structures. This will enable engineers to quickly, precisely, and simply produce a truly constructible 3D bridge model, satisfying any Level of Development (LOD) requirement and guaranteeing the consistency of all deliverables, including the model, drawings, and material lists [6, 7]. Moreover, the Finite Element Model of the bridge for structural analysis is generated from the parametric geometrical model.

2 Parametric Model Development

Following the aim of fostering the design and analysis of bridge projects by developing a parametric model that removed repetitive and time-consuming tasks, the overall workflow is divided into different stages: geometrical model, architectural model, and structural model definition.

2.1 Parametric Geometry Design

One of the initial stages involves identifying the principal parameters that enable the adaptation of the generic model to match the specific project requirements. The selection process relies on engineering criteria to differentiate between essential and less critical parameters. These parameters are: (i) Alignment, (ii) N° of girders and axis distance, (iii) N° and location of cross beams, (iv) Girders length, (v) Girders cross-section, (vi) Slab dimensions, (vii) Bracing systems.

Generally, during the design phase, the input data is the bridge guideline. Then, the algorithm starts creating a guiding curve that passes through three points. It can be described by the quadratic equation:

$$y = ax^2 + bx + c$$

The subsequent steps involve the definition of the bridge transversal section elements through a cluster, which is a collection of Grasshopper components to create cross-beams, diaphragms, and dummy transversal elements. Additionally, it generates the elastic and rigid links needed for structural analysis. Once the transversal elements are defined, they are properly placed along the bridge alignment according to the user's inputs by means of another cluster. Figure 1 presents the two clusters designed to perform the transversal elements definition.

The following step involves the longitudinal beams creation by connecting the transversal elements. The algorithm allows the evaluation of properties changes, such as effective width or cracked cross-sections properties.

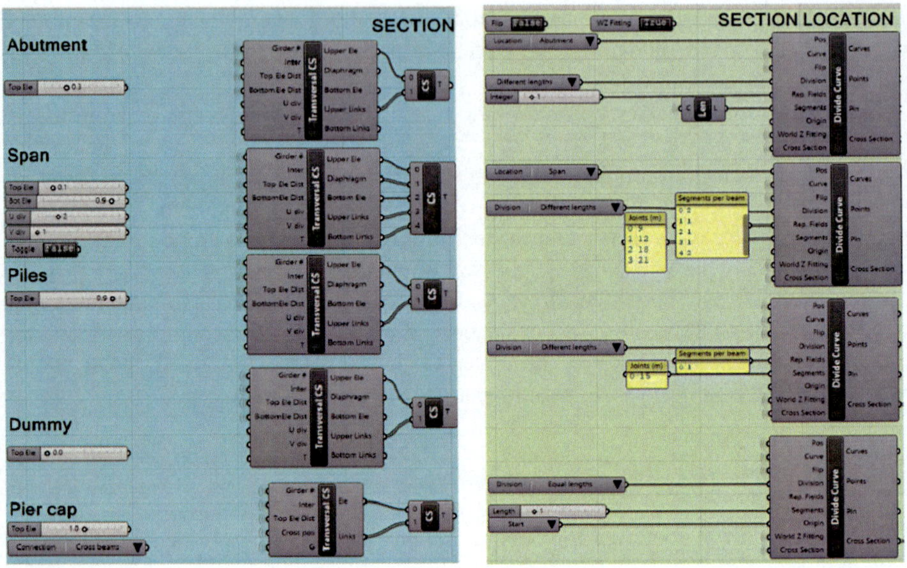

Fig. 1. Clusters of GH components to create and locate bridge transversal elements

Finally, horizontal bracing systems are parametrically defined. Following the objective of a generic model, the script created permits to consider or not the bridge substructure (piers caps and piers), and different support configurations are introduced.

2.2 Architectural Model Development

The next stage of the present workflow encompasses the creation of the 3D global and local definition of the bridge in Tekla Structures. This is performed with the Grasshopper-Tekla Structures live link that allows an automatic transfer of information from Rhinoceros; therefore, the user only needs to modify the parameters in Rhinoceros and the model will be automatically updated in Tekla Structures.

Firstly, the beams and slab are defined from curves, points, and surfaces. For that, Tekla components like Beam, Item and Component are used (Fig. 2(a)). Additionally, connections and details are parametrically introduced to cover most of the configurations. The user can select and modify the components according to the project demands. Figure 2(b) presents two examples of connections designed.

Figure 3(a) presents the algorithm designed for the longitudinal beams' connection; while in (b), some of the parameters, that make the connection parametric, are shown.

2.3 Finite Element Model Definition

In the field of structural analysis, various models are available, ranging from the simplest ones that employ unidimensional elements to more complex ones that introduce brick elements. For this research, the Grid model is chosen, resulting in a network of beam elements situated on the same plane that encompasses both the longitudinal and transverse

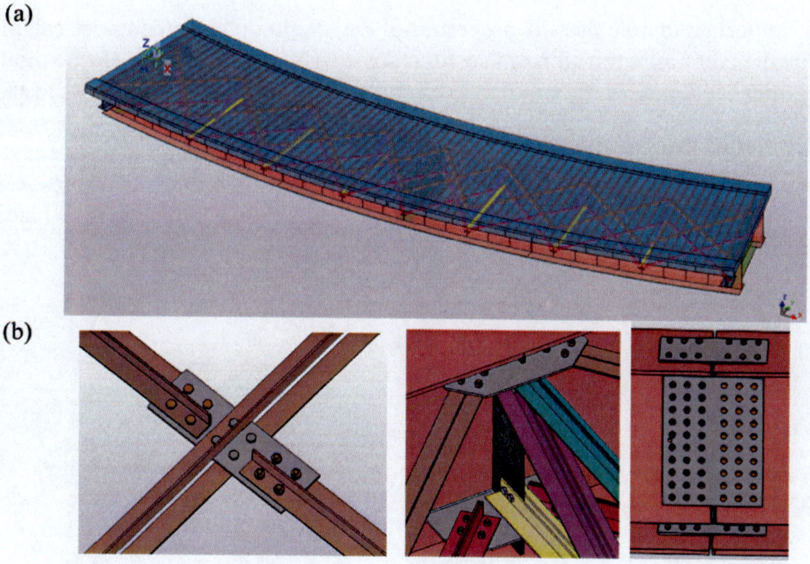

Fig. 2. (a) Global elements; (b) Connections and details in Tekla Structures

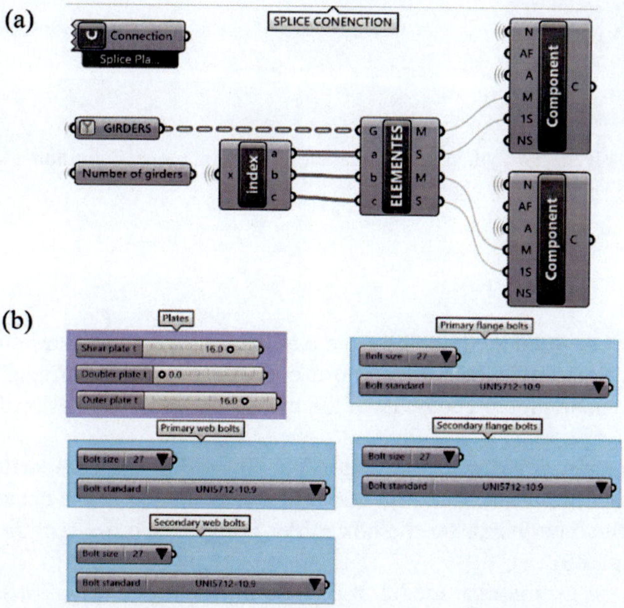

Fig. 3. (a) Splice connection GH script; (b) Part of the connection attributes

stiffness characteristics of the bridge structure. Subsequently, the slab is represented as strips of beam elements, like girders and cross beams.

It is important to note that the properties of the longitudinal elements are calculated based on the composite transformed section, considering the difference elastic modulus of the girders and slab. Furthermore, the torsional constant is determined by the sum of the girder's torsional constant and half of the slab's torsional constant.

Hence, the geometrical information previously established is efficiently converted using a customized Python plugin within the Grasshopper Environment, as presented in Fig. 4. Additionally, a set of predefined load cases and corresponding applied loads are assigned to the elements.

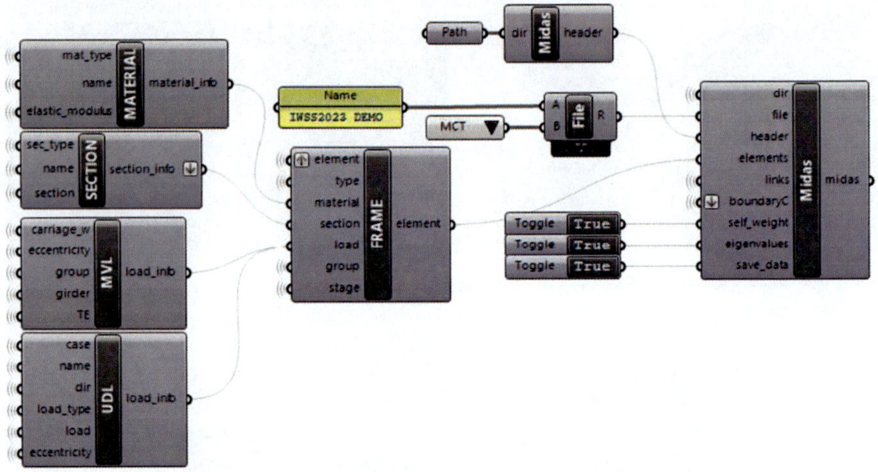

Fig. 4. GhPython components for structural model definition

3 Workflow Application

To illustrate the proposed methodology, the parametric workflow is employed to design both a three-girder composite bridge and a concrete box girder bridge. It begins by establishing the main global geometry of the bridge, the alignment or bridge axis, with a total length of 152, as shown in Fig. 5.

The following step is the cross-section definition. For the composite bridge, the diaphragm elements, presented in Fig. 6(a), are created thanks to the cluster previously presented. On the other hand, for the box girder bridge, the cross-section is defined in Rhinoceros (Fig. 6(b)).

To achieve the geometrical model of the composite bridge (Fig. 7(a)), the primary dimensions of the bridge section are defined, the offsets between the girder and the top and bottom element are linked to the development of the inner inclined trusses, these latter can be modified ed to achieve different connection schemes such as K and inverted K. Following, the sections are positioned along the direction of increasing chainage of the infrastructure ensuring the proper connection of the elements to the main beams using rigid links. Exploiting the geometrical deconstruction of these elements which

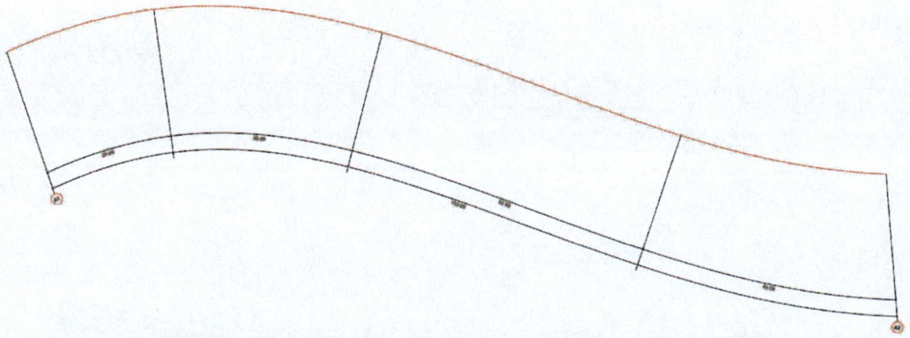

Fig. 5. Bridge alignment

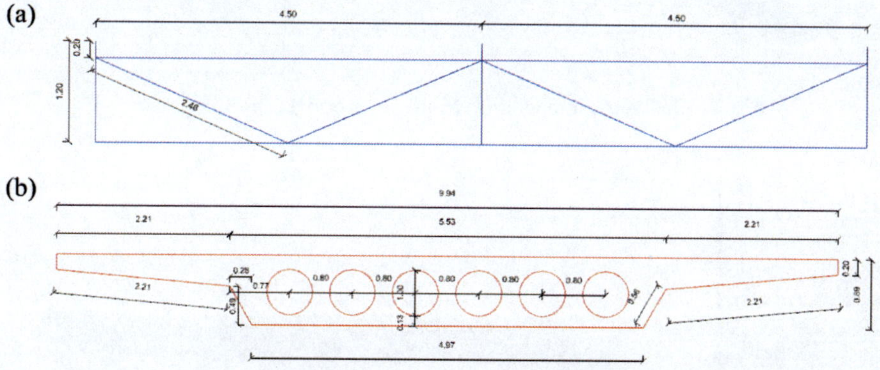

Fig. 6. (a) Composite bridge diaphragm elements; (b) Box girder cross-section

is intuitively realized within Grasshopper framework, the horizontal diaphragms are correctly positioned enabling different setups: crossed brace and diamond configurations. Finally, piers and pier caps are properly defined.

For the box girder, extrusion of the cross section along the alignment is performed (Fig. 7(b)).

Once the correct geometrical setup has been validated in Grasshopper, the categorized native elements undergo a conversion process into a string of codes. To accomplish this, a custom Python plugin is utilized within the Grasshopper environment achieving a seamless conversion to Midas Civil software. This approach establishes a connection between the geometrical settings and various element features, such as the cross-section and the burden loads. After that, the model is ready to be analyzed (Fig. 8).

To address the challenging task of 3D modeling, Tekla Live Link is employed. This integration allows the created elements to be easily transferred and computed within the Tekla Structures environment. Exploiting the connection components, automatic connections are realized using predefined parameters. Which are further customized to comply structural configurations goals. Figure 9 shows the architectural models in Tekla Structures.

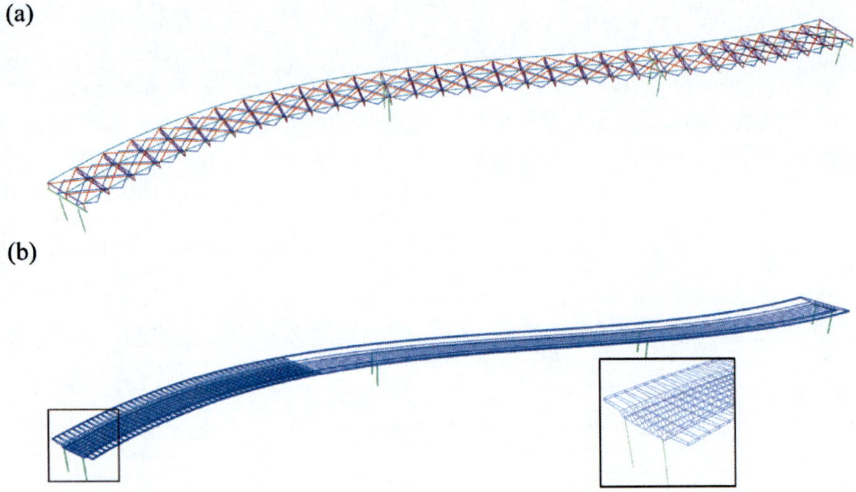

Fig. 7. Geometrical model: (a) Multigirder bridge; (b) Box girder

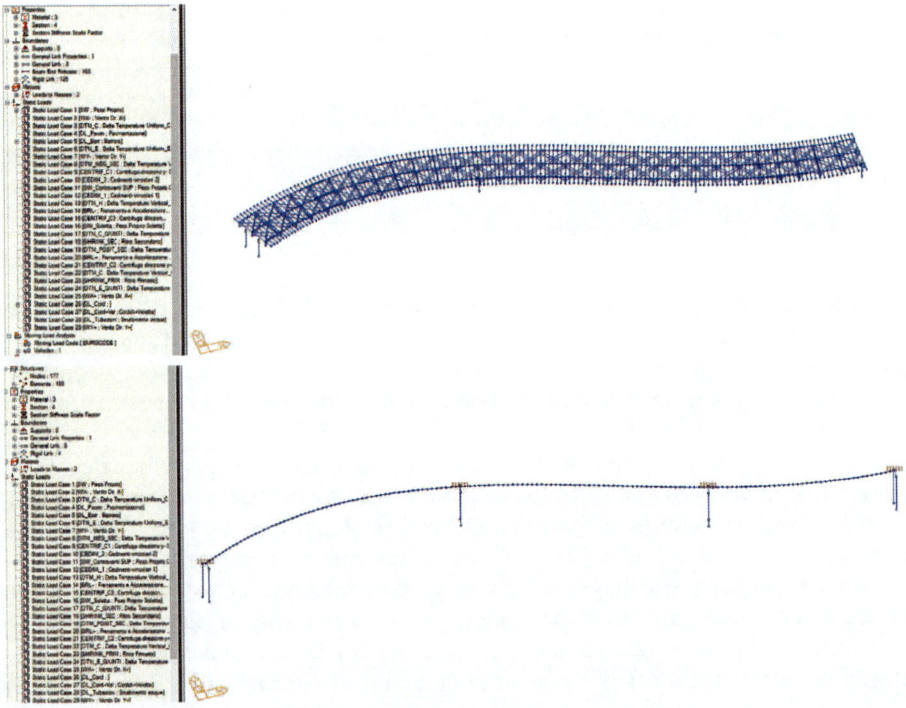

Fig. 8. Bridges structural model in Midas

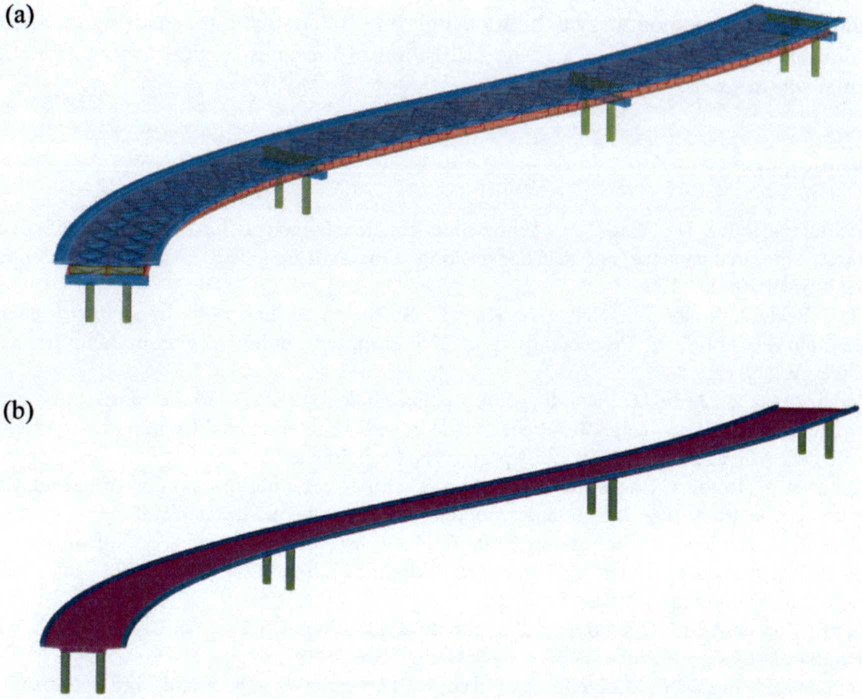

Fig. 9. 3D architectural model: (a) Multi-girder bridge, (b) Box girder bridge

By leveraging Tekla Live Link, the design process becomes more efficient and streamlined, ensuring consistent and accurate results. Overall, this parametric workflow offers an efficient and systematic approach to bridge design. It facilitates the automatic generation of structural elements and their accurate connection, enhancing the overall integrity and performance of the bridge.

4 Conclusion

Grasshopper/Rhinoceros is utilized to implement parametric modeling, resulting in a versatile 3D model in Tekla Structures and enabling the evaluation of the structural response using Midas. As a result, a comprehensive workflow is established, allowing for a resource-efficient design and analysis of geometrically complex bridges. This approach offers numerous advantages over traditional, non-interoperable workflows.

The presented tool offers significant benefits that improve the design process and ensure high-quality outcomes. It enhances productivity and flexibility by enabling users to efficiently test and analyze different alternatives, leading to the discovery of innovative and cost-effective solutions. Furthermore, this algorithm is not limited to composite or concrete box bridges, as it can be applied to other bridge typologies as well.

By adopting this approach, all operations are conducted within a unified interface, facilitating data consolidation, and minimizing the risk of information loss and errors.

Additionally, the methodology exhibits a high level of adaptability, ensuring the incorporation of new technologies and the utilization of the latest digital tools to identify optimal solutions.

References

1. Martínez-Muñoz, D., Martí, J.V., Yepes: Steel-concrete composite bridges: design, life cycle assessment, maintenance, and decision-making. Adv. Civil Eng. **2020** (2020). https://doi.org/10.1155/2020/8823370
2. Ji, Y., Beetz, J., Bonsma, P., Nisbet, N., Katz, C., Borrmann, A.: Integration of parametric geometry into IFC-bridge. In: Proceedings of the 23rd European Conference Forum Bauinformatik, Cork, Ireland (2011)
3. Yalcinkaya, M., Arditi, D.: Building Information Modeling (BIM) and the construction management body of knowledge. In: Proceedings of the IFIP International Conference on Product Lifecycle Management for Society, Springer, Nantes, France (2013)
4. Girardet, A., Boton, C.: A parametric BIM approach to foster bridge project design and analysis. Autom. Construct. **126** (2021). https://doi.org/10.1016/j.autcon.2021.103679
5. Boretti, V., Sardone, L., Bohórquez Graterón, L.A., Masera, D., Marano, G.C., Domaneschi, M.: Algorithm-aided design for composite bridges. Buildings **13**(4), 865 (2023). https://doi.org/10.3390/buildings13040865
6. Konia Company Ltd. Overcoming Complexity and Driving Efficiency in Bridge Design with Parametric Design. Storekonia.com. Accessed 21 Nov 2022
7. Sardone, L., Fiore, A., Manuello, A., Greco, R.: Performative structural design optimization: generative algorithm for a preliminary study of a voided beam. Appl. Sci. **12**(17), 8663 (2022). https://doi.org/10.3390/app12178663

Auxetic Metamaterial and Flagstone Tessellation Patterns Via Convex Airy Stress Functions

Marina Konstantatou(✉)

Specialist Modelling Group, Foster + Partners, London, UK
mk822@cam.ac.uk

Abstract. Auxetic metamaterials are gaining popularity as deployable architectural design systems due to their inherent flexibility, adaptability and programmable nature. The efficiency and constructability of these systems – which comprise of rigid panels and hinges - is interlinked to their geometry and corresponding kinematic behaviour. As such they can be considered as structural design systems in which the structural performance directly defines their architectural conceptual design.

This research paper presents a purely geometrical methodology for developing Auxetic metamaterials based on graphic statics. Specifically, it is shown how by employing reciprocal convex Airy Stress Functions (ASF) and Minkowski sums it is possible to derive in a direct (and noniterative) way non-standard 2D Auxetic patterns with negative Poisson's ratio. Furthermore, this methodology can be applied to cases of convex 3D polyhedral surfaces by combining said reciprocal ASFs into cantellated polyhedral objects. Lastly, it is discussed how this construction relates to origami flagstone tessellations.

Keywords: Graphic Statics · Flagstone tessellations · Auxetics · Metamaterials · Origami · Airy Stress Function · Reciprocal Diagrams

1 Introduction

Graphic statics and specifically the legacy of J.C. Maxwell [1, 2] have recently led to a wealth of contemporary research focusing on structural design, analysis and architectural geometry. In this research paper we will restrict our attention to one of the key concepts described in [1, 2] – that of reciprocal Airy Stress Functions (ASF) and reciprocal form and force diagrams [3]. Specifically, we will show how this relates to the geometry of auxetic meta-material patterns with corner-connected, rigid n-gonal panels. This typology of deployable structures holds the promise of adaptable architectural systems and is an active research field [4].

Auxetic metamaterials have a negative Poisson's ratio. Consequently, they expand when stretched (Fig. 1) – a property that makes them suitable for a plethora of applications, from adaptable wearables to material science, architecture, and structural engineering. To derive non-standard patterns of metamaterials comprising corner-connected rigid panels (Fig. 1.e,f) we will combine reciprocal diagrams in Minkowski Sums (***MS***)

S. Gabriele et al. (Eds.): IWSS 2023, LNCE 437, pp. 619–625, 2024.
https://doi.org/10.1007/978-3-031-44328-2_64

[5] (Fig. 2) in a similar way to applications for discrete stress fields in reinforced concrete [6]. Furthermore, we will incorporate a rotation parameter as in [7] and show how this method results to a bipartite set of n-gonal rigid panels and parallelogram gaps resulting in a kinematic system of an auxetic pattern. Lastly, we will discuss how this method applies also to origami and specifically to flagstone tessellation patterns.

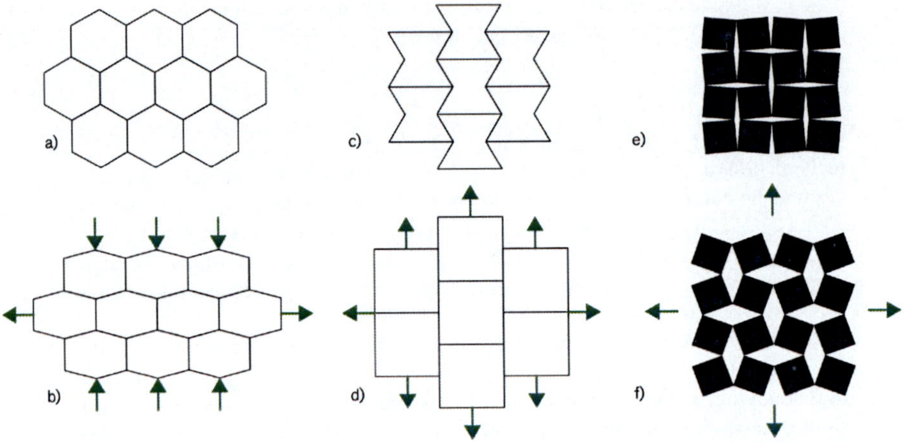

Fig. 1. (a) A honeycomb pattern has a positive Poisson's ratio – consequently when stretched longitudinally it contracts in the perpendicular direction (b). In contrast, an auxetic pattern (c) has negative Poisson's ration, thus, it expands when stretched; a property that might seem counter-intuitive and results in the characteristic 'meta-material' behavior. A special case of auxetic materials – to which we will restrict our attention - is the one comprising rigid panels and parallelogram/rhombic gaps (e,f).

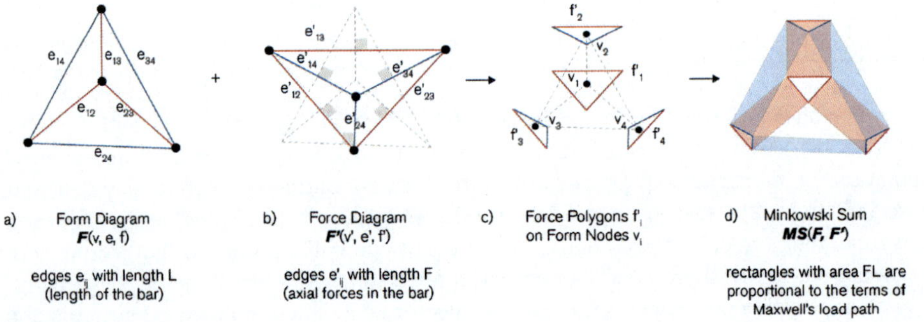

Fig. 2. (a) 2D form diagram ***F***(v,e,f); (b) reciprocal force diagram ***F'***(v',e',f') – note that corresponding form and force edges are perpendicular; (c) Force polygons scaled down with respect to their reciprocal form vertices; (d) Minkowski sum ***MS(F, F')*** combining ***F*** and ***F'*** into ***FL*** rectangles. Adapted from [6].

2 Method and Case Studies

2.1 Auxetics

2-dimensional corner-connected, rigid n-gonal auxetic patterns can be derived as follows:

(a) Given an initial 2-dimensional pattern (form diagram) which is an orthogonal projection of a convex polyhedral ASF (Fig. 3b), its reciprocal force diagram can be derived (Fig. 3a) as described in [3]. Note that the convexity of the ASF ensures that the faces of the form diagram – and hence the rigid panels – are also convex which is necessary for this construction.
(b) With the pair of reciprocal diagrams derived in the previous step, it is possible to position, scale, and rotate each face with respect to its corresponding vertex (Fig. 3e) whilst deleting the geometrical elements corresponding to the tension hoop (Fig. 3f). The remaining faces can be seen as rigid panels (Fig. 3f) which when connected to form a ***MS*** reveal the complete bipartite set of n-gonal rigid panels connected at their corners and the resulting parallelogram gaps (Fig. 3g and h) (which are the *FL* terms of the ***MS*** (Fig. 2d)).
(c) Subsequently the rigid n-gons can be rotated around their reciprocal vertices (Fig. 3h and l) whilst retaining the geometrical constraints of the kinematic system. Also, the proportions of rigid panels and gaps are directly controlled from the scalar terms of the ***MS*** [5].

It should be noted that this method provides a non-iterative way for obtaining this typology of auxetic patterns. This is due to the construction hinging on reciprocal ASFs and therefore, direct global equilibrium and automatic perpendicularity between reciprocal diagrams. What is more, the same construction can be applied directly to the reciprocal ASFs in a 3-dimensional manner. That is to say, the flat faces of one ASF can be positioned with respect to the vertices of its reciprocal ASF. This results to a cantellated polyhedron comprising flat n-gonal faces and parallelogram flat gaps. This method can be applied to any n-gonal pattern which is a result of an orthographic projection of a convex ASF – a complicated example is provided in Fig. 4.

This framework was implemented in terms of Computer-Aided Design (CAD). Specifically, the Rhinoceros software was used and in particular its Grasshopper platform for parametric geometry in conjunction to Python programming language.

2.2 Flagstone Tessellations

As discussed in Lang [8]: 'Origami tessellations are a branch of geometric origami in which an abstract tiling is physically embodied by folding a single uncut sheet of paper in such a way that the folded edges of the paper define a pattern related to the underlying tiling'. Initially, origami tessellations were largely based on regular grids, however; the development of computational algorithms enables the generation of general and irregular patterns [8]. One of the these algorithms, is based on Maxwell's observations of reciprocal diagrams being projections of polyhedra [1][2].

In particular, if an origami tiling has an underlying dual planar graph with corresponding reciprocal edges being perpendicular (Fig. 5), then the origami crease pattern

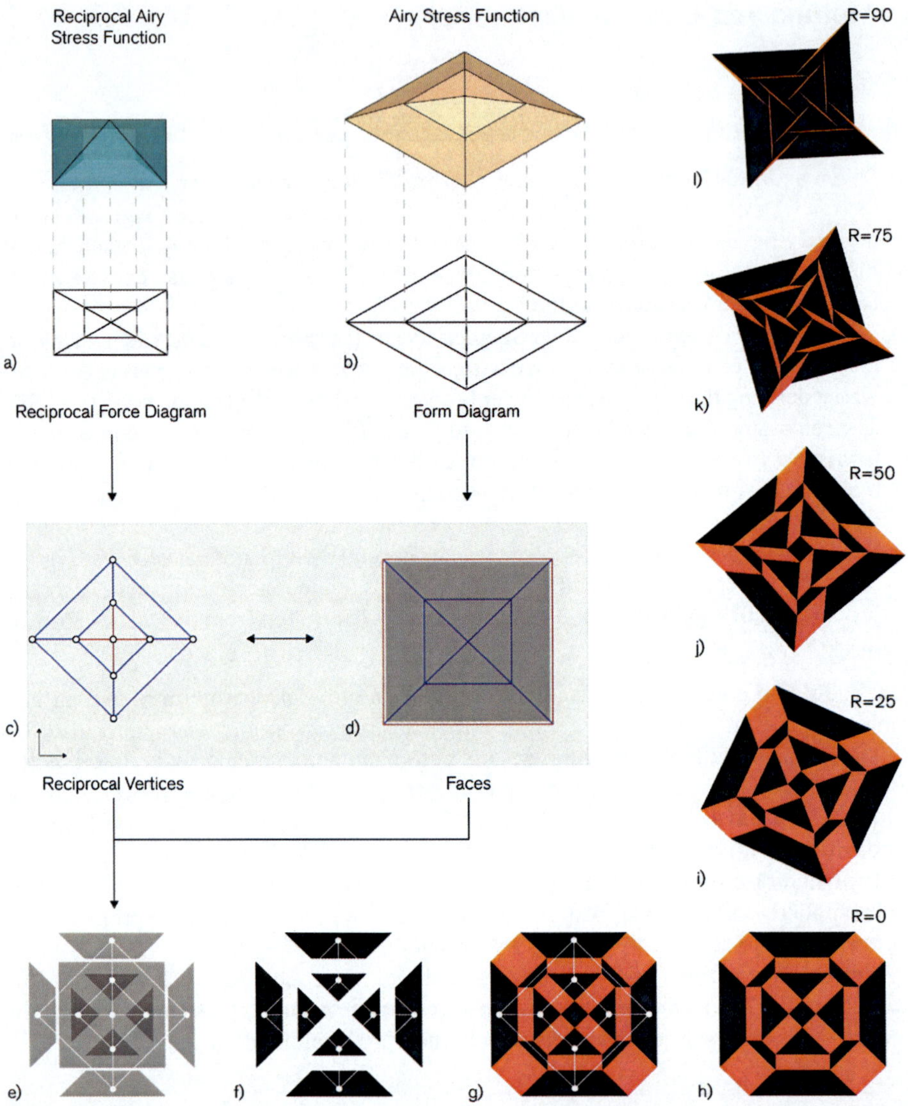

Fig. 3. A reciprocal pair of Airy Stress Functions generates a pair of reciprocal form and force diagrams by orthographic projection (a, b) with vertices mapping to faces (c, d); (e) The vectorial addition of the force faces on the corresponding form vertices; (f) The vertex and face corresponding to the tension ring element are deleted and the first group of rigid panels (in black) is defined from the remaining force faces; (g) Connecting the latter results in a ***MS*** which in turn defines the second group of rigid panels (in black) as the remaining spaces between the orange *FL* rectangles; (h) Resulting bipartite set of orange parallelogram gaps and black rigid panels; (i–l) the gradual rotation of the rigid panels with respect to their corresponding reciprocal vertices reveals the behaviour of the resulting auxetic meta-material with negative Poisson's ratio.

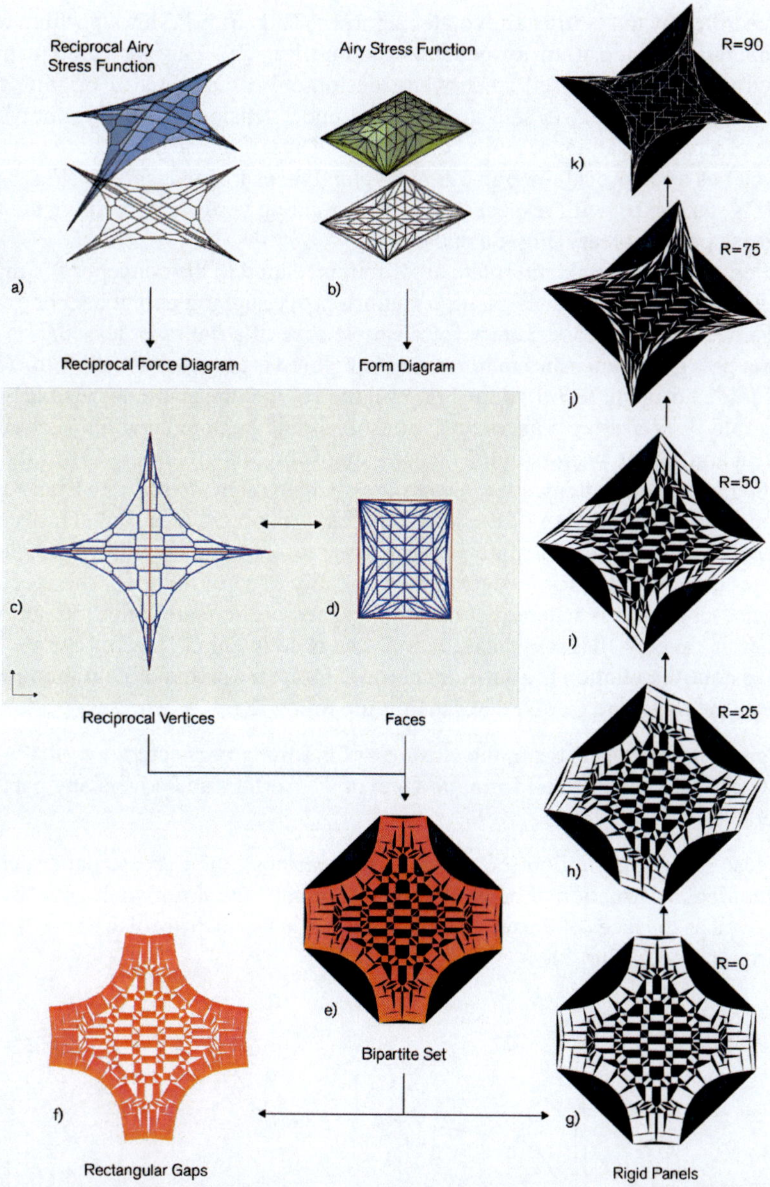

Fig. 4. A reciprocal pair of Airy Stress Functions generates a pair of reciprocal form and force diagrams by orthographic projection (a, b) with vertices mapping to faces (c, d); (e) The resulting ***MS*** which defines the bipartite set of rigid panels (in black) and orange parallelogram gaps (f); (g–k) the gradual rotation of the rigid panels with respect to their corresponding reciprocal vertices reveals the behaviour of the resulting auxetic meta-material with negative Poisson's ratio for a complicated pattern of various n-gons.

can be described by the 'shrink and rotate' algorithm (Fig. 6) [9]. This algorithm was the first computational algorithm for general origami tiling [9]. Lang and Batemam, who developed this approach, used spiderweb (tension-only) diagrams of origami tilings, where the equivalent truss is self-stressed, and under tension internally, implying the existence of a reciprocal figure and a convex ASF. Thus, if the origami tiling is a spiderweb which has a reciprocal diagram, the polygonal faces f can be shrunk (Fig. 5b), and rotated (Fig. 6a and b), with respect to the corresponding vertices v' defining the widths of the crease pattern pleats (Fig. 5d and 6a).

Furthermore, the shrink and rotate algorithm is related to the concept of 'simple flat twists' [9] where the origami tessellation comprises overlapping polygons. For example, in Fig. 6a, we see the crease pattern for a simple case of a flat twist tessellation where the central polygon rotates and in response the colored areas (pleats) will fold. Thus, in the final folded origami tessellation (Fig. 6b), the colored areas are not visible, and the resulting fold lines overlap adjacent polygons. It should be noted that the vectorial sum of the pleat and wedge vectors of the crease pattern (Fig. 6a–c) is zero. When the pleats fold on themselves only the wedge vectors are visible which also add up head-to-tail to zero. Thus, the vectorial sum of the pleat vectors is also zero (Fig. 6c). This relates to the internal tensile forces of a spiderweb (pleat vectors) in static equilibrium, where the force edges of concurring form edges sum vectorially to zero (Fig. 6c). The special case of flat twist tessellations where the visible polygons of the tessellation do not overlap each other, is called a 'flagstone tessellation' and is defined in [9] as follows:

'A flagstone tessellation is a flat-foldable origami crease pattern ***F*** containing a subset f of facets (the flagstone facets) that satisfies the following:

(a) No gaps: in the folded form, the closure of f is simply connected
(b) No overlaps: In the folded form, no facet of ***F*** overlaps and covers any part of any facet in f.

Because of their definition (no gaps and no overlaps), their crease pattern contains the geometrical information of both reciprocal diagrams: the shrunk polygons/ flagstone facets as well as the faces of the reciprocal original tiling, which are also a shrunk version of the original faces (Fig. 5).

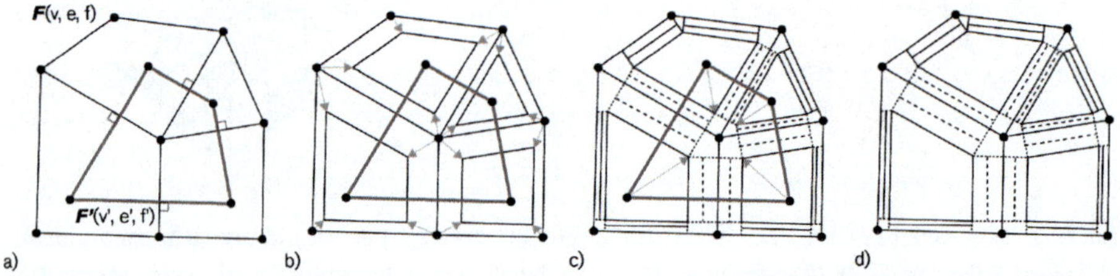

Fig. 5. Flagstone tessellation: (a) Origami tiling ***F*** and its dual ***F'*** with perpendicular reciprocal edges; (b) The faces f shrink with respect to their reciprocal vertices v'; (c) Equivalently faces f' shrink with respect to their reciprocal vertices v; (d) Resulting crease pattern of a flagstone tessellation. Adapted from [9].

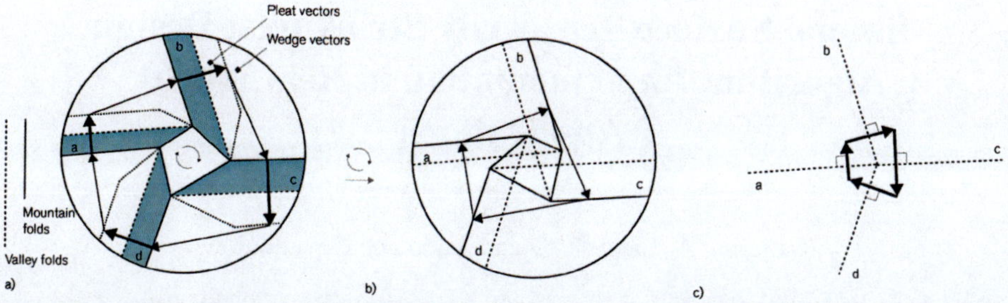

Fig. 6. Shrink and rotate algorithm: (a) Crease pattern for a rotating central polygon; (b) Folded configuration after rotation; (c) The vectorial sum of the pleat vectors is 0. Adapted from [9].

3 Discussion

This research paper discussed the application of graphic statics concepts such as reciprocal ASFs and Minkowski Sums to auxetic meta-materials and flagstone tessellations. Said concepts and implementations originate from J.C. Maxwell's 19th century legacy as well as contemporary literature. This diverse provenance highlights the relevance of fundamental structural engineering theories in conjunction to contemporary CAD advancements for the derivation of geometry-based, visual, and intuitive computational design frameworks.

Acknowledgments. The author would like to thank Professor Allan McRobie and Daniel Piker for fruitful conversations relating to these geometrical constructions over the last years.

References

1. Maxwell, J.C.: On reciprocal figures, frames, and diagrams of forces. Trans. Royal Soc. Edinburgh **7**, 160–208 (1870)
2. Maxwell, J.C.: On reciprocal figures and diagrams of forces. Philos. Mag. J. Sci. **26**, 250–261 (1864)
3. Konstantatou, M., D'Acunto, P., McRobie, A.: Polarities in structural analysis and design: N-dimensional graphic statics and structural transformations. Int. J. Solids Struct. **152–153**, 272–153 (2018)
4. Warisaya, K., Sato, J., Tachi, T.: Freeform auxetic mechanisms based on corner-connected tiles. In: Proceedings of the IASS Symposium, Beijing, China (2022)
5. McRobie, A.: Maxwell and Rankine reciprocal diagrams via Minkowski sums for two-dimensional and three-dimensional trusses under load. Int. J. Space Struct. **31**(2–4), 203–216 (2016)
6. Konstantatou, M., D'Acunto, P., McRobie, A., Schwartz, J.: Unified geometrical framework for the plastic design of reinforced concrete structures. Struct. Concr. **21**, 2320–2338 (2020)
7. McRobie, A., Konstantatou, M., Athanasopoulos, G., Hannigan, L.: Graphic kinematics, visual virtual work and elastographics. R. Soc. Open Sci. **4**, 170202 (2017)
8. Lang, R.J.: Spiderwebs, tilings, and flagstone tessellations. Am. Math. Soc. Origami **6**, 189–200 (2015)
9. Lang, R.J., Bateman, A.: Every spider web has a simple flat twist tessellation. American Mathematical Society – Origami **5**, 455–473 (2011)

Minimal Surface Tensegrity Structures: Design Algorithm Implementation in Real World Applications

Katherine A. Liapi(✉) and Andreana Papantoniou

Department of Architecture, University Campus, University of Patras, 26504 Patras, Greece
kliapi@upatras.gr

Abstract. While the exploration of new configurations of tensegrity structures is expected to contribute to the potential application of the tensegrity concept in building design, their real world implementation poses significant challenges. In this paper, double layer tensegrity networks the two layers of which are minimal surfaces, are considered. Algorithms that address minimal surfaces were developed and were applied and tested during the design and construction of two minimal surface tensegrity structures of 'helical' and 'enneper' geometry. These two structures were displayed as installations in the context of international exhibitions. For the real world implementation of the algorithms, a method that permits the construction of double layer tensegrity structures from the assembly of collapsible tensegrity units was employed. Spatial constraints, structural considerations, and constructability challenges had also to be addressed. The materialization of these structures has proven the validity of the algorithms, while the challenges encountered, can be used to improve both the design algorithms and the construction and assembly method.

Keywords: Tensegrity structures · minimal surfaces · enneper surface · helicoid · design algorithms · installations

1 Introduction

Tensegrities are a special type of tensile structures that offer a viable alternative to conventional and temporary space covering or space enclosure structures. Tensegrity structures, or structures of tensional integrity are composed of isolated compression members kept in equilibrium within a pre-stressed cable network, which, also defines their spatial boundaries.

The morphology of tensegrity structures is uniquely and directly related to their structural and mechanical properties, while their geometric and topological complexity accounts for significant difficulties in their study, and for their limited application in building design. A geometric method and computational processes that generate digital models of double-layer tensegrity networks that occur from the assembly of identical tensegrity prismatic units of a square base has already been developed [1]. The principal geometries addressed are domical, vaulted, and slab configurations. Algorithms that can

S. Gabriele et al. (Eds.): IWSS 2023, LNCE 437, pp. 626–635, 2024.
https://doi.org/10.1007/978-3-031-44328-2_65

generate a developable helical tensegrity structure, as well as spherical and ellipsoidal configurations from non-identical prismatic units of square base have also been developed (Fig. 1). However, our investigation has shown that in the case of structures of ellipsoidal shape, a geometric solution was not possible without using approximation methods [2].

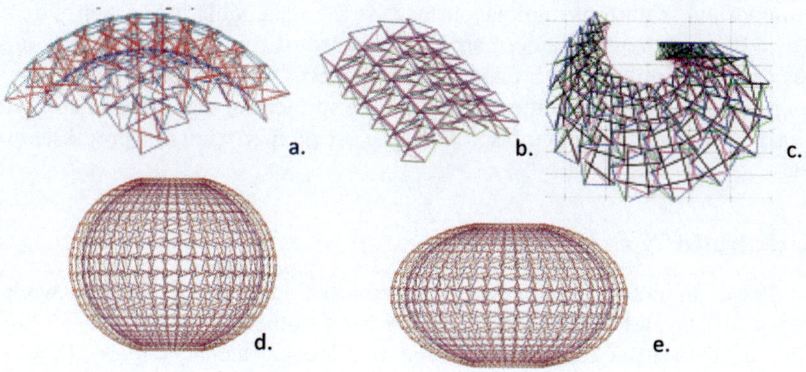

Fig. 1. Double-layer tensegrity structures of: a. domical, b. vaulted, c. developable helical shape, d. spherical and e. ellipsoidal shape

The exploration of new configurations of tensegrity structures is expected to contribute to the expansion of the applications of the tensegrity concept in building design. In this context, the investigation of the geometric configuration of various morphological types of curved tensegrity networks presents a challenging problem, as each type of structure may require a different geometric approach and solution. Among the geometries that were explored, special attention was given to double layer tensegrity networks the two layers of which are minimal surfaces, since, for their geometric solution, the utilization of approximation algorithms was not required.

The validity of the exploration of the geometry of minimal surfaces has been investigated through the design and construction of two full scale structures that were displayed in international exhibits. In the following section a discussion of the geometric processes and the developed algorithms, that were used in their implementation, are presented.

2 Minimal Surface Tensegrity Networks

2.1 Geometric Features

Minimal surface tensegrity networks are double layer tensegrity structures, that are composed of tensegrity units of square base, and their two layers are minimal surfaces.

Minimal surfaces described as *surfaces with zero mean curvature*, can be also defined as *surfaces of the smallest area spanned by a given boundary* [3]. This definition suggests significant potential applications of minimal surfaces in architecture, in particular when their design involves membrane surfaces.

The geometric investigation of minimal surface tensegrity networks has led to a design method that takes into account the constraints determined by the space packing geometry of prismatic tensegrity units of square base, or more precisely, of skewed truncated pyramids of square base, and the special geometric features of minimal surfaces. Accordingly, the overlapping square faces of the adjacent tensegrity units will form a two-square pattern on each layer of the network.

An important feature of minimal surfaces is the rectangularity of their parametric lines (F = 0), and their isogenic mapping. Because of this specific feature, minimal surfaces can be tessellated with planar square patterns. In this regard they present an advantage when compared to other doubly curved surfaces of regular geometry, such as the ellipsoid, as, in their geometric solution , the use of approximation processes are not needed.

2.2 Investigated Types and Algorithms

Three types of minimal surface tensegrity networks, namely tensegrity network of a catenoid, a helical, and an enneper shape have been studied [3].

A catenoid is a surface resulting from the rotation of a catenary curve γ: {x = coshu, y = 0, z = 0} around the z axis. The helicoid is a minimal surface having a helix as its boundary and it is the only ruled minimal surface other than the plane [4] resulting from a straight line that rotates at a constant speed around an axis perpendicular to the line, while at the same time, the line is moving along the axis at a constant speed. An important feature of these two minimal surfaces is that one can transform to the other with a sequence of isometric transformations.

The equations that allow the transformation of one surface into another are the following:

$$x = \cos a \sinh v \sin u + \sin a \cosh v \cos u$$

$$y = -\cos a \sinh v \cos u + \sin a \cosh v \sin u$$

$$z = u \cos a + v \sin a$$

where α = 0 corresponds to the helicoid and a = π/2 to the catenoid. All intermediate states of the transformation process are achieved by giving different values to the variable α, ranging from 0 to π/2. During this transformation, the catenary curves of the catenoid are transformed into the helicoid's spiral lines and the curves on the of v direction of the catenoid are transformed into the straight lines of the helicoid (Fig. 2 a, 2b).

A third minimal surface that was studied is the enneper, which presents a particular morphological interest for architectural applications (Fig. 2c), and is defined by the following parametric equations.

x = u-1/3 u ^ 3 + uv ^ 2
y = -v-u ^ 2 v + 1/3 v ^ 3
z = u ^ 2-v. 2

Algorithmic processes that generate a tessellation of planar square surfaces on the two layers of each studied category or type of minimal surfaces were developed. These

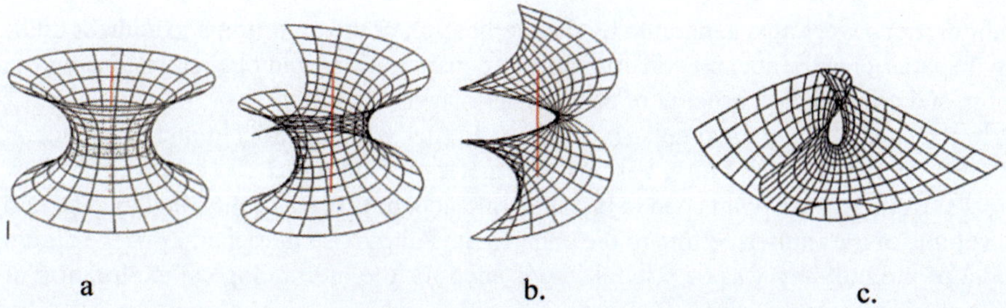

Fig. 2. Minimal surfaces of: a. catenoid, b. helical [transition from catenoid], and c. enneper shape

algorithms integrate the geometric and stability constraints that apply to double layer tensegrity networks composed of prismatic units [5]. The algorithms generate the results in a graphical environment. The Grasshopper visual programming language has been used for the parametric description of the minimal surface tensegrity networks (Fig. 3).

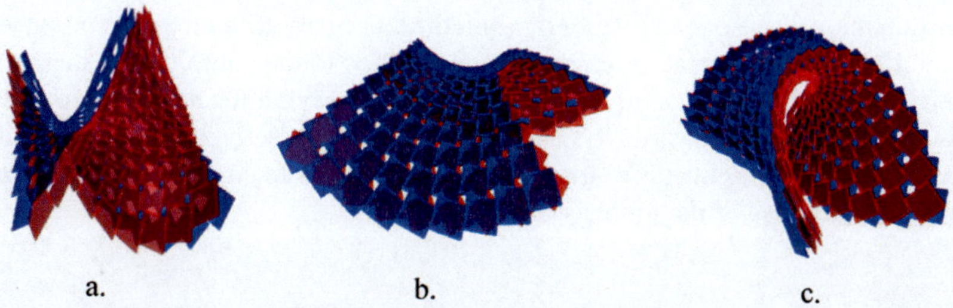

Fig. 3. Minimal surface tensegrity networks of a. helical, b. catenoid, c. enneper shape

3 Minimal Surface Tensegrity Algorithm Implementation

The developed algorithms and computational processes for the generation of tensegrity minimal surface networks were applied and tested during the form exploration and materialization of two tensegrity installation structures, displayed in international exhibitions.

The first tensegrity installation was a helical tensegrity structure that was displayed at Palazzo Mora in Venice, in the context of the 14th Biennale Architecture of Venice, in 2018. The structure was part of the "Time Space Existence" Exhibition that was organized by the European Cultural Center. The second installation was a tensegrity pavilion of enneper shape that was displayed at FIRA Center in Barcelona in 2019. This structure was part of the *Form & Force* International Exhibition, that was organized by the International Association for Shell and Spatial Structures.

In each case, taking into account various constraints such as the dimensions of the space for the ιinstallation, support points and several other parameters, the algorithm were used for the development of digital models of the two minimal surfaces. Alternate configurations were also generated by the application of the algorithms to address additional spatial constraints and constructability considerations and challenges during the design and construction process of each tensegrity installation.

In the case of the helical tensegrity structure that was constructed and displayed at the entry of the Palazzo Mora, various parameters including spatial constraints, view points and suspension points had to be taken into account. The structure had to fit within the volume of the vaulted ceiling of the entry to the Palazzo, so the accuracy of the digital model of the building was of critical importance for the morphological exploration of the studied structure.

Several variations of the helical structure were developed, all of which took into account critical spatial constraints such as door and window swings and relationship to the enclosing surfaces. Besides, strict regulations regarding the suspension points that one could use for hanging the structure within the vaulted ceiling, posed an additional challenge, as for example, one could only use old nails on the upper part of the arched window wood frames (Fig. 4).

Once the final configuration was decided, additional morphological studies followed that resulted in the integration of lightweight wire-mesh surface elements of triangular shape within the volume of each tensegrity unit the tensegrity structure was composed of. These semi-transparent surface elements were attached to the compression members of each unit following a pattern expected to bear visual and symbolic associations to a folk of birds trying to fly out towards the city of Venice; the objective was to highlight and enhance the feeling of lightness of the structure that comes in contrast to the symmetry, rigidity and robustness of the gothic vault.

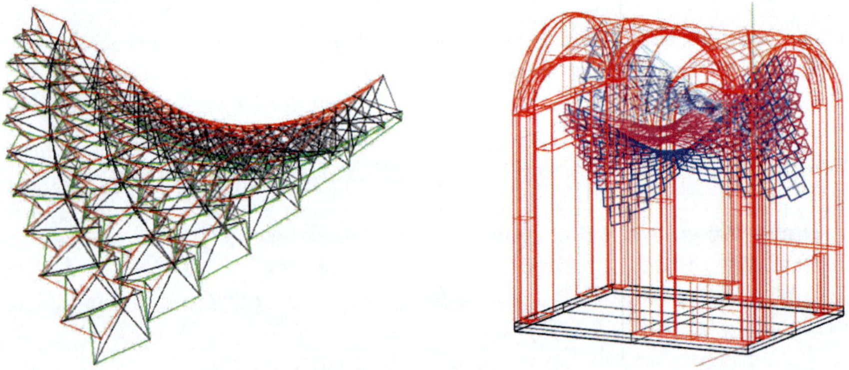

Fig. 4. Helical shape and variations of the helical tensegrity structure within the vaulted ceiling

The visual associations of the integrated surface elements have also dictated the dimension range of the square bases of the tensegrity units that form the structure. (Fig. 5 and 6).

For the *Form & Force* Exhibition, an enneper tensegrity pavilion structure was designed. In this case the structure had to be placed within given boundaries on the museum's floor surface. Inviting the visitors to walk around, and under the pavilion

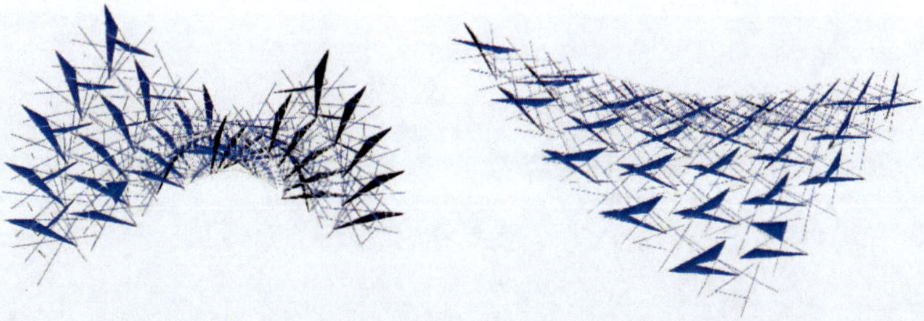

Fig. 5. Integration of surface elements: Morphological studies

Fig. 6. Tensegrity Structure of Helical Shape at Palazzo Mora

structure, in order to explore its provoking geometry and structural concept was one of the main reason for selecting the enneper shape. In its final configuration, the visitor could access the structure from at least two of its sides and observe it from a close and distant view.

The algorithms permitted the development of alternate configurations with variable unit sizes (Fig. 7). In the final configuration of the structure, the 4-sided lightweight surfaces elements that were attached on the four compression members of each unit, allowed the visitors to better perceive the structure's geometric shape. (Fig. 8).

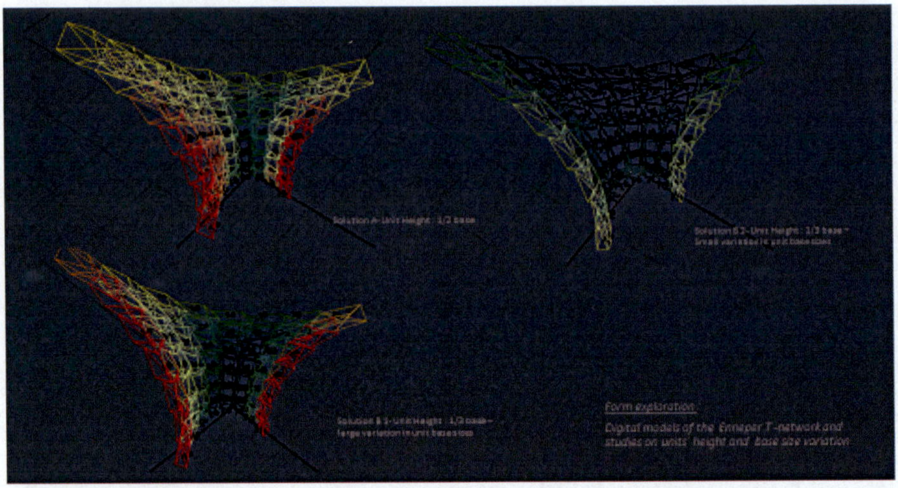

Fig. 7. Alternate configurations of the enneper structure with variable unit sizes

Fig. 8. Tensegrity Structure of enneper shape at Fira Center

4 Assembly and Construction Method

For transferring the virtual models developed in a graphical interface into a real world scenario, a patented method developed by one of the authors, that permits the construction of double layer tensegrity structures from the assembly of collapsible tensegrity units had been used [7], (Fig. 9). The application of the method greatly facilitated the packaging and transportation of the structure to the exhibition destination, as well as the on-site assembly and the erection of the structure.

In addition, a stiffness-based structural analysis method that takes into account the main features of the developed technology, such as the assembly and pre-stress methods, had been developed. This method can be applied to various configurations of double layer tensegrity networks composed of square-base pyramidal units [8].

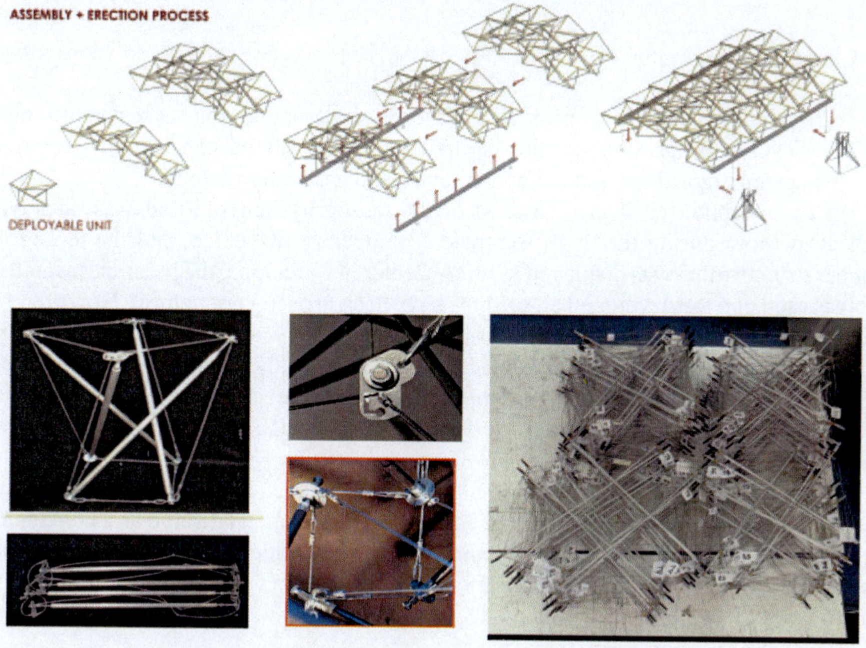

Fig. 9. Tensegrity Unit, Structure, and Method of Construction

Each structure was constructed, pre-assembled and tested at the University of Patras, Greece, in the context of one-week workshops. For the construction of the tensegrity units, aluminum tubular pipes, metal wire mesh and wire rope were used. Special attention was placed to the design and fabrication of the nodes, as they had to permit unit prestress and, at the same time, facilitate the connection of one unit to another.

Once the final shape was decided, the developed algorithms were used to determine the dimensions of all members, the overlapping cable lengths of the adjacent units, as well as the sizes of the surface elements to be added to the structure after the completion of the assembly process.

It was observed that both structures, after assembly, slightly deviated from their initial configuration. In the case of the enepper tensegrity pavilion, once the structure was assembled, as the museum's floor surface should remain intact, a base had to be developed that would constrain the compression members that reached the ground in a fixed position. In addition, after the structure was completed and set in place, a couple of bars, the longest ones, presented a slight deformation due to buckling loads. To address this problem, they were replaced with bars of slightly larger section. It thus become obvious, that, in particular for the enneper tensegrity structure, a stuctural analysis process should have been performed before the construction and assembly of the structure. The stiffness based method that has already been developed, could have been adapted and integrated into the algorithms, to provide more accurate results on the structures' member dimensions and prestress conditions, and eventually, on the final configuration of the structure.

5 Conclusion

The materialization of the two minimal surface tensegrity structures has tested the validity of the developed algorithms, while the challenges encountered, can be used to improve both the design algorithms and the construction and assembly method.

For the morphological investigation of the tensegrity helicoid that was displayed at Palazzo Mora during the 2018 Biennale Architecture in Venice, and the tensegrity enneper structure that was displayed at FIRA Center in Barcelona, the parametric method that was used and the developed algorithms have been proven very helpful. However, for a permanent tensegrity structure, or for structures of larger scale, the integration into the algorithms of the already developed stiffness based structural analysis method applicable to double-layer tensegrity structures composed of square base pyramidal tensegrity units, will be required. In addition, in real world scenarios, special attention need to be given to joint design and materials selection.

As a closing statement it is hoped that the approach taken has led to space structures that provoke innovation in architectural and structural design, and serve as a convincing argument that the developed design method and the embedded mathematical processes account for their rigorous spatial concept and form.

References

1. Liapi, K.A., Kim, J.: A parametric approach to the design of vaulted tensegrity networks. Int. J. Archit. Comput. **2**(2), 245–262 (2004)
2. Liapi, K.A., Papantoniou, K., Nousias, A.: Double-Layer ellipsoidal tensegrity structures: geometric investigation and parametric representation. In: Proceedings of the International Association for Shell and Spatial Structures (IASS) Symposium 2016, Tokyo, Japan, 26 - 30 September 2016
3. Velimirović, T., Radivojević, L.S., Stanković, G.R., Kostić, M.S.: Minimal surfaces for architectural constructions. Facta Universitatis Series: Architecture Civil Eng. **6**(1), 89 – 96 (2008)
4. Carmo, D., Helicoid, T.: In: Fischer, G. (ed.) Mathematical Models from the Collections of Universities and Museums, pp. 44–45. Vieweg, Braunschweig, Germany (1986)

5. Liapi, K., Papantoniou, A.: Square tessellation patterns on double layer minimal surfaces: geometric investigation and design algorithms. eCAADe 36 (2018)
6. Liapi, K.A., Papantoniou, K., Nousias, A.: Morphological exploration of curved tensegrity networks: towards double-layer minimal surface configurations In: Proceedings, IASS 2017, Hamburg, Germany (2017a)
7. Liapi, K.: (inventor). Tensegrity Unit, Structure and Method for Construction, U.S. Patent No 6, 901, 7 June 2005
8. Liu, S,G., Liapi, K.A., Tassoulas, J.L.: The effects of static loading and sup- port conditions on the performance of Double-layer Vaulted Tensegrity Structures IASS 2004

Parametric Analysis of Ribbed Domes

Giulio Mirabella Roberti(✉), Vittorio Paris, and Giuseppe Ruscica

University of Bergamo, Bergamo, Italy
giulio.mirabella@unibg.it

Abstract. This work explores the potential of parameterisation in the design of ribbed domes, specifically focusing on lowered domes. Drawing inspiration from the cross-herringbone pattern employed during the Renaissance, the study compares various configuration of rib patterns. To achieve this, parametric models were constructed, incorporating a limited set of geometrical parameters, allowing for the generation of multiple geometrical configurations. Each ribbed lowered dome configuration was analysed to comprehend its behavior in terms of displacements and stresses. By conducting numerous simulations and comparing the results, the impact of ribs was estimated, considering two different load conditions. This research sheds light on the influence of various parameters on the structural performance of ribbed domes, contributing to advancements in dome design and engineering.

Keywords: Dome · Loxodromes · Ribs · Parametric Analysis · Sangallo

1 Introduction

The term dome denotes a double-curved structure known for stability, often used to cover spaces in monumental buildings. The hemispherical dome is the archetype of all domes. Its geometry is essentially a half sphere, but, within centuries, domes have come in various shapes and sizes [1–3]. Domes characterise spaces and buildings [4], and generally, different domes' shapes follow different cultures [5, 6]; e.g., hemispherical domes were widely adopted in the roman empire [7]. In contrast, onion domes, characterised by a bulbous shape, were commonly found in Eastern Orthodox churches [8]. Domes can be built using various materials; historic domes were mainly made of stone, bricks, modern concrete or lightweight and eco-friendly materials such as recycled plastic [9]. In the 20th and 21st centuries, advancements in building techniques and technology allowed a shift towards more complex shapes, leading to a gradual move away from traditional geometric shapes. During the 21st century architects like Eladio Deste or Pier Luigi Nervi built extreme shells [10–12]. Nowadays, free-form shells are mainly applied to public or monumental spaces.

S. Gabriele et al. (Eds.): IWSS 2023, LNCE 437, pp. 636–643, 2024.
https://doi.org/10.1007/978-3-031-44328-2_66

Generally speaking, shells could exhibit good structural behaviour if subject to defined loads [13]. If well designed, they require very few amounts of material to resist a defined load system, allowing the realisation of remarkably thin structures [14, 15]. The reason is ascribed to the double-curvature shape, which increases the structure's stiffness.

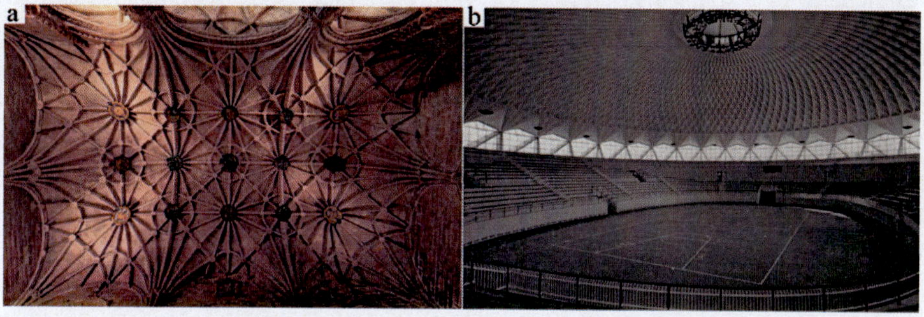

Fig. 1. a) Ribbed vault of Santa Maria de Belém Church, Jerónimos Monastery, Belém, Lisbon. b) Palazzatto dello Sport, Rome (Nervi).

Ribs can characterise domes and shells (see Fig. 1a). Their use in dome construction has been a common practice for centuries and continues to be employed in modern architecture [16]. Historically, ribs were typically made of stone or other durable materials and often decorated and organised in various geometry [17, 18]. E.g., in hemispherical domes, the ribs can be aligned with parallel and meridian or follow loxodromes curves (see Fig. 1b).

The rib's role in domes and shells can be studied from different perspectives; ribs were often adopted to facilitate the construction [17], e.g., for cross vaults, where ribs were built at first along edges and then used to support the vault. From a structural point of view, the role of ribs in shells is still under study [19, 20]; a systematic determination of their contribution has yet to be achieved, perhaps due to the already complex issue of describing the structural behaviour of free-form shapes.

Within this context, this paper aims to contribute a better understanding of the role of ribs in shells. Following the authors' previous study [21], on using the loxodromic curve in historic shells, the research focuses on the structural evaluation of the ribbed-lowered dome.

2 Geometry and Methodology

Parametric analyses were carried out, three different lowered domes studied, and six different configurations of ribs examined for each. Thus, 18 geometries are the basis of the analyses. These were compared to identical lowered domes without ribs. The geometry of all analysed structures corresponds to a portion of spheres cut at different heights, i.e., the spring of the domes change in inclination concerning the angle φ, (see Fig. 2). Further, defining R as the radius of the sphere and r by Eq. (1), the analyses were carried out considering r constant, as well as dome's thickness t is assumed constant and equal to 10 cm.

$$r = R\sin\varphi \tag{1}$$

A double system of loxodromic curves, one right-handed and the other left-handed, was modeled to describe the trajectories of ribs. Their analytical description is reported in (2), where parameter u affects the frequency and variates in the analyses. The number of ribs investigated in the analyses changes between the values 24, 36, and 48.

$$\Gamma = \begin{cases} x = R\,\sin(\alpha)\cos(u\beta) \\ y = R\,\sin(\alpha)\cos(u\beta) \\ z = R\,\cos(u\beta) \end{cases} \tag{2}$$

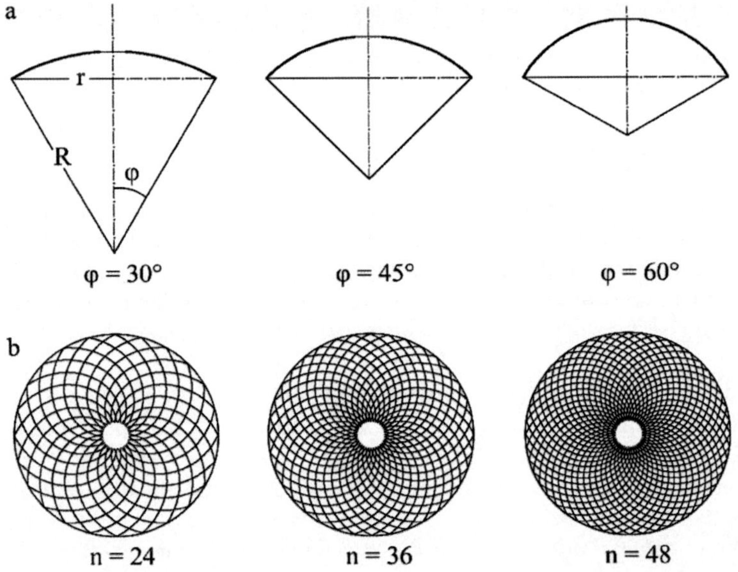

Fig. 2. a) Geometry of lowered domes, b) Loxodromic trajectory of ribs.

Finite Element Method (FEM) [22] was adopted to simulate the behaviour of the lowered ribbed domes. Through FEM, structures were described by finite elements, and for each, a system of equations solved to provide a description of real structures. Indeed, the domes were simulated through two different classes of elements: beam, for the ribs with a cross-section of 10 × 20 cm, and shell, with a thickness of about 10 cm and whose geometries are delimited by the beams. All simulations were implemented in Karamba software [23], which allows real-time interaction between parametric variations and numerical analyses. FEM analyses were executed considering only static loads, linear behaviour, homogeneous material, and effects of the first order. Two different load cases were considered to evaluate the structural performance of the domes. In the first set of simulations (Set A), the structures were loaded solely with their weight, while, in the second set of simulations (Set B), an additional load due to the presence of snow was simulated.

All structures were constrained by radial rollers to ensure realistic boundary conditions during the simulations. These rollers were placed at the lower extremities of the ribs so that their axes were perpendicular to the radial direction R of the domes. Near the crown, an additional rib was modelled at the upper end of the rib; here, beam elements with a cross-section of about 20 × 20 cm were adopted.

3 Results

Three reference domes with a thickness of $t = 10$ cm, a radius of $r = 10$ m, and $\varphi = 30°$, 45°, and 60° were analysed considering only the self-weight to assess the influence of ribs and design parameters (see Fig. 3). The results of simulations run on these three reference structures were confirmed from membrane theory, i.e., maximum displacements occur in the upper portion of the domes and widen due to the horizontal component of the thrust at the base of the structures. Specifically, when φ increases, the maximum displacements decrease consistently near the crown of the domes (maximum displacement $d = 1.42$ - 0.86 - 0.70 cm), while at the base's structure, smaller movements appear (0.09 - 0.12 cm). Compressive stresses show the same behaviour of displacements, and the tensile stresses increase with the growth of φ.

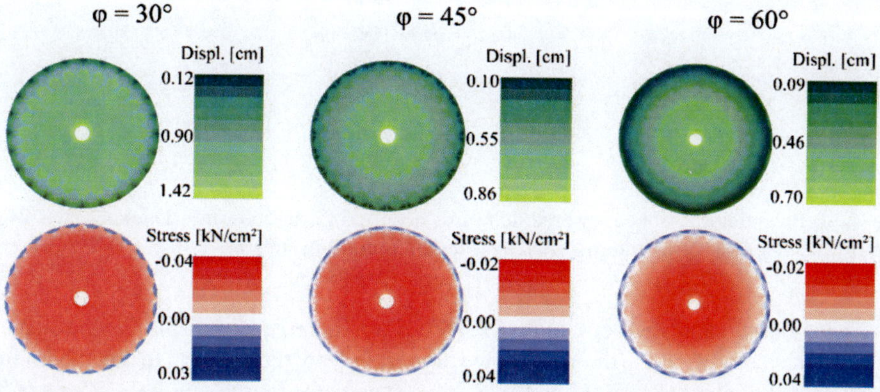

Fig. 3. a) Behaviour of three reference domes ($\varphi = 30°$, 45°, and 60°). Top row displacements, bottom row stresses.

Analyses executed on Set A, i.e., the investigations performed under only self-weight load, show that the presence of ribs increase the magnitude of displacements. Comparing two lowered domes with same thickness t, R and φ, only one ribbed, the bigger displacements are always recorded in the ribbed one. The range of the increments corresponds to 0.27 - 0.49 cm. The minimum increment of displacements occurs when φ is 60°, u is equal to 1 and 24 loxodromes were modelled. The simulations show that the number of ribs impact on displacements. Specifically passing from 24 to 36 and then

48 loxodromes, leads to an increment of displacement. Same behaviour is displayed when the frequency rises (see Table 1 and Fig. 4). Ribbed lowered domes do not highlight substantial differences in term of displacements with respect to the three reference domes, while significant alterations are visible in terms of stresses. The highest stresses are recorded in the ribs where values of compression can reach of about 4.95 kN/cm^2 in the case of $\varphi = 30°$ $u = 1$ and 24 loxodromes, and the maximum tension of about 4.24 kN/cm^2 when $\varphi = 30°$ $u = 2$ and 48. The highest compressive stresses are recorded at the beams placed at the oculus and only at the domes' base tensile appear. Stresses in shell elements do not show significant alteration (see Fig. 4).

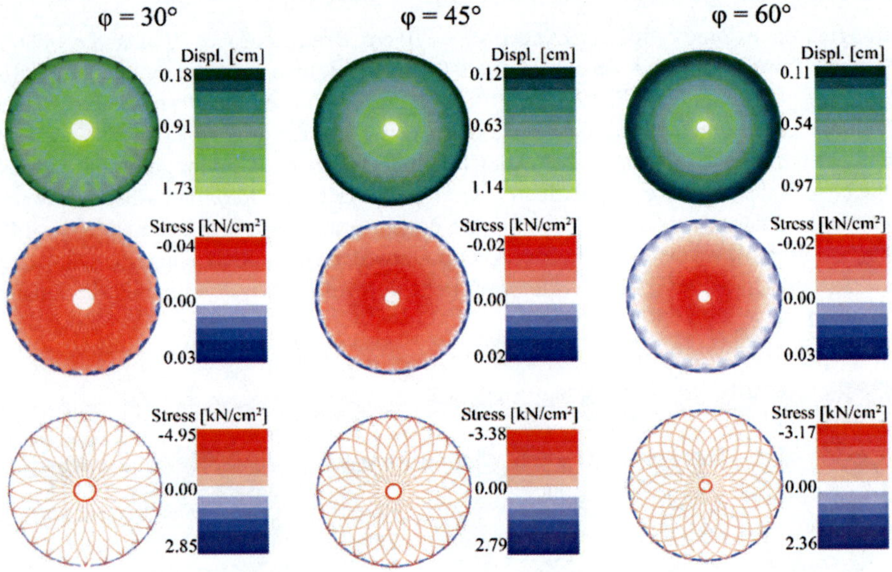

Fig. 4. a) Behaviour of ribbed lowered domes ($\varphi = 30°$, 45°, and 60°, $u = 1$, loxodromes 24). Top row displacements, central row shell stresses, bottom beam stresses.

The simulations carried out under Set B show an increment of displacements, especially, in the analyses of the three reference domes. Indeed, the ribs lead to the maximum displacement of about 2.11, 1.33 and 1.18 cm, respectively, for the $\varphi = 30°, 45°$ and 60°, $u = 1$ and 36 loxodromes, while of about 1.93, 1.25 and 0.91 cm in the case of the three reference domes. This becomes evident normalising the increment of the maximum displacement Δd_{g+s} as describe by Eq. (3). The Δd_{g+s} ranges recorded for lowered ribbed domes are 10 - 18% ($u = 1$) and 10 - 16% ($u = 2$), meanwhile reach the 30 - 36% in case of the three reference domes.

$$\Delta d_{g+s} = \left(\left(d_{(g+s)}^{max} - d_g^{max}\right)/d_g^{max}\right) \cdot 100 \quad (3)$$

With the aim of classifying the ribbed structures, a parameter K was defined as the ratio of weight w above displacement d, see Eq. (4). Values of K are reported in Table 1, they increase with the presence of the ribs and by increasing the number of loxodrome.

$$K = w/d \tag{4}$$

Indeed, fixing φ and increasing the number of loxodromes (24, 36 and 48) the value of K increases and Δd_{g+s} decreases. The simulations executed under Set B allow to highlight the relevance of K parameter. The role of ribs emerges more by comparing the values of Δd_{g+s} with the normalised increment of load Δw_{g+s} described by Eq. (5) that corresponds to the range of 18–27% for the ribbed domes ($\Delta d_{g+s} = 10$–18%), meanwhile, for the three reference domes is about 30–37% ($\Delta d_{g+s} = 30$–36%).

$$\Delta w_{g+s} = \left(\left(w_{(g+s)} - w_g\right)/w_g\right) \cdot 100 \tag{5}$$

Table 1. Values recorded for both set of simulations.

Set A

Reference domes

	$\varphi = 60°$			$\varphi = 45°$			$\varphi = 30°$		
	d [cm]	w [kN]	K [kN/cm]	d [cm]	w [kN]	K [kN/cm]	d [cm]	w [kN]	K [kN/cm]
	0.70	1.0E+03	1.5E+03	0.87	9.1E+02	1.1E+03	1.42	8.3E+02	5.8E+02
Ribbed domes									
u = 1									
24	0.97	1.4E+03	1.4E+03	1.14	1.2E+03	1.1E+03	1.73	1.1E+03	6.3E+02
36	1.04	1.5E+03	1.5E+03	1.22	1.3E+03	1.1E+03	1.87	1.2E+03	6.5E+02
48	1.09	1.7E+03	1.6E+03	1.22	1.5E+03	1.2E+03	1.92	1.3E+03	6.9E+02
u = 2									
24	1.04	1.5E+03	1.4E+03	1.13	1.3E+03	1.1E+03	1.64	1.1E+03	6.7E+02
36	1.17	1.7E+03	1.4E+03	1.24	1.4E+03	1.1E+03	1.66	1.1E+03	6.9E+03
48	1.25	1.9E+03	1.5E+03	1.38	1.6E+03	1.1E+03	1.89	1.3E+03	7.1E+02

Set B

Reference domes

	d [cm]	Δw_{g+s} [%]	Δd_{g+s} [%]	d [cm]	Δw_{g+s} [%]	Δd_{g+s} [%]	d [cm]	Δw_{g+s} [%]	Δd_{g+s} [%]
	0.91	30	30	1.13	34	30	1.94	37	36
Ribbed domes									
u = 1									
24	1.14	22	17	1.33	25	17	2.04	27	18
36	1.19	20	14	1.34	23	10	2.11	25	13
48	1.21	18	11	1.34	20	10	2.12	22	10
u = 2									
24	1.21	20	16	1.31	24	18	1.90	26	16
36	1.32	18	13	1.41	21	14	1.86	24	13
48	1.40	16	12	1.54	19	11	2.09	20	11

4 Conclusions

The ongoing research aims to investigate on the ribs' role in shell structures, and their impact on the structural behavior. Through the paper different geometrical parameters, such as φ, u and number of loxodromes, were investigated to assess the impact of the ribs' loxodrome on lowered domes. The ribs' influence has been recorded in term of stresses and displacements and through these quantities evaluated. All cases studied show that the presence of ribs leads to a concentration of stresses localised in the ribs self, and do not drastically impact on the displacement behaviour. With the same geometrical conditions, same constrains and considering only the self-weight the presence of ribs increases displacements. Despite that, under different load condition, e.g., snow-load, ribbed domes display better behavior showing smaller increments of displacements (Δd_{g+s}). Therefore, according with these preliminary simulations, the ribs provide additional resource to resist to different load condition and with a modest quantity of materials. This research, far to be completed will be pursued, with more numerical simulations investigating dynamic response and the construction of these marvelous structures.

References

1. Ashkan, M., Ahmad, Y.: Persian domes: history, morphology and typologies. Archnet-IJAR Int. J. Archit. Res. **3** (2009)
2. Huerta, F.S.: Oval domes: history, geometry and mechanics. Nexus Netw. J. **9**(2), 211–248 (2007)
3. MacDonald, W.L.: The Pantheon: Design, Meaning, and Progeny. Harvard University Press (2002)
4. Conforti, C.: Lo specchio del cielo. Forme, significati, tecniche e funzioni della cupola dal Pantheon al Novecento. Milano: Electa (2021)
5. Gurrieri, F.: La cupola del Brunelleschi. Firenze: Accademia dell'Iris (2010)
6. Creswell, K.A.C.: Persian domes before 1400 A. D. Burlington Mag. Connoisseurs **26**(142), 146–151+155 (1915)
7. Fletcher, R.: Geometric proportions in measured plans of the pantheon of Rome Nexus Netw. J. **21**(2), 329–345 (2019)
8. Shalunts, G., Haxhimusa, Y., Sablatnig, R.: Architectural style classification of domes. In: Bebis, G. (eds.) et al. Advances in Visual Computing, Lecture Notes in Computer Science. Berlin, Heidelberg: Springer, pp. 420–429 (2012)
9. Talocchino, G.: Design & Construction Criteria for Domes in Low Cost-Housing. University of the Witwatersrand (2005)
10. Iori, T., Poretti, S.: Pier Luigi Nervi: his construction system for shell and spatial structures. In J. Int. Assoc. Shell Spat. Struct. **54**(2–3), 117–126 (2013)
11. Lainez, J.M.C., Verdejo, J.R.J., Macias, B.S.-M., Calero, J.I.P.: The key-role of Eladio Dieste, Spain and the Americas in the evolution from brickwork to architectural form. J. Asian Archit. Build. Eng. **8**(2), 355–362 (2009)
12. Bianchino, G., Costi, D.: Cantiere nervi. SKIRA (2012)
13. Heyman, J.: The stone skeleton. Int. J. Solids Struct. **2**(2), 249–279 (1966)
14. Heyman, J.: On shell solutions for masonry domes. Int. J. Solids Struct. **3**(2), 227–241 (1967)
15. Pini, V., Ruz, J.J., Kosaka, P.M., Malvar, O., Calleja, M., Tamayo, J.: How two-dimensional bending can extraordinarily stiffen thin sheets. Sci. Rep. 6(1) (2016)

16. Block, P., Bayl-Smith, M.K., Schork, T., Bellamy, J., Pigram, D.A.: Ribbed tiled vaulting: innovation through two design-build workshops. gta Verlag (2014)
17. Taín-Guzmán, M., Alonso-Rodríguez, M.Á., Calvo-López, J., Natividad-Vivó, P.: 'Stonecutters' literature and construction practice in Early Modern Gothic: the tracings for a rib vault at the Cathedral of Tui in Galicia. Constr. Hist. **27**, 1–21 (2012)
18. Halpern, A.B., Billington, D.P., Adriaenssens, S.: The ribbed floor slab systems of Pier Luigi Nervi. In: Proceedings of the IASS Symposium 2013, Wroclaw (Poland) (2013)
19. Alexander, K.D., Mark, R., Abel, J.F.: The structural behavior of medieval ribbed vaulting. J. Soc. Archit. Hist. **36**(4), 241–251 (1977)
20. Faleh, S.K., Saleh, I.S., Chkheiwer, A.H.: Comparison study between brick and ribbed dome by using F. E. analysis. In: AIP Conf. Proc., vol. 2660, no. 1, p. 020051 (2022)
21. Mirabella Roberti, G., Ruscica, G., Paris, V.: From the herringbone dome by Sangallo to the Serlio floor of Emy (and beyond). Curved Layer. Struct. **8**(1), 259–270 (2021)
22. Reddy, N.: Introduction to the Finite Element Method, 4th Edition. McGraw-Hill Education (2019)
23. Karamba3D. https://karamba3d.com/

Conceptual Design of Reciprocal Self-stressed Roof Structures with Graphic Statics

Pierluigi D'Acunto[1](✉), Patrick Ole Ohlbrock[1], and Marina Konstantatou[2]

[1] Professorship of Structural Design, TUM School of Engineering and Design, Technical University of Munich, Accisstraße 21, 80333 Munich, Germany
pierluigi.dacunto@tum.de

[2] Specialist Modelling Group, Foster + Partners, Riverside, 22 Hester Rd, SW11 4AN London, UK

Abstract. The ability to work with both form and forces in the structural design process is crucial, particularly during the early design phase, as these are key parameters that impact the static behavior of a structure. Graphic statics offers a powerful geometric framework for the conceptual design of structures in static equilibrium, owing to the interdependence between form and force diagrams. When the form diagram of a self-stressed spatial pin-jointed framework has an underlying planar graph, its corresponding force diagram is reciprocal. It can itself be viewed as a self-stressed framework whose force diagram is the original form diagram. These configurations generally possess inherent architectural elegance and structural performance, making them particularly suitable for lightweight roof structures. This contribution outlines a graphic statics workflow for generating such structures using vector- and polyhedron-based approaches.

Keywords: form-finding · structural design · graphic statics · form and force diagrams · reciprocal diagrams · self-stressed structures

1 Introduction

1.1 Graphic Statics for Structural Design

Graphic statics was developed as an autonomous discipline in the late 19th century by the likes of Rankine [16], Maxwell [12], Culmann [3], and Cremona [2], among others [10]. It comprises a set of geometric methods and techniques that allow for establishing a direct correlation between the form of a structure in static equilibrium and the forces acting on and within it. To this end, graphic statics makes use of two diagrams that are topologically and geometrically interrelated: the form diagram, which represents the geometry of the structure and its external forces, and the force diagram, which portrays a synthetic representation of the equilibrium of the forces at each node of the structure. Starting from the early 20th century, structural designers realized the potential of graphic statics as a novel tool for discovering new structural shapes [7]. In recent years, the field of graphic statics has seen significant progress with the emergence of new digital methods and tools. As a result, a wealth of research has been conducted on extending graphic

S. Gabriele et al. (Eds.): IWSS 2023, LNCE 437, pp. 644–650, 2024.
https://doi.org/10.1007/978-3-031-44328-2_67

statics to the third dimension and integrating it into computational tools for equilibrium-based design. The expansion of graphic statics into 3D has given rise to two distinct approaches for constructing force diagrams [5]: the *polyhedron-based* and the *vector-based*. Both of these approaches provide a means of depicting the forces acting on a structure in 3D space, allowing for a complete understanding of the equilibrium in the structure.

1.2 Vector-Based Approach

According to Maxwell [12], pin-jointed frameworks in static equilibrium can always be represented using force diagrams in which forces are represented by edges, either in the plane or in three-dimensional space. When the form diagram of a self-stressed spatial pin-jointed framework has an underlying 2-planar graph (i.e. a graph that can be drawn on the plane without edge intersections), its corresponding force diagram is reciprocal [5, 15]. It can itself be viewed as a self-stressed framework whose force diagram is the original form diagram. In the case of reciprocal 3D form ***F*** and force ***F**** diagrams (Fig. 1), the two diagrams are related by a 3D duality in the sense that each edge of ***F*** maps to a parallel edge in ***F****. Moreover, this duality maps vertices of ***F*** to closed, possibly skew, polygons in the ***F**** (i.e., force polygons) and closed polygons in ***F*** to vertices in ***F****. Reciprocal form and force diagrams have been referred to as *Cremona Reciprocals* or *Cremona 3D* in the literature, named after the reciprocal diagrams in 2D described by Cremona [2]. Discrete surface structures made of linear elements and tensegrity structures with underlying 2-planar graphs (Fig. 1) are examples of spatial structures that can be represented by reciprocal form and force diagrams based on this approach.

1.3 Polyhedron-Based Approach

The polyhedron-based approach in 3D graphic statics was first proposed by Rankine [16] and further refined by Maxwell [12]. It involves constructing a force diagram (known as Rankine diagram) comprising polyhedral cells instead of polygons. In this approach, ***F*** and ***F**** are related by a 4D duality, which maps vertices into closed polyhedral cells, edges into perpendicular planar faces, and vice versa (Fig. 2). In this case, ***F*** can be mapped to a reciprocal ***F**** only if the underlying graph of ***F*** is 3-planar (able to be drawn in 3D space without face intersections). Although polyhedron-based graphic statics has been known for some time, general procedures for constructing and manipulating polyhedron-based ***F*** and ***F**** have only recently been developed, including iterative reconstruction methods [1, 11] and direct mathematical frameworks based on 4D stress functions [9, 13]. It should be noted that this approach can also be applied to 3D form diagrams that possess 2-planar graphs, such as cases of tensegrities (n-prisms, Jessen's icosahedron) [8, 13, 14].

2 Method

Constructing reciprocal self-stressed structures as vector-based and polyhedron-based form and force diagrams follow a similar procedure, albeit with a crucial distinction. The key difference lies in the dimensionality of planarity of the underlying graphs of the

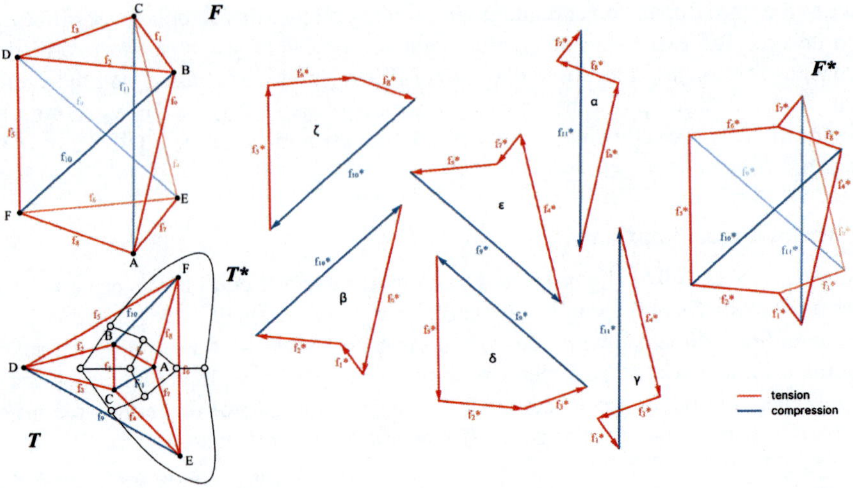

Fig. 1. Top left: 3D form diagram ***F*** of a tensegrity structure with the topology of an octahedron. Bottom left: Underlying graph ***T*** of the 3D form diagram ***F*** and underlying graph ***T**** of the 3D force diagram, which is also the dual graph of ***T***. Center: Six 3D force polygons corresponding to the six nodes of ***F***. Right: 3D force diagram ***F**** reciprocal to ***F***; ***F**** can be constructed by connecting the 3D force polygons so that corresponding vectors are overlapped; ***F**** is a tensegrity structure in equilibrium with the topology of a cube.

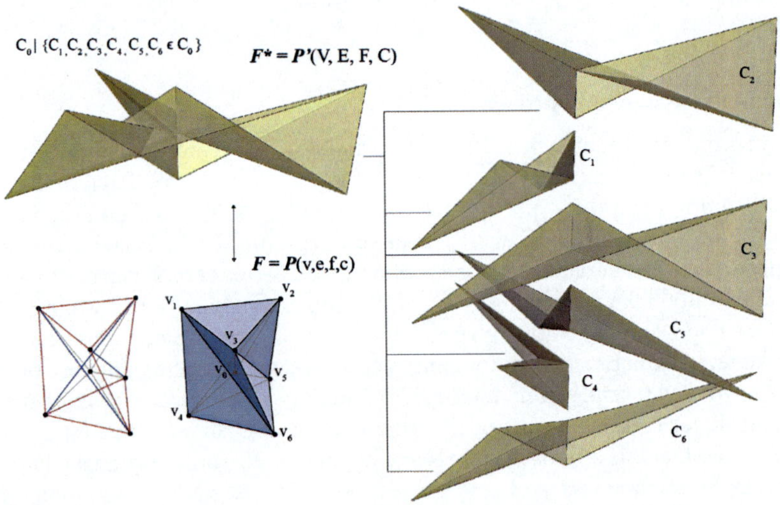

Fig. 2. Bottom left: 3D polyhedral form diagram ***F*** of a tensegrity 3-prism structure, which is topologically a spoked octahedron. Right: Six 3D force polyhedral cells representing the equilibrium of the six external nodes of ***F***. Top Left: 3D polyhedral force diagram ***F**** reciprocal to the 3D form diagram ***F*** with forces represented by faces; ***F**** is a self-stressed structure in equilibrium.

structures: 2-planarity for the vector-based case and 3-planarity for the polyhedron-based case.

2.1 Construction of Vector-Based Reciprocal Structures

The construction of reciprocal vector-based form and force diagrams consists of the following steps based on graph-theory procedures [5] and numerical optimization [4]:

1. Define a preliminary geometry for the 3D form diagram ***F*** as a self-stressed pin-jointed framework. Note that this preliminary geometry is not necessarily already in equilibrium.
2. Assign the boundary and geometric constraints to ***F***.
3. Assign preliminary internal forces to the edges of ***F***.
4. Derive the underlying graph ***T*** of ***F***.
5. Construct the dual graph of ***T***, which is the underlying graph ***T**** of the 3D force diagram ***F****.
6. Construct a first instance of ***F**** as a combination of force polygons representative of the equilibrium of the forces in the nodes of ***F*** and based on the order defined by ***T****.
7. Iteratively adjust the position of the nodes of ***F*** and the magnitude of the internal forces until the equilibrium of the nodes of ***F*** is obtained. This constraint-based optimization enforces that the corresponding edges between ***F*** and ***F**** are parallel, the force polygons in ***F**** are closed, and the boundary conditions in ***F*** are respected.

Note that the same procedure above can be used to generate the 3D form diagram starting from the initial definition of the 3D force diagram.

2.2 Construction of Polyhedron-Based Reciprocal Structures

The construction of reciprocal polyhedron-based form and force diagrams can have the form or force space as a starting point. This is due to the interlink between 3D form and force diagrams and their corresponding 4D stress functions [9]. In the case of the form space, the method consists of the following steps:

1. Define a preliminary geometry for the 3D form diagram ***F***. In this approach, the initial form diagrams should comprise polyhedral cells which: have flat faces, each face belongs to only two cells, and each edge to at least 3 cells. Note that this preliminary geometry is not necessarily already in equilibrium.
2. Assign the *external* points (corresponding to one of the cells) from the set of vertices of ***F*** – the remaining vertices are *internal*.
3. Lift the *internal* vertices to the 4th dimension to obtain the corresponding 4D stress function ***P***. By doing so and imposing cell planarity (all vertices of a cell lying on the same hyper-plane in 4D), the co-ordinates of the *internal* vertices will automatically update, and ***F*** will consequently be in static equilibrium.
4. Map ***P*** to its reciprocal 4D stress function ***P****.
5. Project ***P**** back to 3D space to obtain the reciprocal Rankine ***F****.

6. This construction provides a direct way for adjusting the preliminary geometry of ***F*** and hence for obtaining global static equilibrium in terms of a Rankine force diagram ***F****. In the latter, the closed cells ensure nodal equilibrium for the form diagram whilst the faces are perpendicular to their corresponding form edges, and their surface areas define the axial forces.

The same method described above can be repeated by having as a starting point the Rankine force polyhedron. Note that each one of the form and force diagrams derived in this way can be seen as the 3D self-stressed structure – with the other being its reciprocal Rankine.

3 Case Study

Reciprocal form and force diagrams generally possess inherent architectural elegance and structural performance, making them particularly suitable for lightweight roof structures. This section applies the previously introduced vector-based procedure to the conceptual design of a large-span self-stressed roof structure. The presented design is generated using the *VGS Tool* [6].

4 Vector-Based Reciprocal Roof Structures

The preliminary geometry of the 3D form diagram ***F*** is defined by an undulating compression ring with 36 struts, and an inner tension ring with 36 ties. Radial spokes in tension connect the compression ring on the outside with the tension ring on the inside. The radial spokes are redirected by an additional intermediate compression ring that gives the self-stressed roof a peculiar three-dimensional shape. A series of preliminary values for the internal forces are defined: all the elements in the outer and intermediate compression ring are set to -500kN, all elements in the tension ring to 1′000kN, and the spokes to 10 kN. Moreover, the ranges of possible variation for the positions of the nodes in ***F*** are set to: 1.0m in z- and 0.5m in x- and y-direction. Based on this initial setup, the *VGS Tool* [6] automatically generates the underlying graph ***T*** of the form diagram ***F*** and its dual ***T**** of the force diagram ***F****. The tool then constructs a first instance of ***F**** as a combination of unclosed force polygons. Subsequently, a constraint-based optimization algorithm iteratively adjusts the position of the nodes of ***F*** and the magnitude of the internal forces until the equilibrium of the nodes of ***F*** is obtained. Figures 3 and 4 show the final equilibrium geometries of ***F*** and ***F****. The latter can be regarded as a self-stressed structure in its own right. These diagrams can then be used as initial geometry to design form-active lightweight roof structures.

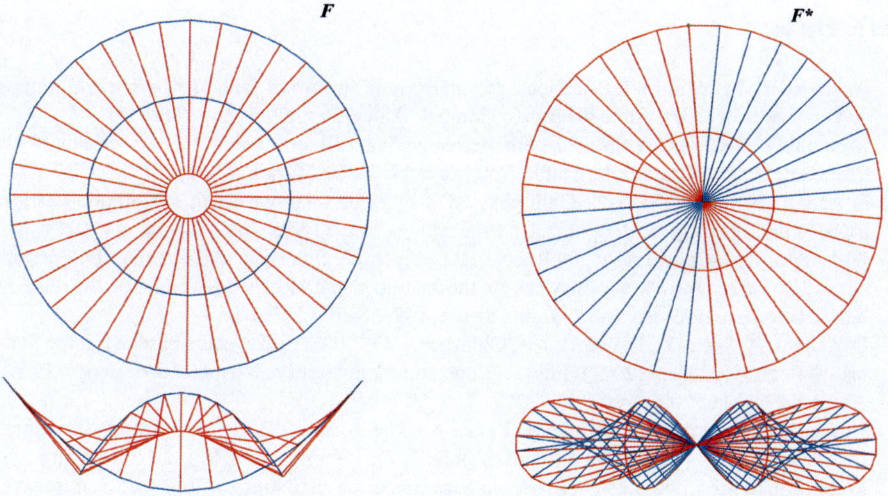

Fig. 3. Left: top and side views of the self-stressed form diagram ***F***; right: top and side views of the reciprocal force diagram ***F****.

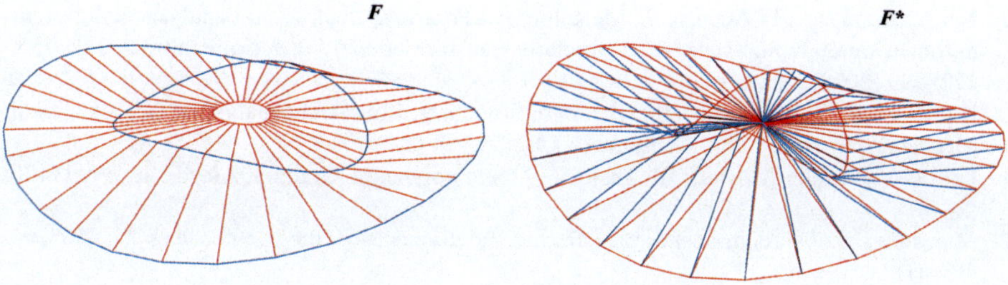

Fig. 4. Left: 3D view of the self-stressed form diagram ***F***; right: 3D view of the reciprocal force diagram ***F****.

5 Conclusion

In this contribution, we have introduced a comprehensive graphic statics procedure that enables the generation of reciprocal form and force diagrams, resulting in self-stressed structures. The proposed workflow encompasses both vector-based and polyhedron-based methods, providing flexibility and adaptability in the design process. By employing graphic statics, designers can effectively work with the interplay of form and forces, ensuring static equilibrium and optimizing structural performance. The presented case study highlights the versatility and advantages of employing reciprocal self-stress configurations in the conceptual design of lightweight roof structures. By showcasing this example, we aim to inspire designers and engineers to explore new possibilities in conceiving aesthetically pleasing and structurally efficient roof structures.

References

1. Akbarzadeh, M., Van Mele, T., Block, P.: On the equilibrium of funicular polyhedral frames and convex polyhedral force diagrams. Comput. Aided Des. **63**, 118–128 (2015)
2. Cremona, L.: Le figure reciproche nella statica grafica, Milano: Tipografia Bernardoni (1872)
3. Culmann, C.: Die Graphische Statik. Von Meyer & Zeller, Zurich (1866)
4. D'Acunto, P., Jasienski, J-P., Ohlbrock, P.O., Fivet, C.: Vector-based 3D graphic statics: transformations of force diagrams. In: Proceedings of the IASS 2017, Hamburg (2017)
5. D'Acunto, P., Jasienski, J.-P., Ohlbrock, P.O., Fivet, C., Schwartz, J., Zastavni, D.: Vector-based 3D graphic statics: a framework for the design of spatial structures based on the relation between form and forces. Int. J. Solids Struct. **167**, 58–70 (2019)
6. D'Acunto, P., Shen, Y., Jasienski, J-P., Ohlbrock, P.O.: VGS Tool: Vector-based Graphic Statics - Release 1.00 Beta (2021). https://github.com/computational-structural-design/VGS.git. Accessed 31 May 2023
7. Huerta, S.: Designing by Geometry: Rankine's Theorems of Transformation of Structures. In: Cassinello, P., Huerta, S., de Prada Poole, J.M., Lampreave, R.S. (eds.) Geometry and Proportion in Structural design, Essays in Ricardo Aroca's Honour, pp. 262–285. Lampreave, Madrid (2010)
8. Konstantatou, M., McRobie, A.: 3D Graphic statics and graphic kinematics for spatial structures, Presented in International Association for Shell and Spatial Structures (IASS), Hamburg (2017)
9. Konstantatou, M., D'Acunto, P., McRobie, A.: Polarities in structural analysis and design: n-dimensional graphic statics and structural transformations. Int. J. Solids Struct. **152–153**, 272–293 (2018)
10. Kurrer, K.-E.: The History of the Theory of Structures, from Arch Analysis to Computational Mechanics. Ernst & Sohn, Germany (2008)
11. Lee, J., Van Mele, T., Block, P.: Disjointed force polyhedra. Comput. Aided Des. **99**, 11–28 (2018)
12. Maxwell, J.C.: On reciprocal figures, frames and diagrams of forces. Phil. Mag. **27**, 250–261 (1864)
13. McRobie, A.: Maxwell and Rankine reciprocal diagrams via Minkowski sums for two-dimensional and three-dimensional trusses under load. Int. J. Space Struct. **31**(2–4), 203–216 (2016)
14. McRobie, A.: Rankine Reciprocals with Zero Bars (2017), https://www.researchgate.net/publication/315027343. Accessed 31 May 2023
15. Micheletti, A.: On generalized reciprocal diagrams for self-stressed frameworks. Int. J. Space Struct. **23**(3), 153–166 (2008)
16. Rankine, W.J.M.: Principle of the equilibrium of polyhedral frames, The London, Edinburgh, and Dublin Philosophical Magazine and Journal of Science XXVII - Fourth Series (CLXXX - February), 92(1864)

A New Sustainable Design Approach for Optimized Structures Under Seismic Loads

Emanuele Crespino[1(✉)], Sigrid Adriaenssens[2], Aguinaldo Fraddosio[1], Carlo Olivieri[2,3], and Mario Daniele Piccioni[1]

[1] Department of Architecture, Construction and Design (ArCoD), Politecnico di Bari, Via Orabona, 4, 70125 Bari (BA), Italy
e.crespino@phd.poliba.it

[2] Department of Civil and Environmental Engineering, Princeton University, Princeton, NJ 08544, USA

[3] Department of Civil Engineering (DICIV), University of Salerno, Via Giovanni Paolo II, 132, 84084 Fisciano, (SA), Italy

Abstract. Large-span structures are essential for various applications, including commercial buildings, train stations, ferry-boat stations, and airports. However, constructing such structures in seismic areas requires the use of a considerable amount of structural materials, resulting in high resource consumption and CO2 emissions. To overcome this problem, a sustainable design strategy for large-span structures in seismic areas using form finding and structural optimization tools is here proposed, starting from a previous work by some of the authors. This strategy is employed for designing an optimized innovative hybrid structural system comprising of a concrete shell coupled with suspended steel multi-floor frame system. The shape of the shell is initially determined by a form finding procedure for different plan forms; a linear finite element analysis under static and dynamic loads allows to select the most performing shell shape in terms of stress and deformation levels. The framed system is then suspended from the selected shell, and the structural layout is optimized by a shell thickness optimization conducted for gravitational loads. The whole structural system, optimized for gravitational loads, reveals efficient performance also under significant seismic actions, indicating a beneficial interaction between the shell and the frame structure. Therefore, the innovative hybrid structural system and the design approach here proposed helps reduce the quantity of structural material usage, minimizing environmental impact and enhancing structural efficiency.

Keywords: Form-finding · Structural optimization · Concrete shells · Sustainable structures · Seismic engineering

1 Introduction

The construction industry is one of the main consumers of resources and energy [12], primarily due to the ever-increasing demand for buildings and infrastructure. This demand is strongly influenced by conventional design approaches and

S. Gabriele et al. (Eds.): IWSS 2023, LNCE 437, pp. 651–661, 2024.
https://doi.org/10.1007/978-3-031-44328-2_68

code requirements that mandate compliance with strict performance criteria for robustness, stiffness, and strength in structures, especially in seismic areas. Indeed, even if seismic codes allow to rely upon structural damage, even severe, as a means of dissipating seismic energy while ensuring human safety, the design of massive, rigid structural elements is inevitably obtained. This leads to substantial resource consumption, CO2 emissions, and material waste throughout the structure's life cycle [4]. instances are separately examined. Structural typology impacts the entire life cycle of a construction, from material selection and construction methods (dry or wet solutions) to maintenance operations, and ultimately, the end-of-life cycle, encompassing demolition, reuse, or recycling. The example of the efficiency of ancient constructions, conceived with a strong connection between shape and structure, indicates a possible direction for the progress of structural design. Renowned designers such as Pier Luigi Nervi [10] and Sergio Musmeci [8] rediscovered this approach in the last century, paving the way for an evolution toward modern and sustainable applications, supported by ongoing research advancement [3,14–16]. In line with sustainable objectives and seismic regulations, this paper presents a conceptual design that couples form-finding and structural optimization tools for obtaining reduced material consumption and optimal performance. The proposed design is founded on the specialization of structural tasks between the two main components of the construction: a reinforced concrete shell that primarily transfers loads (including seismic loads) to the foundations via compression forces; and a steel-framed structure suspended from the shell, acting as a load collector, and having high deformability corresponding to large oscillation periods. Coupling of these two structures synergistically combines the shell's efficiency in resisting seismic forces, with the efficiency of the suspended structure's behavior, similar to that of a steel frame with base isolation. To determine the shell's shape, an algorithm was developed for implementation in Rhinoceros using Grasshopper, a visual programming environment, and Kangaroo 2, a form-finding tool employing the discrete dynamic relaxation method [2,17,20]. Utilizing 3D modeling software for this phase is justified by the ease of code implementation and the versatility of generated shapes. In particular, different shapes are obtained considering different plan forms. The efficiency of each shape is then studied by a seismic analysis, allowing for the choice of the most performing shapes. Seismic actions are defined by horizontal and vertical accelerograms applied in three orthogonal directions. The latter are then integrated with the suspended steel-framed structure, and the thickness of each shell is optimized under gravitational loads for Ultimate Limit States (ULS), according to Italian building regulations [9]. The thickness optimization process is performed by the Method of Feasible Directions (MDF) with the aid of Altair HyperWorks software, to minimize stress and maximum displacements of the shell [21]. This stage of the analysis allows for the selection of the most efficient structural shape, which is then studied by a nonlinear dynamic analysis also under seismic actions, employing the finite element method (FEM) Strand7 software [1,18]

2 Methodology

Form Finding. As illustrated in [20], the Dynamic Relaxation method solves a static problem by applying a pseudo-dynamic process over time. The algorithm simplifies a nonlinear system of equations into a linear and iterative calculation at discrete steps Δt. The continuum structure is discretized into a lattice of elastic bars converging at points called nodes. The accuracy of the form-finding process results depends on the number of elements constituting the lattice, i.e., on the density of the elastic bars lattice. The nodes of the lattice are subjected to an external load P, suitably representing the self-weight of the structures, and the elastic bars are characterized by an axial stiffness EA. The application of nodal loads results in an oscillatory motion that will reach equilibrium downstream of the analysis, due to the contribution of a fictitious kinematic damping. The final equilibrium between external loads and the bar's internal forces corresponds to the minimum potential energy state.

Thickness Optimization. The optimization procedure employed in this study focuses on minimizing the material and thus the weight of the structure, aiming to save material through sizing optimization of the shell thickness. To this aim, an algorithm based on the Method of Feasible Directions (MDF), implemented in Altair "HyperWorks" software, is used. The discrete minimization problem can be outlined as follows:

$$min \frac{\mathbf{t^T a}}{V_i} \tag{1}$$

$$g(t) = \sigma^{VM} - \sigma_{VM,max} \leq 0 \tag{2}$$

$$t_{min} < t < t_{max} \tag{3}$$

where $\frac{\mathbf{t^T a}}{V_i}$ is the objective function f, $\mathbf{a}$ is the area vector, and $\mathbf{t}$ is the thickness vector of the finite elements, V_i is the volume calculated considering t_{min}, $g(t)$ the constraint function, σ^{VM} the Von Mises stress, $\sigma_{VM,max}$ the maximum von Mises stress allowed, t_{min} and t_{max}, minimum and maximum thickness allowed, respectively. If the two constraints are satisfied, the design is considered "feasible. The optimization process proceeds iteratively, defining a variable vector after each iteration $\mathbf{t}^{1+q} = \mathbf{t}^q + \lambda \mathbf{s}^q$, where $\mathbf{s}^q$ is a displacement vector within the feasible domain, and λ is a scalar multiplier (step length). At the q-th iteration; moreover, $\mathbf{s}^q$ must be small enough to contract the objective function (usable descent direction), i.e., $\nabla f^T \cdot \mathbf{s}^q \leq 0$ and without violating the constraints (feasible direction), $\nabla g^T \cdot \mathbf{s}^q \leq 0$. Once $\mathbf{s}^q$ is known, the problem reduces to a linear function in λ to be determined to minimize the objective function [19]. As shown in Fig. 1, at iteration q, the function f intersects the constraint g at a point constituting an optimum. ∇f represents the normal to the objective function f, while ∇g represents the normal to the constraint function g. It is necessary for the angle formed by $\mathbf{s}^q$ with both f and g to be obtuse, to be the scalar products above recalled negative. This condition determines the admissible space for the displacement vector, called the "usable-feasible sector" [19].

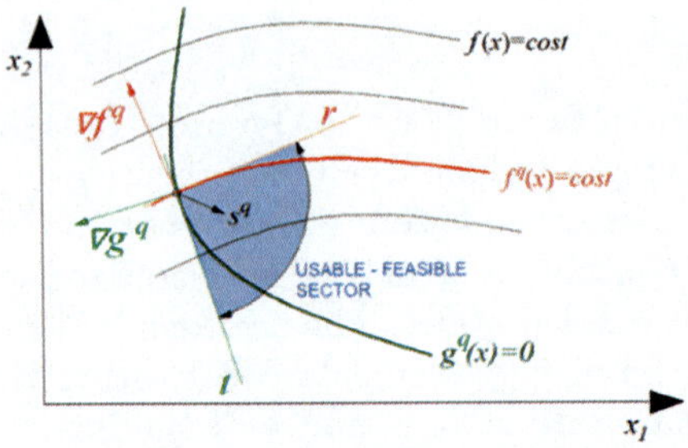

Fig. 1. Feasible direction within the usable feasible sector, generated by the tangents of the objective function f and constraint g at the q-th iteration.

Finite Element Method. The stress and deformation states of the entire structural system are assessed using Strand7 finite element method (FEM) code [1], via both linear and nonlinear analyses. The linear analysis is utilized during the optimization process for the reinforced concrete shell structure, while the nonlinear analysis evaluates the total structural response, including the suspended steel frame. This accounts for geometric non-linearities due to potential large displacement of the steel structure and associated seismic energy dissipation. The shell is modeled in Strand7 with plate-type elements, specifically tri3 finite elements, at a discretization of 2% of the plan dimension. These elements are immune to in-plane distortions, ensuring node position. The selected discretization effectively represents structural behavior without significantly raising computational costs. For seismic analyses, the seismic input is defined for discrete acceleration values over time, using the Newmark constants method [11].

3 Geometries and Materials

The design of a complex structural system under seismic loads requires upfront evaluations on rigidity, robustness, and load bearing capacity [6]. This study analyzes various ground plan shapes to identify the optimal shape that minimizes the shell's deformative and stress states. As demonstrated in Fig. 2, while maintaining plan dimensions (60 m × 20 m), different plan shapes were created by adjusting the curvature of the constrained edges and introducing openings on the short edges. The considered curvatures are elliptical, with minor semi-axes equal to 10% and 20% of the plan's short side L_{min} (2 m and 4 m, respectively). For five plan shapes, openings are also added on the constrained sides to meet architectural needs, providing building access from the sides.

The shell is composed of C50/60 reinforced concrete (Table 1), with the suspended structure made of S355 structural steel. The "residential" module (Fig. 3) is a steel frame with plan dimensions of 30 m × 14 m × 9 m and a total area of 1092 m^2. Positioned 4 m above the ground, the area underneath can accommodate a driveway, railroad, or river passage, depending on the building's purpose. The steel frame was previously designed according to the Italian seismic code

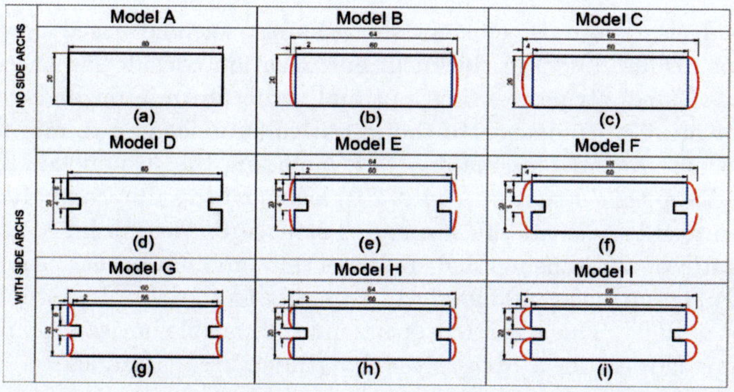

Fig. 2. The investigated geometries with plan dimensions. In blue the starting constrained edge, in red the new edge form. Edge curvature: a) null, b) 10% L_{min}, c) 20% L_{min}, d) null, e) 10% L_{min}, f) 20% L_{min}, g) -10% L_{min}, h) 10% L_{min}, i) 20% L_{min}.

[9], assuming the structure was suspended above. Further details on properties and geometries will be specified in a future paper.

Table 1. Concrete properties.

Concrete C50/60			
f_{ck}	50 MPa	f_{cd}	28 MPa
E	37.640 MPa	ρ	2400 kg/m^3
ν	0.2	ξ	5%

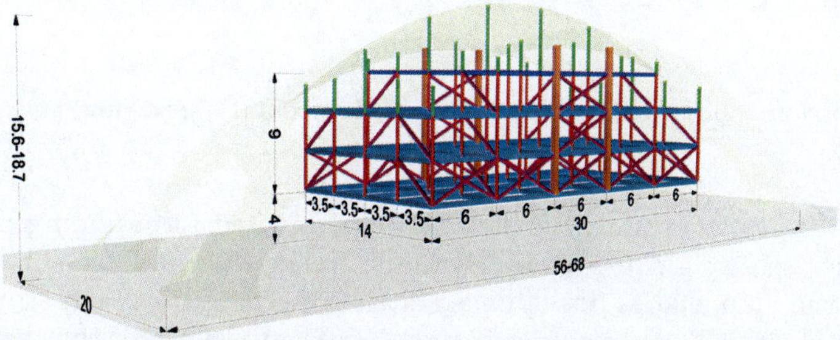

Fig. 3. Structural system, overall plan form goes from 56 m to 68 m, while elevation goes from 15.6 m to 18.7 m due to the different edge curvature.

4 Analysis

This section presents the results of the form-finding process, where each ground plan shell shape (Fig. 2) is subjected to seismic load analysis through linear dynamic analysis. The most efficient shell shapes, selected based on maximum displacement results, undergo thickness optimization considering the load from the suspended steel structure. The optimal structure minimizes both volume and volume variation compared to the initial volume, leading to minimal material usage. A non-linear dynamic analysis, including the suspended steel frame, is conducted on this structure, with displacement results compared to pre-optimization results without the suspended structure. The minimization process follows a static analysis using the fundamental combination at Ultimate State Limit (ULS) [9], with dynamic analyses adhering to the seismic combination for ULS of life safety [9]. This considers spectrum compatible horizontal and vertical accelerograms (Fig. 4) for a high seismicity Italian area (Catanzaro).

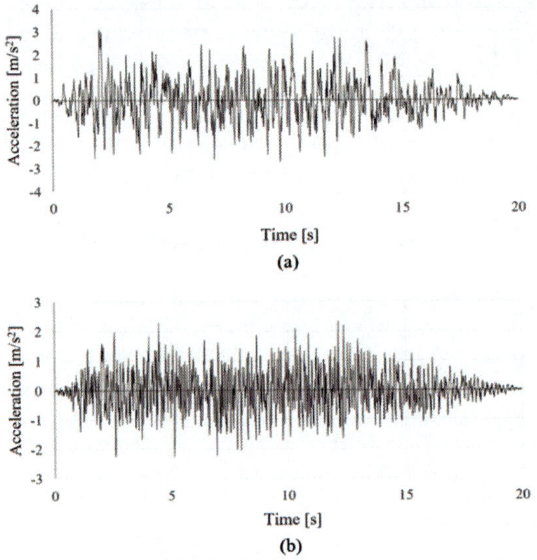

Fig. 4. Spectrum compatible accelerogram: (a) horizontal - spectrum compatible: (b) vertical.

Form Finding. As outlined in Sect. 3, we conducted form-finding of all plan shapes under self-weight using the Dynamic Relaxation method. This involved converting the plan shapes into a dense lattice of elastic bars, and adjusting the parameters of the rods to ensure sufficient height at the top of the shells for the suspended "residential" modules. This form finding was iterative, with the nodal load (distributed evenly) being updated until a vertical node height variation of less than 0.5% was reached. The arrangement of openings induced a local curvature variation (corrugation), enhancing the shell's seismic performance due to its stiffening effect [5]. Figure 5 illustrates the middle surfaces of the shell shapes derived from form finding. The shell thickness was maintained at 0.15 m, thus ensuring a thickness-to-maximum shell span ratio $t/L_{\max} = 0.0025$, comparable to the Algeciras market coverage of $t/L_{\max} = 0.0022$ [13].

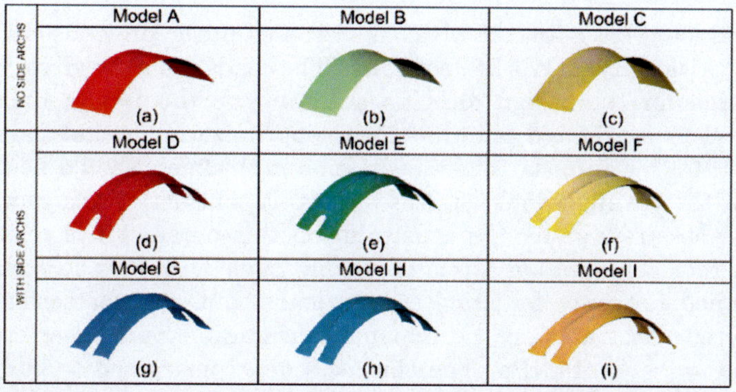

Fig. 5. Shell mid surfaces. Edge curvature: a) null, b) 10% L_{min}, c) 20% L_{min}, d) null, e) 10% L_{min}, f) 20% L_{min}, g) −10% L_{min}, h) 10% L_{min}, i) 20% L_{min}.

Dynamic Analysis of the Shells Alone. All the shells obtained by form finding are then analyzed under seismic loads through a linear dynamic analysis, based on the accelerograms in Fig. 4. Table 2 lists, for each examined model, the frequency of the mode with the highest participating mass, the resulting maximum displacement D_{XYZ}, and the maximum von Mises stress $\sigma_{VM,max}$. From the results, the first four shell models (A, B, C, and D) are the most deformable, with significant seismic displacements. For this reason, these structures will not be considered for subsequent analyses. On the contrary, consistent with the literature [5–7], the corrugation results in a high shell stiffness, consequently reducing maximum displacements.

Table 2. Shells comparison under seismic analysis.

Model	f	Time	D_{XYZ}	$\sigma_{VM,max}$
	[Hz]	[s]	[cm]	[MPa]
A	0.31	18.89	32.6	24
B	0.55	15.64	33.0	72
C	0.72	19.59	17.5	24
D	0.55	13.28	32.4	65
E	1.21	16.05	10.3	47
F	1.46	18.04	16.6	55
G	1.42	16.60	13.9	57
H	1.48	15.07	19.1	63
I	1.77	13.90	17.6	73

Thickness Optimization. The shells corresponding to models E, F, G, H, and I are coupled with the suspended steel structure, and a thickness optimization

(TO) is conducted based on the MFD as discussed in the previous chapter, with the aid of "Altair HyperWorks" software. The loads considered correspond to those required for the design for ULS according to the Italian building code [9]. Essentially, the adopted criterium for the optimization is that the shell must remain light; to this aim, the allowed variation in thickness should be sufficiently limited. For the optimization, von Mises stress is considered as a representative measure of the stress state; the constraint on this parameter is related to the design concrete compression strength f_{cd} in Table 1. Since the structure has to be designed not only for gravitational loads, but also for seismic actions, the considered constraint on the maximum achievable von Mises stress is set to $\sigma_{VM,max} = \frac{f_{cd}}{2} = 14\text{MPa}$. The thickness t is constrained to vary between a minimum of 0.15 m and a maximum of 0.30 m. From the results shown in Table 3, it emerges that shell Model I (Fig. 5i) involves the smallest volume variation. For this case, the variation of thickness is shown in Fig. 6. Therefore, from a sustainable design perspective, the shell model I is the most efficient as it involves the least use of material.

Table 3. TO Comparison: V_i initial volume, $V\%$ variazione percentuale di volume rispetto al volume inziale [%], V_f final volume after thickness optimization.

Model	Vi	Vf	V
	[m3]	[m3]	[%]
E	201	402	100
F	208	269	29.5
G	188	280	49
H	198	245	23.9
I	201	232	15.3

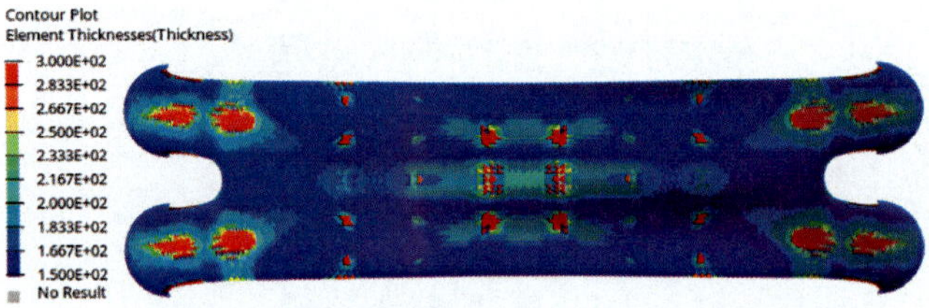

Fig. 6. Thickness variation of Model I.

Dynamic Analysis of the Whole Structural System After Thickness Optimization. The dynamic interaction between the shell and the steel structure for the shell model I is analysed. A nonlinear dynamic analysis is performed, considering as seismic input the accelerograms in in Fig. 4. As can be seen in Fig. 7, despite the addition of the suspended structure, the maximum dynamic displacement of the shell with optimized thickness is about 53% compared to the maximum dynamic displacement of the constant thickness shell in the absence of the suspended structure (0.082 m vs 0.176 m). Moreover, the maximum dynamic displacement of the shell with optimized thickness is approximately 1/820 of the maximum span of the shell (for model I, 68 m). This displacement is compliant with the Italian building code [9], which allows for a maximum displacement between 1/200 and 1/500 of the span.

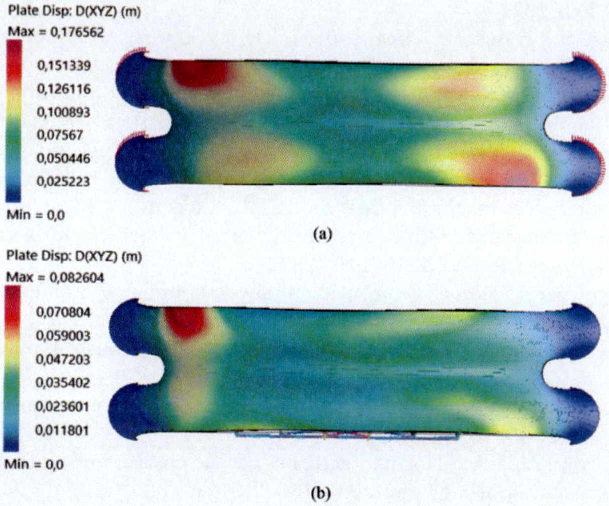

Fig. 7. Maximum resulting displacement of shell model I: a) linear dynamic analysis of shell only, where the maximum value is 0.176 m and occurs at instant $t = \mathbf{13.9}$ s; b) nonlinear dynamic analysis of the whole structural system, where the maximum value is 0.082 m and occurs at instant $t = \mathbf{13.92}$ s.

5 Conclusions

This study explores the structural behavior of nine hybrid concrete-steel structures with different initial shell planforms. The initial shell shapes are determined by form-finding, revealing corrugated shells, created by introducing curvature on the constrained planform edge, as the most efficient in seismic analysis. An ensuing thickness optimization under the gravitational load of the entire hybrid structure facilitates the selection of the most efficient structural shape under such loads. Further non-linear dynamic analysis under seismic loads indicates that this shape also exhibits high efficiency for seismic actions, with acceptable maximum displacement and stress levels. The design format allows for minimal structural material use while maintaining high structural performance under complex load scenarios including gravitational and seismic loads. The interaction between the concrete shell and the suspended structure is noteworthy, reducing displacements in the dynamic phase. Future research will delve deeper into the suspended structure's contribution during the dynamic phase, comparing results of the two-step structural optimization with form-finding that includes the self-weight of the suspended steel frame from the outset.

References

1. Strand7-finite element analysis. https://www.strand7.com/html/brochure.htm. Accessed 20 Feb 2021
2. Adriaenssens, S., Block, P., Veenendaal, D., Williams, C.: Shell Structures for Architecture: Form Finding and Optimization. Routledge, London (2014)
3. Block, P., Ochsendorf, J.: Thrust network analysis: a new methodology for three-dimensional equilibrium. J. Int. Assoc. Shell Spatial Struct. **48**(3), 167–173 (2007)
4. Bruun, E.P., et al.: Three cooperative robotic fabrication methods for the scaffold-free construction of a masonry arch. Autom. Constr. **129**, 103803 (2021)
5. Lai, M., Eugster, S., Reccia, E., Spagnuolo, M., Cazzani, A.: Corrugated shells: an algorithm for generating double-curvature geometric surfaces for structural analysis. Thin-Walled Struct. **173**, 109019 (2022)
6. Lai, M.: Edge-corrugation as structural enhancement of shallow shells. Nexus Netw. J. 1–7 (2023)
7. Lai, M., Reccia, E., Spagnuolo, M., Cazzani, A.M.: Structural behaviour of corrugated shells: a look at the Flaminio dome in Rome
8. Marmo, F., et al.: On the form of the Musmeci's bridge over the Basento river. Eng. Struct. **191**, 658–673 (2019)
9. Mordà, N., Mancini, A.: Norme tecniche per le costruzioni (ntc 2018) d. min. infrastrutture e trasporti 17 gennaio 2018 (2018)
10. Nervi, P.: Aesthetics and Technology in Building: The Twenty-First-Century Edition. University of Illinois Press (2018)
11. Newmark, N.M.: A method of computation for structural dynamics. J. Eng. Mech. Div. **85**(3), 67–94 (1959)
12. Ochsendorf, J.: Sustainable engineering: the future of structural design. In: Structures Congress 2005: Metropolis and Beyond, pp. 1–9 (2005)
13. Ochsendorf, J., Antuña, J.: Eduardo torroja and "cerámica armada". In: First International Congress on Construction History. Instituto Juan de Herrera, Madrid (2003)
14. Ohmori, H., Futai, H., Iijima, T., Muto, A., Hasegawa, Y.: Structural design of office building by computational morphogenesis (structures). AIJ J. Technol. Des. **10**(20), 77–82 (2004)
15. Olivieri, C.: Formerly-math: constrained form-finding through membrane equilibrium analysis in mathematica. Softw. Impacts, 100512 (2023)
16. Olivieri, C., Angelillo, M., Gesualdo, A., Iannuzzo, A., Fortunato, A.: Parametric design of purely compressed shells. Mech. Mater. 103728 (2021)
17. Otter, J.R.H., Cassell, A.C., Hobbs, R.E.: POISSON: dynamic relaxation. Proc. Inst. Civ. Eng. **35**(4), 633–656 (1966)
18. Ridho, B.M.A., Kaewunruen, S.: Failure investigations into interspersed railway tracks exposed to flood and Washaway conditions under moving train loads. Eng. Fail. Anal. **129**, 105726 (2021)

19. Vanderplaats, G.N., Moses, F.: Structural optimization by methods of feasible directions. Comput. Structur. **3**(4), 739–755 (1973)
20. Veenendaal, D.: Evolutionary optimization of fabric formed structural elements. Master's thesis 1223, 1224 (2008)
21. Zhou, M., Pagaldipti, N., Thomas, H., Shyy, Y.: An integrated approach to topology, sizing, and shape optimization. Struct. Multidiscip. Optim. **26**, 308–317 (2004)

A Hybrid Deployable Vault Structure by Coupling Origami Folded Plates with a Scissors – Four Bar Linkage Frame

Evangelia Vlachaki(✉) and Katherine A. Liapi

Department of Architecture, University of Patras, 26504 Patras, Greece
arc935@yahoo.gr

Abstract. In this paper, a novel concept of a lightweight deployable vault structure, consisting of a curved folded plate surface, paired with a retractable frame mechanism is proposed. The retractable frame, in its deployed and constrained configuration, will function as a structural and load-bearing system, whereas the rigid origami folded plate structure will serve as a covering surface. A miura-ori first-level derivative crease pattern, such as the Arc Pattern, will be used for this purpose. The retractable frame consists of scissor bars of different lengths combined with four-bar linkages with specific sliding joints. This frame acts as a motion actuator, as well as a supporting and locking mechanism for the folded plate surface. For the study of the form and kinematic behavior of the proposed vault structure, a parametric model has been developed in the Rhino Grasshopper programming environment and small-scale physical models have been constructed.

Keywords: Deployable Structures · Kinetic Architecture · Structural Origami · Computational Geometry · Parametric Design · Adaptive Architecture · Origami · Scissors · Linkages · Transformable Architecture

1 Introduction

Origami is the ancient Japanese art of paper folding (from "oru" meaning "to fold" and "kami" meaning "paper"), technically defined as the creasing of surfaces into various forms without stretching or cutting. Surfaces are specified as thin sheets of material with significantly high in-plane stiffness compared to out-of-plane stiffness [1]. Among various scientific applications of origami (medical, aerospace, etc.), structural origami has recently gained attention from architects and engineers due to the interrelation between folds and structural performance of surfaces [2]. Through the application of any of the existing thickness accommodation techniques [3], it is possible to fabricate rigid and thick origami structures [4]. These structures are scalable mechanisms composed of stiff plates and hinges that can deploy and retract with a unified and synchronous movement of all creases [5]. As a result, rigid thick origami structures can be effectively used as envelopes in architectural applications that require adaptiveness in response to changing needs in space configuration [6]. Typically, building-scale deployable structures require

S. Gabriele et al. (Eds.): IWSS 2023, LNCE 437, pp. 662–671, 2024.
https://doi.org/10.1007/978-3-031-44328-2_69

a control mechanism that is capable of actuating motion between retracted and deployed configurations. Previous research conducted by the authors [7, 8] has explored the coupling of various geometric forms of folded plate structures with planar scissor linkages, which can function as both structural and load-bearing frames and are also able to control motion. This paper introduces a novel hybrid deployable structure of vaulted geometry.

2 Methodological Approach

The process employed to develop the hybrid deployable structure proposed in the current study involves a series of sequential steps. Firstly, the crease pattern (CP) for the folded plate structure is selected by considering the desired deployed form of the surface. Subsequently, the chosen CP's base geometry and kinematic equations are delineated and parameterized within the Grasshopper visual programming environment. The tapered panels thickness accommodating technique [3] is utilized to add thickness to the origami plates. The following stage involves the development of a retractable linkage frame mechanism that follows the same motion path in space as the folded surface. Subsequently, the relations between the constant and variable parameters of the folded plates and the linkages are established and parameterized. Finally, the developed Grasshopper algorithm is utilized to generate several morphological variations of the proposed hybrid deployable vault, while selected ones are fabricated into small-scale kinetic physical models.

3 Origami Crease Pattern

The rigid-foldable Miura-ori and its derivative crease patterns are widely utilized in engineering and architecture due to their unique property of representing an overconstrained case of a general 4-degree vertex with only one degree of freedom (DOF). Additionally, they possess useful characteristics such as flat-foldability (panels are co-planar when fully folded) and developability (can be folded from a continuous flat sheet) [9]. This research focuses on the Arc Pattern, a Miura-ori first-level derivative crease pattern, as described in reference [10]. This pattern is chosen for its ability to achieve single curvature geometries, such as vault structures. The following subsections describe the base geometry of the CP unit, including its constant and variable parameters relations, as well as the kinematic equations that describe its motion path in space.

3.1 Arc Pattern Unit Geometrical Definition

The basic unit of the Arc Pattern, illustrated in Fig. 1a, is constructed from four identical trapezoidal plates and can be geometrically defined by three constant parameters: side lengths α_1 and b, and sector angle φ. Two more dependent constants, side length α_2 and panel width w, are employed to describe the CP unit. These constants are determined by Eqs. (1) and (2), subject to the condition that $\alpha_1 < \alpha_2$. In the special scenario where $\alpha_1 = 0$, the CP is simplified to triangular plates, commonly referred to as the Yoshimura CP.

Four variable parameters, namely dihedral angles η_A and η_Z, along with edge angles θ_A and θ_Z, play a crucial role in defining a specific folded configuration of a rigid unit.

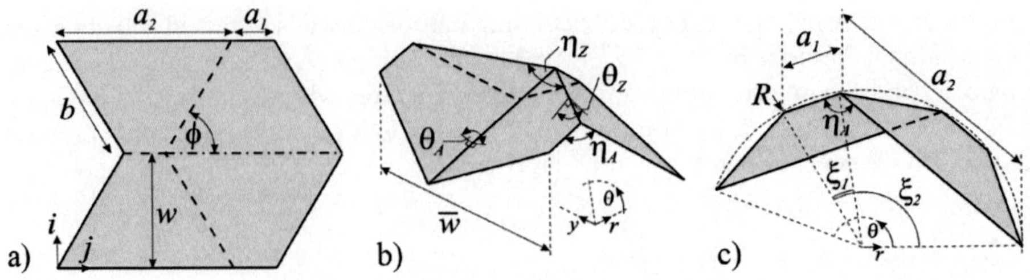

Fig. 1. Unit geometry of the Arc Pattern.

These parameters can be determined by Eqs. (3) - (5). Additionally, the folded width parameter $\overline{w}$, as provided by Eq. (6), is also valuable in this context. The vertices of a partially folded Arc Pattern unit lie along an arc profile characterized by a radius R, which can be calculated using Eqs. (7) - (9). Here, ξ_1 and ξ_2 represent the central angles formed by creases α_1 and α_2, respectively.

$$a_2 = a_1 + 2b\cos\varphi \tag{1}$$

$$w = b\sin\varphi \tag{2}$$

$$(1 + \cos\eta_z)(1 - \cos\eta_A) = 4\cos^2\varphi \tag{3}$$

$$\cos\eta_A = \sin^2\varphi\cos\theta_Z - \cos^2\varphi \tag{4}$$

$$\cos\eta_Z = \sin^2\varphi\cos\theta_A + \cos^2\varphi \tag{5}$$

$$\overline{w} = b\sin(\eta_Z/2) = wsin(\eta_Z/2)/\sin\varphi \tag{6}$$

$$\xi_1 + \xi_2 = 2(\pi - \eta_A) \tag{7}$$

$$R = a_1/[2\sin(\xi_1/2)] \tag{8}$$

$$R = a_2/[2\sin(\xi_2/2)] \tag{9}$$

3.2 Arc Pattern Motion Path

Given a rigid Arc Pattern tessellation with m straight and n zigzag CP lines (rows and columns), the precise location of any vertex can be determined using a 3D cylindrical coordinate system. This coordinate system, with its specified orientation and origin as illustrated in Fig. 1c, provides the necessary framework for accurately defining the location of each vertex in any folded configuration. By utilizing kinematic Eqs. (10) and

(11), it is possible to spatially define every $V_{i,j}$ vertex of the CP, where i = 0,1,…,m-1 and j = 0,1,…,n-1.

$$V_{i,j} = (x, y, z) = (R\cos\theta, ib\sin(\eta_Z/2), R\sin\theta) \tag{10}$$

$$\theta = \begin{cases} j(\xi_1 + \xi_2)/2 & \text{for even } i \text{ and even } j \\ j(\xi_1 + \xi_2)/2 + (\xi_2 - \xi_1)/2 & \text{for even } i \text{ and odd } j \\ (j-1)(\xi_1 + \xi_2)/2 + \xi_2 & \text{for odd } i \text{ and even } j \\ (j-1)(\xi_1 + \xi_2)/2 + \xi_1 + (\xi_2 - \xi_1)/2 & \text{for odd } i \text{ and odd } j \end{cases} \tag{11}$$

To summarize, an Arc Pattern tessellation can be geometrically and kinematically defined using fifteen parameters. Seven of these parameters ($\alpha_1, \alpha_2, b, \varphi, w, m, n$) are considered pattern constants, and remain unchanged regardless of the folded state of the CP. The remaining eight parameters ($\theta_A, \theta_Z, \eta_A, \eta_Z, \overline{w}, \xi_1, \xi_2, R$) are referred to as pattern variables, as they alternate for every different folded configuration of the same CP. The spatial deployment of a rigid tessellation is illustrated in Fig. 2 for a CP with constants $\alpha_1 = 1.4$, $b = 3.2$, $\varphi = 30°$, $m = 13$, $n = 6$, and a sequence of variable edge angles $\theta_A = 180°, 125°, 75°, 20°$.

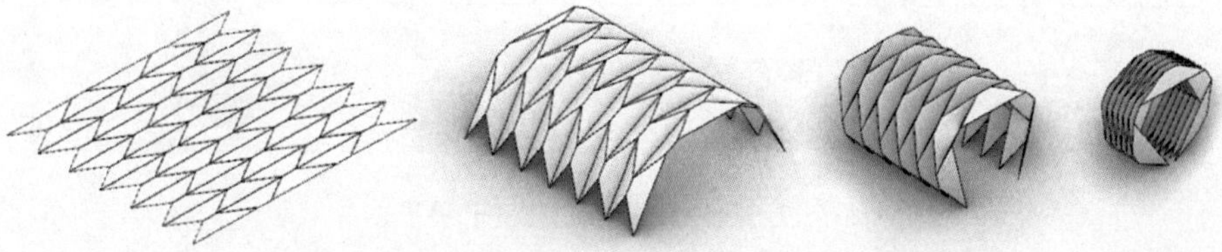

Fig. 2. Folding sequence of the Arc Pattern tessellation.

4 Algorithm Development

In order to facilitate the analysis of the form and kinematics of the vault structure, an algorithm was developed within the Rhino Grasshopper visual programming environment. The following subsections will provide a comprehensive description of the entire process, encompassing the generation of the Arc Pattern tessellation, the incorporation of plate thickness, and the creation of the complementary linkages frame mechanism. The developed parametric model generates a hybrid vault structure and also offers a visual representation of its motion in space.

4.1 Folded Plate Surface Parametric Model

The first part of the algorithm involves the creation of the thickened folded plate surface. The initial inputs consist of the constant parameters of the Arc Pattern: side lengths a_1 and b, as well as sector angle φ. Additionally, a variable parameter θ_A is provided as input through a slider, allowing control over the folding motion of the CP from retracted to deployed configurations. The *"Arc Pattern Eqs"* cluster, depicted in Fig. 3, calculates the rest of the constant and variable parameters, as discussed in Sect. 3, for a single CP unit. These calculations employ Eqs. (1) - (9).

The next cluster within the parametric model, referred to as the *"Arc Pattern Mesh"* cluster (Fig. 3), integrates the following additional inputs: the starting point of the tessellation $V_{0,0}$, the number of rows and columns m and n, and the desired plate thickness t. This cluster generates a mesh representation of the CP by utilizing Eqs. (10) and (11), while also incorporating thickness to the individual plates.

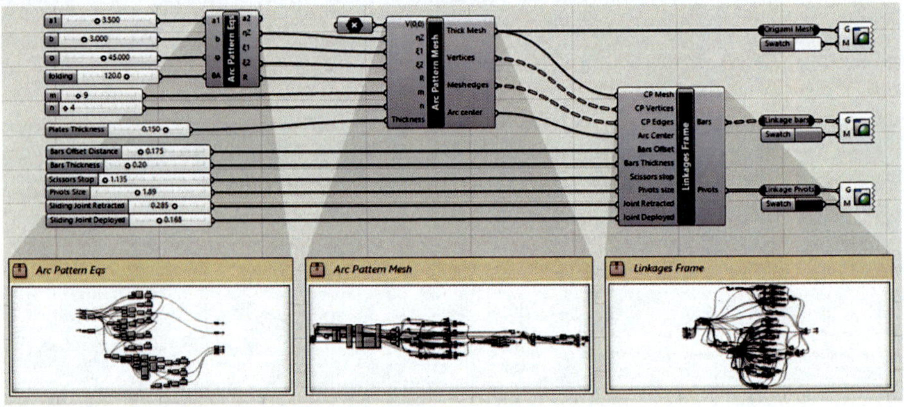

Fig. 3. Grasshopper algorithm preview.

Thickness Accommodation. The chosen technique for accommodating thickness in this study is the layered tapered panels technique [3]. This technique has been selected primarily because it preserves the kinematics of a zero-thickness model while also being easily fabricable using laser cutting or 2-axis CNC routing methods.

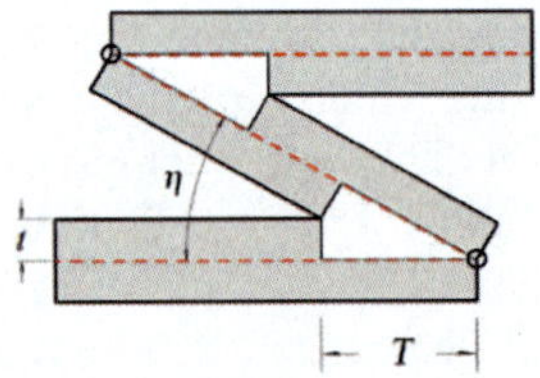

Fig. 4. Detail of the layered tapered panels technique.

In order to achieve the desired total thickness, an equal amount of thickness is added to each side of the zero-thickness surface, which is represented by red dashed lines in Fig. 4. To prevent self-intersection, the necessary taper length T for each panel in the CP is determined using Eq. (12), where t is the half thickness of the panel, and η

the dihedral angle (η_A / η_Z) of the embedded zero-thickness plates at the maximum retracted configuration. It should be noted that to maintain a non-zero panel thickness, the maximum fold angle of the thick origami mesh must not reach 180°. By utilizing the above-mentioned technique, thickness is applied to both sides of the Arc Pattern tessellation, as shown in a darker shade in Fig. 5 for both the convex (top) and concave (bottom) side of the surface.

$$\mathrm{T} = t\cot(\eta/2) \tag{12}$$

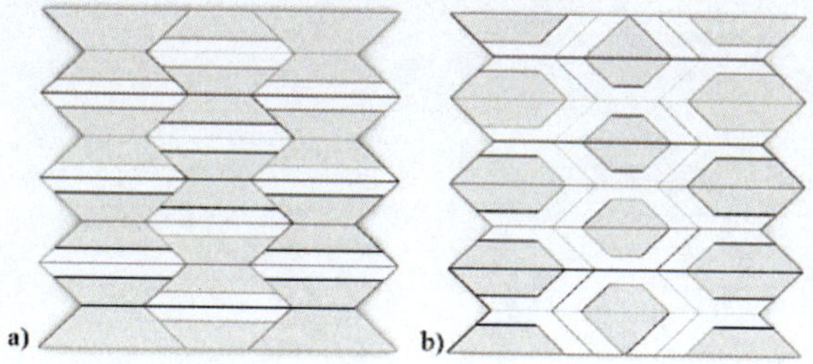

Fig. 5. Thickness accommodation on the a) convex and b) concave sides of the CP.

4.2 Structural Frame Parametric Model

Observations suggest that constraining the boundary edges of folded surfaces enhances their stability and load-carrying capacity [2]. This can be accomplished by incorporating a complementary mechanism that follows the same deployment path as the edge vertices of the folded surfaces. A crucial requirement for this mechanism is its ability to be constrained or locked into particular configurations. Previous studies conducted by the authors [7, 8] have examined planar scissor linkages as a suitable mechanism for this purpose. However, due to the variable curvature of the Arc Pattern during deployment, a different type of linkage is needed for the curved edges of this specific CP. Therefore, a modified scissors-four bar linkage is proposed for the retractable structural frame of the developed vault. This frame comprises two pairs of different linkages, one oriented lengthwise (rectilinear) and the other widthwise (curvilinear).

The "Linkages Frame" cluster of the algorithm, as shown in Fig. 3, receives input regarding the CP geometry and the essential additional parameters required for constructing the frame, such as the thickness of the bars and offset distance. The geometric construction of the two distinct types of linkages that constitute the developed retractable frame is described in the following sections:

Rectilinear Linkages Pair. In the case of the lengthwise edge vertices pair of the Arc Pattern [($V_{0,0}$, $V_{1,0}$, …, $V_{(m-1),0}$) and ($V_{0,(n-1)}$, $V_{1,(n-1)}$, …, $V_{(m-1),(n-1)}$)], a translational scissors linkage of two different lengths is employed (as shown in Fig. 6b). The longer length of the scissors linkage corresponds to the side length b of the Arc Pattern, and it is aligned with the lengthwise opposite edges of the CP. The intentionally shorter length of the scissors linkage is chosen to "stop" the unfolding of the surface, effectively restricting its mobility. This design choice ensures that the folded plate surface achieves and maintains the anticipated space covering configuration.

Curvilinear Linkages Pair. For the widthwise edge vertices pair of the Arc Pattern $[(V_{0,0}, V_{0,1}, \ldots, V_{0,(n-1)})$ and $(V_{(m-1),0}, V_{(m-1),1}, \ldots, V_{(m-1),(n-1)})]$, a novel scissors- four bar linkage with sliding joints is utilized (as shown in Fig. 6a, 7b). This linkage is specifically designed to conform to the geometric characteristics of the folding Arc Pattern and is composed of bars with lengths equivalent to the CP side lengths α_1 and α_2. The pivots located at the intersecting points of the linkages are designed with sliding capability to accommodate the variable curvature of the widthwise opposing edges of the Arc Pattern tessellation.

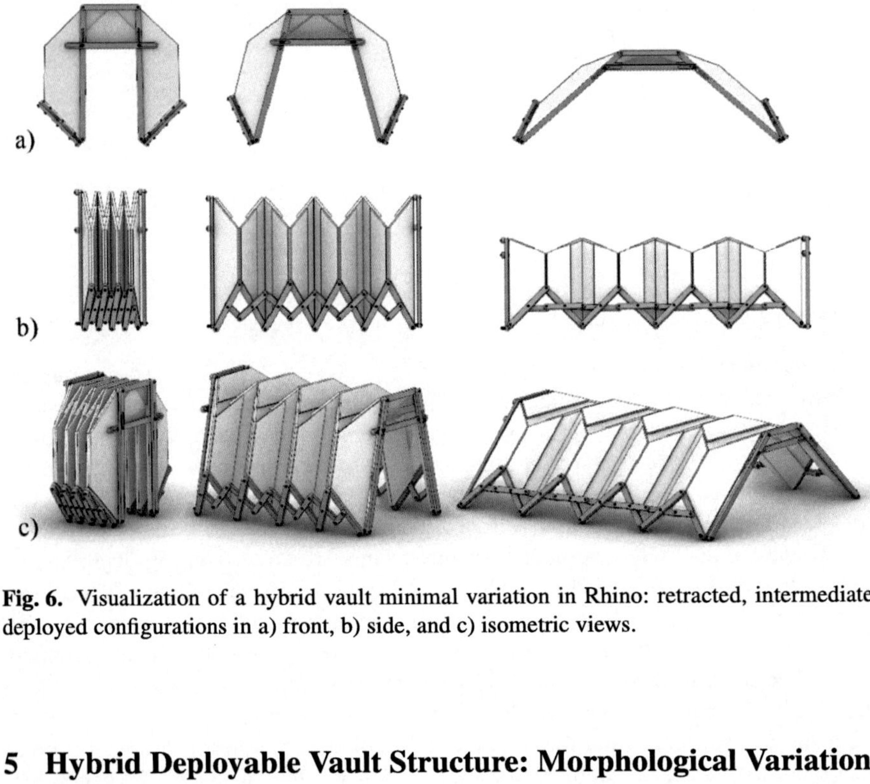

Fig. 6. Visualization of a hybrid vault minimal variation in Rhino: retracted, intermediate and deployed configurations in a) front, b) side, and c) isometric views.

5 Hybrid Deployable Vault Structure: Morphological Variations

The parametric model developed in Grasshopper enables the generation of diverse morphological variations of the proposed vault structure by manipulating the various parameters of the folded plate surface and the structural frame. By analyzing the resulting morphologies, important considerations such as deployability, compactness, stability, and aesthetic qualities can be evaluated. By iteratively refining the design through this process, it becomes possible to identify and select vault variations that best align with the specific requirements of architectural applications. This approach facilitates a more informed decision-making process and enables the creation of structures that effectively meet the desired criteria.

The key parameters that significantly influence the geometry of the proposed vault structure are the initial input parameters of the Arc Pattern, namely α_1, b, and φ, along with the tessellation rows and columns counts, m and n. These parameters not only determine the form of the vault but also affect the size ratio between its retracted and deployed configurations. For instance, the hybrid vault variation illustrated in Fig. 6 is constructed using the parameter values $\alpha_1 = 2$, $b = 2$, $\varphi = 45°$, $m = 9$, and $n = 4$. This configuration results in a compact design with a minimal number of moving parts. The vault variation depicted in Fig. 7, created using parameter values $\alpha_1 = 1.5$, $b = 1$, $\varphi = 35°$, $m = 41$, and $n = 6$, has a larger span and exhibits greater curvature changes. In the course of experimentation with different parameter values, we observed that, in some cases, self-intersection occurs between the opposing rectilinear edges of the vault at a certain point. Furthermore, it was noted that as the number of n counts or the values of φ angle increase, the likelihood of self-intersection instances also increases.

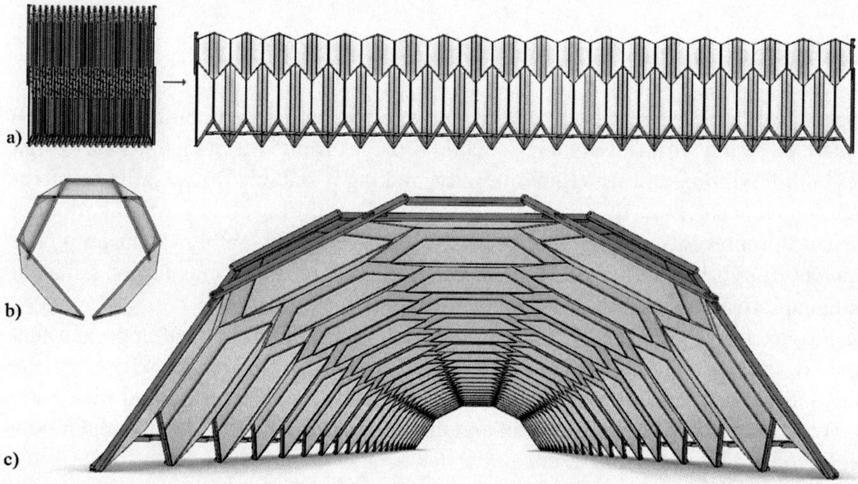

Fig. 7. Visualization of a hybrid vault large variation in Rhino: a) side view of retracted and deployed configurations, b) front view of retracted configuration, and c) perspective view of deployed configuration

A small-scale kinetic physical model of a chosen vault variation has been constructed to assess the stability and the kinematic behavior of the structure, as well as to provide a tangible representation of the proposed hybrid structure. The physical model, depicted in Fig. 8, utilizes plywood linkages for the structural frame and 2 mm cardboard plates attached to both sides of a standard 80 gsm paper for the folded surface. The mechanism exhibits the expected kinematic behavior, validating the developed parametric model. Furthermore, the physical model has been tested and proven capable of supporting weights that are at least three times its own.

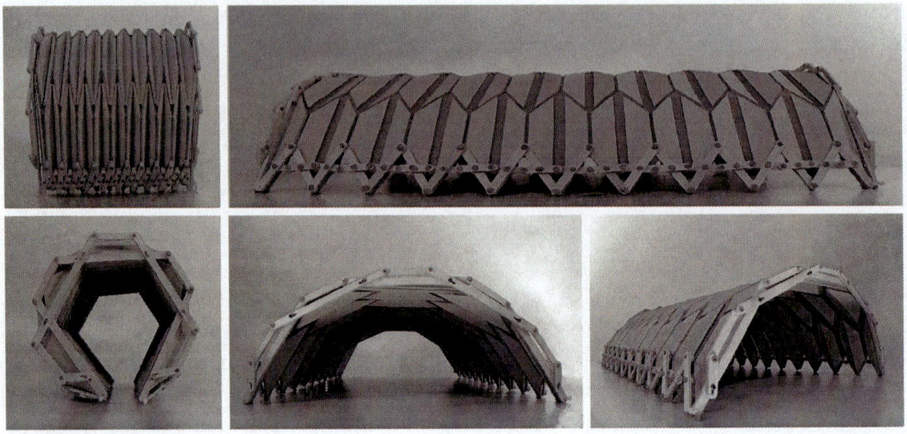

Fig. 8. Small-scale kinetic physical model of the proposed vault in retracted (left) and deployed (right) configurations.

6 Conclusion

This paper introduces a lightweight deployable vault structure that combines a curved folded plate covering surface with a retractable linkages frame mechanism. This design provides both flexibility and structural integrity, making it suitable for versatile architectural solutions that can meet diverse requirements and conditions. It is worth mentioning that this study represents an ongoing research endeavor by the authors focused on the development of hybrid portable and transformable structures of various forms, aimed at promoting adaptive and sustainable architectural practices.

The proposed hybrid structure is developed using a methodological approach that encompasses the geometric and parametric definition of the selected crease pattern, the accommodation of folded plates thickness, the geometric construction and kinematic pairing of the retractable linkages frame, and the fabrication of a kinetic physical model of the interconnected mechanism.

Further research and experimentation can expand upon the findings of this work to explore the practical implementation of the proposed deployable vault structure in real-world scenarios. Important aspects to consider include joint design and materials selection, which play a crucial role in ensuring the structural integrity and functionality of the system. Finite element analysis could be employed to evaluate the stress distribution and deformation under different loading conditions. This would provide valuable insights into the structural performance and potential design optimizations.

References

1. Tachi, T.: Introduction to structural origami. J. Int. Assoc. Shell Spatial Structures: J. IASS (2019). https://doi.org/10.20898/j.iass.2019.199.004
2. Buri, H.U.: Origami - Folded Plate Structures. École Polytechnique Federale De Lau-sanne (2010)

3. Lang, R.J., Tolman, K.A., Crampton, E.B., Magleby, S.P., Howell, L.L.: A review of thickness-accommodation techniques in origami-inspired engineering. Appl. Mech. Rev. American Society of Mechanical Engineers (ASME) **70**(1) (2018). https://doi.org/10.1115/1.4039314
4. Macri, S.: Practical applications of rigid thick origami in kinetic architecture, pp. 1–106 (2015)
5. Tachi, T.: Rigid-foldable thick origami. In: Origami 5: Fifth International Meeting of Origami Science, Mathematics, and Education (2016). https://doi.org/10.1201/b10971-24
6. Osorio, F., Oliveira, S.: Origami folded surfaces: kinetic system behind the folding. Kine[SIS]tem - From Nature to Architectural Matter Conference Proceedings, pp. 28–35 (2017)
7. Vlachaki, E., Liapi, K.A.: Folded surface elements coupled with planar scissor linkages: a novel hybrid type of deployable structures. Curved Layered Struct. **8**(1), 137–146 (2021). https://doi.org/10.1515/cls-2021-0013
8. Vlachaki, E., Liapi, K.: Hybrid deployable structures of miura-ori folded surfaces & translational scissors expandable frames. In: Inspiring the Next Generation - Proceed-ings of the International Conference on Spatial Structures 2020/21 (IASS2020/21-Surrey7), Surrey: Spatial Structures Research Centre of the University of Surrey, pp. 3283–3292 (2021). https://doi.org/10.15126/900337
9. Tachi, T.: Generalization of rigid foldable quadrilateral mesh origami. In: Proceedings of the International Association for Shell and Spatial Structures (IASS) Symposium 2009. Valencia, pp. 2287–2294 (2009)
10. Gattas, J.M., Wu, W., You, Z.: Miura-base rigid origami: parameterizations of first-level derivative and piecewise geometries. J. Mech. Des. Trans. ASME, **135**(11) (2013). https://doi.org/10.1115/1.4025380

Consistent Complementarity Problem Formulation for the Mechanical Modellisation of Spatial Cable–Rib Structures

Giuseppe Cocchetti[1] (✉), Aram Cornaggia[2], Rosalba Ferrari[2], and Egidio Rizzi[2]

[1] Politecnico di Milano, DICA, piazza L. da Vinci 32, I–20133 Milano, Italy
giuseppe.cocchetti@polimi.it

[2] Università degli studi di Bergamo, DISA, viale G. Marconi 5, I–24044 Dalmine (BG), Italy

Abstract. This paper outlines a general mechanical formulation to computationally handle cable–rib structures by a consistent Complementarity Problem modellisation, so far considering potential material non–linearity within the ribs, localised at plastic joints, as per a classical plastic hinge hypothesis in the Limit Analysis of frames. The formulation is also outlined in a prototype self–made implementation, allowing to achieve and display first consistent results on sample test structures, for necessary understanding and control. This shows that the present modellisation concept, and implementation, shall constitute a liable mechanical tool for promising utilisation and end use adoption in different related application scenarios.

Keywords: cable–rib structures · Complementarity Problem formulation · no–compression cables · rulable pretension · structural optimisation

1 Introduction

Cable–rib structures constitute peculiar structural forms with bearing capacity and flexible morphological appearance, according to different aims that are being pursued, for instance in three–dimensional, even large–scale, lattice structures, tensegrity architectural and design forms, deployable configurations like for aerospace satellite antennas, and so on. A comprehensive, and effective, modellisation and structural analysis tool shall be developed, to consistently address the issue of no–compression cables and at potential rulable pretensioning, which may be tuned to achieve appropriate stiffening/resilience and desired overall operational features, along with different employment regimes, like static, dynamic, possibly non–linear (materially and/or geometrically), at quasi–static increasing loading up to potential collapse, etc. Moreover, such a feasibly runnable "Mechanical Solver" may then be coupled to "Optimisation Tools", in nested loops of structural optimisation, toward maximisation of structural performance quest or form–finding, which shall provide fundamental guidelining in design, tuning, experimentation and realisation (see e.g. [1, 2]).

S. Gabriele et al. (Eds.): IWSS 2023, LNCE 437, pp. 672–682, 2024.
https://doi.org/10.1007/978-3-031-44328-2_70

Within such a framework, this contribution aims to present a Complementarity Problem (CP) modellisation, apt to simulate the mechanical behaviour of a cable–rib structure, in which the non–linear holonomic behaviour of the cables is properly accounted for, and the material non–linearity is condensed in plastic joints (as a generalisation of the Plastic Hinge model used in classical Limit Analysis), located at the rib's ends.

Toward that, a rather general step–by–step evolutive elastic–plastic strategy has been successfully developed by the present authors and consistently applied to the analysis of the mechanical response of large–scale 3D truss–frame structures, according to the classical principles of Limit Analysis (LA) [3–5], within the framework of the Theory of Plasticity, with reference to 3D applications to multi–storey frames and metallic truss bridges. This methodology relies on an iterative procedure in which the global elastoplastic matrix of the structure is iteratively updated, as activations of plastic joints evolutively occur, at increasing external load multiplier. In this manner, the true non–linear evolutive elastoplastic behavior of a large–scale structure with several degrees of freedom can exactly be traced until ductile collapse, in the real spirit of LA [4]. Otherwise, the analysis may be supported, or even substituted, by an efficient Direct Method, based on the upper–bound kinematic theorem of LA, to directly achieve the collapse characteristics, by a very fast and efficient iterative procedure [5].

Moreover, in works [1, 2], such a computational procedure has been extended, and applied, to the analysis of large–scale aerospace cable–rib satellite antennas, thereby showing the need of a more general and comprehensive mechanical strategy, for generalised uses of structural analysis and structural performance optimisation. The present work heads on, toward a complete, consistent, algorithmic (i.e. non–empirical) elastic–plastic optimisation of cable–rib structures, assuming that typical principles of LA may constitute a general paradigm and criterion to achieve functional and resilient structures, and by achieving a consistent and efficient handling of no–compression cables and adjusting pretensioning.

The outline of the paper goes as follows. In Sect. 2, a Finite Element formulation of beams and cables is briefly recalled, whereby equilibrium is enforced by the Virtual Work Principle; the elements of the so called "elastic background" are then prepared. In Sect. 3, a non–holonomic model for non–linear joints and a holonomic description of no–compression cables are cast in a piece–wise linear mathematical structure. In Sect. 4, nodal kinematic variables are condensed to form the elastic background of the structure; as a consequence, the whole structural behaviour is governed by internal variables (i.e. "plastic multipliers") within a Complementarity Problem, ruling both the incremental solution (for dissipative joints) and the total solution (for cables). A numerical example is shown in Sect. 5, and some conclusions are briefly listed in Sect. 6.

2 Elastic Background of the Cable–Rib Structure

2.1 Rib Elements with Joints

Let one consider a generic (n–th) beam element for the ribs (see Fig. 1). Potential material non–linearities are assumed to be concentrated at its edges (**A** and **B**, nodes of the beam element). Further, two rigid block segments (**HA** and **VB**) are added, in order to represent the (rigid) joints connecting the various beam elements (and then setting the supports for tensioning cables).

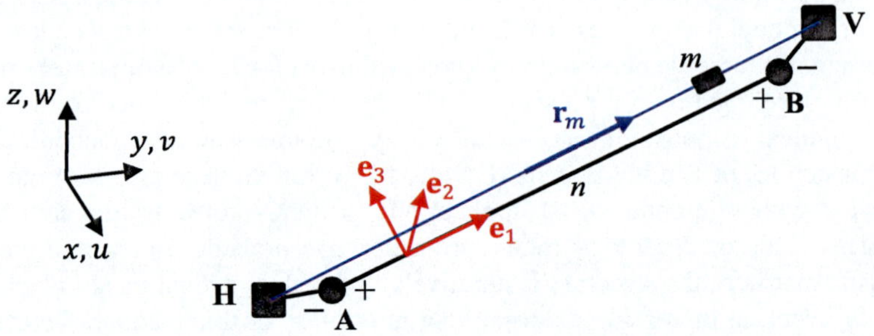

Fig. 1. Beam element (n–th) with non–linear joints at **A** and **B**, rigid segments at nodes **H** and **V** and cable (m–th), connecting rigid joints, with local versor $\mathbf{r}_m$; local and global reference systems are also shown.

Thereby (Fig. 1), a local reference system is described by orthogonal versors $(\mathbf{e}_1, \mathbf{e}_2, \mathbf{e}_3)$, , within global reference (x, y, z), and corresponding displacements (u, v, w).

Relative displacements and rotations across the non–linear joints are defined in (beam) local reference system $(\mathbf{e}_1, \mathbf{e}_2, \mathbf{e}_3)$ and collected in (6×1) vectors $\boldsymbol{\eta}_\mathrm{A}$ and $\boldsymbol{\eta}_\mathrm{B}$.

Kinematic variables $\mathbf{U}_\mathrm{H}$ and $\mathbf{U}_\mathrm{V}$, relevant to the nodes where rigid joints are present, are collected in global structural vector $\mathbf{U}$. Similarly, kinematic discontinuities $\boldsymbol{\eta}_n$ across the non–linear joints are stored in a global vector $\boldsymbol{\eta}$.

2.2 Elastic Cables

Elastic cables, possibly pre–tensioned, may be added to the truss–frame by connecting rigid joints, for instance **H** and **V**, or some other couples of rigid joints, generally not connected through a specific beam element.

When pulled $\left(\Delta L_m^e \geq 0\right)$, the m–th cable is assumed to display a classical linear elastic behaviour in generalised structural variables (with usual meaning of symbols):

$$N_m = \frac{E_m A_m}{L_m} \Delta L_m^e \tag{1}$$

The elastic elongation is expressed as the difference between the total axial elongation and the relative displacement (η_m^c) imposed at the m–th cable–joint (e.g. a tensioner, see Fig. 1):

$$\Delta L_m^e = \begin{bmatrix} -\mathbf{r}_m \\ +\mathbf{r}_m \end{bmatrix}^T \begin{bmatrix} \mathbf{U}_\mathrm{H} \\ \mathbf{U}_\mathrm{V} \end{bmatrix} - \eta_m^c \tag{2}$$

where nodal kinematic variables $\mathbf{U}_\mathrm{H}$ and $\mathbf{U}_\mathrm{V}$ are taken from global vector $\mathbf{U}$ and relative displacements η_m^c are cast in vector $\boldsymbol{\eta}^c$.

2.3 Overall Cable–Rib Structure

The global stiffness matrix of the cable–rib structure ($\mathbf{K}_{\text{glob}}$) can readily be obtained by assembling the stiffness and coupling matrices of beam and cable elements (cables assumed, in this phase, as truss elements). Then, the total strain energy of the structure can be given the following quadratic form in kinematic variables:

$$\mathcal{E}_{str} = \frac{1}{2} \begin{bmatrix} \mathbf{U} \\ \boldsymbol{\eta} \\ \boldsymbol{\eta}^c \end{bmatrix}^T \mathbf{K}_{\text{glob}} \begin{bmatrix} \mathbf{U} \\ \boldsymbol{\eta} \\ \boldsymbol{\eta}^c \end{bmatrix} \tag{3}$$

To express static/kinematic variable coupling, the internal virtual work, relevant to the initial configuration of the structure, can be written as the sum of the following contributions: the virtual work produced by internal (nodal) actions due to the elastic behaviour of the various elements (beams and cables); the virtual work made by internal nodal actions ($\mathbf{M}_n$) applied to the beam non–linear joints, each one referred to the local beam reference system; the virtual work made by internal nodal actions (N_m) applied to the cable–joints, each one referred to the local cable reference system:

$$\delta W_i = \begin{bmatrix} \mathbf{U} \\ \boldsymbol{\eta} \\ \boldsymbol{\eta}^c \end{bmatrix}^T \mathbf{K}_{\text{glob}} \begin{bmatrix} \delta\mathbf{U} \\ \delta\boldsymbol{\eta} \\ \delta\boldsymbol{\eta}^c \end{bmatrix} + \sum_{n=1}^{N_{beams}} \mathbf{M}_n^T \delta\boldsymbol{\eta}_n + \sum_{m=1}^{N_{cables}} N_m \delta\eta_m^c \tag{4}$$

namely:

$$\delta W_i = \begin{bmatrix} \mathbf{U} \\ \boldsymbol{\eta} \\ \boldsymbol{\eta}^c \end{bmatrix}^T \mathbf{K}_{\text{glob}} \begin{bmatrix} \delta\mathbf{U} \\ \delta\boldsymbol{\eta} \\ \delta\boldsymbol{\eta}^c \end{bmatrix} + \mathbf{M}^T \delta\boldsymbol{\eta} + \mathbf{N}^T \delta\boldsymbol{\eta}^c \tag{5}$$

vectors $\mathbf{M}$ and $\mathbf{N}$ being the assemblage of joint actions $\mathbf{M}_n$ and N_m, respectively, and the kinematic set ($\delta\mathbf{U}, \delta\boldsymbol{\eta}, \delta\boldsymbol{\eta}^c$) representing a virtual variation of the (independent) kinematic variables governing the generic structure configuration.

Then, by exploiting the symmetry of the (stiffness) matrix and collecting the variations of the kinematic variables (on the left side of the product form), the internal virtual work can be written, in compact form, as follows:

$$\delta W_i = \begin{bmatrix} \delta\mathbf{U} \\ \delta\boldsymbol{\eta} \\ \delta\boldsymbol{\eta}^c \end{bmatrix}^T \left(\mathbf{K}_{\text{glob}} \begin{bmatrix} \mathbf{U} \\ \boldsymbol{\eta} \\ \boldsymbol{\eta}^c \end{bmatrix} + \begin{bmatrix} \mathbf{0} \\ \mathbf{M} \\ \mathbf{N} \end{bmatrix} \right) \tag{6}$$

2.4 External Loads and Final Equations

To account for external loading, the external virtual work (δW_e) is produced by the following nodal actions, expressed in the global reference system: ($\mathbf{f}_n$) equivalent to concentrated and distributed loads applied along each beam element; ($\mathbf{P}_r$) directly applied to the rigid joints (possibly, including constraint reactions). Thus, the total external virtual work can be written as:

$$\delta W_e = \sum_{n=1}^{N_{beams}} \mathbf{f}_n^T \delta \mathbf{U}_n^+ + \sum_{r=1}^{N_{rigidjoints}} \mathbf{P}_r^T \delta \mathbf{U}_r \tag{7}$$

and also expressed as:

$$\delta W_e = \begin{bmatrix} \delta \mathbf{U} \\ \delta \boldsymbol{\eta} \\ \delta \boldsymbol{\eta}^c \end{bmatrix}^T \begin{bmatrix} \mathbf{F} + \mathbf{P} \\ \mathbf{G} \\ \mathbf{0} \end{bmatrix} \tag{8}$$

in which, $\mathbf{G}$ stores FEM equivalent nodal actions $\mathbf{f}_n$ applied on beam joints; $\mathbf{P}$ and $\mathbf{F}$ collect the nodal actions, $\mathbf{P}_r$ and $\mathbf{f}_n$, respectively acting directly and not directly on rigid nodes.

Now, the set of equilibrium equations, governing the whole cable–rib structural system, is obtained by equating the internal and the external virtual works, for a generic variation of the underlying kinematic variables:

$$\delta W_i = \delta W_e, \quad \forall \delta \mathbf{U}, \quad \forall \delta \boldsymbol{\eta}, \quad \forall \delta \boldsymbol{\eta}^c \tag{9}$$

to get:

$$\begin{bmatrix} \delta \mathbf{U} \\ \delta \boldsymbol{\eta} \\ \delta \boldsymbol{\eta}^c \end{bmatrix}^T \left(\mathbf{K}_{\text{glob}} \begin{bmatrix} \mathbf{U} \\ \boldsymbol{\eta} \\ \boldsymbol{\eta}^c \end{bmatrix} + \begin{bmatrix} \mathbf{0} \\ \mathbf{M} \\ \mathbf{N} \end{bmatrix} - \begin{bmatrix} \mathbf{F} + \mathbf{P} \\ \mathbf{G} \\ \mathbf{0} \end{bmatrix} \right) = 0 \quad \forall \begin{bmatrix} \delta \mathbf{U} \\ \delta \boldsymbol{\eta} \\ \delta \boldsymbol{\eta}^c \end{bmatrix} \tag{10}$$

After some algebraic manipulations, the final governing system shall become:

$$\mathbf{K}_{\text{glob}} \begin{bmatrix} \mathbf{U} \\ \boldsymbol{\eta} \\ \boldsymbol{\eta}^c \end{bmatrix} = \begin{bmatrix} \mathbf{F} + \mathbf{P} \\ \mathbf{G} - \mathbf{M} \\ -\mathbf{N} \end{bmatrix} \tag{11}$$

This (linear) system can be solved for nodal kinematic variables $\mathbf{U}$, nodal actions $\mathbf{M}$ and cable axial forces $\mathbf{N}$, to get the so called "elastic background solution" of the structure.

3 Piece–Wise Linear Modeling of Non–Linear Joints and Cables

The same modeling concept, based on linear complementarity relations, is here meant to apply to non–linear beam joints and to cable joints, as described by piece–wise linear constitutive laws.

3.1 Non–Holonomic Non–Linear Beam Joints

The non–linear (beam) joint constitutive behaviour is described by a non–holonomic piece–wise (three–) linear relationship, both for the domain in the stress (**M**) space (Fig. 2a) and for their hardening/softening evolution (Fig. 2b); λ_r^s are "slack" variables called "plastic multipliers", in analogy with the classical elastic–plastic description of the Theory of Plasticity (see e.g. [3]).

By collecting the relevant relationships in vectors and matrices, the following Linear Complementarity Problem (LCP) allows to describe the activation, and subsequent non–holonomic (irreversible) evolution, of the r-plane (see e.g. [3, 6]):

$$\boldsymbol{\varphi}_r^s = \begin{bmatrix} \varphi_r^{s,1} \\ \varphi_r^{s,2} \\ \varphi_r^{s,3} \end{bmatrix} = [\mathbf{n}_r, \mathbf{0}, \mathbf{0}]^T \mathbf{M}_s - \mathbf{H}_r^s \boldsymbol{\lambda}_r^s - \mathbf{Y}_{0,r}^s \leq \mathbf{0},\ \dot{\boldsymbol{\lambda}}_r^s \geq \mathbf{0},\ (\boldsymbol{\varphi}_r^s)^T \dot{\boldsymbol{\lambda}}_r^s = 0 \tag{12}$$

where

$$\mathbf{H}_r^s = \begin{bmatrix} H_r^{s,1} & -H_r^{s,1} + H_r^{s,2} & -H_r^{s,2} \\ 1 & -1 & 0 \\ 1 & 0 & -1 \end{bmatrix} \tag{13}$$

and

$$\mathbf{Y}_{0,r}^s = \begin{bmatrix} Y_r^{0s} \\ \bar{\lambda}_r^{s,1} \\ \bar{\lambda}_r^{s,2} \end{bmatrix} = \begin{bmatrix} Y_r^{0s} \\ \bar{\lambda}_r^{s,1} \\ \bar{\lambda}_r^{s,1} + \left(Y_r^{0s} + H_r^{s,1}\bar{\lambda}_r^{s,1}\right)/\left(-H_r^{s,2}\right) \end{bmatrix} \tag{14}$$

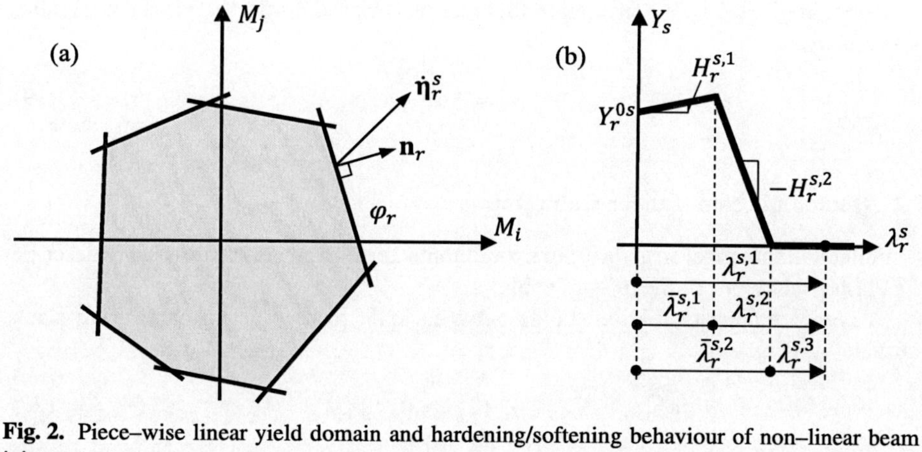

Fig. 2. Piece–wise linear yield domain and hardening/softening behaviour of non–linear beam joints.

The constitutive behaviour of the $s-$ th joint can be obtained by collecting these last relationships for all the yielding planes:

$$\boldsymbol{\varphi}^s = \begin{bmatrix} \varphi_1^s \\ \vdots \\ \varphi_r^s \\ \vdots \end{bmatrix} = \mathbf{n}_s^T \mathbf{M}_s - \mathbf{H}_s \boldsymbol{\lambda}^s - \mathbf{Y}_0^s \leq \mathbf{0}, \quad \dot{\boldsymbol{\lambda}}^s \geq \mathbf{0}, \quad (\boldsymbol{\varphi}^s)^T \dot{\boldsymbol{\lambda}}^s = 0 \tag{15}$$

and, finally, collecting the relationships for all the beam–joints, the general LCP assembly governing the behaviour of all the beam joints can be set as follows:

$$\boldsymbol{\varphi} = \begin{bmatrix} \boldsymbol{\varphi}^1 \\ \vdots \\ \boldsymbol{\varphi}^s \\ \vdots \end{bmatrix} = \mathbf{n}^T \mathbf{M} - \mathbf{H}\boldsymbol{\lambda} - \mathbf{Y}_0 \leq \mathbf{0}, \quad \dot{\boldsymbol{\lambda}} \geq \mathbf{0}, \quad \boldsymbol{\varphi}^T \dot{\boldsymbol{\lambda}} = 0 \tag{16}$$

The constitutive behaviour of each beam joint is completed by the "flow–rules", namely the equations relating kinematic variables $\boldsymbol{\eta}^s$ (relative displacements and rotations at the s–th beam joint) to plastic multipliers $\boldsymbol{\lambda}^s$ in rate form:

$$\dot{\boldsymbol{\eta}}^s = \sum_r \dot{\boldsymbol{\eta}}_r^s = \sum_r \mathbf{V}_r^s \dot{\lambda}_r^{s,1} = \mathbf{V}^s \dot{\boldsymbol{\lambda}}^s \tag{17}$$

and collecting these relations for all beam joints:

$$\dot{\boldsymbol{\eta}} = \mathbf{V}\dot{\boldsymbol{\lambda}} \tag{18}$$

Accordingly, being $\mathbf{V}$ constant, in finite terms (by assuming $\boldsymbol{\lambda} = \mathbf{0}$ at the beginning of the analysis):

$$\boldsymbol{\eta} = \mathbf{V}\boldsymbol{\lambda} \tag{19}$$

3.2 Holonomic Non–Linear Cable Joints

A similar mathematical structure, but in holonomic terms, can also be devised to describe the elastic behaviour of the $m-$ th cable.

A complementarity relation holds between axial force N_m and cable joint slack variable λ_m^c (Fig. 3a):

$$\varphi_m^c = -N_m \leq 0, \quad \lambda_m^c \geq 0, \quad \varphi_m^c \lambda_m^c = 0 \tag{20}$$

Such relations can be collected in vector form:

$$\boldsymbol{\varphi}^c = -\mathbf{N} \leq \mathbf{0}, \quad \boldsymbol{\lambda}^c \geq \mathbf{0}, \quad (\boldsymbol{\varphi}^c)^T \boldsymbol{\lambda}^c = 0 \tag{21}$$

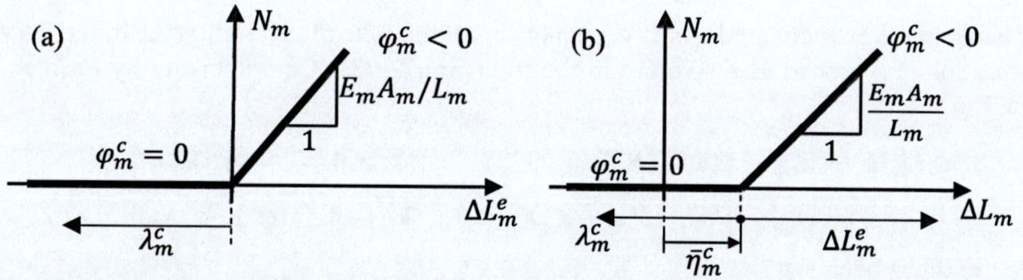

Fig. 3. Constitutive holonomic law of linear elastic cable joints without (a) and with a linear elastic cable with pre–set elongation $\overline{\eta}_m^c$ (b).

For the $m-$ th cable, the relationship between the cable joint relative displacement (η_m^c) and the relevant slack variable (λ_m^c) can be written as:

$$\eta_m^c = \overline{\eta}_m^c - \lambda_m^c \tag{22}$$

$\overline{\eta}_m^c$ being the initial elongation applied to the cable through a tensioner (see Fig. 3b).

Collecting the kinematic variables for all the cables, one gets:

$$\boldsymbol{\eta}^c = \overline{\boldsymbol{\eta}}^c - \boldsymbol{\lambda}^c \tag{23}$$

4 Complementarity Relations Governing the Global Structural Behaviour

The whole set of equations governing the structural constitutive behaviour is represented by Eqs. (16), (19), (21), (23). By replacing the solution to Eqs. (11), Eqs. (19) and (23) in Eqs. (16), for the non–linear beam joints it results:

$$\boldsymbol{\varphi} = \boldsymbol{\varphi}_0 - \widehat{\mathbf{H}}\boldsymbol{\lambda} - \widehat{\mathbf{W}}^c\boldsymbol{\lambda}^c \leq \mathbf{0}, \quad \dot{\boldsymbol{\lambda}} \geq \mathbf{0}, \quad \boldsymbol{\varphi}^T\dot{\boldsymbol{\lambda}} = 0 \tag{24}$$

in which, vectors and matrices are readily defined by prompt comparison.

In an analogous way, a replacement of the solution to Eqs. (11), Eqs. (19) and (23) in Eqs. (21) gives, for the non–linear cable joints:

$$\boldsymbol{\varphi}^c = \boldsymbol{\varphi}_0^c - \left(\mathbf{W}^c\right)^T\mathbf{V}\boldsymbol{\lambda} + \mathbf{Z}^c\boldsymbol{\lambda}^c \leq \mathbf{0}, \quad \boldsymbol{\lambda}^c \geq \mathbf{0}, \quad \left(\boldsymbol{\varphi}^c\right)^T\boldsymbol{\lambda}^c = 0 \tag{25}$$

4.1 Step–by–Step Evolutive Algorithm

Let one consider as "initial" the configuration in which the set of initial elongations $(\overline{\boldsymbol{\eta}}^c)$ is applied. In this configuration, in general terms, it is expected that a purely elastic behaviour of the structure is produced. However, within an optimisation procedure involving the initial elongations as optimisation variables, it may happen that some beam or cable joints are activated. As a consequence, for the sake of generality, the initial configuration is searched as a solution to the following Linear Complementarity Problem (LCP):

$$\begin{cases} \boldsymbol{\varphi} = \boldsymbol{\varphi}_{00} - \widehat{\mathbf{H}}\boldsymbol{\lambda} - \widehat{\mathbf{W}}^{c}\boldsymbol{\lambda}^{c} \leq \mathbf{0}, & \boldsymbol{\lambda} \geq \mathbf{0}, \quad \boldsymbol{\varphi}^{T}\boldsymbol{\lambda} = 0 \\ \boldsymbol{\varphi}^{c} = \boldsymbol{\varphi}_{00}^{c} - (\mathbf{W}^{c})^{T}\mathbf{V}\boldsymbol{\lambda} + \mathbf{Z}^{c}\boldsymbol{\lambda}^{c} \leq \mathbf{0}, & \boldsymbol{\lambda}^{c} \geq \mathbf{0}, \quad (\boldsymbol{\varphi}^{c})^{T}\boldsymbol{\lambda}^{c} = 0 \end{cases} \tag{26}$$

where, it has been set:

$$\boldsymbol{\varphi}_{00} = \widehat{\mathbf{W}}^{c}\overline{\boldsymbol{\eta}}^{c} - \mathbf{Y}_{0}, \quad \boldsymbol{\varphi}_{00}^{c} = -\mathbf{Z}^{c}\overline{\boldsymbol{\eta}}^{c} \tag{27}$$

The solution to LCP Eq. (26) can be split into three complementary subsets.

The first one (A) including the active planes in beam joints, and the active planes in cable joints for which $\boldsymbol{\lambda}'^{c} = \mathbf{0}$:

$$(\mathrm{A})\begin{cases} \boldsymbol{\varphi}' = \mathbf{0}, \boldsymbol{\lambda}' \geq \mathbf{0} \\ \boldsymbol{\varphi}'^{c} = \mathbf{0}, \boldsymbol{\lambda}'^{c} = \mathbf{0} \end{cases} \tag{28}$$

The second subset (B1) collecting the active planes in cable joints for which it results $\boldsymbol{\lambda}''^{c} > \mathbf{0}$:

$$(\mathrm{B1}) \quad \boldsymbol{\varphi}''^{c} = \mathbf{0}, \boldsymbol{\lambda}''^{c} > \mathbf{0} \tag{29}$$

The last subset (B2) listing the inactive planes for all the joints:

$$(\mathrm{B2})\begin{cases} \boldsymbol{\varphi}''' < \mathbf{0}, \boldsymbol{\lambda}''' = \mathbf{0} \\ \boldsymbol{\varphi}'''^{c} < \mathbf{0}, \boldsymbol{\lambda}'''^{c} = \mathbf{0} \end{cases} \tag{30}$$

Then, following [6], the applied loads are increased with a common load factor μ. The rate form of LCP Eqs. (24), restricted to variables belonging to subset (A), is used to capture the solution along the first step:

$$\begin{cases} \dot{\boldsymbol{\varphi}}' = \dot{\boldsymbol{\varphi}}'_{0} - \widehat{\mathbf{H}}'\dot{\boldsymbol{\lambda}}' - \widehat{\mathbf{W}}'^{c}\dot{\boldsymbol{\lambda}}'^{c} \leq \mathbf{0}, & \dot{\boldsymbol{\lambda}}' \geq \mathbf{0}, \quad \dot{\boldsymbol{\varphi}}'^{T}\dot{\boldsymbol{\lambda}}' = 0 \\ \dot{\boldsymbol{\varphi}}'^{c} = \dot{\boldsymbol{\varphi}}'^{c}_{0} - (\mathbf{W}'^{c})^{T}V'\dot{\boldsymbol{\lambda}}' + \mathbf{Z}'^{c}\dot{\boldsymbol{\lambda}}'^{c} \leq \mathbf{0}, & \dot{\boldsymbol{\lambda}}'^{c} \geq \mathbf{0}, \quad (\dot{\boldsymbol{\varphi}}'^{c})^{T}\dot{\boldsymbol{\lambda}}'^{c} = 0 \end{cases} \tag{31}$$

where

$$\dot{\boldsymbol{\varphi}}'_{0} = \mathbf{n}^{T}\dot{\mathbf{M}}_{0}, \quad \dot{\mathbf{M}}_{0} = \mathbf{M}_{0}\dot{\mu} \tag{32}$$

and

$$\dot{\boldsymbol{\varphi}}'^{c}_{0} = -\dot{\mathbf{N}}_{0}, \quad \dot{\mathbf{N}}_{0} = \mathbf{N}_{0}\dot{\mu} \tag{33}$$

Within LCP problem (31), the load amplifier rate is generally assumed to be positive ($\dot{\mu} = +1$); however, if a solution cannot be found, other choices must be assumed ($\dot{\mu} = -1$ or $\dot{\mu} = 0$).

All the expressions are linear within the increment, $\Delta\mu = \dot{\mu}\Delta t$: then, the end of the first step is straightforwardly computed for the minimum value Δt_1 for which the first strictly negative activation function ($\varphi_k < 0$, subset (B2)) and/or the first strictly positive plastic multiplier for cables ($\lambda_m^c > 0$), subset (B1), reach a nil value. After this, above sets (A) and (B) must be updated, and the next step can be tackled following the same procedure.

5 Numerical Example

The CP computational formulation, outlined in the preceding sections, has been self–implemented within a Matlab environment, by relying on previous implementation framings that were made on the side, for 3D truss-frame structures [4, 5] and cable-rib structures with a standard handling of no–compression cables not supported by a CP formulation [1, 2].

First, it has been tested on sample elementary structures, formed by a few ribs and cables, to check if the process of pre–tensioning of the cables and subsequent loading shall correctly have been handled by the CP treatment.

Second, it has been preliminary applied to a prototype structure that has already been analysed with a previous formulation, namely a 3D large–scale cable–rib aerospace antenna substructure of a diameter of about 18 m. Details on the characteristic and geometrical features may be found in [1, 2]. The present procedure reveals complete effectiveness in handling: (a) the first phase of cable pre–tensioning, where characteristic displacements go one way; (b) the subsequent initial linear elastic phase under external loading (uniformly distributed vertical loadings), where characteristic displacements go back the other way, up to first yield; (c) the successive piece-wise linear elastic-plastic response with plastic joint activations, up to plastic collapse (rather near, in this case, to first yield). A typical representative plot reported in Fig. 4 depicts a characteristic load (multiplier)–(normalised) displacement diagram, which elucidates the above-described mechanical phases. Convergence is successfully attained, and expected qualitative features appear to be reproduced.

This shall provide a first demonstration of the effectiveness of the computational procedure, in the considered context. Further results and details may be reported elsewhere, where more room shall be available.

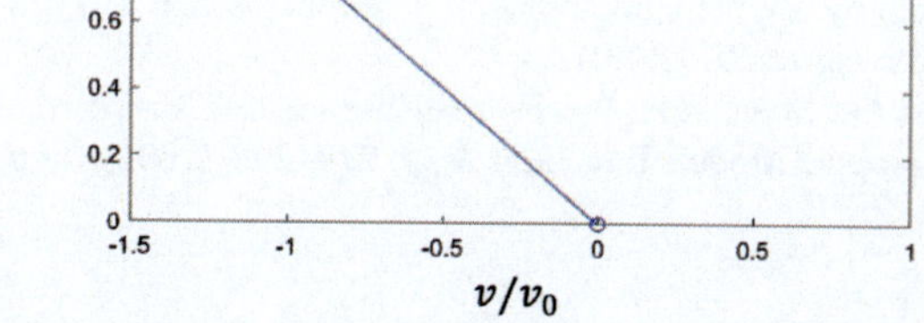

Fig. 4. P-delta piece-wise linear response curve of the antenna.

6 Conclusions

The present work has attempted to outline a comprehensive computational formulation to effectively handle cable–rib structures. As a characteristic feature, the procedure is based on the Complementarity Problem, to express both the behaviour of non–linear plastic joints and of no–compression cables, so that to handle in a unified way, and with a similar strategy, all issues of (material) non–linearity that are involved.

In principle, the numerical strategy could be adopted to simulate different types of cable–rib structures. A first demonstration has been attempted on a specific cable–rib aerospace antenna substructure of a large diameter, which was being handled with a standard formulation for the account of no–compression cables. Initial satisfactory results are recovered. Subsequent attempts may also consider different structural forms, as for instance lattice or cable–net structures and tensegrity structures.

Acknowledgements. The support by ministerial (MUR) funding "Fondi di Ricerca d'Ateneo ex 60%" at the University of Bergamo is gratefully acknowledged.

References

1. Cocchetti, G., Liu, R., Cornaggia, A., Ferrari, R., Rizzi, E.: Elastic–plastic optimisation of a cable–rib satellite antenna. Springer Book on Direct Methods for Limit States of Structures and Materials (DM2022), Eds. Giovanni Garcea and Dieter Weichert, Lecture Notes in Applied and Computational Mechanics 101, contribution earlier presented at DM2022 Workshop, University of Calabria, Arcavacata di Rende, Italy, 28 June 2022, 27 p. Springer Nature (2023). https://doi.org/10.1007/978-3-031-29122-7_3
2. Cocchetti, G., Cornaggia, A., Ferrari, R., Rizzi, E.: Limit Analysis based structural performance optimisation methodology for cable–rib aerospace antennas. *In preparation*
3. Cocchetti, G., Maier, G.: Elastic–plastic and limit-state analyses of frames with softening plastic-hinge models by mathematical programming. Int. J. Solids Struct. **40**(25), 7219–7244 (2003). https://doi.org/10.1016/S0020-7683(03)00363-9
4. Ferrari R, Cocchetti G, Rizzi E. Limit Analysis of a historical iron arch bridge. Formulation and computational implementation. Comput. Struct. **175**, 184–196 (2016). https://doi.org/10.1016/j.compstruc.2016.05.007
5. Ferrari, R., Cocchetti, G., Rizzi, E.: Effective iterative algorithm for the Limit Analysis of truss-frame structures by a kinematic approach. Comput. Struct. **197**, 28–41 (2018). https://doi.org/10.1016/j.compstruc.2017.11.018
6. Cocchetti, G., Maier, G., Shen, X.P.: Piecewise linear models for interfaces and mixed mode cohesive cracks. Comput. Model. Eng. Sci. **3**(3), 279–298 (2002). https://doi.org/10.3970/cmes.2002.003.279

Finite-Friction Effects in Self-standing Symmetric Circular Masonry Arches

Giuseppe Cocchetti[1] and Egidio Rizzi[2(✉)]

[1] Department of Civil and Environmental Engineering, Politecnico di Milano, piazza Leonardo da Vinci 32, 20133 Milano, Italy
giuseppe.cocchetti@polimi.it

[2] Department of Engineering and Applied Sciences, Università degli studi di Bergamo, viale Guglielmo Marconi 5, 24044 Dalmine (BG), Italy
egidio.rizzi@unibg.it

Abstract. This note concerns a general issue, in the mechanics of masonry arches, with reference to symmetric circular geometries, with variable opening, and possible stereotomy with radial joints (to be potentially formed, at failure, within the ideal continuous arch), in a least-thickness condition, under self-weight, namely the role that a finite inherent friction, among the theoretical joints, may play in ruling out the self-standing conditions and the mechanical features at incipient collapse, setting a change from purely-rotational modes to mechanisms that may include sliding. The issue is systematically investigated, by a full analytical derivation, and validated through an original Complementarity Problem/Mathematical Programming formulation, and numerical implementation, reconstructing the complete underlying map of thickness-to-radius ratio versus friction coefficient of all arch states, and corresponding collapse mechanisms. This investigation shall clear the issue, of the theoretical influence of finite friction, in the above-stated setting, and contribute to provide a full understanding of basic aspects in the methodological description, and physical interpretation, of the mechanics of masonry arches, with implications that may come up to appear also in practical terms, once dealing with this traditional and remarkable structures, in real cases, possibly endowed of historical character and architectural value, to be preserved and renewed.

Keywords: (symmetric circular) masonry arches · Couplet-Heyman problem · least thickness · collapse modes · finite friction

1 Introduction

The self-standing (symmetric circular) masonry arch (with ideal radial joints) of least thickness constitutes a fascinating example of structural form optimization

S. Gabriele et al. (Eds.): IWSS 2023, LNCE 437, pp. 683–692, 2024.
https://doi.org/10.1007/978-3-031-44328-2_71

("Couplet-Heyman problem"), which may even be appreciated in related beautiful natural manifestations that likely shall have inspired ancient builders in conceiving and developing this classical form of bearing structure, through, first, empirical rules of construction and, then, attempts in mechanical codifications.

In "modern" times, this has been investigated on a rational basis within the Mechanics of masonry structures, which has settled over the last few centuries, specifically since 1700, after the contributions of Pierre Couplet, coming, since the beginning of 1900, to contemporary thrust-line analyses, from cornerstone Milutin Milankovitch formulation [1] and, in the second half of 1900, to the codifications in the realm of Limit Analysis of masonry constructions, according to the light-breaking studies by Jacques Heyman [2,3]. The latter were based on three classical behavioural assumptions of masonry constructions: no tensile strength; infinite compressive strength; no sliding failure. Specifically, the last stated hypothesis shall rule out the possible onset of sliding within the masonry arch, as e.g. per the presence of a high or even infinite inherent friction, and allows to characterize the main collapse mechanisms of the masonry arch related just to purely-rotational collapse.

That has widely been adopted to analyze the statics of continuous and discrete (symmetric circular) masonry arches under self-weight, with remarkable and fundamental results in terms of kinematical (e.g. angular intrados inner-hinge position from the crown β, of the five-hinge purely-rotational collapse mode), geometrical (e.g. thickness-to-radius ratio $\eta = t/r$) and statical (e.g. non-dimensional horizontal thrust $h = H/[(\gamma t d)\, r]$ or alternatively intrinsic non-dimensional horizontal thrust $\hat{h} = \eta h = H/(\gamma d r^2)$ at given arch characteristics: γ uniform specific self-weight per unit volume; d out-of-plane width) arch features at incipient collapse, in the critical least-thickness self-standing condition.

The whole has been revisited through a comprehensive analysis by the present authors [4–9], where focus was first made on a complete analytical derivation, with separate numerical validations, of the conditions at purely-rotational collapse [4,5,8,9]. Then, on the investigation of the manifestation of possible sliding, friction effects have initially been investigated, in order to inspect the presence of potential sliding failure at reducing friction, namely as linked to a finite friction coefficient [6,7]. This has analytically been revealed for the complete semi-circular arch [6], where all basic underlying equations were derived, and explored, for such a specific case, with first perspectives in terms of additional implications at variable half-angle of embrace of the arch, as further numerically investigated in [7], whereby an original Complementarity Problem/Mathematical Programming (CP/MP) numerical formulation and self-implementation was also developed to the purpose. The latter has also been adopted to systematically illustrate features of both purely-rotational collapse modes [8] and indeed collapse modes involving sliding [7], while independent numerical validations by a Discontinuous Deformation Analysis (DDA) modelization, as per a Discrete Element Model (DEM) approach, have also been provided in [5,8].

Such personal efforts have been framed within a flourishing external literature on the specific subject (whose reference is not pursued in the present short note), where least-thickness conditions of masonry arches have been enquired by different analytical-numerical approaches, aimed at revealing the above-mentioned characteristics, and by modern thrust-line analyses, particularly in view of achieving arch form optimization. Specifically, as per the characterization of purely-rotational collapse, these works have constituted a basic and complementary information concerning the mentioned issues, while, as per the specific investigation on finite-friction effects, various attempts have separately provided a solid basis and reference framework, to be complemented by additional, systematic treatments, to be analytically pursued, whenever possible, as herein sought.

The present paper systematically addresses the issue of reducing friction, at general half-opening angle of the symmetric circular masonry arch from the crown (α), by outlining all the characteristic features and locating all the possible collapse modes that can be manifested. This is developed by a comprehensive analytical approach, where all governing (equilibrium and limit compatibility) equations are derived, and resolved until a very end. All self-standing states of the arch are characterized and separated by the arising collapse modes, in the critical least-thickness/finite-friction condition, by revealing interesting and multi-fold features of such an ideally simple mechanical system. Indeed, and instead, the appearance of even a single, new system parameter (masonry friction angle φ or friction coefficient $\mu = \tan\varphi$) brings in surprising manifestations, in all sought admissible ranges of the physical parameters. The analytically derived states of the masonry arch are then further illustrated through the above-mentioned CP/MP numerical formulation [7], which is also here meant to supply, at the same time, a sort of independent/interrelated numerical validation of the rigorous analytical treatment.

Current reference is here made just to the ideal classical Heyman-like uniform self-weight distribution along the geometrical centreline of the arch [2], although truly referring to the correct least-thickness solution [3], as widely analytically outlined and discussed in [4,5,8,9]. Thereby, the analysis was contextually developed also for the real Milankovitch-like uniform self-weight distribution [1], which constitutes a generalized, though little more complicated, case of mechanical analysis. Thus, at the entrance of finite friction, into the mechanical scene, as an additional physical system parameter, the analysis is here developed only for the classical Heyman self-weight distribution, given the contained differences experienced for the purely-rotational collapse states at infinite friction, though the extension to a Milankovitch-like self-weight treatment shall still be possible, as a generalization of the present analysis, and recorded outcomes.

Presentation flows as follows. Subsequent Sect. 2 outlines the basic underlying equations ruling the present analytical treatment. Section 3 develops the main body of the proposed outcomes, by reporting the various solution instances of the above equations, and illustrating them, based on the variable physical parameters that come into place. Section 4 illustrates the various arch states at incipient collapse, through a separate running of a CP/MP numerical formulation, providing a corresponding, though independent, numerical validation. Finally, closing Sect. 5 gathers the main aspects and achievements of the study and traces down some further research perspectives, in terms of possible generalizations of the present endeavours to other masonry arch forms.

2 Underlying Relations and Analytical Treatment

The underlying governing equations rely on the (correct) Heyman-like treatment of purely-rotational collapse [2–4] and on the introduction of the account of the appearance of finite friction [6]. Assuming the purely-rotational collapse mode (at an infinite or finite supercritical friction) as given from scratch, critical arch characteristics β, η, h can be determined, by equilibrium and (least-thickness) optimality (intrados tangency of the line of thrust) conditions [4,8,9]. At reducing friction, the appearance of joint sliding is ruled by additional equations [6], which sets in a hierarchy of the various collapse modes that can be recorded, at variable half-angle of embrace of the circular arch, and its inherent friction coefficient. Thus, a first approach relies on the assumed purely-rotational collapse mode, and attached discovered analytical solutions, and foresees new (sliding) occurrences based on the value of a reducing (finite) friction coefficient.

This is ruled by the fundamental underlying equations conceptually described as follows [4,6,8,9]:

- for the characterization of purely-rotational collapse, at a high or infinite friction coefficient:
  - equilibrium relation concerning the relative rotational equilibrium of any upper or lower portion of the half-arch (symmetry conditions apply), with respect to the inner intrados hinge forming at the haunch;
  - equilibrium relation concerning the absolute rotational equilibrium of the whole half-arch, with respect to the extrados hinge at the shoulder, involving a dependence on half-opening angle α, just through variable $A = \alpha \cot(\alpha/2)$;
  - limit (truly correct) tangency condition of the line of thrust (locus of pressure points) at the haunch intrados in the least-thickness condition, which can be expressed by a stationary condition;
- for the characterization of sliding onset, at a finite reducing friction coefficient:
  - sliding relation setting the activation of a sliding joint at the shoulder;
  - sliding relation setting the activation of an inner sliding joint, in the limit least-thickness condition, which can be expressed by a stationary condition.

The interplay of the above relations allows to derive a full map of the various manifestations of the considered mechanical system, in terms of admissible self-standing domains, and separating boundaries between them. Primarily, a thickness-to-radius ratio vs. friction coefficient η-μ diagram of the system can be derived, with the representation of all admissible states and collapse types, as illustrated in the next section.

At such a stage, a further, complete analytical treatment may additionally be developed, disregarding the role of any pre-assumed purely-rotational mechanism at infinite friction, as a leading benchmark reference and guideline, in the analysis. Indeed, the previous derivation looks exhaustive by itself but comes from the initial assumption of an underlying Heymanian purely-rotational condition, by further inspecting, then, the possible insertion and consequence of finite-friction effects. Thus, one may wonder if search may be biased and if anything may be missing within that.

Toward deriving all possible self-standing states of the symmetric circular arch, with no a priori hypotheses, a subsequent approach may be set by outlining all equilibrium and limit relations, in terms of classical "line of thrust", and of *"line of friction"* [7]. This is further based on the following two groups (as above) of main 3+2 ruling underlying equations of the physical arch states:

- eccentricity of line of thrust setting passage from shoulder extrados;
- eccentricity of line of thrust setting passage from haunch intrados;
- stationary condition of eccentricity of line of thrust at haunch intrados, as per line-of-thrust tangency condition there;
- eccentricity of line of friction setting passage from shoulder extrados;
- stationary condition of eccentricity of line of friction at inner sliding joint intrados, as per line-of-friction tangency condition there.

The analytical analysis of either set of system equations leads to the same outcomes in terms of least-thickness collapse states and arising characteristic features of the masonry arch states. This provides a full, complete demonstration that no further withstanding or collapse states of the arch shall appear, beyond those that spontaneously come up from a classical (though updated) Heymanian analysis, after the insertion of finite-friction effects.

Actually, to say the truth, in the chronological developments, a part from this a posteriori presentation, such way to proceed was synergically explored together with the previous one (and with the CP/MP numerical formulation later adopted in Sect. 4), and truly helped in corroborating and providing a main guideline, of treatment and derivation, specifically to locate the self-standing states above the multiple point later revealed in the main η-μ plot presented next, and markedly those for the $1 \leq \eta \leq 2$ range representing sliding evolutions and manifestations that may be achieved beyond the limit condition of zero horizontal thrust, at $\eta = 1$, for the validity of purely-rotational collapse [4].

3 Self-standing Domains and Collapse Characteristics

Out of the analytical treatment, the family of contemplated failure mechanisms of the symmetric circular arch comprises the following collapse modes:

- purely-rotational (classical Heymanian one);
- mixed sliding-rotational (with shoulder sliding);
- mixed rotational-sliding (with inner sliding);
- overturning (with a two-block separation at the crown), for infinite μ and $\eta \geq 1$, at zero horizontal thrust;
- overturning (with inner sliding and no block separation at the crown), for finite μ and $\eta \geq 1$, at non-zero horizontal thrust ruled by μ;
- purely-sliding.

Figure 1 depicts the domains of the arch states in the main η-μ plot, at variable half-opening angle α, where four main regions are shaded, with collapse ruled by the following modes: purely-rotational (South-East); purely-sliding (North-West); mixed sliding-rotational (inner "pond" among the above); mixed rotational-sliding, including possible overturning (North-East "basin", until bounded $\eta \leq 2$ and unbounded μ, with manifested asymptotic trend for $\eta \to 1$ as $\mu \to \infty$).

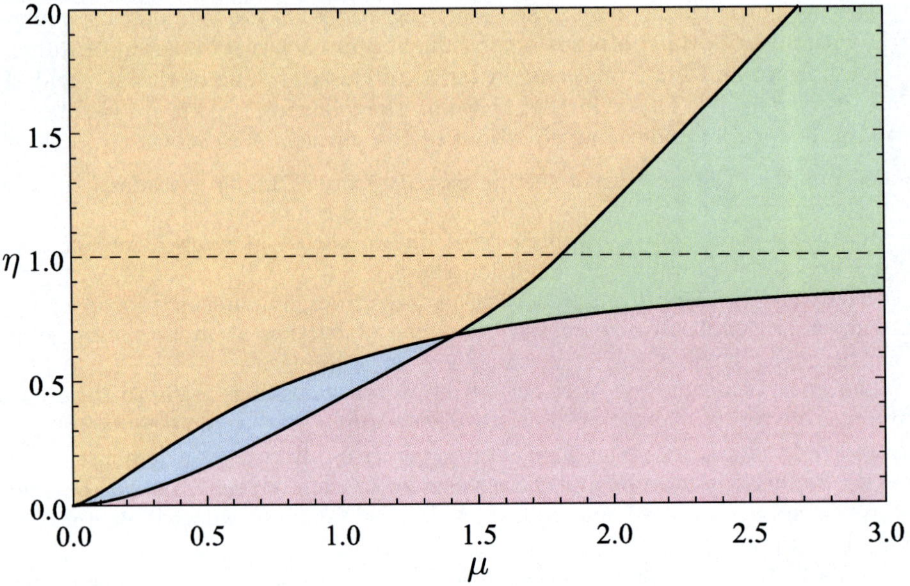

Fig. 1. Main η-μ plot of self-standing and collapse states, for all admissible $0 \leq \eta \leq 2$.

4 Illustration/Validation by a Numerical CP/MP Formulation

A final separate illustration by a CP/MP formulation, as a numerical validation and cross-corroboration of the achieved analytical results is then pursued. For the visual illustration of the characteristic features of the previous rigourous analytical analysis, and also toward independent numerical validation of the achieved analytical predictions, an original CP/MP approach can be adopted [7,8], to deliver and depict all final results corresponding to the self-standing domains and collapse conditions that have earlier been derived on an analytical basis.

As a main resuming outcome, Figs. 2 and 3 report typical masonry arches and collapse modes that may be obtained for characteristic half-opening angles and friction coefficients, all together showing: geometrical features of the arch in the least-thickness condition (black); all-internal line of thrust (red) and line of friction (blue); symmetric associated collapse mode, possibly including sliding (green). Specifically, Fig. 2 reports two different instances of mixed modes, respectively in the "pond" of mixed sliding-rotational modes and in the "basin" of mixed rotational-sliding modes (Fig. 1), while Fig. 3 shows the characteristics of the multiple point in the η-μ plot separating them, where a direct shift from purely-rotational to purely-sliding modes is rejoined. In practice, the four collapse modes depicted in Figs. 2 and 3, represent characteristic samples of the modes that can be seen in the four shaded regions of the η-μ filled plot (for $\eta < 1$).

Further quantitative information should be provided elsewhere, where more room may be available.

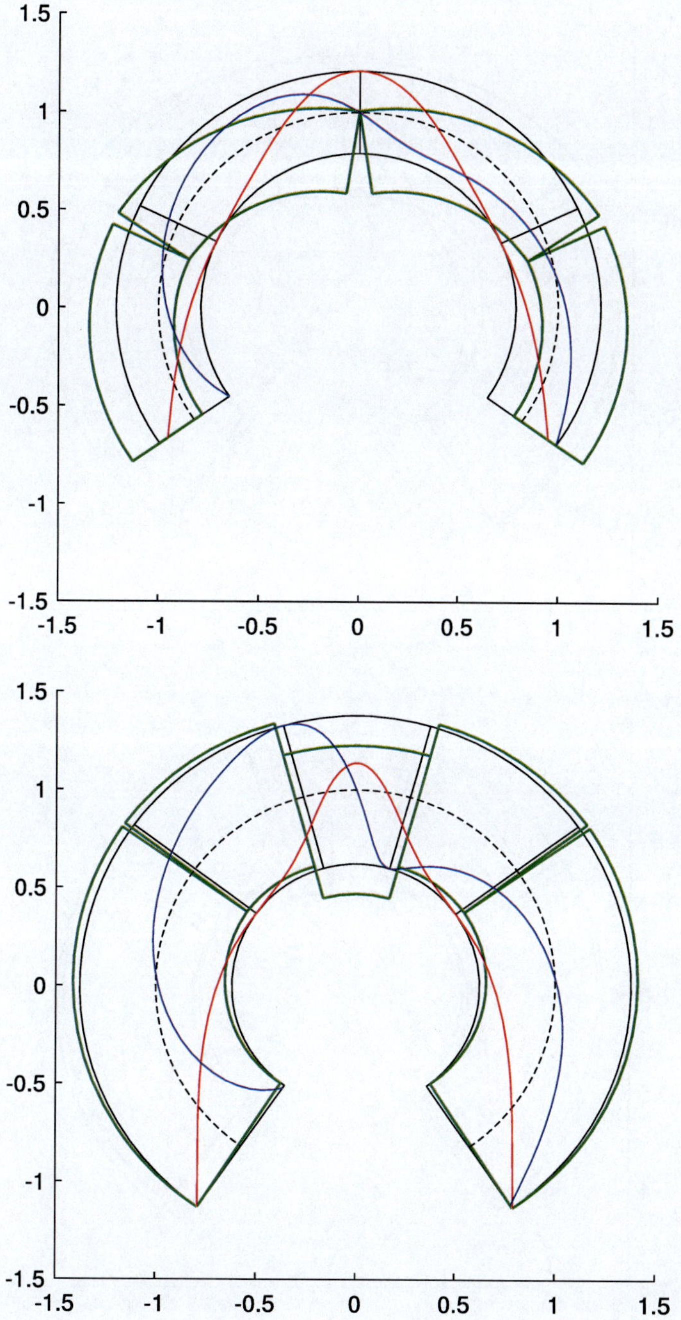

Fig. 2. Least-thickness mixed collapse mechanisms, with lines of thrust and of friction: sliding-rotational, for $\alpha = 125°$, $\mu = 0.9$ ($\varphi \simeq 42°$), with $\eta = 0.4208869235292454$; rotational-sliding, for $\alpha = 145°$, $\mu = 1.6$ ($\varphi \simeq 58°$), with $\eta = 0.7566761737936022$.

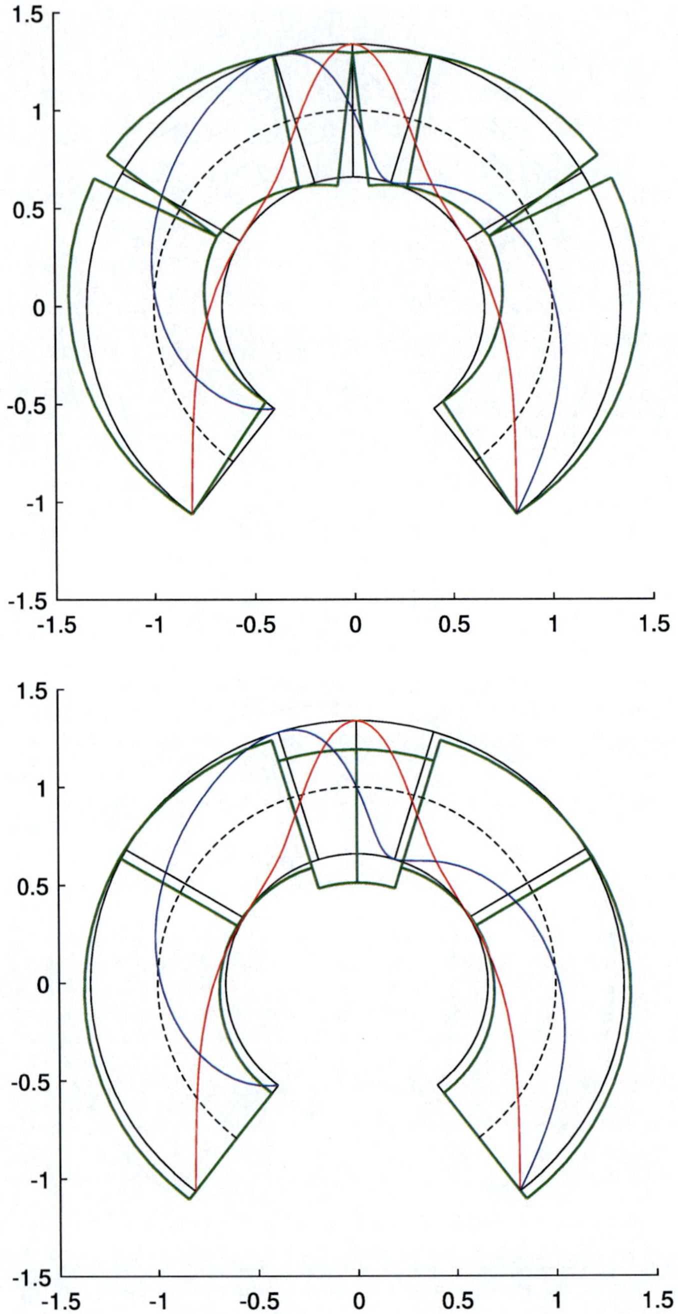

Fig. 3. Least-thickness collapse mechanisms, with (same) lines of thrust and of friction at multiple point with $\mu = 1.415270827756885$ ($\varphi \simeq 54.8°$), $\eta = 0.6796047320988855$, $\alpha = 2.487161163767182 \simeq 142.5°$: purely-rotational (top); purely-sliding (bottom).

5 Conclusions

The present work has addressed the issue of setting the least-thickness optimization condition of self-standing symmetric circular masonry arches of variable half-angle of embrace at finite internal friction. The cardinal equilibrium and limit compatibility conditions have been outlined, in two mutually corresponding ways, and then analytically elaborated, and completely solved in view of exploring the full range of arch behaviour in the physical space of the underlying mechanical parameters and variables.

Specifically, the safe states of the self-standing masonry arch have been recovered, and the domains of arch collapse located, with the contour boundaries separating them and the localization of all the relevant collapse mechanisms, possibly including sliding, as linked to the presence of finite friction.

At given specific self-weight, radius and out-of-plane depth of the symmetric circular masonry arch, the least-thickness self-standing condition of minimum thickness-to-radius ratio is sought, as a structural form-optimization problem, at variable physical parameters, which are the half-opening angle and the underlying friction coefficient (angle) at the theoretical (radial) joints of the continuous arch. This sets, in principle, a rather simple, and fascinating, mechanical system. Once all characteristic solutions are investigated, and delivered, with no a priori limitations on the ideally legitimate mechanical parameters (for instance, thickness-to-radius ratio between 0 and 2, half-opening angle among 0 and π, friction coefficient from 0 to ∞), rather surprising, variegated and intricate features arise, in terms of resulting outcomes and recorded richness of manifestations, in both academic and engineering terms, for the formulated mechanical problem.

The above has been revealed thanks to a complete and systematic analytical approach, up to deliver "exact" solutions for the problem at hand, and then further illustrated and validated by a numerical treatment based on a homemade Complementarity Problem/Mathematical Programming formulation and implementation. While the former is necessarily linked and likely limited to the intrinsic simplicity of the morphological shape and assumed self-weight loading condition of the circular masonry arch, still allowing to recover "exact" solutions, with such rich features, the latter, with consistent adaptations, may then allow for further inspections and applications to other reference configurations of the masonry arch, in view of generalizing the present results to other arch settings, where a complete analytical treatment may become unfeasible, and also in view of the analysis of specific practical cases.

This may lead to further possible developments of the present research investigation, enquiring, for instance, unsymmetric or different geometrical shapes of the masonry arch, issues of stereotomy (i.e. the shape of the stone cuttings of the mutual joints of voussoir arches [3]), effects of boundary conditions, different loading conditions (e.g. including lateral actions), and so on.

However, the considered neat setting of a symmetric circular masonry arch under self-weight shall constitute a main, fundamental reference, once now fully resolved in characteristic features, even at variable finite internal friction, toward a general perception and understanding of the Mechanics of masonry arches.

Acknowledgments. This work has been developed at the University of Bergamo, School of Engineering (Dalmine). The support by ministerial (MUR) funding "Fondi di Ricerca d'Ateneo ex 60%" at the University of Bergamo is gratefully acknowledged.

References

1. Milankovitch, M.: Theorie der Druckkurven. Zeitschrift für Mathematik und Physik **55**, 1–27 (1907)
2. Heyman, J.: The safety of masonry arches. Int. J. Mech. Sci. **11**(4), 363–385 (1969). https://doi.org/10.1016/0020-7403(69)90070-8
3. Heyman, J.: La coupe des pierres. In: Proceedings of the Third International Congress on Construction History, 20–24 May 2009, vol. 2, pp. 807–812. Brandenburg University of Technology, Cottbus, Germany (2009)
4. Cocchetti, G., Colasante, G., Rizzi, E.: On the analysis of minimum thickness in circular masonry arches. Part I: State of the art and Heyman's solution. Part II: Present CCR solution. Part III: Milankovitch-type solution. Appl. Mech. Rev. (ASME) **64**(5), Paper 050802, 1–27 (2011). https://doi.org/10.1115/1.4007417
5. Rizzi, E., Rusconi, F., Cocchetti, G.: Analytical and numerical DDA analysis on the collapse mode of circular masonry arches. Eng. Struct. **60**(February 2014), 241–257 (2014). https://doi.org/10.1016/j.engstruct.2013.12.023
6. Cocchetti, G., Rizzi, E.: Analytical and numerical analysis on the collapse modes of least-thickness circular masonry arches at decreasing friction. Frattura ed Integrità Strutturale - Fracture Struct. Integr. Spec. Issue Fracture Damage Detect. Masonry Struct. **14**(51), 356–375 (2020). https://doi.org/10.3221/IGF-ESIS.51.26
7. Cocchetti, G., Rizzi, E.: Non-linear programming numerical formulation to acquire limit self-standing conditions of circular masonry arches accounting for limited friction. Int. J. Masonry Res. Innov. Spec. Issue 10IMC Masonry Res. Third Millennium Theory Pract. Appl. **5**(4), 569–617 (2020). https://doi.org/10.1504/IJMRI.2020.111806
8. Cocchetti, G., Rizzi, E.: Static upper/lower thrust and kinematic work balance stationarity for least-thickness circular masonry arch optimization. J. Optim. Theory Appl. Spec. Issue Comput. Optim. Struct. Eng. Appl. **187**(3), 707–757 (2020). https://doi.org/10.1007/s10957-020-01772-0
9. Cocchetti, G., Rizzi, E.: Least-thickness symmetric circular masonry arch of maximum horizontal thrust. Arch. Appl. Mech. **91**(6), 2617–2639 (2021). https://doi.org/10.1007/s00419-021-01909-1

Tiled Plates with One Degree of Freedom

Filipe A. dos Santos[1], Antonino Favata[2], Andrea Micheletti[3], Roberto Paroni[4], and Marco Picchi Scardaoni[4](✉)

[1] CERIS-NOVA, Department of Civil Engineering, NOVA School of Science and Technology, Universidade NOVA de Lisboa, Caparica, Portugal
fpas@fct.unl.pt

[2] Department of Structural and Geotechnical Engineering, Sapienza University of Rome, Rome, Italy
antonino.favata@uniroma1.it

[3] Dipartimento di Ingegneria Civile e Ingegneria Informatica, University of Rome Tor Vergata, Rome, Italy
micheletti@ing.uniroma2.it

[4] Dipartimento di Ingegneria Civile e Industriale, Universitá di Pisa, Pisa, Italy
roberto.paroni@unipi.it,marco.picchiscardaoni@ing.unipi.it

Abstract. We investigate a design concept for thin tiled plates having exactly one degree of freedom. A tiled plate is realized as a tessellation composed by rigid tiles hinged to each other along the sides. The plate can deform in just one way, that is, into a predetermined surface. The family of tilings we consider, the monohedral hexagonal one, with either convex or concave tiles, is rich enough to include many noteworthy tessellations. Corresponding plates can approximate surfaces with positive, negative, and null Gaussian curvature, and may exhibit an auxetic behavior. The proposed architecture is highly scalable and easily manufacturable, and it can find applications not only for civil and aerospace engineering purposes, but also for biomechanical scaffolds, energy harvesters, and wearable devices.

Keywords: Tessellations · Plates · Auxetic

1 Introduction

The best-known example of surfaces obtainable by folding an (almost everywhere) rigid thin plate is probably represented by origami. Origami have been deeply studied in recent years in connection to bespoke material and metamaterial properties. For instance, adaptive origami metasurfaces were considered to realize multistable structures [4], robots [10] and active façades [6]. An unfolded origami looks like a tessellated plate, in which adjacent regions (or tiles), delimited by the network of folds, rotate relative to each other around the edges of such network. Given a pattern of folds, i.e., a tessellation of the plane, one can fold a thin plate, say a piece of paper, to achieve surprisingly complex shapes and surfaces [2]. However, there is not a one-to-one correspondence between fold

S. Gabriele et al. (Eds.): IWSS 2023, LNCE 437, pp. 693–698, 2024.
https://doi.org/10.1007/978-3-031-44328-2_72

patterns and target shapes: there can be many different shapes for a given single pattern (see, for instance, [3]).

In this paper, we present the case of a class of tessellated plates which can be designed to take a predetermined deformed shape under any flexural loading condition. In plain words, the tessellated plates we consider possess exactly one deformation mode. In this sense, the aforementioned multiplicity of folding modes of usual origami is ruled out, at the price to consider infinitesimal relative rotations between tiles instead of finite rotations. This feature pave the way for our tiled plates to be employed as shape morphing structures, or elements of those, activable in principle by a single actuator. More in general, they can be used in all applications where the reliability and robustness in achieving a predefined shape is of paramount importance.

Preliminary results were announced in [8], where it was shown that tessellated plates formed by assembling D_2-symmetric convex hexagons can deform into synclastic surfaces with given curvature and bending Poisson ratio when subjected to flexural loads. Here we announce new results, by showing that monohedral, hexagonal tessellations, i.e., those in which all tiles are identical irregular hexagons, either convex or concave, have one deformation mode. Preferentially, we say that the plate has one degree of freedom. Needless to say, the shape of the tile rules the final shape assumed by the corresponding tessellated plate in the deformed configuration.

We focus on the hexagonal case because many examples of hexagonal, 3-valent tessellations can be found in nature. For instance, in [5], it was shown that the hexagonal tessellation of the echinoid shell is a more efficient load-carrying structure if compared to a continuum doubly curved shell of same constitution. Interestingly, the ratio between the principal curvatures in echinoid shells is qualitatively reproduced by the tessellated plates studied in [8] made of similarly shaped tiles.

2 Tiled Kinematics and Main Result

2.1 Kinematics

Consider three tiles $\tau^{(i)}$, $i = 1, 2, 3$, lying on a plane whose normal is $\mathbf{e}$. We assume that the intersection of any of these regions are contained in straight lines that intersect into a point, see Fig. 1. The directions of the intersecting edges are referenced to as by $\mathbf{t}^{(i,j)}$, so that $\mathbf{n}^{(i,j)} = \mathbf{e} \wedge \mathbf{t}^{(i,j)}$ is the unit normal directed from $\tau^{(i)}$ to $\tau^{(i+1)}$.

In [8] it has been shown that

$$\mathbf{W}^{(i)} - \mathbf{W}^{(i+1)} = \varphi^{\{i,i+1\}} \mathbf{e}_3 \otimes \mathbf{n}^{(i,i+1)}$$

where $\mathbf{W}^{(i)}$ is a skew-symmetric tensor representing small rotations around $\mathbf{t}^{(i)}$ and $\varphi^{\{i,i+1\}}$ is the relative angle of rotation between tiles $\tau^{(i)}$ and $\tau^{(i+1)}$ (indices modulo 3). We can write a similar relation for all the three sides merging in the common node. Summing up, and after little manipulation, we get

$$\varphi^{\{1,2\}} \mathbf{t}^{(1,2)} + \varphi^{\{2,3\}} \mathbf{t}^{(2,3)} + \varphi^{\{3,1\}} \mathbf{t}^{(3,1)} = \mathbf{0}. \tag{1}$$

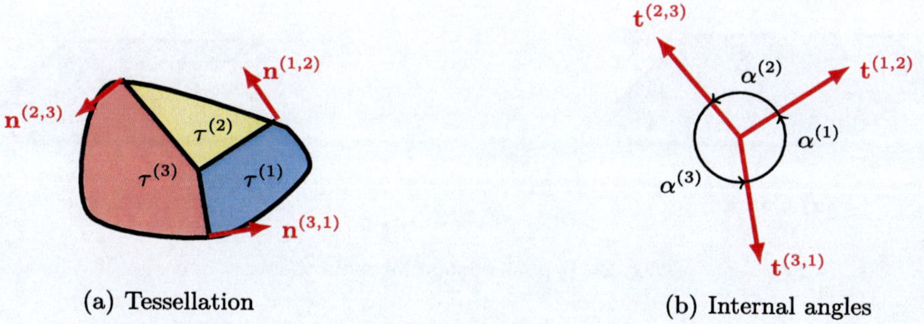

Fig. 1. Minimal example of 3-valent tessellation.

Let $\alpha^{(i+1)}$ be the angle enclosed by $\mathbf{t}^{(i,i+1)}$ and $\mathbf{t}^{(i+1,i+2)}$, and suppose $\alpha^{(i+1)} \neq 0, \pi$. By using $\{\mathbf{t}^{(1,2)}, \mathbf{n}^{(1,2)}\}$ as orthonormal basis, we have

$$\begin{aligned} \mathbf{t}^{(2,3)} &= \cos\alpha^{(2)}\mathbf{t}^{(1,2)} + \sin\alpha^{(2)}\mathbf{n}^{(1,2)}, \\ \mathbf{t}^{(3,1)} &= \cos(\alpha^{(2)} + \alpha^{(3)})\mathbf{t}^{(1,2)} + \sin(\alpha^{(2)} + \alpha^{(3)})\mathbf{n}^{(1,2)}, \end{aligned} \tag{2}$$

and Eq. 1 rewrites as

$$\begin{cases} \varphi^{\{1,2\}} + \varphi^{\{2,3\}} \cos\alpha^{(2)} + \varphi^{\{3,1\}} \cos(\alpha^{(2)} + \alpha^{(3)}) = 0, \\ \varphi^{\{2,3\}} \sin\alpha^{(2)} + \varphi^{\{3,1\}} \sin(\alpha^{(2)} + \alpha^{(3)}) = 0 \end{cases} \tag{3}$$

from which, using elementary trigonometric identities,

$$\varphi^{\{2,3\}} = \varphi^{\{3,1\}} \frac{\sin\alpha^{(1)}}{\sin\alpha^{(2)}}, \qquad \varphi^{\{1,2\}} = \varphi^{\{3,1\}} \frac{\sin\alpha^{(3)}}{\sin\alpha^{(2)}}. \tag{4}$$

We deduce that any two out of three relative rotations can be written in terms of the third.

2.2 Monohedral Hexagonal Tessellations

A tessellation of the plane is *monohedral* if all tiles are congruent. The basic tile is referenced to as *monotile*. The whole tessellation is then obtained by rotations, reflections, and translations of the monotile.

In [1,7], it has been shown that there exist exactly three types of hexagonal, *convex* monotiles. To describe these types let us denote the consecutive angles of the hexagonal monotile by $\alpha_1, \alpha_2, \ldots, \alpha_6$ and the lengths of the sides by $\ell_1, \ell_2, \ldots, \ell_6$, in such a way that ℓ_i and ℓ_{i+1} are the lengths of the sides of the angle α_i (indices modulo 6). In this paper we will deal with the so called Type 1 tiles. An exhaustive analysis of all the types will appear in [9]. With reference to Fig. 2, Type 1 hexagonal monotiles are characterized by $\alpha_1 + \alpha_2 + \alpha_3 = 2\pi$, $\ell_1 = \ell_4$.

(a) Convex (b) Concave

Fig. 2. Type 1 hexagonal monotiles

If at least one of the angles α_i is greater than π, the hexagonal tile is concave (see Fig. 2). Nevertheless, concave tiles, if they fulfill the requirements to belong to one of the aforementioned types, still pave the plane. We point out that it is not known whether there exist more than three types of hexagonal, concave monotiles. It is clear that only concave monotiles whose boundary does not self-intersect are of practical interest. Assuming that all internal angles are different from 0 and π, and that the tessellation is 3-valent, we can prove that the tessellations obtained with Type 1 monotiles (see Fig. 3) have one degree of freedom. The proof relies on Eq. 4, applied at the vertices of the tiles composing the minimum repetitive unit within the tessellation. Hence, one shall verify the compatibility of the assumed known rotation. Details will appear in [9], for the other types of monotiles as well.

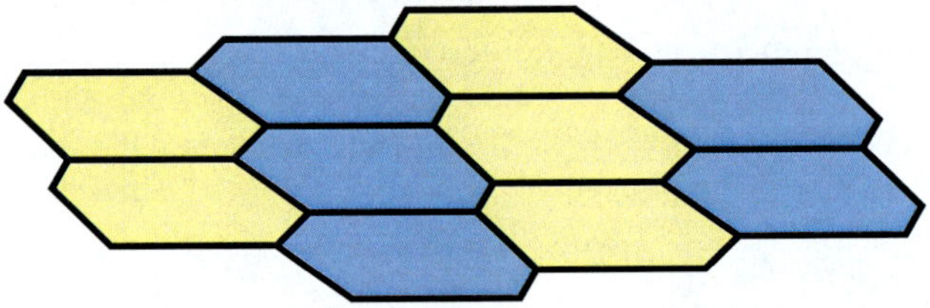

Fig. 3. Example of Type 1 tessellation

Figure 4 shows a parade of surfaces obtained as one degree of freedom deformation of tiled plates generated by Type 1 tiles. It can be appreciated that we can obtain surfaces with positive Gaussian curvature (a), negative Gaussian curvature (b), and zero Gaussian curvature (c).

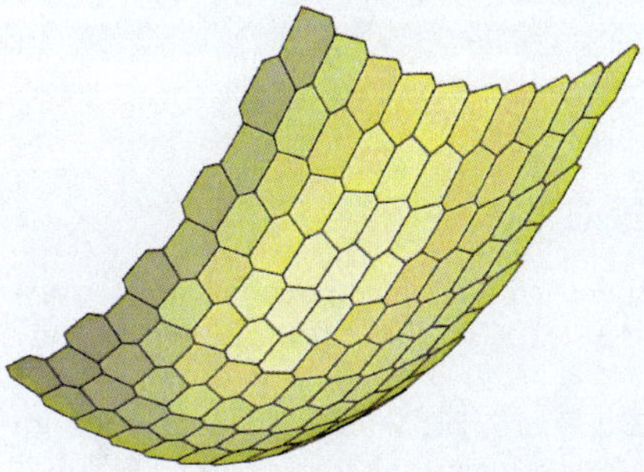

(a) Type 1, convex monotile

(b) Type 1, concave monotile

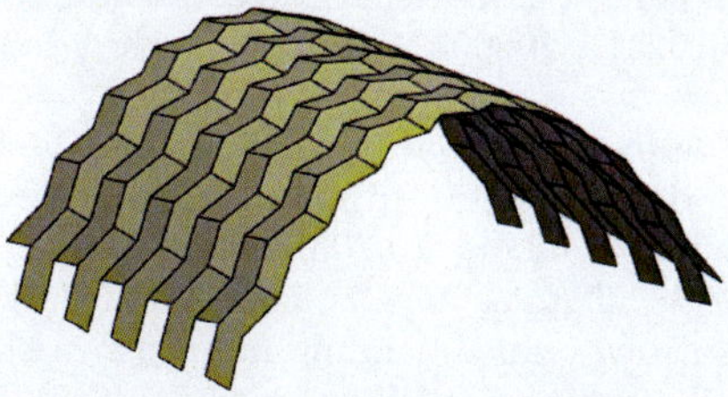

(c) Type 1, concave monotile

Fig. 4. Numerical examples of surfaces obtained as one degree of freedom deformation of tiled plates

3 Conclusions

We presented some new result concerning the possibility to design tiled plates with one degree of freedom. In particular, we described the proof that Type 1 monohedral hexagonal tessellations have one degree of freedom. As far as future perspectives are concerned, one could investigate other well-known tessellations, to see whether they deform in one mode only. For instance, one could consider monohedral, pentagonal tessellations or Voronoi tessellations. Besides, one could study the problem of approximating a target surface with a one DOF tessellated plate, not necessarily with periodic pavings. The knowledge of classes of tessellations fulfilling automatically the one-degree-of-freedom requirement is a fundamental step forward for the design of accurately controllable shape-morphing metasurfaces.

Acnowledgements. FAdS acknowledges the Fundação para a Ciência e a Tecnologia (FCT) in the framework of project UIDB/04625/2020. RP and MPS acknowledge the support from the University of Pisa through the project PRA2022-69 "Advanced modelling of ultra-lightweight materials and structures". AF acknowledges the support from Sapienza University of Rome through the research project 2022 "Mechanics of thin structures and 2D materials: advanced models and new applications". AM acknowledges the funding received in the framework of the project "OPTYMA - Optimized tensegrity metamaterials", grant number E83C22002290005, of the University of Rome Tor Vergata.

The Italian INdAM-GNFM is acknowledged.

References

1. Bollobas, B.: Filling the plane with congruent convex hexagons without overlapping. Ann. Univ. Sci. Budapest **6**, 117–123 (1963). http://annalesm.elte.hu/annales06-1963/Annales_1963_T-VI.pdf
2. Dang, X., Feng, F., Plucinsky, P., James, R.D., Duan, H., Wang, J.: Inverse design of deployable origami structures that approximate a general surface. Int. J. Solids Struct. **234–235**, 111224 (2022). https://doi.org/10.1016/j.ijsolstr.2021.111224
3. Liu, H., Plucinsky, P., Feng, F., James, R.D.: Origami and materials science. Phil. Trans. R. Soc. A **379**(2201), 20200113 (2021). https://doi.org/10.1098/rsta.2020.0113
4. Liu, K., Tachi, T., Paulino, G.: Invariant and smooth limit of discrete geometry folded from bistable origami leading to multistable metasurfaces. Nat. Commun. **10**(1), 4238 (2019). https://doi.org/10.1038/s41467-019-11935-x
5. Marmo, F., Perricone, V., Cutolo, A., Daniela Candia Carnevali, M., Langella, C., Rosati, L.: Flexible sutures reduce bending moments in shells: from the echinoid test to tessellated shell structures. R. Soc. Open Sci. **9**(5), 211972 (2022). https://doi.org/10.1098/rsos.211972
6. Miranda, R., Babilio, E., Singh, N., Santos, F., Fraternali, F.: Mechanics of smart origami sunscreens with energy harvesting ability. Mech. Res. Commun. **105**, 103503 (2020). https://doi.org/10.1016/j.mechrescom.2020.103503 ,http://www.sciencedirect.com/science/article/pii/S009364132030032X
7. Reinhardt, K.: Uber die Zerlegung der Ebene in Polygone. Ph.D. thesis, Universitat Frankfurt am Main (1918). https://gdz.sub.uni-goettingen.de/id/PPN316479497
8. dos Santos, F.A., Favata, A., Micheletti, A., Paroni, R.: Design of auxetic plates with only one degree of freedom. Extreme Mech. Lett. **42**, 101091 (2021). https://doi.org/10.1016/j.eml.2020.101091
9. dos Santos, F.A., Favata, A., Micheletti, A., Paroni, R., Picchi Scardaoni, M.: Programming quadric metasurfaces via infinitesimal origami maps of monohedral hexagonal tessellations (in preparation)
10. Treml, B., Gillman, A., Buskohl, P., Vaia, R.: Origami mechanologic. Proc. Nat. Acad. Sci. **115**, 201805122 (2018). https://doi.org/10.1073/pnas.1805122115

Dimensional and Mechanical Assessment of Gyroid Lattices Produced in Aluminum by Laser Powder Bed Fusion

Silvio Defanti(✉), Mauro Giacalone, Sara Mantovani, and Emanuele Tognoli

Department of Engineering "Enzo Ferrari",
University of Modena and Reggio Emilia, Modena, Italy
silvio.defanti@unimore.it

Abstract. The study investigates the use of aluminum gyroid lattices for structural purposes, with a particular focus on feasibility, dimensional accuracy, and compressive load performance. Gyroid lattice samples were built using laser powder bed fusion technology with AlSi10Mg, utilizing cell sizes of 6, 8, and 12 mm and a wall thickness of 0.5 mm. The compression performance of the samples was tested. The study revealed differences in dimensional accuracy in different directions, which was attributed to the fabrication process. All samples were heavier than expected, with additional materials being proportional to cell size. However, the samples exhibited high compressive strength and stiffness, indicating their potential use in load-bearing applications.

Keywords: Lattice structures · Gyroid · Additive Manufacturing · Laser-Powder Bed Fusion · AlSi10Mg

1 Introduction

Additive Manufacturing (AM) technologies enable the fabrication of complex geometries with increased design flexibility, such as lattice structures [1]. These intricate frameworks have high specific stiffness and strength, as well as large surface-to-volume ratios, making them ideal for a range of applications that were originally in the structural domain but are now expanding into fluid dynamics and heat transfer [2, 3]. Reticular structures, particularly those made of aluminum, open up the possibility of creating lightweight components with tailored mechanical and functional properties. Conventional manufacturing processes often struggle to produce intricate lattice structures due to their complexity and the limitations imposed by subtractive processes [4]. In contrast, AM techniques, particularly Laser-Powder Bed Fusion (L-PBF), enable the fabrication of these complex geometries with relative ease and are the most recommended processes in the metal parts field for producing net-shape components with mechanical behavior equivalent to that of traditionally fabricated parts [5].

However, it should be noted that parts manufactured by AM, especially L-PBF of metals, often exhibit inherent anisotropic behavior. The anisotropy arises from the

S. Gabriele et al. (Eds.): IWSS 2023, LNCE 437, pp. 699–707, 2024.
https://doi.org/10.1007/978-3-031-44328-2_73

layer-by-layer buildup and thermal evolution during fabrication [6]. The manufactured parts exhibit varying roughness, mechanical properties, and dimensional stability along two main directions, parallel to the build-up direction (Z) or to the powder bed (XY plane). Understanding and characterizing this anisotropy is critical to the successful implementation of additively manufactured parts in general, but even more so in lattices, where thin features with different orientations can emphasize the directional properties of the structure as a whole [7].

Another critical aspect in AM is the challenge of overhangs and self-supporting structures [8]. When fabricating lattice geometries, support structures, typically required in L-PBF, must be avoided because they cannot be removed. Therefore, the relationship between the design of the 3D structure and the feasibility of this geometry with a support-free L-PBF process must be thoroughly investigated. It is critical to design lattice reticula with the most self-supporting properties possible and to understand how and to what extent the additive process will change the target geometry in the manufactured part [9]. The self-supporting properties depend on many factors (powder density, particle size distribution, flowability). In addition, the surface roughness and metallurgical quality achieved by L-PBF vary consistently in the upskin and downskin regions, with the latter usually affected by the presence of additional material in the form of partially fused powders and agglomerates, as well as by frequent cracking [10]. Finally, lattice structures test the limits of L-PBF in terms of minimum feature size, since the typical beam diameter or wall thickness are at the lower end of dimensions feasible with laser spots used in industrial machines [11].

Compared to strut-based lattices, gyroid structures present fewer difficulties in AM construction [12]. Their continuous, interconnected network of curved surfaces not only results in high specific strength, stiffness, and energy absorption capability, but also mitigates the problem of overhangs, especially when the cell size is not too large [13].

To ensure straightforward industrial application, accurate modeling and simulation of lattice structures are required, based on reliable experimental data to validate and calibrate simulations using the Finite Element technique (FE) [14]. The objective of this work is to explore the capability of L-PBF to fabricate gyroid lattice structures in aluminum, to investigate the technological feasibility from the point of view of anisotropy, and to collect experimental data to generate charts of prompt access for subsequent computational simulations.

2 Materials and Methods

To investigate the manufacturability and mechanical performance of gyroid structures, cubic samples were designed with 3 different cell sizes and correspondingly different relative density. The thickness was set at 0.5 mm, a value reported as a technological limit in [10]. The basic cell is repeated along the 3 directions to obtain cubes with 8x8x8 cells. Their nominal values are given in Table 1.

Table 1. Nominal values of geometry and relative density of the samples.

Sample type	Cell size [mm]	Sample size [mm]	Nominal relative density (ρ_{rel})
G_6 × 0.5	6	48	0.230
G_8 × 0.5	8	64	0.169
G_12 × 0.5	12	96	0.117

2.1 Samples Manufacturing

Cubic samples were built on an SLM 500 (SLM Solutions Group AG, Lübeck Germany) using AlSi10Mg powder provided by SLM Solutions [15]. Two replicas of every sample type were produced. The process parameters were chosen to obtain a compromise between the manufacturability of thin structures and build-time:

- Layer thickness: 60 μm
- Scan speed: 860 mm/s
- Laser power: 370 W
- Borders-only strategy
- Temperature of the build platform 150 °C
- Argon atmosphere (Oxygen content < 0.1%)

After the production, the entire platform was heat treated at 270 °C for 2 h. The samples were then cut out by electrical discharge machining.

2.2 Characterization

The mass and external dimensions of each sample were respectively measured with an Ohaus PA124C analytical balance (Ohaus Europe GmbH, Switzerland) and a digital caliper, to an accuracy of ±0.02 mm. Then, the experimental relative density was calculated. The comparison of the obtained value with the nominal one gives a first insight into the manufacturing quality of the samples.

A second method of evaluating technological feasibility consisted of viewing one sample per type under a calibrated Nikon SMZ1270 stereomicroscope (Nikon, Japan). The overall aspect of the lattice structures was observed and some quantitative data on the thickness of the cells was collected.

In particular, the thickness of the gyroid was measured from the front (XZ), side (YZ), and top (XY) views, to investigate both the downskin-related defects and the precision of laser consolidation on the XY plane [8] (see Fig. 1). The measured thickness obviously refers only to the outer sample walls. The IT grade was also calculated according to the guidelines of ISO 286-1:2010 [16].

After the measurements, all the samples were tested at quasi-static compression. Since no specific standard for lattice structures was available at the time of testing, compression tests were performed according to ISO 13314-2011 [17]. The compression tests were performed on the samples along their building direction on a Schenck Hydropuls PSB machine equipped with a 250 kN load cell. Tests were performed at an initial strain

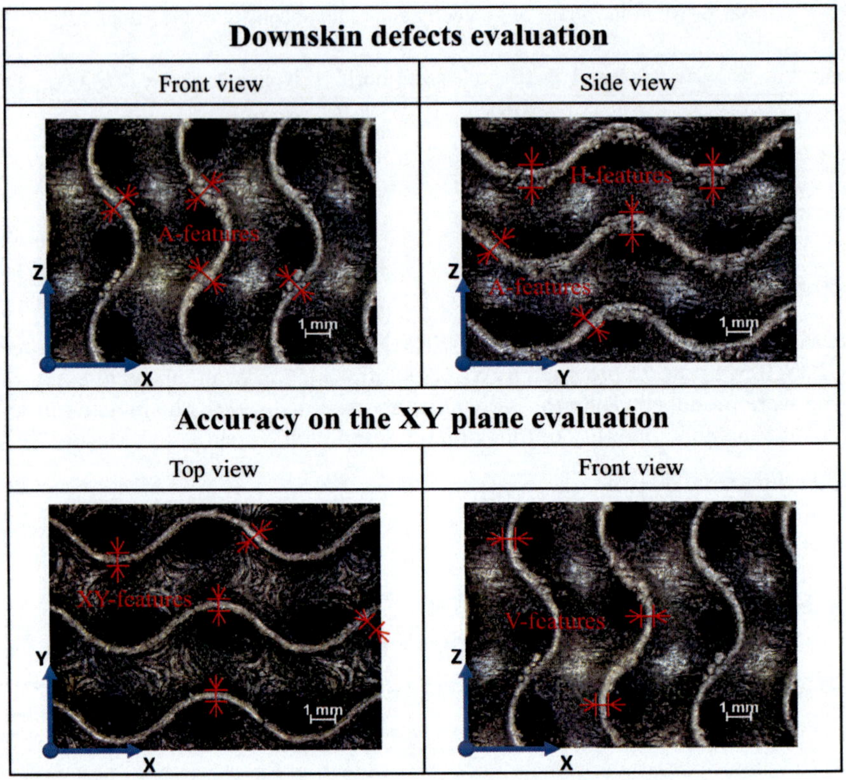

Fig. 1. Representation of the measured features.

rate of 0.0015 s^{-1}; the strain was recorded using the displacement value of the crosshead position encoder, with a correction coefficient used to exclude the deformation of the machine structure.

The loading cycle consists of 3 compression phases similar to the approach of Leary *et al.* [18]:

1. Compression to 1% of strain;
2. Unloading to initial position:
3. Compression to 2% of strain;
4. Unloading to initial position;
5. Compression to 80% of strain.

The following results are obtained from the stress-strain curve of the last compression cycle (step 5) as described in the previous authors contribution [19]:

- Quasi-elastic gradient (E): slope of the stress-strain curve in the linear elastic region.
- First maximum compressive strength (σ_r): compressive stress at the first local maximum of the stress-strain curve.
- Plateau stress (σ_{pl}): integral mean of the stresses between the strain values of 0.25 and 0.4.

- Energy absorption efficiency (W_{eff}): area below the stress-strain curve up to a strain value of 0.5, divided by the product of the maximum compressive stress within the strain range and the magnitude of the strain range.

To compare the performance of the 3 gyroid cells, and to isolate the effect of the difference in density, some normalized-specific values were also calculated:

- Normalized specific stiffness ($E_{n,sp}$), calculated according to Eq. (1).

$$E_{n,sp} = \frac{E}{E_b} \cdot \frac{1}{\rho_{rel,exp}} \tag{1}$$

Where:

- E_b = Young's modulus of the bulk material. For the AlSi10Mg in the stress relieved state, $E_b = 60$ GPa measured in preliminary tests of bulk specimens.
- $\rho_{rel,exp}$ = experimental relative density.

- Normalized specific compressive strength ($\sigma_{r(n,sp)}$), calculated as Eq. (2).

$$\sigma_{r(n,sp)} = \frac{\sigma_r}{\sigma_{ys}} \cdot \frac{1}{\rho_{rel,exp}} \tag{2}$$

Where:

- σ_{ys} = yield stress of the bulk material. For the AlSi10Mg in the stress relieved state, $\sigma_{ys} = 160$ MPa measured in preliminary tests of bulk specimens.

- Normalized specific plateau stress ($\sigma_{pl(n,sp)}$), calculated as Eq. (3).

$$\sigma_{pl(n,sp)} = \frac{\sigma_{pl}}{\sigma_{ys}} \cdot \frac{1}{\rho_{rel,exp}} \tag{3}$$

3 Results

3.1 Manufacturability

The overall aspect of the lattice samples can be observed in Fig. 2. All samples were produced successfully and in good quality. In the top view, it can be seen that the staircase effect does not endanger the wall continuity even in the saddle areas. Even from this initial visual inspection, it is clear that accuracy is critical, especially for the horizontal and angled features, while the vertical geometries benefit from much better feasibility. In particular, the V and XY walls are very uniform and smooth, while for the A features, but especially for the H features, the presence of partially melted powder particles adhering to the downskin areas is marked.

This is confirmed by the results of the measurements on the outer walls, which are shown in Fig. 3 and explained in more detail in Table 2. In most cases, the gyroids are manufactured with thicker walls than designed, with a maximum deviation of 96%.

The only exception is the XY elements, which are slightly thinner than the nominal dimension. It is interesting to note that the deviation of the wall thickness values is poorly related to the cell size, indicating that the process parameters such as layer thickness, diameter of the laser spot and width of the melt pool play a major role. Conversely, a smaller cell leads to more difficult removal of the loose powder and a more compact gyroid structure, where the material adhering to the downskin zones is more likely to affect the adjacent vertical features as well. As a result, different behaviours are observed, *i.e.*, the thickness of H features appears to increase with increasing cell size, while the opposite is true for the other feature (V). The two trends seem to balance each other, leading to a final relative density of the 3 sample types that is slightly higher (from 8% to 13%) compared to the nominal values, with no clear effect of cell size.

Table 2 also shows the IT grades, separately for the measurements affected by the phenomena of overhang and downskin and for those not. As for the first type of geometric features, the lattices can be classified in IT16. Better results are obtained for the V and XY features, whose IT ranges from 14 to 12 in the best case of the largest cell dimension.

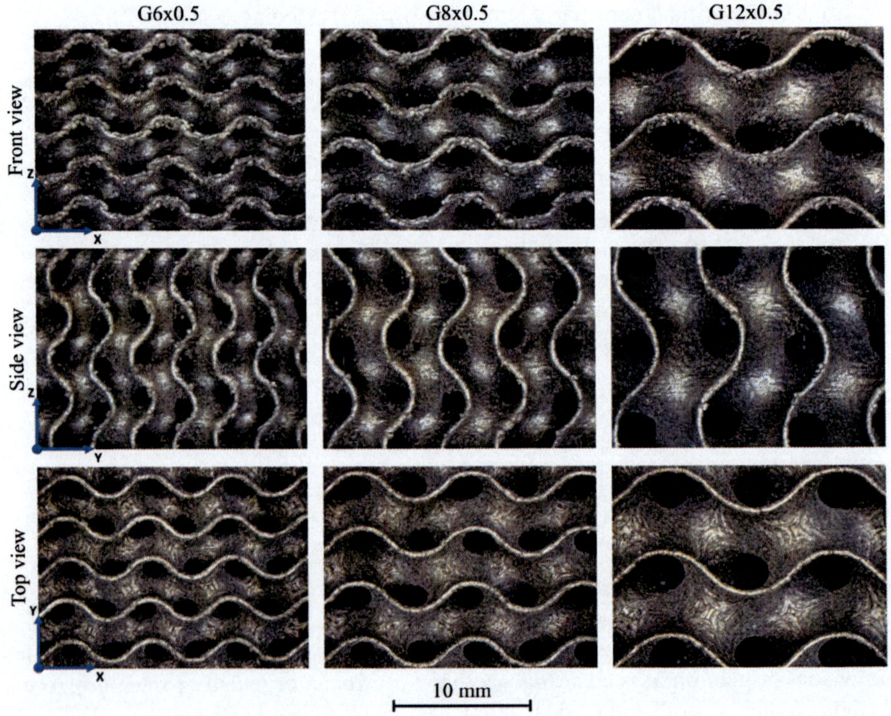

Fig. 2. Front (XZ plane), side (YZ plane) and top (XY plane) view of the 3 gyroid sample types.

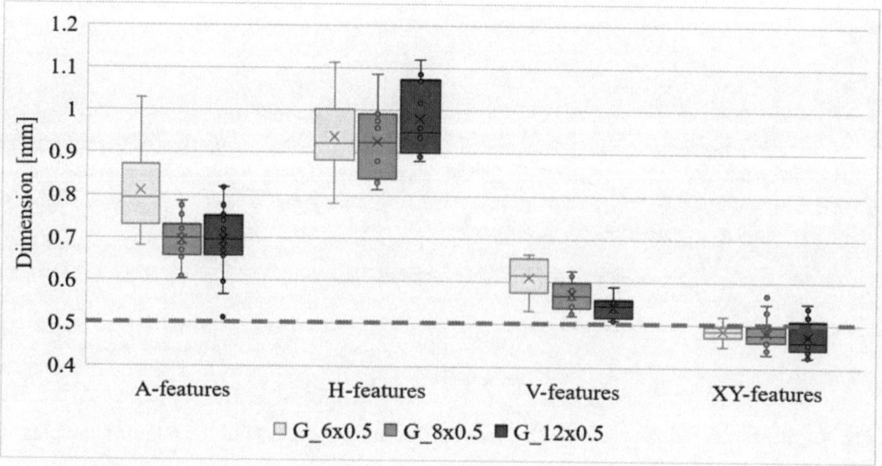

Fig. 3. Results of the thickness measurements of the 3 gyroid types. The dashed line indicates the nominal value.

Table 2. Results of the feasibility analysis in the form of the percentage deviation. The IT grade is also given.

Sample type	Density [Δ%]	Overhang/Downskin effect			XY Accuracy		
		A-features thickness [Δ%]	H-features thickness [Δ%]	IT grade	V-features thickness [Δ%]	XY-features thickness [Δ%]	IT grade
G_6x0.5	+10	+62	+88	16	+22	−3	14
G_8x0.5	+13	+39	+85	16	+14	−3	13
G_12x0.5	+8	+39	+96	16	+8	−5	12

3.2 Compression Tests

The stress-strain curves of the compression tests are shown in Fig. 4. The curves of the same samples type show good repeatability. Table 3 lists the absolute and specific mechanical properties calculated from the curves. The specific modulus is well matched for the 6- and 8-mm cells, while the structure with the largest cell is significantly less stiff than the others. The specific compressive strength and normalized specific plateau stress decrease slightly with increasing cell size. The same happens for the energy absorption efficiency that reaches its maximum for the 6 x 0.5 gyroid.

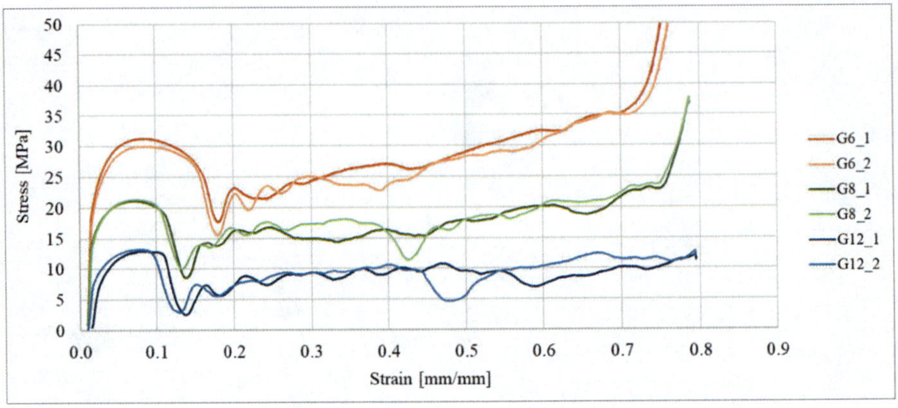

Fig. 4. Stress vs strain curves resulted from the compression tests of the 6 lattice samples.

Table 3. Results of the compressive tests.

Sample type	E [GPa]	$E_{n,sp}$	σ_r[MPa]	$\sigma_{r(n,sp)}$	σ_{pl}[MPa]	$\sigma_{pl(n,sp)}$	W_{eff} [%]
G_6 × 0.5	3.6	0.234	30.7	0.758	24.5	0.605	81
G_8 × 0.5	2.7	0.233	21.3	0.697	16.3	0.533	75
G_12 × 0.5	1.0	0.127	13.1	0.646	9.3	0.457	65

4 Conclusions

Gyroid lattice structures with a thickness of 0.5 mm and cell dimensions ranging from 6 to 12 mm were successfully fabricated by L-PBF of AlSi10Mg alloy. The results confirm the feasibility of gyroid lattice structures with relative densities as low as 0.13 with a mass supplement of about 10% compared to the nominal density. A comprehensive dimensional characterization was performed, allowing evaluation of accuracy and IT degree separately for geometric features affected and unaffected by the overhang/downskin effects. The approach provides insight into the dimensional performance of the process as a function of part orientation in the anisotropic buildup system. The compressive mechanical response was also investigated by calculating both the absolute and specific properties using an approach that quantifies the response of the specimen during subsequent failure steps.

References

1. Chen, L.-Y., Liang, S.-X., Liu, Y., Zhang, L.-C.: Additive manufacturing of metallic lattice structures: unconstrained design, accurate fabrication, fascinated performances, and challenges. Mater. Sci. Eng. R Reports **146**, 100648 (2021). https://doi.org/10.1016/j.mser.2021.100648

2. Mahmoud, D., et al.: Enhancement of heat exchanger performance using additive manufacturing of gyroid lattice structures. Int. J. Adv. Manuf. Technol. **126**, 4021–4036 (2023). https://doi.org/10.1007/s00170-023-11362-9
3. Sélo, R.R.J., Catchpole-Smith, S., Maskery, I., Ashcroft, I., Tuck, C.: On the thermal conductivity of AlSi10Mg and lattice structures made by laser powder bed fusion. Addit. Manuf. **34**, 101214 (2020). https://doi.org/10.1016/j.addma.2020.101214
4. Bassoli, E., Defanti, S., Tognoli, E., Vincenzi, N., Degli Esposti, L.: Design for additive manufacturing and for machining in the automotive field. Appl. Sci. **11**, 7559 (2021). https://doi.org/10.3390/app11167559
5. Froes, F., Boyer, R.: Additive Manufacturing for the Aerospace Industry. Elsevier (2019). ISBN 9780128140628
6. Defanti, S., Bassoli, E.: Repeatability of the fatigue performance of additively manufactured A357.0 under different thermal treatment conditions. Mater. Sci. Eng. A **805**, 140594 (2021). https://doi.org/10.1016/j.msea.2020.140594
7. Wang, Y., et al.: Lattice structure design optimization coupling anisotropy and constraints of additive manufacturing. Mater. Des. **196**, 109089 (2020). https://doi.org/10.1016/j.matdes.2020.109089
8. Bassoli, E., Mantovani, S., Giacalone, M., Merulla, A., Defanti, S.: On the technological feasibility of additively manufactured self-supporting AlSi10Mg lattice structures. Adv. Eng. Mater. **25**, 2201074 (2023). https://doi.org/10.1002/adem.202201074
9. Atzeni, E., Salmi, A.: Study on unsupported overhangs of AlSi10Mg parts processed by Direct Metal Laser Sintering (DMLS). J. Manuf. Process. **20**, 500–506 (2015). https://doi.org/10.1016/j.jmapro.2015.04.004
10. Sola, A., Defanti, S., Mantovani, S., Merulla, A., Denti, L.: Technological feasibility of lattice materials by laser-based powder bed fusion of A357.0. 3D Print. Addit. Manuf. **7**, 1–7 (2020). https://doi.org/10.1089/3dp.2019.0119
11. Chakraborty, A., et al.: In-process failure analysis of thin-wall structures made by laser powder bed fusion additive manufacturing. J. Mater. Sci. Technol. **98**, 233–243 (2022). https://doi.org/10.1016/j.jmst.2021.05.017
12. Yan, C., Hao, L., Hussein, A., Raymont, D.: Evaluations of cellular lattice structures manufactured using selective laser melting. Int. J. Mach. Tools Manuf **62**, 32–38 (2012). https://doi.org/10.1016/j.ijmachtools.2012.06.002
13. Piscopo, G., Salmi, A., Atzeni, E.: On the quality of unsupported overhangs produced by laser powder bed fusion. Int. J. Manuf. Res. **14**, 198–216 (2019). https://doi.org/10.1504/IJMR.2019.100012
14. Yang, E., et al.: Effect of geometry on the mechanical properties of Ti-6Al-4V Gyroid structures fabricated via SLM: a numerical study. Mater. Des. **184**, 108165 (2019). https://doi.org/10.1016/j.matdes.2019.108165
15. SLM Solutions Group AG Al-Alloy AlSi10Mg/EN AC-43000 Material Datasheet
16. International Organization for Standardization ISO 286-1:2010: Geometrical product specifications (GPS) - ISO code system for tolerances on linear sizes - Part 1: Basis of tolerances, deviations and fits **(2010)**
17. ISO ISO 13314-2011: Mechanical testing of metals - Ductility testing - Compression test for porous and cellular metals (2011)
18. Leary, M., et al.: Selective laser melting (SLM) of AlSi12Mg lattice structures. Mater. Des. **98**, 344–357 (2016). https://doi.org/10.1016/j.matdes.2016.02.127
19. Mantovani, S., Giacalone, M., Merulla, A., Bassoli, E., Defanti, S.: Effective mechanical properties of AlSi7Mg additively manufactured cubic lattice structures. 3D Print. Addit. Manuf. (2021). https://doi.org/10.1089/3dp.2021.0176

Stability Assessment of Masonry Arches Deformed by Thermal Distortions

Stefano Galassi, Eloisa Fazzi, and Maria Luisa Satta(✉)

Department of Architecture, University of Florence, Florence, Italy
marialuisa.satta@stud.unifi.it

Abstract. Masonry arches and vaults are widespread in historical buildings and infrastructures. The shape of these structures is a crucial aspect for their equilibrium condition and is responsible for their safety. Their current shape can be different from the design one, because it may have been affected by deformations caused by various phenomena over time, such as load changes, earthquakes, foundation settlements and temperature raising/lowering. It follows that the current safety verification must be done by comparing the load thrust line with the current deformed shape of the arch.

The paper investigates the effect of thermal distortions on the original shape of the arch and the safety condition, through a rigid-block and lumped deformability model. Thermal distortions are assigned at the joints.

The effect of thermal distortions on masonry arches is investigated by assessing the increase or decrease of the load bearing capacity. To this aim, the behavior of an arch case study is investigated by applying an increasing point load and ranging the thermal gradient. A parametric analysis is also performed to assess the effect of the tensile strength of joints.

Keywords: Masonry arches · rigid-block and lumped deformability model · thermal distortions · nonlinear analysis

1 Introduction

Masonry arches and vaults are construction types recurring and widespread in the Italian building landscape and, particularly, in historical buildings belonging to the architectural heritage. Furthermore, there are many highway and railway masonry arch-bridges which are still in use [1]. All these structural elements are daily or occasionally exposed to external actions in addition to loads, such as stresses and strains caused by the movement of supports due to soil settlements or by the increase or decrease of temperature, that may result in the decrease in their load-bearing capacity. In this regard, the Italian NTC 2018 [2] provides for temperature-induced actions as actions to be considered in structural design. Nevertheless, to deal with this topic it is necessary to distinguish between:

- actions induced by temperature, caused by the transition from the minimum winter to the maximum summer temperature and vice versa;
- actions induced by severe thermal shock, such as those caused by fire.

S. Gabriele et al. (Eds.): IWSS 2023, LNCE 437, pp. 708–716, 2024.
https://doi.org/10.1007/978-3-031-44328-2_74

Indeed, the effect of temperature on the structures can be twofold as a function of the temperature attained: 1) the occurrence of additional inner stresses (referred to in the literature as eigenstresses of thermal origin [3]) caused by the thermal deformations prevented by constraints in statically undetermined structures, such as those of masonry; 2) progressive decay of material properties.

While the first point is the result of the effect of both low and high temperatures, the second specifically concerns the action of very high temperatures and depends on the exposure time of structures.

In relation to the 1st point, the Italian NTC 2018 describe the temperature field over the section of a structural element by means of a uniform thermal distortion (calculated as difference between the current average temperature and the initial average temperature at the date of construction) and one that varies linearly along the principal axes of the section. Where temperature is not a fundamental action for safety or functional efficiency of the structure, this standard allows considering, for structural analysis, only the uniform thermal distortion, whose value is provided as a function of the type of the building structure and distinguishes among reinforced concrete structures, prestressed concrete structures and steel structures, but no reference is made to masonry structures.

Because the stresses produced by the thermal action can exceed the tensile strength of the material and cause even macroscopic cracks, thus resulting in a clear source of accelerated degradation of the building, it is necessary to also consider thermal distortions in the analysis of the structure to ensure a proper durability.

In relation to the 2nd point, instead, the exposure of the structure to the very high temperatures caused by fire for some time gives rise to chemical-physical reactions and phase changes that lead to a change in the thermophysical properties of the material even though the bulk density has been shown not to vary. In particular, a significant change in these parameters was found around the temperature of 100 °C, which can be attributed to the phenomenon of evaporation of residual water in the pores of individual constituent materials of masonry (bricks and mortar). It results in a decay of the mechanical properties of masonry, namely the change in the stress-strain constitutive law (i.e. Young's modulus and compression strength) as a function of the reached value of temperature.

The Eurocode 6 [4] shows that for temperatures lower than the environmental one (20 °C) no thermal strains in clay bricks occur while for higher values the strains grow linearly as the temperature increases up to 750°. For even higher temperatures, strains grow at constant stress value. Furthermore, the stress-strain constitutive laws indicate an elastic-brittle behavior of the masonry units, highlighted by the sudden decay which follows the peak strength. Considering different values of temperature, recent studies [5, 6] indicate that as the temperature increases (starting from the environmental one), both the pre- and post-pick Young's modulus decrease according to a parabolic law and the ultimate strain increases according to a linear relation. Furthermore, the peak strength is reached only at the environmental temperature, while for temperatures higher than 20 °C the decay of material leads to a strength reduction that reaches 30% at a temperature of 750°, which is considered the threshold value beyond which the cross section of the structural member is no longer reacting. To deal with this topic also a decay of the stress-strain constitutive law and the reduction of the resisting sections of the structure must be considered.

The aim of this paper is developing a numerical model for the analysis of the effect produced by thermal distortions on masonry arched structures. Nevertheless, models and numerical procedures that are found in the literature are mostly addressed to vertical members (such as walls or piers) with flat surface [7] and, only recently, studies have been conducted on arches and vaults subjected to a high thermal gradient [5, 6, 8]. Neither do current regulations consider structures with curvilinear axis, such as arches and vaults, perhaps due to the lack of adequate testing, which is difficult and expensive. Thus, at present, there is a lack of numerical models for the analysis of masonry arches subject to thermal distortions, which, on the other hand, would be necessary in Italy because of their widespread presence throughout the country.

With the purpose of future extension to the analysis of masonry arches subject to severe thermal shocks produced by fire, the rigid-block and lumped deformability model developed in [9] is adopted and the numerical procedure for the analysis is extended to also consider the effect of thermal distortions. Thermal distortions are introduced in correspondence to the joints. The increase or decrease of the load bearing capacity caused by thermal distortions is proven by the numerical analysis of a reference arch case study. The effect of joint tensile strength is also investigated.

2 Mechanical Model

The rigid-block and lumped deformability model developed in [9] is adopted. The masonry arch is modelled by a set "*n*" rigid blocks linked in-between and to the soil by "*n* + *1*" deformable mortar joints (Fig. 1a). Joints are modelled in a discrete way, by a set of two axial links arranged orthogonally to the mid-surface of the joints and two links along the mid-surface of the joints (Fig. 1b).

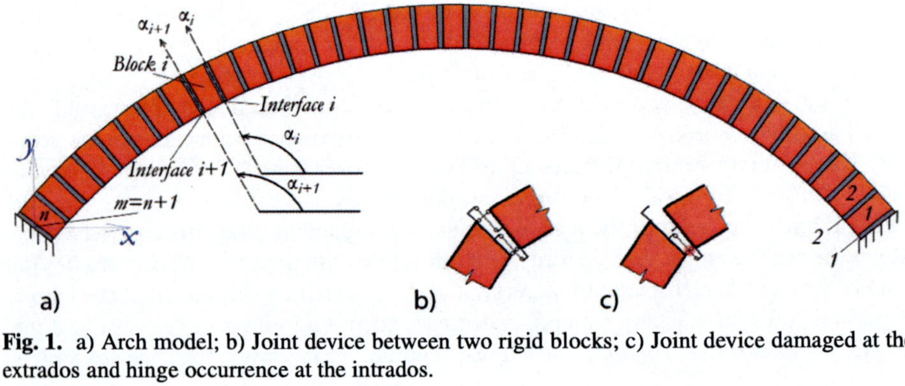

Fig. 1. a) Arch model; b) Joint device between two rigid blocks; c) Joint device damaged at the extrados and hinge occurrence at the intrados.

Axial links transfer the axial force and the tangential ones transfer the shearing force. Deformability is concentrated in correspondence to the joints and is a function of the Young's modulus assigned to the interface links.

A numerical procedure for the analysis of the arch model is conceived, based on a system of two matrix equations expressing the equilibrium and the elastic-kinematic compatibility conditions of the structure:

$$\begin{cases} AX + F + \lambda\overline{F} = 0 \\ A^t U + KX = D_T + \Delta \end{cases} \quad (1)$$

In Eq. 1, matrix A and its transpose A^t are the equilibrium and kinematic matrices, respectively. Matrix K is the diagonal deformability matrix populated by entries of the type $\frac{\ell_k}{EA_k}$ which are defined, for the kth link of the reference joint shown in Fig. 1b, by the ratio between the lenght ℓ_k of the link and its axial stiffness (defined by the Young's modulus E of the joint and the cross-sectional area A_k of the link). Vectors X and U are the static and kinematic unknowns of the problem, the former one collecting the forces in the axial and tangential links of all joints and the latter one collecting the horizontal, vertical and rotational displacements of the block centroids.

Two types of external actions acting on the structures are considered: loads and thermal distortions. F and $\overline{F}$ are the vectors of dead and variable loads, respectively, the latter one increased by the factor λ. Vector D_T is the thermal distortion vector, which is assigned to the axial links of the joints causing their elongation or shortening as a function of an increase or decrease in temperature.

These external actions can cause damage to the structure consisting in cracks in correspondence to the joints. Figure 1c shows a joint damaged by an extrados crack which causes the opening of an intrados hinge at the joint. The occurrence of cracks implies that the material nonlinearity must be considered in the numerical procedure. To consider the material nonlinearity, two failure criteria can be assigned to the interface links, related to the axial force (traction and compression) and shearing force, at analyst's choice as a function of the type of analysis to be performed:

$$\begin{cases} S_n \cdot \tilde{\sigma}^{(-)} \leq X_n \leq S_n \cdot \tilde{\sigma}^{(+)} \\ X_s \leq \mathrm{f} \cdot X_n \end{cases} \quad (2)$$

where X_n and X_s are the sub-vectors of vector X collecting the axial and shearing forces in the axial and tangential links, $\tilde{\sigma}^{(-)}$ and $\tilde{\sigma}^{(+)}$ are the threshold value of the compression and tensile stress causing material failure, S_n is the sub-vector of vector S collecting the cross-sectional areas of axial links and f is the friction factor.

The first failure criterion related to the axial links implies an elastic-brittle behavior in traction and an elastic-plastic behavior in compression. The second criterion, which concerns the shearing links and is based on the Coulomb's sliding-friction behavior, implies an elastic-brittle behavior of the tangential links.

To deal with these failure criteria, in Eq. 1 the distortion vector Δ is introduced. This vector is aimed at nullifying the tensile forces in the axial links greater than the tensile threshold values $S_n \cdot \tilde{\sigma}^{(+)}$ and, concurrently, the associated tangential links in such a way as to develop a crack and turn the original joint device into an ideal rotational hinge device (refer to Fig. 1c). Likewise, the compression forces in the axial links lower than the compression threshold values $S_n \cdot \tilde{\sigma}^{(-)}$ are reduced to this value by the distortion vector. Finally, to address the possible sliding of blocks along the joints, distortions can also be assigned to the tangential links with the purpose of nullifying the shearing forces greater than the friction force $\mathrm{f} \cdot X_n$ and to transform the original joint device of Fig. 1b into a "sliding support"-type constraint.

In the following paper, since the thermal distortions acting together with the external loads are only the effect of a small change in ambient temperature, it is assumed that they can cause tensile failure of the joints but not crushing failure, since the joint elongation caused by compression forces occurs in the linear-elastic field in this temperature range. For this reason, Heyman's assumptions can still be adopted and, therefore, only the tensile failure criterion in Eq. 2 is used in the analysis of the case study in Sect. 3.

The static solution of Eq. 1, obtained by exploiting the Somigliana Distortions Method (SDM) formulated in [9] under the failure criteria chosen by the analyst, is shown in Eq. 3:

$$X = X_0 + \left(I - A^t\left(AK^{-1}A^t\right)^{-1}AK^{-1}\right)\Delta \tag{3}$$

where, the first addendum, explicated in detail in Eq. 4, is the linear-elastic solution:

$$X_0 = -K^{-1}A^t\left(AK^{-1}A^t\right)^{-1}\left(F + \lambda\overline{F}\right) + K^{-1}\left(I - A^t\left(AK^{-1}A^t\right)^{-1}AK^{-1}\right)D_T \tag{4}$$

In Eq. 4 the first addendum provides the forces in the joint links caused by the external loads $F + \lambda\overline{F}$ and the second addendum the forces caused by the thermal distortions D_T.

At the beginning of the analysis $\Delta = 0$ because, for very small values of λ, the arch is assumed to be undamaged.

During the analysis, the value of λ is increased and, in correspondence to each value, the procedure iteratively computes vector X (Eq. 3) and:

1) extracts from vector X the sub-vector X_n and checks the first failure criterion in Eq. 2. If some entries do not meet the tensile failure criterion, distortions are introduced in the axial links subject to traction and the associated tangential links to nullify the tensile and shearing forces; if some entries do not meet the compression failure criterion, distortions are introduced in the axial links subject to compression to reduce the compression forces to the threshold values;
2) extracts from vector X the sub-vector X_s and checks the second failure criterion in Eq. 2 relating to the shearing force; if some entries of this vector do not meet the failure criterion because exceed the friction force, distortions are introduced in the tangential links to nullify the shearing forces.

The algorithm converges when all entries of vector X respect the failure criteria in Eq. 2. Non convergence implies the collapse of the arch.

At each step of the analysis, the algorithm updates the geometrical configuration of the arch. To update it, the vector of displacements of block centroids is computed. To this aim, the second equation of Eq. 1 is partitioned in such a way as to isolate entries which respect the failure criteria of Eq. 2 (subscript "g") from those which do not (subscript "w"). By doing so, a system of two matrix equations is obtained:

$$\begin{cases} A_g^t U + K_g X_g = D_{Tg} \\ A_w^t U = D_{Tw} + \Delta_w \end{cases} \tag{5}$$

Solving it yields the vector of the displacements of the block centroids (Eq. 6) and the width of the cracks measured in correspondence to the axial links of mortar that failed (Eq. 7):

$$U = -A_g\left(A_g^tA_g\right)^{-1}K_gX_g + A_g\left(A_g^tA_g\right)^{-1}D_{Tg} \tag{6}$$

$$\Delta_w = A_w^tU - D_{Tw} \tag{7}$$

3 Case Study and Preliminary Results

In this section a circular arch, with a 300 cm span constituted by 40 rigid blocks with 25 cm depth and height is analyzed; the arch is characterized by 18 KN specific weight, Young's modulus E = 8000 MPa and a point load of 1KN acting at ¼ of the span.

A preliminary analysis was carried out to evaluate the collapse factor of the incremental load, under the assumption of zero tensile stregth of the material and ambient temperature. As well-known in the literature, the incremental load produced the activation of a four-hinged collapse mechanism, causing cracks at joints 1I-12E-29I and joint 37E at collapse (Fig. 2). The collapse load, therefore, corresponds to the activation value of the mechanism, that for the case study is equal to 11.49.

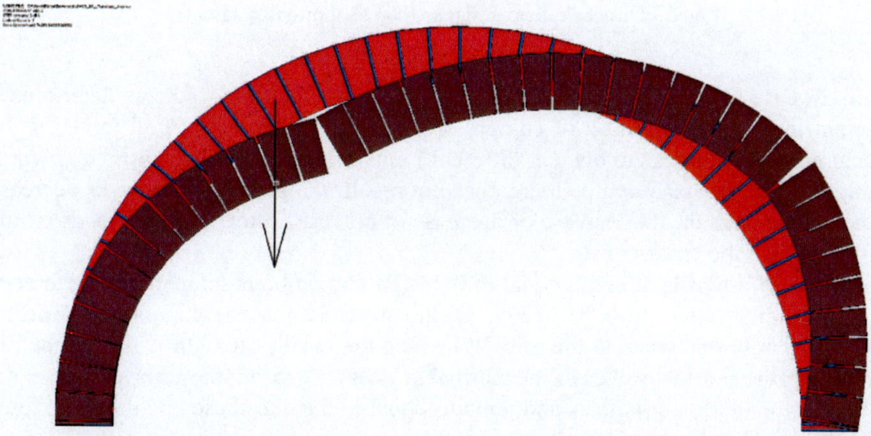

Fig. 2. Collapse mechanism of the arch analyzed under the assumption of no tensile strength and ambient temperature condition.

Then a thermal gradient was inserted into the model in correspondence to the joints.

The assignement of a negative thermal distortion to the structure did not allow finding a collapse factor; strains induced in the material by the assigned thermal distortion determine tensile stresses in all joints, which crack simultaneously. Therefore, it is clear that the no-tension model is not suitable to describe the behavior of the arch when subject to thermal gradients.

The variation of the collapse factor, for a fixed non-zero value of tensile strength of the joints, was then studied by increasing and decreasing the thermal gradient with steps equal to 5 °C. Tenesile strength vaules equal to 0.05, 0.1 and 0.2 MPa were considered.

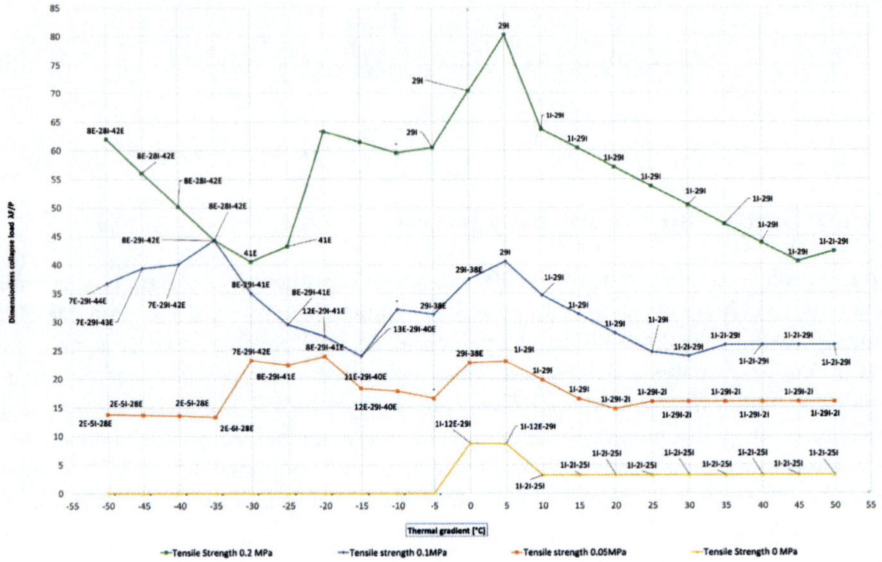

Fig. 3. Dimensionless collapse load vs thermal gradients.

In all case studies (Fig. 3) it can be observed that thermal gradients determine a different internal distribuition of compressive and tensile stresses in all joints of the structure, causing cracks to open at different joints on a case by case basis. It is worth noting that a decrease in temperature does not result in a gradual increase or decrease in the collapse factor: the increase or decrease of colapse factor values depends on the joints at which the cracks open.

In case of a tensile strength equal to 0.2 MPa and ambient temperature (reference thermal gradient equal to 0 °C in Fig. 3), the structure reaches the collapse load by developing only one crack at the joint 29I, since the tensile strength is not reached in any other joint. For low vaules of the thermal gradients (∓ 5 °C) the crack pattern is not altered by the thermal gradients and remains equal to the one caused by the point force and self weight only.

As the thermal gradient decreases in a range between −10 °C and −20 °C, the effect of the thermal gradient produces an optimal state of internal stress, since the tensile strength of the material is not exceeded at any interface and no crack develops. In particular, when the thermal gradient reaches the value of −20 °C, an extremly favorable internal stress state is determined so that the maximum load factor is reached. Collapse occurs due to instant failure of many joints.

By assigning a thermal gradient in the range −25 °C to −30 °C, the occurrence of a crack is determined again, although in a different joint; further decreases of temperature, i.e. negative tehermal gradients, induce relevant tensile stresses affecting the structure more than the stresses determined by loads and the maximum load factor is reached with the formation of three cracks correponding to joints (8E-28I-42E) where the tensile stress exceeds the treshold value.

As the temperature increases with thermal gradients from 10 °C to 45 °C, a progressive decrease of the collapse load, which corresponds to a crack pattern marked by cracks at joints 1I-29I, is observed. Further increasing temperature with a thermal gradient of 50 °C, a change in the crack pattern occurs, affecting the joints 1I-2I-29I, which provides an increase of the collapse load.

The maximum collapse factor is obtained in correspondence to a very low thermal gradient (+5 °C), which does not change the crack pattern caused by loads at ambient temperature.

Denoted by P the weight of a rigid block, in Fig. 3 the dimensionless collapse loads $\frac{\lambda\overline{F}}{P}$ are plotted against thermal gradients, along with the indication of the number and position of the cracked joints (E = extrados, I = intrados).

Also, in cases of lower tensile strength values, i.e. 0.1 and 0.05 MPa, it is observed that different values of the thermal gradient cause cracks to open at always different joints; furthermore the number of cracked joints varies between 0 and 3. In particular, in the same figure it is observed that temperature increases/decreases above a certain threshold do not change the value of the collapse factor, which preserves a constant value.

4 Conclusions

In this paper the effect of thermal distortions on the behavior of masonry arches is investigated. To investigate it, a mechanical rigid-block and lumped deformability model is used. Thermal distortions are assigned to the deformable joints.

The effect of thermal distortions on masonry arches is investigated by assessing the increase or decrease of the load bearing capacity. To this aim, a circular arch subjected to an increasing point load applied at a quarter of the span was analyzed, considering different values of the thermal gradient and different values of the tensile strength of the joints.

Numerical results show that different values of the thermal gradient correspond to different values of the collapse load. However, it does not seem possible to predict a law of variation of the collapse loads as a function of the increase or decrease of the thermal gradient, since different values of the thermal gradient cause different crack patterns which affect the collapse loads. Consequently, it is also not possible to determine a priori whether a positive or negative thermal gradient produces beneficial or harmful effects, as it depends on case by case. For this reason, the development of a numerical model to predict the effect of thermal gradients on masonry arches is needed. The mechanical model proposed here is a first attempt to address this topic and needs to be improved and calibrated in depth.

References

1. Brencich, A., Morbiducci, R.: Masonry arches: historical rules and modern mechanics. Archit. Herit. **1**, 165–189 (2007)
2. Consiglio Superiore dei Lavori Pubblici, Norme Tecniche per le Costruzioni, Italia (2018)

3. Froli, M.: On the eigenstresses of thermal origin in thin walled beams. Costruzioni Metalliche **4**, 229–248 (1993)
4. UNI EN1996-1-2, Eurocode 6: Design of Masonry Structures—Part 1–2: General— Structural Fire Design (2005)
5. Burello, N.S.: Archi in muratura soggetti ad elevati gradienti termici, Politecnico di Torino, B.Sc. Tesis
6. Canevari, E.: Masonry arches in fire conditions, Politecnico di Torino, B.Sc. Tesis (2020)
7. Nadjai, A., O'Garra, M., Ali, F., Laverty, D.: A numerical model for the behaviour of masonry under elevated temperatures. Fire Mater. **27**, 163–182 (2003)
8. Annecchini, A.: Le verifiche di resistenza al fuoco di elementi di arco in condizioni di incendio. http://conference.ing.unipi.it/vgr2016/images/papers/121.pdf
9. Galassi, S., Zampieri, P.: A new automatic procedure for nonlinear analysis of masonry arches subjected to large support movements. Eng. Struct.Struct. **276**, 115359 (2023)

Parametric Modelling in BIM for Structural Analysis of Masonry Vaults

F. Trovatelli[1](✉), A. P. Falcao[2], R. Machete[2], M. Tanganelli[1], and R. Bento[2]

[1] Department of Architecture, University of Florence, Piazza Brunelleschi 6, 50121 Florence, Italy
francesco.trovatelli@unifi.it

[2] CERIS, Instituto Superior Técnico, Universidade de Lisboa, Av. Rovisco Pais, 1049-001 Lisbon, Portugal

Abstract. The use of BIM for historical heritage structures has started to be widely diffused and referred as H-BIM. These models are usually aimed at the management and conservation of monuments; however, these can also contain the information necessary for the characterisation of a numerical structural model. This paper is focused on the interoperability between H-BIM and numerical models applied to masonry vaults, structures that typically require local safety assessments. Different degree of geometry subdivision for structural analysis of masonry vaults, has been identified and related to various numerical modelling approaches and HBIM classification. Based on this, an automatic parametric procedure specific for lunette vaults has been developed. This allows to obtain automatic three-dimensional geometric models that can be efficiently implemented in the H-BIM environment and to obtain and manage the geometric characteristics as input to the numerical modeling. Finally, the application of the methodology to the National Palace of Sintra is presented.

Keywords: Building Information Modeling · Masonry Vaults · Structural analysis · Finite Element Modeling · Parametric modeling

1 Introduction

The BIM methodology applied to historic buildings has reached a high level of diffusion in the international context [1]. H-BIM models, through a centralized database, can provide all the knowledge required for the application of structural analysis methodologies, not only the geometric characteristics, but also the whole spectrum of knowledge and diagnostics [2].

However, BIM software have limitations in the structural field, especially for historic masonry buildings, which are characterized by complex and irregular geometries and large sizes. As a result, simplified models and external software are needed for in-depth structural checks.

S. Gabriele et al. (Eds.): IWSS 2023, LNCE 437, pp. 717–726, 2024.
https://doi.org/10.1007/978-3-031-44328-2_75

In particular, masonry vaults due to their complex geometry represent a challenge in both H-BIM [3] and structural modeling. Those are characterized by damages that typically depend on settlement or seismic actions and require local safety assessments.

Structural analysis of masonry vaults is a widely debated topic in the scientific literature, and various modeling techniques have been proposed [4]. The adoption of an approach and the reliability of the results depend on the level of knowledge acquired. Therefore, it is essential to conduct a specific field campaign to obtain the necessary information for calibrating calculation models, including the historical-constructive characteristics of the parts, mechanical properties of materials, presence of pre-existing crack frameworks, and survey of the masonry pattern.

To address these challenges, a digital modeling methodology for masonry vaults has been proposed specifically for structural analysis.

Three different geometric modelling methods have been identified depending on the grade of subdivision of the structural parts specific to masonry vaults. These are connected to various numerical modelling strategies depending on the objectives of structural assessments required. Based on these, a generative algorithmic modelling for lunetted masonry vaults has been developed. The definition of specific input parameters allows geometric models to be interpreted by both BIM and calculation software, enabling the application of different numerical modelling strategies.

Finally, the procedure has been applied to the case study of the lunetted vault present in the entrance portico of the National Palace of Sintra. The methodology provides a promising approach to digital modeling of masonry vaults for structural analysis, overcoming the limitations of the common tools used in a BIM approach.

2 H-BIM Geometrical Modeling of Complex Historical Structure Background

In the case of historic masonry buildings, BIM modeling typically involves creating a three-dimensional model from a point cloud through the modeling of building elements referred as the Scan-to-BIM procedure.

While data collection and pre-processing systems are becoming increasingly automated, the most significant critical aspect of the Scan-to-BIM procedure is the modeling methods [5]. BIM software designed for the Architecture, Engineering, and Construction (AEC) sector have limited tools for modeling existing and historic buildings that are typically characterized by high complexity on multiple levels, including geometric, material, historical, and construction.

As an alternative to the time-consuming classical modeling methodology, various research groups have proposed several specific alternatives for different objectives. These methodologies can be summarized into three main groups [6]: reverse modeling, direct modeling, and generative modeling. Reverse modeling involves generating a polygonal surface mesh from the processed point cloud, while direct modeling involves extracting a series of profiles from the point cloud and using them as a reference for the creation of complex surfaces, usually NURBS [9]. Generative modeling, on the other hand, involves generating a model through geometric operations algorithms developed in visual scripting environments, based on specific input parameters.

The above strategies lead to non-parametric models but partially solve the problem of geometric complexity. Generative modelling using visual scripting environments such as Dynamo or Grashopper is growing in popularity [9], particularly in the case of masonry vaults. Algorithmic modeling provides the advantage of automatically generating a geometric model specific to the individual object, and with appropriate adaptations, these models can be used to conduct structural analyses through the finite element method.

3 The Degree of Geometry Subdivision

The development of a BIM model for a historical building requires a preliminary assessment to define the objectives of the model to be realised, depending on the model's aims and available resources. This can be done by targeting a specific level of detail (LOD) that the model requires and the level of information (LOI) that the object contains or is connected to [10].

Similarly, as mentioned above, structural analyses of masonry vaults require models that can be characterized by varying degrees of geometric complexity.

Typically, the most significant structural components are the vault shell, the presence of backfill and upper infill, the masonry pattern, and the potential presence of ribs. Based on these, three different degrees of subdivision (DoS) of the geometry with increasing complexity have been identified: zero, 1, and 2 (see Fig. 1). These start with the degree zero (DoS0), where the geometry is represented as a unique generic volume, and end with degree 2 (DoS2), where the vault is divided into the blocks and all construction materials.

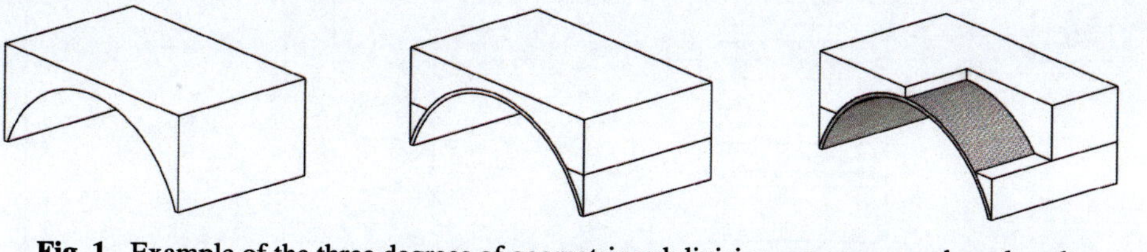

Fig. 1. Example of the three degrees of geometric subdivision on a masonry barrel vault.

Each degree of subdivision is better suited for different analyses and modeling strategies. For instance, a model with a DoS0 is intended for use in global analysis models, where the accuracy of the object and the differentiation of the parts is less significant. In contrast, a model with a DoS1 is subdivided into the most significant structural elements. This model of intermediate geometric simplification is useful in evaluating the local behavior of the vault in relation to the context while considering external connections appropriately. Finally, the DoS2 involves fine modeling, where the geometry is discretized into the stone blocks that compose the vaulted system. This aims at local evaluations of the complete vaulted system or portions requiring specific in-depth studies.

The different degrees of subdivision can be referred to as different levels of detail (LOD) in the HBIM environment. For example, a model with a DoS0 can represent a LOD100. On the other hand, a model with a DoS2 may be useful for higher LODs where it is necessary to catalog individual construction elements for restoration operations (e.g., degradation maps, material maps, etc.).

4 A Methodology for Parametric Modeling of Masonry Vaults

The different DoS discussed in the previous section can be achieved through various modelling methodologies and extended to most kinds of vaulted structures. While BIM tools can be used to achieve a geometry with a DoS0, the other degrees, require more advanced modelling approach, such as generative modelling. This approach covers a broad range of geometric complexities and enables the development of models based on different solutions.

A methodology employing automatic generative modeling procedures for masonry vaulted structures has been proposed (see Fig. 2). The main objective of the procedure is to obtain geometric models intended for two different purposes: one designed for the BIM environment and the other for structural analyses. Generative geometry allows for the processing of both models in parallel, calibrating geometries specifically for BIM objectives and various numerical modelling strategies. The finite element method (FEM) has been adopted as the analysis method.

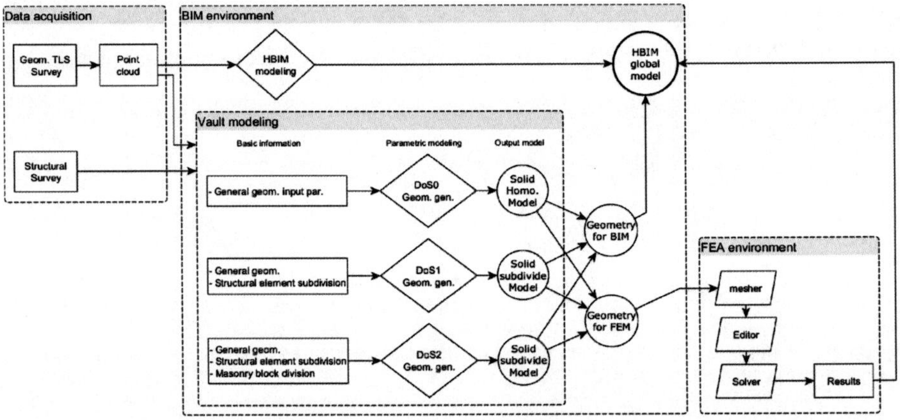

Fig. 2. Proposed methodology for BIM and structural modelling of masonry vaults.

In summary the methodology involves three stages: i) Data acquisition: specific input parameters derived from geometric and structural survey operations, should be extracted from the point cloud through local measurements or automatic procedures [11]. ii) Vault modelling: based on the input parameters, the vault is modelled using specific geometry generator algorithms that provide two outputs – one for the global HBIM model and one for structural analysis. Iii) FEA: the model is processed for discretization, calibration, and analysis procedures by external software.

4.1 A Generative Modelling Procedure for Lunette Masonry Vaults

Based on the discussed methodology a generative algorithmic modelling approach specific for lunette masonry vaults aimed to obtain models of DoS1 has been developed. Lunette vaults are characterised by a main vault, usually barrel or pavilion, with openings on the sides (lunettes) that provide natural light to interior spaces but constitute structural weak elements.

Dynamo [12], the Revit's [13] internal visual scripting environment, has been adopted. The developed algorithm consists of the following phases: definition of input parameters, definition of plan guidelines, geometry generation, assembly, and output (Fig. 3).

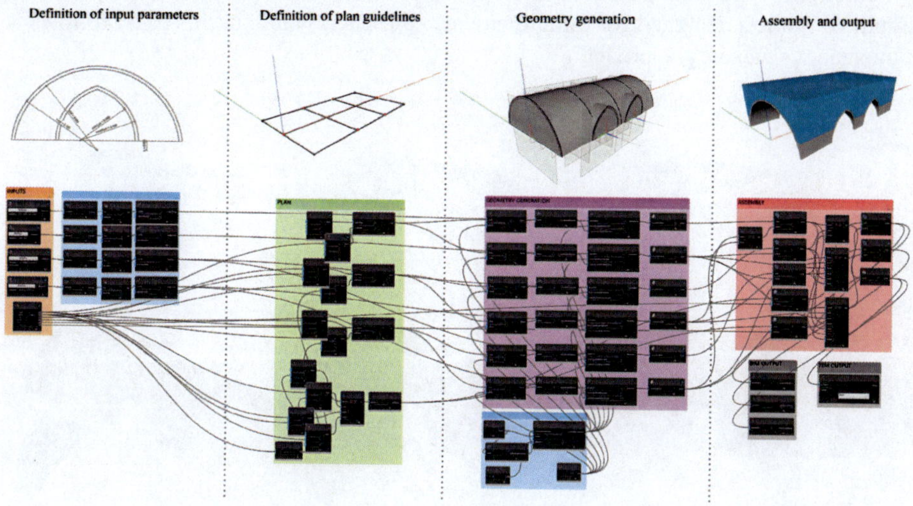

Fig. 3. Dynamo code developed to achieve a DoS1 on a generic lunette vault model.

The extrusion profiles of the main vault and lunettes need to be defined as input parameters. These can be defined parametrically through code, directly in Dynamo, or using elements drawn in the BIM environment. In the latter case, the possibility of adopting different vault profiles is extended. Dimensional parameters are also required for the definition of vault and lunette dimensions, positions of the extrusion axes, height of the backfill and height of the floor.

The procedure uses tools such as extrusion in specific directions of profiles to create solid or shell geometries. These are assembled through various Boolean operations. This methodology, although simplified, allows adaptation to various case studies and the possibility of easily modifying blocks of code according to different needs.

The use of Dynamo allows an easy exchange of information with the Revit model. The geometry can be easily exported and transferred to the comprehensive H-BIM model as a non-parametric object that is subdivided into the required structural part. The HBIM model works as centralized database, collecting not only geometric information as well as reports, related to surveys conducted on site. These models can also be used to aggregate other historic information pertaining to the site and will inform future interventions and analyses. Regarding the structural model, it is possible to transfer geometries from Dynamo to the commercial finite element analysis software Abaqus [14] through the SAT extension; a spatial ACIS solid modeling format.

Finite element models require geometries without intersections or overlaps between parts. The use of generative procedures in this context enables the control of part definition and avoid geometric definition problems typically encountered in direct export from BIM software. In addition, it makes it possible to manage all the model synthesis and simplification operations necessary for the development of structural models. As an example, Fig. 4 shows several geometric models developed using a generative procedure and discretized into various types of finite elements adopting the Abaqus meshing tool. It is noteworthy that minor modifications to the code allow the parallel development of simplified models designed for finite element modelling, which in this case, follows a homogeneous masonry approach.

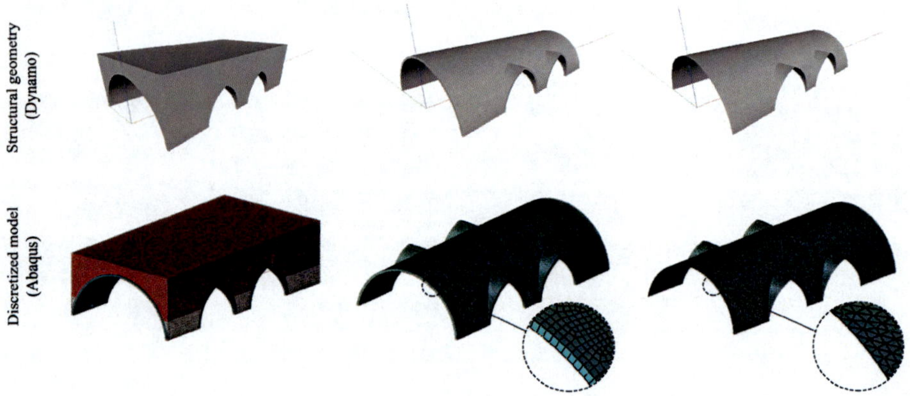

Fig. 4. Example of a DoS1 lunette vault model and different FEM modelling strategies connected.

5 Case Study: Sintra National Palace

The Palace of Sintra is one of most significant royal residence of Portugal. It is an unreinforced masonry building that has been subject to a series of architectural transformations over the centuries, that have conferred a high geometric and material complexity that distinguishes this fascinating structure.

The palace has been the subject of several laser scanning survey and diagnostic campaigns [15] that led to the creation of an HBIM model of the entire complex (see Fig. 5a) [16]. The lower floors of the building present various vaulted masonry structures. The most significant for its size and complexity is the lowered barrel lunetted vault that covers the entrance of the main façade (see Fig. 5b).

The vault is 24.3 m long and 3.5 m span; from the arch impost to the upper floor, it measures 2.8 m. The lunettes have a gothic arch section and cross the main barrel only

a) b)

Fig. 5. a) The BIM model rendered in twin motion. b) The lunetted vault case study.

on the inner side. These allow internal opening to some basement rooms that border the vault.

To model the vault, the generative procedure described above has been adopted. Necessary input parameters have been extracted from the point cloud and structural survey operations, including non-destructive ground penetrating radar (GPR) surveys [15] to assess the dimension of non-detectable internal structural elements.

The Dynamo script has been easily adapted to the case study and the resulting geometric model, divided into the structural shell of the vault, backfill, and infill, has been imported into the BIM environment and classified as different structural elements with different characteristics (see Fig. 6a).

As discussed in Sect. 3, utilizing an DoS1 allows for various geometry modeling strategies based on the specific assessments conducted, whether global or local in nature. In this case, a study has been conducted on the structural behavior of the vaulted structure in relation to a vertical settlement of one of the façade pillars. Although the vault currently exhibits no significant crack pattern, the vault-facade system remains a critical area of the building, as demonstrated by its previous collapse during the famous 1755 Lisbon earthquake [17, 18].

A generative process has been used to create a model of the structural portion of the vault as a solid element. This has been combined with the simplified facade portion exported from the global HBIM model.

Numerical modeling has been performed using Abaqus, with a homogenous macro modeling approach representing the masonry as a continuum material. The numerical model has been discretised in 95,292 tetrahedral four-nodes elements (C3D4) of 0.4 m length for the façade and 0.2 m for the vault. The discretised model is shown in Fig. 6b. Contact faces with adjacent structures have been considered as fixed. The backfill and fill have been considered as loads.

The capacity of the masonry structure in presence of a vertical settlement of the façade pillar has been assessed considering the non-linear behavior of the masonry.

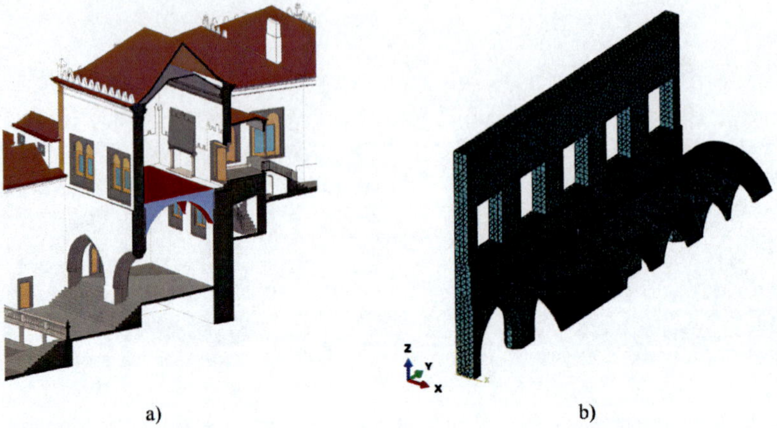

Fig. 6. a) The Vault inserted in the global HBIM model. b) The finite element model of the vault-façade system analysed.

Two different materials have been assigned to the façade and the vault. Mechanical parameters of the facade masonry have been estimated through flat jacks test as reported in [15]. Regarding the vault masonry, in absence of specific investigations, the average values suggested by the Italian standard for the masonry class "solid brick and lime mortar masonry" have been adopted [19].

The nonlinear behaviour has been represented through the concrete damaged plasticity model (CDP), which is based on an isotropic material with a plasticity-damage relation that assigns distinct tensile and compressive strength (f_t and f_c) and different post-peak behaviors with damage parameters (d_c and d_t). Tensile softening has been described through a stress-displacement relation, with an appropriate fracture energy (*Gft*) value calculated according to [20]. Damage parameters d_t and d_c follow a linear evolution law, where the ultimate condition corresponds to 90%. Mechanical parameters adopted for each material are summarized in table 1.

Table 1. Mechanical parameters adopted.

	Young's modulus [MPa]	*Poisson [-]*	*Density [kg/m^3]*	*fc [MPa]*	*ft (fc/20) [MPa]*	*Gft [N/m]*
Façade	2550	0.29	2000	4.04	0.20	13.2
Vault	1400	0.20	1800	3.45	0.18	12.2

The findings of the analysis for a vertical subsidence intensity of 1.3 cm are presented in Fig. 7. The results reveal that the vault and the façade experience localized damage that is linked to the two lateral gothic arches in this configuration. Moreover, damage is observed at the upper floor openings as well.

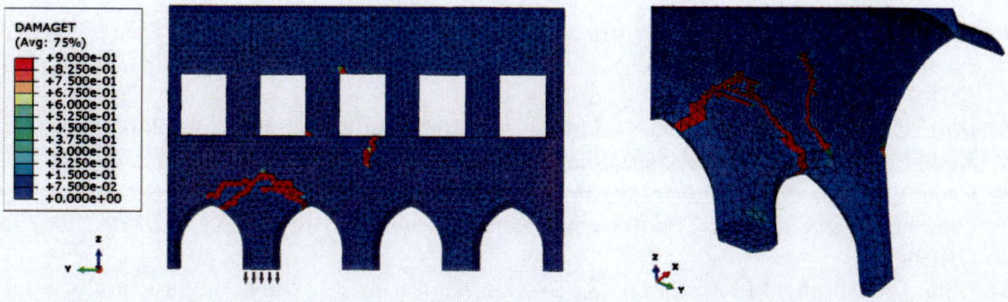

Fig. 7. Tensile damage for a vertical settlement of 1.3 cm intensity of the first pillar.

6 Conclusions

The research presented represents an initial step towards achieving interoperability between BIM environment and structural analysis, for unreinforced masonry structures.

A methodology has been proposed for creating geometries of masonry vaulted structures suitable for both H-BIM and structural analysis, with different levels of detail and application possibilities defined. The generative modeling procedure allows for the generation of geometric models for use in H-BIM and the application of the finite element method.

From experience gained in this research, it can be concluded that the use of generative modeling for masonry vaults allows for the efficient management of both H-BIM and structural models. The proposed procedure for lunette vaults leads to simplified models with lower geometric accuracy compared to direct modeling procedure. However, by appropriately setting the input parameters, a sufficient level of geometric accuracy can be achieved for structural analysis purposes. Moreover, the automation of the process and the parametrization of structural elements allow adaptation of the codes to different case studies, reducing geometric modeling time in both environments. The different classes divided by geometric complexity offer the possibility of applying different structural modeling strategies for vaults. In particular, the procedure demonstrated good interoperability by adopting a finite element macro-modeling approach considering the masonry as a homogeneous material. Future applications could extend the methodology to other numerical modeling strategies, such as FEM micro-modeling or discrete element modeling (DEMs).

References

1. Ewart, I.J., Zuecco, V.: Heritage building information modelling (HBIM): a review of published case studies. In: Mutis, I., Hartmann, T. (eds.) Advances in Informatics and Computing in Civil and Construction Engineering, pp. 35–41 (2019)
2. Cardinali, V., et al.: An oriented H-BIM approach for the seismic assessment of cultural heritage buildings: Palazzo Vecchio in Florence. Buildings **13**(4), 913 (2023)
3. Barazzetti, L., Banfi, F., Brumana, R., Gusmeroli, G., Previtali, M., Schiantarelli, G.: Cloud-to-BIM-to-FEM: Structural simulation with accurate historic BIM from laser scans. Simul. Model. Pract. Theory **57**, 71–87 (2015)

4. D'Altri, A.M., et al.: Modeling strategies for the computational analysis of unrein-forced masonry structures: review and classification. Arch. Comput. Methods Eng. **27**(4), 1153–1185 (2020)
5. Dore, C., Murphy, M.: Current state of the art historic Building Information Modelling. Int. Arch. Photogramm. Remote Sens. Spat. Inf. Sci. **XLII-2/W5**, 185–192 (2017)
6. Capone, M., Lanzara, E.: Scan-to-BIM vs 3D ideal model HBIM: parametric tools to study domes geometry. Int. Arch. Photogramm. Remote Sens. Spat. Inf. Sci. **XLII-2/W9**, 219–226 (2019)
7. Dore, C., Murphy, M., McCarthy, S., Brechin, F., Casidy, C., Dirix, E.: Structural simulations and conservation analysis – historic building information model (HBIM). Int. Arch. Photogramm. Remote Sens. Spat. Inf. Sci. **XL-5/W4**, 351–357 (2015)
8. Barazzetti, L.: Parametric as-built model generation of complex shapes from point clouds. Adv. Eng. Inform. **30**(3), 298–311 (2016)
9. Angjeliu, G., Cardani, G., Coronelli, D.: Digital modelling and analysis of masonry vaults. Int. Arch. Photogramm. Remote Sens. Spat. Inf. Sci. **XLII-2/W11**, 83–89 (2019)
10. Brumana, R., Banfi, F., Cantini, L., Previtali, M., Della Torre, S.: HBIM level of detail-geometry-accuracy and survey analysis for architectural preservation. Int. Arch. Photogramm. Remote Sens. Spat. Inf. Sci. **XLII-2/W11**, 293–299 (2019)
11. Angjeliu, G., Cardani, G., Coronelli, D.: A parametric model for ribbed masonry vaults. Autom. Constr. **105**, 102785 (2019)
12. Dynamo, an Open-Source Computational Design Platform. dynamobim.org
13. Autodesk Revit, BIM Software. www.autodesk.com/products/revit/
14. ABAQUS, Theory and user's manuals. Hibbit, Karls son and Sorensen, Pawtucket (2018)
15. Ponte, M., Bento, R., Vaz, S.D.: A multi-disciplinary approach to the seismic assessment of the national palace of Sintra. Int. J. Archit. Herit. **15**(5), 757–778 (2021)
16. Godinho, M., Machete, R., Ponte, M., Falcão, A.P., Gonçalves, A.B., Bento, R.: BIM as a resource in heritage management: an application for the National Palace of Sintra. Portugal. J. Cult. Herit. **43**, 153–162 (2020)
17. I.A.N.T.T., Almoxarifado dos Reais Paços de Sintra (1784–87). In: Núcleo Extraído do Conselho da Fazenda, Casa das Obras e Paços Reais, no 91, 92 e 93, cx.21
18. Mendonça, J.J.M.: História Universal dos Terramotos, Lisboa, p.150 (1758)
19. Circ. n. 7 del 21/01/2019. MIT. Istruzioni per l'applicazione dell'Aggiornamento delle Norme Tecniche per le Costruzioni di cui al decreto ministeriale 17 gennaio 2018, Rome, Italy
20. Lourenço, P.B.: Recent advances in masonry modelling: micromodelling and homogenization. In: Computational and Experimental Methods in Structures, pp. 251–294. Imperial College Press (2009)

Geometry-Driven Stock-Constrained Truss Design via Equilibrium-Based Structural Models

Tao Sun[1,2(✉)], Marina Konstantatou[2], Corentin Fivet[3], and Pierluigi D'Acunto[1]

[1] Professorship of Structural Design, TUM School of Engineering and Design, Technical University of Munich, Accisstr. 21, 80798 Munich, Germany
tao.sun@tum.de
[2] Specialist Modelling Group, Foster + Partners, Battersea, UK
[3] Swiss Federal Institute of Technology Lausanne (EPFL), Lausanne, Switzerland

Abstract. This paper presents a geometry-driven approach to form-finding with reused stock elements. Our proposed workflow uses a K-mean algorithm to cluster stock elements and incorporate their geometrical values early in the form-finding process. A feedback loop improves the reuse rate over multiple iterations, and a best-fit heuristic algorithm is used to examine the reuse states. A detailed example and a case study demonstrate the efficacy of the method. This results in a form-finding method with reused stock elements; hence, promoting sustainable design and reducing environmental impact in the built environment.

Keywords: graphic statics · form-finding · reuse · pin-jointed structure · machine learning

1 Introduction

1.1 Environmental Impact of the Built Environment and Circular Economy

The construction industry has a significant adverse environmental impact with its embodied carbon [1]. The United Nations Environment Programme (UNEP) has identified sustainability in the built environment as a critical issue for addressing climate change [2]. As such, there is a growing demand for sustainable design approaches that reduce the environmental impact of buildings and infrastructure. To end this, the concept of the circular economy [3] has been introduced. A circular economy aims to prolong the commission of the product throughout multiple life cycles [4]. In opposition to recycling, which is the common strategy to reprocess obsolete built materials, reuse can further reduce the environmental impact by demanding less energy consumption for reprocessing [5]. However, this is not a blanket conclusion. To examine if the impact reduction by reusing outweighs the one by recycling, a life cycle assessment (LCA) should be carried out [6].

S. Gabriele et al. (Eds.): IWSS 2023, LNCE 437, pp. 727–736, 2024.
https://doi.org/10.1007/978-3-031-44328-2_76

1.2 Reuse in Structural Design

Incorporating reused components in structural design can reduce the demand for new materials and optimize the use of existing resources. However, this also introduces unique challenges since members' geometrical and mechanical properties of the original construction will influence the resulting new structure [7]. Hence, this process can be seen as a reverse of a conventional structural design approach [8]. Several studies have focused on reuse in structural design: in 2017, Gorgolewski et al. Presented the concept of "Form follows Availability" [12]. Brütting et al. Introduced a method for designing structures through reuse by matching existing stock elements to the desired form [8]; they subsequently investigated the environmental impact minimization of reticular structures through reuse and optimization [6]; and then combined these methods with form-finding to develop a stock-constrained truss design workflow [13]. However, in these research studies, reuse occurs in the later stages of the design process, which may not fully capitalize on the potential of considering stock element characteristics at the early design phase.

1.3 Proposed Approach

This research paper proposes a geometry-driven approach that introduces a K-mean algorithm for clustering stock elements. This is used to inform the form-finding process by incorporating geometrical features of the stock in the early stage of the design process. The workflow gradually introduces more clusters through a feedback loop, improving the reuse rate over multiple iterations before eventually stabilizing it at a maximum level. The approach also incorporates Phoenix3D [14], a toolkit for reusing structural members using a best-fit heuristic algorithm, to examine and visualize the states of reusing.

1.4 Structural Form-Finding

The proposed approach makes use of the Combinatorial Equilibrium Modelling (CEM) method for form-finding purposes. CEM is a form-finding method [9] and a ready-to-use computational toolkit based [10] on vector-based graphic statics [11]. This method allows for designing structural systems by representing the equilibrium of forces through geometric relationships and user-defined constraints [15]. In the CEM, the edges of a structure are categorized as trail edges or deviation edges. A trail is a sequence of trail edges connected at nodes that do not form a loop. It represents the shortest load path for external forces applied to nodes to reach the closest support. A deviation is an edge connecting nodes on two different trails without being part of either of them. The designer assigns a combinatorial state of tension or compression for each trail and deviation edge, in addition to metric properties such as edge lengths for trail edges and internal force magnitudes for deviation edges. Using this information, an equilibrium state of the structure is established [9].

1.5 Design of Optimum Truss Structures from New and Reused Components

Previous research on stock-constrained form-finding by Brütting et al. Has consolidated into Phoenix3D, an interactive computational tool designed to create structures from a stock of reclaimed components and new components. It incorporates two methods for stock-constrained design: 1) discrete optimization based on Mixed-Integer Linear Programming (MILP), and 2) a heuristic approach. Both methods are combined with Life Cycle Assessment to design structures with the least environmental impact ***EI***. This measure encompasses the life cycle assessment of its components, starting from sourcing materials through manufacturing and all the way to assembly, including transportation. ***EI*** is expressed in terms of greenhouse gas emissions measured in units of equivalent carbon (kgCO_2eq) [14].

2 Method

Our proposed workflow has two parallel sequences for stock members and new structural elements and consists of six main steps, as shown in the flow chart (see Fig. 1). To illustrate the method, we will use the stock-constrained form-finding of a dome as a simple example. Note that in this paper, the term *element* refers to the bars from the stock, while the term *member* refers to the virtual bars in the form-found structure.

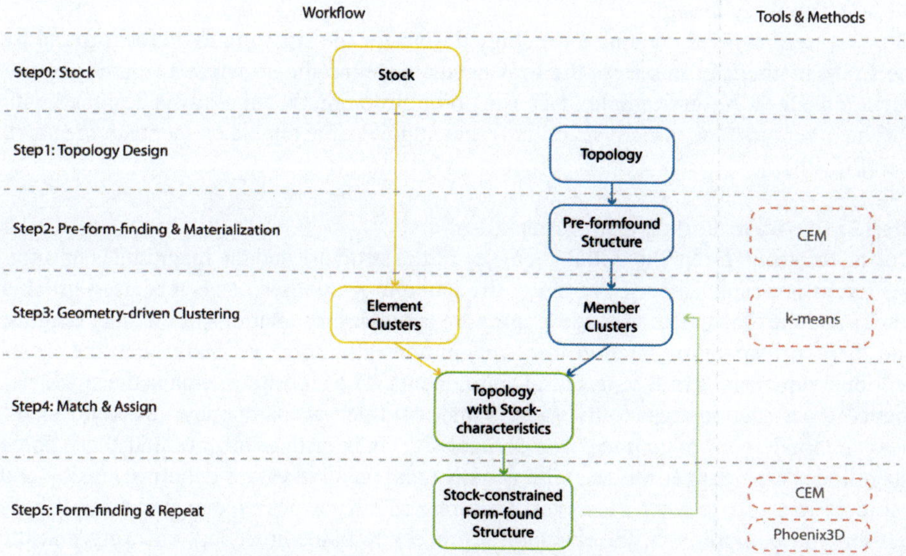

Fig. 1. Flowchart of the proposed method.

2.1 Design Workflow

Step 0: Definition of the Stock
Since this approach is considered an extension of the previous work of Brütting et al., the example is also inspired by one of their case studies for consistency.

The transmission towers of Swissgrid AG are to be decommissioned, resulting in a stock of 565 members per tower, coloured for different cross-sections (see Fig. 2a).

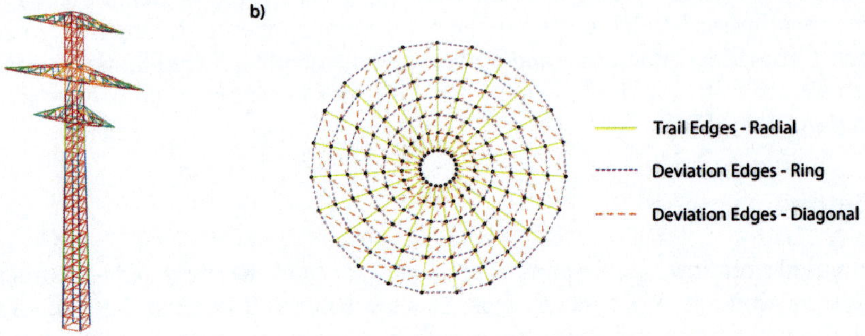

Fig. 2. a) Stock; b) Topology.

Step 1: Topology Design
The first step involves creating a topology design for the structure to be form-found by the CEM method. In this step, the user needs to define the structure's trail edges and deviation edges. As an example of a basic dome structure, the topology is designed with 506 members in total, which is similar to the number of available elements in the stock (see Fig. 2b).

Step 2: Pre-form-finding and Materialisation
During the second step, the initial geometry of the structure and the magnitudes of internal forces are established. Leveraging the inherent properties of vector-based graphic statics and the CEM method, edge lengths are actual representations of member lengths, and their corresponding internal force magnitudes align proportionally with the cross-sectional dimensions of the structural components. The algorithm remaps the stock elements' cross-section areas to the same numerical intervals as the new structural members' internal force magnitudes for comparison. It is critical to note that these force magnitudes carry signed values, with negative and positive values denoting tension and compression, respectively. However, assuming that the stock elements can resist both tension and compression, the remapping process remains unsigned, ensuring that the dimensioning of elements is solely dependent on their inherent capacity. In this step, the effect of buckling is not considered.

Step 3: Geometry-Driven k-means Clustering
In the third step, the stock elements and the structural members are separately clustered into the same number of groups using a k-means algorithm. This is only based on their two geometrical features: their lengths and cross-sectional areas (determined by force magnitudes). The algorithm requires the input of dataset $\boldsymbol{E}$ containing the length $\boldsymbol{len}$ and the cross-sectional areas $\boldsymbol{cs}$, the number of clusters $\boldsymbol{n_k}$. The output $\boldsymbol{I_k}$ is a list of the same length as $\boldsymbol{E}$ that consists of the clustering result. The algorithm initializes $\boldsymbol{n_k}$

cluster centroids randomly, assigns each data point in $\boldsymbol{E}$ to its nearest centroid based on the distance metric, and recalculates the centroids based on the current data points assigned to that cluster. This process is repeated until the centroids no longer change significantly, fulfilling the convergence criterion (see Fig. 3). After the clustering, the algorithm will then compute each cluster's geometric characteristics by calculating the mean length and force magnitude of all the members/elements within every cluster.

Input: $\boldsymbol{E}$ (a dataset containing length ***len*** and cross-section ***cs*** of stock elements)
$\boldsymbol{n_k}$ (number of clusters)
Output: $\boldsymbol{I_k}$ (a list of indicex indicating to which cluster $\boldsymbol{e_i}$ in $\boldsymbol{E}$ belongs to)

1. Initialize:
C_k <- a list of $\boldsymbol{n_k}$ randomly initialized centroids ($\boldsymbol{len_{random}}$, $\boldsymbol{cs_{random}}$)
$\boldsymbol{I_k}$ <- an empty list of size $|\boldsymbol{E}|$

2. Repeat until convergence:
2.1 For each data point $\boldsymbol{e_i}$ in $\boldsymbol{E}$:
2.1.1 Calculate the distance $\boldsymbol{d_{1\dots n}}$ between $\boldsymbol{e_i}$ and $c_{1\dots n}$ in C_k
2.1.2 Find the closest c_j to $\boldsymbol{e_i}$ by comparing $\boldsymbol{d_{1\dots n}}$
2.1.3 Assign $\boldsymbol{e_i}$ to the closest c_j and update $\boldsymbol{I_k}$ accordingly

2.2 For each c_i in C_k:
2.2.1 Calculate the new centroid position based on the current $\boldsymbol{I_k}$
2.2.2 Update C_k

3. Return $\boldsymbol{I_k}$

Fig. 3. Pseudo code of the k-mean algorithm.

Step 4: Match and Assign

In this step, a custom match and assign algorithm is implemented to heuristically find the best possible correspondence between clusters of members and elements by comparing the Euclidean distance between them. The algorithm operates as follows:

1. The input data includes the geometric characteristics of the stock element clusters $\boldsymbol{C_e}$; geometric characteristics of the structural member clusters $\boldsymbol{C_m}$; the clustering result indicating to which cluster a member belongs $\boldsymbol{I_k}$; the features of members attained from the pre-form-finding $\boldsymbol{F}$.
2. $\boldsymbol{C_e}$ is sorted in descending order based on the absolute value of the stock element's force magnitude so that the elements with higher structural demand will find a match first.
3. $\boldsymbol{C_m}$ is sorted in descending order with each index attached forming a tuple into a list $\boldsymbol{TC_m}$ based on the structure members' cross-section so that each set of features in $\boldsymbol{C_m}$ remains indexable by their original indices before sorting.
4. The algorithm initializes an empty list for storing added indices, a list of assigned force magnitudes ***Mag***, and a list of assigned edge lengths ***Len*** as well as a dictionary $\boldsymbol{M}$ to store the match result between stock elements and structure elements clusters.
5. For each set of features $\boldsymbol{tc_{mi}}$ in $\boldsymbol{TC_m}$ and $\boldsymbol{c_{ei}}$ in $\boldsymbol{C_e}$, the algorithm calculates the Euclidean distance $\boldsymbol{d}$ between them. If $\boldsymbol{d}$ is smaller than the current minimum distance $\boldsymbol{d_{min}}$ and the index $\boldsymbol{j}$ of $\boldsymbol{c_e}$ is not in the list of added indices, $\boldsymbol{d_{min}}$ is updated, and $\boldsymbol{j}$ is stored.

6. The match is found in the case of $\boldsymbol{d_{min}}$. And then $\boldsymbol{i}$, the index of $\boldsymbol{tc_{mi}}$ and the corresponding $\boldsymbol{j}$ is added to the match dictionary (see Fig. 4). By sequentially storing the corresponding values into ***Mag*** and ***Len*** based on ***M***, the clustered stock characteristic values are assigned to the structure members for the next step.

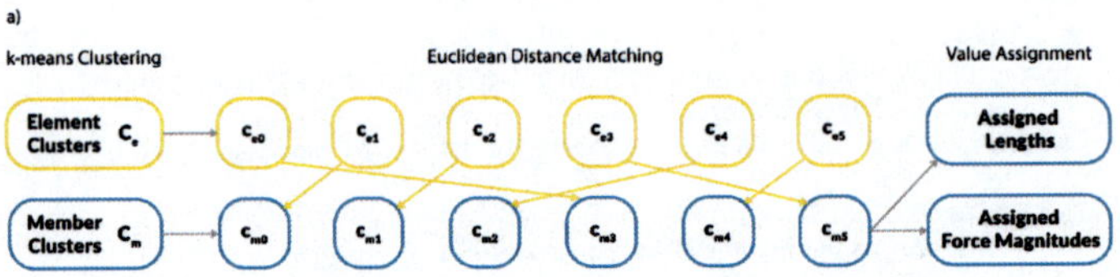

Fig. 4. Diagram of the match and assign algorithm.

Step 5: Form-Finding and Iteration

The final step is the actual form-finding with CEM. The values, including the edge length and the edge force magnitudes, are read out from the assignment lists attained in the previous step, and then they are used as input for CEM form-finding. The new form-found structure is then examined with the same stock by the toolkit Phoenix3D and then used as the input for further iterations through the loop until the convergence criteria are reached. In our example, the form-finding ended after 4 interactions, resulting in a reduction of the environmental impact from 2722 $kgCO_2eq$ to 2041 $kgCO_2eq$ and an increase in the reuse rate from 65.9% to 88.1%. (see Fig. 5).

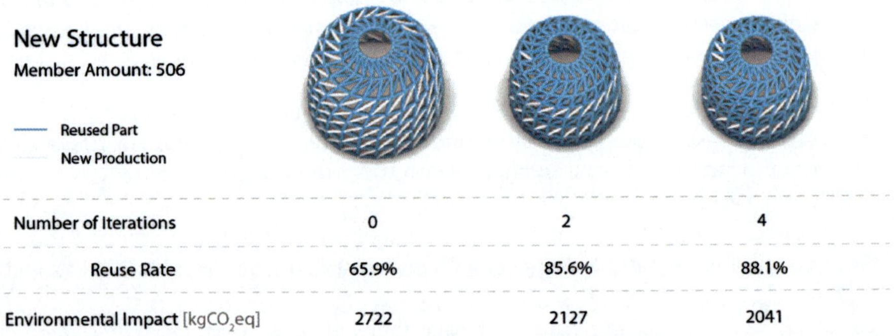

Fig. 5. Form-finding iterations and results of the dome.

2.2 Feedback Loop and Gradual Incrementation of the Number of Clusters

By integrating a feedback loop into the workflow and incrementally increasing the number of clusters, we establish an iterative procedure that enhances adaptive capacity and flexibility through continuous refinement. The main idea behind this approach is to guide the form-finding with the geometrical characteristics of the stock elements by coarse-grained clustering of the stock elements at the beginning so the process is not overwhelmed by the largely various stock elements. Consequently, as the number of clusters increases through each iteration, it is observed how the design space becomes more and more refined and better incorporates the various characteristics of the stock elements. This approach also enhances computational efficiency. Increasing granularity can help manage computational complexity, as smaller cluster sizes are computationally less demanding. This allows the designers to quickly attain an intuition of the potential result of the stock-constrained form-finding and then refine the design via more iterations only when desired.

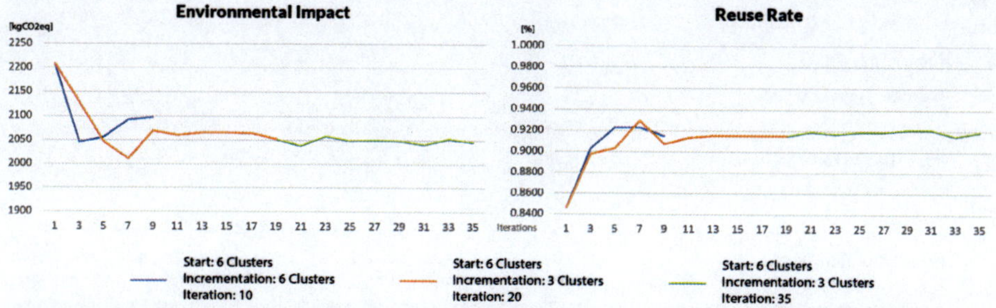

Fig. 6. Line charts for comparing different configurations of the number of clusters and incrementation.

The key thus lies in the number of iterations and the size of incrementation. In our example, the process starts with 6 clusters and increments 6 in each of the 4 iterations. (Fig. 6) In our experiments, we observed that smaller incrementation with more iterations yields similar results to larger incrementation with fewer iterations, indicating a trade-off between granularity and computational effort. Both reuse rate and environmental impact stabilize after a certain number of iterations, suggesting convergence towards an optimized solution within the scope of the proposed method.

3 Case Study

This brief case study aims to explore the application of the method proposed earlier in Sect. 2 for form-finding a funnel-shaped pavilion. Different from the first example, the triangulation is added after form-finding and it is not included in the topology diagram. This structure contains 736 members, which notably exceeds the number of available stock elements by 30% (see Fig. 7a):

The objective of this case study is to minimize the environmental impact ***EI*** (Sect. 1.5). Despite the challenge posed by the increased number of members relative to stock elements, the implemented workflow successfully induced a consistent reduction in ***EI*** throughout each iteration. The results of this study indicate the effectiveness and adaptability of the proposed method under the constraints of stock availability. Indeed, our workflow facilitated a 6.9% increase in the reuse rate coupled with a substantial reduction in ***EI*** by 962 kgCO_2eq. (see Fig. 7b).

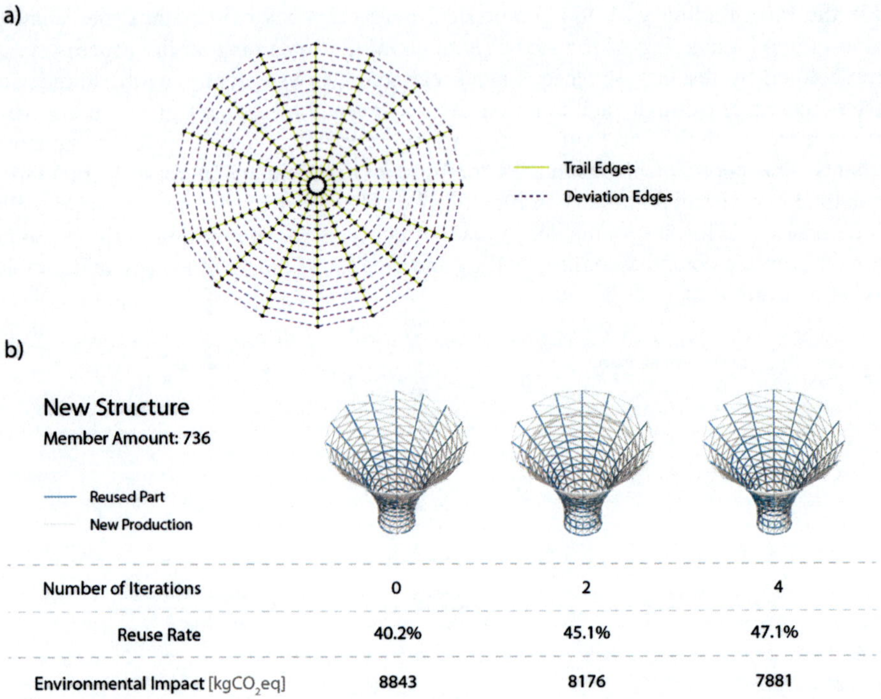

Fig. 7. a) Topology; b) Form-finding iterations and results of the pavilion.

4 Conclusion

By incorporating stock element characteristics into the form-finding process and iteratively refining the design through a feedback loop, our method enables the reuse of materials and components, reducing the environmental impact of structures. The form-finding example in Sect. 3, as well as the structure in Sect. 4 (see Fig. 8), demonstrate the effectiveness of the proposed geometry-driven approach. However, there are limitations and challenges associated with this approach, which also define a range of potential topics for future investigation:

- The proposed approach requires an initial number of clusters, and this method currently initializes with an arbitrary hard-coded number, which may not be optimal for every case. Future research could explore methods for dynamically determining the initial number of clusters based on certain stock characteristics.
- The proposed method frequently falls short of achieving a 100% reuse rate once the results stabilize. To optimize the reuse rate, it is imperative to complement this approach with another method, such as the listing provided in [9].
- The examples show that the final geometry could differ greatly from the initial geometry. Better control of the geometry during the form-finding process is necessary.

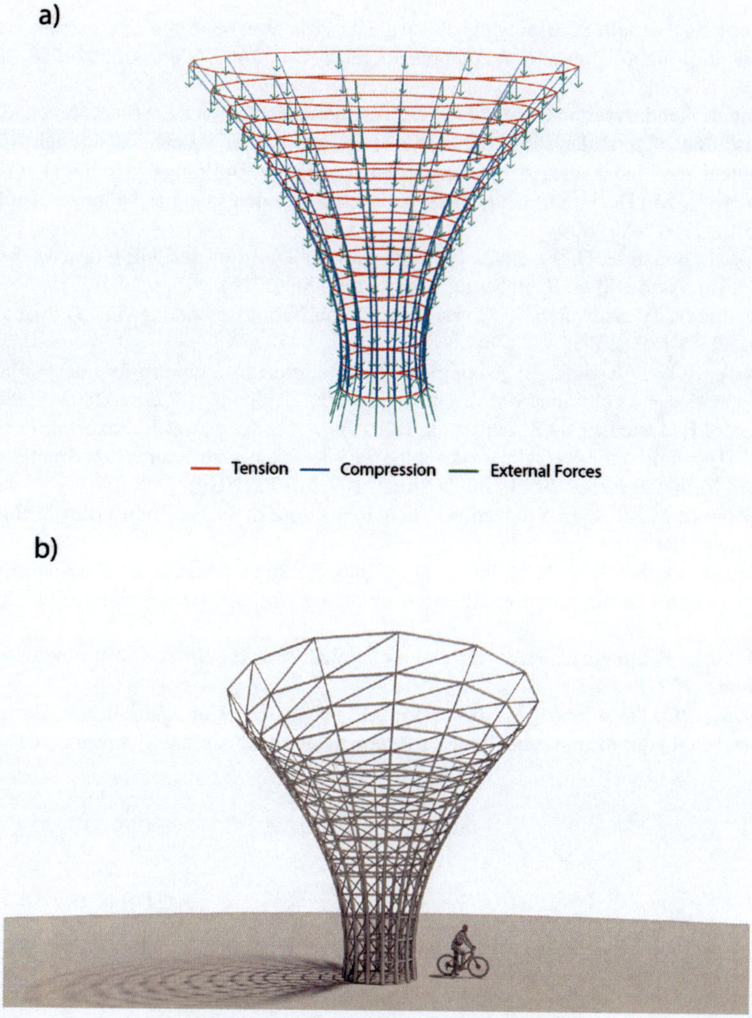

Fig. 8. a) Structural model; b) Visualization

References

1. Dräger, P., Letmathe, P.: Value losses and environmental impacts in the construction industry – Tradeoffs or correlates? J. Cleaner Prod. (2022)
2. UN Environment. https://www.unep.org/news-and-stories/press-release/co2-emissions-buildings-and-construction-hit-new-high-leaving-sector. Accessed 12 Mar 2023
3. Sartori, I., Hestnes, A.G.: Energy use in the life cycle of conventional and low-energy buildings: a review article. Energ. Buildings **39**(3), 249–257 (2007)
4. Brütting, J., Desruelle, J., Senatore, G., Fivet, C.: Design of truss structures through reuse. Structures **18**, 128–137 (2019)

5. Iacovidou, E., Purnell, P.: Mining the physical infrastructure: opportunities, barriers and interventions in promoting structural components reuse. Sci. Total. Environ. **557–558**, 791–807 (2016)
6. Brütting, J., Vandervaeren, C., Senatore, G., Temmerman, N., Fivet, C.: Environmental impact minimization of reticular structures made of reused and new elements through life cycle assessment and mixed-integer linear programming. Energ. Buildings **215**, 109827 (2020)
7. Gorgolewski, M.: Designing with reused building components: some challenges. Build. Res. Inf. **36**(2), 175–188 (2008)
8. Brütting, J., Senatore, G., Fivet, C.: Form follows availability – designing structures through reuse. J. Int. Assoc. Shell Spat. Struct. **60**(4), 257–265 (2019)
9. Ole Ohlbrock, P., Schwartz, J.: Combinatorial equilibrium modeling. Int. J. Space Struct. **31**(2–4), 177–189 (2016)
10. Ohlbrock, P.O., D'Acunto, P.: A computer-aided approach to equilibrium design based on graphic statics and combinatorial variations. Comput. Aided Des. **121**, 102802 (2020)
11. D'Acunto, P., Jasienski, J.-P., Ohlbrock, P.O., Fivet, C., Schwartz, J., Zastavni, D.: Vector-based 3D graphic statics: a framework for the design of spatial structures based on the relation between form and forces. Int. J. Solids Struct. **167**, 58–70 (2019)
12. Gorgolewski, M.: Resource Salvation: The Architecture of Reuse. John Wiley & Sons Ltd., Chichester (2018)
13. Brütting, J., Ohlbrock, P.O., Hofer, J., D'Acunto, P.: Stock-constrained truss design exploration through combinatorial equilibrium modeling. Int. J. Space Struct. **36**(4), 253–269 (2021)
14. Warmuth, J., Brütting, J., Fivet, C.: Computational Tool for Stock-Constrained Design of Structures (2021)
15. Ohlbrock, P.O., D'Acunto, P., Jasienski, J.P., Fivet, C.: Constraint-driven design with combinatorial equilibrium modelling. Proceedings of IASS Annual Symposia (2017)

Stability Assessment of Masonry Arches Reinforced with Fiber-Reinforced Composite Materials Under Large Deformation

Valerio Alecci[1], Stefano Galassi[1], Fausto Mistretta[2], Gianfranco Stipo[1,1(✉)], and Mario De Stefano[1]

[1] Department of Architecture, University of Florence, Florence, Italy
gianfranco.stipo@unifi.it
[2] Department of Civil, Environmental Engineering and Architecture, University of Cagliari, Cagliari, Italy

Abstract. Preservation and retrofitting of masonry arches and vaults belonging to the architectural heritage is a difficult task. Currently, the most widespread technique consists of applying fiber-reinforced composite materials to the intrados of these structures. The result is an increase in load-bearing capacity and a reduction in the width of cracks in the joints between blocks, although large deformations are experienced. The effectiveness of this technique relies on the bond between the reinforcement and the masonry substrate. To account for it, in this paper a delamination model is proposed. The arch is modelled by rigid blocks assembled with elastic-cracking interfaces and the reinforcement by means of deformable links connecting, in series, the intrados' midpoints of adjacent blocks. A nonlinear numerical procedure to investigate the bond behavior is proposed. The delamination process is checked, step by step, searching for both the tensile forces in the mortar joints that cause the occurrence of cracks and the shearing and peeling forces at the intrados' midpoint of each block that cause delamination. The deformed and cracked shape of the arch with the delaminated reinforcement is provided.

Keywords: Masonry arches · rigid-block and lumped deformability model · reinforcement · delamination model · peeling force · nonlinear analysis

1 Introduction

Masonry buildings are sensitive to seismic actions and the reinforcement of masonry structures is a topic of great interest to architects and engineers. The challenge becomes more important when dealing with historic buildings, where knowledge of the structural organization of the building and the mechanical properties of materials plays a fundamental role [1]. In the last decades, traditional techniques to strengthen and retrofit masonry structures have been replaced by innovative systems and materials. Fiber reinforced composite materials bonded at the masonry structure are the most widespread for

S. Gabriele et al. (Eds.): IWSS 2023, LNCE 437, pp. 737–746, 2024.
https://doi.org/10.1007/978-3-031-44328-2_77

their lightness, high tensile strength, high Young's modulus and stiffness, rapid installment in the structure, reversibility of the intervention, and so on. Two families of fiber reinforced composites are used, namely the Fiber Reinforced Polymers (FRPs) and the Fiber Reinforced Cementitious Matrices (FRCMs). The main difference between these two families consists in the use of an organic matrix (namely an epoxy one) in the former case and an inorganic matrix (namely a mortar or cement one) in the latter case to embed the strengthening fibers (carbon, glass, basalt, PBO, steel). Therefore, FRCMs-based systems are preferable because the inorganic matrix is more compatible with the masonry substrate, preserves the breathability of the masonry and prevents moisture problems and is less susceptible to detachment from masonry caused by heating due to thermal jumps or fire.

Prevalent applications of composite materials are columns confinement to increase the load-bearing capacity and ductility [2–6], strengthening of masonry walls to increase the shearing and eccentric axial loading capacity [7, 8], establishment of roof and floor bands to improve the box-like behavior of the building in seismic areas and to prevent overturning of wall macro-elements [9], the installation at the extrados and/or intrados of arches and vaults to prevent the opening of hinges at joints between masonry units [10]. In all cases, the goal of the application is to provide the masonry with tensile strength and increase the ductility.

The effectiveness of these strengthening systems depends largely on their ability to maintain bonding with the substrate. In the literature it is ascertained that debonding occurs at the fiber-matrix interface in FRCMs systems and at the reinforcement-masonry interface in FRPs systems. In the latter case, debonding involves a thin layer of masonry, which has a lower tensile strength than the epoxy resin. This phenomenon is known as ripping.

Therefore, many authors have studied this phenomenon and have proposed CMLs to predict the bond-slip behavior of FRPs and FRCMs, exploiting the experimental results from different test test-ups [11].

In the case of curved substrates, such as arches and vaults, the problem is more complex because the tangential forces at the bonding interface responsible for sliding combine with the peeling forces, which are arranged orthogonally to the curved interface. If the strengthening intervention is applied at the intrados, these peeling forces are tractions responsible for the typical delamination failure.

In this paper a delamination model is proposed to investigate this topic. A rigid-block and lumped deformability model of the arch is adopted [12]. The intrados strengthening system is conceived as a set of deformable links connecting, in series, the intrados' midpoints of adjacent blocks. A non-linear numerical procedure is proposed to analyze the structure, considering both the material and geometry nonlinearity. The procedure computes the axial and shearing forces in the mortar joints of the arch, the tensile forces in the reinforcing sheet and the peeling and shearing forces at the bonding interface and allows detection of hinge openings caused by tensile failure of mortar joints, as well as monitoring the delamination process of the sheet from the arch substrate.

2 Numerical Procedure

The arch is modelled as an assemblage of rigid blocks (Fig. 1a), linked in-between by deformable mortar joints conceived as discrete devices composed of 4 links orthogonal to the mid-surface of the joints to transfer the axial load and 4 links along the mid-surface to transfer the shearing force. The axial links are positioned at the extrados and intrados points of the mid-surface of the joints and at the upper and lower points of the central core (Fig. 1b,c). The strengthening system is modelled as a set of additional deformable links that connect the mid-points of the intrados edge of the blocks in series (ref. to Fig. 1a).

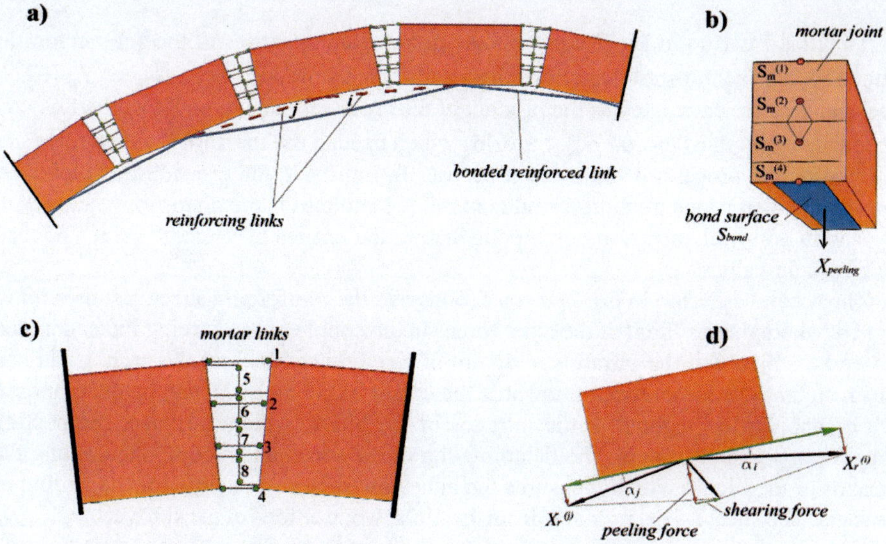

Fig. 1. a) Reinforced arch model; b) detail of a block: mortar joint and cross-sectional area of the four axial and tangential links; c) discrete device of the mortar joint; d) decomposition of the reinforcing link axial forces into the shearing and peeling forces.

To analyze the structure, the equilibrium equation and the elastic-kinematic equation are formulated in matrix form (Eq. 1):

$$\begin{cases} AX + \alpha F = 0 \\ A^t U + KX = \Delta \end{cases} \tag{1}$$

where A is the equilibrium matrix and A^t, its transpose, is the kinematic matrix, K is the deformability matrix of the mortar and the reinforcing links and X the vector of the forces within the links, U is the vector of the displacements of the block centroids, F and α are the vector of the loads acting on the blocks of the arch and the load factor, respectively.

To address the possible failure modes of the reinforced arch, vector X is partitioned in the sub-vector X_{nc}, X_{nt}, X_s, populated by entries relating to the compression, tensile and shearing forces in the mortar joints, and X_r, populated by entries relating to the axial forces in the reinforcing sheet.

To detect the opening of hinges at the mortar joints of the arch and monitor the delamination phase of the reinforcing sheet from the intrados of the arch, two failure criteria are established (Eq. 2):

$$\begin{cases} X_{nt} \leq \overline{X}_{nt} \\ X_{peeling} = \mathrm{X}_r^{(i)} \sin\alpha_i + \mathrm{X}_r^{(j)} \sin\alpha_j \leq \overline{X}_{bond} \end{cases} \tag{2}$$

The first criterion in Eq. 2 concerns the mortar joints: entries of the sub-vector X_{nt} cannot exceed the threshold values of the tensile force stored in vector $\overline{X}_{nt} = S_m \cdot \overline{\sigma}_{nt}^{(m)}$, whose entries are calculated as the product of the cross-sectional area S_m of the link and the tensile strength of mortar $\overline{\sigma}_{nt}^{(m)}$. Entries which exceed the threshold tensile force are nullified by the procedure and a crack occurs. To nullify it, the procedure exploits the Somigliana distortions method formulated in [12] and builds the distortion vector Δ in Eq. 1 with non-null entries in correspondence to the entries of vector X which exceed the threshold values.

The second criterion in Eq. 2, instead, concerns the reinforcing sheet. Let us refer to Fig. 1d, where the resultant of the axial forces in the couple of reinforcing links, denoted as *i* and *j*, adhered at the intrados midpoint of a reference block of the arch, is shown. This resultant is a force directed towards the center of the arch and can be decomposed in a component orthogonal to the intrados of the block, which measures the peeling force $X_{peeling}$ responsible for the delamination of the sheet, and a component along the intrados of the block, which measures the adhesion force responsible for the sliding of the sheet. The sheet *i*–*j* delaminates from the block when entries of the sub-vector $X_{peeling}$ exceed the threshold values of the debonding force stored in vector $\overline{X}_{bond} = S_{bond} \cdot \overline{\sigma}_t^{(b)}$, whose entries are calculated as the product of the bond surface S_{bond} of the sheet under the block and the tensile strength of the bond interface $\overline{\sigma}_t^{(b)}$. It is worth noting that the reinforcement-masonry interface should be assumed as the "bond interface" and the tensile strength $\overline{\sigma}_t^{(b)}$ of the masonry unit should be considered in FRPs systems while, in the case of FRCMs systems, the fiber-matrix interface should be assumed as the "bond interface" and the tensile strength $\overline{\sigma}_t^{(b)}$ of the composite matrix should be considered.

The solution of Eq. 1 is given in Eq. 3, which provides the forces in the mortar and the reinforcing sheet:

$$X = -K^{-1}A^tR^{-1}\alpha F + K^{-1}\left(I - K^{-1}A^tR^{-1}A\right)\Delta \tag{3}$$

where $R = AK^{-1}A^t$ is the stiffness matrix of the structure.

At the beginning of the analysis $\Delta = 0$ because, for very small values of α, the arch is assumed to be undamaged and in its original geometrical configuration.

During the analysis, the value of α is increased and, in correspondence to each value of α, the procedure iteratively:

1) computes vector X (Eq. 3); extracts from vector X the sub-vector X_{nt} and checks the first failure criterium in Eq. 2; if some entries do not meet the failure criterion, distortions are introduced in the axial links and the associated tangential links to nullify the tensile and shearing forces such that cracks develop. It is worth noting that, the failure of three axial links and associated tangential links in a mortar joint develops the formation of a hinge at the intersection of the still working axial and tangential links;
2) extracts from vector X the sub-vector X_r and checks the second failure criterion in Eq. 2 relating to the reinforcing sheet; computes the vector $X_{peeling}$; if some entries of this vector do not meet the failure criterion because exceeds the corresponding entry of vector $\overline{X}_{bond}$, , the pair of reinforcing links that caused the peeling force at the intrados interface of the block where they are connected are deleted from the model; then they are replaced by a new link joining the previous block of the arch to the subsequent block, with a length equal to the sum of the lengths of the removed links (Fig. 1a). The removal of a reinforcing link requires resizing of matrix A with one less column and matrix K with one less row and one less column, respectively.

The algorithm converges when there is no longer any tensile force in the mortar joints or peeling force in the reinforcing links that exceed the threshold values established by Eq. 2. Non convergence implies the collapse of the arch.

At each step of the analysis, the algorithm updates the geometrical configuration of the arch. To update it, the vector of displacements of block centroids is computed. To this aim, the second equation of Eq. 1 is partitioned in such a way as to isolate entries which respect the failure criteria of Eq. 2 (subscript "g") from those which do not (subscript "w"). By doing so, a system of two matrix equations is obtained:

$$\begin{cases} A_g^t U + K_g X_g = 0 \\ A_w^t U = \Delta_w \end{cases} \tag{4}$$

Solving it yields the vector of the displacements of the block centroids (Eq. 5) and the width of the cracks measured in correspondence to the axial links of mortar that failed (Eq. 6):

$$U = -A_g\left(A_g^t A_g\right)^{-1} K_g X_g \tag{5}$$

$$\Delta_w = A_w^t U \tag{6}$$

3 Experimental Campaign and Numerical Predictions

3.1 Experimental Trials

With the purpose of validating the numerical model, experimental trials were performed at the Official Materials and Structures Testing Laboratory of the University of Florence (Italy). The experiments were carried out on half-scaled segmental masonry arches subjected to a monotonically increasing vertical displacement applied at the keystone, impressed through a hydraulic jack. To build the scaled arches, 112 bricks of dimensions $2.5 \times 4.6 \times 10\ cm^3$ were fabricated and a non-commercial lime-based mortar was used for joints. The joints were wedge-shaped with the intrados thickness of 0.8 cm.

The geometry of the arch models is shown in Fig. 2. Each model was fabricated with 112 bricks arranged by alternating, on the front plane, one solid brick with two half bricks and was characterized by an 86.6-cm internal radius, 10-cm width and 10-cm thickness. Accordingly, it results in symmetric segmental arches with effective internal radius and span of 150 cm and 43.2 cm, respectively, and an angle of embrace of 120°. The right and left imposts of the specimens, inclined by 30°, were not directly placed on the testing machine but fixed on bricks arranged in such a way as to simulate the top of two fixed piers.

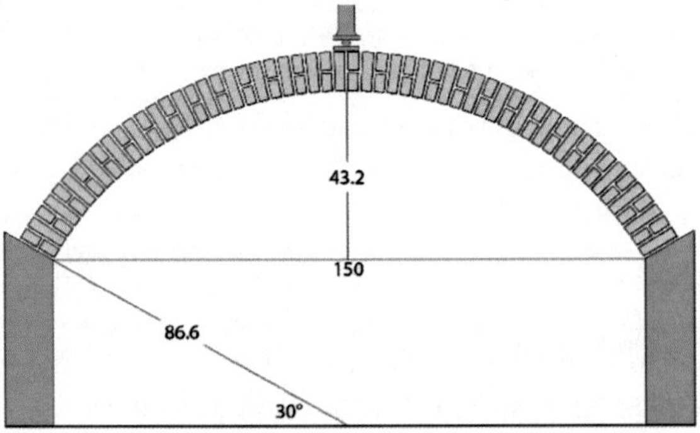

Fig. 2. Geometry of the arch models.

The system adopted to strengthen the arches was the X Mesh Gold, the FRCM (Fiber Reinforced Cementitious Matrix) system provided by Ruredil, consisting of a PBO net embedded within an inorganic matrix. Due to its very high mechanical performance, this composite can match the performance of traditional FRP with carbon fibers and epoxy binder. Some of the arch models were reinforced by applying a continuous 5-cm wide strip bond at the intrados.

Tests were performed both on URM and RM arches to assess the effectiveness of the reinforcement, the change in the failure mode and the increase of the load-bearing capacity. For the need for conciseness, in this section the results of only one URM and RM arch are reported.

Figure 3 shows the URM arch at collapse. For the peak load of 570 N the five-hinge collapse mechanism was fully developed (Fig. 3a) and, successively, the arch collapsed. This is the typical failure mode of a symmetric URM arch, as established in literature.

Then, the arch reinforced at the intrados was also tested in the same way (Fig. 4). Only when the load reached the value of 1980 N (about 3.5 times higher than the collapse load of the URM arch) two cracks at the key occurred (involving two different joints, one at the intrados and the other at the extrados), while for lower values of the load

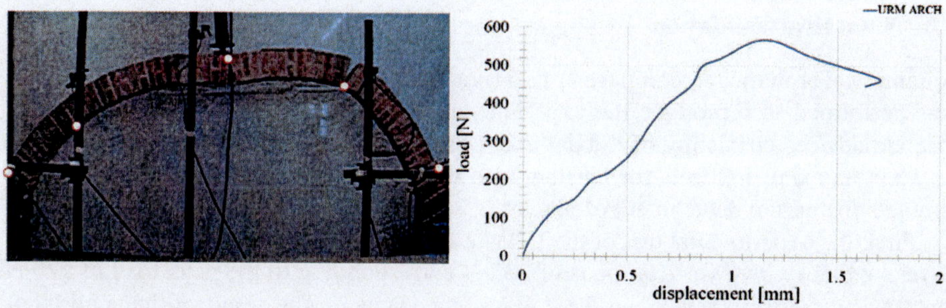

Fig. 3. Experimental trial on the URM arch: a) development of the five-hinge mechanism; b) load-displacement curve.

only microcracks, which could not develop because they were kept closed by the reinforcement, were recorded at the same joints where the hinges opened fully in the URM arch. Already from this evidence, the effectiveness of the reinforcement is ascertained. In correspondence to a load of 2630 N, also the cracks at the haunches occurred, with the composite strengthening still adhered to the intrados of the arch. When the load reached the value of 4320 N, the four hinges at the key and the haunches were fully developed and the composite system initiated the debonding phase caused by the very high tensile forces within it. Starting from the keystone sections, this phenomenon gradually involved the intrados sections of the adjacent bricks as far as beyond the right haunch, where the intrados hinge had occurred (Fig. 4a). The detachment of the reinforcing sheet from the arch occurred by carrying away a thin layer of the brick (ripping). Collapse occurred when the detached sheet arranged following a straight line and worked as an internal truss subjected to tensile force that caused the sudden detachment from the interfaces comprised between the right haunch and the right impost.

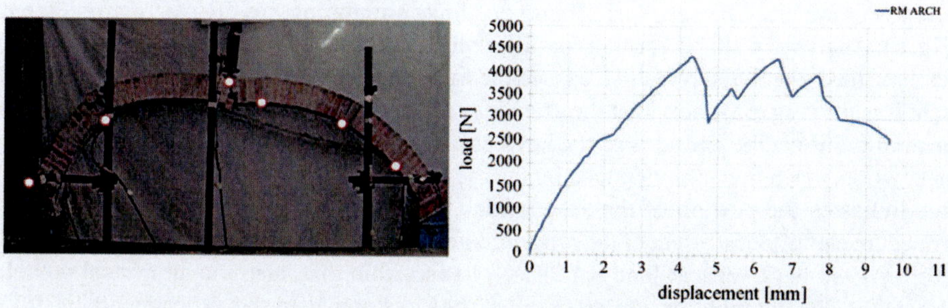

Fig. 4. Experimental trial on the RM arch: a) arch at collapse; b) load-displacement curve.

By comparing the load-displacement curve of the URM and the RM arch (Fig. 3b vs. Fig. 4b) an increase in the load bearing capacity of the arch of 658% and in increase of kinematic ductility of 34% are observed.

3.2 Numerical Validation

A numerical arch model, composed of 56 rigid blocks and 57 deformable mortar joints, was performed to reproduce the experimental arch. The mortar joints were input as discrete devices consisting of 4 axial and 4 associated tangential links, in such a way as when an axial link fails for tension, also the corresponding tangential link fails to simulate the partial detachment of blocks.

First, the experimental test of the URM arch was simulated by applying, at the two keystone blocks, two vertical loads of 285 N corresponding to the peak load of 570 N that, during the trial, developed the collapse mechanism. The result of the analysis of the numerical model is coherent with that of the experimental one since the arch was unstable, with a line of thrust tangent at the extrados of the key and the imposts and slightly lying outside of the arch profile in correspondence to the intrados of the haunch sections (Fig. 5a). To obtain the numerical collapse load, two loads of 100 N were applied at the keystone blocks and the analysis was re-run for increasing values. The numerical peak load obtained was 420N. The result of the arch for the collapse load is shown in Fig. 5b.

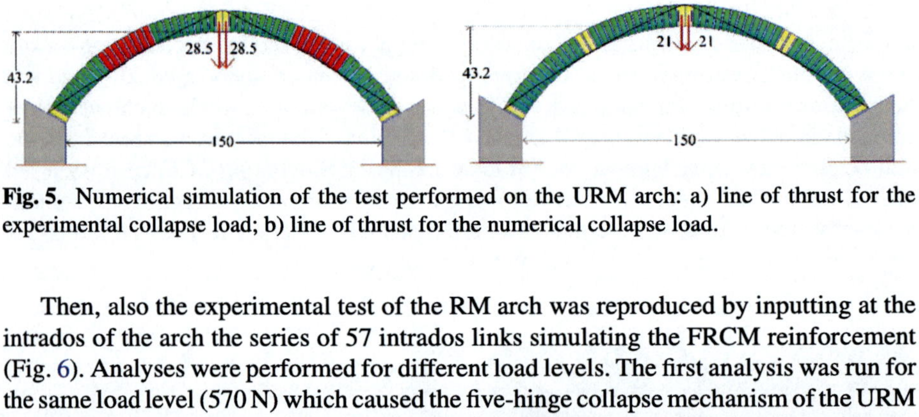

Fig. 5. Numerical simulation of the test performed on the URM arch: a) line of thrust for the experimental collapse load; b) line of thrust for the numerical collapse load.

Then, also the experimental test of the RM arch was reproduced by inputting at the intrados of the arch the series of 57 intrados links simulating the FRCM reinforcement (Fig. 6). Analyses were performed for different load levels. The first analysis was run for the same load level (570 N) which caused the five-hinge collapse mechanism of the URM arch. Results (Fig. 6a) show that the arch does not collapse even if the line of thrust exits the arch profile at the extrados of the keystone joints because the underly reinforcement sheet works in tension, as evidenced by the red color of the sheet itself (conversely, color blue indicates the portion of the sheet which does not work). It demonstrates that the procedure captures the effect of the reinforcement. The second analysis was run for a load level of 1980 N (2 vertical loads of 990 N) to ascertain that, also the numerical model, detects the development of a microcrack at the keystone as in the experimental test. In detail, the numerical value of the microcrack depth was 0.004465 mm at the intrados and 0.002166 mm at the lower boundary of the central core of the 29th joint. The third analysis was run for a load level of 2630 N (2 vertical loads of 1315 N) to ascertain that also in the numerical model the development of microcracks at the haunches occurred. Indeed, the numerical model detected microcracks at the haunches (joints n° 14 and 44) with an extrados width of 0.002788 mm and the microcrack at the keystone enlarged and reached the maximum width of 0.00597 mm at the intrados. The last analysis was performed by inputting the load of 4320 N (2 vertical loads of 2160 N) that, in the experimental model, caused the delamination of the reinforcement. For this load level, the numerical model showed, step by step, the delamination process which also occurred during the trial and the deformed configuration of the arch. Figure 6b and c show an intermediate step of the delamination phase and the step immediately before collapse, with the reinforcement sheet detached up to the right impost, respectively. Snapshot of Fig. 6d is taken at collapse: the sheet detaches also from the right impost and collapse instantly occurs.

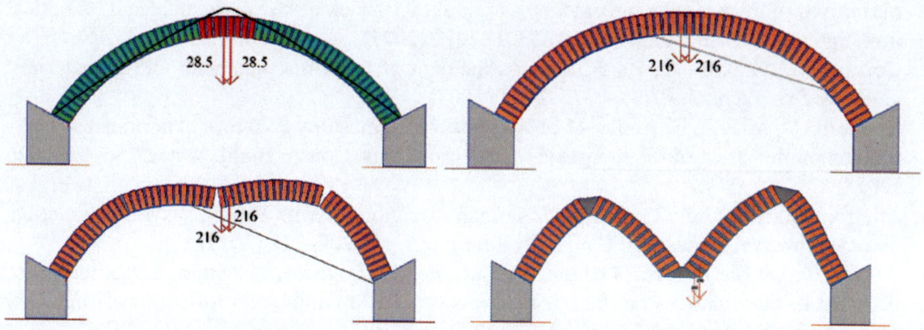

Fig. 6. Numerical simulation of the test performed on the RM arch: a) snapshot for the load level of 570 N; b) snapshot taken during the delamination process; c) snapshot immediately before collapse; d) arch at collapse.

4 Conclusions

The analysis of masonry arches strengthened with fiber reinforced composite materials is the topic investigated in this paper. A rigid-block and lumped deformability model of the arch has been adopted, with the addition of the reinforcing system at the intrados. Since the weak link of the strengthening intervention is the bond between the reinforcement and the substate, a delamination model of the reinforcing system has been conceived. The reinforcing system has been modelled in a discrete manner through a series of deformable links connecting, in series, the intrados' midpoints of adjacent blocks.

A nonlinear numerical procedure is developed and applied to some experimental case studies. The current release of the procedure implements only two failure criteria: tensile failure of mortar joints (causing hinge opening) and delamination of the reinforcing sheet from the masonry arch. However, it is worth noting that by taking advantage of the technique of partitioned matrices used in Sect. 2 (refer to the partitioned vector X), the procedure can be easily extended to consider other failure modes as well, such as sliding of blocks along mortar joints, crushing of masonry, tensile failure of the reinforcing sheet and sliding of the sheet along the intrados profile of the arch where it is bonded.

The reliability of the procedure proposed is ascertained by the good agreement between numerical and experimental results.

Acknowledgements. This research has been funded by the Italian Ministry of University and Research through the program PNRR, Extended Partnership n.3 (PE3) ‘RETURN - Multi-Risk Science for Resilient Communities Under a Changing Climate’, Project Code PE_0000005, CUP: B83C22004820002.

References

1. Alecci, V., Ayala, A.G., De Stefano, M., Marra, A.M., Nudo, R., Stipo, G.: Influence of the masonry wall thickness on the outcomes of double flat-jack test: experimental and numerical investigation. Constr. Build. Mater. **285**, 122912 (2021)
2. Corradi, M., Grazini, A., Borri, A.: Confinement of brick masonry with CFRP materials. Compos. Sci. Technol. **67**(9), 1772–1783 (2007)
3. Cascardi, A., Micelli, F., Aiello, M.: FRCM-confined masonry columns: experimental investigation on the effect of the inorganic matrix properties. Constr. Build. Mater. **186**, 811–825 (2018)
4. Murgo, F.S., Mazzotti, C.: Masonry columns strengthened with FRCM systems: numerical and experimental evaluation. Constr. Build. Mater. **202**, 208–222 (2019)
5. Alecci, V., De Stefano, M., Galassi, S., Luciano, R., Pugliese, D., Stipo, G.: Influence of different mortar matrices on the effectiveness of FRCM composites for confining masonry columns. J. Test. Eval. **51**(2), 20220323 (2022)
6. Alecci, V., De Stefano, M., Galassi, S., Magos, M., Stipo, G.: Confinement of masonry columns with natural lime-based mortar composite: an experimental investigation. Sustainability **13**(24), 13742 (2021)
7. Del Zoppo, M., Di Ludovico, M., Balsamo, A., Prota, A.: Experimental in-plane shear capacity of clay brick masonry panels strengthened with FRCM and FRM composites. J. Compos. Constr. **23**(5) (2019)
8. Bellini, A., Incerti, A., Bovo, M., Mazzotti, C.: Effectiveness of FRCM reinforcement applied to masonry walls subject to axial force and out-of-plane loads evaluated by experimental and numerical studies. Int. J. Architectural Heritage **12**(3), 376–394 (2018)
9. Lourenço, B., Mendes, N., Ramos, L.F., Oliveira, D.V.: Analysis of masonry structures without box behavior. Int. J. Architectural Heritage **5**(4–5), 369–382 (2011)
10. Giresini, L., Sassu, M., Butenweg, C., Alecci, V., De Stefano, M.: Vault macro-element with equivalent trusses in global seismic analyses. Earthquakes Struct. **12**(4), 409–423 (2017)
11. Alecci, V., et al.: Reliability of different test set-ups and influence of mortar mixture on the FRCM-to-brick bond response. J. Test. Eval. **49**(6), 20200656 (2021)
12. Galassi, S., Zampieri, P.: A new automatic procedure for nonlinear analysis of masonry arches subjected to large support movements. Eng. Struct. **276**, 115359 (2023)

Optimization of on Shore Wind Turbine Foundation

Nikolaos Lagaros, Emmanouil Vougioukas, Savvas Triantafillou, Athanasios Stamos, and Charikleia Pappa(✉)

NTUA Greece, Athens, Greece
savtri@mail.ntua.gr, stamthan@central.ntua.gr, pappachara8@gmail.com

Abstract. The current study presents the methodology of the wind turbine foundation optimization. Initially the forces applied on the system wind turbine – foundation are being analyzed. It is presented which phenomena need to be faced due to the applied forces and after implementing the sensitivity analysis, the critical check of the system is identified and proved to be the check against overturn. After following a repetitive procedure, the optimum dimensions and reinforcement of the foundation are being calculated while Eurocodes restrictions are met. At a next phase the foundation is being calculated as beam slab leading to the optimum design and the cost calculation through removal of concrete quantity and insert of rock materials either from deposit areas or by reusing the excavation products. The proposed method is implemented on actual wind turbine showing in numbers the saving in time, cost and environmental footprint. The algorithm can be used for any kind of soil, wind turbine and foundation.

Keywords: optimization · wind turbine · foundation · cost

1 Introduction

Optimization is described as the process through which the objective function is maximized or minimized leading to meeting the target which may be linked to the optimal function or cost reduction. Optimization is met also in the field of civil engineering and it is of high importance as by implementing it, the maximum of the functional behavior of a structural element can be achieved while the cost remains the minimum possible under the studied conditions which follow the regulation restrictions.

The technology of the energy production by the wind turbines is being continuously developed so as higher levels of energy can be produced through a more environmental friendly ways. Plenty of studies have been conducted referring to the hub, the blades or the tower of wind turbine and have proved that their design optimization has high positive effect on the effectiveness of the wind turbine concerning their energy production. Unlikely, the foundation of the wind turbines follows the restrictions by the relevant regulations which aim to the safety of the structure. The foundation design is, in the majority of the cases, provided by the wind turbine manufacturer. As a result, hardly

S. Gabriele et al. (Eds.): IWSS 2023, LNCE 437, pp. 747–757, 2024.
https://doi.org/10.1007/978-3-031-44328-2_78

ever the materials economy is taken into consideration and so does the need of cost minimization.

Taking the afore mentioned into consideration, the need for cost minimization is imposed. In this paper, the optimization of the wind turbines foundations is tackled by introducing a study which leads to the cost minimization while satisfying the regulations restrictions.

More specifically, the loading system applied on the foundation is identified. The relevant forces are the self-weight of the wind turbine, the self-weight of the wind turbine foundation, the soil stresses under the foundation and the lateral ones. The forces lead to the design against overturn, bending, shear and perforation. Initially, the foundation is designed based on the repetitive procedure which sets the cost minimization as objective function and the check against overturn as the crucial check. The second phase of the study includes the design of the foundation as beam slab and the concrete section and reinforcement of each beam are identified. The voids between the beams are filled with rock material which is procured either from deposit areas or by reusing the excavation material in the case of rock foundation ground.

2 Given Data at the Optimization Process

The study developed and analyzed is to be implemented in any type of onshore wind turbine, any kind of soil and any wind conditions. This is the result of inserting as parameters the units that identify the above and thus the algorithm is customized accordingly on each wind turbine and soil type.

The repetitive procedure starts after the height of the turbine over the foundation (H_W), the wings diameter (D_W), the wind turbine tower diameter (D_T) and the weight of the turbine (W_W) are interested. The last are provided by the wind turbine manufacturer. Additionally, the soil conditions should be also given as identified by the internal angle, f, the cohesion factor, c, the special weight of the soil, g_{soil} and the maximum and minimum allowed soil stresses. Last, the wind conditions should be provided including the wind velocity of the area, v_{bo}, that the wind turbine is going to be installed.

The targeted units to be identified are for the 1st phase of the optimization, the width of the foundation, B, the length, L, the height, H, and at the 2nd phase, the width, length and height of the beams that consist the foundation. Following the above, the reinforcement of the foundation as one unit and as beam slab is calculated and adjusted at each repetition circle according to the selected dimensions.

3 Foundation System Identification

The foundation of the wind turbine may be on the mainland or in the sea, distinguishing them between onshore and offshore wind turbines. At the current study the onshore wind turbine is examined and more specifically, the shallow foundation.

The forces applied on the wind turbine are horizontal and vertical. The horizontal force that is applied on the wind turbine and is the cause of overturn of the foundation is the wind (F_W). The calculation of it, is based on the Eurocode restrictions facing the wings of the wind turbine as a mesh.

At the foundation of the wind turbine, the forces applied are the weight of the wind turbine without the foundation (W_W), the self-weight of the foundation (G), the soil passive stresses due to the resistance to the tendency to the overturn of the foundation due to the wind (P_P), and the soil stresses underneath the foundation. Part of these forces contribute to overturn and part for the stability of the foundation. The optimum design results in the balance and stability both for the foundation and the upper structure.

4 Wind Force Calculation

The wind force is the main load that applies on the wind turbine and affects the stability of the foundation. In order to calculate the wind force that applies on the turbine and taking into consideration the difference geometry between the wings and the tower, initially, the influence of the wind on the wind turbine is divided in two parts, one is the at the area of the wings of the turbine and the other at the tower of the wind turbine. The difference lies upon the different geometry of the two parts. The wings are faced as mesh and the tower as a linear element whose diameter is reduced as the height increases.

At the area of the wings, the wind force is actually applied on a mesh, and it is calculated according to Eurocode 1. The velocity inserted is according to the allocation area. At the tower of the wind turbine the wind force is being calculated according to Eurocode 1 and the tower is considered as a circular cylinder with reducing diameter by height.

The wind force at the wings is according to Eurocode 1:

$$\mathrm{Fw1} = \mathrm{cs} * \mathrm{cd} * \mathrm{cf} * \mathrm{qp(ze)} * \mathrm{Aref} \tag{1}$$

The wind force at the tower is according to Eurocode 1:

$$\mathrm{Fw2} = \mathrm{cs} * \mathrm{cd} * \mathrm{cf} * \mathrm{qp(ze)} * \mathrm{Aref} \tag{2}$$

The final wind force calculation applied on the turbine wings and the wind turbine tower is reduced so as to take into consideration that the wings are of aerodynamic shape and not 90° perpendicular to the wind.

Considering that there are two axes at the section of the foundation, x and y, the wind force applied on the wind turbine affects each direction accordingly. The direction of the wind force that applies perpendicular on the wings and at the same time at the column is considered to be the axis x while the direction y is where the force is applied by 0 degrees on the wings so there is practically no force on them, and the only force applied is this at the column of the wind turbine.

As a result, the total wind force that is applied on the wind turbine at axis x is the sum of the two parts of the wind turbine.

$$\mathrm{Fw} = \mathrm{Fw1} + \mathrm{Fw2} \tag{3}$$

while the total force applied at axis y is the force on the column of the wind turbine so

$$\mathrm{Fw} = \mathrm{Fw2} \tag{4}$$

As a conclusion, the horizontal force on the system is fully analyzed and calculated. It should be highlighted that the wind force varies according to the location of the wind turbine installation and is not a default number in the stability equation.

5 Optimization Requirement 1 – Foundation Overturn Check

The forces that are involved at the overturn check are divided into overturn forces and stability forces and their between equilibrium results in the first restriction of the optimization process.

The overturn force is the wind force. The stability forces are the self-weight of the foundation, the self-weight of the wind turbine and the passive stresses from the soil to the foundation surface. The wind force is calculated as described in paragraph 4.

The stability forces are calculated as below:

The self-weigh of the wind turbine is:

$$\mathrm{Ww} = \mathrm{B} * \mathrm{L} * \mathrm{H} * \gamma \mathrm{b} \tag{5}$$

The passive push from the soil is:

$$\mathrm{Pp} = 0.5 * \sigma \mathrm{h} * \mathrm{B} * \mathrm{H} \tag{6}$$

It is the area of the lateral push at the foundation which includes the horizontal stress of the soil, s_h and the surface that they are applied which is the height of the foundation by the length of it, so it is $*H$.

The horizontal stress of the foundation is equal to the coefficient of the passive push of the soil by the vertical stresses of the soil,

$$\sigma \mathrm{h} = \mathrm{Kp} * \sigma \mathrm{v} \tag{7}$$

The coefficient of the passive push is:

$$\mathrm{Kp} = \tan^2(45 + 0.5 * \varphi) \tag{8}$$

where the angle φ is given.

The vertical stress of the soil to the surface of the foundation is equal to

$$\sigma \mathrm{v} = \gamma \mathrm{soil} * \mathrm{HQUOTE}\sigma \nu = \gamma \mathrm{soil} * \mathrm{H} \tag{9}$$

meaning equal to the coefficient for the soil category by the height of the foundation that it is applied.

The above-mentioned forces create moment at the lower right point M of the foundation which is considered the ultimate point to examine the moments as it is the most distanced from the wind turbine wings and foundation axis. This will be the reference point of the moment's equilibrium.

The overturn forces are being separated at axis x and y.

The moment produced by the overturn force is:

Overturn moment around y & x axis:

$$\mathrm{MFwy} = \mathrm{Fwx} * (\mathrm{Hw} + \mathrm{H}) \tag{10}$$

$$\mathrm{MFwx} = \mathrm{Fwy} * (\mathrm{Hw} + \mathrm{H}) \tag{11}$$

it is equal to the wind force multiplied by the height of the turbine plus the height of the foundation.

The moments produced by the stability forces are:

By the self-weight of the foundation: Axis x & y respectively:

$$\mathrm{MGx} = \mathrm{G} * 0.5 * \mathrm{L} \tag{12}$$

$$\mathrm{MGy} = \mathrm{G} * 0.5 * \mathrm{B} \tag{13}$$

which is the self-weight multiplied by the distance from the point M.

By the self-weight of the wind turbine: Axis x & y axis:

$$\mathrm{Mwwx} = \mathrm{Ww} * 0.5 * \mathrm{L} \tag{14}$$

$$\mathrm{Mwwy} = \mathrm{Ww} * 0.5 * \mathrm{B} \tag{15}$$

which is the self-weight of the wind turbine multiplied by the distance of the point M.

By the passive push from the soil: Axis x & y:

$$\mathrm{MPPx} = \mathrm{Pp} * (2 * 0.33 * \mathrm{H}) \tag{16}$$

$$\mathrm{MPPy} = \mathrm{Pp} * (2 * 0.33 * \mathrm{H}) \tag{17}$$

which is the passive push from the soil multiplied by the 2/3 of the vertical surface of the foundation H.

So, in total the sum of the moments at each axis by the stability forces are:

$$\mathrm{SMstab} = \mathrm{MG} + \mathrm{MWW} + \mathrm{MPP} \tag{18}$$

The sum of the stability moments should be greater than the moment by the wind force,

$$\mathrm{SMstab} >= Mww \tag{19}$$

<u>optimization requirement 1</u>

6 Optimization Requirement 2 – Calculation of the Reinforcement of the Foundation

For the specific foundation dimensions that are derived by the 1st optimization requirement and wind force that is applied on the wind turbine and after satisfying the restriction of the stability of the foundation and the safety against overturn, the reinforcement that is needed is calculated.

The final resulted reinforcement should satisfy the minimum and maximum area of the reinforcement for the chosen section of the foundation as one unit. This is the second restriction in the optimization process.

In order to calculate the reinforcement of the foundation, initially the factor μ_{sd} is calculated for both axis x and y taking into consideration the relevant moments.

MFWx & MFWy.

$$\mathrm{Msd} = \mathrm{Med}/(\mathrm{B} * \mathrm{d}^2 * \mathrm{fck}) \quad (20)$$

The moment M_{ed} is the moment that is produced by the wind turbine as calculated above ($Med = Mww$) and d is the static height of the foundation.

The reinforcement needed is calculated for both axis x and y and is equal to

$$\mathrm{ASx} = \mathrm{Medx}/(\mathrm{fyd} * \mathrm{z}) \quad (21)$$

$$\mathrm{ASy} = \mathrm{Medy}/(\mathrm{fyd} * \mathrm{z}) \quad (22)$$

where f_{yd} is the yield point of the steel and z is equal to 0.95*d.

The minimum reinforcement that is required for the foundation against overturn is:

$$\mathrm{Asmin} = 0.15\% * \mathrm{Ac} \quad (23)$$

where

$$\mathrm{Ac} = \mathrm{L} * \mathrm{B} \quad (24)$$

As will be derived below, the minimum reinforcement of the foundation that should be installed according to the Eurocode 1 could be characterized as excessive.

The maximum reinforcement that can be fitted at the foundation against overturn is:

$$\mathrm{ASmax} = 0.04 * \mathrm{Ac} * (\mathrm{L} * \mathrm{B}) \quad (25)$$

The final calculated reinforcement of the foundation should be:

$$\mathrm{Asmin} <= AS\mathrm{x} << AS\mathrm{max} \quad (26)$$

$$\mathrm{Asmin} <= AS\mathrm{y} <= AS\mathrm{max} \quad (27)$$

optimization requirement 2

7 Optimization Requirement 3 – Calculation of Soil Stresses

The foundation chosen after the restrictions followed and the forces and moments applied on it, affect the soil that the wind turbine is founded.

It should be verified that the stresses developed under the foundation for both axis x & y are within the minimum and maximum limits that the soil can accept so as soil failure is avoided.

This leads to the third restriction of the optimization process.

The maximum & the minimum soil stress that is developed is:

$$\sigma\mathrm{maxactual} = (\mathrm{Ww} + \mathrm{G})/\mathrm{R}^2 + \mathrm{Med}/\mathrm{W} \quad (28)$$

$$\sigma \text{minactual} = (\text{Ww} + \text{G})/\text{R}^2 - \text{Med}/\text{W} \tag{29}$$

where W is the resistance moment of the foundation. The foundation that will be selected should produce such soil stress so as:

$$\sigma \text{minactual} >= \sigma \text{min} \tag{30}$$

$$\sigma \text{maxactual} <= \sigma \text{max} \tag{31}$$

optimization requirement 3

8 Optimization Requirement 4 – Calculation of the Strength Against Perforation

The foundation that will be chosen should be secured against the perforation phenomenon. The perforation is faced as the wind turbine tower acts as a gathered load on the foundation leading to the check against it.

This is the fourth restriction of the optimization process.

The maximum load stress that the foundation can accept is:

$$\text{vRDc} = \text{cRdc} * \text{k} * (100 * \rho\text{l} * \text{fck})^0.33 * 2 * \text{d}/\text{a} >= \nu\text{min} * 2 * \text{d}/\text{a} \tag{32}$$

according to Eurocode 2. The acting stress is:

$$\text{Ved} = \text{Ww}/\text{u} * \text{d} \tag{33}$$

according to Eurocode 2. So, it should be:

$$\text{vRDc} > \nu\text{Ed} \tag{34}$$

optimization requirement 4

9 Optimization Requirement 5 – Check Against Bending

The check against bending should satisfy the restriction that the moment due to the wind developed on the foundation is not bigger than the moment the foundation can hold. The check is done both for x and y axis.

$$\text{Med} < M\text{RD} \tag{35}$$

where M_{Ed} is the acting moment and M_{RD} is the moment that the foundation can carry. All calculations are being done according to Eurocode 2.

10 Cost Minimization of Single Foundation

The above four optimization requirements lead to the optimum dimensions B, L & H and reinforcement R in kg of the foundation of the wind turbine.

Having this information, the minimum cost is calculated.

In order to construct the foundation, there are 3 stages which are the excavation, the reinforcement installation and the concrete phase. The cost of each stage and the quantities define the final cost of the whole foundation.

It is considered that: The cost of the excavation is equal to CE = 13€/M3, the cost of the steel reinforcement including material procurement and installation is equal to CR = 1.5€/kg & the cost of the concrete phase including the material procurement and the concreting is equal to CC = 100€/M3.

As a result, the final cost of the foundation will be:

Total Cost:

$$(\mathrm{TC}) = (\mathrm{B} * \mathrm{L} * \mathrm{H}) * \mathrm{CE} + \mathrm{R} * \mathrm{CR} + (\mathrm{B} * \mathrm{L} * \mathrm{H}) * \mathrm{CC} \tag{36}$$

When achieving the minimum volume of the wind turbine foundation and the minimum reinforcement, we lead to the minimum total cost (TC) while satisfying the five restrictions analyzed above and the optimum design will be completed. The total cost compared with current studies without optimization, results in cost effectiveness of 30% for the concrete and 40% for the steel.

11 Beam Slab Foundation

The optimized foundation is studied further and is analyzed as beam slab. The optimized foundation is examined against shear and the excess of shear strength leads to the partial concrete removal forming a unified reinforced concrete with reinforced beams on it and rock materials filling the voids between the beams.

The optimization requirements that were analyzed already, are now applied on each single beam and lead to the optimum section of each beam and its reinforcement. Due to the fact that the minimum and maximum reinforcement requirements are now applied on beam the ratio is lower and this results in reduced need for reinforced concrete. At the same time, the void filling with rock material either by the excavation materials or by supplying them from deposit areas, contribute to the crucial equation which is the check against overturn.

12 Numerical Example

The analyzed methodology is implemented on a wind turbine with the below characteristics:

Wind turbine height Hw = 80m,Wind turbine wings Dw = 52 m, Wind turbine self-weight Ww = 142 tones. The soil characteristics are the internal angle of the soil f, the soil factor $\gamma = 22$KN/m3, the cohesion factor c 120 KN/m^2, the maximum stress of the soil σmax = 750 kPa and the minimum stress of the soil σmin = 0kPa.Wind speed = 33 m/sec. After applying the optimization algorithm, the final dimensions of the foundation after the optimization of the cubic foundation are H = 2.5 m, L = 23 m B = 17 m.

Total Cost (TC) = 87.975.04€.It was calculated with bars of Φ12/40.

At Table 1 the results of the analysis are presented.

Table 1. Analysis results.

Type of foundation	Single foundation	Beam slab foundation
Cost(€)	251903.85	87975.04
Concrete quantity (m3)	675.51	240.59
Steel quantity (kg)	117047.5	27490.08
Rock material quantity (kg)	0	1431750
Stability moment $SM_{STABILITY}$	204270.69	246305.78
Overtun moment $SM_{OVERTURN}$	65903.19	65903.19

It is derived that the cost of the beam slab foundation is reduced by 35% while the concrete and steel quantity are reduced by 36% and 24% respectively. The rock material is inserted and acts for the stability moment which is increased by 20% for the beam slab while the overturn moment remains the same as the location of wind turbine installation remains the same for both cases so as the results are comparable.

As shown in Fig. 1, the cost of the foundation construction including materials procurement and installation decreases linearly as the concrete quantity is reduced and replaced with rocks and the reinforcement is calculated for the beams.

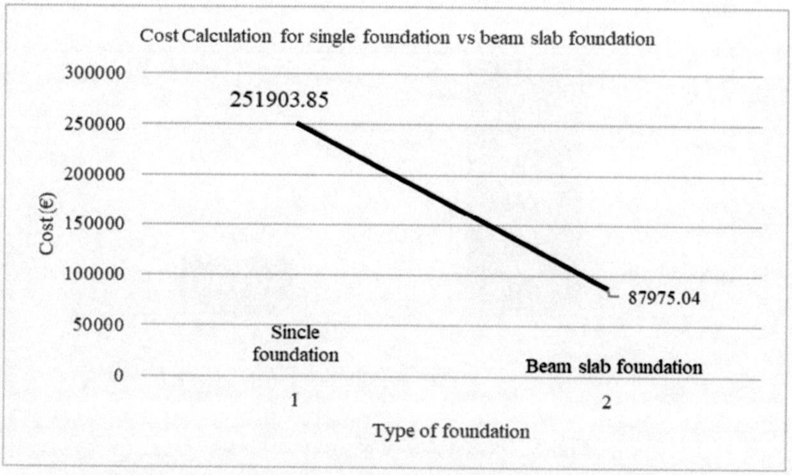

Fig. 1. Cost calculation for singe foundation and beam slab foundation.

The Fig. 2 presents that the stability moment increases for beam slab despite the concrete quantity decreases. This happens as the role of the concrete for stability is taken from the rock materials that contribute to the stability. The Fig. 3 presents the increase in stability moment with the parallel decrease of steel reinforcement. The reinforcement decreases as the section of the beams are lower than the single foundation and the stability moment for each increases.

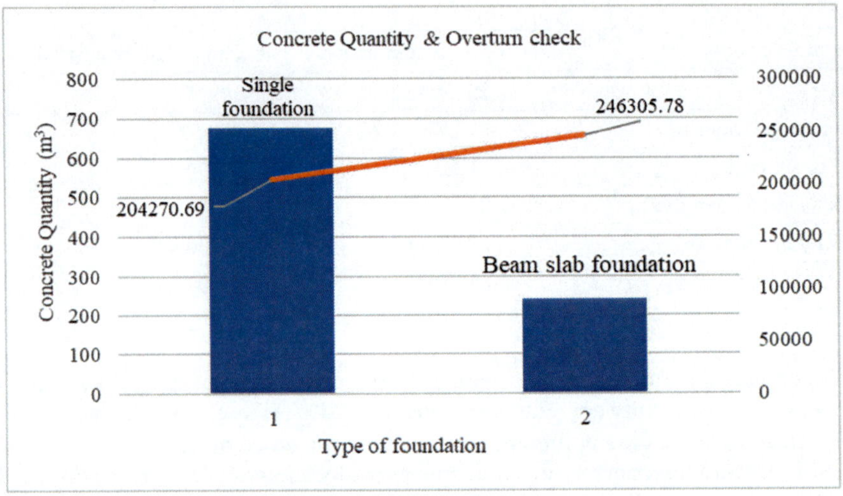

Fig. 2. Concrete quantity & overturn check

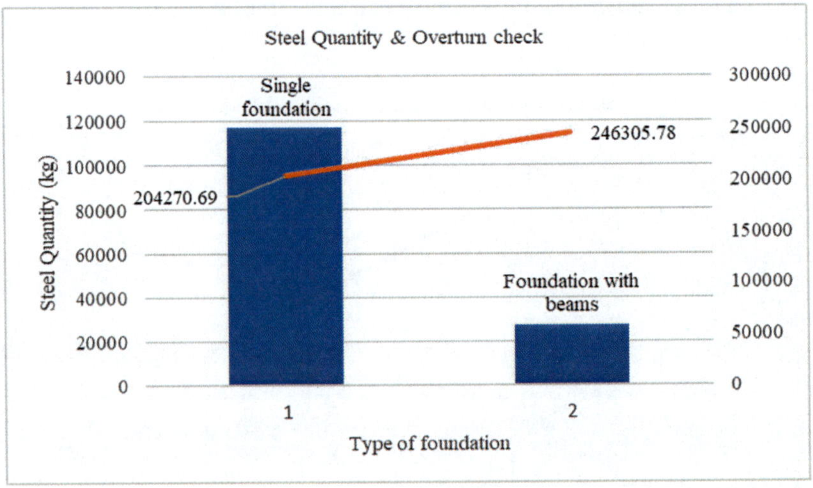

Fig. 3. Steel quantity & overturn check

13 Conclusions

The wind turbine tower and wings have been analyzed and a lot of attempts were made to optimize them, while the wind turbine foundation is provided by the manufacturers and less studies have been published on it so far. The current study presents the methodology to reduce the cost of the onshore wind turbine foundation by analyzing the forces applied and optimizing the structural elements of it. The analysis is implemented on an existing wind turbine and the results verify the expected optimization as the cost is reduced by 35%. The sector of wind turbine foundation may be further analyzed and optimized also for offshore wind turbine where the foundation types are more and need to be separately analyzed.

References

1. Kavvada, M.: Notes Foundations of technical projects, September (2007)
2. Mangal, M., Cheng, J.C.P.: Automated optimization of steel reinforcement in RC building frames using building information modeling and hybrid genetic information, 18 February (2018)
3. Tasios, T., et al.: Notes for concrete structures. Athens (2008)
4. Christaras, B., Chatziaggelou, M.: Laboratory of Technical Geology – Geology department – Aristotelio Univertisy of Thessaloniki, Soil engineering formulas. Thessaloniki (2011)
5. EN 1991–1–4 - Eurocode 1, Loads on constructions. Part 1- 4 General loads – Wind loads. April (2005)
6. ACP 61400–1–202x Wind Energy Generation Systems – Part 1: Design requirements – Modified Adoption of IEC 61400–1, American Clean Power Association Standards Committee, Wessington DC (2005)
7. Sahab, M.G., Ashour, A.F., Toropov, V.: A hybrid genetic algorithm for reinforced concrete flat slab buildings. 31 December (2004)
8. Maragkos, N., Additions Kirtas, E.: Soil Engineering, 2009 (2010)
9. Tasios, T.P., Giannopoulos, P.I., Trezos, K.G., Tsoukantas, S.G.: Reinforced concrete, Athens (2005)
10. Emmanouil, K.: Foundations – Chapter 3 – Shallow foundations, Serres, September (2010)
11. Fardis, M.N.: Lessons of reinforced concrete, Part III, Patra (2015)
12. EN 1992 1–1:2004 - Eurocode 2, Concrete structures design. Part 1–1 General requirements and requirements for buildings, December (2004)
13. Pappa, C., Vougioukas, E., Triantafyllou, S., Lagaros, N.: Detailing design optimization of the wind tower's RC foundation system, June 2022, fib Congress 2022 in Oslo – FIB International (2022)
14. Pappa, C., Vougioukas, E., Lagaros, N.: Detailing design optimization of the wind tower's RC foundation system, October 2022, National Conference of earthquake engineering and technical seismology – Athens Greece (2022)

A Novel Apparatus and Methodology to Characterise the High-Rate Behaviour of Materials Under Complex Loading Conditions

Yuan Xu[1] and Antonio Pellegrino[2](✉)

[1] Lecturer in Mechanical Engineering at Newcastle University, Newcastle upon Tyne NE1 7RU, UK
yuan.xu@newcastle.ac.uk

[2] Lecturer in Mechanical Engineering at University of Bath, North Rd, Claverton Down, Bath BA2 7AY, UK
ap3551@bath.ac.ak

Abstract. More advanced testing methodologies and measurement techniques to identify complex deformation and failure at high strain rates have drawn increasing attention in recent years. This study presents a novel combined tension-torsion split Hopkinson bar (TTHB) system that is conceived to generate a combination of tensile and torsional waves in a single loading case and to measure material data representative of real case impact scenarios. Energy-release mechanism is employed to generate both longitudinal and shear waves practically via the quick release of a bespoke clamp assembly.

The synchronisation of the longitudinal and torsional waves, and the wave rise time, were experimentally assessed. Thin-walled tube specimens made of two metallic materials were utilised to examine the capability of the developed TTHB system. Four-ligament tension-torsion specimens were also used and the results compared to those obtained using the thin-walled tubular specimens.

The capability of the apparatus is demonstrated by the measurement of the failure envelope of standard and additively manufactured materials at high strain rates..

Keywords: Hopkinson Bar · Combined Loading · Tension Torsion · wave synchronisation · ligament specimen · failure envelope · Mises Criterion

1 Introduction

The understanding of the dynamic behaviour of materials subjected to impulsive loading is of paramount importance in the fields of transportation, military, automotive and aerospace industries. The consequences of impact loading due to collision or explosion are generally severe, often threatening the safety of people involved. Hence the desire to understand the dynamic response of engineering materials under loading conditions representative of real case scenarios to design impact resilient structures and materials.

These conditions are generally not uniaxial but deform the materials in multiple directions. Plates and shells, widely used in aerospace and automotive structures, are

S. Gabriele et al. (Eds.): IWSS 2023, LNCE 437, pp. 758–764, 2024.
https://doi.org/10.1007/978-3-031-44328-2_79

typically deformed under non uniform in-plane loading conditions. Researchers in experimental mechanics have made extensive efforts to reproduce in a controlled laboratory environment complex loading scenarios at high rates of strain.

The split Hopkinson bar (SHB) is widely recognized as the prevailing method for assessing the mechanical properties of materials within a controlled laboratory setting, covering a broad range of high strain rates from 10^2 to $10^{4\ s\text{-}1}$. Diverse setups involving tensile and compression Hopkinson bars have been effectively employed to analyze the dynamic response and failure behavior of a vast variety of materials under uniaxial loading conditions. These materials encompass metals, polymers, composites, ceramics, and rocks [1–9].

The most commonly used method to generate a longitudinal stress pulse is through the direct impact of a solid or tubular-like striker bar on the incident Hopkinson bar. This approach produces a sharp stress wave with a rise time in the range of tenths of microseconds, and it can be further enhanced by employing pulse shaping techniques [10]. Additionally, the Split Hopkinson Pressure Bar setup can be modified to determine the dynamic response of granular materials [11–13] and successfully characterize the tensile dynamic response of Dyneema® and Kevlar® fibre yarns [14].

Another significant configuration of the apparatus used for measuring the dynamic response of engineering materials is the Split Hopkinson Torsion Bar (SHToB) [15–17]. In this setup, torsional stress pulses are generated primarily through the elastic energy storage and release mechanism. The most commonly employed method to release torsional elastic energy is by swiftly releasing a clamp, facilitated by the fracture of a notched pin that initially connects the two jaws of the clamp itself [18–20]. Consequently, the rise of the torsional stress wave is greatly influenced by the material of the pin and the geometry of the notch.

In recent years, there has been a growing interest in the advancement of multiaxial Hopkinson bar techniques. Lewis and Goldsmith [21] introduced a method that combines a compressive wave and a torsional wave by utilizing two split bars and employing the impact method and energy-release mechanism mentioned earlier. To simultaneously apply dynamic compression and shear loading on a specimen, Claus et al. [22] made modifications to a compression Kolsky bar [1]. They achieved synchronization between the compression and shear loading by introducing the torsional load near the sample using a torque adapter with a suitable shape. Albertini et al. [23] accomplished the synchronization of a tensile wave and a shear wave in their ground-breaking study by positioning a single clamp in close proximity to the specimen, effectively storing both longitudinal and torsional energies in a single bar. Nevertheless, the data generated with this configuration was relatively limited, and no additional data has been presented since that time. Sato et al. [24] aimed to categorise the performance of adhesive bonds under combined loads using a system of hydraulic actuators with a clamp, such that the sudden release of a clamp would transmit stress waves.

This study introduces a novel split Hopkinson bar system, referred to as the combined tension-torsion split Hopkinson bar (TTHB) system, which is designed to replicate the combination of tension and shear (torsion) under high strain rates in a single loading scenario [25–28]. The system can be equipped with a piezo-driven clamp release for synchronisation and timing of combined direct-shear stress waves [29] and temperature conditioning. The generated results motivate the development of novel traditional and data-driven [30] constitutive models and the assessment of existing ones.

The working principle of this combined Hopkinson bar method and the utilised specimen geometries are elaborated in Sect. 2. Section 3 illustrates the experimental results obtained on additively manufactured Ti-6Al-2Sn-4Zr-2Mo.

2 Experimental Setup and Specimen Geometry

Figure 1 illustrates the schematic of the testing setup. The TTHB system comprises an input bar, an output bar, a specimen positioned between them, a rapid release clamp assembly, a pre-tension loading unit (hydraulic linear actuator), a pre-torsion loading unit (rotary actuator), and a data acquisition system. The input bar and the output bar are both constructed from Al 7075-T6, measuring 2500 mm in length and 20 mm in diameter. The torsion loading unit consists of an electric servo motor coupled with a harmonic drive gear, enabling the generation and storage of torsional elastic strain energy. Harmonic drive gears offer several advantages compared to conventional gearing systems. These advantages encompass a single stage with high reduction ratios, exceptional positional accuracy, absence of backlash, compact size, lightweight design, high torque capacity, and coaxial input and output shafts.

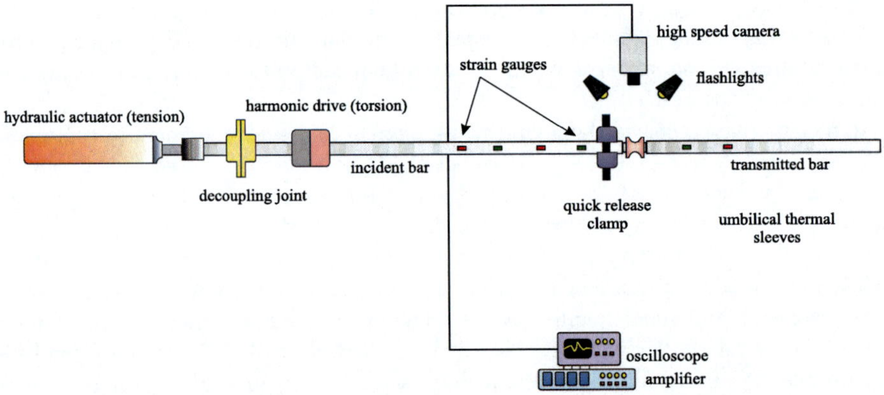

Fig. 1. Schematic of the Tension-Torsion Hopkinson Bar (TTHB) setup.

To prevent the torsional actuator from twisting the shaft of the linear actuator, a decoupling joint is employed to connect the harmonic drive and the hydraulic linear actuator.

In a typical tension-torsion loading experiment, the input bar is clamped mechanically at a short distance from the bar-specimen interface, while the opposite end of the bar is driven to extend until a prescribed force is reached and to rotate until a desired torque is exerted. This process leads to the accumulation of tensile and torsional strain energies in the clamped section of the input bar. On the other hand, the output bar is allowed to slide along and rotate about its axis freely. The stored tensile and torsional energies are released when a notched aluminium pin, securing the clamp tightly to the bar, fails. As a result, shear and axial waves propagate in both forward and backward directions along the bar. Due to the proximity of the clamp to the sample, there is insufficient time for the tension and shear waves, which have different wave speeds, to separate. Consequently, the stress pulses reach the specimen simultaneously.

The data acquisition instrumentation consisted of longitudinal and shear strain gauges installed on both the input and output bars, a high-speed camera that was focused on the specimen, and a series of oscilloscopes and high frequency signal conditioners.

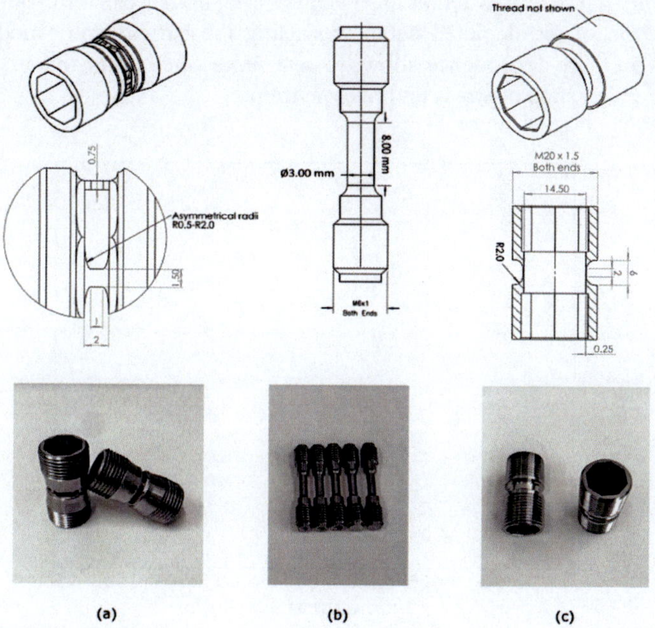

Fig. 2. Tension-Torsion ligament specimens (a), tensile dog-bone specimens (b) and thin-walled tubular specimens (c)

Tension-torsion ligament specimens (Fig. 2a) and thin-walled tubular specimens (Fig. 2c) were fabricated from additively manufactured Ti-6Al-2Sn-4Zr-2Mo to characterise its mechanical response under different stress states including tension, torsion, and their arbitrary combinations in quasi static and dynamic conditions. The specimens have specially shaped ends featuring octagonal slots and external threads, allowing for the transfer of torque and axial load. The input and output bars have octagonal studs that match the slots in the specimen, and they are connected to the specimen using threaded collars. The mechanical response measured in tension on the four ligament specimens was comparable to the one measured on the dog-bone tensile samples (Fig. 2b).

3 Experimental Results

Figure 3 illustrates the partial failure locus of additively manufactured Ti-6Al-2Sn-4Zr-2Mo under various tensile-shear loading conditions in the normal versus shear stress space.

The blue circular markers represent the ultimate stresses measured in quasi static loading conditions while the green circular markers represent the ultimate stresses determined during high-rate experiments.

It is evident that the failure envelope expands radially from the quasi static to the dynamic regime. These results denote an evident rate dependence of the material strength. The interpolation of the depicted datapoints using the parabolically modified Mises criterion with pressure dependence allows to *determine* whether the material fails when subjected to a given state of stress and rate of strain.

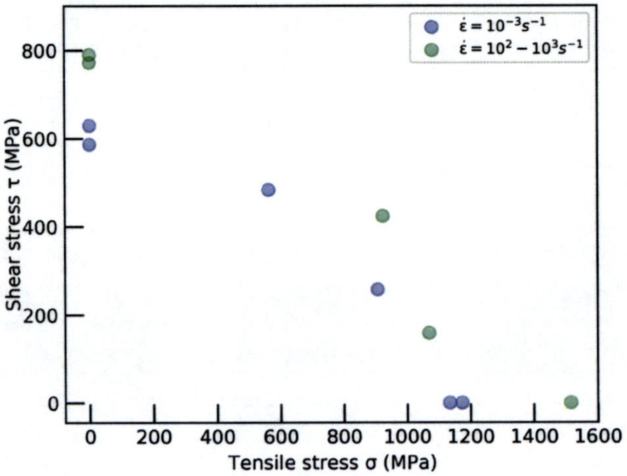

Fig. 3. Quasi static and dynamic ultimate stress locus of additively manufactured Ti-6Al-2Sn-4Zr-2Mo under different direct-shear loading conditions.

4 Conclusions

The main conclusions of this study are summarised below:

We developed a unique Split Hopkinson Tension-Torsion Bar that can generate both tensile and torsional stress waves in a single loading scenario. Using this apparatus, a set of dynamic combined loading experiments were performed on additively manufactured Ti-6Al-2Sn-4Zr-2Mo.

The examination of the high-rate experiments reveals the synchronization of direct and shear stress waves during sample loading.

The experimental results obtained provide an overview of the failure stress locus of Ti-6Al-2Sn-4Zr-2Mo under various combinations of direct and shear stress. These combinations include shear, shear-dominated tension-shear, tension-dominated tension-shear, and plain tension.

The presentation of the quasi-static and dynamic failure loci in the direct-shear stress space aims to motivate the development of novel theoretical constitutive models.

References

1. Kolsky, H.: An investigation of the mechanical properties of materials at very high rates of loading. Proc. Phys. Soc. **62**, 676–700 (1949)
2. Gray, III., G.T.: Classic split-Hopkinson pressure bar testing. In: Kuhn, H., Medlin, D. (eds.) ASM handbook, Mechanical Testing and Evaluation, vol. 8, pp. 1027–1068. ASM International: Materials Park, Ohio (2000)
3. Xu, Y., Dai, F.: Dynamic response and failure mechanism of brittle rocks under combined compression-shear loading experiments. Rock Mech. Rock Eng. **51**, 747–764 (2018)
4. Pellegrino, A., Tagarielli, V.L., Gerlach, R., Petrinic, N.: The mechanical response of a syntactic polyurethane foam at low and high rates of strain. Int. J. Impact EngEng **75**, 214–221 (2015)
5. Chen, W., Ravichandran, G.: Dynamic compressive failure of a glass ceramic under lateral confinement. J. Mech. Phys. Solid **45**(8), 1303–1328 (1997)
6. Xu, Y., Pellegrino, A., Dai, F., Du, H.: Experimental and numerical investigation on the dynamic failure envelope and cracking mechanism of precompressed rock under compression-shear loads. Int. J. Geomech.Geomech. **21**(11), 04021208 (2021)
7. Tagarielli, V.L., Deshpande, V.S., Fleck, N.A.: The high strain rate response of PVC foams and end-grain balsa wood. Compos. B Eng. **39**(1), 83–91 (2008)
8. Xu, Y., Dai, F., Xu, N.W., Zhao, T.: Numerical investigation of dynamic rock fracture toughness determination using a semi-circular bend specimen in split Hopkinson pressure bar testing. Rock Mech. Rock Eng. **49**, 731–745 (2016)
9. Ravichandran, G., Chen, W.: Dynamic failure of brittle materials under uniaxial compression. Experiments in Micromechanics of Fracture Resistant Materials, ed. Kim KS, AMD-130, ASME, New York, pp. 85–90 (1991)
10. Frew, D.J., Forrestal, M.J., Chen, W.: Pulse shaping techniques for testing brittle materials with a split Hopkinson pressure bar. Exp. Mech. **42**(1), 93–106 (2002)
11. Varley, L., Rutherford, M.E., Zhang, L., Pellegrino, A.: The mechanical response of wet volcanic sand to impact loading, effects of water content and initial compaction. J. Dyn. Behav. Mater. **6**, 358–372 (2020)
12. De Cola, F., Pellegrino, A., Barbieri, E., Penumadu, D., Petrinic, N.: Void ratio based representative volume element for modelling the high strain rate behaviour of granular materials. Int. J. Impact Eng. **91**, 46–55 (2016)
13. Pellegrino, A., De Cola, F., Dragnevski, K., Petrinic, N.: The dynamic response of etnean sand and the effect of its impingement on Ti-6Al-4 V alloy. J. Dyn. Behav. Mater. **2**, 438–451 (2016)
14. Zhou, J., et al.: A new technique for tensile testing of engineering materials and composites at high strain rates. Proc. R. Soc. A **475**(2229), 20190310 (2019)
15. Baker, W.W., Yew, C.H.: Strain rate effects in the propagation of torsional plastic waves. J. Appl. Mech. **33**(6), 917–923 (1966)
16. Hartley, K.A., Duffy, J., Hawley, R.H.: The Torsional Kolsky (Split-Hopkinson) Bar. In: 9th ed ASM Handbook, Mechanical Testing, vol. 8, pp. 218–228. American Society for Metals (1985)

17. Gilat, A.: Torsional Kolsky bar testing. In: Kuhn, H., Medlin, D. (eds.) ASM Handbook, Mechanical Testing and Evaluation, vol. 8, pp. 505–515. ASM International: Materials Park, Ohio (2000)
18. Ramesh, K.T.: On the localization of shearing deformations in tungsten heavy alloys. Mech. Mater. **17**(2–3), 165–173 (1994)
19. Peirs, J., Verleysen, P., Degrieck, J.: Experimental study of the high strain rate shear behaviour of Ti6Al4V. Appl. Mech. Mater. **82**, 130–135 (2011)
20. Duffy, J., Campbell, J.D., Hawley, R.H.: On the use of a torsional split Hopkinson bar to study rate effects in 1100-O Aluminum. J. Appl. Mech. **38**, 83–91 (1971)
21. Lewis, J.L., Goldsmith, W.: The dynamic fracture and prefracture response of compact bone by split Hopkinson bar methods. J. Biomech.Biomech. **8**, 27–40 (1975)
22. Claus, B., Chu, J., Beason, M., Liao, H., Martin, B., Chen, W.: Dynamic experiments using simultaneous compression and shear loading. Exp. Mech. **57**, 1359–1369 (2017)
23. Albertini, C., et al.: Material characterization by an innovative biaxial shear experiment at very large strains and at very high strain rates. J. Phys. IV **1**(C3), 435–440 (1991)
24. Sato, C., Ikegami, K.: Strength of adhesively-bonded butt joints of tubes subjected to combined high-rate loads. J. Adhes.Adhes. **70**(1–2), 57–73 (1999)
25. Xu, Y., Zhou, J., Farbaniec, L., Pellegrino, A.: Optimal design, development and experimental analysis of a tension–torsion Hopkinson bar for the understanding of complex impact loading scenarios. Exp. Mech. **63**(4), 773–789 (2023)
26. Xu, Y., et al.: Experimental analysis of the multiaxial failure stress locus of commercially pure titanium at low and high rates of strain. Int. J. Impact EngEng **170**, 104341 (2022)
27. Zhou, J., et al.: The mechanical response of commercially pure copper under multiaxial loading at low and high strain rates. Int. J. Mech. Sci. **224**, 107340 (2022)
28. Xu, Y., Farbaniec, L., Siviour, C., Eakins, D., Pellegrino, A.: The development of split Hopkinson tension-torsion bar for the understanding of complex stress states at high rate. In: Dynamic Behavior of Materials, Volume 1: Proceedings of the 2020 Annual Conference on Experimental and Applied Mechanics, pp. 89–93. Springer International Publishing (2021)
29. Zhou, J., Xu, Y., Farbaniec, L., Pellegrino, A.: Piezo-driven clamp release for synchronisation and timing of combined direct-shear stress waves. Int. J. Impact Eng., 104672 (2023)
30. Tasdemir, B., Pellegrino, A., Tagarielli, V.: A strategy to formulate data-driven constitutive models from random multiaxial experiments. Sci. Rep. **12**(1), 22248 (2022)

Digital Twinning for the Prognosis of Spatial Architectures: Morandi's Underground Pavilion in Turin

Gaetano Miraglia(✉), Valeria Cavanni, Alessio Crocetti, Erica Lenticchia, Valerio Oliva, Linda Scussolini, and Rosario Ceravolo

Politecnico di Torino, 10129 Turin, TO, Italy
gaetano.miraglia@polito.it

Abstract. Concrete spatial architecture was mainly built using techniques that at the time were still experimental and based on design criteria that did not consider seismic actions. The validity of accurate models accounting for such complex structural schemes can be demonstrated, but the latter still would not support a clear comparison with the original predictions.

Different from other Morandi's balanced beam schemes, in the underground Pavilion V of Turin Exhibition Center the main post-tensioned ribs are not parallel beams but are diagonally directed and multiply reciprocally interconnected in order to obtain a spatial structure offering a high overall rigidity and lateral stability, and to contrast the instability of the very thin webs of the main ribs.

The paper focuses on how information from the experimental campaign can help to formulate virtual models for prognostic and diagnostic assessments under different scenarios, such as for example the design of structural health monitoring activities and systems.

Keywords: Experimental campaigns · Gaussian mixture models · Hilbert-Huang Transform · Model parameter calibration · Natural frequency · Sensitivity analysis · Virtual experiment

1 Introduction

One important task to understand the structural behavior of physical systems is to understand the causes and quantify the effects of the variation of its mechanical parameters. Cleary enough, this task cannot be performed on the real structure since this would consist in drastically varying its structural characteristics. However, a virtual model can be designed for this purpose.

A sensitivity analysis activity [1] is aimed at evaluating the effect that the variation of a single mechanical parameter of a numerical model produces on another parameter of the same model (e.g., modal parameter such as natural frequency). The first step in carrying out this type of activity is data collection. In this step, the numerical model is interrogated to generate data on the vibration modes of the system such as the values assumed by the mechanical parameters. The problem of applying this procedure with

S. Gabriele et al. (Eds.): IWSS 2023, LNCE 437, pp. 765–774, 2024.
https://doi.org/10.1007/978-3-031-44328-2_80

modal data is that the individual modes of vibration are ordered in the numerical model in a non-predefined order, which varies with the variation of the value assumed by the mechanical parameter. In other words, to perform a sensitivity analysis of modal data it is necessary to implement a *mode tracking* or *mode coupling* procedure [2]. To achieve this, a reference model, characterized by a set of mechanical reference parameters, is used to generate the reference modal data, which will be used as a comparison term during the sensitivity analysis. This reference model must consider as much as possible the information gathered experimentally, for example through in-situ mechanical tests. To evaluate the coupling between the single reference mode and a set of vibrating modes extracted from a model (to which a different set of tentative mechanical parameters has been attributed), it is possible to use the Modal Assurance Criterion (MAC) [3]. The closer the MAC is to one, the more the single extracted mode will be correlated to the reference mode. The mode to be coupled to the reference mode will be the one with the highest MAC value among those extracted. This procedure allows to track the various reference modes, obtaining, as the mechanical parameters vary, how the order of the mode varies within the mechanical model.

The problem with this procedure is that, sometimes, it can happen that modes with a high MAC can be interpreted incorrectly only because small process errors (e.g., the mesh discretization of the numerical model) affect the MAC estimation. In this context, the *correct* mode to be coupled to the reference one is characterized by a lower MAC value than the mode with the highest MAC. The method proposed in this paper mitigates this misclassification problem through the combined use of: (i) Analogies with time-frequency analysis of the generated data, and (ii) Gaussian processes fitted to data. Section 2 reports the proposed method to alleviate the misclassification error in mode tracking/coupling for sensitivity analysis. In Sect. 3 the methods are applied to Morandi's iconic structure of the Pavilion V of Turin Exhibition Center, and the results of the analysis are critically discussed. Finally, conclusions of the study are drawn in Sect. 4.

2 Method

By interrogating the virtual model, it is possible to define historical series of natural frequency and MAC as a function of the value of the assumed mechanical parameter, which represents the independent variable of the problem, and regulates the natural frequency and MAC variation. In line with the concept of "local" sensitivity analysis, this can be repeated for each mechanical parameter of the model and for each assumed reference mode.

The MAC data, which in fact represent the complement to one of the coupling errors, can be analyzed with signal analysis techniques assuming the considered mechanical parameter as an independent variable, instead of time. In the present work the datum is thus decomposed by Empirical Mode Decomposition (EMD) [4]. The result of the decomposition is a series of Intrinsic Mode Functions (IMFs) and a residual signal, the sum of which provides the starting signal. The IMFs are used for the estimation of the Hilbert-Huang Transform (HHT) [5, 6], and in particular for the estimate of its power spectrum. The peculiarity of the HHT spectrum is that it provides a sparse representation of the signal in the joint parameter / parameter-rate domain. The misclassification errors are thus easily detectable as high-rate components in the Hilbert-Huang domain, and assuming a threshold value of the parameter-rate (calibrated on the observations), it is possible to reject the values of the parameters referred to the non-zero values of the power spectrum, above the threshold. Rejecting these parameter values also means rejecting the associated natural frequency values, and thus correcting the variation of the modal data by rejecting the variation component due to the misclassification error. At this point, it is possible that the natural frequency data is split into two or more classes divided by empty spaces generated by the rejection of the parameter and natural frequency values associated with MAC having high parameter-rate (above threshold) components other than zero in the Hilbert-Huang power spectrum. This leads to a problem in estimating the variation of the modal datum (i.e., natural frequency), as the statistical distribution associated with it could be characterized by several modes (i.e., multivariate distribution). In this situation the variation to be estimated is not that of the entire distribution (which in fact would mean re-introducing a variation component generated by the misclassification error), but rather a combination of the variations due to the single components of the multivariate distribution.

In the present work, in order to estimate the *variance* and the *mean* of the different components of the multivariate distribution, a naive solution has been adopted assuming the gaussianity of each component. In this way it was possible to use the iterative Expectation-Maximization (EM) algorithm [7, 8] to fit a Gaussian mixture model on natural frequency data, already cleaned by the misclassification errors. For each component, the mean and covariance value that maximized the likelihood of the Gaussian process was obtained. It should be noted that not all classes of natural frequency are associated with a correct coupling with the reference mode, and this happens because in the data sampling phase (i.e., virtualization of the Operational Modal Analysis - OMA experiment [9]) it can happen to collect a series of uniform values referred to mechanical parameters which do not produce modes comparable with the reference mode at all. In this situation it is understandable that natural frequency values close to the reference value are certainly more reliable than values far from them. Consequently, in this work it was assumed to estimate the variation of the modal datum as that referred to the component of the Gaussian process having the minimum distance from the reference datum. The Mahalanobis distance [10] was used to estimate the distance between the reference datum (a point) and the components of the Gaussian process (a distribution). It is clear that a procedure of this type can lead to an underestimation of the variation of the modal data, as it could happen that a component of the Gaussian mixture model fit on the data is actually representative of a correct coupling (rejection of meaningful data). To alleviate this problem, it is possible to estimate the variation of the modal datum as relative to the variation of the mechanical parameter (i.e., if a mechanical parameter varies little, the direct consequence is that the modal datum will also have a limited variation). By normalizing the variation of the modal datum to that of the mechanical parameter (also estimated with the procedure previously described using Gaussian mixture models) it is possible to obtain an estimate of the variation not affected by the domain of variation of the mechanical parameter, which must in any case be a domain: (i) composed of a large number of samples in order to reach a statistical significance also after the rejection of some data associated with misclassification errors by the proposed algorithm, (ii) close to the correct values assumed by the parameter (e.g., knowledge of the experimental data or an estimate of it from literature). The obtained variation, in this case estimated as the ratio between the coefficients of variation of the natural frequency and of the mechanical parameter (i.e., Young's modulus, Poisson ratio, and density), can then be corrected by remultiplying its value by the real variation assumed by the mechanical parameters in different purposes of use (e.g., model updating).

3 Application

To demonstrate the proposed methods, the Finite Element (FE) model and the structure of Pavilion V of Turin Exhibition Center is briefly described. A complete description of the Pavilion can be found in [11]. The Pavilion is a perfect example for applying the methods proposed in this paper since its structure consists of a spatial architecture, which from a modal point of view is characterized by innumerable vibration modes in the vertical direction, many of which present a high number of inflections in the mode shape. This large number of inflections (i.e., bubbles modes) can generate a misclassification error because, during an OMA, the eigenvectors are estimated on a small number of Degrees of Freedoms (DoFs). The observation of the eigenvectors on a reduced domain can lead to the highest MAC between a reference mode and a set of modes extracted from a numerical model being related to a mode not really comparable with the reference one. In addition, the Pavilion consists of a jointed three-block structure. The three blocks are thus weakly connected through a perimeter wall linked by connecting rods to the roof beams. In such a situation the modal behavior that most interests one block could easily be misinterpreted with that of another block. Finally, the Pavilion presents vertical and horizontal components of the mode shapes rather coupled, especially for the first modes.

Fig. 1. Pictures of Morandi's Pavilion V (interval views).

This increases the possibility of error in the classification of the modal behavior when it is observed on a small number of DoFs (e.g., a local out-of-plane deformation of the roof slab characterized by horizontal movement could be erroneously interpreted for a diaphragmatic behavior of the same roof in the horizontal direction). Figure 1 reports two pictures of the Pavilion.

The FE model is composed of eight structural components having uniform material characteristics. In the linear elastic field, it contemplates just three material parameters for each component c, i.e., Young's modulus E_c, Poisson ratio ν_c, and density ρ_c. It is composed of both 4-nodes shell and 2-nodes beam finite elements with six DoFs at each node (translation and rotations), and it counts approximately 27'000 elements for a total number of 165'000 DoFs and an average mesh size of 1.5 m. The model is fully restrained at the base. Figure 2 reports the FE model of the Pavilion with components highlighted in different colors, while Table 1 reports the components description.

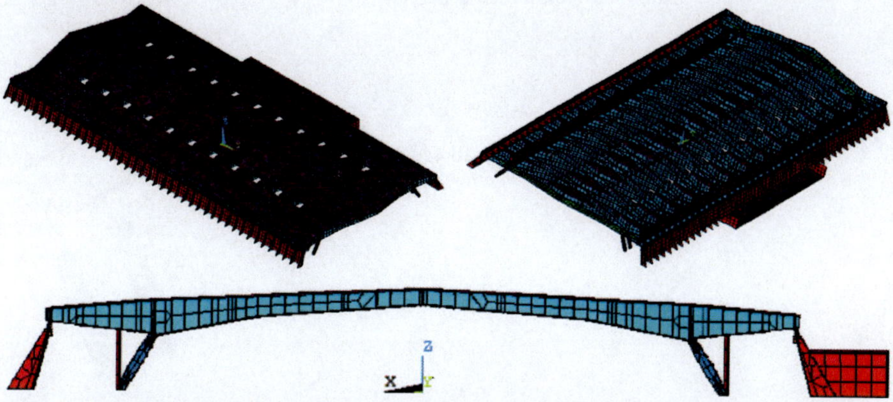

Fig. 2. Finite Element (FE) model of Morandi's Pavilion V (top, bottom, and front view).

The first (first transverse mode #5 at 2.40 Hz) and last (first torsional mode #23 at 5.41 Hz) reference mode used for the analysis are instead reported in Fig. 3 in terms of mode shapes. In addition to the modes reported in Fig. 3, the analysis contemplated also the first #6 and second #7 vertical mode of the slab at 2.62 Hz and 2.75 Hz, respectively, and the first longitudinal mode #21 at 5.02 Hz.

3.1 Results and Discussions

In this section, the results of the analyses are reported and commented on. Figure 4 reports the tracking of the mode number in the FE model for reference mode #5 as a function of the Young's modulus of infill walls. While the modulus remains lower than 2.65e9 Pa, the coupling is quite correct, and contemplates two classes. Between these two classes a misclassification error appears. The same behavior can be perceived in Fig. 5 where the behavior of both natural frequency and MAC of the first transverse mode #5 is depicted as function of the Young's modulus of the infill walls. It is quite easy to understand that misclassification errors can be of different nature; instantaneous-localized errors at 1.30e9 Pa, approximately, and continuous error starting from 2.65e9 Pa, approximately. The first error is rejected with HHT, while the second error is rejected with Gaussian mixture models fitted to natural frequency in this work. The reader, however, can understand that meaningful information is erroneously also rejected (class associated to parameter values lower than 1.30e9 Pa). This error is mitigated by the

Table 1. Description of the components of the Pavilion's numerical model.

FE component, *c*	Description (material)	Representation
1	Roof beams (prestressed concrete)	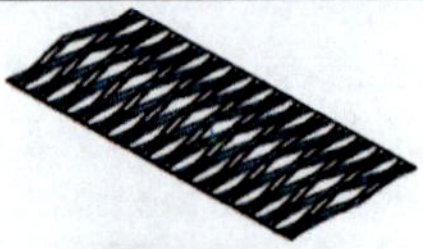
2	Roof slab (reinforced concrete)	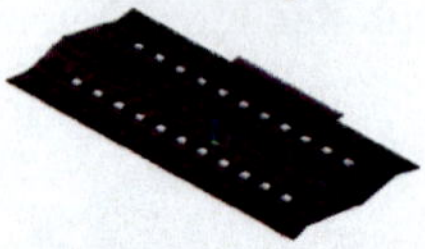
3	Perimetral walls (reinforced concrete)	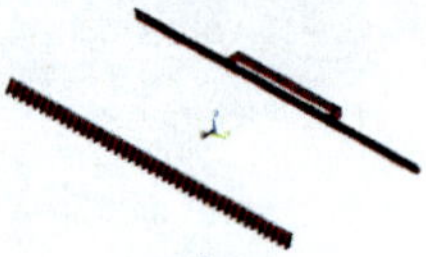
4	Inclined struts (reinforced concrete)	
5	Infill walls (cellular concrete)	
6	Edge beams (reinforced concrete)	
7	Connecting rods and perimetral beams (reinforced concrete)	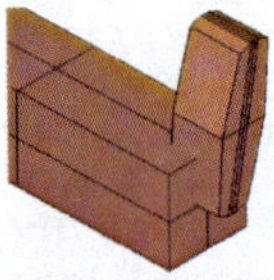
8	Top and bottom edges of inclined struts (steel)	

algorithm by normalizing the natural frequency variation to the variation of the parameter (estimated in the center class, defined between 1.35e9 Pa and 2.65e9 Pa).

Figure 6 reports the same concepts already described with Fig. 5, but is this case the natural frequency and MAC of the first transverse mode #5 is depicted as function of the Young's modulus of the roof beams. Figure 7 instead, introduce a new concept. Here, the HHT of the MAC function reported in Fig. 6 is reported. It is possible to see

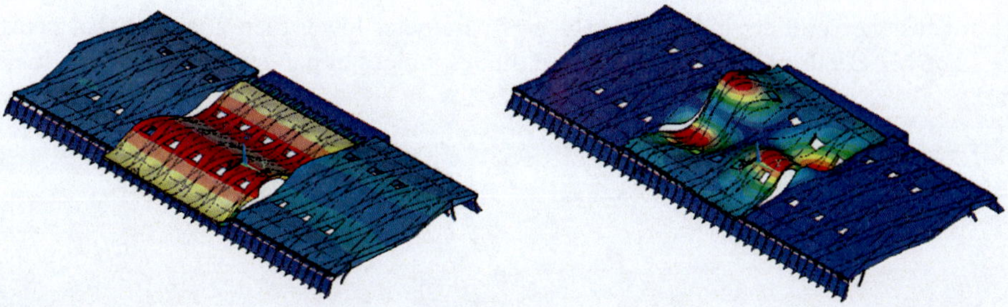

Fig. 3. Mode shapes of the first transverse reference mode #5 at 2.40 Hz, and first torsional reference mode #23 at 5.41 Hz used in the analysis.

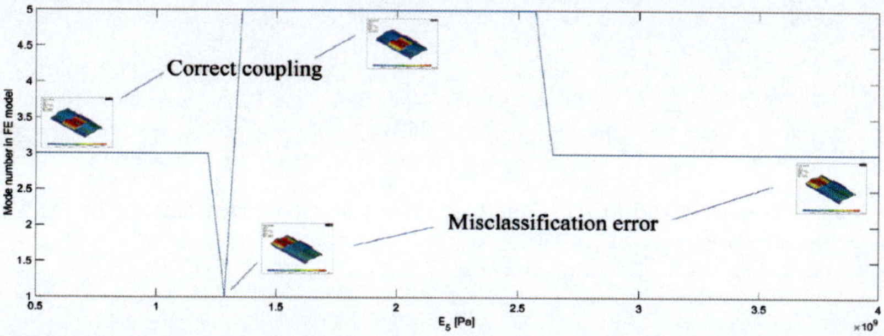

Fig. 4. Mode number in the FE model for the first transverse mode, as a function of the Young's modulus of the infill walls.

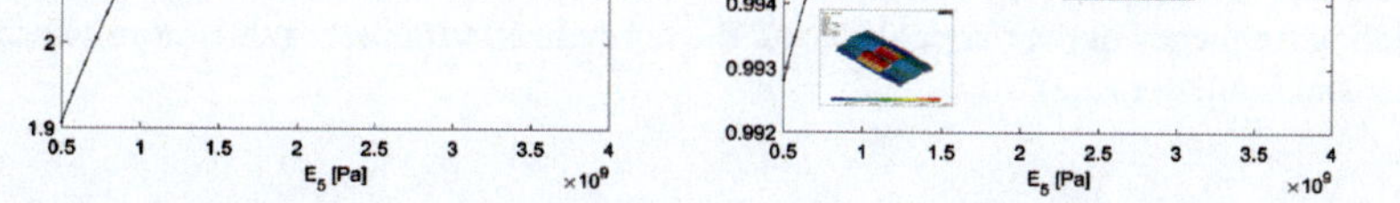

Fig. 5. Natural frequency and MAC of the first transverse mode, as a function of the Young's modulus of the infill walls.

that, on the contrary of Fig. 6, in Fig. 7, the outliers around 26e9 Pa (around 0.3 in the normalized space obtained using the stochastic oscillator indicator rule [12] applied to the parameter domain) are well classified at high parameter rate (threshold manually calibrated and imposed at 0.05 in this analysis). Thus, in this domain is quite easy to reject anomalies with respect to what may be done with MAC function of parameter:

see for instance outliers in Fig. 6 (values of parameter lower than 26e9 Pa) that takes the same MAC values of correct classified modes (values of parameter higher than 26e9 Pa).

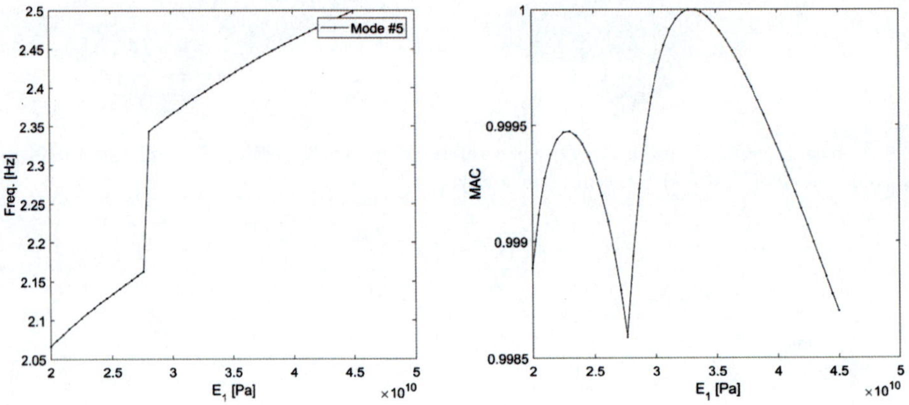

Fig. 6. Natural frequency and MAC of the first transverse mode, as a function of the Young's modulus of the roof beams.

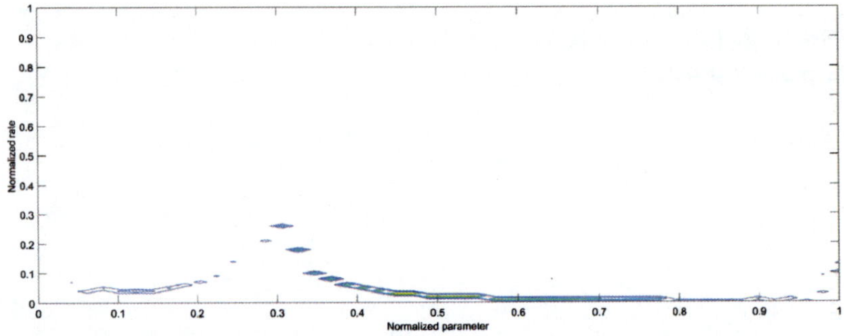

Fig. 7. Hilbert-Huang power spectrum as a function of the parameter rate (normalized to half of the sampling frequency of the parameter), and the normalized parameter value (normalized using the stochastic oscillator rule).

The coefficient of variation (ratio between standard deviation and mean) of the natural frequency is then normalized to the coefficient of variation of the parameter evaluated in the same class (same component) of the natural frequency. Then, these relative variation estimators are re-multiplied to the coefficient of variation of parameter values used for other purposes, such as the variation obtained by the parameters range used in model updating tasks (e.g., assuming a Gaussian prior distribution of the parameters). Figure 8 reports these results in terms of Fraction of Variation (FoV), that is the variation estimators normalized to the sum of all contributions.

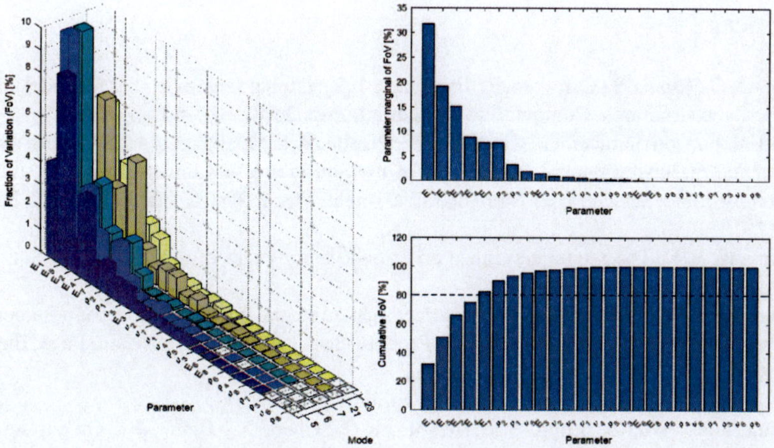

Fig. 8. Result of the sensitivity analysis in terms of the normalized Fraction of Variation (FoV) attributed to each parameter and each mode. On the top right, the marginal over the mode dimension is depicted, while on the bottom right the cumulative sum of the marginal is reported. Following the empirical Pareto's rule, the achievement of 80% of the total variation is highlighted with dashed black line in the cumulative FoV plot.

4 Conclusions

During mode coupling, in sensitivity analysis based on modal data, modes with high MAC value can be incorrectly associated to a target-controlled mode. In this context, the *correct* mode to be coupled to the reference one is characterized by a lower MAC value than the mode with the highest MAC. This problem is very common in architectural spatial structures where vertical and horizontal components of mode shape interact, especially in presence of mode shapes characterized by a huge number of local bubbles, or even worst in presence of interacting diaphragms, as in the case of Morandi's Pavilion V of Turin Exhibition Center. The method proposed in this paper mitigates these misclassification errors through the combined use of "time-frequency" analysis and Gaussian mixture models.

The authors showed that very fast variations of MAC and natural frequency studied as function of the assumed variation of mechanical parameters are most often associated with misclassification errors during the mode coupling process. These fast variations tend to be located at high parameter-rate values in the Hilbert-Huang space. The sparse nature of the Hilbert-Huang power spectrum allows easy identification of the parameter values associated with the misclassification error, allowing error rejection in the calculation of the variance of the natural frequency, which in normal conditions contemplates the presence of a true variance (that one want to estimate) generated by the variation of the mechanical parameter, and a variance component generated by the misclassification (i.e., tracking or coupling error).

References

1. Boscato, G., Russo, S., Ceravolo, R., Fragonara, L.Z.: Global sensitivity-based model updating for heritage structures. Comput. Civ. Infrastruct. Eng. **30**(8), 620–635 (2015)
2. Ubertini, F., Comanducci, G., Cavalagli, N., Pisello, A.L., Materazzi, A.L., Cotana, F.: Environmental effects on natural frequencies of the San Pietro bell tower in Perugia, Italy, and their removal for structural performance assessment. Mech. Syst. Signal Process. **82**, 307–322 (2017)
3. Allemang, R.J.: The modal assurance criterion–twenty years of use and abuse. Sound Vib. **37**(8), 14–23 (2003)
4. Huang, N.E., et al.: The empirical mode decomposition and the Hilbert spectrum for nonlinear and non-stationary time series analysis. Proc. R. Soc. London. Ser. A Math. Phys. Eng. Sci., **454** (1971), 903–995 (1998)
5. Wu, Z.: Hilbert--Huang transform and its applications (2005)
6. Huang, N.E., Wu, Z., Long, S.R., Arnold, K.C., Chen, X., Blank, K.: On instantaneous frequency. Adv. Adapt. Data Anal. **1**(02), 177–229 (2009)
7. Borman, S.: The expectation maximization algorithm-a short tutorial, p. 41. Publ, Submitt (2004)
8. McLachlan, G., Peel, D.: Finite Mix. Model (2000)
9. Civera, M., Sibille, L., Fragonara, L.Z., Ceravolo, R.: A DBSCAN-based automated operational modal analysis algorithm for bridge monitoring. Measurement **208**, 112451 (2023)
10. De Maesschalck, R., Jouan-Rimbaud, D., Massart, D.L.: The mahalanobis distance. Chemom. Intell. Lab. Syst. **50**(1), 1–18 (2000)
11. Ceravolo, R., Lenticchia, E., Miraglia, G., Oliva, V., Scussolini, L.: Modal identification of structures with interacting diaphragms. Appl. Sci. **12**(8), 4030 (2022)
12. Romano, P.K.: Application of the stochastic oscillator to assess source convergence in Monte Carlo criticality calculations (2009)

A Sustainability by Design Lesson Learned from Space

Valentina Sumini[1](✉), Marta Rossi[1], and Amedeo Manuello Bertetto[2]

[1] Politecnico di Milano, Piazza Leonardo da Vinci 22, 20133 Milano, Italy
valentina.sumini@polimi.it

[2] Politecnico di Torino, Corso Duca degli Abruzzi 24, 10129 Torino, Italy

Abstract. Space architecture involves the integration of multiple research disciplines to establish a framework for planning secure human settlements in Low Earth Orbit (LEO), on the Moon, or on Mars. Designing sustainable and safe habitats for space exploration requires a diverse range of skills and knowledge. As humanity enters an era of venturing towards neighboring celestial bodies, NASA's Artemis program envisions establishing permanent settlements at the South Pole of the Moon. These settlements aim to serve as testing grounds for future generations, fostering collaboration in the creation of joint infrastructures akin to the International Space Station's cooperative model—a new paradigm of an "ideal city" in a unique environment. The challenge of designing in extreme space environments is being addressed through innovative approaches such as computational design tools, topology optimization processes, and circular design methodologies.

Keywords: Space Architecture · Sustainability by Design · Computational Design

1 Introduction

Space architecture is an interdisciplinary field that emerges from the convergence of multiple domains, interlacing space sciences, architecture, engineering, robotics, industrial design, medicine, ergonomics, psychology, and art [1]. This dynamic discipline governs the design of habitats tailored for space missions, whether in Low Earth Orbit or on celestial bodies like the Moon or Mars. Prominent manifestations of space architecture include space stations, spacecraft, planetary habitats, and analog test structures [2, 3].

Space architecture requires an engineering mindset that incorporates human factors and architectural principles [4]. It is essential to prioritize human requirements and construct structures capable of sustaining human life both physiologically and psychologically. Design considerations should prioritize safety and comfort, with a human-centered approach being paramount, as future astronauts residing in these settlements will experience prolonged periods of isolation and confinement. Consequently, architectural designs must accommodate human needs and provide spaces that promote comfort and wellbeing.

S. Gabriele et al. (Eds.): IWSS 2023, LNCE 437, pp. 775–783, 2024.
https://doi.org/10.1007/978-3-031-44328-2_81

2 Space Architecture Design Challenges

The interdisciplinary nature of space architecture [5] presents challenges in creating a sustainable and resilient infrastructure to support human missions in LEO, on the Moon, or on Mars. Space architects leverage the technological expertise gained from the space industry to foster a collaborative environment capable of undertaking projects spanning various scales, ranging from urban planning to architecture and interior design. This integration of diverse fields enables the development of comprehensive solutions that can withstand the demands of space exploration.

One of the primary challenges that must be tackled pertains to the correlation between human activities and the resources required to sustain them [6]. Resource supply becomes a critical focal point in this context due to the exorbitant costs associated with transporting resources to outer space. Consequently, it becomes imperative to devise strategies that minimize the reliance on Earth-imported resources and instead foster the production of locally sourced goods. Developing a model that embraces this approach for space applications holds the potential to yield benefits for architecture on Earth, prompting a reframing of how we utilize the planet's resources and encouraging a more conscientious approach. Such an approach aligns with the goals outlined in the United Nations' Sustainable Development Agenda for 2030 [5] and introduces the practical implementation of the Sustainability by Design (SBD) concept.

Undertakings of this nature are characterized by their immense complexity, necessitating the consideration of multiple variables and requirements [7]. However, this challenge can be effectively addressed by leveraging the capabilities of computational design and multi-objective optimization tools. These tools enable the resolution of various objectives, including minimizing terrestrial mass, maximizing the utilization of in situ resources, implementing protective measures against radiation and micrometeoroids, and embracing the principles of sustainability, which should permeate the entire design process [6].

Indeed, the support for a Human Space Mission (HSM) entails a multitude of phases, encompassing habitability requirements and the utilization of In Situ Resources (ISRU). Both technical and human factors hold significant importance in the space architecture design process. Principles rooted in Human Factor Design guide the development of architectural solutions that prioritize human psychological and physiological well-being. Moreover, construction automation stands as a vital aspect that complements and supports human efforts in space endeavors. Lastly, the concept of Sustainability by Design (SBD) enables the creation of structures founded on strategies centered around the principle of 'less is more', a concept that holds potential applicability in terrestrial projects as well.

3 Space Habitats Classification

NASA has defined three primary categories of habitat structures for lunar, Martian, and other celestial body exploration, taking into account mission requirements and reliance on Earth-based materials and technology [3].

Class I habitats involve fully constructing the structures on Earth prior to their departure for outer space. These habitats are pre-integrated and ready-to-use upon lift-off from the ground, eliminating the need for deployment upon arrival at the destination. The concept of pre-integrated structures was initially conceived for the Apollo missions (Fig. 1), wherein habitable spacecraft were entirely assembled on Earth. These structures were designed to provide comprehensive support to astronauts for a duration of 14 days, serving as a viable construction approach for the initial phases of human planetary exploration.

Fig. 1. Apollo Lunar Module (Credit: NASA)

The current International Space Station has been designed and built upon several Class I modules that have been assembled in LEO.

Class II habitats encompass prefabricated structures that necessitate deployment and assembly upon reaching their designated location. This approach offers several advantages, including increased habitable volume while maintaining the same available volume within the transporting vehicle. By deploying and assembling these structures on-site, greater space for human habitation can be achieved, optimizing the efficiency of the available resources. The structures used for habitats can be compactly packaged, occupying a smaller volume compared to their assembled form. This type of habitat design allows for integration and enhancement by combining Earth-made technology with In Situ Resources Utilization (ISRU) constructions. For instance, lunar or Martian regolith (soil) can be utilized to build radiation shields or accommodate additional equipment. Class II habitats serve as an intermediate step in human space exploration as they can provide shelters for longer durations.

There are several noteworthy projects involving Class II habitats for both lunar and Martian exploration. A collaborative effort between ESA, Foster + Partners, Monolite Ltd, Alta SpA, and Scuola Superiore Sant'Anna took place from 2009 to 2015.

They developed a hybrid concept comprising an inflatable structure connected to a pre-integrated section. The project also incorporated a radiation shield, leveraging existing 3D printing technology known as D-Shape and adapting it for space applications [8]. Another example is derived from the NASA 3D Printed Habitat Challenge held in 2018. Hassell Studio, in collaboration with Eckersley O'Callaghan (EOC), presented a project featuring multiple inflatable pods covered by a regolith shell. The regolith shell would be 3D printed prior to the arrival of astronauts [9].

Another notable Class II project is the Moon Village (Fig. 2), a concept jointly developed by ESA, Skidmore, Owings & Merrill, and MIT. This design incorporates inflatable vertical structures that feature three bladder systems connected to a rigid structural element [10]. The research team devised a masterplan for the entire village, with one of the primary objectives being to find a solution for the integration of multiple modules, allowing for incremental growth over time. The Moon Village encompasses various distinct areas based on their functions, including the habitation zone, an activity band, and the energy and transportation zone [11].

Fig. 2. Moon Village by ESA, SOM, MIT [11].

Class III habitats are constructed entirely using resources sourced or produced in situ, representing the most advanced technology for space habitats. However, the feasibility of this approach has yet to be fully demonstrated, and extensive testing and experimentation are required due to the relatively low Technology Readiness Level (TRL) of current examples in this class. This type of technology holds significant potential for future colonization phases, as it would substantially reduce the need for resources transported from Earth [12].

An exemplary instance of a Class III habitat is the Lunar Lantern, developed within the context of Project Olympus by ICON, Search +, and Bjarke Ingels Group (BIG) as part of NASA's Moon to Mars Planetary Autonomous Construction Technologies (MMPACT) initiative. The project envisions a vertical habitat constructed entirely from in situ materials, comprising an inner fabric liner, a 3D-printed vessel, and an outer

Whipple shield [13]. This pioneering endeavor showcases the potential of utilizing local resources for the creation of self-sustaining habitats in space exploration missions.

4 Sustainability by Design

4.1 Decreasing Dependence on Earth's Resources

To facilitate future space exploration missions, reducing reliance on resources from Earth will be imperative. Missions that extend beyond Low Earth Orbit (LEO) will necessitate facilities with increased self-sufficiency. Currently, human space missions aboard the International Space Station (ISS) heavily rely on regular resupplies from Earth. The proximity of the ISS to Earth and the relatively low cost of shipping cargo make this feasible. However, the challenges escalate when considering missions to the Moon, which typically take three days for transportation, or Mars, which requires a journey of approximately 6 to 8 months. Compared to the ISS, shipping cargo to these distant destinations becomes technologically and economically more demanding [14].

To thrive on celestial bodies beyond Earth, it will be crucial to harness and utilize in situ materials. This involves processing resources available on these bodies to produce construction materials, fuel, and even oxygen [15]. NASA has already made significant progress in this regard, successfully extracting oxygen from the thin Martian atmosphere through the Mars Oxygen In-Situ Resource Utilization Experiment (MOXIE). MOXIE has demonstrated the capability to extract 10 g of oxygen per hour [16]. The ultimate objective is to establish a closed-loop system, where resources are efficiently recycled and reused. To achieve this, it will be necessary to incorporate a food production facility that enables the cultivation of plants capable of generating consumable resources and nutrients to support human life. This aspect becomes particularly crucial for Martian missions, where resupply from Earth would be exceedingly challenging. Therefore, astronauts must possess the ability to sustain themselves autonomously throughout the entire duration of the mission.

4.2 Leveraging on Autonomous Construction Systems

Automation plays a fundamental role in construction within space architecture projects as it enables the construction of structures without the need for human presence. Utilizing technology during the preliminary phases is crucial for studying the conditions of future settlement locations and initiating the construction of initial structures. Given the high levels of radiation on Mars and the Moon, it becomes necessary to commence the construction of radiation shielding before human arrival. Additionally, various other operations, such as excavations and soil sintering, can be carried out autonomously to prepare the ground for habitats and the infrastructure connecting the entire system [17].

There are multiple approaches to achieving automated additive construction using lunar or Martian regolith. Numerous technologies have been tested on Earth and can be adapted for space applications. One example involves combining regolith with additives like sulfur, Portland cement, or plastics, employing a cementitious construction method. Sulfur, when combined with regolith, performs well in vacuum conditions, making it a viable solution for space construction. However, transporting additives to Mars or the Moon presents challenges. Therefore, an alternative automated construction method called microwave melting could be utilized. This technique involves melting regolith within a chamber and subsequently extruding it [6]. By leveraging automation and additive manufacturing techniques, construction processes can be optimized for space exploration missions.

4.3 Generating Novel Computational Design Tools

Computational design in architecture entails utilizing computer software, algorithms, and other digital tools to aid in the design and analysis of projects at various scales. This includes employing building information modeling (BIM) software and multi-physics simulation and analysis tools to assist architects and engineers in exploring diverse design options, assessing structural and environmental performance, and optimizing building systems. Furthermore, computational design techniques enable the generation of intricate geometric forms and shapes that may be challenging or even impossible to achieve through traditional design methods, facilitating the exploration of complex systems.

In the field of Space Architecture, mathematical methods such as Multi-Objective Optimization (MOO) are essential for exploring and identifying the best solutions among a range of possible outcomes, where multiple objectives or goals are simultaneously optimized. The multidisciplinary nature of Space Architecture necessitates striking a balance among various objectives, such as minimizing resource usage and mass transfer from Earth, while maximizing protection against micrometeoroids, Galactic Cosmic Rays, and solar flares, as well as ensuring structural performance, energy efficiency, habitable volume, and the utilization of autonomous construction systems [18]. By leveraging computational design and optimization techniques, space architects can address the intricate challenges involved in creating sustainable and efficient space habitats.

In literature, numerous projects leverage computational design techniques in conjunction with Multi-Objective Optimization (MOO) algorithms to guide the form-finding process for lunar or Martian habitats or settlements, addressing a range of essential requirements across different scales. One prominent software tool in this domain is Rhinoceros© [19, 20], a 3D modeling software offering a comprehensive set of modeling tools, including NURBS (Non-Uniform Rational B-Splines), meshes, and surfaces. Rhinoceros© is complemented by Grasshopper3D©, a visual programming platform. Within Grasshopper©, various plugins support form-finding and optimization processes.

For example, Kangaroo Physics© [21] enables dynamic relaxation and multi-physics interactions, facilitating the exploration of complex behaviors. Karamba3D© [22], a parametric structural engineering tool, allows for structural form-finding and optimization, offering accurate analysis of spatial trusses, frames, and shells. Ameba© [23], a topology optimization software based on Bi-directional Evolutionary Structural Optimization (BESO) technology, enables topology optimization processes. Another valuable plugin is Octopus©, which allows users to define design variables and constraints, employing different optimization algorithms to identify optimal solutions that fulfill specified objectives. It also facilitates the visualization and analysis of optimization

Fig. 3. Project "Martian Formicary" of the course "Architecture for Human Space Exploration" at Politecnico di Milano (A.Y. 2021–22) by Prof. Valentina Sumini, T.A. Marta Rossi. Students: Angelova M., Bitik D., Cihan Alkan M., Lichocik K., Marticorena Angela V., Mileni Munari M., Pedrazzini S.

results and trade-offs within the Pareto Front. The final designs generated through these tools (Fig. 3) can be further analyzed and prepared for fabrication.

By employing these computational design techniques and software tools, researchers and practitioners can explore innovative design solutions, optimize performance, and evaluate trade-offs in the development of lunar or Martian habitats and settlements.

5 Conclusions

The emergence, rapid evolution, and continuous expansion of a robust "space economy" globally, focusing on the scientific and economic utilization of outer space, including the Moon and eventually Mars in the coming decades, highlight the necessity for a structured educational framework. Such a framework would encompass the various technological aspects mentioned above and serve as a platform for dedicated learning activities.

The common thread running through all the aforementioned learning activities is the utilization of computational design tools, enabling students from diverse backgrounds and educational levels to communicate through a universal language.

These community tools, often open source, possess captivating features that generate a widespread and enthusiastic interest among younger generations. They perceive these design activities as a fresh canvas, where they can envision their future using innovative tools and approaches.

Furthermore, the imperative to maintain a careful balance and effectively manage limited resources such as energy, water, air, and food reinforces the need for a heightened awareness of sustainability. Applying similar principles and criteria when addressing sustainability concerns on Earth can contribute to guiding humanity towards a secure and prosperous future.

References

1. Adams, C., Arenales, O., Cohen, M.: The Millennium Charter. Space architecture workshop by AIAA DETC Aerospace Architecture Subcommittee, Houston, USA (2002)
2. Häuplik-Meusburger Bannova, S.: Space architecture education for engineers and architects designing and planning beyond earth. San Francisco: Springer, (2016)
3. Howe, A.S., Sherwood, B.: Out of this world the new field of space architecture. (Reston: American Institute of Aeronautics and Astronautics, Inc., (2009)
4. Connors, M.M., Harrison, A.A., Akins, F.R.: Living Aloft Human Requirements for Extended Spaceflight. US Government Printing Office, Washington DC (1985)
5. Transforming our world: the 2030 Agenda for Sustainable Development, United Nations. Accessed 1 Sept 2022. https://sdgs.un.org/2030agenda
6. Mueller, R.P., et al.: Automated additive construction (ACC) for earth and space using In-Situ resources. In: Presented at the Fifteenth Biennial ASCE Aerospace Division International Conference on Engineering, Science, Construction, and Operations in Challenging Environments, Orlando, April 11–15 (2016). https://oro.open.ac.uk/45865/1/3Dadditive_paper_final.pdf
7. Kestelier, X.D., Dini, E., Cesaretti, G., Colla, V., Pambaguian, L.: 3D printing regolith as a construction technique for environmental shielding on the moon. Fabricate 2014: Negotiating Design & Making, pp. 198–205 (2017)
8. Cesaretti, G., Dini, E., Kestelier, X.D., Colla, V., Pambaguian, L.: Building components for an outpost on the Lunar soil by means of a novel 3D printing technology. Acta Astronaut. **93**, 430–450 (2014)
9. NASA 3D Printed Habitat Challenge, Hassell Studio. Accessed 13 Jan 2023. https://www.hassellstudio.com/project/nasa-3d-printed-habitat-challenge
10. Petrov, G.I., et al.: Moon village reference masterplan and habitat design presented at ICES-2019–280. In: 49th International Conference on Environmental Systems, Boston, Massachusetts, 7–11 July (2019). https://ttu-ir.tdl.org/bitstream/handle/2346/84487/ICES-2019-280.pdf
11. Inocente, D., et al.: Master planning and space architecture for a moon village. In: Presented at IAC-2019-D4, 1, 2, 70th International Astronautical Congress (IAC), Washington, DC, 21–25 October (2019)
12. Technology Readiness Level, NASA. Accessed 08 Aug 2022. https://www.nasa.gov/directorates/heo/scan/engineering/technology/technology_readiness_level
13. Yashar, M., et al.: Project Olympus: off-world additive construction for lunar surface infrastructure. In: Presented at ICES-2021–96, 50th International Conference on Environmental Systems, Lisbon, Portugal, 12–15 July (2021). https://ttu-ir.tdl.org/handle/2346/87095
14. Calles, C.: The electrostatic environments of Mars and the Moon. J. Phys. Conf. Ser. (2011)
15. Labeaga-Martínez, N., Sanjurjo-Rivo, M., Díaz-Álvarez, J., Martínez-Frías, J.: Additive manufacturing for a Moon village. Procedia Manuf. **13**, 794–801 (2017)
16. Moxie, NASA. Accessed 13 Jan 2023. https://mars.nasa.gov/mars2020/spacecraft/instruments/moxie/
17. Mueller, R.P., et al.: Automated Additive Construction (ACC) for earth and space using In-Situ resources. In: Proceedings of the Fifteenth Biennial ASCE Aerospace Division International Conference on Engineering, Science, Construction, and Operations in Challenging Environments (Earth & Space 2016), American Society of Civil Engineers (2016)
18. Keane, A.J., Nair, P.B.: Computational Approaches for Aerospace Design: The Pursuit of Excellence. John Wiley & Sons Ltd, Chichester (2005)
19. Ericson, M.: Review: grasshopper algorithmic modeling for rhinoceros 5. J. Soc. Archit. Hist. **76**(4), 580–583 (2017)

20. McNeel, R.: Rhinoceros. Seattle, WA: Robert McNeel & Associates (2018)
21. Kangaroo. Grasshopper, McNeel, Robert, and Associates. https://www.food4rhino.com/app/kangaroo-physics. Accessed 15 Jan 2023
22. Preisinger, Thomas. Karamba3D User Manual. Food4Rhino. https://www.food4rhino.com/en/app/karamba3d?lang=it. Accessed 15 Jan 2023
23. Ameba. Kancloud, https://www.kancloud.cn/woshiyaoyuan0318/ameba_en/582345. Accessed 15 Jan 2023

A Mobile Habitat for Human Lunar Exploration

Valentina Sumini[1,5](✉), Claudio Chesi[1], Daniele Florenzano[1,3], Michela Botti[1,3], Lorenzo Calogero[2,3,3], Ana Carolina Corrêa Caracas[2,3], Alessandro Caruso[1,3,3], Massimo C. Comparini[6], Enrico Ferrone[6], Jeffrey Hoffman[4], Paolo Maggiore[2], Marta Mattioli[1,3], Joseph Paradiso[5], Maria Antonietta Perino[6,6], Marco Portolani[2,3], Angela Rizzo[1,3], and Giulia Signorotto[1,3]

[1] Politecnico di Milano, Piazza Leonardo da Vinci 22, 20133 Milano, Italy
valentina.sumini@polimi.it
[2] Politecnico di Torino, Corso Duca degli Abruzzi 24, 10129 Torino, Italy
[3] Alta Scuola Politecnica, Torino, Italy
[4] Massachusetts Institute of Technology, 77 Massachusetts Avenue, Cambridge, USA
[5] MIT Media Lab, 75 Amherst Street, Cambridge, USA
[6] Thales Alenia Space, Str. Antica di Collegno 253, 10146 Torino, Italy

Abstract. Since the Apollo missions, there have been significant technological advancements and scientific discoveries in robotic exploration of deep space. Currently, NASA's Artemis Program aims to establish human habitation on the Moon, which remains a considerable challenge. To address this, Alta Scuola Politecnica, Thales Alenia Space and MIT have collaborated to design an innovative and adaptable mobile habitat using a holistic multidisciplinary approach for crewed surface exploration missions. The outcome of the Lunar Architecture Design Exploration (LADE) project is a mobile space architecture system that enables human presence on the Moon, supporting medium to long-term missions. This mobile module serves as a crucial component within a more extensive system of hybrid class II and class III shelters, intended for the development of a lunar village. The primary objective is to facilitate the extended stay of four astronauts near the Shackleton crater, strategically located at the South Pole of the Moon, which offers favorable conditions for surface exploration and potential future permanent settlement. The paper presents an in-depth study of the form-finding process and structural analysis adopted for the LADE mobile habitat.

Keywords: Space Architecture · Mobile Habitat · Structural Analysis · Moon

1 Introduction

After the Apollo missions, the exploration of deep space has witnessed significant technological progress and remarkable scientific breakthroughs. Even now, more than 50 years since the extraordinary first Moon landing, NASA is formulating a vision to establish a human presence on the Moon. This new phase of space exploration revolves around the Artemis Program, which brings together international efforts to accomplish these

S. Gabriele et al. (Eds.): IWSS 2023, LNCE 437, pp. 784–790, 2024.
https://doi.org/10.1007/978-3-031-44328-2_82

ambitious objectives. It is worth noting that collaboration among nations in space exploration is not unprecedented, as exemplified by the International Space Station (ISS), a shining example of cooperative ventures that have lasted for over 23 years. The ISS represents the first-ever extraterrestrial society, symbolizing the collective achievements of countries working together in space.

To support the successful human exploration of the Moon and eventual missions to Mars, it is essential to establish a mission framework. From a broader perspective, the project Lunar Architecture Design Exploration (LADE) focuses on initial exploration phases, which are intrinsically connected to the requirement for efficient transportation.

The requirements primarily revolve around ensuring long-term mobility and the well-being of the crew, aligning with the vision of establishing a permanent presence on the Moon. This necessitates the development of a system capable of supporting extended human missions for exploration purposes. Notably, there are significant constraints to consider, including limitations on volume and mass, as well as practical factors such as budgetary considerations and adherence to international law regulations, which are not addressed in this particular article.

In essence, the project aims to design a mobile module that can accommodate four astronauts for a 60-day mission. This module serves as a crucial link between robotic precursory missions for lunar site exploration, specifically the South Pole of the Moon, and the establishment of a human settlement across a network of outposts on the lunar surface.

2 LADE Design

Within the LADE project, a variety of mobile modules are envisioned, each designed to cater to diverse mission objectives and requirements based on their unique functions and internal layout. The underlying concept involves offering a selection of mobile modules that can be interconnected through airlocks, enabling the choice of a specific module for a given mission.

The inspiration behind the LADE project stems from a comprehensive analysis and subsequent interpretation of bees. The inherent characteristics of bees are reimagined in a futuristic context, envisioning a lunar swarm and its interconnected habitat. The project emphasizes the interdependence of both the shelters and the mobile modules, as they are mutually essential for functioning and existence.

The biophilic approach is also applied to the design of mobile modules, where the arrangement of their functions takes inspiration from the body structure of bees. When interconnected, these modules resemble the head, thorax, and abdomen of the bee, with each section serving specific purposes such as navigation, communication, locomotion technologies, and extravehicular activity (EVA) systems in the primary module, and research or storage functions in the secondary module.

Similarly, the method chosen for the movement of mobile modules draws parallels to the coordinated behavior of a bee swarm. These modules form a group of individual units that move together and work towards a shared objective. To ensure order and minimize the need for navigation tools and control equipment within the mobile modules, the project envisions a primary module responsible for navigation and leading the way.

The secondary modules, selected for the mission based on computer vision, artificial intelligence (AI), and fleet learning, follow the primary module. Each secondary module possesses its own mobility system.

This approach offers significant flexibility in achieving the desired performance levels depending on the specific mission requirements. Currently, the primary module focuses on navigation and control, while the secondary modules are dedicated to tasks such as research, storage, or transportation (Fig. 1).

Fig. 1. Render of the exterior.

The primary mobile module's internal layout is structured based on its key functions, divided into three sections: the cockpit, center, and rear. The cockpit is specifically designed for navigation and control purposes, featuring spacious desks equipped with displays, speakers, projectors, and sensors. The chairs are adjustable on rails, allowing astronauts to access all control areas and position themselves in front of the transparent surface of the module's shell. This transparency enables the drivers to have a clear view of the Moon's surface and the surrounding environment.

3 Designing the Module Shell Structure

3.1 General Aspects

The geometry of the LADE module has been determined by the current technological and volumetric constraints coming from the payload launch from Earth and the current limitation of the launcher's capabilities. The geometry of the shell has been therefore designed as a simil-cylindrical shape, the section of which is inscribed within a circle 4.5 m in diameter.

The complexity of the structural system for a mobile lunar habitat has been simplified into two main macro-elements: the independent structure of the mobility system, which includes the structural elements composing the modified rocker-bogie mobility system, providing support to the shell, and the shell structure itself, containing the pressurized habitat space. The shell structure is composed of a panel structure reinforced by a network of lightweight beams in the most stressed areas. Structural optimization was performed to define the most efficient configuration of sections of both beam and shell elements. To this purpose, the geometry has been completely developed within Grasshopper© [1], while the structural analysis has been performed with the Grasshopper© plug-in Karamba3D® [2].

3.2 Definition of Design Loads

Multiple load conditions must be defined when designing a structure that is meant to be operated in space, considering all the different mission phases. A main issue is related to the Moon's gravity, which accounts for roughly 1/6 of Earth's gravity, impacting the calculations of structural dead loads. Another factor to be accounted for is the lack of a consistent atmosphere, which results into the pressurization of the interior spaces to 1 atm, creating an outward directed load on the whole pressurized surface of the shell. Along with these considerations, another crucial aspect from a structural standpoint is the exposure to extremely harsh temperature excursions, which could impose a heavy thermal loading condition on the structures [3].

The design of the shell structure has been based on a simple static analysis, with the purpose of optimizing both the shell shape and cross sections; further research would require performing a dynamic analysis, considering both launch and operating conditions.

3.3 Choice of Materials

In order to limit the module weight, an Al-Li alloy, AA2060-T8, has been selected for the shell structure, which is required to sustain the pressurization load of the interior spaces of the module and its own weight [4]. An additional material, the borosilicate glass, has been selected for the transparent portions of the shell, due to its high resistance to thermal shock and its optimal mechanical properties for space uses [5]. Table 1 provides information on the mechanical properties of such materials.

Table 1. Mechanical properties of structural materials.

Element	Material	Young's modulus [GPa]	Tensile strength [MP]	Compressive strength [MPa]	Poisson ratio
External shell Glass window module	Al-Li alloy AA2060-T8 Borosilicate Glass	76.5 64	476 81.6	470 914.7	0.33 0.20

3.4 Geometrical Configuration

Several requirements have been considered in the search for a suitable geometrical configuration. First of all, due to the conformation of the mobility system [6], a symmetrical shape has been adopted [7]. In addition to this, sharp edges have been avoided, in order to reduce stress concentration and improve manufacturability.

The resulting geometry is composed by 3 portions, which include a central segment, symmetrical with respect to the center of mass of the structure, and two quasi hemispherical shells which host on the one side the airlock dedicated to the physical connection with the secondary module and on the other one the opening nearby the navigation cockpit. These three segments are connected to each other by a transition segment with the cross section composed by two half circular arches and two horizontal beams, in order to expand the horizontal floor surface of the module and increase the usable interior space for the astronauts. Figure 2 shows the parametrically modelled geometry of the shell.

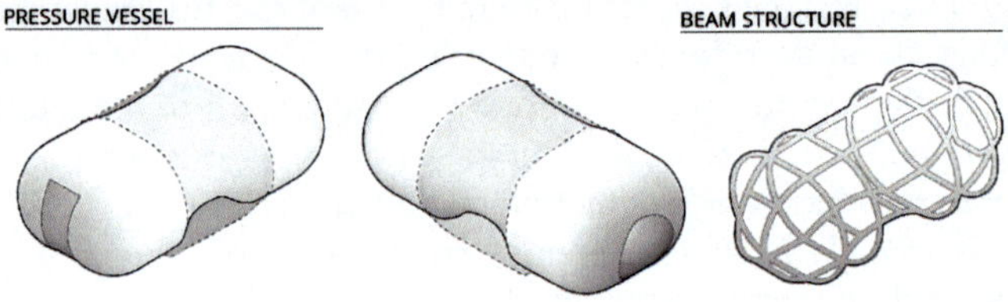

Fig. 2. The shell obtained from a parametric model.

3.5 Structural Analysis and Design Optimization

The design of the pressurized shell has been based on numerical analyses performed by Karamba3D. To achieve this, the computational procedure has included the following steps:

1. Modelling of the shell structure, which has also required the integration of the cockpit opening and the airlock opening. The Rhinoceros® [8] native NURBS [9] geometry has been then transformed into a mesh of variable density to be compatible with the Karamba3D environment.
2. Setting up of the structural model. This step has included the definition of design loads and load combinations, as well as the constraints representing the connections with the rocker-bogie structure; also, the cross sections for both the structural panels and the supporting light beams have been defined. The applied loads include: gravity loads, the operation pressurization load of 1 atm, and additional loads corresponding to the weight of the permanent elements of the structure, in particular the shielding elements and the batteries. For beams, hollow square cross sections with a 10cm side and a thickness of 1.2cm have been adopted, whereas the shell sections, being the object of optimization, are variable. Both structural elements have been applied the same Al-Li alloy as previously described.

3. Lastly, the static analysis of the model, including shell and beam elements, has been conducted. Typical results consist in the graphical representation of stress distributions and displacements. Figure 3 shows displacements occurring on the structure, from which it can be noticed that maximum values occur in the central-lower portion of the structure. It has also to be considered that high displacement values are present in the upper central area of the shell, in the vicinity of the supports. Looking at Fig. 4, the spherical structures show uneven distribution of section utilization at both ends, with a reduced value for the spherical elements. Overall, the displacement and section utilization diagrams show a comparable behavior.

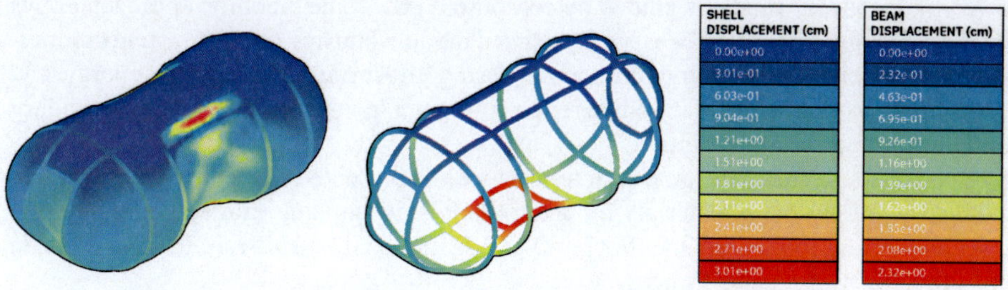

Fig. 3. Displacement levels for both the beams and the shell structure.

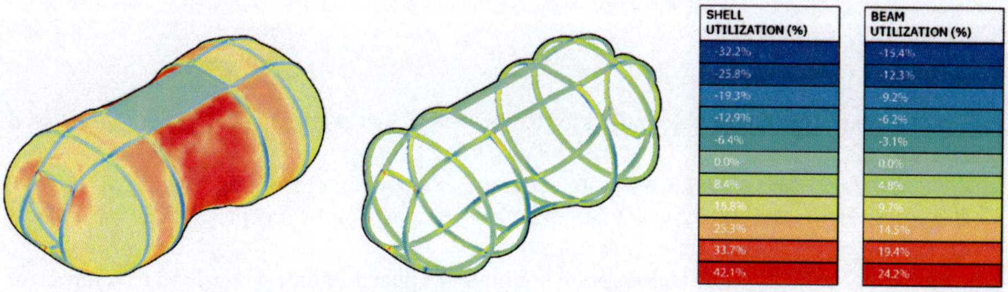

Fig. 4. Section Utilization on the beam and shell structure

3.6 Final Design Choices

The structural analysis of the pressurized shell has revealed the necessity to modify the element cross sections to obtain a more regular response of the structural system to the applied loads.

In particular, two actions have been undertaken to the purpose of improving the structural performance: first, additional segments of supporting light beams have been introduced in highly stressed areas; then the shell portions subject to maximum displacements and with the highest section utilization have been modified increasing the shell thickness by a factor of two, going from the initial value of 1.25 cm to a double layer panel of 2.50 cm.

4 Conclusions

This paper presents a study on a mobile space architecture system designed to enable human presence on the Moon for medium to long-term missions. The LADE research began by identifying the needs and requirements of the mobile habitat and its subsystems through a holistic approach. The objectives include ensuring robust structural support, efficient mobility and speed for lunar surface travel, self-sufficiency in energy under various conditions, hazard shielding, and landing site analysis [10].

Two key concepts served as references for developing a unique solution. The "wagon train" concept involves the primary mobile module leading the way in navigation, followed by secondary modules guided by computer vision. The biophilic approach draws inspiration from the body of bees, with the head module housing navigation and communication systems, the thorax module encompassing life-sustaining functions, energy, and locomotion systems, and the abdomen representing the specific function of the secondary module selected for each mission.

In this paper a focus on some structural aspects is presented since the research aimed to identify cutting-edge materials for the module's design and utilized computational analysis through tools like Grasshopper© and Karamba3D® software to ensure a safe and structurally efficient solution.

LADE, in its entirety, goes beyond being a mere design endeavor for an innovative surface exploration and habitability solution. It strives to establish a firm foundation upon which a new era of enduring Lunar exploration can be built.

References

1. Grasshopper, Algorithmic Modelling for Rhino. https://www.grasshopper3d.com/. Accessed 20 Aug 2022
2. Welcome to Karamba 3D. https://manual.karamba3d.com/. Accessed 04 May 2022
3. Benaroya, H.: Lunar habitats: a brief overview of issues and concepts. Reach **7–8**, 14–33 (2017)
4. Lequeu, P., Smith, K.P., Daniélou, A.: Aluminum-Copper-Lithium Alloy 2050 developed for medium to thick plate. J. Mater. Eng. Perform. **19**(6), 841–847 (2010)
5. Lima, M.M., Monteiro, R.: Characterisation and thermal behaviour of a borosilicate glass. Thermochim. Acta **373**, 69–74 (2001)
6. Li, S., Gao, H., Deng, Z.: Mobility performance evaluation of lunar rover and optimization of rocker-bogie suspension parameters. In: 2nd International Symposium on Systems and Control in Aerospace and Astronautics, Shenzhen, China, 10–12 Dec (2008)
7. Howe, A.S., Sherwood, B., Eds.: Out of This World: The New Field of Space Architecture. American Institute of Aeronautics and Astronautics, Reston, VA (2009)
8. Rhinoceros3D. https://www.rhino3d.com/it/. Accessed 20 Aug 2022
9. Dimas, E., Briassoulis, D.: 3D geometric modelling based on NURBS: a review. Adv. Eng. Softw. **30**(9–11), 741–751 (1999)
10. Florenzano, D., et al.: LADE: a mobile habitat paving the way for sustained lunar exploration. In: Proceedings of 73rd International Astronautical Congress 2022, Sept. 18–22. Paris, France (2023)

The Structural Tile of the Tholos of Delphi, Its Design and Construction Contemporary

Attilio Pizzigoni[1(✉)], Vittorio Paris[1], Matteo Bonasio[2], and Mattia Agazzi[2]

[1] Department of Engineering and Applied Sciences, University of Bergamo, Dalmine, Italy
attilio.pizzigoni@gmail.com, vittorio.paris@unibg.it

[2] FabLab Bergamo, Bergamo, Italy
{matteo.bonasio,mattia.agazzi}@fablabbergamo.it

Abstract. The text explores the fundamental question raised by Edoardo Persico nearly a century ago: Is architecture solely an engineering solution for building problems? It emphasizes the need to combine intuition and demonstration in architecture, drawing inspiration from Dante's works. The study emphasizes the importance of understanding the history and origins of architecture and how seemingly obsolete ideas can become modern again. The research focuses on the Tholos of Athena Pronaia in Delphi, analyzing its structural and architectural form using 3D elaboration. It highlights the interdisciplinary dialogue between contemporary research and classical Greek cosmological vision. The text raises questions about the collapse of ancient Greek temples and proposes innovative solutions. The Tholos of Delphi serves as a timeless icon for studying rational roots and design devices. The research utilizes advanced technologies like 3D printing to create a detailed prototype and evaluate the distribution of elements for balance and structural functionality.

Keywords: Architecture · Tholos · dome · 3D modeling · Greece · tile

1 The Thought

Nearly a century ago, in 1935, Edoardo Persico, in a conference he held in Turin, stated that the "coming architecture" is not an engineering solution for building problems [1]. I believe this remains the fundamental question for those dedicated to the pursuit of advancement in this field, or as Nervi defined it, those who believe architecture to be the "art of construction." The future, whether it be near or distant, of this discipline cannot solely rely on an engineering solution for the construction of buildings.

Edoardo Persico, in envisioning the possible prophecy of a future for architecture, spoke of the unity between intuition and demonstration, drawing upon a figure from Dante's works. He referred to the "coming architecture" as a faith, a substance of hoped-for things, and a demonstration of unseen matters. "Prophecy of Architecture" was the title of his Turin conference, where he analyzed the form finding as a belief in the rationality of humankind and the topic coherence of its history.

A. Pizzigoni—Former professor.

S. Gabriele et al. (Eds.): IWSS 2023, LNCE 437, pp. 791–798, 2024.
https://doi.org/10.1007/978-3-031-44328-2_83

If we are here today discussing technological advancements and innovative forms in the development of architecture, we must also acknowledge the truth, even banal, of the statement that "it is mathematically true that architecture would not exist without its history."

Therefore, if we believe in the evolutionary development of architecture, we must look back to its distant history, its origins. Above all, we must look at a history that moves in an evolutionary and diachronic dimension, where positions that seemed obsolete, lost, and no longer productive often return to being considered modern in different times.

Our current case study, which revolves around the central use of three-dimensional geometry as a conceptual and compositional module, goes beyond analyzing forms. It also considers forces, resistances, and constraints that generate balance and stability in the work. This study becomes a central reference point to guide us towards the future of modern architecture.

The research on the structural roofing tile of the Tholos of Athena Pronaia in Dephi begins with the tradition and events surrounding one of the most emblematic monuments of Western architecture. From the magnificence of its obscure and celebrated design origin to its millennium-long abandonment, and ultimately to its newfound relevance in the history of art and the science of construction.

The analyses conducted on this structural and architectural form, using only the tools of design, reveal a sequence of facts, observations, and correlations that engage in interdisciplinary dialogue with archaeological studies. Here, the cosmological vision of classical Greece from the 5th and 4th centuries BCE merges with the themes of contemporary spatial and constructive research.

How? When? Why did the ancient Greek temples collapse? These questions may appear to be of interest only to philological and archaeological investigations. However, it is evident from the studies and proposals we present here, using 3D elaboration, that all of this can be a source of innovative proposals, not only in terms of typology and form but also tied to a new and ancient constructive conception based on the economy of materials and the forces at play [2, 3]. It is precisely for this reason that they become exemplars of the energy sustainability that humans have always sought [4].

These are questions and answers that we have posed to ourselves while studying a timeless icon of universal architecture: the Tholos of Delphi. Through the study of certain findings related to its vaulted roofing, we have sought to rediscover the rational roots and design devices that have generated and defined these elements and their form. In this research, we have found ourselves traversing paths that have been repeatedly taken during the various renaissances that have marked human history, from 15th–century Florence to 17th-century France. We have discovered that these attentions and urgencies of design are once again relevant to contemporary roofing structures.

We specifically refer to the technologies that are currently at the forefront of structural research, allowing the use of mechanical arms to create structures without the need for temporary supports. Moreover, we aim to imitate the use of materials that require greater technological and energy expenditure, such as concrete and steel.

2 In the Search of the Tile's Model

The original tile, dating back over two thousand years, bears visible signs of age and is not directly accessible to the public as it is housed within a museum. However, through visual analysis and building upon previous studies [2], we were able to observe the symmetry of the element and a cavity on the front side, indicating the presence of an interlocking system for coupling with other tiles. (see Fig. 1).

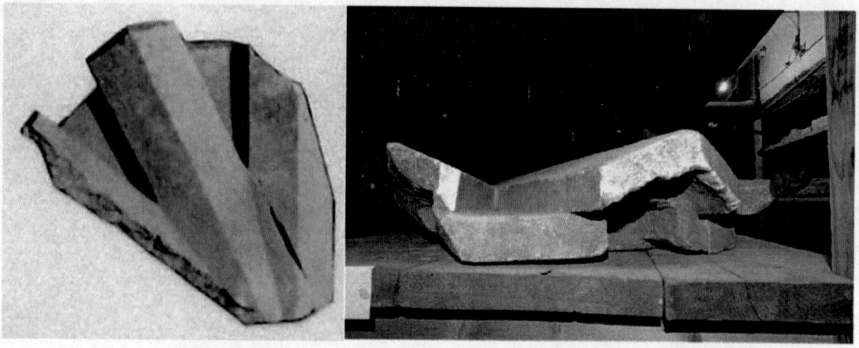

Fig. 1. Original tile in Delphi Archeological Museum. Left, Picture published by G.Roux [5]. Right, back view of the tile.

According with previous researches [2], It has been proceeded to determine the shape of the tile, considering it as the elementary unit of the dome of the Tholos.

The techniques employed in researching the tile's form included 3D modeling in conjunction with 3D printing. The physical models played an essential role in the search of the shape, early digital models were 3D printed or cast. The testing of couplings, interlocking systems, and the physical evaluation of various solutions that led to the proposed model were crucial for the advancement of the research process. (see Fig. 2).

Through a mixed approach (virtual and digital modelling) has been defined a suitable trapezoidal shape which led to interlocking four directions allowing to pile tiles and to build courses (see Fig. 2).

3 The Dome as Aggregation of the Elementary Unit

As described in [3], the presence of a wooden flat roof is not realistic, thus, amongst all conditions considered in the design process, the tile's shape should be able to describe lowered dome with oculus. Thus, the next step in the research was to understand how the elementary unit could be assembly to constitute the roof of the Tholos of Delphi. Therefore, the challenge arises as to how to assemble flat elements to create a circular structure.

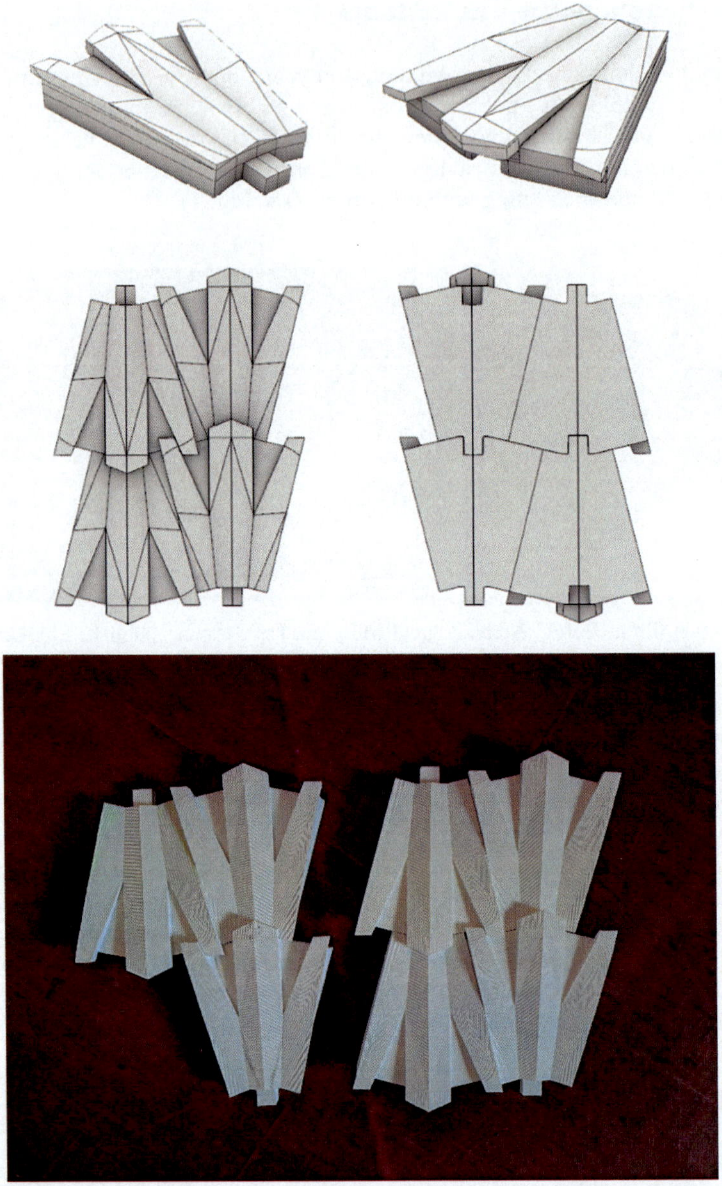

Fig. 2. Virtual and physical modeling of the tile element and his assembling

The study proposed the use of a logarithmic spiral that follows the curvature of the sphere inscribed within the Tholos. This spiral, known as a loxodromic curve, serves as the guiding principle for arranging the courses of tiles. Such geometry was widely adopted in many ancient structures [6], its materialization (tracing) has been widely describe in [7, 8]. Five loxodromic curve were introduced to define the distribution of the

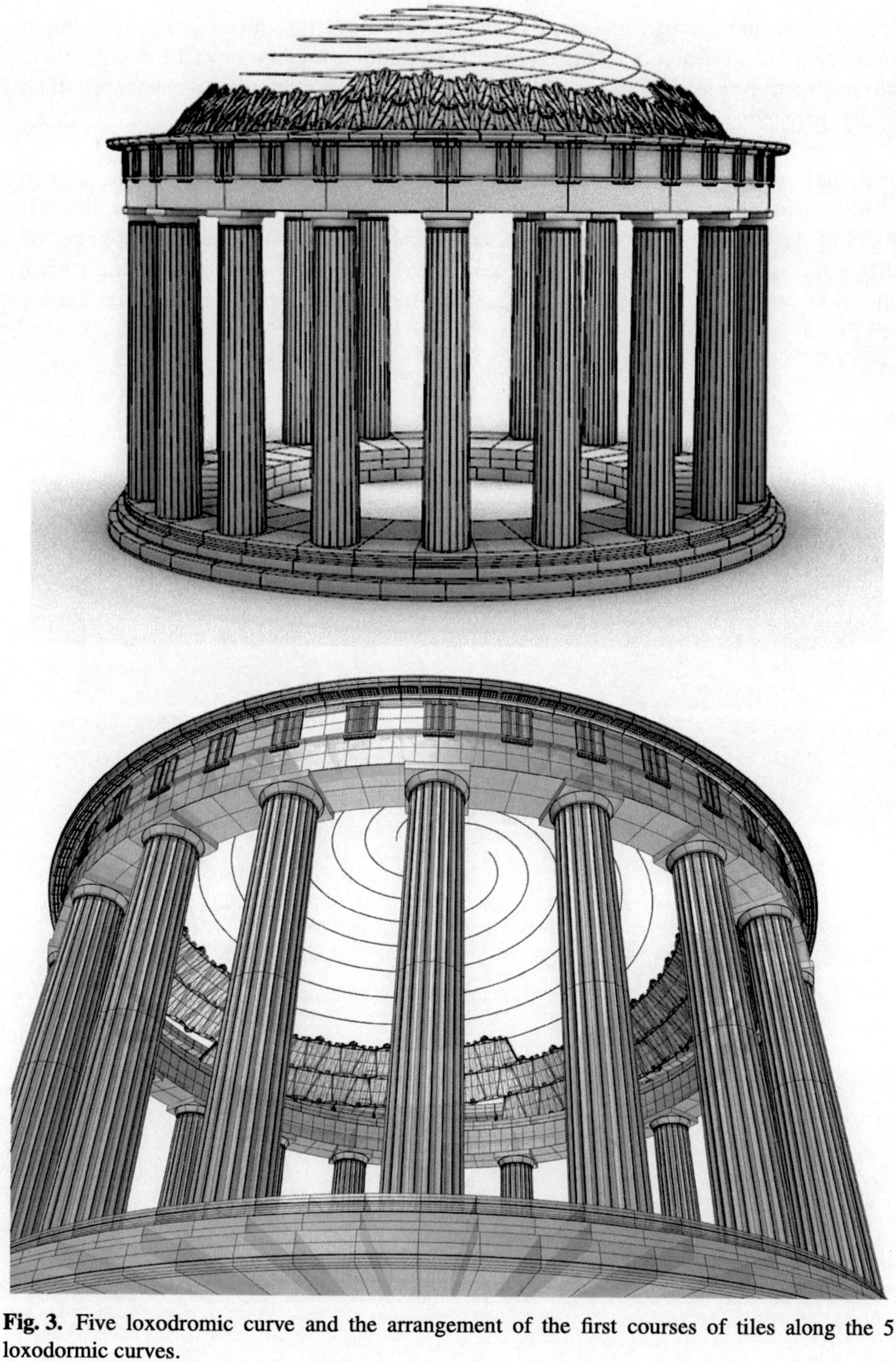

Fig. 3. Five loxodromic curve and the arrangement of the first courses of tiles along the 5 loxodormic curves.

tiles, such The number of curves, and consequently of the tile courses were according to the geometrical construction of the tholos [2], This choice finds confirmation on practical and construction considerations such as the inclination and spacing between them, which almost correspond to the dimension of the tiles modelled.. (see Fig. 3).

As the dome rises, the curvature requires a drastically dimensional variations towards the domes crown (see Fig. 4). Through the analysis of the trapezoidal marble tile and the utilization of virtual modeling, 3D printing, and the study of loxodromes, the first model of the dome of the Athena Pronaia Tholos at Delphi has been developed. The design process is far to be completed (see Fig. 5), but it already shows an alternative, innovative route of traditional interpretation of the roof of the Athena Pronaia Tholos (see Fig. 6).

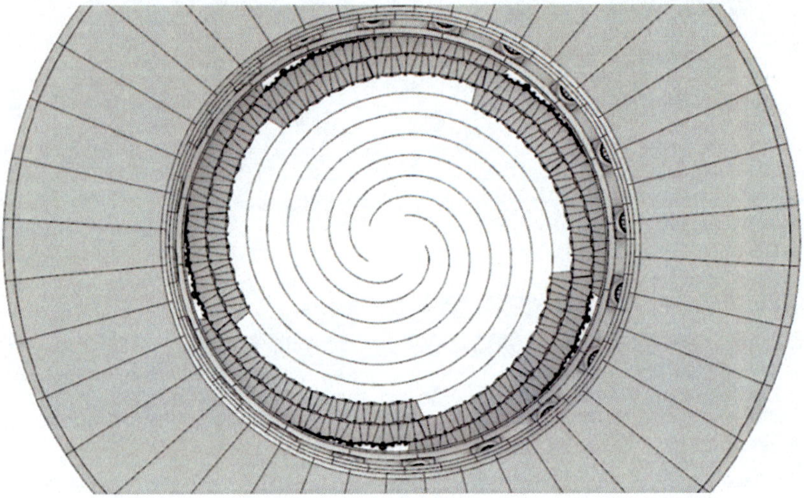

Fig. 4. Bottom-up view of the 5 loxodromic curves.

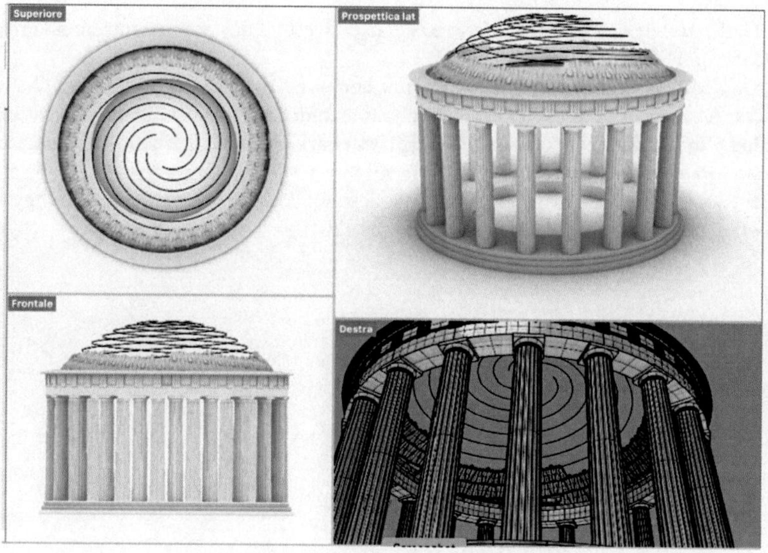

Fig. 5. View of the model of Delphi Tholos.

Fig. 6. View of the model of Delphi Tholos, focus on stone dome.

References

1. Ghirardo, D.: Architecture of Deceit. Perspecta **21**, 110–115 (1984)
2. Pizzigoni, A., Beatini, V., Paris, V.: The Stone Roof. Marble Tholos Tiles in the Sanctuary of Athena Pronaia at Delphi. Stereotomy and Compression Resistant Forms in the Domed Roof Structures of Ancient Greece. Structrual. Mag., vol. 240 (2022)
3. Aita, D., Garavaglia, E., Paris, V., Pizzigoni, A., Sgambi, L.: The stone roof of the Tholos of Athena Pronaia: Structural hypotheses starting from frangements of marble tiles, Florence, Jun. 2023. To be presented CIPA (2023)
4. Pizzigoni, A.: Osservando i marmi bianchi della Tholos di Delfi - Perché sono crollati i templi greci. Christian Marinotti Edizioni, Milano, Italy (2023)

5. Roux, G.: Le toit de la Tholos de Marmaria et la couverture des monuments circulaires grecs. Bull. Corresp. Hell. **76**(1), 442–483 (1952)
6. Conti, G., Sedili, B., Trotta, A.: The loxodromic curves in Architecture. Sci. Philos. **2**(2) (2014)
7. Pizzigoni, A., Paris, V., Ruscica, G.: Herringbone technique: truth and history of a cutting-edge technology. In: Proceedings of IASS Annual Symposia, Boston: International Association for Shell and Spatial Structures. IASS, pp. 1–8 (2018)
8. Jones, B., Sereni, A., Ricci, M.: Building Brunelleschi's dome: a practical methodology verified by experiment. J. Soc. Archit. Hist. **69**(1), 39–61 (2010)

Large Vaulted Structures and Prefabrication. The Case of Pier Luigi Nervi's Hangar in Pontecagnano, Italy

Giulia Neri(✉), Anna Gallo, and Federica Ribera

Department of Civil Engineering (DICIV), University of Salerno, Salerno, Italy
gneri@unisa.it

Abstract. Large-span roof implementation has long been a challenge for designers, who have had to overcome traditional structural engineering limits by introducing creative and innovative solutions. Pier Luigi Nervi revolutionized architecture and engineering with his prefabrication and construction techniques, inspiring professionals globally to explore new materials and solutions. His unique use of reinforced concrete in large-span roofs construction involved prefabricating structural elements and assembling them on-site, leading to faster and more efficient construction. The present research aims to analyse the construction procedures employed for the realisation of the Hangar designed by Nervi in 1939 in Pontecagnano, Italy. By examining original drawings and technical specifications, this research seeks to examine the structural solutions employed in this masterpiece of 20th-century architecture, showcasing Nervi's pioneering work in structural prefabrication and large vaulted structures.

Keywords: Reinforced concrete · Pier Luigi Nervi · Prefabrication · Large-span roof · Vaulted structures

1 Nervi's Hangars Early Experimentation

Pier Luigi Nervi had already established himself as a renowned architect in Italy, primarily known for his achievements in reinforced concrete construction, at the age of less than 40. During this time, Nervi was active in architectural debates and was published in several international architectural and engineering journals. In 1932, Pier Luigi Nervi, relying on his previous professional experiences conducted with Nebbiosi, founded the new company "Ingg. Nervi & Bartoli" in Rome, along with his cousin Giovanni Bartoli, embarking on a parallel experimentation aimed at revolutionizing reinforced concrete structure fabrication methods.

Throughout his multifaceted roles as a theorist, designer, experimenter, technician, construction site organizer, and entrepreneur, Pier Luigi Nervi consistently focused his research on the resolution of the roofing problem [1].

Vaulted architectural structures (domes, parabolic vaults, and geodetic roofing) are regarded as an optimal system for Nervi's personal research on the definition of

S. Gabriele et al. (Eds.): IWSS 2023, LNCE 437, pp. 799–808, 2024.
https://doi.org/10.1007/978-3-031-44328-2_84

large spans with reduced thickness. According to Nervi, these roofing systems are the finest architectural expressions because they embody a perfect synthesis of structural functionality and economic requirements [2].

Nervi first became involved in hangar design around 1930, when he designed a circular reinforced concrete hangar with cantilever beams, supported by V-shaped pillars, and a movable floor, later patented [3]. This initial prototype was later improved in 1932 with the design of a second annular hangar, this time with increased capacity and a structure made up of 32 steel cantilever beams. In contrast to the approach taken for the projects presented at V Milan Triennale in 1933 [4], a distinct and contrasting strategy was employed by him and his company, Nervi & Bartoli, for the competition of 1935 for two aircraft hangars at Ciampino Airport in Rome, in which, a design with a more conventional reinforced concrete structure was proposed [5]. The type of roof of these two hangars was changed from several types of solutions, transitioning from various conventional solutions to an integrated vault-type structure. During the same time period, a competition for two hangars at Orvieto's military airport is held, and Nervi repurposes and refines the structural scheme of Ciampino's project to introduce the so-called "first series of *Aviorimesse*". Nervi's design featured a combination of two structural techniques, utilizing gentle parabolic curves aligned with the ideal shape of a uniformly loaded structure. The Orvieto hangar's roof is a pavilion vault formed by an arrangement of reinforced concrete arched beams, orthogonally crossed with each other and at a 45-degree angle from the perimeter of the hangar. This design and diamond-shape grid pattern, with each rib stiffened by regular crossing ribs, created a hyperstatic network of load paths, where every rib was recruited to resist lateral and vertical loads. This distributed the load onto multiple supports, providing exceptional redundancy and efficiency. Alternatively, the ribs can be seen as a coffering that removed most of the dead weight from a monolithic shell while maintaining its rigidity and carrying capacity. Nervi referred to this as the "geodetic" nature of the structure, highlighting its efficient reinforcing and describing it as "the most economical solution and the one requiring the least steel" [6]. The construction process for the hangar proved more complex than expected, with the cost of formwork being just one contributing factor. Despite the efficiencies achieved through the parabolic shape and lamellar pattern, the joints between the roof shell and buttresses were subject to immense loads, requiring costly reinforcement with steel. Furthermore, due to the monolithic nature of the roof, the entire structure had to be poured as a one-piece, necessitating the use of scaffolding for the entire shell. In 1936, Italy faced sanctions after invading Ethiopia, leading to the proclamation of an autarchy regime by the Fascist government. As a result, the limited supply of steel available in the country was redirected to the war industry. The use of reinforced concrete was thus first restricted and eventually completely banned in 1939, causing a significant change in the development of Italian construction [7]. In the autarchy regime and wartime period, Nervi and his firm, Nervi and Bartoli, embarked on a series of experiments that led to a complete reinvention of the construction method for reinforced concrete structures. Nervi's efforts to improve the efficiency of the first series of hangars were oriented to achieve a monolithic behaviour of the geodetic system with lighter ribs and to investigate the feasibility of prefabricating these elements to speed up the construction processes and eliminate the need for formwork.

Between 1938 and 1942, Nervi had the opportunity to refine and enhance his approach when tasked by the Air Force to construct the "second series of Aviorimesse" [8]. The second series hangars were divided into two types: larger hangars measuring over 100x36 metres built in Orvieto, Orbetello, and Torre del Lago, and smaller hangars measuring 45x55 metres built in Pontecagnano and Marsala.

The hangars designed and constructed by Nervi, unfortunately, fell victim to the ravages of wartime events. However, within this context, the hangar at the Monte Corvino Rovella airstrip, now Pontecagnano, located near Salerno, in southern Italy, assumes a position of remarkable significance as it is the only remaining example of such Nervi's architectural prowess, preserved in good condition and still serving a purpose reminiscent of its original intent [9].

2 Pontecagnano's Hangar: Archival Research

The knowledge of the Pontecagnano hangar follows a path that encompasses several interconnected stages. Bibliographic research, combined with the collection and verification of documentary and iconographic sources, allows for the acquisition of the current state of knowledge, which is then integrated with data derived directly from the work. In particular, the consultation, analysis, and comparison of archival documents enable the identification of unique aspects of the project and define useful directions for further research. Specifically, the Center for Studies and Archives of Communication (CSAC) in Parma houses the archival collection dedicated to Pier Luigi Nervi, which includes paper drawings, sketches, and photographic films, covering a time span from the 1920s to the late 1970s.

Among the folders in the collection, the one titled "Aviorimessa measuring 47x55 meters in Montecorvino, Salerno" [14] contains a series of drawings of the Pontecagnano hangar that are useful for understanding the project's evolution.

The first major issue concerns the chronological placement: in fact, to this day, it is possible to identify a timeframe within which the conception and realization of the work can be encompassed (in 1939, the Aviorimessa is reported as complete), as no document provides an exact date. It is worth noting, in particular, that the design drawings of the Pontecagnano Aviorimessa bear the stamp of the firm "Ingg. Nervi & Bartoli." During the archive consultation, it was deemed useful to investigate further by comparing it with a second folder, also of uncertain dating and named "Aviorimessa with maximum autarchic coefficient" [15]. These are slides, also signed by Nervi and Bartoli, which, with direct reference to the autarchic period, suggest the hypothesis of contemporaneity with the Pontecagnano project, in addition to complementing the analysis and critical interpretation of the work in Salerno.

3 The Project and Structural Solutions

The Aviorimessa of Pontecagnano is the outcome of Nervi's evolving thinking, who conceived several design iterations prior to approving the project's final version. In fact, it is acknowledged that one of Nervi's peculiarities is his propensity to continually advance the project by introducing experimentation in cases where mathematical calculations are inadequate. Through load tests on appropriately constructed models, Nervi manages to make the necessary modifications to ensure the stability of the structure, as well as determine the final form of the building [5,10].

In the case of Pontecagnano, this characteristic is already evident in the floor plans: an initial version includes an interior space with dimensions of 45 m wide and 55 m long, with two symmetrical side corridors. The same symmetry is reflected in the regular succession of exterior buttresses, which, with a spacing of 4.25 m, define the side elevations, while the main facade features an opening of 45 m for aircraft access. A second proposal, named "Aviorimessa m 45x55x8, variant with side appendages", presents a redistribution of the interior spaces with several pencil annotations indicating a study of possible alternative solutions. Actually, the peculiarity of the floor plan lies in an increase in the width of the single interior space, which becomes 47 m, while maintaining the dimension of the main entrance (45 m). The annotation that cancels this possible version (as seen in the top left with the word "annullato") leads to the belief that the final version corresponds to the variant with side appendages, whose internal dimensions remain unchanged, as well as the division of the interior spaces (Fig. 1).

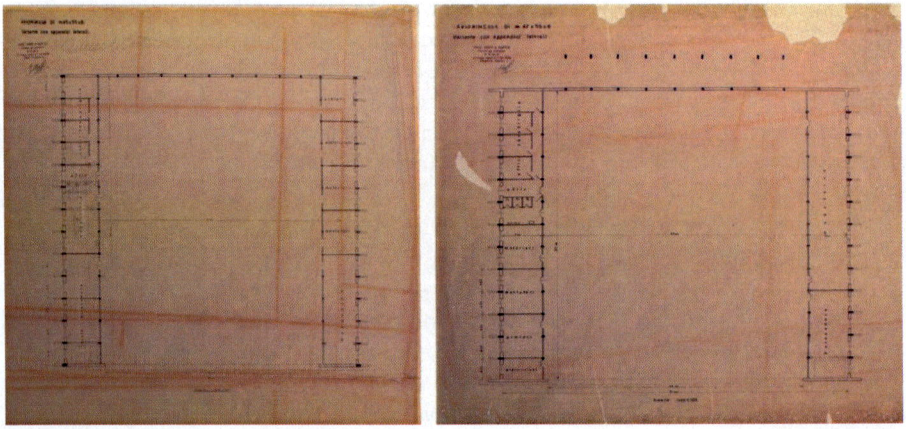

Fig. 1. Project drawings: First version of the floor plan, below: final version with buttresses, CSAC

However, the decisive choice lies in the addition of external buttresses on the rear elevation as well. All these elements can still be observed today, despite the various additions and alterations over time [11]. Therefore, the first question that arises concerns the reasons behind Nervi's decision to introduce a third system of buttresses.

The conducted analysis considers the Aviorimessa as a complex model that, when broken down into its components, allows for greater simplification and identification of the main elements that constitute it (Fig. 2).

The basic scheme resembles that of a barrel vault, which, unlike the earlier hangars with plain crossed structure, is now simplified. The fourteen transverse arches, the system of secondary connecting beams, the appropriately attached roofing "tiles", and not least, the buttress system with slightly inclined foundation plinths, establish a precise functional scheme, ensuring the stability of the entire structure (Fig. 3).

Nervi's meticulous research in addressing the essential aspects related to the nature and purpose of the project is evident. The answer to the previously posed question can be deducted from the fundamental role played by wind load in this specific type of structure. In fact, considering the horizontal thrust generated by the aircraft's movement within the single hall, it becomes imperative to counteract this force by utilizing the nine buttresses strategically positioned at the rear. Notably, Nervi explores an alternative system before ultimately opting for this solution, as evidenced by the archival drawing depicting the specific details of the rear facade (Fig. 4).

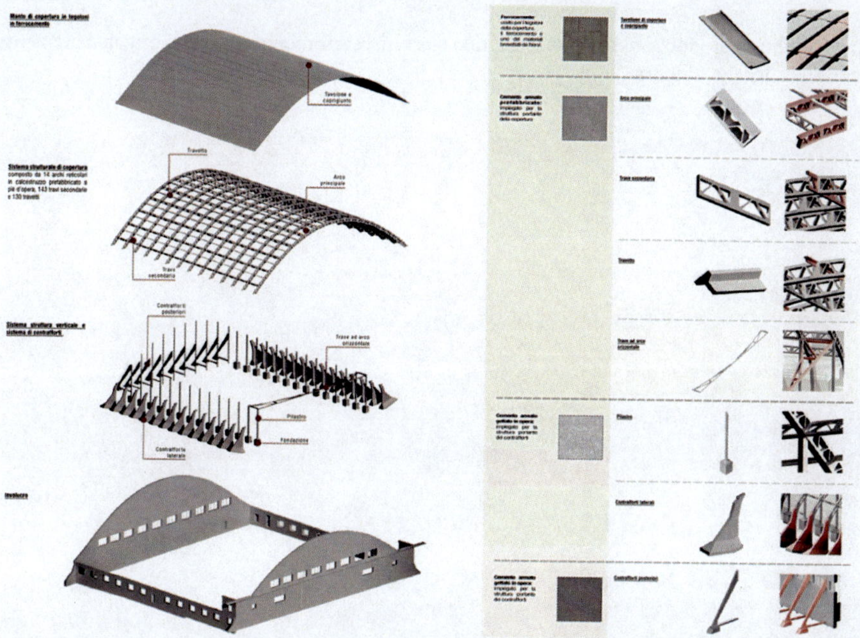

Fig. 2. Axonometric exploded view of the construction system and 3D reconstruction of the structural elements, by the authors

Indeed, Nervi conceives a suspended horizontal reticular beam connected to the structure through a Ø20 steel rod that originates from the junction of the connecting beam, «in order to transfer onto the pillars those horizontal reactions that bring the system of forces that contrast along the arches into balance». In the design documents, the horizontal beam is referred to as the "rib for horizontal stiffening of the tympanums" and thus providing « the sufficient capacity to withstand asymmetrical loads and wind thrust [...]» [12]. However, Nervi does not discard this solution and applies it to the main facade as well. The suspended horizontal reticular beam serves the purpose of accommodating the 45-m opening required for the aircraft's movement. By examining its section at the entrance (Fig. 5), the static scheme resembles that of a simply supported beam subjected

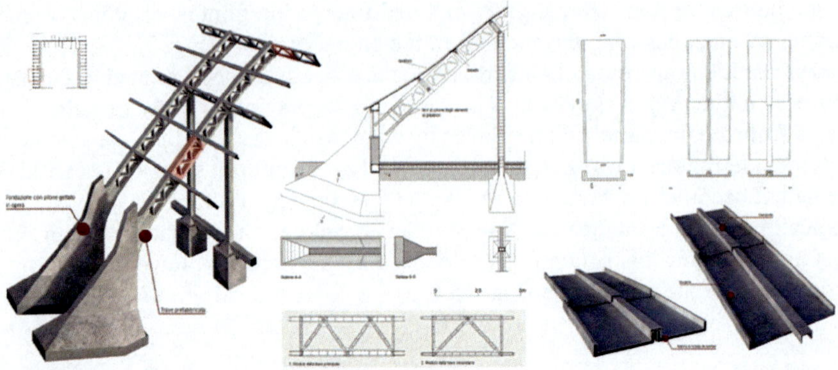

Fig. 3. Graphic reproduction of the construction system, reinforcement of prefabricated elements, and roof tiles, by the authors

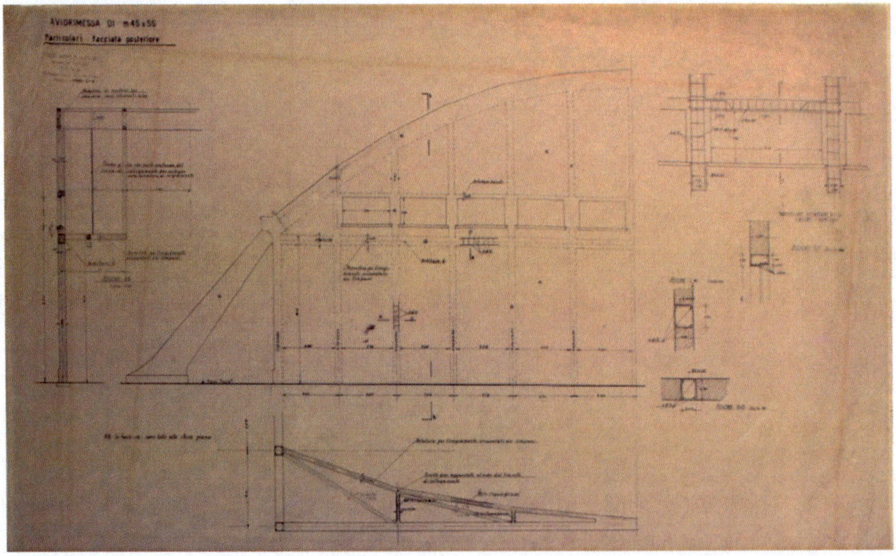

Fig. 4. Project drawings: facade details and reticular beam, CSAC

to wind forces. Simultaneously, a reaction force R is generated and discharged on the lateral part. Further sectioning at the horizontal reticular beam allows identification of the weight force attributed to the beam itself (W), necessitating the inclusion of tension members that ensure the connection to the vaulted structure. Consequently, rise a torsional moment (Mt) that is likely absorbed by the tie rod. It should be noted that these deductions stem from an initial analysis of the design documents, which require further on-site verification using appropriate instrumentation to gain a comprehensive understanding of the type and distribution of reinforcements at various construction elements.

Undoubtedly, it can be agreed that Pontecagnano's hangar serves as a testing ground for Nervi, where he verifies the results of calculations and insights and as a field for experimentation.

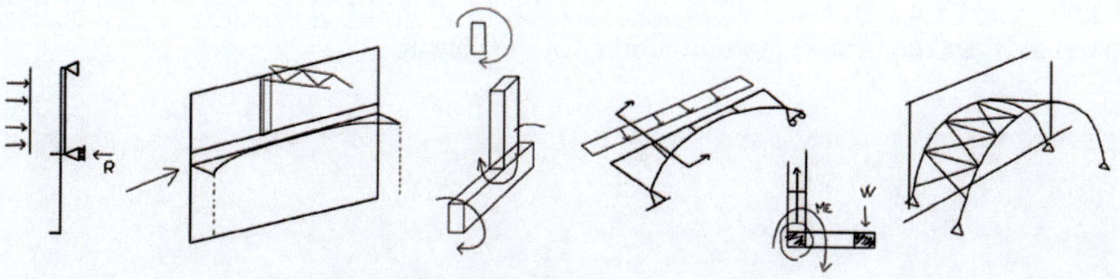

Fig. 5. Static diagrams, by the authors

The significance of wind load in the Aviorimessa project is evident from the documentation of the "Graphical verification of wind load." which, among other things, confirms once again the greatness of Nervi. By defining the data and geometric values (such as rise, relative distances, span, interaxes of the arches, etc.) through graphic construction and calculation of the forces at play, Nervi identifies the maximum negative moment at the impost line of the arch (Fig. 6). Another graphical verification, related to the research of the funicular of the loads, allows to understand the project's approach and to gain new useful information (Fig. 7).

In his analysis, Nervi considers a three-hinged arch and reports the weight values of the various components (arches, joists, roofing) identifying two values of "overload"; considering the actions acting in both directions, he develops the funicular of the loads and identifies the value of the reaction (Ra). One may wonder if Nervi, when indicating

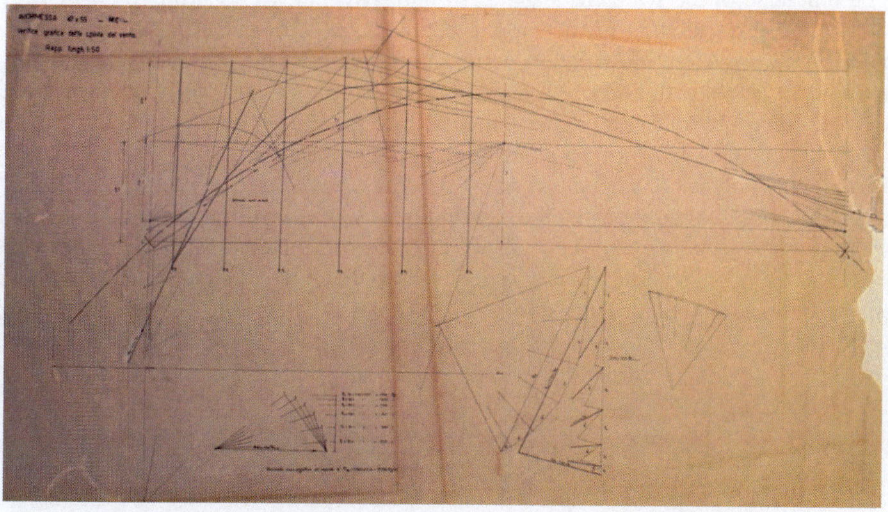

Fig. 6. Project drawings: Graphic verification of wind load, CSAC

the weight of the roofing, had already defined the system of tiles, "*tavelloni*", that we can still appreciate today. These are elements conceived in terms of function, size, material, and method of installation.

It is evident that the function of the system extends beyond mere closure and covering. The entire system creates an infinitely rigid floor within the plane, specifically with a curved axis, and contributes to the stability of the arches.

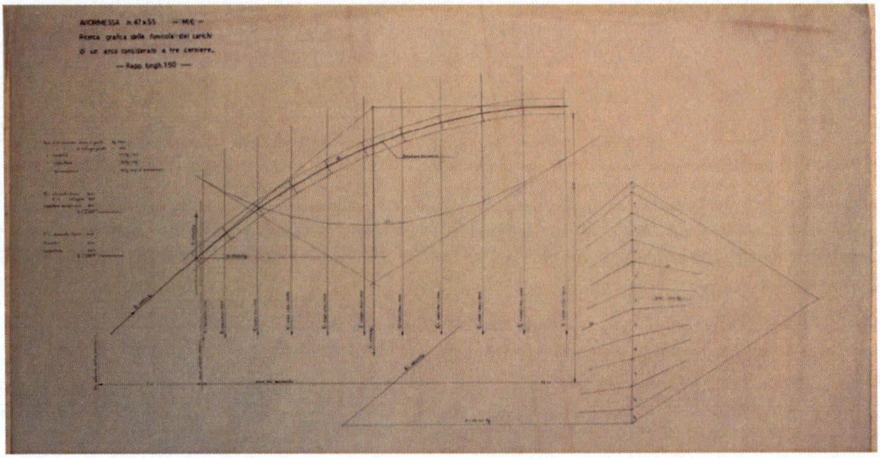

Fig. 7. Project drawings: Graphic investigation of load funicular of a three-hinged arch, CSAC

Therefore, the unresolved inquiry revolves around the technological aspect, specifically concerning the anchoring solution for the "*tavelloni.*" A visual examination of the current state of the structure leads to an initial hypothesis regarding the assembly and subsequent attachment of the roofing elements. It is plausible that Nervi, before considering a dry assembly solution, would have explored the possibility of creating conglomerate joints with anchoring bars to the structure. The various hypotheses arise due to the current lack of compliance between the anchoring system of the tiles and the beams, as indicated in the project's cross-sectional view (Fig. 8).

In conclusion, the decomposition of the model proves to be essential for facilitating a clear and concise reading as well as enabling a more straightforward interpretation. Through this meticulous analysis, we are able to discern the inherent relationship between the various components and their interplay within the entire system. Each individual element assumes a specific and pivotal role, which is integral to the overall proper functioning.

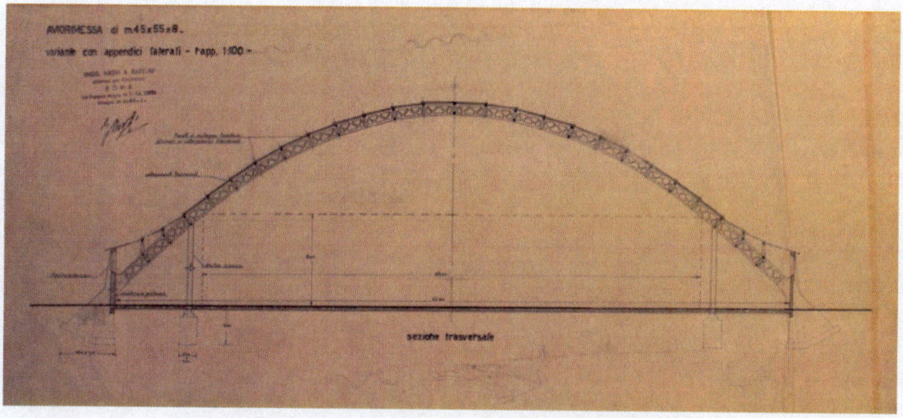

Fig. 8. Project drawings: Cross-section, CSAC

4 Conclusions

Through the implemented projects and exploratory trials conducted in the hangars, Nervi acquires a comprehensive array of construction techniques and design methodologies that will shape his future accomplishments in the post-war era. The subsequent phase of his career, which commences in 1950, garners widespread recognition and acclaim, yet it does not witness any substantial breakthroughs or remarkable architectural innovations of considerable significance [13]. In 1960, Pier Luigi Nervi has become one of the most famous engineers in the world, thanks to the creation of an original, cost-effective, and quickly executable construction method. This construction system, which would later evolve into a true architectural "style", was developed and refined during the construction of the second series of Aviorimesse. It is undeniable that the subsequent works, including those for the 1960 Rome Olympics and the Turin Exhibition's Hall B in 1947, find their origins in these hangars, designed and built under the difficult conditions of autarchic and wartime Italy.

The hangar exemplifies Nervi's pioneering work in structural prefabrication and his mastery of reinforced concrete, showcasing his ability to create efficient, durable, and visually striking structures.

However, despite the initial efforts of analysis, the case of Pontecagnano necessitates a further commitment and a comprehensive approach to address certain unresolved matters. Foremost, it is crucial to undertake a meticulous verification of the existing reinforcements using appropriate instrumentation to ensure their adherence to the original design. Another challenge pertains to the roofing panels and the reconstruction of the stages that culminate in the final solution. In conclusion, the Pontecagnano hangar designed by Pier Luigi Nervi holds great structural significance. Through meticulous archival research and analysis of the construction procedures employed, it becomes evident that Nervi's innovative approach to structural design, particularly in large-span roofs, paved the way for future advancements in the field of architecture and engineering. The Pontecagnano hangar stands as a testament to Nervi's enduring legacy as an influential figure in 20th-century architecture, inspiring generations of professionals to explore new possibilities in structural design and construction.

Acknowledgements. The authors would like to express their sincere gratitude to the CSAC for providing access to their extensive collection and granting permission to utilize the project drawings for this research.

References

1. Poretti, S.: Considerazioni sull'opera di Pier Luigi Nervi. In: Ramazzotti, L., (eds.) Nervi oggi: scritti dalle mostre e dai convegni, Kappa, Rome (1983)
2. Trentin, A., Trombetti, T., Einaudi, R.: La lezione di Pier Luigi Nervi. Mondadori Bruno (2010)
3. Nervi, P.L.: Capannone circolare per aeroplani o veicoli in genere con piattaforma anulare rotante, (Italy, Patent n. 304282) (1932)
4. Sassone, M., Piccoli, E.: Otto aviorimesse in cemento armato. In: Chiorino, C., Olmo, C.M., (eds). Pier Luigi Nervi: architettura come sfida, Silvana Editoriale, Milan (2010)
5. Neri, G.: Capolavori in miniatura: Pier Luigi Nervi e la modellazione strutturale. Silvana (2014)
6. Nervi, P.L.: Concrete and structural form. In: The Structural Engineer, vol. 34, issue 5 (1956)
7. Iori, T., Poretti, S.: The rise and decline of the Italian school of engineering. Front. Matter. Construct. Hist. **33**(2), 85–108 (2018)
8. Nervi, P.: The works of Pier Luigi Nervi. Pref. by Pier Luigi Nervi; introd. by Ernesto N. Rogers; explanatory notes to illus. by Jürgen Joedicke; translation by Ernest Priefert. (Books that matter). New York: F. A. Praeger (1957)
9. Ribera, F., Miccio, G., Apicella, P.: Pier Luigi Nervi a Salerno: l'Hangar ritrovato. In: History of Engineering, Proceedings of the International Conference, vol.2, Cuzzolin, Naples (2014)
10. Argan, G.C.: Pier Luigi Nervi, pp. 12–13. Il Balcone, Milan (1955)
11. Ribera, F., Cucco, P., Gallo, A., Neri, G.: L'eredità di Pier Luigi Nervi in Campania. Esiti di una ricerca. In: Ribera F., Cucco P. (eds) Pier Luigi Nervi e la questione del Moderno. Cultura della conservazione e centralità del progetto. Aracne, Rome (2023)
12. Nervi, P.L.: Un'aviorimessa in cemento armato. Casabella Costruzioni, 124 (1938)
13. Greco, C.: Pier Luigi Nervi: Dai Primi Brevetti Al Palazzo Delle Esposizioni Di Torino 1917–1948. Quart Architektur (2014)
14. CSAC Parma, Fondo Pier Luigi Nervi, Aviorimessa di metri 47x55 di Montecorvino (Salerno), PRA 119, N.id 13252, coll. 140/2, B002741P
15. CSAC Parma, Fondo Pier Luigi Nervi, Aviorimessa a massimo coefficiente autarchico, PRA 803, Progetto 290

Edge-Corrugated Shells: Structural Insights About Nervi's Conception of the Flamino Dome

M. Lai[1](✉), N. L. Rizzi[2], V. Eremeyev[1], E. Reccia[1], and A. M. Cazzani[1]

[1] Dipartimento di Ingegneria Civile Ambientale e Architettura, Università degli Studi di Cagliari, via Marengo 2, Cagliari, Italy
matteolai@unica.it

[2] Dipartimento di Architettura, Università di Roma Tre, via Della Madonna di Monti 40, Roma, Italy

Abstract. The dome designed by P. L. Nervi in 1956 for the roof of the "Palazzetto Dello Sport" in Flaminio district, Rome, has a remarkable feature: the wavy shape of its edge provides relevant enhancements of its structural behaviour. Even though such structure is well-known abroad for its aesthetics, from a Structural Mechanics point of view such shape sparks interest in the optimization of shell structures. Primarily, corrugation enhances the bending stiffness at the edge providing a significant reduction of the dome thickness.

Taking inspiration from Nervi's work, this contribution deals with the mechanical analysis of edge-corrugated shells. Attention is paid to the definition of their mathematical and geometrical descriptions and to the realization of a mesh suitable to perform FE analysis. Then some insights are given about the static and dynamic behaviour of these structures. Therefore, the enhancement provided by the corrugated shape is highlighted by comparison with a non-corrugated shape. It is well known that on the shell edge the effect of bending effects must not be neglected. As a consequence, attention is given to stresses and bending moments.

The other main issue that will be considered is the behaviour in the non-linear field (especially concerning instability aspects such as snap-through phenomena, the effect of imperfections, etc.). Some numerical simulations of the improvement produced by the new shape are shown with different methods.

Keywords: concrete shells · corrugated shells · shells · Nervi · Flaminio dome

1 Flaminio Dome Conception

1.1 Introduction

This paper seeks to approach the topic of structural enhancements provided by clever shape changes in shells. It is recognised that when a metal sheet is corrugated, its strenght increases compared to a plane surface of the identical thickness; this concept is illustrated using a cardboard model [1].

This research originates from P. L. Nervi's conception proposed for the 'Palazzetto dello Sport' dome in Rome's Flaminio district, depicted in Fig. 1 and built on the occasion

S. Gabriele et al. (Eds.): IWSS 2023, LNCE 437, pp. 809–816, 2024.
https://doi.org/10.1007/978-3-031-44328-2_85

Fig. 1. Bird's view of the Flaminio dome, Rome.

of the Rome Olympic games. The architectural design was entrusted to A. Vitellozzi, whereas Nervi was both the structural designer and the owner of the construction company. The peculiar construction system is described in [2, 3], which it made possible to make the dome construction cost cheap, affordable and fast. A renewed look at the history of Italian structural engineering is offered in [4], where Nervi is one of the prominent representatives. He envisaged a shallow spherical dome that is sinuously wave-shaped near the edge. This slight change in shape was not a mere display of mannerism but provided a structural enhancement regarding the bending effect on the edge [5]. The design process was disclosed by Nervi himself in the lectures of construction design at the Architecture faculty of La Sapienza in Rome, which were recorded by Roberto Einaudi and transcribed in [6]:

> "I had the idea for the first time during the war. The reinforced concrete dome was hard to build, very costly. What would happen if we could build it in pieces, like a mosaic? We can make individual prefabricated pieces that can then be joined together with a pour of concrete. But how can we do it for a dome, a three-dimensional surface? All the parts have to fit precisely together. The only way to do it is to prepare a portion of the dome in actual size and construct one of its sectors on the ground.
>
> First this requires in-depth study of the form, as a negative, prepared on the ground. In this way, you can make the necessary changes, even the smallest ones, and shape the silhouette as you like. On this surface a pour is done using reinforced concrete, which when it is ready will become the mother-mold. The forms required will be one or ten, depending on the pace of the construction schedule. They are used to make the final prefabricated parts in ferrocemento, which are then joined together by a pour of concrete and iron bars. The thickness of a shell dome can and should be minimal, but you have to pay particular attention to create a margin of safety in case things [will] go very badly after stripping. The ribbing provides this extra thickness without adding too much weight. These ribs take the form of meridians and parallels, or triangles and diamonds. From a static viewpoint there is great

freedom in the design of the ribs. The dimensions of the individual parts depend on the most easily manageable weight, ranging from one quarter ton to one ton. In any project the work of the builder is essential. The work of the contractor is of great technical value. A project cannot progress without understanding how it can be built."

The stages of construction are summarized in Fig. 2, using historical pictures. Moreover, pieces of information about the recent dome refurbishment are discussed in [7], which describes the philological restoration of the dome using original materials and where the later inconsistent workings were removed.

Not only were Nervi's books a formidable resource for a deep understanding of the construction techniques for his patented material *ferrocemento*, but today it stands as a testimony to the design and construction process and their transformation across World War II [8].

The more general framework regarding the theory of shells is treated in [9, 10], attention to concrete shells design is drawn in [11–15], instead for the structural details specification used at that time one should refer to [16].

The study about corrugation used for civil engineering application are reported. A structural analysis of the vault of Eladio Dieste has been carried out in [17], where the effect of the corrugation against the seismic action is evaluated. Structural analyses of edge-corrugated dome is addressed in the [18–21]. [22] shows an interesting application in tunnel engineering, where instead of using the traditional concrete building techniques the use of corrugated steel plate is proposed.

1.2 Geometric Generation of the Model

The corrugated shape is reproduced using a mathematical description of the geometry, reporting for different parametric equation, that differ each one for the function that controls the edge-corrugation (*i.e.*, the Heaviside or the hyperbolic function) the procedure adopted is accurately described in [19] and in [23] several type of parametric equations are reported, which are useful to tackle different problems.

Afterwards, an algorithm was developed inside the modeler Rhinoceros and in its graphical development environment Grasshopper, that can be used to set the appropriate shape resembling the Flaminio dome shape. Then, the created geometric object can be easily imported in a FE software, see Fig. 3.

2 Structural Analyses of Edge Corrugated Shells

It is well known that for spherical shells, the bending effect on the edge should be considered carefully; in fact, the membrane solution is not representative of the actual stress. The theoretical solution to account for this issue relies on the study of [24]. Nonetheless, some practical solutions to provide stiffness are usually adopted: thickness increasing around the edge and building a ring beam. Nervi's proposal increased the stiffness of the section only with a clever change of shape, where the effect of bending moments is significant.

Fig. 2. A sequence of the construction stages: (a) the temporary support above which the wooden template is constructed; (b) a bricklayer sketches the line of thesingle pieces formwork using a pencil; (c) construction of a formwork upon the wooden spherical template; (d) concrete single pieces of the spherical dome; (e) workers handle a dome concrete piece; (f) maturation of the concrete pieces; (g) storage of the dome pieces; (h) a detailed view of the pieces positioning over the scaffolding; (i) overview of the construction site, a crane moves the concrete pieces over the scaffolding.

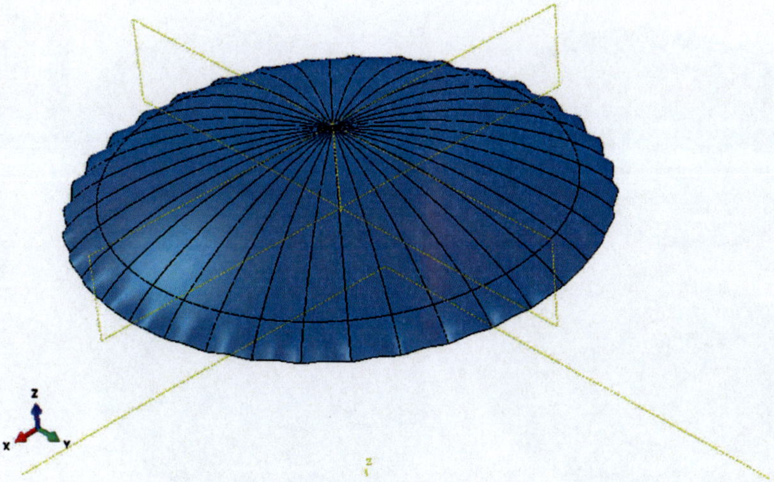

Fig. 3. Edge-corrugate surface model of Flaminio dome.

The static relevance of the edge-corrugation has been thoroughly discussed in [19] for an inward pressure load, whereas [21] considers the dead load and the anti-symmetrical load. For shallow shells (i.e., structures whose rise-to-span ratio is low), the pressure load could be used as a rough approximation of the self-weight. It has been shown that the effect of the edge-corrugation in the case of a pressure load can significantly reduce the stress, owing to a decrease in the magnitude of the bending moment in the meridian direction. Moreover, the ultimate bearing capacity has been investigating considerin the material non-linearity.

The instability issue on concrete shallow spherical shells is tackled in [25, 26] with a practical methodology. Some relevant considerations on the effect of linear buckling loads are presented in [21], which shows a relevant decrease in the value of the critical loads. In Fig. 4 the first four eigenshapes are illustrated. Nonetheless, the instability phenomena should be analysed considering the geometrical non-linearities, such as imperfections, settlement displacements and building defects. This approach can provide a forecast of the kind of instability the structure is likely to be subjected to; it can be a local snap-through, a bifurcation or a limit load.

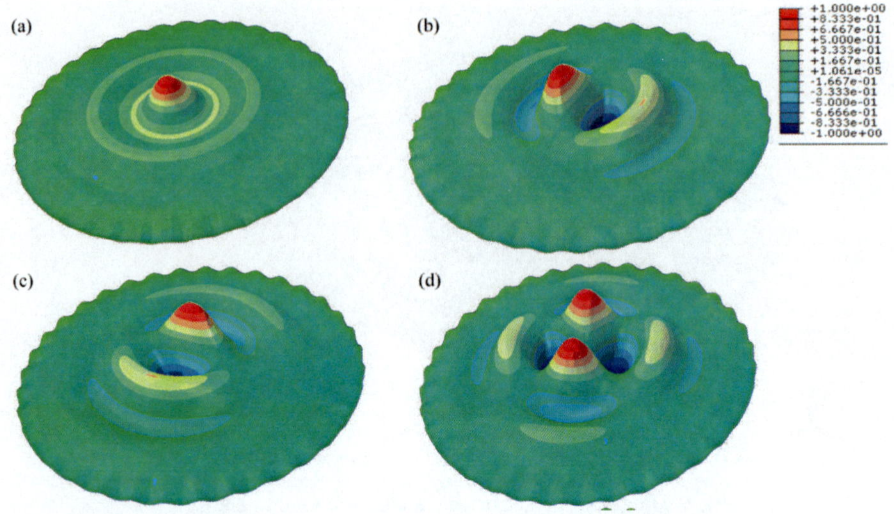

Fig. 4. Eigenshapes of an edge-corrugated shell; 1st, 2nd, 3rd, 4th. The contour plot shows the magnitude of the displacement in z direction.

3 Conclusion

It can be possible to find some comparison results to grapple with the intrinsic difficulties of considering the decoupling of the membrane and bending state using different formulations. Therefore, further developments can be addressed by making use of membrane theory see, for instance, [27–29].

The authors in the forthcoming papers will carry out a multi-parametric analysis, changing the geometrical parameters of the geometry in order to investigate the best set, which can produce an optimal corrugation to enhance the structure's structural strength. In conclusion, the research will consider non-linear buckling as a central issue in the thin-shells design.

Acknowledgment. This research has been developed for the partial fulfilment of the doctoral program of M. Lai at *Scuola di Dottorato in Ingegneria Civile e Architettura*, University of Cagliari.

The financial support of Fondazione di Sardegna through grant Surveying, modelling, monitoring and rehabilitation of masonry vaults and domes *i.e., Rilievo, modellazione, monitoraggio e risanamento di volte e cupole in muratura* (RMMR) (CUP code: F72F20000320007) is grateful acknowledged by M. Lai, E. Reccia, A. Cazzani.

References

1. Siegel, C.: Struttura e forma dell'architettura moderna. CELI, Bologna (1968)
2. Iori, T., Poretti, S.: Pier Luigi Nervi: his construction system for shell and spatial structures. J. Int. Assoc. Shell Spat. Struct. **54**(176–177), 117–126 (2013)

3. Leslie, T.: Carpenter's parametrics: economics, efficiency, and form in Pier Luigi Nervi's concrete designs. J. Int. Assoc. Shell Spat. Struct. **54**(2–3), 107–115 (2013)
4. Iori, T., Poretti, S., (Eds.) SIXXI. Storia dell'ingegneria strutturale in Italia, vol. 1–5. Roma: Gangemi (2014)
5. Nervi, P.L.: Costruire Correttamente. Milano: Hoepli (1955)
6. Trentin, A., Trombetti, T., Einaudi, R., Nervi, P. L. (eds.): La lezione di Pier Luigi Nervi. in Ricerca. Milano: Bruno Mondadori (2010)
7. Iori, T., Sulpizio, R.: La cupola del Palazzetto dello Sport di Pier Luigi Nervi. Ind. Ital. Cem. **857**, 16–23 (2022)
8. Nervi, P.L., Neri, G., Chiorino, M.A., Rossi, A.: Scienza o arte del costruire?: Caratteristiche e possibilità del cemento armato. Milano: CittàStudi (2014)
9. Flügge, W.: Stresses in Shells. Springer, Berlin (1960)
10. Timoshenko, S.P., Woinowsky-Krieger, S.: Theory of plates and shells, 2nd ed. in Engineering societies monographs. New York: McGraw-Hill (1959)
11. Haas, A.M.: Thin Concrete Shells, vol. 1. Wiley, New York (1962)
12. Billington, D.P.: Thin Shell Concrete Structures. McGraw-Hill, New York (1965)
13. Haas, A.M.: Thin Concrete Shells, vol. 2. Wiley, New York (1967)
14. Gioncu, V.: Thin Reinforced Concrete Shells: Special Analysis Problems. Wiley, Bucarest (1979)
15. Calvi, G.M., Nascimbene, R.: Progettare I Gusci. IUSS Press, Pavia (2011)
16. Leonhardt, F., Manning, E.: C.a & c.a.p. calcolo di progetto e tecniche costruttive; Casi speciali di dimensionamento nelle costruzioni c.a. e c.a.p., vol. II. ed. Scienza e Tecnica
17. Michiels, T., Adriaenssens, S., Dejong, M.: Form finding of corrugated shell structures for seismic design and validation using non-linear pushover analysis. Eng. Struct. **181**, 362–373 (2019)
18. Lai, M., Reccia, E., Spagnuolo, M., Cazzani, M.A.: Structural behaviour of corrugated shells: a look at the Flaminio dome in Rome. In: 2nd fib Italy YMG Symposium on Concrete and Concrete Structures, Roma, pp. 87–94 (2021)
19. Lai, M., Eugster, S.R., Reccia, E., Spagnuolo, M., Cazzani, A.: Corrugated shells: an algorithm for generating double-curvature geometric surfaces for structural analysis. Thin-Walled Struct. **173**, 109019 (2022). https://doi.org/10.1016/j.tws.2022.109019
20. Lai, M.: Edge-corrugation as structural enhancement of shallow shells. Nexus Netw. J. (2023). https://doi.org/10.1007/s00004-023-00716-1
21. Lai, M., Zucca, M., Meloni, D., Reccia, E., Cazzani, M.A.: Thin corrugated-edge shells inspired by Nervi's dome: numerical insight about their mechanical behavior. Thin-Walled Struct. vol. to be submitted (2022)
22. Sun, K., Hong, Y., Xu, W., Liu, H., Zhen, Y., Qin, J.: Analysis and prediction of the mechanical behavior of corrugated plate as primary support in tunnels with elastoplastic constitution. Tunn. Undergr. Space Technol. **124**, 104451 (2022). https://doi.org/10.1016/j.tust.2022.104451
23. Lai, M.: Stability analysis of edge corrugated shells. PhD Thesis, Università degli Studi di Cagliari (2022)
24. Geckeler, J.: Über die Festigkeit achsensymmetrischer Schalen. VDI-Verlag, Berlin (1926)
25. Isler, H.: The stability of thin concrete shells. In: Ramm, E. (ed.) Buckling of Shells, pp. 645–672. Springer Berlin Heidelberg, Berlin, Heidelberg (1982). https://doi.org/10.1007/978-3-642-49334-8_23
26. Ramm, E.: Ultimate load and stability analysis of reinforced concrete shells. In: Proceedings of IABSE Colloquium on Computational Mechanics of Concrete Structures (1987)

27. Olivieri, C., Angelillo, M., Gesualdo, A., Iannuzzo, A., Fortunato, A.: Parametric design of purely compressed shells. Mech. Mater. **155**, 103782 (2021)
28. Montanino, A., Olivieri, C., Zuccaro, G., Angelillo, M.: From stress to shape: equilibrium of cloister and cross vaults. Appl. Sci. **11**(9), 3846 (2021)
29. Olivieri, C.: FORMERLY-math: constrained form-finding through membrane equilibrium analysis in Mathematica. Softw. Impacts **16**, 100512 (2023). https://doi.org/10.1016/j.simpa.2023.100512

The Appraisal of The Historic Interventions to the Superstructure Through Time: The Case of Reinforced Concrete

Zeynep Inan Ocak[1] (✉) and Chiara Calderini[2]

[1] Marmara University, Istanbul 34865, Türkiye
zeynepinanocak@gmail.com
[2] DICCA, Universita di Genova, 16145 Genoa, Italy

Abstract. Strengthening interventions are significant actions to pass on heritage buildings to future generations. The study of historic strengthening interventions is worthy of understanding the historical past of the monument and gaining insight into how the building structure and structural damage have been interpreted. This paper aims to analyze historical interventions made on masonry vaults, focusing on the relationships between the deformed geometry of the original structure and the geometry of the intervention. In particular, reinforced concrete interventions made from the late 19th to the mid-20th century will be analyzed, considering the mutual effects of earthquakes, seismic codes, conservation theories and engineering practices. While in the field of conservation, the use of concrete was admitted and encouraged in consolidating monuments to avoid the dangers of dismantling and reinstating the parts to be preserved, in the engineering field, its usage seemed everlasting, given structural design and safety assessment. In this understanding, according to the vault's damage, geometry, and materials, the interventions were with different techniques and in various shapes. Although these interventions do not date back to the first construction of the monuments, they merged into the original structure. Today, the interventions can be defined as the layers showing the evolution of monuments through time.

In this paper, by analyzing the cases from Turkey, an overview of consolidation works of the superstructure through comparing of two cases and an understanding of the concept of the interventions over time and putting forward the current situations to guide the subsequent restorations are presented.

Keywords: Strengthening · reinforced concrete intervention · superstructure · Turkey

1 Introduction

Architectural heritage consists of the initial construction values and their changes through historical time. However, this definition includes two contrasting concepts. The first includes the originality of the monument when it was first built. The second refers to later additions or layers that differ from the original part that occur as a result of interventions made to modify or repair the building after damage. Sometimes, it is

S. Gabriele et al. (Eds.): IWSS 2023, LNCE 437, pp. 817–826, 2024.
https://doi.org/10.1007/978-3-031-44328-2_86

sufficient to repair the cracks and strengthen the structure; in other cases, it is necessary to rebuild some parts. In masonry buildings, where it is almost impossible to distinguish the structural and non-structural parts, reintegrating missing parts is not an easy task. This is particularly true for vaulted masonry structures. Although interventions lead to changes in authenticity, they ensure the survival of monuments and represent the evolution of construction technology and conservation theories over time.

In the past, the problems were mainly technical. On the one hand, there were structural issues related to the load transfer between the new and the old parts; on the other hand, there were building issues related to the geometric compatibility between the new and the old parts (that may have a deformed geometry related to the damage) and to the building operations (scaffoldings, centering etc.). In more recent times (from the XIX century), the development of conservation theories raised new issues. In particular, the necessity of distinguishing the new parts from the original ones was put at the center of the debate. Recognition was seen as a requirement for a good restoration, different materials were used in the reintegration, or surface differences were created between the new and original patterns to identify subsequent additions. From the beginning of the XX century, reinforced concrete has been seen as a way of consolidating masonry structures by distinguishing inherently between original and later additions without any special effort. It was not only clearly recognizable (being a different material) but also structurally efficient and easy to adapt to irregular geometries. However, on the other hand, reinforced concrete also leads to material deterioration and structural problems between reinforced concrete and masonry. There have also been cases where the original part was partly destroyed to construct the reinforced concrete. Despite the negative perception towards reinforced concrete, these interventions do not create confusion or falsification when the initial originality was not known. Considering the pros and cons of reinforced concrete, this study examines how the original geometry was handled and how the monuments were intervened with reinforced concrete.

This study emphasizes that monuments inevitably change over time and discusses the changes in the superstructure of two cases in different contexts in Turkey. One case is the Hagia Sophia, which has been altered over time but still has evidence of its original construction; the other is the Twin Minaret Madrasa at Erzurum, one of the medieval structures of Anatolia. This paper also mentions the perspective on how we should handle reinforced concrete interventions today. In Turkey, reinforced concrete interventions have been dismantled in the recent past on the grounds of not being original material. Instead of preserving, the reconstruction reminiscent of stylistic unity began to emerge without taking into account the documentary value of reinforced concrete as strengthening intervention of the 20th century.

1.1 Methodology

This study compares two cases from different periods and with different historical processes [1]. Analyzing the monuments' changes over time aims to show how the original geometry preserved or and how the reintegration/reconstruction were made. The breadth of knowledge on the topic was gained by conducting research literature reviews and many examples were examined on-site [2–9]. Unless otherwise stated, all photos were taken by the responsible author. Current and old photographs and other records are compared to determine the historical process of the monuments. The dome of Hagia Sophia, one of the cases, shows that authenticity cannot be limited to the original construction features. For this case, the work of Duppel, who studied three-dimensional technical studies on the dome repairs of the monument, was utilized [10]. Twin Minaret Madrasa at Erzurum was selected as a reinforced concrete intervention. Reinforced concrete was deliberately used to restore this monument in the 1930s. It was extensively restored a second time in the 1980s, when iwan vault and the upper floor's smaller vault were reconstructed with reinforced concrete. The project from the 1980s and earlier was prepared using the traditional geometrical measurement methods [11]. It is interesting to study how the restorer determined the shape of the vaults at that time. The studies on the restitution of the building and the vault geometry integrated all the findings into a detailed CAD modelling. The cases are followed by an overview of the interventions that changed the original character of the monument in terms of materials, techniques, and geometry. This assessment is based on UNESCO's world heritage criteria and ICOMOS' conservation principles [12].

2 The Analysis of the Changing of Monuments in time

2.1 First Case: Hagia Sofia

Hagia Sofia, served as the major church of the Byzantine capital was constructed between 532 and 537 by Anthemius of Tralles and Isidorus of Miletus in the reign of Emperor Justinian. It was erected over the previous churches which were demolished due to some reasons. The first church known as having basilica-shaped timber roof in 404 and the second in the Nika Riot 532 were set on fire [6]. It is not easy to briefing the architectural features and construction history of Hagia Sofia in spite of several studies. Considering the topic of the interventions to superstructure, the brief is limited to its dome.

Architectural and Structural Brief. The ground plan of Hagia Sophia measures 70 m × 97 m. The central square area of its nave is covered by a masonry dome with 32 m diameter and 56 m height. The clear spans between the main arches of east and west is 31.19 m, and the opposite direction is 32.55 m (Fig. 1). The dome height from its drum is approximately 14.40 m. The central dome is carried on arches erected on four piers. The transition from the square to the dome's circle area are formed with the pendentives. The superstructure is supported by two semi-domes in an east-west direction. The other sides of the domes, in the north and south direction, are provided by buttresses; some are original, and some are later additions. The construction materials of the dome are brick and pozzolanic mortar like as its whole structure. The dimension of 35–40 cm and a thickness of 4–5 cm of square shaped bricks were used which has had the role of stiffness rather than strength [7]. It's formed with forty brick ribs and have forty windows at its drum level. The mosaics covers the ribs and the shell parts.

Timeline For Changes. The dome of Hagia Sophia exemplifies the fact that monuments do not remain in the form in which they were first built, and vice versa undergo inevitable changes over time. Its change should have taken place in the process of construction itself. Since the time required for the mortar to set, plastic deformations should have started within five years of construction [7]. It means that the monuments original geometry

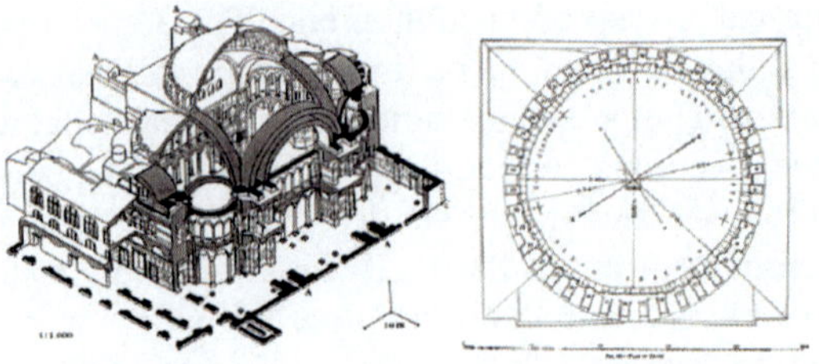

Fig. 1. The isometric view and the plan dome of the Hagia Sophia, isometric view [7].

was inherently starting to change. Today the central dome is a mixture of interventions throughout its long history. The dome has collapsed three times, the first in 558, just 20 years after its completion, the second in 989, and the last in 1346 due to earthquakes (Fig. 2). After the first collapse, the form of the dome altered completely. The new design of the dome and the measurement were undertaken by Isidorus the Younger who was the nephew of one of the original architects. Many chroniclers estimate that the new shape was approximately higher 20 feet than the former one.

After its dome collapsed for the second time in 986, the repair was undertaken by Armenian architect Trdat. In his work by preserving the existing part, the missing part is integrated with the masonry wall, which is thicker than the existing parts. Metal bands were added to connect old and new parts together. In addition, the four windows in western side of the drum were blocked [9]. The devastation of the eastern quarter of the main dome with part of the adjacent semidome in the 14th century was due to the earthquake of 1346; the rebuilding took considerable time (1346 to 1353) because of insufficient finance. Greek G. S. Astras and the Italian G. Peralta were the builders for the reintegration of the collapsed part. After their works, the dome shape and structure still exist today [8]. Although Istanbul was hit by several devastating earthquakes, the dome survived intact. However, the main structure was subjected to comprehensive works and was reinforced with buttresses to control lateral expansion. No significant changes in the dome were made by the Fossati Brothers, except for the removal of the dome's flying buttresses. Instead, a double iron ring was placed around the base of the dome, holding the dome against tensile forces. In addition, the many columns that had deviated from the vertical by about 38 cm, a problem even affecting the superstructure, were stabilized [8]. In the 1950s and 1980s, cement was used in repair work. However, the dome already retains its integrity, so cement was used only as a binder and in the imitation of window frames. In the first decades of 21st century, all cement was replaced with the original material [5].

2.2 The Second Case: The Reinforced Concrete Interventions to the Twin Minarets Madrasah at Erzurum

Architectural and Structural Brief. The Twin Minarets Madrasah, originally named *Çifte Minareli Medrese* at Erzurum, was commissioned by Seljuk Sultan in the 13th century [11]. It has a typical plan scheme of the Anatolian Seljuk Architecture's madrasah, consisting of a court (12x25 m), iwans, and an artwork gate. It is an example of the two-storied madrasah with twin minarets. The student's rooms are nestled on the north and south sides. In the middle of these sides, there are two iwans. The north iwan was arranged as the gate; the south iwan lasted with a mausoleum (Fig. 2).

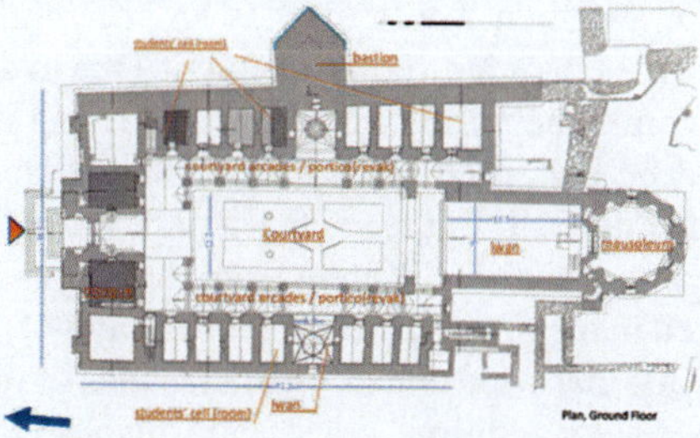

Fig. 2. The ground floor plan of Twin Minaret Madrasah at Erzurum [11]

In madrasah, the major spaces are covered with vaults with two centered pointed profiles except for the masjid, mausoleum, and classrooms covered with domes. Moreover, the superstructure of the mausoleum consisted of two layers: an inner stone dome and a conical dome as a second shell giving a more pleasing external form (Fig. 3). The south vault in front of the mausoleum, no longer existent, has a much larger than the other iwans. It measures 9.2 m and 15.7 m. It is also expected to have been originally more decorative than the others. As the traces of its original geometry, the flat stones and the beginning of a transition element in the section close to the mausoleum can be distinguished in old photographs dated to early 20th century. This space could have a two-centered pointed profile vault and a decorative vault covering an approximately square area. However, much more information is needed to fully understand its geometry and size (Fig. 3).

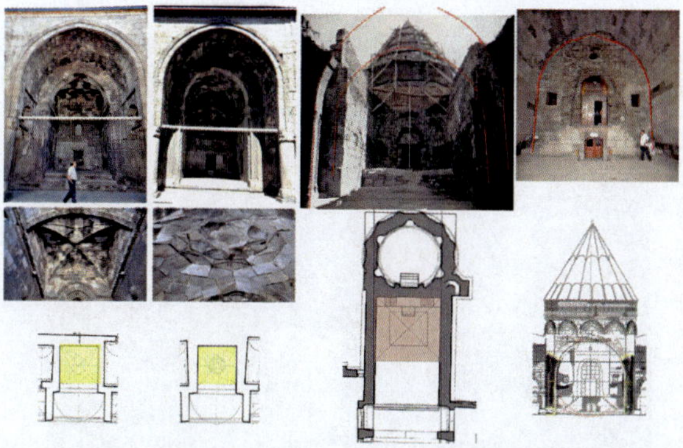

Fig. 3. Comparison of the vaults of the madrasah and possible geometry for south iwan

Timeline For Changes. In the 19th century, many medieval buildings lost their superstructure due to the damage caused by time and the damage and neglect after the First

World War. Moreover, the earthquakes accelerated the damages of the structures. The madrasa exemplifying these conditions was lost its structural integrity. It was an educational institution used until the end of the 18th century. Within the modernization movement, it lost its original function due to the establishment of new modern schools. It was derelict for a while and then it was used as an armory. The monument should have been restored many times in order to survive today. Due to a lack of records, before the 19th century are not comprehensively known. Considering its old photographs, the south iwan's vault would be collapsed before the 1900s [11]. The other iwans were closed with walls to use for deposit. In the southern part, there were rubble stone walls with horizontal timber beams, which is different from the original cut stone. These walls must have been constructed in the previous repair (Figs. 3 and 4). It was understood from the photographs of the 1930s that wide cracks on the façades of the mausoleum occurred due to earthquakes (Fig. 5). In addition, cracks and displacements were seen in the vaults on the ground floor. While the administrative and intellectual interest was rising in cultural heritage in the 1930s, the repair and maintenance works were kicked off with conservation consciousness. The twin madrasa was placed on the selected monuments list for restoration. Its restoration represents the conservation concept and techniques in Turkey. It is a significant case in which the reinforced concrete was used for consolidation deliberately [2]. In this restoration between 1933–35, the conical dome of the mausoleum was dismantled and a new reinforced concrete structural system was built in the inner layer of the structure and covered with stone again as followed in Fig. 4. Reinforced concrete was used by concealing as defined in the regulations and specifications in force at the time [3, 13]. In the 1970s, the reconstruction work of the south iwan vault was started. As explained in architectural features, the south iwan vault no longer exists from the 19th century. Through the examination of the old photographs, the idea of its original geometry can be gleaned from the surviving parts. However, it is unknown how this information was interpreted in the repair decisions. These details were lost or hidden after the restoration, and are difficult to determine again. In the 1980 conservation work, a stone vault with a two-centered pointed profile was reconstructed. It may be a reinforced concrete frame under the masonry stones. But the cement was definitely used in the mortar (Fig. 4).

Fig. 4. First row, the conical dome restoration work in 1933–35; second row; the reconstruction of south iwan vault in 1980s [11, 14].

The intervention in the vaults of the porticoes on the upper floor in the southwest direction of the monument is interesting both in terms of the use of reinforced concrete and in terms of showing a different approach from the other vaults repair. It was determined that the vault was demolished, and the space was covered with a wooden flat ceiling. In the symmetrical structure, the original form of this vault must have been the same as the one on the other side. Although the original geometry of the main vault is unknown, the vault was reconstructed using stone. However, in reconstruction of this small size vault, masonry was not preferred, the reinforced concrete was visibly used.

Since the 1990s, the use of reinforced concrete has been abandoned with the increasing negative effects of cement on masonry structures. In the last ten years, reinforced concrete interventions have been dismantled even if the original structure has been damaged and reconstructions have emerged. This approach was also followed in the madrasa restoration in 2015. However, the reinforced concrete vault and reinforced concrete horizontal beam could not be removed as they could cause structural problems.

The Comparison of The Cases. The historical process of the two monuments provides essential information about both the strengthening techniques and the approach to the survival of the buildings, given that there was no scientific understanding of conservation until the 19th century.

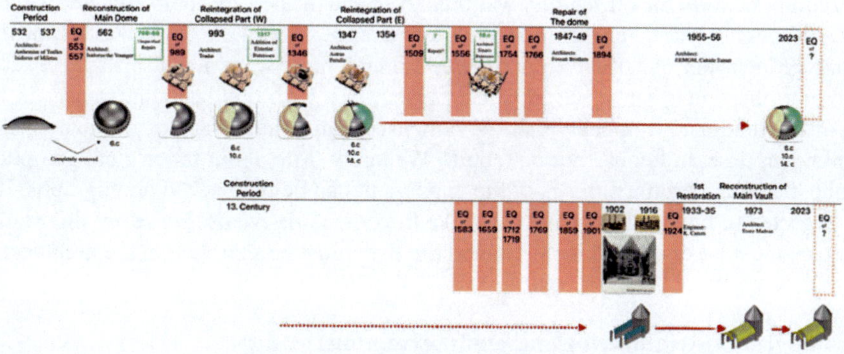

Fig. 5. The comparison of the cases

In the Context of Conservation. Despite the lack of scientific conservation criteria, many monuments have been preserved for historical and religious reasons. Hagia Sophia has always had a universal value as every nation revered it. The fact that a prominent architect in Armenia was called for the restoration of Hagia Sophia is related to the value of this monument. The madrasa was only an educational building with only functional value. It was repaired only to maintain its use. In the 19th century, its use as an armory showed that despite its monumental size, its value was unrelated to religious or historical reasons. While the Madrasa was suffering from neglect, the Italian Fossati Brothers embarked on the restoration of Hagia Sophia with a huge budget in the same years [5]. These examples show that the concept of conservation was not comprehensively established in the Ottoman Empire and not all monuments were treated as cultural assets.

It would only be with the Republic of Turkey that the madrasa would be registered as a cultural property. In this way, it was among the first scientific conservation efforts of the early Republican period.

Strengthening Techniques. Damage is a crucial parameter that determines the scope and scale of interventions. The dome of Hagia Sophia was rebuilt in different geometry within a decade of its initial collapse. Isidorus the Younger reconstructed the new dome in its place higher than the first dome. The architect may have realized that the first dome was weak against the horizontal forces of the earthquake and that increasing its height might be a solution. Another example of completely removing is taken place in the Fatih Mosque where its dome collapsed due to earthquakes in 16th century [5]. Not only its dome but also the whole structure was removed and rebuilt.

In the case of Hagia Sophia, in the repairment of the structure in the 10th century it was intended to reduce the pressure of the skirt band of the dome by blocking the four windows. Interestingly, seven centuries later, the same technique was used in the Ottoman period. Some of the windows in the drum were filled in to stabilize the damaged dome of the Church of Hagia Irene [4]. The dome of Hagia Sophia was destroyed again by an earthquake in the 14th century. This time, the 6th century eastern part was demolished. Looking at the plan of the building, the sides where the main dome was supported by semi-domes were destroyed by earthquakes. In the repair of 14th century, the north and south components of the 6th century was preserved and the new masonry attached to these parts. Although it was taken a homogenous appearance, indeed the dome was composing of mixture interventions of different times. Although there were many earthquakes after this repair, the dome has survived to the present day with this last appearance. In the case of the madrasa, it appears that the collapsed main vault was heavily damaged at an unknown time and could not be rebuilt. We know little about its original geometry. Despite all the uncertainty, traces of the main vault can be read in old photographs. It is also expected to have a decorative vault like the other side iwans. However, this vault's geometry must be examined in detail, and the data must be searched in the archives.

2.3 In Term of Reinforced Concrete Interventions

In the 1940s, Turkey experienced several devastating earthquakes that further structurally deteriorated already neglected and poorly maintained monuments. With conservation consciousness, the restoration practices in Turkey began in the 1930s, but not all monuments were reached in time due to financial difficulties and lack of experts. Despite the need for urgent remedies, many monuments were restored in the 1950s for the first time after a long period of neglect. However, in the further years, due to the increasing number of the many restorations of monuments, the historical research required for reintegration was not realized for all works. Thus, these reasons lead to rapid and unqualified work. Since the availability of traditional materials was scarce and not easy to find craftsmen, the spread of cheap and easily applicable concrete has increased.

The restoration work of the 1930s of madrasa was the first case of consolidation of historic structures with reinforced concrete. But, in further works in the 1970s reinforced concrete was used for reintegration of ruined parts or reconstruction of the missing parts of the madrasa. In either case, the original geometry obtaining data from the monument itself and its historical records and the analogy studies is essential to the demand for constructing the new part. If the knowledge is poor, the new part which is constructed in the traditional technique can be a falsification. Thus, reinforced concrete was a modern way to avoid falsification in spite of its disadvantages.

In Terms of Original Geometry. The intervention for partial demolition is mostly in the form of reintegration. There are many examples of this, as in the case of the dome of Hagia Sophia. However, after the 1950s, a conservation approach emerged in which the new and original parts are distinguished from each other and the intervention is deliberately not made parallel to the original geometry in order to be recognizable. In this approach, the original part is still reintegrated with masonry, but a surface difference is created between the new and the original courses. In the case of the madrasa, this was not followed, and the original part was even partially dismantled and lost. In the examples where the reintegration was of reinforced concrete, the intervention is already directly recognizable due to the difference in materials (Fig. 6).

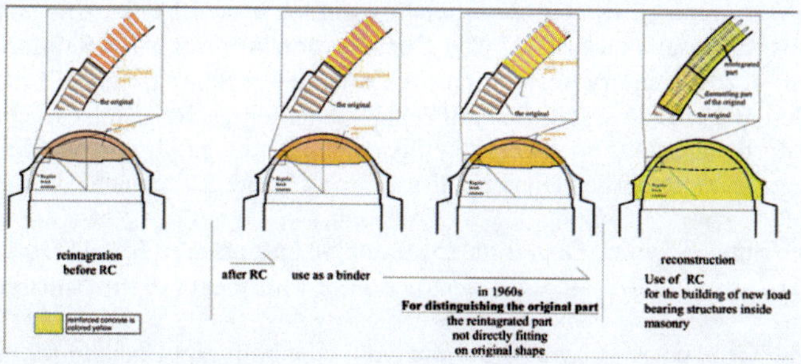

Fig. 6. Variations of interventions depending on the scale and the materials used

3 Conclusion

Each damage contains knowledge about the problem the monuments suffered, how the problem was perceived, and how it was resolved. For example, a collapse may indicate that the building suffered an earthquake or point to another reason that could have led to the collapse. When our assumptions are matched with factual information, we can gain new insights into the monument process and conservation history. These investigations are the key to the unknown chains of the long historical process of any monument. A critical point regarding conservation is how interventions have rectified damages and how human beings have found solutions in the historical process. The historic interventions related to engineering provide essential data showing that monuments were structurally evaluated and kept alive. Subsequent interventions allow the monument to survive and are part of its existence. Therefore, later additions and interventions are as exceptional as their architectural aesthetic qualities.

The cases show that reintegration provides much more detailed knowledge than reconstruction. The dome of Hagia Sophia survived all the damage and that this process is evidenced by its deformed geometry is enthusing over it. It has been intervened with techniques and shapes different from its first original character, these interventions, which are layered and integrated into the whole structure, are also respected. Taking note of the way, even if not convenient and original material and technique, the reinforced concrete interventions should be respected like all later additions. The reinforced concrete interventions are aligned with the principle of "recognition of the intervention", and they ensure by distinguishing these interventions from the original character in terms of transferring knowledge to the future. The original geometries of the missing parts of the monuments can be followed thanks to the reinforced concrete interventions. Reinforced concrete interventions to structures whose original geometry is unknown prevent falsification and create a neutral state.

References

1. This study was prepared by the corresponding author hosted by the University of Genoa from her post-doctoral study titled "The Conservation Approach for Reinforced Concrete Interventions in 20th Century Restorations: A Comparative Study Between Turkey and Italy" article funded by TUBITAK (The Scientific and Technological Research Council of Turkiye)
2. Inan Ocak, Z.: Restorations of anatolian seljuks monumental buildings from the beginning of the 20th century, (Unpublished Doctoral dissertation). Istanbul Technical University, Istanbul (2018)
3. Camilerin ve Bunlara Şumuli Olan Binaların Tamir ve İnşalarına ait Fenni Şartname (the Book of Specifications on the Repair and Construction of Mosques) (1936) Osmanbey Matbaası, İstanbul
4. Ahunbay, Z.: Tarihi çevre koruma ve restorasyon, Istanbul: Yapı Endüstri Merkezi (1996)
5. Coskun, B.S.: İstanbul'daki Anıtsal Yapıların Cumhuriyet Dönemindeki Koruma ve Onarım Süreçleri Üzerine Bir Araştırma, (Unpublished Doctoral dissertation). Mimar Sinan Fine Arts University, İstanbul, Turkey (2012)
6. Mainstone, R.J.: Hagia Sophia: Architecture. Thames and Hudson, Structure and Liturgy of Justinian's Great Church (2006)
7. Cakmak, A., Taylor, R., Durukal, E.: The structural configuration of the first dome of Justinian's Hagia Sophia (A.D. 537–558): an investigation based on structural and literary analysis. Soil Dyn. Earthq. Eng. **29**, 693–698 (2009)
8. Emerson, W., van Nice, R.L.: Haghia Sophia, Istanbul: preliminary report of a recent examination of the structure. Am. J. Archaeol. **47**(4), 403–436 (1943)
9. Maranci, C.: The architect trdat: building practices and cross cultural exchange in Byzantium and Armenia. J. Soc. Archit. Hist. **62**(3), 294–305 (2003)
10. Duppel, C.: Ingenieurwissenschaftliche untersuchungen an der hauptkuppel and den hauptpfeilem der Hagia Sophia in Istanbul, Ph. D. Karlsruhe Institute of Technology (2009)
11. All the documents were obtained from Directorate General of Foundations in Turkey. Its archive is abbreviated as VGMA
12. Icomos charter- principles for the analysis, conservation and structural restoration of architectural heritage (2003). https://www.icomos.org/en/about-the-centre/179-articles-en-francais/ressources/charters-and-standards/165-icomos-charter-principles-for-the-analysis-conservation-and-structural-restoration-of-architectural-heritage
13. https://www.icomos.org/en/167-the-athens-charter-for-the-restoration-of-historic-monuments
14. https://archives.saltresearch.org

Effective Limit Analysis Computational Approaches for the Structural Characterization of Nervi's Palazzetto dello Sport

Rosalba Ferrari[1(✉)], Giuseppe Cocchetti[2], and Egidio Rizzi[1]

[1] Università degli studi di Bergamo, DISA,
viale G. Marconi 5, I-24044 Dalmine (BG), Italy
rosalba.ferrari@unibg.it
[2] Politecnico di Milano, DICA,
piazza L. da Vinci 32, I-20133 Milano, Italy

Abstract. In this work, two formulations, and computational implementations, of Limit Analysis (LA), for large-scale 3D truss-frame structures are employed to efficiently investigate the limit elastoplastic response of the fascinating and well-known Palazzetto dello Sport (1957), located in the Flaminio district of Rome (Italy), designed by Pier Luigi Nervi. To this end, a spatial FEM modeling was first generated, whereby the characteristic inclined trestles and supporting skeleton of the above concrete shell dome were modelled according to the corresponding geometric characteristics and mechanical properties acquired by inspecting original design drawings and relevant bibliographic sources. In performing the structural Limit Analysis, the first algorithm step-by-step traces a fully exact evolutive piece-wise linear elastoplastic response of the structure, up to plastic collapse, by reconstructing the sequence of activation of localised plastic joints. The second algorithm, relying on a kinematic iterative direct approach, determines just the collapse mechanism and associated collapse load multiplier, but in a much shorter computational time, showing a rather impressive performance, in truly precipitating from above on the collapse load multiplier, by rapidly adjusting possible mechanisms to the sought collapse mode, in very few iterations. The performed investigation reveals the ingenuity of Nervi's work, not only with respect to ordinary structural bearing tasks, within the elastic range of regular-service structural response, but also to post-yield performance and ultimate limit load capacity, up to possible ductile failure.

Keywords: structural modeling · large-scale 3D truss-frame structures · computational Limit Analysis (LA) · piece-wise linear evolutive description · direct kinematic method · collapse load multiplier and mechanism

1 Introduction

This paper explores two novel approaches for the computational Limit Analysis of 3D large-scale truss-frame structures, specifically applied to analyze the

S. Gabriele et al. (Eds.): IWSS 2023, LNCE 437, pp. 827–836, 2024.
https://doi.org/10.1007/978-3-031-44328-2_87

renowned *Palazzetto dello Sport* (1957) in Rome's Flaminio district, an architectural masterpiece by Pier Luigi Nervi. The adopted underlying FEM model encompasses approximately 3200 beam finite elements and around 10000 degrees of freedom.

Nervi's globally recognized structures epitomize an exquisite blend of structural efficiency and architectural allure. They stand as true monumental achievements, where static equilibrium harmoniously melds with envisioned structural form. While a considerable literature exists on Nervi and his works, only a few studies have investigated how these remarkable artworks may set their collapse limit in terms of structural response. In this sense, it might be reasonable to anticipate that structures characterized by congenial geometric forms may exhibit an exceptional efficiency not only in their everyday structural tasks, which typically fall within the elastic range, but also in their performance under limit load conditions, nearing collapse. In pursuit of such an investigation objective, the present study explores Nervi's sports complex using two self-made Limit Analysis algorithms. These algorithms were initially developed to comprehensively examine the limit elastic-plastic response of 3D large-scale truss-frame structures under specific loading configurations [1–4].

The first algorithm employed in this study outlines an iterative approach to systematically increase the base forces (live loads) applied to the structure through a common increasing load multiplier. This method effectively traces the "exact" piece-wise linear evolutive elastoplastic response of the structure, considering the hypothesis of plastic joint formation, as a generalization of the plastic hinge concept [1,2]. In each iteration, the computational calibration ensures that, as compared to the previous step, one or more plastic joints located at the ends of the beam finite element are activated. Consequently, the (elastoplastic) stiffness matrix of the structure is updated at each step, precisely reflecting the number and position of the activated plastic joints within the structure.

The second algorithm adopted in this work is based on a kinematic upper-bound approach for computational Limit Analysis, as an efficient "direct" method. Notably, this algorithm stands out for its distinct computational efficiency, as highlighted in [2]. Similar to other kinematic Limit Analysis algorithms, it does not provide a comprehensive characterization of the structural response under increasing loads. However, starting from a collapse configuration randomly defined within the structure, it rapidly identifies the collapse mechanism by converging on the true collapse multiplier in just a few iterations. This unique characteristic renders the algorithm highly suitable for investigating structural collapses, particularly when a significant number of simulations may be required.

The two computational algorithms were employed to conduct an elastoplastic Limit Analysis on a 3D FEM model of the *"Palazzetto dello Sport"*. The model consisted of beam-type finite elements, whereby for the dome part only the articulated rib system that forms the structural skeleton was explicitly considered.

The analyses just focused on a self-weight distribution of the structure, which was simulated by applying concentrated loads to the nodes of the 3D truss-frame model. Despite the significant number of degrees of freedom involved in the computation, both LA methods demonstrated consistent and effective performance. From a computational standpoint, it is in fact shown that the kinematic algorithm rapidly converges onto the collapse state, as also traced by the full evolutive elastoplastic analysis. The kinematic load multiplier quickly approaches the desired collapse load multiplier, while the plastic mechanism efficiently adjusts to the genuine collapse state. Notably, the kinematic algorithm showcases remarkable time-saving advantages, reducing computational requirements by over 98%, as compared to the evolutive elastoplastic analysis. On the other hand, the latter analysis traces a rather ductile piece-wise linear global load-displacement curve.

The present paper is organized as follows. Section 2 briefly introduces the main features of the *Palazzetto dello Sport* (1957). Section 3 recalls the main features of the two LA algorithms employed for the analysis of the structure and the results of the numerical simulations. Brief comments on the various computational aspects and the effectiveness of the performed simulations are concisely pointed out in closing Sect. 4.

2 The Palazzetto dello Sport (1957)

The *Palazzetto dello Sport* (Small Sports Complex) (Fig. 1) is a versatile sports facility situated in the Flamino district of Rome, Italy. It was collaboratively designed by Architect Annibale Vitellozzi and Engineer Pier Luigi Nervi, commissioned by the Italian National Olympic Committee (CONI). While the venue gained prominence during the 1960 s' Summer Olympics in Rome, its original purpose extended beyond hosting events for the Games. The intent was to create a prototype of a moderately sized sports complex, capable of accommodating 4000–5000 spectators, and suitable for a wide range of indoor sporting events. Moreover, the design aimed to facilitate its possible replication in other locations throughout the country.

Fig. 1. Nervi's Palazzetto dello Sport (1957).

The structure boasts a ribbed shell dome with a diameter of 60 m, supported by sloping trestle-style buttresses arranged along its entire perimeter. These buttresses are inclined to match the tangent plane of the shell's surface where they connect. Each of this Y-shaped buttresses displays a vertical support in the spot where they branch off. Taking these elements into account, the total outer perimeter of the dome measures approximately 80 m in diameter. To withstand the dome's thrust, a superficial foundation was chosen. It consists of a prestressed concrete ring beam with an 80 m diameter, positioned at the lower end of the pillars. The sports arena is situated below the ground level and is encompassed by a terraced seating area. Beneath the seating area, service rooms are located, extending towards the external ring that aligns with the buttresses. The structure is enclosed by continuous ribbon windows, spanning the space between the top of the seating area and the ribbed shell dome (Fig. 2).

Fig. 2. Internal view of the *Palazzetto dello Sport*.

The *Palazzetto dello Sport* exemplifies the success of Nervi's system in terms of statics, construction, architecture, and economics. The seamless integration of structural integrity and architectural beauty is evident both inside and outside the building (see Figs. 1 and 2). The complex was completed in just over a year, from July 21, 1956, to September 14, 1957, thanks to the efficient utilization of prefabricated elements for the intricately ribbed roof. This construction approach resulted in a rather cost-effective implementation. Specifically, the ribbed shell dome was divided into individual sections, which were later assembled on-site. Its project encompassed 13 distinct types of slices, referred to as "*tavelloni*" or hollow blocks. Each slice was carefully designed to be easily manufactured and transported by a team of just two workers, owing to their manageable size.

The prefabrication technique employed for the blocks was based on Nervi's concept of "*ferro-cemento*" (steel–concrete) and a clever generational process. Initially, a full-scale wooden template of a section of the canopy was created. This template was placed on the ground and coated with plaster, onto which the block axes were marked. Subsequently, 13 masonry molds were constructed, each producing a prototype block using *ferro-cemento*. The blocks were formed by using a simplified grid of 6 mm metal rods spaced every 15 cm, covered with a 0.6 kg/m^2 layer of mesh. A pliable concrete mix, comprising fine aggregates, was then applied with a trowel onto this lighter steel mesh, supported by a mold, creating a 2.5 cm thick layer. Within this prototype, known as "*nonna*" or granny, a series of identical-shaped blocks called "*mamme*" or mums, with a precise double curvature were generated. Using these *mamme* as templates, multiple teams of workers produced numerous "*figlie*" or daughters, which were ultimately utilized in the construction process. Approximately 30 blocks were manufactured each day. After reassembling the dome mosaic, reinforcement was inserted into the channels formed between the blocks, and casting was performed to finalize the construction. The wedges, which served as disposable formwork, were meticulously finished on the visible inner surface. Subsequently, a series of layers were applied for waterproofing and exterior painting, ensuring both protection and aesthetic appeal [5–7].

3 Numerical Simulations

3.1 Limit Analysis Finite Element Modelization

The two computational LA formulations herein adopted are based on the same underlying hypotheses, as briefly listed here. Potential perfectly-plastic joints have been located at the two edges of each beam finite element composing the model structure; a piece-wise linear yield domain, specifically an uncoupled Rankine-type boxed-form yield domain in the space of the static variables (i.e. beam internal actions), is chosen. Linear kinematics is assumed, namely second-order non-linear geometrical effects are so far neglected and equilibrium is enforced in the initial (unstressed) configuration of the structure. Permanent (i.e. gravity) loads and a set of (base) live loads amplified by a common ("load multiplier") factor are considered. Both formulations have been implemented within MATLAB®, within specific homemade computer programs exploiting 3D beam finite elements.

The evolutive elastoplastic formulation has been developed by applying the "exact" and stepwise holonomic step-by-step analysis method originally described in Cocchetti and Maier [8], then enriched by various original features in order to deal with 3D large-scale truss-frame structures. Such characteristics comprise the treatment of mutual connections by static condensation and Gaussian elimination and the determination of the tangent stiffness formulation through Gaussian elimination (see e.g. Ferrari et al. [1,4] for the computational details).

The direct kinematic algorithm exploits a convenient reshaping of the internal power dissipation description of a beam element, in order to adapt an existing efficient approach originally proposed by Zhang et al. [9], for the LA of continua, within a classical FEM analysis based on beam element discretization (see Ferrari et al. [2] for the computational details).

The numerical model of the *Palazzetto dello Sport* was developed using MATLAB®, based on an analysis of original design drawings to determine the structure's topology and geometric characteristics of its elements. Certain simplifications were made within the FEM formulation. Specifically, a 3D truss-frame model was constructed using beam finite elements built-in at the mutual nodes. However, for the dome, only the articulated rib system that forms the structural skeleton was considered, omitting the explicit representation of the external shell. The FEM model utilized constant cross-section rods with homogenized geometric properties (area, principal moments of inertia, and torsional inertia) between the nodes of the truss-frame structure. The base of the Reinforced Concrete (RC) pillars was fixed using absolute encastres. The elastic mechanical properties assigned to the FEM assembly of the *Palazzetto dello Sport* were characteristic of RC materials, including a Young's modulus (E) of 30000 MPa and a Poisson's ratio (ν) of 0.2 (resulting in a shear modulus (G) of 12500 MPa).

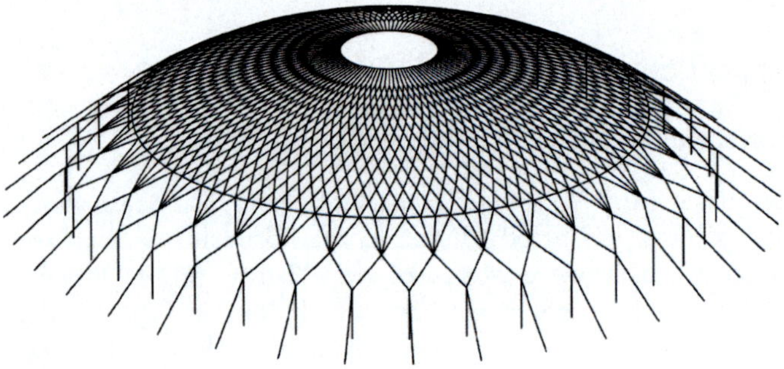

Fig. 3. Assembled complete 3D truss-frame FEM structural model of the *Palazzetto dello Sport.*

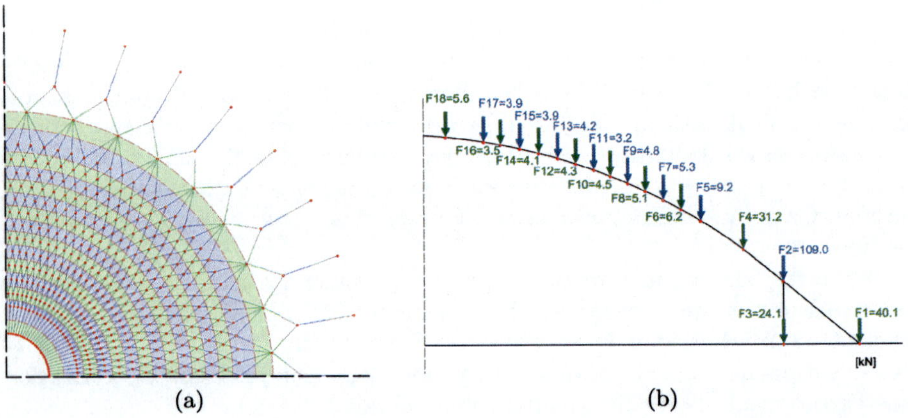

Fig. 4. Self-weight load acting on the nodes of the *Palazzetto dello Sport* FEM model.

The complete 3D truss-frame FEM representation of the *Palazzetto dello Sport*, depicted in Fig. 3, consists of 3204 beam finite elements and 1656 linked nodes, resulting in a total of 9936 degrees of freedom (including node displacements and rotations).

This study focuses on quantifying the plastic collapse load multiplier and the associated collapse mode of the structure, considering only the self-weight of the dome. The self-weight comprises components such as the *ferro-cemento* material, vermiculite insulation, waterproofing layer, and the concrete composing the dome. The estimated weight was applied as concentrated loads on the nodes corresponding to the rib system of the dome (see Fig. 4).

3.2 Numerical Results

The elastoplastic load-carrying capacity of the *Palazzetto dello Sport*, obtained through the aforementioned numerical algorithms of Limit Analysis, is comprehensively summarized in Table 1, with self-matching results, for the two employed methods. Furthermore, Fig. 5–6 illustrate the mechanical elastoplastic response of the structure.

Table 1. Collapse load multipliers for the *Palazzetto dello Sport.*

Method	Collapse load multiplier	Limit state
Evolutive	2.44	First null eigvenvalue of stiffness matrix
Kinematic	2.44	Collapse mechanism with min(λ)

The numerical assessment of the collapse load multiplier, as presented in Table 1, demonstrates a precise alignment between the estimated value obtained through the evolutive analysis (which may be considered as a static evaluation, essentially capturing the earliest singularity of the global tangent stiffness matrix) and the upper-bound value derived from the kinematic method. This correspondence validates the accuracy and reliability of the achieved LA results.

Figure 5 reports the characteristic piece-wise linear load/displacement response curve of the dome obtained from the evolutive algorithm. In particular, the horizontal axis reports the vertical displacement (Δ) of a specific node of interest (the analysis here delivered specifically considers a node belonging to the rib system of the dome); the left vertical axis reports total amplified live load $P = \lambda Q$, where λ is the load multiplier related to the incremental solution of the non-linear elastoplastic analysis. The end point of the P–Δ curve refers to the so-estimated "exact" collapse of the structure in terms of first vanishing minimum eigenvalue of the tangent stiffness matrix of the structure. In this case, the computing time to achieve the collapse solution was about 47265 s. The algorithm has been implemented and run as a non-compiled code within MATLAB®, under a Windows 10 operating system, on a Dell laptop endowed with an Intel Core i7-6500U Processor, clock at 2.50 GHz and 16 GB RAM.

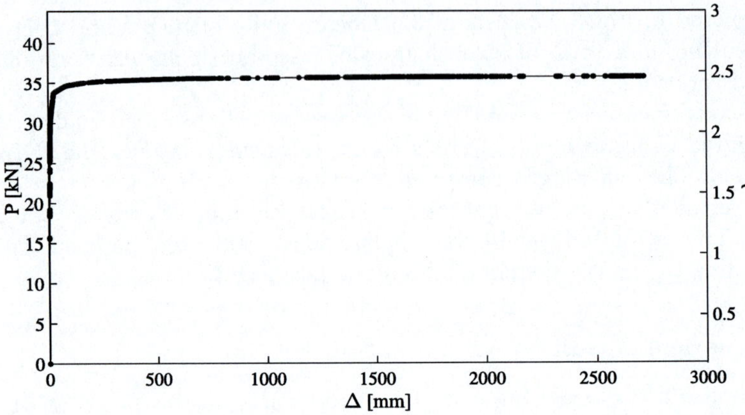

Fig. 5. Representation of the characteristic piece-wise linear load-displacement response curve by the evolutive approach.

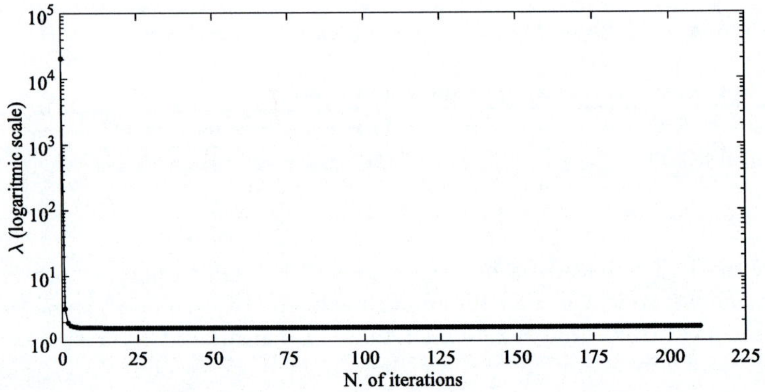

Fig. 6. Collapse load multiplier estimation along the direct kinematic iterative procedure.

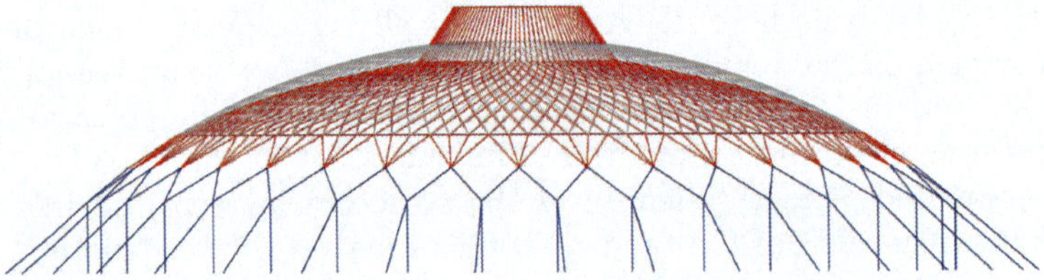

Fig. 7. Estimated collapse mechanism of the *Palazzetto dello Sport*, $\lambda = 2.44$ (undeformed structure in grey).

Figure 6 further illustrates the results related to the kinematic algorithm. It displays the collapse load multiplier computed during the iterative procedure. Remarkably, the collapse mechanism is nearly attained already by the tenth iteration, and in approximately 50 iterations, the kinematic load multiplier already reaches the collapse value. Subsequently, the load multiplier stabilizes, requiring an increasing number of iterations to progress further, until the iterative process ends based on the stopping criteria defined by the user [1,4]. Notably, the direct kinematic method achieved the collapse solution in about 210 iterations, taking approximately 476 s on the same computer platform mentioned earlier, which is about only 1% of the computational time required for the corresponding complete evolutive elastoplastic approach.

Finally, Fig. 7 depicts the configuration of the structure at incipient collapse (namely, the plastic collapse mechanism), as obtained through the implementation of the kinematic Limit Analysis algorithm.

4 Conclusions

This paper presents a comprehensive computational elastoplastic structural analysis implementation, adhering to the principles of Limit Analysis. It focuses on 3D large-scale truss-frame structures. Thereby, it specifically investigates the elastoplastic response of a numerical model that closely resembles iconic Nervi's *Palazzetto dello Sport* (1957). Although the model's geometry is approximated, the analysis aims to offer an initial and forward-looking evaluation of its structural performance in relation to Limit Analysis states.

Two approaches were utilized and compared to uncover the capabilities of the structure regarding plastic collapse limit states. The first approach involved reconstructing the complete evolutive response leading to collapse, tracing the global load-displacement curve. The second approach focused on directly determining the collapse characteristics, such as the collapse load multiplier and the plastic mechanism.

Both Limit Analysis methods consistently and effectively performed, despite the significant number of degrees of freedom involved in the calculations (about 10000). The predictions of plastic collapse aligned remarkably well, with the kinematic iterative algorithm demonstrating an impressive efficiency, by swiftly converging onto the collapse load multiplier and rapidly adjusting the plastic mechanism to the collapse mode in just a few iterations.

These new computational Limit Analysis methodologies shall display the potential to revitalize the application of Limit Analysis in the structural analysis of large-scale structures. They can be employed in the analysis of both innovative design concepts and historically significant structures that shall require preservation. Furthermore, they offer exciting opportunities for structural optimization, allowing for variable structural characteristics and exploring the corresponding variations in elastoplastic response and collapse features through optimization loops.

The present research aims to constitute the starting point of a study focused on the analysis of the limit structural behavior of Nervi's *Palazzetto dello Sport*.

Specifically, it investigates the ductile structural behavior of the sports complex, which undoubtedly was envisioned in the design philosophy of the great designer. Progresses toward achieving this goal may involve enhancing the accuracy of the finite element model and refining the characterization of the plastic behavior of its elements. Additionally, determining the applicable loads in a Limit Analysis scenario will constitute an essential aspect of the research.

Acknowledgments. This study has been developed at the University of Bergamo, Dept. of Engineering and Applied Sciences. The public financial support by ministerial research funds (MIUR) *"Fondi di Ricerca d'Ateneo ex 60%"* at UniBG is truly acknowledged.

References

1. Ferrari, R., Cocchetti, G., Rizzi, E.: Limit Analysis of a historical iron arch bridge. Formulation and computational implementation. Comput. Struct. **175**, 184–196 (2016) https://doi.org/10.1016/j.compstruc.2016.05.007
2. Ferrari, R., Cocchetti, G., Rizzi, E.: Effective iterative algorithm for the Limit Analysis of truss-frame structures by a kinematic approach. Comput. Struct. **197**, 28–41 (2018). https://doi.org/10.1016/j.compstruc.2017.11.018
3. Ferrari, R., Cocchetti, G., Rizzi, E.: New computational algorithms for the Limit Analysis of large-scale 3D truss-frame structures. In: Liu, G.R., Trovalusci, P. (eds.) Proceedings of the 9th International Conference on Computational Methods (ICCM 2018), Rome, Italy, 6–10 August 2018, Paper ID 3387, pp. 206–517. ScienTech Publisher LLC, USA (2018). ISSN: 2374-3948 (online)
4. Ferrari, R., Cocchetti, G., Rizzi, E.: Reference structural investigation on a 19th-century arch iron bridge loyal to design-stage conditions. Int. J. Architectural Heritage **14**(10), 1425–1455 (2020). https://doi.org/10.1080/15583058.2019.1613453
5. Bucur-Horváth, I.V., Săplăcan, R.: Force Lines Embodied in the Building: Palazzetto dello Sport. J. Int. Assoc. Shell Spatial Struct. **54**(2–3), 179–187 (2013)
6. Chiorino, M.A., Chiorino, C.: Pier Luigi Nervi: architecture as challenge. Struct. Archit.: New Concepts Appl. Chall. **13** (2013)
7. Dieli, S.: About the hangar and Pier Luigi Nervi. Ephemer Arts Connection (2012)
8. Cocchetti, G., Maier, G.: Elastic-plastic and limit-state analyses of frames with softening plastic-hinge models by mathematical programming. Int. J. Solids Struct. **40**(25), 7219–7244 (2003)
9. Zhang, P., Lu, M., Hwang, K.: A mathematical programming algorithm for limit analysis. Acta. Mech. Sin. **7**(3), 267–274 (1991)

A Unidirectional Ribbed Slab System with Prefabricated Textile-Reinforced Concrete Shells

Marco Carlo Rampini, Giulio Zani(✉), Matteo Colombo, and Marco di Prisco

Politecnico di Milano, Milan, Italy
giulio.zani@polimi.it

Abstract. The use of innovative high-performance cementitious materials enables the development of advanced structural systems characterized by enhanced lightness, durability, sustainability, and mechanical performance. In this paper, we propose a partially precast unidirectional ribbed floor system consisting of very high-performance fiber-reinforced concrete (VHPFRC) I-beams, textile reinforced concrete (TRC) stay-in-place formworks, and ordinary fiber-reinforced concrete (FRC) finishes. A description of the conceptual design underlying the proposed building system is followed by a depiction of the materials and construction procedures used. Special emphasis is placed on the mechanical behavior of the TRC, both at the material and structural scales. Within this context, the membrane and flexural responses of thin beam specimens are discussed and followed by the experimental analysis of the movable prefabricated shells inspired by Pier Luigi Nervi's ferrocement constructions, subject to transient actions associated with typical construction stages.

Keywords: textile reinforced concrete · stay-in-place formwork · prefabricated shell

1 Introduction

In the ever-evolving world of construction, there is a pressing need for construction solutions that can keep up with the demands of current technical scenarios, which require innovative approaches that can deliver advantages in terms of efficiency, sustainability, speed of implementation, and flexibility. Traditional construction methods are gradually giving way to building industrialization processes, where prefabricated elements and modular systems are adopted to ensure greater precision, reduced environmental impact, and shorter construction times. This paradigm shift has created an increasing demand for innovative solutions that combine characteristics such as lightweight, ease of assembly, cost-effectiveness, and superior mechanical properties.

High-performance cementitious composites offer promising solutions for optimizing structural components [1]. Inspired by Pier Luigi Nervi's groundbreaking work on ribbed slabs, this study further discusses a novel approach for elevated

S. Gabriele et al. (Eds.): IWSS 2023, LNCE 437, pp. 837–846, 2024.
https://doi.org/10.1007/978-3-031-44328-2_88

slabs, using textile-reinforced concrete (TRC) [2,3] as a thin structural shell [4–6]. The TRC serves a dual purpose as a stay-in-place formwork and a finishing layer for the slabs. The selected cement-based material consists of a single layer of alkali-resistant (AR) glass fabric embedded in a fine-grained self-compacting matrix. Automated manufacturing techniques were used to produce two full-scale specimens with similar characteristics. As detailed in [7],these specimens were subjected to load tests, simulating transient conditions such as the presence of a worker during the assembly phase. At the mesoscale level, the composite material was further analyzed through uniaxial tension and flexural tests. The bending tests aimed to evaluate how the position of the reinforcement within the thickness affected the behavior of the TRC. This research contributes to the advancement of lightweight, economical and mechanically efficient solutions for structural elements in the field of building industrialization.

2 Conceptual Design

The proposed slab is represented in Fig. 1 and consists of adjacent structural beams made of very high-performance fiber-reinforced concrete (VHPFRC) [1], with a spacing of 1 m and a span of 6 m. These beams are coupled with TRC stay-in-place formworks. Additionally, a topping layer made of steel fiber-reinforced concrete (SFRC) is introduced to minimize the need for on-site reinforcement placement, unless strictly necessary. The adoption of this design solution brings several significant advantages, including a reduction in overall mass. It also offers enhanced environmental sustainability and increased durability due to the optimized use of innovative materials. Additionally, construction times are considerably reduced compared to traditional cast-in-place solutions [8].

The TRC shell element discussed in this paper is a lost formwork with a square plan dimension of 1 m by 1 m and a depth of approximately 20 cm. Initially, it was conceived with a horizontal walking surface, but this design immediately revealed several issues during transient phases. To address these challenges and improve the structural behavior, alternative geometries were considered. The investigation of alternative solutions was carried out through preliminary numerical modeling, focusing on the following aspects: minimizing principal tensile stresses, ensuring compatibility of the geometry with the selected reinforcement system, reducing potential accumulation of rainwater on the element, and simplifying production procedures.

3 Mechanical Characterization of the TRC

The composite material was made incorporating a layer of alkali-resistant (AR) glass double leno weave fabric into a fine-grained high-performance concrete (HPC). The mix design for the TRC was selected through a procedure aimed at achieving optimal workability in the fresh state. This workability requirement was necessary for the full-scale production process, which involved pouring the matrix into the casting mold through several small openings located on one side. Further details are provided in [7].

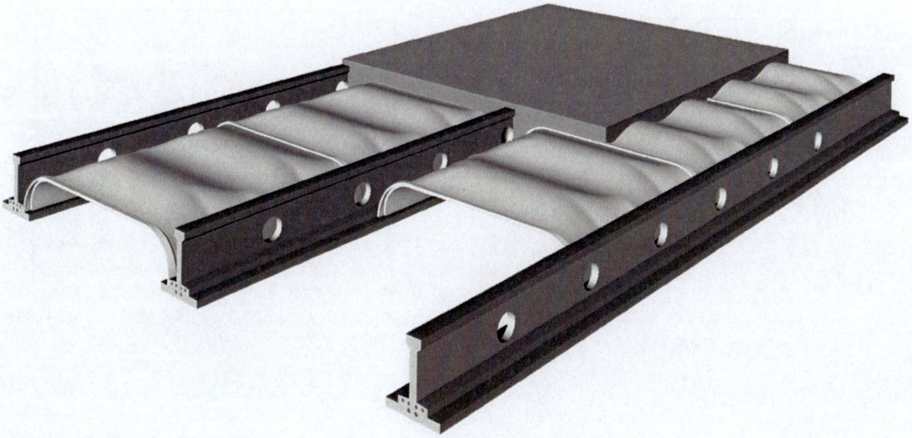

Fig. 1. Scheme of a portion of the proposed lightweight elevated slab composed by VHPFRC I-beams, TRC disposable formworks, and ordinary SFRC concrete.

3.1 Tensile Response

To study the tensile behavior of the material along the two main fabric weaving directions, uniaxial tensile tests were performed on six specimens with a nominal size of 9×70×400 mm^3. Three of the specimens included a fabric oriented in the weft direction, and three in the warp direction. In all cases, the fabric was positioned halfway through the thickness of the sample.

The tests were conducted using a constant displacement control at a set speed of 0.02 mm/s. Steel plates with a thickness of 2 mm (55×70 mm^2 in size) were bonded to the specimens to ensure better distribution of the pressure applied by the grips (about 3.5 bar). Additionally, two LVDT (linear variable differential transformer) sensors were installed, one on the front and one on the back of each specimen, to measure and record the relative crack opening displacements (COD) astride a nominal gauge length of 200 mm. The free length L_0 of the specimens was equal to 300 mm.

The experimental nominal tensile stress (σ_N) vs. normalized stroke displacement (δ/L_0) curves are reported in Fig. 2, in which the individual responses are overlaid in red with the average behavior, both in the warp (Fig. 2a) and in the weft (Fig. 2b) reinforcing directions. Please note that the reported nominal stresses were calculated by dividing the load measured by the load cell by the cross-sectional area of the entire composite, which was determined using the measured values of width and thickness.

Figure 3 shows the crack patterns of all six specimens. It can be observed that all of them displayed multiple cracks before reaching failure, except for the weft-N$_{01}$ specimen. The latter fractured near the steel plate due to the loss of adhesion between the fabric and the matrix, potentially caused by a localized manufacturing defect. As a result, the test was interrupted and the following part of the graph is not plotted.

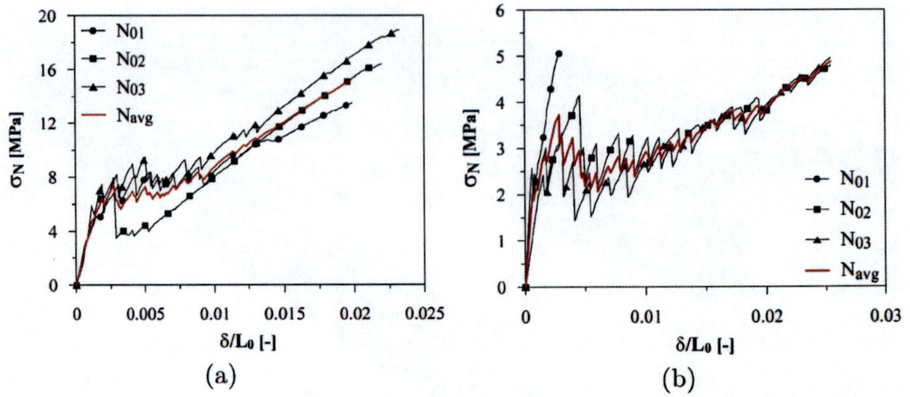

Fig. 2. Tensile tests responses in terms of nominal composite stress vs. normalized displacement in the warp (a) and the weft (b) directions.

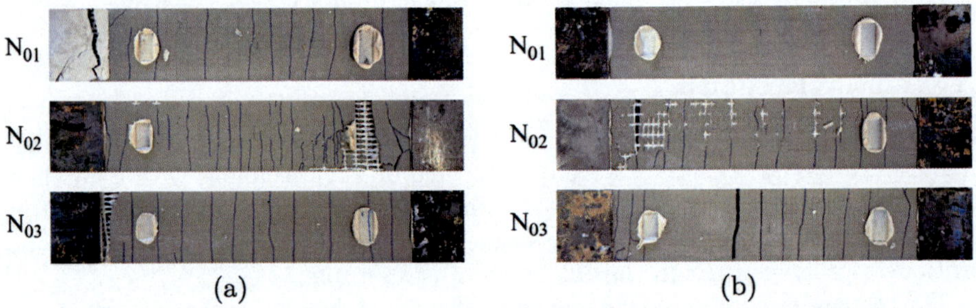

Fig. 3. Crack patterns at the end of the tensile tests in the warp (a) and the weft (b) directions.

It is interesting to note that the TRC achieved a maximum in-plane tensile stress of approximately 15 MPa and 5 MPa, respectively in the warp and in the weft direction. This asymmetry can be attributed to the intrinsic characteristics of the glass fabric, which had a 2×1200 Tex roving with a spacing of 4.9 mm in the warp direction and a 1200 Tex roving with a spacing of 10.1 mm in the weft one.

3.2 Flexural Response

To study the flexural behavior of the TRC along the two principal reinforcing directions, bending tests were conducted on twelve specimens with a nominal size of 15×70×500 mm^3. These specimens were obtained by cutting a larger plate that was cast together with the full-scale shells. Also in this case, six specimens were cut along the weft fabric direction, and six along the warp. However, it is important to mention that the placement of the reinforcing fabric within the plate thickness could not be controlled during the casting process.

Given the possibility of sub-optimal positioning in the full-scale shells as well (the prototypes were made by inserting a fabric sheet into a double fiberglass mold and pouring the matrix through small openings [7]), for each direction it was decided to assess three bending specimens with the fabric placed in the tensile zone and three specimens with the fabric positioned near the top chord of the cross-section.

The tests were performed using a four-point bending scheme, with a distance of 450 mm between supports and a nominal shear span of 150 mm. As the flexural specimens were unnotched, four displacement transducers (LVDT) were arranged as follows: two were positioned on the bottom surface of the specimen, with a nominal gauge length of 200 mm, while two smaller LVDTs were placed on the top surface, spanning a gauge length of 40 mm. The tests were conducted controlling the displacement of the loading knives at a constant speed of 0.002 mm/s. However, for the specimens tested with the fabric positioned near the top edge, a brittle behavior was observed. Therefore, after the initial cracking occurred, the testing rate was increased to 0.004 mm/s.

As expected, the positioning of the AR-glass fabric within the matrix had a notable influence on the flexural response (see Fig. 4, in which only the nominal stress vs. stroke displacement diagrams are shown for brevity). This effect was further amplified by the choice of a statically determinate testing scheme, incapable of allowing the activation of second-order mechanisms.

The warp-N_{01}, warp-N_{02}, and warp-N_{04} specimens, with the fabric positioned in the tensile area of the cross-section (Fig. 4a and Fig. 5), consistently demonstrated the ability to withstand higher loads after the first cracking, thanks to the typical trilinear strain-hardening behaviour of TRC [2]. This provided evidence of the reinforcement effectiveness, leading to approximately 30 MPa of peak flexural stress, which varied depending on the number of intact glass rovings left after the sample cutting phase. On the contrary, specimens warp-N_{03}, warp-N_{05},

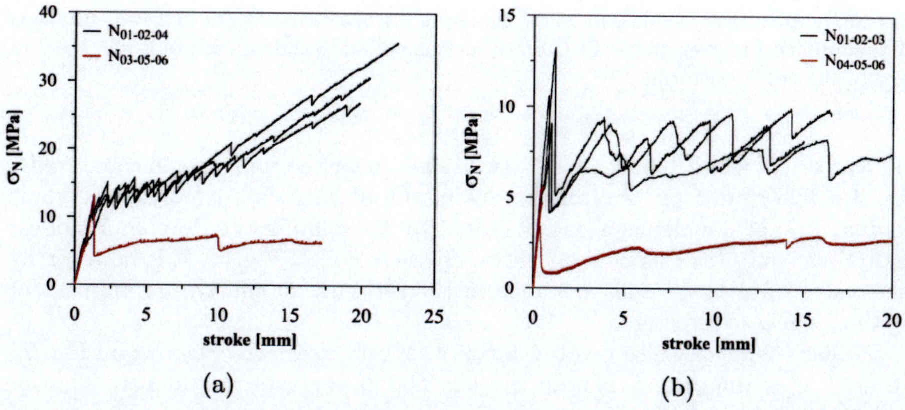

Fig. 4. Nominal flexural stress vs. stroke curves for the (a) warp- and (b) weft-oriented TRC coupons. Cases with the textile positioned upwards are identified in red.

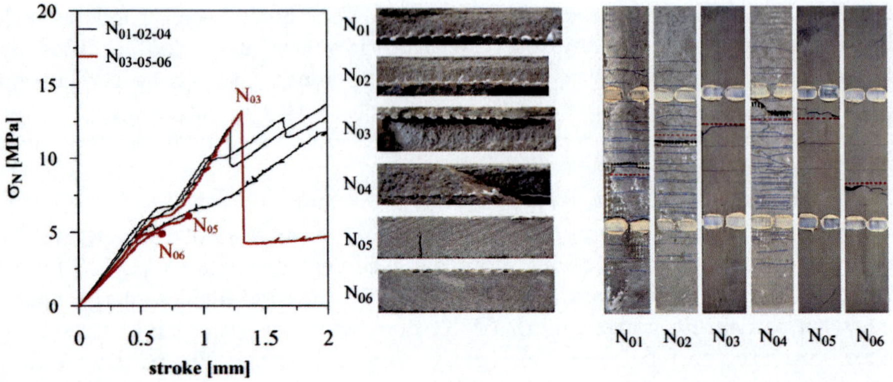

Fig. 5. Detail of the nominal flexural stress vs. stroke responses of the warp-oriented TRC coupons. Pictures of the TRC cross-sections close to the failure point and bottom view of the crack patterns.

Fig. 6. Detail of the nominal flexural stress vs. stroke responses of the weft-oriented TRC coupons. Pictures of the TRC cross-sections close to the failure point and bottom view of the crack patterns.

and warp-N_{06}, which had the fabric positioned near the top edge, encountered a loss of equilibrium after the initial cracking. Only warp-N_{03} demonstrated some residual strength, which can be attributed to its capability to develop a modest flexural capacity (as clearly seen in the cross-section of Fig. 5). It is noteworthy that warp-N_{03} was the sole specimen in the set with an effective reinforcement depth of a few millimeters.

Similar trends can also be observed in the weft direction (Fig. 4b and Fig. 6), although it is important to note the absence of the typical trilinear behavior in the weft-N_{01}, weft-N_{02}, and weft-N_{03}, specimens. Also in this case, the pronounced strain-hardening behavior is not observed due to the lower reinforcement ratio in the weft direction.

The crack patterns on the bottom surfaces of the twelve flexural specimens are depicted in Fig. 5 and Fig. 6. These images provide a clearer insight into the distinct fracture mechanisms – and therefore the different strain energy dissipation potential – that can be inferred based on the location of the AR-glass fabric within the cross-section.

4 Full-Scale Tests Under Centered and Eccentric Loads

Two nominally identical full-scale TRC formworks were tested imposing a constant increase of stroke by means of an electro-mechanical actuator. The purpose of these tests was to evaluate the ultimate capacity of the TRC stay-in-place formworks during the assembly of an elevated slab. The formworks needed to withstand point-loads caused by the movement of workers, which, according to standards, correspond to 1.2 kN distributed over an equivalent footprint of 100×100 mm^2. Two different load positions (centered and eccentric) were considered to investigate the most critical transient conditions. To ensure an accurate introduction of the load, a custom-shaped wooden block was carefully placed on the top surface of the shell. Figure 7 illustrates the two setups, which included six linear potentiometric transducers (PDT) to measure vertical displacements (δ_{1-6}), while a 5 kN piezo-resistive load cell recorded the applied load. Displacements δ_{1-5} were measured under the applied load and at each shell rib, while δ_6 captured the transverse behaviour. Furthermore, to replicate the actual constraints accurately, the TRC formworks were positioned on two segments of VHPFRC I-beams, matching those designed for the elevated slab.

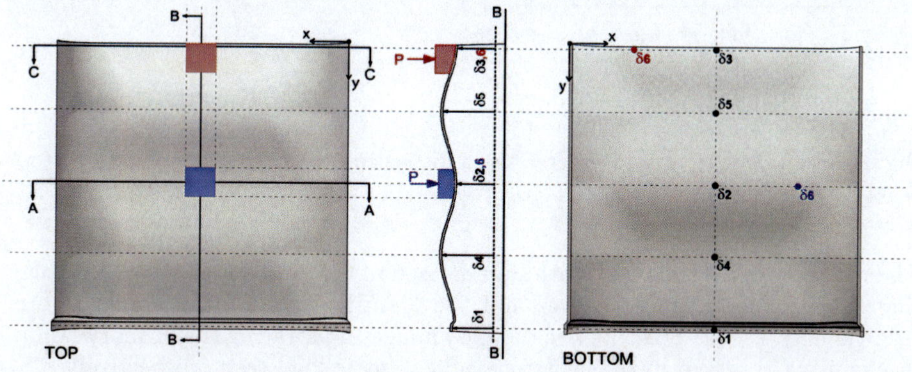

Fig. 7. Test setup and instruments arrangement. Load P and δ_6 positions refer to centered (blue) and eccentric (red) load tests (adapted from [7]). Please note that the glass fabric was arranged with the warp aligned along the x-axis.

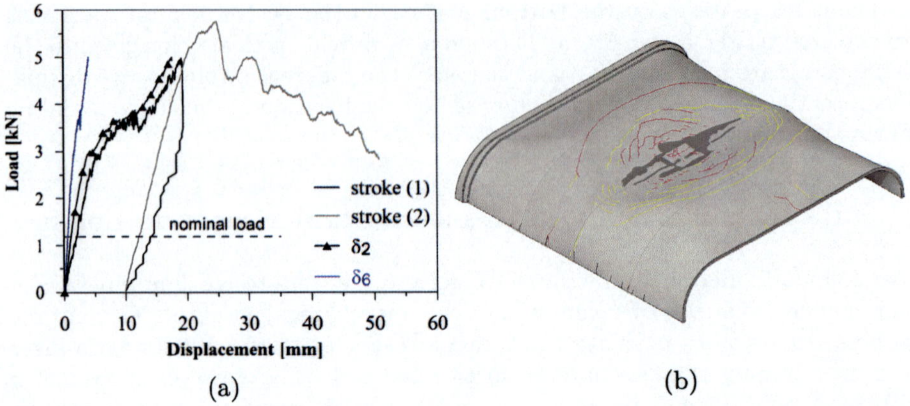

Fig. 8. Centered load test: (a) relevant load-displacement curves and (b) crack pattern at failure. Note that the solid hatch represents the zone in which a complete ejection of mortar was observed.

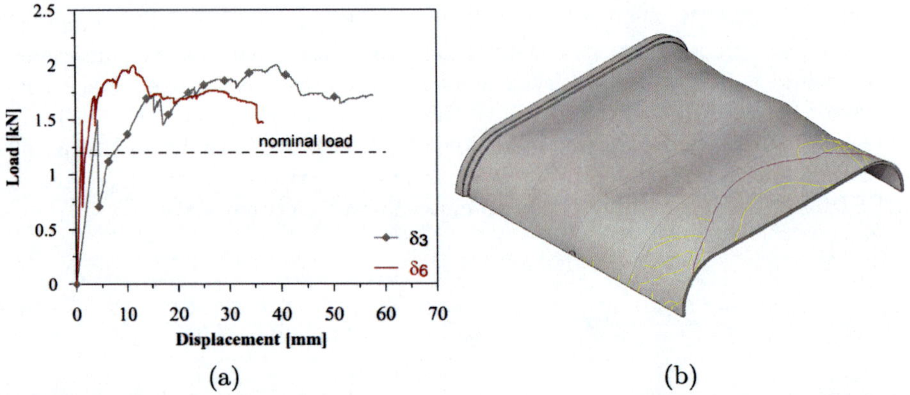

Fig. 9. Eccentric load test: (a) relevant load-displacement curves and (b) crack pattern at failure.

The main experimental results are summarized in Fig. 8 and Fig. 9, which display the load vs. displacement curves and the identified crack patterns at failure. Additionally, Fig. 10 presents two detailed images illustrating the primary failure mechanisms observed in the centered and eccentric load tests, respectively.

When subjected to a centered load, the formwork displayed a load vs. displacement linear response that extended beyond the nominal load of 1.2 kN, as depicted in Fig. 8. This indicates a reversible behavior under operating conditions. At the reaching of the maximum capacity of the load cell (around 5 kN), the test was arrested and visible cracks were marked in magenta. During this initial phase, the damage was primarily concentrated near the loading area, as evidenced by the distribution of cracks and the noticeable difference in vertical displacements δ_1 and δ_6.

After removing the PDTs and load cell, the prototype was reloaded under the same test conditions until failure occurred. The maximum load recorded during the entire test reached approximately 5.8 kN, nearly five times the required one. As visible from the load vs. actuator head displacement (see the stroke displacement recorded in the first and second phases in Fig. 8a), after reaching the peak load, the shell exhibited a softening behavior due to the ejection of HPC layers and the progressive tearing of the glass textile, as observed in Fig. 10a in correspondence to the load footprint. Thanks to the strain-hardening response of TRC and the membrane behavior of the full-scale shell, an extension of the damage was observed, as indicated by the yellow cracks marked at the end of the test (Fig. 8b). It is important to note that no cracks were found near the stiffer overlapping edge on one side of the prototype.

During the eccentric test, the point load was applied to the weakest part of the element to investigate the most severe loading condition. As shown in Fig. 9a, the formwork initially exhibited an elastic response until reaching approximately 1.4 kN, at which point the main crack, highlighted in magenta in Fig. 9b, formed.

Despite its small wall thickness and the inability to rely on shape-related strength phenomena in this area, the specimen demonstrated the activation of membrane behaviors (see Fig. 10), reaching a maximum load of approximately 2 kN. This load is over 1.5 times higher than the required capacity. This phenomenon is further supported by the development of additional cracks (marked in yellow at the end of the test), which indicate the strain-hardening behavior of the material and the presence of tensile stresses in the glass fabric.

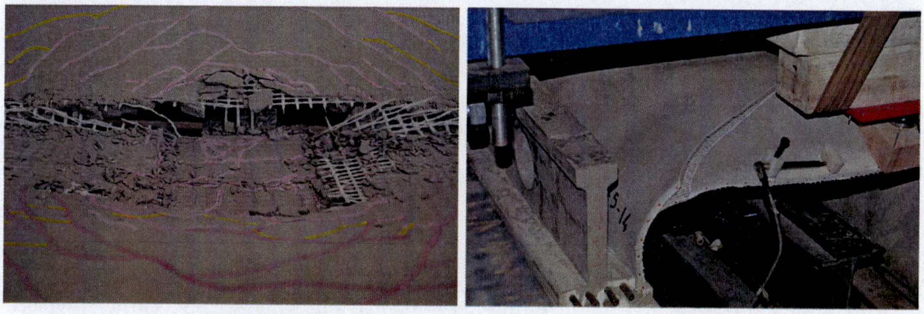

(a) (b)

Fig. 10. Failure mechanisms: (a) detail of the loading point in the centered test and (b) main crack developed in the eccentric load test.

5 Concluding Remarks

Based on the presented findings, the following conclusions can be drawn.

- If properly designed, textile reinforced concrete demonstrates typical trilinear strain-hardening behaviors, which offer membrane and flexural responses capable of withstanding significant post-cracking stresses.
- The positioning of the reinforcing fabric within the TRC cross-section influences its flexural response, with specimens having the fabric placed in the tensile zone demonstrating higher load capacity. In both weaving direction, the position of the fabric within the composite thickness is equally essential for structural purposes.
- Promising results were obtained from full-scale tests conducted on TRC formworks subjected to both centered and eccentric loads. The intricate geometry of the stay-in-place formwork, characterized by a high level of structural redundancy, enables the avoidance of potential local crises arising from incorrect fabric placement within the thickness. As a result, satisfactory outcomes were achieved for both analyzed load combinations and this demonstrates that thin TRC shells are able to ensure adequate safety even during the assembly phases. Of particular significance is the activation of second-order mechanisms, which allow the glass reinforcement to contribute to the overall strength through the development of membrane stresses.

References

1. Zani, G., Rampini, M.C., Colombo, M., di Prisco, M.: Bending behavior of cement-based multi-layered roof elements. Eng. Struct. **190**, 101–115 (2019)
2. Rampini, M. C., Zani, G., Colombo, M., di Prisco, M.: Mechanical behaviour of TRC composites: Experimental and analytical approaches. Appl. Sci. **9**(7), article 1492 (2019)
3. Rampini, M.C., Zani, G., Colombo, M., di Prisco, M.: Stiffness degradation of textile-reinforced mortar under unloading-reloading tensile cycles. J. Compos. Constr. **27**(1), article 04022083 (2023)
4. Sharei, E., Scholzen, A., Hegger, J., Chudoba, R.: Structural behavior of a lightweight, textile-reinforced concrete barrel vault shell. Compos. Struct. **171**, 505–514 (2017)
5. Tsangouri, E., Van Driessche, A., Livitsanos, G., Aggelis, D. G.: Design, casting and fracture analysis of textile reinforced cementitious shells. Dev. Built Environ. **3**, article 100013 (2020)
6. Scholzen, A., Chudoba, R., Hegger, J.: Thin-walled shell structures made of textile-reinforced concrete: part I: structural design and construction. Struct. Concr. **16**(1), 106–114 (2015)
7. Rampini, M. C., Zani, G., Colombo, M., di Prisco, M.: Experimental characterization of TRC disposable formworks. In: BEFIB 2016 - 9th Rilem International Symposium on Fiber Reinforced Concrete, Vancouver (2016)
8. Rampini, M. C., Zani, G., Caverzan, A., Colombo, M., di Prisco, M.: Una proposta di solaio alleggerito con elementi prefabbricati in composito fibrorinforzato. In: Italian Concrete Days - Giornate aicap 2016 Congresso CTE, Rome (2016)

Analysis of Optimal Deployment Strategy for Large Deployable Tensegrity Space Antennas

Muhao Chen[1], Aguinaldo Fraddosio[2], Andrea Micheletti[3], Gaetano Pavone[2(✉)], Mario Daniele Piccioni[2], and Robert E. Skelton[1]

[1] Department of Aerospace Engineering, Texas A&M University, College Station, TX, USA
{muhaochen,bobskelton}@tamu.edu
[2] Department of Architecture, Construction and Design, Polytechnic University of Bari, Bari, Italy
{aguinaldo.fraddosio,gaetano.pavone,mariodaniele.piccioni}@poliba.it
[3] Department of Civil and Computer Science Engineering, University of Rome Tor Vergata, Rome, Italy
micheletti@ing.uniroma2.it

Abstract. This study introduces an integrated approach that merges the design of structure and control to study the deployment strategies for tensegrity structures, particularly in the context of space antennas. First, we establish a nonlinear shape control law for clustered tensegrity structures, the solution turns out to solve a constraint linear algebra equation. Leveraging the symmetric nature of the antenna structure, we designate active actuators for the top and bottom cables of the space antenna while considering the remaining cables as passive. To further reduce the number of actuators required, we employ various clustering strategies for the active actuating cables. Results show that the deployment from the initial state to the predetermined targets is successfully guided by the proposed control law through clustering active cables using different actuation strategies. It is significant to note, however, that the energy cost escalates as more cables are clustered into the deployable antenna structures. In the context of space applications, this scenario emphasizes structure design and control are not independent problems. These insights also offer extensive relevance and can be extrapolated to different deployable tensegrity structures and robotic systems.

Keywords: Space Antenna · Deployable Structure · Tensegrity Structure · Deployment strategy · Integrated Structure and Control Design

1 Introduction

In light of the substantial expenses associated with launching and the constrained size of rocket tanks, the aerospace industry is continuously pursuing lightweight

S. Gabriele et al. (Eds.): IWSS 2023, LNCE 437, pp. 847–856, 2024.
https://doi.org/10.1007/978-3-031-44328-2_89

and deployable alternatives [2]. Recently, the European Space Agency (ESA) commissioned a study to explore breakthrough architectural concepts that offer superior reliability for the deployment of large systems. According to a previous study [7], the application of the new tensegrity-type architectural concept to large space antenna reflectors has shown that it achieves reliability by significantly reducing the number of articulated joints when compared to non-tensegrity architectures. Ganga et al. analyzed the overall feasibility of the deployable system, as well as its design, stiffness, deployment repeatability, and geometric precision by experiments [7]. However, the optimal deployment strategy in terms of structure design and control is still an open question.

Tensegrity structures offer a plethora of cables, which can provide a wide range of options for actuators [4,6]. However, the availability of numerous cables can pose challenges in their practical application [5], such as increased costs for actuators, difficulties in their placement, complicated control, and higher energy requirements. To address this issue, it is necessary to select only a subset of cables as actuators while ensuring that the desired control performance is achieved.

Skelton et al. proposed a systematic design method for sensors and actuators in structural control to minimize the cost of instrumentation [11]. Chen et al. presented the energy-efficient cable-actuation strategy for tensegrity structures [1]. Feng et al. compared the control energy and optimal dynamic performance of clustered tensegrity structures with different actuator placements for clustered cables and struts [12]. Ma et al. presented the dynamics and control law to clustered tensegrity structures [9]. However, most of the research on actuator selection has been conducted on linear systems, whereas tensegrity is a nonlinear system. Limited studies have been conducted on finding energy-efficient control laws with an optimal number of actuators.

This paper presents an energy-effective method for achieving nonlinear shape control in tensegrity structures. The subsequent sections are organized as follows: Sect. 2 introduces the design of deployable tensegrity space antennas. Section 3 outlines the shape control equation for tensegrity structures, derived from nonlinear dynamics formulations, utilizing cable forces as variables with imposed constraints. We also provide solutions to the shape control equation, along with clustering actuation strategies and minimal control energy considerations. In Sect. 4, numerical examples are presented to demonstrate the accuracy and efficiency of the control law and optimal clustering strategy. Finally, Sect. 5 summarizes the study's conclusions.

2 Design of Deployable Tensegrity Space Antennas

2.1 The Tensegrity Antenna Topology

Definition 1 (The Tensegrity Antenna). *Figure. 1 illustrates the structural topology of the deployable tensegrity antenna. The complexity q of the structure is defined by the number of X units in the structure. Figure 2 displays the structure with varying complexities.*

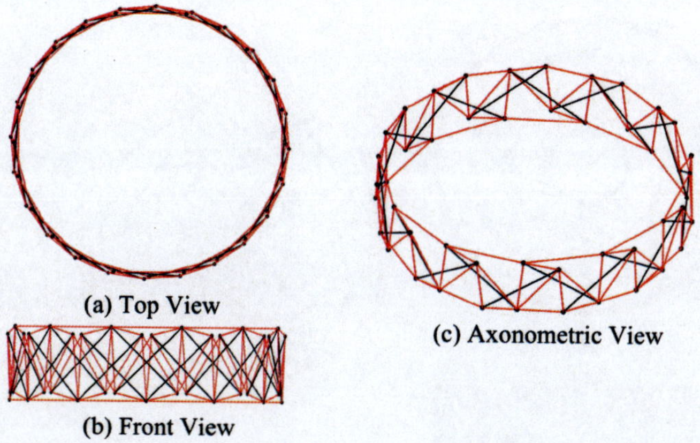

Fig. 1. A tensegrity ring: (a) top, (b) front, and (c) axonometric views with structure complexity of $q = 12$. The thick black lines are bars, and the thin red lines are cables.

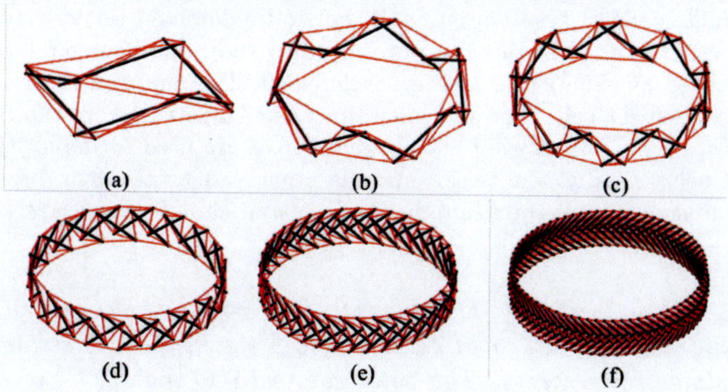

Fig. 2. The tensegrity antenna with different complexities: (a) $q = 3$, (b) $q = 6$, (c) $q = 9$, (d) $q = 12$, (e) $q = 36$, and (f) $q = 72$.

By shrinking the top and bottom cables simultaneously, we can have a deployable antenna. The structure deployment configurations are shown in Fig. 3.

3 Nonlinear Shape Control Law

3.1 Nonlinear Equations of Motion of Full-Order Model

Definition 2 (Nodal Coordinates). *The nodal coordinate vector* $\boldsymbol{n} \in \mathbb{R}^{3n_n}$ *and matrix* $\boldsymbol{N} \in \mathbb{R}^{3\times n_n}$ *in a tensegrity system are:*

$$\boldsymbol{n} = \left[\boldsymbol{n}_1^T \; \boldsymbol{n}_2^T \cdots \boldsymbol{n}_{n_n}^T\right]^T, \tag{1}$$

where $\boldsymbol{n}_i = \left[x_i \; y_i \; z_i\right]^T \in \mathbb{R}^3$ *is the x-, y-, and z-coordinates of the ith* ($i = 1, 2, \cdots, n_n$) *node, and* n_n *is the total number of nodes.*

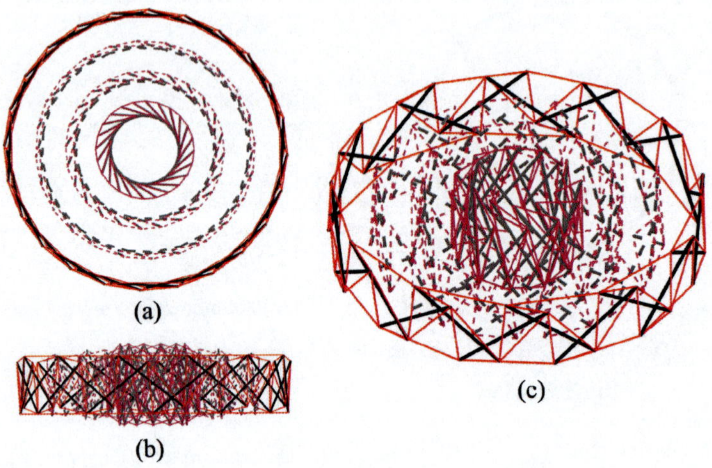

Fig. 3. Deployed, transient, and folded configurations: (a) top view, (b) front view, and (c) axonometric view. The diameters of the circles depicted in the views are 12 m, 9 m, 6 m, and 3 m, arranged from the outermost to the innermost circle. The initial state is represented by the 12 m diameter circle, with black and red lines denoting the bars and cables, respectively. The transient states are illustrated by the 9 m and 6 m diameter circles, where dotted grey and magenta lines are used to depict the bars and cables, respectively. Finally, the target state is portrayed by the 3 m diameter circle, with grey and magenta lines representing the bars and cables, respectively.

Definition 3 (Connectivity Matrices). *The connectivity matrices* $\boldsymbol{C}_s$ *and* $\boldsymbol{C}_b$ *describe the overall pattern of connections between the cables and bars in a tensegrity system, respectively. The total connectivity matrix* $\boldsymbol{C}$ *is obtained by concatenating* $\boldsymbol{C}_s$ *and* $\boldsymbol{C}_b$ *and has dimensions* $\mathbb{R}^{n_e \times n_n}$, *where* $n_e = n_s + n_b$ *is the total number of structural elements. Each row of* $\boldsymbol{C}$ *corresponds to an element* $m = 1, 2, \cdots, n_e$ *and is represented by the vector* $\boldsymbol{C}_m$. *The i-th entry of* $\boldsymbol{C}_m$ *indicates the connection between the two nodes associated with the element m. Specifically, the i-th entry of* $\boldsymbol{C}_m$ *is defined as follows:*

$$\boldsymbol{C}_{mi} = \begin{cases} -1, & i = j \\ 1, & i = k \\ 0, & otherwise \end{cases}, \tag{2}$$

where $j = 1, 2, \cdots, n_n$ *and* $k = 1, 2, \cdots, n_n$.

Theorem 1 (Tensegrity Nonlinear Dynamics). *The tensegrity dynamics with constraints is given by:*

$$\boldsymbol{E}_a^T (\boldsymbol{M}\ddot{\boldsymbol{n}} + \boldsymbol{D}\dot{\boldsymbol{n}} + \boldsymbol{K}\boldsymbol{n}) = \boldsymbol{E}_a^T (\boldsymbol{f}_{ex} - \boldsymbol{g}), \tag{3}$$

where

$$M = \frac{1}{6}(|C|^T \hat{m}|C| + \lfloor |C|^T \hat{m}|C| \rfloor) \otimes I_3, \tag{4}$$

$$K = (C^T \hat{x} C) \otimes I_3, \tag{5}$$

$$g = \frac{g}{2}(|C|^T m) \otimes \begin{bmatrix} 0 \; 0 \; 1 \end{bmatrix}^T, \tag{6}$$

where $M \in \mathbb{R}^{3n_n \times 3n_n}$, $D \in \mathbb{R}^{3n_n \times 3n_n}$, and $K \in \mathbb{R}^{3n_n \times 3n_n}$ are mass, damping, and stiffness matrices, $\hat{\bullet}$ operator converts the vector into a diagonal matrix, $|V|$ gets the absolute value of each element of the matrix V, and $\lfloor V \rfloor$ sets all the off-diagonal elements of a square matrix to zero. $x \in \mathbb{R}^{n_e}$ is the force density (force per unit length) vector of structure members, $f_{ex} \in \mathbb{R}^{3n_n}$ is external forces on the structure nodes, and g is gravity vector (g is the gravity constant, e.g., $g = 9.81\,\mathrm{m/s^2}$ for earth gravity). $E_a \in \mathbb{R}^{3n_n \times n_a}$ (n_a are the number of free nodes) is an orthonormal matrix to extract free nodes n_a from nodal vector n, which satisfies $E_a(:, i) = I_{3n}(:, a_i) \in \mathbb{R}^{3n_n \times n_a}$ and $n_a = E_a^T n$. The value a_i in vector $a = \begin{bmatrix} a_1 \; a_2 \; \cdots \; a_{n_a} \end{bmatrix}^T \in \mathbb{R}^{n_a}$ are the indices of the free nodes in the nodal coordinate vector n. Similarly, vector $b = \begin{bmatrix} b_1 \; b_2 \; \cdots \; b_{n_b} \end{bmatrix}^T \in \mathbb{R}^{n_b}$ are the indices of constrained entries in the nodal coordinate vector n. $E_b(:, i) = I_{3n}(:, b_i) \in \mathbb{R}^{3n_n \times n_b}$ is an orthonormal matrix to abstract constrained nodes n_b from nodal vector n and $n_b = E_b^T n$, The sum of the number of free and fixed nodes is the total number of nodes, and we have $n_a + n_b = 3n_n$.

Proof. Theorem 1 can be derived using the Lagrangian method, and the details are given in [10] and software package [8].

We also provide the statics equations since this part will be used in the later control equations.

Corollary 1 (Tensegrity Statics). *The tensegrity statics with constraints has three equivalent forms:*

$$E_a^T K n = E_a^T A_{1c} x_c = E_a^T A_{2c} t_c = E_a^T (f_{ex} - g), \tag{7}$$

where

$$A_{1c} = A_{2c} \hat{l}_c = \left(C^T \otimes I_3\right) b.d.(H) \hat{l}^{-1} S^T \hat{l}_c. \tag{8}$$

Proof. By neglecting the time derivatives in Theorem 1, we can have $E_a^T K n = E_a^T (f_{ex} - g)$. From Eq. (5), one can get $K n = A_{1c} x_c = A_{2c} t_c$.

3.2 Shape Control Law

Definition 4 (Shape Objectives). *The error between the current position of the structure to the control target is defined as:*

$$e = n_c - \bar{n}_c, \tag{9}$$

where $\boldsymbol{n}_c$ *is the nodes of interest to control,* $\bar{\boldsymbol{n}}_c$ *is the morphing objective. Since the nodal coordinate of interest* $\boldsymbol{n}_c$ *is extracted from the free nodal coordinate* $\boldsymbol{n}_a$*, we have* $\boldsymbol{n}_c = \boldsymbol{E}_c^T \boldsymbol{n}_a$*, where* $\boldsymbol{E}_c$ *is the index matrix.*

Definition 5 (Active and Passive Members). *The active and passive cables are defined as the structure members who actively and passively change their length, satisfying* $n_s = n_{s,act} + n_{s,pas}$*, where* $n_{s,act}$ *and* $n_{s,pas}$ *are the number of active and passive cables, respectively. The force vectors of the active and passive members can be written as follows:*

$$\boldsymbol{t}_{c_{act}} = \boldsymbol{E}_{act}^T \boldsymbol{t}_c, \ \boldsymbol{t}_{c_{pas}} = \boldsymbol{E}_{pas}^T \boldsymbol{t}_c, \tag{10}$$

where $\boldsymbol{E}_{act}$ *and* $\boldsymbol{E}_{pas}$ *are index matrices to separate all the structure members. Since* $\left[\boldsymbol{E}_{act} \ \boldsymbol{E}_{pas}\right]$ *is an orthogonal matrix, we have the following equation:*

$$\boldsymbol{t}_c = \begin{bmatrix} \boldsymbol{E}_{act}^T \\ \boldsymbol{E}_{pas}^T \end{bmatrix}^{-1} \begin{bmatrix} \boldsymbol{t}_{c_{act}} \\ \boldsymbol{t}_{c_{pas}} \end{bmatrix} = \left[\boldsymbol{E}_{act} \ \boldsymbol{E}_{pas}\right] \begin{bmatrix} \boldsymbol{t}_{c_{act}} \\ \boldsymbol{t}_{c_{pas}} \end{bmatrix}. \tag{11}$$

Theorem 2 (Nonlinear Shape Control). *The nonlinear shape control of the tensegrity structure is equivalent to solving the following linear algebra equation:*

$$\boldsymbol{\mu} - \boldsymbol{\Gamma}_{pas} \boldsymbol{t}_{c_{pas}} = \boldsymbol{\Gamma}_{act} \boldsymbol{t}_{c_{act}}, \tag{12}$$

where $\boldsymbol{\mu}$*,* $\boldsymbol{\Gamma}_{act}$*, and* $\boldsymbol{\Gamma}_{pas}$ *are:*

$$\begin{aligned} \boldsymbol{\mu} &= \boldsymbol{E}_c^T \boldsymbol{M}_{aa}^{-1} \boldsymbol{E}_a^T (\boldsymbol{f}_{ex} - \boldsymbol{g} - \boldsymbol{M} \boldsymbol{E}_b \ddot{\boldsymbol{n}}_b - \boldsymbol{D} \dot{\boldsymbol{n}}) \\ &\quad - \ddot{\bar{\boldsymbol{n}}}_c + \boldsymbol{\psi}(\boldsymbol{E}_c^T \dot{\boldsymbol{n}}_a - \dot{\bar{\boldsymbol{n}}}_c) + \boldsymbol{\phi}(\boldsymbol{E}_c^T \boldsymbol{n}_a - \bar{\boldsymbol{n}}_c), \end{aligned} \tag{13}$$

$$\boldsymbol{\Gamma}_{act} = \boldsymbol{E}_c^T \boldsymbol{M}_{aa}^{-1} \boldsymbol{E}_a^T \boldsymbol{A}_{2c} \boldsymbol{E}_{act}, \tag{14}$$

$$\boldsymbol{\Gamma}_{pas} = \boldsymbol{E}_c^T \boldsymbol{M}_{aa}^{-1} \boldsymbol{E}_a^T \boldsymbol{A}_{2c} \boldsymbol{E}_{pas}. \tag{15}$$

Proof. When the nodes of interest reach their targets, the error vector $\boldsymbol{e}$ and its time derivatives should all go to zero. This goal can be expressed as follows [3]:

$$\ddot{\boldsymbol{e}} + \boldsymbol{\psi} \dot{\boldsymbol{e}} + \boldsymbol{\phi} \boldsymbol{e} = \boldsymbol{0}, \tag{16}$$

where $\boldsymbol{\psi}$ and $\boldsymbol{\phi}$ are tune matrices that can adjust the time response of the morphing process. Since $\boldsymbol{E}_c$ is given constants, the time derivatives of the error vector in Eq. (9) are:

$$\dot{\boldsymbol{e}} = \dot{\boldsymbol{n}}_c - \dot{\bar{\boldsymbol{n}}}_c, \ \ddot{\boldsymbol{e}} = \ddot{\boldsymbol{n}}_c - \ddot{\bar{\boldsymbol{n}}}_c. \tag{17}$$

Substitute Eqs. (3) and (9) into Eq. (16), we have:

$$\begin{aligned} &\boldsymbol{E}_c^T \boldsymbol{M}_{aa}^{-1} \boldsymbol{E}_a^T (\boldsymbol{f}_{ex} - \boldsymbol{g} - \boldsymbol{M} \boldsymbol{E}_b \ddot{\boldsymbol{n}}_b - \boldsymbol{D} \dot{\boldsymbol{n}} - \boldsymbol{K} \boldsymbol{n}) \\ &- \ddot{\bar{\boldsymbol{n}}}_c + \boldsymbol{\psi}(\boldsymbol{E}_c^T \dot{\boldsymbol{n}}_a - \dot{\bar{\boldsymbol{n}}}_c) + \boldsymbol{\phi}(\boldsymbol{E}_c^T \boldsymbol{n}_a - \bar{\boldsymbol{n}}_c) = \boldsymbol{0}. \end{aligned} \tag{18}$$

Substitute Eqs. (8) and (11) into Eq. (18), one can have a linear algebra Eqs. (12)–(15).

3.3 Solving for Control Variable

The only unknown variable in Eq. (12) is $\boldsymbol{t}_{c_{act}}$. In most cases, people use cables to control the tensegrity structure, so let us use cable as the control variable as an example. Since cables cannot take compression, and if they do, we should substitute the force to zero, we must add the constraint to this unknown variable $\boldsymbol{t}_{c_{act}} \geq \boldsymbol{0}$. Moreover, the cables should never exceed their yield strength, which can be calculated by the $\sigma_s \boldsymbol{A}_c$. Thus, the problem of solving Eq. (12) becomes a linear algebra problem with inequality constraints.

Theorem 3 (Shape Control Solution). *The solution to Theorem 3.2 is as the following:*

$$\begin{cases} \underset{\boldsymbol{t}_{c_{act}}}{Solve} & \boldsymbol{\Gamma}_{act}\boldsymbol{t}_{c_{act}} = \boldsymbol{\mu} - \boldsymbol{\Gamma}_{pas}\boldsymbol{t}_{c_{pas}} \\ s.t. & \boldsymbol{0} \leq \boldsymbol{t}_{c_{act}} \leq \sigma_s \boldsymbol{A}_c \end{cases}. \tag{19}$$

The resolution of Theorem 3 relies on the characteristics of $\boldsymbol{\Gamma}_{act}$. One favorable feature of tensegrity structures is that they contain an abundant number of cables, resulting in $\boldsymbol{\Gamma}_{act}$ being a matrix with a complete column rank, thereby making at least one solution typically exist. It is worth mentioning that an analytical solution could exist, such as finding solutions from the positive span space in the first equation. However, for different structures, a solution cannot always be guaranteed, but a least square one is always possible, $\min\limits_{\boldsymbol{t}_{c_{act}}} ||\boldsymbol{\mu} - \boldsymbol{\Gamma}_{pas}\boldsymbol{t}_{c_{pas}} - \boldsymbol{\Gamma}_{act}\boldsymbol{t}_{c_{act}}||^2$. We should point out that the least square solution can be arbitrarily bad. Thus, one has to be careful in choosing the target points, control speed, and materials.

3.4 Deployment Strategy Design

In an effort to optimize both the manufacturing process and deployment strategy, we have thoroughly taken advantage of the symmetrical properties of the structure. Specifically, we choose passive cables for the side cables and active cables for the top and bottom cables.

Definition 6 (Clustering Actuation Strategy). *The structure clustering strategy is shown in Fig. 4, where we use the complexity of* $q = 12$ *as an example to illustrate the idea.*

Definition 7 (The Energy Cost). *The total control energy of the system can be computed as follows:*

$$E = \sum_{i=1}^{k} \Delta \boldsymbol{l}_{c_{act},i}^T \boldsymbol{t}_{c_{act},i}, \tag{20}$$

where $\Delta \boldsymbol{l}_{c_{act},i}$ *represents the change in length of the active members at the ith step,* $\boldsymbol{t}_{c_{act},i}$ *denotes the member force in the active members at the ith step, and k is the total time steps of the active actuation process.*

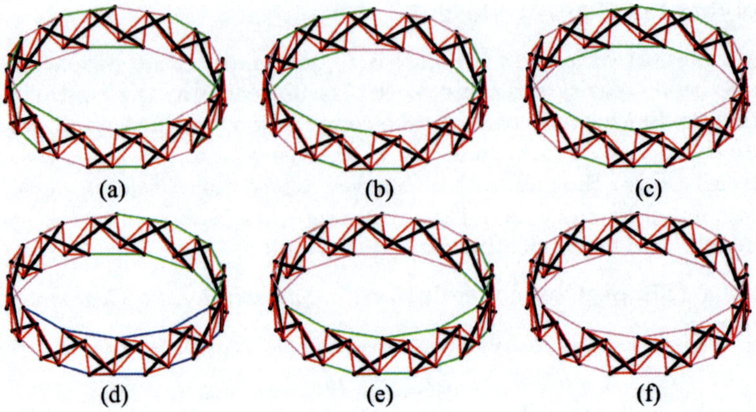

Fig. 4. The clustering strategies involve individual passive cables represented by red cables, as well as active cables found in the top and bottom cables. Within these top and bottom cables, six clustering strategies are employed: (a) non-clustered cables, (b) clustering of every two adjacent cables, (c) clustering of every three adjacent cables, (d) clustering of every four adjacent cables, (e) clustering of every six adjacent cables, and (f) clustering of all twelve adjacent cables.

4 Numerical Results

To demonstrate the proposed method, we choose the antenna complexity to be $q = 12$, as shown in Figs. 1 and 3. The diameter of the ring is 12 m, and the height of ring H is 2.6 m. The end module's height is 2.5m. The damping coefficient is 0.1. The time step is 0.0001 s, and the total simulation time is 20 s. The target node positions are computed from the static deployment process by adjusting the inner ring diameter to be 3 m. The material properties are given in Table 1. In this study, we examine the six cases as shown in Fig. 4. We use the same setups for all six cases except for the clustering strategies. We choose the same control gains for the six cases.

Table 1. Material properties of the structure members.

Properties	Bars	Passive Cables	Active Cables
Material	Steel	Aluminum	UHMWPE
Diameter	16 mm	4 mm	10 mm
Density	7,7870 kg/m^3	2,700 kg/m^3	970 kg/m^3
Young's Modulus	206 Gpa	60 Gpa	120 Gpa
Yield Strength	300 Mpa	110 Mpa	2.7 Gpa

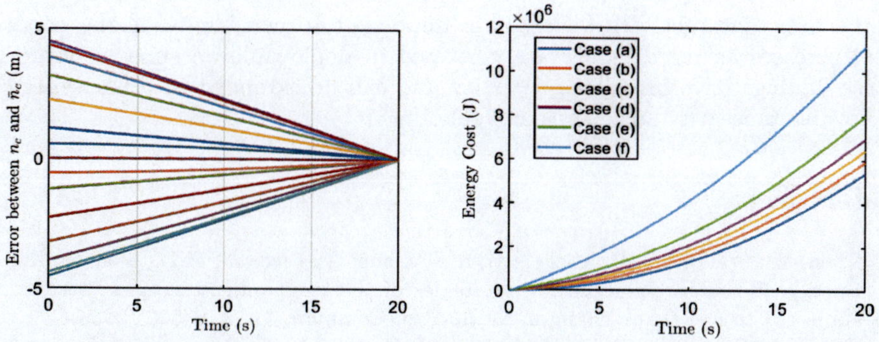

Fig. 5. Left: The performance of control measured by the errors between the current position and the control targets of all the inner nodes of the structure. No significant difference was found between the six clustering strategies in Fig. 4. Right: The control energy cost for the six clustering strategies (a)-(f) in Fig. 4.

As evident from Fig. 5 (left), the control performance, measured by the errors between the current position and control targets, is roughly zero at the final simulation time for all six clustering strategies. This implies that the control law is effective in guiding the structure toward the final targets. However, as we cluster more cables, the energy cost is higher. For example, as shown in Fig. 5 (Right), the energy cost at the final state for the listed six cases are 5.3887×10^6 J, 5.9111×10^6 J, 6.4374×10^6 J, 6.9644×10^6 J, 8.0223×10^6 J, and 1.1183×10^7 J. That is, by clustering cables, one can use fewer actuators in the antenna structure, but the energy cost is higher. One can note that in case (f), where all twelve nearby cables are clustered, the energy consumption is 2.0752 times that of case (a), which involves non-clustered cables. However, this setup results in the saving of 11 actuators.

It is worth noting that energy resources in space applications, such as those derived from solar panels, are fairly restricted. Consequently, in such instances, there must be a careful equilibrium between energy consumption and the number of actuators used. This scenario underlines that, in the context of space applications, the design of the structure and its control are intertwined and should not be considered separate issues.

5 Conclusions

This paper presents an integrated approach that combines structure and control design to study the deployment strategy for tensegrity deployable structures, with a focus on space antennas. We derive a nonlinear shape control law for clustered tensegrity structures by solving a constraint linear algebra equation. The top and bottom cables are selected as active, while the remaining cables are passive. By clustering the active cables using different actuation strategies, the proposed control law effectively drives the structure from its initial configuration to the desired targets. However, it is important to consider that the energy cost increases as more cables are clustered in deployable antenna structures. These findings have broad applicability and can be extended to other types of deployable tensegrity structures and robots.

References

1. Chen, M., Fraddosio, A., Micheletti, A., Pavone, G., Piccioni, M.D., Skelton, R.E.: Energy-efficient cable-actuation strategies of the v-expander tensegrity structure subjected to five shape changes. Mech. Res. Commun. **127**, 104026 (2023)
2. Chen, M., Goyal, R., Majji, M., Skelton, R.E.: Review of space habitat designs for long term space explorations. Prog. Aerosp. Sci. **122**, 100692 (2021)
3. Chen, M., Liu, J., Skelton, R.E.: Design and control of tensegrity morphing airfoils. Mech. Res. Commun. **103**, 103480 (2020)
4. Fraddosio, A., Marzano, S., Pavone, G., Piccioni, M.D.: Morphology and self-stress design of v-expander tensegrity cells. Compos. B Eng. **115**, 102–116 (2017)
5. Fraddosio, A., Pavone, G., Piccioni, M.D.: Minimal mass and self-stress analysis for innovative v-expander tensegrity cells. Compos. Struct. **209**, 754–774 (2019)
6. Fraddosio, A., Pavone, G., Piccioni, M.D.: A novel method for determining the feasible integral self-stress states for tensegrity structures. Curved Layered Struct. **8**(1), 70–88 (2021)
7. Ganga, P.L., Micheletti, A., Podio-Guidugli, P., Scolamiero, L., Tibert, G., Zolesi, V.: Tensegrity rings for deployable space antennas: concept, design, analysis, and prototype testing. Variational Analysis and Aerospace Engineering: Mathematical Challenges for the Aerospace of the Future, pp. 269–304 (2016)
8. Ma, S., Chen, M., Skelton, R.: TsgFEM: tensegrity finite element method. J. Open Source Softw. **7**(75), 3390 (2022)
9. Ma, S., Chen, M., Skelton, R.E.: Dynamics and control of clustered tensegrity systems. Eng. Struct. **264**, 114391 (2022)
10. Ma, S., Chen, M., Skelton, R.E.: Tensegrity system dynamics based on finite element method. Compos. Struct. **280**, 114838 (2022)
11. Skelton, R.E., Li, F.: Economic sensor/actuator selection and its application to flexible structure control. Smart Structures and Materials 2004: Modeling, Signal Processing, and Control (2004)
12. Xiaodong, F., Yaowen, O., Mohammad, S.M.: Energy-based comparative analysis of optimal active control schemes for clustered tensegrity structures. Struct. Control Health Monit. **25**(19) (2018)

Reality-Based 3D Survey and Modeling Supporting Historical Vaulted Structures Studies

Antonia Spanò(✉), Giacomo Patrucco, and Giulia Sammartano

Department of Architecture and Design, Geomatics Lab for Cultural Heritage, Politecnico di Torino, Torino, Italy
antonia.spano@polito.it

Abstract. Geometry has always been a means of proportioning and sizing architectural constructions and has permeated the ways of conceiving buildings from the classical, modern, and contemporary ages, ensuring aesthetic and technical values and thrilling architects and treatise writers. The structural behavior of masonry structures is consequently often aimed at highlighting the close relationship between the geometry and the safety level of structures based on primitive geometries of the architectural heritage, in particular, the arches, the vaulted systems, and the domes.Reality-based 3D surveys and modeling that use laser scanning and photogrammetric methods have important implications for the study of generative geometries, as they are able to allow a direct comparison between the real surfaces surveyed and those designed according to strict geometric principles.The research will highlight the contribution of methods such as point cloud co-registration, deviation analysis, shape recognition and algorithms such as ICP-based (Iterative Closest Points), RANSAC-based to achieve these objectives.

Keywords: TLS (Terrestrial laser Scanning) · Photogrammetry · SfM (Structure from Motion) · generative geometries · C2C (cloud to cloud) comparison

1 Introduction

Since the 2000s, terrestrial laser scanning (TLS) technology has affirmed a crucial role in the structural study of historical built heritage, aimed at static or dynamic assessment in the presence of different stress states and at evaluating the state of conservation. The interest is determined to a large extent by the ability to generate models of built structures being able to predetermine different levels of schematization from reality-based models, mainly three-dimensional, to consequently be able to carry out structural evaluations by assigning simplified geometric contents that respond to the physical reality of the manufactured assets [1–3].

The innovative methods of close-range 3D survey and modeling, based on active or passive sensor systems, i.e. laser scanning or photogrammetry, enable the generation of accurate and highly descriptive three-dimensional models of the surveyed objects [4, 5]. These models, together with the development of hybrid solutions based on the fusion of

S. Gabriele et al. (Eds.): IWSS 2023, LNCE 437, pp. 857–866, 2024.
https://doi.org/10.1007/978-3-031-44328-2_90

methods (see Sect. 2), increasingly accompany the development of conservation projects for the built heritage.

The main goals related to adopting methods capable of providing verifiable metric results of predefined accuracy and overall quality of 3D models are varied and increasingly not independent. In the specific context of the conservation plan of the built heritage, the advantages can be identified above all in the introduction of automation in data collection and the processing, management, production, and communication of the results. The automation of the process development also makes it possible to focus the attention on the reading and interpretation of the information deriving from the 3D survey and modeling results, as well as on the virtual representation and communication systems, the shared dissemination, and the ease of use of the results to wider experts comminity [6].

The field of study of the evaluation and strengthening of architectural structures characterized by traditional construction techniques must undoubtedly face the challenge of an infinite number of different case studies, of different scenarios due to extremely varied systems and construction techniques, different materials, multiplied in different geographical areas that express their own peculiarities. Within the growing awareness of the fragility of assets and at the same time of the risk to which they are subjected for both anthropic and natural reasons [7], as well as the ability to analyze and intervene has increased, trust has developed in interdisciplinary synergic strategies. As a consequence, it is important that the guidelines dedicated to interventions on protected built heritage suggest the necessary levels of knowledge [8]. It is therefore relevant that the different fields of study compare each other and the studies in collaboration are intensified as the need to have the most detailed and in-depth knowledge frameworks is certainly recognized as paramount.

2 Reality-Based 3D Survey and Modeling Methods

The research group of the Geomatics Lab for Cultural Heritage of the Polytechnic of Turin [9] has dedicated several efforts in recent years to focus and coordinate the different challenges that the so-called *rapid mapping* techniques require to adapt technological developments to increasingly specialized 3D documentation needs, capable of providing the primary knowledge bases for inspection, monitoring and assessing damages, risks, anomalies that often afflict historic buildings and historical vaulted structures having complex and irregular geometries.

The advantage that has been fully recognized now to digital photogrammetry and terrestrial laser scanning (TLS) technologies, and above all to their integration, is the high data acquisition productivity. Dense, accurate, and three-dimensional point clouds allow setting up a 3D digital archive from which to select subsequent processing to derive 2.5D digital models such as DEMs, 3D models, and orthophotos [10, 11].

In the framework of photogrammetric systems, also called *image-based* systems, the most significant innovations have come from UAV (Unmanned Aerial Vehicle) systems which have revolutionized survey strategies also in the field of built heritage; in this paper, however, UAVs are of lesser importance as they operate above all in the outdoor aerial environment, while the indoor application of microdrones is more limited, although some devices are able to orient the cameras upwards to detect ceilings and vaulted systems. In the image-based field, which includes a very wide range of terrestrial and low-cost sensors (also involving multi-cameras and spherical lens cameras), it can be said that any sensor capable of recording images, which must be acquired strictly in stereoscopic mode, or videos from which to extract frames characterized by high overlapping, can benefit from the integrated developments of *image matching algorithms* and the *Structure from Motion* (SfM) technique [12, 13].

Photogrammetric softwares now widely use integrated approaches, relating both to the traditional photogrammetry and the new algorithms that develop automation deriving from Computer Vision. In a combined way, *image matching* and the *SfM* technique solve the photogrammetric problem of automatically orienting blocks of stereoscopic images, generating dense and accurate point clouds also in automatic mode, finally benefiting from control points determined by topographic method to optimize the calculation and control the propagation of the error.

As far as range-based solutions are concerned, innovations have been studied and implemented in this sector, which guarantee higher acquisition speeds than the traditional TLS technique, so much so as to develop the so-called mobile mapping systems (MMSs) technology. Excluding the complex mobile mapping systems from cars or other land vehicles, which integrate spherical or multi-cameras with laser profilers and GNSS receivers coordinated with inertial platforms for determining the projection centers, which do not adapt to the indoor situations, LiDAR (Light Detection and Ranging) systems based on SLAM (Simultaneous Localization and Mapping) technology are now remarkably diffused and investigated also in the field of built heritage [14, 15].

The point clouds derived from image or range-based systems or their integration are to all intents and purposes 3D archives, enabling in-depth further analysis of the derived products. Ortho-projection and DEMs have had a crucial role in the study and inquiries of the variegated world of Cultural Heritage, mainly architectural and archaeological but also all other types of heritage since they have made it possible to finalize the investigations of construction systems, architectural elements, materials, surface and structural deformations or distresses, as well as in the monitoring of the state of conservation [16]. The Geomatics research, in addition to the rapid 3D surveying techniques, develops new knowledge and management skills of 3D metric archives, which appear in the first instance in the form of point clouds, and which need to be organized in structured spatial databases from both geometrical and semantic point of view, to give life to the so-called digital twins' scenarios (see Sect. 4).

3 Shapes and Deformations. The Analysis of Generative Geometries

The point clouds generated by both described methods must, first of all, be optimized to identify and remove the portions that have imprecise reconstructions and redundant geometries, and it is necessary to use suitable automatic noise reduction filters (random errors), to obtain the optimization of the final 3D models, ready for subsequent evaluations and interpretations. Without referring to the workflow that led to reaching the results of the 3D survey and modeling about different case studies, the following figures and descriptions discuss a series of issues regarding the shape and possible anomalies found on vaulted structures systems, selected from the rich repertoire of case studies addressed by the Geomatics Laboratory.

The point models and also their conversion into triangulated surfaces (*mesh*) make it possible to address the problems of interpretation of the architectural elements and structures and the relative anomalies, but it is certainly crucial to consider different steps and compare information deriving from an interdisciplinary scenario to address a correct interpretation. The evaluation of deviations analysis from simple geometries of walls or horizontal partitions can be performed through the extraction of section profiles, measuring the angle of inclination that the surface portions assume with respect to planes or axes adopted as references (for example, from the vertical to evaluate the out of plumb, or deviations from horizontal planes which are the premise for the evaluation of any failure). By associating a color scale with the distance variation, it is possible to identify the discrepancies surface-based mapping and the areas that most deviate from the plane (Fig. 1).

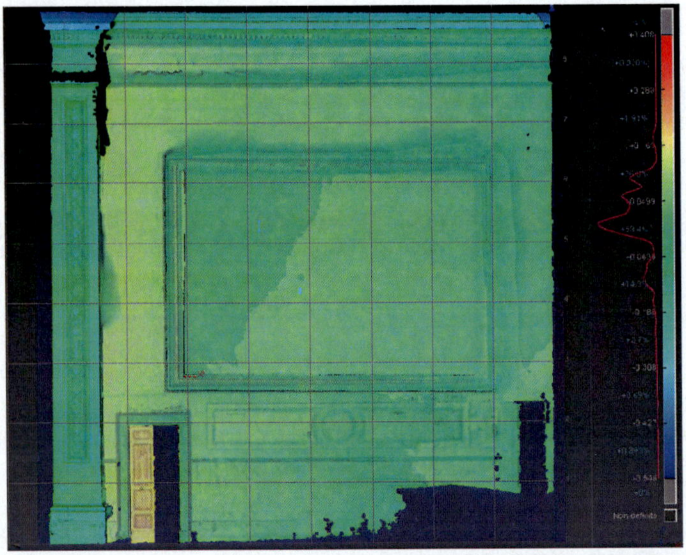

Fig. 1. Analysis of the deviation of a wall from a vertical surface. 63% of the points on the wall substantially coincide with the vertical surface, and, excluding the cornice, about 20% of the surface deviates from the vertical plane by values up to just over 5 cm.

When the surfaces are vaults, the comparison takes place with the surfaces that are estimated to be those identified to guide the project and the construction of the architectural structure. In particular, it is interpreted based on the generative geometry at the basis of the shape. The extraction of section profiles at constant distance, which if applied to the land are contour lines, can largely support the assessment of the deviation from the generative geometry. The following example applied to the central vault of the oval hall of the Stupinigi palace (Filippo Juvarra, 1730) [17] (Fig. 2), represents how the

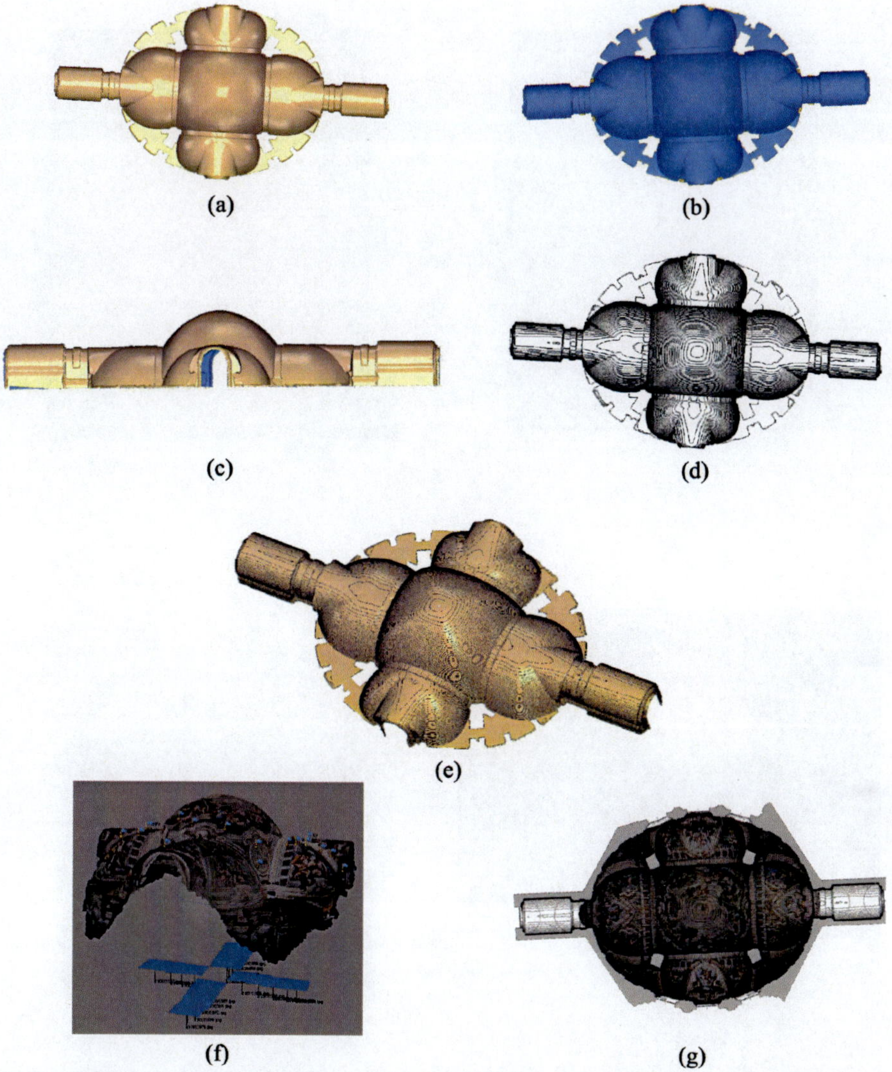

Fig. 2. (a,b) Extrados and intrados surfaces of the central vault of the oval hall of Stupinigi royal residence. (c,d) Extraction of section profiles from horizontal cutting planes; (e) a view of the surveyed surface and overlapping section profiles; (f) a scheme of photogrammetric process and (g) the final orthophoto not highlighting any deformation.

section profiles of the central vault, with a particular regular shape, show appropriate adherence to the geometry of the pavilion vault with a cylindrical surface. On the other hand, the lateral vaults show a yielding in the key point, which had been interpreted as due to deformations due to infiltrations in the roof, leading to restoration works. The

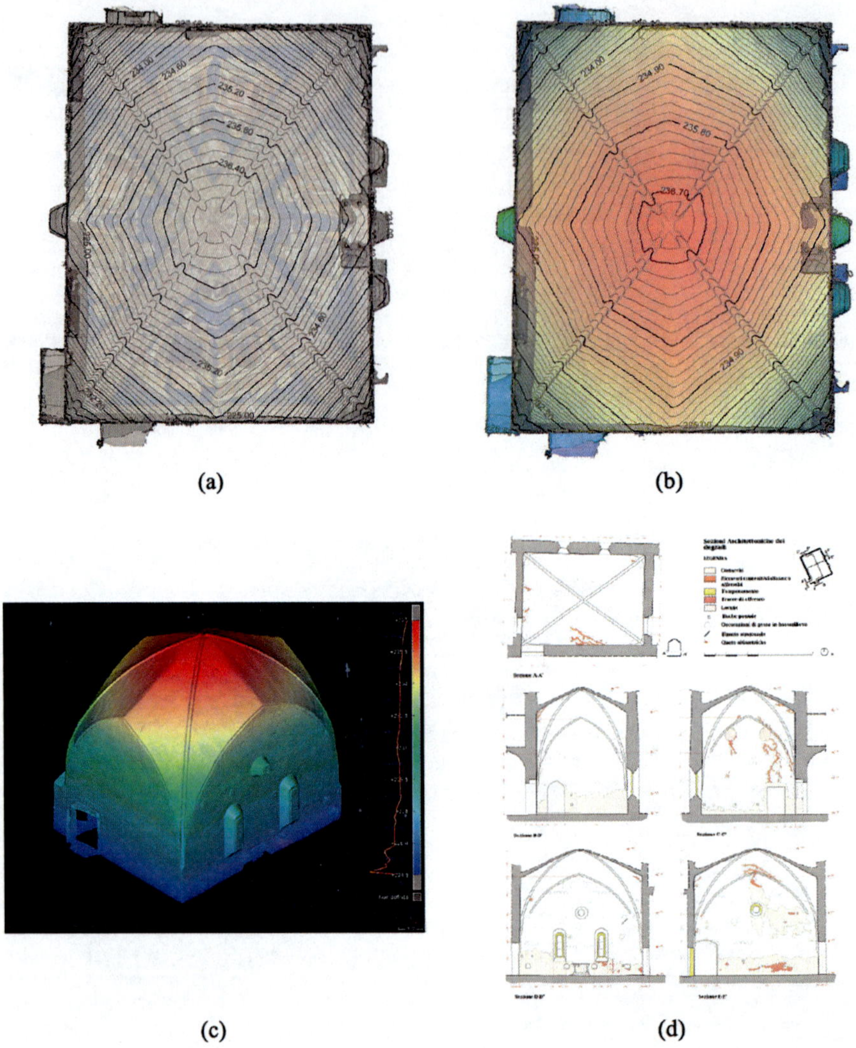

(a) (b) (c) (d)

Fig. 3. DEM analysis and extraction of contour lines to evaluate the geometry and any variations in the shape of the ribbed vault; (a) superposition of the contour lines (equidistance 90 cm) on the orthophoto; (b) superimposition of contours with DSM, (c) DSM calculation of the entire model of the room. (d) Plan projection of the vault and elevations drawings of the walls with identification of the technological and deterioration elements.

orthophoto alone had not highlighted any anomaly of shape, it was the analysis of the surface that revealed the deformations of the cylindrical vaults in key.

Even the vault with ribs, set on ogival arches of the large chapel ("Cappellone") of San Nicola in Tolentino (XIV century, frescoes by Nanni di Bartolo) [18] (Fig. 3), despite the presence of degradations and detachments of the frescoed surface, presents overall a very regular generative geometry of the conic surfaces, as can be seen, both from the uniformity of the section profiles directions, and by the color distribution of the DEM surface.

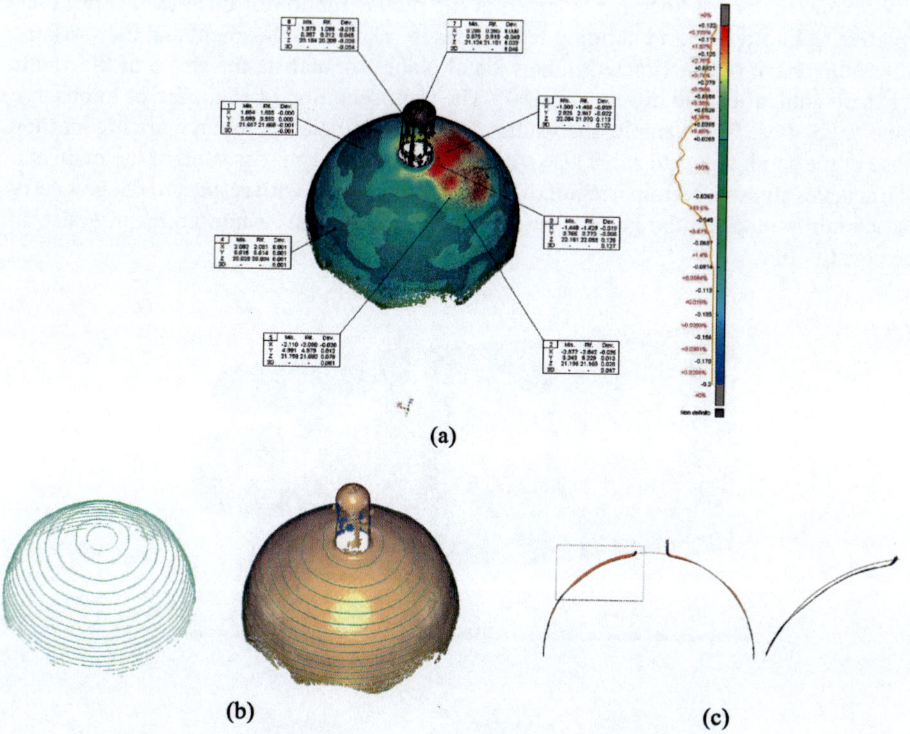

Fig. 4. DSM of the chapel dome. (a) Color mapping of the deviation between the detected geometry and the analytically generated sphere; (b) Analysis of the deviations between the relief surface and the analytical hemisphere using section profiles; (c) Analysis of a section profile and deformation with respect to the spherical generating curve.

The achievable analyses, starting from reality-based techniques, may concern also deviations of the surfaces estimated analytically, such as, for example, comparing a spherical dome with a hemisphere set on the supports. Taking into consideration the Basilica of San Nicola of Tolentino (XVII century) which was subject to the destructive seismic events of 2016 and 2017 in central Italy, the ideal spherical geometry best adapted to the actual geometric surface of the dome of one of the side chapels of the Basilica was defined analytically [19]. This analysis made it possible to obtain a chromatic mapping to identify the areas that present a deviation between the surveyed surface and the reference analytical generative surface (Fig. 4). The crucial point is to evaluate whether the causes of deviations are to be attributed to an imperfect installation. In the present case, also in light of the structural observations of the monitoring team, the post-earthquake settlement is recognized. (max 13 cm). Here more than in other cases, we can say the choice of the profile steps is of fundamental importance based on the investigation's purposes.

Also, the study of the vaulted system of the S. Maria church in Staffarda Abbey, founded in 1135 and transformed in the XV century, featured a comparison among 3D surveys and historical documents aimed at the reconstruction of different cycles, constructive technologies, and building techniques in relation to the client and the workers. 3D profiles have been extracted in horizontal planes to analyse the shape of the vaults to clarify anomalies and differences [20]. The representation of Fig. 5 helps to observe some anomalies. For example, the vaults of the north side of the church are higher than those of the south side and show also a different shape. Each cross vault of the main and lateral naves shows a certain irregularity in the construction with respect to the geometry that certainly inspired the generative curves, which suggests a non-uniformity also in the construction.

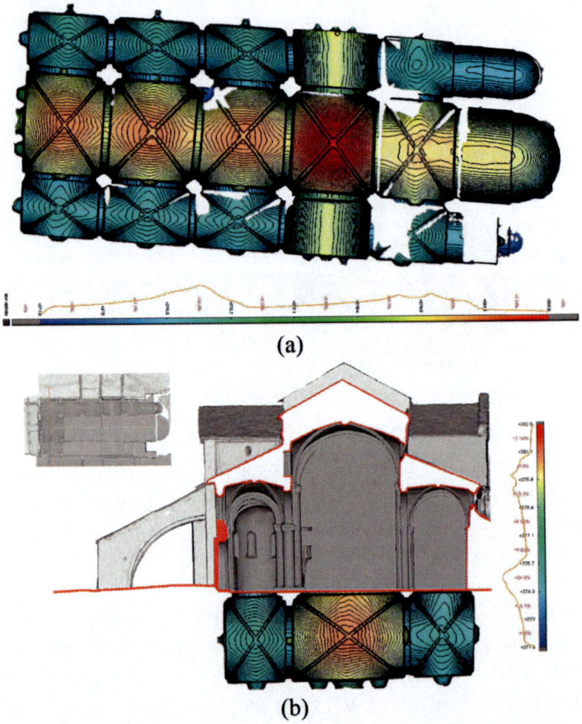

Fig. 5. (a) DSM of the vault systems with contour lines extracted each 20 cm; (b) Vertical transversal section extracted from the 3D model over the first span of the church and corresponding DSM.

4 Conclusions and Perspective

Survey techniques based on laser scanning and photogrammetry have demonstrated in the last two decades their fundamental role in the geometric documentation and interpretation of the historical heritage and in the ability to investigate problems connected to the construction of the factory or to highlight in monitoring processes possible deformations with respect to the initial configuration if associated with structural analysis and historical studies.

An important novelty that has been witnessed in recent years is determined by the development of the archiving of historical architecture models through the so-called HBIM strategy, which aims to establish an intelligent system capable of archiving both the three-dimensional configuration of historical buildings and the connected heterogeneous information. In this scenario, the study and research on generative geometries have undergone a renewed motivation due to technical issues related to the *as-built* approaches [21] which aim to make the transformation of point models into structured models more streamlined and efficient, parametric whenever possible.

References

1. Pieraccini, M., Dei, D., Betti, M., Bartoli, G., Tucci, G., Guardini, N.: Dynamic identification of historical masonry towers through an expeditious and no-contact approach: application to the "Torre del Mangia" in Siena (Italy). J. Cult. Herit. **15**, 1–8 (2014). https://doi.org/10.1016/j.culher.2013.07.006
2. Abbate, E., Invernizzi, S., Spanò, A.: HBIM parametric modelling from clouds to perform structural analyses based on finite elements: a case study on a parabolic concrete vault. Appl. Geomatics **14**, 79–96 (2022). https://doi.org/10.1007/S12518-020-00341-
3. Selvaggi, I., Bitelli, G., Serantoni, E., Wieser, A.: Point cloud dataset and FEM for a complex geometry: the San Luzi bell tower case study. ISPRS – Int. Arch. Photogramm Remote Sens. Spat. Inf. Sci. **XLII-2/W11**, 1047–1052 (2019). https://doi.org/10.5194/isprs-archives-XLII-2-W11-1047-2019
4. Ramos, M.M., Remondino, F.: Data fusion in Cultural heritage: a review. ISPRS – Int. Arch. Photogramm Remote Sens. Spat. Inf. Sci. **XL-5/W7**, 359–363 (2015). https://doi.org/10.5194/isprsarchives-XL-5-W7-359-2015
5. Grussenmeyer, P., Landes, T., Voegtle, T., Ringle, K.: Comparison methods of terrestrial laser scanning, photogrammetry and tacheometry data for recording of cultural heritage buildings. ISPRS – Int. Arch. Photogramm Remote Sens. Spat. Inf. Sci. **37**(B5), 213–218 (2008)
6. Banfi, F., Brumana, R., Stanga, C.: Extended reality and informative models for the architectural heritage: from scan-to-BIM process to virtual and augmented reality. Virtual Archaeol. Rev. **10**, 14 (2019). https://doi.org/10.4995/var.2019.11923
7. Sammartano, G., Avena, M., Fillia, E., Spanò, A.: Integrated HBIM-GIS models for multi-scale seismic vulnerability assessment of historical buildings. Remote Sens. **15**, 833 (2023). https://doi.org/10.3390/rs15030833
8. (2011) DPCM 9/02/2011-Valutazione e riduzione del rischio sismico del patrimonio culturale con riferimento alle norme tecniche per le costruzioni di cui al DM 14/01/2008
9. Lab G4CH - Lab Geomatics for Cultural Heritage. https://g4ch.polito.it/wordpress/
10. Boehler, W., Marbs, A.: 3D scanning and photogrammetry for heritage recording: a comparison. In: Proceedings of 12th International Conference on Geoinformatics, University of Gävle, Sweden, pp. 7–9 (2004)

11. Shan, J., Toth, C.K.: Topographic Laser Ranging and Scanning: Principles and Processing. CRC Press, Boca Raton, FL, USA (2008)
12. Luhmann, T., Robson, S., Kyle, S., Harley, I.: Close Range Photogrammetry: Principles, Techniques and Applications (Vol. 3). Whittles Publishing, Dunbeath, UK (2006)
13. Stavroulaki, M.E., Riveiro, B., Drosopoulos, G.A., Solla, M., Koutsianitis, P., Stavroulakis, G.E.: Modelling and strength evaluation of masonry bridges using terrestrial photogrammetry and finite elements. Adv. Eng. Softw. **101**, 136–148 (2016)
14. Bosse, M., Zlot, R., Flick, P.: Zebedee: design of a spring-mounted 3-D range sensor with application to mobile mapping. IEEE Trans. Robot. **28**, 1104–1119 (2012). https://doi.org/10.1109/TRO.2012.2200990
15. Sammartano, G.: Suitability of 3D dense models from rapid mapping strategies for Cultural Heritage documentation and conservation. Validation of metric and non-metric information extraction from integrated solutions. PhD Thesis, Prof. A. Spanò and Prof. F. Chiabrando supervisors. DAD - Politecnico di Torino (2018)
16. Patias, P.: Overview of applications of close-range photogrammetry and vision techniques in architecture and archaeology. In: Manual of Photogrammetry, 6th Editio. American Society of Photogrammetry and Remote Sensing, pp. 1093–1107 (2013)
17. Chiabrando, F., Lingua, A., Noardo, F., Spanò, A.: 3D modelling of trompe l'oeil decorated vaults using dense matching techniques. ISPRS – Int. Arch. Photogramm Remote Sens. Spat. Inf. Sci. **II–5**, 97–104 (2014). https://doi.org/10.5194/isprsannals-II-5-97-2014
18. Spanò, A., Perri, S.: Ortofoto e modelli 3D ad alta risoluzione per la valutazione del degrado di murature affrescate. Il Cappellone della basilica di San Nicola a Tolentino. Atti & Rassegna Tecnica LXXIII-3 anno **152**, 222–229 (2019)
19. Davide, E., Alessandra, S., Spanò, A.: Nuvole di punti, rappresentazione architettonica e analisi dei modelli 3D ottimizzati per lo studio della basilica di San Nicola a Tolentino. Atti & Rassegna Tecnica LXXIII-3 anno **152**, 206–211 (2019)
20. Beltramo, S., Donadio, E., Spanò, A.: Stratigraphic analyses, historical evidences, and 3D documentation tools. In: Inglese C, Ippolito A (eds.) Analysis, Conservation and Restoration of Tangible and Intangible Cultural Heritage, IGI Global, USA, pp. 273–301 (2019). https://doi.org/10.4018/978-1-5225-6936-7.ch012
21. Sammartano, G., Previtali, M., Banfi, F.: Parametric generation in HBIM workflows for SLAM-based data: discussing expectations on suitability and accuracy. In: Proceedings ARQUEOLÓGICA 2.0 - 9th International Congress & 3rd GEORES - GEOmatics and pREServation, pp. 374–388 (2021). https://doi.org/10.4995/arqueologica9.2021.12155
22. Spanò, A., et al.: Digital twinning for 20th century concrete heritage: HBIM cognitive model for Torino Esposizioni halls. Sensors **23**(10), 4791 (2023). https://doi.org/10.3390/s23104791

Form-finding with Restraint Topology Optimization of a Curved Shell-Supported Footbridge under Vertical and Horizontal Loads

Luigi Fenu[1], Alireza Hosseini[1](✉), Stefano Punzo[1], Bruno Briseghella[2], and Gian Felice Giaccu[3]

[1] University of Cagliari, 09124 Cagliari, Italy
seyeda.hosseini@unica.it
[2] Fuzhou University, Fuzhou, China
[3] University of Sassari, Sassari, Italy

Abstract. The research focused on the design of a curved shell-supported footbridge using a form-finding algorithm and genetic optimization. The bridge was shaped through a parametric design code, which also allows optimization based on finite element structural analysis. The constrained optimization involved a mono-objective approach aided by penalty functions to control the maximum tension utilization of the concrete material. The objective was to find the optimal bridge shape in terms of minimizing displacement under vertical and horizontal loads, with both the topological optimization of the positions of the bridge supports and the optimization of the control points of the Bezier curve describing the form of the curved deck as key parameters. The results provide insights into effective techniques for optimizing the design of curved shell-supported footbridges subjected to earthquake loads.

Keywords: parametric design · form-finding · structural optimization · shell-supported footbridge · seismic design · conceptual design

1 Introduction

Shell-supported bridges are a kind of bridge whose world diffusion is very low with respect to practically any other type of bridge and footbridges except, perhaps, particular kind of footbridges such as tensegrity footbridges [1, 2].

Nevertheless, the few shell-supported bridges that have been constructed are exciting bridges designed by important designers in the history of Structural Engineering and Architecture. The first remarkable example of a shell-supported bridge is the Basento bridge in Potenza, Italy, built in the late '60 s; a concrete road bridge considered the masterpiece of the eminent Italian structural designer Sergio Musmeci [3, 4].

In the same years, a shell-supported road bridge was built in Erdberger (Austria) by Alfred Pauser [5], who, even if not as genius as Musmeci, was an excellent bridge designer (a great number of bridges in Vienna were designed by him) and reputable

S. Gabriele et al. (Eds.): IWSS 2023, LNCE 437, pp. 867–876, 2024.
https://doi.org/10.1007/978-3-031-44328-2_91

professor of "Bridge Design" in Vienna Technical University, where in the Institute for Structural Engineering was considered a leader in innovating research and teaching in bridge design.

After them, for about thirty years, shell-supported bridges were absent in bridge design and construction scene, and only at the beginning of the 2000s, Jiri Strasky participated in an international design competition in Jersey (UK) and another for the Leamouth pedestrian bridge in London [6–8].

In 2004 Fenu et al. [4, 9], following Musmeci's work, after performing the form-finding using a simulated-annealing algorithm, proposed a method of constructing shell-supported footbridges by using a thin membrane made of ferrocement to be set-in-place and which thickness was successively increased with sprayed concrete to obtain the concrete shell supporting the deck. To better exploit the construction material, topology optimization was used to remove the less stressed concrete from the shell by creating holes throughout it [10, 11].

About forty years after the construction of the shell-supported bridges realized by Musmeci and Pauser, in 2008, a shell-supported footbridge in steel with a curved concrete deck, known as the Lichtenlijn bridge, was designed by Laurent Ney and successively realized in Knokke-Heist, Belgium [8], while in 2011 another shell-footbridge was built in Madrid by Hugo Corres Peiretti and collaborators, who designed the Matadero Bridge [12] whose deck is hanged to a concrete shell canopy over it.

More recently, Trumpf Footbridge in Ditzingen (Germany, 2018), a pedestrian bridge whose deck is supported by a steel grid shell [13], was designed by Mike Schlaich and collaborators of sbp, Schlaich Bergermann Partners.

In the knowledge of these authors, since 2018, no other shell bridges have been built, but a model of a shell-supported footbridge has been constructed in the laboratory of ETH in Zurich [14] for experimental studies carried out by Block's Research Group, as part of their research on the shell-bridge construction method to realize them [15].

Besides studying the behaviour and construction methods of shell-bridges, the present research on curved shell-supported footbridges also started from studying the fundamental research on curved bridges carried out by Jörg Schlaich since the '80s of the last century, and that he continued until the first decade of the 2000s [16, 17], by also constructing some of his better masterpieces as a fruit of this research.

Taking into account the properties of ring girders [16, 18], Fenu et al. [19] studied the possibility of obtaining a concrete shell with an anticlastic shape supporting a curved deck, provided that suitable boundary conditions were assigned for the shell form-finding. Curved footbridges supported by an anticlastic shell shaped under vertical loads with different form-finding methods were thus obtained [20, 21], and their static performance was investigated.

The seismic behaviour of curved shell-supported footbridges is herein studied. For this aim, the concrete shell supporting the deck is suitably shaped. Shaping shells to resist the earthquake action better is a problem faced by Michiels during his Ph.D. research at Princeton University [22–24].

This paper faces the problem of shaping a curved-shell-supported footbridge under both vertical and horizontal loads. To this aim, the curvature of the shell top edge supporting the curved deck and parametrically defined through Bezier curves, and the inclined curve of the shell bottom edge with the opposite concavity of its horizontal projection are exploited. The latter curve is obtained as a result of a form-finding procedure performed at each step by means of a genetic algorithm through a topology optimization of the position of the abutment restraints on the ground slope. By also optimizing the deck curvature by varying the parameters of the Bezier curve, the objective of the genetic algorithm is to minimize displacements, provided that concrete stresses, checked through a finite element code, are lower than allowable values.

2 Parametric Design

This study investigates the mechanical performance of a curved shell-supported footbridge with a 40-m-long span, exploiting a variety of shapes and forms. Due to the significant influence of the shell structure's geometry on force amounts and flow patterns, a parametric design methodology is adopted to generate multiple designs by altering several input parameters. Every generated model is analyzed to control the bridge's displacement, stresses, and other properties. Based on the finite element analysis output, a genetic optimization algorithm is utilized to minimize the selected objective function. Generally, the form may be found under self-load, creating a funicular shell system. In this case, the form-finding algorithm defines the shell by incorporating optimized support positions and applying the external load in accordance with the prescribed curvature. Figure 1 shows the flowchart which summarizes the optimization process. Here the integration of the genetic optimization routine with the parametric definition of the design problem, the form-finding algorithm, and the Finite Element Analysis are described.

2.1 Design Variables Definition and Form-Finding

The horizontally curved footbridge geometry is generally determined by defining the deck form as the main constraint, followed by the abutments' positions. Then shell shape is generated through a form-finding algorithm. A horizontal cubic Bézier Curve allows us to investigate the curve shape while maintaining the depth. A particular curve used with four control points defined as P0 P1 P2 P3 defined in Eq. (1)

$$B(t) = P_0(1-t)^3 + 3P_1t(1-t)^2 + 3P_2t^2(1-t) + P_3t^3 \tag{1}$$

where t is a range between 0 to 1, the curve's first (P_0) and the last points (P_3) are stationary points, while P_1 and P_2 act as primary control points that dictate the curve's curvature. These control points are instrumental in determining the direction of the tangent at P_0 and P_3 and are hence regarded as the first shape design variable, as depicted in Fig. 2. Assuming the fixed maximum bent for the bridge (4 m in the middle), the vertical position of P_1 and P_2 are constant. In contrast, their longitudinal position can be altered, resulting in different tangents at the corners of the bridge. As the bridge is symmetric, the x position of P_1 and P_2 is determined based on their distance from the center, which serves as an input parameter for determining the curve shape.

A vertical lofted surface is generated according to the pre-defined trend (x-y) on the z-axis, with the input height as the second design variable -the bridge's height- to obtain

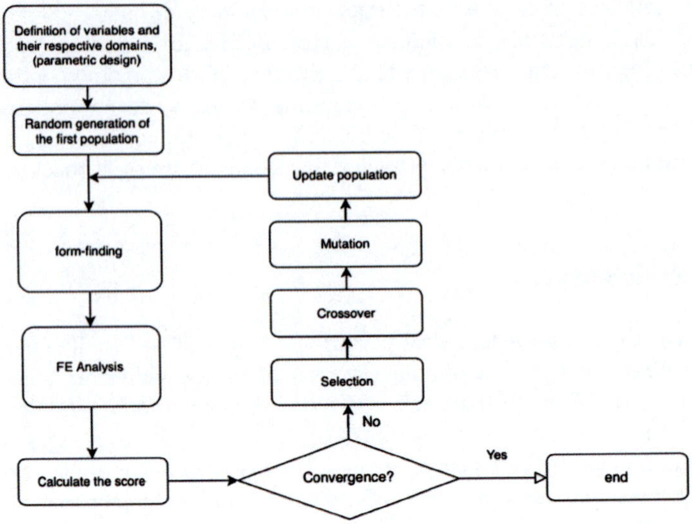

Fig. 1. Design procedure of single objective optimization

the desired shape that is subsequently meshed (Fig. 3). The mesh configuration (19 by 8 quads longitudinal and vertical to the bridge respectively) is defined according to the behavior properties of the reinforced concrete material. The last requirement to start the form-finding procedure is the boundary condition. The position of boundary points at the abutments is defined as two other design variables, particularly the positions of middle-height and bottom supports. Moreover, the top edge -the bridge's deck- is constrained to the x-y surface to maintain its horizontal form during the form-finding (Table 1).

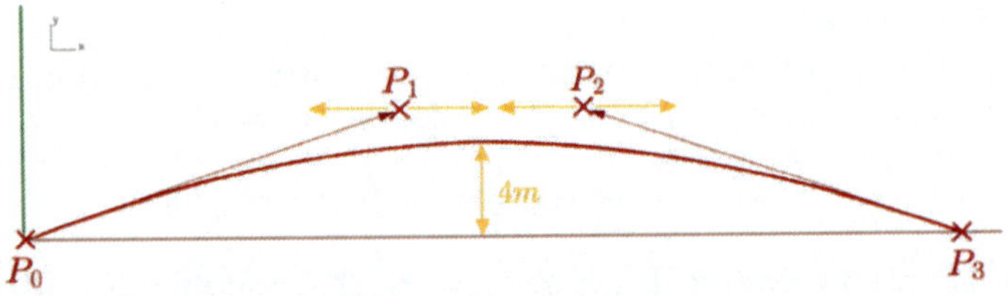

Fig. 2. Curvature introduction

Once the deck geometry and mesh are established, a form-finding procedure is employed using Dynamic Relaxation provided by Kangaroo3d. In this way, a form-active system determines the shape of a form-passive shell structure, ensuring a cohesive and effective design approach [25]. The form-finding algorithm uses a triangular mesh as the initial input and applies dynamic relaxation principles to convert its vertices and edges into nodes and elastic connections (springs). The process ensures the determination of the optimal geometry through dynamic equilibrium methods.

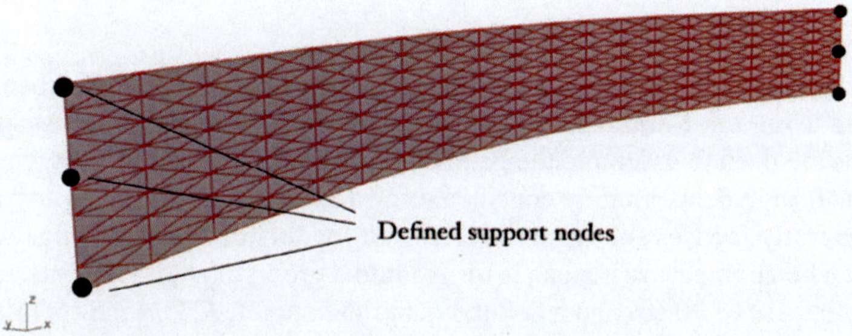

Fig. 3. Initial mesh surface and initial boundary conditions.

Table 1. Design variables and range of variations

Design variables	Range of variation
Edges Bézier tangents	0.35 to 5.0
Bridge height (m)	3 to 5
External force in form-finding (X-dir)	−1 to +1
Position of middle supports (X)	0.0 to 1.0
Position of middle supports (Y)	0.0 to 1.0
Position of middle supports (Z)	(Mid-height −1.0) to Mid-height
Position of bottom supports (X)	0.0 to 5.0
Position of bottom supports (Y)	0 to 6.0

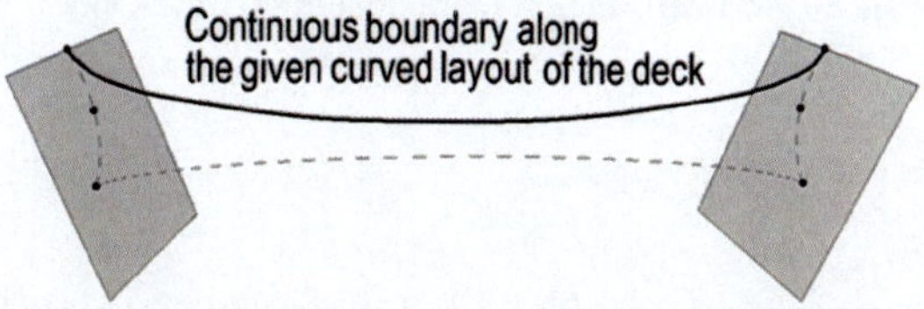

Fig. 4. Initial condition and topography of the ground

Selecting appropriate shell boundary points at the bridge's ends is crucial; in fact, depending on the topography of the slope's ground surface at each end, two or more boundary points may be assigned. However, in this case, only two supports (in addition to the deck connection) are utilized, and the natural morphology of the building site governs their location range variation (Fig. 4). Finally, an external force is applied to the mesh in order to control the shape, particularly in the context of horizontal load application.

2.2 Finite Element Analysis

The form-found mesh is collected to establish the structural model, and a linear analysis is performed using the Karamba3D [26] plugin for each load case. The assigned loads and supports are used to assemble the geometry during each iteration. Leaving apart the layout of the bridge deck, already considered in some previous scientific articles by the authors, this study focuses solely on the effects of loading the shell top edge with a 4 m width deck. The geometry as a mesh is turned into a group of shell elements made from a defined thickness of 30 cm and a defined material concrete C35/45. Supports position are determined through topology optimization, whereas the restriction specification of hinged support is imposed initially.

Concerning vertical loads, the shell self-load and the deck load (20 kN/m) are amplified by 1.35 as permanent loads, as specified by the EC1. In addition, taking into account 5 kN/m^2 live load, amplified by a factor of 1.5, and incorporating a gravity load factor of 1.35. As for horizontal loads, they are accounted for by applying a 20% self-load in perpendicular directions to the bridge, acting in two opposite orientations (Fig. 5).

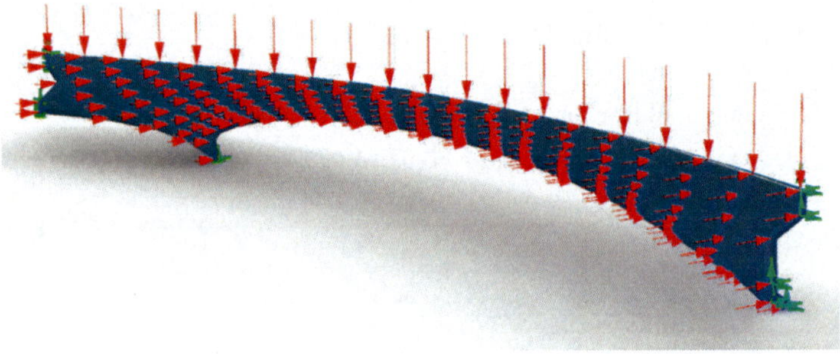

Fig. 5. Vertical and horizontal loading (convex side)

2.3 Optimization

The integration of mathematical algorithms and computational tools has revolutionized the exploration of design spaces, leading to efficient and sustainable solutions. This study utilizes parametric design and Galapagos mono-optimization algorithms to tackle the complex problem of form-finding a shell-supported bridge under bilateral load conditions. The genetic algorithm minimizes the maximum displacement, with a population of 200 individuals (double for the initial generation), and a 75% inbreeding factor results in 'incestuous' couples. Results indicate that an optimized population can be achieved within 20–50 iterations.

It is essential to emphasize that the structural optimization problem in this study involves simultaneous optimization of all design variables, considering both horizontal and vertical load cases. A penalty function is employed to enforce a maximum tension stress limit of 2.2 MPa (using the Rankine criterion), based on the tension limit of concrete C35/45 and safety coefficients of 1.5. The penalty function is applied only when the stress limit is violated, proportionally penalizing the objective function based on the degree of constraint violation.

Regarding horizontal loading, it is crucial for the structure to perform adequately under lateral forces in both directions. Consequently, the optimization algorithm is modified to evaluate constraints for each load combination separately. Thus, the worst load combination is considered to ensure the structure's adequacy in both lateral force directions.

3 Results

The study's results reveal significant structural performance and architectural value. The bridge exhibits an intriguing anticlastic shape characterized by horizontal amd inclined curvature along its length and height (Fig. 6). Notably, the curvature in the z-axis is contingent upon the applied horizontal load. The positions of the supports and other design variables are summarized in Table 2, highlighting the bridge shape that achieved optimal performance in terms of adhering to allowable tensile stresses and limiting displacements. As the bridge is symmetric, the supports position mentioned in Table 2 are only provided for one side of the bridge.

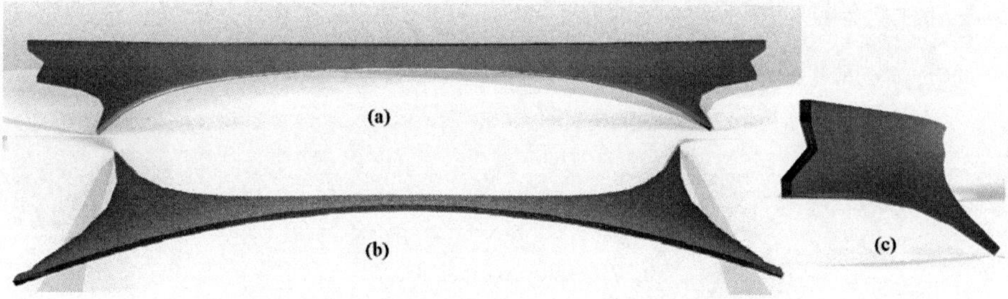

Fig. 6. (a) Front view, (b) Top view, (c) Right view.

The evaluation of principal stresses is utilized to assess the structural performance, enforcing the allowable stresses in tension and compression for unreinforced concrete. The first principal stress is predominantly tensile, constrained to a maximum of 2.2 MPa. While this value may be reached in certain localized regions, particularly in the upper zone of the shell (Fig. 7), local stress concentrations near the restraints are observed.

The distribution of the second principal stress (Fig. 8) is predominantly characterized by compression stresses, spanning over two-thirds of the shell surface, with a maximum value of 8 MPa. These findings confirm that the bridge demonstrates a compression utilization of approximately 35% and a tension utilization of approximately 95%. It is important to note that these utilization values are obtained under design loads, including the horizontal loads.

In facing this problem, the penalty function is an efficient and robust method to control stress levels, and the formula is well-tuned to get the fittest objective function. The impact of the optimal design can also be appreciated in terms of material usage,

Table 2. Design variables, range of variation and the fittest population

Design variables	Range of variation	Optimized value
Edges Bézier tangents	0.35 to 5.0	0.50
Bridge height (m)	3 to 5	4.2
External force in form-finding (X-dir.)	−1 to +1	1
Position of middle supports (X)	0.0 to 1.0	0.28
Position of middle supports (Y)	0.0 to 1.0	0.5
Position of middle supports (Z)	(Mid-height −1.0) to Mid-height	Mid-height −0.25
Position of bottom supports (X)	0.0 to 5.0	5.0
Position of bottom supports (Y)	0 to 6.0	5.1

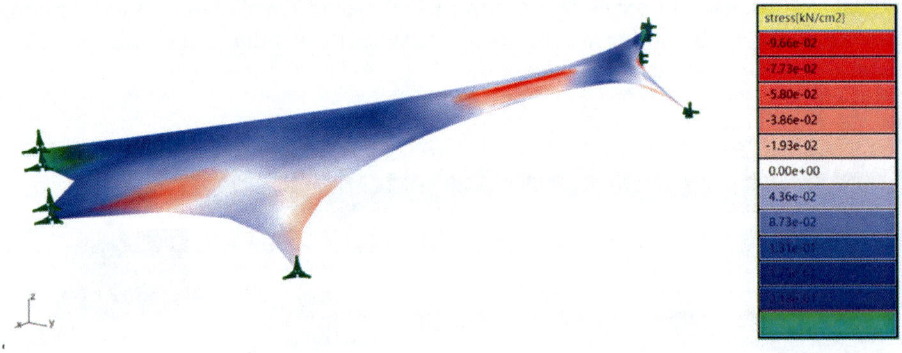

Fig. 7. First principal stresses

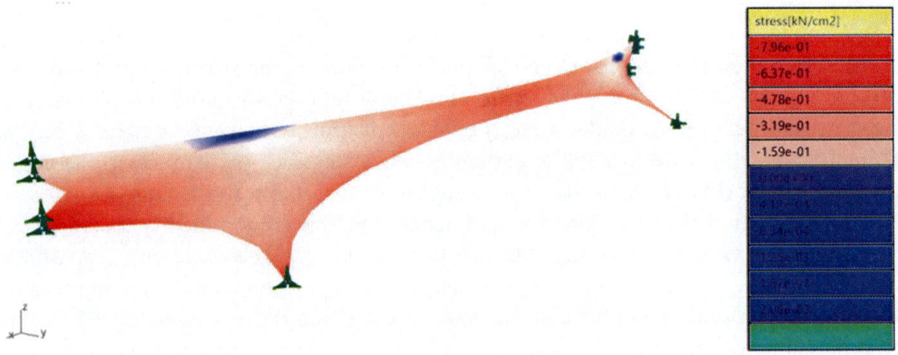

Fig. 8. Second principal stresses

with a limited weight of about 70 tons, despite the research of the best form in terms of minimal displacement. The displacement distribution along the bridge with a maximum deflection of 35.9 mm occurred in the SLS load case at mid-span.

4 Conclusions

This study has shown that the earthquake resistance of a curved shell-supported bridge in concrete is mainly governed by the contemporary work of the top and bottom arched edges of the shell. The prescribed deck curvature demonstrates favorable alignment with transverse horizontal loads applied to the bridge's convex side. Furthermore, the optimization of abutment supports location enables the creation of an inclined arched bottom edge, effectively countering seismic forces acting in the opposite orientation on the bridge's concave side.

Moreover, the findings emphasize the crucial inclusion of horizontal external forces in the form-finding procedure of curved shell-supported bridges. While form-finding can be processed only by restraint location optimization producing anticlastic shells, the introduction of horizontal loads during form-finding substantially enhances the bridge's mechanical performance and reduces tensile stresses in the structure by the factor of two. This enhancement extends beyond earthquake loads and positively impacts the bridge's behavior under vertical load scenarios as well.

References

1. Fenu, L., Congiu, E., Marano, G.C., Briseghella, B.: Shell-supported footbridges. Curved Layer. Struct. **2**, 199–214 (2020)
2. Briseghella, B., Fenu, L., Huang, W., Zordan, T.: Tensegrity bridge with prestressed deck. In: Large Structures and Infrastructures for Environmentally Constrained and Urbanised Areas (2010)
3. Nicoletti, M.: Sergio Musmeci: organicità di forme e forze nello spazio. Testo & Immagine, Torino (1999)
4. Fenu, L., Madama, G.: A method of shaping R/C shells with heuristic algorithms and with reference to Musmeci's work. Stud. e Ric. – Politec. di Milano. Sc. di Spec. Costr. Cem. Armato (25), 199–238 (2004)
5. Kromoser, B., Pachner, T., Tang, C., Kollegger, J., Pottmann, H.: Form finding of shell bridges using the pneumatic forming of hardened concrete construction principle. Adv. Civ. Eng. **2018**, 1–14 (2018). https://doi.org/10.1155/2018/6309460
6. Strasky, J.: Pedestrian bridges utilizing high strength concrete. Int. J. Sp. Struct. **22**(1), 61–79 (2007). https://doi.org/10.1260/026635107781037275
7. Strasky, J.: Bridges Utilizing High-strength concrete. In: 30th Conference of Slovenian Structural Engineers, pp. 15–34 (2008)
8. Adriaenssens, S.: How and why Laurent Ney finds steel structural forms. J. Int. Assoc. Shell Spatial Struct. **61**(1), 39–49 (2020)
9. Fenu, L., Madama, G., Tattoni, S.: On the conceptual design of R/C footbridges with the deck supported by shells of minimal surface. Stud. e Ric. – Politec. di Milano. Sc. di Spec. Costr. Cem. Armato (26), 103–126

10. Briseghella, B., Fenu, L., Feng, Y., Mazzarolo, E., Zordan, T.: Topology optimization of bridges supported by a concrete shell. Struct. Eng. Int. J. Int. Assoc. Bridg. Struct. Eng. **23**(3), 285–294 (2013). https://doi.org/10.2749/101686613X13363929988214
11. Briseghella, B., Fenu, L., Feng, Y., Lan, C., Mazzarolo, E., Zordan, T.: Optimization indexes to identify the optimal design solution of shell-supported bridges. J. Bridg. Eng. **21**(3), 04015067 (2016). https://doi.org/10.1061/(ASCE)BE.1943-5592.0000838
12. Corres-Peiretti, H., Dieste, S., León, J., Pérez, A., Sánchez, J., Sanz, C.: New materials and construction techniques in bridge and building design. In: Fardis, M.N. (ed.) Innovative Materials and Techniques in Concrete Construction: ACES Workshop, pp. 17–41. Springer Netherlands, Dordrecht (2012). https://doi.org/10.1007/978-94-007-1997-2_2
13. Kappelt, H.: Efficient structure: footbridge in Ditzingen. Structure **4**(4), 4–5 (2018)
14. Popescu, M., Reiter, L., Liew, A., Van Mele, T., Flatt, R.J., Block, P.: Building in concrete with an ultra-lightweight knitted stay-in-place formwork: prototype of a concrete shell bridge. Structures **14**(February), 322–332 (2018). https://doi.org/10.1016/j.istruc.2018.03.001
15. Bouten, S., Popescu, M. , Ranaudo, F., Van Mele, T., Block, P.: Design of a funicular concrete bridge with knitted formwork. In: IABSE Congress Ghent – Structural Engineering for Future Societal Needs, pp. 690–699 (2021)
16. Jorg Schlaich, R.B.: Leicht weit/Light Structures. Prestel, Munchen (2004)
17. Schlaich, J.: Conceptual design of light structures. J. Int. Assoc. Shell Spat. Struct. **45**(146), 157–168 (2004)
18. Belluzzi, O.: Scienza delle Costruzioni (In Italian). Zanichelli, Milan (1952)
19. Fenu, L., Briseghella, B., Zordan, T.: Curved shell-supported footbridges. In: IABSE Conference Geneva 2015 Structural Engineering Providing Solution to Global Challenges – Report, no. September 2015, pp. 394–401 (2015). https://doi.org/10.2749/222137815818357430
20. Fenu, L., Congiu, E., Lavorato, D., Briseghella, B., Marano, G.C.: Curved footbridges supported by a shell obtained through thrust network analysis. J. Traffic Transp. Eng. (English Ed.) **6**(1), 65–75 (2019). https://doi.org/10.1016/j.jtte.2018.10.007
21. Congiu, E., Fenu, L., Briseghella, B.: Comparison of form-finding methods to shape concrete shells for curved footbridges. Struct. Eng. Int. **31**(4), 527–535 (2021)
22. Michiels, T.: Form Finding Of Arches And Shell Structures Subjected To Seismic Loading Adviser : Sigrid Adriaenssens. no. May (2018). https://doi.org/10.13140/RG.2.2.18163.22562
23. Michiels, T., Adriaenssens, S., Jorquera-Lucerga, J.J.: Parametric study of masonry shells form-found for seismic loading. J. Int. Assoc. Shell Spat. Struct. **58**(4), 267–275 (2017). https://doi.org/10.20898/j.iass.2017.194.892
24. Michiels, T., Adriaenssens, S., Dejong, M.: Form finding of corrugated shell structures for seismic design and validation using non-linear pushover analysis. Eng. Struct. **181**, 362–373 (2019). https://doi.org/10.1016/j.engstruct.2018.12.043
25. Adriaenssens, S.: Shell Structures for Architecture (2014). https://doi.org/10.4324/9781315849270
26. Preisinger, C.: Linking structure and parametric geometry. Archit. Des. **83**, 110–113 (2013). https://doi.org/10.1002/ad.1564

Parametric Design and Optimization of Reciprocal Frame Bridges

Alireza Hosseini[1], Bruno Briseghella[2], Luigi Fenu[1](✉), Gian Felice Giaccu[3], and Stefano Punzo[1]

[1] University of Cagliari, 09124 Cagliari, Italy
lfenu@unica.it
[2] Fuzhou University, Fuzhou, China
[3] University of Sassari, Sassari, Italy

Abstract. This study explores parametric design and optimization to improve a reciprocal frame bridge's structural efficiency while preserving its historical and architectural significance. The research aims to identify an optimal configuration that minimizes steel requirements for a self-supported bridge while satisfying structural requirements. This objective was achieved by modifying the bridge's geometrical parameters using a genetic algorithm for mono-objective optimization. A finite element structural analysis was conducted to evaluate the maximum stress in the material, with a penalty function used to ensure structural safety. The parametric design software allowed for efficient and precise optimization of the bridge design. The results demonstrate that the proposed optimization method reduces material usage while maintaining the bridge's original structural concept of traditional wooden Chinese bridge, validating the approach's effectiveness for future design of reciprocal frame bridges.

Keywords: reciprocal frame bridge · parametric design · structural optimization · conceptual design

1 Introduction

Reciprocal structures are a type of self-supporting structure with also peculiar architectural value characterized by interconnected elements aimed at creating a rigid and stable structure. These structures used to be constructed for centuries in various cultures; however, with the advent of structural use of materials allowing to have better performant members, have lost their privilege. Nevertheless, reciprocal frame structures have recently seen a resurgence in popularity due to their potential for sustainable construction and for their interesting form.

Not surprisingly, the reciprocal structures are particularly interesting for their ability to be constructed quickly and cover large spans without needing intermediate supports with all elements in bending whose depth is much lower than that of a single beam with the same span length. This makes them suitable for bridges, where spans tend to be longer than in other constructions (Fig. 1) [1].

S. Gabriele et al. (Eds.): IWSS 2023, LNCE 437, pp. 877–887, 2024.
https://doi.org/10.1007/978-3-031-44328-2_92

Fig. 1. Leonardo da Vinci's proposal for temporary bridges. (Sketches by A. E. Piroozfar.)

However, designing a reciprocal frame structure can be challenging due to the complex interlocking between elements, whereby changes to one part of the structure can significantly impact overall stability and performance.

To manage these unusual bridge structures by also maintaining their traditional structural scheme coming from the history of Chinese architecture, parametric design, and optimization techniques were found to be particularly useful, allowing efficient exploration of design options and identification of optimal configurations [2].

In this paper, once the parametric model is defined, an analysis is carried out on an existing bridge to study its behavior. Subsequently, size and parameter optimizations are carried out to find the optimal geometry. The optimization process integrates an optimization algorithm, in particular a genetic algorithm, used to modify the geometric parameters of the bridge while conducting a structural finite element analysis (FEA) to evaluate the maximum stress in the material. Furthermore, penalty functions are used in mono-objective optimization to ensure structural safety, particularly limiting the stresses in members. Penalty functions are mathematical function used to incorporate constraints into the optimization problem through assigning a penalty value to the objective function whenever a constraint is violated. Specifically, in this study, the parametric design environment is Grasshopper, which comes as a plug-in for the 3D modeler Rhinoceros [3], and the FEA is performed with the parametric structural design toolkit Karamba3D, as well as with SAP 2000 for defining the design constraints checking the global stability of the reciprocal structure [4, 5].

2 Parametric Design

In this study, parametric modeling of a reciprocal bridge is at first done, taking inspiration from some historical examples like the Chinese Bridge in Jinze (China), also defined as the Rainbow Bridge, and the Leonardo bridge described in folio 899 of his Codex Atlanticus [1]. In the geometric definition of the bridge, the initial geometry comes from a circular arc with pinned supports. However, in this structural typology, the arch is formed by various linear elements, interlocked with each other and self-supporting; therefore, it is not present in its entirety.

Three design variables, span, height, and width are associated with the circular arc, which defines the general geometry of the bridge to be designed. Once the initial geometry is defined, other design parameters are considered, such as the diameter of the longitudinal members, the longitudinal subdivision of the arc, and the transverse subdivision. Table 1 summarizes all the optimization variables.

Table 1. Design variables

Design variables	Range of variation
Span	Fixed at choice
Height	Fixed at choice
Width	Fixed at choice
X-dir. Subdivisions	4 – 10 (pair numbers)
Y-dir. Subdivisions	2 – 10 (pair numbers)
Diameter of longitudinal elements	10 – 40 cm
Thickness of longitudinal elements	4 mm to radius-length
Thickness of transversal elements	4 mm to radius-length

It is essential to consider that, having defined the dimensions of the model and the sectional characteristics of the longitudinal elements, the geometric section outer diameter of the transverse elements is consequently determined. This is because the available "space" between the crossed elements confines the size of the transverse elements. Hence, to ensure structural continuity, this space must fit the transverse elements' depth (Fig. 2).

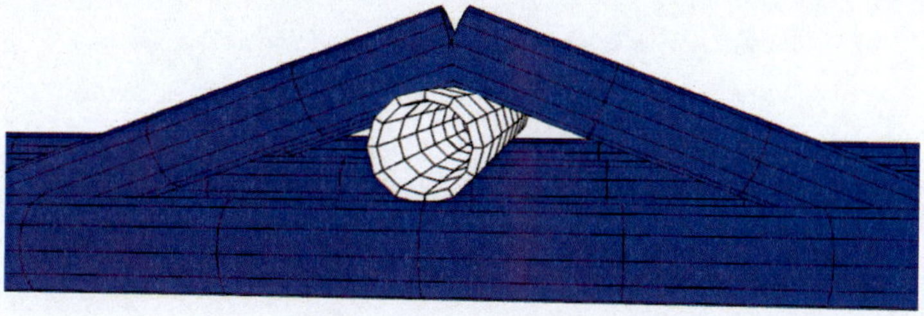

Fig. 2. Transversal element

An interesting morphological aspect of reciprocal frame structures based on superimposed joints is their natural out-of-plane development due to bars lying on the top or the bottom of each other[6]. Such configurations with unaligned elements can be regarded as intrinsically three-dimensional. This needs to introduce rigid elements to be able to simulate their structural behavior; these elements are called: "connectivity elements" in structural modeling.

The midpoints contact between members is modeled by dividing the members at each intersection with the "connectivity elements" representing the above element, and then reconnecting them rigidly in all directions. Furthermore, longitudinal members' start and end points are modeled as hinged to allow rotations in the arch's planes (Fig. 3). Finally, longitudinal and transverse elements are modeled with beam elements, whereas connectivity elements are modeled with rigid links.

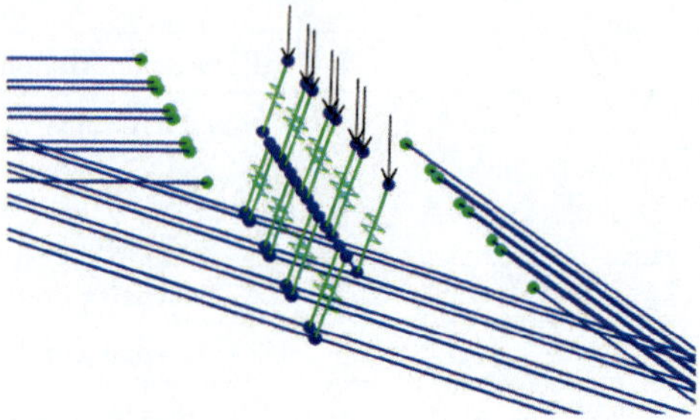

Fig. 3. Connectivity elements and releases (in SAP2000)

Different geometries are shown in Fig. 4, differing in the number of longitudinal subdivisions of the arch, in the number of transversal subdivisions, and in the diameter of the different members, which are some of the design variables listed in Table 1.

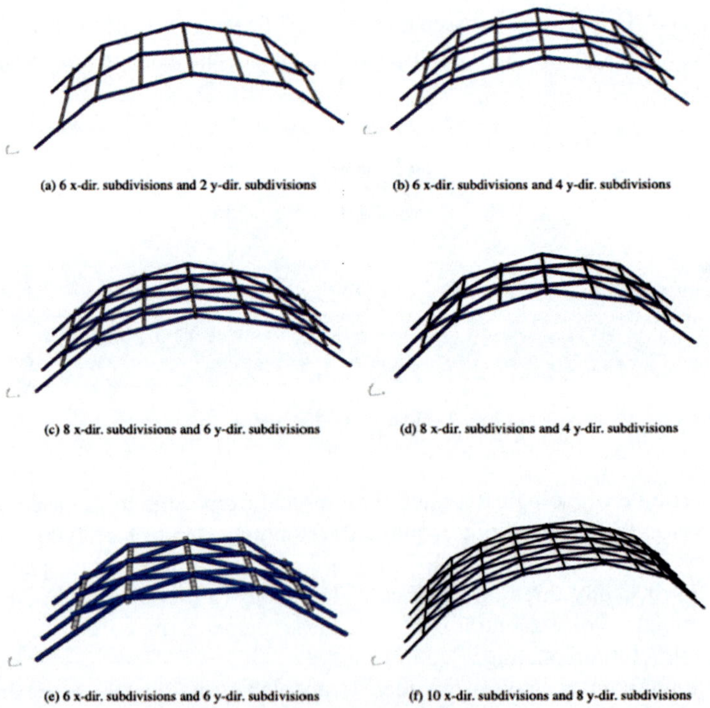

Fig. 4. Primary introduced geometries.

As an example of the versatility of the parametric model, Fig. 4 shows some different model layouts generated for different parameter values. The reciprocal structure bridge under consideration represents the reconstruction of a Chinese Rainbow Bridge, 15m length span, which is located in Jinze, China (Fig. 5) and the parametric model only differ from the prototype for the number of rods, which are doubled [7].

(a) (b)

Fig. 5. (a) Structural Model of Jinze Rainbow Bridge, (b) Jinze Rainbow Bridge

3 Size and Parameter Optimization

The primary objective of this study is to update a structural typology coming from the traditional Architecture of ancient China. To achieve this, an existing bridge is thoroughly examined to understand its behavior. Subsequently, a parametric model is developed, and size and parameter optimization analyses are conducted by generating multiple models. The aim is to find the optimal size of elements and the optimal geometrical shape to create a structure with a minimum weight that complies with modern structure requirements, in terms of maximum allowable stresses and maximum deflection evaluated through a Finite Element Analysis (FEA) under various load conditions.

Two 'loads' combinations were considered during the optimization in accordance with Eurocode 1. Dead load is considered for the structure design, with a safety factor of 1.35 for ULS and 1.00 for SLS. Besides permanent loads, the live load of 5 kN/m^2 is applied with a safety factor of 1.5 for ULS and 1.00 for SLS [8].

The two uniform area loads are initially defined and transformed into point loads at the endpoints of the members. Varying initial geometries correspond to different areas. The area loads are determined once the height, length, and width are assigned. During the optimization process, different geometries result in varying numbers of members and load distribution points, evaluated at each iteration.

Penalty functions were used to perform a simple yet effective single objective constrained optimization. They were used to limit the maximum utilization of each element (Fig. 6), evaluated at the ULS, whose section was constrained to be at least in class 3 and to limit the maximum deflection evaluated at the SLS combination. The optimum layout solution was also constrained to comply with a global stability analysis. In fact, after performing a linear buckling analysis the critical buckling load factor λ was always lower than 6 but constrained to be above 3.

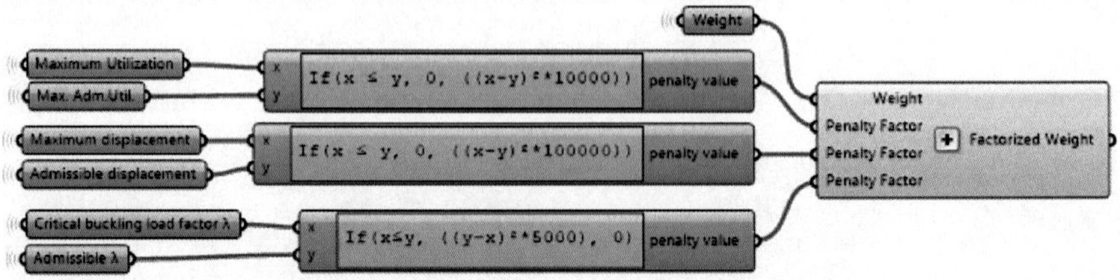

Fig. 6. Factorized weight with penalty functions

Optimization was performed through a genetic algorithm, which turned out to be a powerful tool for optimizing this kind of reciprocal structure. Size and parameter variables of the structure are represented as a set of genes, and the set of candidate solutions are represented as a population of individuals (or genome). These individuals are then evaluated according to the fitness function, and individuals with the highest fitness are selected to reproduce and produce new offspring, which inherit some of the characteristics of their parents. This process is repeated over many generations to reach the optimal solution [9].

4 Results

The outcome demonstrates both satisfactory structural performance and architectural value. By using steel tubes as both longitudinal and transverse members, it is found that the optimal solution has a limited number of elements, consisting of just 8 longitudinal tubular elements and 5 transverse tubular elements with, respectively, 30 and 19 cm in diameter. The optimized values of the design variables are shown in Table 2.

Table 2. Optimal design variables

Design variables	Values
Span	15 m (fixed)
Height	4 m (fixed)
Width	4 m (fixed)
X subdivisions	6
Y subdivisions	2
Diameter of longitudinal elements	30 cm
Thickness of longitudinal elements	5 mm
Thickness of transversal elements	8 mm

Figure 7 shows that, in the reciprocal-structure bridge, two sets of structural members can be identified, a 4-node arch bearing most of the compressions, and a 3-node frame, bearing less loads than the other two arches, but essential for the bridge geometry and the structural scheme. The horizontal members restrain the deformation of the beams of the 4-joint system by means of lap joints and cause the arch beams to deflect upwards (Fig. 9). In addition, note the negative moments along the members of the three-hinged frame because the members of the other frame transmit a slight lift to their ends through the joint.

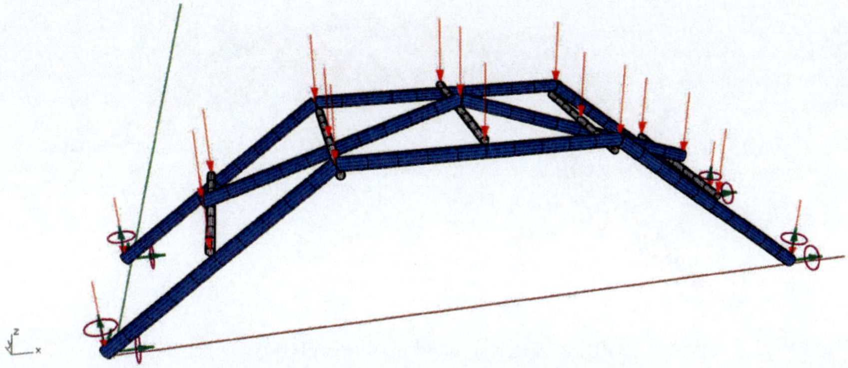

Fig. 7. Optimized geometry of Chinese Bridge

The FEA results demonstrate that the structure behaves like an arch with all the elements in compression, smaller axial forces appear at the top and larger forces at the ends, which are a typical feature of arch behavior with a maximum compressive force of 161 kN (Fig. 8), and therefore less than the allowable force evaluated through a buckling analysis.

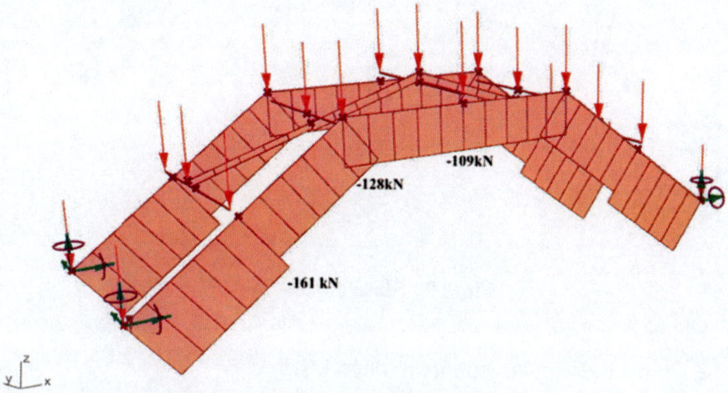

Fig. 8. Normal forces Nx

The maximum value of the moments throughout the optimized structure is found in its lower part of the lateral 4-node frames, and is equal to 48,6 kNm (Fig. 9).

The 3-node frame is characterized by negative moments with maximum value of 23 kNm. In addition, moment values near the ultimate moment are also reached in the transverse and central longitudinal members, underlining the overall and homogeneous utilization of the structure.

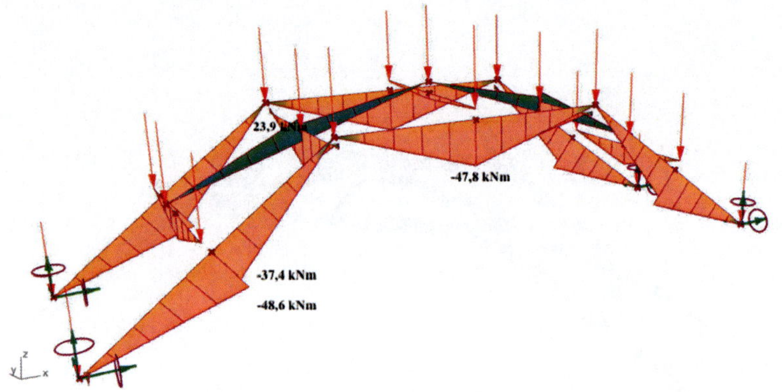

Fig. 9. Internal bending moments My

Figure 10 shows the shear diagram, the shear is higher in the transverse elements also characterized by symmetrical trends and alternately opposite signs. The maximum value of the shear forces is equal to −23 kN.

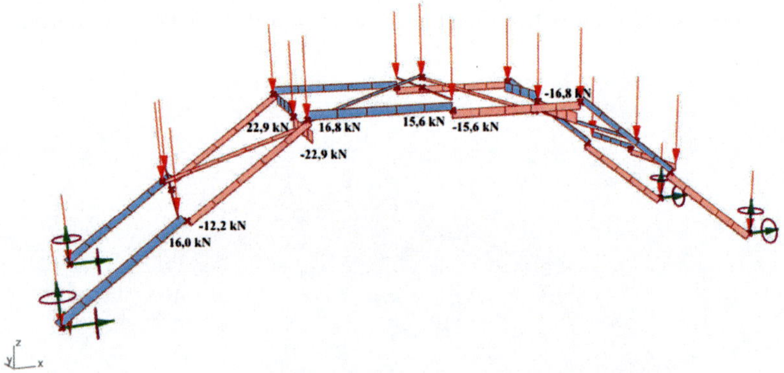

Fig. 10. Shear forces Vy

Figure 11 shows the displacements under SLS.

In the optimization process, the maximum utilization level for each element is constrained and limited up to 1. In fact, the utilization is the ratio between the tensile or compressive strength and the material's maximum stress.

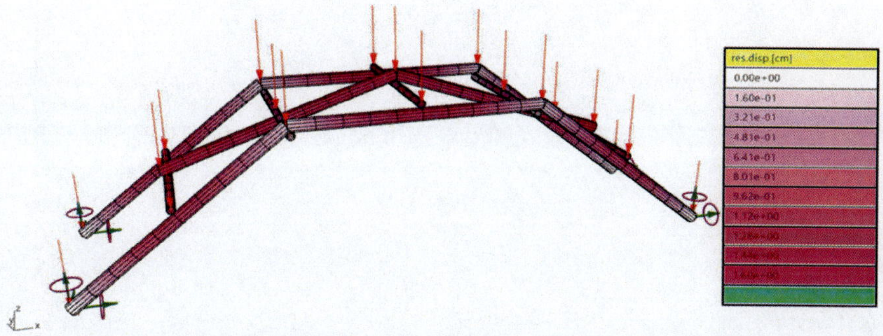

Fig. 11. Displacements (SLS) in axonometric view.

Figure 12 demonstrates that the elements exhibit high utilization under Ultimate Limit State (ULS) condition, both in tensile and compression stresses. This is due to the primary objective of minimizing weight, which correlates with the optimization goal of cost-effectiveness and sustainable construction practices.

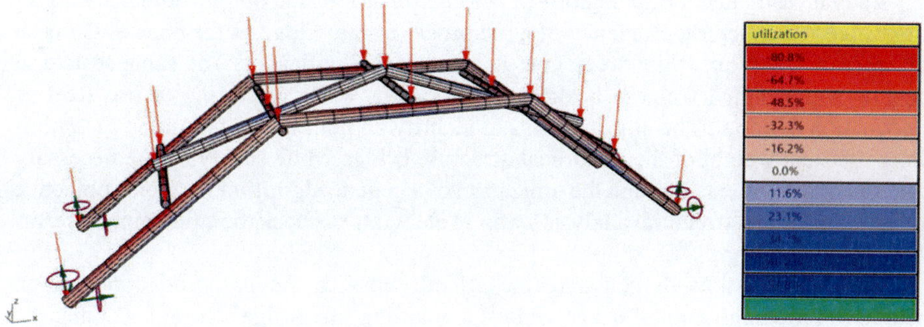

Fig. 12. Utilization of elements in axonometric view.

To check bridge stability, a buckling analysis was first performed, showing that second-order effects have to be considered because the critical buckling load factor λ was quite low, namely 4.8. To take into account second-order effects, a P-Delta analysis was performed by using the FEM software SAP2000 (Fig. 13). It shows that convergence is quickly achieved through obtaining the same bridge displaced shape shown in Fig. 13 after few steps [10].

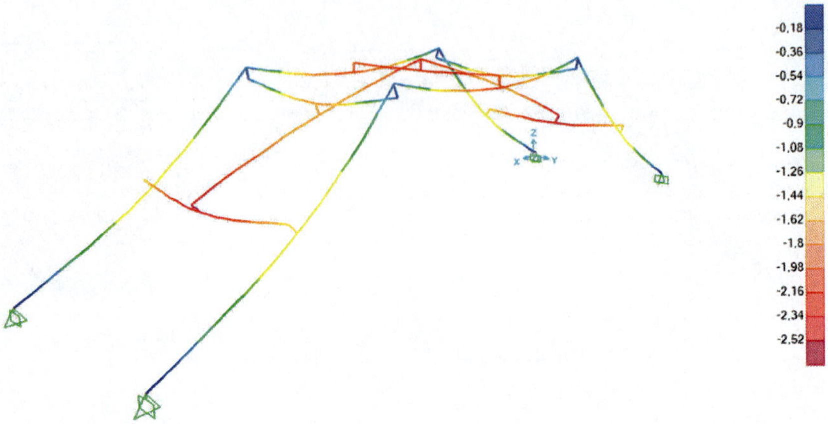

Fig. 13. Displacements after P-Delta analysis (ULS) in axonometric view (in SAP2000).

5 Conclusions

This study investigated the application of parametric design and optimization techniques to enhance the structural efficiency of a reciprocal frame bridge as the ones of Chinese construction tradition, while preserving its historical significance. The same structural scheme of traditional Chinese bridges was adopted, while preferring to use steel as construction material. The objective was to identify an optimal configuration that minimizes the steel weight of the reciprocal-structure bridge while satisfying the necessary structural requirements through the utilization of a genetic algorithm for mono-objective optimization, and a structural analysis by finite elements to check the constraints in terms of structural behavior.

The study's results demonstrate the effectiveness of the proposed optimization method in reducing material usage while maintaining the bridge's structural integrity. The optimized geometry exhibited a compact structure with only 12 elements, efficiently distributing stresses and exhibiting good structural performance in terms of material utilization, stiffness, and stability. The non-linear analysis to check second-order effects confirmed the structural safety of the bridge, despite the high slenderness of the bridge and stress level in its structural elements.

References

1. Larsen, O.P.: Reciprocal Frame Architecture (2008)
2. Robert Woodbury – Elements of Parametric Design. Routledge (2010)
3. Arturo Tedeschi – AAD Algorithms-Aided Design_ Parametric Strategies using Grasshopper. Le Penseur (2014)
4. Preisinger, C., Heimrath, M.: Karamba—a toolkit for parametric structural design. Struct. Eng. Int. **24**(2), 217–221 (2018)
5. Preisinger, C.: Linking structure and parametric geometry. Architectural Des. **83**(2), 110–113 (2013). https://doi.org/10.1002/ad.1564

6. Parigi, D., Pugnale, A.: Three-dimensionality in reciprocal structures: concepts and generative rules. Nexus Netw. J. **16**(1), 151–177 (2014)
7. Liu, D.J.: Ancient Chinese Timber Arch Bridges
8. Eurocode 1: Actions on structures – Part 1–1: General actions – Densities, self-weight, imposed loads for buildings EN.1991.1.1.2002
9. Gandomi, A.H.: Metaheuristic Applications in Structures and Infrastructures. Elsevier Science (2013)
10. CSI: SAP2000 Integrated Software for Structural Analysis and Design. Computers and Structures Inc., Berkeley, California

The Untouchable: An Adaptive Spring-Model Buckling for the Mimosa Pudica Stimuli Response

Fabio Bazzucchi[1,2](✉), Amedeo Manuello Bertetto[2], and Ingrid Maria Paoletti[3]

[1] Senseable City Lab, Massachusetts Institute of Technology, Cambridge, MA, USA
fbaz@mit.edu
[2] Politecnico di Torino, DISEG, Turin, Italy
[3] Politecnico di Milano, DABC, Milan, Italy

Abstract. Mimosa Pudica (MP) possess a hierarchical leaf-folding behavior in response to external disturbances [1]. As observed in other plants [2], such fast movements are mechanically achieved by osmotic regulation of tissue pressure in specific apparatus, pulvini in the case of the MP. Depending on the intensity and location of the stimuli, the triggered kinematics can involve a pair of leaves up to the collapse of the entire branch. Because of this progressive on-off behavior, we modeled the folding system as the linear buckling of a concentrated linear elasticity system, realized by rotational springs and considering a simple time function for the electrochemical signaling. We then described the leaves stimuli adaptive functioning by a modal tuning of the spring stiffnesses. Finally, we proposed a tuned buckling loads to train a neural network to describe a learning model [3] for the equilibrium related MP intelligence and morphological features.

Keywords: snap-through · progressive buckling · mimosa pudica · CNN · plants biomechanics

1 Introduction

From a mechanical point of view, plants are structures that change in size and shape their form [4]. This relentless transformation, other than to sustain growth loads, is often related in being a foundation for plant organs functions. The emerging morphology has a striking complexity in terms of geometry, tissues layout and organs interconnection. Factoring for functional origins on such outcomes is thus a degree of influence assessment rather than a quantitative causal decoupling [5]. Where form and function have at least an injection feature relation (one to more than one), structure and function are instead surjectively related in this matter. Considering the case of reproductive activity, the apparatus can for example build up extra water tissue pressure and vegetal mass by factors of 1.5 and 3 respectively than median values [6], impacting on the stems

S. Gabriele et al. (Eds.): IWSS 2023, LNCE 437, pp. 888–897, 2024.
https://doi.org/10.1007/978-3-031-44328-2_93

wall thickness. On the other hand, to assure tissue integrity during water stress period, the same parameter has to be the lowest possible to curb tissue drying out [7]. Encompassing the metabolic economy of the plant, such plurality requires some sort of a high-order design of the structural arrangement. If evolution and adaptation aggregates inter-generational needs-to-solution shapes, form changes and transformation occurs at individual plant scale as response to external stimuli. In 1811, Knight observed and proved that girth growth of swaying stems was influenced by the direction of bending [8]. Since then, biomechanics research evidenced a plethora of morphological traits that are in fact structural stratagems [9,10]. Furthermore, the sensing apparatus of the plant is based on a mechanical signal translation format [11]. All of this produces growth patterns modifications due to mechanical stimuli, contributing to a unique thigmomorphogenetic [12] final plant shape.

(a) Rest configuration.

(b) Complete folding configuration.

Fig. 1. Mimosa Pudica, before (a) and after (b) the application of a stimulus that caused global folding.

Mimosa Pudica (MP) possesses an additional shape transition layer due to a rapid leaf movement as a response to external threatening disturbances. Such movement is a hierarchical (complete or partial) folding up of the frond, starting from the small leaflets composing the doubly compound leaves [13]. Since observed, this peculiar phenomena sparked the interests of researchers to understand the sensing, signaling and the actuation put in place by the MP. More in details, many interpretations have been provided to understand the conduction of excitation responsible for the progressive kinematic [14,15]. The MP has in fact a complicated and articulated kinematics across the body plant and the parts that constitutes it, depending on the position, intensity and extent of the stimulus. In Fig. 1a and Fig. 1b the MP morphology before (a) and after (b) a complete folding up is reported. It is important to note that partial folding (Fig. 2e and 2l) is not a intermediate state of the kinematics path but a *solved* response state to a determinate stimulus. Figure 2a,e and 2f,l show in fact in frames the side view and top view of a leaf folding when the pinnae gets touched (red circle). This means that exists a hierarchical system in the modulation and actuation of this progressive folding. And behind it, a convoluted electro-physiology signaling apparatus supports such movements, by regulating turgor in in special organs of the plant (pulvini) through osmotic ion flow [16–19]. How this system regulates the pulvini actuation in an adaptive scheme is still unclear, even if the MP suggests to possess a rudimentary shape memory outfit [3]. Encompassing all the limitations eloquently explained by Niklas [4] in applying mechanics to living organism functions, we provide a novel interpretation of the MP leaf movements, using a concentrated elasticity buckling model for pulvini and thin beams for stems. Buckling morphogenesis has been successfully applied in biomechanics for natural shape origin interpretation [20] and plant morphology [5]. This because the solution set of stability eigenforms (vectors) produces shapes of minimum energy, an ubiquitous principle for natural forms [21,22]. Moreover, elastic buckled shapes produces non-incremental modifications of the initial geometry under external forces, guaranteeing still the uniqueness of the solution. These are necessary conditions to model rapid and large movements (often reversible) that are in fact stimuli responses. Finally, because eigenforms are non-dimensional, metrics and indicators in the modal space are sufficient to identify behavior and self-similarity across scales and hierarchies. This approach has been successfully applied to reproduce physical intelligence sensations [23,24] and response behavior identification, for example in MEMS and meta-materials [25].

2 Basic Anatomy of the MP Leaf Movement Organs

In Fig. 3a is represented the essential anatomy of the MP vegetal body. The petiole articulates itself from the branch (stem) through the primary pulvinus, a pad-like swelling of the stem [1]. From the petiole the leaf extends into two to four pinnae, each of them with a central rachis connected by a secondary pulvinus. Lastly, each pinna is subdivided into ten tow twenty oppositely-arranged pars of pinnulas with a tertiary individual pulvinus joining the pinnula with the rachis. To each of these pulivini corresponds the same hierarchical folding movement. This sequential is important because no primary folding can occur if the higher order ones have not materialized yet. In addition to what show in Fig. 2, for example, the nyctinastic (darkness response) folding commonly involves solely the tertiary and secondary pulvini.

Resulting movement is a change in the orientation angle of the connected elements, from tendentiously upward and spread to downward and clutched. The motor tissue is confined to specialized cells of the pulvini, occurring through an osmotic pressure variation between the upper and lower halves, as hypothesized

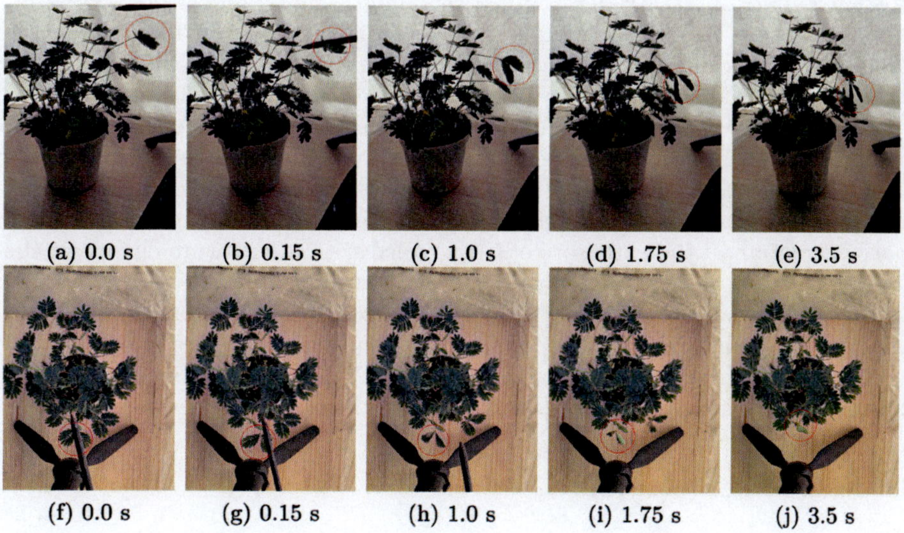

Fig. 2. Mimosa Pudica, before (a) and after (b) the application of a stimulus that caused global folding. (Color figure online)

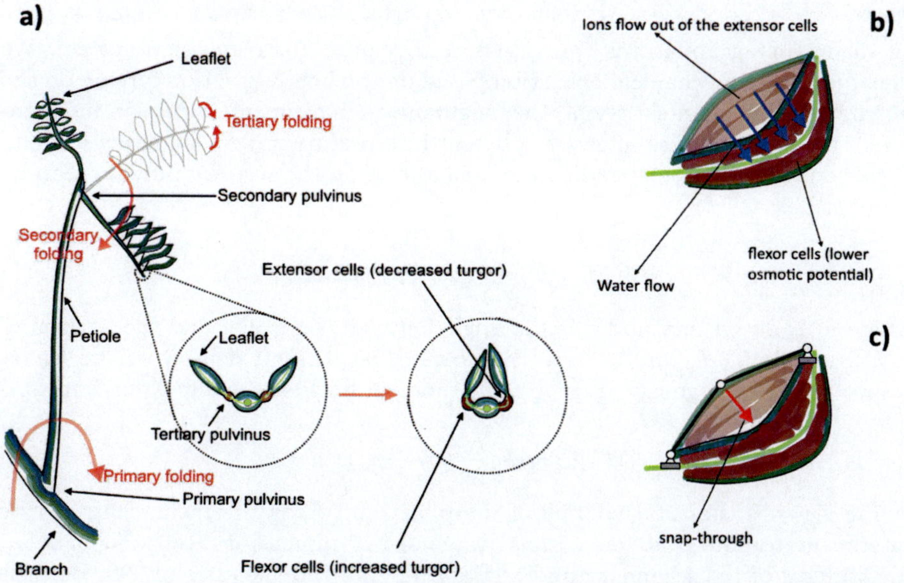

Fig. 3. a) MP essential anatomy. b) Tertiary pulvinus detail, water flows from the upper sensitive half to the lower one. c) simple mechanical modeling od the curvature change of the upper half of the tertiary pulvinus

by Pfeffer in 1912 [1]). In detail, after signaling, osmosis regulating ions[1] concentration varies in the upper half of the pulvinus, causing a flaccidity of the tissue (extensor cells) due to the outgoing water flow. Consequently, turgor on the lower half walls increases by water pressure and volume change coupling, producing a curved stress that fold the pinnulae. Because the secondary and primary pulvini produce an opposite rotation than the pinnulae ones, the corresponding halves of pulvini are inverted as firstly observed by Dutrochet [1]. In Fig. 3b a schematic detail of this mechanism is presented. It is worth to note that contraction of the extensor cells is coupled with a curvature change of the upper half. From a mechanical point of view, the change in wall pressure translates itself into an unbalanced bending action on the joint that the pulvinus represents. We thus decided to model this symmetry breaking mechanism with a snapping three hinged arch for the extensor cells, mimicking the hemispherical compression collapse of the topmost portion of the pulvinus, as in Fig. 3c.

3 Mechanical Modeling

Extensive research has been devoted to the mechanical modeling of the MP, starting directly from Hooke's writings [1]. We present a novel snap-through modeling of the pulvinus, that stands for the elastically concentrated module of the MP body arrangement. Snapping modeling of osmotic pressure change plants has been successfully applied to the Venus Flytrap insect capturing movement [2]. However, contrary to the Venus Flytrap, the stimulus is not applied directly on the snapping apparatus (pulvinus), and require a transmission system. We then modeled a mechanical actuation based on the bending action forcing on the pulvinus. Other than satisfying the thigmonastic movement (direction independent), this solution also allows a coherent interpretation of non-contact stimuli. Considering Fig. 4a, the bending moment acting on the tertiary pulvinus can be expressed as

$$M = Vg\frac{L_f}{2}(\rho_f - \rho_a)\sin\phi, \tag{1}$$

where ρ_a is air density and ϕ is the angle between the leaflet and the horizontal X axis, while V, L_f and ρ_f are the volume, length and density of the leaflet respectively. Analogously (Fig. 4b), the moment on the secondary pulvinus is

$$M = V_{TOT}g\frac{L_r}{2}(\rho_f - \rho_a)\sin\phi, \tag{2}$$

where V_{TOT} is the total volume of the pinna and L_r is the rachis length. It is interesting to note that for nominal value of MP dimensions, self-weight bending stiffness of the secondary and tertiary pulvini is in the ratio of 300. Because inertia follows a fourth root of the diameter, this value is anatomically compensated by other means, such as tissue anisotropy and epidermal wrinkles [26,27]. Under the light of this, we can consider this stiffness exerted by external walls

[1] Manily chloride and potassium [18].

and it can be approximated by membrane $K_R(=\frac{EA}{L})$ stiffness of a three hinged arch (Fig. 5a). This model is particularly useful to simulate compressive buckling strength of shallow curved structures due to sudden curvature inversion [28].

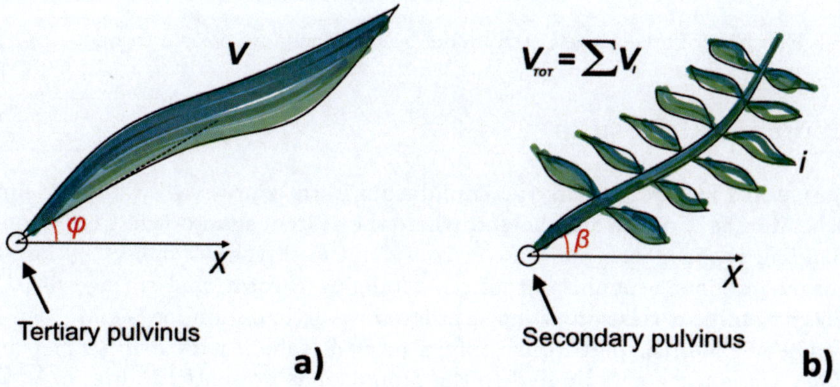

Fig. 4. a) Leaflet configuration. b) Bending moment by the pinna on the secondary pulvinus.

The system snaps down also if loaded by a bending moment m at the crown hinge. Considering h as the arch height, the vertical reaction F to the generalized coordinate u is

$$F(u) = \frac{K_R}{L}(u-h)(u-2h)u. \tag{3}$$

By vertical equilibrium equation,

$$F(u) = 2N_R \sin(\frac{h-u}{L}), \tag{4}$$

where N_R is the axial force in the rod. At the same time, horizontal equilibrium imposes that the N_R torque produced by m is oriented as shown in 5b, giving

$$2N_R = m\frac{\cos\frac{h-u}{L}}{L}. \tag{5}$$

Therefore the relation between m and u can be expressed as

$$m(u) = \frac{1}{2}\frac{K_R(u-h)(u-2h)u}{\cos\frac{h-u}{L}\sin\frac{h-u}{L}}. \tag{6}$$

which is nonlinear and possess limit points corresponding to the vertical snap-through. Solution of 6 cannot be obtained in closed form without trigonometric linearization. We present in the next section a numerical solution by FEM modeling of the presented structure.

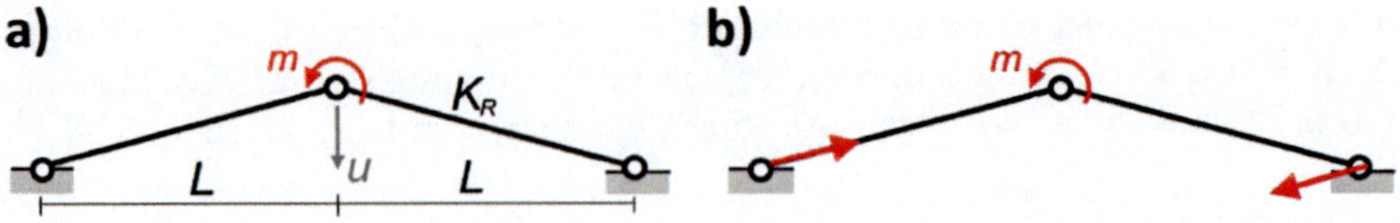

Fig. 5. a) Three-hinged arch model b) Balancing axial force torque.

4 Numerical Results

The curve in Fig. 6 represents the equilibrium path expressed by 6. First limit point lie at circa $\frac{1}{2}$ of the arch height, where the system snap-down. Corresponding bending moment is what we can consider the threshold value m_{T_i} for the i-th order pulvinar stimulus actuation. Modeling a rotational spring with the presented nonlinear relation, a tuned progressive articulation for the MP can be derived. For example, perturbing only a pair of leaflets with a $m > m_{T_3}$ but $m < m_{T_2}$ the folding gets limited to the pinnulae, as presented in Fig. 2. At the same time, if the vertical component of the related bending moment produces a vertical displacement in the other pulvini of the pinna above the threshold value, the response propagate up to that along the rachis, adopting a simple elastica profile for the deform. Under the same line of reasoning, a concentrated stimuli on the rachis with a $m > m_{T_3}$ and $m > m_{T_2}$ would produce a collapse also of the secondary pulvinus. We simulated this behavior on a FEM model, as shown in Fig. 7a. Each element has been modeled with a 3-node nonlinear thin beam, using nonlinear spring connection as in 6. The same response, in term of folding, can be obtained by a force applied to the peripheral leaflets (Fig. 7b) or by a the same vertical resultant spread on four leaflets, with a corresponding pulvinar bending moment $m > m_{T_3}$ and $m < m_{T_2}$.

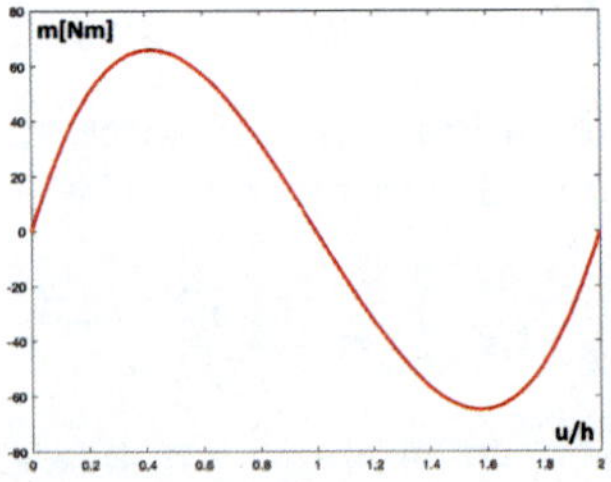

Fig. 6. Equilibrium path - x axis normalized.

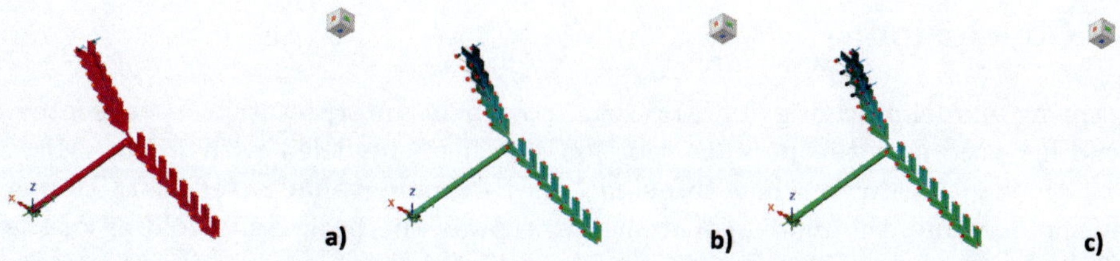

Fig. 7. a) outline of the FEM model. b) 2 leaflets stimuli propagated to 4 c) 4 leaflets stimuli

Finally, because described action follow the relations 1 and 2, a stimuli can be produced by air pressure forcing. Other than wind blow, such model correctly interprets the defensive folding of the MP when decelerating after a fall. In Fig. 8 the frames of a fall decelerating demonstration is presented. Produced stimulus results into a global folding of the MP.

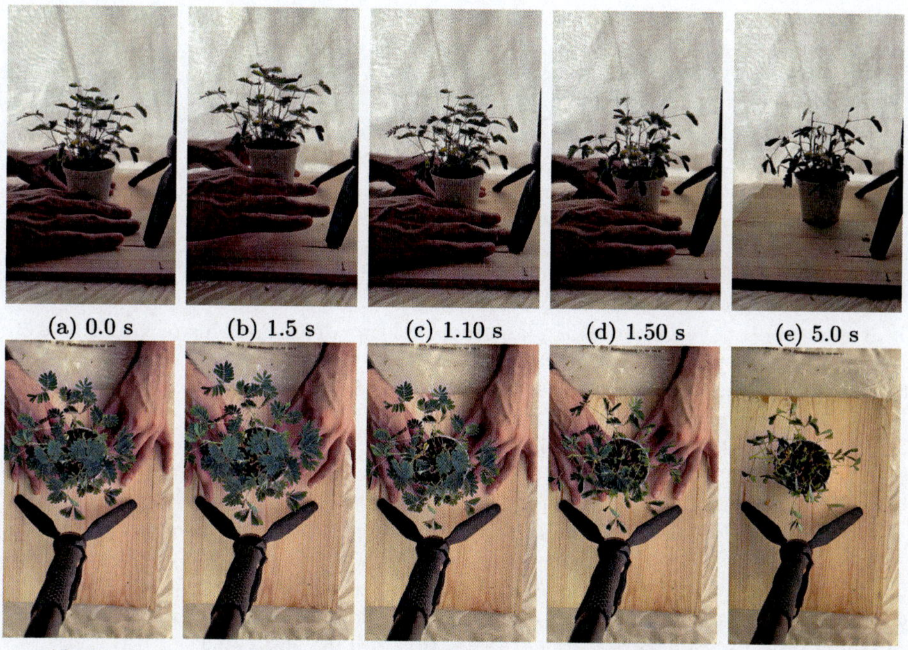

(a) 0.0 s (b) 1.5 s (c) 1.10 s (d) 1.50 s (e) 5.0 s

(f) 0.0 s (g) 1.5 s (h) 1.10 s (i) 1.50 s (j) 5.0 s

Fig. 8. Mimosa Pudica, before (a) and after (b) the application of a stimulus that caused global folding.

5 Conclusion

A spring model buckling for MP leaf movement interpretation is presented. Bending action on the pulvinar portion has been modeled with a Von Mises arch that snaps down when the signaling threshold value is reached. Tuning thresholds value, the model coherently interprets the progressive folding of the MP body elements. Using bending action, inherently thigmonastic movements can be reproduced with different vertical action that produce the same eigenforms. Encompassing these conclusion, we may suggest that *The Unbendable* could serve as a more appropriate nickname for the MP. The presented results, could be useful in the biomechanic field and bio-inspired structures, particuarly in smart facades and deployable structures design.

Acknowledgments. No plant has been hurt for this research. Thanks to Eng. Berellini for letting us play with her Mimmy, a MP whose devotion and discipline taught us all something.

References

1. Weintraub, M.: Leaf movements in mimosa pudica L. New Phytol. **50**(3), 357–382 (1952)
2. Sachse, R., et al.: Snapping mechanics of the Venus flytrap (Dionaea muscipula). Proc. Natl. Acad. Sci. **117**(27), 16035–16042 (2020)
3. Gagliano, M., Renton, M., Depczynski, M., Mancuso, S.: Experience teaches plants to learn faster and forget slower in environments where it matters. Oecologia **175**(1), 63–72 (2014). https://doi.org/10.1007/s00442-013-2873-7
4. Niklas, K.J.: Plant Biomechanics: An Engineering Approach to Plant form and Function, University of Chicago Press (1992)
5. Kaplan, D.R.: The science of plant morphology: definition, history, and role in modern biology. Am. J. Bot. **88**(10), 1711–1741 (2001)
6. Bidhendi, A.J., Geitmann, A.: Methods to quantify primary plant cell wall mechanics. J. Exp. Bot. **70**(14), 3615–3648 (2019)
7. Schopfer, P.: Biomechanics of plant growth. Am. J. Bot. **93**(10), 1415–1425 (2006)
8. Knight, T.A.: XI. On the causes which influence the direction of the growth of roots. by TA knight, Esq. FRS in a letter to the right hon. sir Joseph banks, bart. KBPR S, Philosophical Transactions of the Royal Society of London **101**(1811), 209–219
9. Geitmann, A., Niklas, K., Speck, T.: Plant biomechanics in the 21st century (2019)
10. Moulia, B.: Plant biomechanics and mechanobiology are convergent paths to flourishing interdisciplinary research. J. Exp. Bot. **64**(15), 4617–4633 (2013)
11. Coutand, C.: Mechanosensing and thigmomorphogenesis, a physiological and biomechanical point of view. Plant Sci. **179**(3), 168–182 (2010)
12. Jaffe, M., Forbes, S.: Thigmomorphogenesis: the effect of mechanical perturbation on plants. Plant Growth Regul. **12**, 313–324 (1993)
13. Braam, J.: In touch: plant responses to mechanical stimuli. New Phytol. **165**(2), 373–389 (2005)
14. Snow, R.: Conduction of excitation in stem and leaf of mimosa pudica. Proc. R. Soc. Lond. Ser. B. **96**(678), 349–374 (1924)

15. Houwink, A.L.: The conduction of excitation in mimosa pudica. Recueil des travaux botaniques neerlandais **32**(1), 51–91 (1935)
16. Volkov, A.G.: Plant Electrophysiology, Springer (2006). https://doi.org/10.1007/978-3-540-37843-3
17. Volkov, A.G., Foster, J.C., Ashby, T.A., Walker, R.K., Johnson, J.A., Markin, V.S.: Mimosa pudica: electrical and mechanical stimulation of plant movements. Plant Cell Environ. **33**(2), 163–173 (2010)
18. Volkov, A.G., Foster, J.C., Markin, V.S.: Signal transduction in mimosa pudica: biologically closed electrical circuits. Plant Cell Environ. **33**(5), 816–827 (2010)
19. Hagihara, T., Mano, H., Miura, T., Hasebe, M., Toyota, M.: Calcium-mediated rapid movements defend against herbivorous insects in mimosa pudica. Nat. Commun. **13**(1), 6412 (2022)
20. Nelson, C.M.: On buckling morphogenesis. J. Biomech. Eng. **138**(2), 021005 (2016)
21. Thomson, J.A.: On growth and form (1917)
22. Canham, P.B.: The minimum energy of bending as a possible explanation of the biconcave shape of the human red blood cell. J. Theor. Biol. **26**(1), 61–81 (1970)
23. Sitti, M.: Physical intelligence as a new paradigm. Extreme Mech. Lett. **46**, 101340 (2021)
24. Lobel, T.: Sensation: The New Science of Physical Intelligence, Simon and Schuster (2016)
25. Xu, Y., Huang, G.: Modal sensitivity analysis of acoustic metamaterials for structural damage detection. Int. J. Mech. Sci. **259**, 108571 (2023)
26. Sleboda, D.A., Geitmann, A., Sharif-Naeini, R.: Multiscale structural anisotropy steers plant organ actuation. Curr. Biol. **33**(4), 639–646 (2023)
27. Gibson, L.J.: The hierarchical structure and mechanics of plant materials. J. R. Soc. Interface **9**(76), 2749–2766 (2012)
28. Bazzucchi, F., Manuello, A., Carpinteri, A.: Interaction between snap-through and Eulerian instability in shallow structures. Int. J. Non-Linear Mech. **88**, 11–20 (2017)

Vision, Perception and Intuition in Structural Analysis

Fabio Bazzucchi[1,2](✉)

[1] Senseable City Lab, Massachusetts Institute of Technology, Cambridge, MA, USA
fbaz@mit.edu

[2] Politecnico di Torino, DISEG, Department of Structural, Geotechnical and Building Engineering, Corso Duca Degli Abruzzi, 24, Turin 10128, Italy

Abstract. Forces geometric visualization and representation in one of the key aspects in Structural Mechanics and its advancement went along with the progress of the subject itself. From graphic statics to modern Computer Vision (CV) systems, there is a constant and significant component of the visual factor when compared to other disciplines in the STEM panorama. In this work, I present the experience carried on in the MSCA FORSEES to try to frame and quantify this visual factor. I will then suggest a new perspective to the problem of forces identification (and related stress and strain) in solids based on some recent discoveries in Neuroscience. The analogy that represents the first step of this exploration is between the Free Energy Principle (FEP) [1] and the Elastic Potential Energy, by defining a structure behavior as the inner agent to sustain structural integrity. Limiting the FEP to the visual sense only [2], we will look through the indissoluble link that lies within the relation of a form and its structural behavior, circling back to the base of the Theory of Elasticity. Final step will be the application to common methods in engineering practice and teaching of AI-based visual tools for structural analysis.

Keywords: free energy principle · visual cognition · FEMA · rapid visual screening · intuitive structural analysis

1 Introduction

Professor Baldacci's book [3] on Mechanics of Solids famously begins by introducing the stress and strain tensors as models of the mind. They represents a generalization of the empirical observations in a purely mental construct, that when examined in its ideal dimension, produces relations inaccessible for the experience but deductible through inference. As in a perfect Kantian critique approach, these models are bounded and controlled by experience on the relations that they produce, and not on their logical framework [4,5]. Cauchy stress tensor formulation is in fact axiomatic [6] and based on the mathematical institution of the stress entity $\mathbf{t}_n$. In a specific point of a continuum solid in equilibrium, it is always possible to define the ratio

S. Gabriele et al. (Eds.): IWSS 2023, LNCE 437, pp. 898–905, 2024.
https://doi.org/10.1007/978-3-031-44328-2_94

$$\mathbf{t}_n = \lim_{A_n \to \infty} \frac{\mathbf{R}_n}{A_n}, \tag{1}$$

where $\mathbf{R}_n$ is the resulting force on the small area A_n identified by the normal n. Tetrahedral decomposing of $\mathbf{t}_n$ provide the special components and the terms of the stress tensor. The extension of it to the whole solid domain has two crucial assumptions: *(i)* the limit expressed by 1 exists for every small value of A_n, sacrificing atomic or sub-atomic behavior characterization in favor of the experience representation. And *(ii)*, the properties observed at infinitesimal scale are the same that the ones observed in the solid macrocosms, implying that structural analysis is indeed a phenomenological analysis.

This line of reasoning nimbly overcome the debate between Rationalists and Empiricist, decoupling reality from its elemental representation (*à la* Mach) and mathematical nesting. The elision of this unnecessary harmony, highlighted by Poincaré [7] in a visionary way, resonates with current Machine Learning (ML) and Artificial Intelligence (AI) methodologies. By training models on a large number of certified phenomenological observations, the solving agent infers the results of a new experience, and their fitness is only assessed on the empirical validity of them. We, for example, define a good Neural Network (NN) the one that has a better statistical accuracy on the data, not the one that has an inherent reality representation [8,9]. This quasi-metaphysical epistemology finds even more reverberation in the Consciousness studies and Mind modeling [10–12]. It is in fact necessary, not for a relativism but for a centrality of the mind, to pass from reality to reality mental representation. To do so, the concepts of Markov Blanket (MB) and Intelligence have to be introduced [13–15]. For the scope of this work, we can see a MB as the boundary that encompass and separates the Intelligence from all the rest, i.e. the reality. On its surface, the MB contains the organs for reality perception and interaction. Under the light of this, the Intelligence creates and stores a mental image of the reality through the filter of the sensing and interacting apparatuses. If we consider the example of the human vision sense, vision is the task performed by the MB dedicated organs while perception is the feedback processed knowledge by the brain through cognition [16]. However, the brain is capable of fast MB interactive responses without a significant cognitive work, and relying of some presets and constructs of the Intelligence: intuition [17].

Circling back to the initial problem of structural analysis, we can say that a mind model such the stress tensor is not a construct built by the Intelligence through active (sensing and interacting) perception, but a construct produced by pure cognition. The inference that the intelligence may perform are the relations of this construct, that has the sole boundary of creating coherent results with consensus empirical observation. In other words, we may assume that the use of axiomatic mind models forces the constructivism of the mind and the development of intuitive schemes. This hypothesis is easily verifiable for trusses deform drawings, in where few axioms help to predict the deformation of structural systems. Visual analysis had in fact a major importance in the heroic stage of structural analysis, from internal action representation to the immense work in Graphic Statics [18,19]. Finite Element Method stopped the advancements and the discourse around such approaches in favor of a more rationalist approach, but they are now going through a revival phase. Apart from their use in form finding [20], Computer Vision (CV) is displaying all is power in recognition, detection and classification tasks. In this work, the potential of CV in structural analysis is presented and connected with intuitive-like solutions approach.

2 Structural Integrity and FEP

A well renowned and solid model for brain inference is Karl Friston Free Energy Principle (FEP) [16,21,22]. For what concerns the scope of this work, we can use the FEP intrinsic derivation of the best mental image of the reality through a minimization of the free energy, thus the maximizion of the likelihood between the sensed form and the interactive form. The free energy can be expressed as follow

$$F(s,\mu) = E_q[-\log p(s,\psi|m)] - H[q(\psi|\mu)] = \\ [-\log p(s|m)] + D_{KL}[q(\psi|\mu)||p(s,\psi|m)] \geq -\log p(s|m).$$

Excluding unnecessary details for the purpose of this work, the expression represents a Bayesian approach that measure the difference between the cognitive load (energy) E_q to gain a knowledge of a set p with a set of unknown states q. H stands thus as an entropy measure. Higher is the free energy F, lower is the likelihood and therefore the accurate representation of the reality. Because p and q (term D is called divergence) are conditional states, a large number of observation does not necessarily implies a lower free energy. It is a powerful tool and its validity has been proven in many fields.

Curiously, this equation formally resembles the Minimal Potential Energy for Stability Analysis of solids. By minimizing, in fact, the work of the internal and external states the equilibrium configuration is obtained. Because in this case the two sates are mutual by the force-displacement duplicity, there is no divergence term and the solution is unique. Therefore, in the case of external forces applied to a MB equipped with an Intelligence capable to mutually react, the MB itself assume the minimum energy surface configuration, implying that the scope of the intelligence is the structural integrity. Friston brilliantly describe how the FEP is directly connected with the emergence of life and its fundamental role of resisting [23]. As previously stated, the role of a MB is to separate the intelligence from the rest, therefore a fracture in MB would translate into a destruction of the agent. If we imagine a drop of ink in a water glass, it would immediately dissolve through the act of random fluctuations of the fluid after few seconds. If anyway, the drop could possess an intelligence apt to sense the random fluctuations and respond in order to maintain a structural integrity, it would pulse and look alive to the eye of an external observer, and doted with a metabolism proportional to the resisting work (Fig. 1). Capitalizing on this example, we are interested to know if the drop can learn how to react towards external fluctuations by feeding the intelligence with a numbered of trained solutions. Moreover, creating axiomatic constraints on the same intelligence if the same solutions can be achieved in an intuition-like scheme.

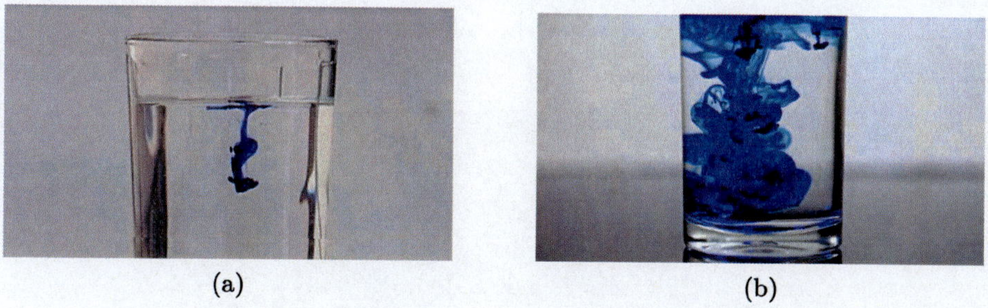

Fig. 1. a) drop of ink b) advance dissolving [23]

3 CauShee Module

CauShee (Computer automated Structural analysis solver) is one of the research project developed during the FORSEES MSCA Action, and it's based on the approach described above. The system uses CV to solve simple structural arrangements as the one in Fig. 2. A CNN based on a standard ResNet50 has been trained to classify loading type (red), restraints (yellow) and geometries (cyan). The same procedure has been used to train the CNN for internal action drawings. Pix2pix layers have been added to have an interpolating drawing result. In Fig. 3, the bending moment is shown in red. At the same time, an intuitive-like solution has been implemented using the FEP. In details, it is possible to reduce the system entropy by discarding all the non-admissible solution by the compatibility equations, from all the possible system state. In fact, we can compute the disorder (D) by

$$D = \sum (p_i - P_{P_Y}), \tag{2}$$

with

$$P_Y = E(\omega_{i-1}), \tag{3}$$

where E is the expected value of the compatible solutions P_Y. Order metric O is instead

$$O = 1 - \frac{C_D}{C_I} \tag{4}$$

in which C_I is the number of indeterminate solution. Therefore, by imposing for example that solutions with non null bending moment at the pin extremity are determinate and not to be computed, the entropy of the system decreases as the free energy. A quantize distribution of possible internal actions points $p_x = p(x_1, x_2, ..., x_n)$ has been considered, using Shannon formula

$$H(x) = -\sum p_i \log p_i. \tag{5}$$

In Fig. 3 the green line represents the solution of this approach.

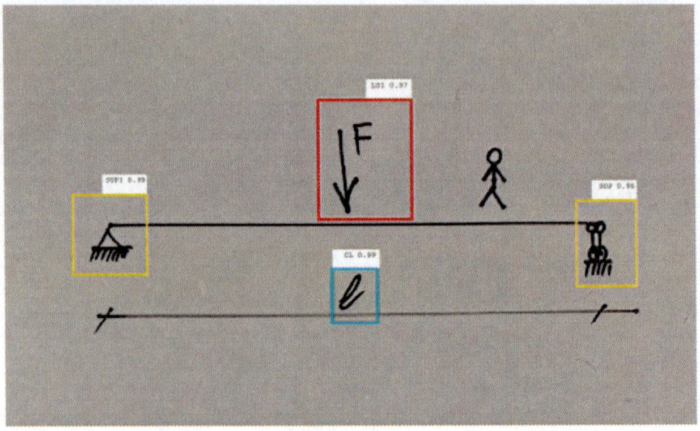

Fig. 2. Structural arrangement data classification

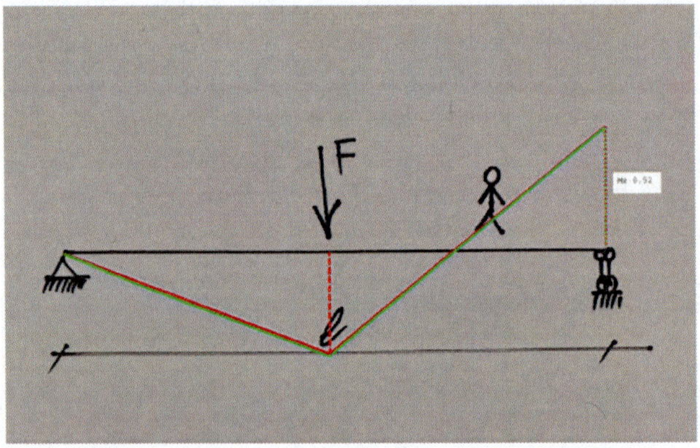

Fig. 3. Bending moment solution with CNN (red) and intuitive FEP approach (green).

4 Vulneracities Module

In this section the results of another research module of the FORSEES MSCA Action are presented. Vulneracities is a project dedicated to assess seismic vulnerability of urban morphologies through visual analysis. The scientific base model is the FEMA Rapid Visual Screening (RSV) implemented more than 30 years ago for large scenario inspection [24]. Developed for the US building stock, it consists on a 2 level survey that identifies the building structural typology (among 16 classes) and assigns a risk basic score. This score can be subjected to modifiers with the presence/absence of structural peculiarities. For example, the soft-story can knock-down the basic score of a concrete building by half. Capitalizing on the CV potential, a DCNN (Deep Convolutional Neural Network) has been trained to perform this classification task. A database of 5m buildings has been used extracting facades from Google Street View API. For details see [25]. In Fig. 4 the activation layer maps of a Densenet-201 have been reported. The first row represent the classification performed using the classical CV training, while the bottom row has been obtained by the use of the FEP approach. In this case, the reduction of the entropy has been obtained by imposing some causal eclusion, i.e. the presence of bow window automatically excludes the structural typologies that are not wooden structures bases (D) or masonry (C). Both the network rightfully predicted the class D, but the intuitive scheme used the window as discriminant, as a deployed engineer would do.

The same results are presented in Fig. 5. In this case the building typology to classify was a masonry wall arrangement (C). Again, it is possible to note that the intuitive solution detected succesfully the building using the peculiarity of the window profile.

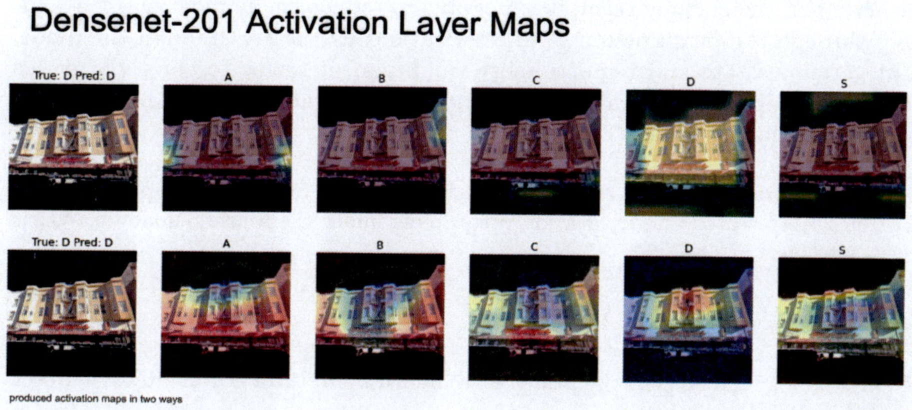

Fig. 4. Activation map for a five building typology classification. First row is the classical ML approach, while bottom row is for the intuitive solution.

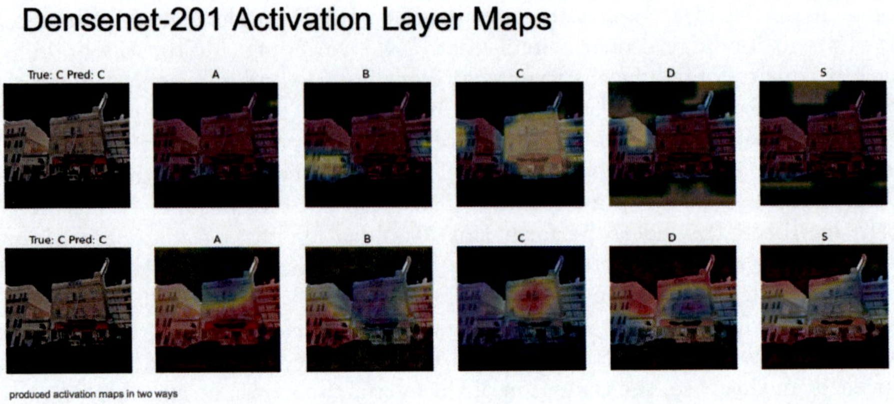

Fig. 5. Activation map for a five building typology classification. First row is the classical ML approach, while bottom row is for the intuitive solution.

5 Conclusion

The paper presents the use of axiomatic schemes for structural analysis tasks in the modern use of Computer Vision and visual analysis. The Free Energy Principle has been use to frame an intuition term inside the Bayesan approach. By acting on the entropy term, it is possible to reduce the number of open solution, through the introduction of idiosyncratic constructs. Although the model is at germinal state and requires more validation, it seems to correctly grasps both the theoretical and practice-based boundaries that experience in structural analysis has consolidated.

Acknowledgments. This project has received funding from the European Union's Horizon 2020 research and innovation programme under the Marie Sklodowska-Curie grant agreement No 845064.

References

1. Friston, K.: The free-energy principle: a unified brain theory? Nat. Rev. Neurosci. **11**(2), 127–138 (2010)
2. Cavanagh, P.: Visual cognition. Vision. Res. **51**(13), 1538–1551 (2011)
3. Baldacci, R.: Scienza delle costruzioni, Utet (1970)
4. Kant, I., Meiklejohn, J.M.D., Abbott, T.K., Meredith, J.C.: Critique of pure reason, JM Dent London (1934)
5. Bridgman, P.W.: The Logic of Modern Physics, vol. 3, Macmillan (1927)
6. Love, A.E.H.: A Treatise on the Mathematical Theory of Elasticity, University Press (1927)
7. Poincaré, H.: The value of science: essential writings of Henri Poincaré, Modern library (2012)
8. Wilkinson, J., et al.: Time to reality check the promises of machine learning-powered precision medicine. Lancet Digital Health **2**(12), e677–e680 (2020)

9. Wheeler, G.: Machine epistemology and big data. In: The Routledge Companion to Philosophy of Social Science, Routledge, pp. 341–349 (2016)
10. Horgan, T., Kriegel, U.: Phenomenal epistemology: what is consciousness that we may know it so well?, Philosophical. Issues **17**, 123–144 (2007)
11. Harman, W.: The scientific exploration of consciousness: towards an adequate epistemology. J. Conscious. Stud. **1**(1), 140–148 (1994)
12. Hill, C.: Modality, modal epistemology, and the metaphysics of consciousness. The Architecture of the Imagination, pp. 205–235 (2006)
13. Hipólito, I., Ramstead, M.J., Convertino, L., Bhat, A., Friston, K., Parr, T.: Markov blankets in the brain. Neurosci. Biobehav. Rev. **125**, 88–97 (2021)
14. Bruineberg, J., Dołęga, K., Dewhurst, J., Baltieri, M.: The emperor's new markov blankets. Behav. Brain Sci. **45**, e183 (2022)
15. Clark, A.: How to knit your own markov blanket (2017)
16. Friston, K., Sengupta, B., Auletta, G.: Cognitive dynamics: from attractors to active inference. Proc. IEEE **102**(4), 427–445 (2014)
17. Kahneman, D., Tversky, A.: On the study of statistical intuitions. Cognition **11**(2), 123–141 (1982)
18. Cremona, L.: Le figure reciproche nella statica grafica, U. Hoepli (1879)
19. Capecchi, D., Ruta, G., Capecchi, D., Ruta, G.: La statica grafica in italia, La scienza delle costruzioni in Italia nell'Ottocento: Un'analisi storica dei fondamenti della scienza delle costruzioni, pp. 239–286 (2011)
20. Van Mele, T., Block, P.: Algebraic graph statics. Comput. Aided Des. **53**, 104–116 (2014)
21. Parr, T., Friston, K.J.: The anatomy of inference: generative models and brain structure. Front. Comput. Neurosci. **12**, 90 (2018)
22. Kirchhoff, M., Parr, T., Palacios, E., Friston, K., Kiverstein, J.: The markov blankets of life: autonomy, active inference and the free energy principle. J. R. Soc. Interface **15**(138), 20170792 (2018)
23. https://www.youtube.com/watch?v=NIu$_d$JGyIQI&t=1s
24. F. E. M. A. (US), Rapid visual screening of buildings for potential seismic hazards: A handbook, Government Printing Office (2017)
25. Bazzucchi, F., Klimenko, N., Sun, M., Rathore, P., Duarte, F., Ratti, C.: Building seismic vulnerability assessment through deep learning at street level imagery. Computer Aided Civil and Infrastructure Engineering

Author Index

S. Gabriele et al. (Eds.): IWSS 2023, LNCE 437, pp. 907–910, 2024.
https://doi.org/10.1007/978-3-031-44328-2